OFDM and
MC-CDMA

for Broadband Multi-user
Communications, WLANs
and Broadcasting

OFDM and MC-CDMA

for Broadband Multi-user Communications, WLANs and Broadcasting

**L. Hanzo, M. Münster,
B.J. Choi** and **T. Keller**

*All of
University of Southampton, UK*

IEEE PRESS

IEEE Communications Society, Sponsor

WILEY

Other Wiley Editorial Offices

John Wiley & Sons, Inc., 111 River Street, Hoboken, NJ 07030, USA

Jossey-Bass, 989 Market Street, San Francisco, CA 94103-1741, USA

Wiley-VCH Verlag GmbH, Boschstr. 12, D-69469 Weinheim, Germany

John Wiley & Sons Australia Ltd, 33 Park Road, Milton, Queensland 4064, Australia

John Wiley & Sons (Asia) Pte Ltd, 2 Clementi Loop #02-01, Jin Xing Distripark, Singapore 129809

John Wiley & Sons (Canada) Ltd, 22 Worcester Road, Etobicoke, Rexdale, Ontario, Canada M9W 1L1

Wiley also publishes its books in a variety of electronic formats. Some content that appears in print may
not be available in electronic books.

IEEE Communications Society, Sponsor COMMS-S Liason to IEEE Press, Mostafa Hashem Sherif

Library of Congress Cataloging-in-Publication Data

OFDM and MC-CDMA for broadband multi-user communications, WLANs and
     broadcasting / L. Hanzo... [et al.].
        p. cm.
     Includes bibliographical references and index.
     ISBN 0-470-85879-6
       1. Wireless communications Systems. 2. Wireless LANs. 3. Wavelength division
     multiplexing. 4. Code division multiple access. 5. Orthogonalization methods. I. Hanzo,
     Lajos, 1952-
     TK5103.2.O33 2003
     621.382—dc21

                                                              2003050188

British Library Cataloguing in Publication Data

A catalogue record for this book is available from the British Library

ISBN 0-470-85879-6

Typeset from pdf files supplied by the author.
Printed and bound in Great Britain by Antony Rowe, Chippenham, Wiltshire.
This book is printed on acid-free paper responsibly manufactured from sustainable forestry in which at
least two trees are planted for each one used for paper production.

We dedicate this monograph to the numerous contributors of this field,
many of whom are listed in the Author Index

Contents

III Advanced Topics: Channel Estimation and Multi-user OFDM Systems 479

About the Authors

Lajos Hanzo received his degree in electronics in 1976 and his doctorate in 1983. During his career in telecommunications he has held various research and academic posts in Hungary, Germany and the UK. Since 1986 he has been with the Department of Electronics and Computer Science, University of Southampton, UK, where he holds the chair in telecommunications. He co-authored 10 books totalling 8000 pages on mobile radio communications, published about 450 research papers, organised and chaired conference sessions, presented overview lectures and has been awarded a number of distinctions. Currently he heads an academic research team, working on a range of research projects in the field of wireless multimedia communications sponsored by industry, the Engineering and Physical Sciences Research Council (EPSRC) UK, the European IST Programme and the Mobile Virtual Centre of Excellence (VCE), UK. He is an enthusiastic supporter of industrial and academic liaison and he offers a range of industrial courses. Lajos is also an IEEE Distinguished Lecturer of both the Communications as well as the Vehicular Technology Society and a Fellow of the IEE. For further information on research in progress and associated publications please refer to http://www-mobile.ecs.soton.ac.uk

Matthias Münster was awared the Dipl. Ing. degree by the RWTH Aachen, Germany and after graduation he embarked on postgraduate research at the University of Southampton, where he completed his PhD in mobile communications on 2002. His areas of interest include adaptive multiuser OFDM transmission, wideband channel estimation, multiuser detection and a range of related signal processing aspects. During his PhD research he contributed over a dozen various research papers and following the completion of his thesis he returned to his native Germany, where he is currently involved in the development of sophisticated signal processing algorithms.

Byoung-Jo Choi received his BSc and MSc degrees in Electrical Engineering from KAIST, Korea, in 1990 and 1992, respectively. He has been working for LG Electronics, Korea, since January 1992, where he was involved in developing the KoreaSat monitoring system, Digital DBS transmission system and W-CDMA based Wireless Local Loop (WLL) system. He was awarded the PhD degree in Mobile Communications at the University of Southampton, UK, where he was a postdoctoral research assistant from 2001 to 2002. He is a recipient of the British Chevening Scholarship awarded by the British Council, UK. His current research interests are related to mobile communication systems design with emphasis on adaptive modulation aided OFDM, MC-CDMA and W-CDMA.

Thomas Keller studied Electrical Engineering at the University of Karlsruhe, Ecole Superieure d'Ingenieurs en Electronique et Electrotechnique, Paris, and the University of Southampton. He graduated with a Dipl.-Ing. degree in 1995. Between 1995 and 1999 he had been with the Wireless Multimedia Communications Group at the University of Southampton, where he completed his PhD in mobile communications. His areas of interest include adaptive OFDM transmission, wideband channel estimation, CDMA and error correction coding. He recently joined Ubinetics, Cambridge, UK, where he is involved in the research and development of third-genertion wireless systems. Dr. Keller co-authored two monographs and about 30 various research papers.

Other Wiley and IEEE Press Books on Related Topics [1]

- R. Steele, L. Hanzo (Ed): *Mobile Radio Communications: Second and Third Generation Cellular and WATM Systems*, John Wiley and IEEE Press, 2nd edition, 1999, ISBN 07 273-1406-8, 1064 pages

- L. Hanzo, W. Webb, and T. Keller, *Single- and Multi-Carrier Quadrature Amplitude Modulation: Principles and Applications for Personal Communications, WLANs and Broadcasting*, John Wiley and IEEE Press, 2000, 739 pages

- L. Hanzo, F.C.A. Somerville, J.P. Woodard: *Voice Compression and Communications: Principles and Applications for Fixed and Wireless Channels*; IEEE Press and John Wiley, 2001, 642 pages

- L. Hanzo, P. Cherriman, J. Streit: *Wireless Video Communications: Second to Third Generation and Beyond*, IEEE Press and John Wiley, 2001, 1093 pages

- L. Hanzo, T.H. Liew, B.L. Yeap: *Turbo Coding, Turbo Equalisation and Space-Time Coding*, John Wiley and IEEE Press, 2002, 751 pages

- J.S. Blogh, L. Hanzo: *Third-Generation Systems and Intelligent Wireless Networking: Smart Antennas and Adaptive Modulation*, John Wiley and IEEE Press, 2002, 408 pages

- L. Hanzo, C.H. Wong, M.S. Yee: *Adaptive wireless transceivers: Turbo-Coded, Turbo-Equalised and Space-Time Coded TDMA, CDMA and OFDM systems*, John Wiley and IEEE Press, 2002, 737 pages

- L. Hanzo, L-L. Yang, E-L. Kuan and K. Yen: *Single- and Multi-Carrier CDMA: Multi-User Detection, Space-Time Spreading, Synchronisation, Standards and Networking*, John Wiley and IEEE Press, June 2003, 1060 pages

---

[1]For detailed contents and sample chapters please refer to http://www-mobile.ecs.soton.ac.uk

Acknowledgments

We are indebted to our many colleagues who have enhanced our understanding of the subject, in particular to Prof. Emeritus Raymond Steele. These colleagues and valued friends, too numerous to be mentioned, have influenced our views concerning various aspects of wireless multimedia communications. We thank them for the enlightenment gained from our collaborations on various projects, papers and books. We are grateful to Steve Braithwaite, Jan Brecht, Jon Blogh, Marco Breiling, Marco del Buono, Sheng Chen, Peter Cherriman, Stanley Chia, Joseph Cheung, Sheyam Lal Dhomeja, Dirk Didascalou, Lim Dongmin, Stephan Ernst, Peter Fortune, Eddie Green, David Greenwood, Hee Thong How, Ee Lin Kuan, W. H. Lam, C. C. Lee, Xiao Lin, Chee Siong Lee, Tong-Hooi Liew, Vincent Roger-Marchart, Jason Ng, Michael Ng, M. A. Nofal, Jeff Reeve, Redwan Salami, Clare Somerville, Rob Stedman, David Stewart, Jürgen Streit, Jeff Torrance, Spyros Vlahoyiannatos, William Webb, Stephan Weiss, John Williams, Jason Woodard, Choong Hin Wong, Henry Wong, James Wong, Lie-Liang Yang, Bee-Leong Yeap, Mong-Suan Yee, Kai Yen, Andy Yuen, and many others with whom we enjoyed an association.

We also acknowledge our valuable associations with the Virtual Centre of Excellence (VCE) in Mobile Communications, in particular with its chief executive, Dr Walter Tuttlebee, and other leading members of the VCE, namely Dr Keith Baughan, Prof. Hamid Aghvami, Prof. Ed Candy, Prof. John Dunlop, Prof. Barry Evans, Prof. Peter Grant, Dr Mike Barnard, Prof. Joseph McGeehan, Prof. Steve McLaughlin and many other valued colleagues. Our sincere thanks are also due to the EPSRC, UK for supporting our research. We would also like to thank Dr Joao Da Silva, Dr Jorge Pereira, Dr Bartholome Arroyo, Dr Bernard Barani, Dr Demosthenes Ikonomou, Dr Fabrizio Sestini and other valued colleagues from the Commission of the European Communities, Brussels, Belgium.

We feel particularly indebted to Denise Harvey for her skilful assistance in correcting the final manuscript in LaTeX. Without the kind support of Mark Hammond, Sarah Hinton, Zöe Pinnock and their colleagues at the Wiley editorial office in Chichester, UK this monograph would never have reached the readers. *Finally, our sincere gratitude is due to the numerous authors listed in the Author Index — as well as to those whose work was not cited owing to space limitations — for their contributions to the state of the art, without whom this book would not have materialised.*

Lajos Hanzo, Matthias Münster, Byuong-Jo Choi and Thomas Keller
Department of Electronics and Computer Science
University of Southampton

Chapter 1

Introduction[1]

1.1 Motivation of the Book

Whilst the concept of Orthogonal Frequency Division Multiplexing (OFDM) has been known since 1966 [1], it only reached sufficient maturity for employment in standard systems during the 1990s. OFDM exhibits numerous advantages over the family of more conventional serial modem schemes [2], although it is only natural that it also imposes a number of disadvantages. The discussion of the associated design tradeoffs of OFDM and Multi-Carrier Code Division Multiple Access (MC-CDMA) systems constitutes the topic of this monograph and in this context our discussions include the following fundamental issues:

1) A particularly attractive feature of OFDM systems is that they are capable of operating without a classic channel equaliser, when communicating over dispersive transmission media, such as wireless channels, while conveniently accommodating the time- and frequency-domain channel quality fluctuations of the wireless channel.

   Explicitly, the channel SNR variation versus both time and frequency of an indoor wireless channel is shown in a three-dimensional form in Figure 1.1 versus both time and frequency, which suggests that OFDM constitutes a convenient framework for accommodating the channel quality fluctuations of the wireless channel, as will be briefly augmented below. This channel transfer function was recorded for the channel impulse response of Figure 1.2, by simply transforming the impulse response to the frequency domain at regular time intervals, while its taps fluctuated according to the Rayleigh distribution.

   These channel quality fluctuations may be readily accommodated with the aid of subband-adaptive modulation as follows. Such an adaptive OFDM (AOFDM) modem is characterised by Figure 1.3, portraying at the top a contour plot of the above-mentioned

[1] *OFDM and MC-CDMA for Broadband Multi-user Communications WLANS and Broadcasting.*
L. Hanzo, Münster, T. Keller and B.J. Choi,
©2003 John Wiley & Sons, Ltd. ISBN 0-470-85879-6

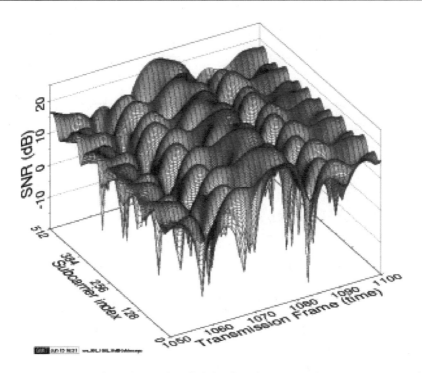

Figure 1.1: Instantaneous channel SNR for all 512 subcarriers versus time, for an average channel SNR of 16 dB over the channel characterised by the channel impulse response (CIR) of Figure 1.2.

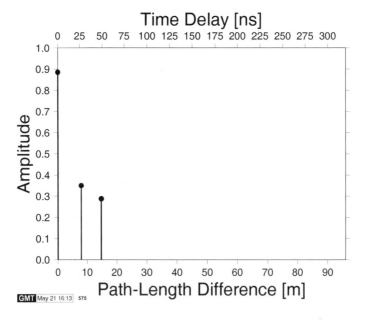

Figure 1.2: Indoor three-path WATM channel impulse response.

wireless channel's signal-to-noise ratio (SNR) fluctuation versus both time and frequency for each OFDM subcarrier. We note at this early stage that these channel quality fluctuations may be mitigated with the aid of frequency-domain channel equalisation, as will be detailed throughout the book, but nonetheless, they cannot be entirely eradicated.

More specifically, as can be seen in Figure 1.1, that when the channel is of high quality — as for example in the vicinity of the OFDM symbol index of 1080 — the sub-band-adaptive modem considered here for the sake of illustration has used the same modulation mode, as the identical-throughput conventional fixed-rate OFDM modem in all subcarriers, which was 1 bit per symbol (BPS) in this example, as in conventional Binary Phase Shift Keying (BPSK). By contrast, when the channel is hostile — for example, around frame 1060 — the sub-band-adaptive modem transmitted zero bits per symbol in some sub-bands, corresponding to disabling transmissions in the low-quality sub-bands. In order to compensate for the loss of throughput in this sub-band, a higher-order modulation mode was used in the higher quality sub-bands.

In the centre and bottom subfigures of Figure 1.3 the modulation mode chosen for each 32-subcarrier sub-band is shown versus time for two different high-speed wireless modems communicating at either 3.4 or 7.0 Mbps, respectively, again, corresponding to an average throughput of either 1 or 2 BPS.

However, these adaptive transceiver principles are not limited to OFDM transmissions. In recent years the concept of intelligent multi-mode, multimedia transceivers (IMMT) has emerged in the context of a variety of wireless systems [2–7]. The range of various existing solutions that have found favour in already operational standard systems has been summarised in the excellent overview by Nanda *et al.* [5]. *The aim of these adaptive transceivers is to provide mobile users with the best possible compromise amongst a number of contradicting design factors, such as the power consumption of the hand-held portable station (PS), robustness against transmission errors, spectral efficiency, teletraffic capacity, audio/video quality and so forth [4].*

2) Another design alternative applicable in the context of OFDM systems is that the channel quality fluctuations observed, for example, in Figure 1.1 are averaged out with the aid of frequency-domain spreading codes, which leads to the concept of Multi-Carrier Code Division Multiple Access (MC-CDMA). In this scenario typically only a few chips of the spreading code are obliterated by the frequency-selective fading and hence the chances are that the spreading code and its conveyed data may still be recoverable. The advantage of this approach is that in contrast to AOFDM-based communications, in MC-CDMA no channel quality estimation and signalling are required. Therefore OFDM and MC-CDMA will be comparatively studied in Part II of this monograph. Part III will also consider the employment of Walsh-Hadamard code-based spreading of each subcarrier's signal across the entire OFDM bandwidth, which was found to be an efficient frequency-domain fading counter-measure capable of operating without the employment of adaptive modulation.

3) A further techique capable of mitigating the channel quality fluctuations of wireless channels is constituted by space-time coding, which will also be considered as an attractive anti-fading design option capable of attaining a high diversity gain. Space-time

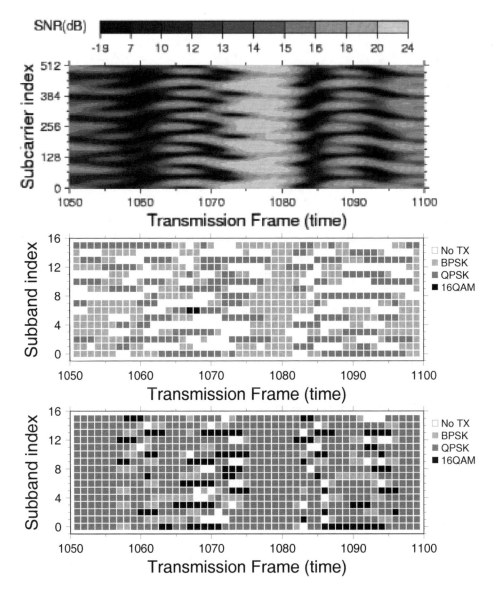

Figure 1.3: The micro-adaptive nature of the sub-band-adaptive OFDM modem. The top graph is a contour plot of the channel SNR for all 512 subcarriers versus time. The bottom two graphs show the modulation modes chosen for all 16 32-subcarrier sub-bands for the same period of time. The middle graph shows the performance of the 3.4 Mbps sub-band-adaptive modem, which operates at the same bit rate as a fixed BPSK modem. The bottom graph represents the 7.0 Mbps sub-band-adaptive modem, which operated at the same bit rate as a fixed QPSK modem. The average channel SNR was 16 dB.

coding employs several transmit and receive antennas for the sake of achieving diversity gain and hence an improved performance.

4) By contrast, in Part III of the book we employ multiple antennas at the base-station for a different reason, namely for the sake of supporting multiple users, rather than to achieving transmit diversity gain. This is possible, since the users' channel impulse responses (CIR) or channel transfer functions are accurately estimated and hence these channel transfer functions may be viewed as unique user signature sequences, which allow us to recognise and demultiplex the transmissions of the individual users, in a similar fashion to the unique user-specific spreading codes employed in CDMA systems. We note, however, that this technique is only capable of reliably separating the users communicating within the same bandwidth, if their CIRs are sufficiently different. This assumption is typically valid for the uplink, although it may have a limited validity, when the base station receives from mobile stations in its immediate vicinity. By contrast, different techniques have to be invoked for downlink multi-user transmissions.

Our intention with the book is:

1) First, to pay tribute to all researchers, colleagues and valued friends, who contributed to the field. Hence this book is dedicated to them, since without their quest for better transmission solutions for wireless communications this monograph could not have been conceived. They are too numerous to name here, hence they appear in the author index of the book. Our hope is that the conception of this monograph on the topic will provide an adequate portrayal of the community's research and will further fuel this innovation process.

2) We expect to stimulate further research by exposing open research problems and by collating a range of practical problems and design issues for the practitioners. The coherent further efforts of the wireless research community is expected to lead to the solution of the range of outstanding problems, ultimately providing us with flexible wireless transceivers exhibiting a performance close to information theoretical limits.

1.2 Orthogonal Frequency Division Multiplexing History

1.2.1 Early Classic Contributions

The first OFDM scheme was proposed by Chang in 1966 [1] for dispersive fading channels. During the early years of the evolution of OFDM research the contributions due to the efforts of Weinstein, Peled, Ruiz, Hirosaki, Kolb, Cimini, Schüssler, Preuss, Rückriem, Kalet *et al.* [1, 8–20] have to be mentioned. As unquestionable proof of its maturity, OFDM was standardised as the European digital audio broadcast (DAB) as well as digital video broadcast (DVB) scheme. It constituted also a credible proposal for the recent third-generation mobile radio standard competition in Europe. Finally, OFDM was recently selected as the high performance local area network's (HIPERLAN) transmission technique as well as becoming part of the IEEE 802.11 Wireless Local Area Network (WLAN) standard.

The system's operational principle is that the original bandwidth is divided into a high number of narrow sub-bands, in which the mobile channel can be considered non-dispersive. Hence no channel equaliser is required and instead of implementing a bank of sub-channel modems they can be conveniently implemented with the aid of a single Fast Fourier Transformer (FFT), as it will be outlined in Chapter 2.

These OFDM systems - often also termed frequency division multiplexing (FDM) or multi-tone systems - have been employed in military applications since the 1960s, for example by Bello [21], Zimmermann [8], Powers and Zimmerman [22], Chang and Gibby [23] and others. Saltzberg [24] studied a multi-carrier system employing orthogonal time–staggered quadrature amplitude modulation (O-QAM) of the carriers.

The employment of the discrete Fourier transform (DFT) to replace the banks of sinusoidal generators and the demodulators was suggested by Weinstein and Ebert [9] in 1971, which significantly reduces the implementation complexity of OFDM modems. In 1980, Hirosaki [20] suggested an equalisation algorithm in order to suppress both intersymbol and intersubcarrier interference caused by the channel impulse response or timing and frequency errors. Simplified OFDM modem implementations were studied by Peled [13] in 1980, while Hirosaki [14] introduced the DFT-based implementation of Saltzberg's O-QAM OFDM system. From Erlangen University, Kolb [15], Schüssler [16], Preuss [17] and Rückriem [18] conducted further research into the application of OFDM. Cimini [10] and Kalet [19] published analytical and early seminal experimental results on the performance of OFDM modems in mobile communications channels.

More recent advances in OFDM transmission were presented in the impressive state-of-the-art collection of works edited by Fazel and Fettweis [25], including the research by Fettweis *et al.* at Dresden University, Rohling *et al.* at Braunschweig University, Vandendorp at Loeven University, Huber *et al.* at Erlangen University, Lindner *et al.* at Ulm University, Kammeyer *et al.* at Bremen University and Meyr *et al.* [26, 27] at Aachen University, but the individual contributions are too numerous to mention. Important recent references are the books by van Nee and Prasad [28] as well as by Vandenameele, van der Perre and Engels [29].

While OFDM transmission over mobile communications channels can alleviate the problem of multipath propagation, recent research efforts have focused on solving a set of inherent difficulties regarding OFDM, namely the peak-to-mean power ratio, time and frequency synchronisation, and on mitigating the effects of the frequency selective fading channel. These issues are addressed below with reference to the literature, while a more in-depth treatment is given throughout the book.

1.2.2   Peak-to-mean Power Ratio

It is plausible that the OFDM signal - which is the superposition of a high number of modulated sub-channel signals - may exhibit a high instantaneous signal peak with respect to the average signal level. Furthermore, large signal amplitude swings are encountered, when the time domain signal traverses from a low instantaneous power waveform to a high power waveform, which may results in a high out-of-band (OOB) harmonic distortion power, unless the transmitter's power amplifier exhibits an extremely high linearity across the entire signal level range. This then potentially contaminates the adjacent channels with adjacent channel interference. Practical amplifiers exhibit a finite amplitude range, in which they can be considered almost linear. In order to prevent severe clipping of the high OFDM signal peaks - which is the main source of OOB emissions - the power amplifier must not be driven to saturation and hence they are typically operated with a certain so-called back-off, creating a certain "head room" for the signal peaks, which reduces the risk of amplifier saturation and OOB emission. Two different families of solutions have been suggested in the literature, in order to mitigate these problems, either reducing the peak-to-mean power ratio, or improving the amplification stage of the transmitter.

More explicitly, Shepherd [30], Jones [31], and Wulich [32] have suggested different coding techniques which aim to minimise the peak power of the OFDM signal by employing different data encoding schemes before modulation, with the philosophy of choosing block codes whose legitimate code words exhibit low so-called crest factors or peak-to-mean power envelope fluctuation. Müller [33], Pauli [34], May [35] and Wulich [36] suggested different algorithms for post-processing the time domain OFDM signal prior to amplification, while Schmidt and Kammeyer [37] employed adaptive subcarrier allocation in order to reduce the crest factor. Dinis and Gusmão [38–40] researched the use of two-branch amplifiers, while the clustered OFDM technique introduced by Daneshrad, Cimini and Carloni [41] operates with a set of parallel partial FFT processors with associated transmitting chains. OFDM systems with increased robustness to non-linear distortion have been proposed by Okada, Nishijima and Komaki [42] as well as by Dinis and Gusmão [43]. These aspects of OFDM transmissions will be treated in substantial depth in Part II of the book.

1.2.3 Synchronisation

Time and frequency synchronisation between the transmitter and receiver are of crucial importance as regards the performance of an OFDM link [44, 45]. A wide variety of techniques have been proposed for estimating and correcting both timing and carrier frequency offsets at the OFDM receiver. Rough timing and frequency acquisition algorithms relying on known pilot symbols or pilot tones embedded into the OFDM symbols have been suggested by Claßen [26], Warner [46], Sari [47], Moose [48], as well as Brüninghaus and Rohling [49]. Fine frequency and timing tracking algorithms exploiting the OFDM signal's cyclic extension were published by Moose [48], Daffara [50] and Sandell [51]. OFDM synchronisation issues are the topics of Chapter 5.

1.2.4 OFDM/CDMA

Combining multi-carrier OFDM transmissions with code division multiple access (CDMA) allows us to exploit the wideband channel's inherent frequency diversity by spreading each symbol across multiple subcarriers. This technique has been pioneered by Yee, Linnartz and Fettweis [52], by Chouly, Brajal and Jourdan [53], as well as by Fettweis, Bahai and Anvari [54]. Fazel and Papke [55] investigated convolutional coding in conjunction with OFDM/CDMA. Prasad and Hara [56] compared various methods of combining the two techniques, identifying three different structures, namely multi-carrier CDMA (MC-CDMA), multi-carrier direct sequence CDMA (MC-DS-CDMA) and multi-tone CDMA (MT-CDMA). Like non-spread OFDM transmission, OFDM/CDMA methods suffer from high peak-to-mean power ratios, which are dependent on the frequency domain spreading scheme, as investigated by Choi, Kuan and Hanzo [57]. Part II of the book considers the related design trade-offs.

1.2.5   Decision-Directed Channel Estimation

In recent years numerous research contributions have appeared on the topic of channel transfer function estimation techniques designed for employment in single-user, single transmit antenna-assisted OFDM scenarios, since the availability of an accurate channel transfer function estimate is one of the prerequisites for coherent symbol detection with an OFDM receiver. The techniques proposed in the literature can be classified as *pilot-assisted*, *decision-directed* (DD) and *blind* channel estimation (CE) methods.

In the context of pilot-assisted channel transfer function estimation a subset of the available subcarriers is dedicated to the transmission of specific pilot symbols known to the receiver, which are used for "sampling" the desired channel transfer function. Based on these samples of the frequency domain transfer function, the well-known process of interpolation is used for generating a transfer function estimate for each subcarrier residing between the pilots. This is achieved at the cost of a reduction in the number of useful subcarriers available for data transmission. The family of *pilot-assisted* channel estimation techniques was investigated for example by Chang and Su [82], Höher [58, 66, 67], Itami *et al.* [71], Li [74], Tufvesson and Maseng [65], Wang and Liu [77], as well as Yang *et al.* [73, 78, 84].

By contrast, in the context of Decision-Directed Channel Estimation (DDCE) all the sliced and remodulated subcarrier data symbols are considered as pilots. In the absence of symbol errors and also depending on the rate of channel fluctuation, it was found that accurate channel transfer function estimates can be obtained, which often are of better quality, in terms of the channel transfer function estimator's mean-square error (MSE), than the estimates offered by pilot-assisted schemes. This is because the latter arrangements usually invoke relatively sparse pilot patterns.

The family of *decision-directed* channel estimation techniques was investigated for example by van de Beek *et al.* [61], Edfors *et al.* [62, 69], Li *et al.* [68], Li [80], Mignone and Morello [64], Al-Susa and Ormondroyd [72], Frenger and Svensson [63], as well as Wilson *et al.* [60]. Furthermore, the family of *blind* channnel estimation techniques was studied by Lu and Wang [79], Necker and Stüber [83], as well as by Zhou and Giannakis [76]. The various contributions have been summarized in Tables 1.1 and 1.2.

In order to render the various DDCE techniques more amenable to use in scenarios associated with a relatively high rate of channel variation expressed in terms of the OFDM symbol normalized Doppler frequency, linear prediction techniques well known from the speech coding literature [85, 86] can be invoked. To elaborate a little further, we will substitute the CIR-related tap estimation filter - which is part of the two-dimensional channel transfer function estimator proposed in [68] - by a CIR-related tap prediction filter. The employment of this CIR-related tap prediction filter enables a more accurate estimation of the channel transfer function encountered during the forthcoming transmission time slot and thus potentially enhances the performance of the channel estimator. We will be following the general concepts described by Duel-Hallen *et al.* [87] and the ideas presented by Frenger and Svensson [63], where frequency domain prediction filter-assisted DDCE was proposed. Furthermore, we should mention the contributions of Tufvesson *et al.* [70, 88], where a prediction filter-assisted frequency domain pre-equalisation scheme was discussed in the context of OFDM. In a fur-

Year	Author	Contribution
'91	Höher [58]	Cascaded 1D-FIR channel transfer factor interpolation was carried out in the frequency- and time-direction for frequency-domain PSAM.
'93	Chow, Cioffi and Bingham [59]	Subcarrier-by-subcarrier-based LMS-related channel transfer factor equalisation techniques were employed.
'94	Wilson, Khayata and Cioffi [60]	Linear channel transfer factor filtering was invoked in the time-direction for DDCE.
'95	van de Beek, Edfors, Sandell, Wilson and Börjesson [61]	DFT-aided CIR-related domain Wiener filter-based noise reduction was advocated for DDCE. The effects of leakage in the context of non-sample-spaced CIRs were analysed.
'96	Edfors, Sandell, van de Beek, Wilson and Börjesson [62]	SVD-aided CIR-related domain Wiener filter-based noise reduction was introduced for DDCE.
	Frenger and Svensson [63]	MMSE-based frequency-domain channel transfer factor prediction was proposed for DDCE.
	Mignone and Morello [64]	FEC was invoked for improving the DDCE's remodulated reference.
'97	Tufvesson and Maseng [65]	An analysis of various pilot patterns employed in frequency-domain PSAM was provided in terms of the system's BER for different Doppler frequencies. Kalman filter-aided channel transfer factor estimation was used.
	Höher, Kaiser and Robertson [66, 67]	Cascaded 1D-FIR Wiener filter channel interpolation was utilised in the context of 2D-pilot pattern-aided PSAM
'98	Li, Cimini and Sollenberger [68]	An SVD-aided CIR-related domain Wiener filter-based noise reduction was achieved by employing CIR-related tap estimation filtering in the time-direction.
	Edfors, Sandell, van de Beek, Wilson and Börjesson [69]	A detailed analysis of SVD-aided CIR-related domain Wiener filter-based noise reduction was provided for DDCE, which expanded the results of [62].
	Tufvesson, Faulkner and Maseng [70]	Wiener filter-aided frequency domain channel transfer factor prediction-assisted pre-equalisation was studied.
	Itami, Kuwabara, Yamashita, Ohta and Itoh [71]	Parametric finite-tap CIR model-based channel estimation was employed for frequency domain PSAM.

Table 1.1: Contributions to channel transfer factor estimation for single-transmit antenna-assisted OFDM.

Year	Author	Contribution
'99	Al-Susa and Ormondroyd [72]	DFT-aided Burg algorithm-assisted adaptive CIR-related tap prediction filtering was employed for DDCE.
	Yang, Letaief, Cheng and Cao [73]	Parametric, ESPRIT-assisted channel estimation was employed for frequency domain PSAM.
'00	Li [74]	Robust 2D frequency domain Wiener filtering was suggested for employment in frequency domain PSAM using 2D pilot patterns.
'01	Yang, Letaief, Cheng and Cao [75]	Detailed discussions of parametric, ESPRIT-assisted channel estimation were provided in the context of frequency domain PSAM [73].
	Zhou and Giannakis [76]	Finite alphabet-based channel transfer factor estimation was proposed.
	Wang and Liu [77]	Polynomial frequency domain channel transfer factor interpolation was contrived.
	Yang, Cao and Letaief [78]	DFT-aided CIR-related domain one-tap Wiener filter-based noise reduction was investigated, which is supported by variable frequency domain Hanning windowing.
	Lu and Wang [79]	A Bayesian blind turbo receiver was contrived for coded OFDM systems.
	Li and Sollenberger [80]	Various transforms were suggested for CIR-related tap estimation filtering-assisted DDCE.
	Morelli and Mengali [81]	LS- and MMSE-based channel transfer factor estimators were compared in the context of frequency domain PSAM.
'02	Chang and Su [82]	Parametric quadrature surface-based frequency domain channel transfer factor interpolation was studied for PSAM.
	Necker and Stüber [83]	Totally blind channel transfer factor estimation based on the finite alphabet property of PSK signals was investigated.

Table 1.2: Contributions to channel transfer factor estimation for single-transmit antenna-assisted OFDM.

ther contribution by Al-Susa and Ormondroyd [72], adaptive prediction filter-assisted DDCE designed for OFDM has been proposed upon invoking techniques known from speech coding, such as the Levinson-Durbin algorithm or the Burg algorithm [85, 89, 90] in order to determine the predictor coefficients.

In contrast to the above-mentioned single-user OFDM scenarios, in a multi-user OFDM scenario the signal received by each antenna is constituted by the superposition of the signal contributions associated with the different users or transmit antennas. Note that in terms of the multiple-input multiple-output (MIMO) structure of the channel the multi-user single-transmit antenna scenario is equivalent, for example, to a single-user space-time coded (STC) scenario using multiple transmit antennas. For the latter a Least-Squares (LS) error channel estimator was proposed by Li *et al.* [91], which aims at recovering the different transmit antennas' channel transfer functions on the basis of the output signal of a specific reception antenna element and by also capitalising on the remodulated received symbols associated with the different users. The performance of this estimator was found to be limited in terms of the mean-square estimation error in scenarios, where the product of the number of transmit antennas and the number of CIR taps to be estimated per transmit antenna approaches the total number of subcarriers hosted by an OFDM symbol. As a design alternative, in [92] a DDCE was proposed by Jeon *et al.* for a space-time coded OFDM scenario of two transmit antennas and two receive antennas.

Specifically, the channel transfer function[2] associated with each transmit-receive antenna pair was estimated on the basis of the output signal of the specific receive antenna upon *subtracting* the interfering signal contributions associated with the remaining transmit antennas. These interference contributions were estimated by capitalising on the knowledge of the channel transfer functions of all interfering transmit antennas predicted during the $(n-1)$-th OFDM symbol period for the n-th OFDM symbol, also invoking the corresponding remodulated symbols associated with the n-th OFDM symbol. To elaborate further, the difference between the subtraction-based channel transfer function estimator of [92] and the LS estimator proposed by Li *et al.* in [91] is that in the former the channel transfer functions predicted during the previous, i.e. the $(n-1)$-th OFDM symbol period for the current, i.e. the n-th OFDM symbol are employed for both symbol detection *as well as* for obtaining an updated channel estimate for employment during the $(n+1)$-th OFDM symbol period. In the approach advocated in [92] the subtraction of the different transmit antennas' interfering signals is performed in the frequency domain.

By contrast, in [93] a similar technique was proposed by Li with the aim of simplifying the DDCE approach of [91], which operates in the time domain. A prerequisite for the operation of this parallel interference cancellation (PIC)-assisted DDCE is the availability of a reliable estimate of the various channel transfer functions for the current OFDM symbol, which are employed in the cancellation process in order to obtain updated channel transfer function estimates for the demodulation of the next OFDM symbol. In order to compensate for the channel's variation as a function of the OFDM symbol index, linear prediction techniques can be employed, as it was also proposed for example in [93]. However, due to the estimator's recursive structure, determining the optimum predictor coefficients is not as straightforward as for the transversal FIR filter-assisted predictor as described in Section 15.2.4 for the single-user DDCE.

---

[2]In the context of the OFDM system the set of K different subcarriers' channel transfer factors is referred to as the channel transfer function, or simply as the channel.

Year	Author	Contribution
'99	Li, Seshadri and Ariyavisitakul [91]	The LS-assisted DDCE proposed exploits the cross-correlation properties of the transmitted subcarrier symbol sequences.
'00	Jeon, Paik and Cho [92]	Frequency-domain PIC-assisted DDCE is studied, which exploits the channel's slow variation versus time.
	Li [93]	Time-domain PIC-assisted DDCE is investigated as a simplification of the LS-assisted DDCE of [91]. Optimum training sequences are proposed for the LS-assisted DDCE of [91].
'01	Mody and Stüber [94]	Channel transfer factor estimation designed for frequency-domain PSAM based on CIR-related domain filtering is studied.
	Gong and Letaief [95]	MMSE-assisted DDCE is advocated which represents an extension of the LS-assisted DDCE of [95]. The MMSE-assisted DDCE is shown to be practical in the context of transmitting consecutive training blocks. Additionally, a low-rank approximation of the MMSE-assisted DDCE is considered.
	Jeon, Paik and Cho [96]	2D MMSE-based channel estimation is proposed for frequency-domain PSAM.
	Vook and Thomas [97]	2D MMSE based channel estimation is invoked for frequency domain PSAM. A complexity reduction is achieved by CIR-related domain-based processing.
	Xie and Georghiades [98]	Expectation maximization (EM) based channel transfer factor estimation approach for DDCE.
'02	Li [99]	A more detailed discussion on time-domain PIC-assisted DDCE is provided and optimum training sequences are proposed [93].
	Bölcskei, Heath and Paulraj [100]	Blind channel identification and equalisation using second-order cyclostationary statistics as well as antenna precoding were studied.
	Minn, Kim and Bhargava [101]	A reduced complexity version of the LS-assisted DDCE of [91] is introduced, based on exploiting the channel's correlation in the frequency-direction, as opposed to invoking the simplified scheme of [99], which exploits the channel's correlation in the time-direction. A similar approach was suggested by Slimane [102] for the specific case of two transmit antennas.
	Komninakis, Fragouli, Sayed and Wesel [103]	Fading channel tracking and equalisation were proposed for employment in MIMO systems assisted by Kalman estimation and channel prediction.

Table 1.3: Contributions on channel transfer factor estimation for multiple-transmit antenna assisted OFDM.

A comprehensive overview of further publications on channel transfer factor estimation for OFDM systems supported by multiple transmit antennas is provided in Table 1.3.

1.2.6 Uplink Detection Techniques for Multi-User SDMA-OFDM

Combining adaptive antenna-aided techniques with OFDM transmissions was shown to be advantageous for example in the context of suppressing co-channel interference in cellular communications systems. Amongst others, Li, Cimini and Sollenberger [129–131], Kim, Choi and Cho [132], Lin, Cimini and Chuang [133] as well as Münster *et al.* [134] have investigated algorithms designed for multi-user channel estimation and interference suppression.

The related family of Space-Division-Multiple-Access (SDMA) communication systems has recently drawn wide reseach interests. In these systems the L different users' transmitted signals are separated at the base-station (BS) with the aid of their unique, user-specific spatial signature, which is constituted by the P-element vector of channel transfer factors between the users' single transmit antenna and the P different receiver antenna elements at the BS, upon assuming flat-fading channel conditions such as those often experienced in the context of each of the OFDM subcarriers.

A whole host of multi-user detection (MUD) techniques known from Code-Division-Multiple-Access (CDMA) communications lend themselves also to an application in the context of SDMA-OFDM on a per-subcarrier basis. Some of these techniques are the Least-Squares (LS) [113, 119, 127, 135], Minimum Mean-Square Error (MMSE) [105–108, 110, 113, 117, 121, 135–137], Successive Interference Cancellation (SIC) [104, 109, 113, 117, 119, 124, 126, 128, 135, 137], Parallel Interference Cancellation (PIC) [125, 135] and Maximum Likelihood (ML) detection [112, 114–118, 120, 123, 135, 137]. A comprehensive overview of recent publications on MUD techniques for MIMO systems is given in Tables 1.4 and 1.5.

1.2.7 OFDM Applications

Due to their implementational complexity, OFDM applications have been scarce until quite recently. Recently, however, OFDM has been adopted as the new European digital audio broadcasting (DAB) standard [11, 12, 138–140] as well as for the terrestrial digital video broadcasting (DVB) system [47, 141].

For fixed-wire applications, OFDM is employed in the asynchronous digital subscriber line (ADSL) and high-bit-rate digital subscriber line (HDSL) systems [142–145] and it has also been suggested for power line communications systems [146, 147] due to its resilience to time dispersive channels and narrow band interferers.

More recently, OFDM applications were studied within the European 4th Framework Advanced Communications Technologies and Services (ACTS) programme [148]. The ME-DIAN project investigated a 155 Mbps wireless asynchronous transfer mode (WATM) network [149–152], while the Magic WAND group [153, 154] developed a wireless local area network (LAN). Hallmann and Rohling [155] presented a range of different OFDM systems that were applicable to the European Telecommunications Standardisation Institute's (ETSI) recent personal communications oriented air interface concept [156].

Year	Author	Contribution
'96	Foschini [104]	The concept of the BLAST architecture was introduced.
'98	Vook and Baum [105]	SMI-assisted MMSE combining was invoked on an OFDM subcarrier basis.
	Wang and Poor [106]	Robust sub-space-based weight vector calculation and tracking were employed for co-channel interference suppression, as an improvement of the SMI-algorithm.
	Wong, Cheng, Letaief and Murch [107]	Optimization of an OFDM system was reported in the context of multiple transmit and receive antennas upon invoking the maximum SINR criterion. The computational was reduced by exploiting the channel's correlation in the frequency direction.
	Li and Sollenberger [108]	Tracking of the channel correlation matrix' entries was suggested in the context of SMI-assisted MMSE combining for multiple receiver antenna assisted OFDM, by capitalizing on the principles of [68].
'99	Golden, Foschini, Valenzuela and Wolniansky [109]	The SIC detection-assisted V-BLAST algorithm was introduced.
	Li and Sollenberger [110]	The system introduced in [108] was further detailed.
	Vandenameele, Van der Perre, Engels and de Man [111]	A comparative study of different SDMA detection techniques, namely that of MMSE, SIC and ML detection was provided. Further improvements of SIC detection were suggested by adaptively tracking multiple symbol decisions at each detection node.
	Speth and Senst [112]	Soft-bit generation techniques were proposed for MLSE in the context of a coded SDMA-OFDM system.
'00	Sweatman, Thompson, Mulgrew and Grant [113]	Comparisons of various detection algorithms including LS, MMSE, D-BLAST and V-BLAST (SIC detection) were carried out.
	van Nee, van Zelst and Awater [114–116]	The evaluation of ML detection in the context of a Space-Division Multiplexing (SDM) system was provided, considering various simplified ML detection techniques.
	Vandenameele, Van der Perre, Engels, Gyselinckx and de Man [117]	More detailed discussions were provided on the topics of [111].

Table 1.4: Contributions on multi-user detection techniques designed for multiple transmit antenna assisted OFDM systems.

Year	Author	Contribution
'00	Li, Huang, Lozano and Foschini [118]	Reduced complexity ML detection was proposed for multiple transmit antenna systems employing adaptive antenna grouping and multi-step reduced-complexity detection.
'01	Degen, Walke, Lecomte and Rembold [119]	An overview of various adaptive MIMO techniques was provided. Specifically, pre-distortion was employed at the transmitter, as well as LS- or BLAST detection were used at the receiver or balanced equalisation was invoked at both the transmitter and receiver.
	Zhu and Murch [120]	A tight upper bound on the SER performance of ML detection was derived.
	Li, Letaief, Cheng and Cao [121]	Joint adaptive power control and detection were investigated in the context of an OFDM/SDMA system, based on the approach of Farrokhi *et al.* [122].
	van Zelst, van Nee and Awater [123]	Iterative decoding was proposed for the BLAST system following the turbo principle.
	Benjebbour, Murata and Yoshida [124]	The performance of V-BLAST or SIC detection was studied in the context of backward iterative cancellation scheme employed after the conventional forward cancellation stage.
	Sellathurai and Haykin [125]	A simplified D-BLAST was proposed, which used iterative PIC capitalizing on the extrinsic soft-bit information provided by the FEC scheme used.
	Bhargave, Figueiredo and Eltoft [126]	A detection algorithm was suggested, which followed the concepts of V-BLAST or SIC. However, multiple symbols states are tracked from each detection stage, where - in contrast to [117] - an intermediate decision is made at intermediate detection stages.
	Thoen, Deneire, Van der Perre and Engels [127]	A constrained LS detector was proposed for OFDM/SDMA, which was based on exploiting the constant modulus property of PSK signals.
'02	Li and Luo [128]	The block error probability of optimally ordered V-BLAST was studied. Furthermore, the block error probability is also investigated for the case of tracking multiple parallel symbol decisions from the first detection stage, following an approach similar to that of [117].

Table 1.5: Contributions on detection techniques for MIMO systems and for multiple transmit antenna assisted OFDM systems.

1.3   Outline of the Book

- **Chapter 2**: In this chapter we commence our detailed discourse by demonstrating that OFDM modems can be efficiently implemented by invoking the Fourier transform or the fast Fourier Transform (FFT). A number of basic OFDM design issues are discussed in an accessible style.

- **Chapter 3**: The BER performance of OFDM modems achievable in AWGN channels is studied for a set of different modulation schemes in the subcarriers. The effects of amplitude limiting of the transmitter's output signal, caused by a simple clipping amplifier model, and of finite resolution D/A and A/D conversion on the system performance are investigated. Oscillator phase noise is considered as a source of intersubcarrier interference and its effects on the system performance are demonstrated.

- **Chapter 4**: The effects of time-dispersive frequency-selective Rayleigh fading channels on OFDM transmissions are demonstrated. Channel estimation techniques are presented which support the employment of coherent detection in frequency selective channels. Additionally, differential detection is investigated, and the resultant system performance is compared, when communicating over various channels.

- **Chapter 5**: We focus our attention on the time and frequency synchronisation requirements of OFDM transmissions and the effects of synchronisation errors are demonstrated. Two novel synchronisation algorithms for frame and OFDM symbol synchronisation are suggested and compared. The resulting system performance over fading wideband channels is examined.

- **Chapter 6**: Based on the results of Chapter 4, the employment of adaptive modulation schemes is suggested for duplex point-to-point links over frequency-selective time-varying channels. Different bit allocation schemes are investigated and a simplified sub-band adaptivity OFDM scheme is suggested to alleviate the associated signalling constraints. A range of blind modulation scheme detection algorithms are also investigated and compared. The employment of long-block-length convolutional turbo codes is suggested for improving the system's throughput and the turbo coded adaptive OFDM modem's performance is compared using different sets of parameters. Then the effects of using pre-equalisation at the transmitter are examined, and a set of different pre-equalisation algorithms is introduced. A joint pre-equalisation and adaptive modulation algorithm is proposed and its BER and throughput performance are studied.

- **Chapter 7**: The adaptive OFDM transmission ideas of Chapter 6 are extended further, in order to include adaptive error correction coding, based on redundant residual number system (RRNS) and turbo BCH codes. A joint modulation and code rate adaptation scheme is presented.

- **Chapter 8**: The discussions of **Part II** of the book commence by a rudimentary comparison of OFDM, CDMA and MC-CDMA in Chapter 8.

- **Chapter 9**: Since the properties of spreading sequences are equally important in both multicarrier CDMA and in DS-CDMA, the basic properties of various spreading sequences are reviewed in Chapter 9.

- **Chapter 10**: The basic characterisation of spreading codes provided in Chapter 9 is followed by Chapter 10, analysing the achievable performance of both single- and multi-user detected MC-CDMA.

- **Chapter 11**: One of the main problems associated with the implementation of multi-carrier communication systems is their high Peak-to-Mean Envelope-Power Ratio (PMEPR), requiring highly linear power amplifiers. Hence the envelope power of the MC-CDMA signal is analysed in Chapter 11. Several orthogonal spreading sequences are examined, in order to assess their ability to maintain low PMEPR of the transmitted signal. The effects of reduced PMEPR are investigated in terms of the BER performance and the out-of-band frequency spectrum.

- **Chapter 12**: An adaptive modulation assisted MC-CDMA technique is investigated using both analysis as well as simulation-based studies. Various techniques of optimising the modem mode switching thresholds are compared and the Lagrangian optimisation method is shown to provide the best solution. This method results in an SNR-dependent set of switching thresholds, because at high SNRs the activation of high-throughput modem modes may be promoted by lowering their respective activation thresholds, while still maintaining the target BER.

- **Chapter 13** deals with three different types of reduced complexity despreading aided MC-CDMA detection techniques that can be used at the receiver. Their BER performance as well as their achievable computational complexity reduction is characterised.

- **Chapter 14**: In **Part III** of the book our discussions are dedicated to the detailed design of channel transfer function estimation and to the conception of multi-user OFDM systems. We commence our discourse with the portrayal of two-dimensional pilot symbol-assisted channel estimation techniques in Chapter 14.

- **Chapter 15**: Our discussions in Chapter 15 continue by reviewing the concepts of 1D- and 2D-MMSE-based Decision-Directed Channel Estimators (DDCE) contrived for single-user OFDM, which was analysed for example by Edfors *et al.* [62, 69, 157], Sandell [158] and Li *et al.* [68]. Based on the observation that in the context of DDCE the most recently received OFDM symbol is equalisd based on a potentially outdated channel transfer function estimate, which was generated during the previous OFDM symbol period, the CIR-related tap *estimation filters* invoked by the 2D-MMSE DDCE proposed by Li *et al.* [68] are substituted by CIR-related tap *prediction filters*. For this configuration we derive the channel predictor's average *a priori* estimation MSE observed in the frequency-domain, as it will be shown in Section 15.2.4.6. In the context of our performance assessments provided in Section 15.3 two methods are compared against each other for evaluating the Wiener prediction filter's coefficients, namely the robust approach as advocated by Li *et al.* [68] and the adaptive approach as proposed by Al-Susa and Ormondroyd [72], based on the Burg algorithm known from speech coding. As a third and most promising alternative, the RLS algorithm could be employed, as proposed in the context of our discussions in Chapter 16.

  Following our in-depth performance assessment of the CIR-related tap prediction-assisted DDCE, both in the context of uncoded and turbo-coded scenarios, combining channel transfer factor prediction-assisted DDCE with AOFDM is proposed in Section 15.4. As a result of the channel transfer factor prediction, AOFDM is rendered attractive also in scenarios having a relatively high OFDM symbol-normalized Doppler frequency. These contributions were published in [159] and [160].

- **Chapter 16** commences in Section 16.4 with an in-depth discussion of Li's Least-Squares (LS) assisted DDCE, which was contrived for space-time coded OFDM systems [91], or more generally for OFDM systems employing multiple transmit antennas encountered in the context of multi-user SDMA-OFDM scenarios. Our contribution in this section is the provision of a sophisticated mathematical description, which provides the standard LS-related solution of the associated estimation problem. In the context of these discussions a necessary condition for the existence of the LS-channel estimates is identified. In contrast to Li's discussions in [91], which provided an expression for the estimation MSE based on sample-spaced CIRs, here we also derived expression for the estimation MSE assuming the more realistic scenario of a non-sample-spaced CIR.

  Motivated by the potentially excessive complexity of the LS-assisted DDCE [91] and by its limitation in terms of the number of users or transmit antennas supported, we focused our attention in Section 16.5 on the investigation of Parallel Interference Cancellation (PIC) assisted DDCE. We found that the PIC operations can either be performed in the CIR-related domain - as proposed by Li [93] - or in the frequency-domain as proposed by Jeon *et al.* [92]. We found that the frequency-domain PIC appears more advantageous in terms of its lower complexity. In contrast to the contributions of Li [93] and Jeon *et al.* [92], in Section 16.5 we provide an in-depth mathematical analysis with respect to a number of key points. Based on identifying the estimator's recursive structure, expressions are derived for the *a posteriori*- and *a priori* estimation MSE and conditions are provided for the estimator's stability. Furthermore, an iterative procedure is devised for the off-line calculation of the *a priori* predictor coefficients. In order to provide an improved flexibility for the PIC-assisted DDCE with respect to variations of the channel's statistics and in order to increase its resilience to impulsive noise imposed by the erroneous symbol decisions that may be encountered in the DDCE, the RLS algorithm is adopted for the task of predictor coefficient adaptation. Part of this work was published in [161] or it was accepted for publication [162].

- **Chapter 17**: Various linear- and non-linear multi-user detection techniques are compared against each other in terms of their applicability to the problem of detecting the subcarrier-based vectors of the L different users' transmitted symbols encountered in a multi-user SDMA-OFDM system. Specifically, the Least-Squares Error (LSE) and Minimum Mean-Square Error (MMSE) schemes are studied in Section 17.2, while the Successive Interference Cancellation (SIC) scheme and its derivatives, such as M-SIC and partial M-SIC, as well as the Parallel Interference Cancellation (PIC) and the Maximum Likelihood (ML) detection algorithms are considered in Section 17.3. These sections include the different techniques' mathematical derivation, as well as their performance and complexity analysis.

  Specifically, in the context of SIC detection our investigations of the effects of error propagation occurring through the different detection stages should be emphasised. Motivated by the results of these investigations an improved metric is proposed for the generation of soft-bit values suitable for turbo-decoding. Furthermore, our contributions contrived for hard-decision based turbo-decoding assisted PIC for SDMA-OFDM should be mentioned here, which was published in [163].

Based on the observation that SIC detection as the second-best performing detection approach after ML detection exhibits a potentially high complexity, further investigations are conducted in Section 17.4 for the sake of enhancing the performance of both MMSE and PIC detection. Specifically, in Section 17.4.1 we discuss the employment of adaptive modulation in the context of an SDMA-OFDM scenario, which is hence termed SDMA-AOFDM. This scheme will be shown to be effective in the context of an almost "fully-loaded" system, where the L number of users supported approaches the P number of reception antennas employed at the base station. This novel scheme was disseminated in [134].

However, the use of the SDMA-AOFDM scheme is restricted to duplex transmission scenarios having a reverse link, such as those in Time Division Duplexing (TDD). An enhancement of the MMSE- and PIC-assisted detection schemes' performance was achieved with the aid of orthogonal Walsh-Hadamard spreading codes, which will be investigated in Section 17.4.2. These advances were published in [160].

- **Chapter 18**: In the penultimate chapter we provide a few application examples based on powerful turbo-coded OFDM systems invoked for the wireless transmission of video telephone signals. In the first system design example turbo-coded adaptive OFDM video transmissions are considered, while in the second system various multi-user detection schemes are invoked and their achievable transmission integrity and video performance is quantified as a function of the number of users supported.

- **Chapter 19**: In this chapter we offer detailed conclusions and highlight a range of further research problems.

1.4 Chapter Summary and Conclusion

Here we conclude our brief introduction to OFDM and the review of its evolution since its conception by Chang in 1966 [1]. Numerous seminal contributions have been reviewed in chronological order in Tables 1.1–1.5, highlighting the historical development of the subject. These contributions reflect the state of the art at the time of writing in the context of the various OFDM system components, outlining a number of open research topics. Let us now embark on a detailed investigation of the topics introduced in this chapter.

Throughout this monograph we endeavour to highlight the range of contradictory system design trade-offs associated with the conception of OFDM and MC-CDMA systems. We intend to present the material in an unbiased fashion and sufficiently richly illustrated in terms of the associated design trade-offs so that readers will be able to find recipes and examples for solving their own particular wireless communications problems. In this rapidly evolving field it is a challenge to complete a timely, yet self-contained treatise, since new advances are discovered at an accelerating pace, which should find their way into a timely monograph. Our sincere hope is that you, the readers, will find the book a useful source of information, but above all a catalyst for further research.

Part I

OFDM System Design

Chapter 2

Introduction to Orthogonal Frequency Division Multiplexing[1]

2.1 Introduction

In this introductory chapter we examine orthogonal frequency division multiplexing (OFDM) as a means of counteracting the channel-induced linear distortions encountered when transmitting over a dispersive radio channel. The fundamental principle of orthogonal multiplexing originates from Chang [1], and over the years a number of researchers have investigated this technique [8–10, 13–18, 22–24]. Despite its conceptual elegance, until recently its deployment has been mostly limited to military applications due to implementational difficulties. However, it has recently been adopted as the new European digital audio broadcasting (DAB) standard, and this consumer electronics application underlines its significance as a broadcasting technique [11, 12, 19, 138, 139].

In the OFDM scheme of Figure 2.1 the serial data stream of a traffic channel is passed through a serial-to-parallel convertor which splits the data into a number of parallel channels. The data in each channel is applied to a modulator, such that for N channels there are N modulators whose carrier frequencies are f_0, f_1, ..., f_{N-1}. The difference between adjacent channels is Δf and the overall bandwidth W of the N modulated carriers is $N\Delta f$.

These N modulated carriers are then combined to give an OFDM signal. We may view the serial-to-parallel convertor as applying every Nth symbol to a modulator. This has the effect of interleaving the symbols into each modulator, e.g. symbols S_0, S_N, S_{2N}, ... are applied to the modulator whose carrier frequency is f_1. At the receiver the received OFDM signal is demultiplexed into N frequency bands, and the N modulated signals are demodulated. The baseband signals are then recombined using a parallel-to-serial convertor.

In the more conventional serial transmission approach [164], the traffic data is applied directly to the modulator transmitting at a carrier frequency positioned at the centre of the

[1]*OFDM and MC-CDMA for Broadband Multi-user Communications WLANS and Broadcasting.*
L.Hanzo, Münster, T. Keller and B.J. Choi,
©2003 John Wiley & Sons, Ltd. ISBN 0-470-85879-6

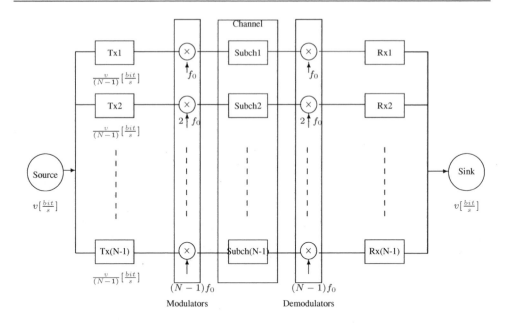

Figure 2.1: Simplified block diagram of the orthogonal parallel modem

transmission band $f_0, \ldots, f_{N-1}$, ie at $(f_{N-1} + f_0)/2$. The modulated signal occupies the entire bandwidth W. When the data is transmitted serially, the effect of a deep fade in a mobile channel is to cause a burst of transmission errors, if the fade extends over the duration of several bits. By contrast, during an N-symbol duration period of the conventional serial system, each of the N number of OFDM subchannel modulators carries only one symbol, each of which has an N times longer duration. Hence an identical-duration channel fade would only affect a fraction of the duration of each of the extended-length subcarrier symbols transmitted in parallel. Therefore the OFDM system may be able to recover all of the partially fading-contaminated N subcarrier symbols. Thus, while the serial system exhibits an error burst, no errors or few errors may occur using the OFDM approach.

A further advantage of OFDM is that because the symbol period has been increased, the channel's delay spread becomes a significantly shorter fraction of a symbol period than in the serial system, potentially rendering the system less sensitive to channel-induced dispersion, than the conventional serial system.

A disadvantage of the OFDM approach portrayed in Figure 2.1 is its increased complexity in comparison to a conventional serial modem, which is a consequence of employing N modulators and transmit filters at the transmitter and N demodulators and receive filters at the receiver. We will demonstrate in Section 2.3 that the associated complexity can be reduced by the employment of the discrete Fourier transform (DFT). When the number of subcarriers is high, the system's complexity may be further reduced by implementing the DFT with the aid of the Fast Fourier transform (FFT), again, as we will demonstrate in Section 2.3. The simple conceptual reason for this is that – as it will be argued in the context of Figure 2.2 in the next section – the OFDM sub-channel modulators use a set of harmonically related sinusoidal and cosinusoidal sub-channel carriers, just like the basis functions of the DFT.

2.2 Principles of QAM-OFDM

The simplest version of the basic OFDM system has N sub-bands, each separated from its neighbour by a sufficiently large guard band in order to prevent interference between signals in adjacent bands. However, the available spectrum can be used much more efficiently if the spectra of the individual sub-bands are allowed to overlap. By using coherent detection and orthogonal sub-band tones the original data can be correctly recovered.

In the system shown in Figure 2.2, the input serial data stream is rearranged into a sequence $\{d_n\}$ of N QAM symbols at baseband. Each serial QAM symbol is spaced by $\Delta t = 1/f_s$ where f_s is the serial signalling or symbol rate. At the nth symbol instant, the QAM symbol $d(n) = a(n) + jb(n)$ is represented by an in-phase component $a(n)$ and a quadrature component $b(n)$. A block of N QAM symbols is applied to a serial-to-parallel convertor and the resulting in-phase symbols $a(0)$, $a(1)$, ..., $a(N-1)$, and quadrature symbols $b(0)$, $b(1)$, ..., $b(N-1)$ are applied to N pairs of balanced modulators. The quadrature components $a(n)$ and $b(n)$, $n = 0, 1, ..., N-1$, modulate the quadrature carriers $\cos w_n t$ and $\sin w_n t$, respectively. Notice that the signalling interval of the sub-bands in the parallel system is N times longer than that of the serial system giving $T = N\Delta t$, which corresponds to an N-times lower signalling rate. The sub-band carrier frequencies $w_n = 2\pi n f_0$ are spaced apart by $w_0 = 2\pi/T$.

The modulated carriers $a(n)\cos w_n t$ and $b(n)\sin w_n t$ when added together constitute a QAM signal, and since $n = 0, 1, ..., N-1$, we have N QAM symbols at RF, where the nth QAM signal is given by:

$$
\begin{aligned}
X_n(t) &= a(n)\cos w_n t + b(n)\sin w_n t \\
&= \gamma(n)\cos(w_n t + \psi_n)
\end{aligned}
\tag{2.1}
$$

where we have

$$
\gamma(n) = \sqrt{a^2(n) + b^2(n)}
\tag{2.2}
$$

and

$$
\psi_n = \tan^{-1}\left(\frac{b(n)}{a(n)}\right) \quad n = 0, 1, ..., N-1.
\tag{2.3}
$$

By adding the outputs of each sub-channel signal $X_n(t)$ whose carriers are offset by $w_0 = 2\pi/T$ we obtain the FDM/QAM signal in the form of

$$
D(t) = \sum_{n=0}^{N-1} X_n(t).
\tag{2.4}
$$

This set of N FDM/QAM signals is transmitted over the mobile radio channel. At the receiver this OFDM signal is demultiplexed using a bank of N filters to regenerate the N QAM signals. The QAM baseband signals $a(n)$ and $b(n)$ are recovered and turned into serial form $\{d_n\}$. Recovery of the data ensues using the QAM baseband demodulator and differential decoding.

In theory, such a system is capable of achieving the maximum transmission rate of

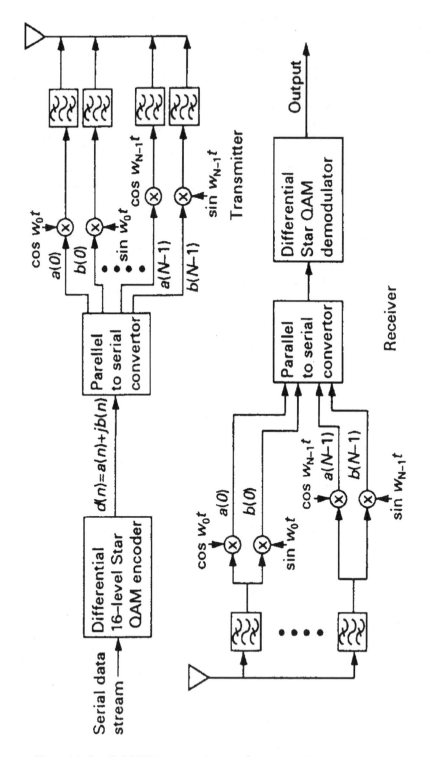

Figure 2.2: Detailed OFDM system schematic ©Hanzo, Webb, Keller, 2000, [164]

$log_2 Q \ bits/s/Hz$, where Q is the number of QAM levels. In practice, there is some spectral spillage due to adjacent frequency sub-bands which reduces this efficiency. Spectral spillage due to the sub-bands at the top and bottom of the overall frequency band requires a certain amount of guard space between adjacent users. Furthermore, spectral spillage between OFDM sub-bands due to the imperfections of each of the sub-band filters requires that the sub-bands be spaced further apart than the theoretically required minimum amount, decreasing spectral efficiency. In order to obtain the highest efficiency, the block size should be kept high and the sub-band filters made to meet stringent specifications.

One of the most attractive features of this scheme is that the bandwidth of the sub-channels is very narrow when compared to the communications channel's coherence bandwidth. Therefore, flat-fading narrowband propagation conditions apply. The sub-channel modems can use almost any modulation scheme, and QAM is an attractive choice in some situations.

2.3   Modulation by Discrete Fourier Transform [165, 166]

A fundamental problem associated with the OFDM scheme described is that in order to achieve high resilience against fades in the channels we consider, the block size, N, has to be of the order of 100, requiring a large number of sub-channel modems. Fortunately, it can be shown mathematically that taking the discrete Fourier transform (DFT) of the original block of N QAM symbols and then transmitting the DFT coefficients serially is exactly equivalent to the operations required by the OFDM transmitter of Figure 2.2. Substantial hardware simplifications can be made with OFDM transmissions if the bank of sub-channel modulators/demodulators is implemented using the computationally efficient pair of inverse fast Fourier transform and fast Fourier transform (IFFT/FFT).

The modulated signal $m(t)$ is given by

$$m(t) = \Re \left\{ b(t) e^{j 2\pi f_0 t} \right\}, \tag{2.5}$$

where $b(t)$ is the equivalent baseband information signal and f_0 is the carrier frequency, as introduced in Equations 2.6, 2.9 and 2.10. Using the rectangular full-response modulation elements $m_T(t - kT) = rect \frac{(t-kT)}{T}$ "weighted" by the complex QAM information symbols $X(k) = I(k) + jQ(k)$ to be transmitted, where $I(k)$ and $Q(k)$ are the quadrature components, the equivalent baseband information signal is given by:

$$b(t) = \sum_{k=-\infty}^{\infty} X(k) m_T(t - kT), \tag{2.6}$$

where k is the signalling interval index and T is its duration. On substituting Equation 2.6 into Equation 2.5 we have:

$$m(t) = \Re \left\{ \sum_{k=-\infty}^{\infty} X(k) m_T(t - kT) e^{j 2\pi f_0 t} \right\}. \tag{2.7}$$

Without loss of generality let us consider the signalling interval $k = 0$:

$$m_0(t) = m(t) rect \frac{t}{T}, \tag{2.8}$$

where adding the modulated signals of the sub-channel modulators yields

$$m_0(t) = \sum_{n=0}^{N-1} m_{0n}(t). \tag{2.9}$$

Observe that the stream of complex baseband symbols $X(k)$ to be transmitted can be described both in terms of in-phase $I(k)$ and quadrature phase $Q(k)$ components, as well as by magnitude and phase. In the case of a square-shaped QAM constellation [167] $X(k) = I(k) + jQ(k)$ might be a more convenient formalism; for the star QAM constellation introduced in [167], $X(k) = |X(k)|e^{j\Phi(k)}$ appears to be more attractive.

If $X_n = X_{n,k=0}$ is the complex baseband QAM symbol to be transmitted via sub-channel n in signalling interval $k = 0$, then

$$m_0(t) = \begin{cases} \sum_{n=0}^{N-1} \Re \left\{ X_{n,0} e^{j2\pi f_{0,n}t} \right\} & \text{for } |t| < \frac{T}{2}, \\ 0 & \text{otherwise.} \end{cases} \tag{2.10}$$

Bearing in mind that $m_0(t)$ is confined to the interval $|t| < \frac{T}{2}$ we drop the sampling interval index $k = 0$ and simplify our formalism to

$$m_0(t) = \sum_{n=0}^{N-1} \Re \left\{ X_n e^{j2\pi f_{0n}t} \right\}. \tag{2.11}$$

When computing the $\Re$ part using the complex conjugate we have:

$$\begin{aligned} m_0(t) &= \sum_{n=0}^{N-1} \frac{1}{2} \left\{ X_n e^{j2\pi f_{0n}t} + X_n{}^* e^{-j2\pi f_{0n}t} \right\} \\ &= \sum_{n=-(N-1)}^{N-1} \frac{1}{2} X_n e^{j2\pi f_{0n}t}, \end{aligned} \tag{2.12}$$

where for $n = 0, \ldots, N-1$ we have:

$$X_{-n} = X_n{}^*, \quad X_0 = 0, \quad f_{0(-n)} = -f_{0n}, \quad f_{00} = 0.$$

This conjugate complex symmetric sequence is shown in Figure 2.3 in case of the square 16-QAM constellation of [167], where both the I and Q components can assume values of ± 1 and ± 3.

We streamline our formalism in Equation 2.12 by introducing the Fourier coefficients F_n

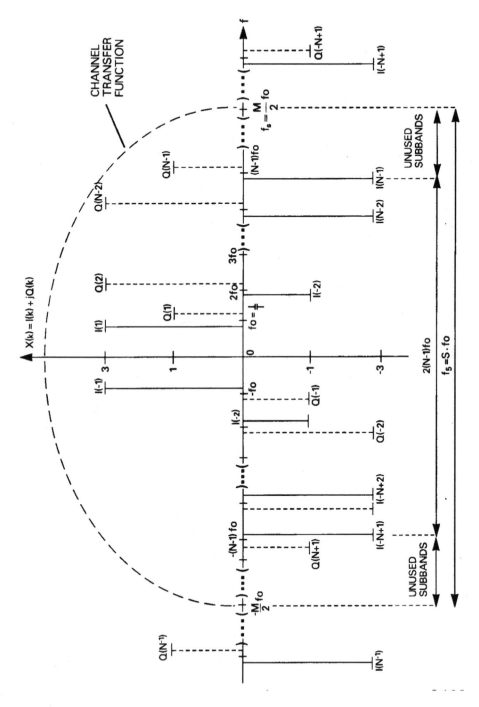

Figure 2.3: Conjugate complex symmetric square 16-QAM transmitted sequence $X(k) = I(k) + jQ(k)$

given by

$$
F_n = \begin{cases} \frac{1}{2}X_n & if & 1 \leq n \leq N-1 \\ \frac{1}{2}X_n^* & if & -(N-1) \leq n \leq -1 \\ 0 & if & n = 0 \end{cases} \tag{2.13}
$$

emphasising that the Fourier coefficients of a real signal are conjugate complex symmetric. Then Equation 2.12 can be rewritten as:

$$
m_0(t) = \sum_{n=-(N-1)}^{N-1} F_n \cdot e^{j2\pi f_{0n}t}. \tag{2.14}
$$

Observe that Equation 2.14 already bears close resemblance to the DFT. Assuming that the sub-channel carriers take values of $f_{0n} = nf_0, n = 0, \ldots, N-1$, where again $f_0 = 1/T$ represents the subcarrier spacing chosen to be the reciprocal of the sub-channel signalling interval, then the total one-sided bandwidth is $B = (N-1)f_0$.

So far the sub-channel modulated signals $m_{0n}(t)$ have been assumed to be continuous functions of time within the parallel signalling interval $k = 0$. In order to assist sampled, time-discrete processing by the DFT within the signalling interval $k = 0$ we introduce the discretised time $t = i\Delta t$, where $\Delta t = 1/f_s$ is the reciprocal of the sampling frequency f_s which must be chosen in accordance with Nyquist's sampling theorem to adequately represent $m_0(t)$. Recall furthermore from the previous section that $\Delta t = 1/f_s$ is the original serial QAM symbol spacing and f_s is the serial QAM symbol rate. Then we have:

$$
m_0(i\Delta t) = \sum_{n=-(N-1)}^{N-1} F_n e^{j2\pi n f_0 i\Delta t}. \tag{2.15}
$$

The Nyquist criterion is met if

$$
f_s > 2(N-1)f_0, \tag{2.16}
$$

where for practical reasons $f_s = Mf_0$ is assumed, implying that the sampling frequency f_s is an integer multiple of the subcarrier spacing f_0, with $M > 2(N-1)$ being a positive integer, implying also that $f_s = 1/\Delta t = Mf_0$, i.e. $f_0\Delta t = 1/M$. Again, these frequencies are portrayed in Figure 2.3. Bearing in mind that the spectrum of a sampled signal repeats itself at multiples of the sampling frequency f_s with a periodicity of $M = f_s/f_0$ samples, and exploiting the conjugate complex symmetry of the spectrum, for the Fourier coefficients F_n we have:

$$
F_n = \begin{cases} F_{n-M} = F_{M-n}^* & if \left(\frac{M}{2}+1\right) \leq n \leq M-1 \\ 0 & if \ N-1 < n \leq \frac{M}{2}. \end{cases} \tag{2.17}
$$

Observe that the frequency region $(N-1)f_0 < f_n \leq \frac{M}{2}f_0$ represents the typical unused transition band of the communications channel, where it exhibits significant amplitude and group delay distortion as suggested by Figure 2.3. However, in the narrow sub-bands the originally wide band frequency selective fading channel can be rendered flat upon using a high number of sub-channels.

The set of QAM symbols can be interpreted as a spectral domain sequence, which due to its conjugate complex symmetry will have a real IDFT pair in the time domain, representing the real modulated signal, which can be written as:

$$m_0(i\Delta t) = \sum_{n=0}^{M-1} F_n e^{j\frac{2\pi}{M}ni}, \; i = 0 \ldots M - 1. \tag{2.18}$$

This is the standard IDFT that can be computed by the IFFT if the transform length M is an integer power of 2. The rectangular modulation elements $m_T(t) = rect\frac{t}{T}$ introduced in Equation 2.6 have an infinite bandwidth requirement since in effect they rectangularly window the set of orthogonal basis functions constituted by the carriers $rect\frac{t}{T} \cdot \cos w_{0l}t$ and $rect\frac{t}{T} \cdot \sin w_{0l}t$. It is possible to use a time domain raised cosine pulse instead of the $rect\frac{t}{T}$ function [9], but this will impose further slight impairments on the time domain modulated signal which is also exposed to the hostile communications channel.

Observe that the representation of $m_0(t)$ by its $\Delta t = 1/f_s$-spaced samples is only correct if $m_0(t)$ is assumed to be periodic and bandlimited to $2(N-1)f_0$. This is equivalent to saying that $m_0(t)$ can only have a bandlimited frequency domain representation if in the time domain it expands from $-\infty$ to ∞. Due to sampling at a rate of $f_s = 1/\Delta t$ the spectral lobes become periodic at the multiples of f_s, but if the Nyquist sampling theorem is observed, no aliasing occurs. In order to fulfil these requirements, the modulating signal $m_0(t)$ derived by IFFT from the conjugate complex symmetric baseband information signal X_l has to be quasi-periodically extended before transmission via the channel, at least for the duration of the channel's memory. The effects of bandlimited transmission media will be discussed in the following section.

Having provided a justification for employing the FFT for carrying out modulation of all subcarriers in a single step, let us now consider the FFT-based QAM/FDM modem's schematic portrayed in Figure 2.4. The bits provided by the source are serial/parallel converted in order to form the n-level Gray coded symbols, N of which are collected in TX buffer 1, while the contents of TX buffer 2 are being transformed by the IFFT in order to form the time domain modulated signal. The digital-to-analogue (D/A) converted, low-pass filtered modulated signal is then transmitted via the channel and its received samples are collected in RX buffer 1, while the contents of RX buffer 2 are being fast Fourier transformed for deriving the demodulated signal. The twin buffers are alternately filled with data to allow for the finite FFT-based demodulation time. Before the data is Gray decoded and passed to the data sink, it can be decontaminated from the effects of the dispersive channel in the frequency domain, since despite using a high number of subchannels and hence a long subchannel symbol duration, the subchannel symbols may experience some dispersion in case of supporting extremely high transmission rates. In simple terms this operation may be carried out by inserting known pilot symbols into the frequency domain OFDM signal. These pilot subcarriers will allow us to sample and reconstruct the channel's complex frequency domain transfer function. Provided that frequency domain transfer function is sampled at a sufficiently high frequency, which is higher than the channel's Nyquist sampling frequency, we will show in Chapter 3 that the channel's frequency domain transfer function can be reliably recovered at the receiver, despite the contaminating effects of the Additive White Gaussian Noise (AWGN). Naturally, if the vehicular speed is high and hence the channel's Doppler frequency is high, a high "sampling frequency", i.e. high pilot subcarrier density has to be used, which reduces the achievable effective date throughput.

2.4   Transmission via Bandlimited Channels

The DFT/IDFT operations assume that the input signal is periodic in both time and frequency domain with a periodicity of M samples. If the modulated sample sequence of Equation 2.18 is periodically repeated and transmitted via the lowpass filter (LPF) preceding the channel, the channel is excited with a continuous, periodic signal. However, in order not to waste precious transmission time and hence channel capacity we would like to transmit only one period of $m_0(t)$ constituted by M samples. Assuming a LPF with a cut-off frequency of $f_c = 1/2\Delta t = f_s/2$ and transmitting only one period of $m_0(i\Delta t)$, the channel's input signal becomes:

$$
\begin{aligned}
m_{0,LPF}(t) &= m_0(i\Delta t) * \frac{1}{\Delta t}\frac{\sin(\pi t/\Delta t)}{\pi t/\Delta t}\\
&= m_0(i\Delta t) * \frac{1}{\Delta t}sinc\frac{\pi t}{\Delta t},
\end{aligned} \tag{2.19}
$$

where the LPF's impulse response is given by the *sinc* function and hence $m_{0,LPF}(t)$ has an infinite time domain duration. One period of the periodic modulated signal $m_{0,p}(i\Delta t)$, is given by

$$
m_0(i\Delta t) = m_{0,p}(i\Delta t)rect\frac{t}{T}. \tag{2.20}
$$

The convolution in Equation 2.19 can be written as:

$$
m_{0,LPF}(t) = \sum_{i=0}^{M-1} m_0(i\Delta t)\frac{1}{\Delta t}sinc\frac{\pi(t-i\Delta t)}{\Delta t}. \tag{2.21}
$$

In the spectral domain this is equivalent to writing

$$
M_{0,LPF}(f) = M_0(f)rect\frac{f}{f_c}, \tag{2.22}
$$

where $M_0(f) = FFT\{m_0(i\Delta t)\}$ and $H_{LPF}(f) = rect\frac{f}{f_c}$ is the LPF's frequency domain transfer function. Transforming Equation 2.20 into the frequency domain yields:

$$
\begin{aligned}
M_0(f) &= FFT\{m_{0,p}(i\Delta t)rect\frac{t}{T}\}\\
&= M_{0,p}(f) * \frac{1}{f_0}sinc\frac{\pi f}{f_0},
\end{aligned} \tag{2.23}
$$

where $M_{0,p}(f)$ is the frequency domain representation of $m_{0,p}(i\Delta t)$, which is convolved with the Fourier transform of the $rect\frac{t}{T}$ function. Now $M_0(f)$ of Equation 2.23 is lowpass

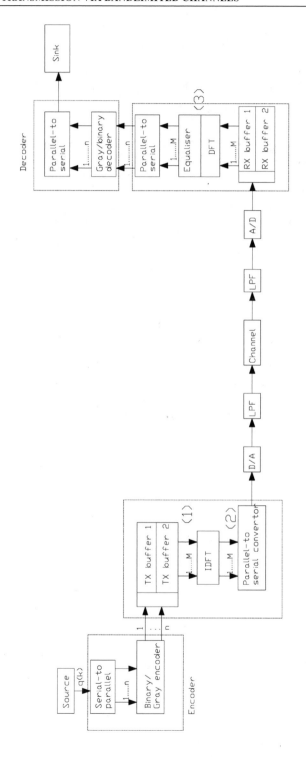

Figure 2.4: FFT-based OFDM modem schematic ©Hanzo, Webb, Keller 2000, [164]

filtered according to Equation 2.22, giving:

$$
\begin{aligned}
M_{0,LPF}(f) &= M_0(f) rect\frac{f}{f_c} \\
&= \left[M_{0,p}(f) * \frac{1}{f_0} sinc\frac{\pi f}{f_0} \right] rect\frac{f}{f_c}.
\end{aligned}
\tag{2.24}
$$

So the effect of the time domain truncation of the periodic modulated signal $m_{0,p}(i\Delta t)$ to a single period as in Equation 2.20 manifests itself in the frequency domain as the convolution of Equation 2.23, generating the infinite bandwidth signal $M_0(f)$. When $M_0(f)$ is lowpass filtered according to Equation 2.24, it becomes bandlimited to f_c, and its Fourier transform pair $m_{0,LPF}(t)$ in Equation 2.19 has an infinite time domain support due to the convolution with $sinc(\pi t/\Delta t)$. This phenomenon results in interference due to time domain overlapping between consecutive transmission blocks, which can be mitigated by quasi-periodically extending $m_0(i\Delta t)$ for the duration of the memory of the channel before transmission. At the receiver only the unimpaired central section is used for signal detection.

In order to portray a practical OFDM scheme and to aid the exposition here and in the previous section, in Figure 2.5 we plotted a few characteristic signals.

Figure 2.5(a) shows the transmitted spectrum using $M = 128$ and rectangular 16-QAM having $I, Q = \pm 1, \pm 3$.

In contrast to Figure 2.3, where the I and Q components were portrayed next to each other in order to emphasise that they belong to the same frequency, here the I and Q components associated with the same frequency are plotted with the same spacing as adjacent sub-bands. In Figure 2.5(a) there are 64 legitimate frequencies between 0 and 4 kHz, corresponding to 128 lines. Observe, however, that for frequencies of 0-300 Hz and 3.4-4.0 kHz we have allocated no QAM symbols. This is because this signal was transmitted after conjugate complex extension and IFFT over the M1020 CCITT telephone channel simulator, which has a high attenuation in these frequency slots. The real modulated signal after IFFT is plotted in Figure 2.5(b), which is constituted by 128 real samples. At the receiver this signal is demodulated by FFT in order to derive the received signal $\tilde{X}(k)$, which is then subjected to hard decision delivering the sequence $\hat{X}(k)$ seen in Figure 2.5(c). Finally, Figure 2.5(d) portrays the error signal $\Delta(k) = X(k) - \tilde{X}(k)$, where large errors can be observed towards the transmission band edges due to the M1020 channel's attenuation. The interpretation of these characteristic signals will be further developed in Section 2.6, as further details of the OFDM system are unravelled.

2.5   Generalised Nyquist Criterion

Using frequency division or orthogonal multiplexing (OFDM) two different sources of interference can be identified [168–172]. Intersymbol interference (ISI) is defined as the cross-talk between signals within the same sub-channel of consecutive FFT frames, which are separated in time by the signalling interval T. Inter channel interference (ICI) is the cross-talk between adjacent sub-channels or frequency slots of the same FFT frame. Since the effects of these interference sources and their mitigation methods are similar, it is convenient to introduce the term multidimensional interference (MDI) [170].

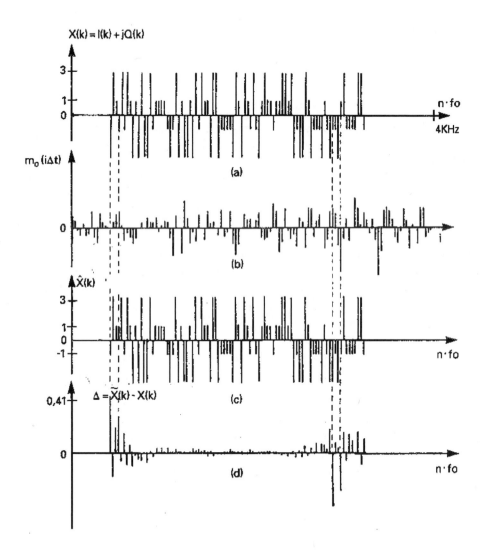

Figure 2.5: Characteristic OFDM signals for $M = 128$ and 16-QAM

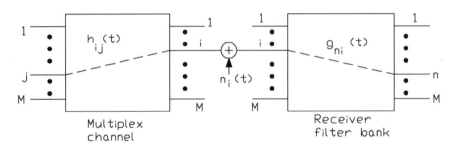

Figure 2.6: Multidimensional interference model [169] ©IEEE, 1975, Van
Etten

In order to describe MDI we assume the linear system model depicted in Figure 2.6, which
is the concatenation of the M-sub-channel multiplex transmission medium to the M-sub-
channel receiver filter bank. The channel's impulse response between its input j and output
i is denoted by $h_{ij}(t)$, while that of the receiver filter bank is $g_{ni}(t)$. With the assumption
of this linear system model the output signal of the receiver filter is the superposition of the
system responses due to all input signals $1, \ldots, M$. Our objective is to optimise the linear
receiver in order to minimise the effects of AWGN and MDI, which will reduce the BER of
the receiver.

Assuming a sub-channel noise spectral density of N_i and receiver filter transfer functions
of $G_{ni}(f)$ between the input i and output n, the noise variance at output n becomes:

$$\sigma_n^2 = \sum_{i=1}^{M} N_i \int_0^\infty G_{ni}^2(f)df, \qquad (2.25)$$

which can also be expressed in terms of $g_{ni}(t)$ using Parseval's theorem:

$$\sigma_n^2 = \sum_{i=1}^{M} N_i \int_0^\infty g_{ni}^2(\tau)d\tau. \qquad (2.26)$$

Following the approaches in [170–172] the noise variance can be minimised as a function of
the receiver filter impulse response $g_{ni}(t)$. Supposing that the sub-channel input signals a_j
$j = 1, \ldots, M$ can take any value from the mutually independent, equiprobable set of L-ary
alphabet, then the channel's input vector in the kth signalling interval is given by:

$$\mathbf{u^T(k)} = \left[a_1^k, a_2^k, \ldots, a_M^k \right]. \qquad (2.27)$$

Observe that the number of possible input vectors is L^M.

¿From Figure 2.6 we can see that the system's response at the receiver output can be
computed through convolution of the channel and receiver filter impulse responses. The input
signal of sub-channel j generates an output at every receiver filter input $i = 1, \ldots, M$, each
of which gives rise to a response at the nth receiver filter output. In order to derive the
nth output, the contributions must be summed first fixing the jth sub-channel input over the
receiver filter input index i. Then, since the input signal vector $\mathbf{u^T(k)}$ excites all sub-channel

inputs $j = 1, \ldots, M$, summing over the nth index j gives the output signal of the receiver filter:

$$s_n{}^k(t) = \sum_{j=1}^{M} a_j{}^k \sum_{i=1}^{M} \int_0^\infty g_{ni}(\tau) h_{ij}(t - \tau) d\tau. \qquad (2.28)$$

At the received sampling instants $(t_s + lT)$ we have

$$s_n{}^k(t_s + lT) = \sum_{j=1}^{M} a_j{}^k \sum_{i=1}^{M} \int_0^\infty g_{ni}(\tau) h_{ij}(t_s + lT - \tau) d\tau. \qquad (2.29)$$

The optimisation problem now is to find the system transfer functions, which minimise the noise variance at the output of the nth receiver filter under the constraint of constant $s_n{}^k(t_s + lT)$, for all l and k. Using the technique of variational calculus, the following functional must be minimised [169]:

$$F_n = \sigma_n^2 - 2 \sum_{k=1}^{L} \sum_{l}^{M} \lambda_{nkl} s_n{}^k(t_s + lT), \qquad (2.30)$$

where λ_{nkl} is the so-called Lagrange multiplier. On substituting Equations 2.26 and 2.29 into Equation 2.30 we have:

$$\begin{aligned} F_n &= \sum_{i=1}^{M} N_i \int_0^\infty g_{ni}^2(\tau) d\tau \\ &- 2 \sum_{k=1}^{L} \sum_{l}^{M} \lambda_{nkl} \sum_{j=1}^{M} a_j{}^k \sum_{i=1}^{M} \int_0^\infty g_{ni}(\tau) h_{ij}(t_s + lT - \tau) d\tau. \end{aligned}$$

The minimisation [169] of F_n in Equation 2.5 yields:

$$g_{ni}(t) = \frac{1}{N_i} \sum_{k=1}^{L} \sum_{l}^{M} \lambda_{nkl} \sum_{j=1}^{M} a_j{}^k h_{ij}(t_s + lT - t). \qquad (2.31)$$

Assuming identical noise spectral density in all sub-channels, i.e. that $N_i = N$ for all i and introducing the shorthand

$$c_{njl} = \frac{1}{N} \sum_{k=1}^{L} a_j{}^k \lambda_{nkl}, \qquad (2.32)$$

then Equation 2.31 can be simplified to:

$$g_{ni}(t) = \sum_{j=1}^{M} \sum_{l} c_{njl} h_{ij}(t_s + lT - t). \qquad (2.33)$$

As seen from Equation 2.33, the set of optimum receiver filters $g_{ni}(t)$ depends on the sub-channel impulse responses $h_{ij}(t)$. The receiver filter impulse responses $g_{ni}(t)$ from the ith filter input to all filter outputs have to be matched to the responses $h_{ij}(t)$ from all sub-channel inputs $j = 1, \ldots, M$ to the ith subchannel output or received filter input, as detailed in section 4.5.2.

In order to derive the matched receiver filter response $g_{ni}(t)$ according to Equation 2.33, we have to superimpose the responses of all receiver filters that are matched to the OFDM sub-channel outputs due to the jth OFDM sub-channel input. The summation over the index l weighted by the coefficients c_{njl} in Equation 2.33 represents the ISI due to the T-spaced h_{ij} contributions from previous sampling instants.

Using our general MDI model in Figure 2.6, the overall impulse response at the sampling instant (lT) between the jth OFDM sub-channel input and the nth receiver filter output can be denoted by $f_{nj}(lT)$, $n, j = 1, \ldots, M$ or in matrix form:

$$
\mathbf{F_l} = \begin{bmatrix}
f_{11}(lT) & f_{12}(lT) & \cdots & f_{1M}(lT) \\
f_{21}(lT) & f_{22}(lT) & \cdots & f_{2M}(lT) \\
\vdots & & & \\
f_{M1}(lT) & f_{M2}(lT) & \cdots & f_{MM}(lT)
\end{bmatrix} .
\tag{2.34}
$$

The total accumulated MDI at the nth receiver filter output due to both ISI and ICI can be defined [169] as:

$$
MDI_n = \frac{|f_{nn}(0)| - \sum_l \sum_{i=1}^{M} |f_{ni}(l)|}{f_{nn}(0)} .
\tag{2.35}
$$

It is plausible that the condition of MDI-free transmission is met if there is no cross-talk amongst the sub-channels, i.e. the matrix of Equation 2.34 is the diagonal identity matrix for any arbitrary sampling instant (lT). In formal terms this means that for $n, j = 1, 2, \ldots, M$, $l = 0, \pm 1, \pm 2, \ldots, \pm \infty$ the generalised Nyquist criterion proposed by Schnidman [168] must be satisfied:

$$
\sum_l f_{nj}(lT) = \begin{cases} 1 & \text{if } n = j \\ 0 & \text{otherwise.} \end{cases}
\tag{2.36}
$$

If Equation 2.36 is satisfied, the MDI term in Equation 2.35 becomes zero and the generalised Nyquist criterion can be fulfilled by the appropriate choice of the coefficients c_{njl}.

Quite clearly, the generalised Nyquist criterion in Equation 2.36 requires not only that the conventional Nyquist criterion shall be met, i.e. $f_{nn}(lT) = \delta_l$, $l = 0, 1, \ldots, \infty$, where δ_l is the Kronecker delta, in order to render the ISI from other signalling intervals zero. It also requires that all the other $M(M - 1)$ sub-channels with their matched receiver filters shall have a zero-crossing in their impulse responses, i.e. for $n, j = 1, 2, \ldots, M$ and $l = 0, \pm 1, \ldots, \pm \infty$ we have that

$$
F_{nj}(lT) = \delta_{nj} \delta_l
\tag{2.37}
$$

shall be satisfied, where

$$
\delta_{nj} = \begin{cases} 1 & \text{if } n = j \\ 0 & \text{otherwise.} \end{cases}
\tag{2.38}
$$

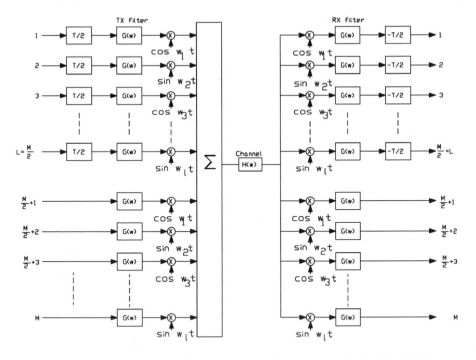

Figure 2.7: Filter-bank implementation of the FDM modem [14] ©IEEE, 1981, Hirosaki

Therefore, the generalised Nyquist criterion requires the equivalent folded-in baseband transfer characteristic introduced in Chapter 4 to be an ideal lowpass characteristic for $n = j$ and zero otherwise.

2.6 Basic OFDM Modem Implementations

A historic OFDM modem implementation was proposed by Hirosaki [14] and is shown in Figure 2.7. This implementation is based on staggered QAM (SQAM) or offset QAM (OQAM), where the quadrature components are delayed by half a signalling interval with respect to each other in order to reduce the signal envelope fluctuation [164]. The system is constituted by M synchronised baseband channels operating with a Baud rate of $f_0 = 1/T$. The baseband modulating signals of sub-channels i and $(i + M/2)$ are amplitude modulated onto the carriers $f_i = [f_0 + (i - 1)f_0]$, $i = 1, \ldots, M/2$, with the carrier suppressed. This implies that the sub-channels are spaced according to the Baud rate of $f_0 = 1/T$. Then the sum of the sub-channel signals i and $(i + M/2)$, $i = 1, \ldots, M/2$ form the ith SQAM sub-channel, where the in-phase and quadrature phase modulating signals are shifted by $T/2$ with respect to each other. The transmit and receive filters $G(\omega)$ are identical square-root raised-cosine Nyquist filters and an equaliser implementation is also proposed in [14].

However, Hirosaki's conceptually simple OFDM implementation may become implementationally prohibitive in terms of complexity and cost [14], especially for a high number of sub-channels. Weinstein [9] suggested the digital implementation of OFDM subcarrier

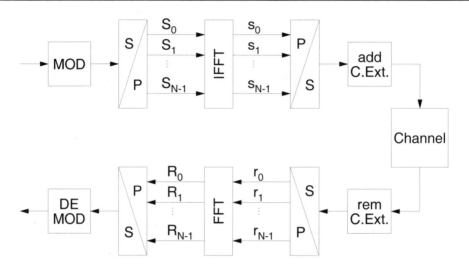

Figure 2.8: Schematic of N-subcarrier OFDM transmission system

modulators/demodulators based on the discrete Fourier transform (DFT).

The DFT and its more efficient implementation, the fast Fourier transform (FFT) are typically employed in practice in the baseband OFDM modulation/demodulation process, as can be seen in the schematic of Figure 2.8. The serial data stream is mapped to data symbols with a symbol rate of $1/T_s$, employing a general phase and amplitude modulation scheme, and the resulting symbol stream is demultiplexed into a vector of N data symbols S_0 to S_{N-1}. The parallel data symbol rate is $1/N \cdot T_s$, ie. the parallel symbol duration is N times longer than the serial symbol duration T_s. The inverse FFT (IFFT) of the data symbol vector is computed and the coefficients s_0 to s_{N-1} constitute an OFDM symbol. The s_n are the time domain samples of the OFDM symbol and are transmitted sequentially over the channel at a symbol rate of $1/T_s$. At the receiver, a spectral decomposition of the received time domain samples r_n is computed employing an N-tap FFT, and the recovered data symbols R_n are restored in serial order and demultiplexed.

If we assume that the bandwidth of the OFDM spectrum is finite, then simple Fourier theory dictates that the corresponding time domain signal has an infinite duration. The underlying assumption upon invoking the IFFT for modulation is that although N frequency domain samples produce N time domain samples, the time domain signal is assumed to be periodically repeated, theoretically, for an infinite duration. In practice, however, it is sufficient to repeat the time domain signal periodically for the duration of the channel's memory, i.e. for a duration that is comparable to the length of the CIR. Hence, for transmission over time dispersive channels, each time domain OFDM symbol is extended by the so-called cyclic extension (C. Ext. in Figure 2.8) or a guard interval of N_g samples duration, in order to overcome the inter-OFDM symbol interference due to the channel's memory.

The samples of the cyclic extension are copied from the end of the time domain OFDM symbol, generating the transmitted time domain signal $(s_{N-N_g-1}, \dots, s_{N-1}, s_0, \dots, s_{N-1})$ depicted in Figure 2.9. At the receiver, the samples of the cyclic extension are discarded. Clearly, the need for a cyclic extension in time dispersive environments reduces

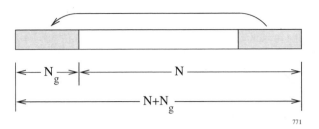

Figure 2.9: Stylised plot of N-subcarrier OFDM time domain signal with a cyclic extension of N_g samples

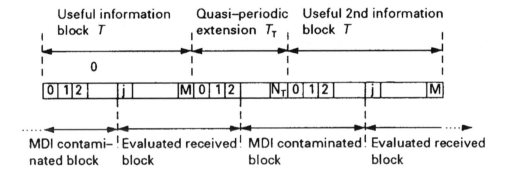

Figure 2.10: Reducing MDI by quasi-periodic block extension

the efficiency of OFDM transmissions by a factor of $N/(N + N_g)$. Since the duration N_g of the necessary cyclic extension depends only on the channel's memory, OFDM transmissions employing a high number of carriers N are desirable for efficient operation. Typically a guard interval length of not more than 10% of the OFDM symbol's duration is employed.

2.7 Cyclic OFDM Symbol Extension

As we have seen in Section 2.4, the response of the low pass filtered OFDM channels to a block of modulated signal is theoretically infinite. In Section 2.5 we introduced the generalised Nyquist criterion, which allowed infinite-length system responses, as long as their periodic zero-crossings yielded no MDI. If, however, the zero-crossings or the sampling instants fluctuate, the MDI soon becomes prohibitively high.

An alternative method of combating the previously characterised multi-dimensional interference (MDI) is to transmit quasi-periodically extended time domain blocks, i.e. OFDM symbols, after modulation by IFFT [15]. The length of the added quasi-periodic extension depends on the memory length of the channel, in other words, on the length of the transient response of the channel to the quasi-periodic excitation constituted by the modulated signal. The method is best explained with reference to Figure 2.10. Every block of length-T modulated signal segment is quasi-periodically extended by a length of T_T transient duration simply repeating N_T samples of the useful information block. Then the total sequence length becomes $(M + N_T)$ samples, corresponding to a duration of $(T + T_T)$. Trailing and leading samples of this extended block are corrupted by the channel's transient response, hence the receiver is instructed to ignore the first j number of samples of the received block and also disregard $(M + N_T - j)$ trailing samples. Only the central M number of samples are demodulated by FFT at the receiver, which are essentially unaffected by the channel's transient response, as seen in Figure 2.10.

The number of extension samples N_T required depends on the length of the channel's transient response and the number of modulation levels. If the number of modulation levels is high, the maximum acceptable MDI due to channel transients must be kept low in order to maintain sufficient noise margins before the data is corrupted. This then requires a longer quasi-periodic extension, i.e. N_T must be also higher. It must also be appreciated that this MDI-corrupted extension actually wastes channel capacity as well as transmitted power. However, if the useful information blocks are long, i.e. $M \geq 128$, the extension length can be kept as low as 10% of the useful information block length.

2.8 Reducing MDI by Compensation [15]

2.8.1 Transient System Analysis

Following Kolb's [15] and Schüssler's [16] approach, we can describe the OFDM transmission system by using input and output vectors $\mathbf{u}(k)$ and $\mathbf{y}(k)$, respectively, that are related by

$$\mathbf{y}(k) = \mathbf{S}\mathbf{u}(k),\tag{2.39}$$

where the convolution matrix $\mathbf{S}$ [173] is given by:

$$\mathbf{S} = \begin{bmatrix} h_0 & 0 & \cdots & \cdots & 0 \\ h_1 & h_0 & \cdots & \cdots & 0 \\ h_2 & h_1 & h_0 & \cdots & 0 \\ \vdots & \vdots & \vdots & & \vdots \end{bmatrix}.\tag{2.40}$$

Due to causality the vectors $\mathbf{u}(k)$ and $\mathbf{y}(k)$ are of infinite length for positive sample indices only. Let us assume furthermore that the input vector $\mathbf{u}(k)$ represents $L \geq 2$ periods of an M-sample periodic sequence and that the channel's impulse response is shorter than M samples. Then, for example, for $L = 3$ periods and a channel impulse response length of $n = M - 1$ samples the convolution in Equation 2.39 can be written as:

$$\begin{bmatrix} \mathbf{y}_1 \\ \mathbf{y}_2 \\ \mathbf{y}_3 \\ \mathbf{y}_4 \end{bmatrix} = \begin{bmatrix} \mathbf{S}_0 & 0 & 0 \\ \mathbf{S}_1 & \mathbf{S}_0 & 0 \\ 0 & \mathbf{S}_1 & \mathbf{S}_0 \\ 0 & 0 & \mathbf{S}_1 \end{bmatrix} \begin{bmatrix} \mathbf{u} \\ \mathbf{u} \\ \mathbf{u} \end{bmatrix},\tag{2.41}$$

where

$$
\mathbf{S}_0 = \begin{bmatrix}
h_0 & 0 & \cdots & \cdots & 0 \\
h_1 & h_0 & \cdots & \cdots & 0 \\
h_2 & h_1 & h_0 & \cdots & 0 \\
\vdots & \vdots & \vdots & & \vdots \\
h_n & h_{n-1} & h_{n-2} & \cdots & h_0
\end{bmatrix}
\tag{2.42}
$$

and

$$
\mathbf{S}_1 = \begin{bmatrix}
0 & h_n & h_{n-1} & \cdots & h_1 \\
0 & 0 & h_n & \cdots & h_2 \\
\vdots & \vdots & \vdots & & \vdots \\
0 & 0 & 0 & \cdots & h_n \\
0 & 0 & 0 & \cdots & 0
\end{bmatrix}.
\tag{2.43}
$$

Upon exciting our transmission channel with a periodic signal, its response is constituted by a transient response $y_{in}(k)$ plus a stationary periodic response $y_{st}(k)$. The stationary response has the periodicity M of the excitation, while the channel's transient response dies down after $n = M - 1$ samples.

The system's response during the first period of M samples is then given by:

$$
\begin{aligned}
\mathbf{y}_1 &= \mathbf{y}_{in} + \mathbf{y}_{st} = \mathbf{S}_0 \mathbf{u} \\
&= [h_0 u_0; (h_1 u_0 + h_0 u_1); (h_2 u_0 + h_1 u_1 + h_0 u_2); \\
&\quad \ldots ; (h_n u_0 + h_{n-1} u_1 + h_{n-2} u_2 + \ldots + h_0 u_n)]^T,
\end{aligned}
\tag{2.44}
$$

where we also exploited Equation 2.41 and $[\ ]^T$ means the transpose of $[\ ]$. During the second cycle of the excitation the transient does not affect the system's response since $u < M$, and hence by the help of Equation 2.41 we get:

$$
\begin{aligned}
\mathbf{y}_2 &= \mathbf{y}_{st} = [\mathbf{S}_1 \mathbf{S}_0][\mathbf{u} \mathbf{u}]^T \\
&= [\mathbf{S}_1 + \mathbf{S}_0]\mathbf{u} = \mathbf{H} \mathbf{u},
\end{aligned}
\tag{2.45}
$$

where the matrix $\mathbf{H} = [\mathbf{S}_1 + \mathbf{S}_0]$ is computed from Equations 2.42 and 2.43 as follows:

$$
\mathbf{H} = \mathbf{S}_1 + \mathbf{S}_0 = \begin{bmatrix}
h_0 & h_n & h_{n-1} & \cdots & h_1 \\
h_1 & h_0 & h_n & \cdots & h_2 \\
\vdots & & & & \\
h_n & h_{n-1} & h_{n-2} & \cdots & h_0
\end{bmatrix}.
\tag{2.46}
$$

Then the system's response for the second cycle from Equation 2.46 can be written as:

$$\mathbf{y}_2 = \begin{bmatrix} h_0 u_0 + h_n u_1 + \ldots + h_1 u_n \\ h_1 u_0 + h_0 u_1 + \ldots + h_2 u_n \\ \vdots \\ h_n u_0 + h_{n-1} u_1 + \ldots + h_0 u_n \end{bmatrix}. \tag{2.47}$$

Similarly, the system's response during the third period of the excitation is unaffected by transients, and hence given by:

$$\mathbf{y}_3 = \mathbf{H} \mathbf{u}. \tag{2.48}$$

The switch-off transient $\mathbf{y}_4 = \mathbf{y}_{out}$ of the system describes the output signal after the excitation died down. From Equation 2.41 we get:

$$\begin{aligned} \mathbf{y}_{out} = \mathbf{y}_4 &= \mathbf{S}_1 \mathbf{u} = \mathbf{S}_1 \mathbf{u} + \mathbf{S}_0 \mathbf{u} - \mathbf{S}_0 \mathbf{u} \\ &= (\mathbf{S}_1 + \mathbf{S}_0) \mathbf{u} - \mathbf{S}_0 \mathbf{u}. \end{aligned} \tag{2.49}$$

Combining Equations 2.45 and 2.46 gives:

$$\mathbf{y}_{out} = \mathbf{y}_4 = \mathbf{y}_{st} - (\mathbf{y}_{in} + \mathbf{y}_{st}) = -\mathbf{y}_{in}, \tag{2.50}$$

which implies that the trailing transient is the "inverse" of the leading transient.

Based on the previous analysis our aim is to transmit each information block only once $(L = 1)$ and remove the effects of MDI by cancellation, rather than reduce the system's effective transmission rate by the factor $M/(M + N_T)$, where N_T is the number of samples used for the quasi-periodic transmission block expansion. The kth received signal vector $\mathbf{y}^k$ is constituted by an initial transient due to the trailing transient $\mathbf{y}_{out}^{k-1}$ of the $(k-1)$th transmitted vector, a stationary response $\mathbf{y}_{st}^k$ due to the kth transmitted vector plus the initial transient $\mathbf{y}_{in}^{k-1}$ due to the kth transmitted vector:

$$\mathbf{y}^k = \mathbf{y}_{out}^{k-1} + \mathbf{y}_{st}^k + \mathbf{y}_{in}^k. \tag{2.51}$$

Equation 2.50, for the trailing transient from block $(k-1)$, gives us:

$$\mathbf{y}_{out}^{k-1} = -\mathbf{y}_{in}^{k-1}, \tag{2.52}$$

which can be substituted in Equation 2.51 to yield:

$$\mathbf{y}^k = \mathbf{y}_{in}^k + \mathbf{y}_{st}^k - \mathbf{y}_{in}^{k-1}. \tag{2.53}$$

2.8.2 Recursive MDI Compensation

Our objective is now to determine the transmitted vector $\mathbf{u}^k$, which can be inferred if $\mathbf{y}^k$ is known. Hence the compensation of MDI ensues as follows:

(1) Initially a known preamble sequence $\mathbf{u}_0$ is transmitted, sending $L \geq 2$ repetitions to determine the channel's impulse response, i.e. the matrix $\mathbf{H}$. This is possible using Equations 2.45 and 2.46, where the system's response is in its stationary phase during the second cycle of the periodic preamble. The system's impulse response can be conveniently measured by transmitting a specific preamble $\mathbf{u}_0$, which is constituted by a "white" spectrum with the real part set to one and the imaginary part set to zero. This input signal modulates the sub-channel modulators, where the modulated signal after IFFT becomes the Kronecker delta. When this preamble signal excites the transmission system, its response is the impulse response itself. Once the impulse response is known, the matrices $\mathbf{H}$, $\mathbf{S}_0$ and $\mathbf{S}_1$ are also known.

(2) Since the system responses $\mathbf{y}_1^0$ and $\mathbf{y}_2^0$ due to the preamble $\mathbf{u}_0$ during the first two periods are now known, the system's initial transient can be computed from Equations 2.45 and 2.46 using:

$$\mathbf{y}_1^0 = \mathbf{y}_{in}^0 + \mathbf{y}_{st}^0 = \mathbf{S}_0 \mathbf{u} \tag{2.54}$$

and

$$\mathbf{y}_2^0 = \mathbf{y}_{st}^0 = [\mathbf{S}_1 + \mathbf{S}_0]\mathbf{u} \tag{2.55}$$

as follows:

$$\mathbf{y}_{in}^0 = \mathbf{y}_1^0 - \mathbf{y}_{st}^0 = \mathbf{y}_1^0 - \mathbf{y}_2^0 = -\mathbf{S}_1 \mathbf{u}. \tag{2.56}$$

With the knowledge of this initial transient response the trailing effects corrupting the consecutive blocks can be recursively compensated.

(3) The preamble $\mathbf{u}_0$ is followed by the useful information blocks $\mathbf{y}^k$, $k = 1, 2, \ldots, \infty$, transmitted only once $(L = 1)$. Upon receiving the system's response $\mathbf{y}^k$ due to $\mathbf{u}^k$, $k = 1, 2, \ldots, \infty$, the trailing transient of the previous block can be subtracted as follows:

$$\mathbf{y}^k - \mathbf{y}_{out}^{k-1} = \mathbf{y}^k + \mathbf{y}_{in}^{k-1}. \tag{2.57}$$

When considering Equations 2.51 and 2.45 we get:

$$\mathbf{y}^k - \mathbf{y}_{out}^{k-1} = \mathbf{y}_{st}^k + \mathbf{y}_{in}^k = \mathbf{S}_0 \tilde{\mathbf{u}}^k, \tag{2.58}$$

where we used $\tilde{\mathbf{u}}^k$ which is different from $\mathbf{u}^k$ due to channel effects. The estimated transmitted signal $\tilde{\mathbf{u}}^k$ can be recovered using Equations 2.57 and 2.58:

$$\tilde{\mathbf{u}}^k = \mathbf{S}_0^{-1}(\mathbf{y}^k - \mathbf{y}_{out}^{k-1}) = \mathbf{S}_0^{-1}(\mathbf{y}^k + \mathbf{y}_{in}^{k-1}). \tag{2.59}$$

From Equation 2.49 we have $\mathbf{y}_{out}^{k-1} = \mathbf{S}_1 \mathbf{u}^{k-1}$ that can be substituted in Equation 2.59, giving

$$\tilde{\mathbf{u}}^k = \mathbf{S}_0^{-1}(\mathbf{y}^k - \mathbf{S}_1 \mathbf{u}^{k-1}), \tag{2.60}$$

which is an explicit formula for the estimated transmitted vector $\tilde{\mathbf{u}}^k$ in terms of the received vector $\mathbf{y}^k$, the previously recovered transmitted vector $\mathbf{u}^{k-1}$ and the matrices $\mathbf{S}_0$ and $\mathbf{S}_1$, which depend only on the system's impulse response.

(4) Now the compensated vector $\tilde{\mathbf{u}}^k$ in Equation 2.60 can be demodulated by FFT and the demodulated signal $\tilde{\mathbf{U}}^k$ is subjected to hard decisions, which are represented as $D\{\ \}$, giving the recovered information signal in the following form:

$$\mathbf{U}^k = D\{\tilde{\mathbf{u}}^k\} = D\{FFT[\tilde{\mathbf{u}}^k]\}. \tag{2.61}$$

(5) Since $\mathbf{u}^k$ is needed in Equation 2.60 for the next recursive compensation step,

$$\mathbf{u}^k = IFFT[\mathbf{U}^k] \tag{2.62}$$

is computed to conclude the compensation process.

This method bears a strong resemblance to the conventional partial response technique, allowing signals belonging to adjacent signalling intervals to overlap. The controlled ISI can then be recursively compensated, if the channel can be considered slowly changing.

2.9   Decision-Directed Adaptive Channel Equalisation

In this section it will be assumed that the MDI was removed by compensation, quasi-periodic block extension or by obeying the generalised Nyquist criterion. The linear distortions introduced by the unequalised channel transfer function $H(f)$ can be removed by estimating $H(if_0)$ and then dividing the received signal spectrum by $H(if_0)$, before hard decision decoding is performed using the function $D\{\ \}$.

In the previous section we mentioned how $H(if_0)$ can be measured using a preamble having a real-valued "white" spectrum. By setting all real spectral lines to unity and all imaginary lines to zero in the preamble data frame, after modulation by the IFFT the transmitted signal is the Kronecker delta. Therefore the received signal is the channel's impulse response, which after demodulation by FFT gives the channel's frequency domain transfer function $H(if_0)$. The received signal's spectrum after demodulation by FFT becomes:

$$\tilde{X}_i = H(if_0)X_i, \ i = 1, \ldots, \ M, \tag{2.63}$$

where X_i is the ith transmitted spectral line at frequency if_0. After equalisation by dividing the received spectral line $\tilde{X}_i$ with the estimated channel transfer function $H(if_0)$ and taking hard decisions, we obtain the recovered sequence:

$$\hat{X}_i = D\left\{\frac{\tilde{X}_i}{H(if_0)}\right\}, \ i = 1, \ldots, \ M. \tag{2.64}$$

If the recovered sequence $\hat{X}_i$ is error-free, it lends itself to the recursive recomputation of the channel's frequency response in order to cope with slowly time-varying transmission media [15]. The updated transfer function is given by:

$$H_a(if_0) = \frac{\tilde{X}_i}{\hat{X}_i} = \frac{\tilde{X}_i}{D\left\{\frac{\tilde{X}_i}{H(if_0)}\right\}}, \ i = 1, \ldots, \ M. \tag{2.65}$$

In order to retain robustness for transmissions over channels having high bit error rate a leaky algorithm can be introduced to generate a weighted average of the previous and current transfer function using the leakage factor β in the following fashion:

$$H_{\text{adaptive}}(if_0) = \beta H_a^k(if_0) + (1 - \beta)H_a^{k-1}(if_0). \tag{2.66}$$

The leakage factor β is a parameter determined by the prevailing channel bit error rate.

An interesting aspect of our OFDM scheme is that if the channel varies slowly, then using differential coding between corresponding sub-channels of consecutive OFDM transmission frames removes the requirement for a channel equaliser as long as the difference is computed before a hard decision takes place [9]. This is due to the fact that spectral lines of the same sub-channel or same frequency will suffer the same attenuation due to the channel's linear distortion. Hence the effect of channel attenuation and phase shift drops out before hard decision takes place. Similar arguments can be exploited between adjacent lines of the same OFDM transmission frame as well, if the channel's transfer characteristic is sufficiently smooth.

If the frequency domain transfer function $H(if_0)$ is more erratic as a function of frequency or time, a number of known pilot tones can be included in the transmitted spectrum. These pilots facilitate the more accurate estimation and equalisation of the channel transfer function. This technique was proposed by Cimini for wide-band, frequency-selective multipath mobile channels in reference [10] and it will be widely used throughout the forthcoming chapters. In case of narrowband Rayleigh-fading mobile radio channels the fading is "frequency flat fading", hence each transmitted frequency component suffers the same attenuation and phase shift. In this case time domain pilot symbols employed as in the Pilot Symbol Assisted Modulation (PSAM) schemes of Chapter 10 in [164] are useful in supporting the tracking of the Rayleigh fading envelope [174]. A detailed discourse on sophisticated two-dimensional pilot symbol-assisted channel estimation techniques will be provided in Chapter 14.

2.10 OFDM Bandwidth Efficiency

In the OFDM system each symbol to be transmitted modulates an assigned carrier of a set of wide-sense orthogonal basis functions and these modulated sub-channel signals are superimposed for transmission via the communications channel. The received signal can be demodulated for example by the correlation receiver described in the previous section or by FFT.

A set of suitable wide-sense orthogonal basis functions of gradually increasing length T, $3T$ and $5T$, similar to those used in our OFDM schemes, is depicted in Figure 2.11 for a one-carrier, three-carrier and five-carrier system, respectively [175, 176].

Both their time domain waveforms and stylised spectra are shown in Figure 2.11. In a simplistic approach here we assume that the signal spectra can be band limited to the bandwidth of its main spectral lobe, as suggested by the figure.

Using an essentially serial system with one carrier, as seen in Figure 2.11(a), the minimum bandwidth required is $f_B = 1/T$ and the bandwidth efficiency is $\eta = 1$ Bd/Hz because the spectrum of this pulse is represented by the sinc function whose first zero is at $f_B = 1/T$. The three-carrier system of Figure 2.11(b) expands the length of the basis functions to $3T$, thereby reducing the bandwidth requirement to $B = \frac{2}{3}f_B$, giving $\eta = 1.5$ Bd/Hz. This is

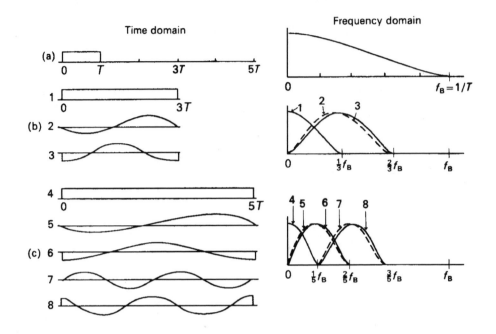

Figure 2.11: OFDM carriers, their stylised spectra and bandwidth requirement: (a) one-carrier system, (b) three-carrier system, (c) five-carrier system [175] ©Springer, 1969, Harmuth

because the rectangularly windowed sin and cos spectra are represented by the convolution of a tonal spectral line and a frequency domain sinc function describing the spectrum of the rectangular time domain window. The five-carrier scheme using basis functions of $5T$ length further reduces the bandwidth to $B = \frac{3}{5}f_B$ and increases the spectral efficiency to 1.67 Bd/Hz.

Similarly, the approximate bandwidth of a $(2M + 1)$-carrier system using an impulse as well as M sine and M cosine carriers of length $(2M + 1)T$ becomes

$$B = \frac{M+1}{2M+1}\frac{1}{T},\tag{2.67}$$

yielding a bandwidth efficiency of

$$\eta = \frac{2M+1}{M+1}\ \text{Bd/Hz}.\tag{2.68}$$

When $M \to \infty$, we have $\lim_{M\to\infty} \eta = 2$ Bd/Hz, which for a typical value of $M = 64$ gives $\eta = 129/65 = 1.98$ Bd/Hz. The interferences caused by the above band limitation are given in closed form in [176] for a variety of carriers, but for the more attractive scenarios using a higher number of carriers it can only be estimated by simulation studies.

2.11   Chapter Summary and Conclusion

In this chapter we have investigated orthogonal multiplexing as an attractive means of transmitting high-rate information over highly dispersive mobile radio channels. The essential premise of orthogonal multiplexing is that of dividing the serial input data stream into a high number of parallel streams and to transmit these low-rate parallel streams simultaneously. This offered two main advantages. First, by increasing the subchannels' symbol duration – owing to the low sub-channel signalling rate – the probability of the subchannel symbols being completely corrupted in a fade was reduced. Second, by decreasing the bandwidth of each subchannel, the need for a channel equaliser was diminished. The modulation scheme of each of these orthogonal subchannels could be either binary or multilevel, and both were considered.

However, orthogonal multiplexing also has a number of disadvantages. It requires an increase in the complexity of the transmitting and receiving equipment, and may also engender an increased transmission delay. It will typically also require pilot symbols or some other similar redundancy such that the receiver can be informed of how each individual channel has been attenuated and phase-rotated by the mobile radio channel.

A number of sections were concerned with the issue of multidimensional interference, whereby interference might occur not only between consecutive symbols of the same sub-channel, but also from one sub-channel to the adjacent sub-channels, and methods to ameliorate this were suggested. Following this rudimentary introduction to the basic concepts of OFDM, in the forthcoming chapters we will delve into the various design issues of OFDM more deeply, considering a range of associated research and implementational aspects.

Chapter **3**

OFDM Transmission over Gaussian Channels[1]

High data rate communications over additive white Gaussian noise (AWGN) channels are limited not only by noise, but especially with increasing symbol rates, often more significantly by the intersymbol interference (ISI) due to the memory of the dispersive wireless communications channel. Explicitly, this channel memory is caused by the dispersive channel impulse response (CIR) due to the different length propagation paths between the transmitting and the receiving antennae. This dispersion effect could theoretically be measured by transmitting an infinitely short impulse and "receiving" the channel impulse response itself. On this basis, several measures of the effective duration of the impulse response can be calculated, one being the delay spread. The multipath propagation of the channel manifests itself by different echos of possibly different transmitted symbols overlapping at the receiver, which leads to error rate degradation.

This effect occurs not only in wireless communications, but also over all types of electrical and optical waveguides, although for these media the relative time differences are comparatively small, mostly due to multimode transmission or incorrect electrical or optical adaption at interfaces.

In wireless communications systems the duration and the shape of the channel impulse response depend heavily on the propagation environment of the communications system in question. While indoor wireless networks typically exhibit only short relative delays, outdoor networks, like the Global System for Mobile (GSM) [164] can face maximum delays spreads in the order of 15 μs.

As a general rule, the effects of ISI on the transmission error statistics are negligible, as long as the delay spread is significantly shorter than the duration of one transmitted symbol. This implies that the symbol rate of communications systems is practically limited by the channel's memory. For higher symbol rates, there is typically significant deterioration of the

[1] *OFDM and MC-CDMA for Broadband Multi-user Communications WLANS and Broadcasting.*
L.Hanzo, Münster, T. Keller and B.J. Choi,
©2003 John Wiley & Sons, Ltd. ISBN 0-470-85879-6

system's error rate performance.

If symbol rates exceeding this limit are to be transmitted over the channel, mechanisms must be implemented in order to combat the effects of ISI. Channel equalisation techniques can be used to suppress the echos caused by the channel. To perform this operation, the channel impulse response must be estimated. Significant research efforts were invested into the development of such channel equalisers, and most wireless systems in operation use equalisers to combat ISI.

There is, however, an alternative approach towards transmitting data over a multipath channel. Instead of attempting to cancel the effects of the channel's echoes, orthogonal frequency division multiplexing (OFDM) [164] modems employ a set of harmonically related carriers in order to transmit information symbols in parallel over the channel. Since the system's data throughput is the sum of all the parallel channels' throughputs, the data rate per sub-channel is only a fraction of the data rate of a conventional single-carrier system having the same throughput. This allows us to design a system supporting high data rates, while maintaining symbol durations much longer than the channel's memory, thus circumventing the need for channel equalisation.

3.1 Orthogonal Frequency Division Multiplexing

3.1.1 History

Frequency division multiplexing (FDM) or multi-tone systems have been employed in military applications since the 1960s, for example by Bello [21], Zimmermann [8], Powers and Zimmermann [22], and others. Orthogonal frequency division multiplexing (OFDM), which employs multiple carriers overlapping in the frequency domain, was pioneered by Chang [1, 23]. Saltzberg [24] studied a multicarrier system employing orthogonal time-staggered quadrature amplitude modulation (O-QAM) on the carriers.

The use of the discrete Fourier transform (DFT) to replace the banks of sinusoidal generators and the demodulators was suggested by Weinstein and Ebert [9] in 1971, which significantly reduces the implementation complexity of OFDM modems. In 1980, Hirosaki [20] suggested an equalisation algorithm in order to suppress both intersymbol and intersubcarrier interference caused by the channel impulse response or timing and frequency errors. Simplified OFDM modem implementations were studied by Peled [13] in 1980, while Hirosaki [14] introduced the DFT-based implementation of Saltzberg's O-QAM OFDM system. From Erlangen University, Kolb [15], Schüßler [16], Preuss [17] and Rückriem [18] conducted further research into the application of OFDM. Cimini [10] and Kalet [19] published analytical and early seminal experimental results on the performance of OFDM modems in mobile communications channels.

More recent advances in OFDM transmission are presented in the impressive state-of-the-art collection of works edited by Fazel and Fettweis [25], including the research by Fettweis *et al.* at Dresden University, Rohling *et al.* at Braunschweig University, Vandendorp at Loeven University, Huber *et al.* at Erlangen University, Lindner *et al.* at Ulm University, Kammeyer *et al.* at Bremen University and Meyr *et al.* [26, 27] at Aachen University, but the individual contributions are too numerous to mention.

While OFDM transmission over mobile communications channels can alleviate the problem of multipath propagation, recent research efforts have focused on solving a set of inherent difficulties regarding OFDM, namely the peak-to-mean power ratio, time and frequency synchronisation, and on mitigating the effects of the frequency selective fading channel. These issues are addressed below in more depth.

3.1.1.1 Peak-to-Mean Power Ratio

It is plausibe that the OFDM signal – which is the superposition of a high number of modulated sub-channel signals – may exhibit a high instantaneous signal peak with respect to the average signal level. Furthermore, large signal amplitude swings are encountered, when the time domain signal traverses from a low instantaneous power waveform to a high power waveform, which may results in a high out-of-band (OOB) harmonic distortion power, unless the transmitter's power amplifier exhibits an extremely high linearity across the entire signal level range, as discussed in [2]. This then potentially contaminates the adjacent channels with adjacent channel interference. Practical amplifiers exhibit a finite amplitude range, in which they can be considered near-linear. In order to prevent severe clipping of the high OFDM signal peaks, which is the main source of OOB emissions, the power amplifier must not be driven into saturation and hence they are typically operated with a certain so-called back-off, creating a certain "headroom" for the signal peaks, which reduces the risk of amplifier saturation and OOB emission. Two different families of solutions have been suggested in the literature, in order to mitigate these problems, either reducing the peak-to-mean power ratio, or improving the amplification stage of the transmitter.

More explicitly, Shepherd [30], Jones [31], and Wulich [32] suggested different coding techniques which aim to minimise the peak power of the OFDM signal by employing different data encoding schemes before modulation, with the philosophy of choosing block codes whose legitimate code words exhibit low so-called crest factors or peak-to-mean power envelope fluctuation. Müller [33], Pauli [34], May [35] and Wulich [36] suggested different algorithms for post-processing the time domain OFDM signal prior to amplification, while Schmidt and Kammeyer [37] employed adaptive subcarrier allocation in order to reduce the crest factor. Dinis and Gusmão [38–40] researched the use of two-branch amplifiers, while the clustered OFDM technique introduced by Daneshrad, Cimini and Carloni [41] operates with a set of parallel partial FFT processors with associated transmitting chains. OFDM systems with increased robustness to non-linear distortion have been proposed by Okada, Nishijima and Komaki [42] as well as by Dinis and Gusmão [43].

3.1.1.2 Synchronisation

Time and frequency synchronisation between the transmitter and receiver is of crucial importance as regards the performance of an OFDM link [44, 45]. A wide variety of techniques have been proposed for estimating and correcting both timing and carrier frequency offsets at the OFDM receiver. Rough timing and frequency acquisition algorithms relying on known pilot symbols or pilot tones embedded into the OFDM symbols have been suggested by Claßen [26], Warner [46], Sari [47], Moose [48], as well as Brüninghaus and Rohling [49]. Fine frequency and timing tracking algorithms exploiting the OFDM signal's cyclic extension were published by Moose [48], Daffara [50] and Sandell [51].

3.1.1.3   OFDM/CDMA

Combining OFDM transmissions with code division multiple access (CDMA) allows us to exploit the wideband channel's inherent frequency diversity by spreading each symbol across multiple subcarriers. This technique has been pioneered by Yee, Linnartz and Fettweis [52], by Chouly, Brajal and Jourdan [53], as well as by Fettweis, Bahai and Anvari [54]. Fazel and Papke [55] investigated convolutional coding in conjunction with OFDM/CDMA. Prasad and Hara [56] compared various methods of combining the two techniques, identifying three different structures, namely multicarrier CDMA (MC-CDMA), multicarrier direct sequence CDMA (MC-DS-CDMA) and multi-tone CDMA (MT-CDMA). Like non-spread OFDM transmission, OFDM/CDMA methods suffer from high peak-to-mean power ratios, which are dependent on the frequency domain spreading scheme, as investigated by Choi, Kuan and Hanzo [57].

3.1.1.4   Adaptive Antennas

Combining adaptive antenna techniques with OFDM transmissions was shown to be advantageous in suppressing co-channel interference in cellular communications systems. Li, Cimini and Sollenberger [129–131], Kim, Choi and Cho [132], Lin, Cimini and Chuang [133] as well as Münster *et al.* [134] have investigated algorithms for multi-user channel estimation and interference suppression.

3.1.1.5   OFDM Applications

Due to their implementational complexity, OFDM applications have been scarce until quite recently. Recently, however, OFDM has been adopted as the new European digital audio broadcasting (DAB) standard [11, 12, 138–140] as well as for the terrestrial digital video broadcasting (DVB) system [47, 141].

For fixed-wire applications, OFDM is employed in the asynchronous digital subscriber line (ADSL) and high-bit-rate digital subscriber line (HDSL) systems [142–145] and it has also been suggested for power line communication systems [146, 147] due to its resilience to time dispersive channels and narrowband interferers.

More recently, OFDM applications were studied within the European 4th Framework Advanced Communications Technologies and Services (ACTS) programme [148]. The MEDIAN project investigated a 155 Mbps wireless asynchronous transfer mode (WATM) network [149–152], while the Magic WAND group [153, 154] developed a wireless local area network (LAN). Hallmann and Rohling [155] presented a range of different OFDM-based system proposals that were applicable to the European Telecommunications Standardisation Institute's (ETSI) recent personal communications oriented air interface concept [156].

3.2   Choice of the OFDM Modulation

Modulation of the OFDM subcarriers is analogous to the modulation in conventional serial systems. The modulation schemes of the subcarriers are generally quadrature amplitude modulation (QAM) or phase shift keying (PSK) [164] in conjunction with both coherent and non-coherent detection. Differential coded star-QAM (DSQAM) [164] can also be used. If

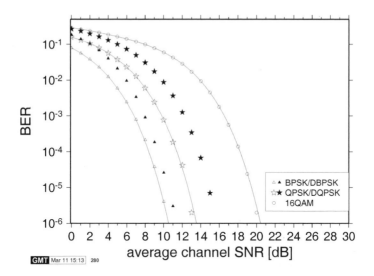

Figure 3.1: BER versus SNR curves for the OFDM modem in AWGN channel using BPSK, DBPSK, QPSK, DQPSK, and 16-QAM. The lines indicate the theoretical performance of the coherently detected modulation schemes in a serial modem over AWGN channels

coherently detected modulation schemes are used, then the reference phase of the OFDM symbol must be known, which can be acquired with the aid of pilot tones [174] embedded in the spectrum of the OFDM symbol, as will be discussed in Chapter 4. For differential detection the knowledge of the absolute subcarrier phase is not necessary, and differentially coded signalling can be invoked either between neighbouring subcarriers or between the same subcarriers of consecutive OFDM symbols.

3.3 OFDM System Performance over AWGN Channels

As the additive white Gaussian noise (AWGN) in the time domain channel corresponds to AWGN of the same average power in the frequency domain, an OFDM modem's performance in an AWGN channel is identical to that of a serial modem. Analogously to a serial system, the bit error rate (BER) versus signal-to-noise rate (SNR) characteristics are determined by the frequency domain modulation scheme used. In Figure 3.1 the simulated BER versus SNR curves for binary phase shift keying (BPSK), differential BPSK (DBPSK), quaternary phase shift keying (QPSK), differential QPSK (DQPSK) and coherent 16-quadrature amplitude modulation (16-QAM) are shown, together with the theoretical BER curves of serial modems, as derived in [177]:

$$p_{e,BPSK}(\gamma) = Q(\sqrt{2\gamma}), \tag{3.1}$$

$$p_{e,QPSK}(\gamma) = Q(\sqrt{\gamma}), \tag{3.2}$$

$$p_{e,16QAM}(\gamma) = 0.75 \cdot Q\left(\sqrt{\frac{\gamma}{5}}\right) 0.25 \cdot Q\left(\sqrt{\frac{\gamma}{5}}\right), \tag{3.3}$$

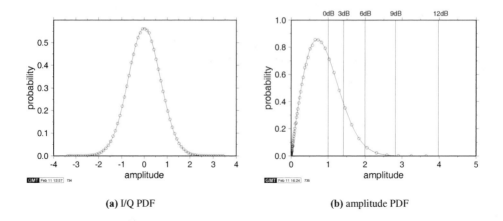

(a) I/Q PDF   **(b)** amplitude PDF

Figure 3.2: Statistics of OFDM time domain signal: (a) amplitude histogram of the I component of
a 256-subcarrier OFDM signal (markers) obeying a Gaussian PDF (continuous line) with
$\sigma = 1/\sqrt{2}$; (b) two-dimensional amplitude histogram (markers) following a Rayleigh
PDF (continuous line) with $\sigma = 1/\sqrt{2}$. The vertical lines in (b) indicate the corresponding
powers above mean in decibels

where γ is the SNR and the Gaussian $Q()$-function is defined as

$$Q(y) = \frac{1}{\sqrt{2\pi}} \int_y^\infty e^{-x^2/2} dx = \text{erfc}\left(\frac{y}{\sqrt{2}}\right).$$

It can be seen from Figure 3.1 that the experimental BER performance of the OFDM modem
is in very good accordance with the theoretical BER curves of conventional serial modems in
AWGN channels.

3.4   Clipping Amplification

3.4.1   OFDM Signal Amplitude Statistics

The time domain OFDM signal is constituted by the sum of complex exponential functions,
whose amplitudes and phases are determined by the data symbols transmitted over the different carriers. Assuming random data symbols, the resulting time domain signal exhibits an
amplitude probability density function (PDF) approaching the two-dimensional or complex
Gaussian distribution for a high number of subcarriers. Figure 3.2(a) explicitly shows that the
measured amplitude histogram of the in-phase (I) component of a 256-subcarrier OFDM signal obeys a Gaussian distribution with a standard deviation of $\sigma = 1/\sqrt{2}$. The mean power
of the signal is $2\sigma^2 = 1$. As the amplitude distribution in both the in-phase and the quadrature phase (Q) component is Gaussian, the two-dimensional amplitude histogram follows a
complex Gaussian, or in other words, a Rayleigh distribution [177] with the same standard
deviation. The observed amplitude histogram of the 256-subcarrier OFDM signal and the
corresponding Rayleigh probability density function are depicted in Figure 3.2(b). The verti-

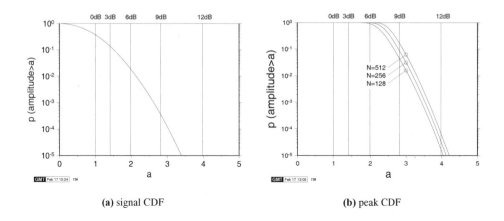

(a) signal CDF **(b)** peak CDF

Figure 3.3: OFDM time domain signal: probability of (a) the instantaneous signal amplitude and (b) the peak amplitude per OFDM symbol being above a given threshold value. The lines indicate the corresponding power levels above the mean. The peak value CDF depends on the number of subcarriers, and curves are shown for 128, 256 and 512 carriers

cal grid lines in the figure indicate the relative amplitude above the mean value in decibels. It can be seen that unlike full response serial modulation schemes – which have a more limited range of possible output amplitudes – the parallel modem's output signal exhibits strong amplitude fluctuations. Note that the standard deviation σ of the probability density functions depicted in Figure 3.2(a) is independent of the number of subcarriers employed, since the mean power of the signal is normalised to 1.

The probability of the instantaneous signal amplitude being above a given threshold, the cumulative density function (CDF), is depicted in Figure 3.3(a). It can be noted that the probability of the signal amplitude exceeding the 6 dB mark is about 1.1%, while the 9dB mark is exceeded with a probability of about $3.5 \cdot 10^{-4}$. The signal amplitude is higher than the average by 10.6 dB with a probability of 10^{-5}.

3.4.2 Clipping Amplifier Simulations

In order to evaluate the effects of a non-linear amplifier on the performance of an OFDM system, simulations have been conducted employing a simple clipping amplifier model. This clipping amplifier limits the amplitude of the transmitted signal to a given level, without perturbing the phase information. This amplitude limitation of the time domain signal affects both the received symbols on all subcarriers in the OFDM symbol, as well as the frequency domain out-of-band emissions, and therefore increases the interference inflicted on adjacent carriers.

In Figure 3.3(a) the probability of the instantaneous signal amplitude exceeding a given level was shown. As amplitude limitation in the time domain affects all the subcarriers in the OFDM symbol, the probability of the time domain peak amplitude per OFDM symbol period being clipped by the amplifier is the probability of at least one of the N time domain samples exceeding a given amplitude limit. This clipping probability for a given maximal amplifier amplitude a is displayed in Figure 3.3(b). It can be observed that the clipping probability per

OFDM symbol is dependent on the number of subcarriers employed.

Given the information in Figure 3.3(b), the necessary amplifier back-off for an OFDM transmitter can be determined. If the acceptable clipping probability per OFDM symbol is 10^{-5}, then the necessary amplifier back-off values would be 12.1 dB, 12.3 dB, and 12.5 dB for a 128, 256, and 512 subcarrier OFDM modem, respectively. If a clipping probability of 10^{-4} is acceptable, then these back-off values can be reduced by about 0.6 dB.

3.4.2.1   Introduction to Peak-Power Reduction Techniques

Two main types of peak-to-mean power ratio reduction techniques have been investigated in the literature, which rely on either introducing redundancy in the data stream or on post-processing the time domain OFDM signal before amplification, respectively. In this section we provide a rudimentary introduction to the problem and provide some quantitative results characterising the performance impairments imposed by the peak-to-mean power ratio fluc-tutations. We will return to the subject by adopting a more advanced approach, where a more detailed exposure of the associated issues will be provided in Part II, namely in the context of Chapter 11 of the book.

Shepherd [30], Jones [31], and Wulich [32] suggest different coding techniques which aim to ensure that only low peak power OFDM symbols are chosen for transmission, whilst excluding the transmission of the specific combinations of modulating bits that are known to result in the highest peak factors. These schemes can be viewed as simple k out of n block codes. Depending on the tolerable peak to mean power ratio, the set of acceptable OFDM symbols is computed and the data sequences are mapped onto OFDM symbols from this set only. These techniques reduce the data throughput of the system, but mitigate the severity of clipping amplification.

Müller [33] proposes techniques based on generating a set of OFDM signals by multiply-ing the modulating data vector in the frequency domain with a set of different phase vectors known to both the transmitter and the receiver, before applying the IDFT. The transmitter will then choose the resulting OFDM symbol exhibiting the lowest peak factor and transmits this together with the chosen phase vector's identification. This technique reduces the aver-age peak factors and requires only a low signalling overhead, but cannot guarantee a given maximum peak factor.

Another group of peak amplitude limiting techniques is based on modifying the time do-main signal, where its amplitude exceeds the given peak amplitude limit. Straightforward clipping falls into this category, but induces strong frequency domain out-of-band emissions. In order to avoid the spectral domain spillage of hard clipping, both multiplicative and addi-tive time domain modifications of the OFDM signal have been investigated. Pauli [34] pro-poses a multiplicative correction of the peak values and their adjacent samples using smooth Gaussian functions for limiting out-of-band emissions. However, multiplying the time do-main signal with the time-varying amplitude limiting function introduces intersubcarrier in-terference.

A similar technique, using additive instead of multiplicative amplitude limiting, is de-scribed by May [35]. A smoothly varying time domain signal is added to the time domain signal, which has been optimised for low out-of-band emissions. As the DFT is a linear op-eration, adding a correction signal in the time domain will overlay the frequency domain data symbols with the DFT of the correction signal, introducing additional noise.

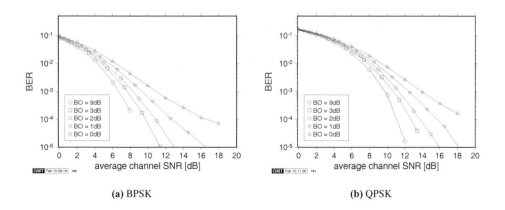

(a) BPSK   (b) QPSK

Figure 3.4: Influence of amplitude clipping on BER performance of an OFDM modem

As an alternative solution, Wulich [36] suggests measuring the peak amplitude in each OFDM time domain symbol and scaling the amplitude for the whole OFDM symbol, so that the peak value becomes constant from symbol to symbol. At the receiver, an amplitude correction similar to pilot-assisted scaling over fading channels has to be performed. The advantage of this technique is that the system's throughput is not reduced by introducing redundant information, nor is the orthogonality of the subcarriers impaired. As the OFDM symbol energy is fluctuating with the peak power exhibited by the different OFDM symbols, however, the BER performance of the system is impaired.

A further approach to reducing the amplifier linearity requirements, and which relies on the findings of Figure 3.3(b), is to split an OFDM symbol into groups of subcarriers, which are processed by IFFT, amplified and transmitted over separate transmit chains. This system, referred to as clustered OFDM, has been proposed by Daneshrad and Cimini [41], and combines the advantages of reduced peak power in each transmitter chain with additional spatial diversity from the use of multiple transmit antennas, which can be exploited by coding or OFDM-CDMA techniques.

3.4.2.2 BER Performance Using Clipping Amplifiers

If lower back-off values are chosen and no peak power reduction techniques are employed in the OFDM transmission system, then the clipping amplifier will influence both the performance of the OFDM link as well as the out-of-band emissions. The effects of the amplifier back-off value on the BER performance in an AWGN channel for coherently detected 512-subcarrier BPSK and QPSK OFDM transmission are depicted in Figure 3.4.

The investigated amplifier back-off values range from 9 dB down to 0 dB. According to Figure 3.3(b), an amplifier back-off of 9 dB results in a clipping probability per OFDM symbol of about 17%, while an amplifier back-off of 0 dB results in nearly certain clipping of every transmitted OFDM symbol.

It can be observed in Figure 3.4 for both BPSK and QPSK that for an amplifier back-off value of 9 dB the BER performance is indistinguishable from the non-limited case, which was portrayed in Figure 3.1. For smaller back-off values, however, the effects of clipping on

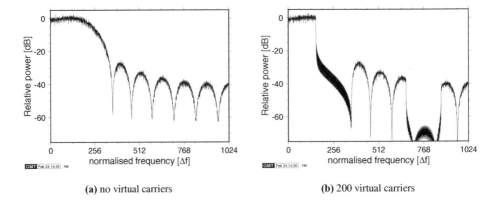

(a) no virtual carriers   (b) 200 virtual carriers

Figure 3.5: Spectrum of a 512-subcarrier OFDM signal using raised-cosine Nyquist pulse shaping filter with an excess bandwidth of $\alpha = 0.35$: (a) no virtual subcarriers, (b) 2×100 subcarriers at the edges of the spectrum. The characteristic sinc-signal shaped curve of this figure is a consequence of rectangular windowing used in the spectral estimation. This effect can be mitigated by smoother windowing.

the BER performance are more severe. A 3 dB amplifier back-off results in a SNR penalty of about 1.3 dB at a BER of 10^{-4}, and a back-off of 0 dB results in more than 8 dB SNR penalty at the same BER.

The BER performance of the OFDM modem can be improved slightly if the receiver's decision boundaries are adjusted according to the amplitude loss in each subcarrier due to the overall signal power loss, which was observed by O'Neill and Lopes [178]. In fading channels this amplitude adjustment would be performed in conjunction with the channel estimation.

3.4.2.3 Signal Spectrum with Clipping Amplifier

The simulated spectrum of a 512-subcarrier OFDM signal is shown in Figure 3.5. Figure 3.5(a) depicts the spectrum of the transmitted signal, when all 512 subcarriers are used for data transmission. A raised cosine Nyquist filter with an excess bandwidth of $\alpha = 0.35$ was used for pulse shaping, resulting in a relatively slow fall-off of the power spectral density outside the FFT bandwidth of $\pm 256 \Delta f$. In order to avoid adjacent channel interference due to spectral overlapping, an adjacent carrier must be placed at a frequency distance considerably higher than the data symbol rate. If a tighter spectral packing of adjacent carriers is important, then pulse shaping filters with a steeper frequency transfer function have to be employed, which are more complex to implement. Since in this scenario all the subcarriers are used for data transmission, the time domain samples are uncorrelated and therefore the spectrum in Figure 3.5(a) is equivalent to that of a serial modem transmitting at the same sample rate and employing the same pulse shaping filter.

Figure 3.5(b) depicts the signal spectrum of a 512-subcarrier modem employing 200 virtual subcarriers at the edges of the bandwidth, which are effectively disabled and carry no energy. Although the same pulse shaping filter was employed as in Figure 3.5(a), the re-

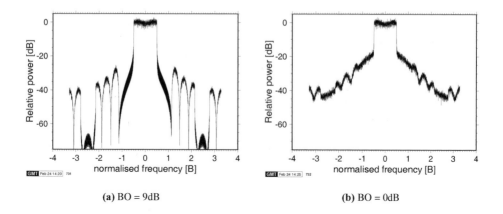

(a) BO = 9dB   (b) BO = 0dB

Figure 3.6: Power spectra of a 512-subcarrier OFDM modem employing 200 virtual carriers and Nyquist filtering with an excess bandwidth of 0.35. The frequencies are normalised to the signal bandwidth of $B = 312 \cdot \Delta f$: (a) 9 dB amplifier back-off (BO), (b) 0 dB amplifier back-off

sulting spectral shape is nearly rectangular, with a drop of approximately 20 dB within one subcarrier distance Δf outside the signal bandwidth. Comparing this spectrum with Figure 3.5(a) it can be observed that the overall shape of the spectrum is determined by the pulse shaping filter, while the employment of disabled virtual subcarriers cuts out parts of the spectrum. It can be seen from the spectrum that by employing virtual subcarriers a very tight power spectrum can be obtained with a simple pulse shaping filter. This rectangular signal spectrum allows tight packing of adjacent carriers, therefore enhancing the spectral efficiency of an OFDM system.

The effects of a clipping amplifier on the OFDM signal spectrum are demonstrated in Figure 3.6. Specifically, Figure 3.6(a) shows the spectrum corresponding to the system employed in the context of Figure 3.5(b), but with the amplifier back-off set to 9 dB. The frequency axis is normalised to the signal bandwidth of $B = 312 \cdot \Delta f$ due to the disabled carriers and the power spectral density is normalised to the average signal power in the signal band. Comparison of Figure 3.6(a) and 3.5(b) shows no discernible change in the spectral domain.

Reducing the amplifier back-off results in increased out-of-band emissions. Figure 3.6(b) shows the averaged power spectral density of the same system with an amplifier back-off of 0 dB. In this case, the spectral spillage is apparent directly outside the signal band.

Since the amplitude limiting of the transmitted time domain signal is due to the power amplifier, no bandlimiting filtering is performed after the amplitude distortion. Li and Cimini [179] proposed employing a hard limiter and a lowpass filter before the non-linear amplification stage. In this case, the out-of-band emissions can be reduced, while the BER penalty due to hard limiting the time domain signal still applies.

In order to evaluate the amount of adjacent channel interference caused by non-linear amplification of the OFDM signals, the total interference power in a frequency domain window of width B having a variable frequency offset was integrated and normalised by the in-band power for varying values of amplifier back-off. Figure 3.7(b) shows the results of this integration for amplifier back-off values up to 9 dB and adjacent carrier separations from B to

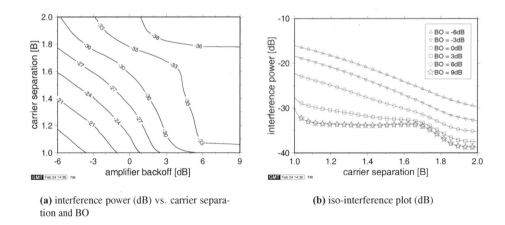

(a) interference power (dB) vs. carrier separation and BO

(b) iso-interference plot (dB)

Figure 3.7: Integrated adjacent channel interference power (dB) in a signal bandwidth of B for different values of normalised carrier separation and amplifier back-off

$2B$. A carrier separation of B corresponds to directly adjacent signal bands, and such a set-up would approach the maximal possible system throughput of 2 symbols/s/Hz or 2 Baud/Hz.

It can be observed from the figure that the integrated interference levels are indistinguishable for amplifier back-off values of 6 dB and 9 dB. The integrated interference level in this case is below -30 dB with directly adjacent carriers and below -33 dB for all carrier separation values above $1.03B$. The drop in interference for carrier separations above $1.7B$ is due to the frequency domain image of the virtual carriers that can be seen in Figure 3.6(a) at frequencies between $2.2B$ and $3.8B$. For a back-off value of 3 dB the interference is approximately 1 dB worse than for the higher back-off values for carrier separations above $1.5B$, and this difference is up to 5 dB for smaller carrier separations.

For a back-off value of 0 dB, which corresponds to the spectrum shown in Figure 3.6(b), the integrated interference is below -22.5 dB for all carrier separations. The figure also shows two graphs for negative amplifier back-off values for comparison, which exhibit considerably higher interference values. Since negative back-off values imply amplifier clipping below the nominal mean power of the signal, this would not be a likely case at the transmitter's power amplifier.

3.4.3   Clipping Amplification – Summary

We have seen that for OFDM transmission a considerable amplifier back-off is necessary, in order to maintain the BER and spectral characteristics of the ideal OFDM system. This complicates the transmitter design and leads to inefficient power amplifier applications.

Although a 512-subcarrier OFDM modem is capable of producing a peak-to-average power ratio of up to 27 dB, we have seen that much lower amplifier back-off values result in acceptable system performance. A back-off of 12.5 dB results in only one of 10^5 OFDM symbols suffering clipping at all, and we have seen for a 6 dB back-off that both the BER performance as well as the averaged spectrum are indistinguishable from the ideally amplified scenario.

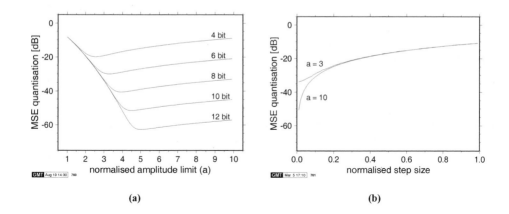

Figure 3.8: Mean square quantisation error for linearly spaced ADC at OFDM receiver: (a) quantisation noise for different amplitude limits normalised to the average input signal amplitude σ versus ADC resolution; (b) quantisation noise for amplitude limits of $a = 3$ and $a = 10$ versus the linear step size

3.5 Analogue-to-Digital Conversion

We saw in Figure 3.2 that the time domain OFDM signal follows a two-dimensional or complex Gaussian distribution. The distribution of signal amplitudes determines the necessary parameters of the analogue-to-digital convertor (ADC) at the receiver. We have investigated the effects of the receiver input ADC on the BER performance of an OFDM system in an AWGN channel environment.

Since an ADC transforms the continuous analogue input signal to discrete-valued digital information, the original input signal is not preserved perfectly. As the dynamic range of an ADC is limited, there are amplitude clipping effects if the input signal exceeds the ADC's maximal input amplitude.

A linear analogue-to-digital convertor quantises the input signal into a set of n equidistant output values, which span an interval with clipping values of $-a$ and a. In this case, the ADC error stems from quantisation error due to the finite step size of $2a/n$ for input values within the dynamic range $[-a, a]$, as well as from amplitude clipping for input values outside this range.

Taking into account the Gaussian nature of the I and Q components of the OFDM signal, the mean square ADC error can be determined for different ADC set-ups. Figure 3.8(a) shows the mean square ADC error normalised to the average input signal power versus clipping amplitude for different quantisation resolution values. For this and the following figures, the input signal is modelled by a white Gaussian noise function with a standard deviation σ. The mean square quantisation error is relative to the mean input power σ^2, and the amplitude limit a is in units of σ.

The mean square ADC error value can be interpreted as the mean power of an additional random noise superimposed on the quantised signal. This ADC noise is not Gaussian, hence estimating the BER performance degradation with the Gaussian noise approximation is inadequate.

It can be seen from the figure that a small dynamic range of $2a$ results in high ADC noise irrespective of the number of ADC bits, which is due to predominant clipping. For $a = 1$, the normalised ADC noise power is about -8 dB, again independently of the resolution of the ADC. For higher values of a, the total ADC noise is dominated by the effects of increasing step size. Depending on the AGC's given resolution, there is an optimal value of a, for which the ADC noise, constituted by clipping and granular noise, is minimal.

In order to separate the effects of the amplitude limitation and the quantisation step size, BER simulations have been conducted with a fixed amplitude limiter and a range of ADC step sizes. The linear ADC step size is given for a range of ADC resolutions b and two amplitude limits a in Table 3.1. The effects of clipping on the OFDM performance have been discussed in the context of non-linear amplification, and Figure 3.4 showed that an amplifier back-off of 9 dB results in no BER performance degradation due to amplifier clipping in an AWGN channel for a 512-subcarrier OFDM modem. Setting the amplitude limits of the simulated ADC to $a = 3$ corresponds to a 9.5 dB dynamic range, hence the amplitude clipping contribution to the ADC noise does not significantly impair the modem's BER performance.

$a\backslash b$	4	5	6	7	8	9	10	11	12
3	0.375	0.188	0.094	0.0469	0.0234	0.0117	0.00586	0.00293	0.00146
10	0.625	0.313	0.156	0.0781	0.039	0.0195	0.00977	0.00488	0.00244

Table 3.1: Linear ADC step size for amplitude limits of $a = 3$ and $a = 10$ and ADC resolutions from $b = 4$ bits/sample to $b = 12$ bits/sample

The average ADC noise power for the normalised amplitude limits of $a = 3$ and $a = 10$ is depicted in Figure 3.8(b). The amplitude limit of $a = 3$ results in a residual ADC noise of about -34 dB due to amplitude clipping, while the amplitude clipping noise contribution for a limit of $a = 10$ is below -50 dB. For step sizes above 0.2, which corresponds to ADC resolutions of 5 bits or less, the gap between the two curves is closed, and the noise power is dominated by the quantisation noise.

The BER performance of a 512-subcarrier OFDM modem employing coherently as well as differentially detected QPSK is depicted in Figure 3.9. These experiments have been conducted for a maximal amplitude of $a = 3$ with step sizes from 0.1 to 0.5, which roughly corresponds to ADC resolutions between 6 and 3 bits. In both graphs the curve without markers represents the modem's BER performance in the absence of ADC quantisation and clipping noise.

It can be observed in both figures that the SNR penalty at a BER of 10^{-4} with an ADC step size of 0.1 is 0.15 dB and 0.2 dB for QPSK and DQPSK, respectively. This resolution corresponds to 60 steps in the range from -3 to 3, and therefore would require an ADC of 6-bit resolution. For higher step sizes the SNR penalty is more severe; a step size of 0.5 results in a SNR penalty of about 2.8 dB and 3.2 dB for coherent and differential detection, respectively. Let us now consider the effects of oscillator noise.

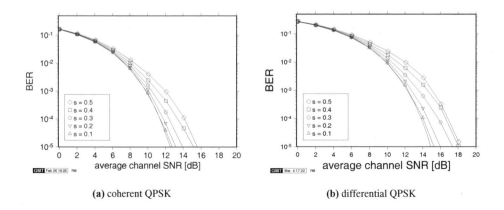

(a) coherent QPSK   (b) differential QPSK

Figure 3.9: BER versus channel SNR for a QPSK modem with coherent and differential detection using different normalised receiver ADC resolution values in an AWGN channel. The maximal normalised amplitude is $a = 3$

3.6 Phase Noise

The presence of phase noise is an important limiting factor for an OFDM system's performance [44, 45, 180], and depends on the quality and the operating conditions of the system's RF hardware. In conventional mobile radio systems around a carrier frequency of 2GHz the phase noise constitutes typically no severe limitation, however, in the 60GHz carrier frequency, 225MHz bandwidth system considered in Section 4.1.1 its effects were less negligible and hence had to be investigated in more depth. Oscillator noise stems from oscillator inaccuracies in both the transmitter and receiver and manifests itself in the baseband as additional phase and amplitude modulation of the received samples [181]. The oscillator noise influence on the signal depends on the noise characteristics of the oscillators in the system and on the signal bandwidth. It is generally split in amplitude noise $A(t)$ and phase noise $\Phi(t)$, and the influence of the amplitude noise $A(t)$ on the data samples is often neglected. The time domain functions $A(t)$ and $\Phi(t)$ have Gaussian histograms, and their time domain correlation are determined by their respective long-term power spectra through the Wiener-Khintchine theorem.

If the amplitude noise is neglected, imperfect oscillators are characterised by the long-term power spectral density (PSD) $N_p(f')$ of the oscillator output signal's phase noise, which is also referred to as the phase noise mask. The variable f' represents the frequency distance from the oscillator's nominal carrier frequency in a bandpass model, or equivalently, the absolute frequency in the baseband. An example of this phase noise mask for a practical oscillator is given in Figure 3.10(a). If the phase noise PSD $N_p(f')$ of a specific oscillator is known, then the variance of the phase error $\Phi(t)$ for noise components in a frequency band $[f_1, f_2]$ is the integral of the phase noise spectral density over this frequency band as in [181]:

$$\bar{\Phi}^2 = \int_{f_1}^{f_2} \left(\frac{2N_p(f')}{C} \right) df', \tag{3.4}$$

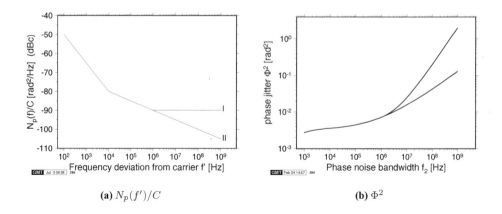

(a) $N_p(f')/C$ **(b)** Φ^2

Figure 3.10: Phase noise characterisation: (a) spectral phase noise density (phase noise mask), (b) integrated phase jitter for two different phase noise masks

where C is the carrier power and the factor 2 represents the double-sided spectrum of the phase noise. The phase noise variance $\bar{\Phi}^2$ is also referred to as the integrated phase jitter, which is depicted in Figure 3.10(b).

3.6.1   Effects of Phase Noise

The phase noise contribution of both the transmitter and receiver can be viewed as an additional multiplicative effect of the radio channel, like fast and slow fading. In serial modulation schemes phase noise manifests itself as random phase errors of the received samples. The effects of this additional random phase component on the received samples depend on the modulation scheme employed, and as expected, they are more pronounced in differential detection schemes than in coherently detected arrangements invoking carrier recovery. The carrier recovery, however, is affected by the phase noise, which in turn degrades the performance of a coherently detected scheme.

For OFDM schemes, multiplication of the received time domain signal with a time-varying channel transfer function is equivalent to convolving the frequency domain spectrum of the OFDM signal with the frequency domain channel transfer function. Since the phase noise spectrum's bandwidth is wider than the subcarrier spacing, this results in energy spillage into other sub-channels and therefore in intersubcarrier interference, an effect which will be quantified below.

3.6.2   Phase Noise Simulations

In our studies two different models of the phase noise in an OFDM communications system have been investigated: the simple band-limited white phase noise model, which is solely based on the value of the integrated phase jitter of Equation 3.4, and a coloured phase noise model, which also takes into account the phase noise mask of Figure 3.10(a). Both of these models will be described below.

3.6.2.1 White Phase Noise Model

The simplest way of modelling the effect of phase noise in simulations is to assume uncorrelated Gaussian distributed phase errors with a standard deviation of $\bar{\Phi}$ at the signal sampling instants. This corresponds to the phase noise exhibiting a uniform PSD $N_p(f') = (\bar{\Phi})^2 / B$ throughout the signal bandwidth B.

As the phase noise is assumed to be white, the integrated phase jitter, or the variance of the phase noise given in Equation 3.4 constitutes sufficient information for generating the phase error signal $\Phi(t)$. Clearly, since no correlation between the noise samples is assumed, this is the worst-case scenario with the highest possible differences of $\Phi(t)$ between two consecutive samples.

Simulations have been performed for both a 512-subcarrier OFDM modem, as well as for a serial modem for comparison. The serial modem's BER performance curves are depicted in Figure 3.11 for BPSK and QPSK employing both coherent and differential detection, for integrated phase jitter values of $\bar{\Phi}^2 = 0.05 \; rad^2$ to $\bar{\Phi}^2 = 0.25 \; rad^2$. For coherently detected BPSK, as shown in Figure 3.11(a), white phase noise of $\bar{\Phi}^2 = 0.05 \; rad^2$ already causes a measurable SNR penalty of about 0.6 dB at a BER of 10^{-4}, while a phase jitter of $0.1 \; rad^2$ results in a SNR penalty of 1.8 dB at this BER. For phase jitter values above $0.1 \; rad^2$, residual bit error rates of $6 \cdot 10^{-5}$, $5 \cdot 10^{-4}$ and $2 \cdot 10^{-3}$ were observed at phase jitters of 0.15, 0.2 and $0.25 \; rad^2$, respectively.

3.6.2.1.1 Serial Modem The serial QPSK modem's performance is more vulnerable to the effects of phase noise than that of BPSK, as is shown in Figure 3.11(a) and 3.11(b). For all simulated values of $\bar{\Phi}^2$, the bit error rate exceeds 10^{-4}. For a phase jitter of $0.25 \; rad^2$, the observed residual bit error rate is 6%. The curves for differential detection of BPSK and QPSK are shown in Figure 3.11(c) and 3.11(d), respectively. Since for differentially detected PSK schemes both the received modulated symbol as well as the reference phase are corrupted by phase noise, the phase noise effects are severe. The DBPSK modem exhibits residual bit errors of more than 10^{-4} for all simulated phase jitter values above $0.05 \; rad^2$, while the DQPSK modem's BER performance is worse than 0.9% for all simulated values of $\bar{\Phi}^2$.

3.6.2.1.2 OFDM Modem While for serial modems the phase noise results in phase errors of the received samples, the FFT modem's performance is affected by the inter-subcarrier interference caused by the convolution of the phase noise spectrum with the OFDM spectrum. The average amount of power spillage from one subcarrier into the rest of the OFDM spectrum observed for different values of integrated phase jitter is depicted in Figure 3.12(a). It can be seen that the white phase noise results in uniform spreading of inter-subcarrier interference over all subcarriers, and a loss of signal power in the desired subcarrier number 256. It can be noted that doubling the variance of the phase noise results approximately in doubling the interference power in the other subcarriers, as demonstrated by the 3 dB differences seen in Figure 3.12(a). The received signal power in the central subcarrier drops accordingly.

The total interference power received in each subcarrier is the sum of all active subcarriers' contributions, which is shown in Figure 3.12(b) for a 512-subcarrier OFDM modem. It can be seen that the signal energy drops while the interference energy rises for rising values of the integrated phase jitter, with the signal-to-interference ratio (SIR) dropping to about 1.5,

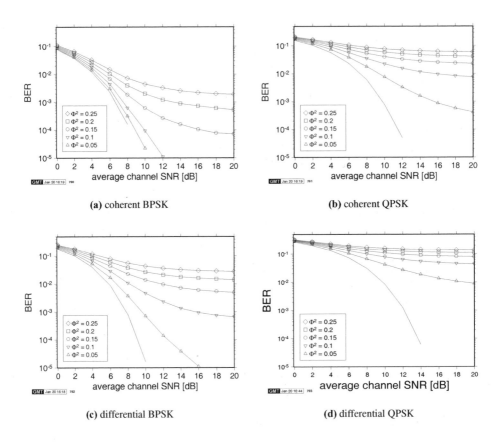

Figure 3.11: Bit error rate versus channel SNR for a serial modem in an AWGN channel and exposed
to white phase noise for integrated phase jitter values of $0.05\ rad^2$ to $0.25\ rad^2$: (a)
BPSK, (b) QPSK, (c) DBPSK, (d) DQPSK

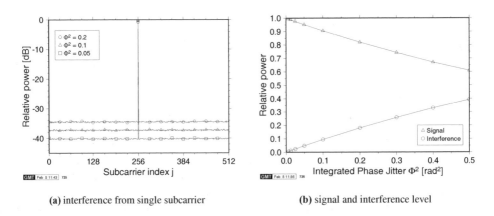

Figure 3.12: The effect of white phase noise on OFDM: (a) power spillage due to white phase noise,
(b) average signal and interference levels for one subcarrier

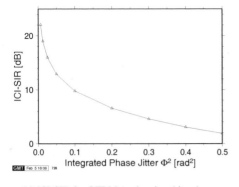

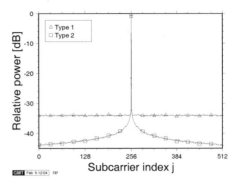

(a) ICI-SIR for OFDM modem in white phase noise

(b) interference from single subcarrier, coloured phase noise

Figure 3.13: OFDM in phase noise channel: (a) ICI-SIR for OFDM modem in white phase noise channel; (b) average signal and interference levels for one subcarrier in the coloured phase noise channels corresponding to the phase noise masks shown in Figure 3.10(a)

which corresponds to 2 dB for $\bar{\Phi}^2 = 0.5$. For small values of phase jitter up to approximately 0.1, the interference power rises linearly with the phase jitter, and more slowly for higher values of phase jitter. The interchannel-interference (ICI) experienced in each subcarrier for the white phase noise channel depending on the phase jitter is shown in Figure 3.13(a). The SIR is above 10 dB for phase jitter values below approximately 0.1, and 20 dB for values phase jitter values of 0.01. This is in accordance to the linear relationship of interference power and phase jitter for small values observed above.

Figure 3.14 shows the BER performance results observed for a 512-subcarrier OFDM modem in the white phase noise channels. For coherently detected OFDM BPSK the BER performance is generally worse than for the serial modem, characterised in Figure 3.11(a). For differential detection, however, the OFDM modem's BER performance is better than the serial modem's depicted in Figure 3.11(c). This different behaviour can be explained by the different effects of the phase noise on the different systems. It has been observed in Figure 3.12(b) that for the investigated values of the integrated phase jitter the mean interference power in each subcarrier of an OFDM modem increases approximately linearly with the mean square of the phase error in the serial modem. For the same values of noise variance, coherent serial BPSK is more robust to the phase errors in the received serial symbols than DBPSK, while the noise-like interference observed in the OFDM system is less detrimental to the differentially detected BPSK than to coherent BPSK.

For both differential and coherent QPSK the OFDM modem performs better than the serial modem, as can be seen comparing Figure 3.14(b) with 3.11(b) and 3.14(d) with 3.11(d). There is, however, considerable degradation of the BER performance in both cases. Having considered the worst-case scenario of uncorrelated white phase noise, let us now focus our attention on the more practical case of coloured phase noise sources.

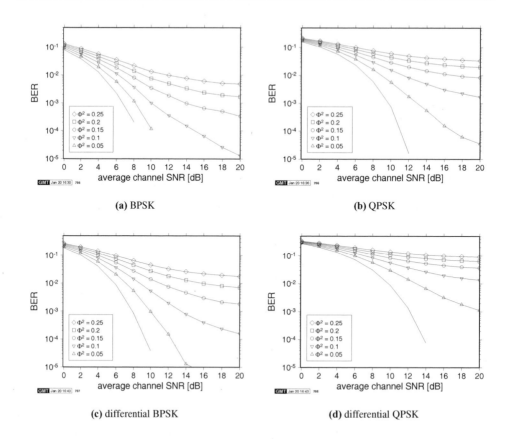

Figure 3.14: Bit error rate versus channel SNR for a 512 subcarrier OFDM modem in an AWGN channel and white phase noise for different values of integrated phase jitter of $0.05\ rad^2$ to $0.25\ rad^2$: (a) BPSK, (b) QPSK, (c) DBPSK, (d) DQPSK (compared with Figure 3.11)

3.6.2.2   Coloured Phase Noise Model

The integral $\bar{\Phi}^2$ of Equation 3.4 characterises the long-term statistical properties of the oscillator's phase and frequency errors due to phase noise. In order to create a time domain function satisfying the standard deviation $\bar{\Phi}^2$, a white Gaussian noise spectrum was filtered with the phase noise mask $N_p(f')$ depicted in Figure 3.10(a), which was transformed into the time domain. A frequency resolution of about 50 Hz was assumed in order to model the shape of the phase noise mask at low frequencies, which led to a FFT transform length of $2^{22} = 4194304$ samples for the frequency range of Figure 3.10(a).

The resulting time domain phase noise channel data is a stream of phase error samples, which were used to distort the incoming signal at the receiver. The double-sided phase noise mask used for the simulations is given in Table 3.2. Between the points given in Table 3.2, a log-linear interpolation is assumed, as shown in Figure 3.10(a). As the commercial oscillator's phase noise mask used in our investigations was not specified for frequencies beyond 1 MHz, two different cases were considered for frequencies beyond 1 MHz: (I) a

f'[Hz]	100	1k	10k	100k	1M
N_p/C[dB]	-50	-65	-80	-85	-90

Table 3.2: Two-sided phase noise mask used for simulations. f' = frequency distance from carrier, N_p/C = normalised phase noise density

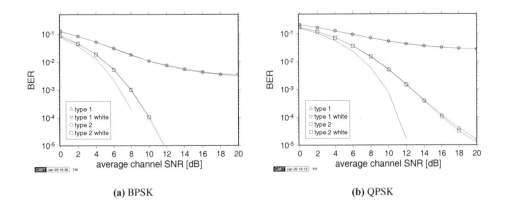

(a) BPSK(b) QPSK

Figure 3.15: Bit error rate versus channel SNR for a 512-subcarrier OFDM modem in the presence of phase noise. Type 1 represents the coloured phase noise channel with the phase noise mask depicted in Figure 3.10(a) assuming a noise floor of 90 rad^2/Hz, while Type 2 is the channel without phase noise floor. The curves designated "white" are the corresponding white phase noise results. The lines without markers give the corresponding results in the absence of phase noise

phase noise floor at -90 dB, and (II) an $f^{-1/2}$ law. Both of these extended phase noise masks are shown in Figure 3.10(a). The integrated phase jitter has been calculated using Equation 3.4 for both scenarios, and the value of the integral for different noise bandwidths is depicted in Figure 3.10(b).

For the investigated 155 Mbits/s wireless ATM (WATM) system's [149–152] double-sided bandwidth of 225 MHz, the integration of the phase noise masks results in phase jitter values of $\bar{\Phi}^2 = 0.2303\ rad^2$ and $\bar{\Phi}^2 = 0.04533\ rad^2$ for the phase noise mask with and without noise floor, respectively.

The simulated BER performance of a 512-subcarrier OFDM system with a subcarrier distance $\Delta f = 440\ kHz$ over the two different phase noise channels is depicted in Figure 3.15 for coherently detected BPSK and QPSK. In addition to the BER graphs corresponding to the coloured phase noise channels described above, graphs of the modems' BER performance over white phase noise channels with the equivalent integrated phase jitter values were also plotted in the figures. It can be observed that the BER performance for both modulation schemes and for both phase noise masks is very similar for the coloured and the white phase noise models.

The interference caused by one OFDM subcarrier in the two coloured phase noise channels is depicted in Figure 3.13(b). It can be seen that the Type 1 channel exhibits a virtually white interference spectrum, very similar to the white phase noise channels depicted in Figure

3.12(a). Only the subcarriers directly adjacent to the signal bearer show higher interference influence than in the white phase noise channel. This is due to the Type 1 phase noise density given in the phase noise mask; it is non-flat only for frequencies below 1 MHz, which corresponds to approximately 2.3 subcarrier distances at a separation of 440 kHz. It can be observed in Figure 3.13(b) that higher interference was measured in the two subcarriers adjacent to the signal carrier. If the interference caused by all subcarriers is combined, then all but the very closest subcarriers have equal contributions to the interference signal, resulting in interference like Gaussian noise.

The Type 2 channel, which exhibits no noise floor in the phase noise mask, results in interference that is dependent on the distance from the interfering subcarrier. The summation of the interference is dominated by the interference contribution of the carriers in close vicinity, hence the resulting interference is less Gaussian than for the Type 1 channel.

The simulated BER results shown in Figure 3.15 show virtually indistinguishable performance for the modems in both the coloured and the white phase noise channels. Only a slight difference can be observed for QPSK between the Type 1 and Type 2 phase noise masks, where the corresponding white phase noise results in a better performance than the coloured noise. This difference can be explained with the interference being caused by fewer interferers compared to the white phase noise scenario, resulting in a non-Gaussian error histogram.

3.6.3   Phase Noise – Summary

Phase noise, like all time-varying channel conditions experienced by the time domain signal, results in intersubcarrier interference in OFDM transmissions. If the bandwidth of the phase noise is high compared to the OFDM subcarrier spacing, then this interference is caused by a high number of contributions from different subcarriers, resulting in a Gaussian noise interference. Besides this noise inflicted upon the received symbols, the signal level in the subcarriers drops by the amount of energy spread over the adjacent subcarriers.

The integral over the phase noise mask, termed phase jitter, is a measure of the signal-to-interference ratio that can be expected in the received subcarriers, if the phase noise has a wide bandwidth and is predominantly white. The relationship between the phase jitter and the SIR is shown in Figure 3.13(a). For narrowband phase noise this estimation is pessimistic.

3.7   Chapter Summary and Conclusion

In this chapter we discussed some of basic components of OFDM modems as well as a number of design factors, which influence the achievable performance. Specifically, in Section 3.3 we characterised the OFDM modem's performance, when communicating over conventional wireline type AWGN channels using various modulation schemes. We have shown in Section 3.4 that a particular characteristic of the OFDM signal is that owing to the presence of a high number of randomly modulated subcarrier signals the power of the composite modulated signal of an OFDM symbol may exhibit a high fluctuation. The corresponding signal PDFs were portrayed in Subsection 3.4.1. In order to avoid the spillage of out-of-band non-linear distortion products into the adjacent transmission bands, high-linearity power amplifiers are required for the transmission of OFDM signals. Furthermore, for the sake of reducing the chances that the highest signal peaks are clipped by the power amplifier, the input signal level is typically reduced by a certain amount by invoking a technique, which is often referred to as amplifier back-off, as was discussed in Section 3.4.2. The related issues of peak factor reduction techniques will be covered in more detail in Chapter 11. The effects of the ADC conversion accuracy on the achievable BER were quantified in Section 3.5. The chapter was concluded in Section 3.6 by considering the effects of phase noise on the OFDM system's BER. We are now ready to discuss the performance of OFDM modems communicating over dispersive fading channels, which constitutes the topic of the next chapter.

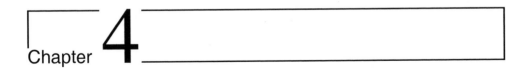

Chapter 4

OFDM Transmission over Wideband Channels[1]

Orthogonal frequency division multiplexing modems were originally conceived in order to transmit data reliably in time dispersive or frequency selective channels without the need for a complex time domain channel equaliser. In this chapter, the techniques employed for the transmission of quadrature amplitude modulated (QAM) OFDM signals over a time dispersive channel are discussed and channel estimation methods are investigated.

4.1 The Channel Model

The channel model assumed in this chapter is that of a finite impulse response (FIR) filter with time-varying tap values. Every propagation path i is characterised by a fixed delay τ_i and a time-varying amplitude $A_i(t) = a_i \cdot g_i(t)$, which is the product of a complex amplitude a_i and a Rayleigh fading process $g_i(t)$. The Rayleigh processes g_i are independent from each other, but they all exhibit the same normalised Doppler frequency f_d', depending on the parameters of the simulated channel.

The ensemble of the p propagation paths constitutes the impulse response

$$h(t, \tau) = \sum_{i=1}^{p} A_i(t) \cdot \delta(\tau - \tau_i) = \sum_{i=1}^{p} a_i \cdot g_i(t) \cdot \delta(\tau - \tau_i), \tag{4.1}$$

which is convolved with the transmitted signal.

All the investigations carried out in this chapter were based on one of the system and channel models characterised below. Each of the models represents a framework for a class of similar systems, grouped into three categories: wireless asynchronous transfer mode (WATM), wireless local area networks (WLAN) and a time division multiple access (TDMA) OFDM

---

[1]*OFDM and MC-CDMA for Broadband Multi-user Communications WLANS and Broadcasting.*
L.Hanzo, Münster, T. Keller and B.J. Choi,
©2003 John Wiley & Sons, Ltd. ISBN 0-470-85879-6

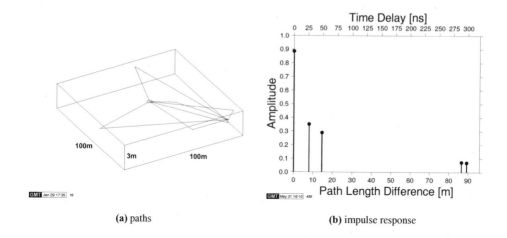

(a) paths   (b) impulse response

Figure 4.1: The 60GHz, 225MHz bandwidth WATM channel: five-path model and resulting impulse response

form of a personal communications type scheme, in a similar framework to the universal mobile telecommunication system (UMTS).

4.1.1   The Wireless Asynchronous Transfer Mode System

The wireless asynchronous transfer mode (WATM) system parameters used for our investigations follow closely the specifications of the Advanced Communications Technologies and Services (ACTS) Median system, which is a proposed wireless extension to fixed-wire ATM-type networks. In the Median system, the OFDM FFT length is 512, and each symbol is padded with a cyclic prefix of length 64. The sampling rate of the Median system is 225 Msamples/s, and the carrier frequency is 60 GHz. The uncoded target data rate of the Median system is 155 MBps.

4.1.1.1   The WATM Channel

The WATM channel employed here is a pessimistic model of the operating environment of an indoor wireless ATM network similar to that of the ACTS Median system. For our simulations, we assumed a vehicular velocity of about 50 km/h or 13.9 m/s, resulting in a normalised Doppler frequency of $f'_d = 1.235 \cdot 10^{-5}$. The impulse response was determined by simple ray tracing in a warehouse-type environment of 100 m × 100 m × 3 m, which is shown schematically in Figure 4.1(a). The resulting impulse response, shown in Figure 4.1(b) exhibits a maximum path delay of 300 ns, which corresponds to 67 sampling intervals. The five taps of the impulse response were derived assuming free space propagation, using the inverse second power law.

The Fourier transform of this impulse response leads to the static frequency domain channel transfer function depicted in Figure 4.2(a). Note that the central frequency of the bandwidth is in FFT bin 0, hence the spectrum appears wrapped around in this graph. Throughout this work the subcarrier index 0 contains the central frequency, with the subcarriers 1 to $N/2$

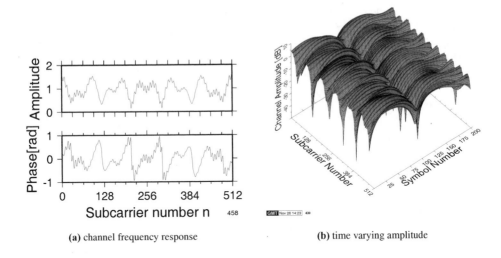

(a) channel frequency response   **(b)** time varying amplitude

Figure 4.2: WATM channel: (a) unfaded frequency domain channel transfer function $H(n)$; (b) time-varying channel amplitude for 200 consecutive OFDM symbols

spanning the positive relative frequencies, and the subcarriers $N/2 - 1$ to $N - 1$ containing the negative relative frequency range. Since the impulse response is assumed to be real, the resultant channel transfer function is conjugate complex symmetric around the central frequency bin.

If each of the impulses in Figure 4.1(b) is faded according to a Rayleigh fading process with the normalised Doppler frequency f'_d, then the resulting time-varying impulse response will lead to a time-varying frequency domain channel transfer function, which is depicted in Figure 4.2(b). In this figure, the amplitude of the channel transfer function has been plotted over the bandwidth of the OFDM system for 200 consecutive OFDM symbol time slots.

Figure 4.2(b) reveals a relatively high correlation of the channel transfer function amplitude both along the time and the frequency axes. The correlation in the frequency domain is dependent on the impulse response; the longer the tap delays, the lower the correlation in the transfer function. The similarity of the channel transfer functions along the time axis stems from the slowly changing nature, i.e. the low normalised Doppler frequency, of the assumed narrowband channels. The impulse response changes only little between consecutive OFDM symbol intervals; the correlation between neighbouring channel transfer functions is therefore affected by the Doppler frequency of the constituting narrowband fading channels relative to the duration of one OFDM symbol. For this graph, the impulse response was kept constant for the duration of each OFDM symbol. We will show in Section 4.2.2 that the notion of defining a frequency domain channel transfer function is problematic in environments exhibiting rapid changes of the impulse response.

As the channel impulse response (CIR) depicted in Figure 4.1(b) is of the line-of-sight (LOS) type with one dominant path, fading of this LOS path results in an amplitude variation across the whole signal bandwidth. This is seen in Figure 4.2(b) at around symbol number 100, where a substantial amplitude fade across all subcarriers occurs.

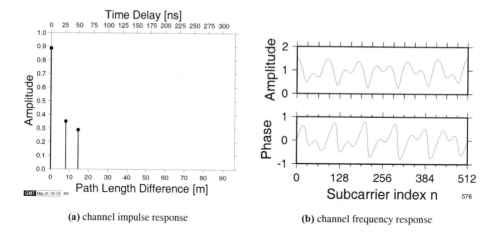

(a) channel impulse response   **(b)** channel frequency response

Figure 4.3: Short WATM channel: (a) impulse response, (b) unfaded frequency domain channel transfer function $H(n)$

4.1.1.2 The Shortened WATM Channel

Since the WATM channel discussed above is the worst-case scenario for an indoor wireless high data rate network, we have introduced a truncated version of the original WATM channel impulse response depicted in Figure 4.1(b), by only retaining the first three impulses. This reduces the total length of the impulse response, with the last path arriving at a delay of 48.9 ns due to the reflection with an excess path length of about 15 m with respect to the line-of-sight path, which corresponds to 11 sample periods. Omitting the two long delay pulses in the impulse response does not affect the total impulse response energy significantly, lowering the power by only 0.045 dB, hence no renormalisation of the impulse response was necessary. The resulting impulse response exhibits a root mean squared (RMS) delay spread of $1.5276 \cdot 10^{-8}$s, and it is shown in Figure 4.3(a). The resulting frequency domain transfer function for the unfaded short WATM impulse response is given in Figure 4.3(b), and comparison with Figure 4.2(a) reveals a close correspondence with the original WATM channel transfer function. As the first three low delay paths are the same for both channel models, the general shape of the channel transfer function is very similar; the last two paths in the WATM impulse response result in fast ripple on top of the dominant lower frequency components of the channel transfer function of Figure 4.2(a), which is absent in the shortened channel of Figure 4.3(b).

4.1.2 The Wireless Local Area Network

The main wireless local area network (WLAN) system parameters we opted for were loosely based on the high performance local area network (HIPERLAN) system [182], leading to a sampling rate of 20 MHz and a carrier frequency of 17 GHz. The assumed data symbols utilise a 1024-point FFT and a cyclic extension of 168 samples.

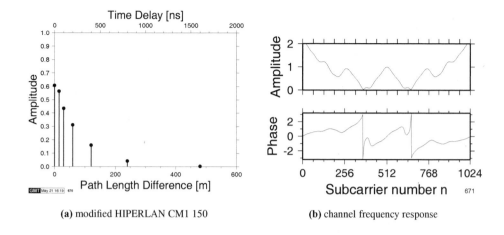

(a) modified HIPERLAN CM1 150

(b) channel frequency response

Figure 4.4: WLAN channel: (a) modified HIPERLAN CM1 150 ns impulse response, (b) unfaded frequency domain channel transfer function $H(n)$

4.1.2.1 The WLAN Channel

The WLAN channel's impulse response is based on the HIPERLAN CM1 impulse response with a RMS delay spread of 150 ns, as described by Tellambura *et al.* [182]. The second and third path of the original impulse response were combined to achieve a symbol-spaced impulse response pattern. The resulting impulse response is shown in Figure 4.4(a). The corresponding frequency domain channel transfer function is displayed in Figure 4.4(b). Assuming a worst-case vehicular velocity of 50km/h, fading channel simulations were conducted employing Rayleigh fading narrowband channels with a normalised Doppler frequency of $f'_d = 3.94 \cdot 10^{-5}$.

4.1.3 UMTS System

The set of parameters used, for the UMTS-type investigations was based on a version of the ACTS FRAMES Mode 1 proposal [183], resulting in a sampling rate of 2.17 MHz, a carrier frequency of 2 GHz, and a bandwidth of 1.6 MHz. A 1024-point FFT OFDM symbol with 410 virtual subcarriers is employed, in order to comply with the spectral constraints. The data segment of the OFDM symbol is padded with a 168-sample cyclic extension.

4.1.3.1 The UMTS Type Channel

A discretised COST 207 bad urban (BU) impulse response, as described in the COST 207 final report [184], was chosen for the UMTS-type channel. The symbol-spaced discretised impulse response employed for our investigations is depicted in Figure 4.5(a). This impulse response results in a strongly frequency selective channel, as shown in Figure 4.5(b). The area shaded in grey shows the location of the virtual subcarriers in the OFDM spectrum. Since no signal is transmitted in this range of frequencies, the channel transfer function for the virtual subcarriers does not affect the modem's performance. For simulations under Rayleigh fading

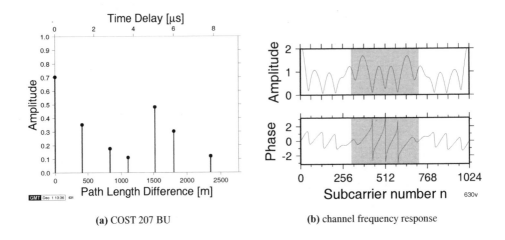

(a) COST 207 BU   (b) channel frequency response

Figure 4.5: UMTS channel: (a) COST 207 BU impulse response, (b) unfaded frequency domain channel transfer function $H(n)$

	f_c	$1/T_s$	$f_{d,max}$	$f'_{d,max}$	τ_{max}	RMS(τ)
UMTS	2 GHz	2.17 MHz	87.9 Hz	$4.05 \cdot 10^{-5}$	7.83 μs	3.28 μs
WLAN	17 GHz	20 MHz	787 Hz	$3.94 \cdot 10^{-5}$	1.6 μs	0.109 μs
WATM	60 GHz	225 MHz	2278 Hz	$2.345 \cdot 10^{-5}$	300 ns	34.3 ns
WATM - short	-"-	-"-	-"-	-"-	48.9 ns	16.9 ns

Table 4.1: Carrier frequency f_c, sample rate $1/T_s$, maximal Doppler frequency $f_{d,max}$, normalised maximal Doppler frequency $f'_{d,max}$, maximal path delay τ_{max}, and channel RMS delay spread RMS(τ) of the various system frameworks

conditions, a carrier frequency of 1.9 GHz and a vehicular velocity of 50 km/h were assumed, leading to a normalised Doppler frequency of $f'_d = 4.0534 \cdot 10^{-5}$.

4.2 Effects of Time Dispersive Channels on OFDM

The effects of the time variant and time dispersive channels on the data symbols transmitted in an OFDM symbol's subcarriers are diverse. First, if the impulse response of the channel is longer than the duration of the OFDM guard interval, then energy will spill over between consecutive OFDM symbols, leading to inter-OFDM symbol interference. We will not investigate this effect here, as the length of the guard interval is generally chosen to be longer than the longest anticipated channel impulse response.

If - as a consequence of a low Doppler frequency - the Rayleigh-fading CIR taps fluctuate only slowly compared to the duration of an OFDM symbol, then an essentially time-invariant CIR response can be associated with each transmitted OFDM symbol. Naturally, all of the Rayleigh-fading CIR tap values change gradually over the duration of a number of consecutive OFDM symbols. Viewing this phenomenon in the frequency domain, the channel transfer function of a specific OFDM symbols can be considered time-invariant for the duration of one OFDM symbol, while in a number of consecutive OFDM symbol durations it changes gradually. In this case, the frequency domain effects of the channel result in a time-invariant, but frequency-dependent multiplicative linear distortion of the frequency domain received symbols. This phenomenon is analogous to the effects of a multiplicative, non-dispersive time domain Rayleigh-fading channel envelope in the context of a serial modem, which does not result in inter-symbol interference. A low-Doppler scenario associated with slowly fluctuating Rayleigh-faded CIR taps will be investigated in Section 4.2.1.

By contrast, when the Rayleigh-fading CIR taps fluctuate rapidly owing to a high vehicular speed and hence impose CIR tap fluctuations associated with a high Doppler frequency, the system experiences Inter-subcarrier Interference (ICI). In simple tangible terms this CIR-tap fluctuation may be interpreted in the frequency domain as a frequency domain channel transfer function change during the reception of the OFDM symbol. When communicating over a channel exhibiting no linear distortions, the individual subchannels have a sinc-function shaped frequency response. All of the sinc-shaped subchannel spectra exhibit zero-crossings at the other subcarrier frequencies and the individual subchannel spectra are orthogonal to each other. This ensures that the subcarrier signals do not interfere with each other, when communicating over perfectly distortionless channels, as a consequence of their orthogonality. However, the orthogonality of the sinc-shaped subchannel spectra may be destroyed by the channel, when the Rayleigh-fading CIR taps change substantially during the transmission of an OFDM symbol and hence the corresponding frequency domain channel transfer function changes during this time interval. The quantitative effects of the ICI will be studied in Section 4.2.2.

4.2.1 Effects of the Stationary Time Dispersive Channel

Here a channel is referred to as stationary, if the Rayleigh-fading CIR taps do not fluctuate significantly over the duration of an OFDM symbol, but do change over longer periods of time. In this case, the time domain convolution of the transmitted time domain signal with the channel impulse response corresponds simply to the multiplication of the spectrum of the signal with the channel frequency transfer function $H(f)$:

$$s(t) * h(t) \longleftrightarrow S(f) \cdot H(f), \qquad (4.2)$$

where the channel's frequency domain transfer function $H(f)$ is the Fourier transform of the impulse response $h(t)$:

$$h(t) \longleftrightarrow H(f). \qquad (4.3)$$

Since the information symbols $S(n)$ are encoded into the amplitude of the transmitted spectrum at the subcarrier frequencies f_n, the received symbols $r(n)$ are the product of the transmitted symbol and the channel's frequency domain transfer function $H(n)$ plus the additive complex Gaussian noise samples $n(n)$:

$$r(n) = S(n) \cdot H(n) + n(n). \qquad (4.4)$$

4.2.2   Non-Stationary Channel

A channel is classified here as non-stationary, if the Rayleigh-fading CIR taps change significantly over the duration of an OFDM symbol, as a consequence of experiencing a high Doppler frequency. In this case, the frequency domain transfer function is time-variant during the transmission of an OFDM symbol and this time-variant frequency domain transfer function leads to the loss of orthogonality between the OFDM symbol's sinc-shaped subcarrier spectra. The amount of intersubcarrier interference imposed by this phenomenon depends on the rate of change, i.e. the Doppler frequency governing the fluctuation speed of the Rayleigh-fading CIR taps.

The simplest environment that may be considered for studying the effects of non-stationary channels is the narrowband channel, whose impulse response consists of a single Rayleigh-fading path. If the amplitude of this path changes as a function of time owing to fading, then the received OFDM symbol's spectrum will be the original OFDM spectrum convolved with the time-variant channel transfer function experienced during the transmission of the OFDM symbol. As this short-term channel transfer function changes between the consecutive transmission bursts, we will investigate the effects of the time-variant narrowband channel averaged over a high number of transmission bursts.

Since the ICI is caused by the variation of the channel impulse response during the transmission of each OFDM symbol, we introduce the "OFDM symbol normalised" Doppler frequency F_d:

$$F_d = f_d \cdot NT_s = f'_d \cdot N, \qquad (4.5)$$

where N is the FFT length, $1/T_s$ is the sampling rate, f_d is the Doppler frequency characterising the fading channel and $f'_d = f_d \cdot T_s$ is the conventional normalised Doppler frequency.

f'_d / N	16	32	64	128	256	512	1024	2048	4096
5×10^{-5}	-	-	0.0032	0.0064	0.0128	0.0256	0.0512	0.1024	0.2048
1×10^{-4}	-	0.0032	0.0064	0.0128	0.0256	0.0512	0.1024	0.2048	-
2×10^{-4}	0.0032	0.0064	0.0128	0.0256	0.0512	0.1024	0.2048	-	-

Table 4.2: OFDM symbol normalised Doppler frequency of $F_d = f_d NT_s = f'_d N$ as a function of the FFT length between 16 and 4096, when using the set of system parameters employed in the experiments characterising the non-stationary transmission scenario

The BER performance for the OFDM modem configurations shown in Table 4.2 was determined by simulation and the simulation results for BPSK are given in Figure 4.6. Figure 4.6(a) depicts the BER performance of an OFDM modem employing BPSK with perfect narrowband fading channel estimation, where it can be observed that for any given value of F_d the different FFT lengths and channels behave similarly. For an F_d value of 0.0256, a residual bit error rate of about $2.8 \cdot 10^{-4}$ is observed, while for $F_d = 0.1024$ the residual BER is about 0.37%.

Figure 4.6(b) shows the performance of an OFDM system without the assumption of perfect phase recovery. Instead, differential detection is employed at the receiver, and, as in Figure 4.6(a), there is good correspondence between the BER curves for the same value of F_d. It can be seen that for F_d values from 0.0128 on, there is already a visible performance

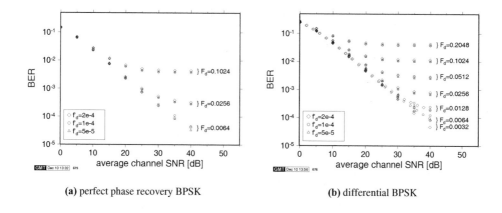

(a) perfect phase recovery BPSK   **(b)** differential BPSK

Figure 4.6: BPSK OFDM modem performance in a fading narrowband channel for normalised Doppler frequencies of $f'_d = 5 \cdot 10^{-5}$, $1 \cdot 10^{-4}$ and $2 \cdot 10^{-4}$ and FFT lengths between 16 and 4096. The FFT length for a given normalised Doppler frequency f'_d and an OFDM symbol normalised Doppler frequency F_d can be obtained from Table 4.2.

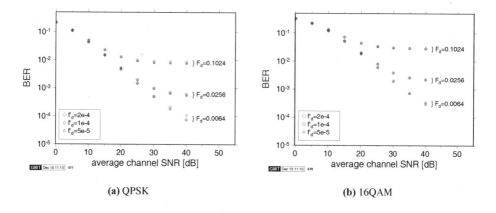

(a) QPSK   **(b)** 16QAM

Figure 4.7: OFDM modem performance assuming perfect phase recovery in a fading narrowband channel with normalised Doppler frequencies of $f'_d = 5 \cdot 10^{-5}$, $1 \cdot 10^{-4}$ and $2 \cdot 10^{-4}$ and FFT lengths between 16 and 2048 symbols. The FFT length for a given normalised Doppler frequency f'_d and an OFDM symbol normalised Doppler frequency F_d can be obtained from Table 4.2.

penalty at a SNR value of 40 dB. The highest observed residual BER is 4.2% for $f_d = 0.2048$.

Higher-order modulation schemes, such as QPSK and 16-QAM, are more affected by the intersubcarrier interference caused by the time-varying narrowband fading channel characteristics, as shown in Figure 4.7. Again, the correspondence between different set-ups resulting in the same value of F_d is very good. Since the BER performance is limited by the intersubcarrier interference instead of noise, the BER curves of Figures 4.6(a), 4.7(a) and 4.7(b) do not exhibit the expected fixed SNR shift of 3 dB between BPSK and QPSK and 9.5 dB between BPSK and 16-QAM, as experienced for noise-only limited systems. From the BER curves and residual BER values of Figures 4.6 and 4.7, it can be deduced that the effects of the intersubcarrier interference for $F_d = 0.1024$ and $F_d = 0.0256$ are equivalent to the effects of noise at SNR values of $\gamma = 17$ dB and $\gamma = 28$ dB, respectively. These values are long-term averages and appear to be valid for all the modulation schemes used.

4.2.2.1 Summary of Time-variant Channels

Time variant channels impose a major influence on the BER performance of OFDM systems. The intersubcarrier interference caused by the time-varying nature of the transmission channel limits the attainable bit error rates for the UMTS and the WLAN scenarios, where normalised Doppler frequencies of around $4 \cdot 10^{-5}$ and FFT lengths of 1024 are assumed, leading to $F_d \approx 0.04$. Therefore, an error floor of about 10^{-3} has to be faced by these systems, even if perfect channel estimation is assumed.

4.2.3   Signalling over Time-Dispersive OFDM Channels

Analogously to the case of serial modems in narrowband fading channels, the amplitude and phase variations inflicted by the channel's frequency domain transfer function $H(n)$ upon the received symbols will severely affect the bit error probabilities, where different modulation schemes suffer to different extents from the effects of the channel transfer function. Coherent modulation schemes rely on the knowledge of the symbols' reference phase, which will be distorted by the phase of $H(n)$. If such a modulation scheme is to be employed, then this phase distortion has to be estimated and corrected. For multilevel modulation schemes, where also the magnitude of the received symbol bears information, the magnitude of $H(n)$ will affect the demodulation. Clearly, the performance of such a system depends on the quality of the channel estimation.

A simpler approach to signalling over fading channels is to employ differential modulation, where the information is encoded in the difference between consecutive symbols. Differential phase shift keying (DPSK) employs the phase of the previously received symbol as phase reference, encoding information in the phase difference between symbols. DPSK is thus only affected by the differential channel phase distortion between two consecutive symbols, rather than by the channel phase distortion's absolute value.

4.3   Channel Transfer Function Estimation

Frequency domain channel estimation algorithms generate the channel transfer function estimates $\hat{H}(n)$ for subsequent correction of the received symbols prior to demodulation. The accuracy of the algorithm influences the total system performance to a great extent, especially for systems employing multilevel modulation and coherent detection. We have investigated several different wideband channel estimation techniques, which can be split into two groups: those operating on the spectrum of the received OFDM symbol and those employing time domain correlation algorithms.

4.3.1 Frequency Domain Channel Transfer Function Estimation

Frequency domain channel estimation algorithms exploit the knowledge of the pilot subcarrier positions in the frequency domain representation of the OFDM symbols, that is after the receiver's FFT-based demodulation stage. At this stage, the frequency domain channel transfer function $H(n)$ can be estimated by using known frequency domain pilot subcarriers embedded in the OFDM symbol's spectrum. These pilot subcarriers facilitate the sampling of the frequency domain channel transfer function $H(n)$, provided that the corresponding sampling frequency is higher than the Nyquist frequency required for the aliasing-free representation of the channel's transfer function at the Doppler frequency encountered. Our treatment of channel transfer function estimation is conceptual in this chapter. A deeper discussion on transfer function estimation will be provided in Chapters 14-16.

4.3.1.1 Pilot Symbol-Assisted Schemes

Pilot symbol-assisted modulation (PSAM) schemes obtain a channel transfer function estimate on the basis of known frequency domain pilot symbols that are interspersed with the transmitted data symbols [174]. Conventionally, PSAM schemes have been utilised in narrowband fading environments for sampling the time domain fading envelope of the channel in the context of conventional serial modems. For each received pilot subcarrier the corresponding channel transfer function value is estimated as the quotient of the received and the expected subcarrier value. The channel transfer function estimate for the information-bearing subcarrier positions of the OFDM symbol is derived from these pilot-based channel transfer function estimates by means of interpolation between them. A range of different interpolation techniques were comparatively studied by Torrance et al. [185] in the context of serial modems communicating over Rayleigh fading narrowband channels, ultimately favouring linear interpolation as a consequence of its low complexity and good performance. We will address the issues of choosing an interpolation scheme in the context of OFDM transmissions over fading wideband channels below.

In parallel modems, PSAM can be utilised for estimating the frequency domain channel transfer function $H(n)$ in a time dispersive environment, if no inter-subcarrier interference is present. Accordingly, n_p pilot symbols P_i are transmitted in the subcarriers with indices $p_i, i = 1, \ldots, n_p$ within the total OFDM symbol bandwidth of N subcarriers. At the receiver, the channel transfer function $\hat{H}(p_i)$ at the pilot subcarriers is estimated from the received samples $r(p_i)$:

$$\hat{H}(p_i) = r(p_i)/P_i. \tag{4.6}$$

In a second step, the values of the channel transfer function are estimated for the unknown data symbols by interpolation using the $\hat{H}(p_i)$ values of Equation 4.6. Clearly, the placement of the pilots and the interpolation technique will influence the quality of the channel estimation. In this chapter we will investigate two different interpolation techniques, linear and lowpass interpolation, both with a varying number of pilot subcarriers.

4.3.1.1.1 Linear Interpolation for PSAM
The simplest interpolation technique that can be used to estimate the channel transfer function for the subcarriers between two neighbour-

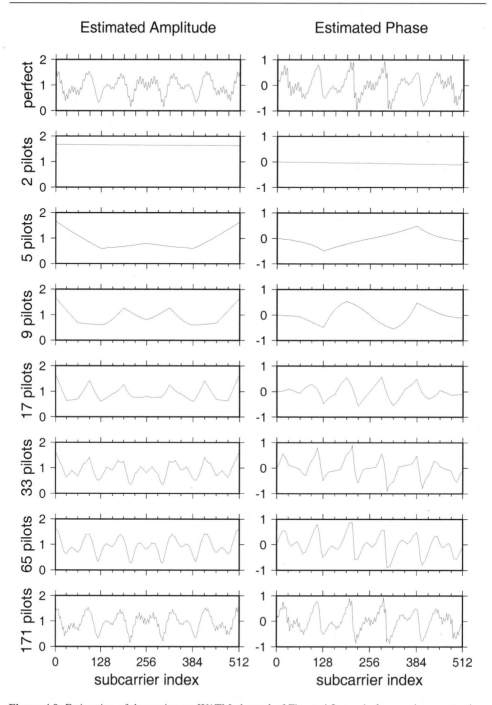

Figure 4.8: Estimation of the stationary WATM channel of Figure 4.2, a noiseless environment using linear interpolation between pilots employing $2\ldots171$ pilots per 512-subcarrier OFDM symbol

ing channel estimation samples $\hat{H}(p_i)$ and $\hat{H}(p_{i+1})$ is a linear function:

$$\hat{H}(n) = \hat{H}(p_i) + \frac{\hat{H}(p_{i+1}) - \hat{H}(p_i)}{p_{i+1} - p(i)} \cdot (n - p_i) \quad \text{for} \quad p_i \leq n \leq p_{i+1}. \quad (4.7)$$

Figure 4.8 shows the amplitude and the phase of the stationary WATM channel's frequency domain transfer function $H(n)$ for different numbers of equidistant pilot symbols in the frequency domain in an ideal noiseless environment. It can be seen that the channel estimation accuracy improves with increasing number of pilot tones in the spectrum. The number of pilot tones in the OFDM spectrum necessary to sample the channel transfer function can be determined on the basis of the sampling theorem as follows.

The frequency domain channel's transfer function $H(f)$ is the Fourier transform of the channel impulse response $h(t)$. Each of the impulses in the impulse response will result in a complex exponential function $e^{-(j2\pi\tau/T_s)f}$ in the frequency domain, depending on its time delay τ. In order to sample this contribution to $H(f)$ according to the sampling theorem, the maximum pilot spacing Δp in the OFDM symbol is:

$$\Delta p \leq \frac{N}{2\tau/T_s} \Delta f. \quad (4.8)$$

In the case of the WATM channel impulse response with a maximum delay of $\tau/T_s = 67$, the resulting pilot spacing in the OFDM symbol is three subcarriers, requiring 171 pilot carriers per 512-subcarrier OFDM symbol. If the channel estimation is to resolve only the effects of the first three dominant paths with a maximum delay of $\tau/T_s = 11$ in Figure 4.1(b), then the minimal pilot spacing can be increased from 3 to 23 subcarriers.

Inspection of Figure 4.8 underlines these points: in order to resolve the full detail of the original transfer function, as depicted at the top of the figure, 171 pilots must be used. Furthermore, even using 171 pilots in the OFDM symbol, the estimated curve is not a very close match of the original. This is due to the linear interpolation algorithm used.

Nonetheless, the estimations performed with 33 and 65 pilots, respectively, show very similar levels of detail. Both schemes can appropriately sample the effects of the first three paths, but they are undersampled for capturing information concerning the two long-delay paths. If the pilot spacing is higher than the calculated number of 23, then the accuracy of the estimation declines further.

The shortened WATM channel, as characterised in Figure 4.3(a), can be sampled adequately utilising any pilot spacing of less than 23 subcarriers, which for our 512-subcarrier system corresponds to a minimum number of 23 pilots per OFDM symbol. Figure 4.9 shows the linearly interpolated channel estimation values for PSAM channel estimation using 2, ... , 171 pilots in the 512-subcarrier OFDM symbol for the noiseless shortened WATM channel. Again, it can be seen that the accuracy of the channel transfer function estimation improves with the number of pilots, but that a total of 171 pilots is necessary to yield a channel estimation that is indistinguishable from the perfect channel estimation curve in this scale. As 33 and 65 pilot subcarriers already exceed the 23 pilots required to sample the frequency domain channel transfer function according to the Nyquist sampling theorem, it becomes clear that the estimation errors are due to the linear interpolation algorithm and that – for optimal channel estimation accuracy – substantial oversampling of the channel has to

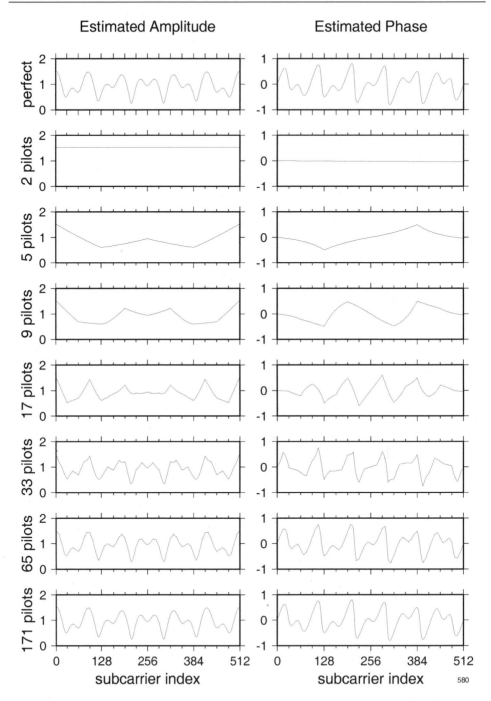

Figure 4.9: Estimation of the stationary shortened WATM channel of Figure 4.3, a noiseless environment using linear interpolation between pilots employing 2 ... , 171 pilots per 512-subcarrier OFDM symbol

be employed, thus deteriorating the system's bandwidth efficiency.

4.3.1.1.2 Ideal Lowpass Interpolation for PSAM The perfect interpolator for the measured channel transfer function samples $\hat{H}(p_i)$ is the ideal lowpass filter with a cut-off frequency of $1/\Delta p$, where Δp is the spectral distance between consecutive pilot subcarriers. Unlike the linear interpolation, this requires equidistant pilot placement in the OFDM symbol.

While an ideal rectangular filter transfer function is impossible to implement perfectly in the time domain due to its infinitely long impulse response, we employed an ideal rectangular frequency domain filter based on a FFT/IFFT operation. A perfect FFT-based lowpass filter can be implemented if its bandwidth coincides with a FFT bin of the chosen Fourier transform, which limits the set of possible pilot numbers. In our case, a 512-point FFT/IFFT operation was employed, therefore allowing for sets of 2^n with $1 \leq n \leq 9$ pilots.

For comparison with the linear interpolation, the same set of pilot distances Δp has been employed for the experiments. Since the lowpass interpolator does not require a pilot in the last subcarrier, the corresponding number of pilots is reduced by one for the lowpass interpolator, when compared to the linear interpolator.

Figure 4.10 gives an overview of the estimation accuracy of the PSAM scheme with ideal lowpass interpolation in a noiseless environment. Again, the resolution of the estimation depends on the number of pilots per OFDM symbol. The estimated transfer functions for 32 and for 64 pilots are identical and exactly correspond to the effect of the first three impulses in $h(t)$. In order to resolve $H(f)$ more finely, at least 170 pilot subcarriers must be employed. Closer inspection of the estimated $\hat{H}(n)$ for 170 pilots reveals relatively high errors in the estimation. This is due to non-optimal lowpass filtering; the optimum bandwidth of the lowpass filter corresponding to $\Delta p = 3\Delta f$ does not line up with the filter's FFT bins, hence the passband is too wide. This leads to considerable estimation errors, especially at the highest subcarrier indices, since there is no pilot symbol at the last subcarrier. If a perfect lowpass interpolation scheme was to be used over the WATM channel, then a pilot distance of $\Delta p = 2\Delta f$ would have to be used, resulting in 256 pilot subcarriers per 512-subcarrier OFDM symbol.

The equivalent estimated channel transfer functions for the case of the shortened WATM channel are depicted in Figure 4.11. It can be seen from the figure that the channel estimation is indistinguishable from the perfect case for the 32-pilot scenario and the 64-pilot scenario. Since the number of pilots necessary for the perfect sampling of this channel is 23, which corresponds to a pilot spacing of 23, the channel estimation utilising less than 23 pilots yields inaccurate results. If 170 pilot subcarriers are employed in each OFDM symbol, then the same effect as observed for the full-length WATM channel applies; because the frequency resolution of the FFT employed for the lowpass filter implementation is insufficient, the filter passband does not correspond exactly to the pilot frequency.

4.3.1.1.3 Summary The Nyquist sampling theorem applies to the channel estimation in the frequency domain utilising pilot subcarriers in the OFDM symbol, but delivers only a lower bound for the number of pilots necessary, which is valid for perfect lowpass interpolation. If linear interpolation is employed, then oversampling beyond the Nyquist rate is necessary, in order to achieve high estimation accuracy. The number of pilot subcarriers that has to be used is dependent on the modulation scheme employed for signalling over the

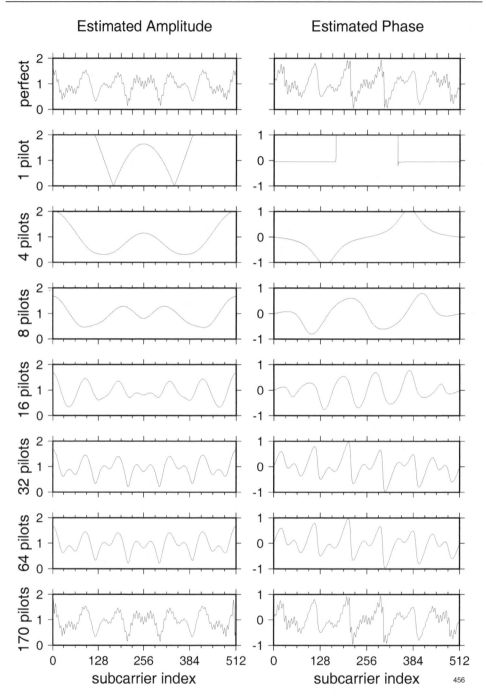

Figure 4.10: Estimation of the stationary WATM channel of Figure 4.2, a noiseless environment using ideal lowpass interpolation between pilots employing 2, ... , 171 pilots per 512-
subcarrier OFDM symbol

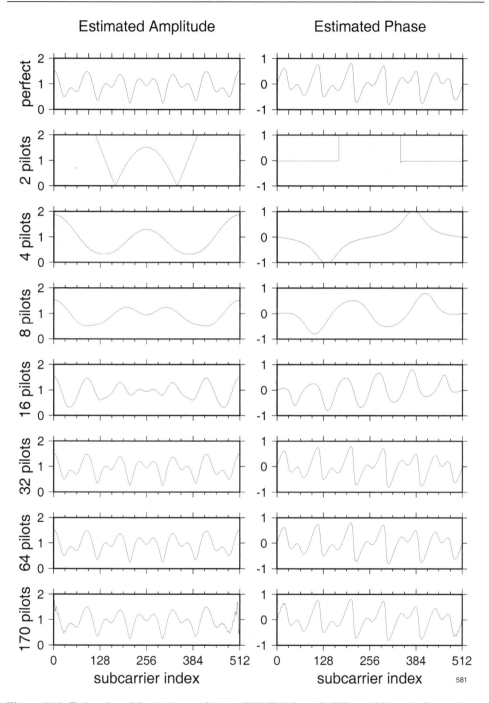

Figure 4.11: Estimation of the stationary shortened WATM channel of Figure 4.3, a noiseless environment using ideal lowpass interpolation between pilots employing 2, ... , 171 pilots per 512-subcarrier OFDM symbol

subcarriers.

The ideal lowpass interpolator yields better estimation accuracy than the linear interpolator for a given number of pilots, but the choice of pilot sets is limited by the implementation
of the lowpass filter. If a FFT/IFFT based filter implementation is chosen, then only a limited
set of possible pilot distances is available. This, in the WATM channel, leads to the need for
a substantially higher number of pilots than would be necessary for the more flexible linear
interpolator.

4.3.2   Time Domain Channel Estimation

An alternative approach to channel estimation in OFDM transmission systems is to estimate
the channel directly from the time domain received signal. Analogous to wideband estimation
in serial modems, a training sequence in the transmitted data stream can be employed in order
to perform a correlation-based impulse response estimation. The corresponding frequency
domain channel transfer function can be computed by FFT from this impulse response.

Again, our treatment of channel transfer function estimation was conceptual in this chapter. A deeper discussion on channel transfer function estimation will be provided in Chapters 14–16.

4.4   System Performance

The various channel estimation techniques discussed above have been studied in a perfectly
noiseless environment and without considering the effects of the channel estimation upon the
system's bit error rate performance. In this section, we will consider the attainable performance of data transmission systems over the time dispersive channel, for both non-fading and
stationary channels.

4.4.1   Static Time-Dispersive Channel

The static time dispersive channel exhibits a time invariant channel impulse response, which
inflicts no intersubcarrier interference as described in Section 4.2.2. Therefore, the OFDM
subcarriers are independent and each subcarrier n corresponds to an AWGN channel with a
signal-to-noise ratio of γ_n. If the average channel SNR is γ, then the sub-channel SNR values
γ_n depend on the magnitude of the frequency domain channel transfer function $H(n)$:

$$\gamma_n = \gamma \cdot |H(n)|^2. \tag{4.9}$$

If we assume perfect channel estimation, corresponding to $\hat{H}(n) = H(n)$, then the bit error
rate $p_{e,n}$ for each subcarrier n depends exclusively on the sub-channel SNR γ_n, which allows
the calculation of the overall system bit error rate p_e by simply averaging over the sub-channel
SNR values:

$$p_e = \frac{1}{N_u} \sum_n p_{e,n} = \frac{1}{N_u} \sum_n p_e(\gamma_n), \tag{4.10}$$

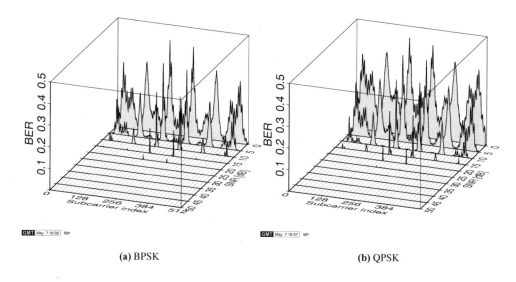

(a) BPSK   (b) QPSK

Figure 4.12: Simulated BER per subcarrier over the static WATM channel of Figure 4.1(b) with perfect channel estimation for (a) coherent BPSK transmission and (b) coherent QPSK transmission. The symmetry of the BER curves indicates that the unfaded CIR was real.

where N_u is the number of subcarriers used for data signalling and the use of the same modulation scheme for all subcarriers is assumed.

4.4.1.1 Perfect Channel Estimation

Perfect channel estimation is the best-case scenario for OFDM transmission over time dispersive channels, as the performance is only limited by the SNR of the subcarriers, not by the reference phase and amplitude estimation errors in the receiver. Simulations have been performed for both coherent and non-coherent modulation schemes, where the frequency domain channel transfer function $H(n)$ was calculated at the receiver by applying the Fourier transform to the perfectly known noiseless impulse response of the channel, as depicted in Figure 4.1(b). The resulting channel transfer function is shown in Figure 4.2(a).

Figure 4.12(a) shows the measured BER per subcarrier for different levels of average OFDM SNR for the simulated 512-subcarrier modem over the WATM channel employing coherent binary phase shift keying (BPSK) in the subcarriers. It is apparent that the bit error probability varies significantly between different subcarriers, from about 2% in good subchannel conditions, up to 40% in the deep fades of the channel transfer function $H(f)$ at an average OFDM SNR of 0 dB. At an average SNR of 5 dB, groups of virtually error-free subcarriers can be observed, interspersed with bundles of carriers exhibiting high bit error probabilities. At 15 dB, only a very small number of carriers experience measurable transmission error rates at all.

The recorded bit error rates in the simulation correspond very closely to the theoretical results obtained by calculating the sub-channel SNR γ_n for each subcarrier index n following Equation 4.9 and evaluating the appropriate bit error probability function for the chosen modulation scheme. Specifically, the BER over the AWGN channel was calculated employ-

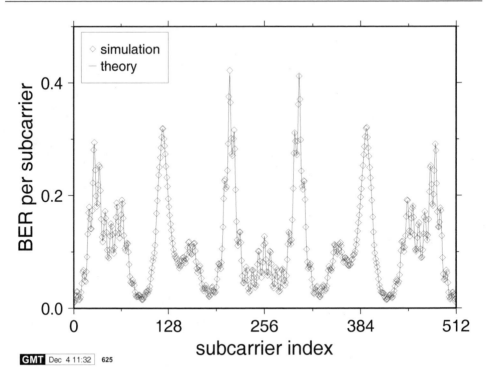

Figure 4.13: Simulated and theoretical BER per subcarrier for coherently detected BPSK transmission with perfect channel estimation over the static WATM channel of Figure 4.1(b) at an average OFDM SNR of 0 dB

ing Equations 3.1 to 3.3 For coherent modulation, the bit error probabilities $p_e(\gamma)$ for BPSK, QPSK and 16-QAM are depicted in Figure 3.1, and Figure 4.13 shows good correspondence between the obtained theoretical bit error rates and the simulation results.

The corresponding graph for QPSK, as shown in Figure 4.12(b), is similar to the BPSK case, only exhibiting higher bit error probabilities. Again, the bit error rate per subcarrier varies strongly with the channel transfer function, which is due to the variation in sub-channel SNR.

The OFDM bit error probability averaged over all subcarriers versus the channel SNR for BPSK, QPSK and 16-QAM transmission over the non-fading WATM channel is shown in Figure 4.14(a). The QPSK curve is shifted to the right by 3 dB compared to the BPSK performance curve, which is in accordance with the performance in a narrowband channel, as demonstrated in Section 3.3. The same observation can be made for the performance difference between 16-QAM and the other signalling schemes.

Although the relative BER performance relationship between the different modulation schemes corresponds to the BER results found in the narrowband AWGN channel, the absolute bit error rate values depend on the channel transfer function. Figure 4.14(b) depicts the corresponding BER versus channel SNR graph for the shortened WATM channel of Figure 4.3. The relative BER performance relationship between the different modulation schemes is the same as over the other channels, but the shape of the BER curve is different from that

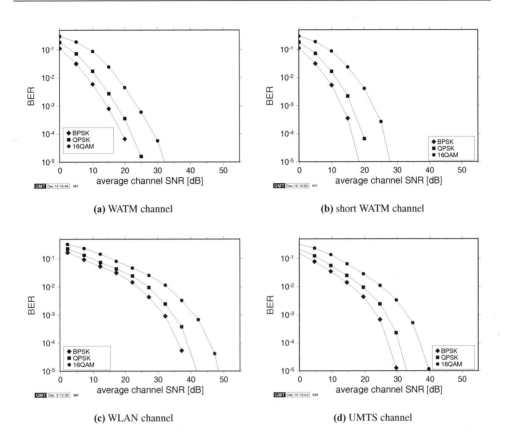

(a) WATM channel

(b) short WATM channel

(c) WLAN channel

(d) UMTS channel

Figure 4.14: Bit error rate versus channel SNR for static (a) WATM channel (Figure 4.1(b)), (b) short-ened WATM channel (Figure 4.3), (c) WLAN channel (Figure 4.4), (d) UMTS channel (Figure 4.5(a)); perfect channel estimation and coherent detection

of the WATM channel. Comparing the frequency domain channel transfer function of the shortened WATM channel of Figure 4.3(b) and that of the original WATM channel of Figure 4.2(a) it can be seen that the fading is less severe in the shortened WATM channel, leading to better BER performance at high SNR values.

The UMTS channel characterised in Figure 4.5(b) exhibits even stronger fading in the frequency domain than the two WATM channels, and consequently the BER performance is worse than that of both the WATM systems. At a BER of 10^{-3} the required SNR for the corresponding modulation schemes is about 9 dB higher for the UMTS system than for the WATM system.

4.4.1.2 Differentially Coded Modulation

Differentially coded modulation enables the receiver to detect the data symbols without knowledge of the reference phase, therefore no channel estimation is required, if differential phase shift keying (DPSK) is used for modulating the subcarriers. Since the transmitted

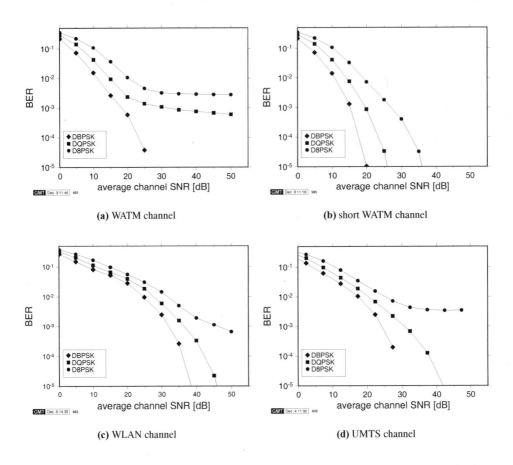

(a) WATM channel   **(b)** short WATM channel

(c) WLAN channel   **(d)** UMTS channel

Figure 4.15: Bit error rate versus channel SNR for static (a) WATM channel (Figure 4.1(b)), (b) shortened WATM channel (Figure 4.3), (c) WLAN channel (Figure 4.4), (d) UMTS channel
(Figure 4.5(a)), using differential detection

information is encoded in the phase difference between two consecutive subcarriers, there
needs to be one phase reference subcarrier per OFDM symbol, typically in the the. first subcarrier.

While the absolute phase of the frequency domain channel transfer function $H(n)$ at
subcarrier n does not affect the reception, the phase change of $H(n)$ between two consecutive
subcarriers does influence the reception of the symbols. Note that the BER performance of
both the WATM system in Figure 4.15(a) as well as that of the UMTS system in Figure
4.15(d) exhibit a residual bit error rate for high SNR levels for certain modulation schemes.
These noise-independent errors are caused by channel effects and are investigated for the
UMTS system below.

In differential phase shift keying, the information to be transmitted is mapped to the phase
difference between two consecutively received symbols. As all the symbols transmitted over
the wideband OFDM channel suffer different phase rotations from the channel's frequency

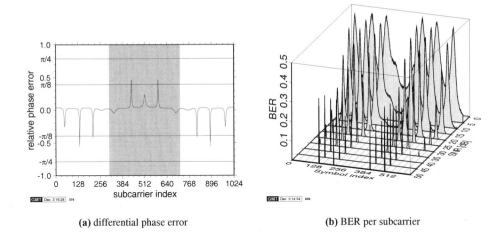

(a) differential phase error   (b) BER per subcarrier

Figure 4.16: Differential modulation over the static UMTS channel of Figure 4.5. (a) Phase difference between consecutive frequency domain samples of the frequency domain channel transfer function $H(n)$; the grey area designates the virtual subcarriers, which are not used for data transmission. (b) Simulated BER per subcarrier in static UMTS channel for D8PSK without channel estimation

domain transfer function $H(n)$, the channel phase rotation difference $(\angle H(n) - \angle H(n-1))$ between subcarrier $(n-1)$ and subcarrier n offsets the receiver's decision boundaries for detected signals. Figure 4.16(a) shows the phase difference $(\angle H(n) - \angle H(n-1))$ between adjacent subcarriers for the UMTS channel of Figure 4.5. The grey area in the graph marks the positions of the virtual subcarriers, which are not used for transmission, and therefore do not have an impact on the system's performance. The horizontal lines at $\pm\pi/8$ and $\pm\pi/4$ mark the receiver's decision boundaries for DQPSK and D8PSK, respectively. If the channel transfer function's differential phase shift crosses the appropriate decision boundary for a given subcarrier index, then the data symbol transmitted over this subcarrier suffers from residual bit errors. In our example of Figure 4.16(a), there are four visible peaks comprising actually 6 out of 612 data-bearing subcarriers crossing the $\pi/8$ decision boundary for D8PSK. As the data is grey mapped on the PSK symbols, exactly one residual bit error will occur on each of the six subcarriers, resulting in a residual bit error rate of $6/1836 = 3.268 \cdot 10^{-3}$. Figure 4.16(b) gives an overview of the resulting BER per subcarrier for the data-bearing subcarriers for SNR values from 0 dB up to 50 dB. It can be seen that the residual bit errors are concentrated in four bursts, comprising the six symbols with subcarrier indices 136, 137, 312, 401, 477, and 478. At 50 dB SNR the symbols in the outer two bursts have a bit error rate of $1/3$, corresponding to one bit error per D8PSK symbol. The bit error rate of the inner two error bursts is lower, at about 23%, but is rising with higher SNR values to $1/3$. This rising BER with higher SNR values for the two inner error bursts is explained by the channel transfer function shown in Figure 4.16(a). The channel induces exactly one bit error for the symbols in each of the corresponding subcarriers in the absence of noise. Since the phase error due to the channel's phase rotation is very close to the decision 8-DPSK boundary, the presence of noise can influence the reception towards the correct decision, therefore reducing

System	FFT length	Max. delay $[T_s]$	max. Δp $[\Delta f]$	min. N_p
WATM	512	67	3	171
short WATM	512	11	23	22
WLAN	1024	32	16	64
UMTS	1024	17	30	34

Table 4.3: Theoretical maximum pilot distance and minimal number of pilots for the WATM, short WATM, WLAN and UMTS systems

the long-term bit error rate.

In addition to residual bit errors, the decision boundary offset caused by the differential phase shift of the channel transfer function as depicted in Figure 4.16(a) will also compromise the noise sensitivity of the subcarriers that are not directly affected by residual errors.

4.4.1.3   Pilot Symbol-Assisted Modulation Aided Channel Transfer Function Estimation

As discussed in Section 4.3.1.1, pilot symbol-assisted modulation (PSAM) schemes acquire a channel transfer function estimate $\hat{H}(n)$ by sampling $H(n)$ with the aid of known pilot symbols embedded in the transmission burst and interpolating between the received pilots. We have seen that the minimal number of pilots needed to sample the channel transfer function follows the sampling theorem and that the maximal pilot distance in the OFDM symbol is given by Equation 4.8. Furthermore, we have seen that this minimal number of pilots only applies to the ideal lowpass filter interpolation algorithm, and more pilot symbols have to be employed, if linear rather than lowpass interpolation is to be used, therefore trading receiver complexity for transmission overhead.

Here we will characterise the performance of the four different OFDM systems in their respective non-fading channel environments, noting that we will return to this topic is substantially more detail, adopting an analytically motivated approach in Chapters 14 and 15. Each system was investigated using a range of pilot numbers. Using Equation 4.8, we can predict the number of pilot symbols necessary for each scenario, if ideal interpolation is assumed, and these numbers are given in Table 4.3. The performance evaluation for all these scenarios was carried out for 8, 16, 32 and 64 pilots with ideal lowpass interpolation and with the equivalent 9, 17, 33 and 65 pilot sets for linear interpolation. We will concentrate on our most robust and least robust modulation schemes, namely BPSK and 16-QAM.

Figure 4.17 shows the BER performance for all the discussed systems in their respective non-fading channels employing PSAM-16-QAM with ideal lowpass interpolation as modulation scheme. It can be seen in Figure 4.17(a) that the WATM system faces high residual bit error rates for all sets of pilot symbols investigated. This is expected, because the number of pilot subcarriers employed in these simulations was lower than the minimum number of 171 stated in Table 4.3. Therefore, the channel transfer function is not adequately sampled in the frequency domain and the resulting channel estimation is of insufficient accuracy for coherently detecting 16-QAM. Note that the residual bit error rate drops with increasing number of pilots employed for the channel estimation, but that the curves for 32-pilot and 64-pilot PSAM exhibit virtually equal performance. Figure 4.17(a) reveals that the estimated channel

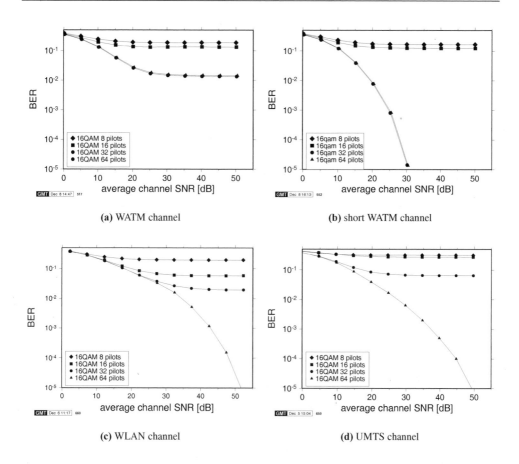

Figure 4.17: Bit error rate versus channel SNR for static (a) WATM channel (Figure 4.1(b)), (b) short-ened WATM channel (Figure 4.3), (c) WLAN channel (Figure 4.4), (d) UMTS channel (Figure 4.5(a)), PSAM-16QAM with ideal lowpass interpolation

transfer function for these two cases is the same, owing to the twin-burst structure of the WATM impulse response of Figure 4.1(b). Therefore, increasing the number of pilots from 32 to 64 will not increase the system's performance, since this does not allow the modem to resolve the channel effects due to the two paths around 90 m.

According to Table 4.3, the short WATM channel can be adequately sampled with 22 pilot symbols equidistantly interspersed with the data symbols. This is in accordance with the system's performance curves in Figure 4.17(b). The BER curves for 32 and 64 subcarriers exhibit essentially identical performance, while for 8 and 16 pilot symbols the residual bit error rate is above 10%. For the WLAN channel, at least 64 pilots have to be employed in order to sample all the frequency components of $H(f)$. We can see in Figure 4.17(c) that all but the 60 pilot symbol systems exhibit residual bit error rates and that the BER performance increases with an increased number of pilots.

If the more robust BPSK modulation scheme is used in the subcarriers, then the effects·

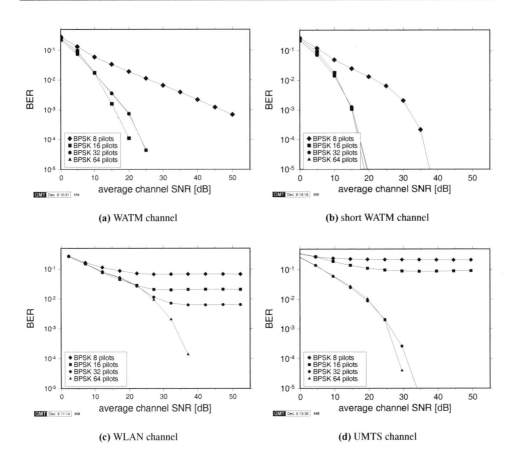

(a) WATM channel

(b) short WATM channel

(c) WLAN channel

(d) UMTS channel

Figure 4.18: Bit error rate versus channel SNR for static (a) WATM channel (Figure 4.1(b)), (b) short-ened WATM channel (Figure 4.3), (c) WLAN channel (Figure 4.4), (d) UMTS channel (Figure 4.5(a)), PSAM-BPSK with ideal lowpass interpolation

of inaccurate channel estimation are less pronounced. Figure 4.18 gives an overview of the BER performance for BPSK modems in the four operational environments, employing a perfect lowpass filter interpolator. The BER performance curves for the WATM system, shown in Figure 4.18(a), exhibit no residual errors. Like the 16-QAM curves, the performance is identical, irrespective of whether 32 or 64 pilot symbols are employed. Unlike in the 16-QAM case, however, the performance is best for 16-pilot subcarriers employed in the OFDM symbol. This oddity can be explained comparing the channel phase estimation errors for 16 and for 32 subcarriers, which are depicted in Figure 4.19. The phase estimation error for the WATM channel transfer function estimated with 16 pilot symbols is shown in Figure 4.19(a), which exhibits high fluctuations of the phase error of up to 0.7 radians. The phase estimation error for 32 pilot symbols, as shown in Figure 4.19(b), while substantially smaller on average, exhibits a peak error value of about 1.0 rad, which is closer to the BPSK decision boundary of $\pm\pi/2$ than the maximum error of the 16-pilot estimation. For higher SNR values the overall

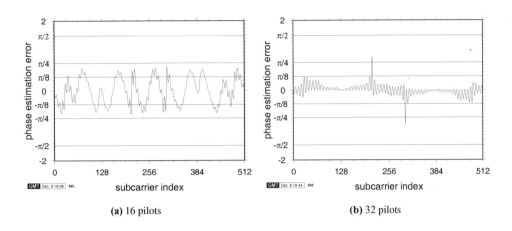

(a) 16 pilots

(b) 32 pilots

Figure 4.19: Channel transfer function phase estimation error for the WATM channel of Figure 4.1(b) employing ideal lowpass filter interpolator: (a) 16 pilot symbols, (b) 32 pilot symbols

BER will be dominated by the bit errors occurring in the subcarriers with the highest phase estimation errors, therefore leading to the 16-pilot system performing better than the overall more accurate 32 and 64 pilot symbol estimation systems.

The WATM system's performance over the short WATM channel of Figure 4.3 is depicted in Figure 4.18(b). No residual bit error rates are observed and the BER performance with 32 and 64 pilots is identical. The 16-pilot BER curve is close to the 32- and 64-pilot results, and the modem performs about 18 dB worse with 8 pilot symbols. Both the WLAN as well as the UMTS channels exhibit residual bit errors for 8 and 16 pilot symbols and show an improving BER performance with increasing numbers of pilots. The WLAN system, whose performance is shown in Figure 4.18(c), performs without residual errors only when using 64 pilot symbols. For the UMTS channel, however, 32-pilot channel estimation results in a BER performance comparable to that of the 64-pilot estimation.

If the linear interpolation algorithm is employed instead of the ideal lowpass filter, then the BER performance of all the systems investigated is degraded. This is in accordance with the channel estimation experiments, as evidenced for example by the WATM channel estimation plots in Figures 4.8 and 4.10. The BER performance curves for 16-QAM transmission employing linearly interpolated PSAM are given in Figure 4.20, and it can be seen that for all modems, the performance is worse than that associated with the lowpass interpolation curves shown in Figure 4.17. The least differences between the two interpolation algorithms are visible for the WATM modem. In both cases, 64-pilot symbols are insufficient for combating residual bit errors and the residual bit error rates for both schemes are similar. In the short UMTS channel, the differences are more obvious. While the performance of the 32- and 64-pilot estimation systems was the same for lowpass interpolation, the linear interpolation algorithm shows residual errors for 32 pilot symbols. Compared to the lowpass interpolation, the BER results for the linear interpolator using 64 pilots are about 3 dB worse.

The BER performance for the WLAN system, as depicted in Figures 4.20(c) and 4.17(c), shows relatively minor differences between lowpass and linear interpolation. The residual bit error rates for 8, 16 and 32 pilots are similar, while for 64 pilots there is a SNR loss of about

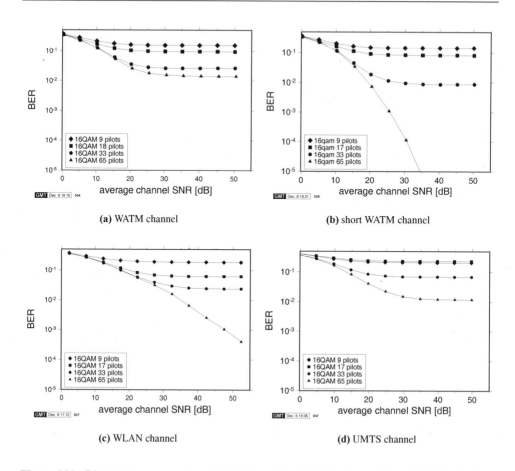

(a) WATM channel

(b) short WATM channel

(c) WLAN channel

(d) UMTS channel

Figure 4.20: Bit error rate versus channel SNR for static (a) WATM channel (Figure 4.1(b)), (b) short-ened WATM channel (Figure 4.3), (c) WLAN channel (Figure 4.4), (d) UMTS channel (Figure 4.5(a)), using PSAM 16-QAM with linear interpolation

5 dB for the linearly interpolated system at a BER of 10^{-3}. The UMTS, whose performance curves are plotted in Figure 4.20(d), exhibits a residual BER of about 1% for 64 pilots.

For BPSK-PSAM transmission over the OFDM systems, the difference in performance between lowpass and linear interpolation algorithms are much less apparent than for 16-QAM. Figure 4.21 gives an overview of the BER performance of the simulated systems in their respective environments using BPSK with linear interpolation-based channel estimation. The BER performance of the WATM system, as depicted in Figure 4.21(a), does not vary significantly with the number of pilots employed. This differs from the curves observed for the lowpass interpolator in Figure 4.18(a), where much higher bit error rates were observed, if only 8 pilots were employed for the channel estimation. Both the linear as well as the lowpass interpolator results show best performance for 17 and 16 pilot symbols, respectively. Over the short WATM channel, systems employing either interpolator exhibit an essentially identical performance for all but 8 and 9 pilots, respectively. In the latter case, there is a performance

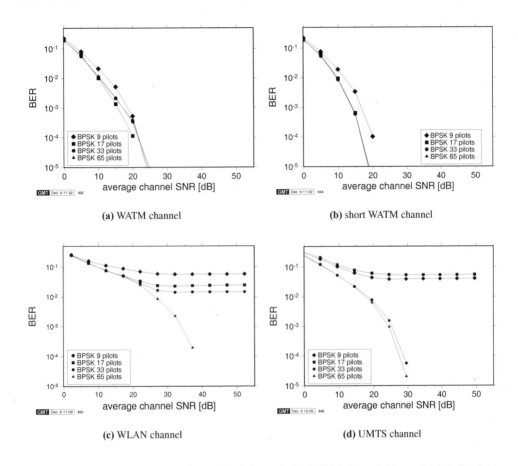

Figure 4.21: Bit error rate versus channel SNR for static (a) WATM channel (Figure 4.1(b)), (b) short-ened WATM channel (Figure 4.3), (c) WLAN channel (Figure 4.4), (d) UMTS channel (Figure 4.5(a)), using PSAM-BPSK with linear interpolation

difference of 13-dB SNR for a BER of 10^{-3} in favour of the linear interpolation. For the WLAN and the UMTS, the BER performance with lowpass and linear interpolation is fairly similar, the difference being slightly lower residual bit error rates for the linearly interpolated channel estimation.

4.4.2 Slowly Varying Time-Dispersive Channel

The slowly varying time-dispersive channel is characterised by an impulse response that is time-varying, but which is assumed constant for the duration of one OFDM symbol. Hence we also refer to this channel model as an OFDM symbol time-invariant channel. This stationary property prevents intersubcarrier interference and renders the wideband channel estimation possible. In this case, the channel can be viewed as a succession of static impulse responses generated upon fading the constituting paths. The rate of change for each of the paths' fading is described by the Doppler frequency of the fading channel.

As we saw in Section 4.2.2, the time-varying nature of the channel transfer function will introduce intersubcarrier interference depending on the symbol normalised Doppler frequency $F_d = f_d \cdot T_s \cdot N$. Performance degradations can be observed even for small values of f_d. We will investigate the validity of the assumption of the stationary channel for our four scenarios in the case of perfect channel estimation.

4.4.2.1   Perfect Channel Estimation

Perfect channel estimation is the best-case scenario for transmission over wideband channels and the system performance in these circumstances constitutes an upper bound to the achievable performance with realistic channel estimation algorithms. As the results for perfect channel estimation are not influenced by sub-optimal channel estimation, the effects of the time-varying channel on the transmitted data symbols can be studied and the validity of the static channel assumption for the four operating frameworks, namely WATM, short WATM, WLAN and UMTS, can be verified.

Perfect channel estimation of the time-varying wideband channel was achieved by taking a snapshot of the channel impulse response in the centre of the OFDM symbol and calculating the corresponding frequency domain channel transfer function from this impulse response by the help of the FFT. This transfer function does not fully characterise the channel, unlike in the static channel case, since it does not give any information concerning the intersubcarrier interference and only takes into account the time-varying nature of the impulse response. Nonetheless, the achieved channel estimation constitutes a best-case benchmark for realistic channel estimation algorithms.

In order to isolate the influence of the time domain variations of the channel during the transmission of an OFDM symbol, experiments have been conducted for all four operating frameworks of Table 4.1 with two different models of the fading channel. First, the channel impulse response was updated after every transmitted sample, resulting in a realistic model for a time-varying wideband channel. The second set of experiments employed a stationary channel model, where the impulse response was kept constant for the duration of each transmitted OFDM symbol. The resulting BER results from these experiments are shown in Figure 4.22. A BER performance degradation due to the time-varying channel can be observed for all the systems. The BER degradation is low for the WATM system over both the WATM and the shortened WATM channel with $F_d = 0.0063$, exhibiting a residual BER of under $2 \cdot 10^{-4}$ for 16-QAM. For the WLAN system and the UMTS, whose F_d values are about 0.04, the BER degradation is much more evident. In both cases, the residual BER for 16-QAM transmission is about $6 \cdot 10^{-3}$ and for BPSK the BER residual is $7.4 \cdot 10^{-4}$. Clearly, for the UMTS and WLAN at the given vehicular speed, the channel cannot be assumed stationary.

4.4.2.2   Pilot Symbol Assisted Modulation Summary

In order to investigate the effects of imperfect channel estimation on the system's BER performance in wideband fading channels separately from the effects of time domain variations of the impulse response, the PSAM experiments have been conducted in the stationary channel. Therefore, the observed performance degradation compared to the perfect channel estimation results depicted in Figure 4.22 are caused by incorrect amplitude and phase estimations.

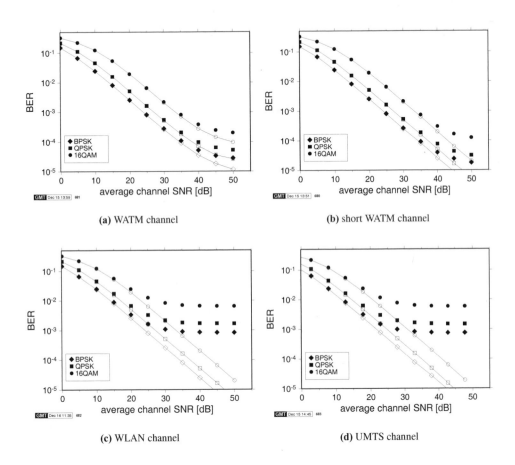

Figure 4.22: Bit error rate versus channel SNR for fading (a) WATM channel (Figure 4.1(b)), (b) shortened WATM channel (Figure 4.3), (c) WLAN channel (Figure 4.4), (d) UMTS channel
(Figure 4.5(a)), using coherent detection and perfect channel estimation. The open symbols correspond to the stationary channel model and the filled symbols correspond to the
continuously fading channel model

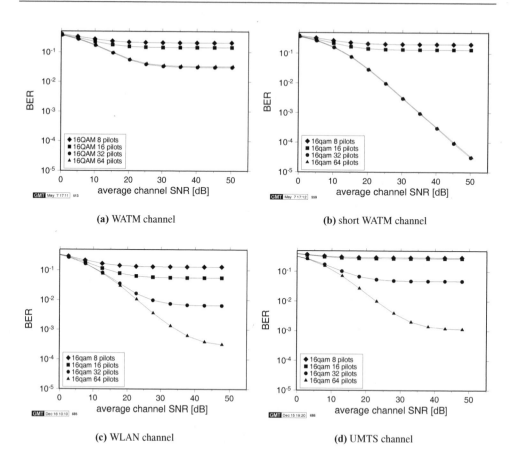

Figure 4.23: Bit error rate versus channel SNR for fading (a) WATM channel (Figure 4.1(b)), (b) short-ened WATM channel (Figure 4.3), (c) WLAN channel (Figure 4.4), (d) UMTS channel (Figure 4.5(a)), using coherent detection and pilot symbol assisted channel estimation with ideal lowpass filter interpolation for 16-QAM

The experimental results for 16-QAM transmission employing PSAM in conjunction with ideal lowpass and linear interpolation techniques are given in Figures 4.23 and 4.24, respectively. The estimation accuracy of the PSAM interpolation algorithms depends on each faded impulse response actually encountered by the modem, with some impulse amplitude combinations resulting in very poor estimation quality for the linear interpolation algorithm.

4.5 Intersubcarrier Interference Cancellation

4.5.1 Motivation

In the previous subsection it was argued that OFDM transmission is sensitive against the fading-induced CIR-tap variations incurred during an OFDM symbol period. This phe-

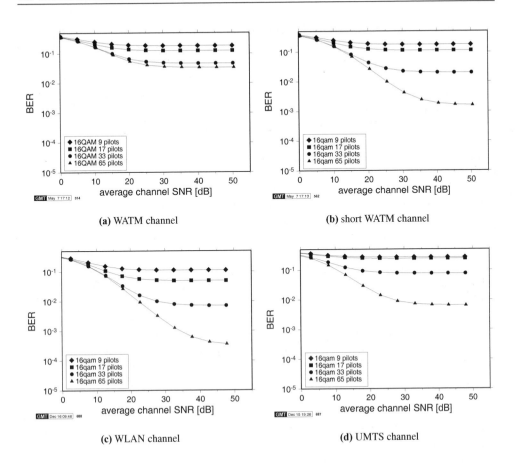

(a) WATM channel

(b) short WATM channel

(c) WLAN channel

(d) UMTS channel

Figure 4.24: Bit error rate versus channel SNR for fading (a) WATM channel (Figure 4.1(b)), (b) short-ened WATM channel (Figure 4.3), (c) WLAN channel (Figure 4.4), (d) UMTS channel (Figure 4.5(a)), using coherent detection and pilot symbol-assisted channel estimation with linear interpolation for 16-QAM

nomenon leads to ICI and hence in this subsection we consider various techniques that may be invoked for mitigating the effects of ICI. ICI can be considered as an additional noise contribution, which linearly depends on the signal power and thus it potentially limits the achievable BER at relatively high SNRs. Here we propose a time domain CIR tap variation estimator, which relies on both the received signal as well as on tentative symbol decisions. Our simulation results obtained in the context of a system employing decision-directed channel-prediction demonstrate that with the aid of the proposed algorithm, ICI cancellation can be performed even in transmission scenarios associated with high Doppler frequencies. We will demonstrate that this may be achieved without inserting additional time domain pilot symbols.

To elaborate a little further, the ICI contribution imposed on each subcarrier can be viewed as additional noise, which becomes - assuming a sufficiently high number of subcarriers - near-Gaussian distributed as a consequence of the central limit theorem. However, in contrast to the effects of AWGN, the ICI imposed on the different subcarriers is correlated. Since the ICI power is a linear function of the signal power, the AWGN power may be exceeded by the ICI power, when the transmitted signal power is high. Several schemes have been proposed in the literature for reducing the ICI noise power [186–188]. For example, Hasholzner *et al.* [186] propose to model the ICI generation by means of a Multiple-Input Multiple-Output (MIMO) system. Equalised and potentially less ICI contaminated subcarrier signals are obtained by a weighted superposition of the signal received in each subcarrier with the contributions from its neighbours. Adaptation of the frequency domain MIMO equaliser weights is achieved by means of the Recursive Least Squares (RLS) algorithm upon invoking the difference between the complex signals at the symbol detector output and at the equaliser output. A drawback of this approach is the RLS algorithm's limited CIR tracking capability under rapid fading channel conditions. An algorithm which follows a similar strategy was proposed in [187]. Here the equaliser weights are obtained by directly inverting the channel matrix describing the MIMO system inflicting the ICI. This approach is equivalent to de-coupling the subcarriers. However, the dimension of the channel matrix is in most cases excessive for direct inversion. Hence, the inversion of multiple lower-dimensional sub-matrices hosting only a limited subset of complex ICI transfer factors between the subcarrier to be decontaminated and its nearest neighbours was proposed in both [186] and [187]. Since the rate of change of the CIR tap values is limited by the channel's Doppler frequency, the ICI is is predominantly restricted to neighbouring subcarriers. The estimation of the channel matrix was achieved in [187] upon regularly inserting time domain pilot symbols into the OFDM symbol stream. More specifically, the channel sounding symbol was a Dirac-impulse like time domain pilot tone surrounded by zero-samples, where the length of each zero-segment was identical to that of the OFDM symbol's cyclic prefix. For the simulations presented in [187], a propagation scenario having a relatively small delay spread was assumed, since otherwise the additional channel-sounding overhead would have been prohibitive. In the context of the Wireless Asynchronous Transfer Mode (WATM) system model, 512 subcarriers and a 64-sample guard interval were assumed [2]. Using the channel sounding symbols proposed in [187], this would impose a pilot overhead of 25%, which is excessive. A potential solution requiring no additional pilot information for estimating the CIR tap's variation will be detailed in Sections 4.5.3 of our contribution. Another approach which we would like to mention is that of Hutter *et al.* [188], where a reduction of the ICI was achieved upon exploiting the correlation between the ICI contributions imposed on adjacent subcarriers. In contrast to the algorithms outlined in [186, 187], in [188] no explicit knowledge concerning the frequency domain MIMO channel transfer factors between the different subcarriers was invoked. An estimate of the short-term channel correlation matrix was obtained by evaluating the vector of error signals between the sliced and remodulated symbol and the received symbol for each subcarrier, followed by multiplication with its Hermitean transpose. A refinement was achieved through averaging with the corresponding tentative estimate associated with the previous OFDM symbol period. The rest of this section is organized as follows. In Section 4.5.2 the ICI generation procedure will be highlighted, followed by the evaluation of the potentially achievable ICI reduction. The philosophy of an attractive CIR tap variation estimator will be detailed in Section 4.5.3, while the ICI canceller is described in Section 4.5.4. The system's BER performance is characterised in Section 4.5.5 and we will draw our conclusions in Section 19.

4.5.2 The Signal Model

It has been demonstrated for example in [187] that the signal $x[n, k]$ received in the k-th subcarrier of the n-th OFDM symbol is a weighted superposition of the complex symbol $s[n, k]$ assigned to this subcarrier at the transmitter, with the contributions $s[n, \acute{k}]$ from all other subcarriers $\acute{k} \neq k$:

$$x[n, k] = \sum_{\acute{k}=0}^{N-1} \left(\sum_{m=0}^{M-1} H_m[n, k - \acute{k}]e^{-j\frac{2\pi}{N}m\acute{k}} \right) s[n, \acute{k}] + n[n, k], \tag{4.11}$$

where:

$$H_m[n, k - \acute{k}] = \frac{1}{N} \sum_{t=0}^{N-1} h_m[n, t]e^{-j\frac{2\pi}{N}t(k-\acute{k})}, \tag{4.12}$$

$$n[n, k] = \sum_{t=0}^{N-1} \acute{n}[n, t]e^{-j\frac{2\pi}{N}tk}. \tag{4.13}$$

Specifically, $H_m[n, k - \acute{k}]$ defined in Equation 4.12 denotes the $(k - \acute{k})$-th frequency bin of the DFT associated with the complex time domain fading signal $h_m[n, t]$ of the m-th CIR tap during the n-th OFDM symbol period - normalised to the number of subcarriers N. Furthermore, $n[n, k]$ is the k-th DFT bin of the complex time domain AWGN noise process $\acute{n}[n, t]$, again for the n-th OFDM symbol period. The term $M - 1$ represents the maximum multipath-delay imposed by the channel, given as an integer multiple of the sampling interval duration T_s. It should be noted that in Equation 4.11 the contributions for $k \neq \acute{k}$ constitute the ICI inflicted by the neighbouring subcarriers upon the k-th subcarrier. In [188, 189] the variance of the ICI has been evaluated, employing the standard definition $\sigma_{ICI}^2 = E\{x[n, k]x^*[n, k]|_{k \neq \acute{k}}\}$ with the aid of Equation 4.11, where $E\{\}$ denotes the expectation. Similarly, we can evaluate the variance of the residual ICI, $\sigma_{ICI,resid}^2$ at the output of an ideal ICI cancellation scheme which is capable of suppressing the undesired ICI contributions in the range of $\{k - sym, \ldots, k + sym\}\backslash\{k\}$ around the subcarrier k, as follows:

$$\frac{\sigma_{ICI,resid}^2}{\sigma_s^2} = 1 - \frac{1}{N^2}\left[(1 + 2 \cdot sym)N + \right.$$
$$+ 2\sum_{\Delta n=1}^{N-1}(N - \Delta n)\frac{sin\left((sym + \frac{1}{2})\frac{2\pi}{N}\Delta n\right)}{sin\left(\frac{\pi}{N}\Delta n\right)} \cdot$$
$$\left. \cdot J_0\left(2\pi T_s f_D \Delta n\right)\right], \tag{4.14}$$

where a normalization to the signal variance σ_s^2 has been performed. In Equation 4.14, $J_0()$ denotes the zero-order Bessel function of the first kind, assuming Jakes fading [131], T_s is the sampling interval duration and f_D is the maximum Doppler frequency of the channel. It should be noted that by setting the one-sided ICI cancellation range sym equal to

$matrix\_size$	1	3	5
$\sigma^2_{ICI,cancel}/\sigma^2_s\|_{lin}$	0.012896	0.005008	0.003063
$reduction\|_{dB}$		-4.108	-2.135

Table 4.4: *Residual* ICI variance normalised to the signal variance for an OFDM symbol normalised Doppler frequency of $f_d T_f = 0.1$ in the context of the indoor WATM channel environment using $N = 512$ and $N_g = 64$.

zero, Equation 4.14 degenerates into the corresponding expression for the ICI noise variance given in [188, 189]. We have evaluated Equation 4.14 for channel sub-matrix sizes of 1, 2 and 4, where the sub-matrix size is related to the one-sided ICI cancellation range according to $matrix\_size = 1 + 2 \cdot sym$. Hence, a sub-matrix size of 1 corresponds to the case of no ICI cancellation. Here we employed the parameters of the indoor WATM system model [2]. Specifically $N = 512$ and $f_d T_f = 0.1$ was assumed as a worst-case scenario, where T_f denotes the frame duration, which is related to the sampling interval duration T_s by $T_f = (N + N_g)T_s$, with N_g as the number of guard samples. The results are listed in Table 4.4, where we have also calculated the differential ICI variance reduction in terms of dB, achieved by increasing the ICI cancellation range. In accordance with [187] the highest improvement is achieved by employing ICI cancellation with a one-sided range of 1 instead of no cancellation. Specifically this results in an ICI reduction of about $4.1dB$. A prerequisite for attaining this performance is the availability of perfect channel- and error-free symbol knowledge, so that the ICI contributions can be perfectly reconstructed and subtracted from the received signal. Upon invoking Equation 4.14 and by considering the first two elements of the Taylor approximation of $J_0(x)$, which is $\tilde{J}_0(x) = 1 - \frac{1}{4}x^2 \ \forall \ x \ll 1$ [89], we also confirm that in the range of $f_D T_f$ of interest, the ICI variance is proportional to the square of the Doppler frequency f_D: $\sigma^2_{ICI,red} = c \cdot f_D^2 \sigma_s^2$, where c is the proportionality constant inherent in Equation 4.14. We conclude furthermore that the maximum achievable ICI reduction is independent of the Doppler frequency f_D. In the next section we will embark on a description of the system model.

4.5.3 Channel Estimation

In Section 4.5.1 we outlined that a prerequisite for the cancellation of CIR variation induced ICI considered here is the availability of an estimate of the variation of each CIR tap during a specific OFDM symbol period. In contrast to [187], our aim is to avoid using dedicated time domain pilot symbols while also circumventing the employment of an iterative MIMO equaliser weight update strategy. **Explicitly, our strategy is to obtain tentative symbol decisions without using ICI cancellation in a first iteration, followed by an estimation of the CIR tap variations on the basis of these tentative symbol decisions and by also capitalising on the knowledge of the received time domain signal. This is followed by frequency domain ICI cancellation and a final iteration to obtain symbol decisions on the basis of a potentially less ICI-contaminated signal.** Hence, our aim in this section is to develop the CIR tap variation estimator. The received time domain signal $r[n, t]$, $t \in \{0, \ldots, N - 1\}$ within the FFT window of the n-th OFDM symbol period is given by the convolution of the time-variant CIR tap values $h_m[n, t]$, where $m \in \{0, \ldots, M - 1\}$ denotes the tap index, with the output $t[n, t]$ of the OFDM modulator, yielding:

$$r[n, t] = \sum_{m=0}^{M-1} h_m[n, t]t[n, t - m] + \acute{n}[n, t]. \tag{4.15}$$

Here we follow the philosophy that the evolution of each CIR tap value during the FFT window of an OFDM symbol period can be linearly approximated, which was also advocated in [187], resulting in:

$$\tilde{h}_m[n, t] = \tilde{c}_m[n] \cdot t + \tilde{b}_m[n], \tag{4.16}$$

where $\tilde{c}_m[n]$ and $\tilde{b}_m[n]$, $m \in \{0, \ldots, M-1\}$ are the CIR tap variation estimator coefficients to be determined. Hence, a cost-function given by the aggregate mean-square error between the received time domain signal and the appropriately synthesized signal can be defined as follows:

$$C[n] = \sum_{t=0}^{N-1} \left| r[n, t] - \sum_{m=0}^{M-1} (\tilde{c}_m[n]t + \tilde{b}_m[n])t[n, t - m] \right|^2. \tag{4.17}$$

In order to determine the estimator coefficients $\tilde{c}_m[n]$ and $\tilde{b}_m[n]$ for all $m \in \{0, \ldots, M-1\}$, a suitable approach is to minimize Equation 4.17 using standard optimisation techniques, namely by evaluating the partial derivatives with respect to the desired variables:

$$\frac{\partial C[n]}{\partial \tilde{b}_m[n]} = \sum_{t=0}^{N-1} \left\{ -r^*[n, t]t[n, t - m] + \right. \tag{4.18}$$

$$\left. + \sum_{\acute{m}=0}^{M-1} (\tilde{c}_{\acute{m}}^* t + \tilde{b}_{\acute{m}}^*) \cdot t[n, t - m]t^*[n, t - \acute{m}] \right\},$$

and

$$\frac{\partial C[n]}{\partial \tilde{c}_m[n]} = \sum_{t=0}^{N-1} \left\{ -r^*[n, t] \cdot t \cdot t[n, t - m] + \right. \tag{4.19}$$

$$\left. + \sum_{\acute{m}=0}^{M-1} (\tilde{c}_{\acute{m}}^* t + \tilde{b}_{\acute{m}}^*) \cdot t \cdot t[n, t - m]t^*[n, t - \acute{m}] \right\}.$$

At the optimum point the partial derivatives $\partial C[n]/\partial \tilde{b}_m$ and $\partial C[n]/\partial \tilde{c}_m$ are zero for all $m \in \{0, \ldots, M-1\}$. Hence, a system of equations can be established assuming knowledge of the received signal $r[n, t]$ and that of the transmitted signal $t[n, t]$, yielding estimates for the desired estimator coefficients upon inverting a matrix of dimensions $2M \times 2M$. In [91] it was proposed - assuming a similar estimation problem - to identify the path delays exhibiting the highest signal power and hence only to incorporate these so-called significant taps in the estimation process. This could be achieved with the aid of dedicated OFDM training symbols employed in the context of decision-directed channel estimation [131, 159]. The advantage would potentially be a significant size reduction of the matrix to be inverted. In the next section we will outline a range of different ICI cancellation schemes applicable to our system.

4.5.4   Cancellation Schemes

We commence with a brief review of the most prominent ICI-cancellation schemes employed. Let us recall Equation 4.11, which established a relationship between the symbols $s[n, \acute{k}]$ transmitted in different subcarriers $\acute{k}$ and the complex symbol $x[n, k]$ received on the k-th subcarrier, as a function of the subcarrier coupling factors and that of the channel noise. This relation could also be expressed in matrix form, potentially leading to an $N \times N$ frequency domain MIMO channel matrix, where N denotes the number of subcarriers. Fully compensating for the effects of ICI would require as alluded to in Section 4.5.2 the inversion of the channel matrix, which is impractical in terms of computational complexity. Hence, a more practical solution is to exploit the band structure of the channel matrix, which is a direct consequence of the sharply decaying ICI influence as a function of the frequency domain separation of subcarriers. This is a ramification of the relatively low Doppler frequency in comparison to the OFDM symbol's bandwidth. Therefore we perform ICI cancellation only for a limited subset of the subcarriers, namely within a range of sym around the subcarrier considered [186, 187]. Hence, for the k-th subcarrier a simplified relation between the transmitted symbol $s[n, k]$ and the received symbol $x[n, k]$ - assuming a cancellation range of $sym = 1$ is given by [187]:

$$\mathbf{x}[k] = \mathbf{H}[k]\mathbf{s}[k] + \mathbf{n}_r[k], \tag{4.20}$$

where

$$\mathbf{x}[k] = (x[k-1], x[k], x[k+1])^T \tag{4.21}$$
$$\mathbf{s}[k] = (s[k-1], s[k], s[k+1])^T \tag{4.22}$$
$$\mathbf{n}_r[k] = (n_r^k[k-1], n_r^k[k], n_r^k[k+1])^T, \tag{4.23}$$

and

$$\mathbf{H}[k] = \begin{pmatrix} H[k-1, k-1] & H[k-1, k] & 0 \\ H[k, k-1] & H[k, k] & H[k, k+1] \\ 0 & H[k+1, k] & H[k+1, k+1] \end{pmatrix}. \tag{4.24}$$

For notational convenience we have omitted the OFDM symbol index n. The vector $\mathbf{n}_r[k]$ of residual noise contributions requires further explanation. Each of its elements $n_r^k[x]$ is constituted by the sum of the AWGN in the x-th subcarrier and the residual ICI taking into account the partial ICI cancellation. For example, following from the structure of the matrix $\mathbf{H}[k]$ defined in Equation 4.24, the residual ICI contribution in the $(k-1)$-th subcarrier is identical to the original contribution minus the contribution from the k-th subcarrier. By contrast, the residual ICI in the k-th subcarrier is constituted by the difference between the original contribution in this subcarrier minus the interference due to the $(k-1)$-th and the $(k+1)$-th subcarrier. The elements of the matrix $\mathbf{H}[k]$ can be directly inferred from Equation 4.11, specifically from the term in round brackets. Equation 4.20 suggests several approaches for recovering the symbol transmitted on the k-th subcarrier. The authors of [187] proposed the direct inversion of the partial channel matrix $\mathbf{H}[k]$, which will be referred to during our comparative study as the **zero-forcing (ZF)** solution. Hence, the signal at the output of the combiner is given by:

$$\tilde{\mathbf{s}}_{ZF}[k] = \mathbf{H}^{-1}[k]\mathbf{x}[k], \tag{4.25}$$

where finally only the desired element in the center of the vector $\tilde{\mathbf{s}}_{ZF}[k]$ is retained. An exception is given by the first and the last vector of recovered symbols, where the first $sym + 1$ elements and the last $sym + 1$ elements are retained [187], respectively. A well-known disadvantage of the ZF solution is the associated potential noise amplification. A more attractive, but also more complex approach is the **minimum mean-square error (MMSE)** solution, which was also advocated in [188]:

$$\tilde{\mathbf{s}}_{MMSE}[k] = \left(\mathbf{H}^H[k]\mathbf{H}[k] + \left(\frac{\sigma_{n,\Sigma}^2}{\sigma_s^2} + dl \right) \mathbf{I} \right)^{-1} \mathbf{H}^H[k]\mathbf{x}[k]. \tag{4.26}$$

It should be noted that in contrast to [188] we have not taken into account the correlation between the residual ICI associated with the subcarriers encompassed by the partial MIMO channel matrix $H[k]$, since the merit of doing so appears to be relatively limited. The residual noise variance $\sigma_{n,\Sigma}^2$ in Equation 4.26 is given by the sum of the AWGN noise variance σ_n^2 and the residual ICI variance $\sigma_{ICI,cancel}^2$, where in the context of our simulations we have employed the estimated noise variance given in the second column of Table 4.4. Furthermore, in Equation 4.26, dl denotes a diagonal loading constant, which is often employed in the literature [188] for facilitating the inversion of ill-conditioned matrices. In the context of our simulations in Section 4.5.5 we experimentally found a value of $dl = 0.055$ to operate best for BPSK and $dl = 0$ for QPSK. A third approach to ICI cancellation, which - in contrast to the previously mentioned ZF- and MMSE solutions - requires the knowledge of the transmitted symbols, is that of **subtractive cancellation**, which is defined here simply by the subtraction of the reconstructed from the received signal. This can be expressed as:

$$\tilde{\mathbf{s}}_{MMSE}[k] = r[k] - \sum_{\substack{\acute{k}=k-sym \\ \acute{k} \neq k}}^{k+sym} H[k, \acute{k}]\hat{s}[\acute{k}], \tag{4.27}$$

where $\hat{s}[\acute{k}]$ denotes the tentative symbol decision of the $\acute{k}$-th subcarrier symbol obtained during the first iteration. In next section the above ICI cancellation schemes will be characterised in conjunction with the CIR tap variation estimator introduced in Section 4.5.3.

4.5.5 ICI Cancellation Performance

In this section we will assess the performance of the proposed system using a decision-directed channel estimator similar to that proposed in [131] for obtaining initial symbol decisions. In [131] Wiener filtering of the CIR taps in the time domain is employed for obtaining a potentially less noise-contaminated estimate of the channel transfer function for the current timeslot, which is then employed as an initial channel estimate during the next timeslot. This implies assuming the invariance of the channel transfer function between two consecutive OFDM symbols. By contrast, we employ a Wiener prediction filter in order to compensate for the CIR tap variations actually incurred in the high-Doppler propagation scenarios considered here. As outlined in [159], the conceptual difference between the interpolation and the prediction filter resides in the structure of the co-variance vector, which is part of the Wiener equation. In our simulations we employed a decision-directed channel estimator capitalising on an 8-tap CIR predictor, a training period of 32 OFDM symbols and a training block length of two OFDM symbols. It should be noted that the training period or equivalently the time domain distance between two training blocks is a crucial parameter, especially at low SNRs, where error propagation is known to deteriorate the decision-directed channel estimator's performance. The training block length has two implications. Firstly, at low SNRs error propagation extending over the training period is experienced as the result of using past CIR estimates in the process of CIR tap prediction. Secondly, the CIR tap predictor is required to deliver sufficiently accurate channel estimates already during the reception of the first information-bearing OFDM symbol. It should be noted that the proposed CIR tap variation estimator could also be employed in a system, which invokes for example frequency domain pilot symbols for obtaining a tentative CIR estimate, in order to assist in the tentative initial demodulation required. Our simulations have been conducted in the context of the 512-subcarrier indoor WATM channel [2] employing a worst-case Doppler scenario of $f_D T_f = 0.1$. Our BER results for BPSK modulation are portrayed in Figure 4.25, while those for QPSK modulation are depicted in Figure 4.26. In both figures we have also plotted the system's performance without ICI cancellation. We considered both, a "frame-invariant" fading scenario, where the fading envelope was kept constant during an OFDM symbol period in order to avoid the generation of ICI, and a "frame-variant" fading scenario. We observe that even in the context of "frame-invariant" fading, where no ICI was generated, the BER performance is limited at high SNRs, which is attributed to the imperfections of the CIR predictor. Amongst the different ICI cancellation schemes the MMSE canceller exhibits the best performance. Specifically, in the context of BPSK the BER is reduced by a factor of 5, while in conjunction with QPSK a reduction by a factor of 3 is observed. The BER difference between these schemes can be explained by pointing out that the minimum distance between signal points hosted by a QPSK constellation are located closer to each other than in the BPSK constellation and hence a higher vulnerability against channel estimation errors and channel noise can be observed. Another phenomenon, which requires further explanation is that for BPSK the ZF-inversion performs worse, than subtraction based cancellation. By contrast, for QPSK the situation was reversed. An explanation is that in the case of BPSK, when relatively reliable first-iteration symbol decisions are available, the subtractive combiner does not encounter the problem of noise amplification, which was associated with the ZF-inversion based solution. Again, by contrast, for QPSK more erroneous symbol decisions are employed in the subtractive cancellation process and hence the ZF-inversion, which does not capitalize directly on tentative symbol decisions is advantageous.

4.5.6   Conclusions on ICI Cancellation

In this section we have demonstrated the feasibility of CIR tap variation-induced ICI cancellation without requiring additional time domain pilot symbols as in [187] or without necessitating iterative weight update techniques as in [186]. We employed a two-stage detection

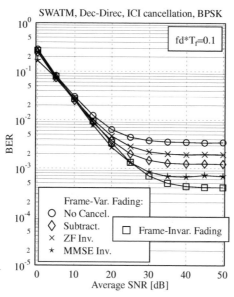

Figure 4.25: BER performance of a system employing 8-tap Wiener filter prediction assisted decisions-directed channel estimation and two-stage ICI cancellation according to Sections 4.5.3 and 4.5.4 by using **BPSK modulation**.

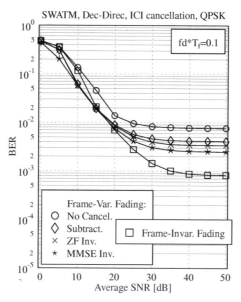

Figure 4.26: BER performance of a system employing 8-tap Wiener filter prediction assisted decisions-directed channel estimation and two-stage ICI cancellation according to Sections 4.5.3 and 4.5.4 by using **QPSK modulation**.

technique, where during the first iteration tentative symbol decisions are provided on the basis of the initially ICI-contaminated received signal. These initial estimates are then employed to estimate the CIR tap variations incurred during the current OFDM symbol period. Following the process of ICI cancellation, the potentially less ICI-impaired received signal is employed in a second demodulation iteration to provide symbol decisions. As a result, the BER was reduced by a factor of five in the context of BPSK and a factor of three in the context of QPSK, when using decision-directed channel estimation as a basis for first-stage detection.

4.6   Chapter Summary and Conclusion

The performance of OFDM transmission over time-varying and time-dispersive channels is mainly limited by two factors, namely the long-term shape and the rate of time variability of the channel's impulse response. The maximum impulse response dispersion determines the rate of frequency domain fluctuation of the channel transfer function, which manifests itself as frequency domain fading across the bandwidth of the OFDM symbol. The effects of this frequency domain channel transfer function fading can be combated with the aid of similar PSAM methods employed in the frequency domain as known from the time domain fading envelope estimation technique employed in conventional narrowband serial modems. Similarly to conventional serial time domain transmission schemes, frequency domain PSAM requires the transmission of known pilot symbols and hence increases the system's overhead.

As an alternative solution, differential detection of the received data symbols may be invoked, where we exploit that the adjacent subcarriers of an OFDM typically experience similar frequency domain attenuations and phase rotations. Similar statements may be made also for the subcarriers at the same frequency of consecutive OFDM symbols. Hence the effects of channel transfer function fluctuations may be eliminated, when the adjacent subcarriers are used as the reference required for differential detection. We have to note, however, that differential detection typically shows a 3 dB SNR performance loss compared to coherently detected schemes. For both differential detection and PSAM the channel's impulse response determines the achievable performance. High-dispersion channels result in rapid fading in the frequency domain, either requiring more pilots per OFDM symbol for PSAM or degrading the achievable BER performance of differential detection.

Variations of the channel's impulse response, caused by fast fading of the constituting paths, result in interference between the OFDM symbol's subcarriers. This intersubcarrier interference is low, if the channel varies only insignificantly during an OFDM symbol, but causes severe performance degradation, if the rate of the channel variation is not much slower than the OFDM symbol rate. The effect of intersubcarrier interference therefore limits the maximum number of subcarriers employed in an OFDM system, depending on the channel's Doppler frequency and the robustness of the modulation scheme employed. Finally, the chapter was closed by proposing an ICI cancellation technique, which was capable of substantially improving the achievable system performance.

In the next chapter we will study a range of time and frequency synchronisation issues encountered in the context of OFDM transmission schemes.

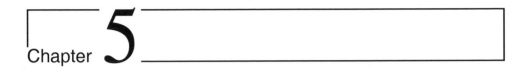

Chapter **5**

OFDM Time and Frequency Domain Synchronisation[1]

In this chapter we will investigate the effects of time and frequency domain synchronisation errors on the performance of an OFDM system, and two different synchronisation algorithms will be presented for time domain burst-based OFDM communications systems.

5.1 System Performance with Frequency and Timing Errors

The performance of the synchronisation subsystem, in particular the accuracy of the frequency and timing error estimations, is of major influence on the overall OFDM system performance. In order to investigate the effects of carrier frequency and time domain FFT window alignment errors, a series of investigations has been performed over different channels.

5.1.1 Frequency Shift

Carrier frequency errors result in a shift of the received signal's spectrum in the frequency domain. If the frequency error is an integer multiple n of the subcarrier spacing Δf, then the received frequency domain subcarriers are shifted by $n \cdot \Delta f$. The subcarriers are still mutually orthogonal, but the received data symbols, which were mapped to the OFDM spectrum, are in the wrong position in the demodulated spectrum, resulting in a bit error rate of 0.5.

If the carrier frequency error is not an integer multiple of the subcarrier spacing, then energy is spilling over between the subcarriers, resulting in loss of their mutual orthogonality. In other words, interference is observed between the subcarriers, which deteriorates the bit

---

[1] *OFDM and MC-CDMA for Broadband Multi-user Communications WLANS and Broadcasting.*
L.Hanzo, Münster, T. Keller and B.J. Choi,
©2003 John Wiley & Sons, Ltd. ISBN 0-470-85879-6

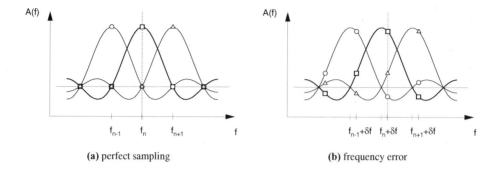

(a) perfect sampling   **(b)** frequency error

Figure 5.1: Stylised plot of OFDM symbol spectrum with sampling points for three subcarriers. The symbols on the curves signify the contributions of the three subcarriers to the sum at the sampling point: (a) no frequency offset between transmitter and receiver, (b) frequency error δf present

error rate of the system. The amount of this intersubcarrier interference can be evaluated by investigating the spectrum of the OFDM symbol.

5.1.1.1 The Spectrum of the OFDM Signal

The spectrum of the OFDM signal is derived from its time domain representation transmitted over the channel. A single OFDM symbol in the time domain can be described as:

$$u(t) = \left[\sum_{n=0}^{N-1} a_n e^{j\omega_n \cdot t} \right] \times \text{rect}\left(\frac{t}{N \cdot T_s} \right), \tag{5.1}$$

which is the sum of N subcarriers $e^{\omega_n \cdot t}$, each modulated by a QAM symbol a_n and windowed by a rectangular window of the OFDM symbol duration T_s. The Fourier transform of this rectangular window is a frequency domain sinc function, which is convolved with the Dirac delta subcarriers, determining the spectrum of each of the windowed complex exponential functions, leading to the spectrum of the nth single subcarrier in the form of

$$A_n(\omega) = \frac{\sin(N \cdot T_s \cdot \omega/2)}{N \cdot T_s \cdot \omega/2} * \delta(\omega - \omega_n).$$

Replacing the angular velocities ω by frequencies and using the relationship $N \cdot T_s = 1/\Delta f$, the spectrum of a subcarrier can be expressed as:

$$A_n(f) = \frac{\sin(\pi \frac{f - f_n}{\Delta f})}{\pi \frac{f - f_n}{\Delta f}} = \text{sinc}\left(\frac{f - f_n}{\Delta f} \right).$$

The OFDM receiver samples the received time domain signal, demodulates it by invoking the FFT and in the case of a carrier frequency shift, it generates the sub-channel signals in the frequency domain at the sampling points $f_n + \delta f$, which are spaced from each other by the subcarrier spacing Δf and misaligned by the frequency error δf. This scenario is shown in Figure 5.1. Figure 5.1(a) shows the sampling of the subcarrier at frequency f_n at the optimum frequency raster, resulting in a maximum signal amplitude and no intersubcarrier interference. If the frequency reference of the receiver is offset with respect to that of the transmitter by a frequency error of δf, then the received symbols suffer from intersubcarrier interference, as depicted in Figure 5.1(b).

The total amount of intersubcarrier interference experienced by subcarrier n is the sum of the interference amplitude contributions of all the other subcarriers in the OFDM symbol:

$$I_n = \sum_{j,j \neq n} a_j \cdot A_j(f_n + \delta f).$$

Since the QAM symbols a_j are random variables, the interference amplitude in subcarrier n, I_n, is also a random variable, which cannot be calculated directly. If the number of interferers is high, however, then the power spectral density of I_n can be approximated with that of a Gaussian process, according to the central limit theorem. Therefore, the effects of the intersubcarrier interference can be modelled by additional white Gaussian noise superimposed on the frequency domain data symbols.

The variance of this Gaussian process σ_{ISI_n} is the sum of the variances of the interference contributions,

$$\sigma_{ISI_n}^2 = \sum_{j,j \neq n} \sigma_{a_j}^2 \cdot |A_j(f_n + \delta f)|^2 .$$

The quantities $\sigma_{a_j}^2$ are the variances of the data symbols, which are the same for all j in a system that is not varying the average symbol power across different subcarriers. Additionally, because of the constant subcarrier spacing of Δf, the interference amplitude contributions can be expressed more conveniently as:

$$A_j(f_n + \delta f) = A_j n(\delta f) = sinc((n - j) + \frac{\delta f}{\Delta f}).$$

The sum of the interferer powers leads to the intersubcarrier interference variance expression:

$$\sigma_{ISI}^2 = \sigma_a^2 \cdot \sum_{i=-N/2-1}^{N/2} \left| sinc(i + \frac{\delta f}{\Delta f}) \right| . \tag{5.2}$$

The value of the intersubcarrier interference (ISI) variance for FFT lengths of $N = 64$, 512 and 4096 and for a range of frequency errors δf is shown in Figure 5.2. It can be seen that the number of subcarriers does not influence the ISI noise variance for OFDM symbol lengths of more than 64 subcarriers. This is due to the rapid decrease of the interference amplitude with increasing frequency separation, so that only the interference from close subcarriers contributes significantly to the interference load on the subcarriers.

In order to investigate the accuracy of the Gaussian approximation, simulations were conducted and histograms of the measured interference amplitude were produced for QPSK and 16-QAM modulation of the subcarriers. The triangles in Figure 5.3 depict the histograms of ISI noise magnitudes recorded for a 512-subcarrier OFDM modem employing QPSK and 16-

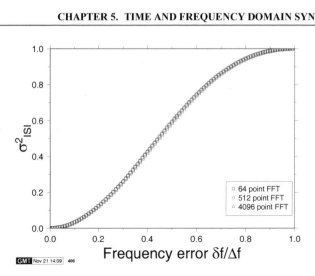

Figure 5.2: Intersubcarrier interference variance due to a frequency shift δf FFT lengths of $N = 64$, 512 and 4096 for normalised frequency errors $\delta f / \Delta f$ between 0 and 1

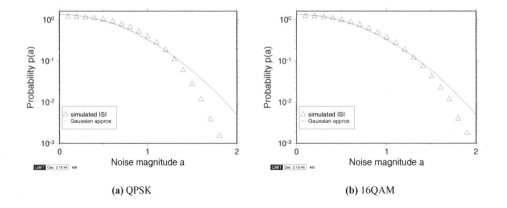

(a) QPSK   (b) 16QAM

Figure 5.3: Histogram of the ISI magnitude for a simulated 512-subcarrier OFDM modem using QPSK or 16-QAM for $\delta f = 0.3\Delta f$; the line represents the Gaussian approximation having the same variance

QAM in a system having a frequency error of $\delta f = 0.3\Delta f$. The continuous line drawn in the same graph is the corresponding approximation of the histogram by a Gaussian probability density function (PDF) of the variance calculated using Equation 5.2. It can be observed that the Gaussian curve is a reasonable approximation for both histograms in the central region, but that for the tails of the distributions the Gaussian function exhibits high relative errors. The histogram of the interference caused by the 16-QAM signal is, however, closer to the Gaussian curve than the QPSK interference histogram.

However, the frequency mismatch between the transmitter and receiver of an OFDM system not only results in intersubcarrier interference, but it also reduces the useful signal am-

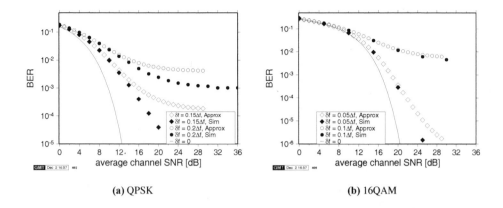

(a) QPSK   (b) 16QAM

Figure 5.4: The effect of intersubcarrier interference due to frequency synchronisation error on the
BER over AWGN channels: (a) bit error probability versus channel SNR for frequency
errors of $0.15\Delta f$ and $0.2\Delta f$ for a QPSK modem. (b) BER versus channel SNR for frequency errors of $0.05\Delta f$ and $0.1\Delta f$ for a 16-QAM modem. In both graphs, the filled
symbols are simulated BER results, while the open symbols are the predicted BER curves
using the Gaussian intersubcarrier interference model

plitude at the frequency domain sampling point by a factor of $f(\delta f) = \mathrm{sinc}(\delta f/\Delta f)$. Using
this and σ_{ISI}^2, the theoretical influence of the inter-subcarrier interference, approximated by
a Gaussian process, can be calculated for a given modulation scheme in an AWGN channel.
In the case of coherently detected QPSK, the closed-form expression for the BER $P_e(\gamma)$ at a
channel SNR γ is given [177] by:

$$P_e(\gamma) = Q(\sqrt{\gamma}),$$

where the Gaussian $Q()$-function is defined as

$$Q(y) = \frac{1}{\sqrt{2\pi}} \int_y^\infty e^{-x^2/2} dx = \mathrm{erfc}\left(\frac{y}{\sqrt{2}}\right).$$

Assuming that the effects of the frequency error can be approximated by white Gaussian
noise of variance σ_{ISI}^2 and taking into account the attenuated signal magnitude $f(\delta f) = sinc(\pi\delta f/\Delta f)$, we can adjust the equivalent SNR to:

$$\gamma' = \frac{f(\delta f) \cdot \sigma_a^2}{\sigma_{ISI}^2 + \sigma_a^2/\gamma},$$

where σ_a^2 is the average symbol power and γ is the real channel SNR. Comparison between
the theoretical BER calculated using γ' and simulation results for different frequency errors
δf are shown in Figure 5.4(a). While for both frequency errors the theoretical BER using
the Gaussian approximation fits the simulation results well for channel SNR values of up
to 12 dB, the predictions and the simulation results diverge for higher values of SNR. The
pessimistic BER prediction is due to the pronounced discrepancy between the histogram

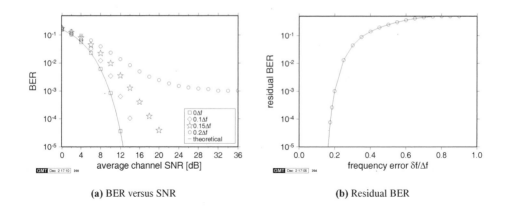

(a) BER versus SNR   (b) Residual BER

Figure 5.5: Bit error rate versus channel SNR performance for a non-pilot-assisted QPSK OFDM modem in an AWGN channel: (a) bit error rate versus signal-to-noise ratio plot for different constant frequency errors, (b) plot of residual bit error rate versus the frequency error

and the Gaussian curve in Figure 5.3 at the tail ends of the amplitude histograms, since for high noise amplitudes the Gaussian model is a poor approximation for the intersubcarrier interference.

The equivalent experiment, conducted for coherently detected 16-QAM, results in the simulated and predicted bit error rates depicted in Figure 5.4(b). For 16-QAM transmission, the noise resilience is much lower than for QPSK, hence for our experiments smaller values of δf have been chosen. It can be observed that the Gaussian noise approximation is a much better fit for the simulated BER in a 16-QAM system than for a 4-QAM modem. This is in accordance with Figure 5.3, where the histograms of the interference magnitudes were depicted.

5.1.1.2   Effects of Frequency Mismatch on Different Modulation Schemes

In order to investigate the effects of frequency mismatch on different modulation schemes, a series of simulations wwa conducted employing both coherently and differentially detected, as well as pilot symbol assisted QPSK systems. Figures 5.5(a), 5.6(a) and 5.7(a) show the performance of the QPSK, pilot symbol-assisted QPSK (PSA-QPSK) and differential QPSK (DQPSK) OFDM schemes, respectively. As a benchmark, the BER performance of the equivalent serial modulation scheme is plotted [177] as a line on all the graphs, which also represents the performance that is achieved by the parallel modem, when $\delta f = 0$.

5.1.1.2.1   Coherent Modulation   Figure 5.5(a) reveals the BER performance degradation due to increasing the carrier frequency offset δf. The adjacent sub-channel interference effects are considerable, even for small frequency errors. It is clear that for a carrier frequency offset of $\delta f = 0.2 \cdot \Delta f$, the BER reaches a residual value of about 10^{-3} at approximately 26 dB of SNR. The minimal possible BER for different values of carrier frequency mismatch are plotted in Figure 5.5(b). For relative frequency errors above $0.18 \cdot \Delta f$, the attainable bit error rate is worse than 10^{-4}.

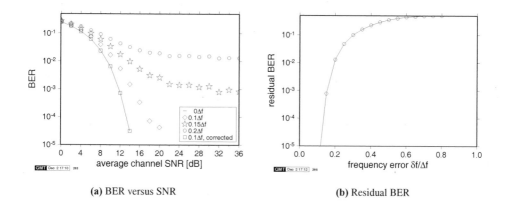

Figure 5.6: Bit error rate versus channel SNR performance for a 3-pilot-assisted QPSK OFDM modem in an AWGN channel: (a) bit error rate versus signal-to-noise ratio plot for different fixed frequency errors, (b) plot of residual bit error rate versus the frequency error

5.1.1.2.2 Pilot Symbol Assisted Modulation Figure 5.6(a) shows the bit error rate performance of the pilot-assisted QPSK system over an AWGN channel, which is consistently worse than that of an equivalent coherently demodulated system without pilot assistance. This is in accordance with the performance of PSAM schemes over AWGN channels, where PSAM systems are generally handicapped by errors in the channel estimation, which is unnecessary for non-fading channels. Since the frequency error mismatch results in additional noise in the received frequency domain pilots, the quality of the channel estimation deteriorates. This reduces the BER performance of a PSAM system compared to a coherently demodulated system refraining from using pilots.

The bit error rate curves given in the figure were computed for three pilot subcarriers per 512-subcarrier OFDM symbol, which were invoked to mitigate the effects of the channel's fading envelope in a time or frequency fading environment. The corresponding relationship between the residual bit error rate and the frequency error is given in Figure 5.6(b). In the simulations presented here linear interpolation was employed between the pilots. Clearly, the BER performance of the system depends on the number of pilots employed as well as on the interpolation method, and these effects are discussed in Section 4.3.1.1 in more depth.

5.1.1.2.3 Differential Modulation The corresponding simulation results for differentially encoded QPSK are shown in Figure 5.7. Again, the impact of the intersubcarrier interference is severe even for small relative frequency errors. Figure 5.7(b) shows that a frequency error of only $0.12\Delta f$ results in a BER residual of 10^{-4}.

5.1.1.2.4 Frequency Error - Summary A frequency error in an OFDM system results in a shift of the received frequency domain symbols relative to the receiver's raster. This leads to intersubcarrier interference, whose nature is noise-like, owing to the great number of contributing interfering subcarriers. Because of the frequency domain aliasing, the variance of this interference is constant for all subcarriers in the OFDM symbol, if all subcarriers carry the same average power. The variance of the interference can be computed and used to model

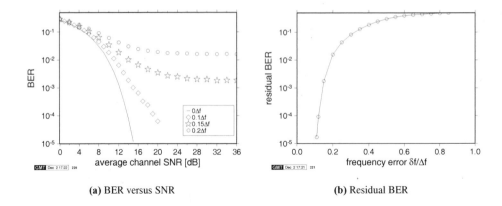

(a) BER versus SNR   (b) Residual BER

Figure 5.7: Bit error rate versus channel SNR performance for differential QPSK modulation scheme
in AWGN channel. The OFDM FFT length is 512. (a) Bit error rate versus signal-to-noise
ratio plot for different fixed frequency errors, (b) plot of residual bit error rate versus the
frequency error

the interference effects by additional white noise superimposed on the data symbols. Depending on the modulation scheme employed, this model typically yields a good estimation
of the actual BER.

Different modulation schemes are affected differently by the presence of frequency errors
in the system, analogously to their performance in purely AWGN environments. Coherent
detection suffers the least penalty, followed by differential detection and PSAM schemes. Let
us now consider the effects of time domain synchronisation errors.

5.1.2   Time Domain Synchronisation Errors

Unlike frequency mismatch, as discussed above, time synchronisation errors do not result in
intersubcarrier interference. Instead, if the receiver's FFT window spans samples from two
consecutive OFDM symbols, inter-OFDM symbol interference occurs.

Additionally, even small misalignments of the FFT window result in an evolving phase
shift in the frequency domain symbols, leading to BER degradation. Initially, we will concentrate on these phase errors.

If the receiver's FFT window is shifted with respect to that of the transmitter, then the
time shift property of the Fourier transform, formulated as:

$$f(t) \quad \longleftrightarrow \quad F(\omega)$$
$$f(t - \tau) \quad \longleftrightarrow \quad e^{-j\omega\tau} F(\omega)$$

describes its effects on the received symbols. Any misalignment τ of the receiver's FFT
window will introduce a phase error of $2\pi\Delta f\tau/T_s$ between two adjacent subcarriers. If the
time shift is an integer multiple m of the sampling time T_s, then the phase shift introduced
between two consecutive subcarriers is $\delta\phi = 2\pi m/N$, where N is the FFT length employed.
This evolving phase error has a considerable influence on the BER performance of the OFDM

system, clearly depending on the modulation scheme used.

5.1.2.1 Coherent Demodulation

Coherent modulation schemes suffer the most from FFT window misalignments, since the reference phase evolves by 2π throughout the frequency range for every sampling time misalignment. Clearly, this results in a total loss of the reference phase, and hence coherent modulation cannot be employed without phase correction mechanisms, if imperfect time Synchronisation has to be expected.

5.1.2.2 Pilot Symbol-Assisted Modulation

Pilot symbol assisted modulation (PSAM) schemes can be employed in order to mitigate the effects of spectral attenuation and the phase rotation throughout the FFT bandwidth. Pilots are interspersed with the data symbols in the frequency domain and the receiver can estimate the evolving phase error from the received pilots' phases.

This operation is performed with the aid of the wideband channel estimation discussed in Section 4.3.1.1, and the number of pilot subcarriers necessary for correctly estimating the channel transfer function depends on the maximum anticipated time shift τ. Following the notion of the frequency domain channel transfer function $H(n)$ introduced in Chapter 4, the effects of phase errors can be written as:

$$H(f) = e^{-j2\pi f\tau}. \tag{5.3}$$

Replacing the frequency variable f by the subcarrier index n, where $f = n\Delta f = n/(NT_s)$ and normalising the time misalignment τ to the sampling time T_s, so that $\tau = m \cdot T_s$, the frequency domain channel transfer function can be expressed as:

$$H(n) = e^{-j2\pi \frac{nm}{N}}. \tag{5.4}$$

The number of pilots necessary for correctly estimating this frequency domain channel transfer function $H(n)$ is dependent on the normalised time delay m. Following the Nyquist sampling theorem, the distance Δp between two pilot tones in the OFDM spectrum must be less than or equal to half the period of $H(n)$, so that

$$\Delta p \le \frac{N}{2m}. \tag{5.5}$$

The simulated performance of a 512-subcarrier 16-QAM PSAM modem in the presence of a constant timing error of $\tau = 10T_s$ in an AWGN channel is depicted in Figure 5.8 for both of the PSAM interpolation algorithms investigated in Section 4.3.1.1. Following Equation 5.5, the maximum acceptable pilot subcarrier distance required for resolving a normalised FFT window misalignment of $m = \tau/T_s = 10$ is $\Delta p = N/20 = 512/20 = 25.6$, requiring at least 20 pilot subcarriers equidistantly spaced in the OFDM symbol. We can see in both graphs of Figure 5.8 that the bit error rate is 0.5 for both schemes if less than 20 pilot subcarriers are employed in the OFDM symbol. For pilot numbers above the required minimum of 20, however, the performance of the ideal lowpass interpolated PSAM scheme does not vary with the number of pilots employed, while the linearly interpolated PSAM

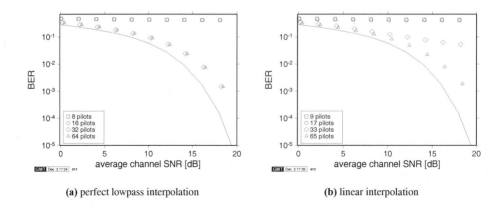

(a) perfect lowpass interpolation   (b) linear interpolation

Figure 5.8: Bit error rate versus channel SNR performance for 16-level PSA-QAM in an AWGN channel for different pilot subcarrier spacings in the presence of a fixed FFT window misalignment of $\tau = 10T_s$. The OFDM FFT length is 512. (a) PSAM interpolation using ideal
lowpass interpolator, (b) PSAM using linear interpolator. In both graphs, the line marks the
coherently detected 16-QAM performance in the absence of both FFT window misalignment and PSAM

scheme needs higher numbers of pilot subcarriers for achieving a similar performance to the
lowpass interpolator scheme. The continuous lines in the graphs show the BER curve for a
coherently detected 16-QAM OFDM modem in the absence of timing errors, while utilising
no PSAM. The BER penalty of PSAM in a narrowband AWGN channel as well as the performance differences between the two PSAM schemes are in accordance with the results found
in Section 4.3.1.1.

5.1.2.3 Differential Modulation

Differential encoding of OFDM symbols can be implemented both between corresponding subcarriers of consecutive OFDM symbols or between adjacent subcarriers of the same
OFDM symbol. We found the latter more advantageous in TDMA environments and hence
this principle is employed here. The BER performance of differentially encoded modulation
schemes is affected by the phase shift $\delta\phi$ between adjacent subcarriers introduced by timing
errors and this influence can be evaluated for example for DPSK systems, as will be shown
below.

The two-dimensional probability density function of a noisy phasor in polar coordinates
is calculated in Appendix 5.6 and is given by Equation 5.49, if we assume the transmitted
phase to be zero. Integration of this function over the magnitude r gives the phase probability
function $p_\phi(\phi)$:

$$p_\phi(\phi) = \int_0^\infty \frac{r}{2\pi\sigma^2} \cdot e^{-(r^2+\mathcal{A}^2-2r\mathcal{A}\cos\phi)/2\sigma^2} dr, \qquad (5.6)$$

where $\mathcal{A}$ is the amplitude of the noiseless phasor. For differential phase modulation, the error
of the difference between the phases of two consecutive symbols, $\Phi = \phi_k - \phi_{k-1}$, determines

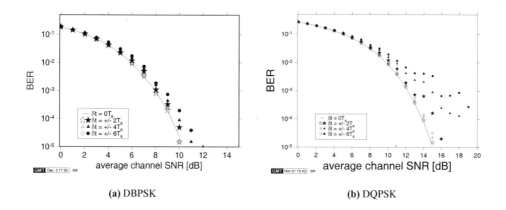

(a) DBPSK   (b) DQPSK

Figure 5.9: Bit error rate versus SNR over AWGN channels for a 512-subcarrier OFDM modem employing DBPSK and DQPSK, respectively. Positive time shifts imply time-advanced FFT window or delayed received data

the symbol error rate (SER). The PDF $p_\Phi(\Phi)$ of this difference can be expressed by a variable transform, which results in the following integral:

$$p_\Phi(\Phi) = \int_{-\pi}^{\pi} p_\phi(\psi - \Phi) \cdot p_\phi(\psi)d\psi, \tag{5.7}$$

where ψ is an auxiliary variable. The received symbol will be demodulated correctly, if the difference error $\Delta\phi$ is within the decision boundaries of the PSK constellation. For M-DPSK, the symbol error rate (SER) is given by:

$$SER = 1 - \int_{-\pi/M}^{\pi/M} p_\Phi(\Phi)d\Phi. \tag{5.8}$$

If there is an FFT window misalignment induced phase shift $\delta\phi$ between consecutive symbols, then the integration limits in Equation 5.8 are biased by this shift:

$$SER = 1 - \int_{-\pi/M+\delta\phi}^{\pi/M+\delta\phi} p_\Phi(\Phi)d\Phi. \tag{5.9}$$

Simulations have been performed for a 512-subcarrier OFDM system, employing DBPSK and DQPSK for different FFT window misalignment values. The BER performance curves for timing errors up to six sampling intervals are displayed in Figure 5.9. Note that one sample interval misalignment represents a phase error of $2\pi/512$ between two consecutive samples, which explains why the BER effects of the simulated positive timing misalignments marked by the hollow symbols are negligible for DBPSK. Specifically, a maximum SNR degradation of 0.5 dB was observed for DQPSK.

Positive FFT window time shifts correspond to a delayed received data stream and hence all samples in the receiver's FFT window belong to the same quasi-periodically extended

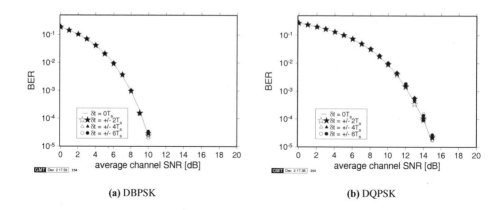

(a) DBPSK   (b) DQPSK

Figure 5.10: Bit error rate versus SNR over AWGN channels for a 512-subcarrier OFDM modem employing a postamble of 10 symbols for DBPSK and DQPSK, respectively. Positive time shifts correspond to time-advanced FFT window or delayed received data

OFDM symbol. In the case of negative time shifts, however, the effects on the bit error rate are much more severe due to inter OFDM-symbol interference. Since the data is received prematurely, the receiver's FFT window contains samples of the forthcoming OFDM symbol, not from the cyclic extension of the wanted symbol.

This non-symmetrical behaviour of the OFDM receiver with respect to positive and negative relative timing errors can be mitigated by adding a short postamble, consisting of copies of the OFDM symbol's first samples. Figure 5.10 shows the BER versus SNR curves for the same offsets, while using a 10-sample postamble. Now, the behaviour for positive and negative timing errors becomes symmetrical. Clearly, the required length of this postamble depends on the largest anticipated timing error, which adds further redundancy to the system. This postamble can be usefully employed, however, to make an OFDM system more robust to time misalignments and thus to simplify the task of the time domain FFT window synchronisation system.

5.1.2.3.1   Time Domain Synchronisation Errors - Summary   Misalignment of the receiver's FFT window relative to the received sample stream leads to possible inter-OFDM symbol interference as well as to an evolving shift of the reference phase throughout the received frequency domain OFDM symbol. While the effects of inter-OFDM symbol interference can be mitigated for moderate misalignments by appending a cyclic postamble to the OFDM symbol, the phase errors in the frequency domain make it impossible to use coherently detected modulation schemes without phase recovery methods. Instead, differentially detected schemes can be employed, which nonetheless suffer from performance degradation due to the phase errors. Alternatively, pilot symbol-assisted channel estimation schemes can be employed in conjunction with coherent detection.

5.2 Synchronisation Algorithms

The results of Section 5.1 indicate that the accuracy of a modem's time and frequency domain synchronisation system dramatically influences the overall BER performance. We have seen that carrier frequency differences between the transmitter and the receiver of an OFDM system will introduce additional impairments in the frequency domain caused by intersubcarrier interference, while FFT window misalignments in the time domain will lead to phase errors between the subcarriers. Both of these effects will degrade the system's performance and have to be kept to a minimum by the synchronisation system.

In a TDMA-based OFDM system, the frame synchronisation between a master station – in cellular systems generally the base station – and the portable stations has also to be maintained. For these systems, a reference symbol marking the beginning of a new time frame is commonly used. This added redundancy can be exploited for both frequency synchronisation and FFT window alignment, if the reference symbol is correctly chosen.

In order to achieve synchronisation with a minimal amount of computational effort at the receiver, while also minimising the amount of redundant information added to the data signal, the synchronisation process is normally split into an acquisition phase and a tracking phase, if the characteristics of the random frequency and timing errors are known. In the acquisition phase, an initial estimate of the errors is acquired, using more complex algorithms and possibly a higher amount of synchronisation information in the data signal, whereas later the tracking algorithms only have to correct for small short-term deviations.

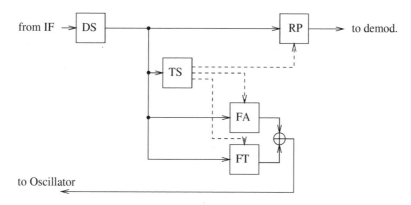

Figure 5.11: Block diagram of the synchronisation system: **DS**-downsampling and clock recovery, **TS**-time synchronisation, **FA**-frequency acquisition, **FT**-frequency tracking, **RP**-remove cyclic extension prefix

The block diagram of a possible synchronisation system is shown in Figure 5.11. The down sampling and clock recovery module DS has to determine the optimum sampling instant. The time synchronisation TS controls the frequency acquisition FA, the frequency tracking FT as well as the time domain alignment of the FFT window, which is carried out by the "remove prefix" or RP block. These operations will be detailed during our further discourse.

At the beginning of the synchronisation process neither the frequency error nor the timing misalignment are known, hence synchronisation algorithms must be found that are sufficiently robust to initial timing and frequency errors.

5.2.1   Coarse Frame and OFDM Symbol Synchronisation Review

Coarse frame and symbol synchronisation algorithms presented in the literature all rely on additional redundancy inserted in the transmitted data stream. The pan-European digital video broadcasting (DVB) system uses a so-called null symbol as the first OFDM symbol in the time frame. No energy is transmitted [190], during the null symbol and it is detected by monitoring the received baseband power in the time domain, without invoking FFT processing. Claßen [26] proposed an OFDM synchronisation burst of at least three OFDM symbols per time frame. Two of the OFDM symbols in the burst would contain synchronisation subcarriers bearing known symbols along with normal data transmission carriers, but one of the OFDM symbols would be the exact copy of one of the other two, thus resulting in more than one OFDM symbol synchronisation overhead per synchronisation burst. For the ALOHA environment, Warner [46] proposed the employment of a power detector and subsequent correlation-based detection of a set of received synchronisation subcarriers embedded in the data symbols. The received synchronisation tones are extracted from the received time domain signal using an iterative algorithm for updating the synchronisation tone values once per sampling interval. For a more detailed discussion on these techniques the interested reader is referred to the literature [26, 46], while for a variety of further treatises to the contributions by Mammela and his team [191–193].

5.2.2   Fine Symbol Tracking Review

Fine symbol tracking algorithms are generally based on correlation operations either in the time or in the frequency domain. Warner [46] and Bingham [194] employed frequency domain correlation of the received synchronisation pilot tones with known synchronisation sequences, while de Couasnon [195] utilised the redundancy of the cyclic prefix by integrating over the magnitude of the difference between the data and the cyclic extension samples. Sandell [51] proposed exploiting the auto-correlation properties of the received time domain samples imposed by the cyclic extension for fine time domain tracking.

5.2.3   Frequency Acquisition Review

The frequency acquisition algorithm has to provide an initial frequency error estimate, which is sufficiently accurate for the subsequent frequency tracking algorithm to operate reliably. Generally the initial estimate must be accurate to half a subcarrier spacing. Sari [47] proposed the use of a pilot tone embedded into the data symbol, surrounded by zero-valued virtual subcarriers, so that the frequency-shifted pilot can be located easily by the receiver. Moose [48] suggested a shortened repeated OFDM symbol pair, analogous to his frequency tracking algorithm to be highlighted in the next section. By using a shorter DFT for this reference symbol pair, the subcarrier distance is increased and thus the frequency error estimation range is extended. Claßen [26, 27] proposed using binary pseudo-noise (PN) or so-called CAZAC training sequences carried by synchronisation subcarriers, which are also employed for the frequency tracking. The frequency acquisition, however, is performed by a search for the training sequence in the frequency domain. This is achieved by means of frequency domain correlation of the received symbol with the training sequence.

5.2.4 Frequency Tracking Review

Frequency tracking generally relies on an already established coarse frequency estimation having a frequency error of less than half a subcarrier spacing. Moose [48] suggested the use of the phase difference between subcarriers of repeated OFDM symbols in order to estimate frequency deviations of up to one-half of the subcarrier spacing, while Claßen [26] employed frequency domain synchronisation subcarriers embedded into the data symbols, for which the phase shift between consecutive OFDM symbols can be measured. Daffara [50] and Sandell [51] used the phase of the received signal's auto-correlation function, which represents a phase shift between the received data samples and their repeated copies in the cyclic extension of the OFDM symbols.

Following the above brief literature survey, we will investigate two different synchronisation algorithms, both making use of a reference symbol marking the beginning of a new time frame. This limits the use of both algorithms to systems whose channel access scheme is based on time division multiple access (TDMA) frames.

5.2.5 Time- and Frequency Domain Synchronisation Based on Auto-correlation

Both the frequency and the time domain synchronisation control signals can be derived from the received signal samples' cyclic nature upon exploiting the OFDM symbols' cyclic time domain extension by means of correlation techniques. A range of symbol timing and fine frequency tracking algorithms were proposed by Mandarini and Falaschi [196]. Originally, Moose [48] proposed a synchronisation algorithm using repeated data symbols, and methods for the frequency error estimation using the cyclic extension of OFDM symbols were presented by Daffara et al. [50] and Sandell et al. [51]. The frequency acquisition and TDMA frame synchronisation proposed here are based on similar principles, employing a dedicated reference symbol exploited in the time domain [197].

No added redundancy in the data symbols and no *a priori* knowledge of the synchronisation sequences constituting the reference symbol are required, since only the repetitive properties of the OFDM symbols and those of the reference symbol (REF) in the proposed time division duplex (TDD) frame structure seen in Figure 5.13 are exploited. All the processing is carried out in the time domain, hence no FFT-based demodulation of the reference symbol is necessary.

5.2.6 Multiple Access Frame Structure

The proposed 64-slot TDMA/TDD frame structure is depicted at the top of Figure 5.12, which is constituted by null symbol, a reference symbol and 62 data symbols. Let us initially consider the role of the reference symbol.

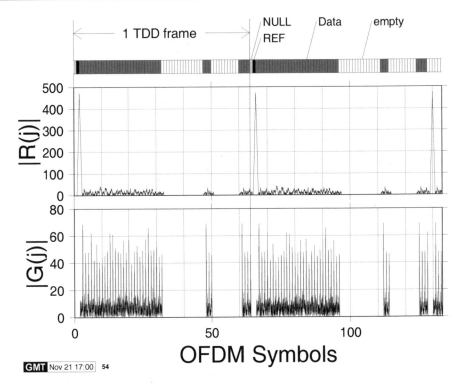

Figure 5.12: Time synchronisation: plots of the correlation terms $R(j)$ and $G(j)$ from Equation 5.10 and 5.11 for two consecutive 64-slot TDD frames under perfect channel conditions. The peaks indicate the correct TDD frame and OFDM symbol synchronisation instants, respectively

5.2.6.1 The Reference Symbol

The reference symbol shown in Figure 5.13 was designed to assist in the operation of the synchronisation scheme and consists of repetitive copies of a synchronisation pattern SP of N_s pseudorandom complex samples. The synchronisation algorithm at the receiver needs no knowledge of the employed synchronisation pattern, hence this sequence could be used for channel-sounding training sequences or for base station identification signals. Note therefore that there are three hierarchical periodic time domain structures in the proposed framing scheme: the short-term intrinsic periodicity in the reference symbol of Figure 5.13, the medium-term periodicity associated with the quasi-periodic extension of the OFDM symbols and the long-term periodicity of the OFDM TDMA/TDD frame structure, repeating the reference symbol every 64 OFDM symbols, as portrayed in Figure 5.12. The long-term reference symbol periodicity is exploited in order to maintain OFDM frame synchronisation, while the medium-term synchronism of the cyclic extension assists in the process of OFDM symbol synchronisation. A detailed discussion of this figure will be provided during our further discourse. Let us begin with the macroscopic structure.

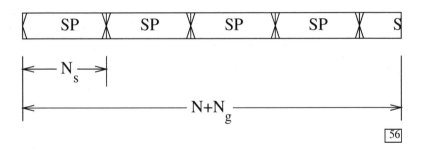

Figure 5.13: Reference symbol consisting of consecutive copies of a synchronisation pattern (SP) in the time domain

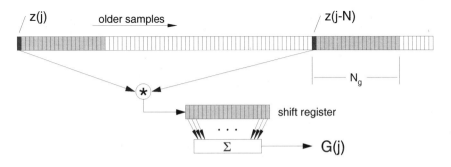

Figure 5.14: Schematic plot of the computation of the correlation function $G(j)$. The grey area represents the memory of the shift register

5.2.6.2 The Correlation Functions

The synchronisation algorithms rely on the evaluation of the following correlation functions $G(j)$ and $R(j)$, where j is the index of the most recent input sample:

$$G(j) = \sum_{m=0}^{N_g-1} z(j-m) \cdot z(j-m-N)^* \tag{5.10}$$

$$R(j) = \sum_{m=0}^{N+N_g-N_s-1} z(j-m) \cdot z(j-m-N_s)^*, \tag{5.11}$$

where $z(j)$ are the received complex signal samples, N is the number of subcarriers per OFDM symbol, N_g is the length of the cyclic extension and N_s is the periodicity within the reference symbol, as seen in Figure 5.13. The asterisk * denotes the conjugate of a complex value.

$G(j)$ is used for both frequency tracking and OFDM symbol synchronisation, expressing the correlation between two sequences of N_g samples length spaced by N in the received sample stream, as shown in Figure 5.14. The second function, $R(j)$, is the corresponding expression for the reference symbol, where the period of the repetitive synchronisation pattern is N_s, as seen in Equation 5.11 and Figure 5.13. In this case, $(N_g + N - N_s)$ samples are taken into account for the correlation computation, and they are spaced by a distance of N_s samples. Having defined the necessary correlation functions for quantifying the time and frequency synchronisation error, let us now concentrate on how the synchronisation algorithms rely on their evaluation.

5.2.7   Frequency Tracking and OFDM Symbol Synchronisation

In this section we consider details of the frequency tracking and OFDM symbol synchronisation algorithms, which make use of $G(j)$, as defined by Equation 5.10.

5.2.7.1   OFDM Symbol Synchronisation

The magnitude of $G(j_{max})$ is maximum, if $z(j_{max})$ is the last sample of the current OFDM symbol, since then the guard samples constituting the cyclic extension and their copies in the current OFDM symbol are perfectly aligned in the summation windows. Figure 5.12 shows the simulated magnitude plots of $G(j)$ and $R(j)$ for two consecutive WATM frames, with $N = 512$ and $N_g = N_s = 50$, under perfect channel conditions. The observed correlation peaks of $|G(j)|$ can be easily identified at the last sample of an OFDM symbol. The amplitudes of the correlation peaks fluctuate, since the transmitted OFDM data symbols differ. The correlation peak magnitude is equal to the energy contained in the N_g samples of the cyclic extension and averages 50 for our system with $N_g = 50$ and an average sample power of unity.

The simulated accuracy of the OFDM symbol synchronisation in an AWGN channel is characterised in Figure 5.15. Observe in the figure that for SNRs in excess of about 7 dB the histogram is tightly concentrated around the perfect estimate, typically resulting in OFDM symbol timing estimation errors below $\pm 20 T_s$. However, as even slightly misaligned time domain FFT windows cause phase errors in the frequency domain, this estimation accuracy is insufficient. In order to improve the OFDM symbol timing synchronisation, the estimates must be lowpass filtered. Let us now concentrate on the issues of fine frequency tracking.

5.2.7.2   Frequency Tracking Studies

A carrier frequency error of δf results in an evolving phase error $\Psi(j)$ of the received samples $z(j)$:

$$\Psi(\delta f, j) = 2\pi \delta f \cdot j \cdot T_s \tag{5.12}$$

$$= 2\pi \frac{j \delta f}{N \Delta f}. \tag{5.13}$$

Clearly, the phase error difference between two received time domain samples $z(j_1)$ and $z(j_2)$ is a function of the frequency error and their time delay, and is given by $\Psi(\delta f, j_2) - \Psi(\delta f, j_1) = \Psi(\delta f, |j_2 - j_1|)$. If the original phase difference between the two received

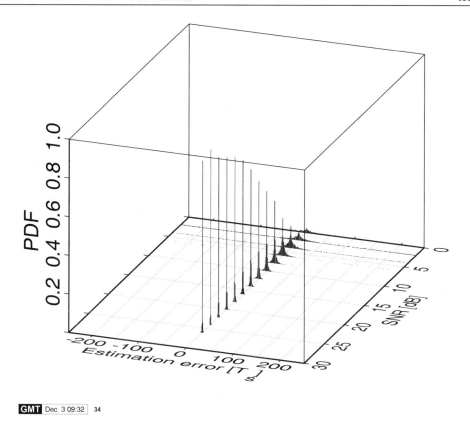

Figure 5.15: Histogram of the symbol timing estimation errors normalised to the sample interval T_s in an AWGN channel for $N = 512$ and $N_g = 50$ with no lowpass filtering of the estimates

symbols $z(j_1)$ and $z(j_2)$ is known, and all other phase distortion is absent, then the phase difference error can be used to determine the frequency error δf.

Since the time domain samples of the cyclic extension or the guard interval are known to be a copy of the last N_g data samples of the OFDM symbol, the frequency error can be estimated using each of these N_g pairs of identical samples. In order to improve the estimation accuracy when exposed to noise and other channel impairments, averaging can be carried out over the N_g estimates.

The phase of $G(j)$ at $j = j_{max}$ equals the averaged phase shift between the guard time samples and the corresponding data samples of the current OFDM symbol. Since the corresponding sample pairs are spaced by N samples, rearranging Equation 5.13 leads to the fine frequency error estimation δf_t given by:

$$\delta f_t = \frac{\Delta f}{2\pi} \cdot \angle G(j_{max}).$$ (5.14)

Because of the 2π ambiguity of the phase, the frequency error must be smaller than $\Delta f/2$. Therefore, the initial frequency acquisition must ensure a rough frequency error estimate with an accuracy of better than $\Delta f/2$, if the proposed fine frequency tracking is used.

Assuming perfect estimation of the position j_{max} of the correlation peak $G(j_{max})$, as we have seen in Figure 5.12, the performance of the fine frequency error estimation in an AWGN environment is shown in Figure 5.16(a). Observe that for AWGN SNR values above 10 dB, the estimation error histogram is concentrated on errors below about $0.02\Delta f$, where Δf is the subcarrier spacing.

Having resolved the issues of OFDM symbol synchronisation and fine frequency tracking, let us now focus our attention on the aspects of frequency acquisition and frame synchronisation.

5.2.8 Frequency Acquisition and Frame Synchronisation Studies

Our proposed frequency acquisition and frame synchronisation techniques are based on the same algorithms as the fine frequency and OFDM symbol synchronisation. However, instead of using the medium-term periodicity of the cyclic extension of the OFDM data symbols, the dedicated reference symbol with shorter cyclic period $N_s < N$ is exploited in order to improve the frequency capture range, as will be highlighted below.

5.2.8.1 Frame Synchronisation Studies

Similarly to the OFDM symbol synchronisation, the magnitude of $R(j)$ in Equation 5.11 and Figure 5.12 is maximum when the periodic synchronisation segments SP of length N_s of the reference symbol shown in Figure 5.13 perfectly overlap. Again, the magnitude of $R(j)$ for two simulated TDD frames is shown at the top of Figure 5.12. The OFDM frame timing is synchronised with the peak of $R(j)$, which can additionally be taken into account for the OFDM symbol synchronisation. The peak height is constant under perfect channel conditions, owing to the fixed reference symbol.

5.2.8.2 Frequency Acquisition Studies

The frequency acquisition algorithm uses the same principle as the frequency tracking scheme of Section 5.2.7.2. Specifically, the phase of $R(j)$ at the last sample of the reference symbol j_{max} contains information on the frequency error:

$$\angle R(j_{max}) = 2\pi \cdot \delta f_a \cdot N_s \cdot T_s = 2\pi \cdot \delta f_a \cdot \frac{N_s}{N \cdot \Delta f} \tag{5.15}$$

leading to

$$\delta f_a = \frac{N}{N_s} \frac{\Delta f}{2\pi} \cdot \angle R(j_{max}). \tag{5.16}$$

Since the spacing between the sample pairs used in the computation of $R(j)$ is smaller than in the case of $G(j)$, which was used for the frequency tracking ($N_s < N$), the maximum detectable frequency error is now increased from $\Delta f/2$ to $N/N_s \cdot \Delta f/2$, where Δf is the subcarrier spacing of the OFDM symbols. The simulated performance of the frequency acquisition algorithm for $N = 512$, $N_s = 50$ and perfect time synchronisation is shown in Figure 5.16(b) for transmissions over an AWGN channel. As seen in the figure, the scheme

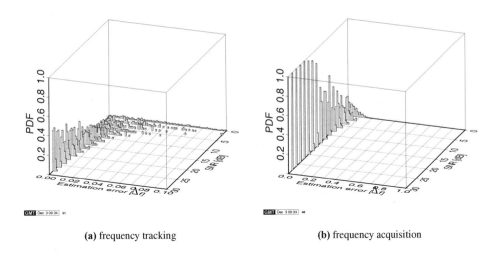

(a) frequency tracking   **(b)** frequency acquisition

Figure 5.16: Histogram of the simulated frequency estimation error for the frequency acquisition and frequency tracking algorithms using $N = 512$ and $N_s = N_g = 50$ over AWGN channels

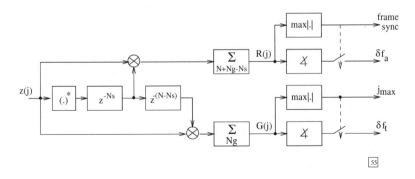

Figure 5.17: Block diagram of the synchronisation algorithms

maintains an acquisition error below $\Delta f/2$ even for SNR values down to 0 dB, which exceeds the system's operational specifications.

5.2.8.3 Block Diagram of the Synchronisation Algorithms

To summarise our previous elaborations, Figure 5.17 shows the detailed block diagram of the synchronisation algorithms. The received samples are multiplied with the complex conjugate of the delayed input sequences and summed up over $(N + N_g - N_s)$ and N_g samples, respectively. The magnitude maxima of the two sequences $G(j)$ and $R(j)$ are detected and they trigger the sampling of the phase estimates $\angle G$ and $\angle R$ in order to derive the two frequency error estimates δf_a and δf_t. Let us now describe an alternative frequency synchronisation technique, which can be used as a benchmark in order to assess the potential of our previously described technique.

5.2.9   Frequency Acquisition Using Frequency Domain Pilots

The algorithm described in this section has been proposed by Mandarini and Falaschi [196]. In contrast to the reference symbol of Figure 5.13, the technique advocated by Mandarini and Falaschi relies on the reference symbol shown in Figure 5.18 containing a set of pilot tones, which is transmitted once per TDD frame. The reference symbol is processed in the frequency domain, after demodulation by FFT.

The idea of the pilot tone algorithm is that a frequency mismatch between the transmitter and the receiver will result in a shift of the pilot tones in the received spectrum. This shift is measured at the receiver by searching for the FFT frequency bin with the maximum amplitude, resulting in an estimation accuracy of half a subcarrier spacing. This estimation is improved further in a second stage by considering the relative amplitudes of the adjacent FFT bins.

5.2.9.1   The Reference Symbol

The reference symbol is transmitted once at the beginning of every TDD frame, as depicted in Figure 5.12. This reference symbol consists of a set of M pilot tones spanning the OFDM signal's bandwidth, spaced from each other by ΔN subcarriers or by a frequency gap of $\Delta F = \Delta N \cdot \Delta f$. Accordingly, the maximum detectable frequency error is $\Delta F/2 = \Delta N/2 \cdot \Delta f$ and hence the higher the number of pilots, the lower the frequency capture range. A stylised plot of the reference symbol in the frequency domain is shown in Figure 5.18.

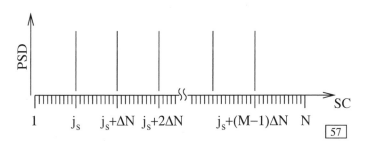

Figure 5.18: Stylised plot of the power spectral density (PSD) of the pilot-tone-based reference symbol using M pilot tones in the N subcarrier (SC) spectrum, spaced by $(\Delta N - 1)$ blank subcarriers

In order to keep the overall OFDM symbol energy constant, the total energy of the reference symbol is set equal to the average OFDM symbol energy and split equally between the M pilot tones.

5.2.9.2   Frequency Acquisition

The frequency acquisition algorithm estimates the frequency error by searching for the position of the pilot tones in the spectrum of the received reference symbol. In order to accomplish this, the received time domain signal of the reference symbol is demodulated by FFT. In order to minimise the influence of noise, the sum of the received power spectral amplitudes over all

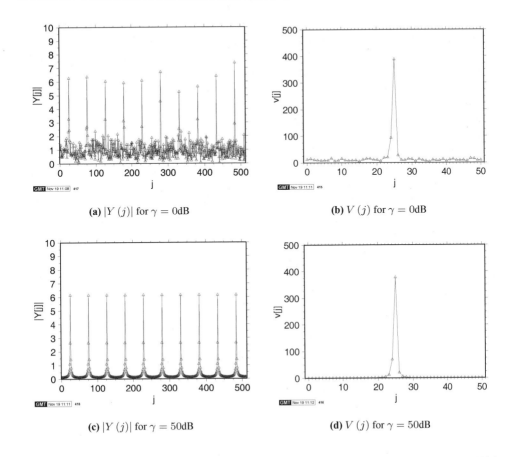

(a) $|Y(j)|$ for $\gamma = 0$dB

(b) $V(j)$ for $\gamma = 0$dB

(c) $|Y(j)|$ for $\gamma = 50$dB

(d) $V(j)$ for $\gamma = 50$dB

Figure 5.19: Magnitude of the received frequency domain samples $|Y(j)|$ and averaged samples $V(j)$ for SNR values of $\gamma = 0$ dB and $\gamma = 50$ dB for a 512-subcarrier OFDM modem employing 10 pilot tone subcarriers in the reference symbol. The simulated frequency mismatch between the transmitter and the receiver was $\delta f = 0.3\Delta f$

the M frequency ranges depicted in Figure 5.18 is then calculated as follows:

$$V(j) = \sum_{m=0}^{M-1} |Y(j + j_s + m\Delta N)|^2 \quad \text{for} \quad -\frac{\Delta N - 1}{2} \le j \le \frac{\Delta N - 1}{2}, \quad (5.17)$$

where $Y(j)$ represents the frequency domain samples of the received demodulated reference symbol. Observe that for the sample position identified by j $V(j)$ corresponds to the superposition of the powers of the ΔN-spaced $Y(j)$ samples, one from each of the M frequency ranges. In Figure 5.19 simulated values for both $|Y(j)|$ and $V(j)$ are given for SNR values of 0 dB and 50 dB, respectively, for a 10-pilot reference symbol in a 512-subcarrier OFDM system with $\delta f = 0.3\Delta f$ frequency error. It can be seen that the positions of the pilot tones in the received spectrum are easily determined even for SNR values of 0 dB, thanks to the high power of the pilot tones relative to the average data symbol power. The total energy of the reference is 512, leading to a pilot tone amplitude of $\sqrt{512/10} = 7.15$. Because of the subcarrier energy spillover caused by the frequency error, the amplitude of the peaks of $Y(j)$ in the figure is lowered to about 6.2. Accordingly, the peak value of $V(j)$ is about 390 instead of 512, with the rest of the received energy located in the adjacent FFT bins.

The frequency error is determined by estimating the position of the highest peak in $V(j)$ over the range of j given in Equation 5.17. First, the index j_{max} of the maximum value of $V(j)$ is found. This gives a rough estimate of the frequency error with a frequency resolution of the subcarrier spacing Δf. As can be observed in Figure 5.19, this rough frequency error estimation is very noise resilient, resulting in a reliable detection of the frequency error. The accuracy of this estimation, however, is insufficient for low BER modem operation. If a subsequent frequency error tracking algorithm like the proposed time domain correlation-based approach is employed in the system, then the $0.5\Delta f$ estimation accuracy can be an adequate starting point for the tracking algorithm. If a better estimate is needed, however, then the amplitudes in the neighbouring bins around the peak value in $V(j)$ can be exploited to refine the first estimate, as follows.

Since the frequency error is not generally an integer multiple of the resolution Δf of the FFT, a better frequency error estimate can be derived by determining the position of the peak value of $V(j)$ around j_{max}. The proposed algorithm exploits the amplitude of $V(j)$ at $j = j_{max}$ and the two adjacent values $V(j_{max}\pm 1)$ in order to derive the following quantities:

$$\rho_1 = \frac{\sqrt{V(j_{max}+1)}}{\sqrt{V(j_{max})}} \tag{5.18}$$

$$\rho_{-1} = \frac{\sqrt{V(j_{max}-1)}}{\sqrt{V(j_{max})}}, \tag{5.19}$$

which represent the normalised frequency domain pilot values at the positions adjacent to the $V(j)$ peak at $j = j_{max}$. Their difference $(\rho_1 - \rho_{-1})$ is computed and the rough frequency error estimate of $j_{max} \cdot \Delta f$ is corrected by the help of the estimated deviation d as follows:

$$d = \frac{\rho_1 - \rho_{-1}}{2} \tag{5.20}$$

$$\delta f = \Delta f \cdot (j_{max} + \text{sgn}\{d\} \cdot \sqrt{|d|}), \tag{5.21}$$

where the value of $\sqrt{|d|}$ is the estimation of the true peak position between the peak $V(j_{max})$ and the higher of the two adjacent values of $V(j_{max} \pm 1)$.

Clearly, the number of pilot tones included in the reference symbol, M, determines the maximal frequency deviation that can be detected using this algorithm, which is given by $\delta f_{max} = \Delta f \cdot \lfloor N/2M \rfloor$, where $\lfloor x \rfloor$ means the highest integer number smaller or equal to x. Not only the frequency capture range, but also the peak-to-average power ratio or crest factor (CF) of the reference symbol in the time domain depends on the number of pilot tones used. If the phase of all the M pilot tones is equal, then the upper bound for the resulting crest factor equals M, which would be the result of all M sinusoids adding constructively to a peak value. Experimental results in Figure 5.20(b) show that the actually measured CF for pilot symbols with a varying number of pilot tones is very close to the stated upper bound. For comparison, the histogram of the simulated CF values for pseudorandom OFDM data symbols is shown in Figure 5.20(b). This PDF shows that the average CF value for a

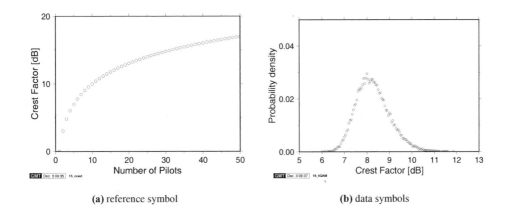

(a) reference symbol   **(b)** data symbols

Figure 5.20: Characterisation of the coarse frequency synchronisation using the reference symbol of
Figure 5.18: (a) crest factor versus number of pilot tones for non-optimised reference
symbols, (b) histogram of simulated data symbol crest factors for a 512-subcarrier, 4-
QAM OFDM system

512-subcarrier 4-QAM OFDM system is about 8.5 dB, although peak values up to 12 dB
have been occasionally observed. For small numbers of pilot tones, the crest factor of the
reference symbol is therefore not worse than that of the average data symbols.

5.2.9.3 Performance of the Pilot-Based Frequency Acquisition in AWGN Channels

A series of simulations was performed in order to characterise the performance of the pilot
tone-based frequency acquisition algorithm in an AWGN environment. For different fixed
frequency errors and noise levels, the frequency error estimation was invoked and the difference between the estimated and the actual error was recorded. Figure 5.21 shows the
histograms of the frequency mismatch estimation errors for 10, 20, 30 and 40 pilots and for a
range of SNR values in the case of an actual frequency error of $\delta f = 0.0$.

Inspection of the figure reveals that the estimation accuracy is not sensitive to the number
of pilot tones used in the reference symbol; in fact, the differences in the simulated histograms
are hardly visible. The estimation accuracy is significantly better than $0.5\Delta f$ throughout the
simulated SNR range and hence meets the requirements of the subsequent fine frequency
synchronisation for all simulated signal-to-noise ratios. It can be observed, however, that
there is a bifurcation of the estimation error histogram which leads to two distinct probability
maxima on both sides of the true value, with a low probability for a perfect estimation. This
behaviour can be explained by the fine peak location estimation, which exhibits a high noise
sensitivity for low δf values and hence small differences in the noise samples on either side
of the correlation peak lead to considerable estimation errors.

The gap between the maxima closes with increasing SNR, but even at 20 dB there are frequency estimation errors of up to 8% of the subcarrier spacing, which underlines the need for
subsequent fine frequency synchronisation. Alternatively, averaging the estimated frequency
errors for several estimations would improve the estimation accuracy, but it would slow down
the system's response time to frequency deviations. The histogram is symmetric around the

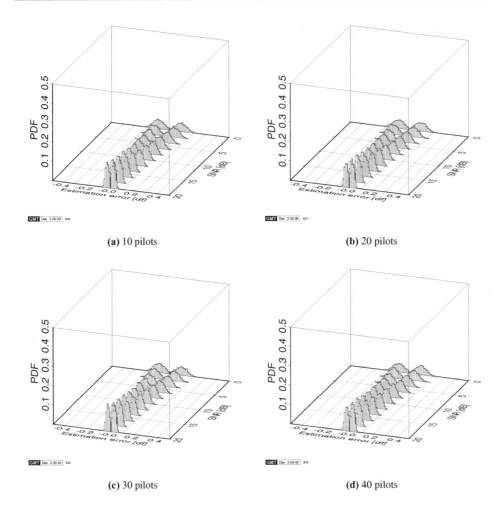

Figure 5.21: Histograms of simulated frequency mismatch estimation error for a 512-subcarrier OFDM modem in a narrowband AWGN channel for 10, 20, 30 and 40 pilot tones. The simulated frequency error was $\delta f = 0.0$

perfect estimation value, hence a longer-term averaging would yield the correct frequency error estimate.

The estimation accuracy of the pilot tone frequency acquisition algorithm depends on the actual frequency error. In order to illustrate this, frequency estimation error histograms for a given frequency mismatch of $\delta f = 0.3\Delta f$ are shown in Figure 5.22, and they exhibit rather different characteristics from the previous case with $\delta f = 0$. The histograms are offset with respect to the perfect estimation value, hence averaging over multiple estimations will result in a residual estimation error. Subsequent fine frequency sychronisation is therefore necessary for low SNR values . For low noise levels, however, the estimation accuracy is greatly improved with respect to the earlier case. For SNRs in excess of 10 dB, the estimation

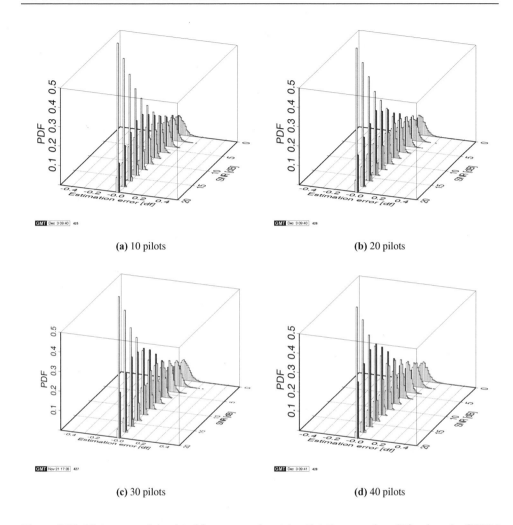

(a) 10 pilots

(b) 20 pilots

(c) 30 pilots

(d) 40 pilots

Figure 5.22: Histograms of simulated frequency mismatch estimation error for a 512-subcarrier OFDM modem in a narrowband AWGN channel for 10, 20, 30 and 40 pilot tones. The simulated frequency error was $\delta f = 0.3$

errors do not exceed 5% of the subcarrier spacing, and will therefore not affect the system performance to a great extent. It can be observed that the asymmetry of the estimation errors is dependent on the number of pilots employed.

The frequency deviation estimation error histogram for $\delta f = 0.5\Delta f$ is shown in Figure 5.23, where it can be observed that in this case the histogram is split into two distinct parts, and that each of the groups is centred at 8% of the subcarrier distance above and below the correct value. Again, fine frequency synchronisation would be necessary to ensure optimal system performance.

In order to evaluate the residual estimation errors, Figure 5.24 depicts the estimated versus the actual frequency error for a perfect, noiseless channel. For frequency errors close to

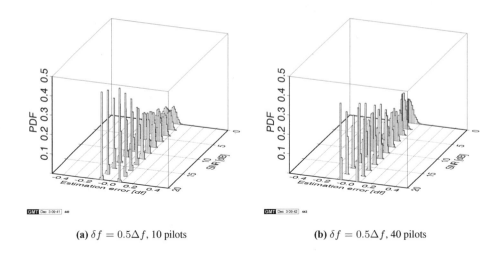

(a) $\delta f = 0.5\Delta f$, 10 pilots   (b) $\delta f = 0.5\Delta f$, 40 pilots

Figure 5.23: Histograms of simulated frequency mismatch estimation error for a 512-subcarrier OFDM modem in a narrowband AWGN channel for 10 and 40 pilot tones. The simulated frequency error was $\delta f = 0.5\Delta f$

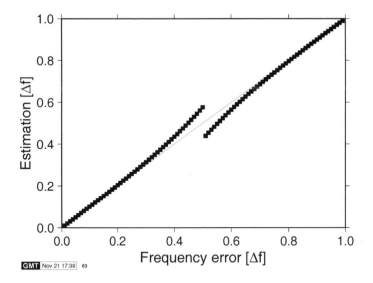

Figure 5.24: Estimated versus actual frequency error for the pilot tone-based coarse frequency synchronisation algorithm in a perfect noiseless channel

an integer multiple of Δf the estimation accuracy is very good, but values close to $(n + 1/2)\Delta f$ are estimated with errors of up to 8% of the subcarrier spacing. This observation is in accordance with the results of the above experiments, which showed a residual estimation error of about 1% of the carrier separation at $\delta f = 0.3\Delta f$ and of $\pm 8\%$ at $\delta f = 0.5\Delta f$.

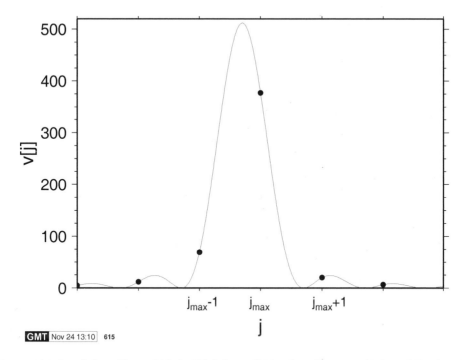

Figure 5.25: Detail from Figure 5.19(d): $V(j)$ for an SNR value of $\gamma = 50$ dB for a 512-subcarrier OFDM modem employing 10 pilot tones in the reference symbol. The simulated frequency mismatch is $0.3\Delta f$. The symbols represent the simulated $V(j)$ values, while the continuous line corresponds to the sinc^2 approximation of Equation 5.23

5.2.9.4 Alternative Frequency Error Estimation for Frequency Domain Pilot Tones

In order to improve the frequency error dependent performance of the frequency domain pilot-based frequency synchronisation algorithm outlined above, an alternative peak position estimation has been investigated in order to enhance the algorithm based on Equation 5.21. As we saw earlier, the continuous spectrum of each OFDM subcarrier follows a sinc function centred around the subcarrier frequency. In the presence of a frequency error between the transmitter and receiver, the receiver's sampling raster in the frequency domain is not aligned with the received spectrum's maximum and nulls, but shifted by the frequency error δf.

The spectrum of the received reference symbol – which contains M pilot tones at the frequencies $(j_s + m\Delta N) \cdot \Delta f$ with $0 \le m \le (M-1)$ – in the presence of a frequency error δf in a noiseless environment can be expressed as:

$$Y(j) = \sqrt{\frac{N}{M}} \cdot \sum_{m=0}^{M-1} \text{sinc}\left(j - j_s - m\Delta N + \frac{\delta f}{\Delta f}\right). \qquad (5.22)$$

The factor $\sqrt{N/M}$ is the amplitude of each pilot tone, ensuring that the overall energy of the reference symbol is equal to the average OFDM symbol energy. If the frequency distance ΔN between two consecutive pilot tones is sufficiently large so that the received spectra of the different pilot tones do not significantly overlap, then the vector $V(j)$, as defined by Equation 5.17, can be approximated as:

$$V(j) \quad = \quad \sum_{m-0}^{M-1} |Y(j + j_s + m\Delta N)|^2 \tag{5.23}$$

$$\approx \quad N \cdot \text{sinc}^2 \left(j + \frac{\delta f}{\Delta f} \right). \tag{5.24}$$

Figure 5.25 shows the simulated values of $V(j)$ from Figure 5.19(d) in greater detail around the peak, along with the continuous line corresponding to the sinc approximation of Equation 5.23. It can be seen that there is a very good correspondence between the simulated and the approximated values for the pilot spacing of $M = 50$ employed in this case.

Using the sinc^2 approximation of Equation 5.23, the values for $V(j_{max})$, $V(j_{max} + 1)$ and $V(j_{max} - 1)$ can be expressed in terms of the normalised fine frequency error estimation $\nu = \delta f/\Delta f - j_{max}$ as follows:

$$V(j_{max}) \quad \approx \quad N\text{sinc}^2(\nu), \tag{5.25}$$
$$V(j_{max} + 1) \quad \approx \quad N\text{sinc}^2(1 + \nu), \quad \text{and} \tag{5.26}$$
$$V(j_{max} - 1) \quad \approx \quad N\text{sinc}^2(-1 + \nu). \tag{5.27}$$

Then the following normalised terms can be defined:

$$\rho_1 \quad = \quad \frac{\sqrt{V(j_{max} + 1)}}{\sqrt{V(j_{max})}} \tag{5.28}$$

$$\approx \quad \frac{\sqrt{N}\,|\sin(\pi(1 + \nu))|}{|\pi(1 + \nu)|} \cdot \frac{|\pi\nu|}{\sqrt{N}\,|\sin(\pi\nu)|} \tag{5.29}$$

$$= \quad \frac{|\nu|}{|1 + \nu|} \quad \text{and} \tag{5.30}$$

$$\rho_{-1} \quad = \quad \frac{\sqrt{V(j_{max} - 1)}}{\sqrt{V(j_{max})}} \tag{5.31}$$

$$\approx \quad \frac{\sqrt{N}\,|\sin(\pi(-1 + \nu))|}{|\pi(-1 + \nu)|} \cdot \frac{|\pi\nu|}{\sqrt{N}\,|\sin(\pi\nu)|} \tag{5.32}$$

$$= \quad \frac{|\nu|}{|-1 + \nu|}. \tag{5.33}$$

The value d, defined as half the difference between ρ_1 and ρ_{-1} in Equation 5.20, can therefore be approximated as:

$$d = \frac{\rho_1 - \rho_{-1}}{2} \approx \frac{1}{2} \left(\frac{|\nu|}{|1 + \nu|} - \frac{|\nu|}{|-1 + \nu|} \right). \tag{5.34}$$

Solving Equation 5.34 for ν values smaller than one subcarrier distance yields:

$$\nu = \begin{cases} -\sqrt{\dfrac{d}{d+1}} & \text{for} \quad -1 < \nu < 0 \quad \text{(or} \quad (d > 0)) \\[2mm] \sqrt{\dfrac{d}{d-1}} & \text{for} \quad 0 \le \nu < 1 \quad \text{(or} \quad d \le 0) \end{cases} . \tag{5.35}$$

An alternative pilot-based frequency synchronisation algorithm can therefore be contrived by calculating the peak position estimate ν from d, as defined in Equation 5.20, using Equation 5.35 and replacing the δf estimate in Equation 5.21 by:

$$\delta f = \Delta f \cdot (j_{max} + \nu). \tag{5.36}$$

A series of simulations were conducted in order to investigate the performance of this modified peak position estimation algorithm in noisy conditions. All the investigations have been performed for a 512-subcarrier system employing a 10-pilot reference symbol in a narrowband Gaussian white noise channel.

Histograms of the estimation errors for fixed frequency errors of $\delta f = 0$, $\delta f = 0.3\Delta f$ as well as for $\delta f = 0.5\Delta f$ are given in Figure 5.26. In all cases, the estimation accuracy was better than $0.5\Delta f$, therefore allowing the subsequent use of the OFDM data symbol-based tracking algorithm of Section 5.2.7. Comparison of Figure 5.26(a) with the corresponding results for the original peak position estimation algorithm in Figure 5.21(a) reveals a similar performance for both algorithms. This was expected, since the modified estimation terms $\sqrt{d/(d-1)}$ and $-\sqrt{d/(d+1)}$ are close to the term $\sqrt{d}$ in Equation 5.21 of the original algorithm for small values of d.

For a frequency error of $\delta f = 0.3\Delta f$, as depicted in Figure 5.26(b), the estimation accuracy was dramatically improved, when compared to the case of $\delta f = 0$. This was in accordance with the results achieved using the original peak position estimation, as shown in Figure 5.22(a). Although both schemes achieved accurate estimation results, the modified algorithm of Equation 5.36 exhibited a lower estimation bias and it was slightly more accurate than the original algorithm of Equation 5.21.

Imposing the third simulated frequency error of $\delta f = 0.5\Delta f$ revealed the different behaviour of the two investigated algorithms. While the original version of the algorithm exhibited a split histogram with systematic estimation errors of about 8% of the subcarrier distance, as can be seen in Figure 5.23(a), the modified algorithm delivered a significantly more accurate frequency error estimate. The estimation accuracy of the modified algorithm in a noiseless environment is demonstrated in Figure 5.26(d). There are no systematic estimation errors for the whole frequency error range that the peak position estimation is expected to handle.

5.3 Comparison of the Frequency Acquisition Algorithms

Both the frequency acquisition algorithms examined achieve the minimum requirements of an estimation accuracy better than half a subcarrier spacing for AWGN channel SNR levels down to 0 dB and hence ensure reliable operation of the subsequent fine frequency synchronisation algorithm.

There is, however, a difference in their absolute accuracy. While the pilot tone algorithm

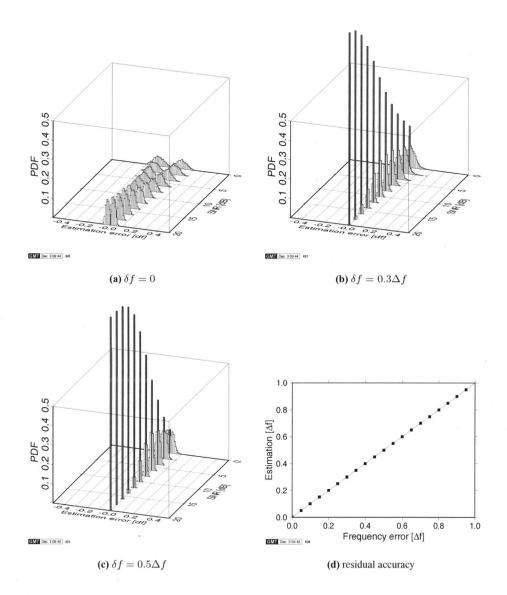

(a) $\delta f = 0$

(b) $\delta f = 0.3\Delta f$

(c) $\delta f = 0.5\Delta f$

(d) residual accuracy

Figure 5.26: (a, b, c): Histograms of the simulated frequency mismatch estimation error for a 512-subcarrier OFDM modem in a narrowband AWGN channel for 10 pilot tones employing the alternative peak position estimation algorithm. The simulated frequency error was $\delta f = 0.3$. (d) Estimated versus actual frequency error for pilot tone-based frequency synchronisation employing the alternative peak position estimation algorithm

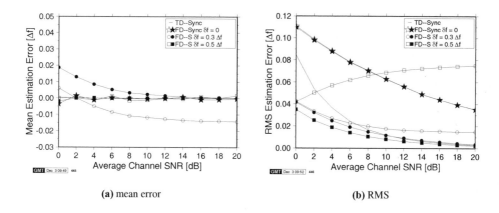

(a) mean error   (b) RMS

Figure 5.27: Comparison of the frequency acquisition algorithms over AWGN channels: (a) mean estimation error, (b) RMS estimation error. TD = time domain estimation algorithm, FD = frequency domain estimation algorithm employing a 10-pilot reference symbol. The open symbols show the performance associated with the original peak position estimation of Equation 5.21, while the closed symbols signify the alternative peak position estimation algorithm of Equation 5.36 for given frequency errors δf of 0, 0.3 and 0.5 times the subcarrier distance Δf

of Section 5.2.9 suffers from residual estimation errors for certain ranges of frequency errors, the correlation-based algorithm of Section 5.2.5 delivers the same estimation accuracy throughout its frequency capture range. Values for the long-term mean and RMS estimation error are shown in Figure 5.27. The solid lines without markers in the two graphs are the mean and RMS error curves for the time domain correlation algorithm, which are valid for all frequency errors within the frequency capture range of the algorithm ($\pm 5.12\Delta f$ for $N = 512$ and $N_s = 50$). The curves with markers correspond to the pilot tone synchronisation system employing a 10-pilot reference symbol, for frequency errors of 0, 0.3, and 0.5 times the subcarrier distance Δf. For these frequency error values, curves are drawn for the original as well as for the modified peak position estimation algorithm. The performance of both the pilot tone algorithms is clearly dependent on the frequency error, both in terms of the mean as well as the RMS errors.

The mean estimation error curves in Figure 5.27(b) show the long-term averaged estimation errors for the different schemes. While none of the algorithms exhibits mean errors above 0.5% of the subcarrier distance for frequency deviations of 0 or 0.5% Δf over the SNR range investigated, frequency errors of $0.3\Delta f$ cause both the pilot tone estimation algorithms to deliver biased estimates. The amount of this mean estimation error depends on the SNR, and the noise level dependency of the estimation is the same for both the original as well as for the alternative peak search algorithm. The residual estimation error in the absence of noise is different for the two peak position estimators, however. While the alternative estimation algorithm delivers a non-biased estimation for high SNR values, the original algorithm exhibits an estimation error of about 1.5% of the subcarrier distance. This is in accordance with Figure 5.24, which shows the residual estimation errors for the original peak position estimation. It is interesting to note that the original peak search algorithm does not result in a biased non-zero mean error estimate for frequency errors of $\delta f = 0.5\Delta f$, although the instantaneous estimation errors at this point are at a maximum, as can be seen from Figures 5.24 and 5.23(a). This is because of the symmetry of the histogram, which will result in a non-biased mean estimation value. For all other frequency errors between 0 and $0.5\Delta f$, however, the histogram of estimation errors will be asymmetrical and the estimation will therefore be biased, as seen for example in Figure 5.22.

The modified peak search algorithm of Section 5.2.9.4, too, delivers biased estimations for frequency errors between 0 and $0.5\Delta f$, as seen in Figure 5.27. Since these are due to the noise affecting the estimation, there is no residual estimation error in noiseless environments. For SNR values in excess of 8 dB, the mean estimation error is smaller than 0.5% of the subcarrier distance for $\delta f = 0.3\Delta f$. The time domain frequency synchronisation algorithm of Section 5.2.5 does not exhibit any significant bias in the estimation error for the range of SNR values investigated. Therefore, long-term averaging of the time domain algorithm's estimates will result in correct frequency error estimation at the cost of reduced agility.

Figure 5.27(b) depicts the RMS estimation error curves for the investigated synchronisation algorithms for the same set of frequency errors. It can be seen that for both frequency domain pilot-based error estimation algorithms the estimation quality varies greatly with the frequency error to be estimated. The original peak position estimation algorithm exhibits residual RMS values of $0.015\Delta f$ and about $0.8\Delta f$ for frequency errors δf of $0.3\Delta f$ and $0.5\Delta f$, respectively. For $\delta f = 0.3\Delta f$ this corresponds to the bias of 1.5% of the subcarrier distance, which has been observed in Figure 5.27(a). For $\delta f = 0.5\Delta f$, this residual estimation error can be observed in the histogram, as shown in Figure 5.23(a). Although the long-term mean of the estimation is the correct value, each single estimation exhibits an error of about 8% of the subcarrier distance. This value corresponds to the residual level of RMS estimation error for $\delta f = 0.5\Delta f$ in Figure 5.27(b).

The frequency domain pilot-based algorithm employing the alternative peak position estimation of Section 5.2.9.4 does not exhibit residual RMS estimation errors. The estimation accuracy does, however, vary with the frequency error to be estimated. For $\delta f = 0$, its performance is virtually the same as that of the original peak position estimation algorithm, with high estimation errors for the whole AWGN SNR range. At $\delta f = 0.3\Delta f$ and $0.5\Delta f$, the estimation accuracy is much better, at RMS estimation errors of around 1% of the subcarrier distance for SNR values in excess of 12 dB and 10 dB, respectively.

The time domain correlation-based algorithm's accuracy does not depend on the frequency error to be estimated. Its performance is comparable to that of the frequency domain algorithm employing the alternative peak search algorithm of Section 5.2.9.4 for frequency errors δf of $0.3\Delta f$ or $0.5\Delta f$ for SNR values above 10 dB, but it performs significantly better at $\delta f = 0$.

Comparing frequency error with the estimation accuracy of the different algorithms it becomes clear that all the algorithms would satisfy the estimation accuracy of $0.5\Delta f$, necessary if subsequent fine frequency tracking mechanisms are used data symbol by data symbol. If no fine frequency tracking is to be employed, however, then the constraints on the estimation accuracy are significantly tighter. We saw in Section 5.1.1.2 that, depending on the modulation scheme used, frequency errors of 5–10% of the subcarrier spacing lead to degradation of the system's SNR performance. In this case – and if no lowpass filtering of the estimated value is to be employed – the time domain correlation-based synchronisation algorithm of Section 5.2.5 would be the only applicable algorithm, since its accuracy is consistently high

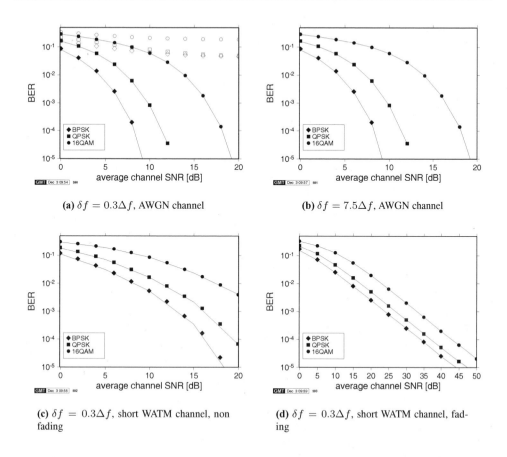

(a) $\delta f = 0.3\Delta f$, AWGN channel

(b) $\delta f = 7.5\Delta f$, AWGN channel

(c) $\delta f = 0.3\Delta f$, short WATM channel, non fading

(d) $\delta f = 0.3\Delta f$, short WATM channel, fading

Figure 5.28: BER versus channel SNR performance curves for the 512-subcarrier OFDM system in the presence of fixed frequency errors. The lines indicate the performance for perfectly corrected frequency error, and the hollow symbols show the performance for uncorrected frequency errors. The filled symbols indicate simulations with our proposed frequency error estimation using the time domain correlation technique. The short WATM channel impulse response is shown in Figure 4.3(a).

and its estimation is unbiased throughout the frequency error range.

If the pilot tone reference symbol- based frequency synchronisation algorithms are to be employed without subsequent frequency tracking, then the estimation accuracy has to be improved using lowpass filtering techniques of the estimates. Let us now consider the system's BER performance when subjected to frequency synchronisation errors.

5.4 BER Performance with Frequency Synchronisation

In order to investigate the effects of the proposed frequency synchronisation algorithm on an OFDM modem, a series of experiments was conducted. The synchronisation algorithm chosen was the time domain correlation algorithm of Section 5.2.5, because of its frequency error independent estimation accuracy. We modelled a system employing one reference symbol and one data symbol per 64-slot TDMA frame. The frequency error estimates were not lowpass filtered, instead every estimate was used directly to correct the frequency error of the subsequent data symbol. Coherently detected BPSK, QPSK and 16-QAM were employed as modulation schemes in the subcarriers and the BER performance of the modem was investigated in both narrowband AWGN as well as in fading and non-fading wideband channels. In all the simulations, a 512-subcarrier OFDM scheme employing no virtual carriers and perfect wideband channel estimation was used.

The BER performance of the simulated modems in all the channels is given in Figure 5.28. Figure 5.28(a) depicts the BER versus channel SNR for BPSK, QPSK and 16-QAM in an AWGN channel with a frequency error of $\delta f = 0.3\Delta f$. The white symbols in the graph portray the BER performance of an OFDM modem employing no frequency synchronisation. It can be seen that the frequency error results in heavy intersubcarrier interference, which manifests itself by a high residual bit error rate of about 5% for BPSK and QPSK and about 20% for 16-QAM. The lines in the graph characterise the performance of the modem in the absence of frequency errors. The black markers correspond to the BER recorded with the frequency synchronisation algorithm in operation. It can be seen that the performance of the modem employing the proposed frequency synchronisation algorithm of Section 5.2.5 is nearly indistinguishable from the perfectly synchronised case. In Figure 5.28(b), the modem's BER curves for an AWGN channel at a frequency error of $7.5\Delta f$ are depicted. Since the synchronisation algorithm's accuracy does not vary with varying frequency errors, the modem's BER performance employing the proposed synchronisation algorithm at $\delta f = 7.5\Delta f$ is the same as at $\delta f = 0.3\Delta f$. The BER for the non-synchronised modem is, however, 50% and the corresponding markers are off the graph.

The synchronised modem's BER performance in wideband channels is given in Figure 5.28(c) and 5.28(d). The impulse response used in these investigations was the short WATM impulse response, which is depicted in Figure 4.3(a). Perfect knowledge of the channel impulse response was assumed for perfect phase and amplitude correction of the data symbols with coherent detection. Again, the BER curves for both the non-fading and the fading channels show a remarkable correspondence between the ideal performance lines and the performance of the synchronised modems. In all the investigated environments, the modem's performance was unaffected by the estimation accuracy of the time domain reference symbol synchronisation algorithm.

5.5   Chapter Summary and Conclusion

In this chapter, the effects of frequency and timing errors in OFDM transmissions have been characterised. While frequency errors result in frequency domain inter-subcarrier interference, timing errors lead to time domain inter-OFDM symbol interference and to frequency domain phase rotations.

In order to overcome the effects of moderate timing errors, a cyclic postamble and the use of pilot symbol-assisted modulation or differential detection was proposed. Different frequency and timing error estimation algorithms were portrayed, and their performance was investigated. A new combined frequency and timing synchronisation algorithm based on a dedicated reference symbol exploited in the time domain was proposed and the system's performance employing this algorithm was investigated in AWGN and fading channels. The solution of Section 5.2.5 was finally advocated for implementation in a 34Mbit/s real-time demonstration testbed.

5.6 Appendix: Theoretical Performance of OFDM Synchronisation Algorithms

5.6.1 Frequency Synchronisation in an AWGN Channel

The correlation operation used in the frequency synchronisation algorithm is based on the conjugate complex multiplication of the noisy input sample with its noisy and phase-shifted copy. To derive the theoretical performance of this algorithm, the influence of the noise on the phase of this product must be known. Throughout the calculations, we will assume no frequency and time invariant phase errors, in order to simplify the notation.

5.6.1.1 One Phasor in AWGN Environment

5.6.1.1.1 Cartesian Coordinates Every received sample is a superposition of the transmitted phasor and two statistically independent quadrature noise samples; if the received phasor is real and of magnitude $\mathcal{A}$, the noisy signal is described by the complex stochastic variable Z:

$$Z \ = \ \mathcal{A} + N \tag{5.37}$$
$$= \ X + jY \tag{5.38}$$

As the two quadrature noise processes are statistically independent, the joint probability density function (PDF) of the two quadrature components x and y is given by the product of two one-dimensional Gaussian PDFs:

$$p_{x,y}(x,y) \ = \ p(x) \cdot p(y) \tag{5.39}$$
$$= \ \left(\frac{1}{\sqrt{2\pi}\sigma}\right)^2 \cdot e^{-(x-\mathcal{A})^2/2\sigma^2} \cdot e^{-y^2/2\sigma^2} \tag{5.40}$$
$$= \ \frac{1}{2\pi\sigma^2} \cdot e^{-((x-\mathcal{A})^2+y^2)/2\sigma^2} \tag{5.41}$$

5.6.1.1.2 Polar Coordinates The stochastic variable Z can be expressed in polar coordinates:

$$Z \ = R \cdot e^{j\Phi} \tag{5.42}$$
$$with R \ = \sqrt{X^2 + Y^2} \tag{5.43}$$
$$\Phi \ = \tan^{-1}\frac{Y}{X} \tag{5.44}$$

To derive the PDF in polar coordinates $p_{r,\phi}(r,\phi)$ from the expression in Cartesian coordinates $p_{x,y}(x,y)$, given in Equation 5.41, the variable transform $(x,y) \Rightarrow (r,\phi)$ is performed:

$$x(r,\phi) = r \cdot \cos\phi \qquad (5.45)$$

$$y(r,\phi) = r \cdot \sin\phi \qquad (5.46)$$

The determinant J of the corresponding Jacobean matrix is:

$$J = \begin{vmatrix} \frac{\partial x(r,\phi)}{\partial r} & \frac{\partial y(r,\phi)}{\partial r} \\ \frac{\partial x(r,\phi)}{\partial \phi} & \frac{\partial y(r,\phi)}{\partial \phi} \end{vmatrix} = \begin{vmatrix} \cos\phi & \sin\phi \\ -r \cdot \sin\phi & r \cdot \cos\phi \end{vmatrix} = r \qquad (5.47)$$

This leads to the probability density function of one noisy phasor in polar coordinates:

$$p_{r,\phi}(r,\phi) = |J| \cdot p_{x,y}(x(r,\phi), y(r,\phi)) \qquad (5.48)$$

$$= \frac{r}{2\pi\sigma^2} \cdot e^{-(r^2 + \mathcal{A}^2 - 2r\mathcal{A}\cos\phi)/2\sigma^2} \qquad (5.49)$$

5.6.1.2   Product of Two Noisy Phasors

5.6.1.2.1   Joint Probability Density
To derive the probability density function of the product of two noisy phasors, we will assume that both of the phasors are stochastic variables with a PDF given in Equation 5.49, with the same signal amplitude $\mathcal{A}$ and zero signal phase. The new stochastic variable Π represents the product of the two noisy phasors Z_1 and Z_2:

$$\Pi = \Xi * e^{j\Psi} = Z_1 \cdot Z_2 \qquad (5.50)$$

The joint probability density function $p_\Pi(\Xi = \xi, \Psi = \psi)$ is derived from Equation 5.49 using complex multiplication:

$$\Xi = R_1 \cdot R_2 \qquad (5.51)$$

$$\Psi = \Phi_1 + \Phi_2 \qquad (5.52)$$

Two auxiliary variables must be introduced in order to solve the system. We choose $\Lambda = R_2$ and $\Omega = \Phi_2$. The resulting transformation functions and their inverse functions are:

$$\left. \begin{array}{rcl} \xi &=& r_1 \cdot r_2 \\ \psi &=& \phi_1 + \phi_2 \\ \lambda &=& r_2 \\ \omega &=& \phi_2 \end{array} \right\} \Leftrightarrow \left\{ \begin{array}{rcl} r_1 &=& \xi/\lambda \\ \phi_1 &=& \psi - \omega \\ r_2 &=& \lambda \\ \phi_2 &=& \omega \end{array} \right. \qquad (5.53)$$

The determinant J of the corresponding Jacobean matrix is:

$$J = \begin{vmatrix} \frac{\partial r_1}{\partial \xi} & \frac{\partial r_2}{\partial \xi} & \frac{\partial \phi_1}{\partial \xi} & \frac{\partial \phi_2}{\partial \xi} \\ \frac{\partial r_1}{\partial \psi} & \frac{\partial r_2}{\partial \psi} & \frac{\partial \phi_1}{\partial \psi} & \frac{\partial \phi_2}{\partial \psi} \\ \frac{\partial r_1}{\partial \lambda} & \frac{\partial r_2}{\partial \lambda} & \frac{\partial \phi_1}{\partial \lambda} & \frac{\partial \phi_2}{\partial \lambda} \\ \frac{\partial r_1}{\partial \omega} & \frac{\partial r_2}{\partial \omega} & \frac{\partial \phi_1}{\partial \omega} & \frac{\partial \phi_2}{\partial \omega} \end{vmatrix} = \begin{vmatrix} 1/\lambda & 0 & 0 & 0 \\ 0 & 1 & 0 & 0 \\ \xi \ln \lambda & 0 & 1 & 0 \\ 0 & 0 & 0 & 1 \end{vmatrix} = 1/\lambda \tag{5.54}$$

The resulting probability density function depending on the four new variables is:

$$p_\Pi(\xi, \psi, \lambda, \omega) = |J| \cdot p_Z(r_1 = \xi/\lambda, \phi_1 = \psi - \omega, r_2 = \lambda, \phi_2 = \omega) \tag{5.55}$$

The two probability density functions $p_Z(r1, \phi_1)$ and $p_Z(r2, \phi_2)$ are statistically independent, therefore

$$p_Z(r_1, \phi_1, r_2, \phi_2) = p_Z(r1, \phi_1) \cdot p_Z(r2, \phi_2) \tag{5.56}$$

Using the Equations 5.49 and 5.56 in 5.55, we find the expression for the PDF of the product of two noisy phasors as a function of the four transformed variables ξ, ψ, λ and ω:

$$p_\Pi(\xi, \psi, \lambda, \omega) = \frac{\xi}{\psi} \cdot \frac{1}{4\pi^2\sigma^4} \cdot e^{-\frac{\xi^2/\lambda^2 + \lambda^2 + 2\mathcal{A}^2 - 2\mathcal{A}(\lambda\cos\omega + \xi/\lambda\cos(\psi-\omega)))}{2\sigma^2}} \tag{5.57}$$

Eliminating the auxiliary variables λ and ω by integrating Equation 5.57 yields:

$$p_\Pi(\Xi = \xi, \Psi = \psi) = \int_0^\infty \int_{-\pi}^\pi p_\Pi(\xi, \psi, \lambda, \omega) d\omega d\lambda \tag{5.58}$$

5.6.1.2.2 Phase Distribution

The distribution of the phase of Π is obtained by integrating Equation 5.58 over the amplitude ξ:

$$p_\Pi(\Psi = \psi) = \int_0^\infty p_\Pi(\xi, \psi) d\xi \tag{5.59}$$

5.6.1.2.3 Numerical Integration

The integrals in Equations 5.58 and 5.59 cannot be solved analytically, therefore numerical integration has to be employed to determine the probability density functions $p_\Pi(\xi, \psi)$ and $p_\Pi(\psi)$. This numerical integral has been evaluated for a range of different signal-to-noise (SNR) values γ:

$$\gamma = \frac{\mathcal{A}}{2\sigma^2} \tag{5.60}$$

Simulations have been performed for the same SNR values to verify the expressions 5.58 and 5.59. The simulated histograms have been employed to perform a χ^2–test with the numerical results. The χ^2 value gives a distance measure between an expected probability distribution

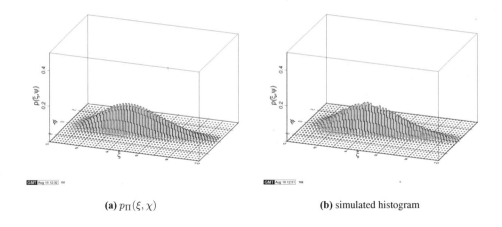

(a) $p_\Pi(\xi, \chi)$ (b) simulated histogram

Figure 5.29: Product of two noisy phasors with $\mathcal{A} = 2$ and $\gamma = 6$dB: (a): numerical solution of Equation 5.58, (b): histogram of 100000 simulated products

SNR	χ^2	Confidence level
0	3.959337e-1	> 99%
2	3.344832e-1	–"–
4	3.012187e-1	–"–
6	2.110520e-1	–"–
8	1.835500e-1	–"–
10	1.718824e-1	–"–

Table 5.1: χ^2 values and confidence levels for $p_\Pi(\xi, \psi)$ for different SNR values

and a N-binned set of experimental data:

$$\chi^2 = \sum_{i=0}^{N-1} \frac{(p(i) - h(i))^2}{p(i)} \tag{5.61}$$

where $h(i)$ is the value of the ith data bin, and $p(i)$ its expected value. As an example, Figure 5.6.1.2.3 illustrates both the numerical evaluation of Equation 5.58 and the histogram of the simulation for a SNR value of 6dB and a phasor amplitude of $\mathcal{A} = 2$. The figures reflect a high degree of correspondence between the numerical and the simulation results. This is confirmed by a quantitative measure, the χ^2 test, which yields a confidence-level in excess of 99% when testing the hypothesis that the simulated results are of the same distribution as given in Equation 5.58. Table 5.1 displays the values of χ^2 and the resulting confidence level $\Gamma(2100, \chi^2)$ for SNR values between 0 and 10dB. The numerically integrated phase probability density function Equation 5.59 and the histogram of the simulated values are shown in Figure 5.30(a). Figure 5.30(b) depicts the phase PDF for different SNR values.

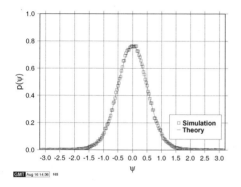

(a) phase PDF, theory and simulation

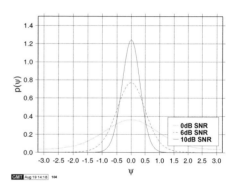

(b) phase PDFs for different SNR values

Figure 5.30: Product of two noisy phasors: phase probability density functions. (a): theoretical PDF and simulated histogram of the phase distribution for SNR = 6dB (b): theoretical PDFs for different SNR values

Chapter 6

Adaptive Single- and Multi-user OFDM Techniques[1]

6.1 Introduction

Steele and Webb [198] proposed adaptive modulation for exploiting the time-variant Shannonian channel capacity of fading narrowband channels, and further research was conducted at Osaka University by Sampei *et al.* [199], at the University of Stanford by Goldsmith and Chua [200], by Pearce, Burr and Tozer at the University of York [201], Lau and McLeod at the University of Cambridge [202], and at Southampton University [203,204]. The associated principles can also be invoked in the context of parallel modems, as it has been demonstrated by Kalet [19], Czylwik [205] as well as by Chow, Cioffi and Bingham [206]. Our treatment of adaptive modulation is conceptual in this chapter. We will revisit this topic in more depth in Chapter 12, where numerous analytical expressions will be derived for characterising the achievable BER perfomance as well as for determining the modem mode switching thresholds. For a more detailed treatise on adaptive modulation please refer to [7, 207–209], while a detailed analytical insight into this topic will be provided in Chapter 12 for the readers seeking an indepth mathematical treatment.

6.1.1 Motivation

We saw in Figure 4.12 that the bit error probability of different OFDM subcarriers transmitted in time dispersive channels depends on the frequency domain channel transfer function. The occurrence of bit errors is normally concentrated in a set of severely faded subcarriers, while in the rest of the OFDM spectrum often no bit errors are observed. If the subcarriers that will exhibit high bit error probabilities in the OFDM symbol to be transmitted can be identified and excluded from data transmission, the overall BER can be improved in exchange for a

[1]*OFDM and MC-CDMA for Broadband Multi-user Communications WLANS and Broadcasting.*
L.Hanzo, Münster, T. Keller and B.J. Choi,
©2003 John Wiley & Sons, Ltd. ISBN 0-470-85879-6

slight loss of system throughput. As the frequency domain fading deteriorates the SNR of certain subcarriers, but improves that of others above the average SNR value, the potential loss of throughput due to the exclusion of faded subcarriers can be mitigated by employing higher-order modulation modes on the subcarriers exhibiting high SNR values.

In addition to excluding sets of faded subcarriers and varying the modulation modes employed, other parameters such as the coding rate of error correction coding schemes can be adapted at the transmitter according to the perceived channel transfer function. This issue will be addressed in Chapter 7.

Adaptation of the transmission parameters is based on the transmitter's perception of the channel conditions in the forthcoming time slot. Clearly, this estimation of future channel parameters can only be obtained by extrapolation of previous channel estimations, which are acquired upon detecting each received OFDM symbol. The channel characteristics therefore have to vary sufficiently slowly compared to the estimation interval.

Adapting the transmission technique to the channel conditions on a time slot by time slot basis for serial modems in narrowband fading channels has been shown to considerably improve the BER performance [210] for time division duplex (TDD) systems assuming duplex reciprocal channels. However, the Doppler fading rate of the narrowband channel has a strong effect on the achievable system performance: if the fading is rapid, then the prediction of the channel conditions for the next transmit time slot is inaccurate, and therefore the wrong set of transmission parameters may be chosen. If, however, the channel varies slowly, then the data throughput of the system is varyies dramatically over time, and large data buffers are required at the transmitters in order to smoothen the bit rate fluctuation. For time critical applications, such as interactive speech transmission, the potential delays can become problematic. A given single-carrier adaptive system in narrowband channels will therefore operate efficiently only in a limited range of channel conditions.

Adaptive OFDM modem channels can ease the problem of slowly time-varying channels, since the variation of the signal quality can be exploited in the time domain and the frequency domain. The channel conditions still have to be monitored based on the received OFDM symbols, and relatively slowly varying channels have to be assumed, since we saw in Section 4.2.2 that OFDM transmissions are not well suited to rapidly varying channel conditions.

6.1.2 Adaptive Techniques

Adaptive modulation is only suitable for duplex communication between two stations, since the transmission parameters have to be adapted using some form of two-way transmission in order to allow channel measurements and signalling to take place.

Transmission parameter adaptation is a response of the transmitter to time-varying channel conditions. In order to efficiently react to the changes in channel quality, the following steps have to be taken:

(1) *Channel quality estimation:* In order to appropriately select the transmission parameters to be employed for the next transmission, a reliable estimation of the channel transfer function during the next active transmit time slot is necessary.

(2) *Choice of the appropriate parameters for the next transmission:* Based on the prediction of the channel conditions for the next time slot, the transmitter has to select the appropriate modulation modes for the subcarriers.

(3) *Signalling or blind detection of the employed parameters:* The receiver has to be informed, which demodulator parameters to employ for the received packet. This information can either be conveyed within the OFDM symbol itself, with the loss of effective data throughput, or the receiver can attempt to estimate the parameters employed by the remote transmitter by means of blind detection mechanisms.

6.1.2.1 Channel Quality Estimation

The transmitter requires an estimate of the expected channel conditions for the time when the next OFDM symbol is to be transmitted. Since this knowledge can only be gained by prediction from past channel quality estimations, the adaptive system can only operate efficiently in an environment exhibiting relatively slowly varying channel conditions.

The channel quality estimation can be acquired from a range of different sources. If the communication between the two stations is bidirectional and the channel can be considered reciprocal, then each station can estimate the channel quality on the basis of the received OFDM symbols, and adapt the parameters of the local transmitter to this estimation. We will refer to such a regime as *open-loop adaptation*, since there is no feedback between the receiver of a given OFDM symbol and the choice of the modulation parameters. A time division duplex (TDD) system in absence of interference is an example of such a system, and hence a TDD regime is assumed to generate the performance results below. Channel reciprocity issues were addressed for example in [211, 212].

If the channel is not reciprocal, as in a frequency division duplex (FDD) system, then the stations cannot determine the parameters for the next OFDM symbol's transmission from the received symbols. In this case, the receiver has to estimate the channel quality and explicitly signal this perceived channel quality information to the transmitter in the reverse link. Since in this case the receiver explicitly instructs the remote transmitter as to which modem modes to invoke, this regime is referred to as *closed-loop adaptation*. With the aid of this technique the adaptation algorithms can take into account effects such as interference as well as non-reciprocal channels. If the communication between the stations is essentially unidirectional, then a low-rate signalling channel must be implemented from the receiver to the transmitter. If such a channel exists, then the same technique as for non-reciprocal channels can be employed.

Different techniques can be employed to estimate the channel quality. For OFDM modems, the bit error probability in each subcarrier is determined by the fluctuations of the channel's instantaneous frequency domain channel transfer function H_n, if no interference is present. The estimate of the channel transfer function $\hat{H}_n$ can be acquired by means of pilot tone based channel estimation, as demonstrated in Section 4.3.1.1. More accurate measures of the channel transfer function can be gained by means of decision directed or time domain training sequence based techniques. The estimate of the channel transfer function $\hat{H}_n$ does not take into account effects, such as co-channel or intersubcarrier interference. Alternative channel quality measures that include interference effects can be devised on the basis of the error correction decoder's soft output information or by means of decision feedback local SNR estimations.

The delay between the channel quality estimate and the actual transmission of the OFDM symbol in relation to the maximal Doppler frequency of the channel is crucial to the adaptive system's performance. If the channel estimate is obsolete at the time of transmission, then poor system performance will result. For a closed-loop adaptive system the delays between channel estimation and transmission of the packet are generally longer than for an open-loop adaptive system, and therefore the Doppler frequency of the channel is a more critical parameter for the system's performance than in the context of open-loop adaptive systems.

6.1.2.2   Parameter Adaptation

Different transmission parameters can be adapted to the anticipated channel conditions, such as the modulation and coding modes. Adapting the number of modulation levels in response to the anticipated local SNR encountered in each subcarrier can be employed, in order to achieve a wide range of different trade-offs between the received data integrity and through-put. Corrupted subcarriers can be excluded from data transmission and left blank or perhaps used for crest factor reduction.

The adaptive channel coding parameters entail code rate, adaptive interleaving and puncturing for convolutional and turbo codes, or varying block lengths for block codes [164]. These techniques can be combined with adaptive modulation mode selection and will be discussed in Chapter 7.

Based on the estimated frequency domain channel transfer function, spectral predistortion at the transmitter of one or both communicating stations can be invoked, in order to partially or fully counteract the frequency selective fading of the time dispersive channel. Unlike frequency domain equalisation at the receiver – which corrects for the amplitude and phase errors inflicted upon the subcarriers by the channel but cannot improve the signal-to-noise ratio in poor quality channels – spectral predistortion at the OFDM transmitter can deliver near-constant signal-to-noise levels for all subcarriers and can be thought of as power control on a subcarrier-by-subcarrier basis.

In addition to improving the system's BER performance in time dispersive channels, spectral predistortion can be employed in order to perform all channel estimation and equalisation functions at only one of the two communicating duplex stations. Low cost, low power consumption mobile stations can communicate with a base station that performs the channel estimation and frequency domain equalisation of the uplink, and uses the estimated channel transfer function for predistorting the down-link OFDM symbol. This set-up would lead to different overall channel quality on the uplink and the downlink, and the superior downlink channel quality could be exploited by using a computationally less complex channel decoder having weaker error correction capabilities in the mobile station than in the base station.

If the channel's frequency domain transfer function is to be fully counteracted by the spectral predistortion upon adapting the subcarrier power to the inverse of the channel transfer function, then the output power of the transmitter can become excessive, if heavily faded subcarriers are present in the system's frequency range. In order to limit the transmitter's maximal output power, hybrid channel predistortion and adaptive modulation schemes can be devised, which would deactivate transmission in deeply faded sub-channels, while retaining the benefits of predistortion in the remaining subcarriers.

6.1.2.3   Signalling the AOFDM Parameters

Signalling plays an important role in adaptive systems and the range of signalling options is summarised in Figure 6.1 for both open-loop and closed-loop signalling, as well as for blind

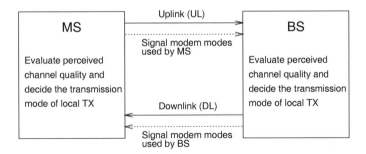

(a) Reciprocal channel, open-loop control

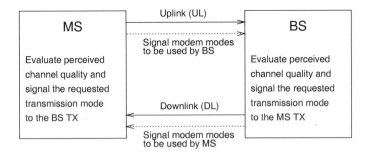

(b) Non-reciprocal channel, closed-loop signalling

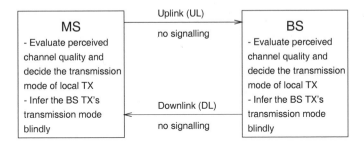

(c) Reciprocal channel, blind modem-mode detection

Figure 6.1: Signalling scenarios in adaptive modems

detection. If the channel quality estimation and parameter adaptation have been performed at the transmitter of a particular link, based on open-loop adaptation, then the resulting set of parameters has to be communicated to the receiver in order to successfully demodulate and decode the OFDM symbol. If the receiver itself determines the requested parameter set to be used by the remote transmitter, the closed-loop scenario, then the same amount of information has to be transported to the remote transmitter in the reverse link. If this signalling information is corrupted, then the receiver is generally unable to correctly decode the OFDM symbol corresponding to the incorrect signalling information.

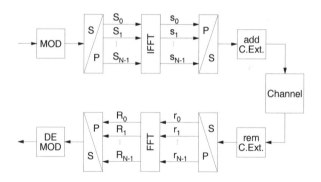

Figure 6.2: Schematic model of the OFDM system

Unlike adaptive serial systems, which employ the same set of parameters for all data symbols in a transmission packet [203, 204], adaptive OFDM systems have to react to the frequency selective nature of the channel, by adapting the modem parameters across the sub-carriers. The resulting signalling overhead may become significantly higher than that for serial modems, and can be prohibitive for subcarrier-by-subcarrier modulation mode adaptation. In order to overcome these limitations, efficient and reliable signalling techniques have to be employed for practical implementation of adaptive OFDM modems.

If some flexibility in choosing the transmission parameters is sacrificed in an adaptation scheme, as in the sub-band adaptive OFDM schemes described below, then the amount of signalling can be reduced. Alternatively, blind parameter detection schemes can be devised, which require little or no signalling information, respectively. Two simple blind modulation scheme detection algorithms are investigated in Section 6.2.6.2 [213].

6.1.3   System Aspects

The effects of transmission parameter adaptation for OFDM systems on the overall communication system have to be investigated in at least the following areas: data buffering and latency due to varying data throughput, the effects of co-channel interference and bandwidth efficiency.

6.2   Adaptive Modulation for OFDM

6.2.1   System Model

The system model of the N-subcarrier orthogonal frequency division multiplexing (OFDM) modem is shown in Figure 6.2 [1]. At the transmitter, the modulator generates N data symbols S_n, $0 \leq n \leq N - 1$, which are multiplexed to the N subcarriers. The time domain samples s_n transmitted during one OFDM symbol are generated by the inverse fast Fourier transform (IFFT) and transmitted over the channel after the cyclic extension (C. Ext.) has been inserted. The channel is modelled by its time-variant impulse response $h(\tau, t)$ and AWGN. At the receiver, the cyclic extension is removed from the received time domain samples, and the data samples r_n are fast Fourier transformed, in order to yield the received frequency domain data symbols R_n.

The channel's impulse response is assumed to be time invariant for the duration of one OFDM symbol, therefore it can be characterised for each OFDM symbol period by the N-point Fourier transform of the impulse response, which is referred to as the frequency domain channel transfer function H_n. The received data symbols R_n can be expressed as:

$$R_n = S_n \cdot H_n + n_n,$$

where n_n is an AWGN sample. Coherent detection is assumed for the system, therefore the received data symbols R_n have to be defaded in the frequency domain with the aid of an estimate of the channel transfer function H_n. This estimate $\hat{H}_n$ can be obtained by the use of pilot subcarriers in the OFDM symbol, or by employing time domain channel sounding training sequences embedded in the transmitted signal. Since the noise energy in each subcarrier is independent of the channel's frequency domain transfer function H_n, the "local" signal-to-noise ratio SNR in subcarrier n can be expressed as

$$\gamma_n = |H_n|^2 \cdot \gamma,$$

where γ is the overall SNR. If no signal degradation due to intersubcarrier interference (ISI) or interference from other sources appears, then the value of γ_n determines the bit error probability for the transmission of data symbols over the subcarrier n.

The goal of adaptive modulation is to choose the appropriate modulation mode for transmission in each subcarrier, given the local SNR γ_n, in order to achieve a good trade-off between throughput and overall BER. The acceptable overall BER varies depending on other systems parameters, such as the correction capability of the error correction coding and the nature of the service supported by this particular link.

The adaptive system has to fulfil these requirements:

(1) Channel quality estimation,

(2) Choice of the appropriate modulation modes.

(3) Signalling or blind detection of the modulation modes.

We will examine these three points with reference to Figure 6.1 in the following sections for the example of a 512-subcarrier OFDM modem in the shortened WATM channel of Section 4.1.1.2.

6.2.2 Channel Model

The impulse response $h(\tau, t)$ used in our experiments was generated on the basis of the symbol-spaced impulse response shown in Figure 6.3(a) by fading each of the impulses obeying a Rayleigh distribution of a normalised maximal Doppler frequency of $f_d' = 1.235 \cdot 10^{-5}$, which corresponds to the WLAN channel experienced by a modem transmitting at a carrier frequency of 60 GHz with a sample rate of 225 MHz and a vehicular velocity of 50 km/h. The complex frequency domain channel transfer function H_n corresponding to the unfaded impulse response is shown in Figure 6.3(b).

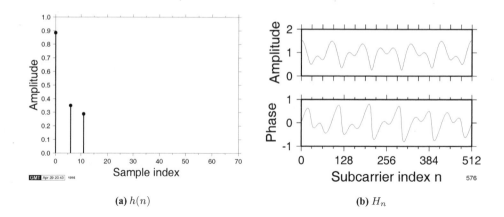

(a) $h(n)$ **(b)** H_n

Figure 6.3: WATM wideband channel: (a) unfaded symbol spaced impulse response, (b) the corresponding frequency domain channel transfer function

6.2.3   Channel Transfer Function Variations

The most convenient setting for an adaptive OFDM (AOFDM) system is a time division duplex (TDD) system in a slowly varying reciprocal channel, allowing open-loop adaptation. Both stations transmit an OFDM symbol in turn, and at each station the most recent received symbol is used for the channel estimation employed for the modulation mode adaptation for the next transmitted OFDM symbol. The channel estimation on the basis of the received symbol can be performed by PSAM (see Section 4.3.1.1), or upon invoking more sophisticated methods, such as decision-directed channel estimation. Initially, we will assume perfect knowledge of the channel transfer function during the received time slot.

6.2.4   Choice of the Modulation Modes

The two communicating stations use the open-loop predicted channel transfer function acquired from the most recent received OFDM symbol, in order to allocate the appropriate modulation modes to the subcarriers. The modulation modes were chosen from the set of binary phase shift keying (BPSK), quadrature phase shift keying (QPSK), 16-quadrature amplitude modulation (16-QAM), as well as "no transmission", for which no signal was transmitted. These modulation modes are denoted by M_m, where $m \in (0, 1, 2, 4)$ is the number of data bits associated with one data symbol of each mode.

In order to keep the system complexity low, the modulation mode is not varied on a subcarrier-by-subcarrier basis, but instead the total OFDM bandwidth of 512 subcarriers is split into blocks of adjacent subcarriers, referred to as sub-bands, and the same modulation scheme is employed for all subcarriers of the same sub-band. This substantially simplifies the task of signalling the modem mode and renders the employment of alternative blind detection mechanisms feasible, which will be discussed in Section 6.2.6.

Three modulation mode allocation algorithms were investigated in the sub-bands: a fixed threshold controlled algorithm, an upper bound BER estimator and a fixed throughput adaptation algorithm.

	l_0	l_1	l_2	l_4
speech system	$-\infty$	3.31	6.48	11.61
data system	$-\infty$	7.98	10.42	16.76

Table 6.1: Optimised switching levels for adaptive modulation over Rayleigh fading channels for the "speech" and "data" systems, shown in instantaneous channel SNR [dB] (from [214])

6.2.4.1   Fixed Threshold Adaptation Algorithm

The fixed threshold algorithm was derived from the adaptation algorithm proposed by Torrance for serial modems [210]. In the case of a serial modem, the channel quality is assumed to be constant for all symbols in the time slot, and hence the channel has to be slowly varying in order to allow accurate channel quality prediction. Under these circumstances, all data symbols in the transmit time slot employ the same modulation mode, chosen according to the predicted SNR. The SNR thresholds for a given long-term target BER were determined by Powell optimisation [214]. Torrance assumed two uncoded target bit error rates: 1% for a high data rate "speech" system, and 10^{-4} for a higher integrity, lower data rate "data" system. The resulting SNR thresholds l_n for activating a given modulation mode M_n in a slowly Rayleigh fading narrowband channel for both systems are given in Table 6.1. Specifically, the modulation mode M_n is selected if the instantaneous channel SNR exceeds the switching level l_n.

As noted before, a more detailed analytically motivated discussion on the optimisation of the modem mode switching thresholds will be provided in Chapter 12. We will show that the multi-dimensional Powell optimisation technique may be replaced by a single-dimensional Lagrangian optimisation procedure, which is capable of further improving the achievable system throughput. More explicitly, the optimisation may be rendered one-simensional, since the swithching thresholds of the high-throughput modem modes are actually at a constant distance from the activation threshold of the lowest-throughput BPSK mode. We will show in Chapter 12 that having determined this threshold, the remaining activation thresholds become readily available.

This adaptation algorithm originally assumed a constant instantaneous SNR over all of the block's symbols, but in the case of an OFDM system in a frequency selective channel the channel quality varies across the different subcarriers. For sub-band adaptive OFDM transmission, this implies that if the sub-band width is wider than the channel's coherence bandwidth [164], then the original switching algorithm cannot be employed. For our investigations, we have therefore employed the lowest quality subcarrier in the sub-band for the adaptation algorithm based on the thresholds given in Table 6.1. The performance of the 16 sub-band adaptive system over the shortened WATM Rayleigh fading channel of Figure 4.3 is shown in Figure 6.4.

Adjacent or consecutive time slots have been used for the uplink and downlink slots in these simulations, so that the delay between channel estimation and transmission was rendered as short as possible. Figure 6.4 shows the long-term average BER and throughput of the studied modem for the "speech" and "data" switching levels of Table 6.1 as well as for a subcarrier-by-subcarrier adaptive modem employing the "data" switching levels. The results show the typical behaviour of a variable throughput AOFDM system, which constitutes a

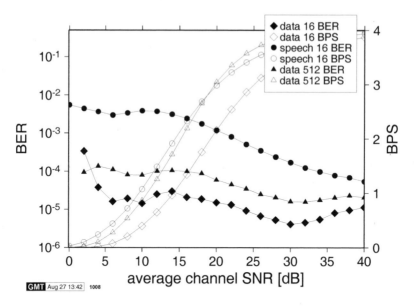

Figure 6.4: BER and BPS throughput performance of the 16 sub-band, 512 subcarrier switching level adaptive OFDM modem employing BPSK, QPSK, 16-QAM and "no transmission" over the Rayleigh fading time dispersive channel of Figure 4.3 using the switching thresholds of Table 6.1

trade-off between the best BER and best throughput performance. For low SNR values, the system achieves a low BER by transmitting very few bits and only when the channel conditions allow. With increasing long-term SNR, the throughput increases without significant change in the BER. For high SNR values the BER drops as the throughput approaches its maximum of 4 bits per symbol, since the highest-order constellation was 16-QAM.

It can be seen from the figure that the adaptive system performs better than its target bit error rates of 10^{-2} and 10^{-4} for the "speech" and "data" systems, respectively, resulting in measured bit error rates lower than the targets. This can be explained by the adaptation regime, which was based on the conservative principle of using the lowest quality subcarrier in each sub-band for channel quality estimation, leading to a pessimistic channel quality estimate for the entire sub-band. For low values of SNR, the throughput in bits per data symbol is low and exceeds the fixed BPSK throughput of 1 bit/symbol only for SNR values in excess of 9.5 dB and 14 dB for the "speech" and "data" systems, respectively.

The upper bound performance of the system with subcarrier-by-subcarrier adaptation is also portrayed in the figure, shown as 512 independent sub-bands, for the "data" optimised set of threshold values. It can be seen that in this case the target BER of 10^{-4} is closely met over a wide range of SNR values from about 2 dB to 20 dB, and that the throughput is considerably higher than in the case of the 16 sub-band modem. This is the result of more accurate subcarrier-by-subcarrier channel quality estimation and fine-grained adaptation, leading to better exploitation of the available channel capacity.

Figure 6.5 shows the long-term modulation mode histograms for a range of channel SNR values for the "data" switching levels in both the 16 sub-band and the subcarrier-by-subcarrier

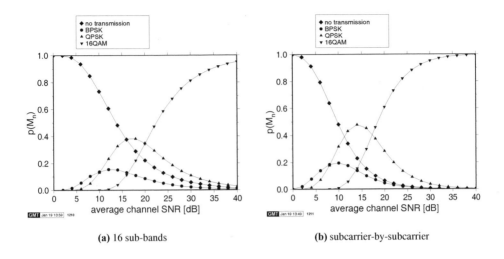

(a) 16 sub-bands   **(b)** subcarrier-by-subcarrier

Figure 6.5: Histograms of modulation modes versus channel SNR for the "data" switching level adaptive 512 subcarrier, 16 sub-band OFDM modem over the Rayleigh fading time dispersive channel of Figure 4.3 using the switching thresholds of Table 6.1

adaptive modems using the switching thresholds of Table 6.1. Comparison of the graphs shows that higher-order modulation modes are used more frequently by the subcarrier-by-subcarrier adaptation algorithm, which is in accordance with the overall throughput performance of the two modems in Figure 6.4.

The throughput penalty of employing sub-band adaptation depends on the frequency domain variation of the channel transfer function. If the sub-band bandwidth is lower than the channel's coherence bandwidth, then the assumption of constant channel quality per sub-band is closely met, and the system performance is equivalent to that of a subcarrier-by-subcarrier adaptive scheme.

6.2.4.2 Sub-band BER Estimator Adaptation Algorithm

We saw above that the fixed switching level based algorithm leads to a throughput performance penalty, if used in a sub-band adaptive OFDM modem, when the channel quality is not constant throughout each sub-band. This is due to the conservative adaptation based on the subcarrier experiencing the most hostile channel in each sub-band.

An alternative scheme taking into account the non-constant SNR values γ_j across the N_s subcarriers in the jth sub-band can be devised by calculating the expected overall bit error probability for all available modulation modes M_n in each sub-band, which is denoted by $\bar{p}_e(n) = 1/N_s \sum_j p_e(\gamma_j, M_n)$. For each sub-band, the mode having the highest throughput, whose estimated BER is lower than a given threshold, is then chosen. While the adaptation granularity is still limited to the sub-band width, the channel quality estimation does not include only the lowest quality subcarrier, which leads to an improved throughput.

Figure 6.6 shows the BER and throughput performance for the 16 sub-band adaptive OFDM modem employing the BER estimator adaptation algorithm in the Rayleigh fading

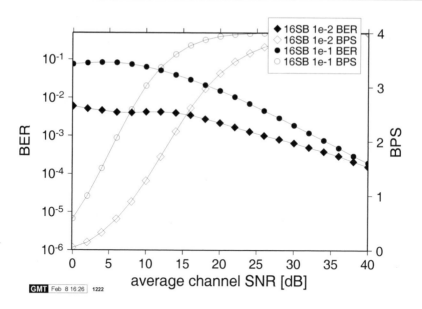

Figure 6.6: BER and BPS throughput performance of the 16 sub-band, 512 subcarrier BER estimator
adaptive OFDM modem employing BPSK, QPSK, 16-QAM and "no transmission" over
the Rayleigh fading time dispersive channel of Figure 4.3

time dispersive channel of Figure 4.3. The two sets of curves in the figure correspond to
target bit error rates of 10^{-2} and 10^{-1}, respectively. Comparing the modem's performance
for a target BER of 10^{-2} with that of the "speech" modem in Figure 6.4, it can be seen
that the BER estimator algorithm results in significantly higher throughput while meeting the
BER requirements. The BER estimator algorithm is readily adjustable to different target bit
error rates, which is demonstrated in the figure for a target BER of 10^{-1}. Such adjustability
is beneficial when combining adaptive modulation with channel coding, as will be discussed
in Section 6.2.7.

6.2.5 Constant Throughput Adaptive OFDM

The time-varying data throughput of an adaptive OFDM modem operating with either of
the two adaptation algorithms discussed above makes it difficult to employ such a scheme
in a variety of constant rate applications. Torrance [210] studied the system implications of
variable-throughput adaptive modems in the context of narrowband channels, stressing the
importance of data buffering at the transmitter, in order to accommodate the variable data
rate. The required length of the buffer is related to the Doppler frequency of the channel, and
a slowly varying channel – as required for adaptive modulation – results in slowly varying
data throughput and therefore the need for a high buffer capacity. Real-time interactive audio
or video transmission is sensitive to delays, and therefore different modem mode adaptation
algorithms are needed for such applications.

    The constant throughput AOFDM scheme proposed here exploits the frequency selectiv-

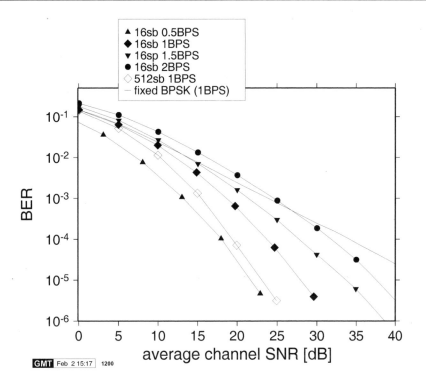

Figure 6.7: BER performance versus SNR for the 512 subcarrier, 16 sub-band constant throughput adaptive OFDM modem employing BPSK, QPSK, 16-QAM and "no transmission" in the Rayleigh fading time dispersive channel of Figure 4.3 for 0.5, 1, 1.5 and 2 bits per symbol (BPS) target throughput

ity of the channel, while offering a constant bit rate. Again, sub-band adaptivity is assumed, in order to simplify the signalling or the associated blind detection of the modem schemes.

The modulation mode allocation of the sub-bands is performed on the basis of a cost function to be introduced below, based on the expected number of bit errors in each sub-band. The expected number of bit errors $e_{n,s}$, for each sub-band n and for each possible modulation mode index s, is calculated on the basis of the estimated channel transfer function $\hat{H}$, taking into account the number of bits transmitted per sub-band and per modulation mode, $b_{n,s}$.

Each sub-band is assigned a state variable s_n holding the index of a modulation mode. Each state variable is initialised to the lowest-order modulation mode, which in our case is 0 for "no transmission". A set of cost values $c_{n,s}$ is calculated for each sub-band n and state s:

$$c_{n,s} = \frac{e_{n,s+1} - e_{n,s}}{b_{n,s+1} - b_{n,s}} \tag{6.1}$$

for all but the highest modulation mode index s. This cost value is related to the expected increase in the number of bit errors, divided by the increase of throughput, if the modulation mode having the next higher index is used instead of index s in sub-band n. In other words,

Equation 6.1 quantifies the expected incremental bit error rate of the state transition $s \rightarrow s+1$ in sub-band n.

The modulation mode adaptation is performed by repeatedly searching for the block n having the lowest value of c_{n,s_n}, and incrementing its state variable s_n. This is repeated until the total number of bits in the OFDM symbol reaches the target number of bits. Because of the granularity in bit numbers introduced by the sub-bands, the total number of bits may exceed the target. In this case, the data is padded with dummy bits for transmission.

Figure 6.7 gives an overview of the BER performance of the fixed throughput 512-subcarrier OFDM modem over the time dispersive channel of Figure 4.3 for a range of target bit numbers. In Figure 6.7 the curve without symbols represents the performance of a fixed BPSK OFDM modem over the same channel and where the modem transmits, i.e. 1 bit over each data subcarrier per OFDM symbol. The diamond shapes give the performance of the equivalent throughput adaptive scheme for the 16 sub-band arrangement filled shapes and for the subcarrier-by-subcarrier adaptive scheme open shapes. It can be seen that the 16 sub-band adaptive scheme yields a significant improvement in BER terms for SNR values above 10 dB. The SNR gain for a bit error rate of 10^{-4} is 8 dB compared to the non-adaptive case. Subcarrier-by-subcarrier adaptivity increases this gain by a further 4 dB. The modem can readily be adapted to the system requirements by adjusting the target bit rate, as shown in Figure 6.7. Halving the throughput to 0.5 BPS, the required SNR is reduced by 6 dB for a BER of 10^{-4}, while increasing the throughput to 2 BPS deteriorates the noise resilience by 8 dB at the same BER.

6.2.6 AOFDM Mode Signalling and Blind Detection

The adaptive OFDM receiver has to be informed of the modulation modes used for the different sub-bands. This information can either be conveyed using signalling subcarriers in the OFDM symbol itself, or the receiver can employ blind detection techniques in order to estimate the transmitted symbols' modulation modes, as seen in Figure 6.1.

6.2.6.1 Signalling

The simplest way of signalling the modulation mode employed in a sub-band is to replace one data symbol by an M-PSK symbol, where M is the number of possible modulation modes. In this case, reception of each of the constellation points directly signals a particular modulation mode in the current sub-band. In our case, for four modulation modes and assuming perfect phase recovery, the probability of a signalling error $p_s(\gamma)$, when employing one signalling symbol, is the symbol error probability of QPSK. Then the correct sub-band mode signalling probability is:

$$(1 - p_s(\gamma)) = (1 - p_{b,QPSK}(\gamma))^2,$$

where $p_{b,QPSK}$ is the bit error probability for QPSK:

$$p_{b,QPSK}(\gamma) = Q(\sqrt{\gamma}) = \frac{1}{2} \cdot \mathrm{erfc}\left(\sqrt{\frac{\gamma}{2}}\right),$$

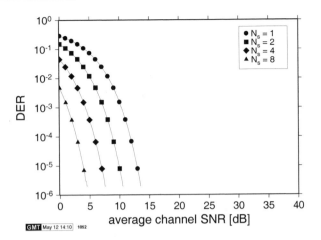

Figure 6.8: Modulation mode detection error ratio (DER) if signalling with maximum ratio combining is employed for QPSK symbols in an AWGN channel for 1, 2, 4 and 8 signalling symbols per sub-band, evaluated from Equation 6.2

which leads to the expression for the modulation mode signalling error probability of

$$p_s(\gamma) = 1 - \left(1 - \frac{1}{2} \cdot \mathrm{erfc}\left(\sqrt{\frac{\gamma}{2}}\right)\right)^2 .$$

The modem mode signalling error probability can be reduced by employing multiple signalling symbols and maximum ratio combining of the received signalling symbols $R_{s,n}$, in order to generate the decision variable R'_s prior to decision:

$$R'_s = \sum_{n=1}^{N_s} R_{s,n} \cdot \hat{H}^*_{s,n},$$

where N_s is the number of signalling symbols per sub-band, the quantities $R_{s,n}$ are the received symbols in the signalling subcarriers, and $\hat{H}_{s,n}$ represents the estimated values of the frequency domain channel transfer function at the signalling subcarriers. Assuming perfect channel estimation and constant values of the channel transfer function across the group of signalling subcarriers, the signalling error probability for N_s signalling symbols can be expressed as:

$$p'_s(\gamma, N_s) = 1 - \left(1 - \frac{1}{2} \cdot \mathrm{erfc}\left(\sqrt{\frac{N_s \gamma}{2}}\right)\right)^2 . \tag{6.2}$$

Figure 6.8 shows the signalling error rate in an AWGN channel for 1, 2, 4 and 8 signalling symbols per sub-band, respectively. It can be seen that doubling the number of signalling subcarriers improves the performance by 3 dB. Modem mode detection error ratios (DER) below 10^{-5} can be achieved at 10 dB SNR over AWGN channels if two signalling symbols

are used. The signalling symbols for a given sub-band can be interleaved across the entire
OFDM symbol bandwidth, in order to benefit from frequency diversity in fading wideband
channels.

As seen in Figure 6.1, blind detection algorithms aim to estimate the employed modulation mode directly from the received data symbols, therefore avoiding the loss of data capacity due to signalling subcarriers. Two algorithms have been investigated, one based on
geometrical SNR estimation and another incorporating error correction coding.

6.2.6.2   Blind Detection by SNR Estimation

The receiver has no *a priori* knowledge of the modulation mode employed in a particular
received sub-band and estimates this parameter by quantising the defaded received data symbols $R_n/\hat{H}_n$ in the sub-band to the closest symbol $\hat{R}_{n,m}$ for all possible modulation modes
M_m for each subcarrier index n in the current sub-band. The decision-directed error energy
e_m for each modulation mode is calculated:

$$e_m = \sum_n \left(R_n/\hat{H}_n - \hat{R}_{n,m} \right)^2$$

and the modulation mode M_m which minimises e_m is chosen for the demodulation of the
sub-band.

The DER of the blind modulation mode detection algorithm described in this section for
a 512-subcarrier OFDM modem in an AWGN channel is depicted in Figure 6.9. It can be
seen that the detection performance depends on the number of symbols per sub-band, with
fewer sub-bands and therefore longer symbol sequences per sub-band leading to a better
detection performance. It is apparent, however, that the number of available modulation
modes has a more significant effect on the detection reliability than the block length. If
all four legitimate modem modes are employed, then reliable detection of the modulation
mode is only guaranteed for AWGN SNR values of more than 15-18 dB, depending on the
number of sub-bands per OFDM symbol. If only M_0 and M_1 are employed, however, the
estimation accuracy is dramatically improved. In this case, AWGN SNR values above 5-7 dB
are sufficient to ensure reliable detection. The estimation accuracy could be improved by
using the estimate of the channel quality, in order to predict the modulation mode, which is
likely to have been employed at the transmitter. For example, at an estimated channel SNR
of 5 dB it is unlikely that 16-QAM was employed as the modem mode and hence this *a priori*
knowledge can be exploited, in order increase our confidence in the corresponding decision.

Figure 6.10 shows the BER performance of the fixed threshold "data" 16 sub-band adaptive system in the fading wideband channel of Figure 4.3 for both sets of modulation modes,
namely for (M_0, M_1) and (M_0, M_1, M_2, M_4) with blind modulation mode detection. Erroneous modulation mode decisions were assumed to yield a BER of 50% in the received
block. This is optimistic, since in a realistic scenario the receiver would have no knowledge of the number of bits actually transmitted, leading to loss of synchronisation in the data
stream. This problem is faced by all systems having a variable throughput and not employing
an ideal reliable signalling channel. This problem must be mitigated by data synchronisation
measures.

It can be seen from Figure 6.10 that while blind modulation mode detection yields poor

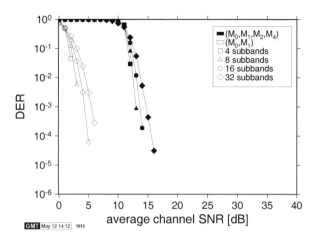

Figure 6.9: Blind modulation mode detection error ratio (DER) for 512-subcarrier OFDM systems employing (M_0, M_1) as well as for (M_0, M_1, M_2, M_4) for different numbers of sub-bands in an AWGN channel

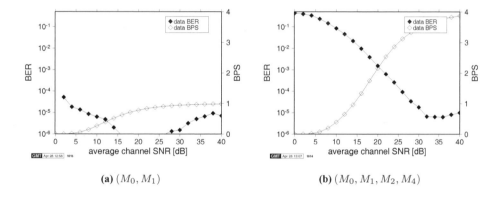

(a) (M_0, M_1) **(b)** (M_0, M_1, M_2, M_4)

Figure 6.10: BER and BPS throughput performance of a 16 sub-band, 512 subcarrier adaptive OFDM modem employing (a) no transmission (M_0) and BPSK (M_1), or (b) (M_0, M_1, M_2, M_4), both using the data-type switching levels of Table 6.1 and the SNR-based blind modulation mode detection of Section 6.2.6.2 over the Rayleigh fading time dispersive channel of Figure 4.3

performance for the quadruple-mode adaptive scheme, the twin-mode scheme exhibits BER results consistently better than 10^{-4}.

6.2.6.3 Blind Detection by Multi-Mode Trellis Decoder

If error correction coding is invoked in the system, then the channel decoder can be employed to estimate the most likely modulation mode per sub-band. Since the number of bits per

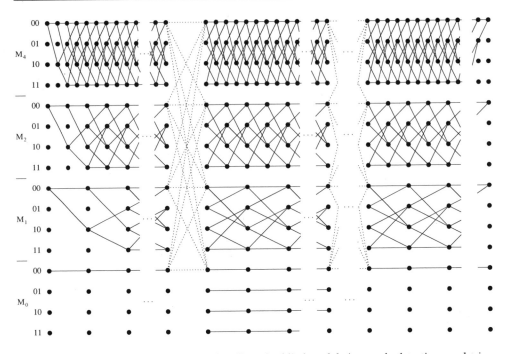

Figure 6.11: Schematic plot of the parallel trellises for blind modulation mode detection employing convolutional coding. In this example, a four-state 00-terminated convolutional encoder was assumed. The dotted lines indicate the intersub-band transitions for the 00 state, and are omitted for the other three states

OFDM symbol is varies in this adaptive scheme, and the channel encoder's block length is therefore not constant, for the sake of implementational convenience we have chosen a convolutional encoder at the transmitter. Once the modulation modes to be used are decided upon at the transmitter, the convolutional encoder is employed to generate a zero-terminated code word having the length of the OFDM symbol's capacity. This code word is modulated on the subcarriers according to the different modulation modes for the different sub-bands, and the OFDM symbol is transmitted over the channel.

At the receiver, each received data subcarrier is demodulated by all possible demodulators, and the resulting hard decision bits are fed into parallel trellises for Viterbi decoding. Figure 6.11 shows a schematic sketch of the resulting parallel trellis if 16-QAM (M_4), QPSK (M_2), BPSK (M_1), and "no transmission" (M_0) are employed, for a convolutional code having four states. Each sub-band in the adaptive scheme corresponds to a set of four parallel trellises, whose inputs are generated independently by the four demodulators of the legitimate modulation modes. The number of transitions in each of the trellises depends on the number of output bits received from the different demodulators, so that the 16-QAM (M_4) trellis contains four times as many transitions as the BPSK and "no transmission" trellises. Since in the case of "no transmission" no coded bits are transmitted, the state of the encoder does not change. Therefore, legitimate transitions for this case are only horizontal ones.

At sub-band boundaries, transitions are allowed between the same state of all the parallel

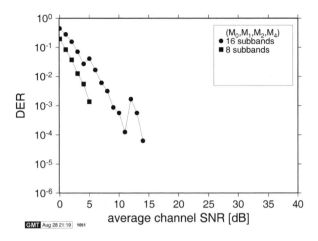

Figure 6.12: Blind modulation mode detection error ratio (DER) using the parallel trellis algorithm of Section 6.2.6.3 with a $K = 7$ convolutional code in an AWGN channel for a 512-subcarrier OFDM modem

trellises associated with the different modulation modes. This is not a transition due to a received bit, and therefore preserves the metric of the originating state. Note that in the figure only the possible allowed transitions for the state 00 are drawn; all other states originate the equivalent set of transitions. The initial state of the first sub-band is 00 for all modulation modes, and since the code is terminated in 00, the last sub-band's final states are 00.

The receiver's Viterbi decoder calculates the metrics for the transitions in the parallel trellises, and once all the data symbols have been processed, it traces back through the parallel trellis on the surviving path. This backtracing commences at the most likely 00 state at the end of the last sub-band. If no termination was used at the decoder, then the backtracing would start at the most likely of all the final states of the last block.

Figure 6.12 shows the modulation mode detection error ratio (DER) for the parallel trellis decoder in an AWGN channel for 16 and 8 sub-bands, if a convolutional code of constraint length 7 is used. Comparison with Figure 6.9 shows considerable improvements relative to the BER estimation-based blind detection scheme of Section 6.2.6.2, both for 16 and 8 sub-bands. Higher sub-band lengths improve the estimation accuracy by a greater degree than has been observed for the BER estimation algorithm of Figure 6.9. A DER of less than 10^{-5} was observed for AWGN SNR values of 6 dB and 15 dB in the 8 and 16 sub-band scenarios, respectively. The use of stronger codes could further improve the estimation accuracy, at the cost of higher complexity.

6.2.7 Sub-band Adaptive OFDM and Turbo Channel Coding

Adaptive modulation can reduce the BER to a level where channel decoders can perform well. Figure 6.13 shows both the uncoded and coded BER performance of a 512-subcarrier OFDM modem in the fading wideband channel of Figure 4.3, assuming perfect channel estimation. The channel coding employed in this set of experiments was a turbo coder [215] with a data block length of 1000 bits, employing a random interleaver and 8 decoder iterations. The log-

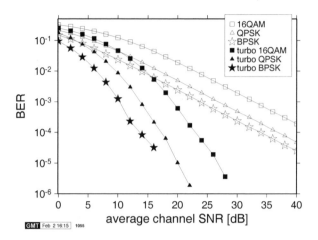

Figure 6.13: BER performance of the 512-subcarrier OFDM modem in the fading time dispersive channel of Figure 4.3 for both uncoded and half-rate turbo-coded transmission, using 8-iteration log-MAP turbo decoding, 1000-bit random interleaver, and a constraint length of 3

MAP decoding algorithm was used [216]. The constituent half-rate convolutional encoders were of constraint length 3, with octally represented generator polynomials of $(7, 5)$ [164]. It can be seen that the turbo decoder provides a considerable coding gain for the different fixed modulation schemes, with a BER of 10^{-4} for SNR values of 13.8 dB, 17.3 dB and 23.2 dB for BPSK, QPSK and 16-QAM transmission, respectively.

Figure 6.14 depicts the BER and throughput performance of the same decoder employed in conjunction with the adaptive OFDM modem for different adaptation algorithms. Figure 6.14(a) shows the performance for the "speech" system employing the switching levels listed in Table 6.1. As expected, the half-rate channel coding results in a halved throughput compared to the uncoded case, but offers low BER transmission over the channel of Figure 4.3 for SNR values of down to 0 dB, maintaining a BER below 10^{-6}.

Further tuning of the adaptation parameters can ensure a better average throughput, while retaining error-free data transmission. The switching level-based adaptation algorithm of Table 6.1 is difficult to control for arbitrary bit error rates, since the set of switching levels was determined by an optimisation process for uncoded transmission. Since the turbo codec has a non-linear BER versus SNR characteristic, direct switching level optimisation is an arduous task. The sub-band BER predictor of Section 6.2.4.2 is easier to adapt to a channel codec, and Figure 6.14(b) shows the performance for the same decoder, with the adaptation algorithm employing the BER prediction method having an upper BER bound of 1%. It can be seen that the less stringent uncoded BER constraints when compared to Figure 6.14(a) lead to a significantly higher throughput for low SNR values. The turbo-decoded data bits are error-free, hence a further increase in throughput is possible while maintaining a high degree of coded data integrity.

The second set of curves in Figure 6.14(b) show the system's performance, if an uncoded target BER of 10% is assumed. In this case, the turbo decoder's output BER is below 10^{-5}

for all the SNR values plotted, and shows a slow decrease for increasing values of SNR. The throughput of the system, however, exceeds 0.5 data bits per symbol for SNR values of more than 2 dB.

6.2.8 Effects of the Doppler Frequency

Since the adaptive OFDM modem employs the most recently received OFDM symbol in order to predict the frequency domain transfer function of the reverse channel for the next transmission, the quality of this prediction suffers from the time variance of the channel transfer function between the uplink and downlink time slots. We assume that the time delay between the uplink and downlink slots is the same as the delay between the downlink and uplink slots, and we refer to this time as the frame duration T_f. We normalise the maximal Doppler frequency f_d of the channel to the frame duration T_f, and define the frame normalised Doppler frequency F'_d as $F'_d = f_d \cdot T_f$. Figure 6.16 depicts the fixed switching level (see Table 6.1) modem's BER and throughput performance in bits per symbol (BPS) for values of F'_d between $7.41 \cdot 10^{-3}$ and $2.3712 \cdot 10^{-1}$. These values stem from the studied WATM system with a time slot duration of 2.67 μs and up/downlink delays of 1, 8, 16 and 32 time slots at a channel Doppler frequency of 2.78 kHz. As mentioned in Section 4.1.1.1, this corresponds to a system employing a carrier frequency of 60 GHz, a sampling rate of 225 Msamples/s and a vehicular velocity of 50 km/h or 13.$\bar{8}$ m/s.

Figure 6.16(a) shows the BER and BPS throughput of the studied modems in a framework with consecutive uplink and downlink time slots. This corresponds to $F'_d = 7.41 \cdot 10^{-3}$, while the target bit error rates for the speech and data system are met for all SNR values above 4 dB, and the BER performance is generally better than the target error rates. This was explained above with the conservative choice of modulation modes based on the most corrupted subcarrier in each sub-band, resulting in lower throughput and lower bit error rates for the switching level based sub-band adaptive modem.

Comparing Figure 6.16(a) with the other performance curves, it can be seen that the bit error rate performance for both the speech and the data system suffer from increasing decorrelation of the predicted and actual channel transfer function for increasing values of F'_d. In Figure 6.16(b) an 8 time slot delay was assumed between uplink and downlink time slots, which corresponds to $F'_d = 5.928 \cdot 10^{-2}$, and therefore the BER performance of the modem was significantly deteriorated. The "speech" system still maintains its target BER, but the "data" system delivers a BER of up to 10^{-3} for SNR values between 25 and 30 dB. It is interesting to observe that the delayed channel prediction mainly affects the higher-order modulation modes, which are employed more frequently at high SNR values. This explains the shape of the BER curve for the "data" system, which is rising from below 10^{-4} at 2 dB SNR up to 10^{-3} at 26 dB SNR. The average throughput of the modem is mainly determined by the statistics of the estimated channel transfer function at the receiver, and this is therefore not affected by the delay between the channel estimation and the packet transmission.

6.2.9 Channel Transfer Function Estimation

All the adaptive modems above rely on the estimate of the frequency domain channel transfer function, both for equalisation of the received symbols at the receiver, as well as for the modem mode adaptation of the next transmitted OFDM symbol. Figure 6.15 shows the

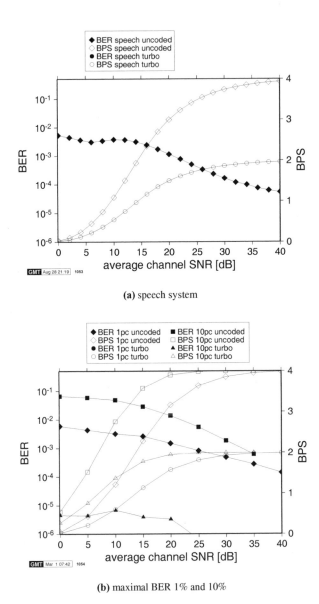

(a) speech system

(b) maximal BER 1% and 10%

Figure 6.14: BER and BPS throughput performance of 16 sub-band, 512 subcarrier adaptive turbo coded and uncoded OFDM modem employing (M_0, M_1, M_2, M_4) for (a) speech-type switching levels of Table 6.1 and (b) a maximal estimated sub-band BER of 1% and 10% over the channel of Figure 4.3. The turbo-coded transmission over the speech system and the 1% maximal BER system are error-free for all examined SNR values and therefore the corresponding BER curves are omitted from the graphs, hence the lack of black circles on (a) and (b)

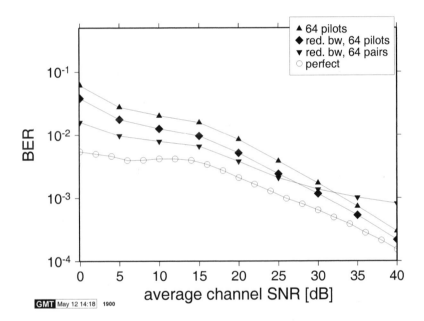

Figure 6.15: BER versus channel SNR performance for the 1% target BER adaptive 16 sub-band, 512 subcarrier OFDM modem employing pilot symbol assisted channel transfer function estimation over the channel of Figure 4.3

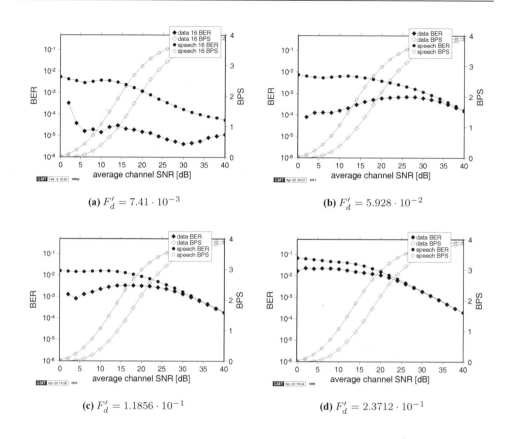

(a) $F'_d = 7.41 \cdot 10^{-3}$

(b) $F'_d = 5.928 \cdot 10^{-2}$

(c) $F'_d = 1.1856 \cdot 10^{-1}$

(d) $F'_d = 2.3712 \cdot 10^{-1}$

Figure 6.16: BER and BPS throughput performance of 16 sub-band, 512 subcarrier adaptive OFDM modem employing (M_0, M_1, M_2, M_4) for both data-type and speech-type switching levels with perfect modulation mode detection and different frame normalised Doppler frequencies F'_d over the channel of Figure 4.3. The triangular markers in (a) show the performance of a subcarrier-by-subcarrier adaptive modem using the data-type switching levels of Table 6.1 for comparison

BER versus SNR curves for the 1% target BER modem, as presented above, if pilot symbol-assisted channel estimation [174] is employed instead of the previously used delayed, but otherwise perfect, channel estimation.

Comparing the curves for perfect channel estimation and for the 64-pilot lowpass interpolation algorithm, it can be seen that the modem falls short of the target bit error rate of 1% for channel SNR values of up to 20 dB. More noise resilient channel estimation algorithms can improve the modem's performance. If the passband width of the interpolation lowpass filter (see Section 4.3.1.1) is halved, which is indicated in Figure 6.15 as the reduced bandwidth (red. bw.) scenario, then the BER gap between the perfect and the pilot symbol assisted channel estimation narrows, and a BER of 1% is achieved at a SNR of 15 dB. Additionally, employing pairs of pilots with the above bandwidth-limited interpolation scheme further improves the modem's performance, which results in BER values below 1% for SNR values above 5 dB. The averaging of the pilot pairs improves the noise resilience of the channel estimation, but introduces estimation errors for high SNR values. This can be observed in the residual BER in the figure.

Having studied a range of different AOFDM modems, let us now embark on a system design study in the context of an adaptive interactive speech system.

6.3 Adaptive OFDM Speech System[2]

6.3.1 Introduction

In this section we introduce a bidirectional high-quality audio communications system, which will be used to highlight the systems aspects of adaptive OFDM transmissions over time dispersive channels. Specifically, the channel-coded adaptive transmission characteristics and a potential application for joint adaptation of modulation, channel coding and source coding are studied.

The basic principle of adaptive modulation is to react to the anticipated channel capacity for the next OFDM symbol transmission burst, by employing modulation modes of different robustness to channel impairments and of different data throughput. The trade-off between data throughput and integrity can be adapted to different system environments. For data transmission systems, which are not dependent on a fixed data rate and do not require low transmission delays, variable-throughput adaptive schemes can be devised that operate efficiently with powerful error correction coders, such as long block length turbo codes [218]. Real-time audio or video communications employing source codecs, which allow variable bit rates, can also be used in conjunction with variable-rate adaptive schemes, but in this case block-based error correction coders cannot be readily employed.

Fixed-rate adaptive OFDM systems, which sacrifice a guaranteed BER performance for the sake of a fixed data throughput, are more readily integrated into interactive communications systems, and can coexist with long block-based channel coders in real-time applications.

For these investigations, we propose a hybrid adaptive OFDM scheme, based on a multi-mode constant throughput algorithm, consisting of two adaptation loops: an inner constant throughput algorithm, having a bit rate consistent with the source and channel coders, and an outer mode switching control loop, which selects the target bit rate of the whole system from a set of distinct operating modes. These issues will become more explicit during our further discourse.

6.3.2 System Overview

The structure of the studied adaptive OFDM modem is depicted schematically in Figure 6.17. The top half of the diagram is the transmitter chain, which consists of the source and channel coders, a channel interleaver decorrelating the channel's frequency domain fading, an adaptive OFDM modulator, a multiplexer adding signalling information to the transmit data, and an IFFT/RF OFDM block. The receiver seen at the bottom of the figure consists of a RF/FFT

[2]T. Keller, M. Münster, L. Hanzo: "A Turbo-coded Burst-by-burst Adaptive Wideband Speech Transceiver", ©IEEE JSAC, 2000, November 2000, Vol. 18, No. 11, pp. 2363-2372 [217]. Details concerning the source codec and its performance are discussed in the above publication and in [6].

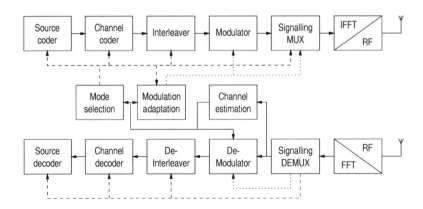

Figure 6.17: Schematic model of the multimode adaptive OFDM system

OFDM receiver, a demultiplexer extracting the signalling information, an adaptive demodulator, a de-interleaver/channel decoder and the source decoder. The parameter adaptation linking the receiver and transmitter chain consists of a channel estimator, and the throughput mode selection as well as the modulation adaptation blocks.

The open-loop control structure of the adaptation algorithms can be seen in the figure: the receiver's operation is controlled by the signalling information that is contained in the received OFDM symbol, while the channel quality information generated by the receiver is employed to determine the remote transmitter's matching parameter set by the modulation adaptation algorithms. The two distinct adaptation loops distinguished by the dotted and dashed lines are the inner and outer adaptation loops, respectively. The outer adaptation loop controls the overall throughput of the system, which is chosen from a finite set of predefined modes, so that a fixed delay decoding of the received OFDM data packets becomes possible. This outer loop controls the block length of the channel encoder and interleaver, and the target throughput of the inner adaptation loop. The operation of the adaptive modulator, controlled by the inner loop, is transparent to the rest of the system. The operation of the adaptation loops is described in more detail below.

6.3.2.1   System Parameters

The transmission parameters have been adopted from the TDD mode of the UMTS system of Section 4.1.3, with a carrier frequency of 1.9 GHz, a time frame and time slot duration of 4.615 ms and 122 μs, respectively. The sampling rate is assumed to be 3.78 MHz, leading to a 1024-subcarrier OFDM symbol with a cyclic extension of 64 samples in each time slot. For spectral shaping of the OFDM signal, there are a total of 206 virtual subcarriers at the bandwidth boundaries.

The 7 kHz bandwidth PictureTel audio codec [6][3] has been chosen for this system because of its good audio quality, robustness to packet dropping and adjustable bit rate. The channel encoder/interleaver combination is constituted by a convolutional turbo codec [215] employing block turbo interleavers in conjunction with a subsequent pseudorandom channel

---

[3]See http://www.picturetel.com

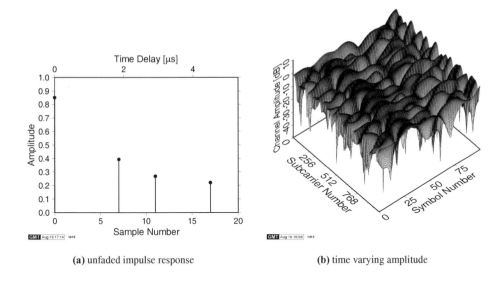

(a) unfaded impulse response   (b) time varying amplitude

Figure 6.18: Channel for PictureTel experiments: (a) unfaded channel impulse response (b) time-varying channel amplitude for 100 OFDM symbols

interleaver. The constituent half-rate recursive systematic convolutional (RSC) encoders are of constraint length 3, with octal generator polynomials of $(7, 5)$ [164]. At the decoder, 8 iterations are performed, utilising the so-called maximum *a posteriori* (MAP) [216] algorithm and log-likelihood ratio soft inputs from the demodulator.

The channel model consists of a four-path COST 207 typical urban impulse response [184], where each impulse is subjected to independent Rayleigh fading with a normalised Doppler frequency of $2.25 \cdot 10^{-6}$, corresponding to a pedestrian scenario with a walking speed of 3 mph.

The unfaded impulse response and the time- and frequency-varying amplitude of the channel transfer function are depicted in Figure 6.18.

6.3.3 Constant Throughput Adaptive Modulation

The constant throughput adaptive algorithm attempts to allocate a given number of bits for transmission in subcarriers exhibiting a low BER, while the use of high BER subcarriers is minimised. We employ the open-loop adaptive regime of Figure 6.1, basing the decision concerning the next transmitted OFDM symbol's modulation scheme allocation on the channel estimation gained at the reception of the most recently received OFDM symbol by the local station. Sub-band adaptive modulation [219] – where the modulation scheme is adapted not on a subcarrier-by-subcarrier basis, but for sub-bands of adjacent subcarriers – is employed in order to simplify the signalling requirements. The adaptation algorithm was highlighted in Section 6.2.5. For these investigations we employed 32 sub-bands of 32 subcarriers in each OFDM symbol. Perfect channel estimation and signalling were used.

6.3.3.1   Constant-Rate BER Performance

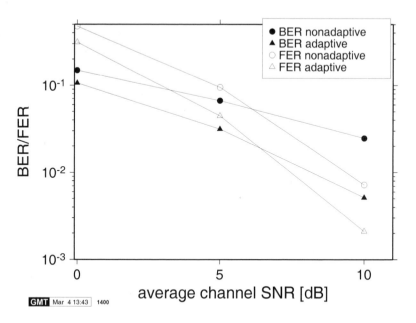

Figure 6.19: BER and FER performance for the fixed-throughput adaptive and non-adaptive OFDM
modems in the fading time dispersive channel of Section 4.1.3 for a block length of 578
coded bits

Figure 6.19 characterises the fixed-throughput adaptive modulation scheme's performance under the channel conditions characterised above, for a block length of 578 coded
bits. As a comparison, the BER curve of a fixed BPSK modem transmitting the same number of bits in the same channel, employing 578 out of 1024 subcarriers, is also depicted. The
number of useful audio bits per OFDM symbol was based on a 200-bit target data throughput,
which corresponds to a 10 kbps data rate, padded with 89 bits, which can contain a checksum for error detection and high-level signalling information. Furthermore, half-rate channel
coding was used.

The BER plotted in the figure is the hard decision-based bit error rate at the receiver
before channel decoding. It can be seen that the adaptive modulation scheme yields a significantly improved performance, which is also reflected in the frame error rate (FER). This FER
approximates the probability of a decoded block containing errors, in which case it is unus-
able for the audio source decoder and hence it is dropped. This error event can be detected
by using the checksum in the OFDM data symbol.

As an example, the modulation mode allocation for the 578 data bit adaptive modem at an
average channel SNR of 5 dB is given in Figure 6.20(a) for 100 consecutive OFDM symbols.
The unused sub-bands with indexes 15 and 16 contain the virtual carriers, and therefore do
not transmit any data. It can be seen that the constant throughput adaptation algorithm of
Section 6.2.5 allocates data to the higher quality subcarriers on a symbol-by-symbol basis,
while keeping the total number of bits per OFDM symbol constant. As a comparison, Figure

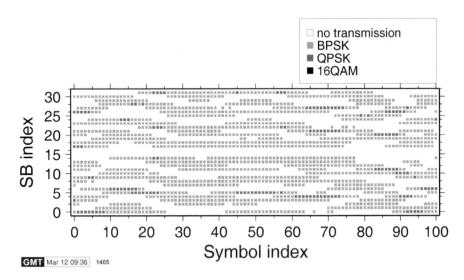

(a) 578 data bits per OFDM symbol

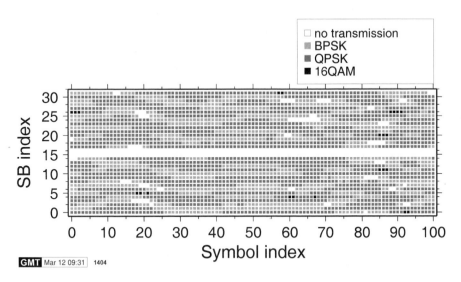

(b) 1458 data bits per OFDM symbol

Figure 6.20: Overview of modulation mode allocation for fixed-throughput adaptive modems over the fading time dispersive channel of Figure 6.18(b) at 5 dB average channel SNR

6.20(b) shows the equivalent overview of the modulation modes employed for a fixed bit rate of 1458 bits per OFDM symbol. It can be seen that in order to meet this increased throughput target, hardly any sub-bands are in "no transmission" mode, and overall higher-order modulation schemes have to be employed.

6.3.4   Multimode Adaptation

While the fixed-throughput adaptive algorithm described above copes with the frequency domain fading of the channel, there is a medium-term variation of the overall channel capacity due to time domain channel quality fluctuations as indicated in Figure 6.18(b). While it is not straightforward to employ powerful block-based channel coding schemes, such as turbo coding, in variable-throughput adaptive OFDM schemes for real-time applications like voice or video telephony, a multimode adaptive system can be designed that allows us to switch between a set of different source and channel codecs as well as transmission parameters, depending on the overall channel quality. We have investigated the use of the estimated overall BER at the output of the receiver, which is the sum of all the $e(j, s_j)$ quantities of Equation 6.1 after adaptation. On the basis of this expected bit error rate at the input of the channel decoder, the probability of a frame error (FER) must be estimated and compared with the estimated FER of the other modem modes. Then the mode having the highest throughput exhibiting an estimated FER of less than 10^{-6} – or alternatively the mode exhibiting the lowest FER – is selected and the source encoder, the channel encoder and the adaptive modem are set up accordingly.

We have defined four different operating modes, which correspond to unprotected audio data rates of 10, 16, 24, and 32 kbps at the source codec's interface. With half-rate channel coding and allowing for checksum and signalling overhead, the number of transmitted coded bits per OFDM symbol was 578, 722, 1058 and 1458 for the four modes, respectively.

6.3.4.1   Mode Switching

Figure 6.21 shows the *observed* FER for all four modes versus the unprotected BER that was *predicted* at the transmitter. The predicted unprotected BER was discretised into intervals of 1%, and the channel-coded FER was evaluated over these BER intervals. It can be seen from the figure that for estimated protected BER values below 5% no frame errors were observed for any of the modes. For higher estimated unprotected BER values, the higher throughput modes exhibited a lower FER than the lower throughput modes, which was consistent with the turbo coder's performance increase for longer block lengths. A FER of 1% was observed for a 7% predicted unprotected error rate for the 10 kbps mode, and BER values of 8% to 9% were allowed for the longer blocks, while maintaining a FER of less than 1%

For this experiment, we assumed the best-case scenario of using the actual measured FER statistics of Figure 6.21 for the mode switching algorithm rather than estimating the FER on the basis of the estimated uncoded BER. In this case, the previously observed FER corresponding to the predicted overall BER values for the different modes were compared, and the mode having the lowest FER was chosen for transmission. The mode switching sequence for the first 500 OFDM symbols at 5 dB channel SNR over the channel of Figure 6.18(b) is depicted in Figure 6.22. It can be seen that in this segment of the sequence 32 kbps transmission is the most frequently employed mode, followed by the 10 kbps mode. The intermediate

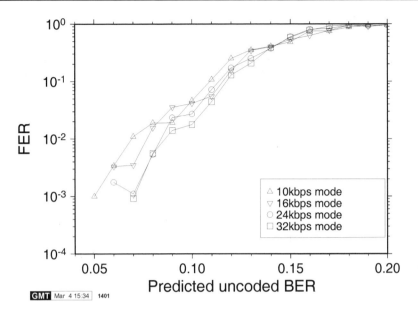

Figure 6.21: Frame error rate versus the predicted unprotected BER for 10 kbps, 16 kbps, 24 kbps and 32 kbps modes

modes are mostly transitory, as the improving or deteriorating channel conditions make it necessary to switch between the 10 kbps and 32 kbps modes. This behaviour is consistent with Table 6.2, for the Switch-I scheme.

6.3.5 Simulation Results

The comparison between the different adaptive schemes will be based on a channel SNR of 5 dB over the channel of Figure 6.18(b), since the audio codec's performance is unacceptable for SNR values around 0 dB, and as the adaptive modulation is most effective for channel SNR values below 10 dB.

6.3.5.1 Frame Error Results

The audio experiments [217] have shown that the audio quality is acceptable for frame dropping rates of about 5%, and that the perceived audio quality increases with increasing throughput. Table 6.2 gives an overview of the frame error rates and mode switching statistics of the system for a channel SNR of 5 dB over the channel of Figure 6.18(b). It can be seen that for the fixed modes the FER increases with the throughput, from 4.45% in the 10 kbps mode up to 18.65% for the 32 kbps mode. This is because the turbo codec performance improves for longer interleavers, the OFDM symbol had to be loaded with more bits, hence there was a higher unprotected BER. The time-variant bit rate mode-switching schemes, referred to as Switch-I and Switch-II for the four- and three-mode switching regimes used, deliver frame dropping rates of 4.44% and 5.58%, respectively. Both these FER values are acceptable for

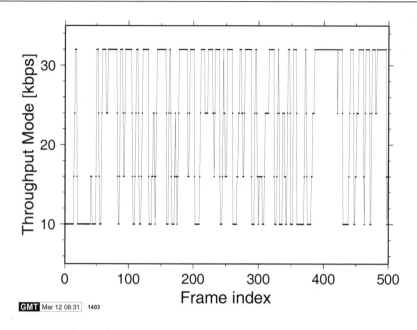

Figure 6.22: Mode switching pattern at 5 dB channel SNR over the channel of Figure 6.18(b)

Scheme	FER [%]	Rate-10kbps [%]	Rate-16kbps [%]	Rate-24kbps [%]	Rate-32kbps [%]
Fixed-10kbps	4.45	95.55	0.0	0.0	0.0
Fixed-16kbps	5.58	0.0	94.42	0.0	0.0
Fixed-24kbps	10.28	0.0	0.0	89.72	0.0
Fixed-32kbps	18.65	0.0	0.0	0.0	81.35
Switch-I	4.44	21.87	13.90	11.59	48.20
Switch-II	5.58	0.0	34.63	11.59	48.20

Table 6.2: FER and relative usage of different bit rates in the fixed bit rate schemes and the variable schemes Switch-I and Switch-II (successfully transmitted frames) for a channel SNR of 5 dB over the channel of Figure 6.18(b)

the audio transmission. It can be seen that upon incorporating the 10 kbps mode in the switching regime Switch-I of Table 6.2, the overall FER is lowered only by an insignificant amount, while the associated average throughput is reduced considerably.

6.3.5.2 Audio Segmental SNR

Figure 6.23 displays the cumulative density function (CDF) of the segmental SNR (SEGSNR) [164] obtained from the reconstructed signal of an audio test sound for all the modes of Table 6.2 discussed above at a channel SNR of 5 dB over the channel of Figure 6.18(b).

Focusing our attention on the figure, we can draw a whole range of interesting conclusions. As expected, for any given SEGSNR it is desirable to maintain as low a proportion as possible of the audio frames' SEGSNRs below a given abscissa value. Hence we conclude that the best SEGSNR CDF was attributable to the Switch-II scheme, while the worst performance was observed for the fixed 10 kbps scheme. Above a SEGSNR of 15 dB the CDFs of the fixed 16, 24 and 32 kbps modes follow our expectations. Viewing matters from a different perspective, the Switch-II scheme exhibits a SEGSNR of less than 20 dB with a probability of 0.8, compared to 0.95 for the fixed 10 kbps scheme.

Before concluding we also note that the CDFs do not have a smoothly tapered tail, since for the erroneous audio frames a SEGSNR of 0 dB was registered. This results in the step-function-like behaviour associated with the discontinuities corresponding to the FER values in the FER column of Table 6.2.

6.4 Pre-equalisation

We have seen above how the receiver's estimate of the channel transfer function can be used by the transmitter in order to dramatically improve the performance of an OFDM system by adapting the subcarrier modulation modes to the channel conditions. For sub-channels exhibiting a low signal-to-noise ratio, robust modulation modes were used, while for sub-carriers having a high SNR, high throughput multi-level modulation modes can be used. An alternative approach to combating the frequency selective channel behaviour is to apply pre-equalisation to the OFDM symbol prior to transmission on the basis of the anticipated channel transfer function. The optimum scheme for the power allocation to the subcarriers is Shannon's water-pouring solution [220]. We will investigate a range of related topics in this section.

6.4.1 Motivation

As discussed above, the received data symbol R_n of subcarrier n over a stationary time dispersive channel can be characterised by:

$$R_n = S_n \cdot H_n + n_n,$$

where S_n is the transmitted data symbol, H_n is the channel transfer function of subcarrier n, and n_n is a noise sample.

The frequency domain equalisation at the receiver, which is necessary for non-differential detection of the data symbols, corrects the phase and amplitude of the received data symbols using the estimate of the channel transfer function $\hat{H}_n$ as follows:

$$R'_n = R_n/\hat{H}_n = S_n \cdot H_n/\hat{H}_n + n_n/\hat{H}_n.$$

If the estimate $\hat{H}_n$ is accurate, this operation defades the constellation points before decision. However, upon defading, the noise sample n_n is amplified by the same amount as the signal, therefore preserving the SNR of the received sample.

Pre-equalisation for the OFDM modem operates by scaling the data symbol of subcarrier n, S_n, by a predistortion function E_n, computed from the inverse of the anticipated channel transfer function, prior to transmission. At the receiver, no equalisation is performed, hence

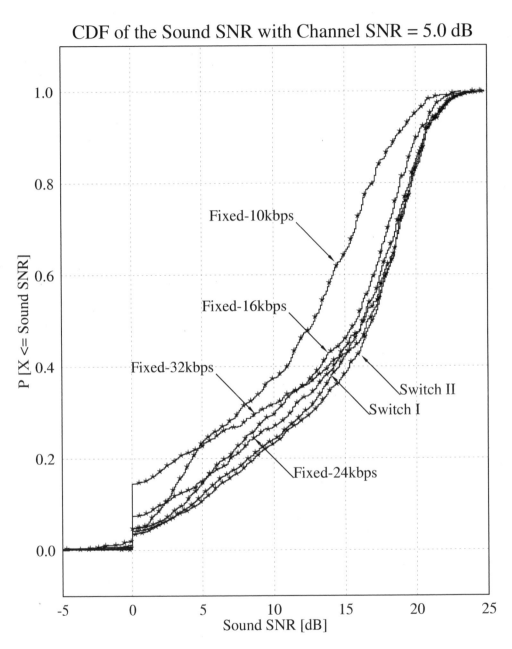

Figure 6.23: Typical CDF of the segmental SNR of a reconstructed audio signal transmitted over the fading time dispersive channel of Section 4.1.3 at a channel SNR of 5 dB (from [217])

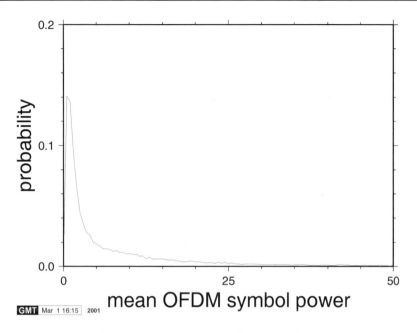

Figure 6.24: OFDM symbol energy histogram for 512-subcarrier 16-QAM with full channel inversion over the short WATM channel of Figure 4.3

the received symbols can be expressed as:

$$R_n = S_n \cdot E_n \cdot H_n + n_n.$$

Since no equalisation is performed, there is no noise amplification at the receiver. Similarly to the adaptive modulation techniques illustrated above, pre-equalisation is only applicable to a duplex link, since the transmitted signal is adapted to the specific channel conditions perceived by the receiver. As in other adaptive schemes, the transmitter needs an estimate of the current frequency domain channel transfer function, which can be obtained from the received signal in the reverse link, as seen in Figure 6.1.

The simplest choice of the pre-equalisation transfer function E_n is the inverse of the estimated frequency domain channel transfer function, $E_n = 1/\hat{H}_n$. If the estimation of the channel transfer function is accurate, then perfect channel inversion would result in an AWGN-like channel perceived at the receiver, since the anticipated time- and frequency-dependent behaviour of the channel is precompensated at the transmitter. The BER performance of such a system, accordingly, is identical to that of the equivalent modem in an AWGN channel with respect to the received signal power.

Since the pre-equalisation algorithm amplifies the power in each subcarrier by the corresponding estimate of the channel transfer function, the transmitter's output power fluctuates in an inverse fashion with respect to time variant channel. The fades in the frequency domain channel transfer function can be deep, hence the transmit power in the corresponding subcarriers may be high. Figure 6.24 shows the histogram of the total OFDM symbol energy

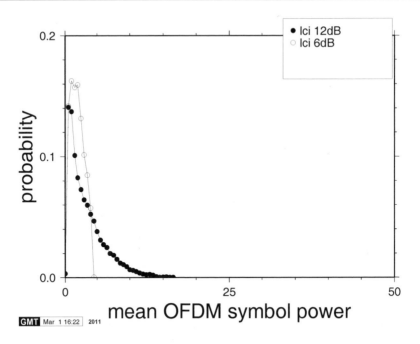

Figure 6.25: OFDM symbol energy histogram for 512-subcarrier 16-QAM transmission with limited
dynamic range channel inversion (lci) over the WATM channel of Figure 4.3.

at the transmitter's output for the short WLAN channel of Figure 4.3 in conjunction with
"full channel inversion", normalised to the fixed average output energy. It can be seen that
the OFDM symbol energy fluctuates widely, with observed peak values in excess of 55. The
long-term mean symbol energy was measured to be 22.9, which corresponds to an average
output power increase of 13.6 dB. This naturally would impose unacceptable constraints on
the required perfectly linear dynamic range of the power amplifier. Hence in practice only
limited dynamic scenarios can be considered.

Explicitly, in order to limit the associated transmit power fluctuations, the dynamic range
of the pre-equalisation algorithm can be limited to a value l, so that the following relations
apply:

$$E_n \;=\; a_n \cdot e^{-j\phi_n}, \text{with} \tag{6.3}$$

$$\phi_n \;=\; \angle \hat{H}_n \quad \text{and} \tag{6.4}$$

$$a_n \;=\; \begin{cases} \left|\hat{H}_n\right| & \text{for } \left|\hat{H}_n\right| \leq l \\ l & \text{otherwise.} \end{cases} \tag{6.5}$$

Limiting the values of E_n to the value of l does not affect the phase of the channel pre-
equalisation. Depending on the modulation mode employed for transmission, reception of
the symbols affected by the amplitude limitation is still possible, perhaps for phase shift keying. Multilevel modulation modes exploiting the received symbol's amplitude will be affected
by the imperfect pre-equalisation. The associated mean OFDM symbol power histogram is

shown in Figure 6.25. Given, for example, a maximum allowed amplification factor of 6 and 12 dB, the normalised transmitted OFDM symbol power in the figure should be limited to 4 and 16, respectively. However, in practice even higher values may be observed, which is due to the OFDM symbol's energy fluctuation as a function of the specific data sequence, if multilevel modulation modes are used. In order to circumvent the above peak-to-mean envelope fluctuation problems, it is in practical terms more attractive to combine pre-equalisation with sub-band blocking, which is the topic of the next section.

6.4.2   Pre-equalisation with Sub-band Blocking

Limiting the maximal amplification of the subcarriers leads to a reduced BER performance compared to the full channel inversion, but the system performance can be improved by identifying the subcarriers that cannot be fully pre-equalised and by disabling subsequent transmission in these subcarriers. This "blocking" of the transmission in certain subcarriers can be seen as adaptive modulation with two modulation modes, which introduces the problem of modulation mode signalling. As was discussed in the context of Figure 6.1 for the adaptive modulation modems above, this signalling task can be solved in different ways, namely by blind detection of blocked subcarriers, or by transmitting explicit signalling information contained in the data block. We have seen above that employing sub-band adaptivity – rather than subcarrier-by-subcarrier adaptivity – simplifies both the modem mode detection as well as the mode signalling, at the expense of a lower system throughput. In order to keep the system's complexity low and to allow for simple modem mode signalling or blind detection, we will assume a 16 sub-band adaptive scheme here.

Analogously to the adaptive modulation schemes above, the transmitter decides for all subcarriers in each sub-band, whether to transmit data or not. If pre-equalisation is possible under the power constraints, then the subcarriers are modulated with the pre-equalised data symbols. The information on whether or not a sub-band is used for transmission, and this is signalled to the receiver.

Since no attempt is made to transmit in the sub-bands that cannot be power-efficiently pre-equalised, the power conserved in the blank subcarriers can be used to "boost" the data-bearing sub-bands. This scheme allows for a more flexible pre-equalisation algorithm than the fixed threshold method described above; here is a summary:

(1) Calculate the necessary transmit power p_n for each sub-band n, assuming perfect pre-equalisation.

(2) Sort sub-bands according to their required transmit power p_n.

(3) Select sub-band n with the lowest power p_n, and add p_n to the total transmit power. Repeat this procedure with the next lowest power, until no further sub-bands can be added without the total power $\sum p_j$ exceeding power limit l.

Figure 6.26 depicts the 16-QAM BER performance over the short WATM channel of Figure 4.3. The BER floor stems from the channel's time-variant nature, since there is a delay between the channel estimation instant and the instant of transmission. The average throughput figures for the 6 dB and 12 dB symbol energy limits are 3.54 and 3.92 bits per data symbol, respectively. It can be noted that the BER floor is lower for $l = 6$ dB than for $l = 12$ dB. This

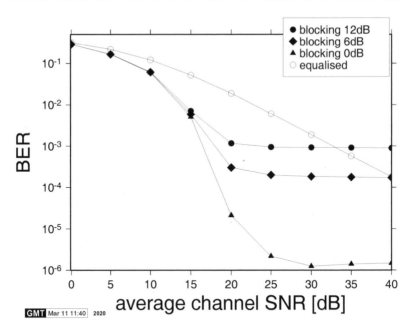

Figure 6.26: BER performance of the 512-subcarrier 16-QAM OFDM modem over the fading short
WATM channel of Figure 4.3 employing 16 sub-band pre-equalisation with blocking and
a delay of 1 time slot between the instants of perfect channel estimation and reception.
Note that the transmit power is not shown in this figure

is because the effects of the channel variation due to the delay between the instants of channel estimation and reception in the faded subcarriers on the equalisation function are more
dramatic than in the higher quality subcarriers. The lower the total symbol energy limit l, the
smaller the number of low quality subcarriers used for transmission. If the symbol energy
is limited to 0 dB, then the BER floor drops to $1.5 \cdot 10^{-6}$ at the expense of the throughput,
which attains 2.5 BPS. Figure 6.27 depicts the mean OFDM symbol energy histogram for this
scenario. It can be seen that, compared with the limited channel inversion scheme of Figure
6.25, the allowable symbol energy is more efficiently allocated, with a higher probability of
high energy OFDM symbols. This is the result of the flexible reallocation of energy from
blocked sub-bands, instead of limiting the output power on a subcarrier-by-subcarrier basis.

6.4.3 Adaptive Modulation with Spectral Predistortion

The pre-equalisation algorithms discussed above invert the channel's anticipated transfer
function, in order to transform the resulting channel into a Gaussian-like non-fading channel, whose SNR is dependent only on the path loss. Sub-band blocking has been introduced
above, in order to limit the transmitter's output power, while maintaining the near-constant
SNR across the used subcarriers. The pre-equalisation algorithms discussed above do not
cancel out the channel's path loss, but rely on the receiver's gain control algorithm to automatically account for the channel's average path loss.

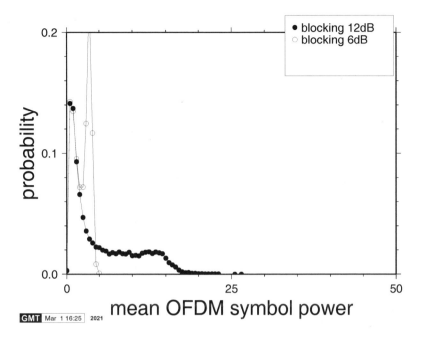

Figure 6.27: OFDM symbol energy histogram for 512 subcarrier, 16 sub-band pre-equalisation with blocking over the short WATM channel of Figure 4.3 using 16-QAM. The corresponding BER curves are given in Figure 6.26

We have already shown that maintaining Gaussian channel characteristics is not the most efficient way of exploiting the channel's time variant capacity. If maintaining a constant data throughput is not required by the rest of the communications system, then a fixed BER scheme in conjunction with error correction coding can assist in maximising the system's throughput. The results presented for the target BER adaptive modulation scheme in Figure 6.14(b) showed that, for the particular turbo coding scheme used, an uncoded BER of 1% resulted in error-free channel coded data transmission, and that for an uncoded target BER of 10% the turbo decoded data BER was below 10^{-5}. We have seen that it is impossible to exactly reach the anticipated uncoded target BER with the adaptive modulation algorithm, since the adaptation algorithm operates in discrete steps between modulation modes.

Combining the target BER adaptive modulation scheme and spectral pre-distortion allows the transmitter to react to the channel's time and frequency variant nature, in order to fine-tune the behaviour of the adaptive modem in fading channels. It also allows the transmitter to invest the energy that is not used in "no transmission" sub-bands into the other sub-bands without affecting the equalisation at the receiver.

The combined algorithm for adaptive modulation with spectral predistortion described here does not intend to invert the channel's transfer function across the OFDM symbol's range of subcarriers, it is therefore not a pure pre-equalisation algorithm. Instead, the aim is to transmit a sub-band's data symbols at a power level which ensures a given target SNR at the receiver, that is constant for all subcarriers in the sub-band, which in turn results in the

Target BER	10^{-4}	1%	10%
SNR(BPSK)[dB]	8.4	4.33	-0.85
SNR(QPSK)[dB]	11.42	7.34	2.16
SNR(16QAM)[dB]	18.23	13.91	7.91

Table 6.3: Required target SNR levels for 1% and 10% target BER for the different modulation schemes over an AWGN channel

required BER. Clearly, the receiver has to anticipate the different relative power levels for the different modulation modes, so that error-free demodulation of the multilevel modulation modes employed can be ensured.

The joint adaptation algorithm requires the estimates of the noise floor level at the receiver as well as the channel transfer function, which includes the path loss. On the basis of these values, the necessary amplitude of E_n required to transmit a data symbol over the subcarrier n for a given received SNR of γ_n can be calculated as follows:

$$|E_n| = \frac{\sqrt{N_0 \cdot \gamma_n}}{\left|\hat{H}_n\right|},$$

where N_0 is the noise floor at the receiver. The phase of E_n is used for the pre-equalisation, and hence:

$$\angle E_n = -\angle \hat{H}_n.$$

The target SNR, of subcarrier n, γ_n, is dependent on the modulation mode that is signalled over the subcarrier, and determines the system's target BER. We have identified three sets of target SNR values for the modulation modes, with uncoded target BER values of 1% and 10% for use in conjunction with channel coders, as well as 10^{-4} for transmission without channel coding. Table 6.3 gives an overview of these levels, which have been read from the BER performance curves of the different modulation modes in a Gaussian channel.

Figure 6.28 shows the performance of the joint predistortion and adaptive modulation algorithm over the fading time-dispersive short WATM channel of Figure 4.3 for the set of different target BER values of Table 6.3, as well as the comparison curves of the perfectly equalised 16-QAM modem under the same channel conditions. It can be seen that the BER achieved by the system is close to the BER targets. Specifically, for a target BER of 10%, no perceptible deviation from the target has been recorded, while for the lower BER targets the deviations increase for higher channel SNRs. For a target BER of 1%, the highest measured deviation is at the SNR of 40 dB, where the recorded BER is 1.36%. For the target BER of 10^{-4}, the BER deviation is small at 0 dB SNR, but at an SNR of 40 dB the experimental BER is $2.2 \cdot 10^{-3}$. This increase of the BER with increasing SNR is due to the rapid channel variations in the deeply faded subcarriers, which are increasingly used at higher SNR values. The light grey curve in the figure denotes the system's performance if no delay is present between the channel estimation and the transmission. In this case, the simulated BER shows only very little deviation from the target BER value. This is consistent with the behaviour of the full channel inversion pre-equalising modem.

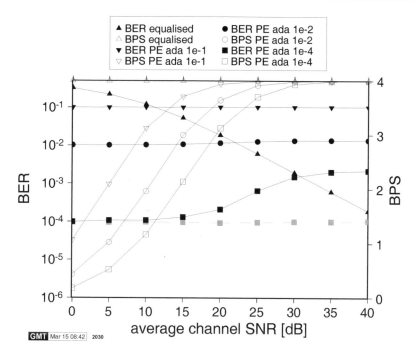

Figure 6.28: BER performance and BPS throughput of the 512 subcarrier, 16 sub-band adaptive OFDM
modem with spectral predistortion over the Rayleigh fading time dispersive short WATM
channel of Figure 4.3, and that of the perfectly equalised 16-QAM modem. The light grey
BER curve gives the performance of the adaptive modem for a target BER of 10^{-4} with
no delay between channel estimation and transmission, while the other results assume 1
time slot delay between uplink and downlink.

6.5   Comparison of the Adaptive Techniques

Figure 6.29 compares the different adaptive modulation schemes discussed in this chapter.
The comparison graph is split into two sets of curves, depicting the achievable data throughput for a data BER of 10^{-4} highlighted for the fixed throughput systems in Figure 6.29(a),
and for the time-variant throughput systems in Figure 6.29(b).

The fixed throughput systems, highlighted in black in Figure 6.29(a), comprise the non-
adaptive BPSK, QPSK and 16-QAM modems, as well as the fixed-throughput adaptive
scheme, both for coded and uncoded applications. The non-adaptive modems' performance
is marked on the graph as diamonds, and it can be seen that the uncoded fixed schemes require the highest channel SNR of all examined transmission methods to achieve a data BER
of 10^{-4}. Channel coding employing the advocated turbo coding schemes dramatically improves the SNR requirements, at the expense of half the data throughput. The uncoded fixed-
throughput (FT) adaptive scheme, marked by filled triangles, yields consistently worse data
throughput than the coded (C-) fixed modulation schemes C-BPSK, C-QPSK and C-16QAM,
with its throughput being about half the coded fixed scheme's at the same SNR values. The
coded FT (C-FT) adaptive system, however, delivers very similar throughput to the C-BPSK

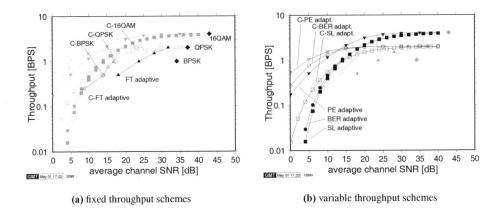

(a) fixed throughput schemes   **(b)** variable throughput schemes

Figure 6.29: BPS throughput versus average channel SNR for non-adaptive and adaptive modulation as well as for pre-equalised adaptive techniques, for a data bit error rate of 10^{-4}. Note that for the coded schemes the achieved BER values are lower than 10^{-4}. (a) Fixed-throughput systems: coded (C-) and uncoded BPSK, QPSK, 16-QAM, and fixed-throughput (FT) adaptive modulation. (b) Variable-throughput systems: coded (C-) and uncoded switching level adaptive (SL), target BER adaptive (BER) and pre-equalised adaptive (PE) systems. Note that the separately plotted variable-throughput graph also shows the light grey benchmark curves of the complementary fixed-rate schemes and vice versa

and C-QPSK transmission, and can deliver a BER of 10^{-4} for SNR values down to about 9 dB.

The variable throughput schemes, highlighted in Figure 6.29(b), outperform the comparable fixed throughput algorithms. For high SNR values, all uncoded schemes' performance curves converge to a throughput of 4 bits/symbol, which is equivalent to 16-QAM transmission. The coded schemes reach a maximal throughput of 2 bits/symbol. Of the uncoded schemes, the "data" switching level (SL) and target BER adaptive modems deliver a very similar BPS performance, with the target-BER scheme exhibiting slightly better throughput than the SL adaptive modem. The adaptive modem employing pre-equalisation (PE) significantly outperforms the other uncoded adaptive schemes and offers a throughput of 0.18 BPS at an SNR of 0 dB.

The coded transmission schemes suffer from limited throughput at high SNR values, since the half-rate channel coding limits the data throughput to 2 BPS. For low SNR values, however, the coded schemes offer better performance than the uncoded schemes, with the exception of the "speech" SL adaptive coded scheme, which is outperformed by the uncoded PE adaptive modem. The poor performance of the coded SL scheme can be explained by the lower uncoded target BER of the "speech" scenario, which was 1%, in contrast to the 10% uncoded target BER for the coded BER and PE adaptive schemes. The coded PE adaptive modem outperforms the target-BER adaptive scheme, thanks to its more accurate control of the uncoded BER, leading to a higher throughput for low SNR values.

It is interesting to observe that for the given set of four modulation modes, the uncoded PE adaptive scheme is close in performance to the coded adaptive schemes, and that for SNR values of more than 14 dB, it outperforms all other studied schemes. It is clear, however, that the coded schemes would benefit from higher-order modulation modes, which would allow these modems to increase the data throughput further when the channel conditions allow. Before concluding this chapter, in the next section let us consider the generic problem of optimum power and bit allocation in the context of uncoded OFDM systems.

6.6 A Fast Algorithm for Near-Optimum Power and Bit Allocation in OFDM Systems [221]

6.6.1 State of the Art

In this section the problem of efficient OFDM symbol-by-symbol based power and bit allocation is analysed in the context of highly dispersive time-variant channels. A range of solutions published in the literature is reviewed briefly and Piazzo's [221] computationally efficient algorithm is discussed in somewhat more detail.

When OFDM is invoked over highly frequency selective channels, each subcarrier can be allocated a different transmit power and a different modulation mode. This OFDM symbol-by-symbol based "resource" allocation can be optimised with the aid of an algorithm which, if the channel is time variant, has to be repeated on an OFDM symbol-by-symbol basis. Some of the existing algorithms [19, 222] are mainly of theoretical interest due to their high complexity. Amongst the practical algorithms [205, 206, 221, 223, 224] the Hughes-Hartog algorithm (HHA) [224–226] is perhaps best known, but its complexity is somewhat high, especially for real-time OFDM symbol-by-symbol based applications at high bit rates. Hence the HHA has stimulated extensive research for computationally more efficient algorithms [205, 206, 221, 223, 224]. The most efficient appears to be that of Lai *et al.* [224], which is a fast version of the HHA and that of Piazzo [221].

6.6.2 Problem Description

Piazzo [221] considered an OFDM system using N subcarriers, each employing a potentially different modulation mode and transmit power. Below we follow the notation and approach proposed by Piazzo [221]. The different modes use different modem constellations and thus carry a different number of bits per subcarrier, ranging from 1 to I bits per subcarrier, corresponding to BPSK and 2^I-ary QAM. We denote the transmit power and the number of bits allocated to subcarrier k ($k = 0, \ldots, N - 1$) by p_k and b_k, respectively. If $b_k = 0$, subcarrier k is allocated no power and no bits, hence it is disabled. The total transmit power is $P = \frac{1}{N} \sum_{k=0}^{N-1} p_k$ and the number of transmitted bits per OFDM symbol is $B = \sum_{k=0}^{N-1} b_k$. The i-bit modulation mode is characterised by the function $R_i(S)$, denoting the SNR required at the input of the detector, in order to achieve a target bit error rate (BER) equal to S. Finally, we denote the channel's power attenuation at subcarrier k by a_k, and the power of the Gaussian noise by P_N, so that the SNR of subcarrier k is $r_k = p_k/(a_k \cdot P_N)$.

We consider the problem of minimising the transmit power for a fixed target BER of S and for a fixed number of transmitted bits B per OFDM symbol. We impose an additional constraint, namely that the BER of every carrier has to be equal to S. This constraint simplifies the problem, while producing a system close to the unconstrained optimum system [206, 224, 226], while [222] considers an unconstrained system. Furthermore, from an important practical point of view, it produces a near-constant BER at the input of the channel decoder, if FEC is used, which maximises the achievable coding gain, since the channel does not become overwhelmed by the plethora of transmission errors, which would be the case for a more bursty error statistics without this constraint. In order to satisfy this constraint, the power transmitted on subcarrier k has to be $p_k = P_N a_k R_{b_k}(S)$, and the total power to be minimised is given by the sum of the N subcarriers' powers across the OFDM symbol:

$$P = \frac{P_N}{N} \sum_{k=0}^{k=N-1} a_k \cdot R_{b_k}(S). \tag{6.6}$$

We now state a property of the optimum system. Namely, in the optimum system if a subcarrier has a lower attenuation than another one – i.e. it exhibits a higher frequency domain transfer function value and hence experiences a higher received SNR – then it must carry at least as many bits as the lower SNR subcarrier. More explicitly:

$$a_k < a_h \;\Rightarrow\; b_k \geq b_h. \tag{6.7}$$

The above property in Equation 6.7 can be readily proven. Let us briefly consider a system, which does not satisfy Equation 6.7, where for subcarriers k and h the above condition is violated and hence we have $a_k < a_h$ and $b_k = i_1 < b_h = i_2$. In other words, although the attenuation a_k is lower than a_h, $i_1 < i_2$. Consider now a second system, where the lower attenuation subcarrier was assigned the higher number of bits, i.e. $b_k = i_2$ and $b_h = i_1$. Since the required SNR for maintaining the target BER of S is lower for a lower number of bits, i.e. we have $R_{i_1}(S) < R_{i_2}(S)$, upon substituting these SNR values in Equation 6.6 we can infer that the second system requires a lower total power P per OFDM symbol for maintaining the target BER S. Thus the first system is not optimum in this sense.

Equation 6.7 states a necessary condition of optimality, which was also exploited by Lai *et al.* in [224], but it can be exploited further, as we will demonstrate below. From now on, we consider the channel's transfer function or attenuation vector sorted in the order of $a_0 \leq a_1 \leq a_2 \ldots$, which simplifies our forthcoming discussions.

6.6.3 Power and Bit Allocation Algorithm

Piazzo's algorithm [221] solves the above resource allocation problem for the general system by repeatedly solving the problem for a simpler system. Explicitly, the simpler system employs only two modulation modes, those carrying J and $J - 1$ bits. This system can be termed the twin-mode system (TMS). On the basis of Equation 6.7 and since the channel's frequency domain attenuation vector was sorted in the order of $a_0 \leq a_1 \leq a_2 \ldots$, for the optimum TMS (OTMS) the OFDM subcarriers will be divided in three groups:

(1) Group J comprises the first or lowest attenuation OFDM subcarriers using a J-bit modulation mode.

(2) Group 0 is constituted by the last or highest attenuation OFDM subcarriers transmitting zero bits.

(3) Group $(J-1)$ hosts the remaining OFDM subcarriers using a $(J-1)$-bit modem mode.

In order to find the OTMS – minimising the required transmit power of the OFDM symbol for a fixed target BER of S and for a fixed number of transmitted bits B per OFDM symbol – we initially assign all the B bits of the OFDM symbol to the highest quality, i.e. lowest attenuation, group J. This of course would be a sub-optimum scheme, leaving the medium quality subcarriers of group $J - 1$ unused, since even the highest quality subcarriers would require an excessive SNR, i.e. transmit power, for maintaining the target BER, when transmitting B bits per OFDM symbol. We note furthermore that the above bit allocation may require padding of the OFDM symbol with dummy bits if J is not an integer divisor of B.

Following the above initial bit allocation, Piazzo suggested performing a series of *bit reallocations, reducing the transmit power upon each reallocation*. Specifically, in each power and bit reallocation step we move the $J \cdot (J - 1)$ bits allocated to the last, i.e. highest attenuation or lowest quality, $(J - 1)$ OFDM subcarriers of group J to group $(J - 1)$. For example, if 1 bit/symbol BPSK and 2 bit/symbol 4-QAM are used, then we move $2 \cdot 1 = 2$ bits, which were allocated to the highest attenuation 4-QAM subcarrier to two BPSK modulated subcarriers. The associated trade-off is that while previously the lowest quality subcarrier of class J had to carry 2 bits, it will now be conveying only 1 bit and additionally the highest quality and previously unused subcarrier has to be assigned 1 bit. This reallocation was motivated by the fact that before reallocation the lowest quality subcarrier of class J would have required a higher power for meeting the target BER requirement of S upon carrying 2 bits, than the regime generated by the reallocation step.

In general, for the sake of performing this power-reducing bit reallocation we have to add J subcarriers to group $(J - 1)$. Hence we assign the last, i.e. lowest quality, $(J - 1)$ OFDM subcarriers of group J and the first, i.e. highest quality, unused subcarrier to group $(J - 1)$. Based on Equation 6.6 and upon denoting the index of the last, i.e. lowest quality, subcarrier of group J before reallocation by M_J and the index of the first, i.e. highest quality, unused subcarrier before reallocation by M_0, the condition of successful power reduction after the tentative bit reallocation can be formulated. Specifically, the bit reallocation results in a system using less power, if the sum of the subcarriers' attenuations carrying J bits weighted by their SNR $R_J(S)$ required for the J-bit modem mode for maintaining the target BER of S is higher than that of the corresponding constellation after the above bit reallocation process, when an extra previously unused subcarrier was invoked for transmission. This can be expressed in a more compact form as:

$$R_J(S) \sum_{k=0}^{J-2} a_{M_J - k} > R_{J-1}(S)(a_{M_0} + \sum_{k=0}^{J-2} a_{M_J - k}). \tag{6.8}$$

If Equation 6.8 is satisfied, the reallocation is performed and another tentative reallocation step is attempted. Otherwise the process is terminated, since the optimum twin-mode power and bit allocation scheme has been found.

According to Piazzo's proposition [221] the above procedure can be further accelerated. Since the attenuation vector was sorted, we have $a_{M_J - k} \approx a_{M_J}$ in Equation 6.8. Upon replacing $a_{M_J - k}$ by a_{M_J}, after some manipulations we can reformulate Equation 6.8, i.e. the condition for the modem mode allocation after the bit reallocation to become more efficient as:

$$K_J(S)a_{M_J} - a_{M_0} > 0, \tag{6.9}$$

where $K_J(S) = (J-1)(\frac{R_J(S)}{R_{J-1}(S)} - 1)$ and $K_J(S) > 0$ holds, since $R_J(S) > R_{J-1}(S)$. Piazzo denoted the values of M_J and M_0 after m reallocation steps by $M_J(m)$ and $M_0(m)$. Since initially all the bits were allocated to group J, we have for the index of the last sub-carrier of group J at the commencement of the bit reallocation steps $M_J(0) = \lceil B/J \rceil - 1$, while for the index of the first unused subcarrier is $M_0(0) = \lceil B/J \rceil$, where $\lceil x \rceil$ is the small-est integer greater than or equal to x. Furthermore, since in each bit reallocation step the last $(J-1)$ OFDM subcarriers of group J and the first subcarrier of group 0 are moved to group $(J-1)$, after m reallocations we have $M_J(m) = \lceil B/J \rceil - 1 - (J-1)m$ and $M_0(m) = \lceil B/J \rceil + m$. Upon substituting these values in Equation 6.9 the left-hand side becomes a function of m, namely $f(m) = K_J(S)a_{M_J(m)} - a_{M_0(m)}$. Because the frequency domain channel transfer function's attenuation vector was ordered and since $K_J(S) > 0$, hence it is readily seen that $f(m)$ is a monotonically decreasing function of the reallocation index m. Therefore the method presented above essentially attempts to find the specific value of the reallocation index m, for which we have $f(m) > 0$ and $f(m+1) < 0$. In other words, when we have $f(m+1) < 0$, the last reallocation step resulted in a power increase, not a decrease, hence the reallocation procedure is completed.

The search commences from $m = 0$ and increases m by one at each bit reallocation step. In order to accelerate the search procedure, Piazzo replaced the linearly incremented search by a logarithmic search. This is possible, since the $f(m)$ function is monotonically decreasing upon increasing the reallocation index m. Piazzo [221] stipulated the search range by commencing from the minimum value of the reallocation index m, namely from $m_0 = 0$.

The maximum value, denoted by m_1, is determined by the number of OFDM subcarriers N or by the number of bits B to be transmitted per OFDM symbol, as will be argued below. There are two limitations, which determine the maximum possible number of reallocation steps. Namely, the reallocation steps have to be curtailed when there are no more bits left in the group of subcarriers associated with the J-bit modem mode group or when there are no more unused carriers left after iteratively invoking the best unused carrier from the group of disabled carriers. These limiting factors, which determine the maximum possible number of bit reallocation steps, are discussed further below.

Recall that at the commencement of the algorithm all the bits were assigned to the sub-carrier group associated with the J-bit modem mode and hence there were $\lceil B/J \rceil$ subcar-riers in group J. Upon reallocating the $J \cdot (J-1)$ bits allocated to the last, i.e. high-est attenuation or lowest quality, $(J-1)$ OFDM subcarriers of group J to group $(J-1)$ until no more bits were left in the subcarrier group associated with the J-bit modem mode naturally constitutes an upper limit for the maximum number of reallocation steps m_1, which is given by $\lceil B/J \rceil/(J-1)$. Again, the other limiting factor of the maxi-mum number of bit reallocation steps is the number of originally unused carriers, which was $N - \lceil B/J \rceil$). Hence the maximum possible number of reallocations is given by $m_1 = min(\lfloor \lceil B/J \rceil/(J-1) \rfloor; N - \lceil B/J \rceil)$, where $\lfloor x \rfloor$ is the highest integer smaller than or equal to x.

The accelerated logarithmic search proposed by Piazzo [221] halves the maximum possi-ble range at each bit reallocation step, by testing the value of $f(m)$ at the centre of the range and by updating the range accordingly. Piazzo's proposed algorithm can be summarised in a compact form as follows [221]:

Algorithm 1 $OTMS(B, S, J, N, a_k)$

(1) Initialise $m_0 = 0$, $m_1 = min(\lfloor \lfloor B/J \rfloor / (J-1) \rfloor, N - \lfloor B/J \rfloor)$.
(2) Compute $m_x = m_0 + \lfloor m_1 - m_0 \rfloor / 2$.
(3) If $f(m_x) \geq 0$ let $m_0 = m_x + 1$; else let $m_1 = m_x$.
(4) If $m_1 = m_0 + 1$ goto step 5); else goto step 2).
(5) Stop. The number of carriers in group J is $N_J = \lfloor B/J \rfloor - m_0 \cdot (J-1)$.

When the algorithm is completed, the value N_J becomes known and this specifies the number of OFDM subcarriers in the group J associated with the J-bit modem mode.

Having generated the optimum twin-mode system, Piazzo also considered the problem of finding the optimum general system (OGS) employing OFDM subcarrier modulation modes carrying $1, ..., I$ bits. The procedure proposed initially invoked Algorithm 1 in order to find the optimum twin-mode system carrying a total of B bits per OFDM symbol using the I-bit and the $(I-1)$-bit per subcarrier modulation modes. At the completion of Algorithm 1 we know N_I, the number of OFDM subcarriers carrying I bits. These subcarriers are now confirmed. These OFDM subcarriers as well as the associated $I \cdot N_I$ bits can now be eliminated from the resource allocation problem, and the optimum system transmitting the remaining $B - I \cdot N_I$ bits of the remaining $(N - N_I)$ subcarriers can be sought, using subcarrier modulation modes transmitting $(I-1), (I-2), ...$ bits.

Again, Algorithm 1 can be applied to this new system, now using the modulation modes with $(I-1)$ and $(I-2)$ bits per subcarrier and repeating the procedure. After each application of Algorithm 1 a new group of subcarrier is confirmed. Piazzo's general algorithm can be summarised in a compact form as follows [221]:

Algorithm 2 $OGS(B, S, I, N, a_k)$
(1) Initialise $\hat{B} = B$, $\hat{N} = N$, $\hat{a}_k = a_k$, $J = I$.
(2) Perform $OTMS(\hat{B}, S, J, \hat{N}, \hat{a}_k)$ to compute N_J.
(3) If $J = 2$, let $N_1 = \hat{B} - 2 \cdot N_2$ and stop.
(4) Remove the first N_J carriers from $\hat{a}_k$, let $\hat{B} = \hat{B} - J \cdot N_J$, $\hat{N} = \hat{N} - N_J$, $J = J - 1$ and goto step 2).

When the algorithm is completed, the values N_i specifying the number of OFDM subcarriers conveying i bits, become known for all the legitimate modes carrying $i = I, (I-1), ..., 1$ bits per subcarrier. Hence we know the number of bits allocated to subcarrier k ($k = 0, \dots, N-1$) expressed in terms of the b_k values as well as the associated minimum power requirements. Hence the system is specified in terms of $p_k = P_N a_k R_{b_k}(S)$. In closing it is worthwhile noting that the algorithm can be readily modified to handle the case where the two modes of the twin-mode system carry J and $K < J - 1$ bits.

Having considered the above near-optimum power- and bit allocation algorithm, in the next section we will consider a variety of OFDM systems supporting multiple users and invoking adaptive beam-steering.

6.7 Multi-User AOFDM

4

---

[4]This section is based on M. Münster, T. Keller and L. Hanzo, "Co-Channel Interference Suppression Assisted Adaptive OFDM in Interference Limited Environments", ©IEEE, VTC'99, Amsterdam, NL, 17-19 Sept. 1999.

6.7.1   Introduction

Signal fading as well as co-channel interference are known to have a severe impact on the system performance in multi-cellular mobile environments. Adaptive modulation as a method of matching the system to fading induced variations of the channel quality was originally proposed for single carrier transmission, but its potential was also soon discovered in the context of multicarrier transmissions, with the aim of concentrating the throughput on subcarriers least affected by frequency selective fading [219]. On the other hand, adaptive antenna array techniques have been shown to be effective in reducing co-channel interference at the receiver side [136,227], when supporting multiple users.

This particular application of smart antennas is often referred to as **Space Division Multiple Access** (SDMA), which exploits the unique, user-specific "spatial signature" of the individual users for differentiating amongst them. In simple conceptual terms one could argue that both a conventional CDMA spreading code and the Channel Impulse Response (CIR) affect the transmitted signal similarly - they are namely convolved with it. Hence, provided that the CIR is accurately estimated, it becomes known and certainly unique, although - as opposed to orthogonal Walsh-Hadamad spreading codes, for example - not orthogonal to the other users' CIRs. Nonetheless, an accurately estimated CIR may be used for uniquely identifying users after channel estimation and hence for supporting several users within the same bandwidth. Provided that a powerful multiuser detector is available, one can support even more users than the number of antennas. Hence this method enhances the achievable spectral efficiency directly. These techniques will be discussed in great detail in Chapter 17, hence in this introductory part of the book only a rudimentary exposure of the basic concepts is offered.

It is worth noting that the family of Multiple Input Multiple Output (MIMO) systems [104] is also closely related to the SDMA arrangements discussed here, since they also employ multiple antennas. However, in contrast to the SDMA arrangements, MIMOs are typically invoked not for the sake of supporting multiple users, but for increasing the throughput of a wireless system in terms of the number of bits per symbol that can be transmitted by a single user in a given bandwidth at a given integrity.

One of the most prominent schemes that may be used for detecting multiple users with the aid of SDMA is the so-called Sample Matrix Inversion (SMI) technique, which has recently drawn wide interest [105,130,131]. We commence our discussions in the next section with a brief description of a system amalgamating adaptive modulation and co-channel interference suppression. Initial performance results will be presented in Subsection 6.7.3 assuming perfect knowledge of all channel parameters, while in Subsection 6.7.4 the problem of channel parameter estimation will be addressed by means of orthogonal pilot sequences, leading to our conclusions.

6.7.2   Adaptive Transceiver Architecture

6.7.2.1   An Overview

The transceiver schematic is shown in Figure 6.30, where the receiver employs a multiple-antenna assisted front end. The signal received by each individual antenna element is fed to an FFT block, and the resulting parallel received OFDM symbols are combined on a subcarrier-by-subcarrier basis. The combining is accomplished on the basis of the weight vector, which

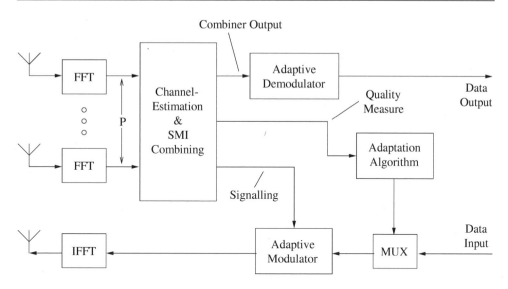

Figure 6.30: Schematic structure of the adaptive transceiver with interference suppression at the receiver

has been obtained by solving the Wiener equation, constituting the core of the sample matrix inversion algorithm [136, 227]. After combining, the signal is fed into the adaptive demodulator of Figure 6.30, which delivers the output bits in the form of soft-decision information to an optional channel decoder. The demodulator operates in one of a set of four modes, namely "no transmission", BPSK, QPSK and 16-QAM. Since an interfered channel cannot be considered to constitute a reciprocal system, the modem mode adaptation operates in a closed-loop fashion, where each of the receivers instructs the remote transmitter as to the required set of modulation modes for the next AOFDM symbol, which is necessary for maintaining a given target Bits per Symbol (BPS) performance. On reception of a packet, the adaptation algorithm computes the set of modulation modes to be employed by the remote transmitter for the next transmitted AOFDM symbol on the basis of a channel quality measure, namely the subcarrier SNR, which can be estimated by the interference suppression algorithm. The set of requested modulation modes is signalled to the remote receiver along with the next transmitted AOFDM symbol, which is then used by the remote transmitter in its next transmission.

6.7.2.2 The Signal Model

The $P \times 1$ vector of complex signals, $\mathbf{x}[n, k]$, received by the antenna array in the k-th subcarrier of the n-th OFDM symbol is constituted by a superposition of the independently faded signals associated with the desired user and the L undesired users plus the Gaussian noise at the array elements:

$$
\begin{aligned}
\mathbf{x}[n,k] &= \mathbf{d}[n,k] + \mathbf{u}[n,k] + \mathbf{n}[n,k], \quad \text{with} \\
\mathbf{d}[n,k] &= \mathbf{H}^{(0)}[n,k]s_0[k] \\
\mathbf{u}[n,k] &= \sum_{l=1}^{L} \mathbf{H}^{(l)}[n,k]s_l[k],
\end{aligned}
\tag{6.10}
$$

where $\mathbf{H}^{(l)}[n,k]$ for $l = 0, \ldots, L$ denotes the $P \times 1$ vector of complex channel coefficients between the l-th user and the P antenna array elements. We assume that the vector components $H_m^{(l)}[n,k]$ for different array elements m or users l are independent, stationary, complex Gaussian distributed processes with zero-mean and different variance $\sigma_l^2, l = 0, \ldots, L$. The variable $s_l[n,k]$ – which is assumed to have zero-mean and unit variance – represents the complex data of the l-th user and $\mathbf{n}[n,k]$ denotes the aforementioned $P \times 1$ vector of additive white Gaussian noise contributions with zero mean and variance σ^2 [130].

6.7.2.3   The SMI Algorithm

The idea behind minimum mean-square error (MMSE) beamforming [136] is to adjust the antenna weights, such that the power of the differential signal between the combiner output and a reference signal – which is characteristic of the desired user – is minimised. The solution to this problem is given by the well-known Wiener equation, which can be directly solved by means of the SMI algorithm in order to yield the optimum weight vector $\mathbf{w}[n,k]$ of dimension $P \times 1$. Once the instantaneous correlation between the received signals, which is represented by the $P \times P$ matrix $\mathbf{R}[n,k]$, and the channel vector $\mathbf{H}^{(0)}[n,k]$ of the desired user are known, the weights are given by [105, 130, 136]:

$$
\mathbf{w}[n,k] = (\mathbf{R}[n,k] + \gamma \mathbf{I})^{-1} \mathbf{H}^{(0)}[n,k],
\tag{6.11}
$$

where γ represents the so-called diagonal augmentation factor [130]. Assuming knowledge of all channel parameters and the noise variance σ^2, the correlation matrix can be determined by:

$$
\begin{aligned}
\mathbf{R}[n,k] &\triangleq E_c\{\mathbf{x}[n,k]\mathbf{x}^H[n,k]\} \\
&= \mathbf{R}_d[n,k] + \mathbf{R}_u[n,k] + \mathbf{R}_n[n,k], \quad \text{with} \\
\mathbf{R}_d[n,k] &= \mathbf{H}^{(0)}[n,k]\mathbf{H}^{(0)^H}[n,k] \\
\mathbf{R}_u[n,k] &= \sum_{l=1}^{L} \mathbf{H}^{(l)}[n,k]\mathbf{H}^{(l)^H}[n,k] \\
\mathbf{R}_n[n,k] &= \sigma^2 \mathbf{I},
\end{aligned}
\tag{6.12}
$$

which is a superposition of the correlation matrices $\mathbf{R}_d[n,k]$, $\mathbf{R}_u[n,k]$ and $\mathbf{R}_n[n,k]$ of the desired and undesired users as well as of the array element noise, respectively. The combiner output can now be inferred from the array output vector $\mathbf{x}[n,k]$ by means of:

$$
y[n,k] = \mathbf{w}^H[n,k]\mathbf{x}[n,k].
\tag{6.13}
$$

The Signal-to-Noise Ratio (SNR) at the combiner output – which is of vital importance for the modulation mode adaptation – is given by [136]:

$$
SNR = \frac{E\{|\mathbf{w}^H[n,k]\mathbf{d}[n,k]|^2\}}{E\{|\mathbf{w}^H[n,k]\mathbf{n}[n,k]|^2\}}
$$

$$
= \frac{\mathbf{w}^H[n,k]\mathbf{R}_d[n,k]\mathbf{w}[n,k]}{\mathbf{w}^H[n,k]\mathbf{R}_n[n,k]\mathbf{w}[n,k]}
$$

(6.14)

and correspondingly the Signal-to-Interference+Noise Ratio (SINR) is given by [136]:

$$
SINR = \frac{\mathbf{w}^H[n,k]\mathbf{R}_d[n,k]\mathbf{w}[n,k]}{\mathbf{w}^H[n,k](\mathbf{R}_u[n,k]+\mathbf{R}_n)\mathbf{w}[n,k]}.
$$

(6.15)

Equation 6.12 is the basis for our initial simulations, where perfect channel knowledge has been assumed. Since in a real environment the receiver does not have perfect knowledge of the channel, its parameters have to be estimated, a problem which we will address in Section 6.7.4 on an OFDM symbol-by-symbol basis by means of orthogonal pilot sequences.

6.7.2.4 The Adaptive Bit-assignment Algorithm -

The adaptation performed by the modem is based on the choice between a set of four modulation modes, namely 4, 2, 1 and 0 bit/subcarrier, where the latter corresponds to "no transmission". The modulation mode could be assigned on a subcarrier-by-subcarrier basis, but the signalling overhead of such a system would be prohibitive. Hence, we have grouped adjacent subcarriers into "sub-bands" and assign the same modulation mode to all subcarriers in a sub-band. Note that the frequency domain channel transfer function is typically not constant across the subcarriers of a sub-band, hence the modem mode adaptation will be suboptimal for some of the subcarriers. The Signal-to-Noise Ratio (SNR) of the subcarriers will be shown to be in most cases an effective measure for controlling the modulation assignment. The modem mode adaptation is hence achieved by calculating in the first step for each sub-band and for all four modulation modes the expected overall sub-band bit error rate (BER) by means of averaging the estimated individual subcarrier BERs. Throughout the second step of the algorithm – commencing with the lowest modulation mode in all sub-bands – in each iteration the number of bits/subcarrier of that sub-band is increased, which provides the best compromise in terms of increasing the number of expected bit errors compared to the number of additional data bits accommodated, until the target number of bits is reached.

6.7.2.5 The Channel Models

Simulations have been conducted for the indoor Wireless Asynchronous Transfer Mode (WATM) channel impulse response (CIR) of Section 4.1 [219]. This three-path impulse response exhibits a maximal dispersion of 11 time domain OFDM samples, with each path faded according to a Rayleigh distribution of a normalised maximal Doppler frequency of $f_d' = 1.235 \cdot 10^{-5}$, where the normalisation interval was the OFDM symbol duration. This model corresponds to the channel experienced by a mobile transmitting at a carrier frequency of 60 GHz with a sampling rate of 225 MHz and travelling at a vehicular velocity of 50 km/h. An alternative channel model, which we considered in our simulations is a Wireless Local Area Network (WLAN) model associated with a seven-path impulse response having a maximal dispersion of 32 samples. However, for this more dispersive and higher Doppler frequency channel adaptive modulation has turned out to be less effective due to its significantly increased normalised Doppler frequency of $f'_d = 3.935 \cdot 10^{-5}$ corresponding to a carrier frequency of 17 GHz, sampling rate of 20 MHz and vehicular velocity of 50 km/h.

6.7.3   Simulation Results - Perfect Channel Knowledge

6.7.3.1   General Remarks

In our initial simulations we assumed that the receiver had perfect knowledge of all channel parameters, which enabled the estimation of the correlation matrix required by the SMI algorithm upon using Equation 6.12. Furthermore, we initially assumed that the receiver was capable of signalling the modulation modes to the transmitter without any additional delay. Throughout our discussions we will gradually remove the above idealistic assumptions. In all simulations we assumed a partitioning of the 512-subcarrier OFDM symbol's total bandwidth into 16 equal-sized 32-subcarrier sub-bands. This has been shown to provide a reasonable compromise between signalling overhead and performance degradation compared to a subcarrier-by-subcarrier based modulation mode assignment.

6.7.3.2   Two-Branch Maximum–Ratio Combining

Initial simulations were conducted in the absence of co-channel interference. In this scenario the SMI equations take the form of the MMSE maximum-ratio combiner, resulting in a high diversity gain even with the minimal configuration of only two reception elements. Adaptive modulation was performed on the basis of the estimated SNR of each subcarrier, which is given by Equation 6.14. Since due to diversity reception the dramatic fades of the channel frequency response have been mitigated, the performance advantage of adaptive modulation is more modest, as illustrated by Figure 6.31 for the "transmission frame-invariant" WATM channel, for which the fading profile is kept constant for the OFDM symbol duration, in order to avoid inter-subcarrier interference. For the equivalent simulations in the "transmission frame-invariant" WLAN channel environment we observed a more distinct performance gain due to adaptive modulation, which is justified by the higher degree of frequency selectivity introduced by the WLAN channel's seven-path impulse response.

6.7.3.3   SMI Co-Channel Interference Suppression

In these simulations we considered first of all the case of a single dominant co-channel interferer of the same signal strength as the desired user. It is well known that if the total number of users – whose signals arrive at the antenna array – is less or equal to the number of array elements, the unwanted users are suppressed quite effectively. Hence, for our modulation adaptation requirements we can assume that $SNR \approx SINR$, which enables us to use the algorithm described in Section 6.7.2, on the basis of the SNR estimated with the aid of Equation 6.14. Figure 6.32 illustrates the impact of adaptive modulation in the WATM channel environment under the outlined conditions. At a given SNR the performance gain due to adaptive modulation decreases with an increasing bitrate, since the higher bitrate imposes a more stringent constraint on the modulation mode assignment, invoking a higher number of

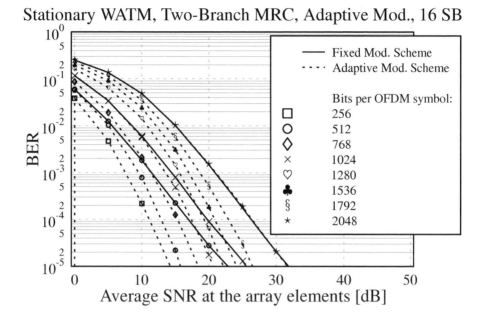

Figure 6.31: BER of 16-sub-band AOFDM modem with two-branch maximum-ratio combining in a "frame-invariant" indoor WATM environment, assuming perfect channel knowledge and zero-delay signalling of the modulation modes

low-SNR subcarriers. Upon comparing Figure 6.32 and 6.33 we observe that AOFDM attains a significantly higher SNR gain in the presence of co-channel interference, than without interference. As alluded to in the previous section this is because under co-channel interference the SMI scheme exploits most of its diversity information extracted from the antenna array for suppressing the unwanted signal components, rather than mitigating the frequency domain channel fades experienced by the wanted user. For decreasing values of the Interference-to-Noise Ratio (INR) at the antenna array output, the system performance will gradually approach the performance observed for the MRC system. In order to render our investigations more realistic in our next experiment we allow a continuous, i.e. "frame-variant" fading across the OFDM symbol duration. The system performance corresponding to this scenario is illustrated in Figure 6.33. At low SNRs the observed performance is identical to that recorded in Figure 6.32 for the "frame-invariant" channel model, whereas at high SNRs we experience a residual BER due to inter-subcarrier interference. Again, for a low number of bits per OFDM symbol the adaptive scheme is capable of reducing the "loading" of subcarriers with low SNR values, which are particularly impaired by inter-subcarrier interference. Hence AOFDM exhibits a BER improvement in excess of an order of magnitude. So far we have assumed that the receiver is capable of instantaneously signalling the required modulation modes for the next OFDM symbol to the transmitter. This assumption cannot be maintained in practice.

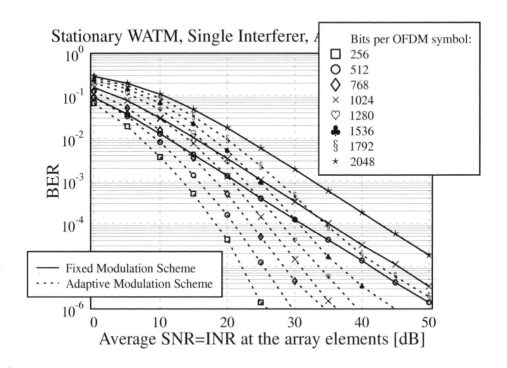

Figure 6.32: BER of 16-sub-band AOFDM modem with two-branch SMI and 2 users in a "frame-invariant" indoor WATM environment, assuming perfect channel knowledge and zero-delay signalling of the modulation modes

Here we assume a time division duplexing (TDD) system with identical transceivers at both ends of the link, which communicate with each other using adjacent uplink and downlink slots. Hence we have to account for this by incorporating an additional delay of at least one OFDM symbol, while neglecting the finite signal processing delay. Simulation results for this scenario are depicted in Figure 6.34. We observe that the performance gain attained by adaptive modulation is reduced compared to that associated with the zero-delay assumption in Figure 6.33. This could partly be compensated for by channel prediction. When employing a higher number of array elements, the performance gain achievable by adaptive modulation will mainly depend on the number of users and their signal strength. If the number of users is lower than the number of array elements, or if the interferers are predominantly weak, the remaining degrees of freedom for influencing the array response are dedicated by the SMI scheme to providing diversity for the reception of the wanted user and hence adaptive modulation proves less effective. If the number of users exceeds the number of array elements, the system becomes incapable of suppressing the undesired users effectively, resulting in a residual BER at high SNRs due to the residual co-channel interference. Since for a relatively high number of users the residual interference exhibits Gaussian-noise like characteristics,

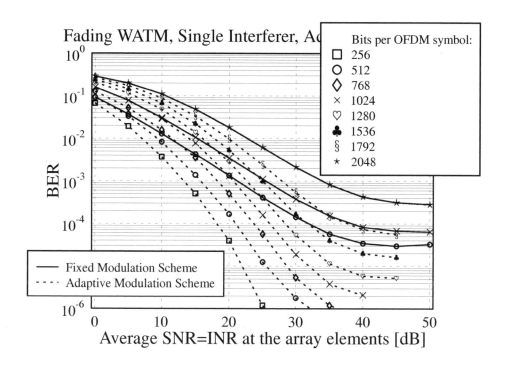

Figure 6.33: Performance results of Figure 6.32 repeated for a fading indoor WATM environment

code/bit	0	1	2	3
0	1	1	1	1
1	1	1	-1	-1
2	1	-1	-1	1
3	1	-1	1	-1

Table 6.4: Orthogonal Walsh codes with a length of 4 bit

the SINR given by Equation 6.15 could be a suitable measure for performing the modulation mode assignment. By contrast, for a low number of interferers it is difficult to predict the impact on the system performance analytically. A possible approach would be to use the instantaneous number of errors in each sub-band (e.g. at the output of a turbo decoder) as a basis for the modulation assignment, which constitutes our future work. Let us now consider the issues of channel parameter estimation.

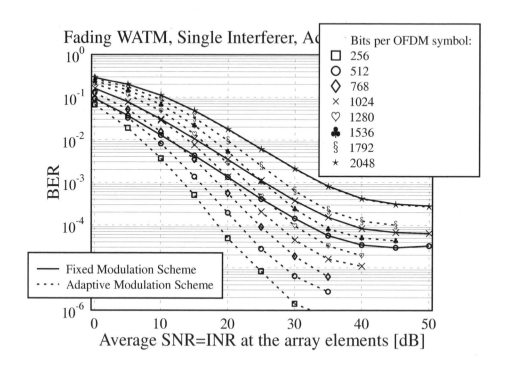

Figure 6.34: Performance results of Figure 6.33 repeated for one OFDM symbol delayed signalling

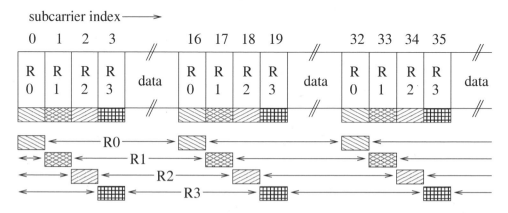

Figure 6.35: Pilot arrangement in each OFDM symbol for a reference length of 4 bit and a group distance of 16 subcarriers; interpolation is performed between pilots associated with the same bit position within the reference sequence

6.7.4   Pilot-based Channel Parameter Estimation

6.7.4.1   System Description

Vook and Baum [105] have proposed SMI parameter estimation for OFDM by means of orthogonal reference sequences carrying pilot slots, which are transmitted over several OFDM
symbol durations. This principle can also be applied on an OFDM symbol-by-symbol basis,
as required for adaptive modulation. Upon invoking the idea of pilot-based channel estimation by means of sampling and low-pass interpolating the channel transfer function, we
replace each single pilot subcarrier by a group of pilots, which carries a replica of the user's
unique reference sequence. This is illustrated in Figure 6.35 for a reference sequence having
a length of 4 bit, and for a pilot group distance of 16 subcarriers, which corresponds to the
frequency required for sampling the WATM channel's transfer function. The corresponding
4-bit orthogonal Walsh code-based reference sequences are listed in Table 6.4. Each of these
4 bits is assigned using BPSK to one of the 4 pilots in a pilot-group. The complex signal
received by the m-th antenna in a pilot subcarrier at absolute index k and local index i within
the reference sequence is constituted by a contribution of all users, each of which consists of
the product of the Walsh code value associated with the user at bit position i of the reference
sequence of Table 6.4 and the complex channel coefficient between the transmitter and the
m-th antenna. MMSE lowpass interpolation is performed between all pilot symbols of the
same relative index i within the k-spaced pilot blocks – as seen in Figure 6.35 – in order
to generate an interpolated estimate of the reference for each subcarrier. An estimate of the
channel vector $\hat{\mathbf{H}}^{(0)}[n,k]$ and the correlation matrix $\hat{\mathbf{R}}[n,k]$ for the k-th subcarrier of the
n-th OFDM symbol is then given by [105, 136, 227]:

$$\hat{\mathbf{H}}^{(0)}[n,k] = \frac{1}{N}\sum_{i=0}^{N-1} r^{(0)*}(i)\mathbf{x}_{LP}[n,k](i) \tag{6.16}$$

$$\hat{\mathbf{R}}[n,k] = \frac{1}{N}\sum_{i=0}^{N-1} \mathbf{x}_{LP}[n,k](i)\mathbf{x}_{LP}^{H}[n,k](i), \tag{6.17}$$

where $r^{(0)}(i)$ denotes the i-th value of the reference sequence associated with the desired
user, $\mathbf{x}_{LP}[n,k](i)$ represents the low-pass interpolated received signal at sequence position i
and N denotes the total reference length.

6.7.4.2   Simulation Results -

The performance of this scheme is characterised by the simulation results presented in Figure 6.36. Compared to the results presented in Figure 6.34 for "perfect channel knowledge",
in Figure 6.36 we observe that besides the reduced range of supported bitrates there is an
additional performance degradation, which is closely related to the choice of the reference
length. Specifically, there is a reduction in the number of useful data subcarriers due to
the pilot overhead, which reduces the "adaptively" exploitable diversity potential. Second,
a relatively short reference sequence results in a limited accuracy of the estimated channel
parameters – an effect which can be partly compensated for by a technique referred to as
diagonal loading [130]. However, the effect of short reference sequences becomes obvious
for a higher number of antenna elements, since more signal samples are required, in order

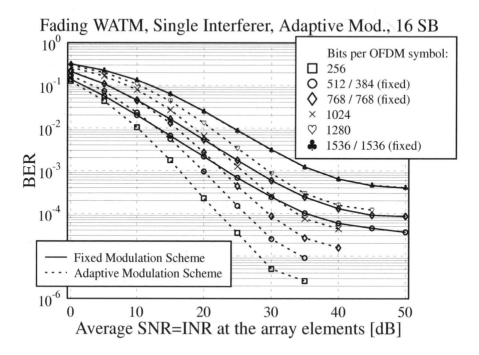

Figure 6.36: BER of 16-sub-band adaptive OFDM modem with two-branch SMI and 2 users in a fading indoor WATM environment, with pilot -based channel parameter estimation and one OFDM symbol delayed signalling of the modulation assignment using a diagonal loading of $\gamma = 1.0$

to yield a reliable estimate of the correlation matrix. Hence our scheme proposed here is attractive for a scenario having 2–3 reception elements, where the interference is due to 1–2 dominant interferers and an additional Gaussian noise-like contribution of background interferers, which renders the SINR of Equation 6.15 to be an effective measure of channel quality. In conclusion, the proposed adaptive array-assisted AOFDM scheme resulted typically in an order of magnitude BER reduction due to employing adaptive modulation.

6.8 Chapter Summary and Conclusion

A range of adaptive modulation and spectral predistortion techniques have been presented in this chapter, all of which aim to react to the time and frequency dependent channel transfer function experienced by OFDM modems in fading time dispersive channels. It has been demonstrated that by exploiting the knowledge of the channel transfer function at the transmitter, the overall system performance can be increased substantially over the non-adaptive case. It has been pointed out that the prediction of the channel transfer function for the next transmission time slot and the signalling of the parameters are the main practical problems in the context of employing adaptive techniques in duplex communications.

The channel prediction accuracy is dependent on the quality of the channel estimation at the receiver, as well as on the temporal correlation of the channel transfer function between the uplink and downlink time slots. Two-dimensional channel estimation techniques [131, 228] can be invoked in order to improve the channel prediction at the receivers.

It has been demonstrated that sub-band adaptivity instead of subcarrier-by-subcarrier adaptivity can significantly decrease the necessary signalling overhead, with a loss of system performance that is dependent on the channel's coherence bandwidth. We have seen that sub-band adaptivity allows the employment of blind detection techniques in order to minimise the signalling overhead. Further work on blind detection algorithms as well as new signalling techniques is needed to improve the overall bandwidth efficiency of adaptive OFDM systems.

Pre-equalisation or spectral predistortion techniques have been demonstrated to significantly improve an OFDM system's performance in time dispersive channels, while not increasing the system's output power. It has been shown that spectral predistortion can integrate well with adaptive modulation techniques, improving the system's performance significantly. We saw in Figure 6.1 that a data throughput of 0.5 bits/symbol has been achieved at 0 dB average channel SNR with a BER of below 10^{-4}.

An adaptive power and bit allocation algorithm was also highlighted and the performance of AOFDM in a beam-forming assisted multi-user scenario was characterised. Our treatment of adaptive modulation was conceptual in this chapter. We will revisit this topic in more depth in Chapter 12, where numerous analytical expressions will be derived for characterising the achievable BER perfomance as well as for determining the modem mode switching thresholds. For a more detailed treatise on adaptive modulation, please refer to [7, 207–209].

Chapter 7

Block-Coded Adaptive OFDM [1] [2]

T. Keller, T-H. Liew and L. Hanzo

7.1 Introduction

7.1.1 Motivation

We saw in Chapter 6 that adaptive modulation techniques can be used in order to achieve
a given target level of bit error rate (BER) in a duplex communications link by choosing
the appropriate modulation modes for the instantaneous channel conditions. The various
adaptive modem mode signalling regimes were portrayed in Figure 6.1. Employing recursive
systematic convolutional (RSC) codes as the constituent codes in turbo coding in conjunction
with a turbo decoder allowed us to dramatically improve the effective data throughput for a
given target bit error rate of 10^{-4} at low SNR values. The performance comparison in Figure
6.29 showed that the observed throughput for the coded, target BER adaptive modem was
higher than in the uncoded case for SNR values up to 18 dB. For higher SNR values, however,
the redundant information overhead of the FEC limits the system's effective data throughput
if only a limited set of modulation modes is employed by the adaptation algorithm. In the
case of the half-rate convolutional coding-based turbo code employed in Section 6.2.7, half
the data bandwidth was unnecessarily absorbed at high SNR values.

In this chapter, we will investigate the employment of variable rate channel coding
schemes within an OFDM transmission system, in order to improve the system's through-

[1] *OFDM and MC-CDMA for Broadband Multi-user Communications WLANS and Broadcasting.*
L.Hanzo, Münster, T. Keller and B.J. Choi,
©2003 John Wiley & Sons, Ltd. ISBN 0-470-85879-6

put at high SNR values. Analogously to the adaptive modulation modes discussed in Chapter 6, in addition to the modem modes, in this chapter the code rate is also adapted in response to the time and frequency dependent estimated channel conditions. We note, however, that this variable code rate scheme can operate along with time-invariant non-adaptive modulation or in conjunction with additional adaptive OFDM (AOFDM) modulation techniques, as will be demonstrated here.

Similarly to the adaptive modulation schemes [207] discussed in Chapter 6, adaptive coding [208] relies on the same fundamental principles of channel quality estimation, parameter adaptation and signalling of the modem parameters employed. Like adaptive modulation, it can operate in open and closed adaptation loops, which were portrayed in Figure 6.29. In this chapter, we will concentrate on the upper bound performance study of adaptive error correction coding for OFDM modems, and assume perfect channel estimation and signalling.

7.1.2 Choice of Error Correction Codes

The behaviour of the adaptive modems in conjunction with long-block-length convolutional coding-based turbo coding [207] was studied in Chapter 6. We saw that the error correction performance of the turbo decoder was high, resulting in output bit error rates of below 10^{-4} for input bit error rates of up to 1%. The disadvantage of this RSC-based turbo codec was the long block length of 2000 bits, which prevented code rate adaptation on a short-term basis. Radically shortening the interleaving block length of turbo codes results in significantly reduced performance, as does additional puncturing of the code word for code rate adaptation.

Similarly to the adaptive modulation algorithms, the adaptive error correction codec has to be able to vary its code rate according to the frequency dependent channel transfer function observed for the OFDM system in a time dispersive channel. Ideally, the error correction capability of the code would be adjustable for each data bit's expected BER independently. For our experiments, short block length codes of less than 100 bits per code word were employed, in order to allow flexible adaptation of code parameters, while delivering reasonable error protection for the data bits.

Two different block codes were investigated for adaptive OFDM transmission, namely variable-length redundant residue number system (RRNS) based codes [208, 229, 230] and turbo BCH codes [208, 231, 232].

7.2 Redundant Residue Number System Codes

Residue number system (RNS) [208,229,230] based algorithms have been studied in the context of digital filtering, spectral analysis, correlation and matrix operations as well as in image processing [233–236]. Until quite recently RRNSs have only been proposed for fault-tolerant processing of signals, for example, in digital arithmetics. RRNSs represent each operand with the aid of a set of residues. The residues are generated as the remainders upon dividing the operand to be represented with the aid of the residues by each of the so-called moduli of the RRNS, which have to be relative primes, i.e. do not have a common divisor. Among others, the following two advantages accrue for RRNS-based processing [237]: they have the ability to use carry-free arithmetic, since all residue-based operations can be processed independently of each other. This is amenable to high-speed parallel processing. A related

RRNS property is the lack of ordered significance among residue digits, which implies that as long as a sufficient number of residues was retained, in order to unambiguously represent the results of the computations, any erroneous residue digit can be discarded without affecting the result.

Error detection and correction algorithms based on the RNS have been proposed by Szabo and Tanaka [229], and by Watson and Hastings [230], which exploited the properties of the redundant residue number system (RRNS) [208]. More recently, a computationally efficient procedure was described in [238] to correct a single error. In [239], the procedure was extended to correct double errors as well as simultaneously correcting single errors and detecting multiple errors. Efficient soft-decision multiple error correcting algorithms were suggested in [240]. Furthermore, a RNS-based M-ary modulation scheme has been proposed and analysed in [241], while an RRNS-based CDMA system was the topic of [242].

RRNS(n, k) error correction codes [208] are akin to the well-known family of Reed-Solomon (RS) codes [208], and both families achieve the so-called maximum minimum distance of $d = n - k + 1$, provided that the moduli of the RRNS code obey certain conditions. It can also be readily shown that the RRNS code's weight distribution can be approximated by the weight distribution of RS codes if all the moduli assume values close to their average value [238,239,243]. For further details concerning the construction, coding theory, encoding and decoding of RRNS codes, the interested reader is referred to the literature [238–243].[3] Based on the above arguments a similar coding performance is achieved by an identical rate RS code and RRNS code, provided that they both use the same number of bits per symbol or bits per residue. However, their encoding and decoding algorithms are distinctly different [208,240], requiring further research into their implementation, coding theory and application. Since in [208, 240] the authors have developed a soft-decision decoding algorithm, here we favoured RRNS codes, with the intention of stimulating further research in this interesting novel field. Typically there exists a larger variety of RRNS codes for a given code rate and code word length than in the family of RS codes.

The RRNS codes employed in our investigations are systematic, which means that k of the n code residues contain the original data bits, and the additional $(n - k)$ redundant residues can be employed for error correction at the decoder. The error correction capability of the code is $t = \lfloor \frac{n-k}{2} \rfloor$ residues [240].

The code rate, and accordingly the error correction capability of the code, can be readily varied by transmitting only a fraction of the generated redundant residues. If the channel conditions are favourable, then only the systematic information-bearing residues are transmitted, resulting in a unity rate code with no added redundancy and no error correction capability. Upon transmitting two redundant residues with the data bits, the resulting code can correct one residue error for a code rate of $\frac{n}{n+2}$. More of the redundant residues can be transmitted, lowering the code rate and improving the code's error resilience at the cost of a lower effective information throughput, when the channel quality degrades.

In our investigations RRNS codes employing 8 bits per residue were chosen. Three or six systematic information-bearing residues – corresponding to 24 or 48 useful data bits per code word – and up to six redundant residues were used. The code parameters for these codes are shown in Table 7.1. As can be seen from the table, the code rates vary from 0.33 to 0.75. The uncoded case, corresponding to a $(3, 3)$ code, is not shown in the table.

---

[3]A range of related papers can be found at http://www-mobile.ecs.soton.ac.uk/lly/

Code	(5, 3)	(7, 3)	(9, 3)	(8, 6)	(10, 6)	(12, 6)
Sys. residues	3	3	3	6	6	6
Red. residues	2	4	6	2	4	6
Correction capability t	1	2	3	1	2	3
Code rate	0.6	0.43	0.33	0.75	0.6	0.5
Data bits per word	24	24	24	48	48	48
Red. bits per word	16	32	48	16	32	48

Table 7.1: RRNS codes employed in our investigations, each using 8-bit residues

7.2.1 Performance in an AWGN channel

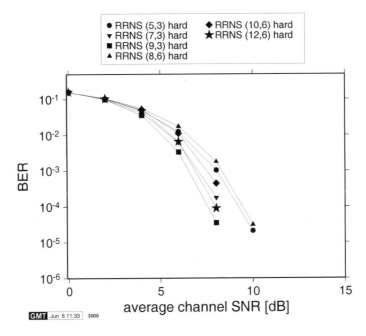

Figure 7.1: BER performance for RRNS-coded QPSK transmission in AWGN channel using hard decision decoding

The BER performance of an OFDM system employing QPSK and the RRNS codes of Table 7.1 is depicted in Figure 7.1. It can be seen that the relative BER performance of the different codes is largely in line with their respective code rates. The $(9, 3)$ code, with a code rate of 0.33, exhibits the strongest error correction properties; the $(8, 6)$ code, with a code rate of 0.75, is the weakest code of the set. Comparing the performance of the $(5, 3)$ code with that of the $(10, 6)$ code, both having a code rate of 0.6, shows that the longer code exhibits a superior performance. The $(12, 6)$ code, having a code rate of 0.5, outperforms the shorter $(7, 3)$ code for SNR values in excess of 6 dB.

7.2.1.1 Performance in a fading time dispersive channel

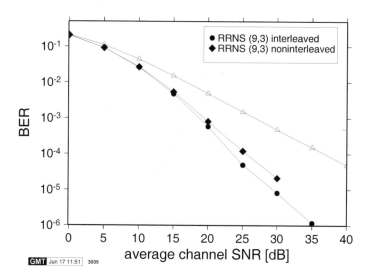

Figure 7.2: BER performance and throughput in bits per symbol for the RRNS-coded OFDM system in the Rayleigh fading shortened WATM channel of Section 4.1.1.2, with 512-FFT QPSK OFDM transmission

Figure 7.2 portrays the BER performance of a RRNS-coded 512-subcarrier modem using a 512-point FFT-based (512-FFT) OFDM modem employing QPSK transmission over the Rayleigh fading short WATM channel described in Section 4.1.1.2. The associated BER curves for uncoded, coded non-interleaved and coded interleaved transmission are shown. The interleaver employed in acquiring the interleaved results was a shortened block interleaver of the structure shown in Figure 7.3. This interleaver structure was chosen since it is sufficiently flexible to vary numbers of residues if adaptive modulation is invoked.

We note that residue-based interleaving has a better performance than bit interleaving, since bit interleaving would increase the probability of residue errors due to spreading bursts of erroneous bits across residues. Since the RRNS decoding algorithm is symbol based, the increased residue error rate would degrade the system's performance.

It can be seen from Figure 7.2 that the coded schemes deliver a gain of about 12 dB in SNR terms. Interleaving of the residues within the OFDM symbol improves the BER performance by about 2 dB.

7.2.1.2 Adaptive RRNS-coded OFDM

Analogously to the adaptive modulation schemes discussed in Chapter 6, the code rate adaptation reacts to the time- and frequency-varying channel conditions experienced in a duplex link. Each station exploits the channel quality information extracted from the last received OFDM symbol for determining the coding parameters of the next transmitted frame.

Although the longer codes exhibit better error correction properties than the shorter codes investigated, we will concentrate on the group of $(9,3)$, $(7,3)$, $(5,3)$ codes and on uncoded

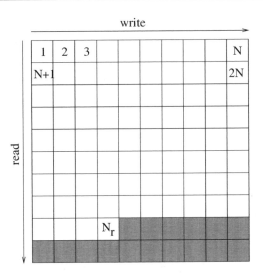

Figure 7.3: Structure of the interleaver employed for the ARRNS system. The interleaver is a square $N \times N$ interleaver for N_r residues, with $N = \lceil\sqrt{N_r}\rceil$; only the first N_r positions are used, where N_r is the number of residues in the OFDM symbol

$(3,3)$ transmission for the variable code rate application, since a short code word length allows for increased flexibility in the adaptation to the channel conditions. These codes exhibit an error correction capability of $t = 3, 2, 1$, and 0 residues per code word, respectively. Correspondingly, four error correction coding modes c are defined, which are shown in Table 7.2.

Mode c	0	1	2	3
n_c	3	5	7	9
k_c	3	3	3	3
t_c	0	1	2	3
R	1	0.6	0.43	0.33

Table 7.2: RRNS coding modes used for the code rate adaptation employing 8-bit residues

The choice of the coding mode for each code word in the AOFDM symbol is determined on the basis of the estimated channel transfer function. As discussed for the adaptive modulation scheme of Section 6.2.4.2, the predicted bit error probabilities p_e are calculated for all bits to be transmitted in an OFDM symbol, based on the estimated subcarrier SNR and the modulation mode to be employed. More explicitly, the expected overall bit error probability for the given OFDM symbol is computed by averaging the individual subcarrier BERs for all legitimate modulation modes M_n, yielding $\bar{p}_e(M_n) = 1/N_s \sum_j p_e(\gamma_j, M_n)$ where γ_j is the subcarrier SNR for $j = 1, \ldots, 512$. The specific subcarrier modem mode allocation with the highest bits per symbol (BPS) throughput, whose estimated BER is lower than the required value, is then confirmed. Currently, however, only QPSK/OFDM is used, which simplifies the associated bit allocation and BER computation procedure. AOFDM will be the topic of Section 7.2.2. This algorithm allows the direct adjustment of the desired maximum BER.

If OFDM is to be employed in conjunction with adaptive RRNS coding, then the number of bits per OFDM symbol and the mapping of bits to subcarriers can change from one OFDM symbol to the next. Explicitly, this implies using a variable bitrate, near-constant BER regime. Hence the RRNS coding scheme adaptation algorithm operates on the basis of the estimated BER, rather than directly relying on the estimated channel transfer function. Once the vector of estimated bit error probabilities $p_e(n)$ for the specific number of bits N_b – which has to be conveyed by the OFDM symbol – is known, the total number of bits to be transmitted is split into blocks of K bits, where K is the number of bits per residue. Again, the error correction capability of the code in each RRNS code word is a given number of residues, not bits. Hence, as argued before, interleaving of bits would increase the residue error rate at the decoder's input and lower the system's performance.

¿From the bit error probability $p_e(n)$ of bit n, $n = 1 \ldots N_b$ – where N_b is the number of bits allocated to the OFDM symbol – the estimated residue error rate $p_r(r)$ for residue index r, which is defined as the proportion of residues in error, for the $N_r = \lfloor N_b/K \rfloor$ number of residues in the OFDM symbol can be calculated as:

$$p_r(r) = 1 - \left(\prod_{n=0}^{K-1} (1 - p_e(r \cdot K + n)) \right). \tag{7.1}$$

The remaining $N_b - K \cdot N_r$ data bits of the OFDM symbol that are not allocated to any residue are filled with padding bits and hence contain no useful data. The mapping of the residues with index r to the RRNS code words is based on the estimated residue error probabilities $p_r(r)$. The interleaver $I(r)$ of Figure 7.3 is used to map the stream of residues to the residue positions in the transmitted OFDM symbol. The interleaver function used in our experiments is defined more explicitly below.

A received code word of the codec mode c_w is irrecoverable if more than t_c of the received residues are in error. The RRNS code word error probability p_w for word w can be calculated as:

$$p_w(w) = p(R_r(w) > t_{c_w}) = 1 - P(R_r(w) \leq t_{c_w}), \tag{7.2}$$

where $R_r(w)$ is the number of residue errors in code word w, and $p_w(w)$ can be calculated from the residue error probabilities $p_r(r)$ as:

$$p_w(w) = 1 - p\left[R_r(w) = 0\right] - p\left[R_r(w) = 1\right] - \ldots - p\left[R_r(w) = t_{c_w}\right]. \tag{7.3}$$

Upon elaborating further:

$$p(R_r(w) = 0) = \prod_{r=0}^{n_{c_w}-1} (1 - p_r(I(r_{0,w} + r))) \tag{7.4}$$

$$p(R_r(w) = 1) = \sum_{r=0}^{n_{c_w}-1} p_r(I(r_{0,w} + r)) \prod_{s=0,s\neq r}^{n_{c_w}-1} (1 - p_r(I(r_{0,w} + s))) \tag{7.5}$$

$$= p(R_r(w) = 0) \cdot \sum_{r=0}^{n_{c_w}-1} \frac{p_r(I(r_{0,w} + r))}{1 - p_r(I(r_{0,w} + r))} \tag{7.6}$$

$$p(R_r(w) = 2) = \frac{1}{2} \cdot p(R_r(w) = 0)$$
$$\cdot \sum_{r=0}^{n_{c_w}-1} \left[\frac{p_r(I(r_{0,w} + r))}{1 - p_r(I(r_{0,w} + r))} \cdot \sum_{s=0,s\neq r}^{n_{c_w}-1} \frac{p_r(I(r_{0,w} + s))}{1 - p_r(I(r_{0,w} + s))} \right] \tag{7.7}$$

$$p(R_r(w) = 3) = \frac{1}{3!} \cdot p(R_r(w) = 0) \cdot \sum_{r=0}^{n_{c_w}-1} \left[\frac{p_r(I(r_{0,w} + r))}{1 - p_r(I(r_{0,w} + r))} \right. \tag{7.8}$$
$$\left. \cdot \sum_{s=0,s\neq r}^{n_{c_w}-1} \left[\frac{p_r(I(r_{0,w} + s))}{1 - p_r(I(r_{0,w} + s))} \cdot \sum_{t=0,t\neq q,r}^{n_{c_w}-1} \frac{p_r(I(r_{0,w} + t))}{1 - p_r(I(r_{0,w} + t))} \right] \right],$$

where $r_{0,w}$ is the index of the first residue in code word w.

The code rate adaptation algorithm calculates the word error probability $p_w(w)$ for the RRNS code word index w for the lowest-power codec mode of Table 7.2, $c = 0$. If the word error probability exceeds a certain threshold, i.e. $p_w(w) > \alpha$ for $c = 0$, then the next stronger coding mode $c = 1$ is selected, and the word error probability is evaluated again. If the new RRNS code word error probability exceeds the threshold α, then the next codec mode is evaluated, until the estimated RRNS code word error probability falls below the threshold α, or until the highest power codec mode is selected.

The parameter α is supplied to the algorithm and it can be used to control the adaptation process, similarly to the target BER in the adaptive modulation algorithm of Section 6.2.4.2.

The effect of different settings of the parameter α on the effective data throughput and BER of a QPSK modem in an AWGN channel is shown in Figure 7.4. The word error rate (WER) threshold α was varied from 10^{-1} up to 10^{-4}, and it can be seen that the system performance varies only in a limited range of SNR values, approximately between 7 and 12 dB, with the WER α. The three RRNS-coded BER curves of Figure 7.1 are easily identified in this figure as segments of the adaptive scheme's BER performance curve together with the BER curve of uncoded QPSK. The different codec modes are easily identifiable in this figure, since in the AWGN channel the channel quality is time invariant, and therefore the same codec mode is chosen for all transmitted RRNS code words at a given SNR point. The choice of the RRNS codec mode clearly depends on the value of the WER α. For all four values of α, the system performance is identical for SNR values up to 6 dB. For higher SNR values, it can be observed that the WER α influences the codec mode switching algorithm. For $\alpha = 10^{-1}$, the codec mode is switched to the next lower-power higher-rate mode of

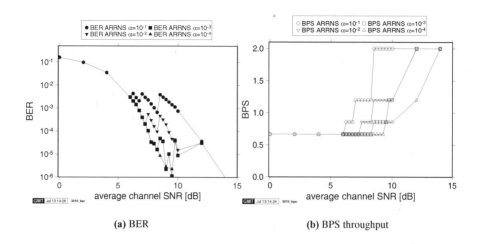

(a) BER   (b) BPS throughput

Figure 7.4: BER performance and throughput in bits per symbol for the adaptive RRNS (ARRNS) coded OFDM system in AWGN channels, with QPSK transmission in the subcarriers

Table 7.2 at SNR values of 6.25, 7 and 8.5 dB, resulting in a BER of about 0.7% after each reconfiguration as seen in Figure 7.4. The maximal throughput of 2 bits/symbol is reached at an SNR of 8.5 dB by the $\alpha = 10^{-1}$ curve. For lower values of α, the codec mode switching is more conservative hence lower bit error rates and throughputs are observed. For $\alpha = 10^{-4}$ the modem employs codec mode 3 of Table 7.2 for SNR values of up to 9.5 dB, reaching mode 0 at 14 dB SNR.

Over fading time dispersive channels the ARRNS codec mode is adapted to the time and frequency variant channel conditions. In order to demonstrate the RRNS code allocation process, in Figure 7.5 we captured the expected subcarrier SNR versus the subcarrier index for one specific OFDM symbol. It can be seen that the channel SNR varies by about 18 dB across the set of subcarriers, which will result in a varying BER across the OFDM symbol. In our illustrative example a specific OFDM symbol was selected that contained at least one RRNS code word corresponding to each of the RRNS codec modes of Table 7.2. In order to augment our understanding, in this example the residues of only one RRNS code word per codec mode are shown in the figure, corresponding to a total of $3 + 5 + 7 + 9 = 24$ residues, signified by vertical bars in the figure. Each vertical bar corresponds to one transmitted residue, and the shading of each bar relates it to one of the four ARRNS code words. The position of a bar gives the residue index r, while its height signifies the expected residue error rate (RER) $p_r(r)$ evaluated from Equation 7.1 for each of the residues of the four chosen ARRNS code words, which depends on the subcarrier SNR experienced at index r.

The channel model used for the fading experiments was described in Section 4.1.1.2. The word error probability threshold α was set to 10^{-1}.

The four code words were chosen so that one of each codec mode of Table 7.2 is demonstrated: the light grey bars mark the position and the residue error probabilities for an uncoded (mode 0) (3, 3) RRNS code word. The positions of the residues in the transmitted OFDM symbol are determined by the residue-based interleaver function. It can be seen that, because

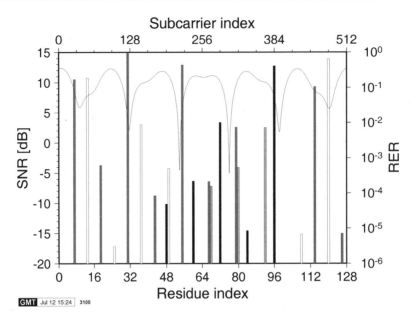

Figure 7.5: Predicted subcarrier SNR (continuous line) for an OFDM symbol and residue error probabilities (vertical bars) for four selected ARRNS code words for a 512-subcarrier OFDM modem employing QPSK in the short WATM channel of Figure 4.3. The average SNR is 10 dB, and $\alpha = 10^{-1}$. Key: light grey = (3, 3) code at indices of 68, 80, 92; black = (5, 3) at 48, 60, 72, 84, 96; white = (7, 3); dark grey (9, 3)

of the good channel quality in the subcarriers used for transmission, all three residues of the uncoded code word exhibit low RER values. The word error probability for this code word is mainly dominated by the highest RER residue transmitted at index 92, which exhibits an error probability of about $7.2 \cdot 10^{-3}$. The resulting word error probability due to all three residues is $7.9 \cdot 10^{-3}$, which is below the threshold of $\alpha = 10^{-1}$, and therefore uncoded transmission was chosen by the algorithm.

The black bars in Figure 7.5 mark the residues for a mode 1 (5, 3) code word, with five transmitted residues, while the white marked residues form a mode 2 (7, 3) code word, consisting of seven transmitted residues. The first of this code word's residues at residue index $r = 0$ is not visible on the figure, as its expected RER is $2.1 \cdot 10^{-7}$.

The dark grey bars of Figure 7.5, represent a mode 3 (9, 3) RRNS code word with 9 transmitted residues. The word error probability for this mode 3 RRNS code word is 11.5%, which exceeds the set threshold. Since there is no stronger code in the set of coder modes of Table 7.2, the highest possible code rate was chosen.

Figure 7.6 shows the BER performance and data throughput of a 512-subcarrier ARRNS-coded OFDM modem employing QPSK in the SWATM channel of Figure 4.3. Figure 7.6(a) depicts the BER and BPS performance for the modem employing no residue interleaving, while for Figure 7.6(b) a residue interleaver obeying the structure introduced in Figure 7.3 was used. This comparison is relevant and necessary, since it is not intuitive as to whether residue interleaving results in performance improvements. This is because the residue inter-

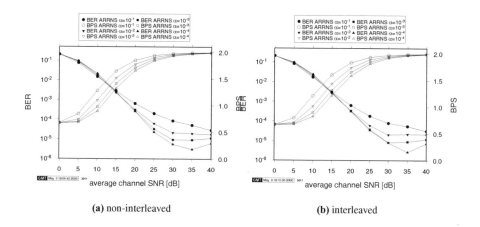

Figure 7.6: BER and BPS throughput versus average channel SNR for ARRNS-coded 512-subcarrier OFDM transmission employing QPSK over the shortened WATM channel of Figure 4.3. The stipulated WER values were $\alpha = 10^{-1}, 10^{-2}, 10^{-3}$ and 10^{-4}

leaving disperses the bursty channel errors across the OFDM subcarriers, which is expected to improve the coding performance of conventional fixed rate coding. We, however, proposed adaptive RRNS-coding, in order to combat the bursty error distribution across the OFDM subcarriers, which was designed to counteract the frequency selective fading and its bursty errors. Hence interleaving and adaptive RRNS coding may disadvantageously interfere with each other, necessitating a comparison of the interleaved and non-interleaved results.

It can be seen that the BER performance is fairly similar for the interleaved and non-interleaved modems. Specifically, although the non-interleaved modem slightly outperforms the interleaved one in BER terms for WERs of $\alpha = 10^{-3}$ and for $\alpha = 10^{-4}$ at SNR values in excess of 25 dB, the BER difference is not significant. Upon comparing the achieved throughput, however, it is clear that the non-interleaved modem offers an average throughput benefit of about 0.1 bit/symbol for all target WERs α in the SNR region up to 25 dB. Since the range of code rates is limited, the BER performance at low values of SNR cannot be significantly lowered. Similar to the adaptive modulation systems of Chapter 6, transmission blocking would have to be introduced in order to guarantee a target bit error rate. Let us now consider the combination of ARRNS coding and AOFDM modulation in the next section.

7.2.2   ARRNS/AOFDM Transceivers

In this section we will demonstrate that upon combining adaptive modulation with adaptive coding, the low SNR performance can be dramatically improved by amalgamating transmission blocking with adaptive error correction coding. We have advocated here the target BER adaptive modulation algorithm of Section 6.2.4.2 due to its high performance and easy adjustability to different target bit error rates.

The transmission parameter adaptation is performed in two steps. First, the modulation modes are allocated to the subcarriers according to the algorithm outlined in Section 6.2.4.2. Recall that for AOFDM the expected overall bit error probability for the given OFDM symbol

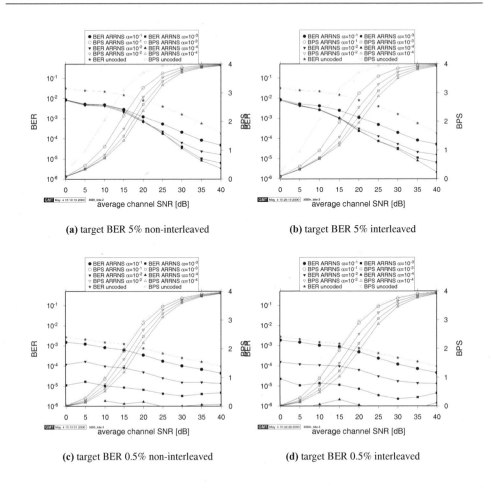

Figure 7.7: BER and BPS throughput versus average channel SNR for hard decision ARRNS-coded 512-subcarrier OFDM transmission employing adaptive modulation over the shortened WATM channel of Figure 4.3: (a, b) uncoded adaptive modulation target BER 5%, (c, d) uncoded target BER 0.5%, (a, c) non-interleaved, (b, d) interleaved. The stipulated WER values were $\alpha = 10^{-1}, 10^{-2}, 10^{-3}$ and 10^{-4}. The light grey curves show the uncoded BER and BPS throughput

was computed by averaging the individual subcarrier BERs for all legitimate modulation modes M_n, yielding $\bar{p}_e(M_n) = 1/N_s \sum_j p_e(\gamma_j, M_n)$ where γ_j is the subcarrier SNR for $j = 1, \ldots, 512$. The specific subcarrier modem mode allocation with the highest bits per symbol (BPS) throughput, whose estimated BER is lower than the required value is then confirmed. According to this step, the number of bits N_b transmitted in the next OFDM symbol and their estimated bit error probabilities $p_e(n)$ are known. On the basis of this, the code rate adaptation algorithm calculates the residue error rates $p_r(r)$ from Equation 7.1, constructs the interleaver $I(r)$ for the correct number of residues and invokes the appropriate codec modes for the RRNS code words, as outlined above.

Figure 7.7 gives an overview of the ARRNS/AOFDM system's BER and throughput performance over the short WATM channel of Figure 4.3. Two target BER values have been stipulated, both with and without interleaving of the transmitted residues. Figure 7.7(b) and 7.7(a) portray the system's performance if a target BER of 5% is assumed for the adaptive modulation with and without interleaving, respectively. As has been observed in Chapter 6, the uncoded BER, represented by the grey curves, is lower than the AOFDM target BER, which is due to the operation of the AOFDM algorithm. It can be seen that the coded BER is below 1% for all simulated modem configurations, and that the SNR gain is much higher than for fixed QPSK transmission in Figure 7.2. The BER performance is limited, however, by the limited error correction capability of the RRNS (9, 3) mode when the SNR is very low.

Lowering the adaptive modulation's target BER to 0.5%, the BER of the system can be influenced over the whole SNR range by varying the WER α. Figure 7.7(d) and 7.7(c) depict the corresponding BER and BPS throughput. For a WER of $\alpha = 10^{-1}$ the achieved BER is better than $2 \cdot 10^{-3}$, while for $\alpha = 10^{-2}$ a BER of $2 \cdot 10^{-4}$ is never exceeded and for $\alpha = 10^{-3}$ a BER of $2 \cdot 10^{-5}$.

Comparing the interleaved performance with the non-interleaved results, it can be seen that the BER performance of comparable modems is fairly similar. The throughput is slightly higher, however, for the non-interleaved systems, demonstrating the efficiency of the hard decision ARRNS/AOFDM schemes in terms of combating the bursty errors of frequency selective fading. Let us now consider the effects of invoking soft decisions.

7.2.3   Soft Decision RRNS Decoding

The performance of the ARRNS codes can be increased by soft decision decoding [240]. Previously, the ARRNS decoder assumed that the outputs of the demodulator were binary hard decision values. However, the transceiver is capable of exploiting soft outputs provided by the demodulator at the receiver. Soft decoding of the ARRNS codes can be implemented by combining the classic Chase algorithm [244] with the hard decision ARRNS decoder.

Figure 7.8 shows the soft decision decoded performance of the AOFDM/ARRNS system. Comparison with Figure 7.7 shows an improved BER performance for the soft decoder. This is especially significant for the target bit error rate of 5%, where the soft decoded AOFDM/ARRNS system achieves BER values of below 0.3% at 0 dB SNR. Since the adaptation algorithm is unchanged, the same throughput is observed for the hard and soft decoded systems. A BER of below 10^{-4} was registered for the 0.5% target BER system for a WER of $\alpha = 10^{-2}$. Under these circumstances the interleaved system exhibits a lower throughput and worse BER performance than the non-interleaved system. Let us now compare the performance results to those of turbo BCH codecs in the next section.

7.3   Turbo BCH Codes

As an alternative to the above ARRNS codes, we have investigated the employment of turbo BCH codes for the adaptive modem. In Section 6.2.7, half-rate convolutional turbo codes were employed in the adaptive modem. However, the convolutional constituent codes can also be replaced by block codes [231, 232], which have been shown, for example by Hagenauer [245, 246] to perform impressively even at near-unity coding rates. Generally, block codes are appropriate for high code rates and convolutional codes for lower code rates, typ-

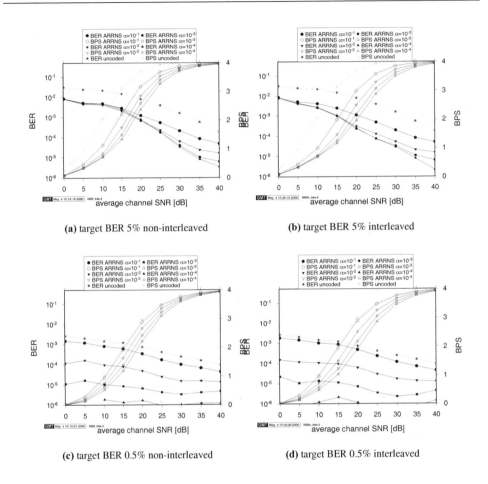

(a) target BER 5% non-interleaved

(b) target BER 5% interleaved

(c) target BER 0.5% non-interleaved

(d) target BER 0.5% interleaved

Figure 7.8: BER and BPS throughput versus average channel SNR for soft decision ARRNS-coded 512-subcarrier OFDM transmission employing adaptive modulation over the shortened WATM channel of Figure 4.3. (a, b) uncoded adaptive modulation target BER 5%, (c, d) uncoded target BER 0.5%, (a, c) non-interleaved, (b, d) interleaved. The stipulated WER values were $\alpha = 10^{-1}, 10^{-2}, 10^{-3}$ and 10^{-4}. The light grey curves show the uncoded BER and BPS throughput. The corresponding hard decision results are plotted in Figure 7.7

ically below 2/3, as argued by Hagenauer *et al.* [245]. Turbo block codes have since been studied in a variety of applications [247–250].

For our adaptive coding system, a turbo (T) code with constituent BCH(31, 26) coders was advocated. The parity bits of the two constituent codes were not punctured, resulting in a TBCH(36, 26) code. The resulting code rate was therefore 0.72. The turbo interleaver employed was a 4×13 bit interleaver, leading to a block size of 52 uncoded or 72 coded bits.

Figure 7.9 shows the BER performance of the TBCH code over an AWGN channel with QPSK transmission using 6 iterations and a 4×13 turbo interleaver. Comparison with the AWGN performance of the hard decision decoded RRNS codes in Figure 7.1 reveals that the TBCH code outperforms all the investigated RRNS codes. The SNR gain of the 0.72 rate

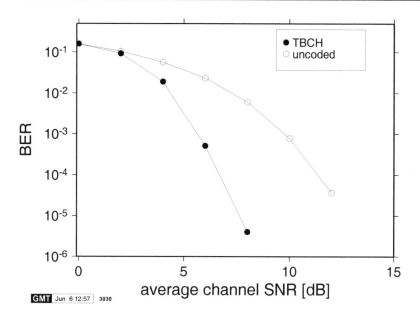

Figure 7.9: BER performance of the 0.72 rate $TBCH(36, 26)$ coded OFDM system over the AWGN channel, employing QPSK transmission in the subcarriers using 6 iterations and a 4×13 turbo interleaver

TBCH code at a BER of 10^{-4} is about 4.5 dB, which is 0.8 dB better than that of the $1/3$ rate $(9, 3)$ RRNS code. We note, however, that the RRNS coding performance can be improved by invoking turbo coding. This issue is the subject of further research.

For transmission over the short WATM channel of Figure 4.3, a block channel interleaver obeying the structure depicted in Figure 7.3 was employed. In the case of the 0.72 rate TBCH(26, 26) coder, the interleaving was performed bit-by-bit, instead of the residue-based interleaving of the RRNS system. Figure 7.10 shows the BER performance of the studied TBCH code over the short WATM channel of Figure 4.3. It can be seen that channel interleaving of the TBCH-coded bits significantly improves the performance of the codec. For a BER of 10^{-4} a SNR gain of 10 dB was recorded for the non-interleaved, while an SNR gain of 16.5 dB was observed for the interleaved system.

Figure 7.11 gives a comparison between the 0.72 rate TBCH performance of Figure 7.10 and the long-block-length half-rate turbo convolutional code of Figure 6.13 in E_b/N_0 terms. It can be seen that the TBCH code with additional channel interleaving performs only about 1 dB worse in E_b/N_0 than the convolutional turbo code at a BER of 10^{-4}.

7.3.1 Adaptive TBCH Coding

Unlike the ARRNS code rate adaptation algorithm, which employed four code rates, the proposed TBCH rate adaptation regime operates as a simple on/off switch. If the channel quality is lower than a threshold, then N_d data bits are encoded into N_c coded bits and transmitted in the bit positions allocated for the code word. If the channel quality is sufficiently high for

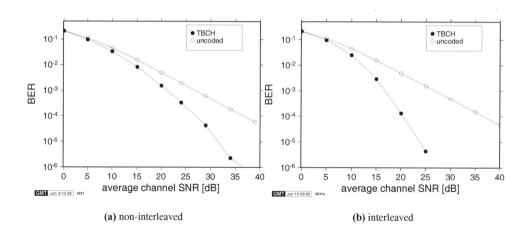

(a) non-interleaved **(b)** interleaved

Figure 7.10: BER versus average channel SNR for the 0.72 rate $TBCH(36, 26)$ coded 512-subcarrier OFDM transmission employing QPSK over the short WATM channel of Figure 4.3 with and without channel interleaving, using 6 iterations and a 4×13 turbo interleaver

uncoded transmission, then N_c data bits are directly mapped into the allocated bit positions.

Since binary BCH codes are employed, the estimated bit error rate in each code word can be used directly as the adaptation criterion. Based on the estimated bit error probability $p_e(n)$, the uncoded bit error probability in the code word w of length N_c can be calculated as:

$$\bar{p}_e(w) = \frac{1}{N_c} \cdot \sum_{n=0}^{N_c} p_e\left(I\left(n + n_0(w)\right)\right),\tag{7.9}$$

where $I()$ is the channel interleaver, while $n_o(w)$ is the index of the first bit of the code word w.

The adaptation algorithm compares the estimated BER in the code word with a bit error rate threshold β, and if $\bar{p}_e(w) \leq \beta$ then N_c data bits are transmitted in the word w. Otherwise, coded transmission is selected.

Figure 7.12 shows the BER and throughput of the ATBCH modem employing QPSK over the short WATM channel of Figure 4.3 for BER thresholds of $\beta = 10^{-2}, 10^{-3}$ and 10^{-4}. It can be seen that for both the interleaved and non-interleaved cases the BER curves follow the trends found in Figure 7.10 for low SNR values before the adaptation algorithm gradually increases the average code rate towards unity, approaching in a throughput of 2 BPS. Comparing the throughput of the interleaved and non-interleaved modems, it can be seen that the non-interleaved scheme yields a slightly higher throughput across the SNR range. Let us now combine ATBCH coding with AOFDM in the next section.

7.3.2 Joint ATBCH/AOFDM Algorithm

As in the ARRNS/AOFDM system of Section 7.2.2, the ATBCH code has been combined with the adaptive modulation scheme selection regime discussed in Section 6.2.4.2. The

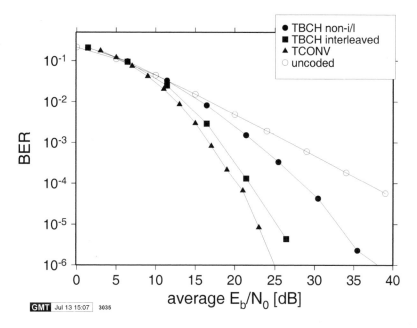

Figure 7.11: Comparison of the BER versus E_b/N_0 performance of the 0.72 rate $TBCH(36, 26)$
coded with and without channel interleaver and the $1/2$ rate turbo convolutional coded
OFDM system with 2000-bit turbo interleaver of Section 6.2.7 (Figure 6.13) employing
QPSK transmission over the short WATM channel of Figure 4.3

BER and BPS throughput performance of the ATBCH/AOFDM system is determined by the
combination of the target BER for the adaptive modulation scheme and the BER threshold
β for the code rate adaptation. Figure 7.13 depicts the resulting system performance for
uncoded AOFDM target BER values of 10% and 1%, and for ATBCH activation threshold
BER values of $\beta = 10^{-2}, 10^{-3}$ and 10^{-4}. For the uncoded AOFDM target BER of 10%,
neither the interleaved nor the non-interleaved modems can reduce the data BER to acceptable
levels for low SNR values. The interleaved system shows a better BER performance, but
exhibits lower throughput than the non-interleaved system. For an uncoded AOFDM target
BER of 1%, the ATBCH decoder can efficiently reduce the data BER below 10^{-4} for $\beta =
10^{-4}$. This meets the requirements of the adaptive modem's BER performance, as set out in
Chapter 6.

7.4 Signalling

Signalling the codec modes employed for the code words of each OFDM symbol requires
a significant amount of bandwidth, analogously to the modulation mode signalling for
AOFDM. The effects of erroneous mode detection at the receiver are the same as for the
modulation mode signalling, resulting in loss of synchronisation in the data stream.

Blind detection of the codec mode at the transmitter in the case of open-loop coding

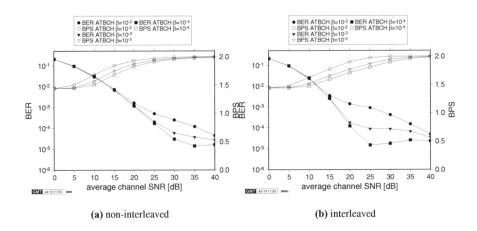

(a) non-interleaved   **(b)** interleaved

Figure 7.12: BER and BPS throughput versus average channel SNR for adaptive TBCH(36, 26) coded
512-subcarrier OFDM transmission employing QPSK over the short WATM channel of
Figure 4.3 with and without channel interleaving

adaptation could be employed for longer block length codes, by tentatively decoding the
received data sequence and exploiting soft outputs of the decoder as a reliability measure. If
decoding of the received sequence increases the reliability measure for the received block's
data bits, then the corresponding coder mode is chosen. However, the short block lengths,
necessitated by the adaptation algorithms discussed in this chapter make it impossible to use
blind detection of the codec mode based on the decoder's soft outputs.

The two different adaptive error correction schemes discussed in this chapter both require
signalling of the codec mode for each of the transmitted code words, but the scenarios are
different. The ARRNS scheme, which employs four codec modes with code words of variable
length between 24 and 72 bits, requires a signalling scheme similar to that of the AOFDM
scheme discussed in Section 6.2.6.1.

The ATBCH scheme, with only two different codec modes and a fixed block length, is
easier to signal. Depending on the required mode detection error rate, and assuming that the
achieved uncoded BER is the uncoded AOFDM target BER, then the number of signalling
bits necessary for majority voting signalling can be calculated. If a target BER of 1% is
assumed, then a 5-bit signalling sequence results in a detection error rate of 10^{-5}. If 7
signalling bits are used, then a DER of $3.5 \cdot 10^{-7}$ can be achieved.

7.5   Chapter Summary and Conclusion

Figure 7.14 shows the throughput of the various adaptive transmission systems studied for
a target data BER of 10^{-4}. The lightly shaded curves represent the variable throughput
systems' performance graphs from Figure 6.29. The ARRNS/AOFDM and ATBCH/AOFDM
systems employed no interleaving and had an AOFDM target BER of 1%. The ARRNS
adaptation algorithm used a WER threshold of $\alpha = 10^{-2}$, while the ATBCH system used a
data BER threshold of $\beta = 10^{-4}$. The pre-equalisation results from Figure 6.29 were omitted

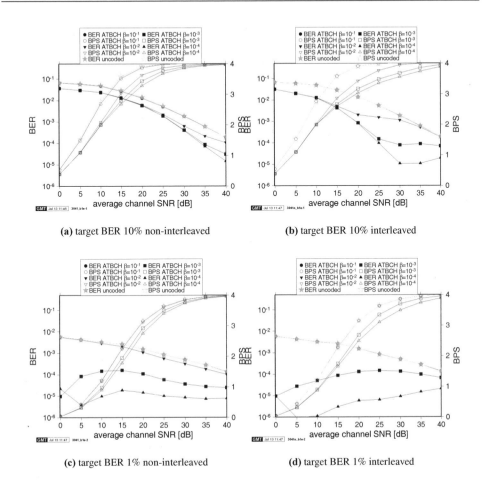

Figure 7.13: BER and BPS throughput versus average channel SNR for ATBCH-coded 512-subcarrier AOFDM over the shortened WATM channel of Figure 4.3: (a, b) uncoded AOFDM target BER 10%, (c, d) target BER 1%, (a, c) non-interleaved, (b, d) interleaved. The stipulated BER threshold values were $\beta = 10^{-1}, 10^{-2}, 10^{-3}$ and 10^{-4}. The light grey curves show the uncoded BER and BPS throughput

in this figure. It should be noted that the above performance figures do not take into account the signalling overhead required for adaptive modulation and adaptive code rates and hence constitute the upper bound performance of the system.

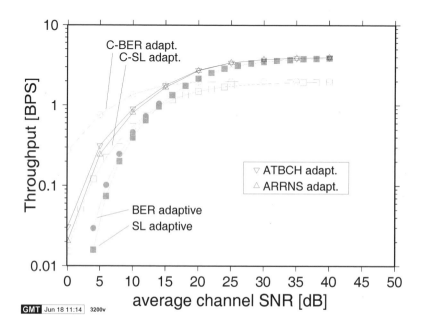

Figure 7.14: BPS throughput versus average channel SNR for AOFDM for a maximum data bit error
rate of 10^{-4}. The light grey curves present the performance of the variable-throughput
adaptive modulation schemes Figure 6.29 with and without convolutional turbo coding.
The variable throughput systems include convolutionally turbo coded (C-) and uncoded
switching level adaptive (SL) and target BER adaptive (BER) systems, as well as the joint
adaptive RRNS/AOFDM and ATBCH/AOFDM systems. The short WATM channel of
Figure 4.3 was employed

Part II

OFDM versus MC-CDMA Systems, Their Spreading Codes and Peak Factor Reduction

Chapter **8**

OFDM versus MC-CDMA[1]

In 1993, a number of hybrid transmission techniques employing an amalgam of Code Division Multiple Access (CDMA) and Orthogonal Frequency Division Multiplexing (OFDM) were proposed [52, 55], which are expected to combine the benefits of pure CDMA and OFDM techniques. The aim of Chapters 8–12 is to examine various aspects of these techniques.

Since multi-carrier CDMA techniques rely on the combination of CDMA and OFDM, the two conventional techniques will be reviewed briefly. The concept of multi-carrier CDMA will be presented next.

8.1 Amalgamating DS-CDMA and OFDM

8.1.1 The DS-CDMA Component

Direct-Sequence (DS) CDMA is a spread-spectrum communication technique. DS-CDMA systems [251, 252] are capable of supporting a multiplicity of users within the same bandwidth by assigning different – typically unique user-specific – codes to different users for their communications, in order to be able to distinguish their signals from each other at the receiver. Spread-spectrum techniques were developed originally for military guidance and communications systems [253]. During the Second World War, radar engineers used spread-spectrum techniques to mitigate intentional jamming and to achieve high resolution ranging. During the late 1970s, the employment of spread-spectrum techniques was proposed for efficient cellular communication [254]. It is interesting to note that [254] addressed most essential issues involved in DS-CDMA cellular communications at such an early stage, although the proposed scheme was based on frequency hopping spread-spectrum communication. For a detailed historical review of spread-spectrum based communications, the reader is referred to the overview papers by Scholtz [253], by Yue [255] or to Chapter 2 of [256] by Simon,

---

[1] *OFDM and MC-CDMA for Broadband Multi-user Communications WLANS and Broadcasting.*
L.Hanzo, Münster, T. Keller and B.J. Choi,
©2003 John Wiley & Sons, Ltd. ISBN 0-470-85879-6

Omura, Scholtz and Levitt.

In CDMA systems, users share the same broad bandwidth all the time, using different spreading codes. This unique feature of CDMA results in a soft capacity limit [254], while conventional Frequency Division Multiple Access (FDMA) and Time Division Multiple Access (TDMA) have hard capacity limits, since they use a finite number of orthogonal resources. The limiting factor of the capacity of CDMA systems is self-interference and Multi-User Interferences (MUI). The effects of these interference sources depend on the channel's characteristics and on the properties of the spreading codes used. Since any method, which reduces the interference is capable of increasing the overall user capacity in CDMA, significant research efforts have been invested in reducing the interference [257]. Several simple techniques, such as spatial isolation of users by cell sectorisation and discontinuous transmission relying on a voice activity detection or on the burstiness of the information source were shown to increase the user capacity [258]. Other sophisticated techniques employing interference cancellation [259], joint detection [260, 261] and adaptive antenna arrays [262] are realistic at the time of writing and offer further capacity gains.

One of the important merits of CDMA in cellular environments is its near-unity frequency reuse factor [258]. In TDMA and FDMA systems, the same frequency could be reused in different cells only beyond a sufficiently hight reuse distance, where the effects of interference between users of the same channel became negligible [2, 263, 264]. By contrast, in CDMA systems, the adjacent cells can use the same frequency in conjunction with unique cell-specific spreading sequences assigned for each cell site. Frequency planning is not required any more, while cell planning still remains an important issue, for example, for reasons of transmit power reduction [265]. An obvious advantage of the universal frequency reuse is the substantial potential increase in user capacity per unit bandwidth in comparison to 7-cell clusters. Another important advantage is the ability to use soft hand-off [258]. When a mobile roams across a cell boundary, its communication channel has to be handed over to a more suitable base station [2, 263]. In conventional multiple access systems often this means an abrupt RF channel change. Hence the communication link typically experiences a short discontinuity, while this RF channel change takes place. This is referred to as hard hand-off or hard hand-over [263]. In CDMA systems all base stations may use the same frequency. So-called Rake receivers [266, 267] are used in the mobile stations for combining the signals from the two base stations involved in the hand-over [265]. This soft hand-off makes seamless communications possible [268]. The soft hand-off also mitigates the multi-path fading effects during the hand-over period with the aid of cell site diversity [265].

In mobile communication the transmitted signal passes through multiple paths generally exhibiting different path lengths. The arrival time of the signals at a receiver spans from several tens of ns to several tens of μs, depending on the environments. The channel impulse characterises the scattering of the transmitted energy in the time domain. The associated time domain description results in Inter-Symbol Interference (ISI), where each impulse response component is exposed to fast Rayleigh fading. In conventional FDMA and TDMA systems complex channel equalisers are required for combating dispersion. As the transmission rate is increased, the number of ISI-contaminated adjacent symbols is proportionately increased [2]. Ironically, CDMA communication systems intentionally use significantly higher transmission rates, or chip rates, in order to take advantage of these scattered multi-path signals. Rake receivers [266, 267] are capable of resolving each delayed signal component with an accuracy of a chip period. Each demodulated component is combined by the Rake receivers in order to make effective use of the channel-induced multi-path diversity. A system having a shorter chip period or wider bandwidth is capable of resolving lower delay differences and hence may benefit from higher-order diversity. This characteristic together with the demands of higher data transfer rates render wide-band CDMA (W-CDMA) more attractive than the existing Pan-American narrow-band CDMA system known as IS-95 [269].

DS-CDMA systems typically suffer from the so-called near/far problem [270, 271]. This problem is more acute in the up-link due to its asynchronous nature, than in the synchronous down-link [265]. The base station's receiver will experience excessive multi-user interferences resulting from the undesirable non-zero asynchronous cross-correlation between the spreading sequences. This is so, even when the asynchronous cross-correlations of the sequences employed are low, due to the effects of the multi-path channel. Thus, a mobile station near the base station should reduce its transmit power so that all the reverse link signals at the base station's receiver can have an equal power, independent of their geographical locations. In open-loop power control schemes [271] a mobile station measures the power of the signal received from the base station for determining its required transmit power, assuming that the transmit path and the receiving path have approximately equal attenuations. Sometimes this is not the case, especially in frequency division duplex (FDD) CDMA systems. Thus closed-loop power control is required. In closed-loop power control schemes [272, 273] the base station sends power adjustment commands to each mobile station based on several measures, such as the received signal strength, the ratio of signal energy per bit to noise density (E_b/N_o) and the Bit Error Ratio (BER). Stability problems may arise in closed-loop power control and hence the global stability of the system has to be ensured [272]. Control latency and power adjustment levels are two key parameters in designing a closed-loop power control system [271].

Although power control is typically capable of tracking the power variation due to slow fading, it is almost impossible to compensate for the effects of fast Rayleigh fading, simply because the actual power variation speed exceeds the power control speed. In this case, interleaving combined with channel coding can support the operation of the power control in attaining the target performance. It is desirable to design the interleaver such that the originally consecutive symbols become separated in time by more than the channel's coherence time at the output of interleaver [271]. This implies that a higher interleaver depth is desirable for mobile stations travelling at lower speed, for example. However, a higher interleaver depth implies a longer processing delay, which is undesirable, for example, in interactive voice communications.

CDMA systems use a significantly higher bandwidth than the modulating signal's bandwidth by spreading the original signal with the aid of high rate spreading sequences. The noise-like spreading sequences play an important role in characterising a CDMA communication system. In frequency selective multi-path fading environments ensuring low off-peak, auto-correlation of the spreading sequences is necessary to reduce the Inter-Symbol Interference (ISI) and to combine the energy scattered in the time domain using Rake receivers. In multi-user environments additionally a low cross-correlation of the different user's sequences is required to distinguish the desired users' signals. The family of various spreading sequences will be reviewed in Chapter 9, while a range of related classic CDMA design aspects are treated in the excellent monographs by Viterbi and Lee [251, 252].

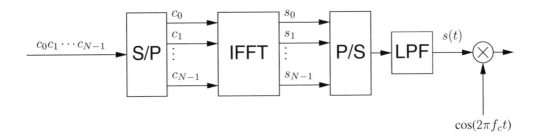

Figure 8.1: OFDM transmitter schematic

8.1.2 The OFDM Component

Orthogonal Frequency Division Multiplexing (OFDM) constitutes a specific form of multi-carrier modulation technique [274]. The basic approach of OFDM [1, 24, 275] is to group serial message symbols and transmit each message symbol on different frequency domain carriers at a reduced signalling rate. The conventional Frequency Division Multiplexing (FDM) technique requires guard bands between adjacent channels, since the receivers make use of band-pass filters to isolate each channel's signal. However, OFDM exploits the orthogonality of the subcarrier signals, although the actual frequency spectra of the different subchannels partially overlap with each other. The receiver in OFDM systems makes use of the orthogonality between the subcarriers to recover the symbols mapped on to a given subcarrier.

The complex baseband equivalent of the OFDM signal $s(t)$ in a symbol duration can be represented as [10]:

$$s(t) = \sum_{k=0}^{N-1} c_k \, e^{j2\pi \frac{k}{T} t} , \tag{8.1}$$

where N is the number of carriers and T is the subchannel signalling interval, while c_k is the symbol modulating the subcarrier k. In the next signalling interval a new set of symbols c_k is transmitted. At the receiver, the received signal is multiplied by $e^{-j2\pi \frac{n}{T} t}$ and integrated over a symbol duration in order to recover c_n. The resultant signal becomes, assuming perfect carrier frequency and symbol time recovery over an ideal channel:

$$\frac{1}{T} \int_0^T s(t) e^{-j2\pi \frac{n}{T} t} dt \;=\; \frac{1}{T} \sum_{k=0}^{N-1} c_k \int_0^T e^{j2\pi \frac{k}{T} t} e^{-j2\pi \frac{n}{T} t} dt = c_n . \tag{8.2}$$

In order to implement directly the transmitter and the receiver of an OFDM system, N oscillators are required.

Weinstein and Ebert [9] presented a method involving the Discrete Fourier Transform (DFT) to perform baseband modulation and demodulation, which spurred the development of OFDM systems with the advent of efficient real-time Digital Signal Processing (DSP) technology. By sampling the modulated signal N times during an OFDM symbol at instants of $t = \frac{m}{N}T$, Equation 8.1 becomes:

$$s\left(\frac{m}{N}T\right) = \sum_{k=0}^{N-1} c_k e^{j2\pi\frac{km}{N}}, \text{ (for } m = 0, 1, \cdots, N-1) . \tag{8.3}$$

Since $s\left(\frac{m}{N}T\right)$ depends only on m, it can be represented as s_m in discrete form, and Equation 8.3 can also be written as:

$$s_m = N \cdot \text{IDFT}(\{c_k\}), \text{ (for } m = 0, 1, \cdots, N-1), \tag{8.4}$$

where IDFT represents the Inverse Discrete Fourier Transform operator. The efficient implementation of the IDFT is the Inverse Fast Fourier Transform (IFFT). The overall structure of the OFDM transmitter is shown in Figure 8.1.

The N message sequences, c_0, c_1, $\cdots$, c_{N-1}, form a frame, which is converted into a parallel form, where c_k is modulating the kth carrier. The IFFT module takes the parallel data and calculates N sampled time domain signals, s_0, s_1, $\cdots$, s_{N-1}, which are low-pass filtered to generate the continuous time domain signal. This baseband time domain signal modulates a single carrier frequency and transmitted. The IFFT eliminates the use of N oscillators and renders the OFDM transmitter implementationally attractive [10].

At the OFDM receiver, the reverse action takes place. The down-converted received signal is sampled at a rate of N/T and converted into a parallel stream of N values. Then, the FFT is applied in order to recover the desired frequency components, i.e. the information symbols, c_k.

In mobile channel environments the transmitted signals experience reflection and refraction [2]. This results in different path delays ranging from several ns to tens of μs, depending on the environments encountered. The receiver has to have a channel equaliser in order to cope with the time dispersion of the received signals. OFDM reduces this requirement by increasing the signalling interval duration. However, it needs a guard time between the OFDM signaling intervals in order to reduce the ISI. Instead of a passive guard space, inserting a quasi-periodically repeated cyclic prefix is used to remove inter-symbol interference [274]. Let us assume that the channel impulse response, $h(t)$, spans $(\nu + 1)\frac{T}{N}$, where $\nu \ll N$ and the corresponding discrete channel impulse response, h_n, $0 \leq n \leq \nu$, is $h\left(\frac{n}{N}T\right)$. Let $s_{i,n}$ be the nth transmitted symbol in the ith frame as defined in Equation 8.4. The cyclic prefix consists of $s_{i,N-\nu}$, $s_{i,N-\nu+1}$, $\cdots$, $s_{i,N-1}$, which is inserted before $s_{i,0}$, $s_{i,1}$, $\cdots$, $s_{i,N-1}$. Then, the sampled baseband received signal, $r_{i,n}$, for

$n = -\nu, -\nu + 1, \cdots, -1, 0, 1, 2, \cdots, N - 1$, becomes

$$
\begin{aligned}
r_{i,-\nu} &= h_0 s_{i,N-\nu} + h_1 s_{i-1,N-1} + h_2 s_{i-1,N-2} + \cdots + h_\nu s_{i-1,N-\nu} \\
r_{i,-\nu+1} &= h_0 s_{i,N-\nu+1} + h_1 s_{i,N-\nu} + h_2 s_{i-1,N-1} + \cdots + h_\nu s_{i-1,N-\nu+1} \\
&\vdots \qquad \vdots \\
r_{i,-1} &= h_0 s_{i,N-1} + h_1 s_{i,N-2} + h_2 s_{i,N-1} + \cdots + h_\nu s_{i-1,N-1} \\
r_{i,0} &= h_0 s_{i,0} + h_1 s_{i,N-1} + h_2 s_{i,N-2} + \cdots + h_\nu s_{i,N-\nu} \\
r_{i,1} &= h_0 s_{i,1} + h_1 s_{i,0} + h_2 s_{i,N-1} + \cdots + h_\nu s_{i,N-\nu+1} \\
&\vdots \qquad \vdots \\
r_{i,\nu} &= h_0 s_{i,\nu} + h_1 s_{i,\nu-1} + h_2 s_{i,\nu-2} + \cdots + h_\nu s_{i,0} \\
&\vdots \qquad \vdots \\
r_{i,N-1} &= h_0 s_{i,N-1} + h_1 s_{i,N-2} + h_2 s_{i,N-3} + \cdots + h_\nu s_{i,N-\nu+1} \; .
\end{aligned}
$$

The first ν samples are discarded and the remaining N samples are processed by FFT in the receiver. The remaining N samples can be represented in a more compact form:

$$
r_{i,n} = \sum_{j=0}^{\nu} h_j s_{i,(n-j \bmod N)} \text{ for } n = 0, \cdots, N - 1 \; . \tag{8.5}
$$

The recovered data symbol, $\hat{c}_{i,k}$, becomes

$$
\hat{c}_{i,k} = \sum_{n=0}^{N-1} r_{i,n} e^{-j 2\pi \frac{n}{N} k} \tag{8.6}
$$

$$
= c_{i,k} \sum_{n=0}^{\nu} h_n e^{-j 2\pi \frac{n}{N} k} \tag{8.7}
$$

$$
= c_{i,k} H_k \; , \tag{8.8}
$$

where H_k is the frequency domain channel transfer function and $c_{i,k} = \sum_{n=0}^{N-1} s_{i,n} e^{-j 2\pi \frac{n}{N} k}$ was used to arrive at Equation 8.7. Equation 8.8 states that the original symbol, $c_{i,k}$, can be recovered without inter-frame interference or inter-subcarrier interference given the knowledge of $\{H_k\}$, which can be estimated.

The frequency spectrum of an OFDM signal using no pulse shaping is presented in Figure 8.2. The required bandwidth is approximately $N \times \frac{1}{T}$, which is the minimum bandwidth imposed by Nyquist's sampling theorem. Since the first spectral sidelobes are only about -17.3dB below the passband level, various pulse shaping and wavelet based transform methods were proposed to suppress the sidelobes [274] which result in adjacent channel interference.

One of the main drawbacks of the above OFDM system is the "peaky" time domain signal, which requires amplifiers having a high dynamic range [31, 276]. This issue will be

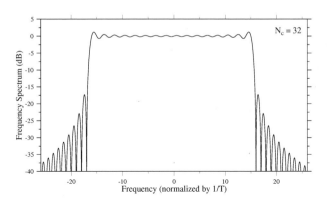

Figure 8.2: Power spectrum of OFDM signal using 32 carriers, without any pulse shaping

discussed in more depth in Chapter 11. Another impediment is that OFDM systems are more sensitive to carrier frequency drift, than single carrier [44, 47] systems.

8.2 Multi-Carrier CDMA

A range of novel techniques combining DS-CDMA and OFDM have been presented in the literature [52, 53, 55, 277, 278].

A DS-CDMA system applies spreading sequences in the time domain and uses Rake receivers to optimally combine the time-dispersed energy in order to combat the effects of multi-path fading. However, in indoor wireless environments the time dispersion is low, on the order of nano seconds, and hence a high chip rate, of the order of tens of MHz, is required for resolving the multi-path components. This implies a high clock-rate, high power consumption as well as implementation difficulties. In order to overcome these difficulties, several techniques have been proposed, which combine DS-CDMA and multi-carrier modulation, such as MC-CDMA [52, 53, 55], MC-DS-CDMA [277] and Multi-Tone CDMA (MT-CDMA) [278]. This overview is mainly based on references [56, 279] by Prasad and Hara, [280] by Scott *et al.*

8.2.1 MC-CDMA

In MC-CDMA, instead of applying spreading sequences in the time domain, we can apply them in the frequency domain, mapping a different chip of a spreading sequence to an individual OFDM subcarrier. Hence each OFDM subcarrier has a data rate identical to the original input data rate and the multicarrier system "absorbs" the increased rate due to spreading in a wider frequency band. The transmitted signal of the ith data symbol of the jth user $s_i^j(t)$ is

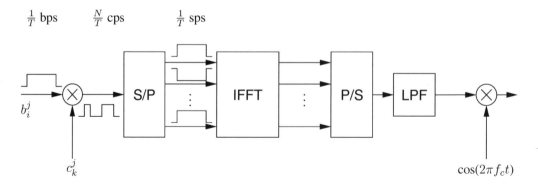

Figure 8.3: Transmitter schematic of MC-CDMA

written as [52, 281]:

$$s_i^j(t) = \sum_{k=0}^{N-1} b_i^j \, c_k^j \, e^{2\pi(f_0 + k f_d)t} \, p(t - iT) \,, \tag{8.9}$$

where

- N is the number of subcarriers

- b_i^j is the ith message symbol of the jth user

- c_k^j represents the kth chip, $k = 0, \cdots, N - 1$, of the spreading sequence of the jth user

- f_0 is the lowest subcarrier frequency

- f_d is the subcarrier separation

- $p(t)$ is a rectangular signalling pulse shifted in time given by:

$$p(t) \triangleq \begin{cases} 1 & \text{for } 0 \le t \le T \\ 0 & \text{otherwise} \,. \end{cases} \tag{8.10}$$

If $1/T$ is used for f_d, the transmitted signal can be generated using the IFFT, as in the case of an OFDM system. The overall transmitter structure can be implemented by concatenating a DS-CDMA spreader and an OFDM transmitter, as shown in Figure 8.3. At the spreader, the information bit, b_i^j, is spread in the time domain by the jth user's spreading sequence, c_k^j, $k = 0, \cdots, N - 1$. In this implementation, high speed operations are required at the output of the spreader in order to carry out the chip-related operations. The spread chips are fed into the serial-to-parallel (S/P) block and IFFT is applied to these N parallel chips. The

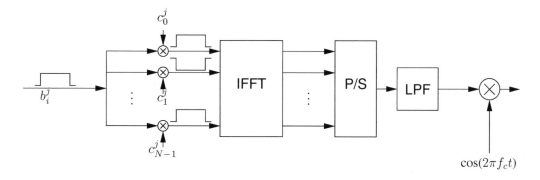

Figure 8.4: Alternative transmitter schematic of MC-CDMA

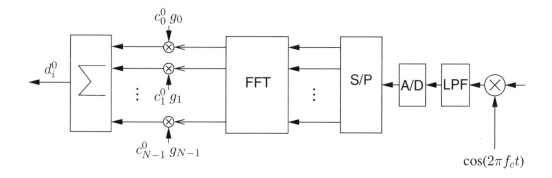

Figure 8.5: Receiver schematic of MC-CDMA

output values of the IFFT in Figure 8.3 are time domain samples in parallel form. After being parallel to serial (P/S) conversion these time domain samples are low-pass-filtered, in order to obtain the continuous time domain signal. The signal modulates the carrier and is transmitted to the receiver.

Figure 8.4 shows another implementation, which removes the time domain spreader. In this implementation, the spreading sequence is applied directly to the identical parallel input bits. Hence, the high speed spreading operation is not required.

The spreading sequences in MC-CDMA separate other users' signal from the desired signal, provided that their spreading sequences are orthogonal to each other. Orthogonal codes have zero cross-correlation and hence they are particularly suitable for MC-CDMA. Walsh codes and orthogonal Gold codes are two well known such codes, which will be examined in Chapter 9.

At the MC-CDMA receiver shown in Figure 8.5 each carrier's symbol, i.e. the corresponding chip c_k^j of user j, is recovered using FFT after sampling at a rate of N/T samples/sec and the recovered chip sequence is correlated with the desired user's spreading code in order to recover the original information, b_i^j. Let us define the ith received symbol at the kth carrier

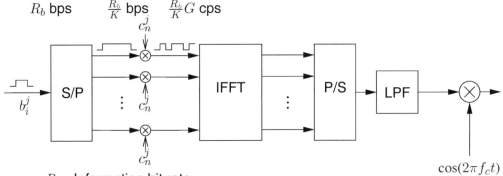

R_b : Information bit rate
K : Number of subcarriers
G : Processing Gain

Figure 8.6: Transmitter schematic of MC-DS-CDMA

in the downlink as:

$$r_{k,i} = \sum_{j=0}^{J-1} H_k\, b_i^j\, c_k^j + n_{k,i} \,, \qquad (8.11)$$

where J is the number of users, H_k is the frequency response of the kth subcarrier and $n_{k,i}$ is the corresponding noise sample. The MC-CDMA receiver of the 0-th user multiplies $r_{k,i}$ of Equation 8.11 by its spreading sequence chip, c_k^0, as well as by the gain, g_k, which is given by the reciprocal of the estimated channel transfer factor of subcarrier k, for each received subcarrier symbol for $k = 0, \cdots, N-1$, and sums all these products, in order to arrive at the decision variable, d_i^0, which is given by:

$$d_i^0 = \sum_{k=0}^{N-1} c_k^0\, g_k\, r_{k,i} \,. \qquad (8.12)$$

Without the frequency domain equalisation of the received subcarrier symbols, the orthogonality between the different users cannot be maintained. Several methods have been proposed for choosing g_k [52, 56, 281]. The associated BER analysis was performed using various equalisation methods over both the Rayleigh channels and Rician channels by Yee and Linnartz [52]. The comparative summary of numerical results for various equalisation strategies was given, for example, by Prasad and Hara [56, 279].

8.2.2 MC-DS-CDMA

In order to transmit high-rate data with sufficient processing gain, the chip rate of the DS-CDMA system should be significantly higher than the practical limit imposed mainly by the processing speed and power consumption of state-of-the-art electronics. In this case, parallel

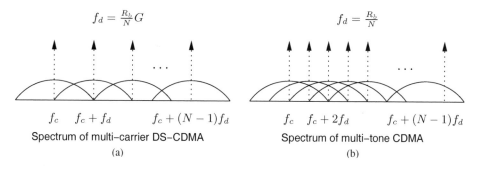

Figure 8.7: Typical power spectra of MC-DS-CDMA and MT-CDMA

transmissions of DS-CDMA signals using the OFDM structure [56, 277] can be a solution. Figure 8.6 shows the transmitter structure of a MC-DS-CDMA system.The N consecutive input bits of the jth user, b_i^j, $i = 0 \cdots N - 1$, are serial-to-parallel converted first. Then, each bit b_i^j is spread by the jth user's spreading sequence in the time domain. The other operations are identical to those of MC-CDMA. In other words, in the MC-CDMA scheme of Figure 8.3 the data bits arriving at a rate of $1/T$ are first spread and hence have a rate of N/T, before reducing the rate again to $1/T$. By contrast, in the MC-DS-CDMA scheme of Figure 8.6 the bit rate of R_b is first reduced to R_b/N and then we produce the rate $R_b G/N$ using spreading.

Figure 8.7(a) shows the typical spectrum of the MC-DS-CDMA signal operated using the schematic of Figure 8.6. The subcarrier separation, f_d, meets the orthogonality condition [56] of:

$$f_d = \frac{R_b}{N}G \, , \tag{8.13}$$

where R_b is the source bit rate, N is the number of subcarriers and G is the processing gain. When N is equal to G, the MC-DS-CDMA spectrum [277] has the same shape, as that of an MC-CDMA system [52]. While the spectra of MC-CDMA signals [52] and MC-DS-CDMA signals [277] exhibit orthogonality between the subcarriers, that of the MT-CDMA scheme of Figure 8.7(b) does not maintain orthogonality between the subcarriers.

The transmitted signal $s_i^j(t)$ of the jth user during the ith signalling interval is written as:

$$s_i^j(t) = \sum_{k=0}^{N-1} \sum_{g=0}^{G-1} b_{k,i}^j c_g^j \, p(t - iT_s - gT_c) \cdot e^{2\pi(f_0 + kf_d)t} \, , \tag{8.14}$$

where $b_{k,i}^j$ is the ith bit of the jth user modulating the kth subcarrier after serial-to-parallel conversion, c_g^j, $g = 0 \cdots G - 1$ represents the jth user's spreading sequence, $T_s = \frac{N}{R_b}$ is the signalling interval, $T_c = \frac{T_s}{G}$ is the chip duration and $p(t)$ is defined in Equation 8.10.

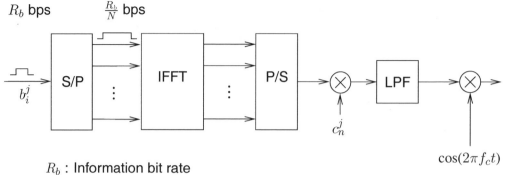

R_b : Information bit rate
N : Number of subcarriers
G : Processing Gain

Figure 8.8: Transmitter schematic of MT-CDMA

8.2.3 MT-CDMA

MT-CDMA is a combined technique employing time domain spreading and a similar multi-carrier transmission scheme to that of the MC-DS-CDMA scheme of Figure 8.6. However, the time domain spreading is applied after the IFFT stage. Figure 8.8 represents a simple block diagram of the transmitter structure. The required operations are the same as in OFDM, but spreading takes place after the IFFT stage. In this way, the system has a multiple access capability. The corresponding power spectrum is shown in Figure 8.7(b). Each subcarrier's spectrum overlaps with other subcarriers' spectra. The subcarrier frequency spacing is given by [56]:

$$f_d = \frac{R_b}{N} \; , \tag{8.15}$$

which does not retain the subcarriers' orthogonality. The main intention of this operation is to increase the processing gain within a given bandwidth.

8.3 Further Research Topics in MC-CDMA

The seminal contributions of [52, 53, 55, 277, 278] have motivated intensive further research in the field of wireless communications [52, 53, 56, 282–325].

The fundamental aim of MC-CDMA research has been the support of high data rate services in hostile wireless environments. As noted before, the main benefit of combining OFDM with DS-spreading is that it is possible to prevent the obliteration of certain subcarriers by deep frequency domain fades. This is achieved by spreading each subcarrier's signal with the aid of a spreading code and thereby increasing the achievable error-resilience, since in case of corrupting a few chips of a spreading code the chances are that the subcarrier signal still may be recovered.

The performance of both OFDM and MC-CDMA can be improved also by frequency hopping, when a fraction of the subcarriers is activated, concentrating for example on activating the high-quality subcarriers. The frequency domain repetition of the same information symbols may also be invoked, where the same symbols are mapped to several subcarriers, although this inevitably results in the reduction of the system's effective throughput [52, 53, 310, 319]. Multicarrier DS-CDMA using adaptive frequency-hopping has been studied in [325], while adaptive subchannel allocation based multicarrier DS-CDMA was the subject of [302]. Yang *et al.* studied constant-weight code aided multicarrier DS-CDMA in conjunction with slow frequency-hopping [285, 286, 314].

8.4 Chapter Summary and Conclusion

In this chapter we commenced by a rudimentary discourse on combining DS-CDMA and OFDM. We have briefly reviewed three types of multi-carrier spread-spectrum schemes combining DS-CDMA and OFDM techniques, namely MC-CDMA [52, 53, 55], MC-DS-CDMA [277] and MT-CDMA [278]. While MC-CDMA employs frequency domain spreading, the other two schemes, i.e. the MC-DS-CDMA and MT-CDMA schemes, employ time domain spreading. Hence, MC-CDMA is capable of exploiting frequency diversity in an explicit manner, since the energy of a symbol is spread over several subcarriers. On the other hand, since all the energy of a symbol is confined to one subcarrier in the MC-DS-CDMA and MT-CDMA schemes, the potential frequency diversity provided by the independently fading different subcarriers cannot be exploited, unless a channel coding scheme is employed in conjunction with cross-subcarrier interleaving [56].

Compared to a MC-DS-CDMA scheme requiring the same frequency band, MT-CDMA is capable of providing a significantly higher spreading factor than that of MC-DS-CDMA, resulting in lower self-interference and in a better suppression of the Multiple Access Interference (MAI) [278]. However, MT-CDMA suffers from inter-subcarrier interference due to the fact that the subcarriers are not orthogonal to each other. Since MC-DS-CDMA is capable of providing backward compatibility with the existing IS-95 DS-CDMA system, a specific form of MC-DS-CDMA has been chosen as one of the Third Generation (3G) mobile communication standards [326].

Prasad and Hara reported that MC-CDMA employing Minimum Mean Square Error Combining (MMSEC) shows the lowest BER among the three above-mentioned multi-carrier spread-spectrum schemes in a downlink scenario [279]. Hence, the performance of MC-CDMA in a downlink scenario will be investigated in more depth in Chapter 10.

The further outline and the rationale of Part II of the book, which entails Chapters 9–12 is highlighted below. Since the properties of spreading sequences are equally important in multicarrier CDMA as in DS-CDMA, the properties of various spreading sequences are reviewed in Chapter 9. One of the main problems associated with the implementation of multicarrier communication systems is their high Peak-to-Mean Envelope-Power Ratio (PMEPR), requiring highly linear power amplifiers. The envelope power of the MC-CDMA signal is analysed in Chapter 11. Several orthogonal spreading sequences are examined, in order to assess their ability to maintain low PMEPR of the transmitted signal. The effects of reduced PMEPR are investigated in terms of the BER performance and the out-of-band frequency spectrum. In Chapter 12 an adaptive modulation assisted MC-CDMA technique is investigated using both analysis as well as simulation-based studies.

Basic Spreading Sequences[1]

A spread spectrum communication system spreads the original information signal using user-specific signature sequences. The receiver then correlates the synchronised replica of the signature sequences with the received signal, in order to recover the original information. Due to the noise-like properties of the spreading sequences, "eavesdropping" is not straightforward. DS-CDMA exploits the code's autocorrelation properties in order to optimally combine the multipath signals of a particular user. By contrast, the different users' codes exibit a low cross-correlation, which can be exploited for seperating each user's signal. MC-CDMA also relies on this cross-correlation property in supporting multi-user communications. The characteristics of the spreading sequences play an important role in terms of the achievable system performance and hence the properties of several sequences will be examined in this chapter.

9.1 PN Sequences

Pseudo-Noise (PN) sequences are binary sequences, which exhibit noise-like properties. Maximal length sequences (m-sequences), Gold sequences and Kasami sequences are well-known PN sequences. Other important PN sequences are considered in [256].

9.1.1 Maximal Length Sequences

Various pseudo-random codes can be generated using Linear Feedback Shift Registers (LFSR). The so-called generator polynomial or LFSR connection polynomial governs all the characteristics of the generator. For a given generator polynomial, there are two ways of implementing LFSRs [256, 327]. The sequence generator shown in Figure 9.1 uses only the LFSR's output bits for feeding information back into several intermediate stages of the shift register, which is attractive for high speed hardware as well as software implementations. The other implementation known as the Fibonacci feedback generator [327], which is shown

---

[1]*OFDM and MC-CDMA for Broadband Multi-user Communications WLANS and Broadcasting.*
L.Hanzo, Münster, T. Keller and B.J. Choi,
©2003 John Wiley & Sons, Ltd. ISBN 0-470-85879-6

in Figure 9.2, is capable of generating several delayed versions of the same sequence at the output of each shift register without any additional logic. Note that the connection position is reversed in Figure 9.2, in order to generate the same sequences as the Galois implementation of [327] as shown in Figure 9.1. Shift-register sequences having the maximum possible

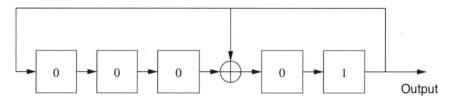

Figure 9.1: Galois Field based feedback implementation of the m-sequence generator. The generator polynomial of $g(x) = 1 + x^2 + x^5$ determines the positions of the non-zero feedback taps.[3]

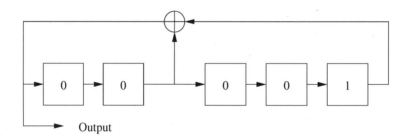

Figure 9.2: Fibonacci feedback implementation of the m-sequence generator. The same generator polynomial was employed as in Figure 9.1.

repetition period of $2^r - 1$ for an r-stage shift register are referred to as maximal-length sequences or m-sequences. A so-called primitive generator polynomial [327] always yields an m-sequence. The m-sequences have three important properties, namely the so-called balance property, the run-length property as well as the shift-and-add property [328]. The periodic autocorrelation function $R_a(k)$ of the sequence $\{a_n\}$ is defined as:

$$R_a(k) \triangleq \frac{1}{N} \sum_{n=0}^{N-1} a_n a_{n+k}, \tag{9.1}$$

where N is the period of the sequence. The periodic autocorrelation function of an m-sequence is given by:

$$R_a(k) = \begin{cases} 1.0 & k = lN \\ -\frac{1}{N} & k \neq lN, \end{cases} \tag{9.2}$$

where l is an integer and N is the period of the m-sequence. The excellent auto-correlation property manifested by the high ratio of these two $R_a(k)$ values is a consequence of the above-mentioned first and third properties, as justified in references [256, 327, 328]. The above-mentioned good autocorrelation features justify the employment of m-sequences for example in the Pan-American IS-95 standard CDMA system [269]. Moreover, as the cross-correlation property of these sequences is relatively poor compared to that of the family of Gold codes, usually the same sequences with different offsets are employed for different users or for different base stations in IS-95 based networks.

9.1.2 Gold Codes

Certain pairs of m-sequences having the same degree can be used for generating the so-called Gold codes by linearly combining two m-sequences associated with different offsets, where the operations are defined over the so-called finite Galois field [256, 327]. Not all pairs of m-sequences yield Gold codes and those which do are referred to as *preferred pairs*. The autocorrelations and cross-correlations of Gold codes may exhibit the legitimate values of $\{-1, -t(m), t(m) - 2\}$, where

$$t(m) = \begin{cases} 2^{(m+1)/2} + 1 & \text{for odd } m \\ 2^{(m+2)/2} + 1 & \text{for even } m . \end{cases} \tag{9.3}$$

Gold codes exhibit lower peak cross-correlations, than m-sequences, as shown in Table 13-2-1 of [274] and hence they differentiate among different users more confidently or distinctively. Furthermore, the cross-correlation function of Gold codes exhibits numerous "-1" values, which is the lowest value among the three possible cross-correlation values. By contrast, Kasami sequences exhibit a lower proportion of "-1"s in their cross-correlation functions, while exhibiting a peak cross-correlation, which is half of that of the Gold codes.

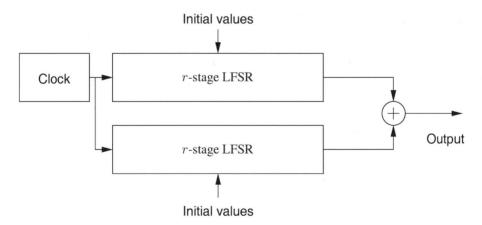

Figure 9.3: Schematic of a Gold code generator

The generation of Gold codes is straightforward. The simple block diagram of a Gold code generator is shown in Figure 9.3. Using two preferred m-sequence generators of degree r, in conjunction with a set of r initial values in the upper LFSR where at least one value is non-zero, 2^r number of different Gold codes can be obtained by changing the set of r initial values of the bottom LFSR according to the range of 0 to $2^r - 1$. Additional Gold sequence can be obtained by setting the contents of the upper LFSR to all zeros, which results in a Gold sequence that is identical to the second m-sequence itself, which is generated from the bottom LFSR of Figure 9.3. Hence in total, $2^r + 1$ Gold codes can be obtained from a Gold code generator, which is characterised by a pair of preferred m-sequences.

9.1.3 Kasami Sequences

Kasami sequences have optimal cross-correlation values, reaching the so-called Welch lower bound [256, 274]. The lower bound on the cross-correlation between any pair of binary sequences of period n in a set of m-sequences is [274]:

$$\phi_{max} \geq n\sqrt{\frac{M-1}{Mn-1}} \, . \tag{9.4}$$

Kasami sequences can be generated by the following procedure [256]. For the m-sequence $\mathbf{a}$, the so-called decimated sequence of $\mathbf{a}$ is obtained by taking every qth bit of $\mathbf{a}$, which is denoted by $\mathbf{a[q]}$. By choosing $q = 2^{r/2} + 1$, where r is the degree of sequence $\mathbf{a}$, having retained every qth bit of $\mathbf{a}$, $\mathbf{a[q]}$ is periodic with the period of $2^{r/2} - 1$. By repeating $\mathbf{a[q]}$ q times, a new sequence $\mathbf{b}$ of length $2^r - 1$ is obtained. With the aid of $\mathbf{a}$ and $\mathbf{b}$, we form a new set of sequences by module-two adding $\mathbf{a}$ and the $2^{r/2} - 2$ number of cyclically shifted versions of the sequence $\mathbf{b}$. Including $\mathbf{a}$ and $\mathbf{b}$, we get a total of $2^{r/2}$ number of sequences. These sequences are the so-called Kasami sequences [256]. The hardware implementation of Kasami sequences is daunting, because the decimation process requires a high clock frequency. Fortunately, the decimated sequence itself is an m-sequence [256] of order $r/2$ and this fact can be exploited for implementing Kasami sequence generators, as shown in Figure 9.4. The m-sequence generated by the r-stage LFSR and the other m-sequence output by

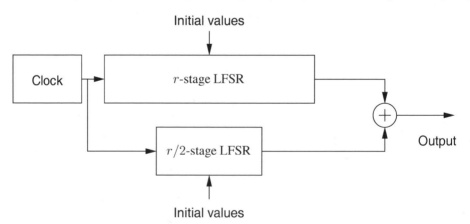

Figure 9.4: Schematic of a Kasami sequence generator

the $r/2$-stage LFSR are added in the binary Galois field in order to form a Kasami sequence.

9.2 Orthogonal Codes

Orthogonal codes have zero cross-correlation. They may appear attractive in terms of replacing PN codes, which have non-zero cross-correlations. However, the cross-correlation value is zero only, when there is no offset between the codes. In fact, they exhibit higher cross-correlations at non-zero offsets, than PN codes. Their autocorrelation properties are usually not attractive either. Nonetheless, orthogonal codes have found application, for example, in perfectly synchronised environments, such as in the down-link of mobile communications.

There are several so-called code expansion techniques that can be used in order to generate orthogonal codes. Probably the Hadamard transform [327] is the best-known technique. A modified Hadamard transform also appeared in the literature [329], which is essentially constituted by the Hadamard transform using a different transform coefficient indexing. Orthogonal Gold codes show reasonable cross-correlation and off-peak autocorrelation values, while providing perfect orthogonality in the zero-offset case. Finally, the multi-rate orthogonal codes of Section 9.2.3 are attractive, since they can provide variable spreading factors depending on the information rate to be supported.

9.2.1 Walsh Codes

Walsh codes are generated by applying the Hadamard transform [327] to a one by one dimensional zero matrix repeatedly. The Hadamard transform is defined as [327]:

$$
\begin{aligned}
\mathbf{H}_1 &= \begin{bmatrix} 0 \end{bmatrix} \\
\mathbf{H}_{2n} &= \begin{bmatrix} \mathbf{H}_n & \mathbf{H}_n \\ \mathbf{H}_n & \overline{\mathbf{H}_n} \end{bmatrix} .
\end{aligned}
\tag{9.5}
$$

This transform gives us a Hadamard matrix, $\mathbf{H}_n$, only for $n = 2^i$, where i is an integer. The Hadamard matrix is a symmetric square-shaped matrix. Each column or row corresponds to a Walsh code of length n. Every row of $\mathbf{H}_n$ is orthogonal to all other rows.

As an example, let us consider the case of $n = 8$. We can generate 8-bit Walsh codes, $\mathbf{H}_8$, applying the transform continuously from $\mathbf{H}_1$ three times repeatedly. The resultant matrix is as follows:

$$
\mathbf{H}_8 =
\begin{bmatrix}
0 & 0 & 0 & 0 & 0 & 0 & 0 & 0 \\
0 & 1 & 0 & 1 & 0 & 1 & 0 & 1 \\
0 & 0 & 1 & 1 & 0 & 0 & 1 & 1 \\
0 & 1 & 1 & 0 & 0 & 1 & 1 & 0 \\
0 & 0 & 0 & 0 & 1 & 1 & 1 & 1 \\
0 & 1 & 0 & 1 & 1 & 0 & 1 & 0 \\
0 & 0 & 1 & 1 & 1 & 1 & 0 & 0 \\
0 & 1 & 1 & 0 & 1 & 0 & 0 & 1
\end{bmatrix}
\tag{9.6}
$$

The first column is the all-zero sequence and the second one is an alternating sequence of "0" and "1". This is true for all Hadamard matrices. Let $\mathbf{H}_8[i]$ and $\mathbf{H}_8[j]$, i, $j = 0 \cdots 7$, be two sequences formed from the ith and the jth rows of $\mathbf{H}_8$. Then, we can verify that $\mathbf{H}_8[i]$ and $\mathbf{H}_8[j]$ are orthogonal to each other. Let us refer to i and j as the indices of $\mathbf{H}_8$.

A possible hardware implementation of the $\mathbf{H}_8$ Walsh code is shown in Figure 9.5. It comprises two T-flipflops, a clock generator, three AND gates denoted as $\otimes$ and one XOR gate denoted as $\oplus$. The clock source generates the $010101\cdots$ sequence, which corresponds to $\mathbf{H}_8[1]$. After the first T-flipflop, the sequence becomes $00110011\cdots$, which is $\mathbf{H}_8[2]$, since the T-flipflop simply halves the clock rate. The output sequence of the second T-flipflop becomes $00001111\cdots$, which represents $\mathbf{H}_8[4]$. In fact, these three sequences, namely $\mathbf{H}_8[1]$, $\mathbf{H}_8[2]$ and $\mathbf{H}_8[4]$, form bases so that the remaining rows of $\mathbf{H}_8$ can be generated by the linear combination of these three sequences. The desired Walsh code index is described by u_2, u_1, u_0, which is a binary bit-based representation of the index. For example, we can set $u_2 = 0$, $u_1 = 1$, $u_0 = 1$, in order to obtain the $\mathbf{H}_8[3]$ sequence. This can

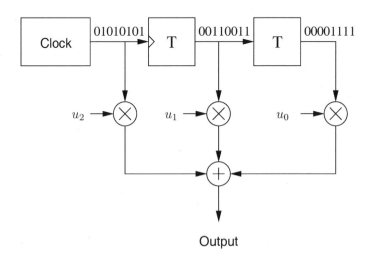

Output

Figure 9.5: Implementation of a Walsh code generator of length$=2^3$. (© 1995 Prentice-Hall. Inc. [Figure 9-8, page 542] [327])

be verified by showing that $\mathbf{H}_8[1] \oplus \mathbf{H}_8[2]$ actually yields $\mathbf{H}_8[3]$. A specific disadvantage of this implementation, however, is that the total delay of generating a specific code is not fixed, it is proportional to the length of the Walsh code to be generated.

9.2.2 Orthogonal Gold Codes

Experiments show that the cross-correlation values of the Gold codes of Section 9.1.2 are "-1" for many code offsets. This suggests that it may be possible to render the cross-correlation values associated with these offsets "0" by attaching an additional "0" to the original Gold codes. In fact, 2^r orthogonal codes can be obtained by this simple zero-padding from a preferred pair of two r-stage LFSR. These codes are referred to as orthogonal Gold codes.

As an example, let us consider a Gold code characterised by a preferred pair of m-sequences, which are denoted by $g_0(x) = 1 + x + x^3$ and $g_1(x) = 1 + x^2 + x^3$ of order three. We can obtain eight Gold codes of length 7 with the aid of the schematic of Figure 9.3 by changing the initial values of the upper LFSR from "000" to "111", while maintaining the bottom LFSR's initial values as "001". We arrive at eight orthogonal Gold codes by attaching a "0" at the tails, which are given below:

$$
\mathbf{G}_8 =
\begin{bmatrix}
1 & 0 & 1 & 1 & 1 & 0 & 0 & 0 \\
0 & 1 & 0 & 1 & 0 & 0 & 0 & 0 \\
1 & 1 & 0 & 0 & 1 & 1 & 0 & 0 \\
0 & 0 & 1 & 0 & 0 & 1 & 0 & 0 \\
1 & 0 & 0 & 0 & 0 & 0 & 1 & 0 \\
0 & 1 & 1 & 0 & 1 & 0 & 1 & 0 \\
1 & 1 & 1 & 1 & 0 & 1 & 1 & 0 \\
0 & 0 & 0 & 1 & 1 & 1 & 1 & 0
\end{bmatrix}
\tag{9.7}
$$

The ith row of $\mathbf{G}_8$, namely $\mathbf{G}_8[i]$, is the ith orthogonal Gold code of length 8. Let us define an n by n dimensional *cross-correlation matrix* $\mathbf{C}(\mathbf{G})$ of an n by n dimensional code matrix $\mathbf{G}$, where each row is a code of length n, defined as $\mathbf{G}\mathbf{G}^T$, and the operations take place after replacing "0" and "1" with "1" and "-1", respectively. Then, the element in the ith row of the jth column, namely $\mathbf{C}_{ij}$, is the cross-correlation between the two codes, $\mathbf{G}[i]$ and $\mathbf{G}[j]$. Let us also define an n by n dimensional *auto-correlation matrix* $\mathbf{A}(\mathbf{G})$, which comprises $\mathbf{A}_{ij}$, the element in the ith row of the jth column, defined as the auto-correlation value of $\mathbf{G}[i]$ with offset j. Now we get the cross-correlation matrices of $\mathbf{H}_8$ in Equation 9.6 and $\mathbf{G}_8$ in Equation 9.7 as:

$$
\mathbf{C}(\mathbf{H}_8) = 8\,\mathbf{I}_8 \tag{9.8}
$$
$$
\mathbf{C}(\mathbf{G}_8) = 8\,\mathbf{I}_8 \,, \tag{9.9}
$$

where $\mathbf{I}_8$ is the 8 by 8 dimensional identity matrix. Equations 9.8 and 9.9 show that there is no difference between the Walsh codes and the orthogonal Gold codes in terms of their cross-correlations. However, their auto-correlation properties are different. The auto-correlation matrix of the Walsh codes of length 8 is shown below:

$$
\mathbf{A}(\mathbf{H}_8) =
\begin{bmatrix}
8 & 8 & 8 & 8 & 8 & 8 & 8 & 8 \\
8 & -8 & 8 & -8 & 8 & -8 & 8 & -8 \\
8 & 0 & -8 & 0 & 8 & 0 & -8 & 0 \\
8 & 0 & -8 & 0 & 8 & 0 & -8 & 0 \\
8 & 4 & 0 & -4 & -8 & -4 & 0 & 4 \\
8 & -4 & 0 & 4 & -8 & 4 & 0 & -4 \\
8 & 4 & 0 & -4 & -8 & -4 & 0 & 4 \\
8 & -4 & 0 & 4 & -8 & 4 & 0 & -4
\end{bmatrix}.
\tag{9.10}
$$

Let us compare (Equation 9.10 with the auto-correlation matrix of the orthogonal Gold codes of length 8 formulated as:

$$
\mathbf{A}(\mathbf{G}_8) = \begin{bmatrix}
8 & 0 & 0 & -4 & 0 & -4 & 0 & 0 \\
8 & 0 & 4 & 0 & 0 & 0 & 4 & 0 \\
8 & 0 & -8 & 0 & 8 & 0 & -8 & 0 \\
8 & 0 & 0 & 4 & 0 & 4 & 0 & 0 \\
8 & 0 & 4 & 0 & 0 & 0 & 4 & 0 \\
8 & -4 & 0 & 0 & 0 & 0 & 0 & -4 \\
8 & 0 & 0 & 4 & 0 & 4 & 0 & 0 \\
8 & 4 & 0 & -4 & -8 & -4 & 0 & 4
\end{bmatrix} . \tag{9.11}
$$

Observe in Equation 9.11 that $\mathbf{A}(\mathbf{H}_8)$ has 24 number of "± 8"s and 16 number of "0"s at the various offsets, while $\mathbf{A}(\mathbf{G}_8)$ has 4 number of "± 8"s and 36 number of "0"s. Both have sixteen "± 4"s. Hence, the orthogonal Gold codes exhibit better characteristics than the Walsh codes in terms of their auto-correlations. We can conclude that orthogonal Gold codes are desirable in some applications, where the codes' auto-correlations have to be low, which is desirable, in order to be able to avoid falsely registering the main peak of the autocorrelation function.

9.2.3   Multi-rate Orthogonal Gold Codes

State-of-the-art wireless communication systems are expected to support multi-rate transmissions. CDMA systems can use multi-rate orthogonal codes for supporting this feature. Despite the terminology "multi-rate", typically a constant physical chip rate is maintained during variable bit rate transmissions and it is the number of chips per information bit, i.e. the spreading factor, which is varied, as it will be described during our further discourse. Generating multi-rate orthogonal codes is fairly straightforward. Commencing from any orthogonal code set, multi-rate codes can be obtained by applying one of the orthogonal transformation techniques described in [327]. Using the Walsh code generator of Figure 9.5 and the orthogonal Gold code generator of Section 9.2.2, one can readily implement multi-rate orthogonal code generators, as shown in Figure 9.6. The clock rate of the Walsh code generator is R_c/L, where R_c is the chip rate and L is the length of the orthogonal Gold code. In other words, the Walsh code generator is clocked once per orthogonal Gold code word. The maximum Walsh code length, n, is given by:

$$
n = \frac{SF_{max}}{L} , \tag{9.12}
$$

where SF_{max} is the maximum spreading factor and the orthogonal Gold code generator is clocked at the chip-rate of R_c, where SF_{max} is given by $SF_{max} = R_c/R_b$, and R_b is the lowest transmission symbol rate of the multi-rate system. The Walsh code index, $(u_{n-1}, \cdots u_1, u_0)_2$, has to be assigned carefully, in order to preserve the orthogonality between the spread multi-rate data symbols [330]. For the lowest-rate bit stream any free Walsh code index can be assigned from the range of 0 to $n - 1$. However, for a $2^k \cdot R_b$-rate data stream, a total of 2^i free Walsh code indices has to be assigned corresponding to the codes $\{W_j\}$, $j = u + \frac{n}{2^i}v$, where u is selected from the set of $\{0, \cdots, \frac{n}{2^i} - 1\}$ and v from the set of $0, \cdots, 2^i - 1$. From the above set of indices only the first assigned index associated

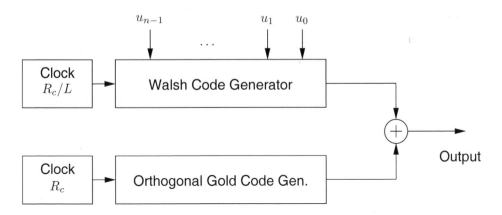

Figure 9.6: A multi-rate orthogonal Gold code generator. L is the length of the orthogonal Gold code, R_c is the chip rate and n is the length of Walsh code.

with $v = 0$ should be actually used.

In order to augment the concept of index assignment, let us now consider a simple example. Let us assume that we have to design a multi-rate CDMA system, which supports two rates, namely 8 kbps and 16 kbps, and that the chip rate, R_c, is 256 kcps. The maximum spreading factor is given by:

$$SF_{max} = R_c/R_b = 256\,kcps/8\,kbps = 32 \,, \tag{9.13}$$

where we used $R_b = 8$kbps as the lowest transmission symbol rate. Let us select $L = 8$, so that we can use $\mathbf{G}_8$ of Equation 9.7. According to Equation 9.12, the maximum Walsh code length n, is $SF_{max}/L = 32/8 = 4$. Thus, we use the 4 by 4 dimensional matrix $\mathbf{H}_4$, given as:

$$\mathbf{H}_4 = \begin{bmatrix} 0 & 0 & 0 & 0 \\ 0 & 1 & 0 & 1 \\ 0 & 0 & 1 & 1 \\ 0 & 1 & 1 & 0 \end{bmatrix}, \tag{9.14}$$

according to the schematic of Figure 9.6 for expanding $\mathbf{G}_8$, in order to generate the multi-rate orthogonal Gold code of dimension 32, $\mathbf{M}_{32}$, by using $\mathbf{G}_8$ of Equation 9.7 in its original form in those positions of $\mathbf{H}_4$ in Equation 9.14, where there is a "0" and its complementary form in the positions, where there is "1", yielding:

$$\mathbf{M_{32}} = \begin{bmatrix} \mathbf{G}_8 & \mathbf{G}_8 & \mathbf{G}_8 & \mathbf{G}_8 \\ \mathbf{G}_8 & \overline{\mathbf{G}_8} & \mathbf{G}_8 & \overline{\mathbf{G}_8} \\ \mathbf{G}_8 & \mathbf{G}_8 & \overline{\mathbf{G}_8} & \overline{\mathbf{G}_8} \\ \mathbf{G}_8 & \overline{\mathbf{G}_8} & \overline{\mathbf{G}_8} & \mathbf{G}_8 \end{bmatrix}. \tag{9.15}$$

Let us now consider the properties of $\mathbf{M}_{32}$. Each row of $\mathbf{M}_{32}$ is orthogonal to other rows.

The codes $\mathbf{M}_{32}[0]$, $\mathbf{M}_{32}[8]$, $\mathbf{M}_{32}[16]$ and $\mathbf{M}_{32}[24]$ are expanded from $\mathbf{G}_8[0]$. In general, $\mathbf{M}_{32}[i]$, $\mathbf{M}_{32}[8 + i]$, $\mathbf{M}_{32}[16 + i]$ and $\mathbf{M}_{32}[24 + i]$, where i spans the range of $0, \cdots, 7$, are generated from $\mathbf{G}_8[i]$ and hence they are not orthogonal, when they are used for multi-rate spreading. This lack of orthogonality is a consequence of the fact that when supporting a high transmission rate, while fixing the chip-rate, the number of chips per data bit has to be reduced. For example, when doubling the bit-rate, the number of chips per bit is halved, which inevitably halves the number of orthogonal sequences available. Hence a double bit-rate user has to be assigned two orthogonal codes, which otherwise could be assigned to two basic-rate users. Figure 9.7 shows this concept more explicitly. Since $\mathbf{H}_4[0]$ and $\mathbf{H}_4[2]$ are not

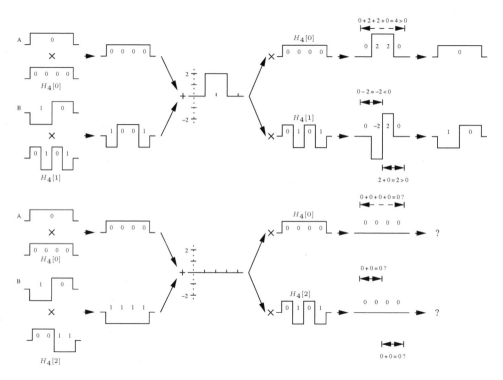

Figure 9.7: A conceptual diagram for multi-rate spreading: "A" transmits at half the rate of the transmission rate of "B". "A;' uses $\mathbf{H}_4[0]$ and "B" $\mathbf{H}_4[1]$ in the upper graph, which results in perfect information recovery at the receiver. In the lower graph, "B" uses $\mathbf{H}_4[2]$, which breaks the orthogonality and makes information recovery impossible.

orthogonal in a multi-rate environment, the receiver may not be able to recover the original symbols from the received signal, as shown in the graph at the bottom of Figure 9.7. In order to prevent this, two free Walsh codes have to be assigned to user "B", transmitting at twice the transmission rate of user "A" although effectively only one of these codes is transmitted. More explicitly, this potential non-orthogonality problem only arises in conjunction with the codes, which have the same generating orthogonal Gold code.

9.3 Chapter Summary and Conclusion

This chapter examined the correlation properties of a set of PN sequences, specifically those of the m-sequences, Gold codes and Kasami codes, as well as the correlation properties of a set of orthogonal sequences, namely those of Walsh codes, orthogonal Gold codes and multi-rate orthogonal Gold codes. The schematic diagram of each sequence generator was presented in Figure 9.1 through Figure 9.6. Based on these schematics, a sequence generator library[4] was constructed for employment in our numerical and simulation studies.

It was observed in Section 9.2.2 that orthogonal Gold codes result in lower auto-correlation values, than those of the Walsh codes studied in Section 9.2.1. The effects of the different autocorrelation values of the spreading sequences on the envelope power of the corresponding MC-CDMA signal employing those sequences will be examined in Section 11.5.2.

Specifically, when studying the structure of multi-rate orthogonal Gold codes in Section 9.2.3, we observed that for suppporting 2^k times higher bit-rate users than the basic transmission rate we have to assign 2^k number of code indices, even though only one is actually used, in order to preserve the code's orthogonality with respect to the lower bit-rate user's spread signal.

---

[4]http://www-mobile.ecs.soton.ac.uk/bjc97r/pnseq/index.html

10

MC-CDMA Performance in Synchronous Environments[1]

In wireless environments the transmitted signals may travel through different propagation paths having different lengths and hence these multi-paths components arrive at the receiver with different delays. This time-dispersive nature of the channel causes Inter-Symbol Interference (ISI) and frequency selective fading [263]. The *RMS delay spread* or multipath spread, τ_{rms}, of the channel determines the amount of ISI inflicted and the gravity of the channel-induced linear distortions. The reciprocal of the delay spread, namely $1/\tau_{rms} = (\Delta f)_c$, is referred to here as the *coherence bandwidth* of the channel [2]. If the total signal bandwidth is wider than $(\Delta f)_c$, the signal experiences *frequency selective* fading [274]. Typically, this is the case, when high rate data is transmitted in a wide signal bandwidth.

In highly frequency selective channels the ISI becomes a major problem in serial modems and hence usually complex channel equalisers are required. In OFDM systems the ISI becomes negligible, as long as a sufficiently high number of guard symbols is introduced [226]. However, due to the frequency selective nature of the channel, each subcarrier has a different bit error ratio (BER). In order to combat this phenomenon several techniques have been used in OFDM, such as error correcting codings in conjunction with frequency domain interleaving [47] and frequency domain adaptive loading [226].

Another traditional method of combating the effect of fading is to involve diversity techniques [274]. The main benefit of MC-CDMA in comparison to other OFDM-based multiple access methods [331] is the inherent provision of frequency diversity. By contrast, a disadvantage of MC-CDMA is the Multi-User Interference (MUI) encountered. These key factors predetermine the performance of MC-CDMA.

When the MC-CDMA signal experiences severe channel fades, the receiver is likely to make a wrong decision concerning the bit carried by the signal. Diversity techniques arrange

---

[1] *OFDM and MC-CDMA for Broadband Multi-user Communications WLANS and Broadcasting.*
L.Hanzo, Münster, T. Keller and B.J. Choi,
©2003 John Wiley & Sons, Ltd. ISBN 0-470-85879-6

for generating several replicas of the signal arriving at the receiver over independent fading paths. From the family of various diversity techniques, frequency diversity, time diversity and antenna diversity are most widely used [274]. Although both MC-CDMA and DS-CDMA use frequency diversity, their receiver structures differ in many aspects. DS-CDMA systems use so-called Rake receivers [266] and the number of fingers in the Rake receiver determines the number of diversity paths exploited. In most cases, the number of fingers in DS-CDMA Rake receivers is limited due to their affordable complexity and size. MC-CDMA systems, on the other hand, use a simpler approach, since they transmit the same information on several subcarriers in order to achieve diversity. The number of subcarriers transmitting the same information determines the order of diversity. Thus, in general, it is easier to achieve a higher-order diversity in MC-CDMA, than in DS-CDMA [332].

However, the maximum achievable order of frequency diversity, L, in a specific channel is approximately given by [274, 333]

$$L \approx \frac{W}{(\Delta f)_c} , \qquad (10.1)$$

where W is the total bandwidth of the channel and $(\Delta f)_c$ is its coherence bandwidth. Hence, spreading a symbol over more than this number of subcarriers is expected to have no more benefits in terms of diversity gain. Instead, as the number of users increases, the MUI will increase significantly, thus, reducing the overall performance of the MC-CDMA system.

In this chapter, the performance of the various diversity techniques and the multi-user interference reduction techniques applicable to MC-CDMA are investigated over a frequency selective indoor fading channel, assuming that all users' signals are synchronous, which is typically the case in the downlink.

10.1   The Frequency Selective Channel Model

A narrow-band fading channel can be modelled as a Rayleigh process associated with a specific Doppler spectrum [2]. This channel model is based on the worst case scenario, assuming that no line-of-sight path is available between the transmitter and the receiver. On the other hand, a wideband fading channel can be modeled as a sum of several differently delayed, independent Rayleigh fading processes. The corresponding channel impulse response is described as

$$h(t, \tau) = \sum_{p=1}^{P} a_p \cdot R_p(t) \cdot \delta(\tau - \tau_p) , \qquad (10.2)$$

where a_p is the normalised amplitude such that $\sum_{p=1}^{P} a_p^2 = 1.0$, $R_p(t)$ is the Rayleigh fading process with $E[R_p^2] = 1.0$ and τ_p is the delay of the p-th path. For the computer simulation purposes, τ_p is given in units of the sampling interval duration.

An indoor channel model is considered for investigating the performance of MC-CDMA in synchronous environments. It is a shortened Wireless Asynchronous Transfer Mode (W-ATM) channel model used for example in [2]. The impulse response and frequency domain response of the channel are shown in Figure 10.1. The impulse response was derived by

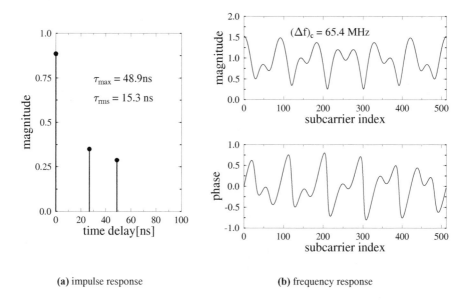

(a) impulse response   **(b)** frequency response

Figure 10.1: The shortened W-ATM channel model, where τ_{max} is the longest delay, τ_{rms} is the RMS delay spread and $(\Delta f)_c$ is the coherence bandwidth of the channel

simple ray-tracing in a warehouse-type room of $100 \times 100 \times 3\text{m}^3$ [2] . The main system parameters and channel parameters of the system investigated are summarised in Table 10.1.

parameters	Shortened W-ATM
Carrier frequency, f_c	60 GHz
Sampling rate, $1/T_s$	225 MHz
BPSK data rate, R_b	155 Mbps
Max. speed of mobile, v_{max}	50 Km/h
Max. Doppler frequency, $f_{d,max}$	2277.8 Hz
Normalised Doppler freq., $f_{n,max}$	$1.23 \cdot 10^{-5}$
Max. delay, τ_{max}	11 samples
RMS delay spread, τ_{rms}	15.3 ns
Coherence bandwidth, $(\Delta f)_c$	65.4 MHz
No. of subcarriers, N	512
No. of guard symbols, N_g	64

Table 10.1: Main system parameters of the considered W-ATM system and the channel parameters

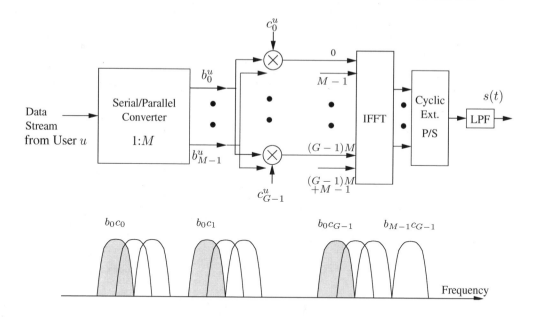

Figure 10.2: MC-CDMA transmitter model and its spectrum for a single user

10.2   The System Model

The MC-CDMA system considered here employs N subcarriers and each user transmits M bits during a signalling interval, leading to the spreading factor or, synonymously, to the processing gain, G, of N/M [56, 333]. Figure 10.2 shows the transmitter structure of the system model. The data stream of user u is converted into M parallel streams and each stream is spread with the aid of orthogonal spreading sequences of Section 9.2, which maps the same bit to G number of subcarriers. The subcarrier spacing between the subcarriers conveying the same bit is set to M, in order to minimise the correlation of the fading of these subcarriers. For simplicity, only a single user is depicted in the figure, but a number of users up to G can be incorporated. The uth user's spreading sequence, c_g^u, $g = 0,\ 1,\ \cdots,\ G-1$, is orthogonal to other users' spreading sequences. The Walsh codes and the orthogonal Gold codes described in Section 9.2 can be used as the spreading sequences. In this model, each user transmits its data at the rate of M/T bps, where T is the frame duration. If the transmission rate can be lowered, the maximum number of users, U, can be increased such that the data streams generated by the M users replace the $1:M$ serial-to-parallel converter in Figure 10.2 [333]. In this case, the total number of users supported can reach the total number of subcarriers, yielding $U = N$. Bearing this in mind, in our discourse, only the configuration shown in Figure 10.2 will be considered.

The complex baseband representation of the transmitted signal, $s(t)$, in a certain signalling interval can be written as:

$$s(t) = \sum_{u=0}^{U-1} \sum_{m=0}^{M-1} \sum_{g=0}^{G-1} \sqrt{E_c} \, b_m^u \, c_g^u \, e^{j2\pi \frac{1}{T}(gM+m)t} \,, \tag{10.3}$$

where

> U is the number of users, which has a maximum of G
> M is the number of bits transmitted per user
> G is the spreading factor or the processing gain given as N/M, where N is the number of subcarriers
> E_c is the energy per subcarrier, or chip, and $E_c = E_b/N$, where E_b is the energy per bit before spreading
> T is the signalling interval, during which M number of bits per user are transmitted and $1/T$ is equal to the spacing between adjacent subcarriers
> $b_m^u \in \{\pm 1\}$ is the mth bit of user u
> $c_g^u \in \{\pm 1\}$ is the gth chip of the uth user's spreading sequence.

10.3   Single User Detection

At the receiver the mth bit of user u is detected independently from the m'th bits, where $m' \neq m$, for all users. Furthermore, assume that there is no inter-subcarrier interference. Thus, without loss of generality, the subscript m can be omitted and the G subcarriers conveying the same bit can be considered for the detection of a bit. In view of Equation 8.8, the demodulated received symbol r_g of the gth subcarrier can be expressed as:

$$r_g = \sum_{u=0}^{U-1} \sqrt{E_c} \, b^u \, c_g^u \, H_g + n_g \,, \tag{10.4}$$

where H_g is the $(gM+m)$th subcarrier's frequency domain channel transfer factor, and n_g is a discrete AWGN process having zero mean and a one-sided power spectral density of N_o. The decision variable of the u'th user's bit, $d^{u'}$, is given for a single user detector as:

$$d^{u'} = \sum_{g=0}^{G-1} q_g \, c_g^{u'} \, r_g \,, \tag{10.5}$$

where q_g is a frequency domain equalisation gain factor, which is dependent upon the diversity combining scheme employed. The decision variable $d^{u'}$ can be expanded with the aid of

Equations 10.4 and 10.5 as:

$$
d^{u'} = \sum_{g=0}^{G-1} q_g \, c_g^{u'} \left(\sum_{u=0}^{U-1} \sqrt{E_c} \, b^u \, c_g^u \, H_g + n_g \right)
$$

$$
= \sqrt{E_c} \, b^{u'} \sum_{g=0}^{G-1} H_g \, q_g + \sqrt{E_c} \sum_{u=0, u \neq u'}^{U-1} b^u \sum_{g=0}^{G-1} c_g^u \, c_g^{u'} \, H_g \, q_g + \sum_{g=0}^{G-1} c_g^{u'} \, n_g \, q_g
$$

$$
= \alpha + \zeta + \eta \,, \tag{10.6}
$$

where α is the desired signal component given by

$$
\alpha = \sqrt{E_c} b^{u'} \sum_{g=0}^{G-1} H_g q_g \,, \tag{10.7}
$$

ζ is the MUI given by

$$
\zeta = \sqrt{E_c} \sum_{u=0, u \neq u'}^{U-1} b^u \sum_{g=0}^{G-1} c_g^u c_g^{u'} H_g q_g \,, \tag{10.8}
$$

and η is the noise component given by

$$
\eta = \sum_{g=0}^{G-1} c_g^{u'} n_g q_g \,. \tag{10.9}
$$

These three signal components predetermine the performance of the single user detector considered.

10.3.1 Maximal Ratio Combining

In Maximal Ratio Combining (MRC) [52, 274] a stronger signal is assigned a higher weight by the diversity combiners, than a weaker signal, since its contribution is more "reliable". The corresponding equalisation gain, q_g, introduced in Equation 10.5 is given as:

$$
q_g = H_g^* \,. \tag{10.10}
$$

This equalisation gain attempts to de-attenuate and de-rotate the fading-induced attenuation and phase rotation. The corresponding user's received signal component, α, is given by:

$$
\alpha = \sqrt{E_c} \, b^{u'} \sum_{g=0}^{G-1} |H_g|^2. \tag{10.11}
$$

On the other hand, the MUI associated with MRC is given by:

$$\zeta = \sqrt{E_c} \sum_{u=0, u \neq u'}^{U-1} b^u \sum_{g=0}^{G-1} c_g^u c_g^{u'} |H_g|^2 . \qquad (10.12)$$

Finally, the noise term, η, of Equation 10.9 can be calculated as:

$$\eta = \sum_{g=0}^{G-1} c_g^{u'} n_g H_g^* . \qquad (10.13)$$

The effect of MRC is equivalent to that of matched filtering, where the filtering is matched to the channel's transfer function. Matched filtering constitutes the optimal receiver, which maximises the SNR at the output of the decision device [274]. The single user performance of MRC has been widely studied for transmission over a Rayleigh fading channel having L-independent propagation paths and the achievable bit error rate, P_e, is given by [274]:

$$P_e = \left[\frac{1}{2}(1 - \mu) \right]^L \sum_{k=0}^{L-1} \binom{L-1+k}{k} \left[\frac{1}{2}(1 + \mu) \right]^k , \qquad (10.14)$$

where μ is defined as

$$\mu \triangleq \sqrt{\frac{\overline{\gamma}_b}{L + \overline{\gamma}_b}} , \qquad (10.15)$$

and $\overline{\gamma}_b$ is the average energy per bit, $\overline{E}_b$, divided by the noise power spectral density, N_o. The associated BER curves for $L = 1, \cdots, 4$ are shown in Figure 10.3. More specifically, the simulation results for a single-user scenario are shown in Figure 10.3(a). The available diversity order, L, is given in Equation 10.1. For the shortened W-ATM channel, we have $225MHz/65.4MHz = 3.44$. Hence, a system having a spreading factor of 4 is expected to reach the diversity-assisted performance enhancement limit. When a spreading factor of 1 is employed corresponding to no spreading, the BER performance of MC-CDMA was identical to the BER of single carrier scheme operating over a Rayleigh fading channel. As the spreading factor was increased, the BER performance was improved. However, as can be seen in Figure 10.3(a), the systems having a Spreading Factor (SF) in excess of 4 did not improve the achievable bit error rate. As the SNR increased, the simulation-based BER approached the theoretical BER associated with $L = 3$ which was evaluated from Equation 10.14 under the assumption of independent fadings over L paths. The discrepancy between the simulation-based and theoretical results was due to the correlated fading over G number of subcarriers.

It is interesting to observe the effects of the number of users on the bit error rate, when MRC is used. Figure 10.3(b) shows our simulation results for the spreading factor of four. As the number of users was increased, the bit error rate increased significantly due to the increased amount of MUI. We can view the frequency domain spreading as a form of repetition coding. Thus, the two-user scenario employing SF = 4 in MC-CDMA corresponds to a half rate encoder, since 2 information bits are transmitted using a total of 4 symbols. If we compare the BER curves of this two-user system with that of the single-user scenario in

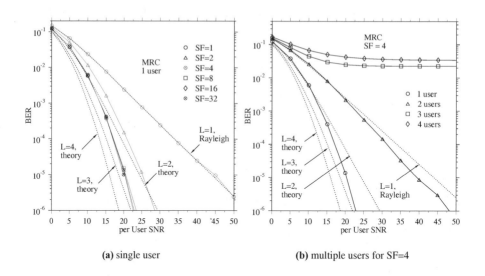

(a) single user   (b) multiple users for SF=4

Figure 10.3: BER of synchronous MC-CDMA for the downlink in conjunction with MRC for transmission over W-ATM channel. The theoretical curves were evaluated from Equation 10.14.

MC-CDMA in conjunction with SF=2, where the code rate can also be regarded as $1/2$, we can conclude that the latter scheme performs significantly better. However, in general, more complex channel coding techniques will perform better than this simple repetition code [274].

In so-called fully loaded conditions when the number of users is equal to the SF, the multi-user interference as opposed to the noise component dominates the system's performance, which becomes poor. The corresponding simulation results are given in Figure 10.4. The BER performance of fully loaded MC-CDMA is worse, than that of OFDM, which corresponds to having SF = 1. In Figure 10.4(b) it can be observed that the BER increased to a certain limit and then decreased across the SNR range spanning from 20dB to 50dB, as the spreading factor was increased. In order to investigate this phenomenon, we have to consider the statistical properties of α and ζ given in Equations 10.11 and 10.12, respectively.

As the total power conveyed by the Channel's Impulse Response (CIR) is normalised such that we have $E\{\sum_{n=0}^{N-1}|h_n|^2\} = 1.0$, the frequency domain channel response satisfies $E\{\sum_{n=0}^{N-1}|H_n|^2\} = N$ according to the discrete form of Parseval's theorem [334], where E denotes the expected value. Hence we can assume that $E\left[\sum_{g=0}^{G-1}|H_g|^2\right]$ becomes G. Thus, the average signal power, $E\left[\alpha^2\right]$ becomes

$$\begin{aligned} E\left[\alpha^2\right] &= E_c \cdot G^2 \\ &= E_b \cdot G\,, \end{aligned}$$

where E_b is the signal energy per bit during the bit interval. If we assume that the absolute value of the subcarrier channel transfer factor H_g obeys an independent identically distributed (iid) Rayleigh process with $E\left[|H_g|^2\right] = 2\sigma^2 = 1$ and $c_g^u c_g^{u'}$ has an equal probability of $+1$ or -1, then Equation 10.12 can be regarded as a zero mean Gaussian process, provided

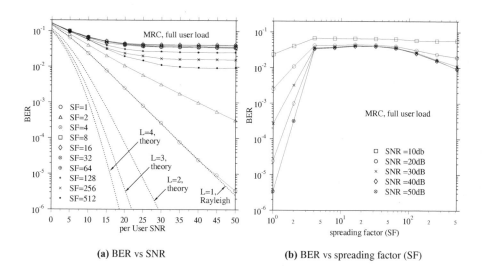

(a) BER vs SNR   **(b)** BER vs spreading factor (SF)

Figure 10.4: BER of the synchronous MC-CDMA downlink using MRC for full user load over the W-ATM channel described in Figure 10.1, where SF is the spreading factor and L represents the diversity order.

that U and G are sufficiently high. The variance, $E\left[\zeta^2\right]$ becomes:

$$
\begin{aligned}
E\left[\zeta^2\right] &= E_c\,(U-1)\,G\,\left(E\left[|H_g|^4\right] - E\left[|H_g|^2\right]^2\right) \\
&= E_c\,(U-1)\,G \\
&= E_b\,(U-1)\,,
\end{aligned}
$$

where $E\left[|H_g|^4\right] = 8\sigma^4 = 2$ is used. In high SNR scenarios supporting a full user load of $U = G$, the bit error probability, P_e, can be approximated as

$$
\begin{aligned}
P_e &\simeq \frac{1}{2}\mathrm{erfc}\left(\sqrt{\frac{E\left[\alpha^2\right]}{E\left[\zeta^2\right]}}\right) \\
&= \frac{1}{2}\mathrm{erfc}\left(\sqrt{\frac{E_b\cdot G}{E_b(U-1)}}\right) \\
&\simeq \frac{1}{2}\mathrm{erfc}(1) \\
&= 0.08\,.
\end{aligned}
$$

This approximate analysis is supported by Figure 10.4. Specifically, at high SNRs the bit error ratio was constant, namely approximately 0.08 for the spreading factors of 8 to 64. Figure 10.4(b) is a plot of the BER versus the spreading factor. As the spreading factor was increased beyond 64, the associated bit error ratios were reduced. This can be interpreted as follows. The MUI defined in Equation 10.12 is reduced, when the channel transfer factors

$\{H_g\}$ are correlated, which is the case for high SFs, when the subcarriers conveying the same bit are close to each other.

In conclusion, we observed that the channel exhibits a limited amount of diversity and the MUI plays a major role in terms of determining the BER, when MRC is used as the diversity combining method.

10.3.2   Equal Gain Combining

Although the spreading codes are orthogonal, due to the differently delayed multi-path components received, their orthogonality is destroyed. The MRC scheme, which is characterised by the equalisation gain given by Equation 10.10, optimally combines these multi-path components in an effort to maximise the SNR, but at the same time it may further impair the orthogonality of the codes. In order to avoid this problem, Equal Gain Combining (EGC) attempts to correct the channel-induced phase rotations [263], leaving the faded magnitudes uncorrected [52]. In this case, the equalisation gain, q_g, is given by:

$$q_g = \frac{H_g^*}{|H_g|}. \tag{10.16}$$

The corresponding three signal components defined in Equations 10.7 to 10.9 are given as:

$$\alpha = \sqrt{E_c}\, b^{u'} \sum_{g=0}^{G-1} |H_g| \tag{10.17}$$

$$\zeta = \sqrt{E_c} \sum_{u=0, u \neq u'}^{U-1} b^u \sum_{g=0}^{G-1} c_g^u c_g^{u'} |H_g| \tag{10.18}$$

$$\eta = \sum_{g=0}^{G-1} c_g^{u'} n_g \frac{H_g^*}{|H_g|}. \tag{10.19}$$

Our simulation results are shown in Figure 10.5. In the single user case shown in Figure 10.5(a), the BER curves approached the theoretical ones, but overall EGC performed slightly worse than MRC. No further improvement was observed for the systems having a SF higher than four, as in the case of MRC. When the number of users was increased for the system employing SF = 4, the performance was degraded but remained better, than that of MRC. This results corroborates the BER comparison reported in [52].

When all users are transmitting, the multi-user interference increases significantly, eroding the benefits of the frequency diversity and the performance becomes worse than that of OFDM, corresponding to SF = 1. The corresponding simulation results are given in Figure 10.6. As in the case of MRC, the bit error ratio curve converged to a BER floor, when the SNR was high. The BER floor was dependent on the spreading factor. The worst BER floor was observed for SF = 16. As the spreading factor increases beyond SF = 16, the BER floor is reduced. This can be interpreted using the similar arguments to those in Section 10.3.1.

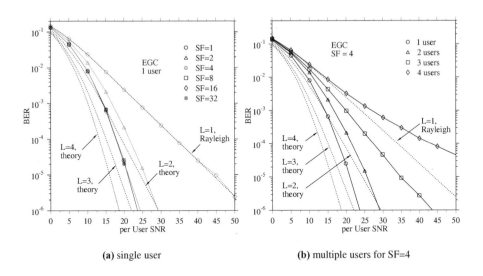

(a) single user

(b) multiple users for SF=4

Figure 10.5: BER performance of the synchronous MC-CDMA downlink using EGC over W-ATM channel, where SF is the spreading factor and L is the diversity order.

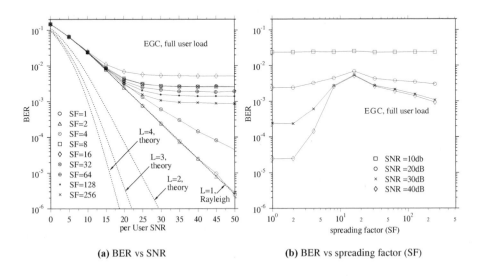

(a) BER vs SNR

(b) BER vs spreading factor (SF)

Figure 10.6: BER performance of the synchronous MC-CDMA downlink using EGC for a full user load, where SF is the spreading factor and L is the diversity order.

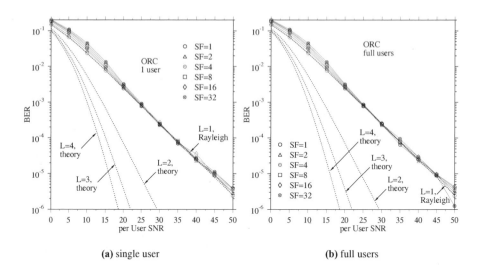

(a) single user   (b) full users

Figure 10.7: BER of the synchronous MC-CDMA downlink using ORC for a single user over the W-ATM channel

10.3.3 Orthogonality Restoring Combining

As argued before, frequency selective fading destroys the orthogonality of the different users' spreading codes. The drawback of the MRC and the EGC scheme is that they inflict MUI due to the above mentioned lack of orthogonality, which is not desirable. If we cancel the effect of the channel transfer function by estimating it and reversing its effects, the orthogonality of the different users can be maintained. This is the aim of Orthogonality Restoring Combining (ORC) or Zero Forcing (ZF) equalisation, which uses the equalisation gain, q_g, given by:

$$q_g = \frac{H_g^*}{|H_g|^2}. \tag{10.20}$$

We have only two components in the decision variable of Equation 10.6 as $\zeta = 0$, which are given by:

$$\alpha = \sqrt{E_c}\, G\, b^{u'} = \sqrt{E_b G}\, b^{u'} \tag{10.21}$$

$$\eta = \sum_{g=0}^{G-1} c_g^{u'}\, n_g\, \frac{H_g^*}{|H_g|^2}\,. \tag{10.22}$$

Figure 10.7 shows our simulation results for the various spreading factors, when ORC is applied in our MC-CDMA scheme. The MC-CDMA system using various spreading factors, i.e. different-order diversity schemes, performed worse than the single-user OFDM scheme associated with SF = 1, when the average per user SNR, $\overline{\gamma}_b$, was less than 25dB. In fact, as the the spreading factor was increased, the performance degraded, since the ORC scheme using perfect channel estimation had already removed the frequency selectivity of the channel transfer function, leaving no room for improvement with the aid of frequency domain diversity. The source of performance degradation in comparison to OFDM was the noise enhancement, particularly at lower SNR values. More specifically, when the fading was severe for a specific subcarrier, the value of n_g/H_g became high, inevitably amplifying both the signal and the noise. As the spreading factor increased, it became more likely that $\{H_g\}$ encountered deep frequency domain fades, leading to an excessive noise enhancement. This effect is less significant at high SNRs. In order to combat to the noise enhancement problem, the technique of Controlled Equalisation (CE) was proposed by Yee and Linnartz [281], which uses a subset of subcarriers for the ORC scheme, associated with the channel transfer factors that are above a certain threshold.

No noticeable differences were observed between the bit error ratios for the various number of users, as shown in Figure 10.7(b). This result is consistent with the fact that no MUI exists, when ORC is employed.

10.4   Multi-User Detection

The detection methods of Section 10.3 exploited single user information only, thus the performance was interference limited except in the case of ORC. Moreover, when using all the available information at the receiver, such as the spreading codes of all users and their CIRs, the performance can be improved with the aid of Multi-User Detection (MUD).

In the third-generation W-CDMA system referred to as IMT2000 [209] a range of powerful optional performance enhancement techniques have been proposed, which are expected to be used only in mature implementations. For example, powerful but complex Multi-user Detectors (MUDs) [335,336] can be used at the base station to achieve a near-single-user performance in the uplink, while supporting a multiplicity of users. The fundamental approach of MUDs, which are also often referred to as multi-user equalisers, accrues from recognising the fact that the nature of the interference is similar, regardless whether its source is dispersive multipath propagation or multi-user interference. In other words, the effects of imposing interference on the received signal by a K-path dispersive channel or by a K-user system are similar. Hence MUDs offer the benefit of jointly counteracting both MUI and multipath interference.

MUDs [335, 336] may be categorised in a number of ways, such as linear versus non-linear, adaptive versus non-adaptive algorithms or burst transmission versus continuous transmission regimes. Excellent summaries of some of these sub-optimum detectors can be found in the monographs by Verdu [335], Prasad [337], Glisic and Vucetic [338] or in the tutorial reviews by Woodward and Vucetic [339], Moshavi [340] as well as Duel-Hallen, Holtzman and Zvonar [341], just to mention a few. Other MAI-mitigating techniques include the employment of interference rejection techniques, which typically impose a lower implementation complexity than MUDs [342].

Adaptive antenna arrays (AAAs) [209] are also capable of mitigating the level of MAI at the receiver by forming a beam in the direction of the wanted user and a null towards the interfering users. It is worth noting, however, that since the angle of arrival of the multipath components becomes more limited owing to the focused beam of the AAAs, which limits the number of received rays and hence the achievable diversity gain. Therefore often alternative fading counter-measures, such as the employment of adaptive modulation [207] become nec-

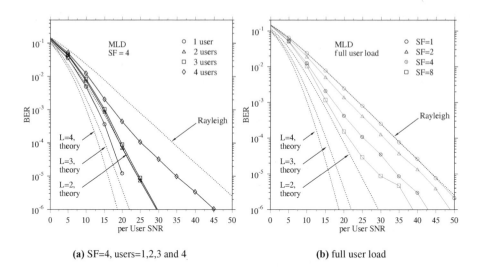

(a) SF=4, users=1,2,3 and 4   (b) full user load

Figure 10.8: BER of MC-CDMA downlink with MLD

essary, further advocating the techniques reviewed in Chapter 12. Research efforts invested in the area of AAAs area include, amongst others, the investigations carried out by Thompson, Grant and Mulgrew [262, 343]; Naguib and Paulraj [344]; Godara [345]; as well as Kohno, Imai, Hatori and Pasupathy [346]. However, the area of adaptive antenna arrays is beyond the scope of this monograph and the reader is referred to the references cited for further discussions. The Maximum Likelihood Detector (MLD) is investigated in the rest this section as an example. For an in-depth study of MUDs the interested reader is referred to [335, 336].

10.4.1   Maximum Likelihood Detection

In this scheme the receiver constructs all possible combinations of the transmitted signals of all users and applies their estimated channel transfer functions in order to generate the expected received signals [333]. Explicitly, the receiver chooses the legitimate transmitted signal, which has the smallest Euclidean distance from the received signal.

Let $\mathbf{C}$ be a $G \times U$-dimensional matrix containing the uth user's spreading code in column u, where $u = 1, 2, \cdots, U$, and $\mathbf{b}$ be a column vector, $(b^0 \ b^1 \ \cdots \ b^{U-1})^T$ comprised of possible source bits combination. If we define the channel's transfer factor matrix, $\mathbf{H}$, as a $G \times G$-dimensional diagonal matrix having H_g at $\mathbf{H_{gg}}$, then the expected received column vector $\hat{\mathbf{r}}$ is given by $\hat{\mathbf{r}} = \mathbf{HCb}$. The Euclidean distance, d, between the actually received signal vector $\mathbf{r}$ containing r_g, $g = 1, 2, \cdots, G$ of Equation 10.4, and the received vector candidate $\hat{\mathbf{r}}$, is given by:

$$d = \| \mathbf{r} - \hat{\mathbf{r}} \|^2 .$$

(10.23)

Our simulation results characterising this MUD are given in Figure 10.8. As can be seen from Figure 10.8(a), the BER curve approached the theoretical BER curve of the single-user scenario for $L = 3$. As the number of users increased, the performance degraded, but nonetheless remained better than the OFDM BER associated with SF = 1. The effect of MUI was considerably reduced compared to the MRC and EGC assisted scenario. Figure 10.8(b) shows the simulation results for the "fully loaded" MC-CDMA system using the MLD. Even in this condition, the BER was lower than that of OFDM. This implies that MC-CDMA systems employing the MLD are superior to the conventional OFDM scheme in terms of their bit error ratio, while providing the same spectral efficiency. In Section 10.3.1 we observed that spreading a bit to more subcarriers than the available diversity order of the channel does not improve the performance of single user detection schemes. However, Figure 10.8(b) suggests that in a multi-user scenario spreading the same bit to more subcarriers than the available diversity order of the channel can improve the bit error ratio due to both the reduced MUI and the better exploitation of the available diversity.

10.5 Chapter Summary and Conclusion

The BER performance of synchronous MC-CDMA operating in the downlink was investigated in Section 10.3 using simulation based studies. In Section 10.3 we employed single-user detectors, namely MRC, EGC and ORC, while in Section 10.4 an MLD multi-user detector operating over the W-ATM channel was used. It was observed in Section 10.3.1 that while MRC is the optimum detector in a single-user scenario, its BER performance was the worst among the three single-user detector investigated in a fully loaded scenario. Neither the EGC nor the ORC-assisted MC-CDMA performed better than OFDM in a fully loaded scenario. However, in the same scenario the MLD multi-user detector-assisted MC-CDMA scheme of Section 10.4.1 exhibited SNR gains of 5dB, 10dB and 15dB for SF = 2, 4 and 8, respectively, in comparison to OFDM, when viewed at the BER of 10^{-6}. We can conclude that the employment of multi-user detectors is essential for MC-CDMA in order to successfully support a multiplicity of simultaneous users.

Chapter 11

Advanced Peak Factor Reduction Techniques [1]

11.1 Introduction

A rudimentary introduction to the peak-factor problems of OFDM was provided in Section 3.4. In this chapter we adopt a more advanced approach and embark on a more detailed discussion of the related problems in the context OFDM and MC-CDMA.

Many practically important absolutely integrable signals can be decomposed into the sum of trigonometric series. These signals have various applications in radars, communications, measurements and so on. They exhibit perfect band-pass nature in the frequency domain. However, they often exhibit a large dynamic range in terms of their amplitude variations in the time domain.

The transmitted signals of multi-carrier modulation systems comprise sums of trigonometric series. These techniques are amenable to transmission over highly time-dispersive channels and they are capable of approaching the Shannonian 2Bd/Hz bandwidth efficiency limit. The main drawback of these techniques is their high peak-to-mean envelope power fluctuation or, synonymously, high crest factor, which is common to the sums of trigonometric series. In simple terms, this statement implies that the higher the number of OFDM/MC-CDMA subcarriers, the higher the dynamic range of their sum. Hence expensive linear high power amplifiers [2] having a large dynamic range are required in order to reduce the nonlinear distortion of the signals. Multi-carrier CDMA systems inevitably inherit this drawback.

Historically, there have been numerous efforts dedicated to reducing these amplitude variations, in order to mitigate the associated amplifier linearity requirements, the spurious out-of-band (OOB) emissions or the required signal level reduction, termed also as amplifier 'back-off', necessary for preventing the amplifier's saturation at high input signal peaks [2]. In this chapter, an overview of past research in this field is given, with an emphasis on several

[1] *OFDM and MC-CDMA for Broadband Multi-user Communications WLANS and Broadcasting.*
L. Hanzo, Münster, T. Keller and B.J. Choi,
©2003 John Wiley & Sons, Ltd. ISBN 0-470-85879-6

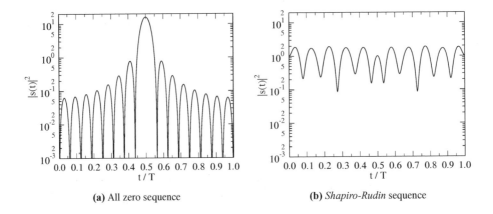

(a) All zero sequence   (b) *Shapiro-Rudin* sequence

Figure 11.1: Examples of normalised envelope power waveforms plotted in a logarithmic scale for 16-carrier BPSK modulated OFDM signals, transmitting (a) the all zero sequence, (b) the Shapiro-Rudin sequence, which will be introduced in Section 11.3.1. The differences in dynamic ranges are quite noticeable.

crest factor reduction techniques applicable to OFDM signals. For MC-CDMA systems, the relationship between the crest factor and the spreading sequences employed will be analysed and the effects of the crest factor on the system performance will be investigated.

As mentioned above, OFDM signals are based on the sum of trigonometric series. OFDM is a parallel transmission technique [2], which uses orthogonal subcarriers modulated by the information symbols. The complex time domain baseband representation, $s(t)$, of the OFDM signal is given by:

$$s(t) = \sqrt{\frac{1}{N}} \sum_{k=1}^{N} c_k e^{j2\pi \frac{k}{T} t} \, , \tag{11.1}$$

where N is the number of subcarriers, T is the OFDM signalling interval and c_k is the information symbol modulating the frequency domain subcarriers. For a specific application of multicarrier transmission the following modulated signal representation is proposed by Schroeder [347]:

$$s'(t) = \sqrt{\frac{2}{N}} \sum_{k=1}^{N} \cos(2\pi \frac{k}{T} t + \theta_k) \, , \tag{11.2}$$

where θ_k is the phase associated with the complex symbol mapped to the kth subcarrier. We can observe that $s'(t)$ of Equation 11.2 corresponds to the real part of $s(t)$ in Equation 11.1, while having the same average power. Both $s(t)$ and $s'(t)$ are normalised signals, having average powers of unity. Depending on the specific values of c_k or θ_k, $s(t)$ or $s'(t)$ may have a high dynamic range. Figure 11.1 shows $|s(t)|^2$ versus time for two representative information sequences $\{c_k\}$ in the context of 16-carrier BPSK modulated OFDM systems. Figure 11.1(a) corresponds to the all zero sequence, whereas Figure 11.1(b) to the so-called

Shapiro-Rudin sequences, which will be introduced in Section 11.3.1. The unmodulated signal associated with the all zero information sequence can readily be visualised, since the associated 16 unity-valued OFDM/BPSK subcarriers represent a 16-subcarrier rectangular frequency domain window function. Hence, the corresponding time domain modulated signal generated by the IFFT [2] is a sinc-shaped function, which exhibits the logarithmic power-envelope versus time function seen in Figure 11.1(a). These figures illustrate that the power envelope versus time function changes dramatically, depending also on the specific codes used for mapping the modulating information sequence on to the subcarriers, while having the same total power.

11.2 Measures of Peakiness

A measure of the amplitude variation of $s(t)$ or $s'(t)$ versus time is required for the sake of a quantitative analysis. Various authors [347–349] have defined the measure as "peak factor", "crest factor" and so on, sometimes using the same terminology associated with different meanings. "Peak-factor (PF_1)", introduced by Schroeder [347], appeared to be the first terminology in various publications. It was defined as the difference between the maximum and minimum amplitudes of a signal divided by its Root-Mean-Square (RMS) value. In reference [347], the "relative peak-factor", defined as $PF_1/(2\sqrt{2})$, was also proposed as a normalised measure, which becomes 1.0 for sine waves. $PF_1/2$ was referred to as the "crest factor (CF_2)" later in reference [348]. In 1981, Greenstein and Fitzgerald [349] used the same term, namely "peak factor (PF_2)", for the peak-to-average power (PAP [350–353]) ratio (PAPR [354–356]), which is also referred to as the peak-to-mean envelope-power ratio (PMEPR [31, 357]). Boyd [358] defined "crest factor (CF_1)" as the ratio of the peak to the RMS amplitude, $||s||_\infty/||s||_2$, which also enjoyed wide-spread use [350, 355].

In this chapter the terminology "crest factor (CF_1)" and "peak factor (PF_2)" will be used, in order to avoid unnecessary confusion, which are defined as the ratio of the peak to the RMS amplitude:

$$CF_1 \triangleq \frac{||s||_\infty}{||s||_2} , \tag{11.3}$$

where $||s||_\infty = \max_t |s(t)|$ and $||s||_2 = \sqrt{\int_T |s(t)|^2 dt}$, while:

$$PF_2 \triangleq \frac{\text{Peak Power}}{\text{Average Power}} = \frac{||s||_\infty^2}{||s||_2^2} , \tag{11.4}$$

which is equivalent to the ratio of the peak power to the average power.

One should note that the crest factors of $s(t)$ and $s'(t)$ are not the same in terms of their values. In fact, the nature of the associated two crest factors is different as well. Let the subscript m represent the index of the sequence applied, as in $s_m(t)$ and $s'_m(t)$. Then, in general, the statement

$$CF_1[s_m(t)] > CF_1[s_n(t)] \rightarrow CF_1[s'_m(t)] > CF_1[s'_n(t)] \tag{11.5}$$

is not true. Below we illustrate this with the aid of a counter-example. Figure 11.2 compares

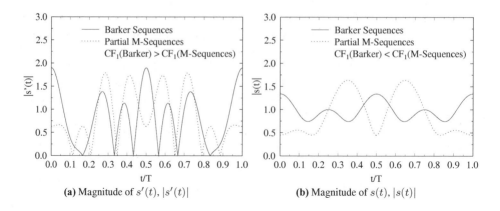

Figure 11.2: A counter-example of Equation 11.5 comparing $|s'(t)|$ and $|s(t)|$ for $\{00010\}$, a Barker sequence, and $\{01001\}$, a subsequence of a 7-bit m-sequence

$|s(t)|$ and $|s'(t)|$ for the data sequences $\{00010\}$ and $\{01001\}$. The former is the so-called Barker sequence, which will be introduced in Section 11.3.5, of length 5 and the latter is the fragment of an m-sequence derived using the octally represented generator polynomial of $g = (013)_0$ [Table 3-5] of [327], with the aid of the initial seed of $(06)_0$). In terms of $|s'(t)|$, the partial m-sequence shows the lower CF$_1$ of 1.79, while the Barker sequence exhibits CF$_1 = 1.90$. However, as shown in Figure 11.2(b), this crest factor relationship is reversed in terms of $|s(t)|$, where CF$_1 = 1.342$ for the Barker sequence and CF$_1 = 1.637$ for the partial m-sequence, when they are applied to $s(t)$. Thus, in general, a rank ordering according to CF$_1$ changes, depending on the application and/or on the specific formulation of the time domain modulated signal, as seen for $s(t)$ and $s'(t)$ in Equations 11.1 and 11.2, noting that for BPSK both equations provide the same result.

11.3   Special Sequences for Reducing Amplitude Variations

There have been numerous efforts dedicated to reducing the envelope variations of $s(t)$ or $s'(t)$, which were reported for example in [347, 351, 354, 357, 359–361]. A specific approach is to encode the information sequences, $\{c_k\}$, of Equation 11.1 or to adjust $\{\theta_k\}$ in Equation 11.2 so that the resultant signal can exhibit low peak factors. In this section, these efforts will be highlighted.

11.3.1   Shapiro-Rudin Sequences

Shapiro [362] and Rudin [359] were interested in trigonometric polynomials having coefficients of ± 1, which correspond to $s(t)$ given in Equation 11.1, as early as in the 1950s. Specifically, Rudin [359] mentioned the existence of the apparent lower bound for $||s||_\infty$

using Parseval's theorem, which is formulated as:

$$\frac{1}{T} \int_T f^2(t)dt = \sum_{k=-\infty}^{\infty} |c_k|^2 , \tag{11.6}$$

where $\{c_k\}$ are the complex frequency domain Fourier series coefficients of the periodic signal $f(t)$. Using Equation 11.6, the average power of $s(t)$ becomes:

$$
\begin{aligned}
\frac{1}{T} \int_T s^2(t)dt &= \sum_{k=-\infty}^{\infty} |c_k|^2 \\
&= \sum_{n=1}^{N} |c_k|^2 \\
&= N .
\end{aligned}
\tag{11.7}
$$

Using the inequality of $\frac{1}{T} \int_T (||s||_\infty)^2 dt \geq \frac{1}{T} \int_T s^2(t)dt$ we may conclude that:[2]

$$||s||_\infty \geq \sqrt{N} . \tag{11.8}$$

Then Rudin [359] introduced the problem of finding an upper bound for $||s||_\infty$, an absolute constant A having the property that for each N one can find $\{c_k; c_k \in (\pm 1)\}$, satisfying:

$$||s||_\infty \leq A\sqrt{N}. \tag{11.9}$$

He also invoked Shapiro's result [362] that $A = \sqrt{2}$ for $N = 2^n$, otherwise $A = 2 + \sqrt{2}$. Their proposed sequences, which satisfy the equality in Equation 11.9 were later referred to as Shapiro-Rudin sequences or phases [358]. This remarkable early work in mathematics was not widely recognised in the communications community until 1981 [349] and it was after 1986 [358], when the work was explained comprehensively. Shapiro-Rudin sequences, later were shown to constitute a specific case of Golay's complementary sequences [361, 363].

In their original formulation [359], Shapiro-Rudin sequences can be constructed using a pair of recursive polynomial relationships, which is given by:

$$
\begin{aligned}
P_0(x) &= Q_0(x) = x , \\
P_{k+1}(x) &= P_k(x) + x^{2^k} Q_k(x) , \\
Q_{k+1}(x) &= P_k(x) - x^{2^k} Q_k(x) ,
\end{aligned}
$$

and the sign of each term x^n forms the individual components of the sequences. Following several recursive steps commencing from $k = 0$ reveals that the sequence is of length 2^k and that the above relation can be stated more directly [358]:

> Starting with the string $p = 11$ and repeatedly concatenating to p a copy of p with its second half negated, we arrive at the set of Shapiro-Rudin sequences.

---

[2]This sets the minimum CF$_1$ to unity.

The first 32-bit sequence is given by

$$1110\ 1101\ 1110\ 0010\ 1110\ 1101\ 0001\ 1101 \ ,$$

where 1 represents a 1 and 0 stands for -1. As mentioned before, Figure 11.1(b) represents the associated power envelope waveform for 16-carrier BPSK OFDM, when Shapiro-Rudin sequences are applied. The peak-factor PF_2 of Equation 11.4 for this case is $1.9\,(2.8\text{dB})$, while the worst case PF_2 associated with using no CF-reduction coding, seen in Figure 11.1(a), is $16\,(12.0\text{dB})$. Shepherd *et al.* [30] proposed half-rate precoding for QPSK OFDM using the Shapiro-Rudin sequences of length 8 for 4 message bits, in order to reduce the peak factor. However, the associated halving of the effective throughput is unacceptable in most practical applications.

11.3.2   Golay Codes

Golay found in 1961 a set of complementary series [363], while he was working on the optical problem of spectrometry. Let $\{a_i\}$ and $\{b_i\}$ be two Golay complementary series of length N, and let their aperiodic autocorrelations A_k and B_k be defined by:

$$A_k \triangleq \sum_{i=1}^{N-k} a_i a_{i+k} \ , \quad k = 0,\ 1,\ \cdots,\ N-1 \tag{11.10}$$

and

$$B_k \triangleq \sum_{i=1}^{N-k} b_i b_{i+k} \ , \quad k = 0,\ 1,\ \cdots,\ N-1 \ , \tag{11.11}$$

where the terminology "aperiodic" implies that the sequences are not used cyclically in the auto-correlation calculation process. Then, the following property holds:

$$A_k + B_k \ = \ 0 \ \forall\ k,\ k \neq 0 \ , \tag{11.12}$$

$$A_0 + B_0 \ = \ 2N \ . \tag{11.13}$$

The aperiodic autocorrelations of the transmitted sequence $s(t)$ given in Equation 11.1 play a major role [347,356,364] in deciding the peak factor of the signal. However, Equations 11.12 and 11.13 do not say anything about the aperiodic autocorrelations of individual series. Golay found several transforms of the sequences, which do not alter their properties, and hence were shown to be invariant in terms of the peak factors.

It was Popović, who showed in 1991 that PF_2 of $s(t)$ employing any Golay complementary series for $\{c_k\}$ is bounded by a value of two (or 3dB) [361]. This is also true when representing $\{c_k\}$ with the aid of complex complementary sequences [365–367]. Popović also noted that *Shapiro-Rudin* sequences belong to a large family of Golay sequences of length 2^k. Davis and Jedwab [357] reported that Golay sequences can be used as a precoder for OFDM in order to bound the PF_2 within a ratio of two (or 3dB) as well as additionally to provide some error correction capability. The code rate, R, is given [357] by:

$$R = \frac{\lfloor \log_2(m!/2) \rfloor + m + 1}{2^m} , \qquad (11.14)$$

where 2^m is the length of the Golay codes used. Figure 11.3 shows the code rate versus the

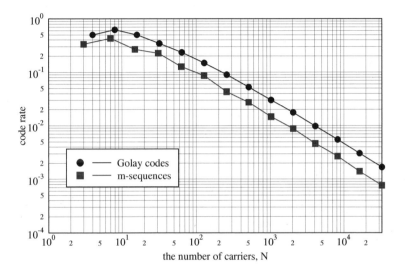

Figure 11.3: The coding rate of Golay sequences and m-sequences used in OFDM

number of carriers. As the number of carriers increases, the code rate becomes too low to be applicable to OFDM. For $N > 16$, the coding rates are below $1/2$, which is unacceptable in most applications and hence it is better to use amplifier-back-off or other error correction codes [368].

11.3.3 M-Sequences

Maximal-length sequences were already discussed in Section 9.1.1, which exhibit randomness properties, also known as the "balance property", "run-length property" and "shift-and-add property" [328]. Their periodic auto-correlations $R_a(k)$, defined as $R_a(k) \triangleq \sum_{i=1}^{N} c_i c_{i+k}$, are given as:

$$R_a(k) = \begin{cases} N & k = lN \\ -1 & k \neq lN , \end{cases} \qquad (11.15)$$

where l is an integer and N is the period of the m-sequence. The excellent auto-correlation property manifested by the high ratio of these two $R_a(k)$ values accrues from the "balance property" and "shift-and-add property", an issue detailed in references [256,327,328]. Since the definition of aperiodic auto-correlation given in Equation 11.10 is similar to that of the periodic auto-correlation provided in Equation 11.15, $s(t)$ of Equation 11.1 employing an m-sequence for c_k may result in a low crest factor. Li and Ritcey [351] proposed a precoding scheme using m-sequences for OFDM, in order to reduce the peak factors, claiming to have achieved very low PF_2 values, which were later shown to be optimistic [352,353]. The CF_1 values of $s'(t)$ and $s(t)$, when encoded by m-sequences, appear in Figures 11.4 and 11.5, respectively, in conjunction with a range of other schemes, which will be described at a later stage in the forthcoming sections. The coding rate R is given by:

$$R = \frac{\lfloor \log_2 N + \log_2 N_p \rfloor}{N},$$ (11.16)

where N_p is the number of primitive polynomials of degree r, given by [328]:

$$N_p = \frac{2^r - 1}{r} \prod_{j=1}^{J} \frac{p_j - 1}{p_j},$$ (11.17)

where $p_j, j = 1, 2, \cdots, J$ are prime factors of the period $N = 2^r - 1$ and J is the number of the prime factors. Figure 11.3 also shows the corresponding code rate R of Equation 11.16. Compared with the code rate of the Golay codes given in Equation 11.14 of Section 11.3.2, m-sequences require a lower code rate, while the peak factors of the Golay codes lie between the those of the worst case and the best case of the m-sequences. This implies that m-sequences are impractical for peak factor reduction motivated coding in most applications.

11.3.4   Newman Phases and Schroeder Phases

In 1965 Newman [360] suggested, a formula for adjusting the phases, while he was aiming for maximising the so-called L^1 norm representing the absolute value of trigonometric polynomials. This Newman phases are given by [360]:

$$\theta_k^{(Newman)} = \frac{\pi(k-1)^2}{N},$$ (11.18)

where the denominator N is the number of subcarriers. An application of $\{\theta_k^{(Newman)}\}$ given in Equation 11.18 for the choice of $\{\theta_k\}$ of $s'(t)$ defined in Equation 11.2 was reported by Boyd [358] together with the achievable CF_1 results.

While Newman constructed the set of low crest factor phases $\{\theta_k^{(Newman)}\}$ purely mathematically, Schroeder [347] used engineering arguments for deriving similarly advantageous, low-CF phases. Based on the observation that Frequency Modulation (FM) signals have constant envelopes and that there is a relationship between the instantaneous frequencies of FM signals and their power spectra, Schroeder derived the phase sequences, $\{\theta_k^{(Schroeder)}\}$, given by [Eq.11] of [347].

$$\theta_k^{(Schroeder)} = \theta_1 - \frac{\pi k(k-1)}{N},$$ (11.19)

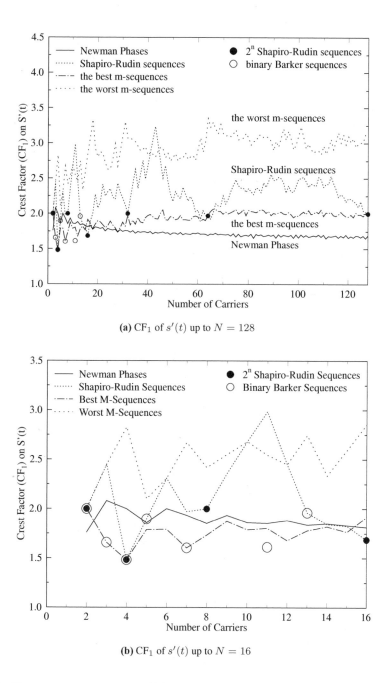

(a) CF_1 of $s'(t)$ up to $N = 128$

(b) CF_1 of $s'(t)$ up to $N = 16$

Figure 11.4: The comparison of CF_1 of $s'(t)$ defined in Equation 11.2 using various CF-reduction schemes

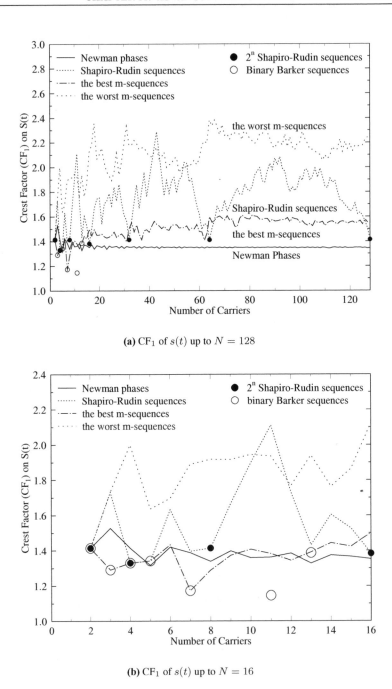

(a) CF_1 of $s(t)$ up to $N = 128$

(b) CF_1 of $s(t)$ up to $N = 16$

Figure 11.5: The comparison of CF_1 of $s(t)$ defined in Equation 11.1 using various CF-reduction schemes

which are similar to the Newman phases of Equation 11.18. Specifically, assuming $\theta_1 = -\frac{\pi}{N}$, and a time shift of $T/(2N)$ together with time reversal yields an $s'(t)$ function identical to that upon using the Newman phases of Equation 11.18. Figure 11.6 shows the $s'(t)$ functions using both Newman phases and Schroeder phases for $N = 16$ number of harmonics. A

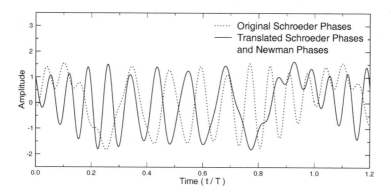

Figure 11.6: The function $s'(t)$ of Equation 11.2 when Newman phases [360] and Schroeder phases [347] are used

time translation of $T/(2N) = T/32$ and time reversal of $s'(t)$ with respect to the original Schroeder phases results the same $s'(t)$ function generated using Newman phases. The CF_1 value in this case is 1.813 (5.17dB), while the worst case is $\sqrt{2N} = 4\sqrt{2}$ (15.05dB).

Greenstein and Fitzgerald [349] proposed a similar phasing scheme in 1981. Narahashi and Nojima [354] also reported a similar phasing scheme, given by:

$$\theta_k^{(Narahashi)} = \frac{(k-1)(k-2)}{N-1}\pi .\qquad(11.20)$$

These phases were derived by setting $\rho_1 = 0$, where ρ_1 is the aperiodic autocorrelation coefficient at a displacement of one, which is given by $\rho_1 = \sum_{i=1}^{N-1} c_i c_{i+1}^*$. The achievable crest factor reduction is similar to that attained with the aid of Newman phases.

11.3.5 Barker Codes

According to Schroeder [Eq.18] of [347][3] the envelope power of $s(t)$ can be written as: as:

$$s^2(t) = 1 + \frac{2}{N}\sum_{k=1}^{N-1} \rho_k \cos(k\omega t) ,\qquad(11.21)$$

---

[3]This relation appeared again in the recent literature [356, 364].

N	Codes	CF_1 for $s(t)$
2	11	2.00
3	110	1.66
4	1110, 1101	1.48
5	11101	1.90
7	1110010	1.60
11	11100010010	1.61
13	1111100110101	1.96

Table 11.1: All the known binary Barker sequences [371]

Sequences	N=2	4	8	16	32	64	128
Binary Barker	2.00	1.48	-	-	-	-	-
Newman phases	1.76	2.00	1.85	1.81	1.76	1.69	1.69
The best partial m-sequences	2.00	1.48	1.73	1.92	1.97	1.98	2.00
Shapiro-Rudin sequences	2.00	1.48	2.00	1.69	2.00	1.97	2.00
The worst partial m-sequences	2.00	2.83	2.42	2.86	3.06	3.35	3.14

Table 11.2: CF_1 values of $s'(t)$ defined in Equation 11.2 for certain sequence length N using various CF-reduction schemes

where ρ_k are the aperiodic autocorrelations defined by $\rho_k = \sum_{i=1}^{N-k} c_i c_{i+k}^*$ and ω is the fundamental angular frequency. It is apparent that a low $\{\rho_k\}$ value is desirable for attaining low peak factors. In this sense, Barker codes [369,370] are optimal. More specifically, Barker codes are defined as arbitrary binary sequences (or complex sequences for generalised Barker codes) of length N, for which the aperiodic autocorrelations, ρ_k, satisfy [369,370]:

$$|\rho_k| \begin{cases} = N & \text{for} \quad k = 0, \\ \leq 1 & \text{for} \quad k = 1, 2, \cdots, N-1, \\ = 0 & \text{for} \quad k \geq N. \end{cases} \tag{11.22}$$

Unfortunately, only few binary Barker sequences [371] are known, which are summarised in Table 11.1, where Table 11.1, 1 represents +1 and 0 corresponds to -1. In the case of $N = 4$, the two codes are also Golay complementary sequences. Other than those listed above, no binary Barker sequences seem to exist [371,372]. However, the generalised Barker sequences [372], which allow c_k to have unit magnitude complex values, exist for an arbitrary value of N.

11.3.6   Comparison of Various Schemes

Having introduced a range of CF reduction techniques, it is interesting to compare their crest factors (CF_1). Figure 11.4 and Table 11.2 show the crest factors (CF_1) of the various schemes, when they are applied to $s'(t)$ of Equation 11.2.

For the range of sequence lengths N that we studied, Newman phases show the lowest

Sequences	N=2	4	8	16	32	64	128
Binary Barker	1.41	1.33	-	-	-	-	-
Newman phases	1.41	1.41	1.34	1.35	1.35	1.35	1.35
The best m-sequences	1.41	1.33	1.29	1.50	1.50	1.60	1.55
Shapiro-Rudin sequences	1.41	1.33	1.41	1.38	1.41	1.41	1.41
The worst m-sequences	1.41	2.00	1.92	2.12	2.18	2.38	2.24

Table 11.3: CF_1 values of $s(t)$ defined in Equation 11.1 for certain sequence lengths N using various CF-reduction schemes

crest factors (CF_1). For N values up to 15 shown in Figure 11.4(b), the best case of partial m-sequences or the binary Barker sequences show the lowest CF_1 values. The partial m-sequences show a large range of CF_1 values, depending on the polynomial used and the initial seed applied. Broadly speaking, Shapiro-Rudin sequences and thus Golay complementary codes exhibit crest factors between those of the worst-case and best-case m-sequences. The crest factors of all the investigated sequences fall well below four (or 6dB), regardless of the sequence length N. Beside the schemes investigated, there is a numerically near-optimal solution [348], which appears to be the best known solution at the time of writing, since its crest factor approaches $\sqrt{2}$.

Figure 11.5 and Table 11.3 show the crest factors (CF_1) of the various CF reduction schemes, when they are applied to $s(t)$ of Equation 11.1. In Section 11.2, we showed that the rank ordering based on CF_1 was not the same for $s(t)$ and $s'(t)$. For $s(t)$ the Barker sequences of Table 11.1 show the best results amongst the binary sequences considered. However, it is interesting to note that the Newman phases of Equation 11.18 exhibit lower CF_1 values than Barker sequences, when considering $N = 5$ and 13. This suggests that some complex sequences, such as generalised Barker sequences [372], may exhibit lower crest factors than those of binary Barker sequences.

11.4 Crest Factor Reduction Mapping Schemes for OFDM

11.4.1 Some Properties of the Peak Factors in OFDM

The normalised complex envelope $s(t)$ of the transmitted waveform of BPSK-modulated OFDM signal can be written as in [352]:

$$s(t) = \frac{1}{\sqrt{N}} \sum_{k=0}^{N-1} c_n \, e^{j2\pi \frac{kt}{T}}, \tag{11.23}$$

where N is the number of carriers, $c_k \in \{-1, +1\}$ is the information bearing frequency domain modulating symbol transmitted by the kth subcarrier and T is the OFDM symbol duration. Applying Parseval's theorem to Equation 11.23, we find that the average power of $s(t)$ is unity. Hence the peak-to-mean envelope power ratio ($PMEPR = PF_2$) is simply given by the maximum value of the envelope power, $|s(t)|^2$, over the OFDM symbol period.

As an example, let us examine the 4-carrier BPSK modulated OFDM scenario. There are

2^4 possible message sequences. The sequences can be represented by their decimal values, from $0, 1, \cdots, 15$, corresponding to the messages of '0000', '0001', $\cdots$, '1111'. Figure 11.7

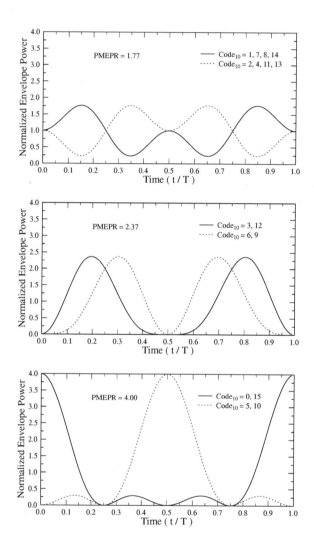

Figure 11.7: Normalised envelope power waveforms, $|s(t)|^2$, for 4-carrier BPSK modulated OFDM signals for the message symbols (top: PMEPR = 1.77, middle: PMEPR = 2.37 and bottom: PMEPR = 4.00), where each frequency domain sample is represented by its corresponding decimal value. For example $Code_{10} = 1$ denotes a message sequence of '0001' and $Code_{10} = 7$ '0111'.

shows the normalised envelope power waveforms, $|s(t)|^2$, of all possible message sequences of 4-carrier BPSK modulated OFDM, where each graph has two waveforms, which have the same PMEPR. All the waveforms in Figure 11.7 have the same average power, namely unity. Furthermore all the waveforms are symmetric to $0.5t/T$, since the power envelopes can be represented by the sums of cosine harmonics, as shown in Equations 11.126, 11.127 and 11.128 in the Appendix. The dotted lines represent the waveforms that can be generated with the aid of shifting the solid lines by $0.5t/T$ and have the same PMEPR as their counterparts shown using solid lines. The waveforms in the top graph are more desirable in CF terms than the others, since they show the lowest power variations and the lowest PMEPR. Eight message symbols corresponding to the decimal notations of 1, 7, 8, 14 and 2, 4, 11, 13, belong to this category. By contrast, the sub-figure at the bottom shows the worst case among the three possible PMEPR values, which is associated with PMEPR = 4.

The power envelope waveforms are dependent on the characteristics of the corresponding message sequences, but some sequences yield exactly the same power envelopes and some others show shifted envelopes. This observation led us to investigate the underlying nature of the sequences employed. Appendix 11.7.2 provides a detailed analysis of the PMEPR for BPSK modulated OFDM signals. The PMEPR of a message symbol is shown to be dependent on its aperiodic autocorrelation values defined in Equation 11.10. Based on this observation, the message symbols can be grouped in terms of their PMEPR. It is shown in Appendix 11.7.2 that irrespective of the number of subcarriers, in BPSK modulated OFDM there are at least four message symbols, which yield an identical PMEPR. These four messages can be obtained by three distinctive transformations of one message. It is also shown that in certain conditions another four messages also yield the same PMEPR and hence form an equivalent class. In general, four or eight message groups belong to a PMEPR group.

For QPSK-modulated OFDM signals, the normalised complex envelope waveform, $s(t)$, can be written as in [276]:

$$s(t) = \frac{1}{\sqrt{2N}} \sum_{k=0}^{N-1} (c_{2k} + jc_{2k+1}) \, e^{j2\pi \frac{kt}{T}}, \qquad (11.24)$$

where N is the number of carriers, c_{2k} and c_{2k+1} are the in-phase symbol and the quadrature phase symbol, respectively, mapped to the kth subcarrier, while $c_i \in \{-1, +1\}$ and T is the OFDM symbol duration. Van Eetvelt, Shepherd and Barton [276] presented an excellent analysis of QPSK modulated multi-carrier systems and showed that 16 messages form a so-called coset, in which messages yield the same PMEPR. They also presented an algorithm, which can be used for finding equivalent classes from a given coset leader and tabulated the equivalence classes up to $N = 5$ carriers. Although their results are useful, Algorithm 2 of [page 91] in [276] does not give the results claimed [Table 4, page 94] in [276] for the 5-carrier case. Table 11.9 shows the correct equivalent groupings for 5-carrier QPSK OFDM.

11.4.2 CF-Reduction Block Coding Scheme

An important observation concerning the peak factor distributions of multi-carrier systems is that there are not many message symbols which yield high peak factors. Figure 11.8 presents the Cumulative Probability Distributions (CDF) for 4-, 6- and 8-carrier OFDM signals. The graph at the left of Figure 11.8 represents BPSK modulated OFDM and the right one is for QPSK modulated OFDM signals. The cumulative distributions presented in Figure 11.8 for

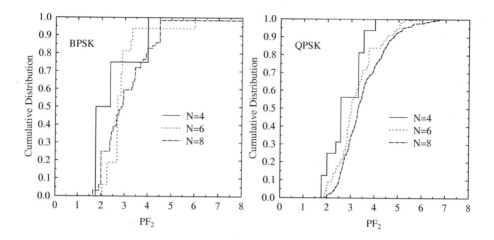

Figure 11.8: Cumulative distributions of the peak factor PF$_2$ of Equation 11.4 for BPSK modulated OFDM (left) and QPSK modulated OFDM (right) using 4, 6 and 8 carriers.

QPSK exhibit twice as many steps, as those related to BPSK, since there are twice as many legitimate symbols for the same number of carriers. The distributions of the real and imaginary amplitudes of $s(t)$ for a high number of subcarriers N were shown to be approximated by the Gaussian distribution [373, 374].

Jones, Wilkinson and Barton [31] proposed a simple block coding scheme for eliminating those message symbols from the coded message symbol set, which yield high peak factors. For example, in the case of 8-carrier BPSK modulated OFDM, the cumulative distribution reaches 50% at a PMEPR of 2.8, which is shown in Table 11.4. This means that the worst case PMEPR can be reduced from 8.0 to 2.8 by employing an (8,7) block code at the cost of a throughput loss of $1/8$. This scheme achieves a low worst case peak factor in a simple manner with the aid of high-rate block codes. By employing lower-rate block codes, even lower peak factors can be obtained [31] at the cost of further effective throughput loss. For example, a (4,3) block code is capable of reducing the PF$_2$ of Equation 11.4 further, down to 2.0 for an 8-carrier BPSK OFDM system. The same method can be used for QPSK modulated OFDM [276, 375]. The main drawback of this scheme is that the encoding and decoding processes rely on lookup tables, since no structured or algorithmic description of the block coding method concerned is known [351].

A half-rate coding method was proposed by Friese [355], which can be regarded as a non-structured block coding method. According to this scheme, the phase difference of a pair of subcarriers is dependent on the message symbol they carry. Thus, there is a degree of freedom in choosing the phase differences between the consecutive pairs of carriers. Friese proposed optimising the phase differences of adjacent subcarriers using the algorithm given in [348]. He reported a considerable reduction in terms of the achievable crest factors. However, the computational complexity inherent in the algorithm hinders its real-time application for a high number of subcarriers N and in the context of high speed data transmission.

PF_2	1.7	1.9	2.0	2.4	2.5	2.6	2.7
PDF	0.031	0.031	0.188	0.109	0.031	0.031	0.031
CDF	0.031	0.063	0.250	0.359	0.391	0.422	0.453
PF_2	2.8	2.9	3.0	3.3	3.4	3.5	3.7
PDF	0.047	0.063	0.031	0.016	0.078	0.031	0.031
CDF	0.500	0.563	0.594	0.609	0.688	0.719	0.750
PF_2	3.8	3.9	4.0	4.2	4.3	4.5	8.0
PDF	0.016	0.031	0.031	0.016	0.016	0.125	0.016
CDF	0.766	0.797	0.828	0.844	0.859	0.984	1.000

Table 11.4: Probability Density Functions (PDF) and Cumulative Distribution Functions (CDF) of the peak factors in BPSK-OFDM using 8 subcarriers: The PF_2 value of Equation 11.4 at CDF=0.5 is 2.8, which is the worst-case peak factor, when an (8,7) nonlinear block code is used for eliminating half of the message codes, yielding higher values of PF_2 than 2.8 or 4.47dB.

Some structured block coding methods appeared in the literature as well. Shepherd, Eetvelt, Millington and Barton [30] proposed the employment of Shapiro-Rudin codes for block coding. Davis and Jedwab [357] proposed a block coding scheme based on Golay codes. However, the code rates of these structured block coding methods were typically low, as seen in Figure 11.3.

11.4.3 Selected Mapping-Based CF-Reduction

In 1996, Bäuml, Fischer and Huber [350] proposed a new mapping algorithm for reducing peak factors in OFDM. One of the advantages of the scheme advocated in [350,376] is that it has a low transmission overhead. In their scheme, the best mapping resulting in the lowest CF is selected from a set of n random mapping attempts, which map $\{c_k\}$ to $\{c_k^{(i)}\}, i = 1, \cdots, n$. As n increases, there is a reduced chance that the selected mapping will exhibit a high worst-case crest factor. The problem is in the context as to select the best mapping efficiently in real time. A crude approach is to calculate all the crest factors associated with the n different tentative mapping attempts, which requires a considerable computational effort.

Eetvelt, Wade and Tomlinson [356] proposed a heuristic measure for identifying possible message codes yielding high crest factors. The corresponding measure was defined as:

$$SF = \frac{|W_H - N|^2 + |R_1|2}{2N^2} , \qquad (11.25)$$

where W_H is the Hamming weight of the message codes and $R_1 = \sum_{i=1}^{N-1} c_{i-1}/c_i$, which has high values for structured sequences exhibiting high crest factors. The smaller SF, the more likely that the message code has low peak factors. Eetvelt, Wade and Tomlinson [356] also proposed a scrambling scheme based on m-sequences for deriving the suggested random mapping method.

Mestdagh and Spruyt [373] analysed a specific version of the "selected mapping" method of [350] in conjunction with fixed peak-clipping in the amplifier. It was shown that there exists an optimal clipping level in terms of the achievable SNR gain for the given D/A or A/D precision and for the given number of random mapping trials used for finding the best mapping.

11.4.4 Partial Transmit Sequences

A different approach within the framework of the mapping concept is to use partial transmit sequences [377]. The problem associated with using the selected mapping based techniques of the previous section is that the selection functions may require n implementationally demanding IFFT operations for arriving at n different mapping choices. Müller and Huber [377] reduced the number of IFFT operations required to $\log_2 \sqrt{n}$, where n is the number of mapping choices available. In their scheme the legitimate information symbol set $\{c_k\}$ is partitioned into V number of sets. The V number of partitioned sets are subjected to IFFT, in order to generate V number of time domain "Partial Transmit Sequences (PTS)" and the transmitted signal is represented as a linear combination of these PTS. The optimisation of the peak factor is performed using V number of independent multiplication factors for each PTS. Allowing $\{\pm 1, \pm j\}$ for each multiplication factor, 4^V legitimate choices are presented and the best one is selected for transmission. Müller and Huber also compared the amplitude distributions associated with their scheme to those of the selected mapping schemes of Section 11.4.3. The authors claimed that their scheme was more efficient in terms of reducing the peak factors than the selected mapping based method of Section 11.4.3 for the same number of IFFT operations, namely when assuming $V = n$, where again V indicates the number of PTS and n is the number of tentative CF-reducing mapping attempts in the context of the selected mapping scheme. However, since their calculation of the peak factors was not based on the continuous time domain signal, but on the discrete values generated by the IFFT, the corresponding peak factors were lower, than the actual peak factors. Furthermore, their assumption of $V = n$ resulted in 4^V number of PTS-based choices, requiring 4^V number of CF calculating operations, while the number of mapping choices for the selected mapping method remained V. In general, comparisons have to be carried out under the assumption of constant computational efforts. In this sense, using effective selection functions [356, 378] for identifying low crest factor messages is of salient importance.

11.5 Peak Factors in Multi-Carrier CDMA

In order to capitalise on the joint benefits of direct-sequence code division multiple access (DS-CDMA) and orthogonal frequency division multiplexing (OFDM), multi-carrier CDMA (MC-CDMA) was proposed [52, 55] and has attracted significant research interests in recent years. DS-CDMA spreads the original spectrum of the information signal over a wide bandwidth and hence even if certain frequencies are faded, it is capable of exploiting frequency diversity using Rake receivers, where the number of fingers in the Rake receiver determines the maximum achievable diversity gain. OFDM, on the other hand, is resilient to frequency selective fading since it effectively converts a high-rate single carrier scheme to numerous parallel low-rate schemes. Furthermore, OFDM asymptotically approaches the theoretically highest 2Bd/Hz Shannonian bandwidth efficiency. One of the main drawbacks of OFDM is that the envelope power of the transmitted signal fluctuates widely, requiring a highly linear RF power amplifier. Since our advocated MC-CDMA system spreads a message symbol across the frequency domain and uses an OFDM transmitter for conveying each spread bit with the aid of each subcarrier, its transmitted signal also exhibits a high CF.

In this section, we analyse the characteristics of the envelope power of the MC-CDMA signal and establish the relationship between the envelope power and certain properties of the spreading sequences employed. The associated crest factors of several spreading sequences are examined and the results are discussed. In order to investigate the effects of the different crest factors on the performance of MC-CDMA systems, the bit error rate and the power spectrum are also studied, employing a specific solid-state transistor amplifier (SSTA) model. Finally, a range of practical solutions are presented for mitigating the crest factor problem in the context of MC-CDMA communication systems.

11.5.1   System Model and Envelope Power

The simplified transmitter structure of a multi-code MC-CDMA system is portrayed in Figure 11.9, where L symbols $\{b_l \mid 0 \leq l < L\}$ are transmitted simultaneously using L orthogonal spreading sequences $\{\mathbf{C}_l \mid 0 \leq l < L\}$, each constituted by N chips according to $\mathbf{C}_l = \{c_l[n] \mid 0 \leq n < N\}$. Since the L number of N-chip parallel spreading codes seen in Figure 11.9 are orthogonal, their superposition can be decomposed into L parallel message symbols at the receiver. Observe in Figure 11.9 that the superposition of the L parallel messages is modulated with the aid of the IFFT block in a single step, as in OFDM. Throughout this section, we use a subscript for the spreading code index, a square bracket $[\cdot]$ for a discrete index to time, frequency or sequence, and a round bracket for a continuous index to time. The number of parallel symbols L mapped to L N-chip orthogonal codes can be adjusted on a frame-by-frame basis, in order to accommodate the channel quality variations and for supporting variable-rate sources. In other words, this structure is suitable for the transmission of variable bit rate (VBR) data as well as constant bit rate (CBR) data. Both base stations and mobile terminals can employ this structure. Various multiple access schemes such as FDMA, TDMA and CDMA can be incorporated for supporting multiple users. Blind transmission-rate detection can readily be employed in the corresponding receivers. When $L = N$, the N-point IFFT of Figure 11.9 conveys N message symbols and hence the spectral efficiency of the multi-code MC-CDMA system is the same as that of OFDM, with the additional advantage of attaining frequency diversity gain, since the information symbols are spread using the N-chip spreading codes.

Although the more attractive family of Quadrature Amplitude Modulation (QAM) schemes [2] can readily be used for generating the information symbols $\{b_l\}$ from the source data bits, here we limit the symbol mapping scheme to M-ary phase shift keying (MPSK) for the rest of this section. Furthermore, we limit our choice of the spreading sequences to those having elements satisfying $|c_l[n]| = 1$. The normalised complex envelope of an MPSK modulated multi-code MC-CDMA signal may be represented for the duration of a symbol period T as:

$$s(t) = \frac{1}{\sqrt{N}} \sum_{l=0}^{L-1} \sum_{n=0}^{N-1} b_l\, c_l[n]\, e^{j2\pi F n \frac{t}{T}}, \tag{11.26}$$

where F is the subcarrier separation parameter [52] invoked for mapping the spread infor-

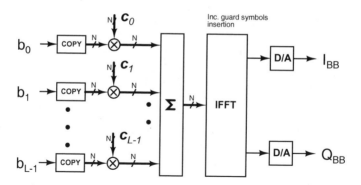

Figure 11.9: MC-CDMA transmitter model: $\mathbf{b}_i$ and $\mathbf{C}_i$ are the i-th message symbol and spreading sequence, respectively.

mation symbol to N subcarriers that are sufficiently far apart in the frequency domain, in order to experience independent fading over different subcarriers. We assume the idealistic condition that the system is operated in a synchronous environment, which is always the case in the downlink scenario. Furthermore, the transmitted signals of all the users undergo the same channel-effects.

The spreading sequences are orthogonal to each other and satisfy:

$$\sum_{n=0}^{N-1} c_i[n]\, c_j^*[n] = N\delta_{i,j}, \tag{11.27}$$

where c^* is the complex conjugate of c and $\delta_{i,j}$ is the Kronecker delta function.

Having established the system model, let us now analyse the envelope power of the signal. Let us define the combined complex symbol mapped to the n-th subcarrier $d[n]$ as:

$$d[n] \triangleq \sum_{l=0}^{L-1} b_l\, c_l[n] . \tag{11.28}$$

Then, following the approach used by Schroeder [347] and Tellambura [364], the envelope power, $|s(t)|^2$, may be expressed as:

$$
\begin{aligned}
|s(t)|^2 &= s(t)s^*(t) \\
&= \left(\frac{1}{\sqrt{N}} \sum_{n=0}^{N-1} d[n]e^{j2\pi F n \frac{t}{T}} \right) \cdot \left(\frac{1}{\sqrt{N}} \sum_{n=0}^{N-1} d[n]^* e^{-j2\pi F n \frac{t}{T}} \right) \\
&= \frac{1}{N} \sum_{n=0}^{N-1} |d[n]|^2 + \frac{2}{N}\mathrm{Re}\left\{ \sum_{n=1}^{N-1} D[n]e^{j2\pi F n \frac{t}{T}} \right\},
\end{aligned}
\tag{11.29}
$$

where $D[n]$ is:

$$D[n] = \sum_{i=0}^{N-n-1} d[i]d^*[i+n] \tag{11.30}$$

and $\text{Re}[\cdot]$ represents the real part of a signal.

The first term in Equation 11.29 is the DC component, which is constant over time, and the second term represents the AC components which has an average of zero over the signalling interval duration of T. Thus, the average power of $s(t)$ is given by the first term in Equation 11.29, which turns out to be L, regardless of the spreading sequences employed, as it is shown below.

The term $|d[n]|^2$ in Equation 11.29 can be expanded as:

$$
\begin{aligned}
|d[n]|^2 &= \left(\sum_{l=0}^{L-1} b_l\, c_l[n]\right) \cdot \left(\sum_{l=0}^{L-1} b_l^*\, c_l^*[n]\right) \\
&= \sum_{l=0}^{L-1} |b_l|^2\, |c_l[n]|^2 + \sum_{l=0}^{L-1} \sum_{l'=0, l'\neq l}^{L-1} b_l b_{l'}^*\, c_l[n]c_{l'}^*[n] \\
&= L + \sum_{l=0}^{L-1} \sum_{l'=0, l'\neq l}^{L-1} b_l b_{l'}^*\, c_l[n]c_{l'}^*[n]. \tag{11.31}
\end{aligned}
$$

Thus, the first term in Equation 11.29, which is the average power during the signalling interval T, becomes:

$$
\begin{aligned}
\frac{1}{N}\sum_{n=0}^{N-1} |d[n]|^2 &= \frac{1}{N}\sum_{n=0}^{N-1} L + \frac{1}{N}\sum_{n=0}^{N-1}\left(\sum_{l=0}^{L-1}\sum_{l'=0, l'\neq l}^{L-1} b_l b_{l'}^*\, c_l[n]c_{l'}^*[n]\right) \\
&= L + \frac{1}{N}\sum_{l=0}^{L-1}\sum_{l'=0, l'\neq l}^{L-1}\left(b_l b_{l'}^* \sum_{n=0}^{N-1} c_l[n]c_{l'}^*[n]\right). \tag{11.32}
\end{aligned}
$$

Considering the orthogonality condition of Equation 11.27 defined for the different spreading codes, the last term in Equation 11.32 becomes zero. Thus, the average transmitted power becomes:

$$\frac{1}{N}\sum_{n=0}^{N-1} |d[n]|^2 = L, \tag{11.33}$$

where again, L is the number of simultaneously used spreading codes.

Let us now concentrate our attention on the second term of Equation 11.29. The n-th harmonic component $D[n]$, of the instantaneous power $|s(t)|^2$, which is defined in Equation 11.30, is constituted by the terms $d[i]d^*[i+n]$, which can be expanded as:

$$d[i]d^*[i+n] = \left(\sum_{l=0}^{L-1} b_l\, c_l[i] \right) \cdot \left(\sum_{l=0}^{L-1} b_l^*\, c_l^*[i+n] \right)$$

$$= \sum_{l=0}^{L-1} |b_l|^2 c_l[i]\, c_l^*[i+n] + \sum_{l=0}^{L-1} \sum_{l'=0, l' \neq l}^{L-1} b_l\, b_{l'}^*\, c_l[i]\, c_{l'}^*[i+n]. \qquad (11.34)$$

Upon substituting Equation 11.34 into Equation 11.30, we arrive at:

$$
\begin{aligned}
D[n] &= \sum_{i=0}^{N-n-1} \left(\sum_{l=0}^{L-1} |b_l|^2\, c_l[i] c_l^*[i+n] + \sum_{l=0}^{L-1} \sum_{l'=0, l' \neq l}^{L-1} b_l b_{l'}^*\, c_l[i] c_{l'}^*[i+n] \right) \\
&= \sum_{l=0}^{L-1} \sum_{i=0}^{N-n-1} c_l[i] c_l^*[i+n] + \sum_{l=0}^{L-1} \sum_{l'=0, l' \neq l}^{L-1} \sum_{i=0}^{N-n-1} b_l b_{l'}^*\, c_l[i] c_{l'}^*[i+n] \\
&= \sum_{l=0}^{L-1} A_l[n] + \sum_{l=0}^{L-1} \sum_{l'=0, l' \neq l}^{L-1} b_l b_{l'}^*\, X_{l,l'}[n], \qquad (11.35)
\end{aligned}
$$

where $A_l[n]$ represents the **aperiodic auto-correlations of the l-th spreading code**, defined by:

$$A_l[n] \triangleq \sum_{i=0}^{N-n-1} c_l[i]\, c_l^*[i+n] \qquad (11.36)$$

and $X_{l,l'}[n]$ represents the **aperiodic cross-correlations between the l-th and l'-th spreading codes**, defined by:

$$X_{l,l'}[n] \triangleq \sum_{i=0}^{N-n-1} c_l[i]\, c_{l'}^*[i+n], \qquad (11.37)$$

where the term "aperiodic" indicates that the correlation is calculated as if the sequences are one-shot sequence followed by all-zero sequences. We introduce $A[n]$ and $X[n]$, which are the collective forms of the correlations, defined as:

$$A[n] \triangleq \sum_{l=0}^{L-1} A_l[n] \text{ for } n \neq 0, \ A[0] \triangleq NL/2 \qquad (11.38)$$

$$X[n] \triangleq \sum_{l=0}^{L-1} \sum_{l'=0, l' \neq l}^{L-1} b_l b_{l'}^*\, X_{l,l'}[n]. \qquad (11.39)$$

The collective aperiodic auto-correlation $A[n]$ is message independent, i.e. independent of b_l, while the collective aperiodic cross-correlation $X[n]$ is message dependent.

Upon applying Equations 11.33, 11.35, 11.38 and 11.39, we can represent the envelope power $|s(t)|^2$ as:

$$|s(t)|^2 = L + \frac{2}{N}\text{Re}\left[\sum_{n=1}^{N-1}(A[n] + X[n])\,e^{j2\pi Fn\frac{t}{T}}\right] \qquad (11.40)$$

$$= \frac{2}{N}\text{Re}\left[\sum_{n=0}^{N-1}(A[n] + X[n])\,e^{j2\pi Fn\frac{t}{T}}\right]. \qquad (11.41)$$

The Fourier transform of $|s(t)|^2$ can be directly derived from Equation 11.41:

$$F\{|s(t)|^2\} = \frac{1}{N}\sum_{n=1}^{N-1}\left[(A[n] + X[n])\,\delta\left(f - Fn\frac{1}{T}\right) + (A^*[n] + X^*[n])\,\delta\left(f + Fn\frac{1}{T}\right)\right]$$

$$= S(f) * S^*(f), \qquad (11.42)$$

where a^* is the complex conjugate of a and $a(t) * b(t)$ is the convolution of $a(t)$ and $b(t)$, defined as:

$$a(t) * b(t) \triangleq \int_{-\infty}^{\infty} a(\zeta)b(t - \zeta)d\zeta. \qquad (11.43)$$

Furthermore, $S(f)$ in Equation 11.42 is the Fourier transform of $s(t)$:

$$S(f) = \frac{T}{\sqrt{N}}\sum_{n=0}^{N-1}d[n]\,\text{sinc}(fT - Fn)e^{-j\pi(fT - Fn)}, \qquad (11.44)$$

where $d[n]$ is the combined complex symbol mapped to the n-th subcarrier given in Equation 11.28 and $\text{sinc}(x)$ is defined as $\sin(\pi x)/(\pi x)$.

Ideally $|s(t)|^2$ should be constant. A trivial case exists for BPSK MC-CDMA with $N = 2$ and $L = 2$, where $\{c_0[n]\} = \{1, 1\}$ and $\{c_1[n]\} = \{1, -1\}$. In this case, both $A[n]$ and $X[n]$ become zero, leaving $|s(t)|^2 = 2$. For other values of N and L, there do not seem to be any ideal choices of the spreading sequences in terms of producing a constant envelope. However, some spreading sequences produce lower envelope fluctuation than others and this issue will be addressed in the next two subsections.

The envelope may also be expressed as:

$$|s(t)|^2 = \sum_{l=0}^{L-1}|s_l(t)|^2 + \text{Re}\left[\sum_{l=0}^{L-1}\sum_{l'=0,l'\neq l}^{L-1}b_lb_{l'}^*\frac{2}{N}\sum_{n=1}^{N-1}X_{l,l'}[n]e^{j2\pi Fn\frac{t}{T}}\right], \qquad (11.45)$$

where $|s_l(t)|^2$ is the envelope power of the lth information bearing signal, which is given by [364]:

$$|s_l(t)|^2 = 1 + \frac{2}{N}\text{Re}\sum_{n=1}^{N-1}A_l[n]e^{j2\pi Fn\frac{t}{T}}, \qquad (11.46)$$

which is a rephrased version of the well-known auto-correlation theorem [379]. Equation 11.45 suggests that the total envelope power is the sum of the power of L single-code MC-CDMA transmit signal envelopes, plus a term that depends on the aperiodic cross-correlations of the lth and l'th spreading codes defined in Equation 11.37 multiplied by the transmitted bits $b_l b_{l'}$. It can be observed in Equation 11.46 that for the single-code case corresponding to $L = 1$ the above cross-correlation terms do not contribute to the envelope power. However, as L increases, the cross-correlation terms given in Equation 11.37 play an increasingly dominant role.

11.5.2   Spreading Sequences and Crest Factors

In the previous section, we observed that the envelope power of a multi-code MC-CDMA signal is characterised by the aperiodic auto-correlations and cross-correlations of the spreading sequences employed. Hence, low values of $A[n]$ and $X[n]$ are desirable in Equation 11.45 and Equation 11.46 in order for a set of spreading codes to exhibit a low crest factor. In this section, we introduce several well-known sequences and investigate their crest factor properties based on their correlation characteristics.

11.5.2.1   Single-code Signal

For a single-code MC-CDMA signal the envelope power is entirely characterised by the aperiodic auto-correlations of the employed spreading sequence, as shown in Equation 11.46 [347, 364]. The Walsh codes of Section 9.2.1 and orthogonal Gold codes (OGold) of Section 9.2.2 are well-known binary orthogonal codes. There exist complex orthogonal sequences as well, such as the family of Frank codes [380–382] and Zadoff-Chu codes [383, 384], which have been applied for multi-carrier CDMA systems [317, 319].

A. Frank codes

A Frank code [380–382] is defined as:

$$s[k + iM] = e^{j2\pi pki/M}, \tag{11.47}$$

where $M^2 = N$ is the length of the codes, $0 \leq k, i < M$, and p is relatively prime with respect to M, i.e. they do not have a common divisor. Here, we set p to 1 for simplicity. A Frank code has perfect periodic auto-correlation, implying that

$$\sum_{n=0}^{N-1} s[n]s^*[n + k] = \begin{cases} N & \text{if } k = 0 \\ 0 & \text{otherwise.} \end{cases} \tag{11.48}$$

Hence, a Frank code and its $N - 1$ circularly shifted versions form a set of N complex orthogonal sequences [380, 381].

Popović [385] as well as Antweiler and Bömer [386] recognised that there exist $\lceil M/2 \rceil$ distinct magnitude sets $\{|A_l[n]|\}$ of aperiodic auto-correlations for the set of N Frank sequences, hence $\lceil M/2 \rceil$ number of corresponding crest factors. In general, from Equation 11.46 we can observe that the envelope power waveform $|s_l'(t)|^2$ having $\{A_l[n]e^{j\alpha n}\}$

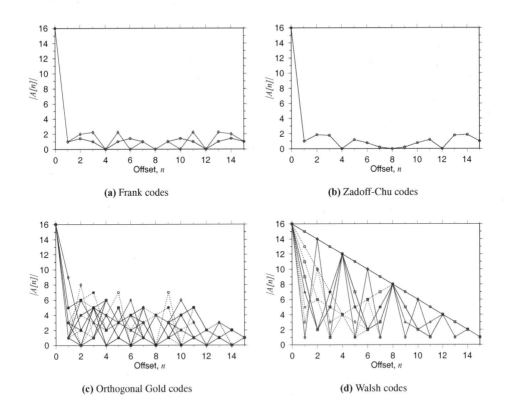

Figure 11.10: The magnitude $|A[n]|$ of the aperiodic auto-correlation functions defined in Equation 11.36 for various spreading sequences having a length of $N = 16$ and for a single code of $L = 1$. It can be shown that there are 2, 1, 16 and 8 sets of magnitudes for the Frank, Zadoff-Chu, orthogonal Gold and Walsh codes, respectively.

instead of $\{A_l[n]\}$ becomes:

$$|s'(t)|^2 = 1 + \frac{2}{N}\text{Re}\sum_{n=1}^{N-1} A_l[n]e^{j\alpha n}\,e^{j2\pi Fn\frac{t}{T}}$$

$$= 1 + \frac{2}{N}\text{Re}\sum_{n=1}^{N-1} A_l[n]e^{j2\pi Fn\frac{1}{T}\left(t+\frac{T\alpha}{2\pi F}\right)}$$

$$= \left|s\left(t + \frac{T\alpha}{2\pi F}\right)\right|^2, \tag{11.49}$$

which is a time-shifted version of the original envelope power waveform. Figure 11.10 depicts the magnitudes $\{|A[n]|\}$ of the auto-correlations of four different classes of spreading sequences having a length of $N = 16$. In case of Frank codes, only two sets of auto-correlation magnitudes $\{|A[n]|\}$ appear in Figure 11.10(a). This was expected, since $M^2 = N$, yielding $M = 4$ for $N = 16$ and according to [385, 386] there exist $\lceil 4/2 \rceil = 2$

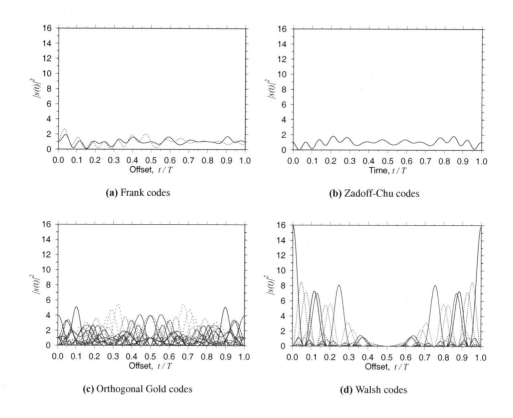

(a) Frank codes

(b) Zadoff-Chu codes

(c) Orthogonal Gold codes

(d) Walsh codes

Figure 11.11: The envelope power $|s(t)|^2$ of single-code MC-CDMA signals employing various spreading sequences with the length N of 16. Only waveforms which are unique with respect to time-shifting were displayed. There are 2, 1, 16 and 8 sets of unique waveforms for Frank, Zadoff-Chu, orthogonal Gold and Walsh codes respectively.

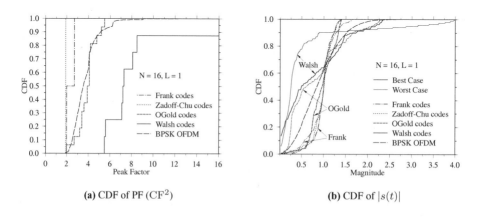

(a) CDF of PF (CF^2)

(b) CDF of $|s(t)|$

Figure 11.12: The crest factor and magnitude cumulative distribution function (CDF) of BPSK MC-CDMA and BPSK OFDM for the spreading factor of $N = 16$ and for the number of simultaneously used codes $L = 1$

magnitude sets. Furthermore, we can observe in Figure 11.10(a) that the off-peak auto-correlation magnitude $|A[n]|$ of Frank codes shows symmetry around $n = 8$. This is be-cause $A[n] = -A[N - n]$ for any code satisfying Equation 11.48 [387]. Frank codes exhibit low off-peak auto-correlations, which is typically quantified in terms of their so-called merit factor (MF) [385–387], defined as:

$$\text{MF} \triangleq A^2[0]/(2 \sum_{n=1}^{N-1} |A[n]|^2).$$ (11.50)

A high value of MF is beneficial in terms of low PF as well as of reliable code-acquisition at the beginning of establishing synchronisation between the transmitter and receiver. Half of the Frank codes exhibit a MF of 8, while we have MF = 4 for the other half of the set of codes. By contrast, all the Zadoff-Chu codes exhibit MF = 6.7. The merit factors of orthogonal Gold codes lie between 0.6 and 1.8, except for one of them, which is 4.0, and those for Walsh codes are below 0.52.

The time domain power envelope waveforms of the four codes characterised in terms of $|A[n]|$ in Figure 11.10 are shown in Figure 11.11. Only the power envelope waveforms are depicted in the Figure which cannot be mapped to each other by time-shifting. Similarly to the observation made in terms of the auto-correlation magnitudes $\{|A[n]|\}$, there are two unique power envelope waveforms for Frank codes. Both waveforms fluctuate less dramatically, than those of the binary OGold and Walsh codes shown in Figures 11.11(c) and 11.11(d). The CDF of the crest factors shows a two-step distribution function for the Frank codes in Figure 11.12(a), since there are two different power envelope waveforms for the MC-CDMA signals employing N-different Frank codes.

B. Zadoff-Chu Codes

The family of Zadoff-Chu code [383, 384] is defined as:

$$s[k] = \begin{cases} e^{j\pi k^2 p/N} & \text{for even } N \\ e^{j\pi k(k+1)p/N} & \text{for odd } N, \end{cases}$$ (11.51)

where N is the length of the code, $0 \leq k < N$ and p is relatively prime with respect to N, and we set p to 1 again. Similarly to Frank codes, Zadoff-Chu codes have perfect periodic auto-correlations. Hence, we can generate $N - 1$ further Zadoff-Chu sequences $s_l[k] = s[k + l]$, $1 \leq l \leq N - 1$, by circularly shifting a Zadoff-Chu code. It can be shown that the family of Zadoff-Chu codes defined in Equation 11.51 is similar to the set of Newman phases [360], Schroeder phases [347], Greenstein phases [349] and Narahashi phases [354], which were introduced in Equations 11.18, 11.19 and 11.20 of Section 11.3.1. It is interesting to note that various researchers have found closely related codes independently, using different methods over a time-span of 30 years.

All the Zadoff-Chu codes of length $N = 16$ showed the same aperiodic correlation mag-nitude sets, as seen in Figure 11.10(b), hence a single unique power envelope waveform is shown in Figure 11.11(b), which becomes also explicit in terms of the crest factor CDF of Figure 11.12(a). This is because the aperiodic auto-correlations of Zadoff-Chu codes are given as:

$$A_l[n] = \sum_{i=0}^{N-n-1} s_l[i]s_l^*[i+n]$$

$$= \sum_{i=0}^{N-n-1} e^{j\pi \frac{(i+l)^2}{N}} e^{-j\pi \frac{(i+l+n)^2}{N}}$$

$$= e^{-j\pi \frac{n^2+2nl}{N}} \sum_{i=0}^{N-n-1} e^{-j\pi \frac{2ni}{N}}$$

$$= e^{-j2\pi \frac{nl}{N}} A_0[n] . \tag{11.52}$$

Substituting Equation 11.52 into Equation 11.46, we have

$$|s_l(t)|^2 = \left| s_0\left(t - \frac{Tl}{NF}\right) \right|^2, \tag{11.53}$$

which states that the lth user's power envelope waveform is a time-shifted version of the first user's power envelope, retaining all the magnitude statistics.

The magnitudes of the aperiodic autocorrelation of Zadoff-Chu codes also exhibit a symmetry around $n = 8$ due to their perfectly periodic auto-correlation function. The crest factor of Zadoff-Chu codes was the lowest among the investigated four spreading sequences. Furthermore, the high peak-to-sidelobe ratio of the auto-correlation function of Figure 11.10(b) is advantageous in terms of reliable code acquisition.

C. Orthogonal Gold codes As was mentioned in Section 9.2.2, orthogonal Gold (OGold) code [388] are derived from binary pseudo-random codes. It was observed in Figure 11.10(c) that each of the 16 OGold code has different auto-correlation magnitude sets $\{|A[n]|\}$. Accordingly, the 16 envelope power waveforms seen in Figure 11.11(c) are all different. However, four codes of the 16 codes happen to have the same crest factor and some further codes exhibit similar crest factors. This explains the various steps in the crest factor distribution of OGold codes in Figure 11.12(a). The crest factor distribution is similar to that of BPSK OFDM using no spreading codes, which correspond to random codes, since all 2^{16} message sequences were encountered. It is interesting to see that some of the orthogonal Gold codes have low crest factors, approaching those of Frank codes and Zadoff-Chu codes. This implies that we can employ a specific code assignment scheme to reduce the crest factor of each uplink signal in synchronous MC-CDMA systems employing orthogonal Gold codes, when the number of simultaneously transmitting mobiles is low and hence low-CF sequences can be invoked. This is however not the case, when a large user population has to be supported using short codes of a limited-sized set.

D. Walsh codes

The family of Walsh codes was introduced in Section 9.2.1. Members of this code family exhibit some structure and this contributes to the high values of their aperiodic auto-correlations seen in Figure 11.10(d). In fact, the all-zero sequence, $\{W_0[k]\}$, which is the first sequence in any Walsh code set, gives us the highest possible auto-correlation magnitude set for $\{|A[n]|\} = N - n$ and hence yields the highest possible crest factor of $\sqrt{N}$ or a peak factor of N. In Figures 11.10(d), 11.11(d) and 11.12(a), we can observe that there are $N/2$

different auto-correlation magnitude values $\{|A[n]|\}$ for the N Walsh codes of length N and hence $N/2$ different unique power envelope waveforms $|s(t)|^2$ and crest factors are observed. This can be explained as follows. Observing Equation 9.6 in Section 9.2.1, we find:

$$W_{2m+1}[k] = W_{2m}[k]\, e^{j\pi k}, \tag{11.54}$$

where $m = 0, \cdots, N/2-1$. In other words, the odd-indexed Walsh sequence can be obtained by changing the sign of every other element. The corresponding aperiodic auto-correlations of Equation 11.36 result in:

$$A_{2m+1}[n] = A_{2m}[n]\, e^{-j\pi n}, \tag{11.55}$$

where $m = 0, \cdots, N/2 - 1$, and hence the envelope power waveform of the $(2m + 1)$-th Walsh code is a time-shifted version of that of the $(2m)$-th Walsh code, retaining all the same magnitude statistics. Figure 11.12(a) shows the 8-step distribution function of the crest factors for the Walsh-spread MC-CDMA signals using $N = 16$. Even the lowest crest factor of the Walsh codes is higher than the highest crest factor of OGold codes.

Figure 11.12(b) depicts the magnitude distributions of single-code MC-CDMA signals employing four different spreading codes, where the "best case" and the "worst case" represent the specific codes of having the lowest crest factor and the highest crest factor, respectively, for the particular class of spreading sequences. Desired magnitude distributions are expected to exhibit a steep CDF curve around the average magnitude, indicating the predominance of the average magnitudes. The Frank, Zadoff-Chu and the best-case OGold sequences exhibited this desirable tendency.

We conclude that the Zadoff-Chu sequences are the most attractive ones from the set of four spreading sequences investigated in the context of synchronous single-code MC-CDMA systems in terms of their low crest factors.

So far, we implicitly considered a synchronous uplink scenario, where different mobiles use different orthogonal spreading codes for transmission using the same set of subcarriers. In a downlink scenario or in a uplink scenario, where different mobiles are assigned non-overlapping sets of subcarriers, we do not need N orthogonal sequences for spreading a message symbol over N subcarriers. We only need one spreading sequence. If this is the case, we are free to choose any sequence producing a low crest factor, such as a poly-phase code based on the Newman phases of Section 11.3.4 [360] or a Golay code of Section 11.3.2 [363], also known as a complementary code. Popović [361] showed that in this context the crest factor of the single-code MC-CDMA signal employing any Golay codes is bounded by 3dB.

11.5.2.2 Shapiro-Rudin-based Spreading Sequences

Multi-code MC-CDMA signals are transmitted by the base station transmitter in the downlink or by mobile terminals in the uplink, when more than one symbols have to be transmitted simultaneously. As we have shown in Section 11.5.1, the power envelope of the transmitted signal depends on the sum of the aperiodic autocorrelations $A[n]$ given in Equation 11.38 and on the modulated sum of the aperiodic cross-correlations $X[n]$ given in Equation 11.39. In this section, we will consider a range of specific special sequences, which exhibit zero or small values of $A[n]$ and $X[n]$.

While studying an absolute bound for a sum of trigonometric polynomials having binary coefficients, Shapiro, then an MSc student, found [359] a pair of binary sequences having a low bound. These sequences are referred to as Shapiro-Rudin sequences, which are defined recursively as:

$$s_0 = c_0 = 1, \tag{11.56}$$

$$s_{n+1} = s_n c_n, \tag{11.57}$$

$$c_{n+1} = s_n \bar{c}_n, \tag{11.58}$$

where the sequence ab $= (a_1 a_2 \cdots a_{N-1} b_1 b_2 \cdots b_{N-1})$ represents a concatenation of sequences a $= (a_1 a_2 \cdots a_{N-1})$ and b $= (b_1 b_2 \cdots b_{N-1})$, and $\bar{a} = (-a_1 - a_2 \cdots - a_{N-1})$ a negation of the original sequence a. Later it was recognised [361] that Shapiro-Rudin sequences constitute a special case of Golay's complementary sequences [363]. A pair of equally long sequences is said to be a complementary pair, if their combined aperiodic auto-correlation defined in Equation 11.38 for $L = 2$ is zero except at zero shift. As we mentioned before, this is the property we want to retain in order to maintain low crest factors.

In our forthcoming discourse, we use $\otimes$ as the aperiodic correlation operator, $\phi_{a,b}$ as the aperiodic cross-correlation sequence $\phi_{a,b}[n] \triangleq (a \otimes b)[n]$, $n = 0, 1, \cdots, N - 1$ and $\Upsilon(a)$ as a specific version of a that was shifted to the left by one, i.e. $\Upsilon(a)[n] = a[n + 1]$, $n = 0, 1, 2, \cdots, N - 1$. We also use $\vec{a}$ to denote the sequence a that was shifted by the length of a, i.e. $\vec{a}[n] = a[n - N]$, $n = N, N + 1, \cdots, 2N - 1$. Furthermore, $\hat{a}$ represents a version of the sequence a which is reverse-ordered, i.e. $\hat{a}[n] = a[N - 1 - n]$, $n = 0, 1, \cdots, N - 1$. Finally, we denote the concatenated sequence of the sequences a and b as ab.

The following relationship regarding the cross-correlation sequence is useful in deriving the crest factor properties of multi-code MC-CDMA signals:

$$a \otimes \vec{b} = \left\{ \sum_{i=0}^{n-1} a_{N-n+i} b_i \mid 0 \leq n < N \right\} \left\{ \sum_{i=0}^{N-n-1} a_i b_{i+n} \mid 0 \leq n < N \right\}$$

$$= \Upsilon(\phi_{b,a}) \, \phi_{a,b} \, . \tag{11.59}$$

In other words, the aperiodic cross-correlation sequence created from the sequence a and the sequence $\vec{b}$, which is the N times right-shifted version of b, is given by the concatenated sequence of $\Upsilon(\phi_{b,a}) = \Upsilon(b \otimes a)$ and $\phi_{a,b} = a \otimes b$. This can be readily verified by the direct calculation of the aperiodic cross-correlation function of a and $\vec{b}$.

The next theorem is required to characterise the crest factor property of MC-CDMA signals employing a pair of Shapiro-Rudin sequences.

Theorem 11.1 *Let* s_n *and* c_n *be a pair of Shapiro-Rudin sequences [359] of length* $N = 2^n$. *With the aid of these sequence, we derive four half-length sequences defined as* $\{u_{n-1}[k] = s_n[2k]\}$, $\{v_{n-1}[k] = s_n[2k+1]\}$, $\{x_{n-1}[k] = c_n[2k]\}$ *and* $\{y_{n-1}[k] = c_n[2k+1]\}$, *where* $0 \leq k < N/2$. *Then,* u *and* v *form a complementary pair, and so do* x *and* y.

For example, we can derive two complementary pairs of length 8, i.e. a total of four sequences, from a pair of complementary sequences of length 16.

Proof: From the definition of the Shapiro-Rudin sequences [359], we can derive the recursive relationships between the new decimated sequences:

$$\mathbf{u}_0 = s_0 \qquad \mathbf{u}_n = \mathbf{u}_{n-1} \, \mathbf{x}_{n-1} \tag{11.60}$$

$$\mathbf{v}_0 = c_0 \qquad \mathbf{v}_n = \mathbf{v}_{n-1} \, \mathbf{y}_{n-1} \tag{11.61}$$

$$\mathbf{x}_0 = s_0 \qquad \mathbf{x}_n = \mathbf{u}_{n-1} \, \overline{\mathbf{x}}_{n-1} \tag{11.62}$$

$$\mathbf{y}_0 = \overline{c}_0 \qquad \mathbf{y}_n = \mathbf{v}_{n-1} \, \overline{\mathbf{y}}_{n-1} \, . \tag{11.63}$$

It will be shown that:

$$\phi_{u_n}[k] + \phi_{v_n}[k] = \phi_{x_n}[k] + \phi_{y_n}[k] = 2^{n+1}\delta[k] \, . \tag{11.64}$$

For $n = 0$, this is trivial. For $n = 1$, $\mathbf{u}_1 = [s_0 s_0]$, $\mathbf{v}_1 = [c_0 \overline{c}_0]$, $\mathbf{x}_1 = [s_0 \overline{s}_0]$ and $\mathbf{y}_1 = [c_0 c_0]$. The sums of the autocorrelations associated with $n = 1$ are given as:

$$\mathbf{u}_1 \otimes \mathbf{u}_1 + \mathbf{v}_1 \otimes \mathbf{v}_1 = [+2 \, +1] \, + \, [+2 \, -1] \, = \, [+4 \, 0] \tag{11.65}$$

$$\mathbf{x}_1 \otimes \mathbf{x}_1 + \mathbf{y}_1 \otimes \mathbf{y}_1 = [+2 \, -1] \, + \, [+2 \, +1] \, = \, [+4 \, 0] \, . \tag{11.66}$$

Therefore Equation 11.64 is true for $n = 0$ and $n = 1$. Let us now assume that Equation 11.64 holds for $n = m - 2$ and $n = m - 1$. Then, $\phi_{u_m}[k] + \phi_{v_m}[k]$ is given as:

$$\begin{aligned}
\phi_{u_m}[k] + \phi_{v_m}[k] &= \mathbf{u}_m \otimes \mathbf{u}_m + \mathbf{v}_m \otimes \mathbf{v}_m \\
&= (\mathbf{u}_{m-1}\mathbf{x}_{m-1}) \otimes (\mathbf{u}_{m-1}\mathbf{x}_{m-1}) + (\mathbf{v}_{m-1}\mathbf{y}_{m-1}) \otimes (\mathbf{v}_{m-1}\mathbf{y}_{m-1}) \\
&= \mathbf{u}_{m-1} \otimes \mathbf{u}_{m-1} + \mathbf{x}_{m-1} \otimes \mathbf{x}_{m-1} + \mathbf{u}_{m-1} \otimes \vec{\mathbf{x}}_{m-1} + \vec{\mathbf{x}}_{m-1} \otimes \mathbf{u}_{m-1} \\
&\quad + \mathbf{v}_{m-1} \otimes \mathbf{v}_{m-1} + \mathbf{y}_{m-1} \otimes \mathbf{y}_{m-1} + \mathbf{v}_{m-1} \otimes \vec{\mathbf{y}}_{m-1} + \vec{\mathbf{y}}_{m-1} \otimes \mathbf{v}_{m-1} \\
&= 2^m \delta[k] + 2^m \delta[k] + \mathbf{u}_{m-1} \otimes \vec{\mathbf{x}}_{m-1} + \mathbf{v}_{m-1} \otimes \vec{\mathbf{y}}_{m-1} \\
&= 2^{m+1}\delta[k] + \mathbf{S}_{m-1} \, ,
\end{aligned} \tag{11.67}$$

where $\mathbf{S}_{m-1} = \mathbf{u}_{m-1} \otimes \vec{\mathbf{x}}_{m-1} + \mathbf{v}_{m-1} \otimes \vec{\mathbf{y}}_{m-1}$. Applying Equation 11.59, $\mathbf{S}_{m-1}$ in the last term of Equation 11.67 can be further expanded as:

$$\begin{aligned}
\mathbf{S}_{m-1} &= \mathbf{u}_{m-1} \otimes \vec{\mathbf{x}}_{m-1} + \mathbf{v}_{m-1} \otimes \vec{\mathbf{y}}_{m-1} \\
&= \Upsilon \left(\phi_{x_{m-1},u_{m-1}} \right) \, \phi_{u_{m-1},x_{m-1}} + \Upsilon \left(\phi_{y_{m-1},v_{m-1}} \right) \, \phi_{v_{m-1},y_{m-1}} \\
&= \Upsilon \left(\mathbf{x}_{m-1} \otimes \mathbf{u}_{m-1} \right) \, \mathbf{u}_{m-1} \otimes \mathbf{x}_{m-1} + \Upsilon \left(\mathbf{y}_{m-1} \otimes \mathbf{v}_{m-1} \right) \, \mathbf{v}_{m-1} \otimes \mathbf{y}_{m-1} \\
&= \Upsilon \left([\mathbf{u}_{m-2} \overline{\mathbf{x}}_{m-2}] \otimes [\mathbf{u}_{m-2} \mathbf{x}_{m-2}] \right) \left([\mathbf{u}_{m-2} \mathbf{x}_{m-2}] \otimes [\mathbf{u}_{m-2} \overline{\mathbf{x}}_{m-2}] \right) \\
&\quad + \Upsilon \left([\mathbf{v}_{m-2} \overline{\mathbf{y}}_{m-2}] \otimes [\mathbf{v}_{m-2} \mathbf{y}_{m-2}] \right) \left([\mathbf{v}_{m-2} \mathbf{y}_{m-2}] \otimes [\mathbf{v}_{m-2} \overline{\mathbf{y}}_{m-2}] \right) \\
&= \Upsilon \left(\mathbf{u}_{m-2} \otimes \vec{\mathbf{x}}_{m-2} + \mathbf{v}_{m-2} \otimes \vec{\mathbf{y}}_{m-2} \right) \left(-\mathbf{u}_{m-2} \otimes \vec{\mathbf{x}}_{m-2} - \mathbf{v}_{m-2} \otimes \vec{\mathbf{y}}_{m-2} \right) \\
&= \Upsilon \left(\mathbf{S}_{m-2} \right) \left(\mathbf{S}_{m-2} \right) \, .
\end{aligned} \tag{11.68}$$

Upon exploiting that $\mathbf{S}_0 = \mathbf{u}_0 \otimes \vec{\mathbf{x}}_0 + \mathbf{v}_0 \otimes \vec{\mathbf{y}}_0 = [0 \, 0]$, we get:

$$\mathbf{S}_{m-1} = \mathbf{0} \tag{11.69}$$

When substituting Equation 11.69 into Equation 11.67, we arrive at:

$$\phi_{u_m}[k] + \phi_{v_m}[k] = 2^{m+1}\,\delta[k]\,. \tag{11.70}$$

Hence, we can conclude that Equation 11.64 holds for any integer $n \geq 0$. ¶

Corollary 11.2 *The autocorrelation* ϕ_s *of any Shapiro-Rudin sequence* s *satisfies that:*

$$\phi_s[2k] = 0 \qquad \text{for any } k \geq 1 \tag{11.71}$$

Proof: Let $\mathbf{u}_{n-1}$ and $\mathbf{v}_{n-1}$ be the even indexed sequence and the odd indexed sequence of $\mathbf{s}_n$, as defined in Theorem 11.1. Then,

$$\begin{aligned}
\phi_{s_n}[2k] &= \phi_{u_{n-1}}[k] + \phi_{v_{n-1}}[k] \\
&= 2^n \delta[k],
\end{aligned} \tag{11.72}$$

as given in Equation 11.64. ¶

We are now ready to introduce a crest factor property of the family of Shapiro-Rudin sequences in the context of MC-CDMA systems.

Theorem 11.3 *If we apply a pair of Shapiro-Rudin sequences [359, 361, 363, 389] as the spreading sequences for a two-code BPSK modulated MC-CDMA transmitter having* $N = 2^m$ *($m = 1, 2, 3, \cdots$) subcarriers, then the crest factor is bounded by 3dB.*

For example, a pair of Shapiro-Rudin sequences of length $N = 8 = 2^3$, $\mathbf{a} = (+1 + 1 + 1 - 1 + 1 + 1 - 1 + 1)$ and $\mathbf{b} = (+1 + 1 + 1 - 1 - 1 - 1 + 1 - 1)$, are orthogonal to each other and hence can be used as spreading sequences for a two-code MC-CDMA scheme.

Proof: Let $\mathbf{s}_m = \mathbf{s}_{m-1}\mathbf{c}_{m-1}$ and $\mathbf{c}_m = \mathbf{s}_{m-1}\bar{\mathbf{c}}_{m-1}$ be a pair of Shapiro-Rudin sequences of length $N = 2^m$, where the pair $\mathbf{s}_{m-1}$ and $\mathbf{c}_{m-1}$ is constituted by a pair of Shapiro-Rudin sequence of length $N/2$. It is well known that Shapiro-Rudin sequences are complementary and hence:

$$A_m^{(2)}[n] = \mathbf{s}_m \otimes \mathbf{s}_m + \mathbf{c}_m \otimes \mathbf{c}_m = 0 \qquad \text{for } n \geq 1, \tag{11.73}$$

where the superscript (2) represents the number of codes used for supporting multiple users or multiple bits per symbol. On the other hand, the aperiodic cross-correlations are given as:

$$\begin{aligned}
\mathbf{s}_m \otimes \mathbf{c}_m &= (\mathbf{s}_{m-1}\mathbf{c}_{m-1}) \otimes (\mathbf{s}_{m-1}\bar{\mathbf{c}}_{m-1}) \\
&= \mathbf{s}_{m-1} \otimes \mathbf{s}_{m-1} - \mathbf{s}_{m-1} \otimes \vec{\mathbf{c}}_{m-1} + \vec{\mathbf{c}}_{m-1} \otimes \mathbf{s}_{m-1} - \mathbf{c}_{m-1} \otimes \mathbf{c}_{m-1}
\end{aligned} \tag{11.74}$$

$$\begin{aligned}
\mathbf{c}_m \otimes \mathbf{s}_m &= (\mathbf{s}_{m-1}\bar{\mathbf{c}}_{m-1}) \otimes (\mathbf{s}_{m-1}\mathbf{c}_{m-1}) \\
&= \mathbf{s}_{m-1} \otimes \mathbf{s}_{m-1} + \mathbf{s}_{m-1} \otimes \vec{\mathbf{c}}_{m-1} - \vec{\mathbf{c}}_{m-1} \otimes \mathbf{s}_{m-1} - \mathbf{c}_{m-1} \otimes \mathbf{c}_{m-1}
\end{aligned} \tag{11.75}$$

The combined aperiodic cross-correlation $X_m^{(2)}[n]$ becomes:

$$
\begin{aligned}
X_m^{(2)}[n] &= b_0 b_1 (\mathbf{s}_m \otimes \mathbf{c}_m) + b_1 b_0 (\mathbf{c}_m \otimes \mathbf{s}_m) \\
&= 2 b_0 b_1 (\mathbf{s}_{m-1} \otimes \mathbf{s}_{m-1}) - 2 b_1 b_0 (\mathbf{c}_{m-1} \otimes \mathbf{c}_{m-1}) \\
&= 4 b_0 b_1 (\mathbf{s}_{m-1} \otimes \mathbf{s}_{m-1}) \\
&= 4 b_0 b_1 A_{m-1}^{(1)}[n] \,,
\end{aligned}
\tag{11.76}
$$

where b_0 and b_1 are the information bits. Applying Equations 11.73 and 11.76 to Equation 11.40, we can establish the relationships between the two envelope powers $|s_m^{(2)}(t)|^2$ and $|s_{m-1}^{(1)}(t)|^2$, one for $L = 2$ and N number of subcarriers, and another one for $L = 1$ and $N/2$ number of subcarriers. First, let us consider the case for $b_0 b_1 = 1$.

$$
\begin{aligned}
\left| s_m^{(2)}(t) \right|^2 &= 2 + \frac{2}{N} \mathrm{Re} \left[\sum_{n=1}^{N/2-1} 4 A_{m-1}^{(1)}[n] e^{j 2\pi F n \frac{t}{T}} \right] \\
&= 2 + 2 \frac{2}{N/2} \mathrm{Re} \left[\sum_{n=0}^{N/2-1} A_{m-1}^{(1)}[n] e^{j 2\pi F n \frac{t}{T}} \right] \\
&= 2 \left| s_{m-1}^{(1)}(t) \right|^2 .
\end{aligned}
\tag{11.77}
$$

When $b_0 b_1 = -1$, $X_m^{(2)}[n]$ becomes $-4 A_{m-1}^{(1)}[n]$, which is equivalent to $4 A_{m-1}^{(1)}[n] e^{j\pi n}$ due to Corollary 11.2. Then the envelope power becomes:

$$
\begin{aligned}
\left| s_m^{(2)}(t) \right|^2 &= 2 + \frac{2}{N} \mathrm{Re} \left[\sum_{n=1}^{N/2-1} 4 A_{m-1}^{(1)}[n] e^{j\pi n} e^{j 2\pi F n \frac{t}{T}} \right] \\
&= 2 + 2 \frac{2}{N/2} \mathrm{Re} \left[\sum_{n=0}^{N/2-1} A_{m-1}^{(1)}[n] e^{j 2\pi F n \frac{1}{T} \left(t + \frac{T}{2F} \right)} \right] \\
&= 2 \left| s_{m-1}^{(1)} \left(t + \frac{T}{2F} \right) \right|^2 .
\end{aligned}
\tag{11.78}
$$

Since the average power of $s_m^{(2)}(t)$ is 2, and the peak factor is defined as the ratio of the maximum value of $s_m^{(2)}(t)$ to the average value, the peak factor remains the same for $s_m^{(2)}(t)$ as in $s_{m-1}^{(1)}(t)$. Considering that the peak factor of $s_{m-1}^{(1)}(t)$ is at most 2 [361], we can conclude that a pair of Shapiro-Rudin sequences produces a crest factor of less than or equal to 3dB for a two-code BPSK modulated MC-CDMA system. ¶

It is interesting to note that the combined cross-correlation of the two constituent codes of a pair of Shapiro-Rudin sequences exhibits the property of so-called sub-complementarity [389]. Sivaswamy introduced sub-complementary sequences in [389]. A pair of sequences of length $N = 2^k N_o$, $k \geq 1$, is called a sub-complementary pair, if the combined aperiodic autocorrelation $A[n]$ of Equation 11.38 is zero for $n > N_o$. A method of constructing sub-complementary pairs was also given in [389] as follows. Let $\mathbf{s}_{m-1}$ be any sequence of length $N/2$. Then, a pair of sequences s_m and $\hat{s}_m$, constructed as [389]:

$$s_m = s_{m-1}s_{m-1} \tag{11.79}$$

$$\hat{s}_m = s_{m-1}\bar{s}_{m-1}, \tag{11.80}$$

is known to form a sub-complementary pair [389]. Sivaswamy's sub-complementary pairs exhibit similar crest factor characteristics to those of Shapiro-Rudin sequences in the context of two-code BPSK modulated MC-CDMA systems.

Theorem 11.4 *If we apply a Shapiro-Rudin sequence [359] of length $N/2$ as the base sequence s_{m-1}, then the sub-complementary pair [389] s_m and $\hat{s}_m$ of Equations 11.79 and 11.80 produces a crest factor less than or equal to 3dB, provided that they are applied in two-code BPSK modulated MC-CDMA systems.*

For example, the sequence $a = (+1+1+1-1+1+1-1+1)$ in the previous example for Theorem 11.3 can be used as the base sequence of a pair of sub-complementary sequences. According to the generation methods defined in Equations 11.79 and 11.80, a pair of sub-complementary sequences s and $\hat{s}$, given as:

$$s = (+1+1+1-1+1+1-1+1+1+1+1-1+1+1-1+1) \tag{11.81}$$

$$\hat{s} = (+1+1+1-1+1+1-1+1-1-1-1+1-1-1+1-1) \tag{11.82}$$

can be used for a two-code MC-CDMA system, since they are inherently orthogonal to each other owing to the Walsh code-like definition of Equations 11.79 and 11.80.

Proof: the aperiodic auto-correlations can be formulated as:

$$s_m \otimes s_m = (s_{m-1}s_{m-1}) \otimes (s_{m-1}s_{m-1})$$
$$= s_{m-1} \otimes s_{m-1} + s_{m-1} \otimes \vec{s}_{m-1} + \vec{s}_{m-1} \otimes s_{m-1} + s_{m-1} \otimes s_{m-1} \tag{11.83}$$

$$\hat{s}_m \otimes \hat{s}_m = (s_{m-1}\bar{s}_{m-1}) \otimes (s_{m-1}\bar{s}_{m-1})$$
$$= s_{m-1} \otimes s_{m-1} - s_{m-1} \otimes \vec{s}_{m-1} - \vec{s}_{m-1} \otimes s_{m-1} + s_{m-1} \otimes s_{m-1}. \tag{11.84}$$

Thus the combined aperiodic auto-correlation $A_m^{(2)}[n]$ defined in Equation 11.38 becomes:

$$A_m^{(2)}[n] = s_m \otimes s_m + \hat{s}_m \otimes \hat{s}_m$$
$$= 4(s_{m-1} \otimes s_{m-1}) = 4A_{m-1}^{(1)}[n], \tag{11.85}$$

where $A_{m-1}^{(1)}[n]$ is the aperiodic auto-correlation of the Shapiro-Rudin sequence s_{m-1}. On the other hand, the aperiodic cross-correlations defined in Equation 11.37 are given as:

$$s_m \otimes \hat{s}_m = (s_{m-1}s_{m-1}) \otimes (s_{m-1}\bar{s}_{m-1})$$
$$= s_{m-1} \otimes s_{m-1} - s_{m-1} \otimes \vec{s}_{m-1} + \vec{s}_{m-1} \otimes s_{m-1} - s_{m-1} \otimes s_{m-1} \tag{11.86}$$

$$\hat{s}_m \otimes s_m = (s_{m-1}\bar{s}_{m-1}) \otimes (s_{m-1}s_{m-1})$$
$$= s_{m-1} \otimes s_{m-1} + s_{m-1} \otimes \vec{s}_{m-1} - \vec{s}_{m-1} \otimes s_{m-1} - s_{m-1} \otimes s_{m-1}. \tag{11.87}$$

The combined aperiodic cross-correlation $X_m^{(2)}[n]$ becomes:

$$X_m^{(2)}[n] = b_0 b_1 (\mathbf{s}_m \otimes \hat{\mathbf{s}}_m) + b_1 b_0 (\hat{\mathbf{s}}_m \otimes \mathbf{s}_m)$$
$$= 0. \tag{11.88}$$

Upon substituting Equations 11.85 and 11.88 in Equation 11.40, we can conclude that:

$$\left| s_m^{(2)}(t) \right|^2 = 2 \left| s_{m-1}^{(1)}(t) \right|^2, \tag{11.89}$$

which is identical to Equation 11.77. Therefore, we reached the same conclusion, as stated in Theorem 11.3, namely that the corresponding crest factor is bounded by 3dB. ¶

It is interesting to see in Equation 11.88 that the combined aperiodic cross-correlation of Sivaswamy's sub-complementary pair vanishes, and hence we conclude that the envelope power waveform does not depend on the message sequences. Considering that a pair of Shapiro-Rudin sequences has zero off-peak combined aperiodic auto-correlation since it is a specific case of complementary pair, the Shapiro-Rudin and the Sivaswamy's sub-complementary families of pairs are dual.

So far we have concentrated our attention on the cases when two symbols had to be transmitted simultaneously. We found that the crest factors of MC-CDMA signals employing the Shapiro-Rudin sequence pairs of Equations 11.57 and 11.58 or the Sivaswamy sub-complementary pairs of Equations 11.79 and 11.79 are bounded by 3dB. Let us now extend our discussions to the scenario where more than two spreading sequences are used to support multiple symbols generated by a single user or for supporting multiple users.

Tseng and Liu [365] introduced a wider range of complementary sets of sequences. A set of sequences of equal length is said to be a complementary set, if the sum of auto-correlations of all the sequences in that set is zero, except for the peak term in zero-shift position [365]. Sivaswamy [389] proposed a method of constructing a complementary set of 2^{n+1} number of sequences of length $2^n N$, where $n \geq 1$ and N is the length of the base complementary pair. When $n = 1$, we can obtain a complementary set of four sequences of length N from a base complementary pair $\mathbf{s}_0$ and $\mathbf{c}_0$ of length $N/2$ using the following method [389]:

$$\mathbf{S}_1 = \mathbf{s}_0 \mathbf{s}_0 \tag{11.90}$$
$$\mathbf{S}_2 = \mathbf{s}_0 \bar{\mathbf{s}}_0 \tag{11.91}$$
$$\mathbf{S}_3 = \mathbf{c}_0 \mathbf{c}_0 \tag{11.92}$$
$$\mathbf{S}_4 = \mathbf{c}_0 \bar{\mathbf{c}}_0 . \tag{11.93}$$

Theorem 11.5 *If we apply a pair of Shapiro-Rudin sequence [359] of length $N/2$ as the base sequences $\mathbf{s}_0$ and $\mathbf{c}_0$, the set of four complementary sequences, given by Equations 11.90 – 11.93, produces a crest factor less than or equal to 3dB, provided that the sequences $\mathbf{S}_1$, $\mathbf{S}_2$, $\mathbf{S}_3$ and $\mathbf{S}_4$ are applied to four-code BPSK modulated MC-CDMA systems.*

For example, a pair of Shapiro-Rudin sequences of length $N/2 = 4$, $\mathbf{a} = (+1 + 1 + 1 - 1)$ and $\mathbf{b} = (+1 + 1 - 1 + 1)$, can be used to generate $\mathbf{S}_1 = (+1 + 1 + 1 - 1 + 1 + 1 + 1 - 1)$, $\mathbf{S}_2 = (+1 + 1 + 1 - 1 - 1 - 1 - 1 + 1)$, $\mathbf{S}_3 = (+1 + 1 - 1 + 1 + 1 + 1 - 1 + 1)$ and $\mathbf{S}_4 = (+1 + 1 - 1 + 1 - 1 - 1 + 1 - 1)$. We can readily verify that they form an orthogonal set and hence these four sequences can be used as the spreading sequences of four-code BPSK modulated MC-CDMA systems.

Proof: Sivaswamy [389] showed that the set of four sequences form a complementary set and hence we have:

$$A[n] = 4N\delta[n]. \tag{11.94}$$

The cross-correlations between all possible combinations of pairs are given as follows:

$$\mathbf{S}_1 \otimes \mathbf{S}_2 = \mathbf{s}_0 \otimes \mathbf{s}_0 - \mathbf{s}_0 \otimes \mathbf{s}_0 - \mathbf{s}_0 \otimes \vec{\mathbf{s}}_0 + \vec{\mathbf{s}}_0 \otimes \mathbf{s}_0 = -\mathbf{s}_0 \otimes \vec{\mathbf{s}}_0 \tag{11.95}$$

$$\mathbf{S}_2 \otimes \mathbf{S}_1 = \mathbf{s}_0 \otimes \mathbf{s}_0 - \mathbf{s}_0 \otimes \mathbf{s}_0 + \mathbf{s}_0 \otimes \vec{\mathbf{s}}_0 - \vec{\mathbf{s}}_0 \otimes \mathbf{s}_0 = \mathbf{s}_0 \otimes \vec{\mathbf{s}}_0 \tag{11.96}$$

$$\mathbf{S}_1 \otimes \mathbf{S}_3 = \mathbf{s}_0 \otimes \mathbf{c}_0 + \mathbf{s}_0 \otimes \mathbf{c}_0 + \mathbf{s}_0 \otimes \vec{\mathbf{c}}_0 + \vec{\mathbf{s}}_0 \otimes \mathbf{c}_0 = 2\mathbf{s}_0 \otimes \mathbf{c}_0 + \mathbf{s}_0 \otimes \vec{\mathbf{c}}_0 \tag{11.97}$$

$$\mathbf{S}_3 \otimes \mathbf{S}_1 = \mathbf{c}_0 \otimes \mathbf{s}_0 + \mathbf{c}_0 \otimes \mathbf{s}_0 + \mathbf{c}_0 \otimes \vec{\mathbf{s}}_0 + \vec{\mathbf{c}}_0 \otimes \mathbf{s}_0 = 2\mathbf{c}_0 \otimes \mathbf{s}_0 + \mathbf{c}_0 \otimes \vec{\mathbf{s}}_0 \tag{11.98}$$

$$\mathbf{S}_1 \otimes \mathbf{S}_4 = \mathbf{s}_0 \otimes \mathbf{c}_0 - \mathbf{s}_0 \otimes \mathbf{c}_0 - \mathbf{s}_0 \otimes \vec{\mathbf{c}}_0 + \vec{\mathbf{s}}_0 \otimes \mathbf{c}_0 = -\mathbf{s}_0 \otimes \vec{\mathbf{c}}_0 \tag{11.99}$$

$$\mathbf{S}_4 \otimes \mathbf{S}_1 = \mathbf{c}_0 \otimes \mathbf{s}_0 - \mathbf{c}_0 \otimes \mathbf{s}_0 + \mathbf{c}_0 \otimes \vec{\mathbf{s}}_0 - \vec{\mathbf{c}}_0 \otimes \mathbf{s}_0 = \mathbf{c}_0 \otimes \vec{\mathbf{s}}_0 \tag{11.100}$$

$$\mathbf{S}_2 \otimes \mathbf{S}_3 = \mathbf{s}_0 \otimes \mathbf{c}_0 - \mathbf{s}_0 \otimes \mathbf{c}_0 + \mathbf{s}_0 \otimes \vec{\mathbf{c}}_0 - \vec{\mathbf{s}}_0 \otimes \mathbf{c}_0 = \mathbf{s}_0 \otimes \vec{\mathbf{c}}_0 \tag{11.101}$$

$$\mathbf{S}_3 \otimes \mathbf{S}_2 = \mathbf{c}_0 \otimes \mathbf{s}_0 - \mathbf{c}_0 \otimes \mathbf{s}_0 - \mathbf{c}_0 \otimes \vec{\mathbf{s}}_0 + \vec{\mathbf{c}}_0 \otimes \mathbf{s}_0 = -\mathbf{c}_0 \otimes \vec{\mathbf{s}}_0 \tag{11.102}$$

$$\mathbf{S}_2 \otimes \mathbf{S}_4 = \mathbf{s}_0 \otimes \mathbf{c}_0 + \mathbf{s}_0 \otimes \mathbf{c}_0 - \mathbf{s}_0 \otimes \vec{\mathbf{c}}_0 - \vec{\mathbf{s}}_0 \otimes \mathbf{c}_0 = 2\mathbf{s}_0 \otimes \mathbf{c}_0 - \mathbf{s}_0 \otimes \vec{\mathbf{c}}_0 \tag{11.103}$$

$$\mathbf{S}_4 \otimes \mathbf{S}_2 = \mathbf{c}_0 \otimes \mathbf{s}_0 + \mathbf{c}_0 \otimes \mathbf{s}_0 - \mathbf{c}_0 \otimes \vec{\mathbf{s}}_0 - \vec{\mathbf{c}}_0 \otimes \mathbf{s}_0 = 2\mathbf{c}_0 \otimes \mathbf{s}_0 - \mathbf{c}_0 \otimes \vec{\mathbf{s}}_0 \tag{11.104}$$

$$\mathbf{S}_3 \otimes \mathbf{S}_4 = \mathbf{c}_0 \otimes \mathbf{c}_0 - \mathbf{c}_0 \otimes \mathbf{c}_0 - \mathbf{c}_0 \otimes \vec{\mathbf{c}}_0 + \vec{\mathbf{c}}_0 \otimes \mathbf{c}_0 = -\mathbf{c}_0 \otimes \vec{\mathbf{c}}_0 \tag{11.105}$$

$$\mathbf{S}_4 \otimes \mathbf{S}_3 = \mathbf{c}_0 \otimes \mathbf{c}_0 - \mathbf{c}_0 \otimes \mathbf{c}_0 + \mathbf{c}_0 \otimes \vec{\mathbf{c}}_0 - \vec{\mathbf{c}}_0 \otimes \mathbf{c}_0 = \mathbf{c}_0 \otimes \vec{\mathbf{c}}_0 \tag{11.106}$$

The combined cross-correlation defined in Equation 11.39, assuming BPSK modulation, is given by:

$$\begin{aligned} X[n] =& b_1 b_2 \left(\mathbf{S}_1 \otimes \mathbf{S}_2 + \mathbf{S}_2 \otimes \mathbf{S}_1 \right) + b_1 b_3 \left(\mathbf{S}_1 \otimes \mathbf{S}_3 + \mathbf{S}_3 \otimes \mathbf{S}_1 \right) \\ &+ b_1 b_4 \left(\mathbf{S}_1 \otimes \mathbf{S}_4 + \mathbf{S}_4 \otimes \mathbf{S}_1 \right) + b_2 b_3 \left(\mathbf{S}_2 \otimes \mathbf{S}_3 + \mathbf{S}_3 \otimes \mathbf{S}_2 \right) \\ &+ b_2 b_4 \left(\mathbf{S}_2 \otimes \mathbf{S}_4 + \mathbf{S}_4 \otimes \mathbf{S}_2 \right) + b_3 b_4 \left(\mathbf{S}_3 \otimes \mathbf{S}_4 + \mathbf{S}_3 \otimes \mathbf{S}_4 \right), \end{aligned} \tag{11.107}$$

where again b_1, b_2, b_3 and b_4 are the four information bits to be transmitted. Considering that $\mathbf{S}_1 \otimes \mathbf{S}_2 + \mathbf{S}_2 \otimes \mathbf{S}_1 = 0$ and $\mathbf{S}_3 \otimes \mathbf{S}_4 + \mathbf{S}_4 \otimes \mathbf{S}_3 = 0$, $X[n]$ can be reduced to:

$$\begin{aligned} X[n] =& b_1 b_3 \left(\mathbf{S}_1 \otimes \mathbf{S}_3 + \mathbf{S}_3 \otimes \mathbf{S}_1 \right) + b_1 b_4 \left(\mathbf{S}_1 \otimes \mathbf{S}_4 + \mathbf{S}_4 \otimes \mathbf{S}_1 \right) \\ &+ b_2 b_3 \left(\mathbf{S}_2 \otimes \mathbf{S}_3 + \mathbf{S}_3 \otimes \mathbf{S}_2 \right) + b_2 b_4 \left(\mathbf{S}_2 \otimes \mathbf{S}_4 + \mathbf{S}_4 \otimes \mathbf{S}_2 \right). \end{aligned} \tag{11.108}$$

First, let us assume that $b_1 b_4 = b_2 b_3$, which implies that $b_1 b_3 = b_2 b_4$ holds as well. It can be shown that eight messages out of the $s^L = 2^4 = 16$ possible 4-bit messages satisfy this condition. In this case, $X[n]$ becomes:

$$X[n] = \pm 4 \left(\mathbf{s}_0 \otimes \mathbf{c}_0 + \mathbf{c}_0 \otimes \mathbf{s}_0 \right). \tag{11.109}$$

Since $\{\mathbf{s}_0, \mathbf{c}_0\}$ constitutes a Shapiro-Rudin pair, the sequences can be written as $\mathbf{s}_0 = \mathbf{ab}$ and $\mathbf{c}_0 = \mathbf{a\overline{b}}$, where $\mathbf{a}$ and $\mathbf{b}$ are also Shapiro-Rudin sequences of length $N/4$. Then, as we observed in Equations 11.74, 11.75 and 11.76, $X[n]$ becomes:

$$X[n] = \pm 16\,\mathbf{a} \otimes \mathbf{a}. \tag{11.110}$$

According to Equation 11.45 the normalised power envelope is given by:

$$\frac{|(s(t)|^2}{4} = 1 + \frac{2}{4N}\mathrm{Re}\left[\sum_{n=1}^{N/4-1} \pm 16\,\mathbf{a} \otimes \mathbf{a}\, e^{j2\pi Fn\frac{t}{T}}\right] \tag{11.111}$$

$$= \begin{cases} 1 + \frac{2}{N/4}\mathrm{Re}\left[\sum_{n=1}^{N/4-1} \mathbf{a} \otimes \mathbf{a}\, e^{j2\pi Fn\frac{t}{T}}\right] & \text{for } b_1 b_4 = 1 \\ 1 + \frac{2}{N/4}\mathrm{Re}\left[\sum_{n=1}^{N/4-1} \mathbf{a} \otimes \mathbf{a}\, e^{j2\pi Fn\frac{1}{T}\left(t+\frac{T}{2F}\right)}\right] & \text{for } b_1 b_4 = -1 \end{cases} \tag{11.112}$$

$$\leq 2\,. \tag{11.113}$$

Let us now assume that $b_1 b_4 = -b_2 b_3$, which also implies that $b_1 b_3 = -b_2 b_4$. Depending on the signs of $b_1 b_4$ and $b_1 b_3$, we get:

$$X[n] = \begin{cases} \pm 4\,\mathbf{s}_0 \otimes \vec{\mathbf{c}}_0 & \text{for } b_1 b_4 = -b_2 b_3 \\ \pm 4\,\mathbf{c}_0 \otimes \vec{\mathbf{s}}_0 & \text{for } b_1 b_4 = b_2 b_3\,. \end{cases} \tag{11.114}$$

Considering that:

$$\mathbf{s}_0 \mathbf{c}_0 \otimes \mathbf{s}_0 \mathbf{c}_0 = \mathbf{s}_0 \otimes \mathbf{s}_0 + \mathbf{c}_0 \otimes \mathbf{c}_0 + \mathbf{s}_0 \otimes \vec{\mathbf{c}}_0 + \vec{\mathbf{c}}_0 \otimes \mathbf{s}_0 = N\delta[n] + \mathbf{s}_0 \otimes \vec{\mathbf{c}}_0\,, \tag{11.115}$$

we can express $\mathbf{s}_0 \otimes \vec{\mathbf{c}}_0$ of Equation 11.114 as $\mathbf{s}_0 \mathbf{c}_0 \otimes \mathbf{s}_0 \mathbf{c}_0$. In a similar manner, $\mathbf{c}_0 \otimes \vec{\mathbf{s}}_0$ can be replaced by $\mathbf{c}_0 \mathbf{s}_0 \otimes \mathbf{c}_0 \mathbf{s}_0$. We note that $\mathbf{s}_1 = \mathbf{s}_0 \mathbf{c}_0$ and $\mathbf{s}_0 \vec{\mathbf{c}}_0$ form a complementary pair of length N and so do $\mathbf{s}_1' = \mathbf{c}_0 \mathbf{s}_0$ and $\vec{\mathbf{c}}_0 \mathbf{s}_0$. Therefore, when $b_1 b_4 = -b_2 b_3$, the normalised power envelope can be formulated as:

$$\frac{|(s(t)|^2}{4} = 1 + \frac{2}{4N}\mathrm{Re}\left[\sum_{n=1}^{N-1} \pm 4\,\mathbf{s}_1 \otimes \mathbf{s}_1\, e^{j2\pi Fn\frac{t}{T}}\right] \tag{11.116}$$

$$= \begin{cases} 1 + \frac{2}{N}\mathrm{Re}\left[\sum_{n=1}^{N-1} \mathbf{s}_1 \otimes \mathbf{s}_1\, e^{j2\pi Fn\frac{t}{T}}\right] & \text{for } b_1 b_4 = 1 \\ 1 + \frac{2}{N}\mathrm{Re}\left[\sum_{n=1}^{N-1} \mathbf{s}_1 \otimes \mathbf{s}_1\, e^{j2\pi Fn\frac{1}{T}\left(t+\frac{T}{2F}\right)}\right] & \text{for } b_1 b_4 = -1 \end{cases} \tag{11.117}$$

$$\leq 2\,. \tag{11.118}$$

In a similar manner, the normalised power envelope associated with $b_1 b_4 = b_2 b_3$ can be shown to be less than or equal to 2. Hence, the corresponding crest factors are always bounded by 3dB. ¶

Figure 11.13 depicts the peak-factors of some of Sivaswamy's complementary sets constituted by $L = 4$ sequences. As shown in Theorem 11.5, there can be three different magnitude distributions for the specific sequence length of $N = 2^n$, depending on the message bits. The marker ○ in Figure 11.13 represents the peak factor of the signal, when $X[k] = \pm 16\,\mathbf{a} \otimes \mathbf{a}$, which is the case for half the number of message bits from the set of 16 possible message bits. The marker □ represents the peak factors, when $X[k] = \pm 4\,\mathbf{s}_0 \otimes \mathbf{c}_0$, and the marker ⋆ when $X[k] = \pm 4\,\mathbf{c}_0 \otimes \mathbf{s}_0$, both of which occur for 4 message bits from the set of

$2^L = s^4 = 16$ possible message bits. We can observe in Figure 11.13 that the peak factors associated with marker $\circ$ are the lowest among the three values when the sequence length N is $2^2 = 4$, $2^4 = 16$ or $2^6 = 64$. An ideal peak factor of 1 was achieved for the sequence length of $N = 1^2 = 4$, representing a constant power envelope. The fact that all the peak factors in Figure 11.13 are bounded by 2 corroborates Theorem 11.5.

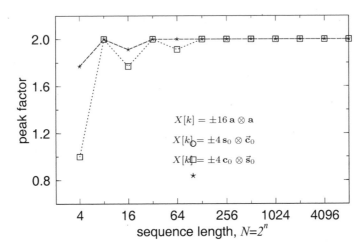

Figure 11.13: The peak-factors of Sivaswamy's complementary set of sequences. The number of simultaneously used sequences is $L = 4$ and the sequence length of $N = 2^n$ is between $2^2 = 4$ and $= 2^{13} = 8192$.

11.5.2.3   Peak Factor Distribution of Multi-Code MC-CDMA

In this section, we investigate the peak factor distribution of multi-code BPSK MC-CDMA. Two binary sequences, namely Walsh codes and orthogonal Gold codes, along with two so-called root-unity poly-phase sequences, namely the so-called Frank codes [381] and Zadoff-Chu codes [383, 384] are considered as the orthogonal spreading codes for our BPSK modulated MC-CDMA scheme. We assumed that the number of subcarriers, N, is 16. However, the results shown in this section apply also to MC-CDMA schemes with $16 \times F$ subcarriers, where the originally adjacent chips are F-subcarriers apart, in order to achieve independent fading on each subcarrier, which improves the achievable diversity gain.

When L number of codes are used simultaneously from the set of N available spreading codes, there are $\binom{N}{L}$ possible choices of selecting a specific set of spreading codes. If there are no other constraints to be satisfied, when selecting the codes, maintaining a low worst case peak factor can be a selection criterion. We can observe the effects of the code selection with the aid of the Cumulative Distribution Functions (CDF) of the worst case peak factors encountered, when choosing the spreading codes. The CDF of a single-code MC-CDMA signal was shown Figure 11.12(a). Figure 11.14 depicts the corresponding CDFs for $L = 2, 4, 8, 15$ number of codes in conjunction with the spreading factor of $N = 16$.   When $L = 2$, there are $\binom{16}{2} = 120$ possible pairs of spreading codes, which we can choose. We can observe in Figure 11.14(a) that the peak factors of the Zadoff-Chu code pairs are in the

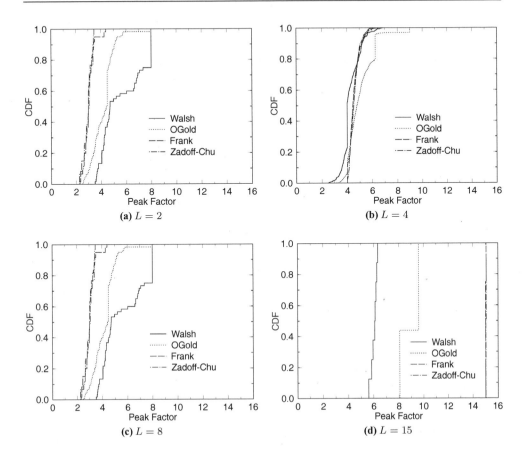

Figure 11.14: Cumulative Distribution Function (CDF) of the maximum PF (CF^2) for multi-code BPSK modulated MC-CDMA signals. The spreading factor was $N = 16$ and the number of simultaneously used codes was $L = 2, 4, 8, 15$. There exists $\binom{N}{L}$ number of possible sets of codes for the given value of L. Walsh codes exhibit desirable peak factor distributions for large L values, i.e. either when supporting a high number of users or transmitting a high number of bits per symbol.

rage of 2.3 to 3.5, depending on the specific pairs selected. The lowest peak factors of the Zadoff-Chu, Frank and OGold codes are similar. A quarter of the Walsh pairs exhibited the highest peak factor of 8, which was extremely undesirable.

However, as the number of codes L increases, the Walsh code sets begin to show more desirable peak factor distributions. This is already evident for $L = 4$, as shown in Figure 11.14(b). The lowest and the highest peak factors were 2.2821 and 6.8595, respectively, for Walsh codes of length $N = 16$ and for $L = 4$ simultaneously employed codes. This implies that the code selection is important in multi-code MC-CDMA, in order to reduce the corresponding peak factor.

For a higher number of simultaneously used codes L, the CF advantage of Walsh codes over the other spreading codes considered becomes more pronounced. When $L = 15$, any

Class	$L = 2$	$L = 4$	$L = 8$	$L = 15$
Walsh	0x0480; 3.512	0x03c0; 2.282	0x23cd; 2.786	0xbfff; 5.602
OGold	0x0202; 2.489	0x0003; 3.161	0xf20e; 4.480	0xfbff; 8.067
Frank	0x0900; 2.307	0x04a4; 4.000	0x4d74; 8.000	0xff7f; 15.000
Zadoff	0x4040; 2.262	0x2174; 4.000	0x44de; 8.000	0xff74; 15.001

Table 11.5: The best code combinations and the corresponding peak factors for $L = 2, 4, 8$ and 15 number of simultaneous codes. The code combinations are represented in hexadecimal form, where the index of each contributing code is indicated by a "1" at the corresponding bit position of the equivalent binary bit pattern. For example, the best Walsh code combination for $L = 2$ is given by $0x0480 = (0000\,0100\,1000\,0000)_2$ having "1" at the 8th and 11th positions, corresponding to these codes from the set of $\binom{16}{2} = 120$ possible combinations. The least significant digit refers the first code in the set of $N = 16$ spreading codes.

combination of Walsh codes exhibited lower peak factors than the best combinations of the other spreading codes investigated, as we can observe in Figure 11.14(d)

The best code combinations are summarised in Table 11.5 together with the corresponding peak factors for $L = 2, 4, 8, 15$ number of simultaneous codes and for the spreading factor of $N = 16$. In Table 11.5 the code combinations are represented by hexadecimal numbers, where the index of each contributing code is indicated by a "1" at the corresponding bit position of the equivalent binary bit pattern. For example, the best Walsh code combination for $L = 2$ is given by the pair of 8th and 11th Walsh codes, exhibiting the lowest peak factor, which is indicated by a "1" at positions 8 and 11 in the binary pattern of "0x0480 = $(0000\,0100\,1000\,0000)_2$", where "0x" is the hexadecimal notation prefix. The least significant bit represents the first code in the set.

Figure 11.15(a) depicts the maximum peak factors evaluated for the set of all 2^L messages using the best code-combinations, taking into account all possible $\binom{N}{L}$ choices. For our BPSK modulated OFDM modem, (N, L) non-linear PF reduction codes [31] associated with the code rate of L/N were used in order to arrive at the maximum peak factors plotted. The PFs plotted were multiplied by N/L in order to normalise them for the sake of maintaining the same average power as the original uncoded MC-CDMA. Walsh-spread MC-CDMA shows the lowest maximum peak factors for $L \geq 3$ among the four MC-CDMA systems studied. In fact, the peak factors of the Walsh-spread MC-CDMA signals approached the practically achievable minimum peak factor for various numbers of simultaneous codes L.

As mentioned before, Jones, Wilkinson and Barton [31] introduced a non-linear block coding scheme for reducing the overall peak factor of OFDM signals, hence we apply the same coding scheme for reducing the peak factors of MC-CDMA signals. Let us eliminate half of the messages which yield high values of peak factors. Then the worst case peak factor corresponds to the median peak factor of the original uncoded MC-CDMA signals. The median peak factor also depends on the spreading sequence combination employed. The best code combination which has the lowest median peak factor may be different from the code combination yielding the lowest maximum peak factor, which was displayed in Table 11.5. Figure 11.15(b) depicts the median peak factors for the best combination of L simultaneous codes for the spreading factor of $N = 16$. We can observe in Figure 11.15(b) that the $15/16$

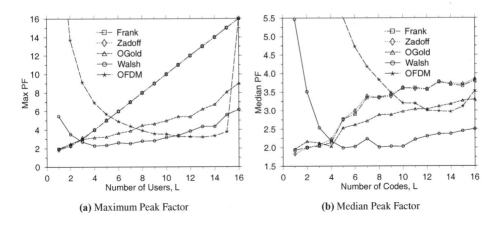

(a) Maximum Peak Factor   **(b)** Median Peak Factor

Figure 11.15: The maximum and median PF (CF^2) when using various spreading codes for BPSK modulated MC-CDMA. A set of nonlinear peak-factor reduction codes [31] of rate L/N were used for BPSK modulated OFDM and the PFs plotted were normalised by N/L for the sake of maintaining the same average power as the MC-CDMA scheme. The spreading factor was $N = 16$.

nonlinear block code used for Walsh spread MC-CDMA reduces the peak factor from 6.18 to 2.50, which is only 1dB higher than the practically achievable minimum peak factor of 3dB. This implies that Walsh-spread MC-CDMA can be viewed as a peak-factor reducing scheme in the context of conventional OFDM, in addition to the inherent diversity gain provided by frequency domain spreading. Figure 11.15(b) suggests that by adaptively applying the specific spreading code family, which exhibits the lowest peak factor for the given number of simultaneous codes L in conjunction with a rate of $(L-1)/L$ crest factor reduction coding, we can reduce the overall peak factor below 2.5 for any L, provided that the other system requirements are not violated by the PF reduction motivated spreading code adaptation.

11.5.3 Clipping Amplifier and BER Comparison

Having investigated the peak factor distributions numerically, let us now study the effects of the peak factor on the achievable BER performance, when the MC-CDMA signal is subjected to non-linear amplification.

11.5.3.1 Non-linear Power Amplifier Model

During our forthcoming investigations, we need a tractable model for the power amplifiers to be used in order to study the effects of nonlinear amplification. Most portable wireless devices use Solid-State-Power-Amplifiers (SSPA), which are referred to as Solid-State-Transistor-Amplifiers (SSTA). The amplitude transfer characteristics or the AM-AM curve $g(x)$ of an SSPA can be modelled as [390–392]:

$$g(x) = \frac{x}{\left(1 + x^{2p}\right)^{1/2p}} \tag{11.119}$$

where x is the input magnitude and p is the smoothness factor during its transition from the linear region to the saturation region. Furthermore, it is assumed that the original phase characteristics of the MC-CDMA signal are retained by the linear-phase amplifier, implying the absence of phase distortion. Figure 11.16 illustrates the AM-AM curves for a few different smoothness factors, p, together with the 1dB compression line. Specifically, the 1dB com-

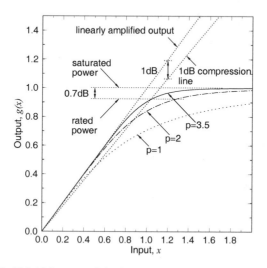

Figure 11.16: AM-AM curves of the SSPA model of Equation 11.119 [390–392]

pression point is defined, as the point on the output versus input curve, where the output level falls 1dB below its expected linearly amplified value. Using simple arithmetic, we can find the 1dB compression point in the form of:

$$\frac{x}{(1 + x^{2p})^{1/2p}} = kx, \tag{11.120}$$

where k is the 1dB compression line's steepness coefficient given as $k = 10^{-1/20}$. The solution of Equation 11.120 for the input level xdB, where the 1dB output power compression is encountered, is given by $x_{1dB} = (k^{-2p} - 1)^{1/(2p)}$. For typical SSPAs, the output power at the 1dB compression point is about 0.7 dB below the saturation power (http://www.vertexepi.com/brochure/broch3.htm). Therefore the smoothness factor p in Equation 11.119 should satisfy $g(x_{1dB}) = k(k^{-2p} - 1)^{1/(2p)} = \alpha$, where α is $10^{-0.7/20}$. The value of the smoothness factor p which satisfies this relation is $p = 3.58$. The corresponding value of the input level, where the 1dB compression is encountered, is denoted by x_{1dB}, which becomes $x_{1dB} = 1.035117539$.

More explicitly, the average output power is typically backed off from the saturation power, in order to reduce the non-linear distortion of the amplified signal. The 1dB compression point is the typically used reference point specified in terms of the input power. Thus, we get a 0.7 dB lower output power, than the saturation power, when the power amplifier is operating at 0dB back-off. Sometimes, we want to investigate the effects of the non-linear amplification on the bit error ratio, as well as on the out-of-band frequency response, which determines the amount of out-of-band spurious emissions inflicted on other communication systems operating in the adjacent frequency band.

11.5.3.2 Clipping Effects on Output Power

We saw in Section 11.5.2.3 that MC-CDMA signals exhibit a fluctuating envelope power and hence when they are subject to non-linear amplification, they produce out-of-band spurious emissions. The extreme peaks of the signal inevitably suffer from so-called "clipping effects", when they enter the saturation region of the amplifier, resulting in the loss of effective transmission power as well as the signal distortion, unless the gain of the amplifier is reduced to a level, where the amplifier's maximum output level is sufficiently high for the signal peaks not to be clipped. We showed in Section 11.5.2.3 that the degree of power envelope variations of a MC-CDMA signal depends on the spreading sequences employed. Therefore, the loss of effective transmission power is also expected to depend on the spreading sequences used. Figure 11.17 depicts the relationship between the input-backoff of the power amplifier and the effective relative output power of the MC-CDMA signals spread by the various spreading sequences employed.

When only one simultaneous spreading code was used, we found that the effective transmission power loss was about 0.7dB for the Frank and Zadoff-Chu spread MC-CDMA signals, while the corresponding losses were 1.3dB, 2.3dB and 4.4dB for a complementary code, a single OGold code and a single Walsh code. A Shapiro-Rudin sequence was used as the complementary code, which was found to exhibit the lowest loss of effective transmission power among the three binary spreading sequences compared. Apart from the effects of signal distortion and the associated spurious out-of-band emissions, a high power loss may be encountered, which is a serious impediment in practical systems.

When two simultaneous spreading codes were used as seen in Figure 11.17(b), the MC-CDMA signal employing a complementary pair of Shapiro-Rudin codes showed the lowest power loss of 1.6dB for the set of five spreading codes investigated. A pair of OGold codes and a pair of Frank codes were next, showing an associated power loss of 1.7dB and 1.8dB, respectively. Finally, the Walsh codes and Zadoff-Chu codes exhibited a high power loss of around 3dB at 0dB input back-off. The spreading code pairs were selected on the basis of maintaining the lowest crest factors for each family of spreading codes. It is interesting to observe that the Frank and Zadoff codes showed nearly identical maximum peak factors, although they exhibited quite a different power loss, as it can be observed in Figure 11.17(b). This is because the power loss depends on the magnitude distribution, rather than on the maximum peak factor of the signal.

When four spreading codes were used simultaneously, as seen in Figure 11.17(c), the effective transmission power losses at 0dB input back-off were 1.3dB for Sivaswamy's complementary set, 1.8dB for OGold codes, 2.0dB for Zadoff-Chu codes, 2.1dB for Frank codes and 2.5dB for Walsh codes. Considering that Walsh codes exhibited the lowest maximum crest factor of 2.28 in Figure 11.15(a) among the spreading codes investigated, except for Sivaswamy's complementary set, the Walsh code related high power loss is surprising. This implies that the Walsh-spread quad-code MC-CDMA signal still exhibits a relatively poor magnitude distribution.

The Walsh-spread MC-CDMA signals using the maximum possible number of simultaneous multi-codes of $L = 16$ showed the lowest power loss of 1.3dB among the four different

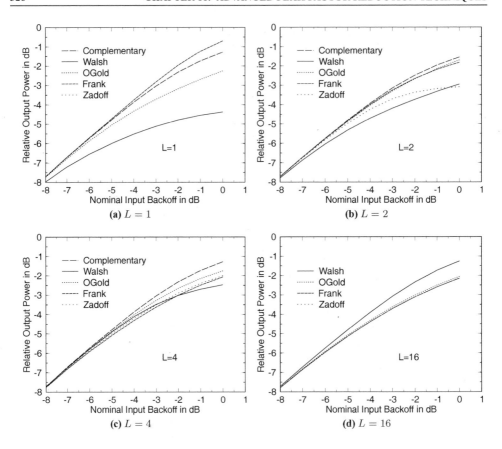

Figure 11.17: The relative output power versus input back-off in MC-CDMA for various spreading codes. The spreading factor was $N = 16$, while the number of simultaneous codes was $L = 1, 2, 4$ and 16. The corresponding curve for OFDM – although not shown in the figures – was almost identical to that of the Frank and the Zadoff-Chu spreading code based MC-CDMA using $L = 16$ simultaneous codes. Shapiro-Rudin codes were used as the complementary codes.

spreading codes investigated in the context of MC-CDMA, as seen in Figure 11.17(d). The desirable characteristic of attaining a low effective power loss was observed for Walsh codes, when the number of simultaneous codes L was higher than 8. The power loss curve for OFDM – although not shown explicitly in the figure – was almost identical to those observed for Zadoff-Chu and Frank code spread MC-CDMA signals, as seen in Figure 11.17(d).

We may conclude that the effective transmission power loss is also an important system performance factor, when choosing spreading sequences for multi-code MC-CDMA transmitters, and hence in addition to the crest factor studies further power-loss related investigations are required. In the fully loaded case, when the number of simultaneous codes is $L = 16$, the Walsh-spread MC-CDMA signal showed a lower power loss than OFDM, underlining another advantage of using spreading in comparison to conventional OFDM employing no spreading.

11.5.3.3   Effects of Clipping on the Bit Error Ratio

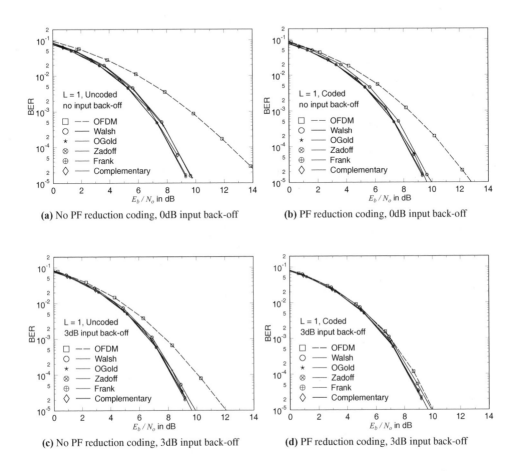

(a) No PF reduction coding, 0dB input back-off

(b) PF reduction coding, 0dB input back-off

(c) No PF reduction coding, 3dB input back-off

(d) PF reduction coding, 3dB input back-off

Figure 11.18: The BER of BPSK modulated MC-CDMA and BPSK modulated OFDM over AWGN channels for a SF of $N = 16$ and for $L = 1$ user.

Simulations were carried out, in order to investigate the effects of non-linear amplification in the context of BPSK modulated MC-CDMA and BPSK modulated OFDM systems over AWGN channels. When a single spreading code was employed for BPSK MC-CDMA, the effects of non-linear amplification were moderate, resulting in an almost unimpaired BPSK bit error probability, as can be observed in Figure 11.18. Since there is little margin for any BER improvement in comparison to a perfectly linearly amplified scenario, neither CF-reduction coding nor input back-off affects the corresponding BER. Only OFDM could achieve some BER improvement with the advent of crest factor reduction coding and when using an input back-off of 3dB.

When the number of simultaneous codes L was increased to two, as seen in Figure 11.19, Walsh-spread BPSK MC-CDMA showed approximately 1dB SNR loss, as can be observed in Figure 11.19, due to the associated non-linear amplification and the high crest factor en-

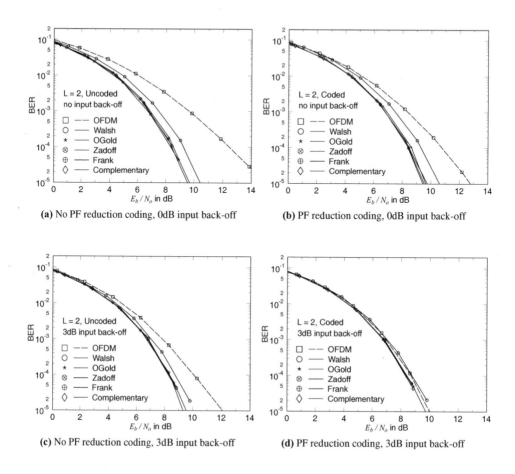

(a) No PF reduction coding, 0dB input back-off

(b) PF reduction coding, 0dB input back-off

(c) No PF reduction coding, 3dB input back-off

(d) PF reduction coding, 3dB input back-off

Figure 11.19: The BER of BPSK modulated MC-CDMA and BPSK modulated OFDM over AWGN channels for a SF of $N = 16$ and for $L = 2$ users.

countered. The BER of Walsh-spread MC-CDMA could not be improved by crest factor reduction coding, since all the $2^L = 2^2 = 4$ number of bit combinations result in the same crest factor distribution. However, involving 3dB input back-off slightly reduced the BER. The SNR values used in Figures 11.18 to 11.21 are based on the measured received signal powers, hence they do not show the effective output power loss encountered at the transmitter side, which was studied in the previous subsection.

When the number of simultaneous codes L was four as portrayed in Figure 11.20, all the MC-CDMA systems exhibited similar bit error rates. The Frank and Zadoff-Chu based systems exhibited a slightly higher BER, than the other MC-CDMA systems, however, the differences became negligible as crest factor reduction coding or the input back-off of 3dB was applied.

Figure 11.21 depicts the BER curves of the investigated MC-CDMA systems, when the number of simultaneous codes L is 16. For a fully loaded MC-CDMA system, only Walsh

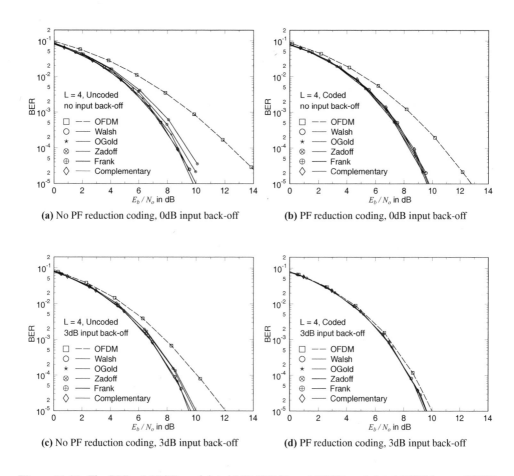

(a) No PF reduction coding, 0dB input back-off

(b) PF reduction coding, 0dB input back-off

(c) No PF reduction coding, 3dB input back-off

(d) PF reduction coding, 3dB input back-off

Figure 11.20: The BER of BPSK modulated MC-CDMA and BPSK modulated OFDM over AWGN channels for a SF of $N = 16$ and for $L = 4$ users.

spreading showed a better BER, than OFDM at an input back-off of 0dB and without crest factor reduction coding. Considering that Walsh-spread MC-CDMA has an approximately 1dB lower loss of effective transmission power than that of OFDM, the SNR difference of about 2dB observed at a given BER is deemed to be a significant benefit of Walsh-spread MC-CDMA from a practical point of view. Both crest factor reduction coding and 3dB input back-off of 3dB had the potential of reducing the BER of Walsh-spread MC-CDMA to a value near the ideal BER associated with using perfectly linear, infinite dynamic range amplification. When these two techniques were combined, the bit error rates of all the investigated systems approached the ideal BER, as we can observe in Figure 11.21(d) .

The BER performance of MMSE Joint-Detection (JD) assisted MC-CDMA [323] and that of single-user OFDM is shown in Figure 11.22 for transmission over Rayleigh channels, where independent subcarrier fading and perfect channel channel transfer function estimation were assumed. The independent subcarrier fading can be achieved in a system, where many

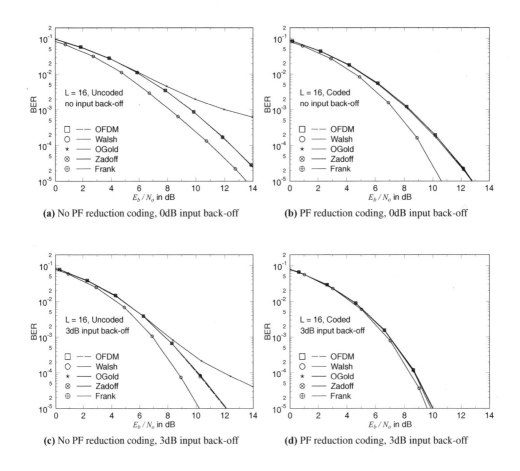

Figure 11.21: The BER of BPSK modulated MC-CDMA and BPSK modulated OFDM over AWGN channels for a SF of $N = 16$ and for $L = 16$ users.

subcarriers are employed in conjunction with frequency domain interleaving, provided that the number of resolvable multi-path components in the channel is higher than or equal to the spreading gain. Figure 11.22(a) shows the performance advantage of MC-CDMA systems over OFDM. MC-CDMA exhibited a significantly better BER performance than OFDM, while maintaining the same bandwidth efficiency. This was achieved, despite supporting $L = 16$ users, while the OFDM scheme supported a single user, which was a benefit of using the MMSE JD assisted receiver. This performance trend is only true, however, when there is sufficient diversity gain in order to compensate for the multiple user interference (MUI) or multi-code interference. It was observed in other simulations not included here that a diversity order of two or three is sufficient for MMSE JD-MC-CDMA to overcome the MUI. We can also observe, however, that JD MC-CDMA is more sensitive to amplitude clipping than OFDM. However, Walsh-spread MC-CDMA was not the best performing scheme over Rayleigh channels. Two of the poly-phase codes, namely the Frank and Zadoff-Chu spread-

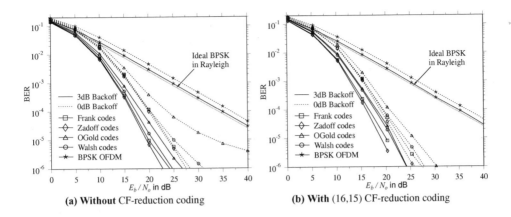

(a) Without CF-reduction coding   **(b) With** $(16,15)$ CF-reduction coding

Figure 11.22: BER versus E_b/N_o performance of BPSK modulated JD MC-CDMA and that of BPSK modulated OFDM when transmitting over Rayleigh channels for a SF of $N = 16$ and for $L = 16$ users. MMSE Joint Detection was used for MC-CDMA and independent fading of each subcarrier was assumed.

ing codes, were more effective, than the binary spreading codes, namely the Walsh and OGold codes, in this environment. Considering that the poly-phase spread signals suffer from higher distortion due to clipping, this result is surprising. It implies that the MUI imposed by the poly-phase spread systems is lower than that of the binary spreading based systems and this MUI difference plays a more dominant role in determining their BER performance than the clipping-induced distortion. The BER performance of the MMSE JD MC-CDMA and OFDM systems using CF-reduction coding is shown in Figure 11.22(b). The BER improvement was around 1dB for the MC-CDMA systems at a BER of 10^{-5} and at a back-off of 3dB, in comparison to the uncoded system. This BER improvement was more noticeable at 0dB back-off, where the effects of amplitude clipping are more dramatic.

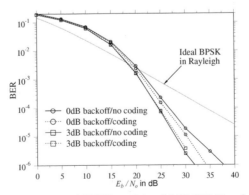

Figure 11.23: The BER of Walsh-spread BPSK modulated MC-CDMA, when a MMSE single-user detector was employed. The crest factor reduction coding increased the bit error ratio at 3dB back-off. The number of users L and the spreading factor N were both 16.

When MMSE single user detection was employed for our MC-CDMA systems, we found that the CF-reduction coding degraded the BER performance at a 3dB back-off for the Walsh-spread system, as can be observed in Figure 11.5.3.3. This implies that the CF-reduction coding is not optimal in reducing the overall BER, since it may select the specific message codes, which incur a high MUI, resulting in a degraded BER performance. This is especially true for the case, when there is little margin for crest factor improvement, such as for example in conjunction with the amplifier having a 3dB back-off.

11.5.3.4   Clipping Effects on Frequency Spectrum

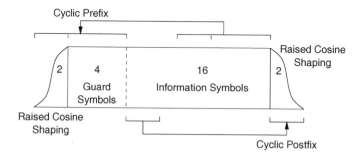

Figure 11.24: The time domain frame structure of MC-CDMA system. The number of subcarriers was $N = 16$ conveying 16 information symbols.

Since the frequency spectrum is a valuable commodity, it has to be effectively shared. The clipping-induced out-of-band emissions result in an increased adjacent channel interference. Multicarrier systems typically exhibit attractive spectral characteristics. In Section 11.5.2.3, we observed that the crest factors of MC-CDMA signals are dependent on the spreading sequences employed and also found that MC-CDMA signals generated with the aid of the appropriate spreading sequences exhibit lower crest factors, than those of OFDM. Therefore, it is expected that MC-CDMA signals will show better spectral characteristics than OFDM signals. The corresponding spectrum is dependent on the time domain transmission frame structure employed. We used a modified version of the frame structure employed in the IEEE 802.11 high-speed Wireless LAN [393] operated in the 5 GHz band. Figure 11.24 depicts the time domain frame structure employed in our reference system. The time domain frame consists of 16 information symbols, which are obtained by applying the IFFT to 16 frequency domain symbols plus six cyclic prefix and two cyclic postfix symbols. The first and the last two symbols are subjected to time domain raised cosine shaping, in order to prevent abrupt time domain signal changes.

The spectra shown in Figure 11.25 were obtained by averaging the individual spectra associated with all the possible messages using the best combinations of spreading sequences in terms of attaining the lowest possible crest factors. Specifically, Figure 11.25(a) displays the spectra of BPSK modulated MC-CDMA signals employing a Walsh code and a Zadoff-Chu code. We can observe that the spectrum of an MC-CDMA signal employing the Zadoff-Chu spreading code exhibits a lower out-of-band power than that of a Walsh code. The spectra corresponding to MC-CDMA signals employing the OGold code, the Frank code and the complementary code were between those associated with the Walsh and Zadoff-Chu spread-

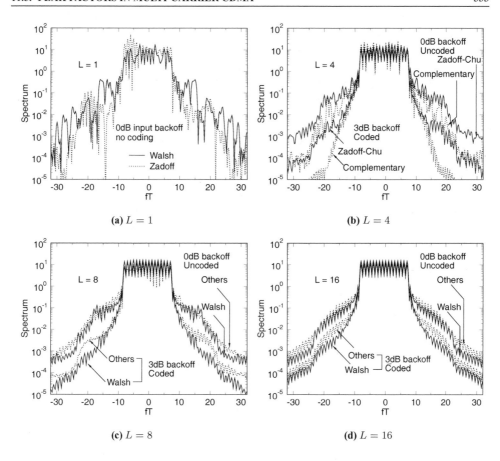

Figure 11.25: The frequency spectrum of BPSK modulated MC-CDMA signals. The number of sub-carriers N was 16 and the non-linear amplifier model of Section 11.5.3.1 was used. The frequency axis was scaled in terms of the normalised frequency fT, where T denotes the bit duration.

ing codes. This observation about the spectra is consistent with the comparison of their corresponding crest factors as we saw in Section 11.5.2.3. When the number of simultaneous codes L was four, the spectrum recorded for a set of four Sivaswamy's complementary sequences showed the lowest out-of-band spurious spillage, which is depicted in Figure 11.25(b) . With the aid of the previously introduced 15/16-rate peak-factor reduction coding and using a 3dB input back-off, the out-of-band spectral spillage was considerably reduced. When the number of simultaneous codes L was higher than eight, the Walsh-spread BPSK modulated MC-CDMA scheme exhibited the lowest out-of-band power spillage for the set of investigated systems, as can be seen in Figure 11.25(c) and Figure 11.25(d). Therefore, we conclude that the Walsh spreading codes are effective in reducing out-of-band spurious emissions as well in reducing the worst case peak factors.

11.5.4   Diversity Considerations

In this section, we investigate the frequency domain symbol magnitude of a multi-code MC-CDMA system employing the Walsh codes and its impact on the achievable diversity gain. The scaled frequency domain symbol magnitude $\sqrt{N}\,|s_n|$ of the n-th subchannel can be given as:

$$\sqrt{N}\,|s_n| = \left| \sum_{l=0}^{L-1} b_l\, c_l[n] \right|, \qquad (11.121)$$

where b_l is the lth user's information bit, c_l is the lth user's spreading sequence, L is the number of simultaneous users and N is the spreading factor. Since the scaled frequency domain symbol magnitude $\sqrt{N}\,|s_n|$ is a sum of an even number of binary values ± 1, the legitimate values are $\{(L - 2i)|i = 0, .., \lfloor L/2 \rfloor\}$. Let us define Q_i as the number of subchannels, which have the magnitude of a $L - 2i$. By definition, the Q_i values are non-negative integers. Additionally, the following two constraints are imposed on $\{Q_i\}$:

$$\sum_{i=0}^{\lfloor L/2 \rfloor} Q_i\,(L - 2i)^2 = NL, \qquad (11.122)$$

$$\sum_{i=0}^{\lfloor L/2 \rfloor} Q_i = N, \qquad (11.123)$$

where Equation 11.122 simply states that the total power in the frequency domain is equal to the average time domain power given in Equation 11.33 and Equation 11.123 implies that the total number of subcarriers is N.

Let us consider an example, where the number of subcarriers is $N = 16$ and the number of simultaneous codes is $L = 2$. Applying Equation 11.122, we get $Q_0 \times 2^2 = 16 \times 2$ and hence $Q_0 = 8$. From Equation 11.123, we have $Q_0 + Q_1 = 16$, and hence $Q_1 = 8$. In this example, we have only one solution for the set of $\{Q_i\}$ values, namely $\{Q_i\} = \{8, 8\}$ for the corresponding magnitude set of $\sqrt{N}|s_n| \in \{2, 0\}$.

Below we provide another example for $N = 16$ and $L = 6$. In this scenario, the possible magnitude set is given by $\{6, 4, 2, 0\}$. For this magnitude set, there exist 10 solutions for $\{Q_i\}$, namely $\{0, 3, 12, 1\}$, $\{0, 4, 8, 4\}$, $\{0, 5, 4, 7\}$, $\{0, 6, 0, 10\}$, $\{1, 0, 15, 0\}$, $\{1, 1, 11, 3\}$, $\{1, 2, 7, 6\}$, $\{1, 3, 3, 9\}$, $\{2, 0, 6, 8\}$ and $\{2, 1, 2, 11\}$ depending on the $L = 6$ number of bits b_l in Equation 11.121.

Now a question arises as to "Which set of Q_i will maximise the achievable diversity gain over N independent fading channels?" or "Which set of Q_i is the worst in this respect?". A seemingly good candidate can be found in the second example for $L = 6$. Specifically, the fifth solution, namely $\{1, 0, 15, 0\}$, does not "waste" any diversity potential, since there is no zero magnitude subcarriers in the frequency domain. By contrast, the other Q_i sets have at least one zero magnitude. For example, the first solution for Q_i given by $\{0, 3, 12, 1\}$ contains a single zero magnitude, while the second solution given by $\{0, 4, 8, 4\}$ has four zero-magnitude subcarriers in the frequency domain.

Having observed that some bit combinations yield a better set of Q_i values, while some do not, we can devise a coding scheme, which eliminates the source bit code words associated with a high number of zero magnitudes in the corresponding Q_i set, in order to more efficiently exploit the frequency diversity potential achievable. However, further investigations are required regarding the statistical distributions associated with the set of Q_i values and the feasibility of joint code construction methods has to be explored with the aim of reducing the worst case peak factor as well as for the sake of increasing the average diversity order.

11.6   Chapter Summary and Conclusion

In contrast to the rudimentary approach to the peak-factor problems of OFDM, which was based on simulations in Section 3.4, in this chapter we adopted a more advanced approach and provided a detailed account of the related problems in the context OFDM and MC-CDMA.

In addition to the more widely recognised benefits of MC-CDMA – such as its substantial frequency diversity gain – we showed in this chapter that MC-CDMA is also capable of mitigating the CF problem. Specifically, in Section 11.5.1 we investigated the relationship between the envelope power and the properties of a range of spreading sequences. It was shown in the context of Equation 11.40 that the envelope power is determined by the sums of the aperiodic autocorrelations defined in Equation 11.36 and by the aperiodic cross-correlations of the spreading sequences defined in Equation 11.37. We showed furthermore in Section 11.5.2.1 and 11.5.2.3 that the former property is more important when supporting a low number of users, L, while the latter plays a more dominant role under highly loaded conditions, i.e. for high values of L.

When we consider that the weighted sums of the aperiodic auto-correlations and the aperiodic cross-correlations defined in Equations 11.38 and 11.39 are bounded by a constant [394], it is apparent that there are no "magic" sequences, which exhibit "good" CF distributions for an arbitrary number of simultaneous codes, namely for all values of L. For example, the family of orthogonal Gold codes is more attractive for low values of L, while Walsh codes are preferred for high L values. The two poly-phase codes considered, namely the Frank and Zadoff-Chu spreading codes, were also found attractive for $L = 1$, while their CF distributions were the least attractive among the codes studied for $L = 1$, as we saw in Section 11.5.2.3.

In Section 11.5.2.2, we found that the crest factors of a two-code BPSK modulated MC-CDMA signal, employing a pair of Shapiro-Rudin sequences or a pair of sub-complementary sequences based on a Shapiro-Rudin pair, are bounded by 3dB. We also showed in Section 11.5.2.2 that the crest factor of the four-code BPSK modulated MC-CDMA signal employing Sivaswamy's set of complementary sequences is also bounded by 3dB. Therefore these sequences are preferred spreading sequences for MC-CDMA in terms of low crest factors, low BERs and attractive spectral characteristics. However, we were unable to find a larger set of such sequences which had more than four codes.

When we set $L = N$, in order to maintain the same spectral efficiency as OFDM, Walsh-spread BPSK modulated MC-CDMA showed a significant peak factor reduction over OFDM in conjunction with the same rate CF-reduction block encoder, upon assuming "fully loaded" conditions. When a 3dB back-off was applied to the transmitter's power amplifier, the Walsh-spread MC-CDMA system showed an approximately 2dB SNR gain at a BER of 10^{-5} in Figure 11.21 in an AWGN environment, when compared to OFDM. When transmitting over a channel exhibiting independent Rayleigh fading over each subcarrier, MMSE JD MC-CDMA employing Zadoff-Chu spreading codes was found to be the best scheme, as seen in Figure 11.22.

11.7 Appendix: Peak-to-mean Envelope Power Ratio of OFDM Systems

11.7.1 PMEPR Analysis of BPSK Modulated OFDM

The normalised complex envelope of the BPSK modulated OFDM signal, $s(t)$, is given by:

$$s(t) = \frac{1}{\sqrt{N}} \sum_{k=0}^{N-1} c_k \, e^{j2\pi \frac{kt}{T}}, \tag{11.124}$$

where N is the number of carriers, $c_k \in \{-1, +1\}$ is the frequency domain information symbol mapped to the kth subcarrier of the OFDM symbol and T is the OFDM symbol duration. The peak-to-mean envelope power ratio (*PMEPR*) of the given frequency domain samples, $c = \{c_0, c_1, \cdots, c_{N-1}\}$ is defined as:

$$PMEPR \triangleq \max_{0 \leq t \leq T} \frac{|s(t)|^2}{\mathrm{E}\left[\,|s(t)|^2\,\right]}, \tag{11.125}$$

where $\mathrm{E}\,[\,\cdot\,]$ denotes a time averaging operator.

Using the same technique as described in [354] and assuming $T = 1.0$, $N|s(t)|^2$ can be represented as:

$$
\begin{aligned}
N|s(t)|^2 &= \left(\sum_{k=0}^{N-1} c_k \, e^{j2\pi \frac{kt}{T}} \right)^2 \\
&= \left(\Re\left[\sum_{k=0}^{N-1} c_k \, e^{j2\pi \frac{kt}{T}} \right] \right)^2 + \left(\Im\left[\sum_{k=0}^{N-1} c_k \, e^{j2\pi \frac{kt}{T}} \right] \right)^2 \\
&= \left(\sum_{k=0}^{N-1} c_k \cos(2\pi kt) \right)^2 + \left(\sum_{k=0}^{N-1} c_k \sin(2\pi kt) \right)^2 \\
&= \sum_{k=0}^{N-1} c_k^2 \cos^2(2\pi kt) + 2 \sum_{k=0}^{N-2} \sum_{i=k+1}^{N-1} c_k c_i \cos(2\pi kt)\cos(2\pi it) \\
&\quad + \sum_{k=0}^{N-1} c_k^2 \sin^2(2\pi kt) + 2 \sum_{k=0}^{N-2} \sum_{i=k+1}^{N-1} c_k c_i \sin(2\pi kt)\sin(2\pi it) \\
&= N + 2 \sum_{k=0}^{N-2} \sum_{i=k+1}^{N-1} c_k c_i \cos(2\pi (i-k)t) \\
&= N + 2 P_o(t), \tag{11.126}
\end{aligned}
$$

where $\Re[x]$ and $\Im[x]$ are the real part and the imaginary part of x respectively, and the AC component of the power envelope of the OFDM signal $P_o(t)$ is defined as:

$$P_o(t) = \sum_{k=0}^{N-2} \sum_{i=k+1}^{N-1} c_k c_i \cos(2\pi(i-k)t). \tag{11.127}$$

The double summations in Equation 11.127 can be replaced with a single summation by combining each term according to its harmonic and $P_o(t)$ becomes:

$$P_o(t) = \sum_{k=1}^{N-1} C_k \cos(2\pi kt), \tag{11.128}$$

which can be physically interpreted as the sum of cosine harmonics weighted by the aperiodic auto-correlation C_k of the frequency domain information bits, where the aperiodic autocorrelation C_k is defined as:

$$C_k = \sum_{i=0}^{N-k-1} c_i c_{i+k}. \tag{11.129}$$

Upon using Equations 11.126 and 11.128 the average power of $s(t)$ becomes:

$$
\begin{aligned}
\mathrm{E}\left[\,|s(t)|^2\right] &= \mathrm{E}\left[\,1 + \frac{2P_o(t)}{N}\,\right] \\
&= 1 + \frac{1}{N}\mathrm{E}\left[\,2P_o(t)\,\right] \\
&= 1. \tag{11.130}
\end{aligned}
$$

As the average power of $s(t)$ is unity, the *PMEPR* in Equation 11.125, when considering its symmetry with respect to the half symbol time, becomes:

$$PMEPR = \max_{0 \le t \le 0.5}\left(1 + \frac{2}{N}P_o(t)\right), \tag{11.131}$$

where $P_o(t)$ is given in Equation 11.128.

11.7.2 PMEPR Properties of BPSK Modulated OFDM

It is readily seen from Equations 11.128, 11.129 and 11.131 that the *PMEPR* is completely characterised by the aperiodic autocorrelations, C_k. For the case of $N = 4$ the aperiodic autocorrelation C_k are given by:

$$
\begin{aligned}
C_1 &= c_0 c_1 + c_1 c_2 + c_2 c_3 \\
C_2 &= c_0 c_2 + c_1 c_3 \\
C_3 &= c_0 c_3.
\end{aligned}
$$

$(Code)_2$	$(Code)_{10}$	C_1	C_2	C_3	$PMEPR$
0000	0	3.0	2.0	1.0	4.00
0001	1	1.0	0.0	-1.0	1.77
0010	2	-1.0	0.0	1.0	1.77
0011	3	1.0	-2.0	-1.0	2.37
0100	4	-1.0	0.0	1.0	1.77
0101	5	-3.0	2.0	-1.0	4.00
0110	6	-1.0	-2.0	1.0	2.37
0111	7	1.0	0.0	-1.0	1.77
1000	8	1.0	0.0	-1.0	1.77
1001	9	-1.0	-2.0	1.0	2.37
1010	10	-3.0	2.0	-1.0	4.00
1011	11	-1.0	0.0	1.0	1.77
1100	12	1.0	-2.0	-1.0	2.37
1101	13	-1.0	0.0	1.0	1.77
1110	14	1.0	0.0	-1.0	1.77
1111	15	3.0	2.0	1.0	4.00

Table 11.6: PMEPR values and aperiodic auto-correlations, C_k, of all the 16 different symbols in a 4-carrier BPSK modulated OFDM system

Table 11.6 presents the values of C_k and the *PMEPR* values for BPSK modulated OFDM in conjunction with $N = 4$. There are six different sets of aperiodic auto-correlation $\{C_k\}$ and three different *PMEPR* values, namely $1.77, 2.37$ and 4.00 in Table 11.6. The message symbols can be grouped according to their *PMEPR* values. Figure 11.26 shows the waveforms of these three groups. Half of the message symbols[4] yield the lowest *PMEPR* of 1.77. The corresponding sets of C_k values in Table 11.6 are $(1, 0, -1)$ and $(-1, 0, 1)$. The power envelope waveforms are shown in the top sub-figure of Figure 11.26. One of the waveforms is a $0.5T$-shifted version of the other one corresponding to the same *PMEPR*. This is also true for all other envelope power waveforms of different *PMEPR* values, as seen in Figure 11.26. This observation can be stated more generally for BPSK modulated OFDM signals, where $P_o(t)$ defined in Equation 11.127 can be represented as:

$$P_o(t) = \Re \left[\sum_{k=1}^{N-1} C_k e^{j2\pi kt} \right]. \tag{11.132}$$

---

[4]These sequences are known as Barker sequences [370].

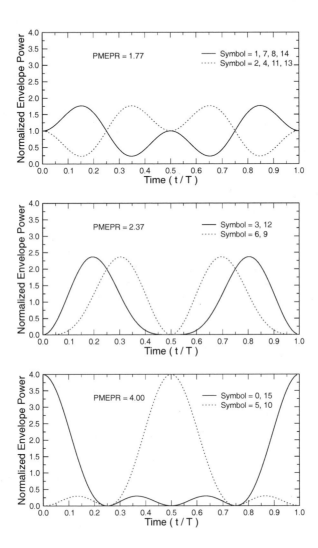

Figure 11.26: Normalised envelope power waveforms, $|s(t)|^2$, of 4-carrier BPSK modulated OFDM signals for various frequency domain bits symbol $\{c_k\}$. The frequency domain symbols $\{c_k\}$ are represented using a decimal notation of the bits, for example Symbol = 7 in the top sub-figure represents $\{0111\}$ and Symbol = 14 $\{1110\}$.

Then, the half symbol-duration time shifted version of $P_o(t)$ by half a symbol time, namely $P'_o(t)$, becomes:

$$
\begin{aligned}
P'_o(t) &= P_o(t + 1/2) \\
&= \Re\left[\sum_{k=1}^{N-1} C_k e^{j2\pi k(t+1/2)}\right] \\
&= \Re\left[\sum_{k=1}^{N-1} C_k e^{j2\pi kt} e^{jk\pi}\right] \\
&= \Re\left[\sum_{k=1}^{N-1} (-1)^k C_k e^{j2\pi kt}\right] \\
&= \Re\left[\sum_{k=1}^{N-1} C'_k e^{j2\pi kt}\right],
\end{aligned}
\tag{11.133}
$$

where C'_k represents the aperiodic auto-correlation values obtained from C_k by changing the sign of every C_k for odd k indices. This can be verified for the 4-carrier case using the C_k values and *PMEPR* values in Table 11.6 together with Figure 11.26.

It is clear that the *PMEPR* is invariant to the message symbol changes, as long as either $\{C_k\}$ does not change or $\{C_k\}$ becomes $\{(-1)^k C_k\}$. This leads to several transformations[5] that can be applied to $\{c_k\}$, which result in the same *PMEPR*, i.e. are invariant in terms of the *PMEPR*. We note that a message symbol set, $\{c_k\}$, can be represented in a polynomial form, defined as:

$$
c(z) \triangleq \sum_{k=0}^{N-1} c_k z^k.
\tag{11.134}
$$

Theorem 11.6 *The following three transforms of $c(z)$ result in the same* PMEPR*s*:

- $T_1[c(z)] = -c(z)$

- $T_2[c(z)] = z^{N-1} c(z^{-1})$

- $T_3[c(z)] = c(-z)$

Proof: It will be shown that $T_1[\bullet]$ and $T_2[\bullet]$ do not change the corresponding aperiodic auto-correlation C_k, while $T_3[\bullet]$ changes the corresponding C_k values to $(-1)^k C_k$. These are sufficient conditions for attaining the same *PMEPR*. The transform $T_1[\bullet]$ replaces each c_i by $c'_i = -c_i$, which corresponds to an exclusive-or operation in terms of their binary representation. Let C'_k represent the new aperiodic auto-correlations after the transform.

---

[5]Having completed these derivations, the author found retrospectively that in 1961 Marcel J. E. Golay mentioned these transforms, while he was investigating the properties of complementarty codes in [363].

Using the definition in Equation 11.129, C_k' becomes:

$$\begin{aligned}
C_k' &= \sum_{i=0}^{N-k-1} c_i' c_{i+k}' \\
&= \sum_{i=0}^{N-k-1} (-c_i)(-c_{i+k}) \\
&= \sum_{i=0}^{N-k-1} c_i c_{i+k} \\
&= C_k.
\end{aligned}$$

This shows that the application of T_1 does not change C_k.

Applying $T_2\,[\bullet]$ to $c(z)$ yields:

$$\begin{aligned}
c'(z) &= T_2\,[c(z)] \\
&= z^{N-1} c(z^{-1}) \\
&= \sum_{k=0}^{N-1} c_k z^{-k} z^{N-1} \\
&= \sum_{k=0}^{N-1} c_k z^{N-k-1}.
\end{aligned}$$

By substituting $N - k - 1$ with i, $c'(z)$ becomes:

$$\begin{aligned}
c'(z) &= \sum_{i=N-1}^{0} c_{N-1-i} z^i \\
&= \sum_{i=0}^{N-1} c_{N-1-i} z^i \, ,
\end{aligned}$$

which corresponds to the reverse ordering of the bits in the OFDM symbol or to a substitution of c_k with $c_k' = c_{N-1-k}$. Then, using Equation 11.129, C_k' becomes:

$$\begin{aligned}
C_k' &= \sum_{i=0}^{N-k-1} c_i' c_{i+k}' \\
&= \sum_{i=0}^{N-k-1} c_{N-1-i} c_{N-1-i-k}.
\end{aligned}$$

By substituting $N - 1 - i - k$ as j, C_k' becomes:

$$
\begin{aligned}
C_k' &= \sum_{j=N-1-k}^{0} c_{j+k} c_j \\
&= \sum_{j=0}^{N-k-1} c_j c_{j+k} \\
&= C_k.
\end{aligned}
$$

Hence the transform T_2 preserves the original C_k.

The result of $T_3[c(z)]$, namely $c'(z)$, becomes:

$$
\begin{aligned}
c'(z) &= \sum_{k=0}^{N-1} c_k(-z)^k \\
&= \sum_{k=0}^{N-1} (-1)^k c_k z^k.
\end{aligned}
$$

This is equivalent to substituting c_k with $c_k' = (-1)^k c_k$. The corresponding aperiodic autocorrelation becomes:

$$
\begin{aligned}
C_k' &= \sum_{i=0}^{N-k-1} c_i' c_{i+k}' \\
&= \sum_{i=0}^{N-k-1} (-1)^i c_i (-1)^{i+k} c_{i+k} \\
&= \sum_{i=0}^{N-k-1} (-1)^{2i+k} c_i c_{i+k} \\
&= \sum_{i=0}^{N-k-1} (-1)^k c_i c_{i+k} \\
&= (-1)^k C_k.
\end{aligned}
$$

This shows that the application of T_3 results in a circular time shift of the envelope power waveform without changing the maximum value of the waveform. This concludes the proof of Theorem 11.6. ¶

By applying the combinations of T_1, T_2 and T_3, we can get eight OFDM message symbols which give the same PMEPR. An example of 4-carrier BPSK modulated OFDM is shown in Table 11.7. The eight codes in the table are equivalent in terms of their PMEPR. A closer look at the combinations of T_1, T_2 and T_3 reveals that there are at least four different codes having the same *PMEPR*. The following theorem states this fact in a formal manner.

Theorem 11.7 *In BPSK modulated OFDM systems at least four message symbols are equiv-*

T	I	T_1	T_2	T_1T_2	T_3	T_3T_1	T_3T_2	$T_3T_2T_1$
$(Code)_2$	0001	1110	1000	0111	0100	1011	0010	1101

Table 11.7: The eight message symbols, which yield the same PMEPR value of 1.77 in a 4-carrier BPSK modulated OFDM system (T_n is defined in Theorem 11.6 and I represents the identical transformation.)

alent in terms of their PMEPR.

Proof: The sufficient and necessary condition for two message symbols, m_1 and m_2, to be identical is that the corresponding polynomial representations, $m_1(z)$ and $m_2(z)$, satisfy

$$m_1(z) - m_2(z) \equiv 0. \tag{11.135}$$

Using this property, $c(z)$, $T_1[c(z)]$, $T_3[c(z)]$ and $T_3T_1[c(z)]$ will be shown to be different from each other. Let us examine $c(z)$ and $T_1[c(z)]$ first, which give:

$$
\begin{aligned}
c(z) - T_1[c(z)] &= c(z) + c(z) \\
&= 2c(z) \\
&\neq 0 \ \forall \, c(z).
\end{aligned}
\tag{11.136}
$$

Thus, $c(z)$ and $T_1[c(z)]$ are always different. In the same way, $c(z)$ and $T_3[c(z)]$ can be shown to be different as follows:

$$
\begin{aligned}
c(z) - T_3[c(z)] &= c(z) - c(-z) \\
&= \sum_{k=0}^{N-1} [c_k z^k - c_k(-z)^k] \\
&= \sum_{k=0}^{N-1} [(1 - (-1)^k)c_k z^k] \\
&= \sum_{k=0}^{\lfloor N/2 \rfloor - 1} [2c_{2k+1} z^{2k+1}] \\
&\neq 0 \ \forall \, c(z).
\end{aligned}
\tag{11.137}
$$

This shows that $c(z)$ and $T_3[c(z)]$ are always different. Furthermore $c(z)$ and $T_3T_1[c(z)]$ are also different, since:

$$
\begin{aligned}
c(z) - T_3T_1[c(z)] &= \sum_{k=0}^{\lfloor (N+1)/2 \rfloor - 1} [2c_{2k} z^{2k}] \\
&\neq 0 \ \forall \, c(z).
\end{aligned}
\tag{11.138}
$$

We note that $T_3[c(z)] - T_1[c(z)] = c(z) - T_3T_1[c(z)]$ and that $c(z)$ and $T_3T_1[c(z)]$ always differ. From the argument so far, we conclude that $c(z)$, $T_1[c(z)]$, $T_3[c(z)]$ and $T_3T_1[c(z)]$

are different from each other. Therefore, there are at least four different message sequences, which have the same PMEPR. ¶

Theorem 11.6 suggests that there may exist eight equivalent message sequences in terms of their *PMEPR*, but theorem 11.7 states that the existence of only four different equivalent message sequences is guaranteed. The following theorem resolves this dilemma.

Theorem 11.8 *If T_2 satisfies any of the following three conditions for any member in the equivalent code set, then the number of the members in the set becomes four.*

- $T_2 \equiv I$

- $T_2 \equiv T_1$ *for even N*

- $T_2 \equiv T_3$ *for odd N*

Proof: If the number of PMEPR-equivalent symbols is less than eight, there exists a number of combined transforms of T_1, T_2 and T_3 defined in Theorem 11.6, which are equivalent in terms of producing the same symbol. It has been argued that I, T_1, T_3 and T_3T_1 are different in the proof of Theorem 11.7. The remaining possibility is that T_2 becomes equivalent to one of I, T_1, T_3 and T_3T_1. Let us consider these cases individually.

Assume that $T_2 \equiv I$. In this case, $c(z) - T_2[c(z)]$ should be 0, which is confirmed below:

$$
\begin{aligned}
c(z) - T_2[c(z)] &= c(z) - z^{N-1}c(z^{-1}) \\
&= \sum_{k=0}^{N-1}(c_k z^k - z^{N-1}c_k z^{-k}) \\
&= \sum_{k=0}^{N-1}(c_k z^k - c_k z^{N-1-k}) \\
&= \sum_{k=0}^{N-1}(c_k - c_{N-1-k})z^k \\
&= 0 \text{ (when } c_k = c_{N-1-k} \ \forall k, \ k \in \{0, 1, \ldots, N-1\}).
\end{aligned}
$$

All codes, which are symmetric with respect to the centre of code meet this condition. These elaborations are valid for arbitrary code lengths, however, in practical OFDM implementations we tend to favour a code length, which is an integer power of two.

Let us now assume that $T_2 \equiv T_1$. In this case, $T_2[c(z)] - T_1[c(z)]$ should be 0, which is confirmed below:

$$
\begin{aligned}
T_2[c(z)] - T_1[c(z)] &= z^{N-1}c(z^{-1}) + c(z) \\
&= \sum_{k=0}^{N-1}(z^{N-1}c_k z^{-k} + c_k z^k) \\
&= \sum_{k=0}^{N-1}(c_k z^{N-1-k} + c_k z^k) \\
&= \sum_{k=0}^{N-1}(c_{N-1-k} + c_k)z^k \\
&= 0 \text{ (when } c_k = -c_{N-1-k} \ \forall k, \ k \in \{0, 1, \ldots, N-1\}).
\end{aligned}
$$

In order for the transforms T_2 and T_1 become identical, c_k should satisfy the above condition, which can be interpreted as requiring odd symmetry of the message sequences in the frequency domain. This condition cannot be met for odd values of N, because there exist a centre symbol, which cannot meet the condition imposed.

Assume that $T_2 \equiv T_3$. Then $T_2[c(z)] - T_3[c(z)]$ becomes:

$$
\begin{aligned}
T_2[c(z)] - T_3[c(z)] &= z^{N-1}c(z^{-1}) - c(-z) \\
&= \sum_{k=0}^{N-1}(z^{N-1}c_k z^{-k} - (-1)^k c_k z^k) \\
&= \sum_{k=0}^{N-1}(c_k z^{N-1-k} - (-1)^k c_k z^k) \\
&= \sum_{k=0}^{N-1}(c_{N-1-k} - (-1)^k c_k)z^k.
\end{aligned}
$$

In order that the transforms T_2 and T_3 become identical, c_k should be equal to $(-1)^k c_{N-1-k}$. Note that this condition cannot be met for even values of N, because it requires $c_0 = c_{N-1}$ and $c_{N-1} = -c_0$ for $k = 0$ and $k = N-1$, respectively.

Finally, let us assume that $T_2 \equiv T_1 T_3$. Then $T_2[c(z)] - T_1 T_3[c(z)]$ becomes:

$$
\begin{aligned}
T_2[c(z)] - T_1 T_3[c(z)] &= z^{N-1}c(z^{-1}) + c(-z) \\
&= \sum_{k=0}^{N-1}(z^{N-1}c_k z^{-k} + (-1)^k c_k z^k) \\
&= \sum_{k=0}^{N-1}(c_k z^{N-1-k} + (-1)^k c_k z^k) \\
&= \sum_{k=0}^{N-1}(c_{N-1-k} + (-1)^k c_k)z^k.
\end{aligned}
$$

In order for the transforms T_2 and $T_1 T_3$ to become equivalent, c_k should be equal to $-(-1)^k c_{N-1-k}$. For even values of N, this requires $c_0 = -c_{N-1}$ and $c_{N-1} = c_0$ for $k = 0$ and $k = N - 1$, respectively. Thus the condition for the equivalence of T_2 and $T_1 T_3$ cannot be met for the even values of N. In case of odd N, $c_{(N-1)/2}$ should be equal to $-c_{(N-1)/2}$, in order to meet the above condition and it is readily seen that this is impossible. Thus, the assumed case never occurs.

Therefore, the first three cases considered above are the only cases, in which eight possible equivalent message sequences yield only four different ones. ¶

Theorem 11.9 *The worst case PMEPR of BPSK modulated OFDM over all the message symbols is N, where N is the number of carriers.*

Proof: From Equations 11.131 and 11.127, the worst case *PMEPR* is encountered when $P_o(t)$ is the maximum over range of $t \in [0, 0.5]$ and over all message symbols. The maximum value of C_k, $C_{k,max}$, is $N - k$ according to Equation 11.129. The time dependent term, $\cos(2\pi k t)$ in $P_o(t)$ of Equation 11.128 should be positive and at its maximum of 1, in order to yield the maximum of $P_o(t)$ by constructively combining all positive C_ks. Considering this argument, the worst case *PMEPR*, $PMEPR_{max}$, becomes:

$$
\begin{aligned}
PMEPR_{max} &= 1 + \frac{2}{N} P_{o,max} \\
&= 1 + \frac{2}{N} \sum_{k=1}^{N-1} C_{k,max} \\
&= 1 + \frac{2}{N} \sum_{k=1}^{N-1} (N - k) \\
&= 1 + \frac{2}{N} \left(N - 1 - \frac{N(N-1)}{2} \right) \\
&= 1 + \frac{2}{N} \frac{N(N-1)}{2} \\
&= N.
\end{aligned}
$$

Thus, the worst case *PMEPR* is N. ¶

Corollary 11.10 *In BPSK modulated OFDM systems the number of message symbols, which yield the worst case PMEPR is four for all N over 1 .*

Proof: The message symbol, which comprises the all zero sequence results in the maximum values of C_k for all k. By applying $T_1 [\bullet]$, $T_3 [\bullet]$ and $T_4 [\bullet]$, we can obtain the other three message symbols yielding the maximum value of C_k. Note that $T_2 \equiv I$ holds for the all zero sequence and corresponds to a degenerated case, yielding only four different OFDM message symbols. ¶

In BPSK modulated OFDM systems having a high number of carriers, it is observed that sometimes more than eight message sequences yield the same PMEPR, which cannot

$(Code)_{10}$	$(Code)_2$	C_1	C_2	C_3	C_4	C_5	PMEPR
4	000100	1	0	1	2	1	2.67
5	000101	-1	2	-1	0	-1	2.67

Table 11.8: Aperiodic auto-correlations evaluated from Equation 11.129 for the message sequences 4 and 5 in 6-carrier BPSK modulated OFDM system

be identified by Theorem 11.6. For example, the message sequences represented by 4 and 5 in a 6-carrier BPSK OFDM system exhibit exactly the same PMEPR value, while they have totally different envelope power waveforms and, thus, Theorem 11.6 does not detect their equivalence in terms of their PMEPR. The modified power envelope waveforms corresponding to the message sequences 4 and 5 in the previously mentioned 6-subcarrier scheme are shown in Figure 11.27. Table 11.8 presents the aperiodic auto-correlations and the

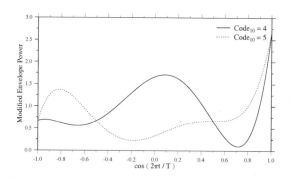

Figure 11.27: Modified power envelope waveforms for the message sequences 4 and 5 in 6-carrier BPSK modulated OFDM systems

PMEPRs for the message sequences 4 and 5. According to Equation 11.126 the power envelope of OFDM signal corresponding to the OFDM message symbol $(Code)_{10} = 4$ and $(Code)_{10} = 5$ is given as $|s(t)|^2 = 1 + 2/N\, P_{o,m}(t)$, where $P_{o,m}(t)$ are given by:

$$P_{o,4}(t) = \cos(2\pi t) + \cos(6\pi t) + 2\cos(8\pi t) + \cos(10\pi t) \tag{11.139}$$

$$P_{o,5}(t) = -\cos(2\pi t) + 2\cos(2\pi t) - \cos(6\pi t) - \cos(10\pi t). \tag{11.140}$$

The harmonics of $\cos(2\pi t)$ in Equations 11.139 and 11.140 can be represented using $\cos(2\pi t)$ only and the parametric polynomial representation is obtained by substituting $\cos(2\pi t)$ by α, which varies between -1 and $+1$ as t changes. In this way, the power envelope of OFDM signal corresponding to the message sequences of 4 and 5 were plotted in Figure 11.27 and their maximum values of $1 + 3P_{o,4}(\alpha)$ and $1 + 3P_{o,5}(\alpha)$ were obtained, which turned out to be exactly the same analytically. Considering that the set of aperiodic auto-correlations $\{C_k\}_5$ for the message sequence 5 shown in Table 11.8 can be obtained by applying T_2 and T_3 defined in Theorem 11.6 to $\{C_k\}_4$ for the message sequence 4, there may be a rule, which governs this phenomenon, although these considerations are beyond the scope of this chapter.

11.7.3 PMEPR Calculation of BPSK Modulated OFDM

In order to determine the value of the PMEPR, we evaluate Equation 11.131 for $t \in \{0, \Delta, 2\Delta, \cdots, K\Delta\}$, where Δ is much less than 0.5 and K is the smallest integer which satisfies $\lfloor 10K\Delta \rfloor \geq 5$. The value of Δ required for attaining a specified accuracy of the PMEPR can be derived using Equation 11.131. Let us consider the worst case scenario, when all the coefficients C_k, $k = 1, \cdots, N-1$ in Equation 11.128 have the maximum possible values and also assume that the derivatives of the cosine functions are all unity. Note that these two worst-case assumptions never occur simultaneously and hence these are very conservative assumptions. With these assumptions, ΔP_o, the worst case error associated with determining $P_o(t)$ and according to Equation 11.131 also the PMEPR, is given by:

$$\Delta P_o = \left| P_o\left(t_{max} + \frac{\Delta}{2}\right) - P_o(t_{max}) \right|$$

$$\leq \frac{\Delta}{2} \sum_{k=1}^{N-1} k \tag{11.141}$$

$$= \frac{\Delta}{4} N(N-1), \tag{11.142}$$

where t_{max} is the exact time instant, when $P_o(t)$ has its maximum value, while $t_{max} + \frac{\Delta}{2}$ is the worst case evaluation time. Using Equations 11.131 and 11.141, the worst case error in associated with determining the PMEPR, namely Δ PMEPR is given by:

$$\Delta PMEPR = \frac{N-1}{2} \Delta. \tag{11.143}$$

As an example, Δ should satisfy the following condition in order to arrive at PMEPR values that are accurate up to two decimal places:

$$\frac{N-1}{2} \Delta \leq 0.005 \tag{11.144}$$

$$\Delta \leq \frac{0.01}{N-1}. \tag{11.145}$$

The choice of Δ is important in order to arrive at the correct PMEPR values, especially in the context of OFDM having a high number of carriers. In [Table 1] of [31], several Peak-Envelope-Power (PEP) values are tabulated up to two decimal places, where according to the above analytical results small amendments are necessary. For example, 7.07 should be 7.08 and 9.45 should be 9.48, when using 20 decimal places in the calculations. In conclusion, since the condition imposed by Equations 11.141 and 11.145 is very conservative, the PMEPR values can be determined according to the required accuracy.

11.7.4   PMEPR Properties of QPSK Modulated OFDM

In QPSK modulated OFDM systems, two message bits are used for modulating each carrier. The normalised complex envelope waveform, $s(t)$, becomes [276]:

$$s(t) = \frac{1}{\sqrt{2N}} \sum_{k=0}^{N-1} \left(c_{2k} + j c_{2k+1} \right) e^{j2\pi \frac{kt}{T}}, \tag{11.146}$$

where N is the number of carriers, c_{2k} and c_{2k+1} are the in-phase symbol and the quadrature phase symbol respectively, modulating the kth carrier of the OFDM symbol, where we have $c_i \in \{-1, +1\}$, and finally, T is the OFDM symbol duration. Equation 11.125 can also be used to define the PMEPR of QPSK modulated OFDM signals. Van Eetvelt, Wade and Tomlinson presented an excellent analysis of QPSK modulated multi-carrier systems and showed that 16 messages form a coset, where the messages belonging to the same coset yield the same PMEPR. For a detailed description we refer to [276].

According to the findings of [276], very large number of message sequences yield exactly the same PMEPR. Algorithm 2 [page 91] [276] was used for generating Table 4 [page 94] [276] for QPSK modulated 5-carrier OFDM, suggesting that 192 message sequences yield a PMEPR value of 2.859, although we found that the application of Algorithm 2 [page 91] [276] resulted in a lower number of such message sequences. In fact, the message symbols represented as 31 and 47 in the associated cosets list in the context of the PMEPR value of 4.095 in Table 4 [page 94] [276] have slightly different PMEPR values, as shown in Figure 11.28, which is likely to be a consequence of the more accurate number representation using 20 effective decimal digits in our calculations. The peak values of the power waveforms found for the message symbols represented by 1, 2, 30, 31, 45 and 47 appear identical in the upper graph of Figure 11.28. In reality, the power envelope waveforms for the messages corresponding to the decimal notations 1, 2, 30, 45 are truly identical except for a time shift and hence have exactly the same PMEPR value. By contrast, the waveforms of the messages represented by 31 and 47 have a different shape and have a slightly different PMEPR value, as shown in the lower graph of Figure 11.28, which is the enlarged version of the marked rectangle seen in the upper graph. The correct characterisation of the 5-carrier QPSK modulated OFDM messages is presented in Table 11.9.

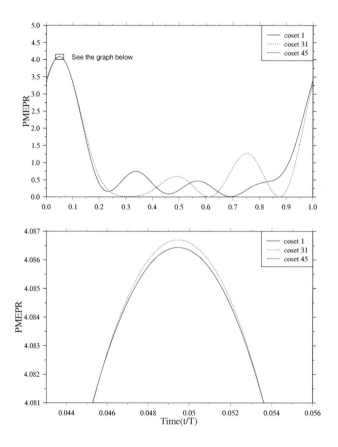

Figure 11.28: Power envelope waveforms of three message sequences corresponding to their decimal notations of 1, 31 and 45 in 5-carrier QPSK modulated OFDM systems (waveforms are appropriately shifted in time, in order to compare their shapes and the peak values)

Number of carriers	Number of equivalence classes	Orders of the equivalence classes	Associated cosets	Associated peak factors
5	23	16	0	5.000
		64	5, 10, 21, 42	4.304
		32	23, 43	4.297
		32	31, 47	4.087
		64	1, 2, 30, 45	4.086
		64	4, 8, 52, 56	3.539
		32	7, 11	3.497
		32	16, 32	3.400
		64	17, 34, 61, 62	3.327
		64	22, 41, 54, 57	2.801
		64	26, 37, 53, 58	2.789
		64	20, 24, 36, 40	2.778
		64	29, 46, 49, 50	2.680
		32	15, 63	2.679
		48	28, 44, 60	2.600
		32	55, 59	2.366
		64	13, 14, 18, 33	2.242
		32	19, 35	2.233
		16	48	2.112
		32	3, 51	2.101
		64	6, 9, 25, 38	1.976
		32	27, 39	1.967
		16	12	1.800

Table 11.9: Re-calculated equivalent cosets of 5-carrier QPSK modulated OFDM signals

Chapter 12

Adaptive Modulation for OFDM and MC-CDMA [1]

12.1 Introduction

Mobile communications channels typically exhibit time-variant channel quality fluctuations [263] and hence conventional fixed-mode modems suffer from bursts of transmission errors, even if the system was designed to provide a high link margin. An efficient approach to mitigating these detrimental effects is to adaptively adjust the transmission format based on the near-instantaneous channel quality information perceived by the receiver, which is fed back to the transmitter with the aid of a feedback channel [395]. This scheme requires a reliable feedback link from the receiver to the transmitter and the channel quality variation versus time should be sufficiently slow for the transmitter to be able to adapt. Hayes [395] proposed transmission power adaptation, while Cavers [396] suggested invoking a variable symbol duration scheme in response to the perceived channel quality at the expense of a variable bandwidth requirement. Since a variable-power scheme increases both the average transmitted power requirements and the level of co-channel interference [397] imposed on other users of the system, instead variable-rate Adaptive Quadrature Amplitude Modulation (AQAM) was proposed by Steele and Webb as an alternative, employing various star-QAM constellations [198, 397]. With the advent of Pilot Symbol-Assisted Modulation (PSAM) [174, 398, 399], Otsuki *et al.* [400] employed square constellations instead of star constellations in the context of AQAM, as a practical fading counter-measure. Analysing the channel capacity of Rayleigh fading channels [401–403], Goldsmith *et al.* showed that variable-power, variable-rate adaptive schemes are optimum, approaching the capacity of the channel and characterised the throughput performance of variable-power AQAM [402]. However, they also found that the extra throughput achieved by the additional variable-power assisted adaptation over the constant-power, variable-rate scheme is marginal for most types

---

[1]*OFDM and MC-CDMA for Broadband Multi-user Communications WLANS and Broadcasting.*
L. Hanzo, Münster, T. Keller and B.J. Choi,
©2003 John Wiley & Sons, Ltd. ISBN 0-470-85879-6

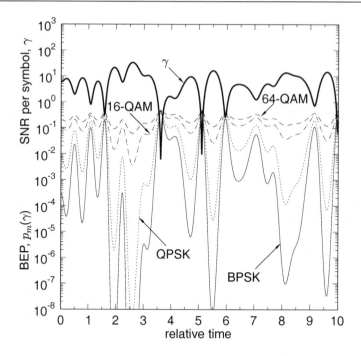

Figure 12.1: Instantaneous SNR per transmitted symbol, γ, in a flat Rayleigh fading scenario and the associated instantaneous bit error probability, $p_m(\gamma)$, of a fixed-mode QAM and an adaptive QAM system. The average SNR is $\bar{\gamma} = 10$dB. The fading magnitude plot is based on a normalised Doppler frequency of $f_N = 10^{-4}$ and for the duration of $100ms$, corresponding to a mobile terminal travelling at the speed of $54km/h$ and operating in the $f_c = 2GHz$ frequency band at the sampling rate of $1MHz$.

of fading channels [402, 404].

12.2 Increasing the Average Transmit Power as a Fading Counter-Measure

The radio frequency (RF) signal radiated from the transmitter's antenna takes different routes, experiencing defraction, scattering and reflections, before it arrives at the receiver. Each multi-path component arriving at the receiver simultaneously adds constructively or destructively, resulting in fading of the combined signal. When there is no line-of-sight component amongst these signals, the combined signal is characterised by Rayleigh fading. The instantaneous SNR (iSNR), γ, per transmitted symbol[2] is depicted in Figure 12.1 for a typical Rayleigh fading using the thick line. The Probability Density Function (PDF) of γ is given

---

[2]When no diversity is employed at the receiver, the SNR per symbol, γ, is the same as the channel SNR, γ_c. In this case, we will use the term "SNR" without any adjective.

as [274]:

$$f_{\bar{\gamma}}(\gamma) = \frac{1}{\bar{\gamma}} e^{\gamma/\bar{\gamma}}, \tag{12.1}$$

where $\bar{\gamma}$ is the average SNR and $\bar{\gamma} = 10$dB was used in Figure 12.1.

The instantaneous Bit Error Probability (iBEP), $p_m(\gamma)$, of BPSK, QPSK, 16-QAM and 64-QAM is also shown in Figure 12.1 with the aid of four different thin lines. These probabilities are obtained from the corresponding bit error probability over an AWGN channel conditioned on the iSNR, γ, which are given as [2]:

$$p_m(\gamma) = \sum_i A_i Q(\sqrt{a_i \gamma}), \tag{12.2}$$

where $Q(x)$ is the Gaussian Q-function defined as $Q(x) \triangleq \frac{1}{\sqrt{2\pi}} \int_x^\infty e^{-t^2/2} dt$ and $\{A_i, a_i\}$ is a set of modulation mode dependent constants. For the Gray-mapped square QAM modulation modes associated with $m = 2, 4, 16, 64$ and 256, the sets $\{A_i, a_i\}$ are given as [2, 405]:

$$
\begin{array}{lll}
m = 2, & \text{BPSK} & \{(1, 2)\} \\
m = 4, & \text{QPSK} & \{(1, 1)\} \\
m = 16, & \text{16-QAM} & \left\{ \left(\frac{3}{4}, \frac{1^2}{5}\right), \left(\frac{2}{4}, \frac{3^2}{5}\right), \left(-\frac{1}{4}, \frac{5^2}{5}\right) \right\} \\
m = 64, & \text{64-QAM} & \left\{ \left(\frac{7}{12}, \frac{1^2}{21}\right), \left(\frac{6}{12}, \frac{3^2}{21}\right), \left(-\frac{1}{12}, \frac{5^2}{21}\right), \left(\frac{1}{12}, \frac{9^2}{21}\right), \left(-\frac{1}{12}, \frac{13^2}{21}\right) \right\} \\
m = 256, & \text{256-QAM} & \left\{ \left(\frac{15}{32}, \frac{1^2}{85}\right), \left(\frac{14}{32}, \frac{3^2}{85}\right), \left(\frac{5}{32}, \frac{5^2}{85}\right), \left(-\frac{6}{32}, \frac{7^2}{85}\right), \left(-\frac{7}{32}, \frac{9^2}{85}\right), \right. \\
& & \left. \left(\frac{6}{32}, \frac{11^2}{85}\right), \left(\frac{9}{32}, \frac{13^2}{85}\right), \left(\frac{8}{32}, \frac{15^2}{85}\right), \left(-\frac{7}{32}, \frac{17^2}{85}\right), \left(-\frac{6}{32}, \frac{19^2}{85}\right), \right. \\
& & \left. \left(-\frac{1}{32}, \frac{21^2}{85}\right), \left(\frac{2}{32}, \frac{23^2}{85}\right), \left(\frac{3}{32}, \frac{25^2}{85}\right), \left(-\frac{2}{32}, \frac{27^2}{85}\right), \left(-\frac{1}{32}, \frac{29^2}{85}\right) \right\}.
\end{array}
\tag{12.3}
$$

As we can observe in Figure 12.1, $p_m(\gamma)$ exhibits high values during the deep channel envelope fades, where even the most robust modulation mode, namely BPSK, exhibits a bit error probability $p_2(\gamma) > 10^{-1}$. By contrast even the error probability of the high-throughput 16-QAM mode, namely $p_{16}(\gamma)$, is below 10^{-2}, when the iSNR γ exhibits a high peak. This wide variation of the communication link's quality is a fundamental problem in wireless radio communication systems. Hence, numerous techniques have been developed to combat this problem, such as increasing the average transmit power, invoking diversity, channel inversion, channel coding and/or adaptive modulation techniques. In this section we will investigate the efficiency of employing an increased average transmit power.

As we observed in Figure 12.1, the instantaneous Bit Error Probability (BEP) becomes excessive for sustaining an adequate service quality during instances, when the signal experiences a deep channel envelope fade. Let us define the cut-off BEP p_c, below which the Quality Of Service (QOS) becomes unacceptable. Then the outage probability P_{out} can be defined as:

$$P_{out}(\bar{\gamma}, p_c) \triangleq \Pr[p_m(\gamma) > p_c], \tag{12.4}$$

where $\bar{\gamma}$ is the average channel SNR dependent on the transmit power, p_c is the cut-off BEP

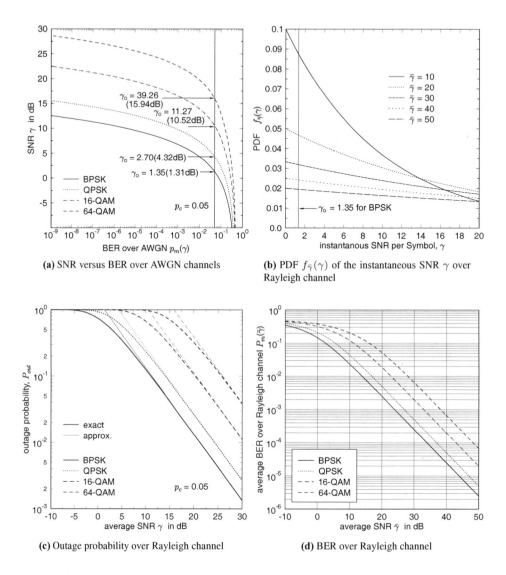

(a) SNR versus BER over AWGN channels

(b) PDF $f_{\bar{\gamma}}(\gamma)$ of the instantaneous SNR γ over Rayleigh channel

(c) Outage probability over Rayleigh channel

(d) BER over Rayleigh channel

Figure 12.2: The effects of an increased average transmit power. (a) The cut-off SNR γ_o versus the cut-off BEP p_c for BPSK, QPSK, 16-QAM and 64-QAM. (b) PDF of the iSNR γ over Rayleigh channel, where the outage probability is given by the area under the PDF curve surrounded by the two lines given by $\gamma = 0$ and $\gamma = \gamma_o$. An increased transmit power increases the average SNR $\bar{\gamma}$ and hence reduces the area under the PDF proportionally to $\bar{\gamma}$. (c) The exact outage probability versus the average SNR $\bar{\gamma}$ for BPSK, QPSK, 16-QAM and 64-QAM evaluated from Equation 12.7 confirms this observation. (d) The average BER is also inversely proportional to the transmit power for BPSK, QPSK, 16-QAM and 64-QAM.

and $p_m(\gamma)$ is the instantaneous BEP, conditioned on γ, for an m-ary modulation mode, given for example by Equation 12.2. We can reduce the outage probability of Equation 12.4 by increasing the transmit power, and hence increasing the average channel SNR $\bar{\gamma}$. Let us briefly investigate the efficiency of this scheme.

Figure 12.2(a) depicts the instantaneous BEP as a function of the instantaneous channel SNR. Once the cut-off BEP p_c is determined as a QOS-related design parameter, the corresponding cut-off SNR γ_o can be determined, as shown for example in Figure 12.2(a) for $p_c = 0.05$. Then, the outage probability of Equation 12.4 can be calculated as:

$$P_{out} = \Pr[\gamma < \gamma_o] \,, \tag{12.5}$$

and in physically integrated terms its value is equal to the area under the PDF curve of Figure 12.2(b) surrounded by the left y-axis and $\gamma = \gamma_o$ vertical line. Upon taking into account that for high SNRs the PDFs of Figure 12.2(b) are near-linear, this area can be approximated by $\gamma_o/\bar{\gamma}$, considering that $f_{\bar{\gamma}}(0) = 1/\bar{\gamma}$. Hence, the outage probability is inversely proportional to the transmit power, requiring an approximately 10-fold increased transmit power for reducing the outage probability by an order of magnitude, as seen in Figure 12.2(c). The exact value of the outage probability is given by:

$$P_{out} = \int_0^{\gamma_o} f_{\bar{\gamma}}(\gamma)\, d\gamma \tag{12.6}$$

$$= 1 - e^{-\gamma_o/\bar{\gamma}} \,, \tag{12.7}$$

where we used the PDF $f_{\bar{\gamma}}(\gamma)$ given in Equation 12.1. Again, Figure 12.2(c) shows the exact outage probabilities together with their linearly approximated values for several QAM modems recorded for the cut-off BER of $p_c = 0.05$, where we can confirm the validity of the linearly approximated outage probability[3], when we have $P_{out} < 0.1$.

The average BER $P_m(\bar{\gamma})$ of an m-ary Gray-mapped QAM modem is given by [2, 274, 406]:

$$P_m(\bar{\gamma}) = \int_0^{\infty} p_m(\gamma)\, f_{\bar{\gamma}}(\gamma)\, d\gamma \tag{12.8}$$

$$= \frac{1}{2} \sum_i A_i \{ 1 - \mu(\bar{\gamma}, a_i) \} \,, \tag{12.9}$$

where a set of constants $\{A_i, a_i\}$ is given in Equation 12.3 and $\mu(\bar{\gamma}, a_i)$ is defined as:

$$\mu(\bar{\gamma}, a_i) \triangleq \sqrt{\frac{a_i \bar{\gamma}}{1 + a_i \bar{\gamma}}} \,. \tag{12.10}$$

In physical terms Equation 12.8 implies weighting the BEP $p_m(\gamma)$ experienced at an iSNR γ by the probability of occurrence of this particular value of γ - which is quantified by its PDF $f_{\bar{\gamma}}(\gamma)$ - and then averaging, i.e. integrating, this weighted BEP over the entire range of γ. Figure 12.2(d) displays the average BER evaluated from Equation 12.9 for the average

---

[3]The same approximate outage probability can be derived by taking the first term of the Taylor series of e^x of Equation 12.7.

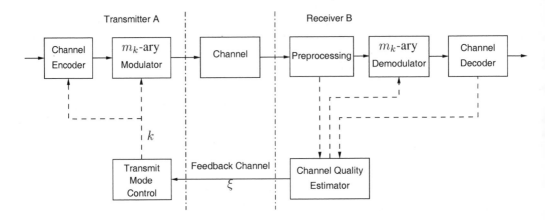

Figure 12.3: Stylised model of a near-instantaneous adaptive modulation scheme

SNR range of $-10\text{dB} \geq \bar{\gamma} \geq 50\text{dB}$. We can observe that the average BER is also inversely proportional to the transmit power.

In conclusion, we studied the efficiency of increasing the average transmit power as a fading counter-measure and found that the outage probability as well as the average bit error probability are inversely proportional to the average transmit power. Since the maximum radiated powers of modems are regulated in order to reduce the co-channel interference and transmit power, the acceptable transmit power increase may be limited and hence employing this technique may not be sufficiently effective for achieving the desired link performance. We will show that the AQAM philosophy of the next section is a more attractive solution to the problem of channel quality fluctuation experienced in wireless systems.

12.3   System Description

A stylised model of our adaptive modulation scheme is illustrated in Figure 12.3, which can be invoked in conjunction with any power control scheme. In our adaptive modulation scheme, the modulation mode used is adapted on a near-instantaneous basis for the sake of counteracting the effects of fading. Let us describe the detailed operation of the adaptive modem scheme of Figure 12.3. First, the channel quality ξ is estimated by the remote receiver B. This channel quality measure ξ can be the instantaneous channel SNR, the Radio Signal Strength Indicator (RSSI) output of the receiver [397], the decoded BER [397], the Signal to Interference-and-Noise Ratio (SINR) estimated at the output of the channel equaliser [407], or the SINR at the output of a CDMA joint detector [408]. The estimated channel quality perceived by receiver B is fed back to transmitter A with the aid of a feedback channel, as seen in Figure 12.3. Then, the transmit mode control block of transmitter A selects the highest-throughput modulation mode k capable of maintaining the target BER based on the channel quality measure ξ and the specific set of adaptive mode switching levels s. Once k is selected, m_k-ary modulation is performed at transmitter A in order to generate the transmitted signal $s(t)$, and the signal $s(t)$ is transmitted through the channel.

The general model and the set of important parameters specifying our constant-power adaptive modulation scheme are described in the next subsection in order to develop the underlying general theory. Then, in Section 12.3.2 several application examples are introduced.

12.3.1 General Model

A K-mode adaptive modulation scheme adjusts its transmit mode k, where $k \in \{0, 1 \cdots K - 1\}$, by employing m_k-ary modulation according to the near-instantaneous channel quality ξ perceived by receiver B of Figure 12.3. The mode selection rule is given by:

$$\text{Choose mode } k \text{ when } \quad s_k \leq \xi < s_{k+1}, \tag{12.11}$$

where a switching level s_k belongs to the set $\mathbf{s} = \{s_k \mid k = 0, 1, \cdots, K\}$. The Bits Per Symbol (BPS) throughput b_k of a specific modulation mode k is given by $b_k = \log_2(m_k)$ if $m_k \neq 0$, otherwise $b_k = 0$. It is convenient to define the incremental BPS c_k as $c_k = b_k - b_{k-1}$, when $k > 0$ and $c_0 = b_0$, which quantifies the achievable BPS increase, when switching from the lower-throughput mode $k-1$ to mode k.

12.3.2 Examples

12.3.2.1 Five-Mode AQAM

A five-mode AQAM system has been studied extensively by many researchers, which was motivated by the high performance of the Gray-mapped constituent modulation modes used. The parameters of this five-mode AQAM system are summarised in Table 12.1. In our inves-

k	0	1	2	3	4
m_k	0	2	4	16	64
b_k	0	1	2	4	6
c_k	0	1	1	2	2
modem	No Tx	BPSK	QPSK	16-QAM	64-QAM

Table 12.1: The parameters of five-mode AQAM system

tigation, the near-instantaneous channel quality ξ is defined as instantaneous channel SNR γ. The boundary switching levels are given as $s_0 = 0$ and $s_5 = \infty$. Figure 12.4 illustrates the operation of the five-mode AQAM scheme over a typical narrow-band Rayleigh fading channel scenario. Transmitter A of Figure 12.3 keeps track of the channel SNR γ perceived by receiver B with the aid of a low-BER, low-delay feedback channel – which can be created for example by superimposing the values of ξ on the reverse direction transmitted messages of transmitter B – and determines the highest-BPS modulation mode maintaining the target BER depending on which region γ falls into. The channel-quality related SNR regions are divided by the modulation mode switching levels s_k. More explicitly, the set of AQAM switching levels $\{s_k\}$ is determined such that the average BPS throughput is maximised, while satisfying the average target BEP requirement, P_{target}. We assumed a target BEP of $P_{target} = 10^{-2}$ in

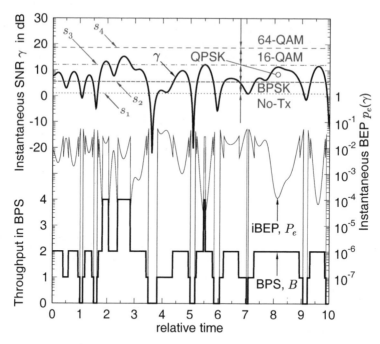

Figure 12.4: The operation of the five-mode AQAM scheme over a Rayleigh fading channel. The instantaneous channel SNR γ is represented as a thick line at the top part of the graph, the associated instantaneous BEP $P_e(\gamma)$ as a thin line at the middle, and the instantaneous BPS throughput $b(\gamma)$ as a thick line at the bottom. The average SNR is $\bar{\gamma} = 10$dB, while the target BER is $P_{target} = 10^{-2}$.

Figure 12.4. The associated instantaneous BPS throughput b is also depicted using the thick stepped line at the bottom of Figure 12.4. We can observe that the throughput varied from 0 BPS, when the "no transmission" (No-Tx) QAM mode was chosen, to 4 BPS, when the 16-QAM mode was activated. During the depicted observation window the 64-QAM mode was not activated. The instantaneous BEP, depicted as a thin line using the middle trace of Figure 12.4, is concentrated around the target BER of $P_{target} = 10^{-2}$.

12.3.2.2   Seven-Mode Adaptive Star-QAM

Webb and Steele revived the research community's interest on adaptive modulation, although a similar concept was initially suggested by Hayes [395] in the 1960s. Webb and Steele reported the performance of adaptive star-QAM systems [397]. The parameters of their system are summarised in Table 12.2.

12.3.2.3   Five-Mode APSK

Our five-mode Adaptive Phase-Shift-Keying (APSK) system employs m-ary PSK constituent modulation modes, where the magnitude of all the constituent constellations' phasors remained constant. The adaptive modem parameters are summarised in Table 12.3.

k	0	1	2	3	4	5	6
m_k	0	2	4	8	16	32	64
b_k	0	1	2	3	4	5	6
c_k	0	1	1	1	1	1	1
modem	No Tx	BPSK	QPSK	8-QAM	16-QAM	32-QAM	64-QAM

Table 12.2: The parameters of a seven-mode adaptive star-QAM system [397], where 8-QAM and 16-QAM employed four and eight constellation points allocated to two concentric rings, respectively, while 32-QAM and 64-QAM employed eight and 16 constellation points over four concentric rings, respectively.

k	0	1	2	3	4
m_k	0	2	4	8	16
b_k	0	1	2	3	4
c_k	0	1	1	1	1
modem	No Tx	BPSK	QPSK	8-PSK	16-PSK

Table 12.3: The parameters of the five-mode APSK system

12.3.2.4 Ten-Mode AQAM

Hole, Holm and Øien [409] studied a trellis coded adaptive modulation scheme based on eight-mode square- and cross-QAM schemes. Upon adding the No-Tx and BPSK modes, we arrive at a ten-mode AQAM scheme. The associated parameters are summarised in Table 12.4.

k	0	1	2	3	4	5	6	7	8	9
m_k	0	2	4	8	16	32	64	128	256	512
b_k	0	1	2	3	4	5	6	7	8	9
c_k	0	1	1	1	1	1	1	1	1	1
Tx	No Tx	BPSK	QPSK	8-Q	16-Q	32-C	64-Q	128-C	256-Q	512-C

Table 12.4: The parameters of the ten-mode adaptive QAM scheme based on [409], where m-Q stands for m-ary square QAM and m-C for m-ary cross QAM.

12.3.3 Characteristic Parameters

In this section, we introduce several parameters in order to characterise our adaptive modulation scheme. The constituent mode selection probability (MSP) $\mathcal{M}_k$ is defined as the probability of selecting the k-th mode from the set of K possible modulation modes, which can be calculated as a function of the channel quality metric ξ, regardless of the specific metric used, as:

$$\mathcal{M}_k = \Pr[\, s_k \leq \xi < s_{k+1} \,] \tag{12.12}$$

$$= \int_{s_k}^{s_{k+1}} f(\xi)\, d\xi\,, \tag{12.13}$$

where s_k denotes the k-th mode switching level and $f(\xi)$ is the probability density function (PDF) of ξ. Then, the average throughput B expressed in terms of BPS can be described as:

$$B = \sum_{k=0}^{K-1} b_k \int_{s_k}^{s_{k+1}} f(\xi)\, d\xi \tag{12.14}$$

$$= \sum_{k=0}^{K-1} b_k \, \mathcal{M}_k\,, \tag{12.15}$$

which in simple verbal terms can be formulated as the weighted sum of the throughput b_k of the individual constituent modes, where the weighting takes into account the probability $\mathcal{M}_k$ of activating the various constituent modes. When $s_K = \infty$, the average throughput B can also be formulated as:

$$B = \sum_{k=0}^{K-1} b_k \int_{s_k}^{s_{k+1}} f(\xi)\, d\xi \tag{12.16}$$

$$= \sum_{k=0}^{K-1} \left[b_k \int_{s_k}^{\infty} f(\xi)\, d\xi - b_k \int_{s_{k+1}}^{\infty} f(\xi)\, d\xi \right] \tag{12.17}$$

$$= b_0 \int_{s_0}^{\infty} f(\xi)\, d\xi + \sum_{k=1}^{K-1} (b_k - b_{k-1}) \int_{s_k}^{\infty} f(\xi)\, d\xi \tag{12.18}$$

$$= \sum_{k=0}^{K-1} c_k \int_{s_k}^{\infty} f(\xi)\, d\xi \tag{12.19}$$

$$= \sum_{k=0}^{K-1} c_k \, F_c(s_k)\,, \tag{12.20}$$

where $c_k \triangleq b_k - b_{k-1}$ and $F_c(\xi)$ is the complementary Cumulative Distribution Function (CDF) defined as:

$$F_c(\xi) \triangleq \int_{\xi}^{\infty} f(x)\, dx\,. \tag{12.21}$$

Let us now assume that we use the instantaneous SNR γ as the channel quality measure ξ, which implies that no co-channel interference is present. By contrast, when operating in a co-channel interference limited environment, we can use the instantaneous SINR as the channel quality measure ξ, provided that the co-channel interference has a near-Gaussian distribution.

In such a scenario, the mode-specific average BER P_k can be written as:

$$P_k = \int_{s_k}^{s_{k+1}} p_{m_k}(\gamma)\, f(\gamma)\, d\gamma \,, \tag{12.22}$$

where $p_{m_k}(\gamma)$ is the BER of the m_k-ary constituent modulation mode over the AWGN channel and we used γ instead of ξ in order to explicitly indicate the employment of γ as the channel quality measure. Then, the average BEP P_{avg} of our adaptive modulation scheme can be represented as the sum of the BEPs of the specific constituent modes divided by the average adaptive modem throughput B, formulated as [410]:

$$P_{avg} = \frac{1}{B} \sum_{k=0}^{K-1} b_k\, P_k \,, \tag{12.23}$$

where b_k is the BPS throughput of the k-th modulation mode, P_k is the mode-specific average BER given in Equation 12.22 and B is the average adaptive modem throughput given in Equation 12.15 or in Equation 12.20.

The aim of our adaptive system is to maximise the number of bits per symbol transmitted, while providing the required QOS. More specifically, we are aiming to maximise the average BPS throughput B of Equation 12.14, while satisfying the average BEP requirement of $P_{avg} \leq P_{target}$. Hence, we have to satisfy the constraint of meeting P_{target}, while optimising the design parameter of s, which is the set of modulation-mode switching levels. The determination of optimum switching levels will be investigated in Section 12.4. Since the calculation of the optimum switching levels typically requires the numerical computation of the parameters introduced in this section, it is advantageous to express the parameters in a closed form, which is the objective of the next section.

12.3.3.1 Closed Form Expressions for Transmission over Nakagami Fading Channels

Fading channels are often modelled as Nakagami fading channels [411]. The PDF of the instantaneous channel SNR γ over a Nakagami fading channel is given as [411]:

$$f(\gamma) = \left(\frac{m}{\bar{\gamma}}\right)^m \frac{\gamma^{m-1}}{\Gamma(m)}\, e^{-m\gamma/\bar{\gamma}} \,, \quad \gamma \geq 0 \,, \tag{12.24}$$

where the parameter m governs the severity of fading and $\Gamma(m)$ is the Gamma function [412]. When $m = 1$, the PDF of Equation 12.24 is reduced to the PDF of γ over a Rayleigh fading channel, which is given in Equation 12.1. As m increases, the fading behaves like Rician fading, and the AWGN channel is encountered, when m tends to ∞. Here we restrict the value of m to be a positive integer. In this case, the Nakagami fading model of Equation 12.24, having a mean of $\bar{\gamma}_s = m\,\bar{\gamma}$, will be used to describe the PDF of the SNR per symbol γ_s in a multiple-antenna based diversity assisted system employing Maximal Ratio Combining (MRC).

When the instantaneous channel SNR γ is used as the channel quality measure ξ in our adaptive modulation scheme transmitting over a Nakagami channel, the parameters defined in Section 12.3.3 can be expressed in a closed form. Specifically, the mode selection probability

$\mathcal{M}_k$ can be expressed as:

$$\mathcal{M}_k = \int_{s_k}^{s_{k+1}} f(\gamma)\, d\gamma \tag{12.25}$$

$$= F_c(s_k) - F_c(s_{k+1})\,, \tag{12.26}$$

where the complementary CDF $F_c(\gamma)$ is given by:

$$F_c(\gamma) = \int_{\gamma}^{\infty} f(x)\, dx \tag{12.27}$$

$$= \int_{\gamma}^{\infty} \left(\frac{m}{\bar{x}}\right)^m \frac{x^{m-1}}{\Gamma(m)}\, e^{-mx/\bar{\gamma}}\, dx \tag{12.28}$$

$$= e^{-m\gamma/\bar{\gamma}} \sum_{i=0}^{m-1} \frac{(m\gamma/\bar{\gamma})^i}{\Gamma(i+1)}\,. \tag{12.29}$$

In deriving Equation 12.29 we used the result of the indefinite integral of [379]:

$$\int x^n e^{-ax}\, dx = -(e^{-ax}/a) \sum_{i=0}^{n} x^{n-i}/a^i\, n!/(n-i)!\,. \tag{12.30}$$

In a Rayleigh fading scenario, i.e. when $m = 1$, the mode selection probability $\mathcal{M}_k$ of Equation 12.26 can be expressed as:

$$\mathcal{M}_k = e^{-s_k/\bar{\gamma}} - e^{-s_{k+1}/\bar{\gamma}}\,. \tag{12.31}$$

The average throughput B of our adaptive modulation scheme transmitting over a Nakagami channel is given by substituting Equation 12.29 into Equation 12.20, yielding:

$$B = \sum_{k=0}^{K-1} c_k\, e^{-ms_k/\bar{\gamma}} \left\{ \sum_{i=0}^{m-1} \frac{(ms_k/\bar{\gamma})^i}{\Gamma(i+1)} \right\}\,. \tag{12.32}$$

Let us now derive the closed form expressions for the mode specific average BEP P_k defined in Equation 12.22 for the various modulation modes when communicating over a Nakagami channel. The BER of a Gray-coded square QAM constellation for transmission over AWGN channels was given in Equation 12.2 and it is repeated here for convenience:

$$p_{m_k,QAM}(\gamma) = \sum_i A_i\, Q(\sqrt{a_i\gamma})\,, \tag{12.33}$$

where the values of the constants A_i and a_i were given in Equation 12.3. Then, the mode specific average BER $P_{k,QAM}$ of m_k-ary QAM over a Nakagami channel can be expressed as shown in Appendix 12.7 as follows:

$$P_{k,QAM} = \int_{s_k}^{s_{k+1}} p_{m_k,QAM}(\gamma) f(\gamma) \, d\gamma \tag{12.34}$$

$$= \sum_i A_i \int_{s_k}^{s_{k+1}} Q(\sqrt{a_i\gamma}) \left(\frac{m}{\bar{\gamma}}\right)^m \frac{\gamma^{m-1}}{\Gamma(m)} e^{-m\gamma/\bar{\gamma}} \, d\gamma \tag{12.35}$$

$$= \sum_i A_i \left\{ -e^{-m\gamma/\bar{\gamma}} Q(\sqrt{a_i\gamma}) \sum_{j=0}^{m-1} \frac{(m\gamma/\bar{\gamma})^j}{\Gamma(j+1)} \right]_{s_k}^{s_{k+1}} + \sum_{j=0}^{m-1} X_j(\gamma, a_i) \Big]_{s_k}^{s_{k+1}} \right\}, \tag{12.36}$$

where $g(\gamma)]_{s_k}^{s_{k+1}} \triangleq g(s_{k+1}) - g(s_k)$ and X_j is given by:

$$X_j(\gamma, a_i) = \frac{\mu^2}{\sqrt{2a_i\pi}} \left(\frac{m}{\bar{\gamma}}\right)^j \frac{\Gamma(j+\frac{1}{2})}{\Gamma(j+1)} \sum_{k=1}^j \left(\frac{2\mu^2}{a_i}\right)^{j-k} \frac{\gamma^{k-\frac{1}{2}}}{\Gamma(k+\frac{1}{2})} e^{-a_i\gamma/(2\mu^2)}$$
$$+ \left(\frac{2\mu^2 m}{a_i\bar{\gamma}}\right)^j \frac{1}{\sqrt{\pi}} \frac{\Gamma(j+\frac{1}{2})}{\Gamma(j+1)} \mu Q\left(\sqrt{a_i\gamma}/\mu\right), \tag{12.37}$$

where, again, $\mu \triangleq \sqrt{\frac{a_i\bar{\gamma}}{2+a_i\bar{\gamma}}}$ and $\Gamma(x)$ is the Gamma function.

On the other hand, the high-accuracy approximated BER formula of a Gray-coded m_k-ary PSK scheme ($k \geq 3$) transmitting over an AWGN channel is given as [413]:

$$p_{m_k,PSK} \simeq \frac{2}{k} \left\{ Q\left(\sqrt{2\gamma} \sin(\pi/2^k)\right) + Q\left(\sqrt{2\gamma} \sin(3\pi/2^k)\right) \right\} \tag{12.38}$$

$$= \sum_i A_i Q(\sqrt{a_i\gamma}), \tag{12.39}$$

where the constants $\{(A_i, a_i)\}$ are given by $\{(2/k, \, 2\sin^2(\pi/m_k)), (2/k, \, 2\sin^2(3\pi/m_k))\}$. Hence, the mode-specific average BEP $P_{k,PSK}$ can be represented using the same equation, namely Equation 12.36, as for $P_{k,QAM}$.

12.4 Optimum Switching Levels

In this section we restrict our interest to adaptive modulation schemes employing the SNR per symbol γ as the channel quality measure ξ. We then derive the optimum switching levels as a function of the target BEP and illustrate the operation of the adaptive modulation scheme. The corresponding performance results of the adaptive modulation schemes communicating over a flat-fading Rayleigh channel are presented in order to demonstrate the effectiveness of the schemes.

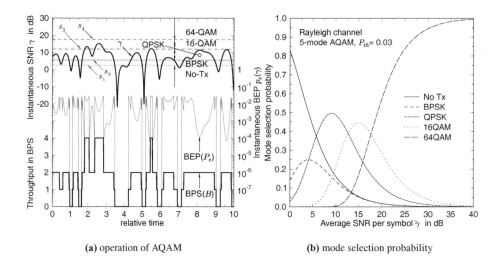

(a) operation of AQAM   **(b)** mode selection probability

Figure 12.5: Various characteristics of the five-mode AQAM scheme communicating over a Rayleigh
fading channel employing the specific set of switching levels designed for limiting the
peak instantaneous BEP to $P_{th} = 3 \times 10^{-2}$. (a) The evolution of the instantaneous
channel SNR γ is represented by the thick line at the top of the graph, the associated
instantaneous BER $p_e(\gamma)$ by the thin line as the trace in the middle and the instantaneous
BPS throughput $b(\gamma)$ by the thick line at the bottom. The average SNR is $\bar{\gamma} = 10$dB. (b)
As the average SNR increases, the higher-order AQAM modes are selected more often.

12.4.1   Limiting the Peak Instantaneous BEP

The first attempt of finding the optimum switching levels that are capable of satisfying various transmission integrity requirements was made by Webb and Steele [397]. They used
the BEP curves of each constituent modulation mode, obtained from simulations conducted
for transmissions over an AWGN channel, in order to find the Signal-to-Noise Ratio (SNR)
values, where each modulation mode satisfies the target BEP requirement [2]. This intuitive
concept of determining the switching levels has been widely used by researchers [400, 404]
since then. The regime proposed by Webb and Steele can be used for ensuring that the instantaneous BEP always remains below a certain threshold BEP P_{th}. In order to satisfy this
constraint, the first modulation mode should be "no transmission". In this case, the set of
switching levels **s** is given by:

$$\mathbf{s} = \left\{ s_0 = 0, \; s_k \mid p_{m_k}(s_k) = P_{th} \; k \geq 1 \right\}. \tag{12.40}$$

Figure 12.5 illustrates how this scheme operates over a Rayleigh channel, using the example
of the five-mode AQAM scheme described in Section 12.3.2.1. The average SNR was $\bar{\gamma} =$
10dB and the instantaneous target BEP was $P_{th} = 3 \times 10^{-2}$. Using the expression given
in Equation 12.2 for p_{m_k}, the set of switching levels can be calculated for the instantaneous
target BEP, which is given by $s_1 = 1.769$, $s_2 = 3.537$, $s_3 = 15.325$ and $s_4 = 55.874$. We
can observe that the instantaneous BEP represented as a thin line by the trace in the middle
of Figure 12.5(a) was limited to values below $P_{th} = 3 \times 10^{-2}$.

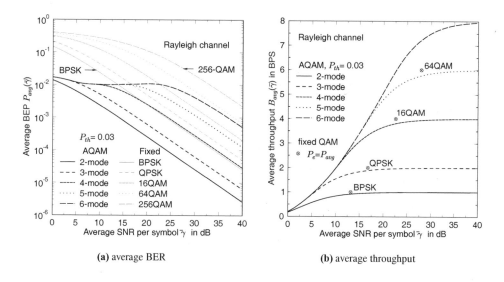

(a) average BER

(b) average throughput

Figure 12.6: The performance of AQAM employing the specific switching levels defined for limiting the peak instantaneous BEP to $P_{th} = 0.03$. (a) As the number of constituent modulation modes increases, the SNR region where the average BEP remains around $P_{avg} = 10^{-2}$ widens. (b) The SNR gains of AQAM over the fixed-mode QAM scheme required for achieving the same BPS throughput at the same average BEP of P_{avg} are in the range of 5dB to 8dB.

At this particular average SNR predominantly the QPSK modulation mode was invoked. However, when the instantaneous channel quality was high, 16-QAM was invoked in order to increase the BPS throughput. The mode selection probability $\mathcal{M}_k$ of Equation 12.26 is shown in Figure 12.5(b). Again, when the average SNR is $\bar{\gamma} = 10$dB, the QPSK mode is selected most often, namely with the probability of about 0.5. The 16-QAM, No-Tx and BPSK modes had the mode selection probabilities of 0.15 to 0.2, while 64-QAM is not likely to be selected in this situation. When the average SNR increases, the next higher order modulation mode becomes the dominant modulation scheme one by one and eventually the highest order of 64-QAM mode of the five-mode AQAM scheme prevails.

The effects of the number of modulation modes used in our AQAM scheme on the performance are depicted in Figure 12.6. The average BER performance portrayed in Figure 12.6(a) shows that the AQAM schemes maintain an average BEP lower than the peak instantaneous BEP of $P_{th} = 3 \times 10^{-2}$ even in the low SNR region, at the cost of a reduced average throughput, which can be observed in Figure 12.6(b). As the number of the constituent modulation modes employed of the AQAM increases, the SNR region, where the average BEP is near constant around $P_{avg} = 10^{-2}$ expands to higher average SNR values. We can observe that the AQAM scheme maintains a constant SNR gain over the highest-order constituent fixed QAM mode, as the average SNR increases, at the cost of a negligible BPS throughput degradation. This is because the AQAM activates the low-order modulation modes or disables transmissions completely, when the channel envelope is in a deep fade, in order to avoid inflicting bursts of bit errors.

Figure 12.6(b) compares the average BPS throughput of the AQAM scheme employing

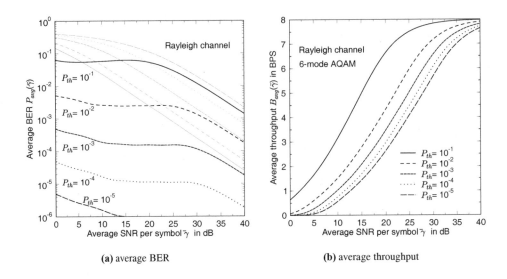

(a) average BER   (b) average throughput

Figure 12.7: The performance of the six-mode AQAM employing the switching levels of Equation 12.40 designed for limiting the peak instantaneous BEP.

various numbers of AQAM modes and those of the fixed QAM constituent modes achieving the same average BER. When we want to achieve the target throughput of $B_{avg} = 1$ BPS using the AQAM scheme, Figure 12.6(b) suggest that 3-mode AQAM employing No-Tx, BPSK and QPSK is as good as four-mode AQAM, or in fact any other AQAM schemes employing more than four modes. In this case, the SNR gain achievable by AQAM is 7.7dB at the average BEP of $P_{avg} = 1.154 \times 10^{-2}$. For the average throughputs of $B_{avg} = 2, 4$ and 6, the SNR gains of the 6-mode AQAM schemes over the fixed QAM schemes are 6.65dB, 5.82dB and 5.12dB, respectively.

Figure 12.7 shows the performance of the six-mode AQAM scheme, which is an extended version of the five-mode AQAM of Section 12.3.2.1, for the peak instantaneous BEP values of $P_{th} = 10^{-1}, 10^{-2}, 10^{-3}, 10^{-4}$ and 10^{-5}. We can observe in Figure 12.7(a) that the corresponding average BER P_{avg} decreases as P_{th} decreases. The average throughput curves seen in Figure 12.7(b) indicate that, as anticipated, the increased average SNR facilitates attaining an increased throughput by the AQAM scheme and there is a clear design trade-off between the achievable average throughput and the peak instantaneous BEP. This is because predominantly lower-throughput, but more error-resilient AQAM modes have to be activated, when the target BER is low. By contrast, higher-throughput but more error-sensitive AQAM modes are favoured, when the tolerable BER is increased.

In conclusion, we introduced an adaptive modulation scheme, where the objective is to limit the peak instantaneous BEP. A set of switching levels designed to meet this objective was given in Equation 12.40, which is independent of the underlying fading channel and the average SNR. The corresponding average BER and throughput formulae were derived in Section 12.3.3.1 and some performance characteristics of a range of AQAM schemes for transmitting over a flat Rayleigh channel were presented in order to demonstrate the effectiveness of the adaptive modulation scheme using the analysis technique developed in Section 12.3.3.1. The main advantage of this adaptive modulation scheme is in its simplicity regarding the design of the AQAM switching levels, while its drawback is that there is no direct relationship between the peak instantaneous BEP and the average BEP, which was used as our performance measure. In the next section a different switching-level optimisation philosophy is introduced and contrasted with the approach of designing the switching levels for maintaining a given peak instantaneous BEP.

12.4.2 Torrance's Switching Levels

Torrance and Hanzo [214] proposed the employment of the following cost function and applied Powell's optimisation method [334] for generating the optimum switching levels:

$$\Omega_T(\mathbf{s}) = \sum_{\bar{\gamma}=0\text{dB}}^{40\text{dB}} \left[10 \log_{10}(\max\{P_{avg}(\bar{\gamma}; \mathbf{s})/P_{th}, 1\}) + B_{max} - B_{avg}(\bar{\gamma}; \mathbf{s})\right], \qquad (12.41)$$

where the average BER P_{avg} is given in Equation 12.23, $\bar{\gamma}$ is the average SNR per symbol, $\mathbf{s}$ is the set of switching levels, P_{th} is the target average BER, B_{max} is the BPS throughput of the highest order constituent modulation mode and the average throughput B_{avg} is given in Equation 12.14. The idea behind employing the cost function Ω_T is that of maximising the average throughput B_{avg}, while endeavouring to maintain the target average BEP P_{th}. Following the philosophy of Section 12.4.1, the minimisation of the cost function of Equation 12.41 produces a set of constant switching levels across the entire SNR range. However, since the calculation of P_{avg} and B_{avg} requires the knowledge of the PDF of the instantaneous SNR γ per symbol, in reality the set of switching levels $\mathbf{s}$ required for maintaining a constant P_{avg} is dependent on the channel encountered and the receiver structure used.

Figure 12.8 illustrates the operation of a five-mode AQAM scheme employing Torrance's SNR-independent switching levels designed to maintain the target average BER of $P_{th} = 10^{-2}$ over a flat Rayleigh channel. The average SNR was $\bar{\gamma} = 10$dB and the target average BEP was $P_{th} = 10^{-2}$. Powell's minimisation [334] involved in the context of Equation 12.41 provides the set of optimised switching levels, given by $s_1 = 2.367, s_2 = 4.055, s_3 = 15.050$ and $s_4 = 56.522$. Upon comparing Figure 12.8(a) with Figure 12.5(a) we find that the two schemes are nearly identical in terms of activating the various AQAM modes according to the channel envelope trace, while the peak instantaneous BEP associated with Torrance's switching scheme is not constant. This is in contrast to the constant peak instantaneous BEP values seen in Figure 12.5(a). The mode selection probabilities depicted in Figure 12.8(b) are similar to those seen in Figure 12.5(b).

The average BER curves, depicted in Figure 12.9(a) show that Torrance's switching levels support the AQAM scheme in successfully maintaining the target average BEP of $P_{th} = 10^{-2}$ over the average SNR range of 0dB to 20dB, when five or six modem modes are employed by the AQAM scheme. Most of the AQAM studies found in the literature have applied Torrance's switching levels owing to the above mentioned good agreement between the design target P_{th} and the actual BEP performance P_{avg} [414].

Figure 12.9(b) compares the average throughputs of a range of AQAM schemes employing various numbers of AQAM modes to the average BPS throughput of fixed-mode QAM arrangements achieving the same average BEP, i.e. $P_e = P_{avg}$, which is not necessarily identical to the target BEP of $P_e = P_{th}$. Specifically, the SNR values required by the fixed mode

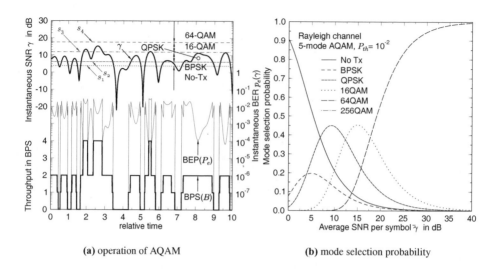

(a) operation of AQAM　　　　　　　　　　　　**(b)** mode selection probability

Figure 12.8: Performance of the five-mode AQAM scheme over a flat Rayleigh fading channel employing the set of switching levels derived by Torrance and Hanzo [214] to achieve the
target average BEP of $P_{th} = 10^{-2}$. (a) The instantaneous channel SNR γ is represented
as a thick line at the top part of the graph, the associated instantaneous BEP $p_e(\gamma)$ as
a thin line at the middle, and the instantaneous BPS throughput $b(\gamma)$ as a thick line at
the bottom. The average SNR is $\bar{\gamma} = 10$dB. (b) As the SNR increases, the higher-order
AQAM modes are selected more often.

scheme in order to achieve $P_e = P_{avg}$ are represented by the markers "$\otimes$", while the SNRs,
where the target average BEP of $P_e = P_{th}$ is achieved, are denoted by the markers "$\odot$".
Compared to the fixed QAM schemes achieving $P_e = P_{avg}$, the SNR gains of the AQAM
scheme were 9.06dB, 7.02dB, 5.81dB and 8.74dB for the BPS throughput values of 1, 2, 4
and 6, respectively. By contrast, the corresponding SNR gains compared to the fixed QAM
schemes achieving $P_e = P_{th}$ were 7.55dB, 6.26dB, 5.83dB and 1.45dB. We can observe
that the SNR gain of the AQAM arrangement over the 64-QAM scheme achieving a BEP of
$P_e = P_{th}$ is small compared to the SNR gains attained in comparison to the lower-throughput
fixed-mode modems. This is due to the fact that the AQAM scheme employing Torrance's
switching levels allows the target BER to drop at a high average SNR due to its sub-optimum
thresholds, which prevents the scheme from increasing the average throughput steadily to
the maximum achievable BPS throughput. This phenomenon is more visible for low target
average BERs, as can be observed in Figure 12.10.

In conclusion, we reviewed an adaptive modulation scheme employing Torrance's switching levels [214], where the objective was to maximise the average BPS throughput, while
maintaining the target average BEP. Torrance's switching levels are constant across the entire
SNR range and the average BEP P_{avg} of the AQAM scheme employing these switching levels shows good agreement with the target average BEP P_{th}. However, the range of average
SNR values, where $P_{avg} \simeq P_{th}$ was limited up to 25dB.

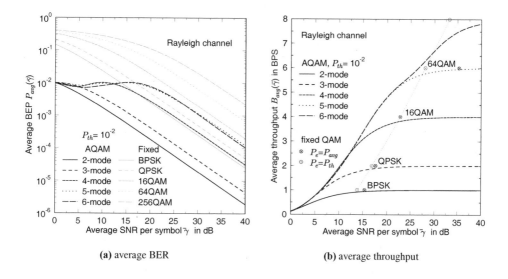

(a) average BER

(b) average throughput

Figure 12.9: The performance of various AQAM systems employing Torrance's switching levels [214] designed for the target average BEP of $P_{th} = 10^{-2}$. (a) The actual average BEP P_{avg} is close to the target BEP of $P_{th} = 10^{-2}$ over an average SNR range which becomes wider, as the number of modulation modes increases. However, the five-mode and six-mode AQAM schemes have a similar performance across much of the SNR range. (b) The SNR gains of the AQAM scheme over the fixed-mode QAM arrangements, while achieving the same throughput at the same average BEP, i.e. $P_e = P_{avg}$, range from 6dB to 9dB, which corresponds to a 1dB improvement compared with the SNR gains observed in Figure 12.6(b). However, the SNR gains over the fixed mode QAM arrangement achieving the target BEP of $P_e = P_{avg}$ are reduced, especially at high average SNR values, namely for $\bar{\gamma} > 25$dB.

12.4.3   Cost Function Optimisation as a Function of the Average SNR

In the previous section, we investigated Torrance's switching levels [214] designed to achieve a certain target average BEP. However, the actual average BEP of the AQAM system was not constant across the SNR range, implying that the average throughput could potentially be further increased. Hence here we propose a modified cost function $\Omega(\mathbf{s}; \bar{\gamma})$, putting more emphasis on achieving a higher throughput and optimising the switching levels for a given SNR, rather than for the whole SNR range [415]:

$$\Omega(\mathbf{s}; \bar{\gamma}) = 10 \log_{10}(\max\{P_{avg}(\bar{\gamma}; \mathbf{s})/P_{th}, 1\}) + \rho \log_{10}(B_{max}/B_{avg}(\bar{\gamma}; \mathbf{s})), \quad (12.42)$$

where $\mathbf{s}$ is a set of switching levels, $\bar{\gamma}$ is the average SNR per symbol, P_{avg} is the average BEP of the adaptive modulation scheme given in Equation 12.23, P_{th} is the target average BEP of the adaptive modulation scheme, B_{max} is the BPS throughput of the highest order constituent modulation mode. Furthermore, the average throughput B_{avg} is given in Equation 12.14 and ρ is a weighting factor, facilitating the above-mentioned BPS throughput enhancement. The first term on the right-hand side of Equation 12.42 corresponds to a cost function, which accounts for the difference, in the logarithmic domain, between the average BEP P_{avg} of

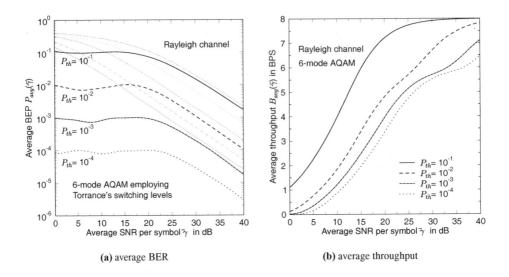

(a) average BER   (b) average throughput

Figure 12.10: The performance of the six-mode AQAM scheme employing Torrance's switching levels [214] for various target average BERs. When the average SNR is over 25dB and the
target average BER is low, the average BEP of the AQAM scheme begins to decrease,
preventing the scheme from increasing the average BPS throughput steadily.

the AQAM scheme and the target BEP P_{th}. This term becomes zero, when $P_{avg} \leq P_{th}$,
contributing no cost to the overall cost function Ω. On the other hand, the second term
of Equation 12.42 accounts for the logarithmic distance between the maximum achievable
BPS throughput B_{max} and the average BPS throughput B_{avg} of the AQAM scheme, which
decreases, as B_{avg} approaches B_{max}. Applying Powell's minimisation [334] to this cost
function under the constraint of $s_{k-1} \leq s_k$, the optimum set of switching levels $s_{opt}(\bar{\gamma})$ can
be obtained, resulting in the highest average BPS throughput, while maintaining the target
average BEP.

Figure 12.11 depicts the switching levels versus the average SNR per symbol optimised in
this manner for a five-mode AQAM scheme achieving the target average BEP of $P_{th} = 10^{-2}$
and 10^{-3}. Since the switching levels are optimised for each specific average SNR value, they
are not constant across the entire SNR range. As the average SNR $\bar{\gamma}$ increases, the switching
levels decrease in order to activate the higher-order mode modulation modes more often in
an effort to increase the BPS throughput. The low-order modulation modes are abandoned
one by one, as $\bar{\gamma}$ increases, activating always the highest-order modulation mode, namely
64-QAM, when the average BEP of the fixed-mode 64-QAM scheme becomes lower, than
the target average BEP P_{th}. Let us define the avalanche SNR $\bar{\gamma}_\alpha$ of a K-mode adaptive
modulation scheme as the lowest SNR, where the target BEP is achieved, which can be
formulated as:

$$P_{e,m_K}(\bar{\gamma}_\alpha) = P_{th} , \tag{12.43}$$

where m_K is the highest order modulation mode, P_{e,m_K} is the average BEP of the fixed-
mode m_K-ary modem activated at the average SNR of $\bar{\gamma}$ and P_{th} is the target average BEP

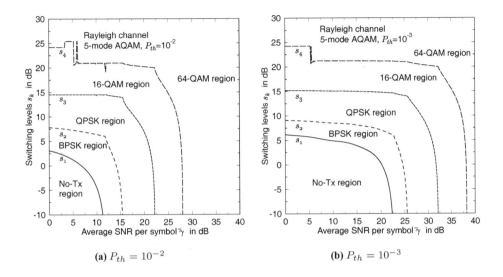

(a) $P_{th} = 10^{-2}$ **(b)** $P_{th} = 10^{-3}$

Figure 12.11: The switching levels optimised at each average SNR value in order to achieve the target average BER of (a) $P_{th} = 10^{-2}$ and (b) $P_{th} = 10^{-3}$. As the average SNR $\bar{\gamma}$ increases, the switching levels decrease in order to activate the higher-order mode modulation modes more often in an effort to increase the BPS throughput. The low-order modulation modes are abandoned one by one as $\bar{\gamma}$ increases, activating the highest-order modulation mode, namely 64-QAM, all the time when the average BEP of the fixed-mode 64-QAM scheme becomes lower than the target average BEP P_{th}.

of the adaptive modulation scheme. We can observe in Figure 12.11 that when the average channel SNR is higher than the avalanche SNR, i.e. $\bar{\gamma} \geq \bar{\gamma}_\alpha$, the switching levels are reduced to zero. Some of the optimised switching level versus SNR curves exhibit glitches, indicating that the multi-dimensional optimisation might result in local optima in some cases.

The corresponding average BEP P_{avg} and the average throughput B_{avg} of the two to six-mode AQAM schemes designed for the target average BEP of $P_{th} = 10^{-2}$ are depicted in Figure 12.12. We can observe in Figure 12.12(a) that now the actual average BEP P_{avg} of the AQAM scheme is exactly the same as the target BEP of $P_{th} = 10^{-2}$, when the average SNR $\bar{\gamma}$ is less than or equal to the avalanche SNR $\bar{\gamma}_\alpha$. As the number of AQAM modulation modes K increases, the range of average SNRs where the design target of $P_{avg} = P_{th}$ is met extends to a higher SNR, namely to the avalanche SNR. In Figure 12.12(b), the average BPS throughputs of the AQAM modems employing the "per-SNR optimised" switching levels introduced in this section are represented in thick lines, while the BPS throughput of the six-mode AQAM arrangement employing Torrance's switching levels [214] is represented using a solid thin line. The average SNR values required by the fixed-mode QAM scheme for achieving the target average BEP of $P_{e,m_K} = P_{th}$ are represented by the markers "⊙". As we can observe in Figure 12.12(b) the new per-SNR optimised scheme produces a higher BPS throughput than the scheme using Torrance's switching regime, when the average SNR $\bar{\gamma} > 20$dB. However, for the range of 8dB $< \bar{\gamma} < 20$dB, the BPS throughput of the new scheme is lower than that of Torrance's scheme, indicating that the multi-dimensional optimisation technique might reach local minima for some SNR values.

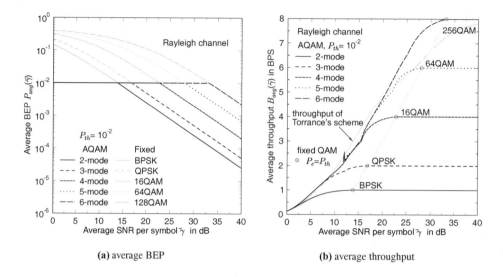

(a) average BEP   **(b)** average throughput

Figure 12.12: The performance of K-mode AQAM schemes for $K = 2, 3, 4, 5$ and 6, employing the switching levels optimised for each SNR value designed for the target average BEP of $P_{th} = 10^{-2}$. (a) The actual average BEP P_{avg} is exactly the same as the target BER of $P_{th} = 10^{-2}$, when the average SNR $\bar{\gamma}$ is less than or equal to the so-called avalanche SNR $\bar{\gamma}_\alpha$, where the average BEP of the highest-order fixed-modulation mode is equal to the target average BEP. (b) The average throughputs of the AQAM modems employing the "per-SNR optimised" switching levels are represented by the thick lines, while that of the six-mode AQAM scheme employing Torrance's switching levels [214] is represented by a solid thin line.

Figure 12.13(a) shows that the six-mode AQAM scheme employing "per-SNR optimised" switching levels satisfies the target average BEP values of $P_{th} = 10^{-1}$ to 10^{-4}. However, the corresponding average throughput performance shown in Figure 12.13(b) also indicates that the thresholds generated by the multi-dimensional optimisation were not satisfactory. The BPS throughput achieved was heavily dependent on the value of the weighting factor ρ in Equation 12.42. The glitches seen in the BPS throughput curves in Figure 12.13(b) also suggest that the optimisation process might result in some local minima.

We conclude that due to these problems it is hard to achieve a satisfactory BPS throughput for adaptive modulation schemes employing the switching levels optimised for each SNR value based on the heuristic cost function of Equation 12.42, while the corresponding average BEP exhibits a perfect agreement with the target average BEP.

12.4.4 Lagrangian Method

As argued in the previous section, Powell's minimisation [334] of the cost function often leads to a local minimum, rather than to the global minimum. Hence, here we adopt an analytical approach to finding the globally optimised switching levels. Our aim is to optimise the set of switching levels, **s**, so that the average BPS throughput $B(\bar{\gamma}; \mathbf{s})$ can be maximised under the constraint of $P_{avg}(\bar{\gamma}; \mathbf{s}) = P_{th}$. Let us define P_R for a K-mode adaptive modulation

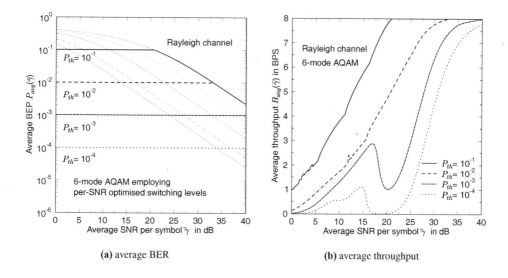

(a) average BER   **(b)** average throughput

Figure 12.13: The performance of six-mode AQAM employing "per-SNR optimised" switching levels for various values of the target average BEP. (a) The average BEP P_{avg} remains constant until the average SNR $\bar{\gamma}$ reaches the avalanche SNR, then follows the average BEP curve of the highest-order fixed-mode QAM scheme, i.e. that of 256-QAM. (b) For some SNR values the BPS throughput performance of the six-mode AQAM scheme is not satisfactory due to the fact that the multi-dimensional optimisation algorithm becomes trapped in local minima and hence fails to reach the global minimum.

scheme as the sum of the mode-specific average BEP weighted by the BPS throughput of the individual constituent mode:

$$P_R(\bar{\gamma}; \mathbf{s}) \triangleq \sum_{k=0}^{K-1} b_k \, P_k \,, \tag{12.44}$$

where $\bar{\gamma}$ is the average SNR per symbol, $\mathbf{s}$ is the set of switching levels, K is the number of constituent modulation modes, b_k is the BPS throughput of the k-th constituent mode and the mode-specific average BEP P_k is given in Equation 12.22 as:

$$P_k = \int_{s_k}^{s_{k+1}} p_{m_k}(\gamma) \, f(\gamma) \, d\gamma \,, \tag{12.45}$$

where again, $p_{m_k}(\gamma)$ is the BEP of the m_k-ary modulation scheme over the AWGN channel and $f(\gamma)$ is the PDF of the SNR per symbol γ. Explicitly, Equation 12.45 implies weighting the BEP $p_{m_k}(\gamma)$ by its probability of occurrence quantified in terms of its PDF and then averaging, i.e. integrating it over the range spanning from s_k to s_{k+1}. Then, with the aid of Equation 12.23, the average BEP constraint can also be written as:

$$P_{avg}(\bar{\gamma}; \mathbf{s}) = P_{th} \iff P_R(\bar{\gamma}; \mathbf{s}) = P_{th} \, B(\bar{\gamma}; \mathbf{s}) \,. \tag{12.46}$$

Another rational constraint regarding the switching levels can be expressed as:

$$s_k \leq s_{k+1} . \tag{12.47}$$

As we discussed before, our optimisation goal is to maximise the objective function $(\bar{\gamma}; \mathbf{s})$ under the constraint of Equation 12.46. The set of switching levels s has $K + 1$ levels in it. However, considering that we have $s_0 = 0$ and $s_K = \infty$ in many adaptive modulation schemes, we have $K - 1$ independent variables in s. Hence, the optimisation task is a $K - 1$ dimensional optimisation under a constraint [416]. It is standard practice to introduce a modified object function using a Lagrangian multiplier and convert the problem into a set of one-dimensional optimisation problems. The modified object function Λ can be formulated employing a Lagrangian multiplier λ [416] as:

$$\Lambda(\mathbf{s}; \bar{\gamma}) = B(\bar{\gamma}; \mathbf{s}) + \lambda \left\{ P_R(\bar{\gamma}; \mathbf{s}) - P_{th} B(\bar{\gamma}; \mathbf{s}) \right\} \tag{12.48}$$

$$= (1 - \lambda P_{th}) B(\bar{\gamma}; \mathbf{s}) + \lambda P_R(\bar{\gamma}; \mathbf{s}) . \tag{12.49}$$

The optimum set of switching levels should satisfy:

$$\frac{\partial \Lambda}{\partial \mathbf{s}} = \frac{\partial}{\partial \mathbf{s}} \left(B(\bar{\gamma}; \mathbf{s}) + \lambda \left\{ P_R(\bar{\gamma}; \mathbf{s}) - P_{th} B(\bar{\gamma}; \mathbf{s}) \right\} \right) = 0 \quad \text{and} \tag{12.50}$$

$$P_R(\bar{\gamma}; \mathbf{s}) - P_t B(\bar{\gamma}; \mathbf{s}) = 0 . \tag{12.51}$$

The following results are helpful in evaluating the partial differentiations in Equation 12.50:

$$\frac{\partial}{\partial s_k} P_{k-1} = \frac{\partial}{\partial s_k} \int_{s_{k-1}}^{s_k} p_{m_{k-1}}(\gamma) f(\gamma) \, d\gamma = p_{m_{k-1}}(s_k) f(s_k) \tag{12.52}$$

$$\frac{\partial}{\partial s_k} P_k = \frac{\partial}{\partial s_k} \int_{s_k}^{s_{k+1}} p_{m_k}(\gamma) f(\gamma) \, d\gamma = -p_{m_k}(s_k) f(s_k) \tag{12.53}$$

$$\frac{\partial}{\partial s_k} F_c(s_k) = \frac{\partial}{\partial s_k} \int_{s_k}^{\infty} f(\gamma) \, d\gamma = -f(s_k) . \tag{12.54}$$

Using Equations 12.52 and 12.53, the partial differentiation of P_R defined in Equation 12.44 with respect to s_k can be written as:

$$\frac{\partial P_R}{\partial s_k} = b_{k-1} p_{m_{k-1}}(s_k) f(s_k) - b_k p_{m_k}(s_k) f(s_k) , \tag{12.55}$$

where b_k is the BPS throughput of an m_k-ary modem. Since the average throughput is given by $B = \sum_{k=0}^{K-1} c_k F_c(s_k)$ in Equation 12.20, the partial differentiation of B with respect to s_k can be written as, using Equation 12.54:

$$\frac{\partial B}{\partial s_k} = -c_k f(s_k) , \tag{12.56}$$

where c_k was defined as $c_k \triangleq b_k - b_{k-1}$ in Section 12.3.1. Hence Equation 12.50 can be evaluated as:

$$\left[-c_k(1 - \lambda P_{th}) + \lambda \left\{b_{k-1} p_{m_{k-1}}(s_k) - b_k p_{m_k}(s_k)\right\}\right] f(s_k) = 0 \text{ for } k = 1, 2, \cdots, K - 1 .$$
$$(12.57)$$

A trivial solution of Equation 12.57 is $f(s_k) = 0$. Certainly, $\{s_k = \infty, k = 1, 2, \cdots, K - 1\}$ satisfies this condition. Again, the lowest throughput modulation mode is "No-Tx" in our model, which corresponds to no transmission. When the PDF of γ satisfies $f(0) = 0$, $\{s_k = 0, k = 1, 2, \cdots, K - 1\}$ can also be a solution, which corresponds to the fixed-mode m_{K-1}-ary modem. The corresponding avalanche SNR $\bar{\gamma}_\alpha$ can obtained by substituting $\{s_k = 0, k = 1, 2, \cdots, K - 1\}$ into Equation 12.51, which satisfies:

$$p_{m_{K-1}}(\bar{\gamma}_\alpha) - P_{th} = 0 .$$
$$(12.58)$$

When $f(s_k) \neq 0$, Equation 12.57 can be simplified upon dividing both sides by $f(s_k)$, yielding:

$$-c_k(1 - \lambda P_{th}) + \lambda \left\{b_{k-1} p_{m_{k-1}}(s_k) - b_k p_{m_k}(s_k)\right\} = 0 \text{ for } k = 1, 2, \cdots, K - 1 .$$
$$(12.59)$$

Rearranging Equation 12.59 for $k = 1$ and assuming $c_1 \neq 0$, we have:

$$1 - \lambda P_{th} = \frac{\lambda}{c_1} \left\{b_0 \, p_{m_0}(s_1) - b_1 p_{m_1}(s_1)\right\} .$$
$$(12.60)$$

Substituting Equation 12.60 into Equation 12.59 and assuming $c_k \neq 0$ for $k \neq 0$, we have:

$$\frac{\lambda}{c_k} \left\{b_{k-1} p_{m_{k-1}}(s_k) - b_k p_{m_k}(s_k)\right\} = \frac{\lambda}{c_1} \left\{b_0 \, p_{m_0}(s_1) - b_1 p_{m_1}(s_1)\right\} .$$
$$(12.61)$$

In this context we note that the Lagrangian multiplier λ is not zero because substitution of $\lambda = 0$ in Equation 12.59 leads to $-c_k = 0$, which is not true. Hence, we can eliminate the Lagrangian multiplier dividing both sides of Equation 12.61 by λ. Then we have:

$$y_k(s_k) = y_1(s_1) \text{ for } k = 2, 3, \cdots K - 1 ,$$
$$(12.62)$$

where the function $y_k(s_k)$ is defined as:

$$y_k(s_k) \triangleq \frac{1}{c_k} \left\{b_k p_{m_k}(s_k) - b_{k-1} p_{m_{k-1}}(s_k)\right\} , \quad k = 2, 3, \cdots K - 1 ,$$
$$(12.63)$$

which does not contain the Lagrangian multiplier λ and hence it will be referred to as the "Lagrangian-free function". This function can be physically interpreted as the normalised BEP difference between the adjacent AQAM modes. For example, $y_1(s_1) = p_2(s_1)$ quantifies the BEP increase, when switching from the No-Tx mode to the BPSK mode, while $y_2(s_2) = 2 p_4(s_2) - p_2(s_2)$ indicates the BEP difference between the QPSK and BPSK modes. This curve will be more explicitly discussed in the context of Figure 12.14. The significance of Equation 12.62 is that the relationship between the optimum switching levels s_k, where $k = 2, 3, \cdots K - 1$, and the lowest optimum switching level s_1 is independent of the underlying propagation scenario. Only the constituent modulation mode related parameters, such as b_k, c_k and $p_{m_k}(\gamma)$, govern this relationship.

Let us now investigate some properties of the Lagrangian-free function $y_k(s_k)$ given in Equation 12.63. Considering that $b_k > b_{k-1}$ and $p_{m_k}(s_k) > p_{m_{k-1}}(s_k)$, it is readily seen that the value of $y_k(s_k)$ is always positive. When $s_k = 0$, $y_k(s_k)$ becomes:

$$y_k(0) \triangleq \frac{1}{c_k} \left\{ b_k p_{m_k}(0) - b_{k-1} p_{m_{k-1}}(0) \right\} = \frac{1}{c_k} \left\{ \frac{b_k}{2} - \frac{b_{k-1}}{2} \right\} = \frac{1}{2} . \qquad (12.64)$$

The solution of $y_k(s_k) = 1/2$ can be either $s_k = 0$ or $b_k p_{m_k}(s_k) = b_{k-1} p_{m_{k-1}}(s_k)$. When $s_k = 0$, $y_k(s_k)$ becomes $y_k(\infty) = 0$. We also conjecture that:

$$\frac{d\,s_k}{d\,s_1} = \frac{y_1'(s_1)}{y_k'(s_k)} > 0 \quad \text{when } y_k(s_k) = y_1(s_1), \qquad (12.65)$$

which states that the k-th optimum switching level s_k always increases, whenever the lowest optimum switching level s_1 increases. Our numerical evaluations suggest that this conjecture appears to be true.

As an example, let us consider the five-mode AQAM scheme introduced in Section 12.3.2.1. The parameters of the five-mode AQAM scheme are summarised in Table 12.1. Substituting these parameters into Equations 12.62 and 12.63, we have the following set of equations.

$$y_1(s_1) = p_2(s_1) \qquad (12.66)$$
$$y_2(s_2) = 2\,p_4(s_2) - p_2(s_2) \qquad (12.67)$$
$$y_3(s_3) = 2\,p_{16}(s_3) - p_4(s_3) \qquad (12.68)$$
$$y_4(s_4) = 3\,p_{64}(s_4) - 2\,p_{16}(s_4) \qquad (12.69)$$

The Lagrangian-free functions of Equations 12.66 through 12.69 are depicted in Figure 12.14 for Gray-mapped square-shaped QAM. As these functions are basically linear combinations of BEP curves associated with AWGN channels, they exhibit waterfall-like shapes and asymptotically approach 0.5, as the switching levels s_k approach zero (or $-\infty$ expressed in dB). While $y_1(s_1)$ and $y_2(s_2)$ are monotonic functions, $y_3(s_3)$ and $y_4(s_4)$ cross the $y = 0.5$ line at $s_3 = -7.34$ dB and $s_4 = 1.82$ dB respectively, as can be observed in Figure 12.14(b). One should also notice that the trivial solutions of Equation 12.62 are $y_k = 0.5$ at $s_k = 0$, $k = 1, 2, 3, 4$, as we have discussed before.

For a given value of s_1, the other switching levels can be determined as $s_2 = y_2^{-1}(y_1(s_1))$, $s_3 = y_3^{-1}(y_1(s_1))$ and $s_4 = y_4^{-1}(y_1(s_1))$. Since deriving the analytical inverse function of y_k is an arduous task, we can rely on a graphical or a numerical method. Figure 12.14(b) illustrates an example of the graphical method. Specifically, when $s_1 = \alpha_1$, we first find the point on the curve y_1 directly above the abscissa value of α_1 and then draw a horizontal line across the corresponding point. From the crossover points found on the curves of y_2, y_3 and y_4 with the aid of the horizontal line, we can find the corresponding values of the other switching levels, namely those of α_2, α_3 and α_4. In a numerical sense, this solution corresponds to a one-dimensional (1-D) root finding problem [Ch 9] of [334]. Furthermore, the $y_k(s_k)$ values are monotonic, provided that we have $y_k(s_k) < 0.5$ and this implies that

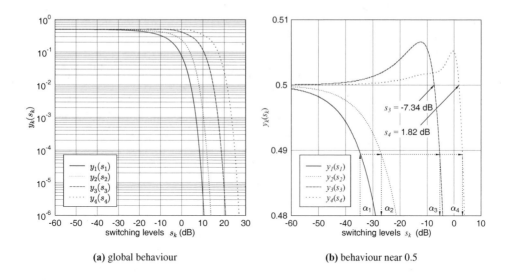

(a) global behaviour **(b)** behaviour near 0.5

Figure 12.14: The Lagrangian-free functions $y_k(s_k)$ of Equation 12.66 through Equation 12.69 for Gray-mapped square-shaped QAM constellations. As s_k becomes lower $y_k(s_k)$ asymptotically approaches 0.5. Observe that while $y_1(s_1)$ and $y_2(s_2)$ are monotonic functions, $y_3(s_3)$ and $y_4(s_4)$ cross the $y = 0.5$ line.

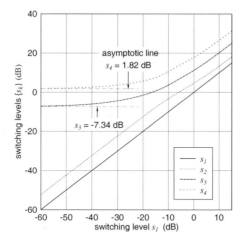

Figure 12.15: Optimum switching levels as functions of s_1. Observe that while the optimum value of s_2 shows a linear relationship with respect to s_1, those of s_3 and s_4 asymptotically approach constant values as s_1 is reduced.

the roots found are unique. The numerical results shown in Figure 12.15 represent the direct relationship between the optimum switching level s_1 and the other optimum switching levels, namely s_2, s_3 and s_4. While the optimum value of s_2 shows a near-linear relationship with respect to s_1, those of s_3 and s_4 asymptotically approach two different constants, as s_1 becomes smaller. This corroborates the trends observed in Figure 12.14(b), where $y_3(s_3)$

and $y_4(s_4)$ cross the $y = 0.5$ line at $s_3 = -7.34$ dB and $s_4 = 1.82$ dB, respectively. Since the low-order modulation modes are abandoned at high average channel SNRs in order to increase the average throughput, the high values of s_1 on the horizontal axis of Figure 12.15 indicate encountering a low channel SNR, while low values of s_1 suggest that high channel SNRs are experienced, as transpires, for example, from Figure 12.11.

Since we can relate the other switching levels to s_1, we have to determine the optimum value of s_1 for the given target BEP, P_{th}, and the PDF of the instantaneous channel SNR, $f(\gamma)$, by solving the constraint equation given in Equation 12.51. This problem also constitutes a 1-D root finding problem, rather than a multi-dimensional optimisation problem, which was the case in Sections 12.4.2 and 12.4.3 . Let us define the constraint function $Y(\bar{\gamma}; \mathbf{s}(s_1))$ using Equation 12.51 as:

$$Y(\bar{\gamma}; \mathbf{s}(s_1)) \triangleq P_R(\bar{\gamma}; \mathbf{s}(s_1)) - P_{th} B(\bar{\gamma}; \mathbf{s}(s_1)), \tag{12.70}$$

where we represented the set of switching levels as a vector, which is the function of s_1, in order to emphasise that s_k satisfies the relationships given by Equations 12.62 and 12.63. More explicitly, $Y(\bar{\gamma}; \mathbf{s}(s_1))$ of Equation 12.70 can be physically interpreted as the difference between $P_R(\bar{\gamma}; \mathbf{s}(s_1))$, namely the sum of the mode-specific average BEPs weighted by the BPS throughput of the individual AQAM modes, as defined in Equation 12.44 and the average BPS throughput $B(\bar{\gamma}; \mathbf{s}(s_1))$ weighted by the target BEP P_{th}. Considering the equivalence relationship given in Equation 12.46, Equation 12.70 reflects just another way of expressing the difference between the average BEP P_{avg} of the adaptive scheme and the target BEP P_{th}. Even though the relationships implied in $\mathbf{s}(s_1)$ are independent of the propagation conditions and the signalling power, the constraint function $Y(\bar{\gamma}; \mathbf{s}(s_1))$ of Equation 12.70 and hence the actual values of the optimum switching levels are dependent on propagation conditions through the PDF $f(\gamma)$ of the SNR per symbol and on the average SNR per symbol $\bar{\gamma}$.

Let us find the initial value of $Y(\bar{\gamma}; \mathbf{s}(s_1))$ defined in Equation 12.70, when $s_1 = 0$. An obvious solution for s_k when $s_1 = 0$ is $s_k = 0$ for $k = 1, 2, \cdots, K - 1$. In this case, $Y(\bar{\gamma}; \mathbf{s}(s_1))$ becomes:

$$Y(\bar{\gamma}; 0) = b_{K-1} (P_{m_{K-1}}(\bar{\gamma}) - P_{th}), \tag{12.71}$$

where b_{K-1} is the BPS throughput of the highest-order constituent modulation mode, $P_{m_{K-1}}(\bar{\gamma})$ is the average BEP of the highest-order constituent modulation mode for transmission over the underlying channel scenario and P_{th} is the target average BEP. The value of $Y(\bar{\gamma}; 0)$ could be positive or negative, depending on the average SNR $\bar{\gamma}$ and on the target average BEP P_{th}. Another solution exists for s_k when $s_1 = 0$, if $b_k \, p_{m_k}(s_k) = b_{k-1} \, p_{m_{k-1}}(s_k)$. The value of $Y(\bar{\gamma}; 0^+)$ using this alternative solution turns out to be close to $Y(\bar{\gamma}; 0)$. However, in the actual numerical evaluation of the initial value of Y, we should use $Y(\bar{\gamma}; 0^+)$ for ensuring the continuity of the function Y at $s_1 = 0$.

In order to find the minima and the maxima of Y, we have to evaluate the derivative of

$Y(\bar{\gamma}; s(s_1))$ with respect to s_1. With the aid of Equations 12.52 to 12.56, we have:

$$
\begin{aligned}
\frac{dY}{ds_1} &= \sum_{k=1}^{K-1} \frac{\partial Y}{\partial s_k} \frac{ds_k}{ds_1} \\
&= \sum_{k=1}^{K-1} \frac{\partial}{\partial s_k} \{P_R - P_{th}\, B\} \frac{ds_k}{ds_1} \\
&= \sum_{k=1}^{K-1} \left\{ b_{k-1}\, p_{m_{k-1}}(s_k) - b_k\, p_{m_k}(s_k) + P_{th}\, c_k \right\} f(s_k) \frac{ds_k}{ds_1} \\
&= \sum_{k=1}^{K-1} \left[\frac{c_k}{c_1} \left\{ b_0\, p_{m_0}(s_1) - b_1\, p_{m_1}(s_1) \right\} + P_{th}\, c_k \right] f(s_k) \frac{ds_k}{ds_1} \\
&= \frac{1}{c_1} \left\{ b_0\, p_{m_0}(s_1) - b_1\, p_{m_1}(s_1) + P_{th} \right\} \sum_{k=1}^{K-1} c_k\, f(s_k) \frac{ds_k}{ds_1} .
\end{aligned}
\tag{12.72}
$$

Considering $f(s_k) \geq 0$ and using our conjecture that $\frac{ds_k}{ds_1} > 0$ given in Equation 12.65, we can conclude from Equation 12.72 that $\frac{dY}{ds_1} = 0$ has roots, when $f(s_k) = 0$ for all k or when $b_1\, p_{m_1}(s_1) - b_0\, p_{m_0}(s_1) = P_{th}$. The former condition corresponds to either $s_i = 0$ for some PDF $f(\gamma)$ or to $s_k = \infty$ for all PDFs. By contrast, when the condition of $b_1\, p_{m_1}(s_1) - b_0\, p_{m_0}(s_1) = P_{th}$ is met, $dY/ds_1 = 0$ has a unique solution. Investigating the sign of the first derivative between these zeros, we can conclude that $Y(\bar{\gamma}; s_1)$ has a global minimum of Y_{min} at $s_1 = \zeta$ such that $b_1\, p_{m_1}(\zeta) - b_0\, p_{m_0}(\zeta) = P_{th}$ and a maximum of Y_{max} at $s_1 = 0$ and another maximum value at $s_1 = \infty$.

Since $Y(\bar{\gamma}; s_1)$ has a maximum value at $s_1 = \infty$, let us find the corresponding maximum value. Let us first consider $\lim_{s_1 \to \infty} P_{avg}(\bar{\gamma}; s(s_1))$, where upon exploiting Equations 12.23 and 12.44 we have:

$$
\lim_{s_1 \to \infty} P_{avg}(\bar{\gamma}; s_k) = \frac{\lim_{s_1 \to \infty} P_R}{\lim_{s_1 \to \infty} B}
\tag{12.73}
$$

$$
= \frac{0}{0} .
\tag{12.74}
$$

When applying l'Hopital's rule and using Equations 12.52 through 12.56, we have:

$$
\frac{\lim_{s_1 \to \infty} P_R}{\lim_{s_1 \to \infty} B} = \frac{\lim_{s_1 \to \infty} \frac{d}{ds_1} P_R}{\lim_{s_1 \to \infty} \frac{d}{ds_1} B}
\tag{12.75}
$$

$$
= \lim_{s_1 \to \infty} \frac{1}{c_1} b_1\, p_{m_1}(s_1) - b_0\, p_{m_0}(s_1)
\tag{12.76}
$$

$$
= 0^+ ,
\tag{12.77}
$$

implying that $P_{avg}(\bar{\gamma}; s_k)$ approaches zero from positive values, when s_1 tends to ∞. Since according to Equations 12.23, 12.44 and 12.70 the function $Y(\bar{\gamma}; s(s_1))$ can be written as

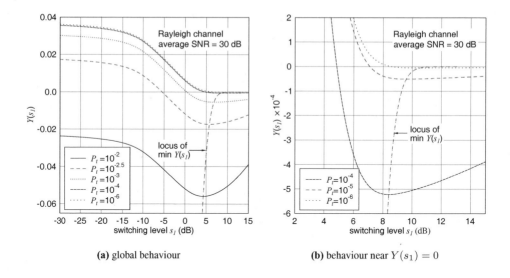

(a) global behaviour   **(b)** behaviour near $Y(s_1) = 0$

Figure 12.16: The constraint function $Y(\bar{\gamma}; \mathbf{s}(s_1))$ defined in Equation 12.70 for our five-mode AQAM scheme employing Gray-mapped square-constellation QAM operating over a flat Rayleigh fading channel. The average SNR was $\bar{\gamma} = 30$ dB and it is seen that Y has a single minimum value, while approaching 0^-, as s_1 increases. The solution of $Y(\bar{\gamma}; \mathbf{s}(s_1)) = 0$ exists, when $Y(\bar{\gamma}; 0) = 6\{p_{64}(\bar{\gamma}) - P_{th}\} > 0$ and is unique.

$B(P_{avg} - P_{th})$, we have:

$$\lim_{s_1 \to \infty} Y(\bar{\gamma}; s_1) = \lim_{s_1 \to \infty} B(P_{avg} - P_{th}) \qquad (12.78)$$

$$= \lim_{s_1 \to \infty} B(0^+ - P_{th}) \qquad (12.79)$$

$$= 0^- , \qquad (12.80)$$

Hence $Y(\bar{\gamma}; \mathbf{s}(s_1))$ asymptotically approaches zero from negative values, as s_1 tends to ∞. From the analysis of the minimum and the maxima, we can conclude that the constraint function $Y(\bar{\gamma}; \mathbf{s}(s_1))$ defined in Equation 12.70 has a unique zero only if $Y(\bar{\gamma}; 0^+) > 0$ at a switching value of $0 < s_1 < \zeta$, where ζ satisfies $b_1 \, p_{m_1}(\zeta) - b_0 \, p_{m_0}(\zeta) = P_{th}$. By contrast, when $Y(\bar{\gamma}; 0^+) < 0$, the optimum switching levels are all zero and the adaptive modulation scheme always employs the highest-order constituent modulation mode.

As an example, let us evaluate the constraint function $Y(\bar{\gamma}; s_1)$ for our five-mode AQAM scheme operating over a flat Rayleigh fading channel. Figure 12.16 depicts the values of $Y(s_1)$ for several values of the target average BEP P_{th}, when the average channel SNR is 30dB. We can observe that $Y(s_1) = 0$ may have a root, depending on the target BEP P_{th}. When $s_k = 0$ for $k < 5$, according to Equations 12.23, 12.44 and 12.70 $Y(s_1)$ is reduced to

$$Y(\bar{\gamma}; 0) = 6(P_{64}(\bar{\gamma}) - P_{th}) , \qquad (12.81)$$

where $P_{64}(\bar{\gamma}$ is the average BEP of 64-QAM over a flat Rayleigh channel. The value of $Y(\bar{\gamma}; 0)$ in Equation 12.81 can be negative or positive, depending on the target BEP P_{th}.

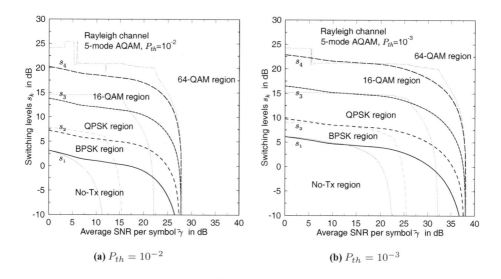

(a) $P_{th} = 10^{-2}$ **(b)** $P_{th} = 10^{-3}$

Figure 12.17: The switching levels for our five-mode AQAM scheme optimised at each average SNR
value in order to achieve the target average BEP of (a) $P_{th} = 10^{-2}$ and (b) $P_{th} = 10^{-3}$
using the Lagrangian multiplier based method of Section 12.4.4. The switching levels
based on Powell's optimisation are represented in thin grey lines for comparison.

We can observe in Figure 12.16 that the solution of $Y(\bar{\gamma}; \mathbf{s}(s_1)) = 0$ is unique, when it exists. The locus of the minimum $Y(s_1)$, i.e. the trace curve of points $(Y_{min}(s_{1,min}), s_{1,min})$,
where Y has the minimum value, is also depicted in Figure 12.16. The locus is always below
the horizontal line of $Y(s_1) = 0$ and asymptotically approaches this line, as the target BEP
P_{th} becomes smaller.

Figure 12.17 depicts the switching levels optimised in this manner for our five-mode
AQAM scheme maintaining the target average BEPs of $P_{th} = 10^{-2}$ and 10^{-3}. The switching
levels obtained using Powell's optimisation method in Section 12.4.3 are represented as the
thin grey lines in Figure 12.17 for comparison. In this case all the modulation modes may
be activated with a certain probability, until the average SNR reaches the avalanche SNR
value, while the scheme derived using Powell's optimisation technique abandons the lower
throughput modulation modes one by one, as the average SNR increases.

Figure 12.18 depicts the average throughput B expressed in BPS of the AQAM scheme
employing the switching levels optimised using the Lagrangian method. In Figure 12.18(a),
the average throughput of our six-mode AQAM arrangement using Torrance's scheme discussed in Section 12.4.2 is represented as a thin grey line. The Lagrangian multiplier based
scheme showed SNR gains of 0.6dB, 0.5dB, 0.2dB and 3.9dB for a BPS throughput of 1, 2, 4
and 6, respectively, compared to Torrance's scheme. The average throughput of our six-mode
AQAM scheme is depicted in Figure 12.18(b) for the several values of P_{th}, where the corresponding BPS throughput of the AQAM scheme employing per-SNR optimised thresholds
determined using Powell's method are also represented as thin lines for $P_{th} = 10^{-1}$, 10^{-2}
and 10^{-3}. Comparing the BPS throughput curves, we can conclude that the per-SNR optimised Powell method of Section 12.4.3 resulted in imperfect optimisation for some values of

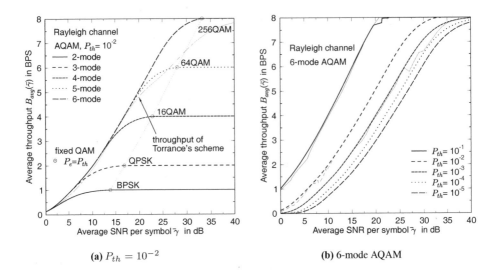

(a) $P_{th} = 10^{-2}$ **(b)** 6-mode AQAM

Figure 12.18: The average BPS throughput of various AQAM schemes employing the switching levels optimised using the Lagrangian multiplier method (a) for $P_{th} = 10^{-2}$ employing two to six-modes and (b) for $P_{th} = 10^{-2}$ to $P_{th} = 10^{-5}$ using six-modes. The average throughput of the six-mode AQAM scheme using Torrance's switching levels [214] is represented for comparison as the thin grey line in figure (a). The average throughput of the six-mode AQAM scheme employing per-SNR optimised thresholds using Powell's optimisation method are represented by the thin lines in figure (b) for the target average BEP of $P_{th} = 10^{-1}$, 10^{-2} and 10^{-3}.

the average SNR.

In conclusion, we derived an optimum mode-switching regime for a general AQAM scheme using the Lagrangian multiplier method and presented our numerical results for various AQAM arrangements. Since the results showed that the Lagrangian optimisation based scheme is superior in comparison to the other methods investigated, we will employ these switching levels in order to further investigate the performance of various adaptive modulation schemes.

12.5   Results and Discussion

The average throughput performance of adaptive modulation schemes employing the globally optimised mode-switching levels of Section 12.4.4 is presented in this section. The mobile channel is modelled as a Nakagami-m fading channel. The performance results and discussion include the effects of the fading parameter m, that of the number of modulation modes, the influence of the various diversity schemes used and the range of Square QAM, Star QAM and MPSK signalling constellations.

12.5.1 Narrow-band Nakagami-m Fading Channel

The PDF of the instantaneous channel SNR γ of a system transmitting over the Nakagami fading channel is given in Equation 12.24. The parameters characterising the operation of the adaptive modulation scheme were summarised in Section 12.3.3.1.

12.5.1.1 Adaptive PSK Modulation Schemes

Phase Shift Keying (PSK) has the advantage of exhibiting a constant envelope power, since all the constellation points are located on a circle. Let us first consider the BER of fixed-mode PSK schemes as a reference, so that we can compare the performance of adaptive PSK and fixed-mode PSK schemes. The BER of Gray-coded coherent M-ary PSK (MPSK), where $M = 2^k$, for transmission over the AWGN channel can be accurately approximated by [413]:

$$p_{MPSK}(\gamma) \simeq \sum_{i=1}^{2} A_i\, Q(\sqrt{a_i\gamma})\,, \tag{12.82}$$

where $M \geq 8$ and the associated constants are given by [413]:

$$A_1 = A_2 = 2/k \tag{12.83}$$
$$a_1 = 2\sin^2(\pi/M) \tag{12.84}$$
$$a_2 = 2\sin^2(3\pi/M)\,. \tag{12.85}$$

Figure 12.19(a) shows the BER of BPSK, QPSK, 8PSK, 16PSK, 32PSK and 64PSK for transmission over the AWGN channel. The differences of the required SNR per symbol, in order to achieve the BER of $p_{MPSK}(\gamma) = 10^{-6}$ for the modulation modes having a throughput difference of 1 BPS are around 6dB, except between BPSK and QPSK, where a 3dB difference is observed.

The average BER of MPSK schemes over a flat Nakagami-m fading channel is given as:

$$P_{MPSK}(\bar{\gamma}) = \int_0^{\infty} p_{MPSK}(\gamma)\, f(\gamma)\, d\gamma\,, \tag{12.86}$$

where the BEP $p_{MPSK}(\gamma)$ for a transmission over the AWGN channel is given by Equation 12.82 and the PDF $f(\gamma)$ is given by Equation 12.24. A closed form solution of Equation 12.86 can be readily obtained for an integer m using the results given in [(14-4-15)] [274], which can be expressed as:

$$P_{MPSK}(\bar{\gamma}) = \sum_{i=1}^{2} A_i \left[\tfrac{1}{2}(1 - \mu_i)\right]^m \sum_{j=0}^{m-1} \binom{m-1+j}{j} \left[\tfrac{1}{2}(1 + \mu_i)\right]^j\,, \tag{12.87}$$

where μ_i is defined as:

$$\mu_i \triangleq \sqrt{\frac{a_i\bar{\gamma}}{2m + a_i\bar{\gamma}}}\,. \tag{12.88}$$

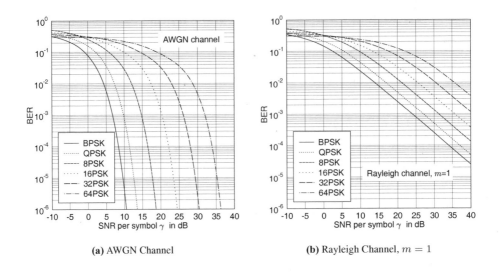

(a) AWGN Channel   (b) Rayleigh Channel, $m = 1$

Figure 12.19: The average BER of various MPSK modulation schemes

Figure 12.19(b) shows the average BER of the various MPSK schemes for transmission over a flat Rayleigh channel, where $m = 1$. The BER of MPSK over the AWGN channel given in Equation 12.82 and that over a Nakagami channel given in Equation 12.87 will be used in comparing the performance of adaptive PSK schemes.

The parameters of our nine-mode adaptive PSK scheme are summarised in Table 12.5 following the definitions of our generic model used for the adaptive modulation schemes developed in Section 12.3.1. The models of other adaptive PSK schemes employing a different number of modes can be readily obtained by increasing or reducing the number of columns in Table 12.5. Since the number of modes is $K = 8$, we have $K + 1 = 9$ mode-switching

k	0	1	2	3	4	5	6	7	8
m_k	0	2	4	8	16	32	64	128	256
b_k	0	1	2	3	4	5	6	7	8
c_k	0	1	1	1	1	1	1	1	1
Tx	No Tx	BPSK	QPSK	8PSK	16PSK	32PSK	64PSK	128PSK	256PSK

Table 12.5: Parameters of a nine-mode adaptive PSK scheme following the definitions of the generic adaptive modulation model developed in Section 12.3.1.

levels, which are hosted by the vector $\mathbf{s} = \{s_k \mid k = 0, 1, 2, \cdots, 8\}$. Let us assume $s_0 = 0$ and $s_8 = \infty$. In order to evaluate the performance of the nine-mode adaptive PSK scheme, we have to obtain the optimum switching levels first. Let us evaluate the "Lagrangian-free" functions defined in Equation 12.63, using the parameters given in Table 12.5 and the BER expressions given in Equation 12.82. The "Lagrangian-free" functions of our nine-mode adaptive PSK scheme are depicted in Figure 12.20. We can observe that there exist two solutions for s_k satisfying $y_k(s_k) = y_1(s_1)$ for a given value of s_1, which are given by the

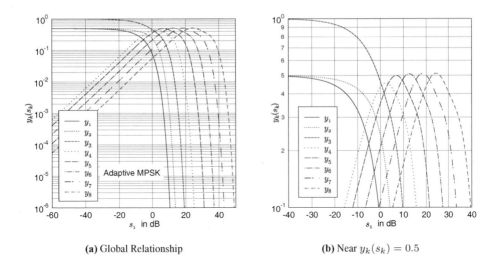

(a) Global Relationship   (b) Near $y_k(s_k) = 0.5$

Figure 12.20: "Lagrangian-free" functions of Equation 12.63 for a nine-mode adaptive PSK scheme. For a given value of s_1, there exist two solutions for s_k satisfying $y_k(s_k) = y_1(s_1)$. However, only the higher value of s_k satisfies the constraint of $s_{k-1} \leq s_k, \ \forall \ k$.

crossover points over the horizontal lines at the various coordinate values scaled on the vertical axis. However, only the higher value of s_k satisfies the constraint of $s_{k-1} \leq s_k, \ \forall \ k$. The enlarged view near $y_k(s_k) = 0.5$ seen in Figure 12.20(b) reveals that $y_4(s_4)$ may have no solution of $y_4(s_4) = y_1(s_1)$, when $y_1(s_1) > 0.45$. One option is to use a constant value of $s_4 = 2.37$dB, where $y_4(s_4)$ reaches its peak value. The other option is to set $s_4 = s_3$, effectively eliminating 16PSK from the set of possible modulation modes. It was found that both policies result in the same performance up to four effective decimal digits in terms of the average BPS throughput BPS.

Upon solving $y_k(s_k) = y_1(s_1)$, we arrive at the relationships sought between the first optimum switching level s_1 and the remaining optimum switching levels s_k. Figure 12.21(a) depicts these relationships. All the optimum switching levels, except for s_1 and s_2, approach their asymptotic limit monotonically, as s_1 decreases. A decreased value of s_1 corresponds to an increased value of the average SNR. Figure 12.21(b) illustrates the optimum switching levels of a seven-mode adaptive PSK scheme operating over a Rayleigh channel associated with $m = 1$ at the target BER of $P_{th} = 10^{-2}$. These switching levels were obtained by solving Equation 12.70. The optimum switching levels show a steady decrease in their values as the average SNR increases, until it reaches the avalanche SNR value of $\bar{\gamma} = 35$dB, beyond which always the highest-order PSK modulation mode, namely 64PSK, is activated.

Having highlighted the evaluation of the optimum switching levels for an adaptive PSK scheme, let us now consider the associated performance results. We are reminded that the average BEP of our optimised adaptive scheme remains constant at $P_{avg} = P_{th}$, provided that the average SNR is less than the avalanche SNR. Hence, the average BPS throughput and the relative SNR gain of our APSK scheme in comparison to the corresponding fixed-mode modem are our concern.

Let us now consider Figure 12.22, where the average BPS throughput of the various

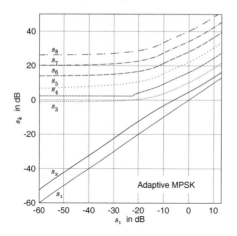

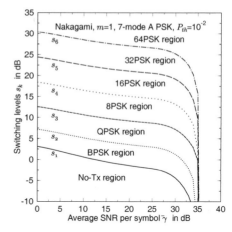

(a) Relationship between the optimum switching levels of a nine-mode PSK scheme

(b) Optimum switching levels of a seven-mode adaptive PSK scheme

Figure 12.21: Optimum switching levels. (a) Relationships between s_k and s_1 in a nine-mode adaptive PSK scheme. (b) Optimum switching levels for a 7-mode adaptive PSK scheme operating over a Rayleigh channel at the target BER of $P_{th} = 10^{-2}$.

adaptive PSK schemes operating over a Rayleigh channel associated with $m = 1$ are plotted, which were designed for the target BEP of $P_{th} = 10^{-2}$ and $P_{th} = 10^{-3}$. The markers "$\otimes$" and "$\odot$" represent the required SNR of the various fixed-mode PSK schemes, while achieving the same target BER as the adaptive schemes, operating over an AWGN channel and a Rayleigh channel, respectively. It can be observed that introducing an additional constituent mode into an adaptive PSK scheme does not make any impact on the average BPS throughput, when the average SNR is relatively low. For example, when the average SNR $\bar{\gamma}$ is less than 10dB in Figure 12.22(a), employing more than four APSK modes for the adaptive scheme does not improve the average BPS throughput. In comparison to the various fixed-mode PSK modems, the adaptive modem achieved the SNR gains between 4dB and 8dB for the target BEP of $P_{th} = 10^{-2}$ and 10dB to 16dB for the target BER of $P_{th} = 10^{-3}$ over a Rayleigh channel. Since no adaptive scheme operating over a fading channel can outperform the corresponding fixed-mode counterpart operating over an AWGN channel, it is interesting to investigate the performance differences between these two schemes. Figure 12.22 suggests that the required SNR of our adaptive PSK modem achieving 1BPS for transmission over a Rayleigh channel is approximately 1dB higher, than that of fixed-mode BPSK operating over an AWGN channel. Furthermore, this impressive performance can be achieved by employing only three modes, namely No-Tx, BPSK and QPSK for the adaptive PSK modem. For other BPS throughput values, the corresponding SNR differences are in the range of 2dB to 3dB, while maintaining the BEP of $P_{th} = 10^{-2}$ and 4dB for the BEP of $P_{th} = 10^{-3}$.

We observed in Figure 12.22 that the average BPS throughput of the various adaptive PSK schemes is dependent on the target BEP. Hence, let us investigate the BPS performances of the adaptive modems for the various values of target BEPs using the results depicted in Figure 12.23. The average BPS throughputs of a nine-mode adaptive PSK scheme are repre-

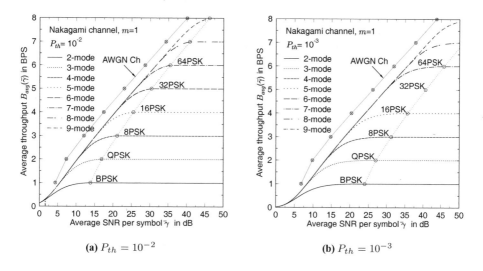

(a) $P_{th} = 10^{-2}$

(b) $P_{th} = 10^{-3}$

Figure 12.22: The average BPS throughput of various adaptive PSK schemes operating over a Rayleigh channel ($m = 1$) at the target BER of (a) $P_{th} = 10^{-2}$ and (b) $P_{th} = 10^{-3}$. The markers "$\otimes$" and "$\odot$" represent the required SNR of the corresponding fixed-mode PSK scheme, while achieving the same target BER as the adaptive schemes, operating over an AWGN channel and a Rayleigh channel, respectively.

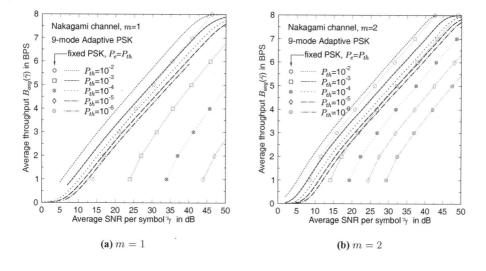

(a) $m = 1$

(b) $m = 2$

Figure 12.23: The average BPS throughput of a nine-mode adaptive PSK scheme operating over a Nakagami fading channel (a) $m = 1$ and (b) $m = 2$. The markers represent the SNR required for achieving the same BPS throughput and the same average BER as the adaptive schemes.

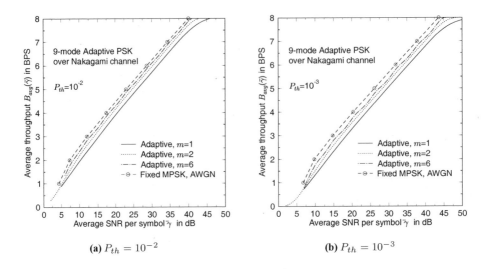

Figure 12.24: The effects of the Nakagami fading parameter m on the average BPS throughput of a nine-mode adaptive PSK scheme designed for the target BER of (a) $P_{th} = 10^{-2}$ and (b) $P_{th} = 10^{-3}$. As m increases, the average throughput of the adaptive modem approaches the throughput of fixed PSK modems operating over an AWGN channel.

sented as various types of lines without markers depending on the target average BERs, while those of the corresponding fixed PSK schemes are represented as various types of lines with markers according to the key legend shown in Figure 12.23. We can observe that the difference between the required SNRs of the adaptive schemes and fixed schemes increases, as the target BER decreases. It is interesting to note that the average BPS curves of the adaptive PSK schemes seem to converge to a set of densely packed curves, as the target BEP decreases to values around $10^{-4} - 10^{-6}$. In other words, the incremental SNR required for achieving the next target BEP, which is an order of magnitude lower, decreases as the target BER decreases. On the other hand, the incremental SNR for the same scenario of fixed modems seems to remain nearly constant at 10dB. Comparing Figure 12.23(a) and Figure 12.23(b), we find that this seemingly constant incremental SNR of the fixed-mode modems is reduced to about 5dB, as the fading becomes less severe, i.e. when the fading parameter becomes $m = 2$.

Let us now investigate the effects of the Nakagami fading parameter m on the average BPS throughput performance of various adaptive PSK schemes by observing Figure 12.24. The BPS throughput of the various fixed PSK schemes for transmission over an AWGN channel is depicted in Figure 12.24 as the ultimate performance limit achievable by the adaptive schemes operating over Nakagami fading channels. For example, when the channel exhibits Rayleigh fading, i.e. when the fading parameter becomes $m = 1$, the adaptive PSK schemes show 3dB to 4dB SNR penalty compared to their fixed-mode counterparts operating over the AWGN channel. Compared to fixed-mode BPSK, the adaptive scheme required only a 1dB higher SNR. As the fading becomes less severe, the average BPS throughput of the adaptive PSK schemes approaches that of fixed-mode PSK operating over the AWGN channel. For the target BEP of $P_{th} = 10^{-3}$, the SNR gap between the BPS throughput curves becomes

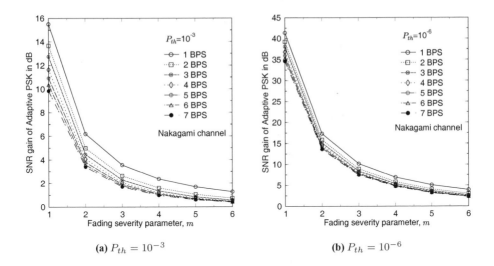

Figure 12.25: The SNR gain of adaptive PSK schemes in comparison to the corresponding fixed-mode PSK schemes yielding the same BPS throughput for the target BER of (a) $P_{th} = 10^{-3}$ and (b) $P_{th} = 10^{-6}$. The performance advantage of employing adaptive PSK schemes decreases, as the fading becomes less severe.

higher. The adaptive PSK scheme operating over the Rayleigh channel required 4dB to 5dB higher SNR for achieving the same throughput compared to the fixed PSK schemes operating over the AWGN channel.

Figure 12.25 summarises the relative SNR gains of our adaptive PSK schemes over the corresponding fixed PSK schemes. For the target BEP of $P_{th} = 10^{-3}$ the relative SNR gain of the nine-mode adaptive scheme compared to BPSK changes from 15.5dB to 1.3dB, as the Nakagami fading parameter changes from 1 to 6. Observing Figure 12.25(a) and Figure 12.25(b) we conclude that the advantages of employing adaptive PSK schemes are more pronounced when

1) the fading is more severe,

2) the target BER is lower, and

3) the average BPS throughput is lower.

Having studied the range of APSK schemes, let us in the next section consider the family of adaptive coherently detected Star-QAM schemes.

12.5.1.2   Adaptive Coherent Star QAM Schemes

In this section, we study the performance of adaptive coherent QAM schemes employing Type-I Star constellations [2]. Even though non-coherent Star QAM (SQAM) schemes are more popular owing to their robustness to fading without requiring pilot symbol-assisted channel estimation and Automatic Gain Control (AGC) at the receiver, the results provided

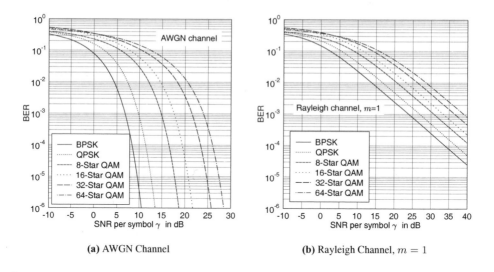

(a) AWGN Channel

(b) Rayleigh Channel, $m = 1$

Figure 12.26: The average BER of various SQAM modulation schemes

in this section can serve as benchmark results for non-coherent Star QAM schemes and the coherent Square QAM schemes.

The BER of coherent Star QAM over an AWGN channel is derived in Appendix 12.8. It is shown that their BER can be expressed as:

$$p_{SQAM}(\gamma) \simeq \sum_i A_i\, Q(\sqrt{a_i\gamma}\,), \tag{12.89}$$

where A_i and a_i are given in Appendix 12.8 for 8-Star, 16-Star, 32-Star and 64-Star QAM. The SNR-dependent optimum ring ratios were also derived in Appendix 12.8 for these Star QAM modems. Figure 12.26(a) shows the BER of BPSK, QPSK, 8-Star QAM, 16-Star QAM, 32-Star QAM and 64-Star QAM employing the optimum ring ratios over the AWGN channel. Comparing Figure 12.19(a) and Figure 12.26(a), we can observe that 16-Star QAM, 32-Star QAM and 64-Star QAM are more power-efficient than 16 PSK, 32 PSK and 64 PSK, respectively. However, the envelope power of the Star QAM signals is not constant, unlike that of the PSK signals. Following an approach similar to that used in Equations 12.86 and 12.87, the average BEP of the various SQAM schemes over a flat Nakagami-m fading channel can be expressed as:

$$P_{SQAM}(\bar{\gamma}) = \sum_i A_i \left[\tfrac{1}{2}(1 - \mu_i)\right]^m \sum_{j=0}^{m-1} \binom{m-1+j}{j} \left[\tfrac{1}{2}(1 + \mu_i)\right]^j, \tag{12.90}$$

where μ_i is defined as:

$$\mu_i \triangleq \sqrt{\frac{a_i\bar{\gamma}}{2m + a_i\bar{\gamma}}}\, . \tag{12.91}$$

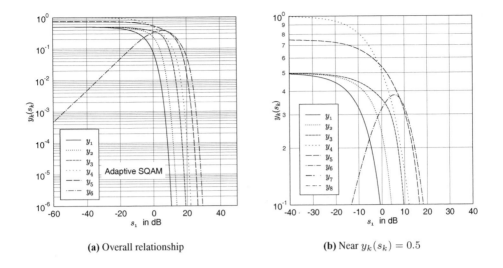

(a) Overall relationship   (b) Near $y_k(s_k) = 0.5$

Figure 12.27: "Lagrangian-free" functions of Equation 12.63 for a seven-mode adaptive Star QAM scheme.

Figure 12.26(b) shows the average BEP of various SQAM schemes for transmission over a flat Rayleigh channel, where $m = 1$. It can be observed that the 16-Star, 32-Star and 64-Star QAM schemes exhibit SNR advantages of around 3.5dB, 4dB, and 7dB compared to 16-PSK, 32-PSK and 64-PSK schemes at a BEP of 10^{-2}. The BEP of SQAM for transmission over the AWGN channel given in Equation 12.89 and that over a Nakagami channel given in Equation 12.90 will be used in comparing the performance of the various adaptive SQAM schemes.

k	0	1	2	3	4	5	6
m_k	0	2	4	8	16	32	64
b_k	0	1	2	3	4	5	6
c_k	0	1	1	1	1	1	1
mode	No Tx	BPSK	QPSK	8-Star	16-Star	32-Star	64-Star

Table 12.6: Parameters of a seven-mode adaptive Star QAM scheme following the definitions developed in Section 12.3.1 for the generic adaptive modulation model

The parameters of a seven-mode adaptive Star QAM scheme are summarised in Table 12.6 following the definitions of the generic model developed in Section 12.3.1 for adaptive modulation schemes. Since the number of modes is $K = 7$, we have $K + 1 = 8$ mode-switching levels hosted by the vector $\mathbf{s} = \{s_k \mid k = 0, 1, 2, \cdots, 7\}$. Let us assume that $s_0 = 0$ and $s_7 = \infty$. Then, we have to determine the optimum values for the remaining six switching levels using the technique developed in Section 12.4.4. The "Lagrangian-free" functions corresponding to a seven-mode Star QAM scheme are depicted in Figure 12.27 and the relationships obtained for the switching levels are displayed in Figure 12.28(a). We

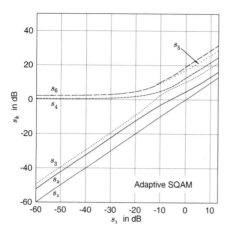

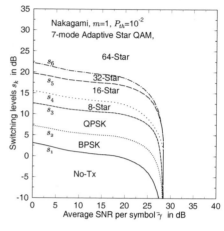

(a) Relationship between the optimum switching levels

(b) Optimum switching levels

Figure 12.28: Optimum switching levels of a seven-mode Adaptive Star QAM scheme. (a) Relationships between s_k and s_1. (b) Optimum switching levels of a seven-mode adaptive Star QAM scheme operating over a Rayleigh channel at the target BER of $P_{th} = 10^{-2}$.

can observe that as seen for APSK in Figure 12.20 there exist two solutions for s_6 satisfying $y_6(s_6) = y_1(s_1)$ for a given value of s_1, when $y_1 \leq 0.382$. However, only the higher value of s_k satisfies the constraint of $s_6 \geq s_5$. When $s_1 \leq 7.9$dB, the optimum value of s_6 should be set to s_5, in order to guarantee $s_6 \geq s_5$. Figure 12.28(b) illustrates the optimum switching levels of a seven-mode adaptive Star QAM scheme operating over a Rayleigh channel at the target BER of $P_{th} = 10^{-2}$. These switching levels were obtained by solving Equation 12.70. The optimum switching levels show a steady decrease in their values, as the average SNR increases, until they reach the avalanche SNR value of $\bar{\gamma} = 28.5$dB, beyond which always the highest-order modulation mode, namely 64-Star QAM, is activated.

Let us now investigate the associated performance results. We are reminded that the average BEP of our optimised adaptive scheme remains constant at $P_{avg} = P_{th}$, provided that the average SNR is less than the avalanche SNR. Hence, the average BPS throughput and the SNR gain of our adaptive modem in comparison to the corresponding fixed-mode modems are our concern.

Let us first consider Figure 12.29, where the average BPS throughput of the various adaptive Star QAM schemes operating over a Rayleigh channel associated with $m = 1$ is shown at the target BEP of $P_{th} = 10^{-2}$ and $P_{th} = 10^{-3}$. The markers "$\otimes$" and "$\odot$" represent the required SNR of the corresponding fixed-mode Star QAM schemes, while achieving the same target BEP as the adaptive schemes, operating over an AWGN channel and a Rayleigh channel, respectively. Comparing Figure 12.22(a) and Figure 12.29(a), we find that the tangent of the average BPS curves of the adaptive Star QAM schemes is higher than that of adaptive PSK schemes. Explicitly, the tangent of the Star QAM schemes is around 0.3BPS/dB, whereas that of the APSK schemes was 0.18BPS/dB. This is due to the more power-efficient constellation arrangement of Star QAM in comparison to the single-ring constellations of

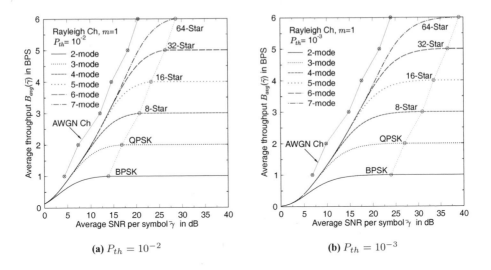

(a) $P_{th} = 10^{-2}$ **(b)** $P_{th} = 10^{-3}$

Figure 12.29: The average BPS throughput of the various adaptive Star QAM schemes operating over a Rayleigh fading channel associated with $m = 1$ at the target BEP of (a) $P_{th} = 10^{-2}$ and (b) $P_{th} = 10^{-3}$. The markers "⊗" and "⊙" represent the required SNR of the corresponding fixed-mode Star QAM schemes, while achieving the same target BER as the adaptive schemes, operating over an AWGN channel and a Rayleigh channel, respectively.

the PSK modulations schemes. In comparison to the corresponding fixed-mode Star QAM modems, the adaptive modem achieved an SNR gain of 6dB to 8dB for the target BEP of $P_{th} = 10^{-2}$ and 12dB to 16dB for the target BEP of $P_{th} = 10^{-3}$ over a Rayleigh channel. Compared to the fixed-mode Star QAM schemes operating over an AWGN channel, our adaptive schemes approached their performance within about 3dB in terms of the required SNR value, while achieving the same target BER of $P_{th} = 10^{-2}$ and $P_{th} = 10^{-3}$.

Since Figure 12.29 suggests that the relative SNR gain of the adaptive schemes is dependent on the target BER, let us investigate the effects of the target BEP in more detail. Figure 12.30 shows the BPS throughput of the various adaptive schemes at the target BEP of $P_{th} = 10^{-2}$ to $P_{th} = 10^{-6}$. The average BPS throughputs of a seven-mode adaptive Star QAM scheme is represented with the aid of the various line types without markers, depending on the target average BERs, while those of the corresponding fixed-mode Star QAM schemes are represented as various types of lines having markers according to the legends shown in Figure 12.30. We can observe that the difference between the SNRs required for the adaptive schemes and fixed schemes increases, as the target BER decreases. The fixed-mode Star QAM schemes require additional SNRs of 10dB and 6dB in order to achieve an order of magnitude lower BER for the Nakagami fading parameters of $m = 1$ and $m = 2$, respectively. However, our adaptive schemes require additional SNRs of only 1dB to 3dB for achieving the same goal.

Let us now investigate the effects of the Nakagami fading parameter m on the average BPS throughput performance of the various adaptive Star QAM schemes by observing Figure 12.31. The BPS throughput of the fixed-mode Star QAM schemes for the transmission

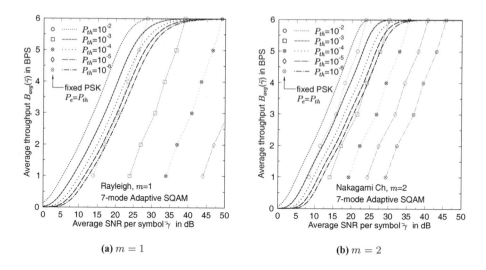

Figure 12.30: The average BPS throughput of a seven-mode adaptive Star QAM scheme operating over a Nakagami fading channel (a) $m = 1$ and (b) $m = 2$. The markers represent the SNR required by the fixed-mode schemes for achieving the same BPS throughput and the same average BER as the adaptive schemes.

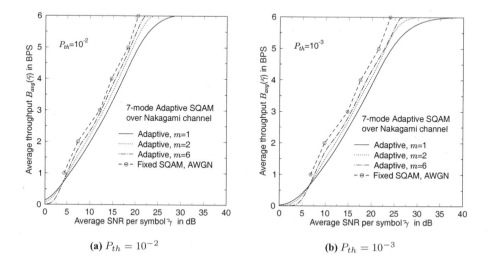

Figure 12.31: The effects of the Nakagami fading parameter m on the average BPS throughput of a seven-mode adaptive Star QAM scheme at the target BER of (a) $P_{th} = 10^{-2}$ and (b) $P_{th} = 10^{-3}$. As m increases, the average throughput of the adaptive modem approaches the throughput of the fixed-mode Star QAM modems operating over an AWGN channel.

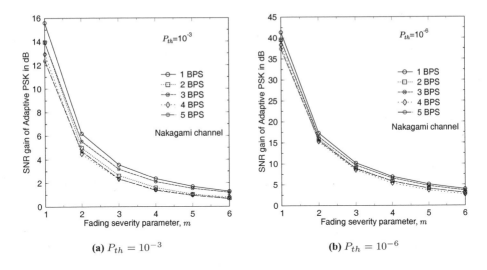

Figure 12.32: The SNR gain of the various adaptive Star QAM schemes in comparison to the fixed-mode Star QAM schemes yielding the same BPS throughput at the target BEP of (a) $P_{th} = 10^{-3}$ and (b) $P_{th} = 10^{-6}$. The advantage of the adaptive Star QAM schemes decreases, as the fading becomes less severe.

over an AWGN channel is depicted in Figure 12.31 as the ultimate performance limit achievable by the adaptive schemes operating over Nakagami fading channels. As the Nakagami fading parameter m increases from 1 to 2 and to 6, the SNR gap between the adaptive schemes operating over a Nakagami fading channel and the fixed-mode schemes decreases. When the average SNR is less than $\bar{\gamma} \le 6$dB, the average BPS throughput of our adaptive schemes decreases, when the fading parameter m increases. The rationale of this phenomenon is that as the channel becomes more and more like an AWGN channel, the probability of activating the BPSK mode is reduced, resulting in more frequent activation of the No-Tx mode and hence the corresponding average BPS throughput inevitably decreases.

The effects of the Nakagami fading factor m on the SNR gain of our adaptive Star QAM scheme can be observed in Figure 12.32. As expected, the relative SNR gain of the adaptive schemes at a throughput of 1 BPS is the highest among the BPS throughputs considered. However, the order observed in terms of the SNR gain of the adaptive schemes does not strictly follow the increasing BPS order at the target BEP of $P_{th} = 10^{-3}$ and $P_{th} = 10^{-6}$, as did for the adaptive PSK schemes of Section 12.5.1.1. Even though the adaptive Star QAM schemes exhibit a higher throughput than the adaptive PSK schemes, the SNR gains compared to their fixed-mode counterparts are more or less the same, showing typically less than 1dB difference, except for the 5 BPS throughput scenario, where the adaptive QAM scheme gained up to 1.3dB more in terms of the required SNR than the adaptive PSK scheme.

Having studied the performance of a range of adaptive Star QAM schemes, in the next section we consider adaptive modulation schemes employing the family of square-shaped QAM constellations.

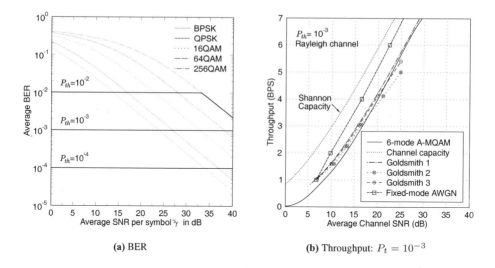

(a) BER   **(b)** Throughput: $P_t = 10^{-3}$

Figure 12.33: The average BER and average throughput performance of a six-mode adaptive Square
QAM scheme operating over a flat Rayleigh channel ($m = 1$). (a) The constant target average BER is maintained over the entire range of the average SNR values up the
avalanche SNR. (b) The average BPS throughput of the equivalent constant-power adaptive scheme is compared to Goldsmith's schemes [404]. The "Goldsmith 1" and "Goldsmith 2" schemes represent a variable-power adaptive scheme employing hypothetical
continuously variable-BPS QAM modulation modes and Square QAM modes, respectively. The "Goldsmith 3" scheme represents the simulation results associated with a
constant-power adaptive Square QAM reported in [404].

12.5.1.3   Adaptive Coherent Square QAM Modulation Schemes

Since coherent Square M-ary QAM (MQAM) is the most power-efficient M-ary modulation
scheme [2] and an accurate channel estimation is facilitated with the advent of Pilot Symbol Assisted Modulation (PSAM) techniques [174, 398, 399], Otsuki, Sampei and Morinaga
proposed employing coherent square QAM as the constituent modulation modes for an adaptive modulation scheme [400] instead of non-coherent Star QAM modulation [397]. In this
section, we study the various aspects of this adaptive square QAM scheme employing the optimum switching levels of Section 12.4.4. The closed form BER expressions of square QAM
over an AWGN channel can be found in Equation 12.2 and that over a Nakagami channel can
be expressed using a similar form given in Equation 12.90. The optimum switching levels of
adaptive Square QAM were studied in Section 12.4.4 as an example.

The average BER of our six-mode adaptive Square QAM scheme operating over a flat
Rayleigh fading channel is depicted in Figure 12.33(a), which shows that the modem maintains the required constant target BER, until it reaches the BER curve of the specific fixed-
mode modulation scheme employing the highest-order modulation mode, namely 256-QAM,
and then it follows the BER curve of the 256-QAM mode. The various grey lines in the figure represent the BER of the fixed constituent modulation modes for transmission over a flat
Rayleigh fading channel. An arbitrarily low target BER could be maintained at the expense
of a reduced throughput.

The average throughput is shown in Figure 12.33(b) together with the estimated channel capacity of the narrow-band Rayleigh channel [401, 402] and with the throughput of several variable-power, variable-rate modems reported in [404]. Specifically, Goldsmith and Chua [404] studied the performance of their variable-power variable-rate adaptive modems based on a BER bound of m-ary Square QAM, rather than using an exact BER expression. Since our adaptive Square QAM schemes do not vary the transmission power, our scheme can be regarded as a sub-optimal policy viewed for their respective [404]. However, the throughput performance of Figure 12.33(b) shows that the SNR degradation is within 2dB in the low-SNR region and within half a dB in the high-SNR region, in comparison to the ideal continuously variable-power adaptive QAM scheme employing a range of hypothetical continuously variable-BPS QAM modes [404], represented as the "Goldsmith 1" scheme in the figure. Goldsmith and Chua [404] also reported the performance of a variable-power discrete-rate and a constant-power discrete-rate scheme, which we represented as the "Goldsmith 2" and "Goldsmith 3" in Figure 12.33(b), respectively. Since their results are based on approximate BER formulas, the average BPS throughput performance of the "Goldsmith 3" scheme is optimistic, when the average SNR γ is less than 17dB. Considering that our scheme achieves the maximum possible throughput for the given average SNR value with the aid of the globally optimised switching levels, the average throughput of the "Goldsmith 3" scheme is expected to be lower, than that of our scheme, as is the case when the average SNR γ is higher than 17dB.

Figure 12.34(a) depicts the average BPS throughput of our various adaptive Square QAM schemes operating over a Rayleigh channel associated with $m = 1$ at the target BEP of $P_{th} = 10^{-3}$. Figure 12.34(a) shows that even though the constituent modulation modes of our adaptive schemes do not include 3, 5 and 7-BPS constellations, the average BPS throughput steadily increases without undulations. Compared to the fixed-mode Square QAM schemes operating over an AWGN channel, our adaptive schemes required an additional SNR of less than 3.5dB, when the throughput is below 6.5 BPS. The comparison of the average BPS throughputs of the adaptive schemes employing PSK, Star QAM and Square QAM modems, as depicted in Figure 12.34(b), confirms the superiority of Square QAM over the other two schemes in terms of the required average SNR for achieving the same throughput and the same target average BEP. Since all these three schemes employ BPSK, QPSK as the second and the third constituent modulation modes, their throughput performance shows virtually no difference, when the average throughput is less than or equal to $B_{avg} = 2$ BPS.

Let us now investigate the effects of the Nakagami fading parameter m on the average BPS throughput performance of the adaptive Square QAM schemes by observing Figure 12.35. The BPS throughput of the fixed-mode Square QAM schemes over an AWGN channel is depicted in Figure 12.35 as the ultimate performance limit achievable by the adaptive schemes operating over Nakagami fading channels. Similar observations can be made for the adaptive Square QAM scheme, like for the adaptive Star QAM arrangement characterised in Figure 12.31. A specific difference is, however, that the average BPS throughput recorded for the fading parameter of $m = 6$ exhibits an undulating curve. For example, an increased m value results in a limited improvement of the corresponding average BPS throughput near the throughput values of 2.5, 4.5 and 6.5 BPS. This is because our adaptive Square QAM schemes do not use 3-, 5- and 7-BPS constituent modems, unlike the adaptive PSK and adaptive Star QAM schemes. Figure 12.36 depicts the corresponding optimum mode-switching levels for the six-mode adaptive Square QAM scheme. The black lines rep-

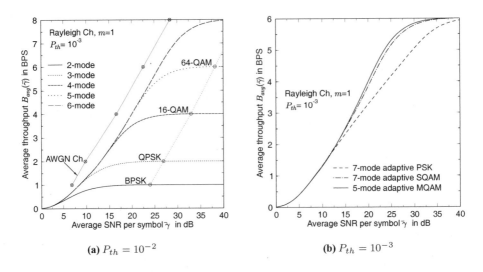

Figure 12.34: The average BPS throughput of various adaptive Square QAM schemes operating over a Rayleigh channel ($m = 1$) at the target BEP of $P_{th} = 10^{-3}$. (a) The markers "$\otimes$" and "$\odot$" represent the required SNR of the corresponding fixed-mode Square QAM schemes achieving the same target BER as the adaptive schemes, operating over an AWGN channel and a Rayleigh channel, respectively. (b) The comparison of the various adaptive schemes employing PSK, Star QAM and Square QAM as the constituent modulation modes.

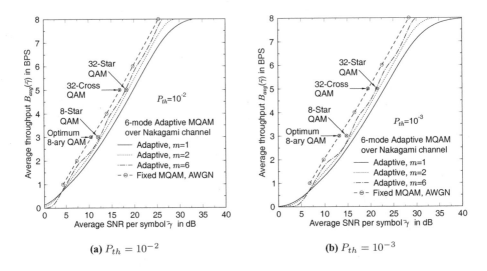

Figure 12.35: The effects of the Nakagami fading parameter m on the average BPS throughput of a seven-mode adaptive Square QAM scheme at the target BEP of (a) $P_{th} = 10^{-2}$ and (b) $P_{th} = 10^{-3}$. As m increases, the average throughput of the adaptive modem approaches the throughput of the corresponding fixed Square QAM modems operating over an AWGN channel.

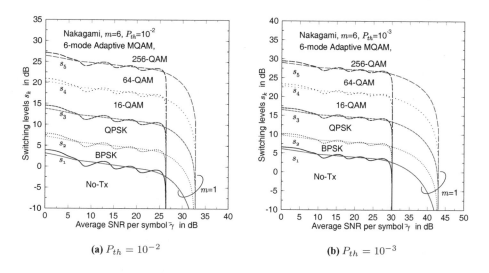

Figure 12.36: The switching levels of the six-mode adaptive Square QAM scheme operating over Nakagami fading channels at the target BER of (a) $P_{th} = 10^{-2}$ and (b) $P_{th} = 10^{-3}$. The bold lines are used for the fading parameter of $m = 6$ and the grey lines are for $m = 1$.

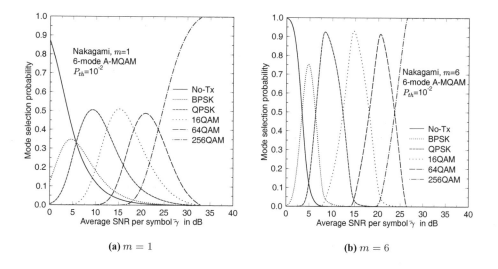

Figure 12.37: The mode selection probability of a six-mode adaptive Square QAM scheme operating over Nakagami fading channels at the target BEP of $P_{th} = 10^{-2}$. When the fading becomes less severe, the mode selection scheme becomes more "selective" in comparison to that for $m = 1$.

resent the switching levels, when the Nakagami fading parameter is $m = 6$ and the grey lines when $m = 1$. In general, the lower the switching levels, the higher the average BPS throughput of the adaptive modems. When the Nakagami fading parameter is $m = 1$, the switching levels decrease monotonically, as the average SNR increases. However, when the fading severity parameter is $m = 6$, the switching levels fluctuate, exhibiting several local minima around 8dB, 15dB and 21dB. In the extreme case of $m \to \infty$, i.e. when operating over an AWGN-like channel, the switching levels would be $s_1 = s_2 = 0$ and $s_k = \infty$ for other k values in the SNR range of $7.3\text{dB} < \bar{\gamma} < 14\text{dB}$, $s_1 = s_2 = s_3 = 0$ and $s_4 = s_5 = \infty$ when we have $14\text{dB} < \bar{\gamma} < 20\text{dB}$, $s_k = 0$ except for $s_5 = \infty$ when the SNR is in the range of $20\text{dB} < \bar{\gamma} < 25\text{dB}$ and finally, all $s_k = 0$ for $\forall k$, when $\bar{\gamma} > 25\text{dB}$, when considering the fixed-mode Square QAM performance achieved over an AWGN channel represented by markers "$\odot$" in Figure 12.35. Observing Figure 12.37, we find that our adaptive schemes become highly "selective", when the Nakagami fading parameter becomes $m = 6$, exhibiting narrow triangular shapes. As m increases, the shapes will eventually converge to Kronecker delta functions.

A possible approach to reducing the undulating behaviour of the average BPS throughput curve is the introduction of a 3-BPS and a 5-BPS mode as additional constituent modem modes. The power-efficiency of 8-Star QAM and 32-Star QAM is insufficient to maintain a linear growth of the average BPS throughput, as we can observe in Figure 12.35. Instead, the most power-efficient 8-ary QAM scheme [page 279]of [274] and 32-ary Cross QAM scheme [page 236] of [2] have the potential of reducing these undulation effects. However, since we observed in Section 12.5.1.1 and Section 12.5.1.2 that the relative SNR advantage of employing adaptive Square QAM rapidly reduces, when the Nakagami fading parameter increases, even though the additional 3-BPS and 5-BPS modes are also used, there seems to be no significant benefit in employing non-square shaped additional constellations.

Again, we can observe in Figure 12.35 that when the average SNR is less than $\bar{\gamma} \leq 6\text{dB}$, the average BPS throughput of our adaptive Square QAM scheme decreases, as the Nakagami fading parameter m increases. As we discussed in Section 12.5.1.2, this is due to the less frequent activation of the BPSK mode in comparison to the "No-Tx" mode, as the channel variation is reduced.

The effects of the Nakagami fading factor m on the relative SNR gain of our adaptive Square QAM scheme can be observed in Figure 12.38. The less severe the fading, the smaller the relative SNR advantage of employing adaptive Square QAM in comparison to its fixed-mode counterparts. Except for the 1-BPS mode, the SNR gains become less than 0.5dB, when m is increased to 6 at the target BEP of $P_{th} = 10^{-3}$. The trend observed is the same at the target BEP of $P_{th} = 10^{-6}$, showing relatively higher gains in comparison to the $P_{th} = 10^{-3}$ scenario.

12.5.2   Performance over Narrow-band Rayleigh Channels Using Antenna Diversity

In the last section, we observed that the adaptive modulation schemes employing Square QAM modes exhibit the highest BPS throughput among the schemes investigated, when operating over Nakagami fading channels. Hence, in this section we study the performance of the adaptive Square QAM schemes employing antenna diversity operating over independent Rayleigh fading channels. The BEP expression of the fixed-mode coherent BPSK scheme can

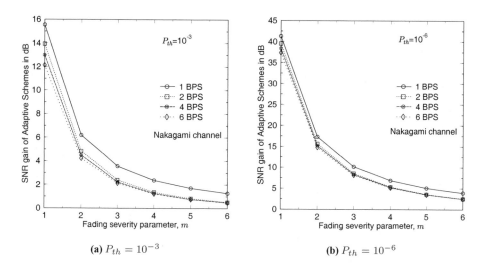

(a) $P_{th} = 10^{-3}$ **(b)** $P_{th} = 10^{-6}$

Figure 12.38: The SNR gain of the six-mode adaptive Square QAM scheme in comparison to the various fixed-mode Square QAM schemes yielding the same BPS throughput at the target BEP of (a) $P_{th} = 10^{-3}$ and (b) $P_{th} = 10^{-6}$. The performance advantage of the adaptive Square QAM schemes decreases, as the fading becomes less severe.

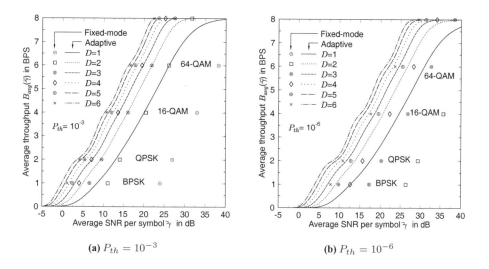

(a) $P_{th} = 10^{-3}$ **(b)** $P_{th} = 10^{-6}$

Figure 12.39: The average BPS throughput of the MRC-aided antenna-diversity assisted adaptive Square QAM scheme operating over independent Rayleigh fading channels at the target average BEP of (a) $P_{th} = 10^{-3}$ and (b) $P_{th} = 10^{-6}$. The markers represent the corresponding fixed-mode Square QAM performances.

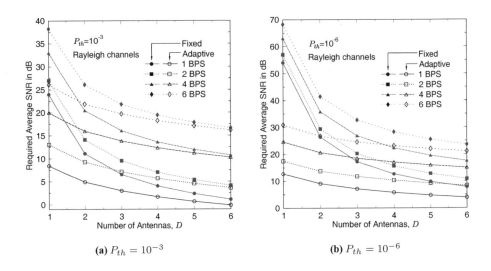

(a) $P_{th} = 10^{-3}$ **(b)** $P_{th} = 10^{-6}$

Figure 12.40: The SNR required for the MRC-aided antenna-diversity-assisted adaptive Square QAM schemes and the corresponding fixed-mode modems operating over independent Rayleigh fading channels at the target average BEP of (a) $P_{th} = 10^{-3}$ and (b) $P_{th} = 10^{-6}$.

be found in [page 781] of [274] and those of coherent Square QAM can be readily extended using the equations in Equations 12.2 and 12.3. Furthermore, the antenna diversity scheme operating over independent narrow-band Rayleigh fading channels can be viewed as a special case of the two-dimensional (2D) Rake receiver analysed in Appendices 12.9 and 12.7. The performance of antenna-diversity assisted adaptive Square QAM schemes can be readily analysed using the technique developed in Section 12.4.4.

Figure 12.39 depicts the average BPS throughput performance of our adaptive schemes employing Maximal Ratio Combining (MRC) aided antenna diversity [Ch 5, 6] of [417] operating over independent Rayleigh fading channels at the target average BEP of $P_{th} = 10^{-3}$ and $P_{th} = 10^{-6}$. The markers represent the performance of the corresponding fixed-mode Square QAM modems in the same scenario. The average SNRs required for achieving the target BEP of the fixed-mode schemes and that of the adaptive schemes decrease, as the antenna diversity order increases. However, the differences between the required SNRs of the adaptive schemes and their fixed-mode counterparts also decrease, as the antenna diversity order increases. The SNRs of both schemes required for achieving the target BEPs of $P_{th} = 10^{-3}$ and $P_{th} = 10^{-6}$ are displayed in Figure 12.40, where we can observe that dual antenna diversity is sufficient for the fixed-mode schemes in order to obtain half of the achievable SNR gain of the six-antenna aided diversity scheme, whereas triple-antenna diversity is required for the adaptive schemes operating in the same scenario. The corresponding first switching levels s_1 are depicted in Figure 12.41 for different orders of antenna diversity up an order of six. As the antenna diversity order increases, the avalanche SNR becomes lower and the switching-threshold undulation effects begins to appear. The required values of the first switching level s_1 are within a range of about 1dB and 0.5dB for the target BEPs of $P_{th} = 10^{-3}$ and $P_{th} = 10^{-6}$, respectively, before the avalanche SNR is reached. This suggests

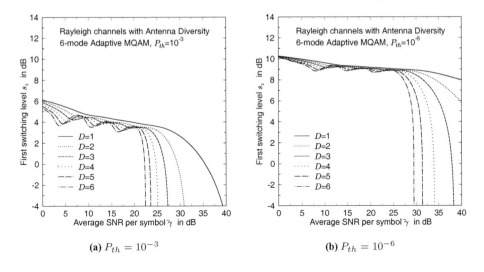

(a) $P_{th} = 10^{-3}$ **(b)** $P_{th} = 10^{-6}$

Figure 12.41: The first switching level s_1 of the MRC-aided antenna-diversity assisted adaptive Square QAM scheme operating over independent Rayleigh fading channels at the target average BEP of (a) $P_{th} = 10^{-3}$ and (b) $P_{th} = 10^{-6}$.

that the optimum mode-switching levels are more dependent on the target BEP, than on the number of diversity antennas.

12.5.3 Performance over Wideband Rayleigh Channels using Antenna Diversity

Wideband fading channels are characterised by their multi-path intensity profiles (MIP). In order to study the performance of the various adaptive modulation schemes, we employ two different MIP models in this section, namely a shortened Wireless Asynchronous Transfer Mode (W-ATM) channel [Ch 20] of [2] for an indoor scenario and a Bad-Urban Reduced-model A (BU-RA) channel [418] for a hilly urban outdoor scenario. Their MIPs are depicted in Figure 12.42. The W-ATM channel exhibits short-range, low-delay multi-path components, while the BU-RA channel exhibits six higher-delay multi-path components. Again, let us assume that our receivers are equipped with MRC Rake receivers [266], employing a sufficiently high number of Rake fingers, in order to capture all the multi-path components generated by our channel models. Furthermore, we employ antenna diversity [Ch 5] [417] at the receivers. This combined diversity scheme is often referred to as a two-dimensional (2D) Rake receiver [page 263] [419]. The BER of the 2D Rake receiver transmission over wideband independent Rayleigh fading channels is analysed in Appendix 12.9. A closed-form expression for the mode-specific average BEP of a 2D-Rake assisted adaptive Square QAM scheme is also given in Appendix 12.7. Hence, the performance of our 2D-Rake assisted adaptive modulation scheme employing the optimum switching levels can readily be obtained.

The average BPS throughputs of our 2D-Rake assisted adaptive schemes operating over the two different types of wideband channel scenarios is presented in Figure 12.43 at the target

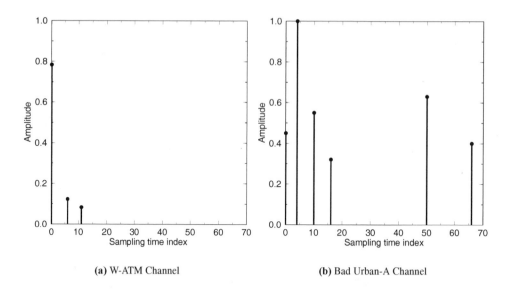

(a) W-ATM Channel   **(b)** Bad Urban-A Channel

Figure 12.42: Multi-path Intensity Profiles (MIPs) of the Wireless Asynchronous Transfer Mode (W-ATM) indoor channel [Ch 20] [2] and that of the Bad-Urban Reduced-model A (BU-RA) channel [418]

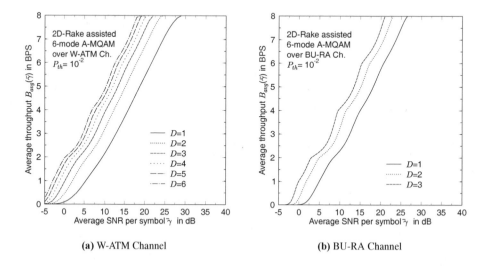

(a) W-ATM Channel   **(b)** BU-RA Channel

Figure 12.43: The effects of the number of diversity antennas on the average BPS throughput of the 2D-Rake assisted six-mode adaptive Square QAM scheme operating over the wideband independent Rayleigh fading channels characterised in Figure 12.42 at the target BEP of $P_{th} = 10^{-2}$. The number of diversity antennas is D.

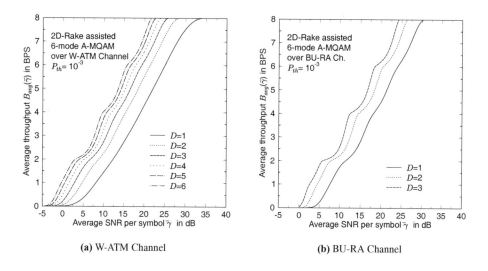

(a) W-ATM Channel   (b) BU-RA Channel

Figure 12.44: The effects of the number of diversity antennas on the average BPS throughput of the 2D-Rake assisted six-mode adaptive Square QAM scheme operating over the wideband independently Rayleigh fading channels characterised in Figure 12.42 at the target BEP of $P_{th} = 10^{-3}$. The number of diversity antennas is D.

BEP of $P_{th} = 10^{-2}$. The throughput performance depicted corresponds to the upper-bound performance of Direct-Sequence Code Division Multiple Access (DS-CDMA) or Multi-Carrier CDMA employing Rake receivers and the MRC-aided diversity assisted scheme in the absence of Multiple Access Interference (MAI). We can observe that the BPS throughput curves undulate, when the number of antennas D increases. This effect is more pronounced for transmission over the BU-RA channel, since the BU-RA channel exhibits six multi-path components, increasing the available diversity potential of the system approximately by a factor of two in comparison to that of the W-ATM channel. The performance of our adaptive scheme employing more than three antennas for transmission over the BU-RA channel could not be obtained owing to numerical instability, since the associated curves become similar to a series of step-functions, which is not analytic in mathematical terms. A similar observation can be made in the context of Figure 12.44, where the target BEP is $P_{th} = 10^{-3}$. Comparing Figure 12.43 and Figure 12.44, we observe that the BPS throughput curves corresponding to $P_{th} = 10^{-3}$ are similar to shifted version of those corresponding to $P_{th} = 10^{-2}$, which are shifted in the direction of increasing SNRs. On the other hand, the BPS throughput curves corresponding to $P_{th} = 10^{-3}$ undulate more dramatically. When the number of antennas is $D = 3$, the BPS throughput curves of the BU-RA channel exhibits a stair-case like shape. The corresponding mode switching levels and mode selection probabilities are shown in Figure 12.45. Again, the switching levels heavily undulate. The mode-selection probability curve of BPSK has a triangular shape, increasing linearly, as the average SNR $\bar{\gamma}$ increases to 2.5dB and decreasing linearly again as $\bar{\gamma}$ increases from 2.5dB. On the other hand, the mode-selection probability curve of QPSK increases linearly and decreases exponentially, since no 3-BPS mode is used. This explains why the BPS throughput curves increase in a near-linear fashion in the SNR range of 0 to 5dB and in a staircase fashion beyond that point. We can

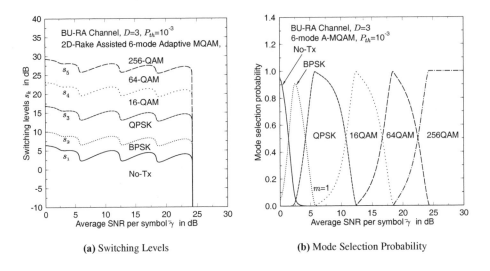

(a) Switching Levels	**(b)** Mode Selection Probability

Figure 12.45: The mode switching levels and mode selection probability of the 2D-Rake-assisted six-mode adaptive Square QAM scheme using $D = 3$ antennas operating over the BU-RA channel characterised in Figure 12.42(b) at the target BEP of $P_{th} = 10^{-3}$.

conclude that the staircase-like shape in the upper range of SNRs is a consequence of the absence of the 3-BPS, 5-BPS and 7-BPS modulation modes in the set of constituent modulation modes employed. As we discussed in Section 12.5.1.3, this problem may be mitigated by introducing power-efficient 8 QAM, 32 QAM and 128 QAM modes.

The average SNRs required for achieving the target BEP of $P_{th} = 10^{-3}$ by the 2D-Rake assisted adaptive scheme and the fixed-mode schemes operating over wide-band fading channels are depicted in Figure 12.46. Since the fixed-mode schemes employing Rake receivers are already enjoying the diversity benefit of multi-path fading channels, the SNR advantages of our adaptive schemes are less than 8dB and 2.6dB over the W-ATM channel and over BU-RA channel, respectively, even when a single antenna is employed. This relative small SNR gain in comparison to those observed over narrow-band fading channels in Figure 12.40 erode as the number of antennas increases. For example, when the number of antennas is $D = 6$, the SNR gains of the adaptive schemes operating over the W-ATM channel of Figure 12.42(a) become virtually zero, where the combined channel becomes an AWGN-like channel. On the other hand, $D = 3$ number of antennas is sufficient for the BU-RA channel to exhibit such a behaviour, since the underlying multi-path diversity provided by the six-path BU-RA channel is higher than that of the tree-path W-ATM channel.

12.5.4   Uncoded Adaptive Multi-Carrier Schemes

The performance of the various adaptive Square QAM schemes has been studied also in the context of multi-carrier systems [2, 420, 421]. The family of Orthogonal Frequency Division Multiplex (OFDM) [226] systems converts frequency selective Rayleigh channels into frequency non-selective or flat Rayleigh channels for each sub-carrier, provided that the number of sub-carriers is sufficiently high. The power and bit allocation strategy of adaptive OFDM

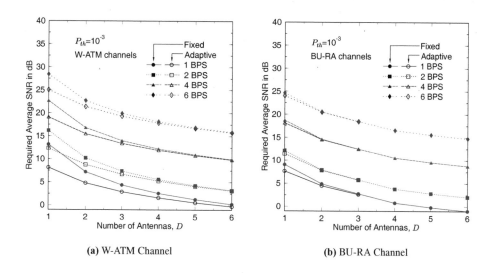

Figure 12.46: The average SNRs required for achieving the target BEP of $P_{th} = 10^{-3}$ by the 2D-Rake assisted adaptive schemes and by the fixed-mode schemes operating over (a) the W-ATM channel and (b) the BU-RA channel

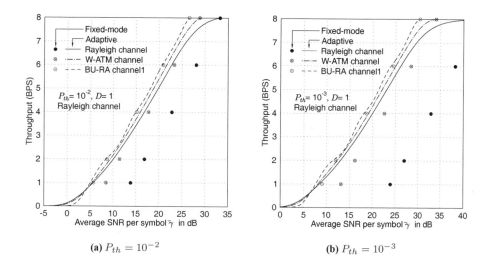

Figure 12.47: The average BPS throughput of the adaptive schemes and fixed-mode schemes communicating over a narrow-band Rayleigh channel, the W-ATM channel and BU-RA channel of Figure 12.42 at the target BEP of (a) $P_{th} = 10^{-2}$ and (b) $P_{th} = 10^{-3}$.

has attracted substantial research interests [2]. OFDM is particularly suitable for combined time-frequency domain processing [421]. Since each sub-carrier of an OFDM system experiences a flat Rayleigh channel, we can apply adaptive modulation for each sub-carrier independently from other sub-carriers. Although a practical scheme would group the sub-carriers into similar-quality sub-bands for the sake of reducing the associated modem mode signalling requirements. The performance of this AQAM assisted OFDM (A-OFDM) scheme is identical to that of the adaptive scheme operating over flat Rayleigh fading channels, characterised in Section 12.5.2.

MC-CDMA [52,55] receiver can be regarded as a frequency domain Rake-receiver, where the multiple carriers play a similar role to that of the time domain Rake fingers. Our simulation results showed that the single-user BER performance of MC-CDMA employing multiple antennae is essentially identical to that of the time domain Rake receiver using antenna diversity, provided that the spreading factor is higher than the number of resolvable multi-path components in the channel. Hence, the throughput of the Rake-receiver over the three-path W-ATM channel [2] and the six-path BU-RA channel [418] studied in Section 12.5.3 can be used for investigating the upper-bound performance of adaptive MC-CDMA schemes over these channels. Figure 12.47 compares the average BPS throughput performances of these schemes, where the throughput curves of the various adaptive schemes are represented as three different types of lines, depending on the underlying channel scenarios, while the fixed-mode schemes are represented as three different types of markers. The solid line corresponds to the performance of A-OFDM and the marker "•" corresponds to that of the fixed-mode OFDM. On the other hand, the dotted lines correspond to the BPS throughput performance of adaptive MC-CDMA operating over wide-band channels and the markers "⊙" and "⊗" to those of the fixed-mode MC-CDMA schemes.

It can be observed that fixed-mode MC-CDMA has the potential to outperform A-OFDM, when the underlying channel provides sufficient diversity due to the high number of resolvable multi-path components. For example, the performance of fixed-mode MC-CDMA operating over the W-ATM channel of Figure 12.42(a) is slightly lower than that of A-OFDM for the BPS range of less than or equal to 6 BPS, owing to the insufficient diversity potential of the wide-band channel. On the other hand, fixed-mode MC-CDMA outperforms A-OFDM, when the channel is characterised by the BU-RA model of Figure 12.42(b). We have to consider several factors, in order to answer, whether fixed-mode MC-CDMA is better than A-OFDM. First, fully loaded MC-CDMA, which can transmit the same number of symbols as OFDM, suffers from multi-code interference and our simulation results showed that the SNR degradation is about 2-4dB at the BER of 10^{-3}, when the Minimum Mean Square Error Block Decision Feedback Equaliser (MMSE-BDFE) [422] based joint detector is used at the receiver. Considering these SNR degradations, the throughput of fixed-mode MC-CDMA using the MMSE-BDFE joint detection receiver falls just below that of the A-OFDM scheme, when the channel is characterised by the BU-RA model. On the other hand, the adaptive schemes may suffer from inaccurate channel estimation/prediction and modem mode signalling feedback delay [404]. Hence, the preference order of the various schemes may depend on the channel scenario encountered, on the interference effects and other practical issues, such as the aforementioned channel estimation accuracy, feedback delays, etc.

12.5.5 Concatenated Space-Time Block Coded and Turbo Coded Symbol-by-Symbol Adaptive OFDM and Multi-Carrier CDMA[4]

In the previous sections we studied the performance of uncoded adaptive schemes. Since a Forward Error Correction (FEC) code reduces the SNR required for achieving a given target BER at the expense of a reduced BPS throughput, it is interesting to investigate the performance of adaptive schemes employing FEC techniques. A variety of FEC techniques have been used in the context of adaptive modulation schemes. In their pioneering work on adaptive modulation, Webb and Steele [397] used a set of binary BCH codes. Vucetic [424] employed various punctured convolutional codes in response to the time-variant channel status. On the other hand, various Trellis Coded Modulation (TCM) [425, 426] schemes were used in the context of adaptive modulation by Alamouti and Kallel [427], Goldsmith and Chua [428], as well as Hole, Holm and Øien [409]. Keller, Liew and Hanzo studied the performance of Redundant Residue Number System (RRNS) codes in the context of adaptive multi-carrier modulation [429, 430]. Various turbo coded adaptive modulation schemes have been investigated also by Liew, Wong, Yee and Hanzo [249, 431, 432]. With the advent of space-time (ST) coding techniques [433–435], various concatenated coding schemes combining ST coding and FEC coding can be applied in adaptive modulation schemes. In this section, we investigate the performance of various concatenated space-time block-coded and turbo-coded adaptive OFDM and MC-CDMA schemes.

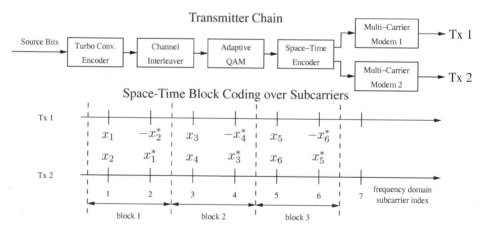

Figure 12.48: Transmitter structure and space-time block encoding scheme

Figure 12.48 portrays the stylised transmitter structure of our system. The source bits are channel coded by a half-rate turbo convolutional encoder [436] using a constraint length of $K = 3$ as well as a random block interleaver having a memory of $L = 3072$ bits. Then, the AQAM block selects a modulation mode from the set of no transmission, BPSK, QPSK, 16-QAM and 64-QAM depending on the instantaneous channel quality perceived by the receiver, according to the SNR-dependent optimum switching levels derived in Section 12.4.4. It is assumed that the perfectly estimated channel quality experienced by receiver A is fed back to transmitter B superimposed on the next burst transmitted to receiver B. The modulation mode

---

[4]This section is based on collaborative research with T.H. Liew [423].

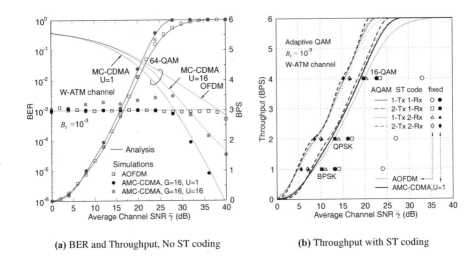

(a) BER and Throughput, No ST coding

(b) Throughput with ST coding

Figure 12.49: Performance of uncoded five-mode AOFDM and AMC-CDMA. The target BER is $B_t = 10^{-3}$ when transmitting over the W-ATM channel [page 474] of [2]. (a) The constant average BER is maintained for AOFDM and single user AMC-CDMA, while "full-user" AMC-CDMA exhibits a slightly higher average BER due to the residual MUI. (b) The SNR gain of the adaptive modems decreases, as ST coding increases the diversity order. The BPS curves appear in pairs, corresponding to AOFDM and AMC-CDMA – indicated by the thin and thick lines, respectively – for each of the four different ST code configurations. The markers represent the SNRs required by the fixed-mode OFDM and MC-CDMA schemes for maintaining the target BER of 10^{-3} in conjunction with the four ST-coded schemes considered.

switching levels of our AQAM scheme determine the average BER as well as the average throughput.

The modulated symbol is now space-time encoded. As seen at the bottom of Figure 12.48, Alamouti's space-time block code [434] is applied across the frequency domain. A pair of the adjacent sub-carriers belonging to the same space-time encoding block is assumed to have the same channel quality. We employed a Wireless Asynchronous Transfer Mode (W-ATM) channel model [page 474] of [2] transmitting at a carrier frequency of 60GHz, at a sampling rate of 225MHz and employing 512 sub-carriers. Specifically, we used a three-path fading channel model, where the average SNR of each path is given by $\bar{\gamma}_1 = 0.79192\bar{\gamma}$, $\bar{\gamma}_2 = 0.12424\bar{\gamma}$ and $\bar{\gamma}_3 = 0.08384\bar{\gamma}$. The Multi-path Intensity Profile (MIP) of the W-ATM channel is illustrated in Figure 12.42(a) in Section 12.5.3. Each channel associated with a different antenna is assumed to exhibit independent fading.

The simulation results related to our uncoded adaptive modems are presented in Figure 12.49. Since we employed the optimum switching levels derived in Section 12.4.4, both our adaptive OFDM (AOFDM) and the adaptive single-user MC-CDMA (AMC-CDMA) modems maintain the constant target BER of 10^{-3} up to the "avalanche" SNR value, and then follow the BER curve of the 64-QAM mode. However, "full-user" AMC-CDMA, which is defined as an AMC-CDMA system supporting $U = 16$ users with the aid of a spreading factor of $G = 16$ and employing the MMSE-BDFE Joint Detection (JD) receiver [261], exhibits a slightly higher average BER, than the target of $B_t = 10^{-3}$ due to the residual Multi-User Interference (MUI) of the imperfect joint detector. Since in Section 12.4.4 we derived the optimum switching levels based on a single-user system, the levels are no longer optimum, when residual MUI is present. The average throughputs of the various schemes expressed in terms of BPS steadily increase and at high SNRs reach the throughput of 64-QAM, namely 6 BPS. The throughput degradation of "full-user" MC-CDMA imposed by the imperfect JD was within a fraction of a dB. Observe in Figure 12.49(a) that the analytical and simulation results are in good agreement, which we denoted by the lines and distinct symbols, respectively.

The effects of ST coding on the average BPS throughput are displayed in Figure 12.49(b). Specifically, the thick lines represent the average BPS throughput of our AMC-CDMA scheme, while the thin lines represent those of our AOFDM modem. The four pairs of hollow and filled markers associated with the four different ST-coded AOFDM and AMC-CDMA scenarios considered represent the BPS throughput versus SNR values associated with fixed-mode OFDM and fixed-mode MMSE-BDFE JD assisted MC-CDMA schemes. Specifically, observe for each of the 1, 2 and 4 BPS fixed-mode schemes that the right-most markers, namely the circles, correspond to the 1-Tx / 1-Rx scenario, the squares to the 2-Tx / 1-Rx scheme, the triangles to the 1-Tx / 2-Rx arrangement and the diamonds to the 2-Tx / 2-Rx scenarios. First of all, we can observe that the BPS throughput curves of OFDM and single-user MC-CDMA are close to each other, namely within 1 dB for most of the SNR range. This is surprising, considering that the fixed-mode MMSE-BDFE JD assisted MC-CDMA scheme was reported to exhibit around 10dB SNR gain at a BER of 10^{-3} and 30dB gain at a BER of 10^{-6} over OFDM [57]. This is confirmed in Figure 12.49(b) by observing that the SNR difference between the ○ and ● markers is around 10dB, regardless of whether the 4, 2 or 1 BPS scenario is concerned.

Let us now compare the SNR gains of the adaptive modems over the fixed modems. The SNR difference between the BPS curve of AOFDM and the fixed-mode OFDM represented by the symbol ○ at the same throughput is around 15dB. The corresponding SNR difference between the adaptive and fixed-mode 4, 2 or 1 BPS MC-CDMA modem is around 5dB. More explicitly, since in the context of the W-ATM channel model [page 474] of [2] fixed-mode MC-CDMA appears to exhibit a 10dB SNR gain over fixed-mode OFDM, the additional 5dB SNR gain of AMC-CDMA over its fixed-mode counterpart results in a total SNR gain of 15dB over fixed-mode OFDM. Hence ultimately the performance of AOFDM and AMC-CDMA becomes similar.

Let us now examine the effect of ST block coding. The SNR gain of the fixed-mode schemes due to the introduction of a 2-Tx / 1-Rx ST block code is represented as the SNR difference between the two right most markers, namely circles and squares. These gains are nearly 10dB for fixed-mode OFDM, while they are only 3dB for fixed-mode MC-CDMA modems. However, the corresponding gains are less than 1dB for both adaptive modems, namely for AOFDM and AMC-CDMA. Since the transmitter power is halved due to using two Tx antennas in the ST codec, a 3dB channel SNR penalty was already applied to the curves in Figure 12.49(b). The introduction of a second receive antenna instead of a second transmit antenna eliminates this 3dB penalty, which results in a better performance for the 1-Tx/2-Rx scheme than for the 2-Tx/1-Rx arrangement. Finally, the 2-Tx / 2-Rx system gives around 3-4dB SNR gain in the context of fixed-mode OFDM and a 2-3dB SNR gain

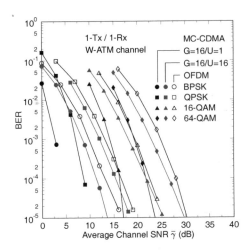

Figure 12.50: Performance of turbo convolutional coded fixed-mode OFDM and MC-CDMA for transmission over the W-ATM channel of [page 474] of [2], indicating that JD MC-CDMA still outperforms OFDM. However, the SNR gain of JD MC-CDMA over OFDM is reduced to 1-2dB at a BER of 10^{-4}.

for fixed-mode MC-CDMA, in both cases over the 1-Tx / 2-Rx system. By contrast, the SNR gain of the 2-Tx / 2-Rx scheme over the 1-Tx / 2-Rx based adaptive modems was, again, less than 1dB in Figure 12.49(b). More importantly, for the 2-Tx / 2-Rx scenario the advantage of employing adaptive modulation erodes, since the fixed-mode MC-CDMA modem performs as well as the AMC-CDMA modem in this scenario. Moreover, the fixed-mode MC-CDMA modem still outperforms the fixed-mode OFDM modem by about 2dB. We conclude that since the diversity-order increases with the introduction of ST block codes, the channel quality variation becomes sufficiently small for the performance advantage of adaptive modems to erode. This is achieved at the price of a higher complexity due to employing two transmitters and two receivers in the ST coded system.

When channel coding is employed in the fixed-mode multi-carrier systems, it is expected that OFDM benefits more substantially from the frequency domain diversity than MC-CDMA, which benefited more than OFDM without channel coding. The simulation results depicted in Figure 12.50 show that the various turbo-coded fixed-mode MC-CDMA systems consistently outperform OFDM. However, the SNR differences between the turbo-coded BER curves of OFDM and MC-CDMA are reduced considerably.

The performance of the concatenated ST block coded and turbo convolutional coded adaptive modems is depicted in Figure 12.51. We applied the optimum set of switching levels designed in Section 12.4.4 for achieving an uncoded BER of 3×10^{-2}. This uncoded target BER was stipulated after observing that it is reduced by half-rate, $K = 3$ turbo convolutional coding to a BER below 10^{-7}, when transmitting over AWGN channels. However, our simulation results yielded zero bit errors, when transmitting 10^9 bits, except for some SNRs, when employing only a single antenna.

Figure 12.51(a) shows the BER of our turbo coded adaptive modems, when a single antenna is used. We observe in the figure that the BER reaches its highest value around the "avalanche" SNR point, where the adaptive modulation scheme consistently activates 64-QAM. The system is most vulnerable around this point. In order to interpret this phenomenon, let us briefly consider the associated interleaving aspects. For practical reasons we have used a fixed interleaver length of $L = 3072$ bits. When the instantaneous channel quality was high, the $L = 3072$ bits were spanning a shorter time-duration during their passage over the fading channel, since the effective BPS throughput was high. Hence the channel errors appeared more bursty, than in the lower-throughput AQAM modes, which conveyed the $L = 3072$ bits over a longer time duration, hence dispersing the error bursts over a longer duration of time. The uniform dispersion of erroneous bits versus time enhances the error correction power of the turbo code. On the other hand, in the SNR region beyond the "avalanche" SNR point seen in Figure 12.51(a) the system exhibited a lower uncoded BER, reducing the coded BER even further. This observation suggests that further research ought to determine the set of switching thresholds directly for a coded adaptive system, rather than by simply estimating the uncoded BER, which is expected to result in near-error-free transmission.

We can also observe that the turbo coded BER of AOFDM is higher than that of AMC-CDMA in the SNR rage of 10-20dB, even though the uncoded BER is the same. This appears to be the effect of the limited exploitation of frequency domain diversity of coded OFDM, compared to MC-CDMA, which leads to a more bursty uncoded error distribution, hence degrading the turbo coded performance. The fact that ST block coding aided multiple antenna systems show virtually error-free performance corroborates our argument.

Figure 12.51(b) compares the throughputs of the coded adaptive modems and the uncoded adaptive modems exhibiting a comparable average BER. The SNR gains due to channel coding were in the range of 0dB to 8dB, depending on the SNR region and on the scenarios employed. Each bundle of throughput curves corresponds to the scenarios of 1-Tx/1-Rx OFDM, 1-Tx/1-Rx MC-CDMA, 2-Tx/1-Rx OFDM, 2-Tx/1-Rx MC-CDMA, 1-Tx/2-Rx OFDM, 1-Tx/2-Rx MC-CDMA, 2-Tx/2-Rx OFDM and 2-Tx/2-Rx MC-CDMA starting from the far right curve, when viewed for throughput values higher than 0.5 BPS. The SNR difference between the throughput curves of the ST and turbo coded AOFDM and those of the corresponding AMC-CDMA schemes was reduced compared to the uncoded performance curves of Figure 12.49(b). The SNR gain owing to ST block coding assisted transmit diversity in the context of AOFDM and AMC-CDMA was within 1dB due to the halved transmitter power. Therefore, again, ST block coding appears to be less effective in conjunction with adaptive modems.

In conclusion, the performance of ST block coded constant-power adaptive multi-carrier modems employing optimum SNR-dependent modem mode switching levels were investigated in this section. The adaptive modems maintained the constant target BER stipulated, whilst maximising the average throughput. As expected, it was found that ST block coding reduces the relative performance advantage of adaptive modulation, since it increases the diversity order and eventually reduces the channel quality variations. When turbo convolutional coding was concatenated to the ST block codes, near-error-free transmission was achieved at the expense of halving the average throughput. Compared to the uncoded system, the turbo coded system was capable of achieving a higher throughput in the low SNR region at the cost of higher complexity. The study of the relationship between the uncoded BER and the corresponding coded BER showed that adaptive modems obtain higher coding gains, than that of fixed modems. This was due to the fact that the adaptive modem avoids burst errors even in deep channel fades by reducing the number of bits per modulated symbol eventually to zero.

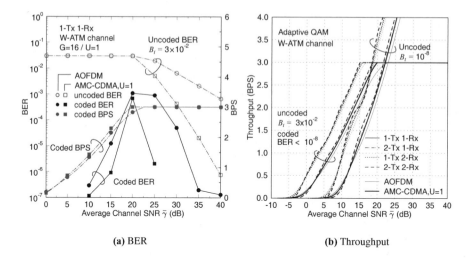

(a) BER   (b) Throughput

Figure 12.51: Performance of the concatenated ST block coded and turbo convolutional coded adaptive OFDM and MC-CDMA systems over the W-ATM channel of [page 474] of [2]. The uncoded target BER is 3×10^{-2}. The coded BER was less than 10^{-8} for most of the SNR range, resulting in virtually error-free transmissions. (a) The coded BER becomes higher near the "avalanche" SNR point, when a single antenna was used. (b) The coded adaptive modems have SNR gains up to 7dB compared to their uncoded counterparts achieving a comparable average BER.

12.6   Chapter Summary and Conclusion

Following a brief introduction to several fading counter-measures, a general model was used to describe several adaptive modulation schemes employing various constituent modulation modes, such as PSK, Star QAM and Square QAM, as one of the attractive fading counter-measures. In Section 12.3.3.1, the closed form expressions were derived for the average BER, the average BPS throughput and the mode selection probability of the adaptive modulation schemes, which were shown to be dependent on the mode-switching levels as well as on the average SNR. After reviewing in Sections 12.4.1, 12.4.2 and 12.4.3 the existing techniques devised for determining the mode-switching levels, in Section 12.4.4 the optimum switching levels achieving the highest possible BPS throughput while maintaining the average target BER were developed based on the Lagrangian optimisation method.

Then, in Section 12.5.1 the performance of uncoded adaptive PSK, Star QAM and Square QAM was characterised, when the underlying channel was a Nakagami fading channel. It was found that an adaptive scheme employing a k-BPS fixed-mode as the highest throughput constituent modulation mode was sufficient for attaining all the benefits of adaptive modulation, while achieving an average throughput of up to $k - 1$ BPS. For example, a three-mode adaptive PSK scheme employing No-Tx, 1-BPS BPSK and 2-BPS QPSK modes attained the maximum possible average BPS throughput of 1 BPS and hence adding higher-throughput modes, such as 3-BPS 8-PSK to the three-mode adaptive PSK scheme resulting in a four-mode adaptive PSK scheme did not achieve a better performance across the 1 BPS throughput range. Instead, this four-mode adaptive PSK scheme extended the maximum achievable BPS throughput by any adaptive PSK scheme to 2 BPS, while asymptotically achieving a throughput of 3 BPS as the average SNR increases.

On the other hand, the relative SNR advantage of adaptive schemes in comparison to fixed-mode schemes increased as the target average BER became lower and decreased as the fading became less severe. More explicitly, less severe fading corresponds to an increased Nakagami fading parameter m, to an increased number of diversity antennas, or to an increased number of multi-path components encountered in wide-band fading channels. As the fading becomes less severe, the average BPS throughput curves of our adaptive Square QAM schemes exhibit undulations owing to the absence of 3-BPS, 5-BPS and 7-BPS square QAM modes.

The comparisons between fixed-mode MC-CDMA and adaptive OFDM (AOFDM) were made based on different channel models. In Section 12.5.4 it was found that fixed-mode MC-CDMA might outperform adaptive OFDM, when the underlying channel provides sufficient diversity. However, a definite conclusion could not be drawn since in practice MC-CDMA might suffer from MUI and AOFDM might suffer from imperfect channel quality estimation and feedback delays.

Concatenated space-time block coded and turbo convolutional-coded adaptive multi-carrier systems were investigated in Section 12.5.5. The coded schemes reduced the required average SNR by about 6dB-7dB at a throughput of 1 BPS achieving near error-free transmission. It was also observed in Section 12.5.5 that increasing the number of transmit antennas in adaptive schemes was not very effective, achieving less than 1dB SNR gain owing to the fact that the transmit power per antenna had to be reduced in order to maintain the same constant total transmit power for the sake of a fair comparison.

12.7 Appendix: Mode Specific Average BEP of an Adaptive Modulation Scheme

A closed form solution for the "mode-specific average BER" of a Maximal Ratio Combining (MRC) receiver using Dth order antenna diversity over independent Rayleigh channels is derived, where the "mode-specific average BER" refers to the BER of the adaptive modulation scheme, while activating one of its specific constituent modem modes. The PDF $f_{\bar{\gamma}}(\gamma)$ of the channel SNR γ is given as [(14-4-13)] [274]

$$f_{\bar{\gamma}}(\gamma) = \frac{1}{(D-1)!\,\bar{\gamma}^D} \gamma^{D-1} e^{-\gamma/\bar{\gamma}} , \quad \gamma \geq 0 , \tag{12.92}$$

where $\bar{\gamma}$ is the average channel SNR. Since the PDF of the instantaneous channel SNR γ over a Nakagami fading channel is given as:

$$f_{\bar{\gamma}}(\gamma) = \left(\frac{m}{\bar{\gamma}}\right)^m \frac{\gamma^{m-1}}{\Gamma(m)} e^{-m\gamma/\bar{\gamma}} , \quad \gamma \geq 0 , \tag{12.93}$$

the following results can also be applied to a Nakagami fading channel with a simple change of variable given as $D = m$ and $\bar{\gamma} = \bar{\gamma}\,/\,m$.

The mode-specific average BEP is defined as:

$$P_r(\alpha, \beta; \bar{\gamma}, D, a) \triangleq \int_\alpha^\beta Q(\sqrt{a\gamma})\, f_{\bar{\gamma}}(\gamma)\, d\gamma \tag{12.94}$$

$$= \int_\alpha^\beta Q(\sqrt{a\gamma})\, \frac{1}{(D-1)!\,\bar{\gamma}^D}\, \gamma^{D-1}\, e^{-\gamma/\bar{\gamma}}\, d\gamma, \tag{12.95}$$

where $Q(x) \triangleq \frac{1}{\sqrt{2\pi}} \int_x^\infty e^{-t^2/2}\, dt$. Applying integration-by-part or $\int u\, dv = uv - \int v\, du$ and noting that:

$$u = Q(\sqrt{a\gamma})$$

$$du = -\frac{\sqrt{a}}{2\sqrt{2\pi\gamma}}\, e^{-a\gamma/2}$$

$$dv = \frac{1}{(D-1)!\,\bar{\gamma}^D}\, \gamma^{D-1}\, e^{-\gamma/\bar{\gamma}}$$

$$v = -e^{-\gamma/\bar{\gamma}} \sum_{d=0}^{D-1} (\gamma/\bar{\gamma})^d\, \frac{1}{d!}$$

Equation 12.94 becomes:

$$P_r(\alpha, \beta; \bar{\gamma}, D, a) = \left[e^{-\gamma/\bar{\gamma}}\, Q(\sqrt{a\gamma}) \sum_{d=0}^{D-1} (\gamma/\bar{\gamma})^d \frac{1}{d!} \right]_\beta^\alpha - \sum_{d=0}^{D-1} I_d(\alpha, \beta), \tag{12.96}$$

where

$$I_d(\alpha, \beta) = \int_\alpha^\beta \frac{\sqrt{a}}{2\sqrt{2\pi}} \frac{1}{d!} (\gamma/\bar{\gamma})^d \frac{1}{\sqrt{\gamma}}\, e^{-a\gamma/(2\mu^2)}\, d\gamma \tag{12.97}$$

and $\mu = \sqrt{\frac{a\bar{\gamma}}{a\bar{\gamma}+2}}$. Let us consider I_d for the case of $d = 0$:

$$I_0(\alpha, \beta) = \int_\alpha^\beta \frac{\sqrt{a}}{2\sqrt{2\pi}} \frac{1}{\sqrt{\gamma}}\, e^{-a\gamma/(2\mu^2)}\, d\gamma. \tag{12.98}$$

Upon introducing the variable $t^2 = a\gamma/\mu^2$ and exploiting that $d\gamma = 2\mu\sqrt{\gamma/a}\, ds$, we have:

$$I_0(\alpha, \beta) = [\mu\, Q(\sqrt{a\gamma}/\mu)]_\beta^\alpha. \tag{12.99}$$

Applying integration-by-part once to Equation 12.97 yields:

$$I_d(\alpha, \beta) = \left[\frac{\mu^2}{\sqrt{2a\pi}} \frac{1}{d!} (\gamma/\bar{\gamma})^d \frac{1}{\sqrt{\gamma}}\, e^{-a\gamma/(2\mu^2)} \right]_\beta^\alpha + \frac{2d-1}{a\bar{\gamma}d}\, \mu^2\, I_{d-1}, \tag{12.100}$$

which is a recursive form with the initial value given in Equation 12.99. For this recursive form of Equation 12.100, a non-recursive form of I_d can be expressed as:

$$
I_d(\alpha, \beta) = \left[\frac{\mu^2}{\sqrt{2a\pi}} \frac{\Gamma(d + \frac{1}{2})}{\bar{\gamma}^d \, \Gamma(d+1)} \sum_{i=1}^{d} \left(\frac{2\mu^2}{a} \right)^{d-i} \frac{\gamma^{i-\frac{1}{2}}}{\Gamma(i + \frac{1}{2})} e^{-a\gamma/(2\mu^2)} \right]_{\beta}^{\alpha}
$$
$$
+ \left[\left(\frac{2\mu^2}{a\bar{\gamma}} \right)^d \frac{1}{\sqrt{\pi}} \frac{\Gamma(d + \frac{1}{2})}{\Gamma(d+1)} \, \mu \, Q(\sqrt{a\gamma}/\mu) \right]_{\beta}^{\alpha} . \tag{12.101}
$$

By substituting $I_d(\alpha, \beta)$ of Equation 12.101 into Equation 12.96, the regional BER $P_r(\bar{\gamma}; a, \alpha, \beta)$ can be represented in a closed form.

12.8 Appendix: BER Analysis of Type-I Star-QAM

The Star Quadrature Amplitude Modulation (SQAM) technique [2], also known as Amplitude-modulated Phase Shift Keying (APSK), employs circular constellations, rather than rectangular constellation as in Square QAM [437]. Although Square QAM has the maximum possible minimum Euclidean distance amongst its phasors given a constant average symbol power, in some situations Star QAM may be preferred due to its relatively simple detector and for its low Peak-to-Average Power Ratio (PAPR) compared to Square QAM [437]. Since differentially detected non-coherent Star QAM signals are robust against fading effects, many researchers analysed its Bit Error Ratio (BER) performance for transmission over AWGN channels [438], Rayleigh fading channels [438, 439] as well as Rician fading channels [440]. The effects of diversity reception on its BER were also studied when communicating over Rayleigh fading channels [441, 442]. The BER of coherent 16 Star QAM was also analysed for transmission over AWGN channels [438] as well as when communicating over *Nakagami-m* fading channels [443]. However, the BER of Star-QAM schemes other than 16-level Star QAM, such as 8, 32 and 64-level Star QAM, has not been studied.

12.8.1 Coherent Detection

The BER of coherent star QAM schemes employing Type-I constellations [437] when communicating over AWGN channels can be analysed using the signal-space method [413, 438, 443, 444]. The phasor constellations of the various Type-I Star QAM schemes are illustrated in Figures 12.52 and 12.57. Let us first consider 8-level Star QAM, which is also referred to as 2-level QPSK [437]. In Figure 12.52(a), a is the radius of the inner ring, while $a\beta$ is the radius of the outer ring. The ring ratio is given by $a\beta/a = \beta$. The three bits, namely b_1, b_2 and b_3, are assigned as shown in Figure 12.52(a), representing Gray coding for each ring using the bits $b_1 b_2$. The third bit, namely b_3 indicates, which ring of the constellation is encountered. The average symbol power is given as:

$$
E_s = \frac{4a^2 + 4a^2\beta^2}{8} = \frac{1}{2}a^2(1 + \beta^2) . \tag{12.102}
$$

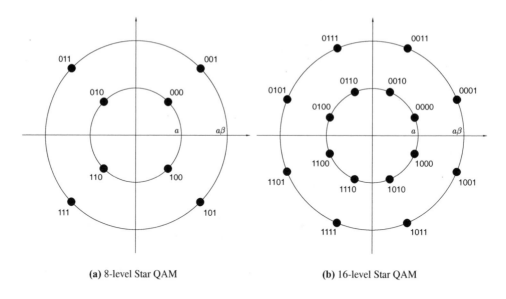

(a) 8-level Star QAM   **(b)** 16-level Star QAM

Figure 12.52: Type-I constellations of Star QAM using two constellation rings.

In order to normalise the constellations so that the average symbol power becomes unity, a should be given as:

$$a = \sqrt{\frac{2}{1 + \beta^2}} \,. \tag{12.103}$$

In terms of the signal space, the modulation scheme with respect to b_3 is an Amplitude Shift Keying (ASK) scheme. The decision rule related to bit b_3 is specified in Figure 12.53(a). The BER of bit b_3 can be expressed as:

$$P_{b_3} = \frac{1}{2} \left[Q\left(\frac{a(\beta - 1)}{2} \sqrt{2\gamma} \right) + Q\left(\frac{a(\beta + 3)}{2} \sqrt{2\gamma} \right) \right]$$

$$+ \frac{1}{2} \left[Q\left(\frac{a(\beta - 1)}{2} \sqrt{2\gamma} \right) - Q\left(a(\beta + 1)\sqrt{2\gamma} \right) \right] \tag{12.104}$$

$$\simeq Q\left(\frac{a(\beta - 1)}{2} \sqrt{2\gamma} \right) \tag{12.105}$$

$$= Q\left(\sqrt{\frac{(\beta - 1)^2}{1 + \beta^2} \gamma} \right) \,, \tag{12.106}$$

where the Gaussian Q-function is defined as $Q(x) = \frac{1}{\sqrt{2\pi}} \int_x^\infty e^{-y^2/2} dy$ and γ is the SNR per symbol. Since bits b_1 and b_2 corresponds to Gray coded QPSK signals, their BER can be

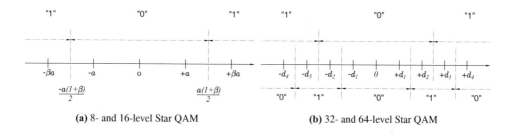

(a) 8- and 16-level Star QAM　　　　　**(b)** 32- and 64-level Star QAM

Figure 12.53: Magnitude-bit decision regions for various Type-I Star QAM constellations.

expressed as:

$$P_{b_1} = P_{b_2} \simeq \frac{1}{2}Q\left(\frac{a}{\sqrt{2}}\sqrt{2\gamma}\right) + \frac{1}{2}Q\left(\frac{a\beta}{\sqrt{2}}\sqrt{2\gamma}\right) \tag{12.107}$$

$$= \frac{1}{2}Q\left(\sqrt{\frac{2}{1+\beta^2}\gamma}\right) + \frac{1}{2}Q\left(\sqrt{\frac{2\beta^2}{1+\beta^2}\gamma}\right). \tag{12.108}$$

Hence, the average BER of an 8-level Star QAM scheme communicating over an AWGN channel can be expressed as:

$$P_8 = \frac{1}{3}P_{b_1} + \frac{1}{3}P_{b_2} + \frac{1}{3}P_{b_3} \tag{12.109}$$

$$\simeq \frac{1}{3}\left[Q\left(\sqrt{\frac{2}{1+\beta^2}\gamma}\right) + Q\left(\sqrt{\frac{2\beta^2}{1+\beta^2}\gamma}\right) + Q\left(\sqrt{\frac{(\beta-1)^2}{1+\beta^2}\gamma}\right)\right]. \tag{12.110}$$

The BER of Equation 12.110 is plotted in Figure 12.54(a) as a function of the ring ratio β for various values of the SNR per symbol γ. We can observe that the BER of 8-level Star QAM reaches its minimum, when the ring ratio is $\beta \simeq 2.4$. This is not surprising, considering that the ring ratio should be $\beta = 1 + \sqrt{2}$ in order to make the Euclidean distances between an inner ring constellation point and its three adjacent constellation points the same. However, the optimum ring ratio β_{opt}, where the BER reaches its minimum is SNR dependent. The optimum ring ratio versus the SNR per symbol is plotted in Figure 12.54(b). It can be observed that when the SNR is lower than 8dB, the optimum ring ratio increases sharply. Since the corresponding BER improvement was, however, less than 0.1dB even at SNRs near 0dB, the fixed ring ratio of $\beta = 1 + \sqrt{2}$ can be used for all SNR values. Figure 12.55(a) compares the BER of 8-level Star QAM and 8-PSK. Observe that 8-level Star QAM exhibits an approximately 1dB SNR performance gain, when the SNR is below 2dB, but above this SNR the SNR gain becomes marginal.

Let us now consider 16-level Star QAM. The corresponding phasor constellation is given in Figure 12.52(b). Since the average symbol power is the same as that of 8-level Star QAM, Equation 12.103 can be used for determining a. The BER analysis for the fourth bit, namely for b_4 is exactly the same as that of 8-level Star QAM and the corresponding value of P_{b_4} is given in Equation 12.106. Since the first three bits, namely b_1, b_2 and b_3, are 8-PSK

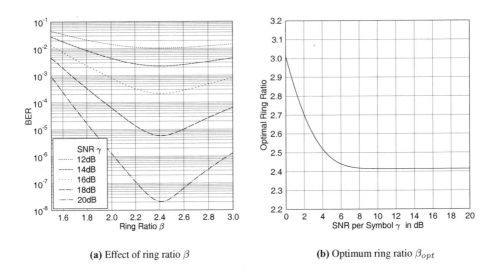

(a) Effect of ring ratio β (b) Optimum ring ratio β_{opt}

Figure 12.54: BER of Gray-mapped 8-level Star QAM for transmission over AWGN channels.

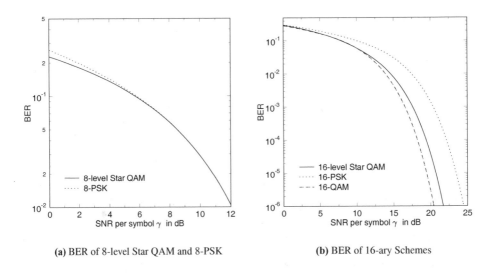

(a) BER of 8-level Star QAM and 8-PSK   (b) BER of 16-ary Schemes

Figure 12.55: BER comparison of 8-ary and 16-ary modulation schemes for transmission over AWGN channels.

modulated, their BER can be expressed as:

$$P_{b_1} = P_{b_2} = P_{b_3} = \frac{1}{2}P_{8PSK}(a^2\gamma) + \frac{1}{2}P_{8PSK}(a^2\beta^2\gamma). \tag{12.111}$$

Lu, Letaief, Chuang and Liou found an accurate approximation of the BER of Gray-coded MPSK, which is given by [413]:

$$P_{MPSK} \simeq \frac{2}{\log_2 M} \sum_{i=1}^{2} Q\left(\sqrt{2\sin^2\left(\frac{2i-1}{M}\pi\right)\gamma}\right), \tag{12.112}$$

where γ is the SNR per symbol. Hence, the BER of Equation 12.111 can be expressed as:

$$P_{b_1} = P_{b_2} = P_{b_3} \tag{12.113}$$

$$\simeq \frac{1}{3}\left[Q\left(\sqrt{\frac{4\sin^2(\pi/8)}{1+\beta^2}\gamma}\right) + Q\left(\sqrt{\frac{4\sin^2(3\pi/8)}{1+\beta^2}\gamma,}\right)\right]$$

$$+ \frac{1}{3}\left[Q\left(\sqrt{\frac{4\beta^2\sin^2(\pi/8)}{1+\beta^2}\gamma}\right) + Q\left(\sqrt{\frac{4\beta^2\sin^2(3\pi/8)}{1+\beta^2}\gamma}\right)\right]. \tag{12.114}$$

Now, the average BER of a 16-level Star QAM scheme for transmission over AWGN channel can be expressed as:

$$P_{16} = \frac{1}{4}P_{b_1} + \frac{1}{4}P_{b_2} + \frac{1}{4}P_{b_3} + \frac{1}{4}P_{b_4} \tag{12.115}$$

$$\simeq \frac{1}{4}\left[Q\left(\sqrt{\frac{4\sin^2(\pi/8)}{1+\beta^2}\gamma}\right) + Q\left(\sqrt{\frac{4\sin^2(3\pi/8)}{1+\beta^2}\gamma,}\right) + Q\left(\sqrt{\frac{4\beta^2\sin^2(\pi/8)}{1+\beta^2}\gamma}\right)\right.$$

$$\left. + Q\left(\sqrt{\frac{4\beta^2\sin^2(3\pi/8)}{1+\beta^2}\gamma}\right) + Q\left(\sqrt{\frac{(\beta-1)^2}{1+\beta^2}\gamma}\right)\right]. \tag{12.116}$$

The BER of Equation 12.116 is plotted in Figure 12.56(a) as a function of the ring ratio β for the various values of the SNR per symbol γ. We can observe that the ring ratio of $\beta \simeq 1.8$ minimises the BER of 16-level Star QAM, when communicating over AWGN channels [438]. This is also expected, since the ring ratio should be $\beta = 1 + 2\cos(3\pi/8) = 1.7654$ in order to render the Euclidean distances between an inner-ring constellation point and its three adjacent constellation points the same. The actual optimum ring ratio β_{opt}, where the BER reaches its minimum is plotted in Figure 12.56(b). As for 8-level Star QAM, even though the optimum ratio is SNR dependent, the difference between the BER corresponding to the optimum ring ratio and that corresponding to the constant ring ratio of $\beta = 1 + 2\cos(3\pi/8)$ is negligible. Figure 12.55(b) compares the BER of 16-level Star QAM, 16-PSK and 16-level Square QAM. We found that the BER performance of 16-level Star QAM is inferior to that of 16-level Square QAM. Viewing the corresponding performance from the perspective of the required SNR per symbol, the former requires an approxmately 1.3dB high SNR to maintain the BER

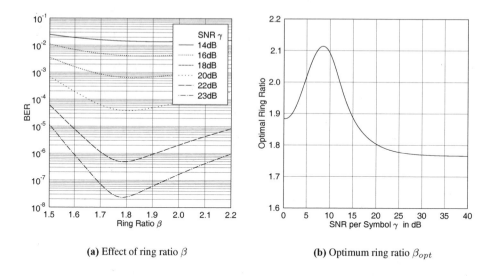

(a) Effect of ring ratio β (b) Optimum ring ratio β_{opt}

Figure 12.56: BER of Gray mapped 16-level Star QAM for transmission over AWGN channels.

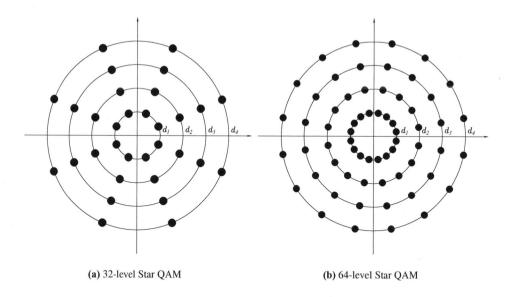

(a) 32-level Star QAM (b) 64-level Star QAM

Figure 12.57: Type-I constellations of Star QAM schemes having four rings.

of 10^{-6}. By contrast, it requires a 2.7dB lower symbol-SNR, than 16-PSK.

Having considered the family of twin-ring constellations, let us focus our attention on two four-ring constellations. The Type-I constellations of 32-level and 64-level Star QAM scheme are depicted in Figure 12.57. The last two bits of a symbol are Gray coded in the "radial direction" and they are four-level ASK modulated. Let us assume that the Gray coding scheme for the bits "b_4b_5" of 32-level Star QAM in the "radial direction" is given as "00", "01", "11" and "10", when viewing it from the inner most ring to the outer rings. The decision regions of these bits were illustrated in Figure 12.53(b). The remaining bits are also Gray coded along each of the four rings and PSK modulated. Let us denote the radius of each ring as d_1, $d_2 = d_1 \beta_1$, $d_3 = d_1 \beta_2$ and $d_4 = d_1 \beta_3$, where β_1, β_2 and β_3 are the corresponding ring ratios of each ring. Since the average power per symbol E_s is given as:

$$E_s = \frac{d_1^2 + d_2^2 + d_3^2 + d_4^2}{4} = \frac{d_1^2}{4}(1 + \beta_1^2 + \beta_2^2 + \beta_3^2),$$
(12.117)

the value of d_1 required for normalising the average power to unity can be expressed as:

$$d_1 = \sqrt{\frac{4}{1 + \beta_1^2 + \beta_2^2 + \beta_3^2}}.$$
(12.118)

Inspecting Figure 12.53(b), the BER of the fourth bit of 32-level Star QAM can be formulated as:

$$
\begin{aligned}
P_{b_4} = &\frac{1}{4}\left[Q\left(\frac{d_2 + d_3 - 2d_1}{2}\sqrt{2\gamma}\right) + Q\left(\frac{d_2 + d_3 + 2d_1}{2}\sqrt{2\gamma}\right)\right] \\
&+ \frac{1}{4}\left[Q\left(\frac{d_3 - d_2}{2}\sqrt{2\gamma}\right) + Q\left(\frac{d_3 + 3d_2}{2}\sqrt{2\gamma}\right)\right] \\
&+ \frac{1}{4}\left[Q\left(\frac{d_3 - d_2}{2}\sqrt{2\gamma}\right) - Q\left(\frac{3d_3 + d_2}{2}\sqrt{2\gamma}\right)\right] \\
&+ \frac{1}{4}\left[Q\left(\frac{2d_4 - d_2 - d_3}{2}\sqrt{2\gamma}\right) - Q\left(\frac{2d_4 + d_2 + d_3}{2}\sqrt{2\gamma}\right)\right] \quad (12.119) \\
\simeq &\frac{1}{4}\left[Q\left(\sqrt{\frac{2(\beta_1 + \beta_2 - 2)^2}{1 + \beta_1^2 + \beta_2^2 + \beta_3^2}\gamma}\right) + 2Q\left(\sqrt{\frac{2(\beta_2 - \beta_1)^2}{1 + \beta_1^2 + \beta_2^2 + \beta_3^2}\gamma}\right)\right. \\
&\left. + Q\left(\sqrt{\frac{2(2\beta_3 - \beta_1 - \beta_2)^2}{1 + \beta_1^2 + \beta_2^2 + \beta_3^2}\gamma}\right)\right].
\end{aligned}
$$
(12.120)

The decision regions depicted in the lower part of Figure 12.53(b) are valid for the fifth bit, namely b_5. The BER of b_5 can be expressed as:

$$
\begin{aligned}
P_{b_5} = {} & \frac{1}{4}\left[Q\left(\frac{d_2 - d_1}{2}\sqrt{2\gamma}\right) + Q\left(\frac{d_2 + 3\,d_1}{2}\sqrt{2\gamma}\right)\right. \\
& \left. -Q\left(\frac{d_3 + d_4 - 2\,d_1}{2}\sqrt{2\gamma}\right) - Q\left(\frac{d_3 + d_4 + 2\,d_1}{2}\sqrt{2\gamma}\right)\right] \\
& +\frac{1}{4}\left[Q\left(\frac{d_2 - d_1}{2}\sqrt{2\gamma}\right) + Q\left(\frac{d_3 + d_4 - 2\,d_2}{2}\sqrt{2\gamma}\right)\right. \\
& \left. -Q\left(\frac{d_1 + 3\,d_2}{2}\sqrt{2\gamma}\right) + Q\left(\frac{d_3 + d_4 + 2\,d_2}{2}\sqrt{2\gamma}\right)\right] \\
& +\frac{1}{4}\left[Q\left(\frac{2\,d_3 - d_1 - d_2}{2}\sqrt{2\gamma}\right) + Q\left(\frac{d_4 - d_3}{2}\sqrt{2\gamma}\right)\right. \\
& \left. -Q\left(\frac{d_1 + d_2 + 2\,d_3}{2}\sqrt{2\gamma}\right) + Q\left(\frac{3\,d_3 + d_4}{2}\sqrt{2\gamma}\right)\right] \\
& +\frac{1}{4}\left[Q\left(\frac{d_4 - d_3}{2}\sqrt{2\gamma}\right) - Q\left(\frac{2\,d_4 - d_1 - d_2}{2}\sqrt{2\gamma}\right)\right. \\
& \left. +Q\left(\frac{d_1 + d_2 + 2\,d_4}{2}\sqrt{2\gamma}\right) - Q\left(\frac{d_3 + 3\,d_4}{2}\sqrt{2\gamma}\right)\right]. \quad (12.121)
\end{aligned}
$$

The expression of the BER P_{b_5} can be accurately approximated as:

$$
\begin{aligned}
P_{b_5} \simeq {} & \frac{1}{4}\left[2\,Q\left(\sqrt{\frac{2(\beta_1 - 1)^2}{1 + \beta_1^2 + \beta_2^2 + \beta_3^2}}\,\gamma\right) + Q\left(\sqrt{\frac{2(\beta_2 + \beta_3 - 2\,\beta_1)^2}{1 + \beta_1^2 + \beta_2^2 + \beta_3^2}}\,\gamma\right)\right. \\
& \left. +Q\left(\sqrt{\frac{2(2\,\beta_2 - 1 - \beta_1)^2}{1 + \beta_1^2 + \beta_2^2 + \beta_3^2}}\,\gamma\right) + 2\,Q\left(\sqrt{\frac{2(\beta_3 - \beta_2)^2}{1 + \beta_1^2 + \beta_2^2 + \beta_3^2}}\,\gamma\right)\right]. \quad (12.122)
\end{aligned}
$$

Let us now find the BER of the PSK modulated bits b_1, b_2 and b_3. Since they are 8-PSK modulated, the BER can be expressed using the results of [413] as:

$$P_{b_1} = P_{b_2} = P_{b_3} = \frac{1}{4}P_{8PSK}(d_1^2\gamma) + \frac{1}{4}P_{8PSK}(d_2^2\gamma) + \frac{1}{4}P_{8PSK}(d_3^2\gamma) + \frac{1}{4}P_{8PSK}(d_4^2\gamma)$$

(12.123)

$$\simeq \frac{1}{6}\left[Q\left(\sqrt{\frac{8\sin^2(\pi/8)}{1+\beta_1^2+\beta_2^2+\beta_3^2}\gamma}\right) + Q\left(\sqrt{\frac{8\sin^2(3\pi/8)}{1+\beta_1^2+\beta_2^2+\beta_3^2}\gamma}\right)\right]$$

$$+ \frac{1}{6}\left[Q\left(\sqrt{\frac{8\beta_1^2\sin^2(\pi/8)}{1+\beta_1^2+\beta_2^2+\beta_3^2}\gamma}\right) + Q\left(\sqrt{\frac{8\beta_1^2\sin^2(3\pi/8)}{1+\beta_1^2+\beta_2^2+\beta_3^2}\gamma}\right)\right]$$

$$+ \frac{1}{6}\left[Q\left(\sqrt{\frac{8\beta_2^2\sin^2(\pi/8)}{1+\beta_1^2+\beta_2^2+\beta_3^2}\gamma}\right) + Q\left(\sqrt{\frac{8\beta_2^2\sin^2(3\pi/8)}{1+\beta_1^2+\beta_2^2+\beta_3^2}\gamma}\right)\right]$$

$$+ \frac{1}{6}\left[Q\left(\sqrt{\frac{8\beta_3^2\sin^2(\pi/8)}{1+\beta_1^2+\beta_2^2+\beta_3^2}\gamma}\right) + Q\left(\sqrt{\frac{8\beta_3^2\sin^2(3\pi/8)}{1+\beta_1^2+\beta_2^2+\beta_3^2}\gamma}\right)\right]. \quad (12.124)$$

Hence, the BER of 32-level Star QAM can be expressed as:

$$P_{32} = \frac{1}{5}(3P_{b_1} + P_{b_4} + P_{b_5}), \quad (12.125)$$

where P_{b_1}, P_{b_4} and P_{b_5} are given in Equations 12.124, 12.120 and 12.122, respectively. The optimum ring ratios of 32-level Star QAM are depicted in Figure 12.58(a). The optimum ring ratios converge to $\beta_1 = 1.77$, $\beta_2 = 2.541$ and $\beta_3 = 3.318$. Note that the first optimum ring ratio is the same as the optimum ring ratio of 16-level Star QAM and the corresponding distances between the second, the third and the fourth rings are approximately equal, as one would expect in an effort to maintain an identical distance amongst the constellations points. Figure 12.58(b) compares the BER of 32-level Star QAM and 32-PSK. We found that the SNR gain of 32-level Star QAM over 32-PSK is 4.6dB at a BER of 10^{-6}.

The BER of 64-level Star QAM can be obtained using the same procedure employed for determing the BER of 32-level Star QAM, considering that now the bits b_1, b_2, b_3 and b_4 are 16-PSK modulated on each ring. The BER of the last two bits, P_{b_5} and P_{b_6} are the same as those given in Equations 12.120 and 12.122, respectively. On the other hand, the BER of the

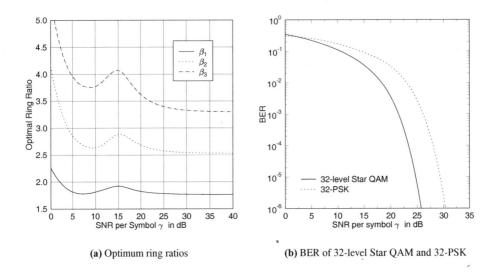

(a) Optimum ring ratios   (b) BER of 32-level Star QAM and 32-PSK

Figure 12.58: BER of 32 Star QAM over AWGN channel.

16-PSK modulated bits of 64-level Star QAM can be expressed as:

$$P_{b_1} = P_{b_2} = P_{b_3} = P_{b_4}$$
$$= \frac{1}{4} P_{16PSK}(d_1^2 \gamma) + \frac{1}{4} P_{16PSK}(d_2^2 \gamma) + \frac{1}{4} P_{16PSK}(d_3^2 \gamma) + \frac{1}{4} P_{16PSK}(d_4^2 \gamma) \quad (12.126)$$
$$\simeq \frac{1}{8} \left[Q\left(\sqrt{\frac{8\sin^2(\pi/16)}{1 + \beta_1^2 + \beta_2^2 + \beta_3^2}\gamma} \right) + Q\left(\sqrt{\frac{8\sin^2(3\pi/16)}{1 + \beta_1^2 + \beta_2^2 + \beta_3^2}\gamma} \right) \right]$$
$$+ \frac{1}{8} \left[Q\left(\sqrt{\frac{8\beta_1^2\sin^2(\pi/16)}{1 + \beta_1^2 + \beta_2^2 + \beta_3^2}\gamma} \right) + Q\left(\sqrt{\frac{8\beta_1^2\sin^2(3\pi/16)}{1 + \beta_1^2 + \beta_2^2 + \beta_3^2}\gamma} \right) \right]$$
$$+ \frac{1}{8} \left[Q\left(\sqrt{\frac{8\beta_2^2\sin^2(\pi/16)}{1 + \beta_1^2 + \beta_2^2 + \beta_3^2}\gamma} \right) + Q\left(\sqrt{\frac{8\beta_2^2\sin^2(3\pi/16)}{1 + \beta_1^2 + \beta_2^2 + \beta_3^2}\gamma} \right) \right]$$
$$+ \frac{1}{8} \left[Q\left(\sqrt{\frac{8\beta_3^2\sin^2(\pi/16)}{1 + \beta_1^2 + \beta_2^2 + \beta_3^2}\gamma} \right) + Q\left(\sqrt{\frac{8\beta_3^2\sin^2(3\pi/16)}{1 + \beta_1^2 + \beta_2^2 + \beta_3^2}\gamma} \right) \right] . \quad (12.127)$$

Hence, the average BER of 64-level Star QAM can be expressed as:

$$P_{64} = \frac{1}{6}(4P_{b_1} + P_{b_5} + P_{b_6}), \quad (12.128)$$

where P_{b_1} is given in Equation 12.127, where P_{b_5} and P_{b_6} are given in Equations 12.120 and 12.122, respectively. The optimum ring ratios of 64-level Star QAM are depicted in

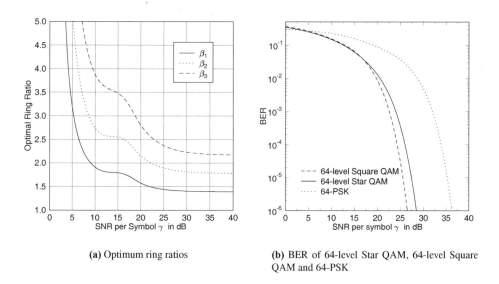

(a) Optimum ring ratios

(b) BER of 64-level Star QAM, 64-level Square QAM and 64-PSK

Figure 12.59: BER of 64-level Star QAM for transmission over AWGN channels.

Figure 12.59(a). The optimum ring ratios converge to $\beta_1 = 1.4$, $\beta_2 = 1.81$ and $\beta_3 = 2.23$. It was observed that the SNR difference between the optimised BER and that employing the asymptotic ring ratio is at most 1dB in the SNR range of 5dB to 15dB. Figure 12.59(b) compares the BER of 64-level Star QAM, 64-level Square QAM and 64-PSK. We found that the SNR gain of 64-level Star QAM over 64-PSK is 7.7dB at the BER of 10^{-6}. The 64-level Square QAM arrangement is the most power-efficient scheme, which exhibits 2dB SNR gain over 64-level Star QAM at the BER of 10^{-6}.

12.9 Appendix: Two-Dimensional Rake Receiver

12.9.1 System Model

The schematic of our Rake-receiver and D-antenna diversity assisted adaptive Square QAM (AQAM) system is illustrated in Figure 12.60. A band-limited equivalent low-pass m-ary QAM signal $s(t)$, having a spectrum of $S(f) = 0$ for $|f| > 1/2W$, is transmitted over time variant frequency selective fading channels and received by a set of D RAKE-receivers. Each Rake-receiver [266, 274] combines all the resolvable multi-path components using Maximal Ratio Combining (MRC). The combined signals of the D-antenna assisted Rake-receivers are summed and demodulated using the estimated channel quality information. The estimated signal-to-noise ratio is fed back to the transmitter and it is used to decide upon the most appropriate m-ary square QAM modulation mode to be used during the next transmission burst. We assume that the channel quality is estimated perfectly and it is available at the transmitter immediately. The effects of channel estimation error and feedback delay on the performance of AQAM were studied, for example, by Goldsmith and Chua [404].

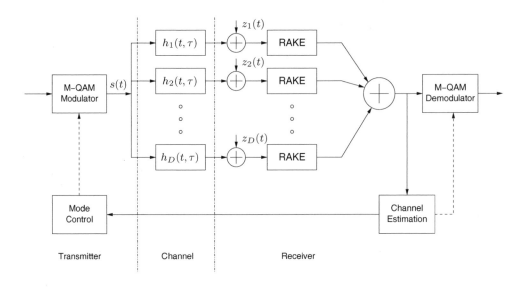

Figure 12.60: Equivalent low-pass model of a D-th order antenna diversity based RAKE-receiver assisted AQAM system.

The low-pass equivalent impulse response of the channel between the transmitter and the d-th antenna may be represented as [274]:

$$h_d(t, \tau) = \sum_{n=1}^{N} h_{d,n}(t)\, \delta\left(\tau - \frac{n}{W}\right) , \qquad (12.129)$$

where $\{h_{d,n}(t)\}$ is a set of independent complex valued stationary random Gaussian processes. The maximum number of resolvable multi-path components N is given by $\lfloor T_m W \rfloor + 1$, where T_m is the multi-path delay spread of the channel [274]. Hence, the low-pass equivalent received signal $r_d(t)$ at the d-th antenna can be formulated as:

$$r_d(t) = \sum_{n=1}^{N} h_{d,n}(t)s\left(t - \frac{n}{W}\right) + z_d(t) , \qquad (12.130)$$

where $z_d(t)$ is a zero mean Gaussian random process having a two-sided power spectral density of $N_o/2$. Let us assume that the fading is sufficiently slow or $(\Delta t)_c \ll T$, where $(\Delta t)_c$ is the channel's coherence time [263] and T is the signalling period. Then, $h_{d,n}(t)$ can be simplified to $h_{d,n}(t) = \alpha_{d,n} e^{j\phi_{d,n}}$ for the duration of signalling period T, where the fading magnitude $\alpha_{d,n}$ is assumed to be Rayleigh distributed and the phase $\phi_{d,n}$ is assumed to be uniformly distributed.

12.9.2 BER Analysis of Fixed-mode Square QAM

An ideal RAKE receiver [266] combines all the signal powers scattered over N paths in an optimal manner, so that the instantaneous Signal-to-Noise Ratio (SNR) per symbol at the RAKE receiver's output can be maximised [274]. The noise at the RAKE receiver's output is known to be Gaussian [274]. The SNR, γ_d, at the d-th ideal RAKE receiver's output is given as [274]:

$$\gamma_d = \sum_{n=1}^{N} \gamma_{d,n} \,, \tag{12.131}$$

where $\gamma_{d,n} = E/N_o\,\alpha_{d,n}^2$ and $\{\alpha_{d,n}\}$ is assumed to be normalised, such that $\sum_{n=1}^{N} \alpha_{d,n}^2$ becomes unity. Since we assumed that each multi-path component has an independent Rayleigh distribution, the characteristic function of γ_d can be represented as [page 802] of [274]:

$$\psi_{\gamma_d}(jv) = \prod_{n=1}^{N} \frac{1}{1 - jv\bar{\gamma}_{d,n}} \,, \tag{12.132}$$

where $\gamma_{d,n} = E/N_o\mathsf{E}[\alpha_{d,n}^2]$. Let us assume furthermore that each of the D diversity channels has the same multi-path intensity profile (MIP), although in practical systems each antenna may experience a different MIP. Under this assumption, $\bar{\gamma}_{d,n}$ in Equation 12.132 can be written as $\bar{\gamma}_n$. The total SNR per symbol, γ, at the output of the demodulator depicted in Figure 12.60 is given as:

$$\gamma = \sum_{d=1}^{D} \gamma_d \,, \tag{12.133}$$

while the characteristic function of the SNR per symbol γ, under the assumption of independent identical diversity channels, can be formulated as:

$$\psi_{\gamma}(jv) = \prod_{n=1}^{N} \frac{1}{(1 - jv\bar{\gamma}_n)^D} \,. \tag{12.134}$$

Applying the technique of Partial Fraction Expansion (PFE) [379], $\psi_{\gamma}(jv)$ can be expressed as:

$$\psi_{\gamma}(jv) = \sum_{d=1}^{D} \sum_{n=1}^{N} \Lambda_{D-d+1,n} \frac{1}{(1 - jv\bar{\gamma}_n)^d} \,. \tag{12.135}$$

Let us now determine the constant coefficients $\Lambda_{d,n}$. Equating Equations 12.134 with 12.135 and substituting $jv = -p$, we have

$$\prod_{i=1}^{N} \frac{1}{(1 + p\bar{\gamma}_i)^D} = \sum_{d=1}^{D} \sum_{i=1}^{N} \Lambda_{D-d+1,i} \frac{1}{(1 + p\bar{\gamma}_i)^d} \,. \tag{12.136}$$

Multiplying by $(1 + p\bar{\gamma}_n)^D$ at both sides, Equation 12.136 becomes:

$$\prod_{\substack{i=1 \\ i \neq n}}^{N} \frac{1}{(1 + p\bar{\gamma}_i)^D} = \sum_{d=1}^{D} \sum_{\substack{i=1 \\ i \neq n}}^{N} \Lambda_{D-d+1,i} \frac{1}{(1 + p\bar{\gamma}_i)^d} + \sum_{d=1}^{D} \Lambda_{d,n} (1 + p\bar{\gamma}_n)^{d-1} . \qquad (12.137)$$

Setting the $(d-1)$th derivatives with respect to p and substituting $p = -1/\bar{\gamma}_n$, we have:

$$\frac{d^{d-1}}{dp^{d-1}} \left[\prod_{\substack{i=1 \\ i \neq n}}^{N} (p\bar{\gamma}_i + 1)^{-D} \right]_{p = -1/\bar{\gamma}_n} = (d-1)! \, \bar{\gamma}_n^{(d-1)} \Lambda_{d,n} . \qquad (12.138)$$

Hence, $\Lambda_{d,n}$ is given as:

$$\Lambda_{d,n} \triangleq \frac{1}{(d-1)! \, \bar{\gamma}_n^{(d-1)}} \, \varphi_{d,n} \left(-1/\bar{\gamma}_n \right) , \qquad (12.139)$$

where $\varphi_{d,n}(x)$ is defined as:

$$\varphi_{d,n}(x) \triangleq \frac{d^{d-1}}{dp^{d-1}} \left[\prod_{\substack{i=1 \\ i \neq n}}^{N} (p\bar{\gamma}_i + 1)^{-D} \right]_{p = x} . \qquad (12.140)$$

Upon setting the derivatives directly, $\varphi_{d,n}(-1/\bar{\gamma}_n)$ of Equation 12.140 can be represented recursively as:

$$\begin{aligned}
\varphi_{1,n}(-1/\bar{\gamma}_n) &= \pi_n^D \\
\varphi_{d,n}(-1/\bar{\gamma}_n) &= \sum_{i=1}^{d-1} \left[C_{d,i} \, D \, \varphi_{d-i,n}(-1/\bar{\gamma}_n) \sum_{\substack{j=1 \\ j \neq n}}^{N} \left(\frac{\bar{\gamma}_n \bar{\gamma}_j}{\bar{\gamma}_n - \bar{\gamma}_j} \right)^i \right] ,
\end{aligned} \qquad (12.141)$$

where π_n is defined as:

$$\pi_n \triangleq \prod_{\substack{i=1 \\ i \neq n}}^{N} \frac{\bar{\gamma}_n}{\bar{\gamma}_n - \bar{\gamma}_i} \qquad (12.142)$$

and the doubly indexed coefficient $C_{d,i}$ of Equation 12.141 can also be expressed recursively as:

$$\begin{aligned}
C_{d,1} &= -1 && \text{for all } d \\
C_{d,d} &= 0 && \text{for } d > 1 \\
C_{d,i} &= -(i-1) C_{d-1,i-1} + C_{d-1,i} && \text{for } d > i .
\end{aligned} \qquad (12.143)$$

The PDF of γ, $f_{\bar{\gamma}}(\gamma)$, can be found by applying the inverse Fourier transform to $\psi_\gamma(j\upsilon)$

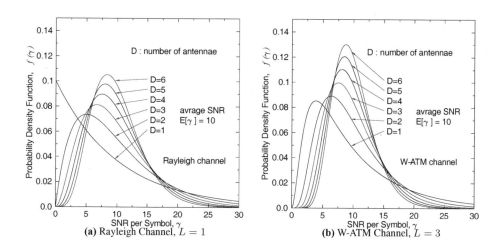

Figure 12.61: PDF of γ given in Equation 12.144 for an average SNR per symbol of $E[\gamma] = 10$dB.

in Equation 12.135, which is given by [page 781, (14-4-13)] of [274]:

$$f_{\bar{\gamma}}(\gamma) = \sum_{d=1}^{D} \sum_{n=1}^{N} \Lambda_{D-d+1,n} \frac{1}{(d-1)! \, \bar{\gamma}_n^d} \gamma^{d-1} e^{-\gamma/\bar{\gamma}_n} . \tag{12.144}$$

Figure 12.61 shows the PDF of the SNR per symbol over both a narrow-band Rayleigh channel and the dispersive Wireless Asynchronous Transfer Mode (W-ATM) channel of [2]. Specifically, the W-ATM channel is a 3-path indoor channel, where the average SNR for each path is given as $\bar{\gamma}_1 = 0.79192\bar{\gamma}$, $\bar{\gamma}_2 = 0.12424\bar{\gamma}$ and $\bar{\gamma}_3 = 0.08384\bar{\gamma}$.

Since we now have the PDF $f_{\bar{\gamma}}(\gamma)$ of the channel SNR, let us calculate the average BEP of m-ary square QAM employing Gray mapping. The average BEP P_e can be expressed as [2,274]:

$$P_e = \int_0^\infty p_m(\gamma) f(\gamma) d\gamma , \tag{12.145}$$

where $p_m(\gamma)$ is the BER of m-ary square QAM employing Gray mapping over Gaussian channels [2]:

$$p_m(\gamma) = \sum_i A_i Q(\sqrt{a_i \gamma}) , \tag{12.146}$$

where $Q(x)$ is the Gaussian Q-function defined as $Q(x) \triangleq \frac{1}{\sqrt{2\pi}} \int_x^\infty e^{-t^2/2} dt$ and $\{A_i, a_i\}$ is a set of modulation mode dependent constants. For the modulation modes associated with

$m = 2, 4, 16$ and 64, the sets $\{A_i, a_i\}$ are given as [2, 405]:

$$
\begin{array}{lll}
m = 2, & \text{BPSK} & \{(1, 2)\} \\
m = 4, & \text{QPSK} & \{(1, 1)\} \\
m = 16, & \text{16-QAM} & \left\{\left(\frac{3}{4}, \frac{1}{5}\right), \left(\frac{2}{4}, \frac{3^2}{5}\right), \left(-\frac{1}{4}, \frac{5^2}{5}\right)\right\} \\
m = 64, & \text{64-QAM} & \left\{\left(\frac{7}{12}, \frac{1}{21}\right), \left(\frac{6}{12}, \frac{3^2}{21}\right), \left(-\frac{1}{12}, \frac{5^2}{21}\right), \left(\frac{1}{12}, \frac{9^2}{21}\right), \left(-\frac{1}{12}, \frac{13^2}{21}\right)\right\}.
\end{array}
$$

$$\tag{12.147}$$

The average BEP of m-ary QAM in our scenario can be calculated by substituting $p_m(\gamma)$ of Equation 12.146 and $f_{\bar{\gamma}}(\gamma)$ of Equation 12.144 into Equation 12.145:

$$
P_{e,m}(\bar{\gamma}) = \int_0^\infty \sum_i A_i Q(\sqrt{a_i \gamma}) f_{\bar{\gamma}}(\gamma) d\gamma \tag{12.148}
$$

$$
= \sum_i A_i P_e(\bar{\gamma}; a_i), \tag{12.149}
$$

where each constituent BEP $P_e(\bar{\gamma}; a_i)$ is defined as:

$$
P_e(\bar{\gamma}; a_i) = \int_0^\infty Q(\sqrt{a_i \gamma}) f_{\bar{\gamma}}(\gamma) d\gamma. \tag{12.150}
$$

Using the similarity of $f_{\bar{\gamma}}(\gamma)$ in Equation 12.144 and the PDF of the SNR of a D-antenna diversity-assisted Maximal Ratio Combining (MRC) system transmitting over flat Rayleigh channels [pp 781] [274], the closed form solution for the component BEP $P_e(\bar{\gamma}; a_i)$ can be expressed as:

$$
P_e(\bar{\gamma}; a_i) = \sum_{d=1}^D \sum_{n=1}^N \frac{1}{\sqrt{2\pi}} \int_0^\infty \int_{\sqrt{2\gamma}}^\infty e^{-x^2/2} \Lambda_{D-d+1,n} \frac{1}{(d-1)! \, \bar{\gamma}_n^d} \gamma^{d-1} e^{-\gamma/\bar{\gamma}_n} \, dx \, d\gamma
$$

$$\tag{12.151}$$

$$
= \sum_{d=1}^D \sum_{n=1}^N \left[\Lambda_{D-d+1,n} \left\{ \tfrac{1}{2}(1 - \mu_n) \right\}^d \sum_{i=0}^{d-1} \binom{d-1+i}{i} \left\{ \tfrac{1}{2}(1 + \mu_n) \right\}^i \right],
$$

$$\tag{12.152}$$

where $\mu_n \triangleq \sqrt{\frac{a_i \bar{\gamma}_n}{2 + a_i \bar{\gamma}_n}}$ and the average SNR per symbol is $\bar{\gamma} = D \sum_{n=1}^N \bar{\gamma}_n$. Substituting $P_e(\bar{\gamma}; a_i)$ of Equation 12.152 into Equation 12.149, the average BEP of an m-ary QAM Rake receiver using antenna diversity can be expressed in a closed form.

Let us consider the performance of BPSK by setting $D = 1$ or $N = 1$. When the number of antennae is one, i.e. $D = 1$, $P_{e,2}(\bar{\gamma})$ is reduced to

$$P_{e,BPSK} = \sum_{n=1}^{N} \Lambda_{1,n} \left\{ \tfrac{1}{2}(1 - \mu_n) \right\} \tag{12.153}$$

$$= \sum_{n=1}^{N} \pi_n \left\{ \tfrac{1}{2}(1 - \mu_n) \right\}, \tag{12.154}$$

which is identical to the result given in [(14-5-28), pp 802] [274]. On the other hand, when the channels exhibit flat fading, i.e. $L=1$, our system is reduced to a D-antenna diversity-based MRC system transmitting over D number of flat Rayleigh channels. In this case, $\Lambda_{D-d+1,n}$ of Equation 12.139 becomes zero for all values of d, except for $\Lambda_{1,1}=1$ when $d = D$, and the average BPSK BEP in this scenario becomes:

$$P_{e,BPSK} = \sum_{d=1}^{D} \left[\Lambda_{D-d+1,1} \left\{ \tfrac{1}{2}(1 - \mu_1) \right\}^d \sum_{i=0}^{d-1} \binom{d-1+i}{i} \left\{ \tfrac{1}{2}(1 + \mu_1) \right\}^i \right] \tag{12.155}$$

$$= \left\{ \tfrac{1}{2}(1 - \mu_1) \right\}^D \sum_{i=0}^{D-1} \binom{D-1+i}{i} \left\{ \tfrac{1}{2}(1 + \mu_1) \right\}^i, \tag{12.156}$$

which is also given in [(14-4-15), pp781] [274].

Chapter 13

Successive Partial Despreading Based Multi-Code MC-CDMA[1]

13.1 Introduction

In the downlink scenario of spread spectrum systems several spreading codes may be simultaneously assigned to one user in order to support a high data throughput [326]. A specific example of this concept, namely a multi-code MC-CDMA scheme, was introduced in Section 11.5.1. When the number of simultaneously transmitted spreading codes L is equal to the spreading factor N, the system is said to be "fully loaded". A fully loaded MC-CDMA scheme was referred to as Walsh Hadamard Transform (WHT) based OFDM in references [445,446]. More explicitly, the bandwidth requirement of the system is increased by a factor of N, when a spreading factor of N is used. However, under fully loaded conditions this system is capable of transmitting N bits of a single user with the aid of employing N parallel codes. Hence a total of N bits are transmitted in an N times higher bandwidth and therefore we may refer to this scheme as a "unity-rate coding arrangement". However, when non-orthogonal spreading codes are used or when the orthogonality of the codes is destroyed by the fading channels, a fully loaded CDMA system suffers from an excessive Multiple Code Interference (MCI). This phenomenon is also referred to as Multiple User Interference (MUI), when the simultaneously transmitted codes are used to support different users, as we observed in Section 10.3. In order to effectively cancel the above-mentioned MCI or MUI, a multi-user detector [447] or a joint detector [323] has to be employed in the receiver. In Section 11.5.3.3 we observed that the BER performance of the Minimum Mean Square Error (MMSE) Joint Detector (JD) assisted fully loaded multi-code MC-CDMA system is better than that of OFDM, when the channel is independently Rayleigh fading over each subcarrier. However, one of the major obstacles of implementing joint-detection receivers has been their high implementational complexity.

---

[1]*OFDM and MC-CDMA for Broadband Multi-user Communications WLANS and Broadcasting.*
L.Hanzo, Münster, T. Keller and B.J. Choi,
©2003 John Wiley & Sons, Ltd. ISBN 0-470-85879-6

In this chapter, we investigate a reduced-complexity despreading technique involved in the context of fully loaded Walsh-spread multi-code systems employing BPSK modulation, based on a Walsh matrix-assisted partitioning. Let us assume that our system model is a BPSK modulated fully loaded multi-code MC-CDMA scheme, which was referred to in [445] as a WHT/OFDM arrangement.

13.2   System Model

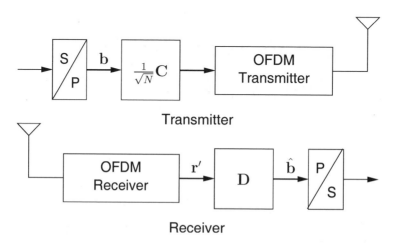

Figure 13.1: A transmitter and a receiver schematic of fully loaded multi-code BPSK modulated MC-CDMA.

The transmitter schematic of the investigated fully loaded multi-code BPSK modulated MC-CDMA scheme is depicted in Figure 13.1. The serial-to-parallel converter gathers N number of different consecutive information bits in order to generate a $N \times 1$ dimensional information bit vector $\mathbf{b}$. Then, an $N \times N$ dimensional Walsh spreading matrix, namely $1/\sqrt{N}\, \mathbf{C}$ is used to transform this bit vector $\mathbf{b}$, in order to generate the $N \times 1$ dimensional frequency domain symbol vector $\mathbf{s}$, which is transmitted using a conventional OFDM transmitter [2]. The $N \times 1$ dimensional frequency domain symbol vector $\mathbf{s}$ can be expressed as:

$$\mathbf{s} = \frac{1}{\sqrt{N}}\, \mathbf{C}\, \mathbf{b}\,, \tag{13.1}$$

where N is the spreading factor. The scaling factor of $1/\sqrt{N}$ was introduced in Equation 13.1 to render the average power of the transformed message the same as that of the unspread symbol.

The OFDM receiver recovers the $N \times 1$ dimensional frequency domain symbol vector $\mathbf{r}$, which was corrupted by fading and contaminated by noise. Various despreading/detection techniques have been developed to recover the original symbol $\mathbf{b}$ from the spread symbol $\mathbf{s}$ [327]. However, all techniques are based on applying the despreading matrix, $\mathbf{D}$. This despreading process can be represented as:

$$\hat{\mathbf{b}} = \mathbf{D}\,\mathbf{r}', \tag{13.2}$$

where $\hat{\mathbf{b}}$ is the estimated $N \times 1$ dimensional binary antipodal message vector, the despreading matrix $\mathbf{D}$ is given as $\mathbf{C}^H$ with the superscript ^{H} indicating the Hermitian [379] of the matrix $\mathbf{C}$ and $\mathbf{r}'$ is the fading-compensated received vector. Since the kth estimated bit $\hat{\mathbf{b}}_k$ is obtained by $\hat{\mathbf{b}}_k = \sum_{n=1}^{N} \mathbf{D}_{kn}\, \mathbf{r}'_n$, N number of multiplications and $N - 1$ number of additions are required to calculate $\hat{\mathbf{b}}_k$, where $k = 1, \cdots, N$, yielding a total number of multiplications given by N^2 and $N^2 - N$ number of additions for detecting all the bits.

The advantage of this Walsh-spreading based scheme is highlighted here in comparison to a conventional OFDM scheme as follows. When a conventional OFDM scheme experiences frequency-selective fading, the specific bits of the N-bit OFDM symbol conveyed by the highly attenuated subcarriers are virtually obliterated by the fading. By contrast, when invoking the above-mentioned Walsh-spreading scheme, each component of the transformed vector $\mathbf{s}$ is dependent on each of the N bits of the vector $\mathbf{b}$. When a specific component of the N-component vector $\mathbf{s}$ is obliterated by a badly faded OFDM subcarrier, the effects of this are spread to all the N bits of the estimated transmitted vector $\hat{\mathbf{b}}$, but they only fractionally contaminate each of the N bits. Hence there is still a high probability that all N bits can be recovered without an error, despite having a corrupted subcarrier. The same approach can be extended to a scenario, where one or several of the components of the vector $\mathbf{s}$ are simply ignored during the detection process, in an effort to reduce the associated detection complexity.

In Section 9.2.1 we observed that Walsh codes are defined recursively using Walsh-Hadamard transform. Exploiting this recursive structure inherent in Walsh codes, we may be able to reduce the number of operations required to despread in Equation 13.2. Hence, we propose an alternative approach to the conventional "full despreading", termed as "successive partial despreading" (SPD), for employment in Walsh-spread systems, where only half of the vector components of the fading compensated received signal $\{r'[k]\}$ are used during the first stage detection, while the other half of the vector components is involved for the final detection only if it is deemed necessary. We will see that the proposed technique requires less than half the number of operations, compared to the conventional despreading method. Three types of SPD detection schemes are introduced in Section 13.3. The proposed schemes are analysed, when communicating over the AWGN channel in terms of their achievable BER performance and their implementational complexity in Section 13.4. In a certain scenario, a time domain impulse noise [448, 449] or a narrow-band jamming/interference [450] in the frequency domain can be a dominant channel impairment source. In this situation some chips of the simultaneous codes have to be discarded, since the magnitude of the impulse noise in the time domain is higher than that of the chip concerned. The same action has to be taken as regards to a particular subcarrier, when the magnitude of the narrow-band jamming signal in the frequency domain is higher than that of the subcarrier. Section 13.5 investigates the performance of the proposed SPD scheme, when one of the symbols modulating a specific subcarrier is unavailable due to the above-mentioned impairments. Section 13.6 concludes with a number of observations and future research directions. Following the above conceptual introduction, let us now elaborate on the associated detection philosophy in more depth.

13.3   The Sequential Partial Despreading Concept

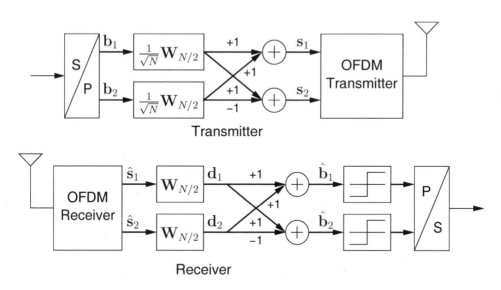

Figure 13.2: Transmitter and receiver schematic of the SPD scheme for fully loaded multi-code BPSK modulated MC-CDMA, where $\mathbf{W}_{N/2}$ is the $N/2$-dimensional Walsh spreading matrix.

The N-dimensional Walsh-Hadamard matrix, $\mathbf{W}_N$, can be represented in its recursive form as [327]:

$$\mathbf{W}_N = \begin{bmatrix} \mathbf{W}_{N/2} & \mathbf{W}_{N/2} \\ \mathbf{W}_{N/2} & \overline{\mathbf{W}}_{N/2} \end{bmatrix} . \tag{13.3}$$

Then the spread symbol of Equation 13.1 can be rewritten with the aid of the $N/2$-dimensional sub-matrices and sub-vectors as:

$$\begin{bmatrix} \mathbf{s}_1 \\ \mathbf{s}_2 \end{bmatrix} = \frac{1}{\sqrt{N}} \begin{bmatrix} \mathbf{W}_{N/2} & \mathbf{W}_{N/2} \\ \mathbf{W}_{N/2} & \overline{\mathbf{W}}_{N/2} \end{bmatrix} \begin{bmatrix} \mathbf{b}_1 \\ \mathbf{b}_2 \end{bmatrix} , \tag{13.4}$$

where $\mathbf{s}_1$ and $\mathbf{s}_2$ are the $N/2 \times 1$ dimensional sub-vectors of $\mathbf{s}$, and $\mathbf{b}_1$ and $\mathbf{b}_2$ are the $N/2 \times 1$ dimensional antipodal sub-vectors of $\mathbf{b}$ having components of ± 1.

The expression in Equation 13.4 can be decomposed into two equations, one for $\mathbf{s}_1$ and the other for $\mathbf{s}_2$ as follows:

$$\mathbf{s}_1 = \frac{1}{\sqrt{N}} \mathbf{W}_{N/2} \mathbf{b}_1 + \frac{1}{\sqrt{N}} \mathbf{W}_{N/2} \mathbf{b}_2 \tag{13.5}$$

$$\mathbf{s}_2 = \frac{1}{\sqrt{N}} \mathbf{W}_{N/2} \mathbf{b}_1 + \frac{1}{\sqrt{N}} \overline{\mathbf{W}}_{N/2} \mathbf{b}_2 . \tag{13.6}$$

The corresponding transmitter and receiver schematics are depicted in Figure 13.2, explicitly showing the decomposed matrix operations. In the proposed SPD scheme, the despreading is performed at the receiver upon multiplying $\mathbf{W}_{N/2}$ by the recovered frequency domain symbol $\hat{\mathbf{s}}_1$ and $\hat{\mathbf{s}}_2$ output by the OFDM receiver, which is expressed as:

$$\mathbf{d}_1 = \mathbf{W}_{N/2}\,\hat{\mathbf{s}}_1 \tag{13.7}$$

$$\mathbf{d}_2 = \mathbf{W}_{N/2}\,\hat{\mathbf{s}}_2 . \tag{13.8}$$

For simplicity, let us assume that $\mathbf{s}_1$ and $\mathbf{s}_2$ have been received without any channel impairments and without any noise for the moment. Then, exploiting that $\mathbf{W}_{N/2}\mathbf{W}_{N/2} = N/2\,\mathbf{I}_{N/2}$, the two $N/2$ dimensional despread symbol vectors of Equations 13.7 and 13.8 can be expressed as:

$$\mathbf{d}_1 = \frac{\sqrt{N}}{2}(\mathbf{b}_1 + \mathbf{b}_2) \tag{13.9}$$

$$\mathbf{d}_2 = \frac{\sqrt{N}}{2}(\mathbf{b}_1 - \mathbf{b}_2) . \tag{13.10}$$

The last operation of the receiver is that of recovering $\mathbf{b}_1$ and $\mathbf{b}_2$ with the aid of:

$$\hat{\mathbf{b}}_1 = \frac{\mathbf{d}_1 + \mathbf{d}_2}{\sqrt{N}} \tag{13.11}$$

$$\hat{\mathbf{b}}_2 = \frac{\mathbf{d}_1 - \mathbf{d}_2}{\sqrt{N}}, \tag{13.12}$$

in order to recover $\hat{\mathbf{b}}$. Explicitly, according to Equations 13.9 and 13.10 in the absence of channel impairments. the detection variables $\mathbf{d}_1$ and $\mathbf{d}_2$ are given by the sum and the difference of the two independent $N/2 \times 1$ dimensional message symbols $\mathbf{b}_1$ and $\mathbf{b}_2$. When the kth bits of the subvectors $\mathbf{b}_1$ and $\mathbf{b}_2$, namely $\mathbf{b}_1[k]$ and $\mathbf{b}_2[k]$, have the same value, $\mathbf{d}_1[k]$ in the numerator of Equation 13.11 is sufficient for the detection of both $\mathbf{b}_1$ and $\mathbf{b}_2$, since $\mathbf{d}_2[k] = \sqrt{N}/2\,(\mathbf{b}_1 - \mathbf{b}_2) = 0$ conveys no information. By contrast, in the presence of channel-impairments $\mathbf{b}_1[k] = \mathbf{b}_2[k]$ is indicated on a probabilistic basis by $\mathbf{d}_2[k] \approx 0$ according to Equation 13.10. On the other hand, when we have $\mathbf{b}_1[k] = -\mathbf{b}_2[k]$, $\mathbf{d}_1[k] = \sqrt{N}/2\,(\mathbf{b}_1 + \mathbf{b}_2)$ becomes zero and $\mathbf{d}_2[k] = \sqrt{N}/2\,(\mathbf{b}_1 - \mathbf{b}_2)$ is required for the correct detection of both bits.

Let us now consider three types of SPD detection schemes, which we referred to as the Type I, Type II, and Type III SPD schemes. They differ in their philosophy of using the despread symbol vectors $\mathbf{d}_1$ of Equation 13.9 and $\mathbf{d}_2$ of Equation 13.10, when generating the decision variables $\hat{\mathbf{b}}_1$ and $\hat{\mathbf{b}}_2$ for the original message vectors of $\mathbf{b}_1$ and $\mathbf{b}_2$. The noise-contaminated PDFs of $\mathbf{d}_1$ and $\mathbf{d}_2$ are plotted in Figure 13.3, which will be consulted throughout our further discourse.

Type I: The decision variables $\hat{\mathbf{b}}_1[k]$ and $\hat{\mathbf{b}}_2[k]$ are identical to those of the conventional despreading detectors, but due to processing two $N/2$-dimensional vectors instead of an N-dimensional vector, the detector requires a reduced number of multiplications and additions, as it will be shown in Section 13.4.1. Upon neglecting the scaling factor of $1/\sqrt{N}$ in Equations 13.11 and 13.12 the associated decision variables are expressed

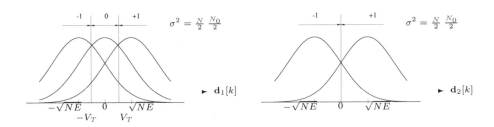

(a) the first stage   (b) the second stage

Figure 13.3: PDF of the detection variables $d_1[k]$ and $d_2[k]$ in the AWGN channel and the first as well as second stage decision region, when the SPD based detection is employed. (a) PDF of the noise contaminated component $d_1[k]$ during the first detection stage. (b) PDF of the noise contaminated component $d_2[k]$ during the second detection stage when $b_1[k] = -b_2[k]$.

as:

$$\hat{b}_1[k] = d_1[k] + d_2[k] \tag{13.13}$$

$$\hat{b}_2[k] = d_1[k] - d_2[k], \tag{13.14}$$

where $d_1[k]$ and $d_2[k]$ are now contaminated by the channel effects and hence they obey the noisy PDFs seen in Figure 13.3.

Type II: If the detector's confidence in $\hat{b}_1[k] = \hat{b}_2[k]$ is high, since we have $|d_1[k]| > V_T$ in Figure 13.3(a), then according to Equation 13.9 we can confidently assume that $\hat{b}_1[k] = \hat{b}_2[k]$. By contrast, if the probability of $\hat{b}_1[k] = \hat{b}_2[k]$ is low, since a low $|d_1[k]|$ value is encountered, which is indicative of $\hat{b}_1[k] = -\hat{b}_2[k]$, then we make the oapposite decision for $\hat{b}_1[k]$ and $\hat{b}_2[k]$, namely $\hat{b}_1[k] = -\hat{b}_2[k]$. This detector has the lowest complexity among the three SPD schemes, as will be shown in Section 13.4.4. The decision variables can be summarised as:

$$\hat{b}_1[k] = \hat{b}_2[k] = d_1[k], \text{ when } |d_1[k]| > V_T, \tag{13.15}$$

$$\hat{b}_1[k] = -\hat{b}_2[k] = d_2[k], \text{ otherwise.} \tag{13.16}$$

Type III: When $d_1[k] \approx 0$, i.e. $|d_1[k]| \leq V_T$ in Figure 13.3(a), the Type III detector uses Type I decision variables of Equations 13.13 and 13.14. Otherwise, it invokes the Type II detector and hence detects $\hat{b}_1[k]$ and $\hat{b}_2[k]$ based on only $d_1[k]$, requiring no further operations. Accordingly, the decision rule can be summarised as:

$$\hat{b}_1[k] = \hat{b}_2[k] = d_1[k], \text{ when } |d_1[k]| > V_T, \tag{13.17}$$

$$\hat{b}_1[k] = d_1[k] + d_2[k], \ \hat{b}_2[k] = d_1[k] - d_2[k], \text{ otherwise.} \tag{13.18}$$

These three types of detectors will be compared in the following sections in terms of their BER performances and their complexities. Although we proposed the above three types of SPD detectors based on the first order decomposition of the Walsh-Hadamard matrix, a recursive application of these SPD schemes using a higher order decomposition is feasible. In fact, the recursive implementation of the Type I despreading detector is known as Fast Walsh-Hadamard Transform [445]. We will observe in Sections 13.4 and 13.5 that the Type II and Type III detectors require a lower number of operations than the Type I detector.

13.4 AWGN Channel

When communicating over the AWGN channel we cannot expect any diversity gain and there is not much benefit in employing spreading system. However, the BER analysis of the proposed successive partial despreading schemes in the AWGN channel is a good starting point for comparing them with the conventional scheme.

In the AWGN channel, the decomposed received vectors, $\hat{s}_1$ and $\hat{s}_2$ of Figure 13.2, can be expressed as:

$$\hat{s}_1 = \sqrt{E}\,s_1 + n_1 \tag{13.19}$$

$$\hat{s}_2 = \sqrt{E}\,s_2 + n_2, \tag{13.20}$$

where E is the energy per received bit, s_1 and s_2 are given in Equations 13.5 and 13.6, and $n_i[k]$, $i = 1$, 2 and $k = 1, \cdots, N/2$ represents the independently and identically distributed (*iid*) zero-mean Gaussian distribution having a variance of $\sigma^2 = N_o/2$. As for the single-carrier DS-CDMA systems, we assume that the PN-based scrambling has already, been removed, leaving Walsh spreading imposed on the original message bits. As for MC-CDMA, some receiver preprocessing was assumed, such as employing an N component FFT to recover the frequency domain symbols from the received time domain symbols. Upon applying the $N/2 \times N/2$-dimensional despreading matrix, $\mathbf{W}_{N/2}$, to $\hat{s}_1$ and $\hat{s}_2$ as expressed in Equations 13.7 and 13.8, we get the the detection variables d_1 and d_2 formulated as:

$$d_1 = \frac{1}{2}\sqrt{NE}\,b_1 + \frac{1}{2}\sqrt{NE}\,b_2 + \eta_1 \tag{13.21}$$

$$d_2 = \frac{1}{2}\sqrt{NE}\,b_1 - \frac{1}{2}\sqrt{NE}\,b_2 + \eta_2\,, \tag{13.22}$$

where each noise component $\eta_i[k]$, $i = 1$, 2 and $k = 1, \cdots, N/2$ is also *iid* (owing to the orthogonality between the rows of the Walsh-Hadamard matrix) and has a zero mean Gaussian distribution with $\sigma^2 = (N/2)(N_o/2)$, since $\eta_i[k]$ is a sum of $N/2$ number of Gaussian random variables $n_i[j]$, $j = 1, \cdots, N/2$ having a variance of $N_o/2$. Having obtained the detection variables d_1 and d_2 in the AWGN channel, let us now investigate the BER performance of the three types of SPD detectors described in Section 13.3.

13.4.1 Type I Detector

As seen in Equations 13.13 and 13.14, in the context of the Type I detector the decision variables become:

$$\hat{\mathbf{b}}_1 = \mathbf{d}_1 + \mathbf{d}_2 = \sqrt{NE}\,\mathbf{b}_1 + \eta_1 + \eta_2 \tag{13.23}$$

$$\hat{\mathbf{b}}_2 = \mathbf{d}_1 - \mathbf{d}_2 = \sqrt{NE}\,\mathbf{b}_2 + \eta_1 - \eta_2 \,. \tag{13.24}$$

The energy per bit becomes N times E after the despreading process, however, the noise variance also becomes N times $N_o/2$. Hence, the bit energy per noise power ratio, γ, does not change. The BER is the same as for the conventional despreading detector or the ideal BPSK detector, which is given by:

$$P_{e,I} = Q(\sqrt{2\gamma}) \,. \tag{13.25}$$

The number of multiplications required by the Type I detector is given by:

$$\begin{aligned} N_I(\times) &= \left(\frac{N}{2}\right)^2 + \left(\frac{N}{2}\right)^2 \\ &= \frac{N^2}{2}, \end{aligned} \tag{13.26}$$

while the number of additions required is given by:

$$\begin{aligned} N_I(+) &= \left(\frac{N}{2} - 1\right)\frac{N}{2} + \left(\frac{N}{2} - 1\right)\frac{N}{2} + N \\ &= \frac{N^2}{2} \,. \end{aligned} \tag{13.27}$$

13.4.2 Type II Detector

Type II detector has the lowest complexity among the three SPD detectors. It uses $\mathbf{d}_1[k]$ to make the first stage decision. If the decision is of low confidence, the detector assumes that $\mathbf{b}_1[k] = -\mathbf{b}_2[k]$ and uses $\mathbf{d}_2[k]$ to decide the specific value of the original bits. The PDF of $\mathbf{d}_1[k]$ defined in Equation 13.21 is illustrated in Figure 13.3(a), when communicating over AWGN channels. The figure shows three decision regions for the first stage, namely region "-1", the low-confidence region denoted by "0" and region "+1". The conditional PDF of $\mathbf{d}_2[k]$ defined in Equation 13.22 when $\mathbf{b}_1[k] = -\mathbf{b}_2[k]$ is also depicted in Figure 13.3(b), where only two decision regions are present, namely those associated with $\mathbf{d}_2[k] = -\sqrt{NE}$ and $\mathbf{d}_2[k] = \sqrt{NE}$. The decision threshold level V_T associated with the first stage detection should lie between 0 and $\sqrt{NE}$, where $\sqrt{NE}$ is the mean of the distribution of $\mathbf{d}_1[k]$ conditioned on $\mathbf{b}_1[k] = \mathbf{b}_2[k]$. If it is too close to zero, the low-confidence region denoted by "0" becomes small and the detector has a problem in confidently detecting the case of $\mathbf{b}_1[k] = -\mathbf{b}_2[k]$. By contrast, if the detection threshold V_T approaches $\sqrt{NE}$, the detection region associated with $\mathbf{d}_1[k] = \pm 1$, i.e. with $\mathbf{b}_1[k] = \mathbf{b}_2[k]$ is reduced to a half, which will lead to a poor bit error rate performance. Hence, it is expected that there exists an optimum threshold level V_T for a given signal power.

First, we assume that $\mathbf{b}_1[k]$ is different from $\mathbf{b}_2[k]$, namely that we had $\mathbf{b}_1[k] = -\mathbf{b}_2[k]$. In this case, the random variable $\mathbf{d}_1[k]$ defined in Equation 13.21 results in $\mathbf{d}_1[k] = \eta_1[k]$, having a zero mean Gaussian PDF, as shown in the middle of Figure 13.3(a). In this scenario, provided that $\mathbf{d}_1[k]$ falls in region "-1" or "+1" of Figure 13.3(a) and hence it is outside the region "0", the Type II detector will make a decision of either $\hat{\mathbf{b}}_1[k] = \hat{\mathbf{b}}_2[k] = -1$ or $+1$, respectively, resulting in a decision error concerning the transmission of one of the two bits. Let us define $P_e(A|B)$ as the conditional BEP associated with this scenario, where the detection variable $\mathbf{d}_1[k]$ falls in the region "A", although it was expected to fall in the region "B" of Figure 13.3(a), where "A" and "B" belong to one of "+1", "0" and "-1". Let us also define $P_e(B)$ as the conditional BEP, when $\mathbf{d}_1[k]$ is expected to fall in the region "B" of Figure 13.3(a). For example, $P_e(+1|0)$ denotes the conditional BEP of the Type II detector associated with the scenario, where $\mathbf{d}_1[k]$ falls in the region "+1" owing to a high positive value of the noise, namely when $\mathbf{d}_1[k] = \eta_1[k] > V_T$, and when we had $\mathbf{b}_1[k] = -\mathbf{b}_2[k]$, corresponding to $\mathbf{d}_1[k] = 0$ in a noiseless environment. More explicitly, the probability of error in the specific scenario of encountering a $\mathbf{d}_1[k]$ value in the "+1" region conditioned on encountering $\mathbf{d}_1[k] = 0$ in the noiseless scenario is denoted by $P_e(+1|0)$, which can be formulated as:

$$P_e(+1|0) = \Pr(\mathbf{d}_1[k] > V_T | \mathbf{b}_1[k] = -\mathbf{b}_2[k]) \cdot \frac{1}{2} \tag{13.28}$$

$$= Q\left(\frac{V_T}{\sigma}\right) \cdot \frac{1}{2}. \tag{13.29}$$

Similarly, the conditional BEP of $P_e(-1|0)$ in the same scenario, except for the difference that $\mathbf{d}_1[k]$ falls in the region "-1" instead of "+1", namely when we have $\mathbf{d}_1[k] = \eta_1[k] < -V_T$, assumes the same value as $P_e(+1|0)$ given in Equation 13.29. Hence, the combined BEP corresponding to the scenario that $\mathbf{d}_1[k]$ falls outside of the region "0", namely when $|\mathbf{d}_1[k]| > V_T$, which is denoted by $P_e(\pm 1|0)$ can be formulated as:

$$P_e(\pm 1|0) = P_e(+1|0) + P_e(-1|0) \tag{13.30}$$

$$= 2 \cdot Q\left(\frac{V_T}{\sigma}\right) \cdot \frac{1}{2} \tag{13.31}$$

$$= Q\left(\frac{V_T}{\sigma}\right), \tag{13.32}$$

where the variance of $\mathbf{d}_1[k]$, and hence also that of the noise component $\eta_1[k]$ is $\sigma^2 = (N/2)(N_o/2)$, as we argued in Section 13.4. On the other hand, when $\mathbf{d}_1[k] = \eta_1[k]$ falls in region "0" as expected, namely when we have $-V_T \leq \mathbf{d}_1[k] = \eta_1[k] \leq +V_T$ owing to a moderate value of the noise, the Type II detector involves the second stage detection variable $\mathbf{d}_2[k]$ in order to arrive at the estimated bits $\hat{\mathbf{b}}_1$ and $\hat{\mathbf{b}}_2$. Since we are considering the scenario of $\mathbf{b}_1[k] = -\mathbf{b}_2[k]$, the detection variable $\mathbf{d}_2[k]$ defined in Equation 13.22 becomes:

$$\mathbf{d}_2[k] = \sqrt{NE}\,\mathbf{b}_1[k] + \eta_2[k], \tag{13.33}$$

and its PDF is depicted in Figure 13.3(b). When $\mathbf{d}_2[k]$ falls in the region "-1", namely when $\mathbf{d}_2[k] \leq 0$, assuming furthermore that $\mathbf{b}_1 = 1$ and hence $\mathbf{b}_2 = -1$ are transmitted, then the Type II detector will result in an erroneous decision concerning both transmitted bits. In the opposite scenario, namely when $\mathbf{d}_2[k]$ falls in the region "+1", the Type II detector will incur no bit errors, provided that $\mathbf{b}_1 = 1$ and hence $\mathbf{b}_2 = -1$ were transmitted. The associated BEP denoted as $P_e(0|0)$, in the scenario when $\mathbf{d}_1[k]$ of Equation 13.21 falls in the region "0", conditioned on the fact that indeed it was expected to fall in the region "0", since $\mathbf{b}_1[k] = -\mathbf{b}_2[k]$ was transmitted, can be formulated as:

$$P_e(0|0) = \Pr(|\mathbf{d}_1[k]| \leq V_T \mid \mathbf{b}_1[k] = -\mathbf{b}_2[k]) \cdot$$
$$\Pr(\{\mathbf{d}_2[k] < 0 \mid \mathbf{b}_1[k] = -\mathbf{b}_2[k] = +1\} \text{ or } \{\mathbf{d}_2[k] \geq 0 \mid \mathbf{b}_1[k] = -\mathbf{b}_2[k] = -1\}) \tag{13.34}$$

$$= \left\{1 - 2Q\left(\frac{V_T}{\sigma}\right)\right\} \left\{Q\left(\frac{\sqrt{NE}}{\sigma}\right) \cdot \frac{1}{2} + Q\left(\frac{\sqrt{NE}}{\sigma}\right) \cdot \frac{1}{2}\right\} \tag{13.35}$$

$$= \left\{1 - 2Q\left(\frac{V_T}{\sigma}\right)\right\} Q\left(\frac{\sqrt{NE}}{\sigma}\right), \tag{13.36}$$

where the variance σ^2 of $\mathbf{d}_1[k]$ and $\mathbf{d}_2[k]$, and hence that of the noise components $\eta_1[k]$ and $\eta_2[k]$ is $\sigma^2 = (N/2)(N_o/2)$, as we argued in Section 13.4. Hence, the bit error probability $P_e(0)$ associated with the case of $\mathbf{b}_1[k] = -\mathbf{b}_2[k]$, namely when $\mathbf{d}_1[k] = 0$ is expected, can be formulated, with the aid of Equations 13.32 and 13.36, as:

$$P_e(0) = P_e(\pm 1|0) + P_e(0|0) \tag{13.37}$$

$$= Q\left(\frac{V_T}{\sigma}\right) + \left\{1 - 2Q\left(\frac{V_T}{\sigma}\right)\right\} Q\left(\frac{\sqrt{NE}}{\sigma}\right)$$

$$= Q(2\sqrt{\gamma}\upsilon) + \{1 - 2Q(2\sqrt{\gamma}\upsilon)\}Q(2\sqrt{\gamma}), \tag{13.38}$$

where γ is the bit energy to noise power ratio, defined as $\gamma = E/N_o$ and υ is the threshold level normalised to $\sqrt{NE}$, defined by $\upsilon = V_T/\sqrt{NE}$.

Let us now assume that the original transmitted bits satisfy $\mathbf{b}_1[k] = \mathbf{b}_2[k]$. In this scenario the detection variable $\mathbf{d}_1[k]$ defined in Equation 13.21 can be expressed as:

$$\mathbf{d}_1[k] = \sqrt{NE}\,\mathbf{b}_1 + \eta_1. \tag{13.39}$$

Specifically, when the bits we have are $\mathbf{b}_1[k] = \mathbf{b}_2[k] = -1$ and $\mathbf{d}_1[k]$ of Equation 13.39 falls in region "+1" of Figure 13.3(a), provided that $\mathbf{d}_1[k]$ is expected to fall in region "-1", it will lead to bit errors for both $\mathbf{b}_1[k]$ and $\mathbf{b}_2[k]$ with a certainty. The associated BEP $P_e(+1|-1)$ can be expressed as:

$$P_e(+1|-1) = \Pr(\mathbf{b}_1[k] = -1 \mid \mathbf{b}_1[k] = \mathbf{b}_2[k]) \cdot \Pr(\mathbf{d}_1[k] > V_T \mid \mathbf{b}_1[k] = \mathbf{b}_2[k] = -1) \tag{13.40}$$

$$= \frac{1}{2} \cdot Q\left(\frac{\sqrt{NE} + V_T}{\sigma}\right). \tag{13.41}$$

Likewise, when the bits encountered are $\mathbf{b}_1[k] = \mathbf{b}_2[k] = +1$ and $\mathbf{d}_1[k]$ of Equation 13.39 falls in region "-1" of Figure 13.3(a), provided that $\mathbf{d}_1[k]$ is expected to fall in region "+1", again both bits will be in error. The conditional BEP associated with this scenario, which is denoted by $P_e(-1|+1)$ has the same value as $P_e(+1|-1)$, as given in Equation 13.41. Hence, the combined BEP denoted by $P_e(\mp 1|\pm 1)$, associated with the scenario where $\mathbf{d}_1[k]$ falls in the opposite region with respect to the expected region, when we have $\mathbf{b}_1[k] = \mathbf{b}_2[k]$, can be expressed as:

$$P_e(\mp 1|\pm 1) = P_e(+1|-1) + P_e(-1|+1) \tag{13.42}$$

$$= Q\left(\frac{\sqrt{NE} + V_T}{\sigma}\right). \tag{13.43}$$

By contrast, when $\mathbf{d}_1[k]$ of Equation 13.39 falls in region "0" of Figure 13.3(a), namely when we have $|\mathbf{d}_1[k]| \leq V_T$, the Type II detector will incorrectly assume that $\mathbf{b}_1[k] = -\mathbf{b}_2[k]$ was transmitted and will use the second stage detection variable $\mathbf{d}_2[k]$ defined in Equation 13.22 in order to carry out a decision. However, regardless of the decision made in the second stage, the incorrect assumption of $\mathbf{b}_1[k] = -\mathbf{b}_2[k]$ made by the Type II SPD detector will remit in a single bit error for the two transmitted bits, yielding a conditional BEP of $1/2$, since we assumed that the same bits are transmitted. The BEP associated with the scenario that $\mathbf{b}_1[k] = \mathbf{b}_2[k] = +1$ are transmitted and that the first stage detection variable $\mathbf{d}_1[k]$ of Equation 13.39 falls in the region "0", which is denoted by $P_e(0|+1)$, can be expressed as:

$$P_e(0|+1) = \Pr(\mathbf{b}_1[k] = +1|\mathbf{b}_1[k] = \mathbf{b}_2[k]) \cdot \Pr(|\mathbf{d}_1[k]| \leq V_T \mid \mathbf{b}_1[k] = \mathbf{b}_2[k]) \cdot \frac{1}{2} \tag{13.44}$$

$$= \frac{1}{2} \cdot \left\{ Q\left(\frac{\sqrt{NE} - V_T}{\sigma}\right) - Q\left(\frac{\sqrt{NE} + V_T}{\sigma}\right) \right\} \frac{1}{2} \tag{13.45}$$

$$= \frac{1}{4} \left\{ Q\left(\frac{\sqrt{NE} - V_T}{\sigma}\right) - Q\left(\frac{\sqrt{NE} + V_T}{\sigma}\right) \right\}. \tag{13.46}$$

Likewise, the BEP associated with the scenario that $\mathbf{b}_1[k] = \mathbf{b}_2[k] = -1$ are transmitted and that the first stage detection variable $\mathbf{d}_1[k]$ of Equation 13.39 falls in the region "0", which is denoted by $P_e(0|-1)$, has the same value as $P_e(0|+1)$ given in Equation 13.46. Hence, the combined BEP associated with the scenario that the first stage detection variable $\mathbf{d}_1[k]$ of Equation 13.39 falls in the region "0", when $\mathbf{b}_1[k] = \mathbf{b}_2[k] = \pm 1$, which is denoted by $P_e(0|\pm 1)$, can be formulated as:

$$P_e(0|\pm 1) = P_e(0|+1) + P_e(0|-1) \tag{13.47}$$

$$= \frac{1}{2} \left\{ Q\left(\frac{\sqrt{NE} - V_T}{\sigma}\right) - Q\left(\frac{\sqrt{NE} + V_T}{\sigma}\right) \right\}. \tag{13.48}$$

Therefore, with the aid of Equations 13.43 and 13.48 the conditional BEP $P_e(\pm 1)$ for the

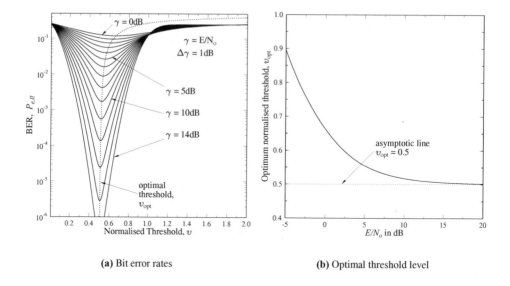

(a) Bit error rates   **(b)** Optimal threshold level

Figure 13.4: BER characteristics of the Type II detector: (a) BER $P_{e,II}$ of the Type II detector versus the normalised threshold level v, for several values of SNR $\gamma = E/N_o$ in terms of dB. It can be observed that there exist optimum threshold levels for given E/N_o values. (b) the optimum normalised threshold levels, v_{opt}, versus E/N_o. v_{opt} asymptotically approaches $1/2$ as E/N_o becomes large.

case of $\mathbf{b}_1[k] = \mathbf{b}_2[k] = \pm 1$ can be summarised as:

$$P_e(\pm 1) = P_e(\mp 1| \pm 1) + P_e(0| \pm 1) \tag{13.49}$$

$$= Q\left(\frac{\sqrt{NE} + V_T}{\sigma}\right) + \frac{1}{2}\left\{Q\left(\frac{\sqrt{NE} - V_T}{\sigma}\right) - Q\left(\frac{\sqrt{NE} + V_T}{\sigma}\right)\right\} \tag{13.50}$$

$$= \frac{1}{2}\left\{Q\left(2\sqrt{\gamma}(1 - v)\right) + Q\left(2\sqrt{\gamma}(1 + v)\right)\right\} . \tag{13.51}$$

¿From Equations 13.38 and 13.51, the total BEP $P_{e,II}$ of the Type II detector may be expressed as:

$$P_{e,II} = \frac{1}{2}\left\{P_e(0) + P_e(\pm 1)\right\} \tag{13.52}$$

$$= \frac{1}{2}Q(2\sqrt{\gamma}v) + \left\{\frac{1}{2} - Q(2\sqrt{\gamma}v)\right\}Q(2\sqrt{\gamma}) + \frac{1}{4}\{Q(2\sqrt{\gamma}(1 - v)) + Q(2\sqrt{\gamma}(1 + v))\}$$

$$= \frac{1}{4}\left[Q\left(2\sqrt{\gamma}(1 - v)\right) + Q\left(2\sqrt{\gamma}(1 + v)\right)\right] + Q\left(2\sqrt{\gamma}\,v\right)\left[\frac{1}{2} - Q\left(2\sqrt{\gamma}\right)\right] + \frac{1}{2}Q\left(2\sqrt{\gamma}\right) . \tag{13.53}$$

The bit error probability $P_{e,II}$ of Equation 13.53 is illustrated in Figure 13.4(a) as a function of both the normalised threshold v and the SNR γ. It can be observed that the threshold levels play an important role in determining the bit error rate. When v is zero, the region "0" in Figure 13.3(a) vanishes. In this case, we see in Equation 13.53 that $P_{e,II}$ becomes $1/4 + 1/2Q(2\sqrt{\gamma}) \approx 1/4$ for $\gamma \gg 0$dB. When v approaches ∞, the region "+1" and "-1" in Figure 13.3(a) vanishes and $P_{e,II}$ becomes $1/4 + 1/2Q(2\sqrt{\gamma})$, which is the same as for the case of $v = 0$.

The optimal threshold can be obtained by finding v_{opt} such that $(\partial P_e/\partial v)\,|_{v=v_{opt}} = 0$. Using the following relation regarding a derivative of the Q-function [379]:

$$\frac{\partial Q(ax+b)}{\partial x} = -\frac{a}{\sqrt{2\pi}}e^{-(ax+b)^2/2}, \tag{13.54}$$

we can derive $\partial P_{e,II}/\partial v$, which can be expressed as:

$$\frac{\partial P_{e,II}}{\partial v} = \frac{\sqrt{\gamma}}{2\sqrt{2\pi}}\left[e^{-2\gamma(1-v)^2} - e^{-2\gamma(1+v)^2} - \{2 - 4Q(2\sqrt{\gamma})\}e^{-2\gamma v^2}\right]. \tag{13.55}$$

Upon equating this expression to zero, the optimum threshold v_{opt} can be expressed as:

$$v_{opt} = \frac{1}{4\gamma}\log_e\left(\mu + \sqrt{1+\mu^2}\right), \tag{13.56}$$

where $\mu = \{1 - 2Q(2\sqrt{\gamma})\}e^{2\gamma}$. The optimum normalised threshold level, v_{opt}, as a function of $\gamma = E_b/N_o$ is illustrated in Figure 13.4(b), which asymptotically approaches $1/2$, as γ grows. When $v_{opt} \simeq 1/2$, $P_{e,II}$ of Equation 13.53 can be simplified to:

$$P_{e,II} \simeq \frac{3}{4}Q\left(\sqrt{\gamma}\right) + \frac{2}{4}Q\left(2\sqrt{\gamma}\right) + \frac{1}{4}Q\left(3\sqrt{\gamma}\right) - Q\left(\sqrt{\gamma}\right)Q\left(2\sqrt{\gamma}\right). \tag{13.57}$$

Having analysed the achievable BEP, let us now examine the complexity of the Type II SPD detector. The probability $P_{II}(0)$ that $\mathbf{d}_1$ falls in the region "0" of Figure 13.3(a), is useful for calculating the number of operations required by the Type II detector, since this quantifies the probability that the Type II detector has to calculate the second stage detection variable $\mathbf{d}_2[k]$ additionally. The probability $P_{II}(0)$ can be formulated as:

$$P_{II}(0) = \Pr(|\mathbf{d}_1| \leq V_T) \tag{13.58}$$

$$\begin{aligned} = \Pr(\mathbf{b}_1 = \mathbf{b}_2 = +1) \cdot \Pr(|\mathbf{d}_1| \leq V_T \mid \mathbf{b}_1 = \mathbf{b}_2 = +1) \\ + \Pr(\mathbf{b}_1 = \mathbf{b}_2 = -1) \cdot \Pr(|\mathbf{d}_1| \leq V_T \mid \mathbf{b}_1 = \mathbf{b}_2 = -1) \\ + \Pr(\mathbf{b}_1 = -\mathbf{b}_2) \cdot \Pr(|\mathbf{d}_1| \leq V_T \mid \mathbf{b}_1 = -\mathbf{b}_2) \end{aligned} \tag{13.59}$$

$$= \frac{1}{4} \left\{ Q\left(\frac{\sqrt{NE} - V_T}{\sigma} \right) - Q\left(\frac{\sqrt{NE} + V_T}{\sigma} \right) \right\}$$

$$+ \frac{1}{4} \left\{ Q\left(\frac{\sqrt{NE} - V_T}{\sigma} \right) - Q\left(\frac{\sqrt{NE} + V_T}{\sigma} \right) \right\}$$

$$+ \frac{1}{2} \left\{ 1 - 2Q\left(\frac{V_T}{\sigma} \right) \right\} \tag{13.60}$$

$$= \frac{1}{2} \{ 1 - 2Q(2\sqrt{\gamma}\upsilon) + Q(2\sqrt{\gamma}(1 - \upsilon)) - Q(2\sqrt{\gamma}(1 + \upsilon)) \}. \tag{13.61}$$

Then, the required number of multiplications, $N_{II}(\times)$, can be expressed as:

$$N_{II}(\times) = \left(\frac{N}{2} \right)^2 + \left(\frac{N}{2} \right)^2 \cdot P_{II}(0), \tag{13.62}$$

where the first term indicates the number of multiplications involved in calculating $\mathbf{d}_1$ expressed in Equation 13.7, while the second term quantifies the number of multiplications involved in determining $\mathbf{d}_2$, weighted by the probability that $\mathbf{d}_2$ is required. Upon substituting Equation 13.61 in Equation 13.62, we have:

$$N_{II}(\times) = \frac{N^2}{8} \{ 3 - 2Q(2\sqrt{\gamma}\upsilon) + Q(2\sqrt{\gamma}(1 - \upsilon)) - Q(2\sqrt{\gamma}(1 + \upsilon)) \}. \tag{13.63}$$

On the other hand, the number of operations required to derive $\mathbf{d}_1$ according to Equation 13.7, $(N/2)(N/2 - 1)$ number of additions are necessitated. Hence, following the approach adopted to calculate $N_{II}(\times)$ above, the number of additions $N_{II}(+)$ required by the Type II detector can be expressed as:

$$N_{II}(+) = \frac{N(N - 2)}{4} (1 + P_{II}(0))$$

$$= \frac{N(N - 2)}{8} \{ 3 - 2Q(2\sqrt{\gamma}\upsilon) + Q(2\sqrt{\gamma}(1 - \upsilon)) - Q(2\sqrt{\gamma}(1 + \upsilon)) \}. \tag{13.64}$$

13.4.3   Type III Detector

In the context of the Type III SPD detector we have to calculate the probability of $\mathbf{d}_1$ falling in the region "0" in Figure 13.3(a), during the first detection stage. Figure 13.3(a) shows the PDF of $\mathbf{d}_1[k]$, where again, the first stage detector has three decision regions denoted by "-1", "0" and "+1". First, let us assume that $\mathbf{b}_1[k] = -\mathbf{b}_2[k]$ are transmitted. Then, according to Equation 13.21 $\mathbf{d}_1[k]$ becomes $\mathbf{d}_1[k] = \eta_1[k]$, which falls in the region "0 ", provided that the noise contribution obéys $|\eta_1[k]| < V_T$. This probability can be expressed as:

$$\Pr(|\eta_1[k]| < V_T) = 1 - 2Q\left(\frac{V_T}{\sigma}\right). \tag{13.65}$$

As described in Section 13.3, when $\mathbf{d}_1[k]$ falls in the region "0" of Figure 13.3(a), the Type III detector uses the Type I detection variables of Equations 13.23 and 13.24, which are repeated here for convenience:

$$\hat{\mathbf{b}}_1 = \mathbf{d}_1 + \mathbf{d}_2 = \sqrt{NE}\,\mathbf{b}_1 + \eta_1 + \eta_2 \tag{13.66}$$

$$\hat{\mathbf{b}}_2 = \mathbf{d}_1 - \mathbf{d}_2 = \sqrt{NE}\,\mathbf{b}_2 + \eta_1 - \eta_2. \tag{13.67}$$

When $\mathbf{b}_1[k] = -1$, $\hat{\mathbf{b}}_1[k]$ of Equation 13.66 becomes $\hat{\mathbf{b}}_1[k] = -\sqrt{NE} + \eta_1 + \eta_2$, which results in a bit error, if $\eta_1 + \eta_2 > \sqrt{NE}$, indicating that the total noise contribution is higher than the signal contribution. The probability of this bit error denoted by $P_e(0|0)$, following the definition given in Section 13.4.2, corresponds to the scenario, where $\mathbf{d}_1[k]$ falls in the region "0", when it was expected to fall in the region "0", can be expressed as:

$$\begin{aligned}
P_e(0|0) &= \Pr(|\eta_1[k]| < V_T \mid \mathbf{b}_1[k] = -\mathbf{b}_2[k] = -1) \cdot \\
&\quad \Pr(\hat{\mathbf{b}}_1[k] > 0 \mid \mathbf{b}_1[k] = -\mathbf{b}_2[k] = -1, |\eta_1[k]| < V_T) \tag{13.68} \\
&= \Pr(|\eta_1[k]| < V_T \mid \mathbf{b}_1[k] = -\mathbf{b}_2[k] = -1) \cdot \\
&\quad \Pr(\eta_1[k] + \eta_2[k] > \sqrt{NE} \mid \mathbf{b}_1[k] = -\mathbf{b}_2[k] = -1, |\eta_1[k]| < V_T) \tag{13.69} \\
&= \Pr(\eta_1[k] + \eta_2[k] > \sqrt{NE}, |\eta_1[k]| < V_T \mid \mathbf{b}_1[k] = -\mathbf{b}_2[k] = -1) \tag{13.70}
\end{aligned}$$

In order to calculate $P_e(0|0)$ in Equation 13.70, we need the joint PDF of η_1 and η_2, denoted as f_{η_1,η_2}. Considering that η_1 and η_2 are independent and identical Gaussian random variables, f_{η_1,η_2} can be expressed as:

$$f_{\eta_1,\eta_2} = \frac{1}{2\pi\sigma^2} e^{-\frac{\eta_1^2 + \eta_2^2}{2\sigma^2}}, \tag{13.71}$$

where $\sigma^2 = (N/2)(N_o/2)$. The integration region associated with the bit error probability of Equation 13.70 is depicted in Figure 13.5(a). Referring to Equation 13.71 and to Figure 13.5(a), we can express the bit error probability as:

$$\begin{aligned}
P_e(0|0) &= \Pr(\eta_1[k] + \eta_2[k] > \sqrt{NE}, |\eta_1[k]| < V_T \mid \mathbf{b}_1[k] = -\mathbf{b}_2[k] = -1) \tag{13.72} \\
&= -\sqrt{NE} + \eta_1[k] + \eta_2[k] > 0, |\eta_1[k]| < V_T\} \tag{13.73} \\
&= \int_{-V_T}^{V_T} \int_{\sqrt{NE}-\eta_1}^{\infty} \frac{1}{2\pi\sigma^2} e^{-(\eta_1^2+\eta_2^2)/(2\sigma^2)} d\eta_2\, d\eta_1 \tag{13.74} \\
&= \int_{-V_T}^{V_T} \frac{1}{\sqrt{2\pi}\sigma} e^{-\eta_1^2/(2\sigma^2)} Q\left(\frac{\sqrt{NE}-\eta_1}{\sigma}\right) d\eta_1 \tag{13.75} \\
&= \frac{1}{\sqrt{2\pi}} \int_{-2\sqrt{\gamma}v}^{2\sqrt{\gamma}v} e^{-\eta^2/2} Q(2\sqrt{\gamma} - \eta) d\eta, \tag{13.76}
\end{aligned}$$

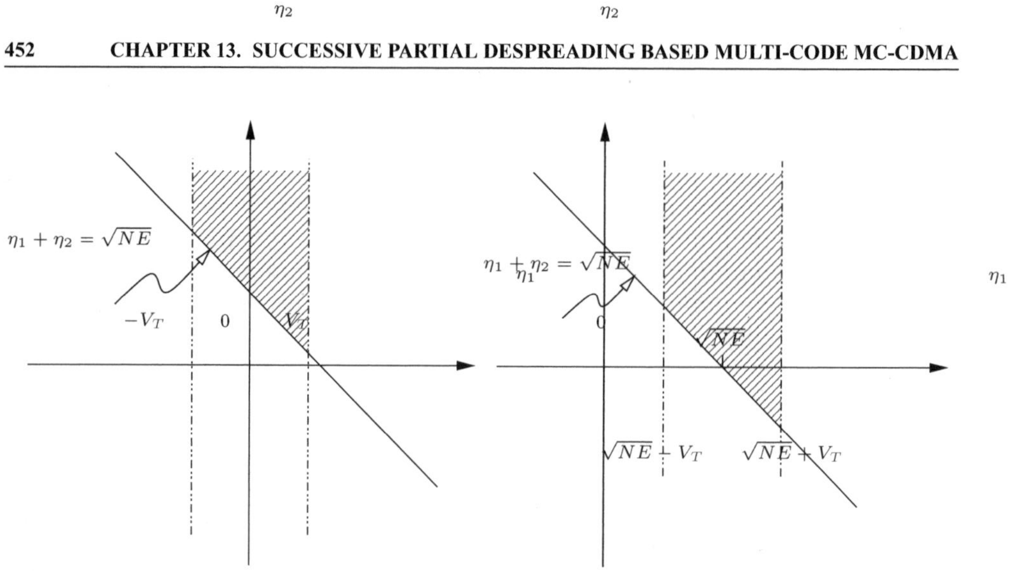

(a) Region 1   (b) Region 2

Figure 13.5: Integration regions for the conditional bit error probabilities associated with the Type III detector: (a) region 1 is for $\mathrm{P}(\eta_1 + \eta_2 > \sqrt{NE}, \; |\eta_1| < V_T)$ and (b) region 2 is for $\mathrm{P}(\eta_1 + \eta_2 > \sqrt{NE}, \; \sqrt{NE} - V_T < \eta_1 < \sqrt{NE} + V_T)$.

where, again, $\gamma = E/N_o$, and $\upsilon = V_T/\sqrt{NE}$. On the other hand, the conditional BEP, when $\mathbf{d}_1[k]$ falls in the region "0" of Figure 13.3(a) and $\hat{\mathbf{b}}_1[k] < 0$, although $\mathbf{b}_1[k] = -\mathbf{b}_2[k] = +1$, has the same value as $P_e(0|0)$ given in Equation 13.76.

Considering the probability that $\mathbf{d}_1[k]$ falls in the regions of "±1" in Figure 13.3(a) despite $\mathbf{b}_1[k] = -\mathbf{b}_2[k]$, which was given in Equation 13.32, the bit error probability for the case of $\mathbf{b}_1[k] = -\mathbf{b}_2[k]$, namely when $\mathbf{d}_1[k]$ is expected to fall in the region "0", can be expressed as:

$$P_e(0) = \mathrm{Q}(2\sqrt{\gamma}\upsilon) + \frac{1}{\sqrt{2\pi}} \int_{-2\sqrt{\gamma}\upsilon}^{2\sqrt{\gamma}\upsilon} e^{-\eta^2/2} \mathrm{Q}(2\sqrt{\gamma} - \eta) d\eta, \tag{13.77}$$

where the first term is the conditional BEP, when $\mathbf{d}_1[k]$ falls in the region of "+1" or "-1", while the second term is the conditional BEP, when $\mathbf{d}_1[k]$ falls in the region "0".

By contrast, when $\mathbf{b}_1[k] = \mathbf{b}_2[k] = -1$ or $+1$, bit errors occur (a) when the first stage detection variable $\mathbf{d}_1[k]$ falls in the region "+1" or "-1" of Figure 13.3(a), or (b) when it falls in the region "0" and the second stage detection variable falls in the region of "+1" or "-1". The probability of case (a), denoted by $P_e(\mp1| \pm 1)$, was already given in Equation 13.43 as:

$$P_e(\mp1| \pm 1) = \mathrm{Q}\left(\frac{\sqrt{NE} + V_T}{\sigma}\right) \tag{13.78}$$

$$= \mathrm{Q}(2\sqrt{\gamma}(1 + \upsilon)), \tag{13.79}$$

where again γ is the SNR per bit and v is the normalised threshold defined as $v = V_T/\sqrt{NE}$.

Let us now consider case (b). The associated conditional BEP is denoted by $P_e(0| \pm 1)$, since this corresponds to the scenario, where $\mathbf{d}_1[k]$ falls in the region "0" of Figure 13.3(a), when $\mathbf{d}_1[k]$ is expected to fall in the region "+1" or "-1", depending on the value of $\mathbf{b}_1[k]$. Specifically, when $\mathbf{b}_1[k] = \mathbf{b}_2[k] = -1$, according to Equation 13.21 the first stage detection variable becomes $\mathbf{d}_1[k] = -\sqrt{NE} + \eta_1[k]$, which falls in the region "0" associated with $|\mathbf{d}_1[k]| \le V_T$, if $(\sqrt{NE} - V_T) < \eta_1[k] < (\sqrt{NE} + V_T)$. In this case a bit error occurs, when $\hat{\mathbf{b}}_1[k] = -\sqrt{NE} + \eta_1[k]$ is positive. Hence, this probability can be expressed as:

$$P_e(0 - 1) = \Pr(|\mathbf{d}_1[k]| < V_T \mid \mathbf{b}_1[k] = \mathbf{b}_2[k] = -1) \cdot$$
$$\Pr(\hat{\mathbf{b}}_1[k] > 0 \mid \mathbf{b}_1[k] = \mathbf{b}_2[k] = -1, |\mathbf{d}_1[k]| < V_T) \qquad (13.80)$$
$$= \Pr(|\mathbf{d}_1[k]| < V_T, \hat{\mathbf{b}}_1[k] > 0 \mid \mathbf{b}_1[k] = \mathbf{b}_2[k] = -1) \qquad (13.81)$$
$$= \Pr(|-\sqrt{NE} + \eta_1[k]| \le V_T, -\sqrt{NE} + \eta_1[k] + \eta_2[k] > 0) \qquad (13.82)$$
$$= \Pr(\sqrt{NE} - V_T \le \eta_1[k] \le \sqrt{NE} + V_T, \eta_1[k] + \eta_2[k] > \sqrt{NE}) \quad (13.83)$$

On the other hand, the conditional BEP $P_e(0| + 1)$ of the Type III detector associated with the scenario, where the first stage detection variable $\mathbf{d}_1[k] = +\sqrt{NE} + \eta_1[k]$ falls in the region "0" of Figure 13.3(a), provided that $\mathbf{d}_1[k]$ is expected to fall in the region "+1", since $\mathbf{b}_1[k] = \mathbf{b}_2[k] = +1$ was transmitted, can be formulated following the same approach as for $P_e(0|-1)$ in Equation 13.80, which results in the same value, as $P_e(0|+1)$ of Equation 13.83. Hence, the combined BEP $P_e(0| \pm 1)$ can be expressed as:

$$P_e(0| \pm 1) = Pr(\mathbf{b}_1[k] = \mathbf{b}_2[k] = -1 \mid \mathbf{b}_1[k] = \mathbf{b}_2[k]) \cdot P_e(0| - 1)$$
$$+ Pr(\mathbf{b}_1[k] = \mathbf{b}_2[k] = +1 \mid \mathbf{b}_1[k] = \mathbf{b}_2[k]) \cdot P_e(0| + 1) \qquad (13.84)$$
$$= \frac{1}{2} P_e(0| - 1) + \frac{1}{2} P_e(0| + 1) \qquad (13.85)$$
$$= \Pr(\sqrt{NE} - V_T \le \eta_1[k] \le \sqrt{NE} + V_T, \eta_1[k] + \eta_2[k] > \sqrt{NE}). \quad (13.86)$$

In order to calculate $P_e(0| \pm 1)$ of Equation 13.86, we need the joint PDF of η_1 and η_2, which is given in Equation 13.71. The integration region, where the bracketed conditions of Equation 13.86 are satisfied, is depicted in Figure 13.5(b). Referring to Equation 13.71 and to Figure 13.5(b), $P_e(0| \pm 1)$ can be expressed as:

$$P_e(0| \pm 1) = \Pr(-\sqrt{NE} + \eta_1[k] + \eta_2[k] > 0, \sqrt{NE} - V_T < \eta_1[k] < \sqrt{NE} + V_T\}$$
$$= \frac{1}{\sqrt{2\pi}} \int_{2\sqrt{\gamma}(1-v)}^{2\sqrt{\gamma}(1+v)} e^{-\eta^2/2} Q(2\sqrt{\gamma} - \eta) \, d\eta . \qquad (13.87)$$

With the aid of Equations 13.79 and 13.87, the BEP associated with $\mathbf{b}_1[k] = \mathbf{b}_2[k]$ can be expressed as:

$$P_e(\pm 1) = Q(2\sqrt{\gamma}(1 + v)) + \frac{1}{\sqrt{2\pi}} \int_{2\sqrt{\gamma}(1-v)}^{2\sqrt{\gamma}(1+v)} e^{-\eta^2/2} Q(2\sqrt{\gamma} - \eta) d\eta . \qquad (13.88)$$

Finally, from Equations 13.77 and 13.88, we can derive the total BEP $P_{e,III}$ of the Type

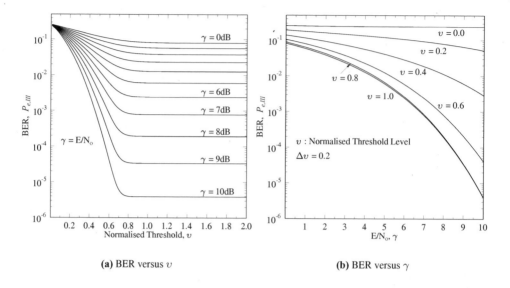

(a) BER versus v (b) BER versus γ

Figure 13.6: The bit error probability $P_{e,III}$ of the Type III detector. As the normalised threshold level v becomes higher, $P_{e,III}$ approaches the BER of the ideal BPSK, which is indistinguishable from the conventional BPSK scheme. However, this is achieved at the expense of increasing computational costs.

III detector, which can be expressed as:

$$P_{e,III} = \frac{1}{2}\left\{Q(2\sqrt{\gamma}v) + Q(2\sqrt{\gamma}(1+v))\right\}$$
$$+ \frac{1}{2\sqrt{2\pi}}\int_{-2\sqrt{\gamma}v}^{2\sqrt{\gamma}v} e^{-\eta^2/2}Q(2\sqrt{\gamma}-\eta)d\eta + \frac{1}{2\sqrt{2\pi}}\int_{2\sqrt{\gamma}(1-v)}^{2\sqrt{\gamma}(1+v)} e^{-\eta^2/2}Q(2\sqrt{\gamma}-\eta)d\eta \,,$$

$$(13.89)$$

which depends on the normalised threshold level, v, as well as on the SNR γ.

The bit error probability $P_{e,III}$ of the Type III detector is depicted against the normalised threshold level v and the bit energy to noise power ratio γ in Figure 13.6. In Figure 13.6(b) we can observe that the BER approaches the ideal BPSK BER curve, as v approaches unity. In fact, the BER curve associated with $v = 1$ was indistinguishable from the ideal BPSK BER curve. This implies that the Type III detector has the potential to reduce the required number of calculations without a noticeable performance degradation.

Having analysed the BEP of the Type III detector communicating over an AWGN channel, let us now consider the complexity of the Type III detector. The first stage detection variable $\mathbf{d}_1$ always has to be evaluated according to Equation 13.7 in the Type III detector, requiring $(N/2)^2$ number of multiplications and $(N/2)(N/2 - 1)$ additions. Additionally, when $\mathbf{d}_1$ falls in the region "0" of Figure 13.3(a) with the probability of $P_{III}(0)$, $\mathbf{d}_2$ has to be evaluated according to Equation 13.8, and then $\hat{\mathbf{b}}_1$ and $\hat{\mathbf{b}}_2$ have to be evaluated according to Equations 13.66 and 13.67, respectively. The former evaluation for $\mathbf{d}_2$ requires the same

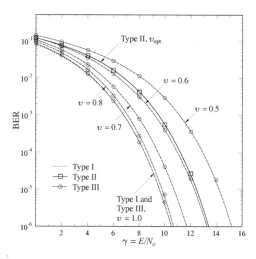

Figure 13.7: The bit error ratio versus γ performance of SPD detectors. The BER performance of the Type I detector is the same as that of a conventional despreading detector or as that of an ideal BPSK detector. The Type II detector exhibits an approximately 3dB disadvantage in terms of the required SNR compared to the Type I detector. The BER curves of the Type III detector show a wide range of variations, depending on the normalised threshold value, v. The markers represent the corresponding simulation results.

number of computations as for $\mathbf{d}_1$. On the other hand, the latter evaluation for $\hat{\mathbf{b}}_1$ and $\hat{\mathbf{b}}_2$ requires N number of additions. It is worth noting that the probability, $P_{III}(0)$, of having a $\mathbf{d}_1[k]$ value in the region "0" of Figure 13.3(a) at the first stage detector is the same as $P_{II}(0)$ given in Equation 13.61. Hence, the expression of the required number of multiplications, $N_{III}(\times)$, is the same as $N_{II}(\times)$ given in Equation 13.63 for the Type II detector. The required number of additions, $N_{III}(+)$, becomes slightly different, which can be expressed as:

$$N_{III}(+) = \frac{N}{2} \cdot \left(\frac{N}{2} - 1\right) + P_{III}(0) \cdot \left\{\frac{N}{2} \cdot \left(\frac{N}{2} - 1\right)\right\} + P_{III}(0) \cdot N \qquad (13.90)$$

$$= \frac{N(N-2)}{4} + \frac{N(N+2)}{8}\{1 - 2Q(2\sqrt{\gamma}v) + Q(2\sqrt{\gamma}(1-v)) - Q(2\sqrt{\gamma}(1+v))\}, \tag{13.91}$$

where the first and the second terms of Equation 13.90 correspond to the number of additions required to evaluate $\mathbf{d}_1$ and $\mathbf{d}_2$, respectively, and the last term of Equation 13.90 is that for evaluating $\hat{\mathbf{b}}_1$ and $\hat{\mathbf{b}}_2$, as mentioned before.

13.4.4 Summary and Discussion

In this section, we analysed the bit error rate of the three different types of SPD detectors in the AWGN channel. The corresponding bit error rate curves are depicted in Figure 13.7. The BER of the Type I detector, as given in Equation 13.25, is identical to that of ideal BPSK.

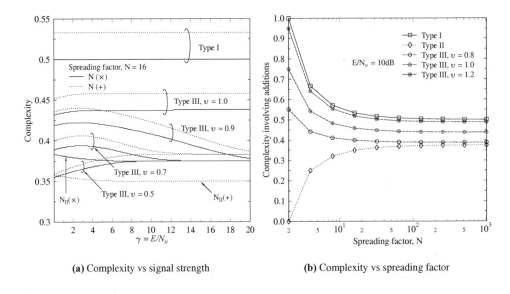

(a) Complexity vs signal strength   (b) Complexity vs spreading factor

Figure 13.8: Relative complexity of SPD detectors to the conventional despreading detector: (a) complexity versus E/N_o for a spreading factor of $N = 16$ (b) complexity versus spreading factor for $E/N_o = 10dB$.

Figure 13.7 shows that the Type II detector has an approximately 3dB SNR disadvantage compared to ideal BPSK and to conventional despreading-based detection, while attaining the same bit error rate performance. The optimum threshold level of Equation 13.56 was assumed for the Type II detector. The BER of the Type III detector varies depending on the threshold levels employed. We can observe that the Type III BER was close to that of the Type II detector, when the normalised threshold level v was set to 0.6 for the Type III detector. At $v = 0.8$ the BER of the Type III detector approached the ideal BER within about 0.2 dB E/N_o difference. The BER of the Type III scheme was indistinguishable from the ideal BER when $v = 1.0$ was used.

Figure 13.8 illustrates the implementational complexity of the three types of SPD detectors in terms of the achievable complexity normalised to the conventional full despreading-based detector. The required number of multiplications and additions is N^2 and $N^2 - N$, respectively, for the conventional full despreading based detector, where N is the spreading factor. The implementational complexity of the Type I, Type II and Type III detectors was given in Equation 13.26, 13.27, 13.63, 13.64 and 13.91. Figure 13.8(a) shows the achievable complexity relative to that of the conventional full despreading detector, which are $N_M(\times)/N^2$ and $N_M(+)/(N(N-1))$, where $M = I, II$ and III for the spreading factor of $N = 16$. We can observe in Figure 13.8(a) that the Type I detectors reduce the number of operations required by about a factor of two, compared to the conventional detectors. The Type II SPD detectors require 35% of the additions and 37.5% of the multiplications necessary for the conventional full despreading detector. The complexity of the Type III detector falls between those of the Type I and the Type II detectors. As the threshold becomes higher, the complexity of the Type III detector increases. As γ becomes higher, $N_{III}(\times)/N^2$ approaches

3/8 for $v < 1$, 7/16 for $v = 1$ and 1/2 for $v > 1$. Figure 13.8(b) shows the reduction of the required number of additions for the three types of SPD detectors as a function of the spreading factor N at $E/N_o = 10$dB. As the spreading factor N increases, the reduction of the required number of additions for the Type I and the Type II detectors approaches 1/2 and 3/8, respectively. The multiplication reduction factors of the three detectors do not vary, as N changes.

13.5 Effects of Impulse Noise or Narrow Band Jamming

As mentioned in Section 13.2, when the received signal is impaired by impulse noise [448, 449] in the time domain or by a narrow-band interferer / tone-jamming [450] in the frequency domain, some of the time domain chips or some of the subcarriers in the frequency domain may be corrupted completely and hence the obliteration of the corrupted symbols may result in a better performance. In this section, we assume that lth time domain chip or frequency domain subcarrier is completely lost and the index l is known to the receiver. We refer to this lost chip or subcarrier signal as the lost Walsh transform-domain symbol.

13.5.1 Conventional BPSK System without Spreading

Since there is no spreading involved, for example as in conventional OFDM, the obliteration of the lth symbol results in a bit error with the probability of 1/2 in conjunction with conventional BPSK schemes. Hence, the average BEP $P_{e,BPSK}$ of conventional BPSK schemes can be expressed as:

$$P_{e,BPSK} = \frac{1}{N}\frac{1}{2} + \frac{N-1}{N}Q(\sqrt{2\gamma}),$$ (13.92)

where the corresponding BER floor is given by $1/(2N)$, since one of the N bits has a 50% BER. The associated bit error rates for the spreading factors of $N = 8, 16, 32$ are depicted in Figure 13.9.

13.5.2 Conventional Despreading Scheme

When strong single-tone jamming is present in a multi-carrier system, the received subcarrier having the same frequency as the tone jammer or the frequency near the jamming tone exhibits a high power owing to the jamming signal, which results in unreliable detection of the symbol conveyed by the subcarrier [451]. In a WHT/OFDM system, this corrupted high magnitude symbol reduces the reliability of detecting the original bit, since the Signal to Interference + Noise Ratio (SINR) is given as:

$$\text{SINR} = \frac{NE}{J + N\sigma^2} = \frac{E}{J/N + \sigma^2},$$ (13.93)

where N is the spreading factor, E is the signal power, J is the jamming power and σ is the noise variance. While a high value of N reduces the adverse effect of jamming, the SINR may be unacceptably low, when the jamming power J is sufficiently high. In this case, we can improve the SINR by discarding the jamming-contaminated symbol during the despreading

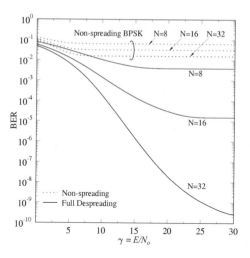

Figure 13.9: The bit error rates of the conventional detectors for transmission over the AWGN channel, when one of N transform-domain symbols is obliterated. The non-spread BPSK scheme has a high irreducible BER of $0.5/N$, while the spread systems have far lower error floor at $1/2^N$.

process. Please note that this approach is equivalent to the frequency domain notch filtering technique proposed, for example, in [452] to suppress the tone-jamming signal. Then the frequency domain received vector, $\hat{\mathbf{s}} = \sqrt{E/N}\, \mathbf{W}_N\, \mathbf{b} + \mathbf{n}$, should be modified to $\hat{\mathbf{s}}'$ such that $\hat{\mathbf{s}}'[l] = 0$ indicating the loss of the lth transform domain symbol. Alternatively, we can model the loss of the lth transform domain symbol by replacing $\mathbf{W}_N$ with $\mathbf{W}'_N$, such that its lth column is replaced by an $N \times 1$-dimensional zero vector. Then, the despread symbol vector $\mathbf{d}$ can be expressed as:

$$\mathbf{d} = \mathbf{W}'_N \hat{\mathbf{s}} \tag{13.94}$$

$$= \sqrt{\frac{E}{N}}\, \mathbf{W}'_N\, \mathbf{W}_N\, \mathbf{b} + \mathbf{W}'_N\, \mathbf{n}\,. \tag{13.95}$$

This equation can be expressed for each vector component $\mathbf{d}[k]$ as:

$$\mathbf{d}[k] = \sqrt{\frac{E}{N}}\, \alpha[k] + \eta[k]\,, \tag{13.96}$$

where the normalised noiseless signal component, $\alpha[k]$, may be expressed using Equation 13.95 as:

$$\alpha[k] = (N-1)\mathbf{b}[k] - \sum_{i=1, i \neq k}^{N} \mathbf{W}_N[k,l]\mathbf{W}_N[l,i]\mathbf{b}[i] \tag{13.97}$$

and the noise component, $\eta[k]$, is given with the aid of Equation 13.95 as:

$$\eta[k] = \sum_{i=1, i \neq l}^{N} \mathbf{W}_N[k, i] \, \mathbf{n}[i] \,, \tag{13.98}$$

where $\eta[k]$ has a zero mean Gaussian distribution with a variance of $(N - 1)N_o/2$, since it is a linear sum of $N - 1$ Gaussian random variables having the variance of $N_o/2$. It is worth mentioning that the second term at the right-hand side of Equation 13.97 is the Multiple Code Interference (MCI) or Multiple User Interference (MUI), which is a consequence of the destroyed orthogonality between the codes owing to the loss of the lth transform-domain symbol.

Let us determine the probability density function (PDF) of $\alpha[k]$, which was defined in Equation 13.97. Assuming that the information bits $\mathbf{b}[i]$ are independent binary random variables, $\alpha[k]$ becomes a binomially distributed random variable, since $\alpha[k]$ is the linear sum of $\mathbf{b}[i]$ [274, 453]. Given a fixed $\mathbf{b}[k]$ value, there are $N - 1$ independent $\mathbf{b}[i]$ values in the N-dimensional vector $\mathbf{b}$. If we choose n number of bits $\mathbf{b}[i]$ out of $N - 1$ such that $\mathbf{b}[i] = \mathbf{W}_N[k, l] \mathbf{W}_N[l, i]$ yielding n number of "+1"s and $N - 1 - n$ number of "-1"s in the summation of the second term in Equation 13.97, then with the aid of Equation 13.97 we have:

$$\begin{aligned} \alpha[k] &= (N - 1)\mathbf{b}[k] - \{n - (N - 1 - n)\} \\ &= (N - 1)(\mathbf{b}[k] + 1) - 2n \,, \end{aligned} \tag{13.99}$$

which becomes either $2(N - 1 - n)$ or $-2n$ depending on the bit $\mathbf{b}[k]$ being $+1$ or -1. The probability of choosing n $\mathbf{b}[i]$'s out of $N - 1$ is given as $\binom{N-1}{n} (1/2^{N-1})$ [453] and the probability of $\mathbf{b}[k]$ being "+1" or "-1" is $1/2$. Considering that $\binom{N-1}{n} = \binom{N-1}{N-1-n}$ [453], we can represent the PDF $f_C(\alpha)$ of $\alpha[k]$ for the conventional despreading scheme as:

$$f_C(\alpha) = \Pr(\mathbf{b}[k] = +1) \cdot \sum_{n=0}^{N-1} \binom{N-1}{n} \frac{1}{2^{(N-1)}} \delta(\alpha - 2(N - 1 - n))$$

$$+ \Pr(\mathbf{b}[k] = -1) \cdot \sum_{n=0}^{N-1} \binom{N-1}{n} \frac{1}{2^{(N-1)}} \delta(\alpha + 2n) \tag{13.100}$$

$$= \frac{1}{2^N} \sum_{n=0}^{N-1} \binom{N-1}{n} \left[\delta(\alpha - 2n) + \delta(\alpha + 2n) \right] \,. \tag{13.101}$$

Figure 13.10 illustrates the PDF of the normalised noiseless signal component α for the spreading factor of $N = 8$. We can observe that two binomial distributions are superimposed, meeting at zero. Since the expected value of $\alpha[k]$ in Equation 13.97 is $E[\alpha[k]] = (N - 1) \mathbf{b}[k]$, the two binomial PDFs in Equation 13.10 are centred at $\pm(N - 1) = \pm 7$.

Having obtained the statistics of the signal component $\alpha[k]$ as well as those of the noise component $\eta[k]$, let us now consider the BEP of the conventional despreading scheme transmitting over AWGN channels, assuming that one transform domain symbol is lost due to a strong tone-jamming signal. When $\alpha[k] = -2n$ and $\mathbf{b}[k] = -1$, the decision variable $\mathbf{d}[k]$

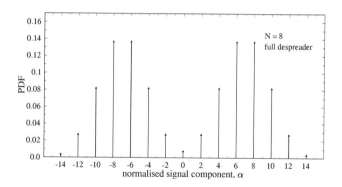

Figure 13.10: The PDF $f_C(\alpha)$ given in Equation 13.101, of the normalised signal component α for the conventional despreading detector when transmitting over AWGN channels under the assumption that one transform-domain is lost due to a strong tone-jamming signal. The spreading factor was $N = 8$, while the PDF is constituted by the superposition of two binomial distributions, as implied by Equation 13.101.

given in Equation 13.96 has the Gaussian distribution having the mean value of $-\sqrt{E/N}\,2n$ and the variance of $\sigma^2 = (N-1)(N_o/2)$. Hence, the bit error probability $P_{e,C}[\alpha = -2n]$ can be formulated as:

$$P_{e,C}[\alpha = -2n] = \Pr(\alpha = -2n) \cdot \Pr(\mathbf{d}[k] > 0 \mid \alpha = -2n) \qquad (13.102)$$

$$= \frac{1}{2^{(N-1)}} \binom{N-1}{n} \cdot Q\left(\frac{\sqrt{E/N}\,2n}{\sigma}\right) \qquad (13.103)$$

$$= \frac{1}{2^{(N-1)}} \binom{N-1}{n} \cdot Q\left(\frac{2n\sqrt{2\gamma}}{\sqrt{N(N-1)}}\right). \qquad (13.104)$$

The BEP corresponding to the scenario of $\mathbf{b}[k] = +1$ can be evaluated in a similar way, yielding the same expression as in Equation 13.104. The probability of $\mathbf{b}[k]$ being $+1$ and -1 is $1/2$ and the BEP given in Equation 13.104 can be used to represent the combined BEP for both scenarios. Referring to Equations 13.101 and 13.104, we can obtain the average BEP $P_{e,C}$ of the conventional despreading detector, which can be expressed as

$$P_{e,C} = \frac{1}{2^{N-1}} \sum_{n=0}^{N-1} \binom{N-1}{n} Q\left(\frac{2n\sqrt{2\gamma}}{\sqrt{N(N-1)}}\right), \qquad (13.105)$$

where γ is the bit energy to noise power ratio, defined as $\gamma = E/N_o$ and N is the spreading factor. The BER floor imposed by the corruption of the lth transform domain symbol is given as $\lim_{\gamma \to \infty} P_{e,C} = 1/2^N$. Figure 13.9 compares the BEP $P_{e,C}$ with that of the non-spreading detectors.

In addition to characterising the detector's performance, we are interested in the required number of operations in the detection process as well. Assuming that the position of the lost transform-domain symbol is known, the number of operations required is given by:

$$N_C(\times) = (N-1)N \qquad (13.106)$$

$$N_C(+) = (N-2)N, \qquad (13.107)$$

where $N_C(\times)$ is the number of multiplications required and $N_C(+)$ is the number of additions required for the conventional despreading detector, since the $N-1$ transform-domain symbols except for the lth symbol, are involved in evaluating N number of decision variables $\mathbf{d}[k]$ according to Equation 13.94.

13.5.3 SPD Detectors

Let us now consider the performance of the various SPD detectors. The received transform-domain symbol vector, $\hat{\mathbf{s}}$, may be expressed in its decomposed form, given by:

$$\hat{\mathbf{s}}_1 = \sqrt{\frac{E}{N}}\mathbf{W}_{N/2}\mathbf{b}_1 + \sqrt{\frac{E}{N}}\mathbf{W}_{N/2}\mathbf{b}_2 + \mathbf{n}_1 \qquad (13.108)$$

$$\hat{\mathbf{s}}_2 = \sqrt{\frac{E}{N}}\mathbf{W}_{N/2}\mathbf{b}_1 + \sqrt{\frac{E}{N}}\overline{\mathbf{W}}_{N/2}\mathbf{b}_2 + \mathbf{n}_2 . \qquad (13.109)$$

Since the position of the corrupted transform-domain symbol, l, is assumed to be known, we can always identify the received sub-vector, which has not been affected by the corrupted transform-domain symbol, which will be used for the first stage detection. Without loss of generality, let us assume that $l = N/2 + l'$, $0 \le l' \le N/2$, i.e. the lower-half received sub-vector $\mathbf{r_2}$ contains the lost symbol at the l'th position. Then, after despreading, the detection variables $\mathbf{d}_1$ and $\mathbf{d}_2$ become:

$$\mathbf{d}_1 = \mathbf{W}_{N/2}\hat{\mathbf{s}}_1 \qquad (13.110)$$

$$= \sqrt{\frac{E}{N}}\frac{N}{2}(\mathbf{b}_1 + \mathbf{b}_2) + \eta_1 , \qquad (13.111)$$

$$\mathbf{d}_2 = \mathbf{W}'_{N/2}\hat{\mathbf{s}}_2 \qquad (13.112)$$

$$= \sqrt{\frac{E}{N}}\mathbf{W}'_{N/2}\mathbf{W}_{N/2}(\mathbf{b}_1 - \mathbf{b}_2) + \eta_2 , \qquad (13.113)$$

where $\mathbf{W}'_{N/2}$ is the modified version of $\mathbf{W}_{N/2}$, such that its l'th column was replaced by a $N/2 \times 1$-dimensional zero vector. Furthermore, η_1 in Equation 13.111 is a zero mean

Gaussian random variable vector. Its covariance matrix, $\text{Cov}(\eta_1)$, can be calculated as:

$$\begin{aligned}
\text{Cov}(\eta_1) &= \text{E}\{\eta_1 \eta_1^T\} \\
&= \text{E}\{\mathbf{W}_{N/2}\, \mathbf{n}_1\, \mathbf{n}_1^T\, \mathbf{W}_{N/2}^T\} \\
&= \mathbf{W}_{N/2}\, \text{E}\{\mathbf{n}_1 \mathbf{n}_1^T\}\, \mathbf{W}_{N/2} \\
&= \mathbf{W}_{N/2}\, \frac{N_o}{2}\, \mathbf{I}_{N/2}\, \mathbf{W}_{N/2} \\
&= \frac{N}{2}\, \frac{N_o}{2}\, \mathbf{I}_{N/2} \, .
\end{aligned} \tag{13.114}$$

Hence, the noise samples $\eta_1[k]$ are *iid* and the associated noise variance is $\sigma_1^2 = (N/2)\,(N_o/2)$, where N_o is the noise power associated with each $\mathbf{n}[i]$. The noise contribution of the second detection variable η_2 in Equation 13.113 is also a zero mean Gaussian random variable vector. The associated covariance matrix $\text{Cov}(\eta_2)$, can be evaluated as:

$$\begin{aligned}
\text{Cov}(\eta_2) &= \text{E}\{\eta_2 \eta_2^T\} \\
&= \text{E}\{\mathbf{W}_{N/2}'\, \mathbf{n}_1\, \mathbf{n}_1^T\, \mathbf{W}_{N/2}'^T\} \\
&= \mathbf{W}_{N/2}'\, \text{E}\{\mathbf{n}_1 \mathbf{n}_1^T\}\, \mathbf{W}_{N/2}'^T \\
&= \mathbf{W}_{N/2}'\, \frac{N_o}{2}\, \mathbf{I}_{N/2}\, \mathbf{W}_{N/2}'^T \\
&= \frac{N_o}{2}\, \mathbf{W}_{N/2}'\, \mathbf{W}_{N/2}'^T \, ,
\end{aligned} \tag{13.115}$$

where the diagonal elements of $\mathbf{W}_{N/2}'\, \mathbf{W}_{N/2}'^T$ have the value of $N/2-1$ and the off diagonal elements are $+1$ or -1 due to the all zero values in the l'th column of $\mathbf{W}_{N/2}'$. Hence, the noise contributions $\eta_2[k]$ are not independent. However, considering that $\sigma_2^2 = (N/2 - 1)\, N_o/2$ in the diagonal elements of $\text{Cov}(\eta_2)$ grows linearly with N, we can conclude that the noise contributions $\eta_2[k]$ become nearly independent for high spreading factors N.

Having investigated the noise statistics, let us now concentrate on the desired signal component. The kth element of the first stage detection variable, $\mathbf{b}_1[k]$, may be rewritten as:

$$\mathbf{d}_1[k] = \sqrt{\frac{E}{N}}\alpha_1[k] + \eta_1[k] \, , \tag{13.116}$$

where the normalised noiseless signal component $\alpha_1[k]$ is defined as:

$$\alpha_1[k] = \frac{N}{2}(\mathbf{b}_1[k] + \mathbf{b}_2[k]) \, . \tag{13.117}$$

Since the first stage detection variable is assumed to be unaffected by the symbol loss, $\alpha_1[k]$ becomes either $\pm N$ or zero with an equal probability of $1/2$.

The modified despreading matrix $\mathbf{W}_{N/2}'$ of Equation 13.113 can be written as $\mathbf{W}_{N/2}' = \mathbf{W}_{N/2}' - \mathbf{L}_{N/2}$, where $\mathbf{L}_{N/2}$ is a $N/2 \times N/2$-dimensional matrix having all zero elements except the l'th column, where the value is the same as that of the despreading matrix $\mathbf{W}_{N/2}$, namely $\mathbf{L}_{N/2}[i, l'] = \mathbf{W}_{N/2}[i, l']$. Hence, the second stage detection variable $\mathbf{d}_2$ of Equation 13.113 can be written as:

$$\mathbf{d}_2 = \sqrt{\frac{E}{N}} \left(\mathbf{W}_{N/2} - \mathbf{L}_{N/2} \right) \cdot \mathbf{W}_{N/2} (\mathbf{b}_1 - \mathbf{b}_2) + \eta_2 \tag{13.118}$$

$$= \sqrt{\frac{E}{N}} \frac{N}{2} (\mathbf{b}_1 - \mathbf{b}_2) - \sqrt{\frac{E}{N}} \mathbf{L}_{N/2} \mathbf{W}_{N/2} (\mathbf{b}_1 - \mathbf{b}_2) + \eta_2 \tag{13.119}$$

The kth element of $\mathbf{d}_2$ in Equation 13.119 can be written as:

$$\mathbf{d}_2[k] = \sqrt{\frac{E}{N}} \, \alpha_2[k] + \eta_2[k] \,, \tag{13.120}$$

where the normalised signal component $\alpha_2[k]$ is defined as:

$$\alpha_2[k] = \frac{N}{2} (\mathbf{b}_1[k] - \mathbf{b}_2[k]) - \sum_{i=1}^{N/2} \mathbf{W}_{N/2}[k, l] \, \mathbf{W}_{N/2}[l, i](\mathbf{b}_1[i] - \mathbf{b}_2[i]) \tag{13.121}$$

$$= \left(\frac{N}{2} - 1 \right) (\mathbf{b}_1[k] - \mathbf{b}_2[k]) - \sum_{i=1, i \neq k}^{N/2} \mathbf{W}_{N/2}[k, l] \, \mathbf{W}_{N/2}[l, i](\mathbf{b}_1[i] - \mathbf{b}_2[i]) \,. \tag{13.122}$$

Following a similar approach to that described in the previous section, we can derive the PDF of $\alpha_2[k]$ of Equation 13.122, which is our next objective. For fixed $\mathbf{b}_1[k]$ and $\mathbf{b}_2[k]$ values, we have $N - 2$ independent random binary variables $\mathbf{b}_1[i]$ and $\mathbf{b}_2[i]$, where $i \neq k$. The probability of choosing n number of specific $\mathbf{b}_1[i]$ values and/or $\mathbf{b}_2[i]$ values out of $N - 2$, such that $\mathbf{b}_1[i] = \mathbf{W}_{N/2}[k, l] \, \mathbf{W}_{N/2}[l, i]$ and $\mathbf{b}_2[i] = -\mathbf{W}_{N/2}[k, l] \, \mathbf{W}_{N/2}[l, i]$, is $\binom{N-2}{n}/2^{N-2}$. Then we have n number of "+1"s and $N - 2 - n$ number of "-1"s in the summation of the second term in the right-hand side of Equation 13.122, yielding:

$$\alpha_2[k] = \left(\frac{N}{2} - 1 \right) (\mathbf{b}_1[k] - \mathbf{b}_2[k]) - \{n - (N - 2 - n)\}$$

$$= \left(\frac{N}{2} - 1 \right) (\mathbf{b}_1[k] - \mathbf{b}_2[k]) + N - 2 - 2n \,. \tag{13.123}$$

Depending on the information bits transmitted, we have either Case 1: $\mathbf{b}_1[k] = \mathbf{b}_2[k]$ or Case 2: $\mathbf{b}_1[k] = -\mathbf{b}_2[k]$. For Case 1, $\alpha_2[k]$ of Equation 13.123 becomes $(N-2)(\mathbf{b}_1[k]+1) - 2n$, which is either $2(N - 2 - n)$ or $-2n$ depending on $\mathbf{b}_1[k]$ being $+1$ or -1, respectively.

Considering that $\binom{N-2}{n} = \binom{N-2}{N-2-n}$, we can represent the PDF of $\alpha_2[k]$, $f_{SPD,1}(\alpha_2)$, as:

$$f_{SPD,1}(\alpha_2) = \Pr(\mathbf{b}_1[k] = +1) \cdot \sum_{n=0}^{N-2} \binom{N-2}{n} \frac{1}{2^{N-2}} \delta(\alpha_2 - 2(N-2-n))$$

$$+ \Pr(\mathbf{b}_1[k] = -1) \cdot \sum_{n=0}^{N-2} \binom{N-2}{n} \frac{1}{2^{N-2}} \delta(\alpha_2 + 2n)$$

$$= \frac{1}{2} \frac{1}{2^{N-2}} \cdot \sum_{n=0}^{N-2} \binom{N-2}{N-2-n} \delta(\alpha_2 - 2n) \tag{13.124}$$

$$+ \frac{1}{2} \frac{1}{2^{N-2}} \cdot \sum_{n=0}^{N-2} \binom{N-2}{n} \frac{1}{2^{N-2}} \delta(\alpha_2 + 2n) \tag{13.125}$$

$$= \frac{1}{2^{N-1}} \sum_{n=0}^{N-2} \binom{N-2}{n} \left\{ \delta(\alpha_2 - 2n) + \delta(\alpha_2 + 2n) \right\} . \tag{13.126}$$

Figure 13.11(a) shows an example of $f_{SPD,1}(\alpha_2)$ for the spreading factor of $N = 8$, which is again constituted by the superposition of two binomial distributions centred at ± 6, because the expected value of $\alpha_2[k]$ is given by the first term of Equation 13.122, which is $\left(\frac{N}{2} - 1\right) (\mathbf{b}_1[k] - \mathbf{b}_2[k]) = 6\,\mathbf{b}_1[k]$.

For Case 2, namely when $\mathbf{b}_1[k] = -\mathbf{b}_2[k]$, $\alpha_2[k]$ of Equation 13.123 becomes $N-2-2n$ with the probability of $\binom{N-2}{n} (1/2^{N-2})$. Hence, we may express the PDF $f_{SPD,2}(\alpha_2)$ of $\alpha_2[k]$ as:

$$f_{SPD,2}(\alpha_2) = \frac{1}{2^{N-2}} \sum_{n=0}^{N-2} \binom{N-2}{n} \delta(\alpha_2 - (N-2-2n)) \tag{13.127}$$

$$= \frac{1}{2^{N-2}} \sum_{n=0}^{N-2} \binom{N-2}{n} \delta(\alpha_2 - 2n) \tag{13.128}$$

An example of $f_{SPD,2}(\alpha_2)$ is depicted in Figure 13.11(b) for the spreading factor of $N = 8$. Since we now have all the necessary statistics of SPD detection variables, we can analyse the bit error rate performance of the three different types of SPD detectors.

13.5.4   Type I Detector

In the Type I SPD detector, according to Equations 13.13 and 13.13 $\hat{\mathbf{b}}_1$ and $\hat{\mathbf{b}}_1$ are given the sum and the difference of $\mathbf{d}_1$ and $\mathbf{d}_2$. Since $\mathbf{d}_1$ and $\mathbf{d}_2$ are given in Equation 13.116 and Equation 13.120, the estimates of the original bits become:

$$\hat{\mathbf{b}}_1[k] = \sqrt{\frac{E}{N}}(\alpha_1[k] + \alpha_2[k]) + \eta_1[k] + \eta_2[k] \tag{13.129}$$

$$\hat{\mathbf{b}}_2[k] = \sqrt{\frac{E}{N}}(\alpha_1[k] - \alpha_2[k]) + \eta_1[k] - \eta_2[k] . \tag{13.130}$$

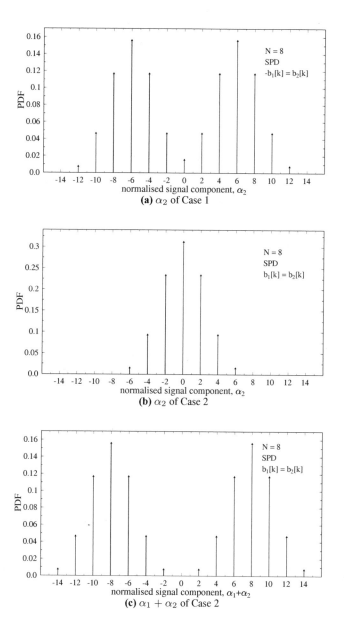

Figure 13.11: The PDF of normalised signal component α_1 and α_2 defined in Equations 13.117 and 13.121, respectively, of the SPD detector for transmission over the AWGN channel, when one transform-domain symbol is obliterated. The spreading factor was $N = 8$. Case 1 is for $-\mathbf{b}_1[k] = \mathbf{b}_2[k]$ and Case 2 is for $\mathbf{b}_1[k] = \mathbf{b}_2[k]$.

Since $\eta_1[k]$ and $\eta_2[k]$ are independent zero-mean Gaussian random variables, the noise term $\eta_1[k]\pm\eta_2[k]$ is also a zero-mean Gaussian random variable with a variance of $(N/2)(N_o/2)+(N/2-1)(N_o/2) = (N-1)(N_o/2)$. We can conclude that the noise statistics of the Type I decision variables are identical to those of the conventional full despreader characterised in Equation 13.98.

For the case of $\mathbf{b}_1[k] = -\mathbf{b}_2[k]$, according to Equation 13.117 $\alpha_1[k]$ becomes zero, and hence according to Equations 13.129 and 13.130 $\sqrt{E/N}\alpha_2[k]$ becomes the mean of the Gaussian random variables $\hat{b}_1$ and $\hat{b}_2$. The PDF of $\alpha_2[k]$ for the case of $\mathbf{b}_1[k] = -\mathbf{b}_2[k]$ was given in Equation 13.126.

On the other hand, when $\mathbf{b}_1[k] = \mathbf{b}_2[k]$, according to Equation 13.117 $\alpha_1[k]$ becomes $N\,\mathbf{b}_1[k] = \pm N$ depending on the value of $\mathbf{b}_1[k]$ and $\alpha_1[k]$ becomes:

$$\alpha_2[k] = -\sum_{i=1,i\neq k}^{N/2} \mathbf{W}_{N/2}[k,l]\,\mathbf{W}_{N/2}[l,i](\mathbf{b}_1[i]-\mathbf{b}_2[i]), \qquad (13.131)$$

while its PDF was given in Equation 13.128. Then, the mean of $\hat{b}_1[k]$ given in Equation 13.129 becomes $\sqrt{E/N}(\alpha_2[k] \pm N)$ depending on the actual value of $\mathbf{b}_1[k]$. In this case, the PDF of the effective normalised signal component α defined by $\alpha = \alpha_1 + \alpha_2[k]$, which becomes $\alpha = \alpha_2[k] \pm N$ according to the above argument, can be derived by introducing the variable $\alpha_2[k] = \alpha - N$ with a probability of $1/2$ and $\alpha_2[k] = \alpha + N$ with a probability of $1/2$ in the expression of the PDF of $\alpha_2[k]$ given in Equation 13.127, yielding:

$$f_{SPD,2}(\alpha) = \frac{1}{2^{N-2}} \sum_{n=0}^{N-2} \binom{N-2}{n} \{\frac{1}{2}\delta(\alpha+N-(N-2-2n)) + \frac{1}{2}\delta(\alpha-N-(N-2-2n))\}$$

$$= \frac{1}{2^{N-1}} \sum_{n=0}^{N-2} \binom{N-2}{n} \{\delta(\alpha + 2(n+1)) + \delta(\alpha - 2(N-1-n))\} \quad (13.132)$$

$$= \frac{1}{2^{N-1}} \sum_{n=0}^{N-2} \binom{N-2}{n} \{\delta(\alpha + 2(n+1)) + \delta(\alpha - 2(n+1))\}, \qquad (13.133)$$

where we used the relationship of $\binom{N-2}{n} = \binom{N-2}{N-2-n}$ to replace $\delta(\alpha - 2(N-1-n))$ of Equation 13.132 with $\delta(\alpha - 2(n+1))$ of Equation 13.133. An example of $f_{SPD,2}(\alpha)$ is given in Figure 13.11(c) for the spreading factor of $N = 8$.

Since $\mathbf{b}_1[k]$ and $\mathbf{b}_2[k]$ are independent, the overall PDF of the effective signal component of the Type I SPD detector can be expressed as the sum of the PDFs given in Equations 13.126

and 13.133, yielding:

$$f_I(\alpha) = \frac{1}{2} f_{SPD,1} + \frac{1}{2} f_{SPD,2'} \tag{13.134}$$

$$= \frac{1}{2^N} \sum_{n=0}^{N-2} \binom{N-2}{n} \{\delta(\alpha - 2n) + \delta(\alpha + 2n)\}$$

$$+ \frac{1}{2^N} \sum_{n=0}^{N-2} \binom{N-2}{n} \{\delta(\alpha - 2(n+1)) + \delta(\alpha + 2(n+1))\} \tag{13.135}$$

$$= \frac{1}{2^N} \sum_{n=0}^{N-2} \binom{N-2}{n} \{\delta(\alpha - 2n) + \delta(\alpha + 2n)\}$$

$$+ \frac{1}{2^N} \sum_{n=1}^{N-1} \binom{N-2}{n-1} \{\delta(\alpha - 2n) + \delta(\alpha + 2n)\} \tag{13.136}$$

$$= \frac{1}{2^N} \sum_{n=0}^{N-1} \binom{N-1}{n} \{\delta(\alpha - 2n) + \delta(\alpha + 2n)\} , \tag{13.137}$$

where we used the relationship of [454]:

$$\binom{N-2}{n-1} + \binom{N-2}{n} = \binom{N-1}{n} . \tag{13.138}$$

We can observe that the PDF $f_I(\alpha)$ given in Equation 13.137 is identical to the PDF of $\alpha[k]$ of the conventional full despreader, which was given in Equation 13.101. Hence we can conclude that the bit error rate performance of the Type I detector for transmission over the AWGN channel with one obliterated transform-domain symbol is identical to that of the conventional despreader, given in Equation 13.105. Figure 13.12(a) shows the BER as a function of E/N_o.

Let us now consider the complexity of the Type I SPD detector. Since $\mathbf{d}_1$ is calculated according to Equation 13.110, the number of operations required is given as:

$$N_{\mathbf{d}_1}(\times) = \frac{N}{2} \cdot \frac{N}{2} = \left(\frac{N}{2}\right)^2 \tag{13.139}$$

$$N_{\mathbf{d}_1}(+) = \left(\frac{N}{2} - 1\right) \cdot \frac{N}{2} . \tag{13.140}$$

On the other hand, $\mathbf{d}_2$ is calculated according to Equation 13.112, exploiting the fact that the l'th transform-domain symbol was obliterated and hence eliminating the unnecessary operations. Hence the number of operations required is given by:

$$N_{\mathbf{d}_2}(\times) = \left(\frac{N}{2} - 1\right) \frac{N}{2} \tag{13.141}$$

$$N_{\mathbf{d}_2}(+) = \left(\frac{N}{2} - 2\right) \frac{N}{2} \tag{13.142}$$

Since the Type I SPD detector only requires additional additions, when evaluating $\hat{\mathbf{b}}_1$ and $\hat{\mathbf{b}}_2$ according to Equations 13.13 and 13.14, respectively, the required number of multiplications $N_I(\times)$ is given as:

$$
\begin{aligned}
N_I(\times) &= N_{\mathbf{d}_1}(\times) + N_{\mathbf{d}_2}(\times) \\
&= \frac{1}{4}(2N^2 - 2N) \\
&= \frac{1}{2}N_C(\times),
\end{aligned}
\tag{13.143}
$$

where $N_C(\times)$ is the number of multiplications required for a conventional full despreading detector given in Equation 13.106. By contrast, the required number of additions $N_I(+)$ involved in the evaluation of Equations 13.13 and 13.14 is given as:

$$
\begin{aligned}
N_I(+) &= N_{\mathbf{d}_1}(+) + N_{\mathbf{d}_2}(+) + N \\
&= \frac{1}{4}(N^2 - 2N + N^2 - 4N + 2N) + \frac{1}{2}N \\
&= \frac{1}{2}N_C(+) + \frac{1}{2}N,
\end{aligned}
\tag{13.144}
$$

where $N_C(+)$ is the number of additions required for a conventional full despreading detector given in Equation 13.106. The achievable complexity reduction is shown in Figure 13.12(b) and will be discussed in more depth in Section 13.5.7.

13.5.5 Type II Detector

In the Type II SPD detector, only $\mathbf{d}_1[k]$ is used at the first stage detection. Recall from Section 13.3 that when $\mathbf{d}_1[k]$ falls in the region "0" of Figure 13.3, $\mathbf{d}_2[k]$ is used for the second stage detection assuming that $\mathbf{b}_1[k] = -\mathbf{b}_2[k]$, as described in Section 13.3 Since we assumed that $\mathbf{d}_1[k]$ is not affected by the obliterated transform-domain symbol, the BER analysis regarding $\mathbf{d}_1[k]$ given in Section 13.4.2 may be used here without any modifications. The analysis regarding $\mathbf{d}_2[k]$ may be modified considering the PDF given in Equation 13.126. Specifically, $P_e(0|0)$ of Equation 13.36 has to be modified, which is the bit error probability, when $\mathbf{d}_1[k]$ falls in the region "0" of Figure 13.3(a), provided that $\mathbf{b}_1[k] = -\mathbf{b}_2[k]$ and, hence $\mathbf{d}_2[k]$ has to be used for the final detection. The BEP $P_e(0|0)$ can be formulated as:

$$
P_e(0|0) = \Pr(\,|\mathbf{d}_1[k]| \le V_T \mid \mathbf{b}_1[k] = -\mathbf{b}_2[k]) \cdot P_{e,2}
\tag{13.145}
$$

$$
= \left\{1 - 2Q\left(\frac{V_T}{\sigma_1}\right)\right\} \cdot P_{e,2},
\tag{13.146}
$$

where $P_{e,2}$ is the average BEP at the second stage detection based on $\mathbf{d}_2[k]$ of Equation 13.120 when $\mathbf{b}_1[k] = -\mathbf{b}_2[k]$. The PDF of the normalised signal component α_2 of $\mathbf{d}_2[k]$ was given in Equation 13.126 and a specific PDF example for the spreading factor of $N = 8$ was illustrated in Figure 13.11(a). Since $\mathbf{d}_2[k]$ is a Gaussian random variable having the mean of $\sqrt{E/N}\,\alpha_2[k]$ and the variance of $\sigma^2 = (N/2 - 1)\,N_o/2$, as argued in

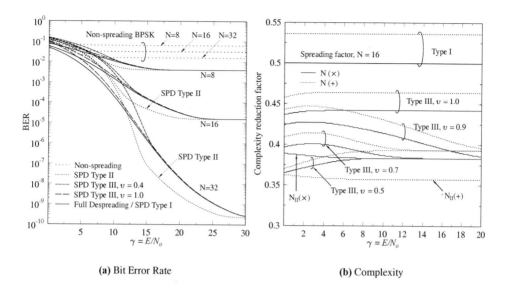

(a) Bit Error Rate

(b) Complexity

Figure 13.12: The performance of various detection schemes for transmission over the AWGN channel, when one of the N transform-domain symbols is obliterated. (a) The non-spread BPSK scheme has a high irreducible BER of $0.5/N$, while the spread systems have far lower error floor at $1/2^N$. The Type II SPD Detector shows a superior performance in the high SNR region compared to the conventional despreading method. (b) The complexity of various SPD detectors normalised to that of conventional despreading detector for the spreading factor of $N = 16$. The Type II SPD detector requires less than 40% of the computations of that for the conventional despreading.

Section 13.5.3, the BEP $P_{e,2}$ can be expressed using Equation 13.126 as:

$$P_{e,2} = \sum_{n=0}^{N-2} \Pr(\alpha = 2n) \cdot Q\left(\frac{2n\sqrt{E/N}}{\sigma_2}\right) \tag{13.147}$$

$$= \frac{1}{2^{N-2}} \sum_{n=0}^{N-2} \binom{N-2}{n} Q\left(\frac{2n\sqrt{E/N}}{\sigma_2}\right) \tag{13.148}$$

$$= \frac{1}{2^{N-2}} \sum_{n=0}^{N-2} \binom{N-2}{n} Q\left(\frac{4n\sqrt{\gamma}}{\sqrt{N(N-2)}}\right). \tag{13.149}$$

Since we now have the modified BEP term $P_e(0|0)$, which is given in Equations 13.146 and 13.149, let us express the average BEP $P_{e,II}$ of the Type II detector, when one transform-domain symbol is obliterated. The BEP of the Type II detector was expressed in Equa-

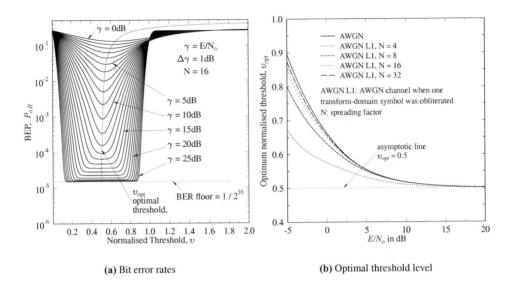

(a) Bit error rates   **(b)** Optimal threshold level

Figure 13.13: Characteristics of the Type II detector: (a) the BEP of the Type II detector, $P_{e,II}$, against the normalised threshold level, v, for several E/N_o values. The spreading factor N is 16. (b) The optimum normalised threshold levels, v_{opt}, versus E/N_o for several values of N. v_{opt} asymptotically approaches $1/2$ as E/N_o becomes large.

tion 13.52 as:

$$P_{e,II} = \frac{1}{2}\{P_e(0) + P_e(\pm 1)\} \tag{13.150}$$

$$P_{e,II} = \frac{1}{2}\{P_e(\pm 1|0) + P_e(0|0) + P_e(\pm 1)\}, \tag{13.151}$$

where $P_e(\pm 1|0)$ and $P_e(\pm 1)$ were given in Equations 13.32 and 13.51, respectively, in Section 13.4.2, while $P_e(0|0)$ was calculated above and was given in Equation 13.146, yielding:

$$P_{e,II} = \frac{1}{4}\left[Q\left(2\sqrt{\gamma}(1-v)\right) + Q\left(2\sqrt{\gamma}(1+v)\right)\right] + Q(2\sqrt{\gamma}\,v)\left[\frac{1}{2} - P_{e,2}\right] + \frac{1}{2}P_{e,2}, \tag{13.152}$$

where $P_{e,2}$ was given in Equation 13.149. The expression for the optimum normalised threshold v_{opt} given in Equation 13.56 is still valid, when using the modified parameter $\mu = \{1 - 2P_{e,2}\}e^{2\gamma}$. The bit error probability $P_{e,II}$ of the Type II SPD detector exhibits the same BER floor at $1/2^N$, as the conventional full despreading detector or the Type I SPD detector. Figure 13.13(a) depicts the bit error probability as functions of the normalised threshold levels used for SNR values of $\gamma = 0\text{dB} \sim 25\text{dB}$ and for the spreading factor of $N = 16$. By comparing Figure 13.4(a) and Figure 13.13(a), we can observe that as the lowest BERs approach the BER floor, the regions of the normalised threshold levels where their BERs are maintained become wide. Figure 13.13(b) shows the optimum threshold levels as a function of E/N_o for several values of the spreading factor N. For low spreading factors, the optimum normalised threshold levels turn out to be lower than those for higher spreading factors. As the spreading factor is increased, the optimum normalised threshold curve approached that valid for the AWGN channel determined without any transform-domain symbol losses.

Let us briefly characterise the computational cost of the Type II SPD detector. Considering that the detector requires $\mathbf{d}_2$ for detection only when $\mathbf{d}_1$ falls in the region "0" of Figure 13.3(a)), the number of multiplications and additions involved in the evaluation of Equations 13.110 and 13.110 can be expressed with the aid of Equations 13.139 to 13.142 as:

$$N_{II}(\times) = N_{\mathbf{d}_1}(\times) + N_{\mathbf{d}_2}(\times)P_{II}(0) = \left(\frac{N}{2}\right)^2 + \frac{N(N-2)}{4}P_{II}(0)\,, \tag{13.153}$$

$$N_{II}(+) = N_{\mathbf{d}_1}(+) + N_{\mathbf{d}_2}(+)P_{II}(0) = \frac{N(N-2)}{4} + \frac{N(N-4)}{4}P_{II}(0)\,, \tag{13.154}$$

where $P_{II}(0)$ is the probability that $\mathbf{d}_1$ falls in the region "0", given in Equation 13.61. Again, Figure 13.12(b) compares the computational costs involved with those of other detection schemes.

13.5.6 Type III Detector

Finally, let us analyse the bit error ratio performance of the Type III SPD detector for transmission over the AWGN channel, when one transform-domain symbol is totally corrupted. The detection method was introduced in Section 13.3 When $\mathbf{d}_1[k]$ of Equation 13.116 falls in the region "0" of Figure 13.3(a), $\hat{\mathbf{b}}_1[k]$ and $\hat{\mathbf{b}}_2[k]$ of Equations 13.129 and 13.130)are used for the detection of $\mathbf{b}_1[k]$ and $\mathbf{b}_2[k]$. Otherwise, the detection process is concluded at the first stage, determining both bits solely based on $\mathbf{d}_1[k]$. The performance analysis given in Section 13.4.3 can be repeated here with some modifications considering the different PDF of $\mathbf{d}_2[k]$ due to the loss of one transform-domain symbol.

First, let us consider the case of $\mathbf{b}_1[k] = -\mathbf{b}_2[k]$ and hence $\mathbf{d}_1[k]$ of Equation 13.116 is expected to fall in the region "0" of Figure 13.3(a) when the noise contribution is not excessive. However, when $\mathbf{d}_1[k]$ of Equation 13.116 falls in the region of "+1" or "-1" of Figure 13.3(a), the Type III SPD detector will assume $\mathbf{b}_1[k] = \mathbf{b}_2[k]$ and will detect one of the bits incorrectly. This error probability, denoted by $P_e(\pm1|0)$ according to our convention introduced in Section 13.4.2, was given in Equation 13.32 and is repeated here for convenience:

$$P_e(\pm1|0) = Q(2\sqrt{\gamma}v)\,. \tag{13.155}$$

It can be inferred from the PDFs seen in Figure 13.3 that a bit error also occurs when $\mathbf{d}_1[k]$ falls in the region "0" and

$$sgn(\hat{\mathbf{b}}_1[k]) \neq \mathbf{b}_1[k] \tag{13.156}$$

$$\Longleftrightarrow sgn(\sqrt{E/N}\alpha[k] + \eta_1[k] + \eta_2[k]) \neq \mathbf{b}_1[k]\,, \tag{13.157}$$

where $sgn(x) = 1$ if $x \geq 0$, otherwise $sgn(x) = -1$ and $\alpha[k]$ is the normalised noiseless signal component defined as $\alpha[k] = \alpha_1[k] + \alpha_2[k]$, which becomes $\alpha_2[k]$ when $\mathbf{b}_1[k] = -\mathbf{b}_2[k]$,

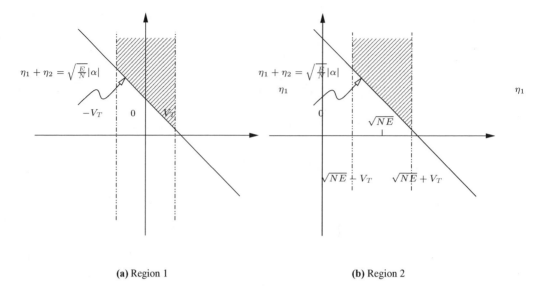

(a) Region 1   (b) Region 2

Figure 13.14: Integration regions for conditional bit error probabilities for the Type III detector in the AWGN channel when one transform-domain symbol was obliterated. (a) Region 1 represents the area satisfying $P(\eta_1 + \eta_2 > \sqrt{\frac{E}{N}}|\alpha|, |\eta_1| < V_T)$ and (b) Region 2 represents that satisfying $P(\eta_1 + \eta_2 > \sqrt{\frac{E}{N}}|\alpha|, \sqrt{NE} - V_T < \eta_1 < \sqrt{NE} + V_T)$.

according to Equations 13.117 and 13.122. For a given positive $\alpha[k]$, the BEP $P_e(\alpha, 0|0)$ can be formulated, following the argument given in Section 13.4.3, as:

$$P_e(\alpha, 0|0) = \Pr(\hat{\mathbf{b}}_1[k] > 0, |\mathbf{d}_1[k]| \leq V_T \mid \mathbf{b}_1[k] = -\mathbf{b}_2[k] = -1). \qquad (13.158)$$

Substituting $\hat{\mathbf{b}}_1[k]$ of Equation 13.129 into Equation 13.158 and noting that $\mathbf{d}_1[k]$ of Equation 13.139 becomes $\mathbf{d}_1[k] = \eta_1[k]$, we have:

$$P_e(\alpha, 0|0) = \Pr(-\sqrt{E/N}\alpha[k] + \eta_1[k] + \eta_2[k] > 0, |\eta_1[k]| < V_T), \qquad (13.159)$$

which can be calculated with the aid of the joint PDF f_{η_1,η_2} of Equation 13.71 having the combined variance of $\sigma^2 = (N-1)N_o/2$ and with the aid of the integration region depicted in Figure 13.14(a), yielding:

$$P_e(\alpha, 0|0) = \int_{-V_T}^{V_T} \int_{\sqrt{E/N}\alpha - \eta_1}^{\infty} \frac{1}{2\pi\sigma_1\sigma_2} e^{-\eta_1^2/(2\sigma_1^2) - \eta_2^2/(2\sigma_2^2)} d\eta_2 \, d\eta_1$$

$$= \int_{-V_T}^{V_T} \frac{1}{\sqrt{2\pi}\sigma_1} e^{-\eta_1^2/(2\sigma_1^2)} Q\left(\frac{\sqrt{E/N}\alpha - \eta_1}{\sigma_2}\right) d\eta_1$$

$$= \frac{1}{\sqrt{2\pi}} \int_{-2\sqrt{\bar\gamma}v}^{2\sqrt{\bar\gamma}v} e^{-\eta^2/2} Q\left(2\sqrt{\bar\gamma}\frac{\alpha}{\sqrt{N(N-2)}} - \sqrt{\frac{N}{N-2}}\eta\right) d\eta. \quad (13.160)$$

Since the random variables α and η in Equation 13.160 have symmetric PDFs, the expression of $P_e(0|0)$ given in Equation 13.160 can be used for a negative $\alpha[k]$. Using the PDF of $\alpha[k]$ given in Equation 13.126, $P_e(\alpha, 0|0)$ can be averaged over $\alpha[k]$, yielding:

$$P_e(0|0) = \sum_{n=0}^{N-2} \Pr(\alpha = 2n) \cdot P_e(\alpha = 2n, 0|0) \quad (13.161)$$

$$= \frac{1}{2^{N-2}} \sum_{n=0}^{N-2} P_e(\alpha, 0|0) \binom{N-2}{n} \delta(\alpha - 2n)$$

$$= \frac{1}{\sqrt{2\pi}} \frac{1}{2^{N-2}} \sum_{n=0}^{N-2} \binom{N-2}{n} \int_{-2\sqrt{\bar\gamma}v}^{2\sqrt{\bar\gamma}v} e^{-\eta^2/2} Q\left(4\sqrt{\bar\gamma}\frac{n}{\sqrt{N(N-2)}} - \sqrt{\frac{N}{N-2}}\eta\right) d\eta. \quad (13.162)$$

From Equations 13.155 and 13.162, we can obtain the total average BEP $P_e(0)$ for the case of $\mathbf{b}_1[k] = -\mathbf{b}_2[k]$, given as $P_e(0) = P_e(\pm 1|0) + P_e(0|0)$.

When $\mathbf{b}_1[k] = \mathbf{b}_2[k]$, there can be two scenarios showing incorrect detection, one at the first stage detector and another at the second stage combined detector. The bit error probability of the fist stage detector was given in Equation 13.43 and remains the same, since the loss of one transform-domain symbol does not affect the first stage detection variable. For a further reference, we repeat the expression of Equation 13.43 here:

$$P_e(\mp 1| \pm 1) = Q\left(2\sqrt{\bar\gamma}(1 + v)\right). \quad (13.163)$$

We obtain the BEP $P_e(0| \pm 1)$ at the second detection stage using similar approaches to those involved in Section 13.4.3. Assuming that $\mathbf{b}_1[k] = \mathbf{b}_2[k] = -1$ is transmitted, according to the Type III detection strategy given in Section 13.3 a bit error occurs, when $\mathbf{d}_1[k] = -\sqrt{NE} + \eta_1[k]$ falls in the region "0" of Figure 13.3(a) and $\hat{\mathbf{b}}_1[k] = \sqrt{E/N}\alpha[k] + \eta_1[k] + \eta_2[k]$ is positive. This error probability can be formulated as:

$$P_e(0| \pm 1) = \Pr(| - \sqrt{NE} + \eta_1[k]| < V_T, \ \sqrt{E/N}\alpha[k] + \eta_1[k] + \eta_2[k] > 0). \quad (13.164)$$

With the aid of the joint PDF of $\eta_1[k]$ and $\eta_2[k]$ given in Equation 13.71 and the corresponding integration region depicted in Figure 13.14(b), the conditional BEP $P_e(\alpha, 0| \pm 1)$ for a fixed value of α can be expressed as:

$$P_e(\alpha, 0| \pm 1) = \int_{\sqrt{NE}-V_T}^{\sqrt{NE}+V_T} \int_{\sqrt{E/N}|\alpha|-\eta_1}^{\infty} \frac{1}{2\pi\sigma_1\sigma_2} e^{-\eta_1^2/(2\sigma_1^2)-\eta_2^2/(1\sigma_2^2)} \, d\eta_2 \, d\eta_1 \quad (13.165)$$

$$= \frac{1}{\sqrt{2\pi}\,\sigma_1} \int_{\sqrt{NE}-V_T}^{\sqrt{NE}+V_T} e^{-\eta_1^2/(2\sigma_1^2)} \, Q\left(\frac{\sqrt{E/N}\,|\alpha|-\eta_1}{\sigma_2}\right) \, d\eta_1, \quad (13.166)$$

where according to Equations 13.114 and 13.115 $\sigma_1^2 = (N/2)(N_o/2)$ and $\sigma_2^2 = (N/2-1)(N_o/2)$. A change of variable according to $\eta_1 = \eta\sigma_1$ in Equation 13.166 yields:

$$P_e(\alpha, 0| \pm 1) = \frac{1}{\sqrt{2\pi}} \int_{2\sqrt{\gamma}(1-v)}^{2\sqrt{\gamma}(1+v)} e^{-\eta^2/2} \, Q\left(2\sqrt{\gamma}\frac{|\alpha|}{N(N-2)} - \sqrt{\frac{N}{N-2}}\,\eta\right) \, d\eta.$$
$$(13.167)$$

Then we can arrive at the average BEP $P_e(0| \pm 1)$ by averaging $P_e(\alpha, 0| \pm 1)$ of Equation 13.167 using the PDF of α given in Equation 13.133:

$$P_e(0| \pm 1) = \sum_{n=0}^{N-2} \Pr(\alpha = 2n) \cdot P_e(\alpha = 2n, 0| \pm 1) \quad (13.168)$$

$$= \frac{1}{\sqrt{2\pi}2^{N-2}} \sum_{n=0}^{N-2} \binom{N-2}{n} \quad (13.169)$$

$$\cdot \int_{2\sqrt{\gamma}(1-v)}^{2\sqrt{\gamma}(1+v)} e^{-\eta^2/2} Q\left(4\sqrt{\gamma}\frac{n+1}{\sqrt{N(N-2)}} - \sqrt{\frac{N}{N-2}}\,\eta\right) \, d\eta.$$

From Equations 13.163 and 13.169, the total average BEP $P_e(\pm 1)$ for the case of $\mathbf{b}_1[k] = \mathbf{b}_2[k]$ can be expressed as $P_e(\pm 1) = P_e(\mp 1| \pm 1) + P_e(0| \pm 1)$.

Using the above analysis, the BEP of the Type III detector for transmission over the AWGN channel, provided that one transform-domain symbol was obliterated, can be summarised as:

$$P_{e,III} = \frac{1}{2} P_e(0) + \frac{1}{2} P_e(\pm 1) \quad (13.170)$$

$$= \frac{1}{2} \{P_e(\pm 1|0) + P_e(0|0) + P_e(\mp 1| \pm 1) + P_e(0| \pm 1)\}. \quad (13.171)$$

Substituting $P_e(\pm 1|0)$, $P_e(0|0)$, $P_e(\mp 1| \pm 1)$ and $P_e(0| \pm 1)$ given in Equations 13.155,

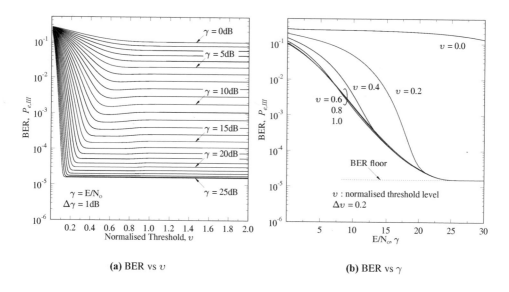

(a) BER vs v

(b) BER vs γ

Figure 13.15: The bit error probability $P_{e,III}$ of the Type III SPD detector for transmission over the AWGN channel, when one transform-domain symbol was obliterated, for the spreading factor of $N = 16$. (a) Bit error rate versus the normalised threshold level v. (b) Bit error rate versus the bit energy to noise power ratio $\gamma = E/N_o$. The BER floor is $1/2^N = 1.53 \times 10^{-5}$.

13.162, 13.163 and 13.169, respectively, into Equation 13.171, we have:

$$
P_{e,III} = \frac{1}{2}Q(2\sqrt{\gamma}v) + \frac{1}{2}Q(2\sqrt{\gamma}(1+v))
$$

$$
+ \frac{1}{2^{N-1}\sqrt{2\pi}} \sum_{n=0}^{N-2} \binom{N-2}{n} \int_{-2\sqrt{\gamma}v}^{2\sqrt{\gamma}v} e^{-\eta^2/2} Q\left(4\sqrt{\gamma}\frac{n}{\sqrt{N(N-2)}} - \sqrt{\frac{N}{N-2}}\eta\right) d\eta
$$

$$
+ \frac{1}{2^{N-1}\sqrt{2\pi}} \sum_{n=0}^{N-2} \binom{N-2}{n} \int_{2\sqrt{\gamma}(1-v)}^{2\sqrt{\gamma}(1+v)} e^{-\eta^2/2} Q\left(4\sqrt{\gamma}\frac{n+1}{\sqrt{N(N-2)}} - \sqrt{\frac{N}{N-2}}\eta\right) d\eta.
$$

$$(13.172)$$

Figure 13.15 shows the bit error rate of the Type III detector for transmission over the AWGN channel, when one transform-domain symbol was obliterated, for the spreading factor of $N = 16$. Comparing Figure 13.15 and Figure 13.6, we can observe that the BER curves in Figure 13.15 become flat at relatively low values of v and at high bit error rates due to the loss of one transform-domain symbol. The irreducible BER can also be observed in Figure 13.15(a) and Figure 13.15(b). The value of this BER floor is not obvious in Equation 13.172. The first two terms certainly vanish, as γ becomes high. We have to consider the integration region depicted in Figure 13.14 in order to identify which terms approach zero. As the signal energy is increased, the integration region becomes smaller, moving away from the origin, except for the third term when $n = 0$, in which case the integration region covers half of the entire η_1-η_2 plane, as υ is increased. Hence, the BER floor becomes $1/2^N$ again. This analysis is corroborated by Figure 13.15 for the spreading factor of $N = 16$, where the BER approached $1/2^{16} = 1.523 \times 10^{-5}$. In the SNR range of about 10-15dB, the bit error rate is slightly increased as the normalised threshold levels becomes higher, widening the region "0". This could not be observed in Figure 13.6 of Section 13.4.3. When no symbol was lost in the AWGN channel, the system always resulted in an improved BER when activating the second stage, because the second stage detector was the ideal detector. However, in the scenario we are considering, the second stage detector processes the impaired detection variable, since the obliterated symbol is assumed to be in the second sub-block. Thus, it is not always better to increase the normalised threshold level in this case. However, as seen in Figure 13.15(b), the BER differences are insignificant.

The Type III SPD detector uses $\mathbf{d}_2$, only when $\mathbf{d}_1$ falls in the region "0" of Figure 13.3(a). Hence, the expression of the number of multiplications required for the Type III SPD detector is identical to that of the Type II detector, which was given in Equation 13.153. However, N number of extra additions are required to compute the sum and the difference of $\mathbf{d}_1[k]$ and $\mathbf{d}_2[k]$ according to Equation 13.18. The probability $P_{III}(0)$ of requiring the computation of $\mathbf{d}_2$ is identical to $P_{II}(0)$ given in Equation 13.61. From the above argument, the number of additions required by the Type III detector can be expressed as:

$$N_{III}(+) = N_{\mathbf{d}_1}(+) + N_{\mathbf{d}_2}(+)P_{III}(0) + N\,P_{III}(0) \tag{13.173}$$

$$= \frac{N(N-2)}{4} + \frac{N^2}{4}P_{III}(0)\,, \tag{13.174}$$

where $N_{\mathbf{d}_1}(+)$ is the number of additions required for computing $\mathbf{d}_1$ given in Equation 13.140 and $N_{\mathbf{d}_2}(+)$ is that for $\mathbf{d}_2$, which has to be evaluated with the probability of $P_{III}(0)$ given in Equation 13.142.

13.5.7 Summary and Discussion

In Section 13.5 we investigated a non-spreading BPSK detector, a full despreading based detector and three different types of SPD detectors in the AWGN channel, when one transform-domain symbol was obliterated. The bit error rate curves of the detectors were depicted in Figure 13.12(a). An uncoded OFDM system can be regarded as a non-spread BPSK system in our context. Non-spread systems have a high irreducible bit error rate of $0.5/N$ in the investigated channel scenario. We can conclude that non-spread systems cannot be used in this channel without powerful error correction mechanisms. All the investigated spreading assisted systems have the same BER floor of $1/2^N$, which is considerably lower than that of the non-spread systems for spreading factor of $N \geq 4$. Among the three different types of SPD detectors the Type II detector exhibited a lower bit error rate than that of the full-despreading aided detectors for bit energy to noise power ratios of $\gamma > 6.5$dB in conjunction with $N = 8$, for $\gamma > 9.0$dB along with $N = 16$ and for $\gamma > 12.0$dB in conjunction with $N = 32$. In Figure 13.12(a) we can observe that the maximum SNR advantage of the Type II SPD detector over a conventional full-despreading based detector is around 3dB, depending on the spreading factor N.

Figure 13.12(b) compares the implementational complexity of the investigated detectors in terms of the required number of operations normalised to that of the conventional full despreading based detector for the spreading factor of $N = 16$. The Type I SPD detectors have about half the complexity of the conventional detector. The Type II and III SPD detectors require a lower number of multiplications and additions than the Type I detector. It is interesting to see that the Type II detectors exhibit a lower BER as well as a lower complexity in the SNR region spanning from 12dB to 30dB in comparison to those of the conventional full despreading based detector.

Although we discussed the performance of our SPD detectors in the context of an AWGN channel contaminated by narrow-band jamming, one of the transform domain symbols may be deliberately obliterated at the receiver. For example the transmitter may specifically choose one of the transform domain symbols in order to reduce the crest factor – an issue, which was studied in Chapter 11 – rather than conveying message-dependent information and then the corresponding transform domain symbol can be discarded at the receiver.

13.6 Chapter Summary and Conclusion

The family of novel SPD detectors was introduced utilising a Walsh spreading matrix in the context of fully loaded multi-code CDMA systems, where the number of simultaneous codes L is equal to the spreading factor N, in an effort to reduce the required number of computations. The associated bit error ratio and the complexity of the detectors were analysed when communicating over an AWGN channel in Section 13.4. Our simulation-based BER results were also presented, in order to corroborate the results of the analyses. It was found in Section 13.4 that some of these detectors show a near ideal bit error rate performance in an AWGN channel, while requiring less than half the number of multiplications and additions, as can be seen in Figure 13.7 and Figure 13.8. When the channel impairments are dominated by time domain impulse noise [455,456] or frequency domain narrow band interference [457, 458], the Type II SPD detection scheme was found to perform better, than the conventional full despreading based detector for the SNR range of 12dB to 30dB for the spreading factor of $N = 32$ in Section 13.5, while requiring less than 40% of the computational efforts of the conventional scheme, as was evidenced by Figure 13.12(b).

Although we observed the potential of the proposed SPD schemes, when communicating over AWGN channels in terms of reducing the computational efforts in comparison to the conventional full despreading based scheme, the performance of the SPD schemes over fading channels has to be investigated in the context of joint detection receivers, in order to confirm the feasibility of this technique in practical systems.

Part III

Advanced Topics: Channel Estimation and Multi-user OFDM Systems

List of General Symbols

B_D: Doppler spread of the channel: $B_D = 2f_D$.

c: Speed of light.

f_c: OFDM carrier frequency.

f_d: Any frequency of the Doppler spectrum.

f_D: Doppler frequency: maximum frequency in the sense of a limitation of the Doppler spectrum.

f_s: OFDM sampling frequency: $f_s = 1/T_s$.

F_D: OFDM symbol duration normalised Doppler frequency: $F_D = f_D T_f$.

$F_{D,K}$: OFDM symbol duration normalised Doppler frequency used in the context of quantifying the effects of ICI: $F_{D,K} = f_D K T_s$.

$h(t, \tau)$: Time-variant channel impulse response (CIR) at time instant t and multipath delay τ.

$H(t, f)$: Time-variant channel transfer function: $H(t, f) = \underset{\tau \to f}{\mathcal{FT}}\{h(t, \tau)\}$.

k: OFDM subcarrier index.

n: OFDM symbol index.

K: Number of OFDM subcarriers.

K_c: Constraint length of the channel code.

K_g: Number of OFDM guard interval samples.

$r_h(\Delta t, \tau_1, \tau_2)$: Auto-correlation function of the CIR: $r_h(\Delta t, \tau_1, \tau_2) = E\{h(t, \tau_1)h^*(t - \Delta t, \tau_2)\} = r_h(\Delta t, \tau_1) \cdot \delta(\tau_1 - \tau_2)$.

$r_h(\Delta t, \tau)$:   Spaced-time auto-correlation function of the CIR for the multipath delay τ, namely: $r_h(\Delta t, \tau) = r_h(\Delta t, \tau_1, \tau_2)|_{(\tau=\tau_1=\tau_2)}$.

$r_h(\tau)$:   Multipath intensity profile or delay power spectrum: $r_h(\tau) = r_h(\Delta t, \tau)|_{(\Delta t=0)}$.

$r_H(\Delta t, \Delta f)$:   Spaced-frequency spaced-time correlation function of the channel: $r_H(\Delta t, \Delta f) = \underset{\tau \to \Delta f}{\mathcal{FT}} \{r_h(\Delta t, \tau)\}$. In the context of 'separability' we have: $r_H(\Delta t, \Delta f) = r_H(\Delta t) \cdot r_H(\Delta f)$.

$r_H(\Delta t)$:   Spaced-time correlation function in the context of 'separability' of $r_H(\Delta t, \Delta f)$.

$r_H(\Delta f)$:   Spaced-frequency correlation function in the context of 'separability' of $r_H(\Delta t, \Delta f)$.

R_c: Coding rate of the channel encoder.

$S_h(f_d, \tau)$:   Scattering function of the channel: $S_h(f_d, \tau) = \underset{\Delta t \to f_d}{\mathcal{FT}} \{r_h(\Delta t, \tau)\}$.

$S_H(f_d, \Delta f)$: $S_H(f_d, \Delta f) = \underset{\Delta t \to f_d}{\mathcal{FT}} \{r_H(\Delta t, \Delta f)\}$.

$S_H(f_d)$:   Doppler power spectrum of the channel: $S_H(f_d) \equiv S_H(f_d, \Delta f)|_{\Delta f=0}$.

T_m: Multipath spread of the channel.

T_s: OFDM sampling period duration: $T_s = 1/f_s$.

T_f: OFDM symbol duration: $T_f = (K + K_g)T_s$.

v: Vehicular speed.

$\mathbf{W}$:   Unitary DFT matrix, where the complex phaser in the i-th row and j-th column is given by $\mathbf{W}|_{i,j} = \frac{1}{\sqrt{K}} e^{-j2\pi \frac{ij}{K}}$: $\mathbf{W} \in \mathbb{C}^{K \times K}$.

$\alpha_n(t)$:   Time-variant attenuation factor of the n-th CIR component.

$\delta(t)$:   Dirac impulse function.

κ_{ICI}:   ICI noise proportionality constant. Note that $\kappa_{\text{ICI}} = f(F_{D,K})$.

ρ: Inverse of the signal to noise ratio: $\rho = \sigma_n^2/\sigma_s^2$.

σ_{AWGN}^2:   Variance of the AWGN.

σ_H^2:   Variance of the frequency-domain channel transfer factors.

σ_{ICI}^2:   Subcarrier domain ICI noise variance: $\sigma_{\text{ICI}}^2 = \kappa_{\text{ICI}} \sigma_s^2$.

σ_n^2:   Total noise variance: $\sigma_n^2 = \sigma_{\text{AWGN}}^2 + \sigma_{\text{ICI}}^2$.

σ_s^2:   Signal variance.

$\tau_n(t)$: Time-variant delay of the n-th CIR component.

ω_D: Doppler angular velocity: $\omega_D = 2\pi f_D$.

ΔF: Subcarrier spacing of the OFDM symbol: $\Delta F = \frac{1}{KT_s}$.

$(\Delta f)_c$: Coherence bandwidth of the channel: $(\Delta f)_c \approx \frac{1}{T_m}$.

$(\Delta t)_c$: Coherence time of the channel: $(\Delta t)_c \approx \frac{1}{B_D}$.

$()|_{opt}$: Index used in the context of the vector of estimator coefficients in order to indicate its optimality in the sense of a minimum MSE.

$()|_{rob}$: Index used in conjunction with auto-correlation matrices or cross-correlation vectors in order to indicate their 'robustness', which is related to the uniform ideally support-limited scattering function.

$()_{unif}$: Channel correlation functions related to a uniform ideally support limited scattering function.

$\tilde{()}$: Index employed to denote variables, vectors and matrices associated with the design of an estimator or predictor. An exception is given by the least-squares channel transfer factor- or CIR-related tap estimates, which are also identified by a superscript of $\tilde{()}$.

$\tilde{()}$: Superscript used in conjunction with the least-squares channel transfer factor and the CIR-related tap estimates. It is also used for indicating that the variable in brackets is associated with the design of the estimator rather than with the original variable, which is employed for characterising the actual channel encountered.

$\lfloor \rfloor$: Largest integer that is not greater than the argument in brackets.

$\lceil \rceil$: Smallest integer that is not less than the argument in brackets.

$||\cdot||_2^2$: Euclidean norm of a vector.

14

Pilot-Assisted Channel Estimation for Single-User OFDM [1]

14.1 Introduction

Recall that in Section 4.3 we provided a rudimentary conceptual introduction to the various aspects of channel transfer function estimation. In this chapter our discussions are substantially deeper and more quantitative. We commence our discourse by providing a slightly more detailed OFDM schematic portrayed in Figure 14.1, which explicitly shows the channel estimation block detailed in Chapters 14 – 16 and its interconnection with the rest of the system components. The OFDM transmitter is shown at the top, while the OFDM receiver is at the bottom of Figure 14.1. Furthermore, a simplified model of the transmission channel, constituted by its time-variant impulse response $h(t, \tau)$ and the Additive White Gaussian Noise (AWGN) contribution $n(t)$ is shown on the right-hand side of Figure 14.1.

In the absence of ISI between the consecutive OFDM symbols the frequency domain effects of the multipath channel manifest themselves as the multiplication of each subcarrier symbol with the channel's transfer factor at this frequency position, plus an AWGN contribution, which can be expressed as:

$$x[n, k] = s[n, k] \cdot H[n, k] + n[n, k], \tag{14.1}$$

where the channel transfer factor $H[n, k]$ is given by:

$$H[n, k] = H(nKT_s, k\Delta F) = \mathcal{FT}_{\tau->f}\Big\{h(nKT_s, \tau)\Big\}\big|_{f=k\Delta F}, \tag{14.2}$$

and $\mathcal{FT}\{\}$ denotes the Fourier Transform. Note that Equation 14.2 is based on the simplified assumption that the CIR is time-invariant for the duration of an OFDM symbol period. For

[1] *OFDM and MC-CDMA for Broadband Multi-user Communications WLANS and Broadcasting.*
L.Hanzo, Münster, T. Keller and B.J. Choi,
©2003 John Wiley & Sons, Ltd. ISBN 0-470-85879-6

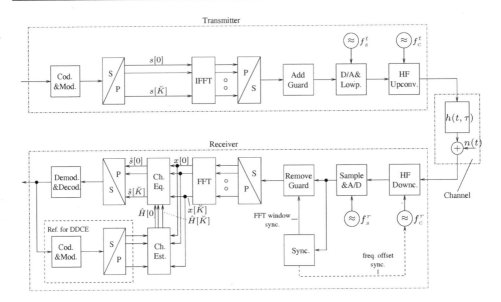

Figure 14.1: Simplified illustration of the OFDM transmitter and receiver, which are linked via the transmission channel. The time domain signal generated in the OFDM transmitter seen at the top is conveyed over the transmission channel, which is characterised by the time-variant impulse response $h(t, \tau)$ plus the AWGN contribution $n(t)$, to the OFDM receiver.

the case of the more realistic scenario of encountering a CIR, which is time-variant during the transmission of a specific OFDM symbol it can be shown that the effective subcarrier channel transfer factors are given as the Fourier transform of the complex CIR tap values, which have been averaged over the duration of the OFDM symbol period. In addition, Inter-Subcarrier Interference (ICI) would be observed as a consequence of the loss of orthogonality between the different subcarrier transfer functions. In order to facilitate the correct detection of the different users' transmitted symbols in the case of coherent detection, the effects of the channel must be removed from the signal $x[n, k]$ associated with the K different subcarriers $k = 0, \ldots, K - 1$. This is the task of the equaliser block ("Ch. Eq.") of Figure 14.1, which performs a normalisation of the signal received by each subcarrier with its associated channel transfer factor estimate. This results in a linear estimate $\hat{s}[n, k]$ of the symbol transmitted on the k-th subcarrier in the following form:

$$\hat{s}[n, k] = \frac{x[n, k]}{\hat{H}[n, k]}, \tag{14.3}$$

where we have $k = 0, \ldots, K - 1$. Following again a parallel-to-serial ("P/S") conversion as seen in Figure 14.1, the linear symbol estimates are demodulated and potentially channel-decoded ("Demod.&Decod."). A prerequisite for obtaining linear symbol estimates according to Equation 14.3 is the availability of the channel transfer factor estimates $\hat{H}[n, k]$, $k = 0, \ldots, K - 1$, which are generated within the channel estimation block ("Ch. Est.") of Figure 14.1. Various approaches are known from the literature for performing the channel estimation task, namely pilot-assisted-, decision-directed and blind methods. While in the pilot-assisted case the channel estimation is performed based on a number of dedicated subcarriers, which employ pilot symbols known both to the receiver and the transmitter, in the decision-directed case the symbol decisions are remodulated and then employed as "pilot symbols". The process of remodulating the subcarrier symbol-decisions is represented by the box at the bottom of Figure 14.1, which is enclosed in dashed lines. In the next section we will summarise some of the main design aspects of OFDM systems, which have been addressed by researchers.

In recent years numerous research contributions have appeared on the topic of channel transfer function estimation for employment in single-user, single transmit antenna-assisted OFDM scenarios, since the availability of an accurate channel transfer function estimate is one of the prerequisites for coherent symbol detection at an OFDM receiver. The techniques proposed in the literature can be classified as pilot-assisted and decision-directed (DD) channel estimation methods.

In the context of pilot-assisted channel transfer function estimation a subset of the available subcarriers is dedicated to the transmission of specific pilot symbols known to the receiver, which are used for "sampling" the desired channel transfer function. Based on these samples of the frequency domain transfer function the well-known process of interpolation is used to generate a transfer function estimate for each subcarrier between the pilots. This is achieved at the cost of a reduction of the number of useful subcarriers available for data transmission. While early publications on pilot-assisted channel estimation for OFDM mainly capitalised on one-dimensional pilot patterns, which span the frequency direction, more recent work on this topic conducted by Höher et al. [58], [66], [67], Edfors et al. [157], Sandell [158] as well as Li et al. [74, 459] invoked the theory of two-dimensional sampling [460, 461], where the pilot pattern is a function of both the subcarrier- and the OFDM symbol index. These investigations were conducted in an effort to reduce the substantial loss in the number of useful subcarriers associated with one-dimensional pilot patterns, since the exploitation of both the frequency and time direction transfer function correlations potentially allows us to reduce the number of pilots required. Two aspects of these schemes, which ultimately influence the channel estimation accuracy are the specific geometry of the pilot pattern used, as well as the specific algorithm applied to estimate the channel transfer factors associated with the data subcarriers.

14.1.1 Classification of Channel Estimation Techniques

In Figure 14.2 we have classified the various 2D-FIR Wiener filter-related channel estimation approaches, which constitute standard techniques known from the literature of multidimensional signal processing. The application of these techniques to channel transfer function estimation of OFDM schemes was investigated by Höher et al. in [66, 67, 462]. A main disadvantage of this technique is the relatively high channel estimator mean-squared error (MSE) in the light of its potentially high complexity.

Höher et al. [66, 67, 462] observed that the specific pilot-assisted channel transfer function samples, which are relatively far apart from the position on the time-frequency plane for which the channel transfer factor is to be estimated, have a moderate influence on the channel transfer factor estimate. Hence, Höher et al. [66, 67, 462] proposed invoking only the channel samples provided by the pilots nearest – according to a specific metric – to the target position

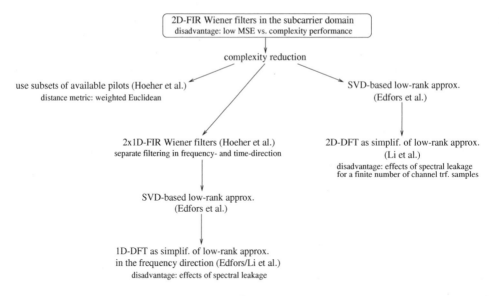

Figure 14.2: Classification of the various 2D-FIR Wiener filter-related channel transfer function estimation approaches.

considered in the process of channel estimation.

A more dramatic complexity reduction was achieved by Höher *et al.* [58,67] upon exploiting the separability[2] of the mobile channel's 2D spaced-time spaced-frequency correlation function in form of the cascaded 1D-FIR filter-based channel estimation technique applied for multi-carrier transmissions. The approach of cascaded 1D-FIR filters constitutes a standard approach often encountered in multi-dimensional signal processing [461], but it has also been shown to be effective in the context of channel estimation for multi-carrier transmissions. The basic idea is to perform the filtering in each of the orthogonal directions of frequency and time, separately. Again, a further complexity reduction can be achieved upon invoking only subsets of the available pilot-assisted channel transfer function samples in the process of channel estimation in each of the orthogonal directions, namely in the frequency and time directions.

A further improvement of the MSE versus complexity trade-off offered by the approach of cascaded 1D-FIR filters was achieved by Edfors *et al.* [69, 157] and Sandell [158] with the aid of Singular Value Decomposition (SVD) based low-rank approximation techniques. More specifically, the pilot-assisted channel samples invoked in the estimation of a specific channel transfer factor are first projected onto a new set of basis vectors, followed by the rank-reduction implemented by retaining only a specific number of the highest energy transform coefficients. The remaining transform coefficients are Wiener filtered, taking into account the transform coefficients' individual variances. In the final stage of processing an inverse transform of the coefficients to the original basis representation is performed. A fundamental characteristic of this type of estimator is that its associated complexity linearly depends on the

---

[2]Separability in this context implies that 1D-FIR filtering may be used first in either the time- or the frequency direction and then in the remaining direction, instead of invoking 2D-FIR filtering in a single step.

number of transform coefficients retained. In the context of the channel scenarios considered by Edfors *et al.* in [69, 157] and Sandell [158] an MSE reduction of a factor of about 1.3-1.5 was achieved, while exhibiting the same complexity as the original FIR Wiener filter-based estimator.

As demonstrated by Van de Beek *et al.* [61], a further potential complexity reduction can be achieved by substituting the unitary linear transform employed in the frequency direction filter related SVD-based low-rank approximation by the computationally less complex DFT. This reduction of the complexity is achieved at the expense of an MSE performance degradation due to leakage effects[3] encountered in the context of non-sample spaced CIRs. As illustrated on the right-hand side of Figure 14.2, low-rank approximation techniques were also applied by Edfors *et al.* [69, 157] and Sandell [158] to the original problem of 2D-FIR Wiener filtering, where again, the case of 2D-DFT-based Wiener filtering can be considered as a simplification of the SVD-based low-rank estimator.

The structure of Chapter 14 is as follows. In Section 14.2 the stochastic channel model and its associated correlation functions are introduced, while in Section 14.4 a short discourse is provided on the topic of two-dimensional signal processing. Based on the Nyquist theorem [90, 177, 463], in Section 14.5 expressions will be provided for the maximum tolerable pilot distances in the context of an infinite-plane rectangular pilot grid. Our discussions on channel transfer function estimation techniques applicable to multi-carrier (MC) transmission systems employing two-dimensional (2D) pilot patterns commence in Section 14.6 with the description and characterisation of 2D-FIR Wiener filter aided estimation. The channel estimation quality metrics in our comparisons will be the mean-square channel estimation error, the system's BER and the associated computational complexity imposed. In an effort to further improve the estimator's MSE performance for a given complexity, cascaded 1D-FIR Wiener filtering is considered in Section 14.7. Our summary and conclusion will be offered in Section 14.8.

14.2 The Stochastic Channel Model

In this section the stochastic channel model employed in our investigations is outlined, noting that a more detailed characterisation of the wireless channel based on the Bello functions can be found for example in /ncitebook:bluebook-v2. The further structure of this section is as follows. In Sections 14.2.1 and 14.3 the model of the time-variant Channel Impulse Response (CIR) is outlined. Its further characterisation is then conducted with reference to Figure 14.3 in terms of the CIR's Auto-Correlation Function (ACF) denoted as $r_h(\Delta t, \tau)$ in Section 14.2.2. Furthermore, the Fourier Transform (FT) of the CIR's ACF with respect to the multipath delay variable τ, namely $r_H(\Delta t, \Delta f)$ – also known as the spaced-time spaced-frequency correlation function – is considered in Section 14.2.3. Further Fourier transforming with respect to the spaced-time variable Δt yields the function $S_H(f_d, \Delta f)$, which is addressed in Section 14.2.4. Finally, the scattering function $S_h(f_d, \tau)$ is obtained by Fourier transforming the CIR's ACF with respect to the spaced-time variable Δt, which is the topic of Section 14.2.5. Our portrayal of the channel model is concluded in Section 14.2.6 by

---

[3]To provide a simple example, if the CIR incurred exhibits a single Dirac impulse at a delay of $0.5T_s$, where T_s denotes the OFDM sampling period duration, then after transforming the channel transfer function samples obtained e.g. with the aid of pilots to the CIR-related domain, all CIR-related taps generated by the IFFT employed in the context of the channel estimation will potentially exhibit a non-zero value. This effect is referred to as "leakage".

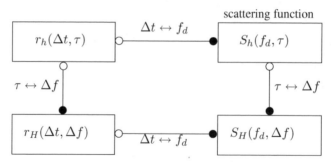

spaced-time spaced-frequency
correlation function

Figure 14.3: Illustration of the Fourier transform relationships between the stochastic channel model's associated correlation functions [177]. Upon setting $\Delta t = 0$, the CIR's auto-correlation function $r_h(\Delta t, \tau)$ is simplified to $r_h(\tau)$, which is known as the multipath intensity profile of the channel. In contrast, when setting $\Delta f = 0$, the Fourier transform of the CIR's auto-correlation function $r_h(\Delta t, \tau)$ with respect to Δt and τ, denoted by $S_H(f_d, \Delta f)$, is simplified to $S_H(f_d)$, which is known as the channel's Doppler power spectrum.

highlighting the conditions for separability of the channel's spaced-time spaced-frequency correlation function $r_H(\Delta t, \Delta f)$.

14.2.1   Model of the Channel Impulse Response

The time-variant channel impulse response CIR is given in terms of its lowpass representation by [177, 464]:

$$h(t, \tau) = \sum_n \alpha_n(t) e^{-j\theta_n(t)} \delta(\tau - \tau_n(t)), \tag{14.4}$$

where $\theta_n(t) = 2\pi f_c \tau_n(t)$ and f_c is the carrier frequency. Furthermore, in Equation 14.4 $\alpha_n(t)$ denotes the time-variant attenuation factor, while $\tau_n(t)$ represents the time-variant delay associated with the n-th CIR path. Please note that since in state-of-the-art communication systems f_c is usually high, relatively small variations of $\tau_n(t)$ result in large changes in the value of the phase $\theta_n(t)$, which in turn impose significant changes on the CIR $h(t, \tau)$. Since the paths' delays are expected to vary randomly as a result of the scatterers' random motion, the CIR tap magnitudes of $h(t, \tau)$ versus time can also be modelled as random processes. More specifically, under these assumptions and for a sufficiently large number of different paths, according to the central limit theorem, each tap of $h(t, \tau)$ can be modelled as a zero-mean complex Gaussian random process with respect to the variable t. Hence, the envelope $|h(t, \tau)|$ is Rayleigh distributed. By contrast, in the presence of static scatterers in addition to the randomly varying scatterers, the mean value of $h(t, \tau)$ is different from zero and hence the envelope $|h(t, \tau)|$ is Ricean distributed [177, 464].

In order to further characterise the above stochastic channel model, the associated Fourier transforms can be defined which are known in the literature as Bello's system functions

[177, 464, 465]. Instead of using the Bello functions to characterise the channel, due to their relevance for the different channel transfer function estimation methods, here we will rather employ the related correlation functions and their associated Fourier transforms. The corresponding relationships are further illustrated in Figure 14.3.

14.2.2 Auto-Correlation Function of the CIR: $r_h(\Delta t, \tau)$

Upon assuming that the process modelling $h(t, \tau)$ is wide-sense-stationary, which implies that the process' mean value is time-invariant and its auto-correlation is only a function of the time-difference, but not of the absolute time instants, the Auto-Correlation Function (ACF) $r_h(\Delta t, \tau_1, \tau_2)$[4] of $h(t, \tau)$ can be defined as [177, 464]:

$$r_h(\Delta t, \tau_1, \tau_2) = E\{h(t, \tau_1)h^*(t - \Delta t, \tau_2)\} \tag{14.5}$$

$$= r_h(\Delta t, \tau_1)\delta(\tau_1 - \tau_2), \tag{14.6}$$

where in the context of Equation 14.6 the assumption of uncorrelated scattering [177, 464] has been invoked. This implies that the amplitudes and phase shifts associated with the different CIR delays of $\tau_1 \neq \tau_2$ are uncorrelated. For $\Delta t = 0$, the function $r_h(\Delta t, \tau_1)$ simplifies to $r_h(0, \tau) \equiv r_h(\tau)$, with $\tau = \tau_1$, which is known as the multipath intensity profile or delay power spectrum [177, 464] of the channel. More explicitly, $r_h(\tau)$ is the average power output of the channel for a given multipath delay τ. The delay range $0 \leq \tau \leq T_m$ across which $r_h(\tau)$ exhibits significant values is referred to as the multipath spread [177, 464] of the channel.

14.2.3 Spaced-Time Spaced-Frequency Correlation Function - Fourier Transform of the CIR's ACF with Respect to the Multipath Delay Variable: $r_H(\Delta t, \Delta f)$

Let us denote the time-variant channel transfer function by $H(t, f)$, which is the Fourier Transform (FT) of the channel impulse response $h(t, \tau)$ with respect to the variable τ, formulated as:

$$H(t, f) = \underset{\tau \to f}{\mathcal{FT}}\{h(t, \tau)\} \tag{14.7}$$

$$= \int_{-\infty}^{\infty} h(t, \tau)e^{-j2\pi f\tau}d\tau. \tag{14.8}$$

Then the auto-correlation of $H(t, f)$, namely $r_H(\Delta t, f_1, f_2)$[5] is given as the Fourier transform of the auto-correlation function $r_h(\Delta t, \tau_1, \tau_2)$ of the CIR $h(t, \tau)$ taken also with respect

---

[4]Note that the definition of the auto-correlation function of $h(t, \tau)$ is different from that of [177, 464], where an arbitrary normalisation factor of $1/2$ was employed, yielding: $r_h(\Delta t, \tau_1, \tau_2) = \frac{1}{2}E\{h(t, \tau_1)h^*(t - \Delta t, \tau_2)\}$.

[5]Note that the definition of the auto-correlation function of $H(t, f)$ is different from that of [177, 464], where an arbitrary normalisation factor of $1/2$ is employed: $r_H(\Delta t, f_1, f_2) = \frac{1}{2}E\{H(t, f_1)H^*(t - \Delta t, f_2)\}$.

to the variable τ [177, 464], which is given by:

$$
\begin{aligned}
r_H(\Delta t, f_1, f_2) &= E\{H(t, f_1)H^*(t - \Delta t, f_2)\} && (14.9) \\
&= \underset{\tau \to \Delta f}{\mathcal{FT}} \{r_h(\Delta t, \tau)\} && (14.10) \\
&\equiv r_H(\Delta t, \Delta f). && (14.11)
\end{aligned}
$$

This relationship is also shown in Figure 14.3. The auto-correlation function $r_H(\Delta t, \Delta f)$ is known as the spaced-frequency, spaced-time correlation function [177, 464] of the channel. Its dependence on the frequency difference $\Delta f = f_1 - f_2$, rather than on the individual frequency values f_1, f_2 is related to the assumption of uncorrelated scattering. For $\Delta t = 0$, which corresponds to considering the channel transfer factors associated with a specific OFDM symbol, the ACF $r_H(\Delta t, \Delta f)$ simplifies to $r_H(0, \Delta f) \equiv r_H(\Delta f)$. Since we have $r_H(\Delta f) = \underset{\tau \to \Delta f}{\mathcal{FT}} \{r_h(\tau)\}$, the reciprocal of the channel's multipath spread T_m can be used to define the coherence bandwidth $(\Delta f)_c$ [177, 464] of the channel, which is the range $-(\Delta f)_c/2 \leq \Delta f \leq (\Delta f)_c/2$ of frequencies over which the channel transfer function exhibits a significant correlation [177, 464], namely:

$$
(\Delta f)_c \approx \frac{1}{T_m}. \tag{14.12}
$$

14.2.4   Fourier Transform of the CIR's ACF with Respect to the Multi-path Delay- and Spaced-Time Variables: $S_H(f_d, \Delta f)$

In order to further characterise the channel, the Fourier transform of $r_H(\Delta t, \Delta f)$ with respect to the difference in time, namely Δt, can be obtained as [177, 464]:

$$
S_H(f_d, \Delta f) = \underset{\Delta t \to f_d}{\mathcal{FT}} \{r_H(\Delta t, \Delta f)\}. \tag{14.13}
$$

This relationship is also shown in Figure 14.3. More specifically, for $\Delta f = 0$ which corresponds to considering a specific subcarrier of consecutive OFDM symbols, each of which fades according to a certain Doppler frequency in the time direction, the function $S_H(f_d, \Delta f)$ simplifies to $S_H(f_d, 0) \equiv S_H(f_d)$, which is known as the *Doppler power spectrum* of the channel. By analogy with the definition of the coherence bandwidth and the related multipath spread, which characterise the correlation properties of the channel for a specific time instant associated with a given OFDM symbol, similar definitions apply to the characterisation of the channel variations as a function of time. More specifically, the frequency range of $-B_D/2 \leq f_d \leq B_D/2$, over which $S_H(f_d)$ exhibits significant values is referred to as the Doppler spread B_D [177, 464] of the channel, while its inverse is known as the *coherence time* [177, 464], which is expressed as:

$$
(\Delta t)_c \approx \frac{1}{B_D}. \tag{14.14}
$$

More explicitly, $(\Delta t)_c$ quantifies the time domain displacement range of $-(\Delta t)_c/2 \leq \Delta t \leq (\Delta t)_c/2$, over which $r_H(\Delta t, \Delta f)$ exhibits significant correlation for $\Delta f = 0$. In this case $r_H(\Delta t, \Delta f)$ simplifies to $r_H(\Delta t, 0) \equiv r_H(\Delta t)$.

14.2.5 Scattering Function - Fourier Transform of the CIR's ACF with Respect to the Time-Delay: $S_h(f_d, \tau)$

A further transform pair can be defined by applying the Fourier transform to $r_h(\Delta t, \tau)$ of Equation 14.6 with respect to the variable Δt [177,464], yielding:

$$S_h(f_d, \tau) = \underset{\Delta t \to f_d}{\mathcal{FT}} \{r_h(\Delta t, \tau)\}, \tag{14.15}$$

which is referred to as the *scattering function* [177,464] of the channel. Furthermore, it follows that $S_H(f_d, \Delta f)$ and $S_h(f_d, \tau)$ also form a Fourier transform pair [177,464], which is expressed as:

$$S_H(f_d, \Delta f) = \underset{\tau \to \Delta f}{\mathcal{FT}} \{S_h(f_d, \tau)\}, \tag{14.16}$$

as seen in Figure 14.3.

14.2.6 Separability of the Channel's Spaced-Time Spaced-Frequency Correlation Function

In this section we will comment on the conditions that have to be satisfied for the sake of maintaining the *separability* [68] of the channel's spaced-time spaced-frequency correlation function $r_H(\Delta t, \Delta f)$. This property will be employed, for example, in the context of our derivation of the cascaded 1D-FIR Wiener filter's coefficients in Section 14.7.1, as well as in the context of our discussions on decision-directed channel estimation (DDCE) invoked in single-user OFDM systems in Chapter 15. Finally, the separability will be also exploited in parallel interference cancellation (PIC) assisted DDCE characterised in the context of multi-user OFDM systems in Section 16.5.

Based on Equation 14.4 and following the arguments of Section 14.2.1, a simplified model of the CIR is given by:

$$h(t, \tau) = \sum_n \gamma_n(t)\delta(\tau - \tau_n), \tag{14.17}$$

where in the absence of static scatterers the time-variant fading factor $\gamma_n(t) \in \mathbb{C}$ obeys a complex Gaussian distribution function. For simplicity we also assume that we have $\tau_{n_1} \neq \tau_{n_2} \ \forall \ n_1 \neq n_2$.

Upon substituting Equation 14.17 describing the time-variant CIR $h(t, \tau)$ into Equation 14.8, the channel's time-variant transfer function $H(t, f)$ is given by:

$$H(t, f) = \sum_n \gamma_n(t)e^{-j2\pi f\tau_n}. \tag{14.18}$$

By further substituting the expression of Equation 14.18 into Equation 14.9 the channel's spaced-time spaced-frequency correlation function $r_H(\Delta t, \Delta f)$ is obtained in the following form:

$$r_H(\Delta t, \Delta f) = E\{H(t_1, f_1)H^*(t_2, f_2)\} \tag{14.19}$$

$$= E\left\{\sum_{n_1} \gamma_{n_1}(t_1)e^{-j2\pi f_1 \tau_{n_1}} \sum_{n_2} \gamma_{n_2}^*(t_2)e^{j2\pi f_2 \tau_{n_2}}\right\} \tag{14.20}$$

$$= \sum_{n} E\{\gamma_n(t_1)\gamma_n^*(t_2)\}e^{-j2\pi \tau_n(f_1 - f_2)}. \tag{14.21}$$

In the context of Equation 14.21 we have exploited that $E\{\gamma_{n_1}(t_1)\gamma_{n_2}^*(t_2)\} = 0 \ \forall \ n_1 \neq n_2$ is satisfied, which follows from the assumption of uncorrelated scattering. Furthermore, in Equation 14.21 the n-th CIR tap's spaced-time correlation function is given by:

$$r_{h,n}(\Delta t) = E\{\gamma_n(t_1)\gamma_n^*(t_2)\} \tag{14.22}$$

$$= \sigma_{h,n}^2 r_{\bar{h},n}(\Delta t), \tag{14.23}$$

where $\sigma_{h,n}^2$ is the n-th CIR tap's variance and $r_{\bar{h},n}(\Delta t)$ is the n-th CIR tap's normalised spaced-time correlation function, which obeys $r_{\bar{h},n}(0) = 1$. Note furthermore that we have $\Delta t = t_1 - t_2$. Upon substituting Equation 14.23 into Equation 14.21 we obtain the following equation for the channel's spaced-time spaced-frequency correlation function:

$$r_H(\Delta t, \Delta f) = \sum_{n} \sigma_{h,n}^2 r_{\bar{h},n}(\Delta t)e^{-j2\pi \tau_n \Delta f}, \tag{14.24}$$

where we have $\Delta f = f_1 - f_2$. For $\Delta t = \Delta f = 0$ the channel transfer factor's variance is obtained, namely $\sigma_H^2 = r_H(0,0) = \sum_n \sigma_{h,n}^2$, which is assumed to be unity [68] in the context of our forthcoming discussions.

Upon stipulating that the same normalised spaced-time correlation function $r_{\bar{h}}(\Delta t)$ is associated with the different CIR taps [68], the spaced-time spaced-frequency correlation function of Equation 14.24 simplifies to:

$$r_H(\Delta t, \Delta f) = r_H(\Delta t) \cdot r_H(\Delta f), \tag{14.25}$$

where

$$r_H(\Delta t) = r_{\bar{h}}(\Delta t). \tag{14.26}$$

and

$$r_H(\Delta f) = \sum_{n} \sigma_{h,n}^2 e^{-j2\pi \tau_n \Delta f}. \tag{14.27}$$

Equation 14.25 reflects the *separability* of the channel's spaced-time spaced-frequency correlation function into the product of the spaced-time correlation function and the spaced-frequency correlation function.

ind. WATM **system**	K	K_g	f_s	f_c
	512	64	225MHz	60GHz
ind. WATM **channel**	N_{path}	$RMS(\tau)$	f_D	F_D
	3	16.9ns	2778Hz	0.0071

Table 14.1: Parameters of the indoor WATM system model [2, 219]

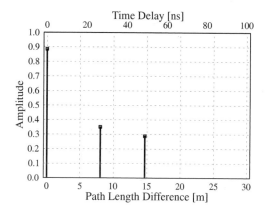

Figure 14.4: Channel impulse response (CIR) of the indoor WATM channel [2, 219]

Following the above rudimentary notes on the properties of the channel transfer function and on the correlation of the CIR, let us now introduce the channel model used in our simulations.

14.3 Channel Model for Monte Carlo Simulations

The simulation results presented throughout our discussions based on Monte Carlo simulations were generated with the aid of the indoor WATM system- and channel model, which will be outlined in the following.

14.3.1 The Indoor WATM Model

The WATM system's parameters used in our investigations follow closely a proposal of the Pan-European Advanced Communications Technologies (ACTS) Median system, which constitutes a wireless extension of wire-line based ATM-type networks [2]. These parameters are listed in Table 14.1. More specifically, the number of OFDM sub-carriers is $K = 512$ and the cyclic prefix exhibits a length of $K_g = 64$ samples. The transmitted signal, which is centred at a carrier frequency of $f_c = 60$GHz is sampled by the receiver at a rate of $f_s = 225$ Msamples/s. Assuming a worst-case vehicular speed of $v = 50$km/h or equivalently $v = 13.9$m/s, the received signal exhibits a maximum Doppler frequency deviation of

$f_D = 2778$Hz from the transmitted carrier frequency [2], which is defined by $f_D = \frac{vf_c}{c}$ [2]. This equals to an OFDM-symbol-normalised Doppler frequency of $F_D = 0.0071$, defined by $F_D = (K + K_g)\frac{f_D}{f_s}$, which is a more adequate metric for characterising the channel's variation versus time. It should be noted that our definition of the OFDM symbol-normalised Doppler frequency F_D is different from the definition presented in [2, 466] in the sense that here the number of guard samples K_g is incorporated as well. Associated with the WATM system model is a three-path Channel Impulse Response (CIR) which exhibits an RMS delay spread of $RMS(\tau) = 0.109\mu s$. The CIR is illustrated in Figure 14.4 [2]. Each of the three paths is independently Rayleigh-faded. In comparison to the original five-path CIR of [2] only the three paths associated with the lowest-delay taps have been retained. This helps to avoid Inter-Symbol Interference (ISI), since the two highest path delays of the original channel model are in excess of the duration of the OFDM symbol's cyclic prefix. Hence, in the following sections we will refer to the three-path CIR as the *indoor WATM channel's-* or, synonymously, the *shortened WATM (SWATM) channel's* CIR.

In the next section we will embark on a short introduction to two-dimensional signal processing, using a vectorial description of the 2D sampling process encountered in the context of the pilot-assisted channel estimation task addressed here.

14.4 Introduction to 2D-Signal Processing

In this section we introduce vector notations encountered during the description of various two-dimensional (2D) signal processing operations [460, 461]. The mathematical convention of denoting vectors by bold faced lower case characters and matrices by bold faced upper case characters will be adopted. The vectors involved are of dimension 2×1, while the matrices are of dimension 2×2. The theorems which will be invoked for the two-dimensional case are also applicable to vector spaces of higher dimension.

In Section 14.4.1 the description of a periodic sequence with the aid of a periodicity matrix is discussed.

14.4.1 Description of a 2D-Sequence by a Periodicity Matrix

The structure of an arbitrary periodic 2D-sequence $x[\mathbf{n}]$ as a function of the integer valued index vector $\mathbf{n} = (n_1 \ n_2)^T$ can be described with the aid of the non-singular and hence invertible 2×2 periodicity matrix $\mathbf{V}$, which fulfils the following condition [460, 461]:

$$x[\mathbf{n} + \mathbf{V}\mathbf{p}] = x[\mathbf{n}], \tag{14.28}$$

implying that the values of the 2D sequence $x[\mathbf{n} + \mathbf{V}\mathbf{p}]$ are identical at the positions corresponding to the displacements of $\mathbf{n} + \mathbf{V}\mathbf{p}$ for any integer vector $\mathbf{p}$. An example is given in Figure 14.5, for which a possible periodicity matrix is given by:

$$\mathbf{V} = \begin{pmatrix} 4 & 2 \\ 1 & 4 \end{pmatrix}, \tag{14.29}$$

where the columns of $\mathbf{V}$ are the basis vectors of the regular structure. Observe the displacements corresponding to the identical elements of the matrix of Equation 14.29 in Figure 14.5,

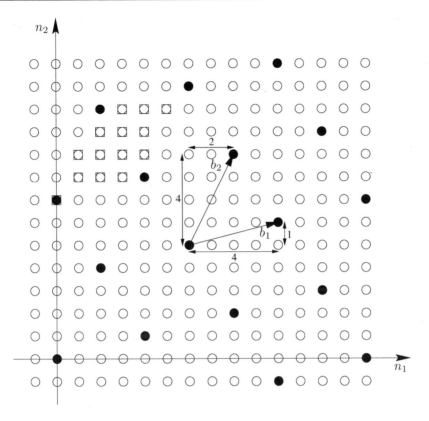

Figure 14.5: Example of a two-dimensional (2D) lattice $x[n]$ generated with the aid of the periodicity matrix V of Equation 14.29.

which are shown in bold. Any integer vector n can be expressed in terms of an arbitrary periodicity matrix V and a unique pair of integer valued vectors k and n_0 as follows:

$$n = k + Vn_0, \qquad (14.30)$$

which is known as the *division theorem of integer vectors* [467–469]. The vector k is related to V by being an element of $\mathfrak{N}(V)$, which is the set of all 2×1 dimensional integer vectors of the form Vx, where x denotes the set of 2×1 dimensional real vectors having components x_i in the range of $0 \leq x_i < 1$. The number of elements contained in $\mathfrak{N}(V)$ is given by $|det V|$, which equals the number of samples associated with one period of $x[n]$.

Most of the signal processing theorems known in the context of one-dimensional signals have their equivalent in both two- and in higher number of dimensions, including the *convolution*- and the *sampling theorem* [460, 461]. In the next section we will embark on the calculation of the maximum tolerable pilot distances in the context of employing a rectangular pilot grid to sample the mobile channel's transfer function versus time and frequency.

14.5   Maximum Pilot-Distances for a Rectangular Pilot Grid

The accuracy of 2D-frequency domain pilot pattern-assisted channel transfer function estimation is influenced by both the 2D geometry and the density of the pilot pattern used. For the case of a rectangular pilot grid represented by the periodicity matrix $\mathbf{V}_{rect}$ [460, 461]:

$$\mathbf{V}_{rect} = \begin{pmatrix} \Delta p_f & 0 \\ 0 & \Delta p_t \end{pmatrix}, \tag{14.31}$$

where Δp_f and Δp_t are the pilot distances in the frequency- and time direction, respectively, the maximum acceptable pilot distances will be determined by applying the Nyquist sampling theorem [177]. The Nyquist theorem states that the minimum sampling frequency required for sampling a band-limited signal in order to permit a perfect reconstruction of the original signal from the discrete set of sample points has to be as twice as high as the highest frequency of the signal. This implies that an ideal rectangular transfer function-based reconstruction filter is required to enable perfect signal reconstruction. In order to allow for a more gradual transition between the filter's passband and the stopband and hence to relax the reconstruction filter specifications, the sampling frequency is usually chosen higher than the minimum required sampling frequency.

14.5.1   Sampling in the Frequency Direction

By capitalising on the Nyquist theorem [463], the maximum tolerable sampling distance $\Delta s_{f,max}$ along the frequency axis normalised to the subcarrier spacing ΔF is given by:

$$\Delta s_{f,max} = \frac{1}{T_m \Delta F}, \tag{14.32}$$

where T_m denotes the multipath spread of the channel, defined in Section 14.2, and the sub-carrier spacing ΔF is given by $\Delta F = 1/(KT_s)$, with K being the number of subcarriers and T_s the time domain sampling period duration. In the idealistic case of a sample-spaced multipath intensity profile, where the multipath spread T_m is given as a multiple of the sampling period duration T_s, we have $T_m = K_0 T_s$. Then Equation 14.32 can be simplified to:

$$\Delta s_{f,max} = \frac{K}{K_0}. \tag{14.33}$$

In order to satisfy the Nyquist theorem, the actual sampling distance Δs_f employed must be lower or equal to the maximum acceptable sampling distance of $\Delta s_{f,max}$, requiring $\Delta s_f \leq \Delta s_{f,max}$. Hence an oversampling factor M_f can be defined as:

$$M_f = \frac{\Delta s_{f,max}}{\Delta s_f}. \tag{14.34}$$

In the more specific case, when the frequency domain channel transfer function samples are provided by the pilot subcarriers, the maximum pilot distance $\Delta p_{f,max}$ along the frequency axis is given by capitalising on Equation 14.32 as:

$$\Delta p_{f,max} = \lfloor \Delta s_{f,max} \rfloor = \left\lfloor \frac{1}{T_m \Delta F} \right\rfloor, \tag{14.35}$$

or in case of a sample-spaced multipath delay profile by exploiting Equation 14.33 as:

$$\Delta p_{f,max} = \left\lfloor \frac{K}{K_0} \right\rfloor. \tag{14.36}$$

In Equations 14.35 and 14.36 the pair of Gauss brackets $\lfloor \rfloor$ denotes the largest integer number, which is not greater than the argument within the brackets. In order to satisfy the Nyquist theorem, the actual pilot distance Δp_f employed must be lower or equal to the maximum sampling distance $\Delta s_{f,max}$, obeying $\Delta p_f \leq \Delta s_{f,max}$ and hence the pilot oversampling factor $M_{p,f}$ can be defined similarly to Equation 14.34 as:

$$M_{p,f} = \frac{\Delta s_{f,max}}{\Delta p_f}. \tag{14.37}$$

In order to give an example, for the $K = 512$ subcarrier indoor WATM channel environment of Section 14.3.1, which exhibits a sample-spaced three-path CIR associated with a multipath spread of $T_m = K_0 T_s$, where $K_0 = 12$, according to Equation 14.36 a maximum pilot distance $\Delta p_{f,max}$ of 42 subcarriers is obtained.

14.5.2 Sampling in the Time Direction

Upon invoking once more the Nyquist theorem [463], a similar expression can be inferred for the maximum acceptable sampling distance $\Delta s_{t,max}$ in the time or OFDM symbol index direction, yielding:

$$\Delta s_{t,max} = \frac{1}{B_D T_f} \tag{14.38}$$

$$= \frac{1}{2F_D}, \tag{14.39}$$

where in Equation 14.38 the variable B_D denotes the Doppler spread of the channel, as defined in Section 14.2. More explicitly, the single-sided equivalent of the channel's Doppler spread is equal to the maximum Doppler frequency of:

$$f_D = \frac{1}{2} B_D = \frac{v f_c}{c}, \tag{14.40}$$

with v being the vehicular speed and f_c the carrier frequency, while c denotes the speed of light. Furthermore, in Equation 14.38 the variable T_f denotes the OFDM symbol duration defined by $T_f = (K + K_g) T_s$ with K being the number of subcarriers and K_g the number of cyclic prefix or guard-interval samples. The product of $F_D = f_D T_f$ is also referred to as the OFDM symbol normalised Doppler frequency F_D of the channel, which is employed in the alternative expression given by Equation 14.39. In order to satisfy the Nyquist theorem, the actual sampling distance Δs_t employed must be lower or equal to the maximum tolerable sampling distance $\Delta s_{t,max}$, requiring $\Delta s_t \leq \Delta s_{t,max}$ and hence analogously to Equation 14.34 the oversampling factor M_t can be defined as:

$$M_t = \frac{\Delta s_{t,max}}{\Delta s_t}. \tag{14.41}$$

Again, in the more specific case, where the channel transfer function samples are provided by the pilot subcarriers, the maximum pilot distance $\Delta p_{t,max}$ along the time direction is given by:

$$\Delta p_{t,max} = \lfloor \Delta s_{t,max} \rfloor = \left\lfloor \frac{1}{B_D T_f} \right\rfloor. \tag{14.42}$$

In order to satisfy the Nyquist theorem, the frequency domain pilot distance Δp_t used must be lower than or equal to the maximum tolerable sampling distance of $\Delta s_{t,max}$, obeying $\Delta p_t \leq \Delta s_{t,max}$ and hence the pilot oversampling factor $M_{p,t}$ can be defined as:

$$M_{p,t} = \frac{\Delta s_{t,max}}{\Delta p_t}. \tag{14.43}$$

In the next section we will embark on a description and assessment of 2D-FIR Wiener filter-aided channel transfer function estimation.

14.6   2D-Pilot Pattern-Assisted 2D-FIR Wiener Filter-Aided Channel Estimation

From a didactic point of view it is useful to commence our discussions on pilot-assisted channel transfer function estimation for OFDM - as employed in the stylised illustration of an OFDM receiver in Figure 14.6 - by describing the 2D-FIR Wiener filtering approach, which simultaneously capitalises on channel transfer function samples collected at individual time and frequency coordinates. The basic principles of 2D-FIR Wiener filter-assisted channel estimation applied in the subcarrier domain were advocated for example in [66, 67, 158, 462, 470] for multi-carrier transmission. In Section 14.6.1 a brief summary of the mathematical concepts of 2D-FIR Wiener filtering will be provided. Furthermore, in Section 14.6.2 the concepts of a robust channel estimator will be outlined, while in Section 14.6.3 a rudimentary strategy for reducing the estimator's complexity will be discussed. This is followed by our MSE performance assessment in Section 14.6.4. After a brief portrayal of the estimator's complexity in Section 14.6.5, our preliminary conclusions will be offered in Section 14.6.6.

14.6.1   Derivation of the Optimum Estimator Coefficients and Estimator MSE for Matched and Mismatched Channel Statistics

The structure of Section 14.6.1 is as follows. In Section 14.6.1.1 an expression is developed for the desired channel transfer factor estimate as a linear combination of the least-squares channel transfer factor estimates associated with the pilot subcarrier positions. Furthermore,

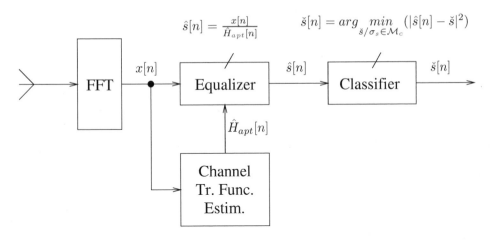

Figure 14.6: Stylised illustration of the OFDM receiver employing pilot-assisted channel estimation. Following FFT processing, the complex frequency domain signal $x[n, k]$ associated with each of the K subcarriers $k = 0, \ldots, K - 1$ is equalised based on the *a posteriori* channel transfer factor estimates generated from the current OFDM symbol's pilot subcarriers, upon assuming 1D-pilot patterns. As a result, the linear estimates $\hat{s}[n, k]$ of the transmitted signals $s[n, k]$ are obtained. These estimates are classified with the aid of the Maximum Likelihood (ML) approach, yielding the complex symbols $\check{s}[n, k]$ that are most likely to have been transmitted. In contrast to this OFDM symbol-by-symbol processing, in the context of the 2D-FIR Wiener filter-based channel estimators employed here the channel estimation, equalisation and classification are performed after the acquisition of an entire block of OFDM symbols. Note furthermore that in the above illustration the subcarrier index k has been omitted from the different variables for the sake of visual clarity.

in Section 14.6.1.2 an equation is provided for the estimator's mean-square error (MSE) conditioned on the knowledge of the channel's spaced-time spaced-frequency correlation function and also conditioned on the estimator's coefficients. The estimator's MSE is minimised by the optimum Wiener solution, which will be outlined in Section 14.6.1.3. Finally, in Section 14.6.1.4 the impact of a mismatch between the statistics of the channel and its estimates used for evaluating the filter coefficients is quantified.

14.6.1.1 Linear Channel Transfer Factor Estimate

To commence our discussions, we assume that in the channel transfer factor estimation process related to a specific subcarrier located on the time-frequency plane a finite number of N_{tap} least-squares channel transfer factor estimates associated with the pilot positions allocated on the 2D-OFDM symbol- and subcarrier plane are invoked. These positions can be arranged in a set $\mathcal{P}_s$:

$$\mathcal{P}_s = \{(\acute{n}_{s1}, \grave{k}_{s1}), (\acute{n}_{s2}, \grave{k}_{s2}), \ldots, (\acute{n}_{sN_{tap}}, \grave{k}_{sN_{tap}})\}. \tag{14.44}$$

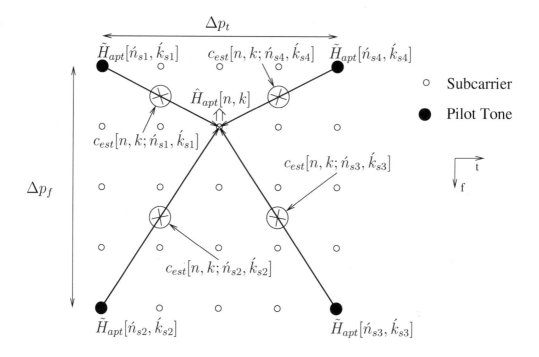

Figure 14.7: Illustration of the process of 2D-FIR Wiener filter-aided channel transfer factor estimation. As shown in Equation 14.47, an estimate $\hat{H}_{apt}[n, k]$ of the channel transfer factor $H[n, k]$ is generated by linear combining of the $N_{tap} = 4$ number of least-squares channel transfer factor estimates $\tilde{H}_{apt}[\acute{n}_s, \acute{k}_s]$ obtained from Equation 14.46 with the aid of the coefficients $c_{est}[n, k; n_s, k_s]$, where $[\acute{n}_s, \acute{k}_s] \in \mathcal{P}_s$, which was defined in Equation 14.44.

Please note that in Equation 14.44 the variable $\acute{n}$ denotes the OFDM symbol index and $\acute{k}$ denotes the subcarrier index, respectively. An *a posteriori* least-squares estimate of the channel transfer factor corrupted by the Additive White Gaussian Noise (AWGN) is given by [67, 158, 470]:

$$\tilde{H}_{apt}[\acute{n}, \acute{k}] \;=\; \frac{x[\acute{n}, \acute{k}]}{s_p[\acute{n}, \acute{k}]} \tag{14.45}$$

$$=\; H[\acute{n}, \acute{k}] + \frac{n[\acute{n}, \acute{k}]}{s_p[\acute{n}, \acute{k}]} \quad \forall \, [\acute{n}, \acute{k}] \in \mathcal{P}_s, \tag{14.46}$$

where according to Equation 14.1, $x[\acute{n}, \acute{k}]$ denotes the received subcarrier signal and $H[\acute{n}, \acute{k}]$ is the channel transfer factor actually experienced by the $\acute{k}$-th subcarrier of the $\acute{n}$-th OFDM symbol. Furthermore, $n[\acute{n}, \acute{k}]$ is the AWGN contribution, $s_p[\acute{n}, \acute{k}]$ is the known pilot symbol and again, $\mathcal{P}_s$ represents the set of pilot positions, which - in contrast to their role in 2D-DFT based channel transfer function estimation [74, 459] - are not necessarily required to be positioned on a periodic grid. From a statistical point of view the channel transfer factor $H[n, k]$ is Rayleigh-distributed having a variance of $\sigma_H^2 = 1$, while the AWGN exhibits zero mean and a variance of σ_n^2. Furthermore, the known pilot symbols $s_p[\acute{n}, \acute{k}]$ exhibit a variance of σ_p^2. A linear estimate $\hat{H}_{apt}[n, k]$ of the channel transfer factor $H[n, k]$ at a specific position $[n, k]$ on the time-frequency plane is given by [66, 67, 470]:

$$\hat{H}_{apt}[n, k] = \sum_{[\acute{n}, \acute{k}] \in \mathcal{P}_s} c_{est}[n, k; \acute{n}, \acute{k}] \cdot \tilde{H}_{apt}[\acute{n}, \acute{k}], \qquad (14.47)$$

where $c_{est}[n, k; \acute{n}, \acute{k}]$ is a filter coefficient which depends both on the position $[n, k]$ of the channel transfer factor to be estimated, as well as on the pilot positions $[\acute{n}, \acute{k}] \in \mathcal{P}_s$. In verbal terms Equation 14.47 implies that the estimate $\hat{H}_{apt}[n, k]$ is generated by an appropriately weighted sum of the pilot-aided least-squares estimates $\tilde{H}_{apt}[\acute{n}, \acute{k}]$ associated with the 2D positions contained in the set $\mathcal{P}_s$. This process is further illustrated in Figure 14.7. Upon invoking vectorial notations, Equation 14.47 can be rewritten as a vector product [471]:

$$\hat{H}_{apt}[n, k] = \mathbf{c}_{est}^H[n, k]\tilde{\mathbf{H}}_{apt}[n, k], \qquad (14.48)$$

where the vector $\tilde{\mathbf{H}}_{apt}[n, k] \in \mathbb{C}^{N_{tap} \times 1}$ of pilot-aided least-squares channel transfer factor estimates is given by:

$$\tilde{\mathbf{H}}_{apt}[n, k] = (\tilde{H}_{apt}[\acute{n}_{s1}, \acute{k}_{s1}], \tilde{H}_{apt}[\acute{n}_{s2}, \acute{k}_{s2}], \dots, \tilde{H}_{apt}[\acute{n}_{sN_{tap}}, \acute{k}_{sN_{tap}}])^T, \qquad (14.49)$$

and the vector $\mathbf{c}_{est}[n, k] \in \mathbb{C}^{N_{tap} \times 1}$ of estimator coefficients is defined by [471]:

$$\mathbf{c}_{est}[n, k] = (c_{est}^*[n, k; \acute{n}_{s1}\acute{k}_{s1}], c_{est}^*[n, k; \acute{n}_{s2}\acute{k}_{s2}], \dots, c_{est}^*[n, k; \acute{n}_{sN_{tap}}, \acute{k}_{sN_{tap}}])^T. \qquad (14.50)$$

In Equation 14.50 the conjugate complex operator is denoted by $()^*$.

14.6.1.2 Estimator MSE and Cost-Function

A prominent approach of determining $\mathbf{c}_{est}[n, k]$ is that of minimising the cost-function $J(\mathbf{c}_{est}[n, k])$, which is defined as the MSE between the "true" channel transfer factor $H[n, k]$ and the estimate $\hat{H}_{apt}[n, k]$ - with respect to $\mathbf{c}_{est}[n, k]$ [66, 67, 470], which is formulated as:

$$J(\mathbf{c}_{est}[n, k]) = \text{MSE}(\mathbf{c}_{est}[n, k]) \qquad (14.51)$$
$$= E\{|H[n, k] - \hat{H}_{apt}[n, k]|^2\}. \qquad (14.52)$$

Upon substituting Equation 14.48 into Equation 14.52 and by evaluating the expectation $E\{\}$ the following expression is obtained [66, 67, 470]:

$$J(\mathbf{c}_{est}[n, k]) = \sigma_H^2 - \mathbf{c}_{est}^T[n, k]\mathbf{r}_{apt}^*[n, k] - \mathbf{c}_{est}^H[n, k]\mathbf{r}_{apt}[n, k] + \\ + \mathbf{c}_{est}^H[n, k]\mathbf{R}_{apt}[n, k]\mathbf{c}_{est}[n, k], \qquad (14.53)$$

where $\mathbf{r}_{apt}[n, k] \in \mathbb{C}^{N_{tap} \times 1}$ denotes the cross-correlation vector between the "true" channel transfer factor $H[n, k]$ at position $[n, k]$ and the pilot-assisted least-squares channel transfer factor estimates $\tilde{H}_{apt}[\acute{n}, \acute{k}]$ at positions $[\acute{n}, \acute{k}] \in \mathcal{P}_s$, defined as [66,67,470]:

$$\mathbf{r}_{apt}[n, k] \;=\; E\{H^*[n, k]\tilde{\mathbf{H}}_{apt}[n, k]\} \tag{14.54}$$

$$=\; E\{H^*[n, k]\mathbf{H}[n, k]\} \tag{14.55}$$

$$=\; \mathbf{r}[n, k], \tag{14.56}$$

and where $\mathbf{H}[n, k] \in \mathbb{C}^{N_{tap} \times 1}$ is the vector of "true" channel transfer factors at the pilot positions, defined similarly to $\tilde{\mathbf{H}}_{apt}[n, k]$ of Equation 14.49. In Equation 14.55 we have exploited the fact that the noise process $n[\acute{n}, \acute{k}]$ is uncorrelated to the channel transfer factor $H[n, k]$, i.e. that $E\{H^*[n, k]n[\acute{n}, \acute{k}]\} = 0$. Hence, as shown in Equation 14.56 the cross-correlation vector $\mathbf{r}_{apt}[n, k]$ is equivalent to the cross-correlation vector $\mathbf{r}[n, k] \in \mathbb{C}^{N_{tap} \times 1}$ of the "true" channel transfer factors at the pilot positions. Returning to Equation 14.53, the symbol $\mathbf{R}_{apt}[n, k] \in \mathbb{C}^{N_{tap} \times N_{tap}}$ represents the auto-correlation matrix of the pilot assisted least-squares channel transfer factor estimates, defined as [66,67,470]:

$$\mathbf{R}_{apt}[n, k] \;=\; E\{\tilde{\mathbf{H}}_{apt}[n, k]\tilde{\mathbf{H}}_{apt}^{H}[n, k]\} \tag{14.57}$$

$$=\; E\{\mathbf{H}[n, k]\mathbf{H}^{H}[n, k]\} + \rho_p \mathbf{I} \tag{14.58}$$

$$=\; \mathbf{R}[n, k] + \rho_p \mathbf{I}, \tag{14.59}$$

and $\mathbf{R}[n, k] \in \mathbb{C}^{N_{tap} \times N_{tap}}$ is the auto-correlation matrix of the "true" channel transfer factors at the pilot positions. Furthermore, $\rho_p = 1/SNR_p = \sigma_n^2/\sigma_p^2$ denotes the reciprocal of the effective subcarrier SNR at the pilot positions.

14.6.1.3   Optimum Estimator Coefficients and Minimum Estimator MSE

Upon invoking standard gradient based optimisation techniques [90,471,472] or by applying the so-called orthogonal projection theorem [90,472] associated with MMSE optimisation problems – both of which will be further outlined in the context of our discussions on *a priori* time direction prediction filtering in Section 15.2.4 - the vector $\mathbf{c}_{est}[n, k]|_{opt}$ of optimum filter coefficients is given by [66,67,90,470–472]:

$$\mathbf{c}_{est}[n, k]|_{opt} \;=\; \mathbf{R}_{apt}^{-1}[n, k]\mathbf{r}_{apt}[n, k] \tag{14.60}$$

$$=\; (\mathbf{R}[n, k] + \rho_p \mathbf{I})^{-1}\mathbf{r}[n, k], \tag{14.61}$$

which is the well-known Wiener solution. In physically tangible terms Equation 14.60 represents a set of N_{tap} number of simultaneous equations delivering the coefficients to be used in Equation 14.48 for determining the channel transfer factor estimate $\hat{H}_{apt}[n, k]$ sought. More specifically, Equation 14.61 is derived from Equation 14.60 by invoking the expressions for $\mathbf{R}_{apt}[n, k]$ and $\mathbf{r}_{apt}[n, k]$ given by Equations 14.59 and 14.56. Upon substituting the vector $\mathbf{c}_{est}[n, k]|_{opt}$ of optimum filter coefficients given by Equation 14.60 into Equation 14.53 we obtain an expression for the estimator's minimum mean-square error (MMSE) at position

$[n,k]$ in the form of [66,67,470]:

$$\text{MMSE}[n,k] \;=\; \text{MSE}(\mathbf{c}_{est}[n,k]|_{opt}) \tag{14.62}$$

$$=\; \sigma_H^2 - \mathbf{c}_{est}^H[n,k]|_{opt}\mathbf{r}[n,k]. \tag{14.63}$$

14.6.1.4 Estimator Coefficients for Mismatched Channel Statistics

In the evaluation of the optimum estimator coefficients according to Equation 14.61, perfect knowledge of the auto-correlation matrix $\mathbf{R}[n,k]$ of the "true" channel transfer factors, of the inverse of the SNR associated with the pilot subcarriers, denoted by $\rho_p = 1/SNR_p$ and of the cross-correlation vector $\mathbf{r}[n,k]$ was assumed. The elements of $\mathbf{R}[n,k]$ and $\mathbf{r}[n,k]$ are defined by the channel's spaced-time spaced-frequency correlation function $r_H[\Delta n, \Delta k] = r_H(\Delta n T_f, \Delta k \Delta F)$, as was discussed in Section 14.2. In general, however, the perfect knowledge of the channel's statistics is not available. Hence, it is more appropriate to stipulate an estimate $\tilde{r}_H[\Delta n, \Delta k]$ of the channel's spaced-time spaced-frequency correlation function and correspondingly that of an auto-correlation matrix $\tilde{\mathbf{R}}[n,k]$ and a cross-correlation vector $\tilde{\mathbf{r}}[n,k]$ related to $\tilde{\mathbf{R}}_{apt}[n,k]$ and $\tilde{\mathbf{r}}_{apt}[n,k]$ according to Equations 14.59 and 14.56, respectively. Furthermore, an estimate of the inverse of the SNR at the pilot positions is given by $\tilde{\rho}_p$. Instead of Equations 14.60 and 14.61 hence a vector $\tilde{\mathbf{c}}_{est}[n,k]|_{opt}$ of sub-optimum filter coefficients is obtained in the form of:

$$\tilde{\mathbf{c}}_{est}[n,k]|_{opt} \;=\; \tilde{\mathbf{R}}_{apt}^{-1}[n,k]\tilde{\mathbf{r}}_{apt}[n,k] \tag{14.64}$$

$$=\; (\tilde{\mathbf{R}}[n,k] + \tilde{\rho}_p \mathbf{I})^{-1}\tilde{\mathbf{r}}[n,k], \tag{14.65}$$

for which the associated estimator MSE is degraded in comparison to the MSE achieved with the aid of the vector $\mathbf{c}_{est}[n,k]|_{opt}$ of optimum filter coefficients invoking the "true" channel statistics, yielding:

$$\text{MSE}(\tilde{\mathbf{c}}_{est}[n,k]|_{opt}) \geq \text{MSE}(\mathbf{c}_{est}[n,k]|_{opt}). \tag{14.66}$$

14.6.2 Robust Estimator Design

Li [74,459] demonstrated in the context of 2D-pilot pattern-assisted 2D-DFT-based Wiener filter channel estimation capitalising on an infinite number of filter taps that the Wiener filter can be designed such that its performance becomes insensitive to the mismatch between the specific channel statistics invoked in the filter calculation and the statistics actually experienced by the OFDM symbols transmitted over the channel. This required the employment of a uniform, ideally support-limited channel scattering function in the calculation of the filter coefficients, whose region of support contained the time-variant channel transfer function's region of support. As a consequence, the robust channel estimator exhibited the same MSE performance – regardless of the specific shape of the support-limited scattering function associated with the channel encountered – as the optimum channel estimator when communicating over a channel having a uniform or constant-valued scattering function. Li [74,459] also demonstrated that the optimum Wiener filter estimator exhibits the worst MSE performance among all channels for transmission over a channel having a uniform or constant-valued scattering function. Hence the robust estimator exhibits the worst-case MSE performance. Please note that in the context of the finite tap- or Finite Impulse Response (FIR) estimators considered here, the invariance of the MSE associated with the robust estimator is only approximate due to the associated imperfections of the filter's transfer function related to employing a finite number of pilot-assisted channel transfer function samples.

The remaining part of Section 14.6.2 exhibits the following structure. In Section 14.6.2.1 the uniform or constant-valued, ideally supported limited scattering function and the related spaced-time spaced-frequency correlation function is expressed in mathematical terms, while in Section 14.6.2.2 the role of the uniform scattering function's associated shift parameter is highlighted. The specific application of the uniform scattering function in the context of robust Wiener filtering is further detailed in Section 14.6.2.3.

14.6.2.1 Uniform Scattering Function and its Associated Spaced-Time Spaced-Frequency Correlation Function

The rectangular *uniform* or constant-valued channel scattering function associated with a multipath spread of T_m and a maximum Doppler frequency f_D, which is also invoked in the calculation of the robust Wiener filter's coefficients is given by:

$$S_{h,unif}(f_d, \tau) = S_{H,unif}(f_d) \cdot r_{h,unif}(\tau), \tag{14.67}$$

where $r_{h,unif}(\tau)$ is the specific normalised uniform multipath intensity profile- or delay power spectrum given by [66, 67, 470]:

$$r_{h,unif}(\tau) = \frac{1}{T_m} \text{rect}\left(\frac{\tau - \tau_{shift}}{T_m}\right) \tag{14.68}$$

and $S_{H,unif}(f_d)$ is the specific normalised uniform Doppler power spectrum given by [470]:

$$S_{H,unif}(f_d) = \frac{1}{2f_D} \text{rect}\left(\frac{f_d}{2f_D}\right). \tag{14.69}$$

In Equation 14.68 the parameter τ_{shift} denotes the displacement of the multipath intensity profile which will be elaborated on in more depth in Section 14.6.2.2. For a value of $\tau_{shift} \geq T_m/2$ the associated multipath intensity profile is causal. As outlined in Figure 14.3 of Section 14.2, the spaced-time spaced-frequency correlation function namely $r_{H,unif}(\Delta t, \Delta f)$ corresponding to the uniform channel scattering function of Equation 14.67 can be obtained upon Fourier transforming Equation 14.67 with respect to the variable τ and inverse Fourier transforming it with respect to the variable f_d:

$$r_{H,unif}(\Delta t, \Delta f) = r_{H,unif}(\Delta t) \cdot r_{H,unif}(\Delta f), \tag{14.70}$$

where the specific uniform spaced-frequency correlation function $r_{H,unif}(\Delta f)$ is given by:

$$r_{H,unif}(\Delta f) = \mathcal{FT}_{\tau \to \Delta f} \{r_{h,unif}(\tau)\} \tag{14.71}$$

$$= \mathrm{sinc}(\pi T_m \Delta f) e^{-j2\pi\tau_{shift}\Delta f}, \tag{14.72}$$

while the specific uniform spaced-time correlation function $r_{H,unif}(\Delta t)$ is given by:

$$r_{H,unif}(\Delta t) = \mathcal{FT}_{f_d \to \Delta t}^{-1} \{S_{H,unif}(f_d)\} \tag{14.73}$$

$$= \mathrm{sinc}(2\pi f_D \Delta t). \tag{14.74}$$

Recall that the sinc-function is defined as $\mathrm{sinc}(x) = \sin(x)/x$ [90, 463].

14.6.2.2 Relevance of the Shift-Parameter

In the context of employing the scattering function of Equation 14.67 or equivalently the spaced-time spaced-frequency correlation of Equation 14.70 in the calculation of the Wiener filter's coefficients,[6] Equation 14.72 suggests setting the shift parameter τ_{shift} of Equation 14.68 equal to zero. As a consequence, the spaced-frequency correlation function $r_{H,unif}(\Delta f)$ becomes real-valued instead of complex-valued, which would be the case for $\tau_{shift} \neq 0$ as can be deduced from Equation 14.72. Hence, the complexity of the filter coefficient calculation and most importantly that of the filtering itself is halved, since the corresponding filter coefficients are real-valued as well. This is achieved at the cost of setting the multipath spread $\tilde{T}_m$ associated with the channel estimator at least twice as high as the specific multipath spread T_m actually observed on the channel. Hence, in order to extract the first repetition of the sampled frequency domain channel transfer function's time domain representation, according to Equation 14.35 the maximum tolerable pilot distance in the frequency direction should be determined as a function of $\tilde{T}_m$ instead of T_m, which implies an increase of the pilot grid density and hence also that of the pilot overhead.

14.6.2.3 Application to a Robust Estimator

In the context of multi-carrier transmission systems employing these channel estimation techniques, the variables Δt and Δf, which denote the relative time and frequency displacements in the argument of the sinc-functions in Equations 14.72 and 14.74 are discretized with the aid of the differential OFDM symbol and subcarrier indices Δn and Δk, respectively, where $\Delta t = T_f \Delta n$ and $\Delta f = \Delta F \Delta k$. More specifically, $T_f = (K + K_g)T_s$ denotes the OFDM symbol duration, where K is the total number of subcarriers, K_g is the number of guard interval samples, and T_s is the sampling period duration, while the factor $\Delta F = 1/(KT_s)$ represents the subcarrier frequency spacing. Hence, the components $\tilde{r}_{apt}[\Delta n, \Delta k]|_{rob}$ of the cross-correlation vector $\tilde{\mathbf{r}}_{apt}[n,k]|_{rob} \in \mathbb{C}^{N_{tap} \times 1}$ and the components $\tilde{R}_{apt}[\Delta n, \Delta k]|_{rob}$ of the auto-correlation matrix $\tilde{\mathbf{R}}_{apt}[\Delta n, \Delta k]|_{rob} \in \mathbb{C}^{N_{tap} \times N_{tap}}$ invoked in the calculation of the filter coefficients for the specific robust channel estimator according to Equation 14.60

---

[6]Also note that in the context of determining the Wiener filter's coefficients, the variables T_m, τ_{shift} and f_D associated with Equations 14.68 and 14.69 are marked by $\tilde{(\,)}$ in order to distinguish them from the true values of the parameters associated with the channel.

are given by:

$$\tilde{r}_{apt}[\Delta n, \Delta k]|_{rob} = r_{H,unif}(T_f \Delta n, \Delta F \Delta k), \qquad (14.75)$$

and:

$$\tilde{R}_{apt}[\Delta n, \Delta k]|_{rob} = r_{H,unif}(T_f \Delta n, \Delta F \Delta k) + \tilde{\rho}_p \delta[\Delta n, \Delta k], \qquad (14.76)$$

respectively, where the operator $\delta[\Delta n, \Delta k]$ denotes the 2D Kronecker delta function [473].

14.6.3   Complexity Reduction

As will be highlighted in Section 14.6.3.1, the number of pilot symbols contained in an estimator input block is potentially excessive, when invoking all associated least-squares channel transfer factor estimates during the estimation of the channel transfer factor at a specific position on the time-frequency plane. Hence, in Section 14.6.3.2 we will use the approach proposed by Höher *et al.* [66,67,462] for selecting a reduced subset constituted by the nearest pilot subcarrier positions during the channel transfer function estimation process.

14.6.3.1   Number of Pilot Subcarriers in the Estimator's Input Block

The N_p number of pilot subcarrier positions contained in an estimator input block of dimensions N_f subcarriers in the frequency direction and N_t OFDM symbols in the time direction can be arranged in a set $\mathcal{P}$ described by:

$$\mathcal{P} = \{(\acute{n}_1, \acute{k}_1), (\acute{n}_2, \acute{k}_2), \dots, (\acute{n}_{N_p}, \acute{k}_{N_p})\}. \qquad (14.77)$$

For the general case of a periodicity matrix $\mathbf{V}$ associated with $|det\mathbf{V}| \neq 0$, an estimate $N_{p\approx}$ of the number of pilot positions hosted by the estimator input block is given by the ratio of the areas associated with the estimator's input block, $N_f N_t$, and that of one period of the pilot grid, given by $|det\mathbf{V}|$:

$$N_{p\approx} = \left\lceil \frac{N_f N_t}{|det\mathbf{V}|} \right\rceil. \qquad (14.78)$$

For the more specific case of a rectangular pilot grid described by the periodicity matrix $\mathbf{V}_{rect}$ we have:

$$\mathbf{V}_{rect} = \begin{pmatrix} \Delta p_f & 0 \\ 0 & \Delta p_t \end{pmatrix}. \qquad (14.79)$$

In this case it is straightforward to exactly determine the total number of pilot positions contained in an estimator input block. More specifically, an estimation block contains $\lceil N_t/\Delta p_t \rceil$ number of OFDM symbols each hosting $\lceil N_f/\Delta p_f \rceil$ pilot positions.

Hence, the total number of pilot positions $N_{p,rect}$ hosted by an estimator input block in the context of a rectangular pilot grid is given by:

$$N_{p,rect} = \left\lceil \frac{N_f}{\Delta p_f} \right\rceil \cdot \left\lceil \frac{N_t}{\Delta p_t} \right\rceil . \tag{14.80}$$

To give an example, for $N_f = K = 512$ subcarriers per OFDM symbol, for $N_t = 64$ OFDM symbols per estimation frame, and for pilot distances of $\Delta p_f = 32$ subcarriers in the frequency direction as well as for $\Delta p_t = 8$ OFDM symbols in the time direction, the total number of pilot subcarrier positions given by $N_{p,rect} = 128$, is excessive to be directly employed in the estimation of all of the channel transfer factors. Hence, a subcarrier selection procedure can be performed, as will be outlined in Section 14.6.3.2.

14.6.3.2 Selection of Subsets of Pilot Subcarriers

As argued in Section 14.6.3.1, the set $\mathcal{P}$ of N_p pilot positions contained in the estimator input block is usually significantly larger than the reduced-size set $\mathcal{P}_s$ of N_{tap} pilot positions defined in Equation 14.44, whose associated least-squares channel transfer factor estimates are invoked in the MMSE estimation of the channel transfer factor at a specific position on the time-frequency plane. This is because the complexity of the 2D-FIR Wiener filter based channel estimator increases linearly with the number of filter taps, and therefore it has to be restricted.

Hence, the selection of a reduced number of N_{tap} out of N_p number of pilot positions to be used to estimate the channel transfer factor of each subcarrier is an important task. As outlined in [67] when using an exhaustive search, a potentially excessive number of $\binom{N_p}{N_{tap}}$ sets of pilot positions would have to be compared in terms of the associated estimator MSE for each subcarrier. In order to circumvent this complexity problem, Höher *et al.* [66, 67] proposed searching for the N_{tap} number of nearest pilot positions indexed by $[\acute{n}, \acute{k}]$ - with respect to a specific distance measure - within a limited range around the target position $[n, k]$ considered on the time-frequency plane. According to [67], a weighted distance measure of the form:

$$d_w = |k - \acute{k}|\tilde{B}_D T_f + |n - \acute{n}|\tilde{T}_m \Delta F \tag{14.81}$$

can be advantageously used, where $\tilde{B}_D T_f$ is the estimated OFDM symbol normalised Doppler spread and $\tilde{T}_m \Delta F$ is the estimated subcarrier spacing normalised multipath spread. However, as it will be argued in Section 14.6.4, the application of the weighted distance-measure of Equation 14.81 is limited to scenarios, which employ a relatively "balanced" pilot grid, further defined in Section 14.6.4.1.

14.6.4 MSE Performance of 2D-FIR Wiener Filtering

In Section 14.6.4 we will concentrate on the performance assessment of the 2D-FIR Wiener filter-based channel estimator, while invoking the mean-square channel transfer factor estimation error as our performance measure. More specifically, while in Section 14.6.4.1 our choice of the simulation parameters is justified, in Section 14.6.4.2 we will characterise the evolution of the estimator's MSE versus frequency and time. Furthermore, the influence of the pilot grid density and that of the number of filter taps on the estimator's MSE will be highlighted in Section 14.6.4.3.

14.6.4.1   Simulation Parameters - Design of the Pilot Grid and the Uniform Scattering Function used in the Calculation of Filter Coefficients

We stipulate an unbounded pilot grid in order to avoid the obfuscating effects of an MSE degradation at the block boundaries due to a lack of pilot subcarriers at these edge positions. More specifically, in our investigations conducted in this section, the rectangular pilot grid described by the periodicity matrix of Equation 14.79 was stipulated. Furthermore, the pilot grid was assumed to be perfectly balanced. This is reflected by the condition that the pilot oversampling factors of $M_{p,f}$, $M_{p,t}$ defined in Equations 14.37 and 14.43, respectively, which are associated with the frequency and time direction, are approximately identical, yielding:

$$M_{p,f} \approx M_{p,t}. \tag{14.82}$$

As our experiments revealed, the pilot grid's "balance" between the time and frequency direction is a prerequisite for a reliable selection of the "nearest" pilot subcarriers according to the weighted distance metric of Equation 14.81. In the context of a strongly "imbalanced" pilot grid significantly more pilot subcarriers would be selected along one of the directions – namely frequency or time direction - associated with the higher oversampling factor. This is undesirable, since due to the lack of surrounding pilots in some cases the estimator would actually have to perform a prediction with the consequence of potentially excessive MSEs.

In our analytical evaluation of the estimator's MSE using Equation 14.63 under perfectly matched channel conditions - where the channel statistics assumed in the filter's calculation were identical to those associated with the channel - a time-variant causal CIR was assumed, which was characterised in terms of its correlation properties by the uniform channel scattering function of $S_{h,unif}(f_d, \tau)$, as defined in Equation 14.67. From our discussions in Section 14.2 we recall that the Wiener filter-based channel estimator exhibits the worst MSE performance for transmission over a channel having a uniform, ideally support-limited scattering function. Hence, upon neglecting the filter's imperfections due to having a finite number of filter taps and due to having a potentially suboptimum selection of the "nearest" pilot subcarriers, the upperbound MSE performance is characterised here.

More specifically, the multipath spread $\tilde{T}_m$ associated with the channel estimator was assumed to be identical to that of the channel, namely to T_m, while both displacement parameters, namely $\tilde{\tau}_{shift}$ and τ_{shift} were set equal to $T_m/2$. This is the scenario of a causal channel, where the first arriving multipath component exhibits a delay of zero. Furthermore, the Doppler spread $\tilde{B}_D$ associated with the channel estimator was also assumed to be identical to that of the channel, namely to B_D.

In our example the multipath spread T_m is given as a multiple of the OFDM sampling period duration T_s, which is formulated as $T_m = K_0 T_s$, where K_0 was set to 16. The number of subcarriers K hosted by the OFDM symbol was assumed to be 512. Hence, the subcarrier spacing normalised multipath spread defined as $T_m \Delta F$, is given by $T_m \Delta F = \frac{K_0}{K} = 0.03125$ for the above example. It is convenient to assume an identical subcarrier spacing normalised multipath spread $T_m \Delta F$ and OFDM symbol normalised Doppler spread defined as $B_D T_f$. Hence we also have $B_D T_f = 0.03125$. Our aim is to avoid the imperfections associated with the sub-optimum selection of pilot subcarrier positions due to an "imbalanced" pilot grid. Following from the relation of $B_D T_f = 2F_D$, the OFDM symbol normalised Doppler frequency

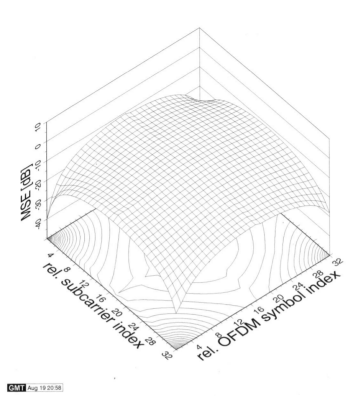

Figure 14.8: Evolution of the MSE exhibited by the 2D-FIR Wiener filter-based channel estimator over one period of the rectangular pilot grid at a channel SNR of 40dB measured at the reception antenna. The same uniform ideally support-limited scattering function obeying Equation 14.67 was associated with both the evaluation of the estimator's filter coefficients and the channel. The subcarrier spacing normalised multipath spread of $T_m \Delta F$ was identical to the OFDM symbol normalised Doppler spread of $B_D T_f$, which was set to $T_m \Delta F = B_D T_f = 0.03125$. The pilot distances of Δp_f, Δp_t employed here were equal to the maximum pilot distances, namely to $\Delta p_f = \Delta p_{f,max} = 32$ and to $\Delta p_t = \Delta p_{t,max} = 32$ as delivered by Equations 14.35 and 14.42, respectively. Hence, the corresponding pilot oversampling factors of $M_{p,f}$, $M_{p,t}$ defined in Equation 14.37 and 14.43, respectively, were both equal to one, $M_{p,f} = M_{p,t} = 1$. The number of FIR filter taps invoked was $N_{tap} = 16$. The results were evaluated from Equation 14.63.

is given by $F_D = 0.015625$. According to Equations 14.35 and 14.42, the maximum tolerable pilot distances of $\Delta p_{f,max}$ and $\Delta p_{t,max}$, respectively, measured in the frequency and time direction in the context of a rectangular pilot grid are given by $\Delta p_{f,max} = \Delta p_{t,max} = 32$.

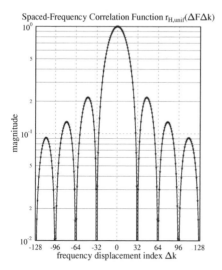
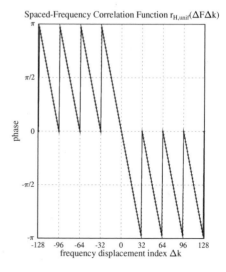

Figure 14.9: Spaced-frequency correlation function $r_{H,unif}(\Delta f)$ according to Equation 14.72 for frequency displacements of $\Delta f = \Delta F \Delta k$ separated into **(a: Left)** magnitude- and **(a: Right)** phase-components, associated with a causal uniform ideally support-limited multipath intensity profile in the context of a subcarrier spacing normalised multipath spread of $T_m \Delta F = 0.03125$. The subcarrier spacing normalised displacement parameter $\tau_{shift}\Delta F$ was set equal to $T_m \Delta F/2$.

14.6.4.2   Evolution of the Estimator MSE over a Period of the Rectangular Pilot Grid

Let us commence our discussion of the 2D-FIR Wiener filter-based estimator's MSE performance with the aid of Figure 14.8, where we have illustrated the evolution of the estimator MSE over a period of the rectangular pilot grid associated with $\Delta p_{f,max} = 32$ and $\Delta p_{t,max} = 32$. The number of FIR filter taps was $N_{tap} = 16$. In Figure 14.8 we observe that upon increasing the distance of the observation point from the pilot positions located at the corners of the square-shaped region, the estimator's MSE is substantially degraded. This is a consequence of the channel's decorrelation, reflected by relatively low values of the channel's sinc-shaped spaced-time spaced-frequency correlation function. The slight imbalance observed in the MSE plot of Figure 14.8 with respect to the relative OFDM symbol index is associated with the imperfections of the pilot selection according to Equation 14.81.

The specific channel's spaced-frequency correlation function associated with the causal uniform ideally support-limited multipath intensity profile of Equation 14.68 for $\tau_{shift} = T_m/2$, has been plotted in Figure 14.9. We observe that for a frequency displacement index of 16, which is half the distance between two adjacent pilot positions, the correlation coefficient becomes as low as 0.6. By contrast, for a two-times oversampled scenario the correlation coefficient at a frequency displacement index of 8, which is half the distance between two adjacent pilot positions would still exhibit a significant value of around 0.9.

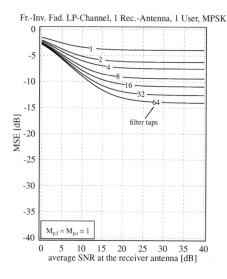
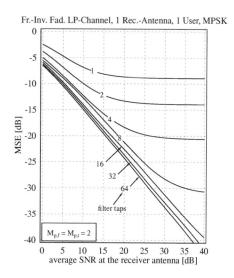

Figure 14.10: MSE exhibited by the 2D-FIR Wiener filter-based channel estimator averaged over one period of the rectangular pilot grid as a function of the SNR at the reception antenna and parameterised with the number of FIR filter taps. The same uniform ideally support-limited scattering function obeying Equation 14.67 was associated with both the evaluation of the filter coefficients and the channel. The subcarrier spacing normalised multipath spread of $T_m \Delta F$ was identical to the OFDM symbol normalised Doppler spread of $B_D T_f$, where we had $T_m \Delta F = B_D T_f = 0.03125$. The pilot sampling distances of Δp_f, Δp_t employed in the context of the **left graph** were equal to the maximum pilot distances, of $\Delta p_f = \Delta p_{f,max} = 32$ and $\Delta p_t = \Delta p_{t,max} = 32$ delivered by Equations 14.35 and 14.42, respectively. This corresponded to unity oversampling factors, namely to $M_{p,f} = M_{p,t} = 1$. By contrast, two-times oversampling expressed as $M_{p,f} = M_{p,t} = 2$ was employed in the context of **the graph on the right**, for which we had $\Delta p_f = \Delta p_t = 16$. The results were evaluated from Equation 14.63.

14.6.4.3 Influence of the Pilot Grid Density and the Number of Filter Taps on the Estimator's MSE

In order to further illustrate the influence of the pilot grid density and that of the number of FIR filter taps on the performance of the 2D-FIR Wiener filter-based channel estimator, we have plotted in Figure 14.10 the estimator's MSE based on Equation 14.63 averaged over one period of the pilot grid. At the left of Figure 14.10 we characterise a maximum-sparsity pilot grid associated with $\Delta p_f = \Delta p_t = 32$, where the oversampling factors of $M_{p,f} = M_{p,t} = 1$ are used. By contrast, at the right we characterize a pilot grid of double density associated with $\Delta p_f = \Delta p_t = 16$, which implies a two-times oversampling, expressed as $M_{p,f} = M_{p,t} = 2$. We note that the scenario associated with the left graph of Figure 14.10 is identical to that of Figure 14.8. Here we observe that when doubling the number of FIR filter taps, the estimator's average MSE decreases by about 2-3dB. This relatively modest MSE reduction is a result of the also modest correlation of the channel transfer function samples employed in the estimation process. By contrast, for the double-density pilot grid, whose associated average MSE is illustrated in the right graph of Figure 14.10, a significant MSE improvement is observed upon increasing the number of FIR filter taps up to a value of 16, which is deemed sufficiently high to exploit most of the channel's predictability or correlation. This is evidenced by the parallel nature of the MSE curves observed, when further increasing the number of filter taps beyond 16. The additional MSE reduction achieved is a result of the more efficient mitigation of the channel noise. By further increasing the pilot oversampling factors of $M_{p,f}$, $M_{p,t}$, the MSE curves would be shifted to lower MSEs, since a higher correlation of the channel transfer factors at the pilot positions with the transfer factors to be estimated permits the Wiener filter to apply more emphasis on the noise mitigation.

14.6.5   Computational Complexity

For the 2D-FIR Wiener filter-based estimator we recall that individual filter coefficient vectors have to be provided for each of the different positions on the time-frequency plane. The number of coefficient vectors to be determined for a specific SNR level can be significantly reduced upon exploiting the periodic structure of the pilot pattern. Here we will assume that the calculation of the coefficient vectors is performed only once before the start of the OFDM symbol transmission and hence it does not contribute to the computational complexity imposed during the demodulation of each OFDM symbol, which will be our main focus.

For each time-frequency point the specific operation of filtering can be described as the computation of the inner-product between the filter's coefficient vector and a vector, which hosts the various pilot-related least-squares channel transfer factor estimates. Upon assuming complex valued filter coefficients, this requires N_{tap} number of complex multiplications and the same number of complex additions. Hence, the total complexity of estimating the channel transfer factor of each time-frequency point associated with the $N_f \times N_t$ estimation block is given by:

$$C_{\text{2D-FIR}}^{(\mathbb{C}*\mathbb{C})} = C_{\text{2D-FIR}}^{(\mathbb{C}+\mathbb{C})} = N_f N_t N_{tap} \qquad (14.83)$$

In case of continuous transmissions, upon neglecting edge effects, the total number of filter coefficient vectors to be pre-calculated for a specific SNR level is given by $|det\mathbf{V}|$, which is the number of time-frequency points contained in one period of the pilot grid. By contrast, in case of the block-based transmission scheme considered here, it is an arduous task to estimate the total number of filter coefficient tuples to be determined, since the associated edge effects would have to be considered in both the time and frequency directions. Note that in cases when a scattering function - which is symmetric with respect to the multipath delay- and Doppler frequency variables is employed - then the resulting filter coefficients are real-valued, and hence only half the number of multiplications and additions is required in the filtering operations.

14.6.6 Summary and Conclusion on 2D-FIR Wiener Filtering

In Section 14.6 2D-FIR Wiener filter-based channel transfer factor estimation was portrayed and assessed. Our discussions commenced in Section 14.6.1 with an analytical description of the 2D-FIR Wiener filter, where the filter was further illustrated in Figure 14.7. This illustration included the representation of the desired channel transfer factor estimate as a weighted superposition of the nearest pilots' least-squares channel transfer factor estimates, which was further discussed in Section 14.6.1.1. It was further argued in Section 14.6.1.2 that a suitable cost-function required for determining the filter coefficients is given by the mean-square error (MSE) between the channel transfer factor to be estimated and the estimator's output signal. The vector of coefficients minimising the MSE-related cost-function was then provided in Section 14.6.1.3, with the aid of the optimum Wiener solution [66,67,90], which is associated with a specific minimum estimation MSE. While the optimum Wiener solution is based on perfect knowledge of the channel's statistics expressed in form of the spaced-time spaced-frequency correlation function $r_H(\Delta t, \Delta f)$ of Section 14.2, as well as relying on perfect knowledge of the SNR at the pilot positions, this knowledge is normally not available. Hence, as discussed in Section 14.6.1.4, the calculation of the filter coefficients has to be based on estimates of these parameters with the concomitant effect of an estimation MSE degradation.

In order to avoid the inconvenience of estimating the channel's statistics based on actual channel measurements, the concepts of a robust estimator design were outlined in Section 14.6.2 as advocated for example by Höher *et al.* [66,67], as well as by Li [74,459]. The approach was to employ a uniform, ideally support-limited channel scattering function in the design of the estimator, regardless of the channel's actual statistics. When using this estimator design, the estimator was rendered relatively insensitive to the specific channel statistics encountered. The concepts of robustness are further elaborated on in the context of our assessment of cascaded 1D-FIR Wiener filtering in Section 14.7.2 and furthermore, in the context of Decision-Directed Channel Estimation (DDCE) invoked in single-user OFDM systems in Section 15 as well as in Parallel Interference Cancellation (PIC) assisted DDCE employed in multi-user OFDM systems in Section 16.5.

Furthermore, in Section 14.6.3 we briefly outlined the approach proposed by Höher *et al.* [66,67] to reduce the dimensionality of the Wiener filter estimation problem, namely by employing only subsets of the set of available least-squares channel transfer factor estimates, based on the weighted distance of their associated pilot positions from the specific position on the time-frequency plane, for which the channel transfer factor is to be estimated.

The 2D-FIR Wiener filter's MSE performance assessment was conducted in Section 14.6.4. We observed in Figure 14.10 that a two-times pilot oversampling is sufficient in conjunction with a finite number of filter taps for removing[7] the MSE floor associated with the scenario of sampling the channel's transfer function at the Nyquist rate. This MSE floor was attributed to the imperfections of the FIR filter, specifically to the finite transition range between the filter's passband and stopband, which is a consequence of having a finite filter order. Note that even for the case of two-times oversampling a substantial N_{tap} number of filter taps is required for achieving a good MSE performance. Given additionally the imperfections associated with the selection of reduced-size subsets of the available pilot subcarrier positions, as described in Section 14.6.3.2, it is useful also to invoke the cascaded

---

[7]The MSE floor is not entirely removed, it is rather reduced to lower MSE values.

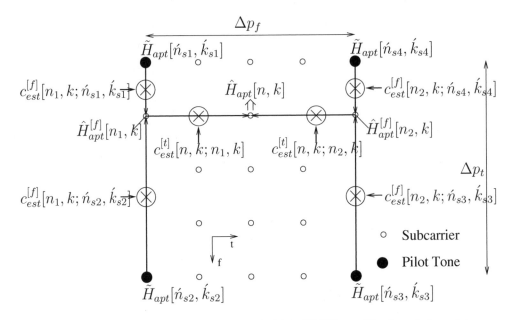

Figure 14.11: Illustration of the process of cascaded 1D-FIR Wiener filter based channel transfer factor estimation. An estimate $\hat{H}_{apt}[n,k]$ of the channel transfer factor $H[n,k]$ is generated in the first step by interpolation in the frequency direction, as outlined in Section 14.7.1.1, for the $(n_1 = \acute{n}_{s1} = \acute{n}_{s2})$-th and the $(n_2 = \acute{n}_{s3} = \acute{n}_{s4})$-th OFDM symbol, associated with the pairs of least-squares channel transfer factor estimates $(\tilde{H}_{apt}[\acute{n}_{s1}, \acute{k}_{s1}], \tilde{H}_{apt}[\acute{n}_{s2}, \acute{k}_{s2}])$ and $(\tilde{H}_{apt}[\acute{n}_{s3}, \acute{k}_{s3}], \tilde{H}_{apt}[\acute{n}_{s4}, \acute{k}_{s4}])$, which results in the k-th subcarriers channel transfer factor estimates $\hat{H}_{apt}^{[f]}[n_1, k]$ and $\hat{H}_{apt}^{[f]}[n_2, k]$. This is followed in a second step, as outlined in Section 14.7.1.2, by interpolation between these estimates in the time direction, which results in the desired estimate $\hat{H}_{apt}[n,k]$.

1D-FIR Wiener filtering approach as proposed by Höher *et al.* [58, 67] in the context of our comparison of different channel transfer function estimation approaches.

14.7   Cascaded 1D-FIR Wiener filtering

The technique of reducing the complexity of a 2D-FIR filter with the aid of two cascaded 1D-FIR filters is a standard approach known from the multi-dimensional signal processing related literature [461]. Its specific application to channel transfer function estimation in the context of multicarrier transmission systems was initially proposed by Höher *et al.* in [58] and later compared to 2D-FIR Wiener filtering in [66, 67, 462]. Here we will only review the basic principles, which can be readily derived from the concepts of 2D-FIR Wiener filter-based channel estimation outlined in Section 14.6.1.

The structure of Section 14.7 is as follows. In Section 14.7.1 the technique of cascaded 1D-FIR Wiener filtering is characterised in mathematical terms using the vector of optimum weights and by quantifying the associated MSE. Our performance assessment commences in Section 14.7.2 with the evaluation of the analytical expressions derived in Section 14.7.1 for the estimator's MSE, while in Section 14.7.3 both the estimator's MSE and the system's BER are evaluated with the aid of Monte Carlo simulations. Finally, following our complexity evaluation in Section 14.7.4, our conclusions will be offered in Section 14.7.5.

14.7.1 Derivation of the Optimum Estimator Coefficients and Estimator MSE for both Matched and Mismatched Channel Statistics

In Section 14.6 2D-FIR Wiener filtering was performed simultaneously both in the frequency and in the time direction. Alternatively, 1D-FIR Wiener filtering could for example be performed first in the frequency direction separately for all OFDM symbols hosting pilot subcarriers, followed by 1D-FIR Wiener filtering in the time direction, separately for all subcarriers. This is further illustrated in Figure 14.11. Alternatively, the filtering could first be performed in the time direction and then in the frequency direction. The notation employed here assumes the former case, namely that of filtering first in the frequency direction.

Our further discussions will be structured as follows. In Section 14.7.1.1 the vector of first-stage filter coefficients together with the associated MSE will be derived under mismatched channel conditions. By contrast, in Section 14.7.1.2 the corresponding vector of second-stage filter coefficients is deduced together with the associated MSE, which at the same time defines the channel transfer function estimator's total MSE. Furthermore, in Section 14.7.1.3 a simplification of the approach of cascaded 1D-FIR Wiener filtering is described with the aim of reducing the number of different sets of filter coefficients to be calculated.

14.7.1.1 First-Stage Channel Estimates - Interpolation in the Frequency Direction

Similar to our discussions of 2D-FIR Wiener filter-aided channel estimation in Section 14.6.1 the structure of this section is as follows. In Section 14.7.1.1.1 the process of generating the linear first-stage channel transfer factor estimates is described, while in Section 14.7.1.1.2 the calculation of the associated filter coefficients is outlined. The associated estimation MSE is then provided in Section 14.7.1.1.3.

14.7.1.1.1 Linear Channel Transfer Factor Estimate
The output value $\hat{H}_{apt}^{[f]}[n,k]$ of the first 1D-FIR Wiener filter - indicated by a superscript of $()^{[f]}$ - in the k-th subcarrier of the n-th OFDM symbol hosting pilot subcarriers is given according to Equation 14.48 by:

$$\hat{H}_{apt}^{[f]}[n,k] = \mathbf{c}_{est}^{[f]H}[n,k]\tilde{\mathbf{H}}_{apt}^{[f]}[n,k], \tag{14.84}$$

where $\tilde{\mathbf{H}}_{apt}^{[f]}[n,k] \in \mathbb{C}^{N_{tap}^{[f]} \times 1}$ is the vector of least-squares channel transfer factor estimates defined according to Equation 14.49 by:

$$\tilde{\mathbf{H}}_{apt}^{[f]}[n,k] = (\tilde{H}_{apt}[n,\acute{k}_{s1}], \tilde{H}_{apt}[n,\acute{k}_{s2}], \ldots, \tilde{H}_{apt}[n,\acute{k}_{sN_{tap}^{[f]}}])^T, \tag{14.85}$$

which have been selected from the n-th OFDM symbol for the first-stage estimation of the specific k-th subcarrier transfer factor. Furthermore, in Equation 14.84, $\mathbf{c}_{est}^{[f]}[n,k] \in \mathbb{C}^{N_{tap}^{[f]} \times 1}$

is the vector of first-stage filter coefficients defined similarly to Equation 14.50 by:

$$\mathbf{c}_{est}^{[f]}[n,k] = (c_{est}^{[f]*}[n,k;n,\acute{k}_{s1}], c_{est}^{[f]*}[n,k;n,\acute{k}_{s2}], \dots, c_{est}^{[f]*}[n,k;n,\acute{k}_{sN_{tap}^{[f]}}])^T. \quad (14.86)$$

14.7.1.1.2 Estimator Coefficients for Mismatched Channel Statistics The vector $\tilde{\mathbf{c}}_{est}^{[f]}[n,k]|_{opt}$ of filter coefficients can be determined upon invoking similar concepts to those outlined in Equation 14.65, namely using the Wiener equation formulated as:

$$\tilde{\mathbf{c}}_{est}^{[f]}[n,k]|_{opt} = \tilde{\mathbf{R}}_{apt}^{[f]-1}[n,k]\tilde{\mathbf{r}}_{apt}^{[f]}[n,k], \quad (14.87)$$

where similarly to Equation 14.64 $\tilde{\mathbf{r}}_{apt}^{[f]}[n,k] \in \mathbb{C}^{N_{tap}^{[f]} \times 1}$ denotes the estimate of the cross-correlation vector $\mathbf{r}_{apt}^{[f]}[n,k]$ defined between the actual channel transfer factor $H[n,k]$ at position $[n,k]$ and the set of pilot assisted least-squares channel transfer factor estimates hosted by the vector $\tilde{\mathbf{H}}_{apt}^{[f]}[n,k]$ of Equation 14.85:

$$\tilde{\mathbf{r}}_{apt}^{[f]}[n,k] = \tilde{E}\{H^*[n,k]\tilde{\mathbf{H}}_{apt}^{[f]}[n,k]\} \quad (14.88)$$

$$= \tilde{E}\{H^*[n,k]\mathbf{H}^{[f]}[n,k]\} \quad (14.89)$$

$$= \tilde{\mathbf{r}}^{[f]}[n,k]. \quad (14.90)$$

In Equation 14.89 $\mathbf{H}^{[f]}[n,k] \in \mathbb{C}^{N_{tap}^{[f]} \times 1}$ represents the vector of true channel transfer factors at the pilot positions, defined similarly to those of the least-squares channel estimates $\tilde{\mathbf{H}}_{apt}[n,k]$ of Equation 14.85.

Furthermore, in Equation 14.87 the variable $\tilde{\mathbf{R}}_{apt}^{[f]}[n,k] \in \mathbb{C}^{N_{tap}^{[f]} \times N_{tap}^{[f]}}$ represents the estimate of the auto-correlation matrix $\mathbf{R}_{apt}^{[f]}[n,k]$ of the pilot assisted least-squares channel transfer factor estimates, defined in harmony with Equation 14.57 as:

$$\tilde{\mathbf{R}}_{apt}^{[f]}[n,k] = \tilde{E}\{\tilde{\mathbf{H}}_{apt}^{[f]}[n,k]\tilde{\mathbf{H}}_{apt}^{[f]H}[n,k]\} \quad (14.91)$$

$$= \tilde{E}\{\mathbf{H}^{[f]}[n,k]\mathbf{H}^{[f]H}[n,k]\} + \tilde{\rho}_p\mathbf{I} \quad (14.92)$$

$$= \tilde{\mathbf{R}}^{[f]}[n,k] + \tilde{\rho}_p\mathbf{I}, \quad (14.93)$$

where $\tilde{\rho}_p = 1/\tilde{\text{SNR}}_p$ denotes the reciprocal of the estimated inverse effective subcarrier SNR at the pilot positions.

14.7.1.1.3 Estimator MSE According to Equation 14.53 the MSE at the first-stage estimator's output – conditioned on $\mathbf{c}_{est}^{[f]}[n,k]$ is hence given by:

$$\text{MSE}^{[f]}(\mathbf{c}_{est}^{[f]}[n,k]) = \sigma_H^2 - \mathbf{c}_{est}^{[f]T}[n,k]\mathbf{r}_{apt}^{[f]*}[n,k] - \mathbf{c}_{est}^{[f]H}[n,k]\mathbf{r}_{apt}^{[f]}[n,k] +$$
$$+ \mathbf{c}_{est}^{[f]H}[n,k]\mathbf{R}_{apt}^{[f]}[n,k]\mathbf{c}_{est}^{[f]}[n,k], \quad (14.94)$$

where $\mathbf{r}_{apt}^{[f]}[n, k]$ and $\mathbf{R}_{apt}^{[f]}[n, k]$ – rather than $\tilde{\mathbf{r}}_{apt}^{[f]}[n, k]$ and $\tilde{\mathbf{R}}_{apt}^{[f]}[n, k]$ of Equations 14.88 and 14.91 are used – where the former two variables are based on perfect knowledge of the channel's statistics, specifically on that of the channel's spaced-frequency correlation function and on the actual value ρ_p of the SNR's reciprocal at the pilot positions. Furthermore, in Equation 14.94 $\sigma_H^2 = E\{H[n, k]H^*[n, k]\}$ denotes the variance of the true channel transfer factors.

As demonstrated in the context of the 2D-FIR Wiener filter-based channel estimation of Section 14.6, in cases where $\tilde{\mathbf{r}}_{apt}^{[f]}[n, k]$ and $\tilde{\mathbf{R}}_{apt}^{[f]}[n, k]$ employed in the filter's calculation are identical to $\mathbf{r}_{apt}^{[f]}[n, k]$ and $\mathbf{R}_{apt}^{[f]}[n, k]$ associated with perfect knowledge of the channel's spaced-frequency correlation function and where the estimate $\tilde{\rho}_p$ is identical to the actual value ρ_p of the reciprocal of the SNR at the pilot positions,[8] Equation 14.94 simplifies to:

$$\text{MMSE}^{[f]}[n, k] = \text{MSE}^{[f]}(\mathbf{c}_{est}^{[f]}[n, k]|_{opt}) \tag{14.95}$$

$$= \sigma_H^2 - \mathbf{c}_{est}^{[f]H}[n, k]|_{opt}\mathbf{r}^{[f]}[n, k]. \tag{14.96}$$

In closing we note again that in contrast to the 2D-FIR Wiener filter of Section 14.6.1, the least-squares channel transfer factor estimates employed in the first filter stage are associated with different subcarriers of the *same* OFDM symbol.

14.7.1.2 Second-Stage Channel Estimates - Interpolation in the Time Direction

Similar to our discussions of the first-stage Wiener filtering in Section 14.7.1.1 the structure of this section is as follows. In Section 14.7.1.2.1 the process of generating the linear first-stage channel transfer factor estimates is described, while in Section 14.7.1.2.2 the calculation of the associated filter coefficients is outlined. The associated estimation MSE is then provided in Section 14.7.1.2.3.

14.7.1.2.1 Linear Channel Transfer Factor Estimate   In the second filter stage – indicated by a superscript of $()^{[t]}$ – which aims at interpolating in the time direction, the estimates $\hat{H}_{apt}^{[f]}[n, k]$ provided by the first filter stage are employed as inputs. Following similar concepts to those used in the first filter stage, the channel transfer factor estimate at the output of the second filter stage at integer position $[n, k]$ of the time-frequency plane is given by:

$$\hat{H}_{apt}^{[t]}[n, k] = \mathbf{c}_{est}^{[t]H}[n, k]\tilde{\mathbf{H}}_{apt}^{[t]}[n, k], \tag{14.97}$$

where $\tilde{\mathbf{H}}_{apt}^{[t]}[n, k] \in \mathbb{C}^{N_{tap}^{[t]} \times 1}$ is the channel transfer factor vector, which hosts a subset of the first filter stage channel estimator's outputs $\hat{H}_{apt}^{[f]}[\acute{n}, k]$ delivered by Equation 14.84 for the k-th subcarrier in the different OFDM symbols, which host pilot subcarriers. This channel transfer factor vector can be expressed as:

$$\tilde{\mathbf{H}}_{apt}^{[t]}[n, k] = (\hat{H}_{apt}^{[f]}[\acute{n}_{s1}, k], \hat{H}_{apt}^{[f]}[\acute{n}_{s2}, k], \dots, \hat{H}_{apt}^{[f]}[\acute{n}_{sN_{tap}^{[t]}}, k])^T. \tag{14.98}$$

---

[8]In this case the vector of optimum coefficients is denoted as $\mathbf{c}_{est}^{[f]}[n, k]|_{opt}$ instead of $\tilde{\mathbf{c}}_{est}^{[f]}[n, k]|_{opt}$.

Furthermore, $\mathbf{c}_{est}^{[t]}[n, k] \in \mathbb{C}^{N_{tap}^{[t]} \times 1}$ in Equation 14.97 is the vector of second-stage filter coefficients defined similarly to Equation 14.86, namely, by:

$$\mathbf{c}_{est}^{[t]}[n, k] = (c_{est}^{[t]*}[n, k; \acute{n}_{s1}, k], c_{est}^{[t]*}[n, k; \acute{n}_{s2}, k], \ldots, c_{est}^{[t]*}[n, k; \acute{n}_{sN_{tap}^{[t]}}, k])^T. \qquad (14.99)$$

14.7.1.2.2 Estimator Coefficients for Mismatched Channel Statistics

Again, the vector $\tilde{\mathbf{c}}_{est}^{[t]}[n, k]|_{opt}$ of filter coefficients can be determined upon invoking the concept of the Wiener filter based estimator, which was also applied in the first stage filter, namely in the context of Equation 14.87, yielding:

$$\tilde{\mathbf{c}}_{est}^{[t]}[n, k]|_{opt} = \tilde{\mathbf{R}}_{apt}^{[t]-1}[n, k]\tilde{\mathbf{r}}_{apt}^{[t]}[n, k], \qquad (14.100)$$

where $\tilde{\mathbf{r}}_{apt}^{[t]}[n, k] \in \mathbb{C}^{N_{tap}^{[t]} \times 1}$ denotes the *estimate* of the cross-correlation vector $\mathbf{r}_{apt}^{[t]}[n, k]$ between the true channel transfer factor $H[n, k]$ at position $[n, k]$ and the subset of first-stage channel transfer factor estimates hosted by the vector $\tilde{\mathbf{H}}_{apt}^{[t]}[n, k]$ of Equation 14.98, which follows a similar philosophy to that of Equation 14.88, yielding:

$$\tilde{\mathbf{r}}_{apt}^{[t]}[n, k] = \tilde{E}\{H^*[n, k]\tilde{\mathbf{H}}_{apt}^{[t]}[n, k]\}. \qquad (14.101)$$

Specifically, the component of $\tilde{\mathbf{r}}_{apt}^{[t]}[n, k]$ associated with the n'-th OFDM symbol's first-stage channel estimator output at the k-th subcarrier, namely with $\hat{H}_{apt}^{[f]}[n', k]$, which was previously defined in Equation 14.84, is represented by:

$$\begin{aligned}
\tilde{r}_{apt}^{[t]}[n, k; n'] &= \mathbf{c}_{est}^{[f]H}[n', k] \cdot \tilde{E}\{H^*[n, k]\tilde{\mathbf{H}}_{apt}^{[f]}[n', k]\} \\
&= \mathbf{c}_{est}^{[f]H}[n', k] \cdot \tilde{E}\{H^*[n, k]\mathbf{H}^{[f]}[n', k]\}. \qquad (14.102)
\end{aligned}$$

More explicitly Equation 14.102 follows from Equation 14.102, since the actual true channel transfer factor $H[n, k]$ is uncorrelated to the AWGN contaminating the least-squares channel transfer factor estimates hosted by the vector $\tilde{\mathbf{H}}_{apt}^{[f]}[n', k]$. Equation 14.102 can be further simplified upon exploiting the separability of the channel's spaced-time spaced-frequency correlation function highlighted in Section 14.2.6, resulting in:

$$\begin{aligned}
\tilde{r}_{apt}^{[t]}[n, k; n'] &= \mathbf{c}_{apt}^{[f]H}[n', k] \cdot \tilde{E}\{H^*[n, k]H[n', k]\} \cdot \tilde{E}\{H^*[n', k]\mathbf{H}^{[f]}[n', k]\} \\
&= \tilde{r}_H^{[t]}[n - n'] \cdot \mathbf{c}_{apt}^{[f]H}[n', k]\tilde{\mathbf{r}}^{[f]}[n', k], \qquad (14.103)
\end{aligned}$$

where $\tilde{r}_H^{[t]}[n - n'] = \tilde{E}\{H^*[n, k]H[n', k]\}$ is the estimate of the channel's spaced-time correlation function, while the estimate of the first stage filter's cross-correlation vector $\tilde{\mathbf{r}}^{[f]}[n, k]$ was defined in Equation 14.90.

Furthermore, in Equation 14.100 the symbol $\tilde{\mathbf{R}}_{apt}^{[t]}[n, k] \in \mathbb{C}^{N_{tap}^{[t]} \times N_{tap}^{[t]}}$ represents the *estimate* of the auto-correlation matrix $\mathbf{R}_{apt}^{[t]}[n, k]$ of the subset of first stage channel transfer factor estimates invoked in the second-stage estimation process for the specific position $[n, k]$

on the time-frequency plane, which is formulated as:

$$\tilde{\mathbf{R}}_{apt}^{[t]}[n,k] = \tilde{E}\{\tilde{\mathbf{H}}_{apt}^{[t]}[n,k]\tilde{\mathbf{H}}_{apt}^{[t]H}[n,k]\}. \tag{14.104}$$

Similarly to the components of the cross-correlation vector $\tilde{\mathbf{r}}_{apt}^{[t]}[n,k]$ in Equation 14.102, the coefficient of $\tilde{\mathbf{R}}_{apt}^{[t]}[n,k]$ associated with the specific n'-th and n''-th OFDM symbols' first-stage channel estimator outputs $\hat{H}_{apt}^{[f]}[n',k]$ and $\hat{H}_{apt}^{[f]}[n'',k]$, respectively, at the k-th subcarrier, which were previously defined in Equation 14.84, is represented by:

$$\begin{aligned}
\tilde{R}_{apt}^{[t]}[n,k;n',n''] &= \mathbf{c}_{est}^{[f]H}[n',k] \cdot \tilde{E}\{\tilde{\mathbf{H}}_{apt}^{[f]}[n',k]\tilde{\mathbf{H}}_{apt}^{[f]H}[n'',k]\} \cdot \mathbf{c}_{est}^{[f]}[n'',k] \\
&= \tilde{r}_{H}^{[t]}[n'-n''] \cdot \mathbf{c}_{est}^{[f]H}[n',k]\tilde{\mathbf{R}}^{[f]}[n,k;n',n'']\mathbf{c}_{est}^{[f]}[n'',k] + \\
&\quad + \delta[n'-n''] \cdot \tilde{\rho}_p \cdot \mathbf{c}_{est}^{[f]H}[n',k]\mathbf{c}_{est}^{[f]}[n'',k]. \tag{14.105}
\end{aligned}$$

Observe again that in the context of Equation 14.105 we have exploited the separability of the spaced-time spaced-frequency correlation function invoked in the calculation of the filter coefficients, where the separability was discussed in Section 14.2.6. More specifically, in Equation 14.105, $\tilde{\mathbf{R}}^{[f]}[n,k;n',n''] \in \mathbb{C}^{N_{tap}^{[t]} \times N_{tap}^{[t]}}$ denotes the cross-correlation matrix between the actual channel transfer factors at the pilot positions invoked in the calculation of the first-stage channel estimates for the k-th subcarrier of the n'-th and the n''-th OFDM symbol, projected onto the n-th OFDM symbol. Furthermore, in Equation 14.105 $\delta[]$ denotes the Kronecker delta and $\tilde{\rho}_p$ is the reciprocal of the estimated effective SNR at the pilot positions.

14.7.1.2.3 Estimator MSE Following again the philosophy of Equation 14.53, the estimator's MSE at the output of the second filter stage – conditioned on $\mathbf{c}^{[t]}[n,k]$ – is given by:

$$\begin{aligned}
\text{MSE}^{[t]}(\mathbf{c}_{est}^{[t]}[n,k]) &= \sigma_H^2 - \mathbf{c}_{est}^{[t]T}[n,k]\mathbf{r}_{apt}^{[t]*}[n,k] - \mathbf{c}_{est}^{[t]H}[n,k]\mathbf{r}_{apt}^{[t]}[n,k] + \\
&\quad + \mathbf{c}_{est}^{[t]H}[n,k]\mathbf{R}_{apt}^{[t]}[n,k]\mathbf{c}_{est}^{[t]}[n,k]. \tag{14.106}
\end{aligned}$$

In contrast to $\tilde{\mathbf{r}}_{apt}^{[t]}[n,k]$ and $\tilde{\mathbf{R}}_{apt}^{[t]}[n,k]$ of Equations 14.101 and 14.104, $\mathbf{r}_{apt}^{[t]}[n,k]$ as well as $\mathbf{R}_{apt}^{[t]}[n,k]$ employed in Equation 14.106 are based on perfect knowledge of the channel's statistics, specifically on that of the channel's spaced-time spaced-frequency correlation function, but again, conditioned on the first stage's *imperfect* filter coefficients. Furthermore, in the calculation of the components of $\mathbf{R}_{apt}^{[t]}[n,k]$, following a similar philosophy to that of Equation 14.105 associated with $\tilde{\mathbf{R}}_{apt}^{[t]}[n,k]$, the actual value ρ_p of the reciprocal of the SNR, recorded at the pilot positions has to be employed.

As outlined in Section 14.7.1.1 in the context of our description of the first-stage 1D-FIR Wiener filter operating in the frequency direction, in cases where $\tilde{\mathbf{r}}_{apt}^{[t]}[n,k]$ and $\tilde{\mathbf{R}}_{apt}^{[t]}[n,k]$ employed in the filter calculation are identical to $\mathbf{r}_{apt}^{[t]}[n,k]$ and $\mathbf{R}_{apt}^{[t]}[n,k]$ associated with the perfect knowledge of the channel's spaced-time correlation function,[9] Equation 14.106

---

[9]In this case the vector of optimum coefficients is denoted as $\mathbf{c}_{est}^{[t]}[n,k]|_{opt}$ instead of $\tilde{\mathbf{c}}_{est}^{[t]}[n,k]|_{opt}$.

simplifies to:

$$\text{MMSE}^{[t]}[n,k] = \text{MSE}^{[t]}(\mathbf{c}_{est}^{[t]}[n,k]|_{opt}) \tag{14.107}$$

$$= \sigma_H^2 - \mathbf{c}_{est}^{[t]H}[n,k]|_{opt}\mathbf{r}^{[t]}[n,k]. \tag{14.108}$$

14.7.1.3 Simplification of the Cascaded 1D-FIR Wiener Filter

In Section 14.7.1.2 we observed that the channel transfer factor estimates provided by the first-stage Wiener filter are employed as the inputs of the second-stage Wiener filter. As a result, the second-stage Wiener filter's coefficients are a function of the first-stage Wiener filter's coefficients, which becomes explicit from the components of the second-stage Wiener filter's cross-correlation vector and auto-correlation matrix, given by Equations 14.102 and 14.105, respectively. In an effort to reduce the number of different sets of filter coefficients to be pre-calculated and stored, a viable approach is to decouple the coefficient calculation of the second-stage filter from that of the first-stage filter. This can be achieved upon assuming that the channel transfer factor estimates provided by the first-stage filter can be represented in the same form as the least-squares channel transfer factor estimates, namely as:

$$\hat{H}_{apt}^{[f]}[n,k] \approx H[n,k] + n_p[n,k], \tag{14.109}$$

where $H[n,k]$ is the "true" channel transfer factor at position $[n,k]$, and $n_p[n,k]$ is an AWGN contribution, which contaminates the transfer factor estimates. Furthermore, we assume that we have $E\{|n_p[n,k]|^2\} = \sigma_p^2$. Note, however, again, that Equation 14.109 is only a simplified model of the signal observed at the output of the first-stage Wiener filter. In fact, upon assuming perfect knowledge of the channel transfer function's correlation in the calculation of the first-stage filter's coefficients, the output of the first-stage filter becomes a scaled version of the true channel transfer factor, which is perturbed by an AWGN contribution having a variance, which is potentially lower than σ_p^2. However, for sufficiently high SNRs the effects of the simplifications imposed here, namely an increase of the estimation MSE compared to that exhibited by the "original" approach of cascaded 1D-FIR filtering diminishes. This will be demonstrated in the context of our performance assessments provided in Section 14.7.3. As a result of the above assumptions the calculation of the second-stage filter's coefficients are decoupled from that of the coefficients of the first-stage filter, with the effect of having to store a potentially lower number of different sets of filter coefficients.

In the next section we will embark on a performance study of the cascaded 1D-FIR Wiener filter-aided channel estimator.

14.7.2 MSE Performance of the Cascaded 1D-FIR Wiener Filter

Our performance assessments in this section are divided into the following parts. In Section 14.7.2.1 the MSE performance exhibited by the cascaded 1D-FIR Wiener filter-based channel estimator is contrasted with that of the 2D-FIR Wiener filter-based channel estimator, which was evaluated in Section 14.6.4. In the context of this comparison we assume that the channel statistics invoked in the evaluation of the filter coefficients are perfectly matched to those of the channel. By contrast, in Section 14.7.2.2 we investigate the case of a mis-adjustment between the channel statistics invoked in the evaluation of the filter coefficients and those of the channel. More specifically, while the channel's statistics are assumed to obey a uniform, ideally support-limited scattering function, as in Section 14.6.2, the statistics used in the evaluation of the filter coefficients, which were also given by a uniform, ideally support-limited scattering function, were varied with respect to the scattering function's region of support. Furthermore, in Section 14.7.2.3 the case of a mismatch between the SNR used in the calculation of the filter coefficients and that measured at the reception antenna will be considered. In Section 14.7.2.4 the effects of OFDM symbol synchronisation errors on the channel estimate are investigated and our recommendations are offered for the dimensioning of the region of support associated with the robust channel estimator's scattering function. Our discussions are concluded in Section 14.7.2.5 by highlighting the effects of a mismatch between the specific shapes of the scattering function invoked in the evaluation of the filter coefficients and that of the channel, while exhibiting an identical region of support.

14.7.2.1 Comparison to 2D-FIR Wiener Filtering

Our discourse on the MSE performance exhibited by the cascaded 1D-FIR Wiener filter assisted channel estimator commences again by considering the scenario of a uniform, ideally $(T_m, 2f_D)$-support-limited scattering function $S_{h,unif}(f_d, \tau)$, obeying Equation 14.67. This scattering function was employed in both the filter coefficient calculation by capitalising on Equations 14.87 and 14.100, as well as in the evaluation of the estimator's MSE based on Equation 14.106. This scenario was also considered in the context of our investigations of 2D-FIR Wiener filter assisted channel estimation in Section 14.6.4. The subcarrier spacing normalised multipath spread $T_m \Delta F$ was identical to the OFDM symbol normalised Doppler spread $B_D T_f$. Specifically, we had $T_m \Delta F = B_D T_f = 0.03125$. Two "balanced" rectangular pilot grid designs were considered, where in the first design the pilot distances in the frequency and time direction were chosen as $\Delta p_f = \Delta p_{f,max} = 32$ and $\Delta p_t = \Delta p_{t,max} = 32$. This resulted in unity oversampling factors of $M_{p,f} = M_{p,t} = 1$, while in the second design the pilot distances were halved to $\Delta p_f = \Delta p_t = 16$, which resulted in a two-times oversampling, expressed as $M_{p,f} = M_{p,t} = 2$. The associated results of our analytical MSE evaluations are portrayed in the left graph of Figure 14.12 for the Nyquist-sampled scenario, while those associated with the doubled sampling grid density, are portrayed in the right graph of Figure 14.12.

Similarly to the 2D-FIR Wiener filter's MSE performance seen in the left graph of Figure 14.10 in conjunction with the lower-density pilot grid, where the pilot density was just sufficiently high for fulfilling the Nyquist theorem [90, 177, 463], the MSE performance of the cascaded 1D-FIR Wiener filter is also relatively poor in this scenario, as illustrated in the left graph of Figure 14.12. The reason for this trend is that due to the FIR filter's limited number of taps, its frequency response exhibits a low-gradient transition between the filter's pass-band and the filter's stop-band. The effect of this is that not only the sampled channel's desired scattering function-related domain segment, but also its attenuated repeated segments are partially retained by the filter in the channel samples' scattering function-related domain representation. By contrast, upon employing a pilot grid using an oversampling factor of two, the associated MSE performance results are plotted in the right graph of Figure 14.12, where the MSE floor is below $-40dB$ for a sufficiently high number of filter taps, such as for

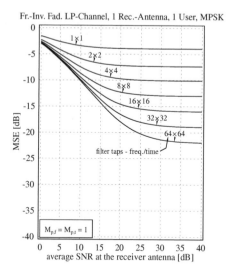
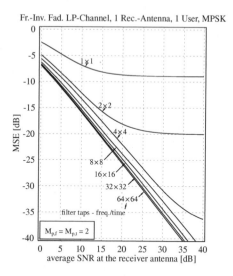

Figure 14.12: MSE exhibited by the cascaded 1D-FIR Wiener filter channel estimator averaged over one period of the rectangular pilot grid as a function of the SNR at the reception antenna and parameterised with the number of FIR filter taps in the frequency and time direction, which is denoted as $N_{tap}^{[f]} \times N_{tap}^{[t]}$. The same uniform ideally support-limited scattering function defined in Equation 14.67 was associated both with the evaluation of the filter coefficients and that of the channel. The subcarrier spacing normalised multipath spread $T_m \Delta F$ was identical to the OFDM symbol normalised Doppler spread $B_D T_f$, where we had $T_m \Delta F = B_D T_f = 0.03125$. The pilot sampling distances Δp_f, Δp_t used in the context of the **left graph** were equal to the maximum pilot distances, namely to $\Delta p_f = \Delta p_{f,max} = 32$ and $\Delta p_t = \Delta p_{t,max} = 32$ delivered by Equations 14.35 and 14.42, respectively, corresponding to the unity oversampling factors, of $M_{p,f} = M_{p,t} = 1$. By contrast, a two-times oversampling, expressed as $M_{p,f} = M_{p,t} = 2$, was used in the context of **the graph on the right**, for which we have $\Delta p_f = \Delta p_t = 16$. The results were evaluated from Equation 14.106.

example upon employing 8 taps for both the time- and frequency direction filters.

In Section 14.7.4 we will provide a complexity analysis of the cascaded 1D-FIR Wiener filter-based channel estimator. Specifically it will be shown that the complexity per estimated channel transfer factor can be quantified as $\frac{N_{tap}^{[f]}}{\Delta p_t} + N_{tap}^{[t]}$ number of complex multiplications and additions, respectively. This is in contrast to the N_{tap} number of complex multiplications and additions per estimated channel transfer factor as demonstrated in Section 14.6.5 for the 2D-FIR Wiener filter-based channel estimator. Upon comparing the MSE results presented in Figure 14.12 to those in Figure 14.10, we observe that the cascaded 1D-FIR Wiener filter-based channel estimator exhibits a significantly better performance, assuming a specific complexity, than the 2D-FIR Wiener filter. This trend was also reported in [67, 470]. For example, upon employing eight taps in each of the cascaded channel estimator's filters, and in conjunction with a pilot distance of $\Delta p_f = \Delta p_t = 16$, as in the more dense pilot grid, on average a complexity of 8.5 complex multiplications and additions per estimated channel transfer factor is imposed by the cascaded 1D-FIR Wiener filter-based channel estimator. As seen in the corresponding right-hand side graph of Figure 14.12, the resultant MSE is below $-40dB$ at an SNR of $40dB$ measured at the reception antenna. By contrast, as illustrated by the MSE curves in the right-hand side graph of Figure 14.10, more than 16 filter taps are required to achieve a similarly low estimator MSE in the context of the 2D-FIR Wiener filter based channel estimator. We note that upon taking into account the cascaded channel estimator's complexity quantified by the expression of $\frac{N_{tap}^{[f]}}{\Delta p_t} + N_{tap}^{[t]}$, the associated MSE could potentially be further reduced by trading off the number of filter taps, $N_{tap}^{[f]}$ and $N_{tap}^{[t]}$, in the frequency and time direction against each other.

14.7.2.2 Misadjustment of Multipath- and Doppler Spread

In the previous section we studied the MSE performance of the cascaded 1D-FIR filter based channel estimator. Specifically, the *same* uniform ideally support-limited scattering function associated with a subcarrier spacing normalised multipath spread of $T_m \Delta F$ and an OFDM symbol normalised Doppler spread of $B_D T_f$ was invoked in both the calculation of the estimator's filter coefficients as well as in the evaluation of the estimator's MSE. By contrast, in this section we will investigate the estimator's MSE under potentially mismatched channel conditions, where the subcarrier spacing normalised multipath spread $\tilde{T}_m \Delta F$ and the OFDM symbol normalised Doppler spread $\tilde{B}_D T_f$, assumed in the filter coefficient calculation were potentially different from the values of $T_m \Delta F$ and $B_D T_f$, associated with the channel encountered. Again, for both the filter coefficient calculation and the channel a uniform ideally support-limited scattering function obeying Equation 14.67 was assumed. While the normalised multipath- and Doppler spread associated with the filter coefficient calculation were fixed to $\tilde{T}_m \Delta F = \tilde{B}_D T_f = 0.03125$, the corresponding values of $T_m \Delta F = B_D T_f$ associated with the channel were varied across a range spanning from 0.01 to 0.1. Furthermore, a fixed SNR of $22.5dB$ was assumed in both the filter coefficient calculation, as well as in the context of the channel. The corresponding analytical MSE simulation results based on Equation 14.106 are portrayed in Figure 14.13. Specifically in the left graph we used a pilot oversampling factor of $M_{p,f} = M_{p,t} = 2$, while in the right graph $M_{p,f} = M_{p,t} = 4$. In both cases we observe in Figure 14.13 that as long as the uniform ideally support-limited scattering function $\tilde{S}_{h,unif}(\tau, f_d)$ associated with the filter calculation, actually envelopes the ideally support-limited scattering function associated with the channel, $S_{h,unif}(\tau, f_d)$, which is the case for $T_m \Delta F = B_D T_f$ values below $\tilde{T}_m \Delta F = \tilde{B}_D T_f = 0.03125$, the estimator's MSE is near-constant, given a sufficiently high number of filter taps. For a relatively low number of filter taps the independence of the MSE of $T_m \Delta F = B_D T_f$ is only approximate due to the imperfections of the filter's scattering function-related domain representation, specifically, due to the relatively wide transitional range between the filter's pass-band and the stop-band. For $T_m \Delta F = B_D T_f$ values in excess of $\tilde{T}_m \Delta F = \tilde{B}_D T_f = 0.03125$, which was associated with the filter coefficient calculation, a significant MSE degradation is observed, since useful components of the desired segment of the sampled channel's scattering function-related domain representation are removed. Please also note that for the higher-order filters characterised in Figure 14.13 the MSE degradation is more significant than for the lower order

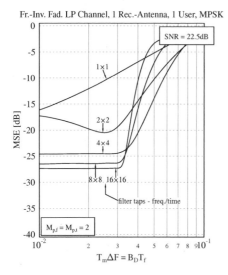
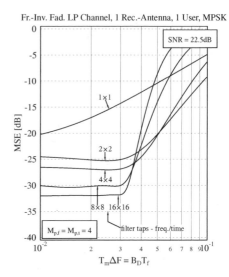

Figure 14.13: MSE exhibited by the cascaded 1D-FIR Wiener filter-based channel estimator averaged over a period of the rectangular pilot grid as a function of the subcarrier spacing normalised multipath spread $T_m\Delta F$ and the OFDM symbol normalised Doppler spread $B_D T_f$ of the channel, which where both set equal to $T_m\Delta f = B_D T_f$, for reasons of simplicity. By contrast, $\tilde{T}_m\Delta F = \tilde{B}_D T_f$ employed in the filter coefficient calculation was fixed to a value of 0.03125. Note that the uniform ideally support-limited scattering function according to Equation 14.67 was associated with both the evaluation of the filter coefficients and the channel. The MSE curves are parameterised with the number of FIR filter taps per frequency- and time direction, namely with $N_{tap}^{[f]} \times N_{tap}^{[t]}$. The SNR at the reception antenna was fixed to a value of $22.5dB$. The pilot sampling distances Δp_f, Δp_t employed in the context of the **left graph** were equal to $\Delta p_f = \Delta p_t = 16$ implying a two-times oversampling, namely $M_{p,f} = M_{p,t} = 2$, while that of the **right graph** were equal to $\Delta p_f = \Delta p_t = 8$ implying a four-times oversampling, namely $M_{p,f} = M_{p,t} = 4$. The results were evaluated from Equation 14.106.

filters, assuming sufficiently high values of $T_m\Delta F = B_D T_f$ compared to $\tilde{T}_m\Delta F = \tilde{B}_D T_f$, namely in comparison to the values associated with the filter coefficient calculation. This can be explained again by noting that for the higher-order filters the boundary between the filter's pass-band and the stop-band is more distinct, resulting in a narrow transitional range.

14.7.2.3　Misadjustment of the SNR

In our previous experiment we have assumed that the SNR of the pilot subcarriers, whose value is denoted by $\text{S\tilde{N}R}_p$ in the context of the filter coefficient calculation, is identical to the corresponding value measured at the reception antenna, denoted by SNR_p. However,

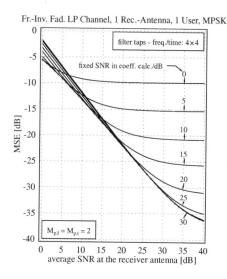
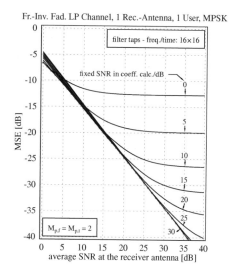

Figure 14.14: MSE exhibited by the cascaded 1D-FIR Wiener filter-based channel estimator averaged over a period of the rectangular pilot grid as a function of the SNR at the reception antenna and parameterised with the SNR employed in the calculation of the filter coefficients. The number of FIR filter taps per frequency and time direction, namely $N_{tap}^{[f]} \times N_{tap}^{[t]}$, employed in the context of the **left graph** is four, while that associated with the **right graph** is 16. The same uniform ideally support-limited scattering function according to Equation 14.67 was associated with both the evaluation of the filter coefficients and the channel. The channel's subcarrier spacing normalised multipath spread $T_m \Delta F$ was identical to the OFDM symbol normalised Doppler spread $B_D T_f$. Specifically, $T_m \Delta F = B_D T_f = 0.03125$ was used. The pilot sampling distances Δp_f, Δp_t employed were equal to $\Delta p_f = \Delta p_t = 16$, corresponding to a two-times oversampling, expressed as $M_{p,f} = M_{p,t} = 2$. The results were evaluated from Equation 14.106.

in a more realistic scenario a misadjustment may be observed between $\tilde{\mathrm{SNR}}_p$ and SNR_p.[10] This will be investigated again in the context of employing the same uniform ideally support-limited scattering function of Equation 14.67 both in the calculation of the filter coefficients as well as in the evaluation of the estimator MSE of Equation 14.106 associated with the specific channel, where $\tilde{S}_h(\tau, f_d) = S_h(\tau, f_d)$. Again, we assume that $T_m \Delta F = B_D T_f = 0.03125$. A two-times pilot oversampling was employed, expressed as $M_{p,f} = M_{p,t} = 2$, in conjunction with frequency- and time direction pilot distances given by $\Delta p_f = \Delta p_t = 16$. The results of our analytical MSE evaluations based on Equation 14.106 and given as a function of the SNR at the reception antenna and parameterised with the SNR employed in the coefficient calculation are presented in Figure 14.14. Specifically, in the left graph 4, while in the right graph 16 FIR filter taps per time and frequency direction were used.

---

[10]Note that here we also assume that the SNR in the data subcarriers is identical to that of the pilot subcarriers, namely $\mathrm{SNR} = \mathrm{SNR}_p$. In other words, no "boosting" of the pilot symbol power is performed.

In order to interpret the evolution of the different MSE curves please recall the mechanisms associated with the Wiener filtering. More specifically, a trade-off has to be found between the mitigation of the AWGN and the exploitation of the correlation between the channel transfer function samples recorded at the different pilot subcarriers. If the SNR at the reception antenna is lower than the SNR assumed in the calculation of the filter coefficients, an MSE degradation is incurred since the channel's correlation is over-weighted in the estimation process. By contrast, if the SNR at the reception antenna is in excess of the SNR assumed in the filter calculation, the channel's correlation along the time direction is not fully exploited and the AWGN mitigation is over-emphasised. As a result, a residual MSE is observed even at high SNRs encountered at the reception antenna. Since the exact SNR at the reception antenna is not normally known in advance, but potentially information about the operating range of SNRs is available, it is advisable to calculate the filter coefficients with respect to the highest SNR in the interval. As suggested by the MSE curves of Figure 14.14, this will assist in reducing the relative average MSE degradation evaluated on a dB scale.

14.7.2.4 Impact of Imperfect Synchronization

So far we have assumed that a uniform ideally support-limited scattering function obeying Equation 14.67 is invoked both in the filter coefficient calculation, as well as in the characterisation of the channel's statistical properties. Specifically, the multipath intensity profile associated with the scattering function was assumed to be causal. This was achieved upon setting the delay variable τ_{shift} of Equation 14.68, characterising the uniform ideally support-limited multipath intensity profile, equal to half of the multipath spread T_m stipulated. This is expressed as $\tau_{shift} = T_m/2$. In the OFDM system a causal CIR and corresponding causal multipath intensity profile is observed by the channel estimator in perfectly synchronised scenarios, where the position of the FFT demodulation window is determined by the lowest-delay multipath component arriving at the receiver. In a more realistic scenario however, the position of the FFT window synchronisation point is described by a probability distribution, centered around the ideal synchronisation point. Hence, it is not advisable to adapt the multipath intensity profile employed in the filter coefficient calculation as tight as possible to the multipath intensity profile associated with the channel. It is a better approach to provide a tolerance range, as it is illustrated in Figure 14.15. Here we have symbolised the multipath intensity profile assumed in the calculation of the filter coefficients by the rectangular shape, while that of the channel is symbolised with the aid of the triangular shape. Figure 14.15 will be considered in more depth during our further discourse.

In our forthcoming analytical MSE evaluations based on Equation 14.106, both the multipath intensity profile associated with the calculation of the filter coefficients as well as that of the channel were assumed to be uniform, ideally support-limited. Again, a subcarrier spacing normalised multipath spread of $T_m\Delta F$ identical to the OFDM symbol normalised Doppler spread of $B_D T_f = 2F_D$ was associated with the channel, which is expressed as $T_m\Delta F = B_D T_f$. Specifically, $F_D = 0.015625$ was used. As in our previous MSE evaluations in this section, identical pilot distances were employed in both the frequency- and time direction. Specifically, we used $\Delta p_f = \Delta p_t = 16$, resulting in a pilot oversampling of two, which is expressed as $M_{p,f} = M_{p,t} = 2$. The OFDM symbol normalised support-range of the Doppler power spectrum or Doppler spread of $\tilde{B}_D T_f = 2\tilde{F}_D$ assumed in the filter coefficients' calculation, was identical to that of the channel, which is given by $B_D T_f = 2F_D$, both

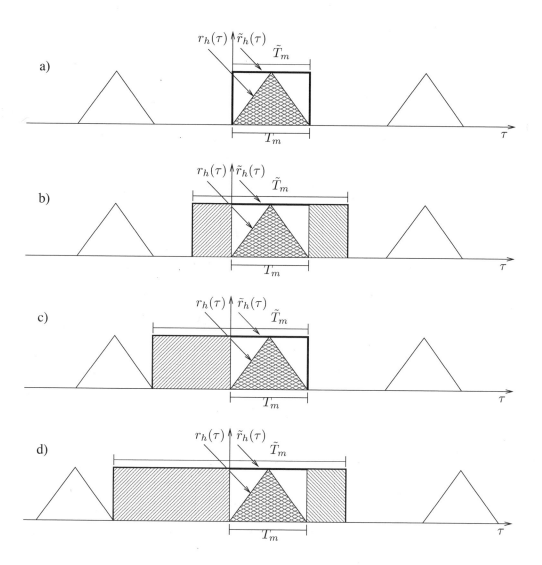

Figure 14.15: Different filter configurations characterised by the spread $\tilde{T}_m$ and displacement $\tilde{\tau}_{shift}$ of the associated multipath intensity profile. More specifically, the multipath intensity profile $\tilde{r}_h(\tau)$ used for the calculation of the filter coefficients is represented by the rectangular shape, while that associated with the channel, $r_h(\tau)$, is indicated by the triangular shape. In arrangements (a) and (b) the displacement parameter $\tilde{\tau}_{shift}$ is given by $T_m/2$, where T_m is the channel's multipath spread, while in arrangements (c) and (d) the displacement parameter τ_{shift} is identical to zero. Note that the filter-related parameters are distinguished from that of the channel by $\tilde{(\ )}$.

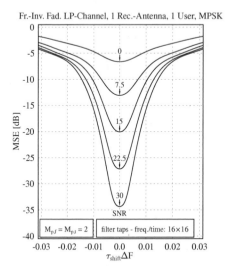

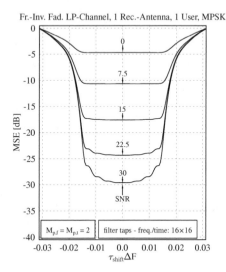

Figure 14.16: MSE exhibited by the cascaded 1D-FIR Wiener filter-based channel estimator averaged over a period of the rectangular pilot grid as a function of the channel's subcarrier spacing normalised displacement $\tau_{shift}\Delta F$, inflicted by FFT window synchronisation errors. The MSE curves are additionally parameterised with the SNR measured at the reception antenna and invoked in the evaluation of the filter coefficients. A uniform, but not necessarily the same ideally support-limited scattering function obeying Equation 14.67 was associated with the evaluation of the filter coefficients. The OFDM symbol normalised Doppler spread associated with the evaluation of the filter coefficients, $\tilde{B}_D T_f$, was identical to that of the channel, namely to $B_D T_f$, which were fixed to a value of 0.015625. The displacement and spread of the multipath intensity profile associated with the evaluation of the filter coefficients and that of the channel were adjusted as outlined in sub-figures (a) and (b) of Figure 14.15, respectively. Specifically, **in both the left and right graphs** the channel's subcarrier spacing normalised multipath spread was $T_m\Delta F = 0.03125$, and the normalised displacement was $\tau_{shift}\Delta F = T_m\Delta F/2 = 0.015625$. In the context of the **left graph** the subcarrier spacing normalised multipath spread- and displacement associated with the evaluation of the filter coefficients was identical to that of the channel, namely we had $\tilde{T}_m\Delta F = 0.03125$ and $\tilde{\tau}_{shift} = 0.015625$. By contrast, in the context of the **right graph**, $\tilde{T}_m\Delta F$ was twice as high, namely we had $\tilde{T}_m\Delta F = 0.0625$, while the normalised displacement $\tilde{\tau}_{shift}$ remained unchanged, namely $\tilde{\tau}_{shift} = 0.03125$. The pilot sampling distances employed were $\Delta p_f = \Delta p_t = 16$, which corresponds to an oversampling factor of two, which is expressed as $M_{p,f} = M_{p,t} = 2$. The results were evaluated from Equation 14.106.

centred around the zero-frequency point. By contrast, the multipath intensity profile used in the filter coefficients' calculation was adjusted with respect to its displacement and spread, as shown in the different scenarios of Figure 14.15, denoted by (a)–(d). Our MSE evaluations were conducted as a function of the displacement of the actual multipath intensity profile associated with the channel encountered - as a result of synchronisation errors - measured from its ideal causal position, where the first arriving multipath component exhibits a zero delay.

In illustration (a) of Figure 14.16 we have characterised the scenario, where the uniform multipath intensity profile invoked in the filter coefficients' calculation exhibits the same spread as the uniform multipath intensity profile associated with the channel encountered. Hence, we have $\tilde{T}_m = T_m$ and $\tilde{\tau}_{shift} = \tau_{shift}$ in the sense of Equation 14.68. The analytical MSE evaluations based on Equation 14.106 corresponding to the subcarrier spacing normalised displacement $\tau_{shift}\Delta F$ of the CIR and to the associated multipath intensity profile are depicted in the left graph of Figure 14.16, where 16 taps per frequency- and time direction filter were employed. The curves are parameterised with the SNR measured at the reception antenna, which was identical to that employed in the filter coefficients' calculation. We observe, that already for relatively small displacements the estimator's MSE is severely degraded. This is due to the misadjustment between the multipath intensity profile invoked in the filter coefficients' calculation and that associated with the channel, encountered, since significant delayed multipath components are removed in the filtering process. Our simulation results, which are not explicitly portrayed here for reasons of space, also revealed that for lower-order filters the degradation of the MSE incurred due to the misadjustment was less severe than for higher-order filters. This is because for the lower-order filters the transitional range between the filter's pass-band and stop-band is wider. Hence, in the case of a displacement of the CIR or multipath intensity profile some of the channel's significant multipath components, which would be outside the range of the perfectly aligned multipath intensity profile associated with the filter coefficients' calculation are conveyed to the filter's output, although potentially attenuated. The problem of MSE degradation due to the misadjustment between the filter's and the channel's multipath intensity profile can be addressed upon incorporating a displacement tolerance into the spread of the multipath intensity profile invoked in the evaluation of the filter coefficients.

This displacement tolerance interval is further illustrated in sub-figure (b) of Figure 14.15. Specifically, the spread $\tilde{T}_m$ of the multipath intensity profile associated with the filter's calculation was assumed to be twice as high as that of the channel T_m, which is expressed as $\tilde{T}_m = 2T_m$. Furthermore, the multipath intensity profiles' displacement in the absence of any external disturbance due to FFT window synchronisation errors was again equal, i.e we had $\tilde{\tau}_{shift} = \tau_{shift} = T_m/2$. The results of our analytical MSE evaluations based on Equation 14.106 and associated with this scenario, as a function of the subcarrier spacing normalised displacement $\tau_{shift}\Delta F$ of the channel's multipath intensity profile are portrayed in the right-hand graph of Figure 14.16. The curves are parameterised with the SNR measured at the reception antenna and invoked in the filter coefficients' calculation. Again, 16 taps were employed in both the time- and frequency direction. We observe that due to the displacement tolerance provided in the spread of the multipath intensity profile associated with the calculation of the filter coefficients, the estimator MSE becomes near-constant for normalised displacements of the channel's multipath intensity profile falling within this tolerance range seen in Figure 14.15(b). In the specific case considered here, the tolerance is equal to $T_m/2$ for both positive and negative displacements. Upon comparing the MSE results portrayed in the left-hand and right-hand graph of Figure 14.16 we observe that the robustness against the displacement of the channel's multipath intensity profile is achieved at the cost of an MSE degradation at all the SNR values considered. This is due to the increased multipath spread associated with the filter's multipath intensity profile, compared to scenario (a) of Figure 14.15, since a higher amount of AWGN is conveyed to the filter's output. Furthermore, at lower MSEs, such as those associated with an SNR of 30dB, the MSE's invariance as a function of the displacement is only approximate. This is because the effects of the filter's imperfections, such as its widened transition range between the filter's pass-band and stop-band become dominant. As a result of these imperfections, not only the desired zero-order segment of the periodic spectrum, but also fractions of its higher-order repetitions are extracted from the sampled channel's output signal.

A disadvantage imposed by the asymmetric shape of the multipath intensity profile associated with the calculation of the filter coefficients, in terms of the associated computational complexity is that the resultant filter coefficients are complex valued. This can be avoided by employing a uniform multipath intensity profile, which is centred around the zero-delay point, as shown in sub-figure (c) of Figure 14.15. Correspondingly, the displacement parameter $\tilde{\tau}_{shift}$ in Equation 14.68 has to be set to zero. Also observe that this filter arrangement does not provide any robustness against a positive-valued displacement of the channel's multipath intensity profile. Upon incorporating the idea of providing a tolerance range in the filter's associated multipath intensity profile, as portrayed in sub-figure (b) of Figure 14.15, also the scenario shown in sub-figure (c) could be rendered robust. This is portrayed in sub-figure (d) of Figure 14.15. In this specific case, in order to extract only the sampled channel's zero-order spectral repetition, the pilot distance would also have to be decreased.

14.7.2.5   Mismatch of the Multipath Intensity Profile's Shape

In our discussions so far we have assumed that a uniform ideally support-limited scattering function defined by Equation 14.67 is invoked both in the evaluation of the filter coefficients, as well as in the characterisation of the channel's properties in the context of the MSE evaluation. To be more realistic, let us now consider in this section the scenario where the channel's multipath intensity profile is derived from the indoor WATM channel model's sample-spaced three-path CIR of Section 14.3.1. Please recall that this specific three-path CIR exhibits a multipath spread of $T_m = K_0 T_s$, where $K_0 = 12$ and T_s is the sampling period duration. Noting that the OFDM symbol's subcarrier spacing is given by $\Delta F = 1/(K T_s)$, where $K = 512$ is the number of OFDM subcarriers, the channel's subcarrier spacing normalised multipath spread can be expressed as $T_m \Delta F = K_0/K = 12/512$. Correspondingly, based on Equation 14.32 the maximum tolerable sampling distance $\Delta s_{f,max}$ in the frequency direction is given by the reciprocal of the subcarrier spacing normalised multipath spread $T_m \Delta F$, namely by $\Delta s_{f,max} = 1/(T_m \Delta F) = 42.66$. Again, upon employing a pilot distance of $\Delta p_f = 32$ subcarriers in the frequency direction, according to Equation 14.37 the pilot oversampling factor becomes $M_{p,f} = \Delta s_{f,max}/\Delta p_f = 1.33$, while in conjunction with a pilot distance of $\Delta p_f = 16$ subcarriers the pilot oversampling factor would be as twice as high, namely $M_{p,f} = 2.66$. Based on this information let us construct a "balanced design", where the pilot oversampling factors $M_{p,f}$ and $M_{p,t}$ associated with the frequency and time direction are identical, as stated in Equation 14.82, which is expressed as $M_{p,f} = M_{p,t}$. By stipulating furthermore identical pilot distances in both the frequency and time direction,

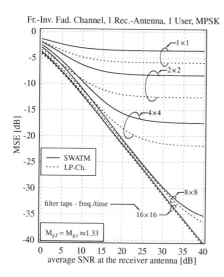
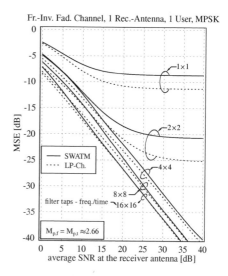

Figure 14.17: MSE exhibited by the cascaded 1D-FIR Wiener filter-based channel estimator averaged over a period of the rectangular pilot grid as a function of the SNR at the reception antenna. The curves are parametrised with the number of FIR filter taps per frequency and time direction, namely by $N_{tap}^{[f]} \times N_{tap}^{[t]}$. A uniform ideally support-limited scattering function obeying Equation 14.67 was associated with the evaluation of the filter coefficients. Two channel models were considered. The MSE performance results associated with the indoor WATM CIR of Figure 14.4 and its related multipath intensity profile are shown in continuous lines, where a Doppler power spectrum obeying Jakes' model [474] was assumed. By contrast, in dashed lines - as a benchmarker - we have plotted the MSE performance results in the context of a channel having a uniform, ideally support limited scattering function obeying Equation 14.67. The subcarrier spacing normalised multipath spread of $\tilde{T}_m \Delta F = T_m \Delta F$ was identical to the OFDM symbol normalised Doppler spread of $\tilde{B}_D T_f = B_D T_f$, where specifically $T_m \Delta F = B_D T_f = 0.03125$ was used. The pilot sampling distances of Δp_f, Δp_t employed in the context of the **left graph** were $\Delta p_f = \Delta p_t = 32$, corresponding to oversampling factors of $M_{p,f} = M_{p,t} = 1.3\dot{3}$. By contrast, in the context of **the graph on the right**, the pilot distances were halved. More specifically, $\Delta p_f = \Delta p_t = 16$ was used, resulting in oversampling factors of $M_{p,f} = M_{p,t} = 2.6\dot{6}$. The results were evaluated from Equation 14.106.

which is expressed as $\Delta p_{p,f} = \Delta s_{p,t}$, the condition of identical oversampling factors is expressed as $B_D T_f = T_m \Delta F$, which is observed from Equations 14.37 and 14.43. Consequently, the OFDM symbol normalised Doppler frequency F_D, which is related to the OFDM symbol normalised Doppler spread $B_D T_f$ by $F_D = (B_D T_f)/2$, should be chosen to be $F_D = 12/1024$. However, in reality, a "balanced design" would be achieved upon appropriately selecting the pilot distances in the frequency and time direction and not by adjusting the channel's normalised Doppler spread as in our simulation set-up.

Our analytical MSE evaluations are portrayed in Figure 14.17 which will be discussed in detail during our further discourse. In the left graph a square-shaped pilot grid of dimensions $\Delta p_f = \Delta p_t = 32$ was used, corresponding to identical pilot oversampling factors of $M_{p,f} = M_{p,t} = 1.3\dot{3}$, while in the right graph a double-density square-shaped pilot grid of dimensions $\Delta p_f = \Delta p_t = 16$ was employed, corresponding to identical time and frequency direction oversampling factors of $M_{p,f} = M_{p,t} = 2.6\dot{6}$. As in our simulations reported in the previous sections, a uniform and ideally support-limited scattering function obeying Equation 14.67 was invoked in the evaluation of the filter's coefficients, where the associated subcarrier spacing normalised multipath spread of $\tilde{T}_m \Delta F$ and the OFDM symbol normalised Doppler spread of $\tilde{B}_D T_f$ were chosen to be identical to that of the channel, namely as $\tilde{T}_m \Delta F = \tilde{B}_D T_f = 12/512$. Note again that the parameters associated with the filter coefficient evaluation are distinguished from those associated with the channel by $\tilde{()}$. The shift parameter $\tilde{\tau}_{shift}$ of the uniform scattering function's associated multipath intensity profile characterised by Equation 14.68 was set to $(K_0 - 1)T_s/2$ instead of $K_0 T_s/2$. Our intention was to stop the SWATM CIR's zero-delay tap coinciding with the filter's left-hand transition between the stop-band and pass-band and thus not to incur the potential MSE degradation, inflicted by the finite-tap filter's imperfections, specifically by its relatively wide transitional band between the filter's stop-band and pass-band. This is particularly important for propagation scenarios characterised by a relatively sparse CIR or by a sparse multipath intensity profile, where a significant fraction of the received energy is associated with a few distinct delays. By contrast in propagation scenarios associated with a near-continuous multipath intensity profile, such as for example the uniform ideally support-limited multipath intensity profile of Equation 14.67, the filter's left-handed transitional band between the stop-band and pass-band and its right-hand transitional band between the pass-band and stop-band have a less adverse influence on the channel estimator's MSE performance. This is because only a relatively limited amount of the channel's low-pass-like energy is located in these transitional ranges. In our MSE evaluations two CIRs and their associated multipath intensity profiles were considered. The main focus of our discussions is on the sample-spaced CIR associated with the SWATM channel model, which was portrayed in Figure 14.4. Furthermore, in the context of these simulations the Doppler power spectrum was given by the Jakes spectrum [474], whose associated spaced-time correlation function is given by the zero-order Bessel function of the first kind [474]. In Figure 14.17 the associated MSE curves evaluated from Equation 14.106 are represented by the continuous lines. As a benchmarker, we have also plotted the MSE performance exhibited by the channel estimator in the context of a uniform, ideally support limited scattering function, again characterised by Equation 14.67. This configuration is identical to that of Section 14.7.2.1. However, by contrast, here we employ a subcarrier spacing normalised multipath spread of $\tilde{T}_m \Delta F = T_m \Delta F = 12/512$ instead of the value of $\tilde{T}_m \Delta F = T_m \Delta F = 16/512$ used in Section 14.7.2.1. As a consequence, the noise reduction in the former scenario is more effective.

We note again, that in both MSE evaluations the uniform, ideally support limited multipath intensity profile of Equation 14.67 was employed in the calculation of the filter's coefficients. Following the same behaviour as observed in the context of our MSE performance assessment in Section 14.7.2.1, it is seen in Figure 14.17 than upon increasing the number of taps associated with each of the cascaded 1D filters, the channel estimator's MSE is reduced. For a sufficiently high number of filter taps the MSE versus SNR curves become straight lines within the relevant SNR range, which indicates that most of the channel's correlation has been exploited. Upon even further increasing the filter order no significant additional MSE reduction is observed. This is, because the pilot-assisted channel transfer function samples, which are spaced further from the position of the specific channel transfer factor to be estimated are less correlated. It is interesting to note that for a sufficiently high filter order the MSE performance observed in a scenario where a uniform, ideally support limited scattering function obeying Equation 14.67 is associated with *both* the evaluation of the filter coefficients, as well as with the actual channel encountered, is identical to the MSE performance observed, when the channel's scattering function is *not ideally* support-limited, although support-limited within the same boundaries, while exhibiting an arbitrary shape. This observation is valid for a so-called "robust" channel estimator [68], which will be further elaborated on in the context of robust decision-directed channel estimation, as discussed in Chapter 15.

14.7.3 MSE and BER Performance Evaluation by Monte Carlo Simulations

In Section 14.7.2 we analytically studied the performance of the cascaded 1D-FIR Wiener filter-based channel estimator in terms of its mean-square estimation error by capitalizing on Equation 14.106. By contrast, in this section we will assess the performance of the cascaded 1D-FIR Wiener filter based channel estimator with the aid of Monte Carlo simulations upon transmitting bursts of OFDM symbols. More specifically, the channel estimation and OFDM symbol demodulation was carried out after the reception of a complete burst- or frame of OFDM symbols. Upon assuming a rectangular pilot grid having a pilot spacing of 16 subcarriers in the frequency direction and 8 OFDM symbols in the time direction, each transmitted frame was chosen to be 57 OFDM symbols long. This allowed us to employ up to 8 taps in the filter operating in the time direction, while also reducing the edge effects in the last OFDM symbol of the frame, as the result of a potential lack of pilot subcarriers in the vicinity of the edges of the frame.

Similarly, in our simulations the boundary effects in the frequency direction were avoided upon exploiting the periodic structure of the channel transfer function and that of the spaced-frequency correlation function associated with the sample-spaced CIR employed here. More specifically, it can be shown that the channel's spaced-frequency correlation function $r_H(\Delta F \Delta k)$ having non-zero values at the subcarrier positions in the context of a sample-spaced CIR can be expressed as:

$$r_H(\Delta F \Delta k) = r_H^*(\Delta F(K - \Delta k)), \tag{14.110}$$

where K denotes the number of subcarriers per OFDM symbol and ΔF denotes the subcarrier spacing. Hence, the least-squares channel transfer factor estimates generated with the aid of the pilot symbols located in the last subcarriers of the OFDM symbol can be used to estimate the channel transfer factors associated with the lowest index subcarriers of the OFDM symbols. Thus, in our simulations the potential edge effects with respect to the frequency direction are avoided. However, the assumption of a sample-spaced CIR is rather idealistic. In a more realistic environment, associated with a non-sample spaced CIR, the edge effects with respect to the subcarrier or frequency direction can potentially be reduced e.g. by adjusting the position and density of the pilot grid appropriately. Furthermore, additional pilot symbols may be transmitted in the last subcarriers of the OFDM symbol, regardless of the specific periodic structure of the pilot grid.

The further structure of Section 14.7.3 is as follows. In Section 14.7.3.1 the simulation parameters will be outlined, while our performance studies commence in Section 14.7.3.2 with a description of the MSE simulation results obtained for the frame-invariant fading indoor WATM channel environment of Section 14.3.1. The corresponding BER simulation results will be presented in Section 14.7.3.3.

14.7.3.1 Simulation Parameters

In the context of our simulations the indoor WATM system and channel model of Section 14.3.1 was employed. Specifically, the number of subcarriers associated with the indoor WATM system's model is $K = 512$ and the indoor WATM channel model's multipath spread T_m - normalised to the sampling period duration T_s - is $K_0 = 12$. Furthermore, the OFDM symbol normalised Doppler frequency was $f_D T_f = 0.007$, which implies a vehicular speed of $50km/h$, given the indoor WATM system's parameters of Section 14.3.1. According to Equation 14.33 the maximum sampling distance in the frequency direction is $\Delta s_{f,max} = 512/12 = 42.6\dot{6}$ subcarriers, while according to Equation 14.38 the maximum sampling distance in the time direction is $\Delta s_{t,max} = 1/(2 \cdot 0.007) \approx 71.43$ OFDM symbols. Hence, upon assuming a pilot distance of 16 subcarriers in the frequency direction, according to Equation 14.37 the corresponding oversampling factor is $M_{p,f} = 2.6\dot{6}$. Similarly, upon assuming a pilot distance of 8 OFDM symbols in the time direction, according to Equation 14.43 the corresponding oversampling factor is $M_{p,t} \approx 8.93$. Consequently, the factor of "imbalance" of the pilot grid is given by the quotient of the oversampling factors, namely $M_{p,t}/M_{p,f} \approx 3.35$. This imbalance may be reduced by increasing the pilot grid density in the frequency direction or by reducing the pilot grid density along the time direction. However, the first option is at the cost of an increased pilot overhead, while the second option would restrict the number of filter taps associated with the filter operating in the time direction. This is, because each frame only included a relatively limited number of 57 OFDM symbols.

Also note that in the calculation of the filter coefficients a uniform ideally support-limited scattering function obeying Equation 14.67 was assumed. The associated expressions describing the uniform multipath intensity profile and the uniform Doppler power spectrum were given by Equations 14.68 and 14.69, respectively. As argued in Section 14.7.2.5, the subcarrier spacing normalised displacement $\tilde{\tau}_{shift}\Delta F$ of the uniform, ideally support-limited multipath intensity profile was set to $(K_0 - 1)\Delta F/2$, rather than to $K_0\Delta F/2$.

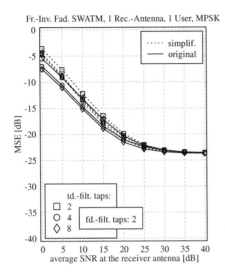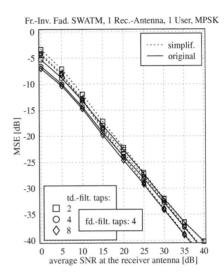

Figure 14.18: MSE exhibited by the cascaded 1D-FIR Wiener filter-based channel estimator, averaged over the subcarriers associated with each of the 57 OFDM symbols contained in a frame, as a function of the SNR at the reception antenna. The curves are parametrised with the number of FIR filter taps per frequency and time direction upon stipulating the indoor WATM system- and channel parameters outlined in Section 14.3.1. Specifically, in the context of the simulation results presented **in the graph on the left**, two taps were associated with the first cascaded filter operating in the frequency direction, while in the context of the simulations presented **in the graph on the right**, four taps were associated with this particular filter. Two different approaches were employed in the calculation of the filter coefficients. Specifically, in Höher's approach [58, 67], denoted by "original", the effects of the first cascaded filter imposed on the initial least-squares channel transfer factor estimates are considered in the evaluation of the second cascaded filter's coefficients, as outlined in Section 14.7. Secondly, a simplified approach denoted by "simplif". was also considered, where in the evaluation of both the first- and the second cascaded filter's coefficients the same SNR, namely that of the initial least-squares channel transfer factor estimates was employed. Hence, the two filters were decoupled. A uniform, ideally support-limited scattering function obeying Equation 14.67 was associated with the evaluation of the filter coefficients. The subcarrier spacing normalised multipath spread used in the calculation of the filter coefficients was identical to that associated with the channel, namely $\tilde{T}_m \Delta F = T_m \Delta F = 12/512 = 0.0234375$. Similarly, the OFDM symbol normalised Doppler spread employed in the calculation of the filter coefficients was also identical to that associated with the channel, namely $\tilde{B}_D T_f = B_D T_f = 2 \cdot 0.007 = 0.014$. In the context of the rectangular pilot grid having pilot distances of $\Delta p_f = 16$ and $\Delta p_t = 8$, the resulting oversampling factors were $M_{p,f} = 2.66$ and $M_{p,t} \approx 8.93$. The subcarrier spacing normalised displacement of the uniform, ideally support-limited multipath intensity profile was equal to $\tilde{\tau}_{shift} \Delta F = 11/1024$.

14.7.3.2 MSE Simulation Results

In order to commence our discussions of the Monte Carlo simulation based results we have plotted in Figure 14.18 the MSE exhibited by the cascaded 1D-FIR Wiener filter-based channel estimator as a function of the SNR at the reception antenna, averaged over a frame of OFDM symbols. The curves are parametrised with the number of taps employed in the context of the first filter operating in the frequency direction. More specifically, in the context of the left graph of Figure 14.18, two taps were associated with this specific filter, while in the context of the right graph of Figure 14.18 a four-tap filter was used. Additionally, the curves are parametrised with the number of taps assigned to the filter operating in the time direction, which ranges between 2 and 8. Furthermore, two different approaches were employed for evaluating the second cascaded filter's associated coefficients. First, the scheme proposed by Höher *et al.* [58,67] - as detailed in Section 14.7.1 - was employed, where the influence of the first cascaded filter on the original least-squares pilot-based channel estimates is taken into account in the calculation of the second cascaded filter's coefficients. In Figure 14.18 this approach is denoted by "original". By contrast, the calculation of the second cascaded filter's coefficients is simplified by assuming that the output signal of the first-stage filter can be modelled similarly to the pilot-assisted least-squares channel transfer factor estimates, namely as described in Section 14.7.1.3.[11] In Figure 14.18 this approach is denoted by "simplif".

In the context of the simulation results presented here the uniform, ideally support limited scattering function of Equation 14.67 invoked in the calculation of the filter's coefficients was adapted to approximate the scattering function associated with the channel encountered, as closely as possible in terms of its multipath delay and Doppler parameters. However, as a result of an imperfect knowledge of these parameters and in order to combat the effects of FFT window synchronisation errors on the channel estimation, the filter's scattering function-related support region should be designed larger than that of the channel, such that the latter is enveloped by the former, as outlined in Section 14.7.2.4. Hence, the channel estimation MSE may potentially become worse than that seen in Figure 14.18, since more noise is conveyed to the output of the channel estimator.

We observe that the channel transfer factor estimates provided by the neighbouring pilot positions of the same OFDM symbol are relatively weakly correlated, as opposed to the channel transfer factor estimates provided by the neighbouring pilot positions in the same subcarrier of the different but adjacent OFDM symbols. Hence, more than two filter taps are required for the filter operating in the frequency direction, in order to reduce the channel decorrelation-induced MSE to values below -40dB. By contrast, upon increasing the number of filter taps associated with the filter operating in the time direction, only a further noise reduction is achieved. With respect to the "original" and "simplified" approaches of evaluating the filter coefficients we observe that the original, computationally more demanding approach, denoted by "original" in Figure 14.18 only provides an advantage in terms of the associated MSE for relatively low SNRs. By contrast, for SNRs in excess of about 20dB, both approaches exhibit the same MSE performance.

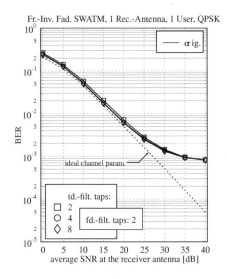
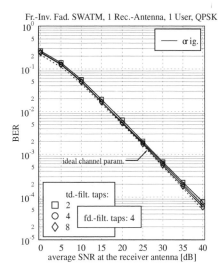

Figure 14.19: BER exhibited by a QPSK modulation assisted system employing the cascaded 1D-FIR
Wiener filter-based channel estimator. The BER results were averaged over the subcarriers associated with each of the 57 OFDM symbols contained in a frame, as a function of
the SNR at the reception antenna. The curves are parametrised with the number of FIR
filter taps per frequency and time direction upon stipulating the indoor WATM system-
and channel parameters outlined in Section 14.3.1. The specific parameters employed
are identical to those of Figure 14.18. In **the graph on the left** the number of taps associated with the filter operating in the frequency direction was two, while in **the graph
on the right** the number of filter taps was four.

14.7.3.3   BER Simulation Results

We have also evaluated the BER performance of a QPSK modulation assisted system employing cascaded 1D-FIR Wiener filter-based channel estimation. The associated BER results are
depicted in Figure 14.19, as a function of the SNR at the reception antenna. The curves are
parametrised with the number of taps associated with each of the two cascaded filters. Again,
in the context of the graph on the left of Figure 14.19 the number of taps associated with the
filter operating in the frequency direction was two, while in the context of the graph on the
right of Figure 14.19 the number of filter taps was four. Note that here we have only plotted
the BER associated with a system employing the original, more computationally demanding
approach in the calculation of the channel estimator's second cascaded filter's coefficients.
This is because no significant BER difference was observed between the two approaches. As
a reference, we have also plotted the BER of a QPSK modulation assisted system capitalising
on perfect channel transfer function knowledge.

---

[11]The advantage of the simplified approach was a potentially lower number of sets of filter coefficients to be
calculated.

When using two taps in the frequency direction filtering, the MSE floor observed in the left graph of Figure 14.18 results in a BER floor at higher SNRs. By contrast, when employing four filter taps, the BER floor is eliminated. Again, the BER improvement achieved upon increasing the order of the filter associated with the time direction is relatively modest. Please observe that the BER achieved with the aid of imperfect, pilot-assisted channel transfer function estimation is within about 1dB of the BER attained upon invoking perfect channel transfer function knowledge - although, only filters of a relatively low order were employed.

14.7.4 Complexity

In the context of the cascaded 1D-FIR Wiener filtering considered here, our complexity estimation will be conducted with respect to two aspects. In Section 14.7.4.1 the computational complexity imposed by the frequency domain filtering operation will be quantified in terms of the number of complex multiplications and additions required. Furthermore, in Section 14.7.4.2 we will evaluate the number of different sets of filter coefficients required for the simplified scenario of decoupled cascaded filters, which determines the storage requirements.

14.7.4.1 Computational Complexity

Again, let us assume employing a rectangular pilot grid using a pilot distance of Δp_f subcarriers in the frequency direction and of Δp_t OFDM symbols in the time direction. Here we also assume that filtering is performed first in the frequency direction in each of the $\lceil N_t/\Delta p_t \rceil$ number of OFDM symbols, which do host pilot tones, where N_t denotes the time direction estimation block dimension and Δp_t the pilot distance in the time direction, respectively. According to the number of taps of the FIR filter operating in the frequency direction this requires $N_{tap}^{[f]}$ number of complex multiplications and additions per time-frequency point. Since each OFDM symbol comprises N_f subcarriers, the total number of complex multiplications and additions in the first filter stage operating in the frequency direction is hence given by:

$$C_{2x1D}^{(\mathbb{C}*\mathbb{C})[f]} = C_{2x1D}^{(\mathbb{C}+\mathbb{C})[f]} = N_f \left\lceil \frac{N_t}{\Delta p_t} \right\rceil N_{tap}^{[f]}, \tag{14.111}$$

where the subscript "$2x1D$" indicates that two cascaded one-dimensional filters are used. In the second stage, filtering is performed in the time direction for all $N_f N_t$ number of time-frequency subcarriers contained in the estimation block. This involves a number of complex multiplications and additions, as quantified by the following equation:

$$C_{2x1D}^{(\mathbb{C}*\mathbb{C})[t]} = C_{2x1D}^{(\mathbb{C}+\mathbb{C})[t]} = N_f N_t N_{tap}^{[t]}, \tag{14.112}$$

where $N_{tap}^{[t]}$ denotes the number of taps of the FIR filter operating in the time direction. The total complexity is hence given by the sum of Equations 14.111 and 14.112:

$$C_{2x1D}^{(\mathbb{C}*\mathbb{C})} = C_{2x1D}^{(\mathbb{C}+\mathbb{C})} = N_f \left(\left\lceil \frac{N_t}{\Delta p_t} \right\rceil N_{tap}^{[f]} + N_t N_{tap}^{[t]} \right). \tag{14.113}$$

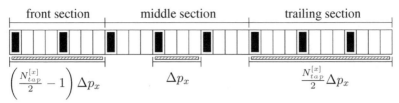

Figure 14.20: Illustration of the sources of individual filter coefficient vectors in a 1D-estimator input block in the context of cascaded 1D-FIR filtering. The corresponding sections are referred to as front-, middle- and trailing section. Note that here we have assumed an even number of filter taps $N_{tap}^{[x]}$ and an estimation block size N_x, which is an integer multiple of the pilot distance Δp_x. The periodically interspersed pilot subcarriers are marked in black.

In the specific case, when the estimator's input blocksize N_t in the time direction is a multiple of the associated pilot distance Δp_t, Equation 14.113 simplifies to:

$$C_{2x1D}^{(\mathbb{C}*\mathbb{C})} = C_{2x1D}^{(\mathbb{C}+\mathbb{C})} = N_f N_t \left(\frac{N_{tap}^{[f]}}{\Delta p_t} + N_{tap}^{[t]} \right) \quad \wedge \ (N_t \bmod \Delta p_t = 0). \qquad (14.114)$$

14.7.4.2 Number of Coefficient Vectors per SNR Level for Decoupled Cascaded 1D-FIR Filters

For the simplified scenario of cascaded 1D-FIR Wiener filters, where the filters are decoupled as described in Section 14.7.1.3, we are also interested in quantifying the complexity imposed in terms of the number of different filter tap vectors to be stored for a specific SNR level. We note that in the context of 2D-FIR filtering N_{tap} number of pilot subcarriers are associated with a specific point on the time-frequency plane, for which a channel transfer factor estimate is required. These N_{tap} pilot subcarriers are the ones that exhibit the minimum distance according to a specific metric to the considered point. Similarly, in the context of the cascaded 1D-FIR filter based estimator considered here, the Euclidean distance metric is employed to select the pilot subcarriers. For a specific 1D-FIR filter, the associated pilot positions must be located within the same OFDM symbol, or in the same subcarrier of different OFDM symbols, depending on whether the operation of filtering takes place in the frequency or in the time direction.

Let us stipulate an even number of filter taps $N_{tap}^{[x]}$ and an even pilot spacing Δp_x, where $x \in \{t/f\}$ throughout our investigations, depending on whether the filtering in the time or frequency direction is considered. This scenario can be viewed in Figure 14.20. Here the partitioning of the 1D-FIR filter input block into a front-, middle- and trailing section has been conducted, in order to motivate our derivation of the number of filter coefficient sets required. Then the number of different filter coefficient vectors to be evaluated and stored per filter direction and per SNR level can be evaluated as follows. Assuming without loss of generality that the 1D-pilot grid observed in a specific filter direction is aligned such that the first pilot symbol is assigned to the first subcarrier of the block, as seen in Figure 14.20 there are $(N_{tap}^{[x]}/2 - 1)\Delta p_x$ unique positions at the left-most side of the block. This front section in Figure 14.20 is followed by a period of positions in the middle of the block, of which only Δp_x number of positions are unique due to the periodicity of the pilot grid. Finally, there are $(N_{tap}^{[x]}/2)\Delta p_x$ number of unique positions at the right end of the block in Figure 14.20, provided that the blocklength is a multiple of the pilot distance. This results in a total number of $N_{tap}^{[x]}$-dimensional coefficient vectors given by:

$$N_{filters}^{[x]} = N_{tap}^{[x]}\Delta p_x. \tag{14.115}$$

If, however, the estimator input blocksize is adjusted such that a pilot symbol is associated with the last position in the block - as was proposed in Section 14.7.3.1 for reducing the so-called edge effects - then the corresponding contribution from the right-most end of the estimation block, seen in Figure 14.20 in terms of the number of unique positions for which filter coefficients have to be calculated, is given by $(N_{tap}^{[x]}/2 - 1)\Delta p_x + 1$. Under these conditions the total number of $N_{tap}^{[x]}$-dimensional filter coefficient vectors to be stored is given by:

$$\acute{N}_{filters}^{[x]} = (N_{tap}^{[x]} - 1)\Delta p_x + 1. \tag{14.116}$$

The scenarios associated with an odd number of filter taps or with an odd pilot distance can be treated similarly. Furthermore, the analysis leading to Equation 14.115 inherently assumed that the channel transfer function estimation blocksizes are sufficiently long, such that all three sections, namely the front, middle and trailing sections of Figure 14.20 can be hosted. This assumption is valid, if the estimation block exhibits a minimum length of:

$$N_x = N_{tap}^{[x]}\Delta p_x \tag{14.117}$$

samples. Upon invoking Equation 14.115 to determine the number of coefficient sets required for filtering in the frequency direction and Equation 14.116 for characterising the filtering in the time direction, under the assumption of hosting pilot symbols in the last OFDM symbol of the estimator's input block, the total number of filter sets is given by:

$$N_{filters} = N_{tap}^{[f]}\Delta p_f + (N_{tap}^{[t]} - 1)\Delta p_t + 1. \tag{14.118}$$

As an example, let $\Delta p_f = 16$, $\Delta p_t = 4$, $N_{tap}^{[f]} = 8$ and $N_{tap}^{[t]} = 4$. Then according to Equation 14.118 the number of different filter coefficient tuples required will be $N_{filters} = 141$. Note again that the above evaluation of the number of filter coefficient vectors is valid only for decoupled cascaded 1D-FIR filters as described in Section 14.7.1.3, where - in contrast to Höher's approach [58, 67] - the influence of the first-stage filter is not considered in the evaluation of the second-stage filter's coefficients.

14.7.5   Summary and Conclusion on Cascaded 1D-FIR Wiener Filtering

In Section 14.7 cascaded 1D-FIR Wiener filtering was analysed in detail. We proceeded in Section 14.7.1 by highlighting the mathematical foundations with respect to the two orthogonal filters' optimum coefficients and the estimator's MSE. A simplification of this approach was alluded to in Section 14.7.1.3, which resulted in a potentially lower number of sets of filter coefficients to be calculated per 2D estimator input block, where decoupled cascaded 1D-FIR filters were employed.

The achievable MSE performance was assessed in Section 14.7.2 by evaluating the analytical expressions derived in Section 14.7.1. Most notably, the MSE curves portrayed in Figure 14.12 of Section 14.7.2 underlined the significantly improved performance versus complexity trade-off of the cascaded 1D-FIR Wiener filter-based estimator compared to that of the 2D-FIR Wiener filter-based estimator portrayed in Section 14.6.4.

Furthermore, in Section 14.7.2.2 the impact of a misadjustment between the multipath- and Doppler spread assumed in the calculation of the filter coefficients and that of the channel was evaluated. We observed that if the channel scattering function's region of support is not entirely enveloped by the channel estimator's associated scattering function's region of support, a potentially excessive performance degradation is incurred. Hence, the robust channel estimator has to be designed for the maximum multipath delay and Doppler frequency encountered by the channel. Our further investigations conducted in Section 14.7.2.4 demonstrated that OFDM symbol synchronisation errors result in a displacement of the channel's associated multipath intensity profile, which must also be accounted for in the estimation of the channel's maximum multipath delay.

In Section 14.7.2.3 we also investigated the effects of a misadjustment of the SNR employed in the calculation of the filter coefficients compared to that at the reception antenna. We observed that if the channel estimator is designed for an SNR, which is lower than that encountered at the reception antenna, an MSE floor is incurred even upon increasing the number of FIR filter coefficients. This is because the channel transfer function's correlation is not optimally exploited.

In Section 14.7.2.5 we concentrated on the robust estimator's relative insensitivity against the variations of the channel scattering function's shape, as long as its region of support is enveloped by the estimator's associated scattering function's region of support. Note that in the context of a non-sample spaced CIR, maintaining this insensitivity requires an infinite number of channel transfer function samples in the estimation process.

MSE and BER performance results based on Monte Carlo simulations were provided in Section 14.7.3. Besides Höher's [58, 67] original cascaded 1D-FIR Wiener filter-based estimator, also a simplified approach was employed, where the orthogonal filters were decoupled. More specifically, the MSE improvement achieved by the first orthogonal filter - compared to the MSE of the initial least-squares channel transfer factor estimates - was not considered in the calculation of the second orthogonal filter's coefficients. As a result, a maximum of about three dB degradation of the estimator's MSE was observed at an SNR of zero dB, while for higher SNRs the MSE curves of the original and simplified approach merged as evidenced by Figure 14.18. We argued in Section 14.7.3 that an advantage of the simplified approach is constituted by a significant reduction of the number of sets of filter coefficients required. Our BER performance results presented for QPSK modulation assisted OFDM transmissions in the context of the "original" cascaded 1D-FIR Wiener filter estimator furthermore revealed an only marginal BER degradation due to the imperfect channel estimation compared to the case of ideal channel knowledge. Note that for channels with a wider multipath spread than that assumed in our simulations, the MSE and BER performance is further degraded, since more noise is injected into the process of channel estimation.

Our discussions were concluded in Section 14.7.4 with an analysis of the cascaded 1D-FIR Wiener filter's complexity.

14.8   Chapter Summary and Conclusion

In Chapter 14 we have portrayed the most prominent channel transfer function estimation techniques designed for single-user OFDM scenarios in terms of their MSE performance, the system's BER and the computational complexity imposed. These techniques exploit the frequency domain channel transfer function's correlation in both the frequency direction, as well as in the time direction by capitalising on FIR Wiener filtering. The Wiener filter solution is the result of an effort to minimise the mean-square error between the true channel transfer factors experienced by the different subcarriers of the OFDM symbols transmitted over the channel and the desired channel transfer factor estimates.

More specifically, the 2D-FIR Wiener filter of Figure 14.7, which simultaneously capitalises on the least-squares channel transfer factor estimates associated with the pilot subcarriers allocated in both the time and the frequency direction, has been studied in Section 14.6. The 2D-FIR filter-based solution was contrasted to the cascaded 1D-FIR Wiener filter of Figure 14.11 in Section 14.7.

Our conclusions drawn for the individual schemes, were offered at the end of the corresponding sections, namely, in Section 14.6.6 for the 2D-FIR approach and in Section 14.7.5 for the cascaded 1D-FIR approach. Hence here we refrain from repeating these conclusions and instead we will argue in favour of one of the schemes. Specifically we found that a disadvantage of the 2D-FIR filter arrangement was that a relatively high number of filter coefficients had to be invoked to remove the MSE floor observed in Figure 14.10 at higher SNRs. By contrast, the best MSE performance versus complexity trade-off was exhibited by the cascaded 1D-FIR filter arrangement of Section 14.7. A simplification of this approach in terms of the number of sets of filter coefficients to be pre-calculated and stored was briefly addressed in Section 14.7.1.3 based on decoupling the two orthogonal filters. The corresponding "best-compromise" MSE results were shown in Figure 14.18.

14.9 List of Symbols Used in Chapter 14

$c_{est}[n, k; \acute{n}, \acute{k}]$:   Estimator coefficient associated with the initial *a posteriori* channel transfer factor estimate at position $[\acute{n}, \acute{k}]$, when estimating the channel transfer factor at position $[n, k]$.

$\mathbf{c}_{est}[n, k]$:   Vector of estimator coefficients associated with the estimation of the channel transfer factor at position $[n, k]$: $\mathbf{c}_{est}[n, k] \in \mathbb{C}^{N_{tap} \times 1}$.

$\mathcal{C}^{(\mathbb{C}*\mathbb{C})}$:   Computational complexity in terms of the number of complex multiplications.

$\mathcal{C}^{(\mathbb{C}+\mathbb{C})}$:   Computational complexity in terms of the number of complex additions.

d_w: Weighted Euclidean distance metric used for selecting the nearest pilot subcarrier positions.

$H[n, k]$:   True channel transfer factor at position $[n, k]$.

$\tilde{H}_{apt}[n, k]$:   Initial *a posteriori* channel transfer factor estimate at position $[n, k]$.

$\tilde{\mathbf{H}}_{apt}[n, k]$:   Vector of initial *a posteriori* channel transfer factor estimates employed for estimating the transfer factor at position $[n, k]$: $\tilde{\mathbf{H}}_{apt}[n, k] \in \mathbb{C}^{N_{tap} \times 1}$.

$\hat{H}_{apt}[n, k]$:   Improved *a posteriori* channel transfer factor estimate at position $[n, k]$.

$J(\mathbf{c}_{est}[n, k])$:   MSE-related cost-function used for optimizing the vector $\mathbf{c}_{est}[n, k]$ of estimator coefficients.

K_0:   Multipath spread T_m normalised to the sampling period duration T_s.

M_f: Oversampling factor in the frequency direction: $M_f = \frac{\Delta s_{f,max}}{\Delta s_f}$.

$M_{p,f}$:   Pilot oversampling factor in the frequency direction: $M_{p,f} = \frac{\Delta s_{f,max}}{\Delta p_f}$.

M_t: Oversampling factor in the time direction: $M_t = \frac{\Delta s_{t,max}}{\Delta s_t}$.

$M_{p,t}$:   Pilot oversampling factor in the time direction: $M_{p,t} = \frac{\Delta s_{t,max}}{\Delta p_t}$.

$\mathrm{MSE}(\mathbf{c}_{est}[n, k])$:   Mean-Square Error (MSE) of the linear estimate at position $[n, k]$ as a function of the coefficient vector $\mathbf{c}_{est}[n, k]$: $MSE(\mathbf{c}_{est}[n, k]) = J(\mathbf{c}_{est}[n, k])$.

$\mathrm{MMSE}[n, k]$:   Minimum Mean-Square Error (MMSE) of the linear estimate at position $[n, k]$ achieved when using the optimum coefficient vector $\mathbf{c}_{est}[n, k] = \mathbf{c}_{est}[n, k]|_{opt}$.

$n[n, k]$:   Additive White Gaussian Noise (AWGN) contribution at position $[n, k]$.

N_{filter}:   Number of filter coefficient sets to be stored.

N_f: Number of subcarriers contained in the estimator input block. Usually we have $N_f = K$.

N_t: Number of OFDM symbols contained in the estimator input block.

N_p: Number of pilot subcarriers contained in the estimator input block which hosts a total of $N_f N_t$ subcarriers.

$N_{p,rect}$:   Number of pilot subcarriers contained in the estimator input block when using a rectangular pilot pattern: $N_{p,rect} = \lceil \frac{N_f}{\Delta p_f} \rceil \cdot \lceil \frac{N_t}{\Delta p_t} \rceil$.

$N_{p\approx}$:   Approximate number of pilot subcarriers contained in the estimator input block.

N_{tap}:   Number of filter taps.

$\mathcal{P}$:   Set constituted by the N_p number of 2D pilot subcarrier positions contained in the estimator's input block.

$\mathcal{P}_s$:   Set constituted by the N_{tap} number of 2D pilot subcarrier positions invoked in the estimation process. Note that $\mathcal{P}_s \subseteq P$.

$\mathbf{r}[n,k]$:   Cross-correlation vector between 'true' channel transfer factor $H[n,k]$ at position $[n,k]$ and the 'true' channel transfer factors $H[\acute{n},\acute{k}]$ at positions $[\acute{n},\acute{k}] \in \mathcal{P}_s$: $\mathbf{r}[n,k] \in \mathbb{C}^{N_{tap}\times1}$.

$\mathbf{r}_{apt}[n,k]$:   Cross-correlation vector between the 'true' channel transfer factor $H[n,k]$ at position $[n,k]$ and the initial *a posteriori* channel transfer factor estimates $\tilde{H}_{apt}[\acute{n},\acute{k}]$ at positions $[\acute{n},\acute{k}] \in \mathcal{P}_s$: $\mathbf{r}_{apt}[n,k] \in \mathbb{C}^{N_{tap}\times1}$.

$\mathbf{R}[n,k]$:   Auto-correlation matrix of the 'true' channel transfer factors at positions $[\acute{n},\acute{k}] \in \mathcal{P}_s$: $\mathbf{R}[n,k] \in \mathbb{C}^{N_{tap}\times N_{tap}}$.

$\mathbf{R}_{apt}[n,k]$:   Auto-correlation matrix of the initial *a posteriori* channel transfer factor estimates $\tilde{H}_{apt}[\acute{n},\acute{k}]$ at positions $[\acute{n},\acute{k}] \in \mathcal{P}_s$: $\mathbf{R}_{apt}[n,k] \in \mathbb{C}^{N_{tap}\times N_{tap}}$.

$s_p[n,k]$:   Complex pilot symbol transmitted at position $[n,k]$.

$\mathbf{V}$:   Periodicity matrix describing the pilot grid: $\mathbf{V} \in \mathbb{Z}^{2\times2}$.

$\mathbf{V}_{rect}$:   Periodicity matrix describing a rectangular pilot grid: $\mathbf{V}_{rect} = \begin{pmatrix} \Delta p_f & 0 \\ 0 & \Delta p_t \end{pmatrix}$.

$\delta[\Delta n, \Delta k]$:   2D Kronecker delta function in the context of integer variables.

Δp_f: Pilot subcarrier distance in the frequency direction.

$\Delta p_{f,max}$:   Maximum pilot subcarrier distance in the frequency direction: $\Delta p_{f,max} = \lfloor \Delta s_{f,max} \rfloor$.

Δp_t: Pilot subcarrier distance in the time direction.

$\Delta p_{t,max}$:   Maximum pilot subcarrier distance in the time direction: $\Delta p_{t,max} = \lfloor \Delta s_{t,max} \rfloor$.

Δs_f: Sampling distance in the frequency direction normalised to the subcarrier spacing ΔF.

$\Delta s_{f,max}$:   Maximum sampling distance in the frequency direction normalised to the subcarrier spacing ΔF: $\Delta s_{f,max} = \frac{1}{T_m \Delta F}$.

Δs_t: Sampling distance in the time direction normalised to the OFDM symbol duration T_f.

$\Delta s_{t,max}$:   Maximum sampling distance in the time direction normalised to the OFDM symbol duration T_f: $\Delta s_{t,max} = \frac{1}{B_D T_f}$.

ρ_p: Inverse of the SNR at the pilot positions: $\rho_p = \sigma_n^2 / \sigma_p^2$.

σ_p^2:   Pilot signal variance.

τ_{shift}:   Displacement of the multipath intensity profile.

$()^{[f]}$:   Index used for denoting the variables associated with the frequency-direction filter in the context of cascaded 1D-FIR Wiener filtering.

$()^{[t]}$:   Index used for denoting the variables associated with the time-direction filter in the context of cascaded 1D-FIR Wiener filtering.

$()[n,k]$:   Parameter associated with the position $[n,k]$.

$()[n,k;\acute{n},\acute{k}]$:   Parameter associated with the estimation of the channel transfer factor at position $[n,k]$, based on information obtained from subcarrier position $[\acute{n},\acute{k}]$.

$()[n,k;n']$:   Parameter associated with the estimation of the channel transfer factor at position $[n,k]$, based on information obtained from the OFDM symbol at position $[\acute{n}]$.

$()[n,k;n',n'']$:   Parameter associated with the estimation of the channel transfer factor at position $[n,k]$, based on information obtained from the OFDM symbols at positions $[n']$ and $[n'']$.

$()_{2D-FIR}$:   Specific variables in the context of 2D-FIR Wiener filtering.

$()_{2x1D}$:   Specific variables in the context of cascaded 1D-FIR Wiener filtering.

Chapter 15

Decision-Directed Channel Estimation for Single-User OFDM[1]

15.1   Introduction

In recent years numerous research contributions have appeared on the topic of channel transfer function estimation techniques designed for employment in single-user, single transmit antenna-assisted OFDM scenarios, since the availability of an accurate channel transfer function estimate is one of the prerequisites for coherent symbol detection at an OFDM receiver. The techniques proposed in the literature can be classified as *pilot-assisted*, *decision-directed* (DD) and *blind* channel estimation (CE) methods.

In the context of pilot-assisted channel transfer function estimation a subset of the available subcarriers is dedicated to the transmission of specific pilot symbols known to the receiver, which are used for "sampling" the desired channel transfer function. Based on these samples of the frequency domain transfer function, the well-known process of interpolation is used for generating a transfer function estimate for each subcarrier residing between the pilots. This is achieved at the cost of a reduction of the number of useful subcarriers available for data transmission. The family of *pilot-assisted* channel estimation techniques was investigated for example by Chang and Su [82], Höher [58, 66, 67], Itami *et al.* [71], Li [74], Tufvesson and Maseng [65], Wang and Liu [77], as well as Yang *et al.* [73, 78, 84].

By contrast, in the context of Decision-Directed Channel Estimation (DDCE) all the sliced and remodulated subcarrier data symbols are considered as pilots. In the absence of symbol errors and also depending on the rate of channel fluctuation, it was found that accurate channel transfer function estimates can be obtained, which are often of better qual-

---

[1]*OFDM and MC-CDMA for Broadband Multi-user Communications WLANS and Broadcasting.*
L.Hanzo, Münster, T. Keller and B.J. Choi,
©2003 John Wiley & Sons, Ltd. ISBN 0-470-85879-6

Year	Author	Contribution
'91	Höher [58]	Cascaded 1D-FIR channel transfer factor interpolation was carried out in the frequency and time direction for frequency domain PSAM.
'93	Chow, Cioffi and Bingham [59]	Subcarrier-by-subcarrier-based LMS related channel transfer factor equalisation techniques were employed.
'94	Wilson, Khayata and Cioffi [60]	Linear channel transfer factor filtering was invoked in the time direction for DDCE.
'95	van de Beek, Edfors, Sandell, Wilson and Börjesson [61]	DFT-aided CIR-related domain Wiener-filter based noise-reduction was advocated for DDCE. The effects of leakage in the context of non-sample-spaced CIRs were analysed.
'96	Edfors, Sandell, van de Beek, Wilson and Börjesson [62]	SVD-aided CIR-related domain Wiener-filter based noise-reduction was introduced for DDCE.
	Frenger and Svensson [63]	MMSE-based frequency domain channel transfer factor prediction was proposed for DDCE.
	Mignone and Morello [64]	FEC was invoked for improving the DDCE's remodulated reference.
'97	Tufvesson and Maseng [65]	An analysis of various pilot patterns employed in frequency domain PSAM was provided in terms of the system's BER for different Doppler frequencies. Kalman filter-aided channel transfer factor estimation was used.
	Höher, Kaiser and Robertson [66,67]	Cascaded 1D-FIR Wiener filter channel interpolation was utilised in the context of 2D-pilot pattern-aided PSAM
'98	Li, Cimini and Sollenberger [68]	An SVD-aided CIR-related domain Wiener filter-based noise-reduction was achieved by employing CIR-related tap estimation filtering in the time direction.
	Edfors, Sandell, van de Beek, Wilson and Börjesson [69]	A detailed analysis of SVD-aided CIR-related domain Wiener filter-based noise-reduction was provided for DDCE, which expanded the results of [62].
	Tufvesson, Faulkner and Maseng [70]	Wiener filter-aided frequency domain channel transfer factor prediction assisted pre-equalisation was studied.
	Itami, Kuwabara, Yamashita, Ohta and Itoh [71]	Parametric finite-tap CIR model based channel estimation was employed for frequency domain PSAM.

Table 15.1: Contributions on channel transfer factor estimation for single-transmit antenna assisted OFDM.

Year	Author	Contribution
'99	Al-Susa and Ormon-droyd [72]	DFT-aided Burg-algorithm assisted adaptive CIR-related tap prediction filtering was employed for DDCE.
	Yang, Letaief, Cheng and Cao [73]	Parametric, ESPRIT-assisted channel estimation was used for frequency domain PSAM.
'00	Li [74]	Robust 2D frequency domain Wiener filtering was suggested for use in frequency domain PSAM using 2D pilot patterns.
'01	Yang, Letaief, Cheng and Cao [75]	Detailed discussions of parametric, ESPRIT-assisted channel estimation were provided in the context of frequency domain PSAM [73].
	Zhou and Giannakis [76]	Finite alphabet-based channel transfer factor estimation was proposed.
	Wang and Liu [77]	Polynomial frequency domain channel transfer factor interpolation was contrived.
	Yang, Cao, and Letaief [78]	DFT-aided CIR-related domain one-tap Wiener filter-based noise-reduction was investigated, which is supported by variable frequency domain Hanning windowing.
	Lu and Wang [79]	A Bayesian blind turbo receiver was contrived for coded OFDM systems.
	Li and Sollenberger [80]	Various transforms were suggested for CIR-related tap estimation filtering assisted DDCE.
	Morelli and Mengali [81]	LS- and MMSE based channel transfer factor estimators were compared in the context of frequency domain PSAM.
'02	Chang and Su [82]	Parametric quadrature surface-based frequency domain channel transfer factor interpolation was studied for PSAM.
	Necker and Stüber [83]	Totally blind channel transfer factor estimation based on the finite alphabet property of PSK signals was investigated.

Table 15.2: Contributions on channel transfer factor estimation for single-transmit antenna assisted OFDM.

ity in terms of the channel transfer function estimator's mean-square error (MSE), than the estimates offered by pilot-assisted schemes. This is because the latter arrangements usually invoke relatively sparse pilot patterns.

The family of *decision-directed* channel estimation techniques was investigated for example by van de Beek *et al.* [61], Edfors *et al.* [62, 69], Li *et al.* [68], Li [80], Mignone and Morello [64], Al-Susa and Ormondroyd [72], Frenger and Svensson [63], as well as Wilson *et al.* [60].

Furthermore, the family of blind channnel estimation techniques was studied by Lu and

Wang [79], Necker *et al.* [83], Petermann, Vogeler, Kammeyer and Boss [475], as well as by Zhou and Giannakis [76]. The various contributions have been summarised in Tables 15.1 and 15.2.

In order to render the various DDCE techniques more amenable to employment in scenarios associated with a relatively high rate of channel variation expressed in terms of the OFDM symbol normalised Doppler frequency, linear prediction techniques well known from the speech coding literature [85, 86] can be invoked. To elaborate a little further, we will substitute the CIR-related tap estimation filter – which is part of the two-dimensional channel transfer function estimator proposed in [68] – by a CIR-related tap prediction filter. The employment of this CIR-related tap prediction filter enables a more accurate estimation of the channel transfer function encountered during the forthcoming transmission timeslot and thus potentially enhances the performance of the channel estimator. We will be following the general concepts described by Duel-Hallen *et al.* [87] and the ideas presented by Frenger and Svensson [63], where frequency domain prediction filter assisted DDCE was proposed. Furthermore, we should mention the contributions of Tufvesson *et al.* [70, 88], where a prediction filter assisted frequency domain pre-equalisation scheme was discussed in the context of OFDM. In a further contribution by Al-Susa and Ormondroyd [72], adaptive prediction filter assisted DDCE designed for OFDM has been proposed upon invoking techniques known from speech-coding, such as the Levinson-Durbin algorithm or the Burg algorithm [85,89,90] in order to determine the predictor coefficients.

Chapter 15 has the following structure. In Section 15.2 the philosophy of the CIR-tap prediction filter-assisted DDCE will be described, while its performance will be studied in Section 15.3, in the context of both the MSE and the BER. In Section 15.4 a system will be described and characterised, which combines decision-directed channel prediction[2] with adaptive modulation since the efficiency of adaptive modulation critically depends on the quality of channel prediction. Our conclusions will be offered in Section 15.5.

15.2   Description

Let us commence our discourse by outlining the structure of Section 15.2. In Section 15.2.1 the concepts of DDCE are outlined with an emphasis on generating the initial *a posteriori* least-squares channel transfer function estimate on the basis of the sliced and remodulated subcarrier symbols. Furthermore, in an effort to improve the initial *a posteriori* least-squares channel estimate, in Section 15.2.2 the concepts of the Karhunen-Loeve Transform (KLT) based 1D-MMSE estimator proposed by Sandell and Edfors *et al.* [62, 69, 157, 158] will be discussed. Based on the structure of this estimator, a 2D-MMSE estimator was proposed by Li *et al.* [68], which potentially capitalises on the availability of an infinite number of previous initial *a posteriori* channel estimates associated with past OFDM symbols. These concepts will be discussed in Section 15.2.3. The improved *a posteriori* channel transfer function estimate calculated for the current OFDM symbol is employed as an *a priori* channel estimate during the demodulation of the OFDM symbol received in the next timeslot. Due to the potentially considerable decorrelation of the channel transfer function incurred between consecutive OFDM symbols, which critically depends on the OFDM symbol nor-

---

[2]In the following we will refer to channel transfer function prediction simply as channel prediction. Furthermore, note that the set of K different subcarriers' channel transfer factors is referred to here as the channel transfer function.

malised Doppler frequency, it was found beneficial to directly obtain an *a priori* estimate of the channel transfer function for the next OFDM symbol upon invoking the MMSE *prediction* techniques to be detailed in Section 15.2.4 instead of the related MMSE *estimation* techniques of Section 15.2.3. Hence, in Section 15.2.4 we will explicitly compare the Finite Impulse Response (FIR) Wiener filter-based estimation- and prediction techniques. A prerequisite for the application of Wiener filtering [90] in the context of channel transfer function estimation or prediction along the time direction is the availability of an estimate of the channel's statistics in the form of the spaced-time correlation function. Note that the various representations of the channel's statistics are outlined in Figure 14.3. In Section 15.2.5 two different approaches are invoked for providing these estimates. Specifically, robust Wiener filtering employing a uniform, ideally support-limited scattering function will be contrasted to a block-adaptive estimation of the channel's statistics, which is combined with the calculation of the Wiener filter coefficients. Numerous techniques are known in the literature for solving the latter problem, such as the Levinson-Durbin algorithm [86] or the Burg algorithm [85, 89]. We will use the latter.

15.2.1 Decision-Directed *A Posteriori* Least-Squares Channel Estimation

The following discussions are with reference to the decision-directed channel transfer function estimation or prediction block seen at the bottom of Figure 15.1, which is the stylised illustration of an OFDM receiver.

An *a posteriori* least-squares estimate $\tilde{H}_{apt}[n, k]$ of the actual channel transfer factor $H[n, k]$ of the k-th subcarrier in the n-th OFDM symbol is obtained upon dividing the complex received OFDM symbol $x[n, k]$ by the subcarrier's sliced symbol $\check{s}[n, k]$ [68], yielding:

$$
\begin{aligned}
\tilde{H}_{apt}[n, k] &= \frac{x[n, k]}{\check{s}[n, k]}, \quad k = 0, \ldots, K - 1 \\
&= H[n, k] \cdot \frac{s[n, k]}{\check{s}[n, k]} + \frac{n[n, k]}{\check{s}[n, k]},
\end{aligned} \tag{15.1}
$$

where $H[n, k]$ denotes the complex Gaussian-distributed channel transfer factor having a variance of σ_H^2, which is unity. Furthermore, $s[n, k]$ represents the complex OFDM symbol transmitted, which exhibits zero mean and a variance of σ_s^2, and finally $n[n, k]$ is the additive noise contribution having a mean value of zero and variance of σ_n^2. The total noise variance σ_n^2 is constituted by the sum of the AWGN process' variance σ_{AWGN}^2 plus the variance of the Gaussian noise-like inter-subcarrier interference (ICI) contribution σ_{ICI}^2 [476]. The latter component can be neglected in fading channels exhibiting a low OFDM symbol normalised Doppler frequency $F_{D,K}$.[3] However, under high-mobility channel conditions - which may be encountered in worst-case scenarios - an estimate of σ_{ICI}^2 has to be provided [476]. Upon assuming error-free symbol decisions, where we have $\check{s}[n, k] = s[n, k]$, the *a posteriori* least-

---

[3]In the context of quantifying the effects of ICI a definition of the OFDM symbol normalised Doppler frequency according to $F_{D,K} = f_D K T_s$ - which is related to the FFT window's duration denoted by $K T_s$ - is more appropriate, than its conventional definition expressed as $F_D = f_D (K + K_g) T_s$ [476]. This conventional definition is rather used to characterise the channel's correlation in the context of investigating channel transfer function estimation schemes.

$$\hat{s}[n] = \frac{x[n]}{\hat{H}_{apr}[n]} \qquad \check{s}[n] = arg \min_{\check{s}/\sigma_s \in \mathcal{M}_c} \left(|\hat{s}[n] - \check{s}|^2 \right)$$

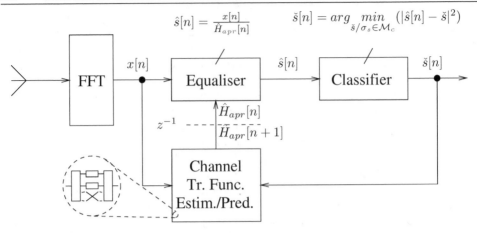

Figure 15.1: Stylised illustration of the OFDM receiver employing decision-directed channel estimation- or prediction. Following FFT processing, the complex frequency domain signal $x[n, k]$ associated with each of the K subcarriers $k = 0, \ldots, K-1$ is equalised based on the *a priori* channel transfer factor estimates generated during the previous OFDM symbol period for employment during the current period. As a result, the linear estimates $\hat{s}[n, k]$ of the transmitted signals $s[n, k]$ are obtained. These estimates are classified with the aid of the Maximum Likelihood (ML) approach, yielding the complex symbols $\check{s}[n, k]$ that are most likely to have been transmitted. These classified symbols $\check{s}[n, k]$ are then employed together with the received subcarrier signals $x[n, k]$ for generating *a priori* channel transfer factor estimates for employment during the $(n + 1)$-th OFDM symbol period. The specific structure of the channel transfer function estimator- or predictor, which is indicated by the stylised illustration at the bottom left corner, will be detailed in the context of Section 15.2. Note that in the above illustration the subcarrier index k has been omitted from the different variables for the sake of visual clarity.

squares channel transfer factor estimate of Equation 15.1 is simplified to:

$$\tilde{H}_{apt}[n, k] = H[n, k] + \frac{n[n, k]}{s[n, k]}, \tag{15.2}$$

which has a mean-square estimation error of:

$$\begin{aligned}
\text{MSE}_{apt}[n, k] &= E\{|H[n, k] - \tilde{H}_{apt}[n, k]|^2\} \\
&= \alpha \cdot \frac{\sigma_n^2}{\sigma_s^2}.
\end{aligned} \tag{15.3}$$

In Equation 15.3 the so-called "modulation-noise enhancement factor" $\alpha = E\{|1/s[n, k]|^2\}$ [477] depends on the modulation mode employed in the k-th subcarrier. For M-PSK modulation for example, we have $\alpha_{\text{M-PSK}} = 1$, while for 16-QAM $\alpha_{\text{16-QAM}} = 17/9$ [69,477].

In order to further reduce the *a posteriori* channel transfer factor estimation error, a one-dimensional (1D) minimum mean-square error (MMSE) channel estimator, which capitalises on the *a posteriori* least-squares channel transfer factor estimates extracted from each OFDM symbol was proposed by Edfors *et al.* [62, 69, 157] as well as Sandell [158]. Its concepts will be outlined in Section 15.2.2. This channel estimation approach was later extended by Li *et al.* [68] to two dimensions, where the *a posteriori* least-squares channel transfer factor estimates of both the current and the previous OFDM symbols are invoked for obtaining an *improved a posteriori* channel transfer function estimate for the current OFDM symbol. The concepts of this estimator will be outlined in Section 15.2.3.

15.2.2 Enhancement of the *A Posteriori* Least-Squares Channel Transfer Factor Estimates by One-Dimensional MMSE Estimation

A one-dimensional (1D) minimum mean-square error (MMSE) channel estimator, which exploits the channel's correlation in the frequency direction was proposed by Edfors *et al.* [62, 69, 157] and Sandell [158], for inferring improved *a posteriori* channel transfer factor estimates $\hat{H}_{apt}[n, k]$ from the initial *a posteriori* least-squares channel transfer factor estimates $\tilde{H}_{apt}[n, \acute{k}]$, $\{k, \acute{k}\} = 0, \ldots, K - 1$. Note that this estimator - in the absence of a rank-reduction, which will be further explained in Section 15.2.2.1 - is the time domain related dual of a 1D-FIR Wiener filter, which invokes all the K least-squares channel transfer factor estimates available in a specific OFDM symbol.

The specific structure of this estimator is outlined in Section 15.2.2.1 with reference to Figure 15.2, where the "Channel Estimator" block of Figure 14.1 is replaced by this schematic. Furthermore, an expression for the estimator's MSE valid under mismatched channel conditions is provided in Section 15.2.2.2.

15.2.2.1 Structure of the 1D-MMSE Channel Estimator

In a first step the vector $\tilde{\mathbf{H}}_{apt}[n] \in \mathbb{C}^{K \times 1}$ of K correlated *a posteriori* subcarrier channel transfer factor estimates $\tilde{H}_{apt}[n, k]$, $k = 0, \ldots, K - 1$ associated with the K subcarriers of the n-th OFDM symbol is subjected to a unitary linear inverse transform $\mathbf{U}^{[f]H}$, specifically the Karhunen-Loeve Transform (KLT) [69, 158], where the matrix $\mathbf{U}^{[f]H}$ used by the KLT is determined from the channel's spaced-frequency correlation matrix $\mathbf{R}^{[f]}$.[4] This transforms the frequency domain least-squares channel transfer factor estimates to a time domain related domain. As a result, the vector $\tilde{\mathbf{h}}_{apt}[n] \in \mathbb{C}^{K \times 1}$ of K uncorrelated CIR-related taps, $\tilde{h}_{apt}[n, l]$, $l = 0, \ldots, K - 1$ is obtained. In matrix notation this can be expressed as:

$$\tilde{\mathbf{h}}_{apt}[n] = \mathbf{U}^{[f]H} \tilde{\mathbf{H}}_{apt}[n]. \tag{15.4}$$

In a second step linear *one-tap* Wiener filtering is performed separately for those CIR-related taps $\tilde{h}_{apt}[n, l]$, for which the variance is significant. The specific value of the l-th

---

[4]The EigenValue Decomposition (EVD) of the Hermitian spaced-frequency correlation matrix $\mathbf{R}^{[f]} = E\{\mathbf{HH}^H\} \in \mathbb{C}^{K \times K}$, where $\mathbf{H} \in \mathbb{C}^{K \times 1}$ is the vector of the different subcarriers' channel transfer factors, is given by $\mathbf{R}^{[f]} = \mathbf{U}^{[f]} \Lambda^{[f]} \mathbf{U}^{[f]H}$, where $\mathbf{U}^{[f]} \in \mathbb{C}^{K \times K}$ is unitary, and $\Lambda^{[f]} \in \mathbb{R}^{K \times K}$ exhibits the form of $\Lambda^{[f]} = diag(\lambda_0^{[f]}, \ldots, \lambda_{K-1}^{[f]})$. The diagonal elements of $\Lambda^{[f]}$ are referred to as the *eigenvalues* of $\mathbf{R}^{[f]}$ [90].

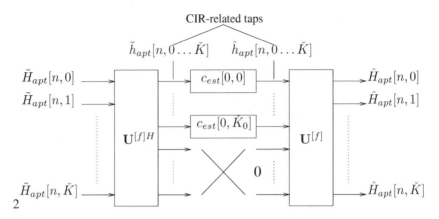

Figure 15.2: 1D-MMSE channel estimator design proposed by Edfors *et al.* [62, 69, 157] replacing the "Channel Transfer Function Estimator" block of Figure 15.1. The set of *a posteriori* least-squares channel transfer factor estimates $\tilde{H}_{apt}[n,k]$, $k = 0,\ldots,K-1$ of Equation 15.1 is subjected to an inverse unitary linear transform $\mathbf{U}^{[f]H}$ which generates a set of uncorrelated CIR-related taps $\tilde{h}_{apt}[n,l], l = 0,\ldots,K-1$. Only the first K_0 coefficients are retained, the rest of them are set to zero. Each of the K_0 coefficients is subjected to time direction one-tap Wiener filtering using the coefficients $c_{est}[0,l]$, $l = 0,\ldots,K-1$ given by Equation 15.5. Finally, a unitary transform $\mathbf{U}^{[f]}$ is applied for conveying the filtered CIR-related taps $\hat{h}_{apt}[n,l], l = 0,\ldots,K-1$ back to the frequency domain, which yields the set of improved *a posteriori* channel transfer factor estimates $\hat{H}_{apt}[n,k]$, $k = 0,\ldots,K-1$. Note that the unitary transform matrix $\mathbf{U}^{[f]}$ is associated with the EVD of the channel's spaced-frequency correlation matrix $\mathbf{R}^{[f]} = \mathbf{U}^{[f]}\Lambda^{[f]}\mathbf{U}^{[f]H}$. For notational simplicity we have defined in the schematic $\check{K}_0 = K_0 - 1$ and $\check{K} = K - 1$.

CIR-related tap's filter coefficient is given by [69]:

$$c_{est}[0,l] = \begin{cases} \dfrac{\lambda_l^{[f]}}{\lambda_l^{[f]} + \frac{\alpha}{\text{SNR}}} & l = 0,\ldots,K_0 - 1 \\ 0 & l = K_0,\ldots,K-1 \end{cases}, \tag{15.5}$$

where $\lambda_l^{[f]}$ is the l-th eigenvalue associated with the EigenValue Decomposition (EVD) of the channel's spaced-frequency correlation matrix $\mathbf{R}^{[f]}$. Since for the set of practical multipath intensity profiles considered the variance $E\{|\tilde{h}_{apt}[n,l]|^2\}$ usually decreases along with increasing the delay index l, a standard complexity reduction procedure [69,477] is to perform time direction filtering individually for each of the CIR-related taps associated with the K_0 number of lowest values of l, while setting the CIR-related taps associated with higher delay or dispersion indices to zero. Again, this reduces the estimator's complexity at the cost of a certain MSE performance degradation due to removing some of the CIR-related taps conveying useful signal components. The filter operation can also be expressed in matrix notation, namely as:

$$\hat{\mathbf{h}}_{apt}[n] = \mathbf{C}_{est}[0]\tilde{\mathbf{h}}_{apt}[n], \tag{15.6}$$

where $\hat{\mathbf{h}}_{apt}[n] \in \mathbb{C}^{K\times 1}$ is the vector of filtered CIR-related tap values and the diagonal-shaped filter matrix $\mathbf{C}_{est}[0] \in \mathbb{R}^{K\times K}$ is given by:

$$\mathbf{C}_{est}[0] = diag(c_{est}[0,0], \ldots, c_{est}[0, K-1]). \tag{15.7}$$

In a last step the vector $\hat{\mathbf{h}}_{apt}[n]$ of filtered CIR-related tap values $\hat{h}_{apt}[n, l], l = 0, \ldots, K-1$ is transformed back to the OFDM symbol's frequency domain representation with the aid of the unitary linear transform $\mathbf{U}^{[f]}$, yielding the vector $\hat{\mathbf{H}}_{apt}[n] \in \mathbb{C}^{K\times 1}$ of MMSE channel transfer factor estimates. Again, this can be formulated in matrix notation as:

$$\hat{\mathbf{H}}_{apt}[n] = \mathbf{U}^{[f]}\hat{\mathbf{h}}_{apt}[n]. \tag{15.8}$$

The 1D-MMSE channel estimator's structure is further illustrated in Figure 15.2.

15.2.2.2   Estimator MSE for Mismatched Channel Conditions

In practice the estimator's spaced-frequency correlation matrix $\tilde{\mathbf{R}}^{[f]}$ and its associated EVD, $\tilde{\mathbf{R}}^{[f]} = \tilde{\mathbf{U}}^{[f]}\tilde{\Lambda}^{[f]}\tilde{\mathbf{U}}^{[f]H}$, as well as the EVD of the channel's spaced-frequency correlation matrix $\mathbf{R}^{[f]}$, namely $\mathbf{R}^{[f]} = \mathbf{U}^{[f]}\Lambda^{[f]}\mathbf{U}^{[f]H}$, are typically different. We often refer to this scenario as a mismatch between the channel statistics assumed and those encountered. For this scenario Edfors *et al.* [69] derived an expression for the MSE of the rank-K_0 estimator seen in Figure 15.2:

$$\overline{\mathrm{MSE}}_{apt}\big|_{K_0}^{N_{tap}^{[t]}=1} = \frac{1}{K}\sum_{l=0}^{K_0-1}\left[v_l^{[f]}(1 - \tilde{c}_{est}[0,l])^2 + \frac{\alpha}{\mathrm{SNR}}\tilde{c}_{est}^2[0,l])\right] + \frac{1}{K}\sum_{l=K_0}^{K-1}v_l^{[f]}, \tag{15.9}$$

where $\tilde{c}_{est}[0,l]$, $l = 0, \ldots, K_0 - 1$ is the set of filter coefficients following the philosophy of Equation 15.5. We note that the coefficients $\tilde{c}_{est}[0,l]$, $l = 0, \ldots, K_0 - 1$ are based on the potentially imperfect estimate $\tilde{\mathbf{R}}^{[f]}$ of the channel's spaced-frequency correlation matrix $\mathbf{R}^{[f]}$ and on the estimate $\tilde{\mathrm{SNR}}$ of the true SNR measured at the reception antenna. Furthermore, in Equation 15.9 the variable $v_l^{[f]}$ denotes the l-th diagonal element of the decomposition associated with the channel's spaced-frequency correlation matrix $\mathbf{R}^{[f]}$ with respect to the unitary transform matrix $\tilde{\mathbf{U}}^{[f]}$ instead of $\mathbf{U}^{[f]}$, which is formulated as $v_l^{[f]} = (\tilde{\mathbf{U}}^{[f]H}\mathbf{R}^{[f]}\tilde{\mathbf{U}}^{[f]})_{[l,l]}$. Also note that if $\tilde{\mathbf{U}}^{[f]} = \mathbf{U}^{[f]}$ holds, we have $v_l^{[f]} = \lambda_l^{[f]}$.

15.2.3   Enhancement of the *A Posteriori* Least-Squares Channel Transfer Factor Estimates by Two-Dimensional MMSE Estimation

Based on the 1D-MMSE channel estimator proposed by Edfors *et al.* [62, 69, 157], which was outlined in Section 15.2.2, a two-dimensional (2D) MMSE channel estimator exploiting the channel's correlation both in the frequency direction as well as in the time direction was proposed by Li *et al.* in [68]. The objective of Li's design was to infer improved *a posteriori* channel transfer factor estimates $\hat{H}_{apt}[n,k]$ from the initial *a posteriori* least-squares channel transfer factor estimates $\tilde{H}_{apt}[n,\acute{k}]$, $\{k, \acute{k}\} = 0, \ldots, K-1$. This can be achieved by capitalising not only on the most recent *a posteriori* estimates

$\tilde{H}_{apt}[n,\acute{k}]$, $\acute{k} = 0, \ldots, K-1$, but also on the *a posteriori* least-squares channel transfer factor estimates $\tilde{H}_{apt}[n-\acute{n},\acute{k}]$, $\acute{n} = 1, \ldots, N_{tap}^{[t]}-1$, $\acute{k} = 0, \ldots, K-1$ of the $(N_{tap}^{[t]}-1)$ number of past OFDM symbols, where $(N_{tap}^{[t]}-1)$ denotes the order of the associated estimation filter. Note that in the absence of a rank-reduction, such as conducted by the scheme portrayed in Figure 15.2, this estimator is the time domain related dual of a frequency domain 2D-FIR Wiener filter, which invokes all least-squares channel transfer factor estimates available in a specific block of $N_{tap}^{[t]}$ number of OFDM symbols.

Our further discourse in Section 15.2.3 evolves as follows. In Section 15.2.3.1 we outline the difference between Li's 2D-MMSE estimator [68] and Edfors's 1D-MMSE estimator, which was the topic of Section 15.2.2. Furthermore, Section 15.2.3.2 highlights Li's discussions on the 2D-MMSE estimator's MSE performance under mismatched channel conditions, with an emphasis on the concepts of robustness or resilience against mismatched channel conditions. Finally, the idea of employing time direction prediction filtering instead of time direction estimation filtering,[5] as advocated by Li *et al.* [68], will be augmented in Section 15.2.3.3.

15.2.3.1   Structure of the 2D-MMSE Estimator

As it was demonstrated by Li *et al.* [68], the two-dimensional channel transfer factor estimation problem can be readily separated into two one-dimensional estimation tasks, which is motivated by the separability of the channel's spaced-time spaced-frequency correlation function originally introduced in Figure 14.3. As noted before, instead of carrying out the channel transfer factor estimation in the frequency domain, the philosophy of rank-reduction in the CIR-related domain, portrayed in Figure 15.2 is used, which results in a reduced complexity.

The structure of Li's 2D-MMSE channel estimator shown in Figure 15.4 is identical to that of the 1D-MMSE channel estimator proposed by Edfors *et al.* [62, 69, 157], as outlined in Section 15.2.2, with the sole difference that $N_{tap}^{[t]}$-tap estimation filters are invoked in the CIR-related domain instead of one-tap filters. More explicitly, filtering of the specific l-th CIR-related tap in the time direction is performed by capitalising on the current value $\tilde{h}_{apt}[n,l]$ and on the $(N_{tap}^{[t]}-1)$ previous values $\tilde{h}_{apt}[n-\acute{n},l]$. This process is illustrated in Figure 15.3. As a result, in the case of estimation filtering [68] an improved *a posteriori* estimate $\hat{h}_{apt}[n,l]$ of $h[n,l]$ is obtained. By contrast, in case of prediction filtering, which will be the topic of Section 15.2.4 an *a priori* estimate $\hat{h}_{apr}[n+1,l]$ of $h[n+1,l]$ is obtained.

15.2.3.2   Estimator MSE for Mismatched Channel Statistics

The salient contribution of Li *et al.* [68] was the investigation of the channel estimator's MSE under mismatched channel conditions, where the spaced-time- spaced-frequency correlation function defined in Figure 14.3 and assumed in the calculation of the channel estimator coefficients actually differed from those of the transmission channel. Note that these investigations were conducted under the assumption of employing the *a posteriori* least-squares channel

---

[5]The basic difference of time direction channel prediction- and estimation filtering is that in the former an *a priori* channel transfer function estimate valid for the next OFDM symbol is obtained, while in the latter an improved *a posteriori* channel estimate is generated for the current OFDM symbol.

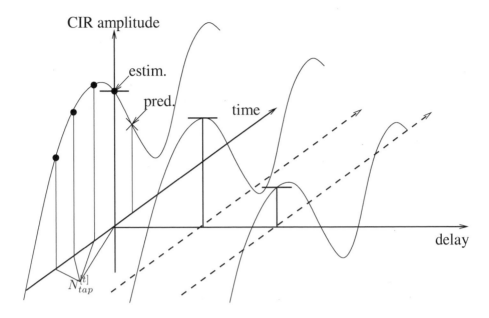

Figure 15.3: Stylised illustration of the estimation and prediction filter, both operating in the CIR-related domain using $N_{tap}^{[t]}$ number of previous *a posteriori* CIR-related tap estimates. In the context of *estimation filtering* improved *a posteriori* tap estimates are generated for the current OFDM symbol period, while in the context *prediction filtering a priori* tap estimates are computed for the next OFDM symbol period.

transfer factor estimates $\tilde{H}_{apt}[n - \acute{n}, \acute{k}]$, $\acute{n} = 1, \ldots, \infty$, $\acute{k} = 0, \ldots, K - 1$ associated with an infinite number of past OFDM symbols in the 2D-MMSE estimation process. Let us now summarise the most significant associated aspects:

1) *MMSE Upper Bound for Channels Band-Limited to* $\omega_D = 2\pi f_D$ *under perfectly matched channel conditions* [68]. From the set of all possible channels band-limited to f_D the channel estimator exhibits the worst MSE for transmission over a channel having an ideally band-limited spaced-time correlation function, with associated Doppler-power spectral density function of:

$$S_{H,unif}(f_d) = \begin{cases} \frac{1}{2f_D}, & \text{if } |f_d| \le f_D \\ 0, & \text{otherwise.} \end{cases} \qquad (15.10)$$

In this case the MMSE averaged over the K subcarriers is given by [68]:

$$\overline{\text{MMSE}}_{apt}\Big|^{N_{tap}^{[t]} \to \infty} = \frac{\rho}{K} \sum_{l=0}^{K-1} \left(1 - \frac{1}{\left(1 + \frac{\pi \lambda_l^{[f]}}{\omega_D \rho}\right)^{\frac{\omega_D}{\pi}}} \right). \qquad (15.11)$$

In Equation 15.11 ω_D is defined as $\omega_D = 2\pi F_D$, where F_D is the OFDM symbol normalised Doppler frequency. Furthermore, the variable ρ is defined as the reciprocal of the average SNR, namely as $\rho = 1/\text{SNR}$, and $\lambda_l^{[f]}, l \in \{0, \ldots, K-1\}$ denotes the l-th eigenvalue[6] of the EVD of the spaced-frequency channel correlation matrix $\mathbf{R}^{[f]}$. Please also note that in contrast to [68] here the first subcarrier is denoted by the index zero.

2) *Spaced-Time Correlation Function Mismatch* [68]. Let us assume that the spaced-frequency correlation function $\tilde{r}_H^{[f]}[\Delta k]$ of the channel estimator matches that of the channel $r_H^{[f]}[\Delta k]$ and that the spaced-time correlation function $\tilde{r}_H^{[t]}[\Delta n]$ of the channel estimator is ideally band-limited to f_D having a PSD function of $S_{H,unif}(f_d)$. Assume furthermore that the spaced-time correlation function $r_H^{[t]}[\Delta n]$ of the channel is associated with an arbitrary PSD function band-limited to f_D. Then the estimator's MSE is given by $\overline{\text{MMSE}}_{apt}|^{N_{tap}^{[t]} \to \infty}$.

3) *Spaced-Frequency Correlation Function Mismatch* [68]. Let us assume that the spaced-time correlation function $\tilde{r}_H^{[t]}[\Delta n]$ of the channel estimator matches that of the channel, namely $r_H^{[t]}[\Delta n]$. Assume furthermore that the eigenvectors of the channel estimator's spaced-frequency correlation matrix $\tilde{\mathbf{R}}^{[f]}$ - which constitute the column vectors of the unitary transform matrix $\tilde{\mathbf{U}}^{[f]}$ - are identical to those of the channel's spaced-frequency correlation matrix $\mathbf{R}^{[f]}$ - which constitute the column vectors of the unitary transform matrix $\mathbf{U}^{[f]}$. Under these assumptions for a channel estimator having spaced-frequency correlation matrix related eigenvalues of [68]:

$$\tilde{\lambda}_l^{[f]} = \begin{cases} \frac{K}{K_0}, & \text{for } 0 \leq l \leq K_0 - 1 \\ 0, & \text{for } K_0 \leq l \leq K - 1 \end{cases} \qquad (15.12)$$

the estimator's MSE is identical for all channels having spaced-frequency correlation matrix related eigenvalues of $\lambda_l^{[f]} = 0$ for $K_0 \leq l \leq K - 1$ and $\sum_{l=0}^{K_0-1} \lambda_l^{[f]} = K$. Note that these conditions can be directly derived from Equation 15.9 associated with the 1D-MMSE channel estimator. As shown in [68, 69], the assumption of $\tilde{\mathbf{U}}^{[f]} \approx \mathbf{U}^{[f]}$ is reasonable, since both the transform matrices $\tilde{\mathbf{U}}^{[f]}$ and $\mathbf{U}^{[f]}$ can be approximated by the DFT matrix $\mathbf{W}$ given by [68]:

$$\mathbf{W} = \frac{1}{\sqrt{K}} \begin{pmatrix} 1 & 1 & \cdots & 1 \\ 1 & e^{-j(2\pi/K)} & \cdots & e^{-j(2\pi(K-1)/K)} \\ \vdots & \cdots & \cdots & \vdots \\ 1 & e^{-j(2\pi(K-1)/K)} & \cdots & e^{-j(2\pi(K-1)(K-1)/K)} \end{pmatrix}. \qquad (15.13)$$

The average MSE of the robust estimator is hence given upon substituting the associated eigenvalues $\tilde{\lambda}_l^{[f]}$, $0 \leq l \leq K - 1$ given by Equation 15.12 into Equation 15.11,

---

[6]Note that in contrast to Li's discussions [68], where the l-th eigenvalue is denoted by d_l, here we follow the conventional notation and denote the l-th eigenvalue by $\lambda_l^{[f]}$.

yielding [68]:

$$\overline{\text{MMSE}}_{apt}\big|_{rob}^{N_{tap}^{[t]} \to \infty} = \frac{K_0 \rho}{K}\left(1 - \frac{1}{\left(1 + \frac{\pi K}{K_0 \omega_D \rho}\right)^{\frac{\omega_D}{\pi}}}\right). \qquad (15.14)$$

The implications of the above assumptions on the estimator's design are illustrated more explicitly in Figure 15.4.

Li *et al.* [68] demonstrated furthermore that the MSE degradation incurred by a robust estimator having a spaced-time spaced-frequency correlation function (seen in Figure 14.3) associated with a uniform, ideally support-limited scattering function as stated in paragraphs 2) and 3) above is marginal compared to a relaxed-specification estimator, which only assumes a spaced-frequency correlation function associated with an ideally support-limited multipath intensity profile, as outlined in paragraph 3) while exactly matching the actual Doppler PSD of the channel.

We emphasise that in the context of the above-mentioned robust estimator design identical CIR-related domain filters are employed for filtering all the different K_0 number of CIR-related taps. We note that the associated estimation filtering could also be performed in the frequency domain by capitalising on the same set of filter coefficients. Hence again, the disadvantage of performing K individual time direction filtering operations related to the different subcarriers, instead of K_0 number of time direction filtering operations for the first K_0 CIR-related taps is the associated higher complexity. The assumption of employing an infinite number of previous *a posteriori* channel transfer function estimates in order to obtain an improved *a posteriori* channel transfer function estimate for the current OFDM symbol is unrealistic. Hence, Li *et al.* [68] also proposed a time domain filter estimator, employing a finite number of previous CIR-related tap estimates. As expected, the MSE of this scheme is lower-bounded by the MSE of an estimator using an infinite number of previous CIR-related tap estimates, which was given in Equation 15.11.

15.2.3.3 Motivation of Time Direction Channel Prediction Filtering

In [68] the improved *a posteriori* channel transfer factor estimates $\hat{H}_{apt}[n, k]$, $k = 0, \ldots, K - 1$ were employed as *a priori* decision-directed channel transfer factor estimates for the demodulation process during the $(n+1)$-th transmission timeslot. More explicitly, we have $\hat{H}_{apr}[n + 1, k] = \hat{H}_{apt}[n, k]$, $k = 0, \ldots, K - 1$, which implies the assumption that the channel transfer function has not changed between the n-th and the $(n + 1) - th$ transmission timeslot. The associated decorrelation-related mean-square channel transfer factor estimation error incurred by this assumption can be quantified for any of the K subcarriers as:

$$
\begin{aligned}
\text{MSE}_{dec}[n, k] &= E\left\{|H[n + 1, k] - H[n, k]|^2\right\} \\
&= 2\left(E\{|H[n + 1, k]|^2\} - \Re\{E\{H[n + 1, k]H^*[n, k]\}\}\right) \quad (15.15) \\
&= 2\left(r_H^{[t]}[0] - \Re\{r_H^{[t]}[1]\}\right). \quad (15.16)
\end{aligned}
$$

Assuming that the channel's Doppler power spectrum obeys Jakes' model [474], the spaced-time channel correlation function is given by $r_{H,J}^{[t]}[\Delta n] = J_0(\Delta n \omega_D)$ [68], where $J_0(x)$ is

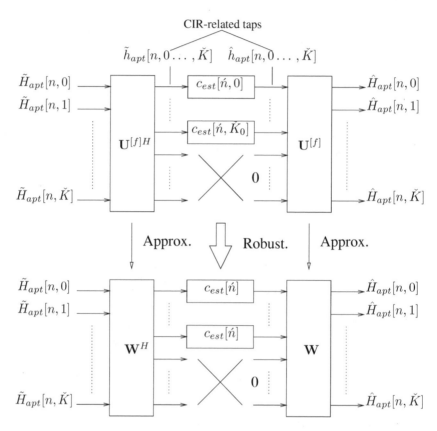

Figure 15.4: 2D-MMSE channel estimator design proposed by Li *et al.* [68] replacing the 1"Channel Estimator" block of Figure 15.1. In contrast to the 1D-MMSE channel estimator design of Edfors *et al.* [69,157], portrayed in Figure 15.2, the 2D-MMSE channel estimator employs CIR-domain-related estimation filters having a filter order higher than one, potentially capitalising on the *a posteriori* least-squares channel transfer factor estimates of several previously received OFDM symbols. In the lower half of this figure, the implications of the assumptions of robustness on the estimator design are illustrated. Specifically, the unitary matrix $\mathbf{U}^{[f]}$ associated with the KLT and hosting the eigenvectors associated with the channel's spaced-frequency correlation matrix $\mathbf{R}^{[f]}$ is substituted by the DFT matrix $\mathbf{W}$ of Equation 15.13. Furthermore identical filters having an impulse response $c[\acute{n}]$, $\acute{n} = 0, \ldots, N_{tap}^{[t]} - 1$ are employed in the context of the identical eigenvalues $\tilde{\lambda}_l^{[f]} = \frac{K}{K_0}$, $l = 0, \ldots, \check{K}_0$ associated with the robust design instead of individual CIR-related tap-specific filters having different impulse responses $c_{est}[\acute{n}, l]$, $\acute{n} = 0, \ldots, N_{tap}^{[t]} - 1$, $l = 0, \ldots, \check{K}_0$ in the context of the individual eigenvalues $\lambda_l^{[f]}$, $l = 0, \ldots, \check{K}_0$ associated with the optimum design. For notational simplicity we have defined in the schematic $\check{K}_0 = K_0 - 1$ and $\check{K} = K - 1$.

F_D	0.007	0.01	0.05	0.1
$\mathrm{MSE}_{dec,J}\big|_{dB}$	-30.14	-27.05	-13.07	-7.05

Table 15.3: Approximate MSE according to Equation 15.19 induced by the channel's decorrelation observed between two successive OFDM symbols in the context of Jakes' model - as a function of the OFDM symbol normalised Doppler frequency $F_D = f_D T_f$.

the zero-order Bessel function of its first kind, which can be approximated by [89, 473]:

$$J_0(x) \approx 1 - \frac{1}{4}x^2, \ x \ll 1. \tag{15.17}$$

Hence, in the context of Jakes' model we obtain for the channel decorrelation-related estimation error:

$$\mathrm{MSE}_{dec,J}[n,k] = 2(1 - J_0(\omega_D)) \tag{15.18}$$

$$\approx \frac{1}{2}\omega_D^2. \tag{15.19}$$

Equation 15.19 has been evaluated for different values of the OFDM symbol normalised Doppler frequency of $F_D = f_D T_f$, which is related to ω_D by $\omega_D = 2\pi F_D$. The corresponding results are summarised in Table 15.3. We observe from Table 15.3 that depending on the value of F_D the channel-decorrelation-related estimation error associated with the assumption of $\hat{H}_{apr}[n+1,k] = \hat{H}_{apt}[n,k]$, $k = 0,\dots,K-1$, may become excessive. This was the motivation for Tufvesson [70, 88] as well as Al-Susa and Ormondroyd [72] to employ prediction filtering instead of estimation filtering along the time direction. The specific structure of the time direction prediction filter will be contrasted against that of the time direction estimation filter of [68] in the next section, namely in Section 15.2.4. The stylised estimation and prediction process were portrayed in Figure 15.3.

15.2.4 MMSE *A Priori* Time Direction Channel Prediction Filtering

In the previous section estimation filtering of the CIR-related taps was employed for enhancing the performance of the decision-directed channel transfer function estimator using the schematic of Figure 15.4. By contrast, in this section we will highlight the structure of the MMSE *a priori* channel predictor, which operates along the time direction [72, 87, 90, 472]. Initially we will *not* impose any of the robustness-related constraints, which were proposed in [68] and summarised in Section 15.2.3 in the context of the two-dimensional channel estimator. Hence, the optimum design in the sense of a minimum channel prediction MSE obeys again the structure seen at the top of Figure 15.4. Most notably, different CIR-related tap prediction filters having an impulse response of $c[\acute{n},l]$, $\acute{n} = 0,\dots,N^{[t]}_{tap}-1$ for each of the l-th CIR-related taps, $l \in \{0,\dots,K_0-1\}$ are invoked for exploiting the correlation of the CIR-related taps along the time direction. In contrast to the discussions of Section 15.2.3, the filters considered here are of finite order. We note that in the context of an optimum channel transfer function predictor design based on the unitary transform matrix $\mathbf{U}^{[f]}$ seen in Figure 15.4, which hosts the eigenvectors of the spaced-frequency correlation matrix $\mathbf{R}^{[f]}$

it would be incorrect to refer to the filtering along the time direction as CIR tap filtering, since by definition the CIR is related to the "frequency-continuous" channel transfer function by the Fourier Transform (FT). By contrast, in the context of the estimator considered here, a finite set of channel transfer function samples is transferred to a CIR-related domain by means of the unitary transform matrix $\mathbf{U}^{[f]}$. However, there is an exceptional case, where the CIR is obtained by subjecting the set of K number of channel transfer factors to the unitary matrix $\mathbf{U}^{[f]}$. This is when the CIR is sample-spaced.[7] Here we will employ the more general terminology of *CIR-related* tap prediction filtering, rather than CIR tap prediction filtering.

The structure of Section 15.2.4 is as follows. In Section 15.2.4.1 the predicted *a priori* value for a specific CIR-related tap is expressed as a linear combination of past *a posteriori* CIR-related tap values. In Section 15.2.4.2 we define the *a posteriori* CIR-related tap values' auto-correlation matrix and cross-correlation vector, which are then employed in the context of Section 15.2.4.3 to derive the Wiener equation with the aid of two different methods, namely the gradient approach [90] and the orthogonality principle [90]. The Wiener equation allows for a direct solution with respect to the desired vector of predictor coefficients, which will be outlined in Section 15.2.4.4. In the same section we also present an expression for the predictor MSE in the CIR-related domain under perfectly matched channel conditions, where the channel statistics invoked in the calculation of the predictor coefficients are identical to those of the channel encountered. Furthermore, in Section 15.2.4.5 the predictor's MSE under mismatched channel conditions will be considered. Finally in Section 15.2.4.6 an expression for the average channel prediction MSE in the frequency domain will be derived.

15.2.4.1   Linear Prediction of the CIR-Related Taps

An *a priori* estimate $\hat{h}_{apr}[n+1, l]$ of the l-th CIR-related tap for the $(n+1)$-th timeslot is given by:

$$\hat{h}_{apr}[n+1, l] = \sum_{\acute{n}=0}^{N_{tap}^{[t]}-1} c_{pre}[\acute{n}, l] \cdot \tilde{h}_{apt}[n - \acute{n}, l], \tag{15.20}$$

where $l = 0, \ldots, K_0 - 1$. In Equation 15.20 the variable $c_{pre}[\acute{n}, l]$ denotes the $\acute{n}$-th coefficient of the $N_{tap}^{[t]}$-tap CIR-related tap predictor and $\tilde{h}_{apt}[n - \acute{n}, l]$ represents the *a posteriori* estimate of the l-th CIR-related tap in the $(n - \acute{n})$-th timeslot, which is related to the *a posteriori* least-squares estimates $\tilde{H}_{apt}[n - \acute{n}, k], k = 0, \ldots, K - 1$ of the channel transfer factors by the unitary transform matrix $\mathbf{U}^{[f]}$, as shown in Equation 15.4. Equation 15.20 can also be expressed in vector notation as:

$$\hat{h}_{apr}[n+1, l] = \mathbf{c}_{pre}^{H}[l]\tilde{\mathbf{h}}_{apt}[n, l], \tag{15.21}$$

where $\mathbf{c}_{pre}[l] \in \mathbb{C}^{N_{tap}^{[t]} \times 1}$ is the vector of CIR-related tap predictor coefficients:

$$\mathbf{c}_{pre}[l] = \left(c_{pre}^{*}[0, l], c_{pre}^{*}[1, l], \ldots, c_{pre}^{*}[N_{tap}^{[t]} - 1, l] \right)^{T}, \tag{15.22}$$

---

[7]Note that in this case the unitary KLT matrix $\mathbf{U}^{[f]}$ is identical to the DFT matrix $\mathbf{W}$.

and $\tilde{\mathbf{h}}_{apt}[n, l] \in \mathbb{C}^{N_{tap}^{[t]} \times 1}$ is the sample vector containing the current and the previous ($N_{tap}^{[t]} -$ 1) *a posteriori* tap estimates at the l-th tap position:

$$\tilde{\mathbf{h}}_{apt}[n, l] = \left(\tilde{h}_{apt}[n, l], \tilde{h}_{apt}[n - 1, l], \ldots, \tilde{h}_{apt}[N_{tap}^{[t]} - 1, l] \right)^T . \tag{15.23}$$

The complex error between the true value $h[n + 1, l]$ of the l-th CIR-related tap with respect to the unitary transform matrix $\mathbf{U}^{[f]}$ and its predicted value $\hat{h}_{apr}[n + 1, l]$ can be expressed as:

$$e_{pre}[n + 1, l] = h[n + 1, l] - \hat{h}_{apr}[n + 1, l]. \tag{15.24}$$

The two most prominent methods of inferring the vector of optimum CIR-related tap predictor coefficients [8] are constituted by the *gradient approach* [90, 472] and the application of the *orthogonality principle* [90, 472].

15.2.4.2 Definition of the CIR-related Taps' Auto-Correlation Matrix and Cross-Correlation Vector

It is convenient to define the auto-correlation matrix $\mathbf{R}_{apt}^{[t]}[l] \in \mathbb{C}^{N_{tap}^{[t]} \times N_{tap}^{[t]}}$ of the CIR-related *a posteriori* tap estimates as well as the cross-correlation vector $\mathbf{r}_{apt}^{[t]}[l] \in \mathbb{C}^{N_{tap}^{[t]} \times 1}$ between the "true" CIR-related tap in the $(n + 1)$-th timeslot and the CIR-related *a posteriori* tap estimates of the $N_{tap}^{[t]}$ number of previous timeslots at the l-th tap position in advance. More specifically, we have:

$$\mathbf{R}_{apt}^{[t]}[l] = E\left\{ \tilde{\mathbf{h}}_{apt}[n, l]\tilde{\mathbf{h}}_{apt}^H[n, l] \right\} \tag{15.25}$$

$$= \mathbf{R}_h^{[t]}[l] + \rho \mathbf{I} \tag{15.26}$$

$$= \lambda_l^{[f]} \mathbf{R}^{[t]}[l] + \rho \mathbf{I}. \tag{15.27}$$

In Equation 15.26 $\mathbf{R}_h^{[t]}[l] = E\{\mathbf{h}[n, l]\mathbf{h}^H[n, l]\} \in \mathbb{C}^{N_{tap}^{[t]} \times N_{tap}^{[t]}}$ denotes the l-th CIR-related tap's spaced-time auto-correlation matrix of the "true" CIR-related taps as hosted by the vector $\mathbf{h}[l] \in \mathbb{C}^{N_{tap}^{[t]} \times 1}$. By contrast, $\mathbf{R}^{[t]}[l]$ of Equation 15.27 denotes the normalised spaced-time auto-correlation matrix,[9] where the normalisation has been performed with respect to the l-th eigenvalue $\lambda_l^{[f]}$ of the spaced-frequency correlation matrix $\mathbf{R}^{[f]}$. Furthermore, the

---

[8]Unless otherwise stated, for the sake of conciseness we will refer to the CIR-related tap predictor coefficients simply as predictor coefficients.

[9]Note that $\mathbf{R}^{[t]}[l] = (1/\lambda_l^{[f]})\mathbf{R}_h^{[t]}[l]$. As a result of the normalisation the main diagonal elements of $\mathbf{R}^{[t]}[l]$ are unity. In case of channels having a separable spaced-time spaced-frequency correlation function $r_H(\Delta t, \Delta f) = r_H(\Delta t) \cdot r_H(\Delta f)$ as outlined in Section 14.2.6, the different CIR-related taps' normalised spaced-time correlation matrices are identical, namely $\mathbf{R}^{[t]} = \mathbf{R}^{[t]}[0] = \ldots = \mathbf{R}^{[t]}[K-1]$ and specifically they contain the samples of the normalised spaced-time correlation function $r_H(\Delta t)$. Specifically we have $\mathbf{R}^{[t]}|_{n_1, n_2} = r_H^{[t]*}(\Delta n T_f)$, $\Delta n = n_1 - n_2$.

cross-correlation vector $\mathbf{r}_{apt}^{[t]}[l] \in \mathbb{C}^{N_{tap}^{[t]} \times 1}$ is defined as:

$$\mathbf{r}_{apt}^{[t]}[l] = E\left\{h^*[n+1,l]\tilde{\mathbf{h}}_{apt}[n,l]\right\} \tag{15.28}$$

$$= \mathbf{r}_h^{[t]}[l] \tag{15.29}$$

$$= \lambda_l^{[f]}\mathbf{r}^{[t]}[l]. \tag{15.30}$$

In Equation 15.29 $\mathbf{r}_h^{[f]}[l] = E\{h^*[n+1,l]\mathbf{h}[l]\} \in \mathbb{C}^{N_{tap}^{[t]} \times 1}$ denotes the l-th CIR-related tap's spaced-time cross-correlation vector. By contrast, in Equation 15.30 $\mathbf{r}^{[t]}[l] \in \mathbb{C}^{N_{tap}^{[t]} \times 1}$ denotes the l-th CIR-related tap's normalised spaced-time cross-correlation vector.[10]

The difference between the CIR-related tap *predictor* considered here and the CIR-related tap *estimator* advocated by Li *et al.* [68] resides in the structure of the cross-correlation vector. Specifically, as seen in Equation 15.28, the *a priori* predictor for the l-th CIR-related tap of the $(n + 1)$-th transmission timeslot capitalises on an estimate of the cross-correlation between the CIR-related tap expected during the $(n + 1)$-th transmission timeslot and that of the current, namely the n-th, as well as the $(N_{tap}^{[t]} - 1)$ previous transmission timeslots. By contrast, the improved *a posteriori* estimator for the l-th CIR-related tap of the current, namely the n-th, transmission timeslot capitalises on the cross-correlation between the CIR-related tap expected during the n-th transmission timeslot and that of the n-th as well as the $(N_{tap}^{[t]} - 1)$ previous transmission timeslots. In the next paragraph Equations 15.27 and 15.30 will be invoked for deriving the optimum predictor coefficients with the aid of the gradient approach and by applying the orthogonality principle.

15.2.4.3   Derivation of the Wiener Equation using the Gradient Approach or the Orthogonality Principle

15.2.4.3.1   Gradient Approach   Upon invoking the complex estimation error expression $e[n + 1, l]$ of Equation 15.24 the squared estimation error is formed, and its mean value is given by:

$$J(\mathbf{c}_{pre}[l]) = \text{MSE}(\mathbf{c}_{pre}[l]) \tag{15.31}$$

$$= E\{|e_{pre}[n+1,l]|^2\} \tag{15.32}$$

$$= \lambda_l^{[f]} - \mathbf{c}_{pre}^T[l]\mathbf{r}_{apt}^{[t]*}[l] - \mathbf{c}_{pre}^H[l]\mathbf{r}_{apt}^{[t]}[l] +$$

$$+ \mathbf{c}_{pre}^H[l]\mathbf{R}_{apt}^{[t]}[l]\mathbf{c}_{pre}[l]. \tag{15.33}$$

Recall that the auto-correlation matrix $\mathbf{R}_{apt}^{[t]}[l]$ and the cross-correlation vector $\mathbf{r}_{apt}^{[t]}[l]$ were defined by Equations 15.25 and 15.28. Furthermore note that $\lambda_l^{[f]} = E\{|h[n+1,l]|^2\}$.

The vector $\mathbf{c}_{pre}[l]|_{opt}$ of optimum predictor coefficients is identified by minimising the value of the cost-function $J(\mathbf{c}_{pre}[l])$. Hence the gradient of $J(\mathbf{c}_{pre}[l])$ given by Equation 15.33 with respect to the conjugate complex vector $\mathbf{c}_{pred}^*[l]$ of predictor coefficients

---

[10]In case of channels having a separable spaced-time spaced-frequency correlation function $r_H(\Delta t, \Delta f) = r_H(\Delta t) \cdot r_H(\Delta f)$ as outlined in Section 14.2.6, the different CIR-related taps' normalised spaced-time cross-correlation vectors are identical, namely $\mathbf{r}^{[t]} = \mathbf{r}^{[t]}[0] = \ldots = \mathbf{r}^{[t]}[K-1]$ and specifically they contain the samples of the normalised spaced-time correlation function $r_H(\Delta t)$. Specifically we have $\mathbf{r}^{[t]}|_{n_1} = r_H^{[t]*}((1+n_1)T_f)$.

is set to zero, which is expressed as:

$$\mathbf{c}_{pre}[l]|_{opt} = arg\,\min_{\mathbf{c}_{pre}[l]} \left(J(\mathbf{c}_{pre}[l]) \right) \quad \longleftrightarrow \quad \left.\frac{\partial J(\mathbf{c}_{pre}[l])}{\partial \mathbf{c}_{pre}^*[l]}\right|_{\mathbf{c}_{pre}[l]|_{opt}} \overset{!}{=} 0. \qquad (15.34)$$

Hence, upon invoking Equation 15.34 together with Equation 15.33 the set of $N_{tap}^{[t]}$ number of Wiener-Hopf equations [90,471,472] is obtained, which is expressed in vectorial notations as:

$$\frac{\partial J(\mathbf{c}_{pre}[l])}{\partial \mathbf{c}_{pre}^*[l]} = -\mathbf{r}_{apt}^{[t]}[l] + \mathbf{R}_{apt}^{[t]}[l]\mathbf{c}_{pre}[l]|_{opt} \overset{!}{=} 0, \qquad (15.35)$$

from which the vector $\mathbf{c}_{pre}[l]|_{opt}$ of optimum predictor coefficients can be inferred.

15.2.4.3.2 Orthogonality Principle

The orthogonality principle of MMSE optimisation implies that the inner product related norm of the difference- or error vector between any vector contained in a vector space S and its projection onto a sub-space spanned by a specific number of data vectors is minimised in the sense of the expectation value, provided that the error vector is orthogonal to the above-mentioned sub-space [90]. Hence, in the context of the optimisation problem considered here we obtain a set of $N_{tap}^{[t]}$ equations:

$$E\left\{ e_{pre}[n+1,l]\tilde{h}_{apt}^*[n+\acute{n},l] \right\} = 0, \quad \acute{n} = -N_{tap}^{[t]} + 1, \dots, 0. \qquad (15.36)$$

Equation 15.36 can be readily transferred into the form of Equation 15.35 by capitalising on the definition of the tap prediction error signal $e_{pre}[n+1,l]$ given by Equation 15.24 as well as on that of the predicted signal given by Equation 15.20 while invoking the definition of the auto-correlation matrix $\mathbf{R}_{apt}^{[t]}[l]$ and that of the cross-correlation matrix $\mathbf{r}_{apt}^{[t]}[l]$ given by Equations 15.27 and 15.30, respectively, yielding:

$$-\mathbf{r}_{apt}^{[t]}[l] + \mathbf{R}_{apt}^{[t]}[l]\mathbf{c}_{pre}[l]|_{opt} \overset{!}{=} 0. \qquad (15.37)$$

15.2.4.4 Optimum Predictor Coefficients and Minimum CIR-Related Domain Predictor MSE

Provided that the inverse of the auto-correlation matrix $\mathbf{R}_{apt}^{[t]}[l]$ exists, which requires that $\mathbf{R}_{apt}^{[t]}[l]$ is of full rank [90], Equations 15.35 or 15.37 can be uniquely solved with respect to the vector $\mathbf{c}_{pre}[l]|_{opt}$ of optimum predictor coefficients associated with the l-th CIR-related tap:

$$\mathbf{c}_{pre}[l]|_{opt} = \mathbf{R}_{apt}^{[t]-1}[l]\mathbf{r}_{apt}^{[t]}[l]. \qquad (15.38)$$

Hence by substituting the vector $\mathbf{c}_{pre}[l]|_{opt}$ of optimum predictor coefficients into the MSE expression of Equation 15.33, an equation is obtained for the minimum MSE (MMSE):

$$\text{MMSE}_{apr}[l] = \text{MSE}_{apr}[l]|_{(\mathbf{c}_{pre}[l]|_{opt})} \tag{15.39}$$

$$= \lambda_l^{[f]} - \mathbf{c}_{pre}^H[l]|_{opt}\mathbf{r}_{apt}^{[t]}[l]. \tag{15.40}$$

15.2.4.5   Optimum Predictor Coefficients for Mismatched Channel Statistics

In the context of deriving the vector $\mathbf{c}_{pre}[l]|_{opt}$ of optimum predictor coefficients given by Equation 15.38 we have implicitly assumed perfect knowledge of the specific spaced-time correlation function associated with the l-th CIR-related tap, which appeared in the form of the auto-correlation matrix $\mathbf{R}^{[t]}[l]$ in Equation 15.27 and in the form of the cross-correlation vector $\mathbf{r}^{[t]}[l]$ in Equation 15.30. We also assumed perfect knowledge of the spaced-frequency correlation matrix $\mathbf{R}^{[f]}$, namely the knowledge of its EVD-related representation $\mathbf{R}^{[f]} = \mathbf{U}^{[f]}\Lambda^{[f]}\mathbf{U}^{[f]H}$ associated with the unitary transform matrix $\mathbf{U}^{[f]}$ and the diagonal matrix $\Lambda^{[f]}$ hosting the eigenvalues $\lambda_l^{[f]}$, $l = 0,\ldots,K-1$. Normally this knowledge is not explicitly available and hence the optimum coefficients are based on the corresponding "estimates" $\tilde{\mathbf{R}}^{[t]}[l]$, $\tilde{\mathbf{r}}^{[t]}[l]$ and $\tilde{\mathbf{R}}^{[f]}$, where $\tilde{\mathbf{R}}^{[f]}$ can be decomposed according to $\tilde{\mathbf{R}}^{[f]} = \tilde{\mathbf{U}}^{[f]}\tilde{\Lambda}^{[f]}\tilde{\mathbf{U}}^{[f]H}$. We also need the *estimates* of the auto-correlation matrix $\tilde{\mathbf{R}}_{apt}^{[t]}[l]$ of the *a posteriori* CIR-related taps, which can be formulated by following the philosophy of Equation 15.27, yielding:

$$\tilde{\mathbf{R}}_{apt}^{[t]}[l] = \tilde{\lambda}_l^{[f]}\tilde{\mathbf{R}}^{[t]}[l] + \tilde{\rho}\mathbf{I}, \tag{15.41}$$

where also a potentially imperfect estimate $\tilde{\rho}$ of the reciprocal value $\rho = 1/\text{SNR}$ of the SNR measured at the reception antenna has been invoked. Similarly, the cross-correlation vector $\tilde{\mathbf{r}}_{apt}^{[t]}[l]$ of *a posteriori* CIR-related tap estimates can be expressed following the philosophy of Equation 15.30:

$$\tilde{\mathbf{r}}_{apt}^{[t]}[l] = \tilde{\lambda}_l^{[f]}\tilde{\mathbf{r}}^{[t]}[l]. \tag{15.42}$$

According to Equation 15.38 the vector $\tilde{\mathbf{c}}_{pre}[l]|_{opt}$ of sub-optimum predictor coefficients is hence given by:

$$\tilde{\mathbf{c}}_{pre}[l]|_{opt} = \tilde{\mathbf{R}}_{apt}^{[t]-1}[l]\tilde{\mathbf{r}}_{apt}^{[t]}[l]. \tag{15.43}$$

Furthermore, the CIR-related domain MSE achieved by this predictor in the context of an arbitrary vector $\tilde{\mathbf{c}}_{pre}[l]$ of coefficients is given similarly to Equation 15.33 by:

$$\text{MSE}_{apr}[l]|_{(\tilde{\mathbf{c}}_{pre}[l])} = v_l^{[f]} - \tilde{\mathbf{c}}_{pre}^T[l]\mathbf{r}_{apt}^{[t]*}[l]|_{\tilde{\mathbf{U}}^{[f]}} - \tilde{\mathbf{c}}_{pre}^H[l]\mathbf{r}_{apt}^{[t]}[l]|_{\tilde{\mathbf{U}}^{[f]}} +$$

$$+ \tilde{\mathbf{c}}_{pre}^H[l]\mathbf{R}_{apt}^{[t]}[l]|_{\tilde{\mathbf{U}}^{[f]}}\tilde{\mathbf{c}}_{pre}[l], \tag{15.44}$$

where the auto-correlation matrix $\mathbf{R}_{apt}^{[t]}[l]|_{\tilde{\mathbf{U}}^{[f]}}$ of the l-th CIR-related tap conditioned on employing the unitary transform matrix $\tilde{\mathbf{U}}^{[f]}$ is defined similarly to Equation 15.27 by:

$$\mathbf{R}_{apt}^{[t]}[l]|_{\tilde{\mathbf{U}}^{[f]}} = v_l^{[f]} \mathbf{R}^{[t]}[l]|_{\tilde{\mathbf{U}}^{[f]}} + \rho \mathbf{I}|. \tag{15.45}$$

By the same token the cross-correlation vector $\mathbf{r}_{apt}^{[t]}[l]|_{\tilde{\mathbf{U}}^{[f]}}$ is defined similarly to Equation 15.30 by:

$$\mathbf{r}_{apt}^{[t]}[l]|_{\tilde{\mathbf{U}}^{[f]}} = v_l^{[f]} \mathbf{r}^{[t]}[l]|_{\tilde{\mathbf{U}}^{[f]}}. \tag{15.46}$$

The variable $v_l^{[f]}$ was defined earlier in Section 15.2.2.2 as the variance of the l-th CIR-related tap, which can be expressed as the l-th diagonal element of the decomposition of $\mathbf{R}^{[f]}$ with respect to the unitary transform matrix $\tilde{\mathbf{U}}^{[f]}$, namely as $v_l^{[f]} = (\tilde{\mathbf{U}}^{[f]H} \mathbf{R}^{[f]} \tilde{\mathbf{U}}^{[f]})_{[l,l]}$. Again, note that if $\tilde{\mathbf{U}}^{[f]} = \mathbf{U}^{[f]}$ holds, we have $v_l^{[f]} = \lambda_l^{[f]}$. Clearly, the predictor's MSE in Equation 15.44 is equal to or higher than the MMSE of Equation 15.40, since the latter assumes perfect knowledge of the channel statistics in the calculation of the predictor coefficients, yielding:

$$\mathrm{MSE}_{apr}[l]|_{(\tilde{\mathbf{c}}_{pre}[l]|_{opt})} \geq \mathrm{MMSE}_{apr}[l]. \tag{15.47}$$

Note again that the predictor MSE values delivered by Equations 15.40 and 15.44 for the cases of matched and mismatched channel statistics, respectively, are valid for the CIR-related domain, observed on a per tap basis. In practice, the average predictor MSE evaluated in the frequency domain is, however, of more importance for the system's BER performance.

15.2.4.6 Average Channel Predictor MSE in the Frequency Domain

Following the notation of [69], below we will characterise the OFDM system's symbol-averaged frequency domain *a priori* estimation MSE. A rank-K_0 predictor is used, where only the first K_0 CIR-related tap values are retained as shown in Figures 15.2 and 15.4, while the CIR-related tap values associated with a higher index are set to zero for the sake of reducing the predictor's computational complexity. The corresponding MSE is given by:

$$\overline{\mathrm{MSE}}_{apr}|_{K_0} = \frac{1}{K} \mathrm{Trace}\left(E\{\mathbf{E}_{pre}[n+1]|_{K_0} \mathbf{E}_{pre}^{H}[n+1]|_{K_0}\} \right), \tag{15.48}$$

where the vector $\mathbf{E}_{pre}[n+1]|_{K_0} \in \mathbb{C}^{K \times 1}$ of frequency domain transfer factor error signals was generated with the aid of the KLT-matrix $\tilde{\mathbf{U}}^{[f]}$ from the vector $\mathbf{e}_{pre}[n+1]|_{K_0} \in \mathbb{C}^{K \times 1}$ of CIR-related domain error signals as follows:

$$\mathbf{E}_{pre}[n+1]|_{K_0} = \tilde{\mathbf{U}}^{[f]} \mathbf{e}_{pre}[n+1]|_{K_0}. \tag{15.49}$$

Recall that $\tilde{\mathbf{U}}^{[f]} \in \mathbb{C}^{K \times K}$ is the matrix of eigenvectors associated with the EVD of the spaced-frequency correlation matrix $\tilde{\mathbf{R}}^{[f]} = \tilde{\mathbf{U}}^{[f]} \tilde{\Lambda}^{[f]} \tilde{\mathbf{U}}^{[f]H}$, invoked in the calculation of the predictor coefficients. The individual components of the vector $\mathbf{e}_{pre}[n+1]|_{K_0}$ of CIR-related domain error signals were given in Equation 15.24, as the difference between the "true" CIR-related tap values with respect to the transform $\tilde{\mathbf{U}}^{[f]}$ and its predictions, where again, in the context of a rank-K_0 predictor, the predicted CIR-related taps for $l = K_0, \ldots, K-1$ are set to zero. Upon substituting Equation 15.49 into Equation 15.48 we obtain:

$$\overline{\mathrm{MSE}}_{apr}|_{K_0} = \frac{1}{K}\mathrm{Trace}\left(\tilde{\mathbf{U}}^{[f]}E\{\mathbf{e}_{pre}[n{+}1]|_{K_0}\mathbf{e}_{pre}^H[n{+}1]|_{K_0}\}\tilde{\mathbf{U}}^{[f]H}\right) \quad (15.50)$$

$$= \frac{1}{K}\mathrm{Trace}\left(E\{\mathbf{e}_{pre}[n{+}1]|_{K_0}\mathbf{e}_{pre}^H[n{+}1]|_{K_0}\}\right), \quad (15.51)$$

where in Equation 15.51 we have exploited that $\mathrm{Trace}\left(\mathbf{U}\mathbf{A}\mathbf{U}^H\right) = \mathrm{Trace}(\mathbf{A})$ for any unitary matrix $\mathbf{U}$ [69, 478]. Note that the first K_0 diagonal elements of the matrix $E\{\mathbf{e}_{pre}[n{+}1]|_{K_0}\mathbf{e}_{pre}^H[n{+}1]|_{K_0}\}$ are equal to the CIR-related tap predictors' MSE values given by Equation 15.44, while the diagonal elements associated with a higher index are equal to the CIR-related tap values' variances - conditioned on employing the unitary transform matrix $\tilde{\mathbf{U}}^{[f]}$. This is, because for these taps associated with indices $l = K_0 \ldots, K-1$ the *a priori* predictions $\hat{h}_{apr}[n{+}1, l]$ are not available and therefore we have $E\{|e_{pre}[n{+}1, l]|^2\} = E\{|h[n+1, l]|^2\} = v_l^{[f]}$. Hence, Equation 15.51 can be reformulated as:

$$\overline{\mathrm{MSE}}_{apr}|_{K_0} = \frac{1}{K}\sum_{l=0}^{K_0-1}\mathrm{MSE}_{apr}[l]|_{(\tilde{\mathbf{c}}_{pre}[l]|_{opt})} + \frac{1}{K}\sum_{l=K_0}^{K-1}v_l^{[f]}. \quad (15.52)$$

Since in general perfect knowledge of the channel statistics is not available, we will highlight in Section 15.2.5 the philosophy of a number of potential strategies that can be invoked for providing estimates of the channel statistics, including Li's approach [68] to render the predictor as robust as possible.

15.2.5 Channel Statistics for *A Priori* Time Direction Channel Prediction Filtering

As we observed in Equation 15.43, evaluating the l-th CIR-related tap's prediction filter coefficients requires an estimate $\tilde{\mathbf{R}}_{apt}^{[t]}[l]$ and $\tilde{\mathbf{r}}_{apt}^{[t]}[l]$ of this tap's auto-correlation matrix and cross-correlation vector, respectively. To elaborate a little further, these correlation-related quantities could be calculated according to Equations 15.27 and 15.30 upon stipulating the availability of an estimate of the spaced-frequency correlation matrix's l-th eigenvalue $\tilde{\lambda}_l^{[f]}$, of the reciprocal of the SNR $\tilde{\rho} = 1/\tilde{\mathrm{SNR}}$ at the reception antenna, and that of the channel's spaced-time correlation function, $\tilde{r}_H^{[t]}[\Delta n]$. These estimates will be provided in Section 15.2.5.1 upon invoking again the concepts of robustness to channel statistics variations, as discussed by Li *et al.* [68] in the context of DDCE. On the other hand, the l-th CIR-related tap's auto-correlation matrix and cross-correlation vector could also be assembled from the auto-correlation function of this tap, which can be estimated on the basis of M_s number of CIR-related tap values. Before continuing our discourse we note that these issues will be further detailed in Section 15.2.5.2.

15.2.5.1 Robust *A Priori* Time Direction Channel Prediction Filtering

While in Section 15.2.5.1.1 we will briefly revisit the fundamental concepts of a robust channel estimator introduced by Li *et al.* in [68], in Section 15.2.5.1.2 we will offer a number of conclusions assisting in the design of a robust channel predictor.

15.2.5.1.1 Review of Robust Channel Estimation
In Section 15.2.4 we highlighted the concept of a robust channel estimator as proposed by Li *et al.* [68]. The conclusion was that an improved *a posteriori* channel transfer function estimator, which capitalises on an *infinite number* of previous initial *a posteriori* channel transfer function estimates can be rendered insensitive to the exact channel statistics. Arriving at such a robust design requires first of all that the spaced-time correlation function $\tilde{r}_H^{[t]}[\Delta n]$ assumed in the calculation of the channel transfer function estimator's coefficients derived for the time direction filter is ideally limited to the frequency-band of ω_D. Second, it was required that the spaced-frequency correlation matrix $\tilde{\mathbf{R}}^{[f]}$ of the channel transfer function estimator can be eigen-decomposed, such that $\tilde{\mathbf{R}}^{[f]} = \tilde{\mathbf{U}}^{[f]} \tilde{\Lambda}^{[f]} \tilde{\mathbf{U}}^{[f]H}$, where the matrix $\tilde{\mathbf{U}}^{[f]}$ of eigenvectors is identical - with tolerable leakage [68] - to the exact matrix $\mathbf{U}^{[f]}$ of eigenvectors associated with the "true" spaced-frequency correlation matrix $\mathbf{R}^{[f]}$ of the channel. We note here that both of the matrices $\tilde{\mathbf{U}}^{[f]}$ and $\mathbf{U}^{[f]}$ are unitary and can in turn be approximated by the DFT matrix $\mathbf{W}$ as we already argued in the context of Section 15.2.3.2. Furthermore, the diagonal matrix $\tilde{\Lambda}^{[f]}$ hosts K_0 identical eigenvalues of $\tilde{\lambda}^{[f]} = K/K_0$ found in the first K_0 diagonal positions of the matrix, while the remaining $(K - K_0)$ number of eigenvalues are equal to zero. Under these conditions the channel estimator's MSE performance is channel-invariant for the specific subset of all channels, which have a spaced-time correlation function $r_H^{[t]}[\Delta n]$ that is limited to the frequency-band of ω_D but not necessarily ideally low-pass shaped and with an arbitrary spaced-frequency correlation function $r_H^{[f]}[\Delta k]$, provided that the related spaced-frequency correlation matrix $\mathbf{R}^{[f]}$ can be eigen-decomposed such that the energy conveyed by the channel is mapped to the K_0 lowest eigenvalues.[11] In this case the *a posteriori* channel estimator's MSE is given by Equation 15.11. It was furthermore demonstrated in [68] and also argued in Section 15.2.4 that in a robust estimator design identical sets of predictor coefficients[12] are associated with the different CIR-related taps' time direction filters. The schematic of the corresponding channel estimator was shown at the bottom of Figure 15.4. This also implies that the estimation filter associated with each CIR-related tap is designed for the same maximum Doppler frequency.

15.2.5.1.2 Design of the Auto-Correlation Matrix and Cross-Correlation Vector of a Robust Channel Predictor
The same concepts of robustness can also be applied to the *a priori* channel predictor advocated here, although in contrast to the *a posteriori* channel estimator of [68] an estimate of the channel transfer function is obtained for the $(n + 1)$-th time slot instead of generating an improved estimate for the current timeslot, namely for the n-th timeslot. More specifically, as highlighted above, identical eigenvalues of $\tilde{\lambda}_l^{[f]} = K/K_0$,

---

[11]Note that in the context of employing the DFT matrix $\mathbf{W}$ of Equation 15.13 as the unitary transform matrix $\tilde{\mathbf{U}}^{[f]}$, this is only true for sample-spaced CIRs.

[12]A prerequisite was the separability of the channel's spaced-time spaced-frequency correlation function, which was discussed in Section 14.2.6.

$l = 0, \ldots, K_0 - 1$ are employed in the calculation of the K_0 number of CIR-related taps' predictor coefficients. Furthermore, the maximum Doppler frequency associated with each of the CIR-related taps' spaced-time correlation functions $\tilde{r}_H^{[t]}[\Delta n]$ is assumed to be identical. Hence identical sets of predictor coefficients are obtained for the different CIR-related taps, which allows us to drop the index l of the specific CIR-related tap considered. Based on Equation 15.41 the auto-correlation matrix $\tilde{\mathbf{R}}_{apt}^{[t]}|_{rob}$ of *a posteriori* CIR-related tap estimates is thus given by:

$$\tilde{\mathbf{R}}_{apt}^{[t]}|_{rob} = \frac{K}{K_0} \tilde{\mathbf{R}}^{[t]}|_{rob} + \tilde{\rho}\mathbf{I}. \tag{15.53}$$

In Equation 15.53 the auto-correlation matrix $\tilde{\mathbf{R}}^{[t]}|_{rob}$ of CIR-related tap estimates is defined in the absence of noise on the basis of the spaced-time correlation function $r_H^{[t]}[\Delta n]|_{rob}$ associated with an ideal low-pass shaped PSD having a cut-off frequency of ω_D, which results in the sinc-shaped correlation function of:

$$\tilde{r}_H^{[t]}[\Delta n]|_{rob} = E\{H[n - n_1]H^*[n - n_2]\} \tag{15.54}$$

$$= \mathrm{sinc}(\Delta n \cdot \omega_D), \tag{15.55}$$

where $\Delta n = n_2 - n_1$ and $\omega_D = 2\pi F_D$. We note that in the case of a slot-by-slot TDD system each downlink transmission time slot is followed by an uplink transmission time slot and vice versa. Hence the effective temporal distance between consecutive OFDM symbols is doubled in comparison to that of a scheme having bursts of multiple consecutive downlink timeslots. This has to be considered in the evaluation of the predictor-related correlation matrices with the aid of Equation 15.55. Upon substituting $\tilde{\mathbf{R}}_{apt}^{[t]}|_{rob}$ and $\tilde{\mathbf{r}}_{apt}^{[t]}|_{rob} = \tilde{\mathbf{r}}^{[t]}|_{rob}$ into Equation 15.43 the set of $N_{tap}^{[t]}$ predictor coefficients associated with the robust CIR-related tap predictor is obtained.

We note furthermore that in case of the finite-order predictors as considered here, the properties of robustness with respect to the specific shape of the channel's Doppler power density spectrum - as highlighted for the case of an infinite-order estimator - are only approximately valid. This will be further elaborated on during our performance assessment of the various techniques in Section 15.3.

15.2.5.2   Adaptive *A Priori* Time Direction Channel Prediction Filtering

In the absence of any *a priori* knowledge concerning the channel's statistics, such as for example the maximum Doppler frequency of the channel, the CIR-related tap predictors' coefficients must be calculated on the basis of a finite set of for example M_s number of past OFDM symbols' *a posteriori* CIR-related tap estimates $\tilde{h}_{apt}[n - \acute{n}, l]$, $\acute{n} = 0, \ldots, M_s - 1$, $l = 0, \ldots, K - 1$. An intuitive method of determining the l-th CIR-related tap's correlation values $\tilde{r}_{\tilde{h}_{apt}}^{[t]}[\Delta n, l]$ constituting the auto-correlation matrix $\tilde{\mathbf{R}}_{apt}^{[t]}[n, l]$ and the cross-correlation vector $\tilde{\mathbf{r}}_{apt}^{[t]}[n, l]$ defined in Equations 15.25 and 15.28, respectively, is given by [89]:

$$\tilde{r}^{[t]}_{\tilde{h}_{apt}}[\Delta n, l] = \frac{1}{M_s - \Delta n} \sum_{\acute{n}=0}^{M_s - \Delta n} \tilde{h}_{apt}[n - \acute{n}, l]\tilde{h}^*_{apt}[(n - \Delta n) - \acute{n}, l], \qquad (15.56)$$

where the calculation is performed for the n-th transmission time slot on the basis of the *a posteriori* CIR-related tap samples of the current plus the previous $M_s - 1$ time slots.

From the literature of predictive speech coding [85] for example, its known that the accuracy of linear prediction is extremely sensitive to the method of estimating the correlation values $\tilde{r}^{[t]}_H[\Delta n, l]$. A number of algorithms have been proposed in the literature for determining the predictor coefficients, such as the Levinson-Durbin algorithm and the related Burg algorithm [85, 89]. The employment of these algorithms in the context of channel transfer function prediction for OFDM has also been investigated by Al-Susa and Ormondroyd in [72]. In the context of employing an adaptive predictor instead of updating the CIR-related tap predictors' coefficients on a block basis, they could also be updated on an OFDM symbol-by-symbol basis with the aid of the RLS-algorithm [90]. This approach will be investigated in the context of PIC-assisted DDCE designed for multi-user OFDM systems in Section 16.6. Let us now embark in the next section on the characterisation of both the robust- and the adaptive channel transfer function predictor in terms of their MSE performance.

15.3 Performance of Decision-Directed Channel Prediction Aided-OFDM Using Fixed Modulation

In the context of our performance assessment presented here the indoor WATM channel and system model of Section 14.3.1 is invoked. Specifically, due to the sample-spaced nature of the indoor WATM channel's associated CIR, the DFT can be employed for perfectly decorrelating the least-squares channel transfer factor estimates with respect to the frequency direction. To elaborate a little further, the EigenValue Decomposition (EVD) of the sample-spaced channel's spaced-frequency correlation matrix given by $\mathbf{R}^{[f]} = \mathbf{U}^{[f]}\Lambda^{[f]}\mathbf{U}^{[f]H}$ is such that the EVD-related unitary transform matrix $\mathbf{U}^{[f]}$ is identical to the DFT matrix $\mathbf{W}$ of Equation 15.13 and the eigenvalues found on the main diagonal of $\Lambda^{[f]}$ are given by the CIR tap variances.

The structure of Section 15.3 is as follows. In Section 15.3.1 we will assess the MSE performance of a decision-directed channel predictor, which adopted the philosophy of the robust approach proposed by Li *et al.* in [68], while in Section 15.3.2 its MSE is contrasted to that of the adaptive decision-directed channel predictor proposed by Al-Susa and Ormondroyd in [72]. Furthermore, in Sections 15.3.3 and 15.3.4 the robust channel predictor's MSE performance, as well as the associated BER performance will be evaluated in the context of a system using no channel coding, upon invoking sliced and hence potentially error-prone symbol decisions in the process of decision-directed channel prediction. By contrast, in Section 15.3.5 we will evaluate the BER performance of a turbo-coded system, which employs robust decision-directed channel prediction.

$K_0, \tilde{K}_0$	Sampling period duration normalised multipath spread
$F_D, \tilde{F}_D$	OFDM symbol normalised Doppler frequency
$\rho, \tilde{\rho}$	Reciprocal of the SNR
$N_{tap}^{[t]}$	Length or range of the CIR-related tap prediction filter
K	Number of subcarriers per OFDM symbol

Table 15.4: Summary of the parameters influencing the MSE performance of the *robust* decision-directed channel predictor, which follows the concepts of the *robust* decision-directed channel estimator of Figure 15.4. Error-free symbol decisions and the absence of ICI are assumed in the corresponding experiments. The parameters associated with the channel predictor are distinguished from those of the channel by $\tilde{(\,)}$.

15.3.1 MSE Performance of a Robust Decision-Directed Channel Predictor in the Context of Error-Free Symbol Decisions

As a consequence of the sample-spaced nature of the indoor WATM channel model's associated CIR employed here, according to our arguments presented concerning the robust channel estimator in Sections 15.2.3.2 and 15.2.5.1.1, an identical MSE performance is expected for any other sample-spaced CIR, provided that the multipath delay spread of the CIR normalised to the sampling period duration, namely $K_0 = T_m/T_s$ is lower than that used in our simulations. This is because for a sample-spaced CIR all of the channel's energy is mapped to the K_0 number of significant CIR-related taps' output by the inverse unitary linear transform employed for generating these taps from the frame of frequency domain channel transfer function samples. By contrast, this is only approximately valid for non-sample-spaced channels, as will be demonstrated in the context of our investigations of DDCE for multi-user OFDM in Chapter 16.

In the context of the simulation results presented in Section 15.3.1 a spaced-time correlation function $r_H^{[t]}[\Delta n]$ as given by Equation 15.55, which follows the concepts of robustness outlined in Section 15.2.5.1.2 was associated with the channel predictor, while a spaced-time correlation function obeying Jakes' model [474] was associated with the specific transmission channel actually encountered.

The further structure of Section 15.3.1 is as follows. Our MSE performance assessment commences in Section 15.3.1.1 by considering a scenario, where the normalised multipath spread $\tilde{K}_0 = \tilde{T}_m/T_s$ and the OFDM symbol normalised Doppler frequency $\tilde{F}_D$, as well as the inverse of the SNR at the reception antenna, namely $\tilde{\rho}$, associated with the channel predictor are identical to that of the channel, namely to K_0, F_D and ρ, respectively. By contrast, in the context of our investigations to be outlined in Section 15.3.1.2 the OFDM symbol normalised Doppler frequency $\tilde{F}_D$ associated with the channel predictor, was fixed, while that associated with the specific channel encountered, namely, F_D, was varied across a range of values. Furthermore, in Section 15.3.1.3 the effects of a mismatch between the inverse of the SNR, $\tilde{\rho}$, associated with the channel predictor, and that of the channel, namely ρ are investigated. In Section 15.3.1.4 we continue our discourse by studying the effects of a misadjustment between the normalised multipath spread, $\tilde{K}_0$, which is associated with the channel predictor, and K_0, associated with the specific channel encountered. Finally, in Section 15.3.1.5 we offer some conclusions.

In Table 15.4 we have summarised again the specific parameters, which influence the MSE performance of the robust decision-directed channel predictor in the context of idealised error-free symbol decisions and in the absence of ICI. Our results related to potentially erroneous, sliced symbol decisions will be presented at a later stage in Section 15.3.3.

15.3.1.1 MSE Performance under Matched Channel Conditions

Initially we will assume employing "robust" prediction filter coefficients, which are perfectly matched to the specific channel conditions encountered. More specifically, the SNR experienced at the reception antenna is identical to that assumed in the calculation of the predictor coefficients, expressed in terms of the effective noise variance as $\rho = \tilde{\rho}$. Furthermore, the OFDM symbol normalised Doppler frequency assumed in the coefficients' calculation was identical to that of the channel, namely we had $F_D = \tilde{F}_D$.

In Figure 15.5 we have portrayed the MSE exhibited by the *a priori* channel predictor as a function of the SNR recorded at the reception antenna and experienced in different propagation scenarios characterised by their specific OFDM symbol normalised Doppler frequency denoted by F_D. Various CIR-related tap prediction filter lengths ranging between one and 64 taps were invoked. We observe that with the aid of a modest prediction filter length of approximately four to eight taps already most of the channel's correlation in the time direction can be exploited. Hence, an even further increase of the prediction filter's length results in no substantial additional channel prediction gain, only in further noise mitigation. This is because a higher-order prediction filter produces its output based on a higher number of input samples and hence averages the additive noise samples more efficiently, which results in a reduced noise variance. This phenomenon is reflected by the parallel nature of the MSE curves corresponding to a relatively high number of CIR-related tap prediction filter coefficients.

Similar conclusions can be drawn from Figure 15.6, where we have portrayed the MSE exhibited by the predictor as a function of the OFDM symbol normalised Doppler frequency for specific values of the SNR. Again, CIR-related tap prediction filter lengths ranging between one and 64 coefficients were invoked. At higher SNRs, where the Wiener filter-based predictor is only required to compensate for the decorrelation of the channel, rather than mitigating the receiver's residual AWGN, a tremendous beneficial impact is observed even in conjunction with prediction filters of a relatively low number of coefficients.

15.3.1.2 MSE Performance under Mismatched Channel Conditions with Respect to the Doppler Frequency

So far the robust channel predictor's *a priori* MSE performance has been investigated in the specific scenario, where its design parameters have been perfectly matched to the channel conditions encountered, as outlined at the beginning of this section. Let us now investigate the impact of a mismatch between the OFDM symbol normalised Doppler frequency F_D observed on the channel and that assumed in the calculation of the predictor coefficients, namely $F_{D,filter} = \tilde{F}_D$. Hence in Figure 15.7 we have portrayed the MSE of the channel predictor as a function of the OFDM symbol normalised Doppler frequency of the channel, while employing predictor coefficients optimised for a specific OFDM symbol normalised Doppler frequency $\tilde{F}_D$, that was potentially different from that of the channel encountered. For comparison we have also plotted the MSE performance in the context of employing the optimum

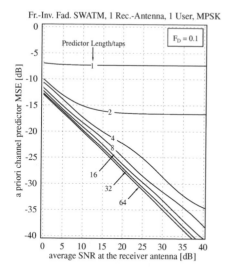

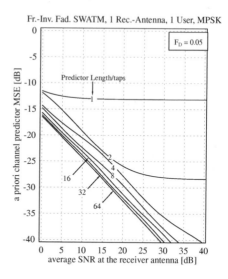

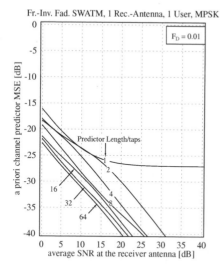

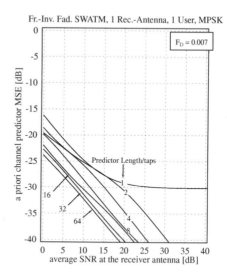

Figure 15.5: Mean-Square Error (MSE) exhibited by the **robust decision-directed _a priori_ channel predictor**, which follows the philosophy of Figure 15.4, as a function of the Signal-to-Noise Ratio (SNR) at the reception antenna. **Error-free symbol decisions** and prediction filter lengths of up to 64 taps were used in the **"frame-invariant"** fading **indoor WATM channel** environment of Figure 14.4 at OFDM symbol-normalised Doppler frequencies of $F_D = 0.1$ (**top left**), 0.05 (**top right**), 0.01 (**bottom left**) and 0.007 (**bottom right**); the variance of the AWGN was $\sigma_n^2 = 2$; the CIR window size was $K_0 = 12$ taps. The results were evaluated from Equation 15.52.

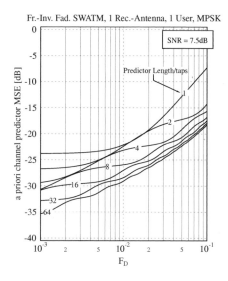

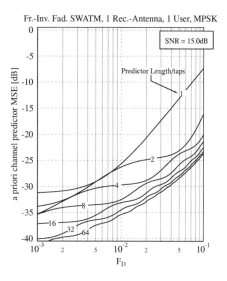

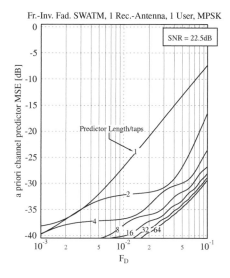

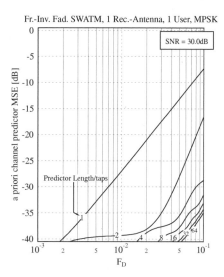

Figure 15.6: Mean-Square Error (MSE) exhibited by the **robust decision-directed** *a priori* **channel predictor**, which follows the philosophy of Figure 15.4, as a function of the OFDM symbol-normalised Doppler frequency F_D. **Error-free symbol decisions** and prediction filter lengths of up to 64 taps were used in the **"frame-invariant"** fading **indoor WATM channel** environment of Figure 14.4 at Signal-to-Noise Ratios (SNRs) of $7.5dB$ (**top left**), $15.0dB$ (**top right**), $22.5dB$ (**bottom left**) and $30.0dB$ (**bottom right**); the variance of the AWGN was $\sigma_n^2 = 2$; the CIR window size was $K_0 = 12$ taps. The results were evaluated from Equation 15.52.

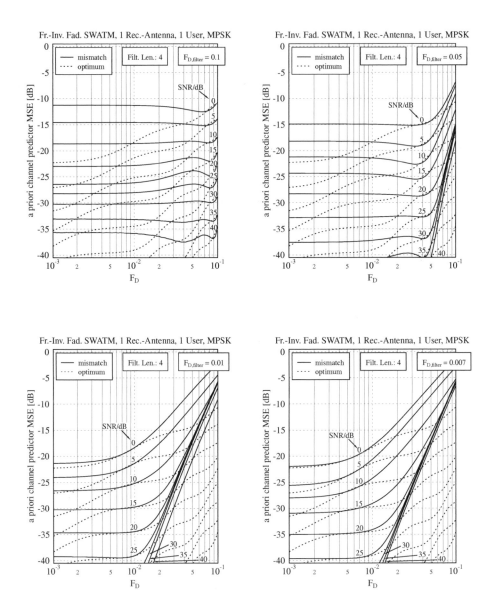

Figure 15.7: Mean-Square Error (MSE) exhibited by the **robust decision-directed *a priori* channel predictor**, which follows the philosophy of Figure 15.4, as a function of the channel's OFDM symbol-normalised Doppler frequency F_D. The results were recorded for different values of the Signal-to-Noise Ratio (SNR) at the reception antenna, using **error-free symbol decisions** and a CIR-related tap prediction filter length of 4 taps in the **"frame-invariant"** fading **indoor WATM channel** environment of Figure 14.4. Fixed OFDM symbol-normalised Doppler frequencies of $F_{D,filter} = \tilde{F}_D$ of 0.1 (**top left**), 0.05 (**top right**), 0.01 (**bottom left**) and 0.007 (**bottom right**) were assumed in the calculation of the CIR-related tap predictor's coefficients; the variance of the AWGN was $\sigma_n^2 = 2$; the CIR window size was $K_0 = 12$ taps. The results were evaluated from Equation 15.52.

"robust" CIR-related tap prediction filter coefficients, which match the channel conditions in terms of both the OFDM symbol normalised Doppler frequency and the SNR measured at the receiver antenna. A fixed CIR-related tap prediction filter length of four taps was assumed. The simulation results have additionally been parametrised with the SNR measured at the receiver antenna, which was assumed to match the SNR employed in the CIR-related tap's prediction filter coefficient calculation. As expected, in Figure 15.7 an MSE degradation is observed in case of encountering mismatched channel Doppler frequencies. An important characteristic of these MSE curves, which motivated the introduction of the terminology of "robust" estimator by Li *et al.* [68] in the context of infinite-order channel estimators is that only a marginal MSE degradation is observed under channel conditions, where the associated OFDM symbol normalised Doppler frequency F_D is lower than the Doppler frequency $\tilde{F}_D$ assumed in the CIR-related tap prediction filter's coefficient calculation, compared to the ideal case, when F_D matches $\tilde{F}_D$. We note that in the context of the finite-order prediction filters considered here, the insensitivity of the MSE to the Doppler frequency encountered is only approximate due to the filters' imperfections imposed by their finite order, which manifests itself in terms of a less selective frequency domain representation. By contrast, if the Doppler value F_D associated with the channel encountered exceeds the specific $\tilde{F}_D$ value associated with the predictor design, an MSE degradation is observed in comparison to the case, when both values are identical. This is, because significant spectral components are removed from the fading signal's frequency domain representation associated with each of the CIR-related taps.

15.3.1.3   MSE Performance under Mismatched Channel SNR Conditions

In the previous paragraph we investigated the influence of a mismatch between the Doppler frequency $\tilde{F}_D$ assumed in the calculation of the CIR-related tap prediction filter's coefficients and the actual Doppler frequency F_D observed on the channel. Similarly, an MSE degradation is observed, if the SNR assumed in the coefficient calculation differs from that measured at the reception antenna. As an example, the MSE degradation incurred has been analytically evaluated for a scenario associated with an OFDM symbol normalised Doppler frequency of $F_D = 0.1$ observed on the channel, which was also assumed in the filter calculation. A filter length of 4 taps was used. The SNR at the reception antenna was varied between 0dB and 40dB, while the SNR invoked in the filter calculation was fixed to specific values. The corresponding results are portrayed in the left graph of Figure 15.8. In order to justify and interpret the evolution of the different curves please recall from Section 15.2.4 the mechanisms associated with the operation of the Wiener prediction filter. More specifically, a trade-off has to be found between the mitigation of the AWGN and the error incurred as a result of the decorrelation of the channel between the transmission of two consecutive OFDM symbols. If the SNR actually measured at the reception antenna is lower than the SNR assumed in the CIR-related tap prediction filter's calculation, an MSE degradation is incurred, since the mitigation of the channel decorrelation is over-weighted, while that of the noise is underestimated. By contrast, if the SNR at the reception antenna is higher than the SNR assumed in the filter calculation, the channel's correlation along the time direction is not optimally exploited and the AWGN mitigation is over-emphasised. As a result a residual MSE is observed even at high channel SNRs.

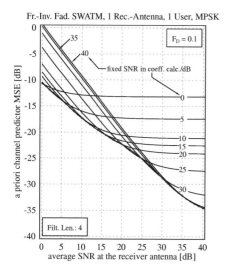
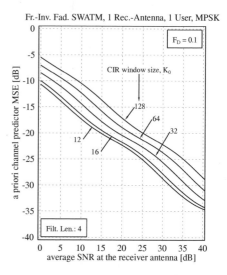

Figure 15.8: **Mean-Square Error (MSE)** exhibited by the **robust decision-directed *a priori* channel predictor**, which follows the concepts of Figure 15.4, as a function of the Signal-to-Noise Ratio (SNR) at the reception antenna. A fixed OFDM symbol-normalised Doppler frequency of $F_D = 0.1$, **error-free symbol decisions** and a CIR-related tap prediction filter length of 4 taps was used in the **"frame-invariant"** fading **indoor WATM channel** environment of Figure 14.4. Fixed SNRs were assumed in the calculation of the predictor coefficients, regardless of the SNR encountered on the channel (**left**) and for perfectly matched conditions as a function of the CIR window size K_0 (**right**); the variance of the AWGN was $\sigma_n^2 = 2$; in the former case characterised in the left-hand side illustration the CIR window size was $K_0 = 12$ taps. The results were evaluated from Equation 15.52.

15.3.1.4 MSE Performance under Mismatched Multipath Spread Conditions

So far we have assumed a fixed CIR window size - or sampling period duration normalised multipath spread - of $K_0 = 12$ CIR-related taps in our analytical evaluations, which - in conjunction with $K = 512$ subcarriers associated with the indoor WATM system model of Section 14.3.1 - resulted in a "filter noise reduction factor" [477] of $K_0/K \approx 0.0234$ or equivalently $K_0/K|_{dB} \approx -16.3 dB$ as a consequence of setting the remaining CIR-related taps to zero. In other systems the maximum CIR-induced delay spread will be potentially different from that of the 12 taps of the indoor WATM channel model and hence K_0 would have to be adjusted accordingly. At this stage we recall that it was demonstrated for example in Equation 15.11 for an idealised channel estimator, which capitalises on an infinite number of previous initial least-squares channel transfer factor estimates that the estimator's MSE achieved is a function of the product $(K_0/K)\rho$, where $\rho = 1/SNR$. Furthermore, this is also the case for the channel predictor considered here, where an increase of the factor K_0/K due to considering a longer CIR can be compensated by increasing the SNR i.e the transmitted signal's power, if this is deemed acceptable in the system considered. Hence, varying the factor K_0/K results in a horizontal shift of the MSE versus SNR curves, where the SNR is associated with the abscissa axis. This is illustrated in the right graph of Figure 15.8 for values of K_0 ranging from 16 to 128 in comparison to $K_0 = 12$ used as a reference and for a fixed number of $K = 512$ subcarriers.

15.3.1.5   Conclusion on the MSE Performance of Robust Decision-Directed Channel Prediction in the Context of Error-Free Symbol Decisions

In Section 15.3.1 the robust decision-directed channel predictor's MSE performance was assessed in the context of error-free symbol decisions as a function of a variety of system parameters. Specifically, in Section 15.3.1.1 we investigated the influence of the number of predictor coefficients and that of the channel's OFDM symbol normalised Doppler frequency on the predictor's MSE versus SNR performance. We found that using four predictor taps was sufficient for exploiting most of the channel's correlation in the time direction. This was concluded on the basis of the quasi-parallel nature of the MSE curves recorded for a higher number of prediction filter taps. Our further investigations in Section 15.3.1.2 were focused on the effects of a misadjustment between the OFDM symbol normalised Doppler frequency of the channel - associated with the Jakes' spectrum [474] - and that associated with the robust channel predictor's uniform Doppler power spectrum invoked in the calculation of the CIR-related tap prediction filter coefficients. We observed even for a relatively modest number of four prediction filter taps that the channel predictor's MSE was relatively insensitive to the OFDM symbol normalised Doppler frequency of the channel, provided that the Doppler frequency encountered was lower than that assumed in the design of the predictor. Furthermore, in Section 15.3.1.3 we evaluated the predictor's sensitivity with respect to a misadjustment of the SNR encountered at the reception antenna and that employed in the calculation of the predictor's coefficients. We found that at SNRs lower than that associated with the design of the predictor, the MSE was degraded, while for SNRs higher than that assumed in the calculation of the predictor, an MSE floor was experienced. This MSE floor was encountered, because the channel's correlation was not optimally exploited. Our investigations were concluded in Section 15.3.1.4 by assessing the influence of the number of significant CIR-related taps on the predictor's MSE performance. We found that for a higher number of taps the MSE versus SNR curves were shifted towards higher SNRs, since more of the AWGN was retained.

15.3.2   MSE Performance of an Adaptive Decision-Directed Channel Predictor in the Context of Error-Free Symbol Decisions

In this section we will assess the MSE performance of the adaptive decision-directed channel predictor briefly addressed in Section 15.2.5.2, where the different CIR-related taps' predictor coefficients are obtained with the aid of the Burg algorithm [85, 89] on the basis of the statistics estimated from actual CIR-related tap samples, as advocated by Al-Susa et al. [72]. We have listed the relevant system parameters in Table 15.5. Recall that in our investigations of the robust channel predictor in Section 15.3.1, we had to explicitly specify the SNR and the OFDM symbol normalised Doppler frequency assumed in the CIR-related tap predictor's calculation as additional parameters, which are potentially different from those actually observed on the channel encountered. By contrast, both of these parameters are implicitly

K_0	Sampling period duration normalised multipath spread
F_D	OFDM symbol normalised Doppler frequency
ρ	Reciprocal of the SNR
$N_{tap}^{[t]}$	Length or range of the CIR-related tap prediction filter
M_s	Number of samples involved in the predictor coefficient calculation
K	Number of subcarriers per OFDM symbol

Table 15.5: Summary of the parameters influencing the MSE performance of the adaptive decision-directed CIR-related tap predictor in the context of error-free symbol decisions and in the absence of ICI.

specified in the context of the adaptive channel prediction process investigated here by the number M_s of the CIR-related tap samples involved in the estimation of the channel's statistics. In addition to the parameters listed in Table 15.5 the particular shape of the power-delay profile will also affect the channel predictor's MSE performance. Please note that during our forthcoming MSE investigations of this section we have ignored the boundary effects related to the predictor's potentially impaired performance during the algorithm's start-up phase, when the required M_s number of channel samples is not yet available.

The further structure of Section 15.3.2 is as follows. In Section 15.3.2.1 we portray the adaptive channel predictor's MSE performance as a function of the SNR encountered at the reception antenna in the context of various OFDM symbol normalised Doppler frequencies, F_D, and additionally parametrised with the number of samples M_s invoked in the Burg algorithm assisted evaluation of the CIR-related tap predictor's coefficients. Based on these MSE performance results, in Section 15.3.2.2 the MSE performance results of Section 15.3.1.1 associated with the robust channel predictor recorded in the context of perfectly matched channel conditions of $K_0 = \tilde{K}_0$ as well as $F_D = \tilde{F}_D$ and $\rho = \tilde{\rho}$ will be compared to those of the adaptive channel predictor. Furthermore, in Section 15.3.2.3 the adaptive channel predictor's MSE performance achieved in conjunction with various sample-spaced negative exponentially decaying multipath intensity profiles will be compared to that associated with a sample-spaced uniform multipath intensity profile. Finally, in Section 15.3.2.4 we will offer our conclusions with respect to the comparison of the robust- and adaptive decision-directed channel predictors.

15.3.2.1   MSE Performance under Matched Channel Conditions as a Function of the Number of Samples invoked in the Predictor Design

In Figure 15.9 we have evaluated the MSE exhibited by the Burg algorithm-assisted decision-directed channel predictor as a function of the SNR recorded at the reception antenna in the context of the indoor WATM channel model of Figure 14.4 using various OFDM symbol normalised Doppler frequencies F_D and a fixed predictor length of 4 taps. The predictor's MSE has been parametrised with the number of time direction CIR-related tap samples M_s considered in the estimation of the individual CIR-related taps' correlation inherent in the Burg algorithm, in form of the spaced-time correlation function associated with each CIR tap's fading process. We observe a reduction of the MSE along with an increasing number of samples M_s invoked. This is due to the increased accuracy of the estimated channel

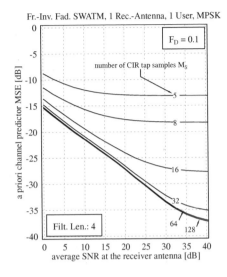

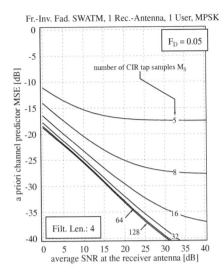

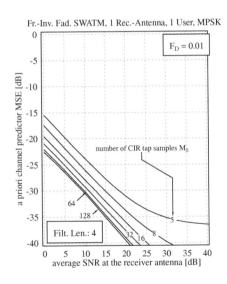

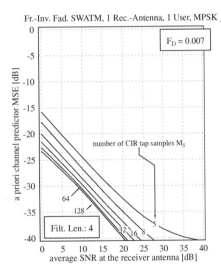

Figure 15.9: **Mean-Square Error (MSE)** exhibited by the **Burg algorithm-assisted decision-directed *a priori* channel predictor** as a function of the Signal-to-Noise Ratio (SNR) at the reception antenna. The results are parametrised with the number of samples M_s invoked in the Burg algorithm. **Error-free symbol decisions** were used and a prediction filter length of 4 taps was employed in the **"frame-invariant"** fading **indoor WATM channel** environment of Figure 14.4. The OFDM symbol-normalised Doppler frequencies used were F_D of 0.1 (**top left**), 0.05 (**top right**), 0.01 (**bottom left**) and 0.007 (**bottom right**). The variance of the AWGN was $\sigma_n^2 = 2$. The CIR window size was $K_0 = 12$ taps.

statistics. However, when the number of channel samples is in excess of about $M_s = 64$, only a marginal further MSE reduction can be observed.

15.3.2.2   MSE Performance in Comparison to that of the Robust Channel Transfer Function Predictor

Upon comparing the MSE curves of Figure 15.9 for a sufficiently high number of time direction CIR-related tap samples M_s with the MSE curves of Figure 15.5, which were associated with the robust channel predictor and were recorded for a prediction filter length of 4 taps, we observe furthermore that the adaptive Burg-algorithm based predictor outperforms the robust approach in terms of the achievable MSE by about 6dB, also depending on the specific channel SNR encountered. The mechanisms responsible for this improvement will be highlighted during our forthcoming discourse in this section. In the context of the simulation results presented here we employed the indoor WATM system model of Section 14.3.1, having a 3-path sample-spaced CIR. The non-zero CIR tap values are associated with the tap indices of 0, 6 and 11. While the robust channel predictor assessed in Section 15.3.1 applied the same CIR-related tap prediction filter coefficients to the different CIR taps, regardless of the power level associated with the specific CIR-related tap, the adaptive predictor performed a more effective suppression of the undesired additive noise associated with the specific estimated CIR-related taps that have a near-zero magnitude. This is a consequence of the spectrally white distribution of the AWGN. As a result, the components of the specific CIR-related tap's estimated cross-correlation vector tend to zero in the absence of a correlated channel signal, provided that a sufficiently high number of samples associated with $M_s \rightarrow \infty$ is used. Hence, for the specific indoor WATM channel model used and for a sufficiently high number of channel samples M_s invoked in the prediction process, the noise reduction factor becomes $3/512$. By contrast, for the same channel model in the context of the robust prediction based approach and for a CIR-related window size of $K_0 = 12$ taps the noise reduction factor is $12/512$. This corresponds to an MSE reduction by a factor of 4, or on a logarithmic scale by 6.02dB. The same sparse nature of the CIR can also be exploited in the context of the robust channel predictor characterised in Section 15.3.1 upon applying a threshold to the estimated CIR-related tap values, below which the corresponding CIR-related taps are forced to zero. This technique was referred to as "significant tap catching" (STC) by Li *et al.* [91] in the context of channel estimation contrived for space-time coded OFDM systems. It should be noted that not only the estimates of the insignificant, low-energy CIR-related taps, but also the estimates of the significant CIR-related taps benefit from the more efficient adaptation of the CIR-related tap prediction filter to the actual channel statistics encountered. From Equation 15.27 it became explicit that in the context of the optimum two-dimensional channel predictor each CIR-related tap is associated with a specific eigenvalue of the channel's spaced-frequency correlation matrix, while in the context of the potentially sub-optimum robust channel predictor an identical eigenvalue of $\tilde{\lambda}_l^{[f]} = K/K_0$ was associated with each CIR-related tap. As a result, the robust channel predictor was rendered insensitive against the specific shape of the channel's multipath intensity profile, while at the same time exhibiting the worst MSE performance of all predictors operating on channels having power-delay profiles confined to a limited CIR window dimension of K_0 taps.

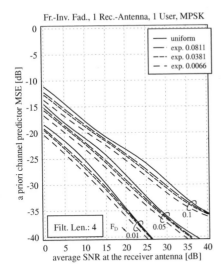

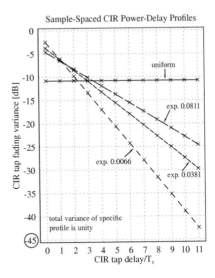

Figure 15.10: **(left): Mean-Square Error (MSE)** exhibited by the **Burg algorithm-assisted decision-directed *a priori* channel predictor** upon invoking a near-infinite number of CIR samples, namely $M_s \rightarrow \infty$ as a function of the Signal-to-Noise Ratio (SNR) recorded at the reception antenna. **Error-free symbol decisions** and a prediction filter length of 4 taps were used for a **"frame-invariant"** fading channel having OFDM symbol-normalised Doppler frequencies of $F_D = 0.1, 0.05$ and 0.01, as well as different power delay profiles. The variance of the AWGN was $\sigma_n^2 = 2$. The CIR window size was $K_0 = 12$ taps. **(right):** Characterisation of the various power delay profiles; the numerical value, which constitutes part of a specific delay profile's identifier is identical to the ratio r defined in Equation 15.58.

15.3.2.3 MSE Performance for Various Multipath Intensity Profiles

The influence of the specific shape of the channel's multipath intensity profile on the adaptive channel predictor's MSE performance has been further illustrated in Figure 15.10. Four different sample-spaced multipath intensity profiles are invoked in our comparisons. The specific multipath intensity trajectories are portrayed in the right-hand side graph of Figure 15.10. The uniform multipath intensity profile serves as a benchmark, while with the aid of the different negative exponentially decaying multipath intensity profiles the influence of a "power compaction" to a few CIR-related taps on the estimator's MSE can conveniently be demonstrated. Here the static negative exponentially decaying amplitude delay profile is defined as:

$$h_{expo}[l] = \beta_{expo} e^{-\alpha_{expo} l}, \qquad 0 \le l \le K_0 - 1, \qquad (15.57)$$

The specific values of the decay factor α_{expo} and the amplitude scaling factor β_{expo} are determined here by the ratio r of the CIR tap amplitude in the K_0-th CIR tap, versus that in the zero-th CIR tap, which is formulated as:

$$r = \frac{h_{expo}[K_0]}{h_{expo}[0]}, \tag{15.58}$$

and by the condition that the sum of the squared CIR tap amplitudes is unity, which implies that no power loss or gain is imposed by the channel, yielding:

$$\sum_{l=0}^{K_0-1} |h_{expo}[l]|^2 \overset{!}{=} 1. \tag{15.59}$$

From Equations 15.58 and 15.57 the value of the decay factor α_{expo} can be determined as:

$$\alpha_{expo} = -\frac{1}{K_0} \ln(r), \tag{15.60}$$

while upon invoking Equations 15.59 and 15.57 the amplitude scaling factor β_{expo} is given by:

$$\beta_{expo} = \sqrt{\frac{1 - e^{-2\alpha_{expo}}}{1 - e^{-2\alpha_{expo}K_0}}}. \tag{15.61}$$

In Figure 15.10 the numerical value, which is part of a specific exponential profile's identifier is identical to the factor r defined by Equation 15.58. As illustrated in the left-hand side graph of Figure 15.10, the worst MSE performance is exhibited by the adaptive channel predictor on a channel having a uniform multipath intensity profile. In this case identical CIR-related tap prediction filters are invoked in the different CIR taps' prediction processes and hence the prediction could also be performed using the same channel transfer factor prediction filter on a subcarrier by subcarrier basis in the frequency domain instead of on a CIR-related tap-by-tap basis in the time domain. By contrast, for the specific negative exponential multipath intensity profiles, which exhibit a non-uniform power distribution across the different CIR taps a reduced MSE is observed. More specifically, the highest performance improvement is achieved for a negative exponentially decaying multipath intensity profile associated with a ratio of $r = 0.0066$, where the highest energy compaction into the first few taps is observed.

15.3.2.4 Conclusion on Adaptive Decision-Directed Channel Prediction in the Context of Error-Free Symbol Decisions

In Section 15.3.2 Burg algorithm-assisted block-adaptive decision-directed channel prediction was investigated in the idealistic scenario of error-free symbol decisions. Our discussions commenced in Section 15.3.2.1 by portraying the effects of the specific number of M_s past time direction CIR-related tap samples invoked in the Burg algorithm on the predictor's MSE. We found that in the context of a four-tap CIR-related tap prediction filter increasing the number M_s of the past time direction CIR-related tap samples beyond $M_s = 64$ does not result in any significant performance improvement. Furthermore, in Section 15.3.2.2 the adaptive channel predictor's MSE was compared against that of the robust predictor on the basis of the results portrayed in Section 15.3.1. These comparisons were further expanded in Section 15.3.2.3 upon invoking various sample-spaced multipath intensity profiles. Recall that the uniform profile is the one associated with the robust CIR-related tap predictor.

In order to elaborate a little further, based on the simulation results presented in Section 15.3.2.3 we conclude that in scenarios associated with a uniform channel multipath intensity profile, where identical prediction filters are applied to the different CIR-related taps, the task of two-dimensional channel transfer function prediction in the frequency domain can be split into that of one-dimensional CIR-related tap windowing in the time domain followed by one-dimensional subcarrier-by-subcarrier based channel transfer factor prediction in the time direction using the same channel transfer factor prediction filter for all subcarriers. However, in order to reduce the computational complexity of the actual implementation, the filtering along the time direction is performed in the time domain, since the number of filtering operations is reduced by filtering K_0 CIR-related taps, rather than K frequency domain channel transfer factors. In this case the MSE of the adaptive channel predictor characterised in Section 15.3.2, while employing the Burg algorithm [85, 89] is identical to that of the robust channel predictor assessed in Section 15.3.1, provided that the Doppler frequency assumed in the design of the robust predictor's time direction filter matches that of the channel encountered and the SNR assumed in the filter's calculation is identical to that experienced at the reception antenna. Otherwise a further MSE degradation is observed for the robust predictor in comparison to the adaptive Burg-algorithm-assisted predictor, as was illustrated in Figures 15.7 and 15.8, respectively. By contrast, in cases where the channel's multipath intensity profile is sharply decaying, an MSE degradation is observed in conjunction with the robust predictor of Section 15.3.1 compared to the adaptive predictor of Section 15.3.2. Recall that the "robust" channel predictor, where the prediction is performed individually for each tap, i.e. on a tap-by-tap basis in the CIR-related domain, is actually identical to two cascaded 1D frequency domain filters, where one of them interpolates the subcarrier channel transfer factors across a given frequency domain OFDM symbol, while the other one provides a predicted channel transfer function for the next OFDM symbol. By contrast, the adaptive channel predictor investigated in Section 15.3.2, where different prediction filters are employed for the different CIR-related taps is actually equivalent to a true $2D$ channel predictor operating in the frequency domain.

15.3.3 MSE Performance of a Robust Decision-Directed Channel Predictor in the Context of an Uncoded System

In Sections 15.3.1 and 15.3.2 we evaluated the MSE of the robust channel predictor and that of the Burg algorithm-assisted adaptive channel predictor in the context of the idealistic scenario of error-free symbol decisions at the receiver. We demonstrated that for transmissions over the channels investigated, which had a sample-spaced impulse response of a maximum dispersion of K_0 samples, the robust channel predictor's MSE upper-bounds the MSE exhibited by the adaptive channel predictor upon invoking a sufficiently high number of time direction CIR samples in the CIR-related tap prediction process. For the OFDM symbol normalised Doppler frequencies considered no significant further *a priori* MSE reduction was observed upon increasing the number of time direction CIR-related tap samples invoked beyond about $M_s = 64$. Motivated by the fact that the MSE performance of the robust channel predictor upper bounds the MSE performance of the adaptive channel predictor provided that a sufficiently high number of error-free CIR-related tap samples is available, in this section we

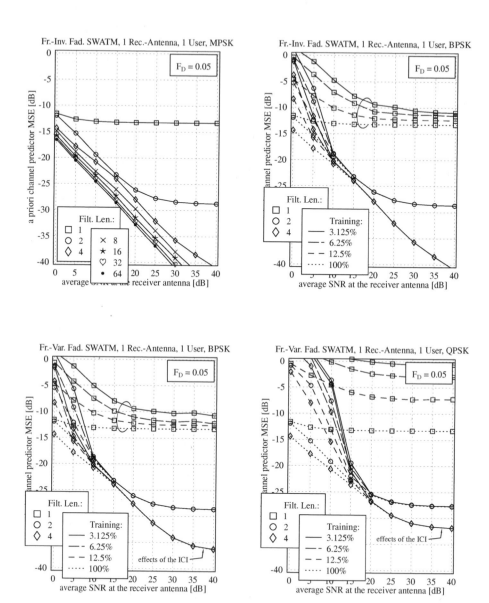

Figure 15.11: **Mean-Square Error (MSE)** exhibited by the **robust decision-directed *a priori* channel predictor** as a function of the SNR encountered at the reception antenna. The results are additionally parametrised with the prediction filter length used. The fading **indoor WATM channel** environment of Figure 14.4 was encountered at an OFDM symbol-normalised Doppler frequency F_D of 0.05; **(top left)**: MSE for various prediction filter lengths of up to 64 taps and for error-free symbol decisions in the context of a frame-invariant fading channel; **(top right)**: MSE for prediction filter lengths of up to 4 taps and potentially error contaminated BPSK symbol decisions in the context of a frame-invariant fading channel; **(bottom left)**: same conditions as in the top right figure, but transmitting over a frame-variant fading channel; **(bottom right)**: same conditions as in bottom left figure, but for QPSK symbol decisions; the variance of the AWGN was $\sigma_n^2 = 2$; the CIR window size was $K_0 = 12$ taps.

will focus our attention on studying the robust channel predictor further, again, in the context of a continuous transmission mode, where the predictor's boundary effects were avoided by always providing a sufficiently high number of least-squares channel transfer factor estimates based on past OFDM symbols. Although the simulation results were obtained in the context of the indoor WATM channel model of Figure 14.4, no specific assumptions were stipulated concerning the shape of the channel's impulse response in the context of the robust channel predictor design. Hence the same performance results would be obtained in the context of any other sample-spaced CIR profile having a maximum dispersion of 11 taps.

The further structure of Section 15.3.3 is as follows. Based on Monte Carlo simulations in Section 15.3.3.1 we will evaluate the MSE performance of the robust channel predictor under frame-invariant fading channel conditions upon assuming error-free symbol decisions. In Section 15.3.3.2 our performance assessments are rendered more realistically by allowing transmission errors to occur in the demodulation process at the receiver. Furthermore, in Section 15.3.3.3 the idealistic assumption of frame-invariant fading is removed in favour of the more realistic frame-variant fading scenario. As a result, Inter-subCarrier Interference (ICI) [476] is encountered. Again, our conclusions will be outlined in Section 15.3.3.4. In contrast to our investigations in Section 15.3.3, where we have employed the predictor MSE as our performance measure, in Section 15.3.4 the system BER will be assessed in both uncoded and coded scenarios in conjunction with various modulation schemes.

15.3.3.1   MSE Performance for a Frame-Invariant Fading Channel and for Error-Free Symbol Decisions

We commence our investigations based on the idealistic assumption of encountering a "frame-invariant" fading channel, where the fading channel's magnitude and phase were kept constant during the OFDM symbol's period, so as to avoid obfuscating the results of channel prediction by the effects of inter-subcarrier interference (ICI) [476]. Furthermore, we assumed the availability of *error-free symbol decisions*. During our discourse we will gradually relax these constraints in favour of more realistic operating conditions. The corresponding MSE simulation results obtained for an OFDM symbol normalised Doppler frequency of $F_D = 0.05$ are portrayed in the top left graph of Figure 15.11, where we have fixed the CIR dispersion to $K_0 = 12$ taps, which is equal to the normalised multipath spread of the CIR associated with the indoor WATM channel model of Figure 14.4. We observe that as expected, upon increasing the prediction filter length from 1 to 64 taps, the MSE performance improves. On the basis of the slope of the MSE curves we also infer - as alluded to in Section 15.3.1 - that most of the channel transfer function's correlation observed in the time direction for consecutive frequency domain OFDM symbols can be exploited with the aid of a prediction filter length of 4 taps. Further increasing the CIR-related tap prediction filter length has the beneficial but modest effect of reducing the influence of the channel noise due to averaging a higher number of additive noise components during the filtering process. As argued before in the context of Figure 15.5, this noise averaging or mitigation process manifests itself in the parallel nature of the MSE curves observed for filter lengths in excess of 4 taps in Figure 15.11.

15.3.3.2   MSE Performance for a Frame-Invariant Fading Channel and Sliced Symbol Decisions

In practice, error-free symbol decisions are normally not available, even if the system is designed for operating at a low BER of say 10^{-5}. To be more realistic, let us now consider the case of potentially error contaminated, *sliced symbol decisions*, where the reference signal is generated by simply remodulating the sliced symbols. Channel coding is not considered at the moment. The corresponding simulation results recorded for the "frame-invariant" fading WATM channel of Figure 14.4 in the context of BPSK modulated OFDM transmissions over a channel that is fading at a rate of $F_D = 0.05$ are illustrated in the top right-hand side graph of Figure 15.11. Observe that we have specified the proportion of pilot-based channel measurement or training information as an additional parameter. More explicitly, training information is employed in form of dedicated OFDM pilot symbols, where all the subcarriers host randomly BPSK-modulated pilot symbols, which are known to the receiver. Thus error propagation extending over a duration longer than the time between consecutive pure pilot-based training OFDM symbols is reduced. It should be noted that a channel sounding training overhead of 3.125% corresponds to assigning pure pilot-based training information to every 32-nd OFDM symbol. Similarly, 6.25% overhead corresponds to a pure pilot-based training period of 16 OFDM symbols and 12.5% corresponds to an 8 OFDM symbol training period, respectively, when assuming the transmission of single, rather than multiple consecutive pure pilot-based training OFDM symbols. In the top right graph of Figure 15.11 the impact of the different training OFDM symbol densities becomes explicit for the case of first-order i.e 1-tap CIR-related tap filtering, which provides relatively inaccurate channel estimates in the context of propagation scenarios having a high OFDM symbol normalised Doppler frequency, such as $F_D = 0.01$. For the more accurate 2- and 4-tap CIR-related tap prediction arrangements, transmitting training OFDM symbols shows the highest impact at SNRs below 10dB, while for higher SNRs the MSE performance curves merge as a result of the increasingly high probability of error-free symbol decisions.

15.3.3.3   MSE Performance for a Frame-Variant Fading Channel and for Sliced Symbol Decisions

In order to further improve the degree of realism in our simulations, let us now focus our attention on the case of a "frame-variant" fading channel. The corresponding simulation results acquired for BPSK modulation are illustrated in the bottom left graph of Figure 15.11. In comparison to the performance exhibited by the decision-directed channel predictor in the idealistic environment of a "frame-invariant" fading channel, the MSE performance becomes more limited, particularly at relatively high SNRs. This is a result of the noise-like ICI contributions [476] induced by the frame-variant fading channel.[13] In order to further illustrate the impact of erroneous symbol decisions on the quality of the predicted channel transfer functions, the *a priori* channel predictor MSE has also been evaluated in the context of the QPSK

---

[13]Note that while the predictor's performance is related to the OFDM symbol normalised Doppler frequency of the channel, namely to $F_D = (K + K_g)T_s f_D$, the variance of the ICI is related to the "FFT window duration normalised" Doppler frequency, namely $F_{D,K} = KT_s f_D$. The quotient of these normalised Doppler frequencies is given by $\frac{F_D}{F_{D,K}} = 1 + \frac{K_g}{K}$. Hence, upon neglecting the effects of error propagation, and for all simulation setups using the same modulation scheme, identical OFDM symbol normalised Doppler frequencies F_D and identical quotients K_g/K the performance will be the same, provided that the noise reduction factor K_0/K is also identical.

modulation scheme used. The corresponding simulation results are portrayed in the bottom right graph of Figure 15.11. In contrast to the corresponding simulation results acquired for BPSK modulation and shown in the bottom left graph of Figure 15.11, the confluence point of the two sets of curves associated with the erroneous symbol decisions and the error-free symbol decisions is found at higher SNRs, namely between 15 and 20dB. Furthermore, we note that the residual MSE inflicted by the ICI at SNRs in excess of about 20dB is increased in the context of QPSK modulation in comparison to BPSK modulation, as it was shown in the bottom left graph of Figure 15.11. Although the variance of the ICI is independent of the specific choice of the modulation scheme used, our results not included here for reasons of space economy suggest that differences can be observed between the subcarrier-based ICI power distributions for the different modulation modes employed. For modulation schemes, which transmit two orthogonal signal components, such as, for example, QPSK or higher-order MPSK modulation schemes, the probability of encountering ICI contributions of a high power is higher than for BPSK. Again, the exact impact of various modulation schemes on the predictor's MSE has not been investigated further, but the existence of a correlation between these two phenomena is conjectured.

At this stage of our discussions it is worth noting that in channel scenarios associated with a high Doppler frequency the specific design of the pilot-based training OFDM symbols has a major influence on the estimator's performance. More specifically, transmitting the same constellation point in each of the subcarriers of the pilot-based training OFDM symbol would result in an excessive amount of ICI as well as in a high crest factor. Hence, in line with standard practice [479], we opted for assigning a randomly BPSK modulated symbol sequence, to each training OFDM symbol, which is also known to the receiver.

On the basis of the simulation results presented so far, we conclude that with the aid of an MMSE CIR-related tap predictor, decision-directed channel estimation can be successfully used even in channel scenarios exhibiting relatively high OFDM symbol normalised Doppler frequencies. After providing our conclusions in the next section, in Section 15.3.4 we will characterise the system's performance in terms of the achievable BER in the context of robust decision-directed channel prediction.

15.3.3.4   Conclusion on the MSE Performance of a Robust Decision-Directed Channel Predictor in the Context of an Uncoded System

In Section 15.3.3 the robust decision-directed channel predictor's MSE performance has been evaluated in the context of an uncoded system upon invoking sliced and hence potentially error-contaminated symbol decisions. In order to reduce the number of different MSE versus SNR curves we employed an OFDM symbol normalised Doppler frequency having a mediocre value of $F_D = 0.05$. From our simulation results portrayed in Figure 15.11 we inferred that for medium SNRs up to about 15dB the predictor's MSE performance deteriorated due to the effects of error propagation. These effects can be mitigated with the aid of dedicated pure pilot-based training OFDM symbols transmitted periodically. We saw in Figure 15.11, that for the frame-variant fading channel scenario and a number of four CIR-related prediction filters taps the channel predictor's *a priori* MSE was limited due to the Gaussian noise like influence of the ICI, rather than as a result of the predictor's imperfections.

15.3.4   BER Performance of an Uncoded System Employing Robust Decision-Directed Channel Prediction

In Figure 15.12 we plotted the BER performance curves corresponding to the "frame-variant" fading channel scenario associated with OFDM symbol normalised Doppler frequencies of $F_D = f_D T_f = 0.1, 0.05$ and 0.01. Different modulation schemes were considered. Specifically, the curves associated with BPSK modulation are shown in the top left graph of Figure 15.12, while those recorded for QPSK modulation and 16QAM are found in the top right graph and in the graph at the bottom of Figure 15.12, respectively. An imperfect reference signal based on sliced and remodulated symbol decisions was assumed in the process of robust decision-directed channel estimation and the CIR-related tap prediction filter length was fixed to four taps. According to our discussions in Section 15.3.3 using four taps was sufficient to exploit most of the channel's correlation in the time direction in the specific propagation environment considered.

The structure of the remaining part of Section 15.3.4 is as follows. The system's BER is analysed in Section 15.3.4.1 in conjunction with BPSK and QPSK. The corresponding BER results related to the more vulnerable 16QAM scheme will be provided in Section 15.3.4.2, where we will specifically focus our attention on the influence of the number of consecutive pure pilot-based training OFDM symbols transmitted periodically. Our preliminary conclusions will be offered in Section 15.3.4.3.

15.3.4.1   BER Performance for BPSK and QPSK

In conjunction with BPSK and QPSK modulation we observe in Figure 15.12 that for propagation scenarios, which exhibit relatively high OFDM symbol normalised Doppler frequencies, such as $F_D = 0.05$ and $F_D = 0.1$, the BER performance is rather limited. This is a result of the ICI imposed by the frame-variant fading. The specific choice of the pure pilot-based OFDM training symbol period plays a more dominant role in the relatively low SNR range below 15dB, where the error propagation encountered deteriorates the performance, also depending on the OFDM symbol normalised Doppler frequency F_D of the channel. As expected, the scenarios associated with a higher Doppler frequency are particularly vulnerable. From our further investigations – which are not explicitly described here for reasons of space – we inferred that for relatively modest values of F_D associated with SNRs close to 0dB the arrangements employing higher order CIR-related tap prediction filters are more susceptible to error propagation, than those employing a lower-order prediction filter. A plausible explanation of this phenomenon is that for higher-order CIR-related tap prediction filters the probability of invoking previous channel estimates potentially based on erroneous symbol decisions is increased. By contrast, in propagation scenarios of relatively high Doppler frequencies, such as $F_D = 0.05$ and $F_D = 0.1$, this effect is compensated by the more accurate channel estimates offered by the higher-order prediction filter. As outlined in Section 15.3.3, this is the result of a better exploitation of the channel's correlation and that of a more efficient noise mitigation due to averaging a higher number of noise components.

15.3.4.2   BER Performance for 16QAM

As shown in the bottom left graph of Figure 15.12, investigations were also conducted employing 16QAM in a system capitalising on robust decision-directed channel estimation. In

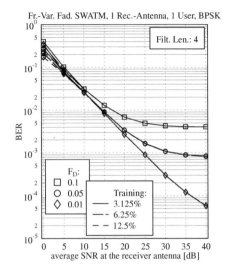

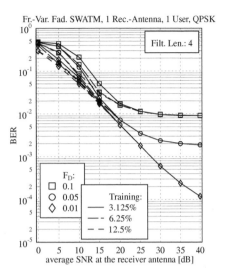

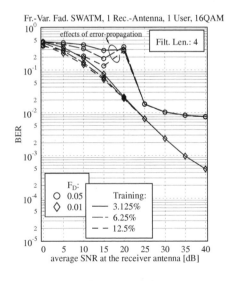

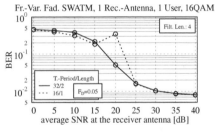

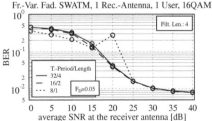

Figure 15.12: **Bit Error Ratio (BER)** exhibited by a system employing **robust decision-directed *a priori* channel prediction** as a function of the SNR encountered at the reception antenna. A fixed CIR-related tap prediction filter length of 4 taps was used and the results were additionally parametrised with the percentage of pure pilot-based OFDM symbol training overhead in a single reception antenna, single user scenario. The **"frame-variant"** fading **indoor WATM channel** environment of Figure 14.4 was encountered at OFDM symbol-normalised Doppler frequencies F_D of 0.1, 0.05 and 0.01 and (**top left**) BPSK, (**top right**) QPSK and (**bottom**) 16QAM modulation schemes were used. The CIR window size was $K_0 = 12$ taps.

the context of the hostile, rapidly fading scenario of $F_D = 0.1$ maintaining high-integrity communication without inflicting excessive BERs became virtually impossible and hence the associated curves have not been plotted in Figure 15.12. This was attributed to the severe error propagation encountered between successive OFDM symbols, despite periodically transmitting dedicated pure pilot-based training OFDM symbols. More specifically, the effect of erroneous symbol decisions is that an inaccurate *a posteriori* channel estimate is generated for the current OFDM symbol, which in turn degrades the quality of the *a priori* channel estimate derived for the next OFDM symbol and thus the associated BER is also degraded. As a result, a potential system instability is observed with the consequence of excessive BERs. Even when benefitting from high SNRs and additionally imposing error-free OFDM symbol decisions in the remodulated DDCE-related reference, a residual BER of $3 \cdot 10^{-2}$ was observed in the context of the rapidly fading propagation scenario of $F_D = 0.1$ as a result of the excessive ICI inflicted by the frame-variant fading channel. In the more benign scenario of $F_D = 0.05$, where the associated BER results are portrayed together with those of the $F_D = 0.01$ scenario in the bottom left graph of Figure 15.12 - after an initial BER improvement recorded for SNRs up to about 15dB - a BER degradation is observed for SNRs up to 20dB. This is followed by a rapid, avalanche-like reduction of the BER to a value of around $8 \cdot 10^{-3}$, within an SNR interval of 5dB. Upon closer inspection we found that two specific mechanisms are responsible for the performance degradation incurred at these intermediate SNRs, which will be outlined below.

First, the AWGN plus ICI-related SNR assumed in the calculation of the CIR-related tap predictor coefficients did not account for the additional impairments resulting in an increased level of "noise" inflicted by erroneous symbol decisions and hence a noise-amplification - also known from zero-forcing (ZF) equalisers - occurred, which resulted in the performance degradation observed. The second design aspect, which influences the system's performance is the specific structure of the block of pure pilot-based training OFDM symbols transmitted. So far, each training block consisted of a single OFDM symbol. Hence, even if the most recently received OFDM symbol was a training symbol, previous erroneously demodulated OFDM symbols within the range of the prediction filter might have deteriorated the MSE of the *a priori* channel estimate for the next OFDM symbol. However, in our previous investigations of Section 15.3.3 this effect was neglected. Our related experiments revealed that this problem requires attention in the context of transmissions employing the relatively vulnerable 16QAM scheme.

In order to further illustrate this effect, we have portrayed in the bottom right graph of Figure 15.12 the BER of a 16QAM-assisted OFDM scheme invoking 4-tap CIR-related tap prediction filtering in the context of an OFDM symbol normalised Doppler frequency of $F_D = 0.05$. The results are parametrised with the number of immediately adjacent pure pilot-based OFDM symbols hosted by a training block, referred to as the training length, as well as with the training period duration between consecutive training blocks, which is assumed to be 32, 16 or, alternatively 8 OFDM symbols long. In the upper graph the training overhead was 6.25%, while in the lower graph it was twice as high, namely 12.5%. From Figure 15.12 we observe that for SNRs of up to about $15dB$ decreasing the training period duration is a more effective measure, than increasing the training block length, while for SNRs between 15- and $25dB$ we encounter the reverse mechanism. More specifically, a training block length of at least two OFDM symbol was required for avoiding the channel predictor instability.

coding rate	$R_c = 1/2$
constraint length	$K_c = 3$
generator polynomial	$(7,5)_8$
number of iterations	4

Table 15.6: Turbo coding parameters

15.3.4.3 Conclusion on the BER Performance of an Uncoded System Employing Robust Decision-Directed Channel Prediction

In Section 15.3.4 we have assessed the uncoded system's BER performance in the context of robust decision-directed channel prediction. Our observation was that with the beneficial assistance of prediction filtering, decision-directed channel estimation can also be supported under rapidly fading channel conditions in conjunction with OFDM symbol normalised Doppler frequencies as high as $F_D = 0.1$. Again, as shown in Section 15.3.4.1 for BPSK and QPSK modulation, the system's BER was limited due to the effects of ICI, rather than as a result of the imperfections of the decision-directed channel predictor. By contrast, it was demonstrated in Section 15.3.4.2 for the more vulnerable 16QAM modulation scheme upon invoking 4-tap prediction filtering that the system's stability could only be guaranteed for the lowest Doppler frequency of $F_D = 0.01$ used in our investigations. For a higher Doppler frequency of $F_D = 0.05$ the system became unstable at intermediate SNRs, which was a consequence of the noise amplification problem incurred. More specifically, the impairments resulting in an increased level of "noise" inflicted by erroneous symbol decisions were not considered in the calculation of the predictor's coefficients. We found that these effects could be mitigated upon increasing the pure pilot-based training block length to two consecutive OFDM symbols per training block. This had the beneficial effect that the error-propagation effects extending beyond the training blocks were mitigated. By contrast, for a Doppler frequency of $F_D = 0.1$ the system's BER was excessive.

In the next section the performance of a system employing both decision-directed channel estimation and turbo channel coding will be investigated, where the process of channel estimation potentially benefits from the more reliable, lower-BER symbol decisions.

15.3.5 BER Performance of a Turbo-Coded System Employing Robust Decision-Directed Channel Prediction

In order to further reduce the BER of the system arrangements studied, invoking turbo channel coding [480, 481] is an attractive option. The turbo coding parameters are listed in Table 15.6.

The structure of the rest of this section is as follows. In Section 15.3.5.1 we will argue that an accurate estimate of the channel's ICI noise variance is required, which is incorporated into the calculation of the subcarrier based SNR. Our investigations of the system's BER performance commence in Section 15.3.5.2 using BPSK modulation, while in Section 15.3.5.3 both QPSK modulation as well as 16QAM are invoked in the context of various OFDM symbol normalised Doppler frequencies. In these investigations the BER performance is presented for a specific configuration employing sliced and remodulated data symbols as a reference in

F_D	0.01	0.05	0.1
κ_{ICI}	0.000105	0.00445	0.01635

Table 15.7: Approximate proportionality constants κ_{ICI} for the indoor WATM system model detailed in Table 14.1 parametrised with the OFDM symbol-normalised Doppler frequency F_D.

the process of decision-directed channel prediction, but refraining from involving the turbo channel decoder in the CIR-related tap prediction process. By contrast, in Section 15.3.5.4 we will discuss the impact of a turbo channel decoded reference signal, where the "source"-related soft-output bits of the turbo decoder were sliced, re-encoded and remodulated. Our conclusions will be offered in Section 15.3.5.5.

15.3.5.1   Influence of the ICI Variance on the Subcarrier SNR

A prerequisite for minimising the BER is the availability of a reliable subcarrier by subcarrier-based SNR estimate, which is employed in the process of soft-bit generation for the turbo channel codec. In propagation scenarios exhibiting a relatively high OFDM symbol normalised Doppler frequency F_D as considered here, the effective additive noise experienced by the complex modulated symbol received in each OFDM subcarrier is constituted by the superposition of the AWGN and the Gaussian noise-like ICI induced by the frame-variant fading channel. Note that the average ICI-induced noise variance σ_{ICI}^2 can be expressed as a function of the signal variance σ_s^2 by means of the proportionality relationship $\sigma_{ICI}^2 = \kappa_{ICI}\sigma_s^2$. Hence estimates of the proportionality constants κ_{ICI} have been obtained with the aid of computer simulations for different values of F_D. The results are listed in Table 15.7.

15.3.5.2   BER Performance for BPSK Modulation in the Context of an Undecoded Reference

The BER results are portrayed in Figure 15.13 assuming BPSK modulation and a training overhead of 3.125% due to transmitting a pure pilot-based training OFDM symbol in every block of 32 OFDM symbols. Here we will characterise the performance observed in the context of a 4-tap CIR-related tap prediction filter, which is capable of exploiting most of the channel's correlation. Hence, it was already highlighted in Section 15.3.3, the effect of a further increase of the predictor's length in the context of error-free symbol decisions would be limited to improving the *a priori* channel estimator's MSE in terms of mitigating the influence of the additive noise due to the weighted averaging of a higher number of noise samples.

We observe that in the context of the more rapidly fluctuating propagation scenarios associated with $F_D = 0.05$ and $F_D = 0.1$ sufficiently low BERs can only be achieved upon invoking prediction filter lengths of at least 2 or 4 taps. For a scenario of $F_D = 0.1$ even in conjunction with a 4-tap CIR-related tap prediction filter a residual BER is observed. This is a consequence of the potentially excessive BER at the input of the turbo decoder, as well as a ramification of the limited accuracy associated with the effective SNR's estimate, which leads to relatively inaccurate soft-bit values to be passed to the turbo-decoder. It should be underlined that the proportionality relationship between the average ICI-induced noise variance

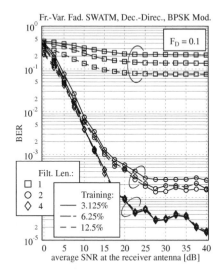

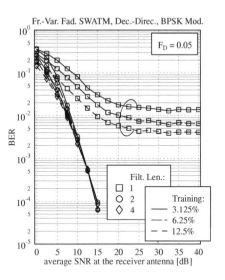

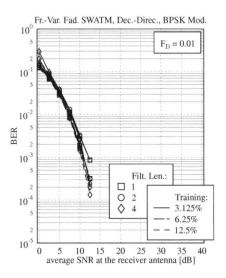

Figure 15.13: **Bit Error Ratio (BER)** exhibited by a **turbo-coded system** employing **robust decision-directed *a priori* channel prediction** with prediction filter lengths of up to 4 taps upon capitalising on potentially error-contaminated, **undecoded symbol decisions**. The **"frame-variant"** fading **indoor WATM channel** environment of Section 14.3.1 was encountered at OFDM symbol-normalised Doppler frequencies F_D of 0.1 (**top left**), 0.05 (**top right**) and 0.01 (**bottom**) and **BPSK modulation** was used. The CIR window size was $K_0 = 12$ taps.

σ_{ICI}^2 and the signal variance σ_s^2 is valid only in terms of their long-term statistics. The instantaneous or short-term ICI variance might substantially differ from its long-term value, since the ICI incurred is strongly dependent on the speed of channel variation experienced during a specific OFDM symbol period. We also note from the BER results not portrayed here due to a lack of space that the BER performance degradation incurred as a result of employing potentially error-contaminated demodulated symbols for channel prediction compared to the case of an ideal reference in terms of error-free symbol decisions is relatively limited even at low SNRs. This is due to the BPSK modulation scheme's relative robustness.

15.3.5.3 BER Performance for QPSK Modulation and 16QAM in the Context of an Undecoded Reference

In the top left corner of Figure 15.14 we have summarised once again our simulation results associated with a BPSK modulation assisted turbo-coded system. By contrast, the simulation results associated with a QPSK modulation- and 16QAM assisted turbo-coded system are portrayed at the top right corner and at the bottom of Figure 15.14, respectively. Again, we have provided simulation results for the case of both an imperfect, potentially error-contaminated reference associated with a pilot-based training overhead of 3.125%, as well as for the idealistic benchmark of an error-free reference, which would be generated by receiving error-free training information in every OFDM symbol.

In the context of QPSK modulation and OFDM symbol-normalised Doppler frequencies of $F_D = 0.01$ and $F_D = 0.05$ we infer that turbo coding is capable of eliminating virtually all transmission errors. By contrast, for the more rapidly fluctuating and hence more ICI-contaminated scenario of $F_D = 0.1$ the BER recorded at the input of the turbo decoder is significantly increased, as illustrated in Figure 15.12. However, compared to the case of BPSK modulation, which was discussed in Section 15.3.5.2, a significant BER performance degradation is incurred due to employing an imperfect reference compared to the idealistic scenario of employing an error-free reference. This is particularly true for the case of the more rapidly fading scenarios of $F_D = 0.05$ and $F_D = 0.1$.

In case of employing 16QAM further restrictions are imposed with respect to the tolerable Doppler frequency. Specifically, in our investigations only the slowly fading scenario of $F_D = 0.01$ allows for virtually error-free transmissions at sufficiently high SNRs. By contrast, for $F_D = 0.05$ a BER floor is observed, which is again attributed to the effects of ICI. The rapidly fading scenario of $F_D = 0.1$ was not considered in the context of 16QAM, since in the uncoded case excessive BERs were observed as a result of the severe error propagation between successive OFDM symbols. Note again in Figure 15.14 the significant BER performance degradation incurred compared to the case of employing an error-free ideal reference.

Similarly to the BER performance results presented in Figure 15.12 for a 16QAM assisted system using no channel coding, investigations were also conducted for the turbo coded system with respect to the specific required pilot-based training period duration and to the training block length necessitated for avoiding the effects of the DDCE's instability at SNRs between 15 and $25dB$. The associated results are illustrated in the graph seen at the bottom right corner of Figure 15.14. Again, for the more rapidly-fading scenario of $F_D = 0.05$, the training block is required to be at least two OFDM symbols long for the sake of avoiding an instability. By contrast, for the more slowly-fading scenario of $F_D = 0.01$, where the effects of an instability are not explicitly observed, a comparison between the curves in the bottom

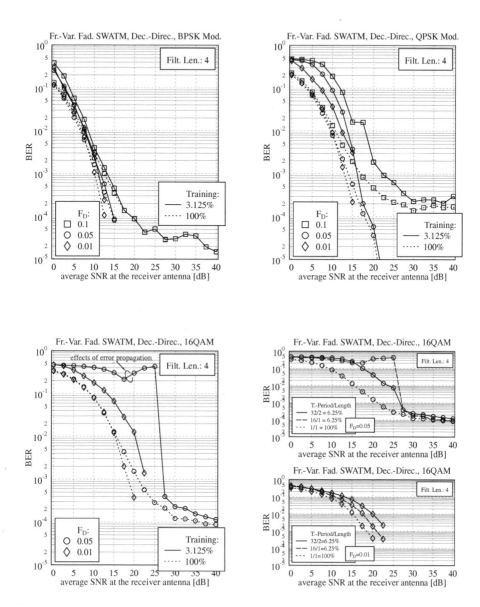

Figure 15.14: Bit Error Ratio (BER) exhibited by a **turbo-coded system** employing **robust decision-directed** *a priori* **channel prediction** using a prediction filter length of 4 taps upon capitalising on potentially error-contaminated, **undecoded symbol decisions**. The **"frame-variant"** fading **indoor WATM channel** environment of Figure 14.4 was encountered at OFDM symbol-normalised Doppler frequencies of $F_D = 0.1, 0.05$ and 0.01 and (**top left**) BPSK, (**top right**) QPSK and (**bottom**) 16QAM modulation schemes were used. The CIR window size was $K_0 = 12$ taps.

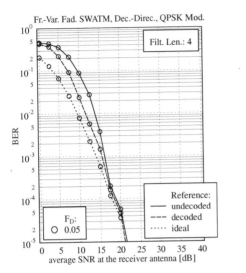

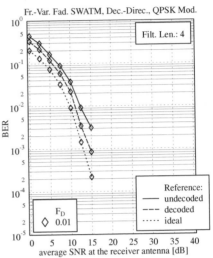

Figure 15.15: **Bit Error Ratio (BER)** exhibited by a **turbo-coded system** employing **robust decision-directed *a priori* channel prediction** using a prediction filter length of 4 taps upon capitalising on an **ideal, undecoded or decoded reference**, respectively. The **"frame-variant"** fading **indoor WATM channel** environment of Figure 14.4 was encountered at OFDM symbol-normalised Doppler frequencies of **(left)** $F_D = 0.05$ and **(right)** $F_D = 0.01$, where QPSK modulation was used. The CIR window size was $K_0 = 12$ taps.

left- and right graph of Figure 15.14 shows that increasing the training block length is not an appropriate measure for improving the system's BER. In this case Figure 15.14 suggests that decreasing the training period duration with the aim of reducing the mean distance between the training OFDM symbols is a more effective measure.

It should also be noted that the systems employing QPSK modulation or 16QAM potentially benefited from the increased channel interleaver length compared to the case of BPSK modulation, since the increased number of bits per modulated subcarrier symbol proportionately increased the number of bits mapped to an OFDM symbol without increasing the system's effective interleaving delay. This was achieved at the cost of an increased decoding complexity at the receiver.

15.3.5.4 BER Performance for QPSK Modulation in the Context of a Decoded Reference

Investigations have also been conducted with respect to employing a turbo decoded reference in the process of decision-directed channel prediction, where the "source"-related soft-output bits of the turbo decoder were sliced, re-encoded and remodulated. In the context of BPSK modulation, which exhibits the highest robustness among the three modulation schemes employed in our simulations, namely BPSK, QPSK and 16QAM, no significant performance advantage was observed when employing a turbo decoded reference instead of an undecoded reference. By contrast, for the more vulnerable QPSK modulation employed in conjunction with a relatively modest training overhead due to transmitting only one training OFDM symbol within every block of 32 OFDM symbols, a notable performance advantage in favour of the turbo decoded reference is observed for intermediate SNRs up to 15dB, where the effects of error propagation are significant. The associated BER performance curves in conjunction with three different scenarios, namely an ideal, an undecoded and a turbo decoded reference are portrayed in Figure 15.15 at the left-hand side for an OFDM symbol normalised Doppler frequency of $F_D = 0.05$ and at the right-hand side for $F_D = 0.05$. Again in contrast, upon invoking 16QAM no BER performance improvement was observed at low OFDM symbol normalised Doppler frequencies, such as $F_D = 0.01$ due to employing a "source"-related soft-output bit based reference. However, for the higher Doppler frequency of $F_D = 0.05$ an excessive BER was incurred, which was caused by the instability of the channel estimator. This is because for some OFDM symbols the turbo decoded reference contained significantly more erroneous subcarrier symbols, than the undecoded or sliced reference, potentially aggravating the effects of error propagation. However, in the context of employing a "source- plus parity"-related soft-output bit based reference, where both the source- and the parity-related soft-output bits of the turbo decoder are sliced and remodulated, a significantly improved performance is expected for all scenarios.

15.3.5.5 Conclusion on the BER Performance of a Turbo-Coded System Employing Robust Decision-Directed Channel Prediction

In Section 15.3.5 the BER of a turbo-coded system employing robust decision-directed channel prediction was evaluated. In our initial simulations conducted with respect to a BPSK-modulated system in Section 15.3.5.2 and with respect to a QPSK- or 16QAM modulated system in Section 15.3.5.3, an undecoded reference was assumed, where the symbols received in different subcarriers were sliced and remodulated without involving turbo decoding- or encoding. More specifically, for the BPSK- and QPSK-modulated systems - given sufficiently high SNRs - error-free transmissions were observed at OFDM symbol normalised Doppler frequencies of $F_D = 0.01$ and $F_D = 0.05$, while for $F_D = 0.1$ the BER curves exhibited a residual value of $2 \cdot 10^{-5}$ and $2 \cdot 10^{-4}$, respectively, which was attributed to the effects of ICI. This was achieved in conjunction with four-tap CIR-related tap prediction filtering. By contrast, for a 16QAM-assisted system error-free transmissions were only observed for the lowest Doppler frequency of $F_D = 0.01$, while for $F_D = 0.05$ a residual BER of 10^{-4} was incurred together with error-propagation and noise-amplification, which were induced by stability problems. Again, these phenomena could potentially be mitigated upon transmitting blocks of training OFDM symbols, rather than a single OFDM symbol. Our further investigations in Section 15.3.5.4 were concerned with the employment of a decoded reference, where the "source"-related soft-output bits of the turbo-decoder were sliced, re-encoded and re-modulated. While the employment of a turbo-decoded reference had no significant impact on the performance of BPSK-assisted transmissions due to the modulation scheme's relative robustness, in the context of QPSK-assisted transmissions a notable BER advantage was observed, when employing only a modest training overhead. Finally, for 16QAM-assisted systems the employment of a turbo-decoded reference was not beneficial, particularly in the context of rapidly fading channel scenarios of e.g. $F_D = 0.05$. Our interpretation of this

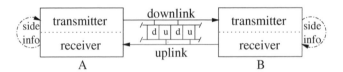

Figure 15.16: Illustration of the Time Division Duplexing (TDD) assisted AOFDM scenario.

phenomenon was that due to generating the DDCE-related reference based on slicing, re-encoding and remodulating the "source"-related soft-output bits of the turbo-decoder only, the detrimental effects of low-SNR subcarriers were smeared over a wider range of sub-carriers. A significant further performance improvement can be achieved upon employing a "source- plus parity"-related soft-output bit based reference, where the source- and parity bits at the output of the turbo-decoder are sliced and remodulated. These beneficial effects will be demonstrated in the context of our investigations on PIC detection-assisted SDMA-OFDM in Section 17.3.2.

15.4 Robust Decision-Directed Channel Prediction-Assisted Adaptive OFDM

Sub-band Adaptive OFDM (AOFDM) has been shown to be an effective method of improving the system's performance in mobile environments, where the sub-bands least affected by frequency selective fading are assigned more bits per subcarrier, than the severely faded sub-bands [2, 219, 482]. The modulation mode assignment to be employed by the remote transmitter A seen in Figure 15.16 in the next downlink timeslot is determined by the local receiver B upon invoking an estimate of the short-term channel quality experienced by the most recently received OFDM symbol. Due to the channel's variation with time, there is a mismatch between the channel quality estimated by receiver B and that actually experienced by the following OFDM symbol transmitted by transmitter A. This potentially limits the achievable performance gain of AOFDM compared to employing a single fixed modulation mode. Hence the application of AOFDM is confined to channel environments exhibiting relatively low Doppler frequencies, if no channel prediction is used. In order to support AOFDM in a broader range of mobility conditions, signal prediction techniques – which are well known from the field of speech-coding [85, 86] for example – can be employed for obtaining a more accurate estimate of the channel quality experienced in the next transmission timeslot on the basis of a weighted sum of that in previous slots. A channel predictor assisted OFDM pre-equalisation scheme was discussed in [70], while prediction assisted decision-directed channel estimation has been proposed in [72]. In our contribution we will study the performance of an AOFDM transceiver, which employs the decision-directed 2D-MMSE channel prediction technique of Sections 15.2.3 and 15.2.4 and AOFDM modulation mode adaptation. It will be demonstrated that this arrangement is advantageous, since potentially both components, namely the process of channel transfer function estimation as well as that of the modulation mode adaptation benefit from the more accurate channel transfer function estimates provided by the predictor.

In Section 15.4.1 we will commence our discussions with an outline of the adaptive

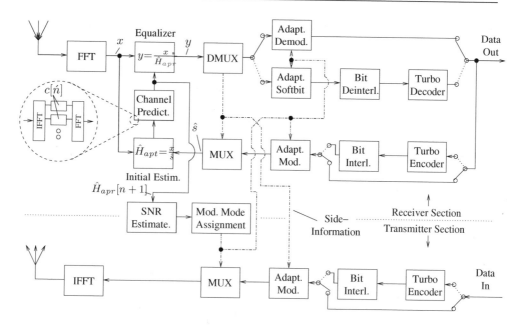

Figure 15.17: Schematic of the sub-band adaptive transceiver employing robust [68] decision-directed channel transfer function prediction. For simplicity the modulation and demodulation of the AOFDM modulation mode related side-information and its associated error protection coding are not shown here. Furthermore, the joint OFDM symbol and subcarrier index $[n, k]$ has been omitted from the different variables associated with the n-th downlink time slot, with the exception of $\hat{H}_{apr}[n + 1]$, which refers to the set of *a priori* subcarrier channel transfer factor estimates for the OFDM symbol received during the $(n + 1)$-th downlink time slot.

transceiver's structure, while in Section 15.4.2 its BER performance is assessed under a variety of channel conditions. Our discussions will be concluded in Section 15.4.3.

15.4.1 Transceiver Structure

The schematic of the adaptive Time Division Duplexing (TDD) OFDM transceiver employed in our simulations is shown in Figure 15.17. The signal received by receiver B of Figure 15.17 from the remote transmitter A in the n-th downlink timeslot is forwarded to a Fast Fourier Transform (FFT) block, followed by the frequency domain equalisation of the complex symbols associated with each of the K subcarriers. Equalisation ensues using the *a priori* channel transfer factor estimates $\hat{H}_{apr}[n, k]$, $k = 0, \ldots , K - 1$ predicted during the $(n - 1)$-th downlink time slot on the basis of the *a posteriori* channel transfer factor estimates $\tilde{H}_{apt}[n - \acute{n}, \acute{k}]$, $\acute{n} = 1, \ldots , N_{tap}^{[t]}$, $\acute{k} = 0, \ldots , K - 1$ of the OFDM symbols transmitted in the previous $N_{tap}^{[t]}$ number of downlink time slots, where each TDD time slot hosts one OFDM symbol. The sub-band modulation mode assignment to be employed by transmitter B of Figure 15.16 for the OFDM symbol transmitted during the next uplink time slot to receiver A is explicitly embedded into the data stream as side-information. As seen in Figure 15.17, the primary data and the AOFDM modem mode signaling streams are separated from each other in the demultiplexer (DMUX) stage of Figure 15.17, followed by adaptive OFDM demodulation of the primary user data. Additionally, turbo coding can be employed in the system, which requires adaptive soft-bit generation at the receiver instead of direct hard-decision based adaptive demodulation, followed by channel-deinterleaving and turbo decoding. The demodulated and turbo decoded data stream is conveyed to the adaptive receiver's output.

Furthermore, the sliced bits are invoked to reconstruct the transmitted OFDM symbol to be used as a reference signal, which allows generating an *a priori* channel estimate for the OFDM symbol received during the $(n+1)$-th downlink time slot. Hence the output bit stream has to be optionally re-encoded and remodulated.

In the proposed arrangement the advantage of employing the 2D-MMSE channel transfer function prediction technique of Section 15.2.4 instead of the 2D-MMSE transfer function estimation process of Section 15.2.3 as proposed by Li *et al.* [68] is two-fold. First, more accurate *a priori* channel estimates are provided for the frequency domain OFDM equalisation at the receiver. Second, the channel quality expressed in terms of the signal-to-noise ratio (SNR) and potentially experienced by an OFDM symbol during the next downlink timeslot can be estimated more reliably. This potentially enhances the performance of our AOFDM scheme in terms of a more accurate modulation mode assignment and more accurate soft-bit values employed in the process of turbo-decoding.

15.4.1.1   Modulation Mode Adaptation

The AOFDM modulation mode adaptation performed by the modem is based on the choice between a set of four modulation modes, namely 4, 2, 1 and 0 bit/subcarrier, where the latter mode corresponds to "no transmission".[14] The AOFDM modulation mode could be in theory assigned on a subcarrier-by-subcarrier basis, but the signalling overhead of such a system would be prohibitive, without significant performance advantages [2]. Hence, we have grouped the adjacent AOFDM subcarriers into "sub-bands" and assigned the same modulation mode to all subcarriers in a sub-band [2,219]. Note that the frequency domain channel transfer function is typically not constant across the subcarriers of a sub-band, hence the modem mode adaptation will be sub-optimal for some of the subcarriers. The modem mode adaptation is achieved on the basis of the *a priori* SNR, SNR_{apr} estimated in each of the K subcarriers for the OFDM symbol hosted by the $(n + 1)$-th downlink time slot, which is formulated as:

$$\hat{\mathrm{SNR}}_{apr}[n+1,k] = |\hat{H}_{apr}[n+1,k]|^2 \frac{\sigma_s^2}{\sigma_n^2}, \tag{15.62}$$

where σ_n^2 is the total noise variance in a subcarrier, given as the sum of the AWGN and the ICI-induced noise variances, expressed as $\sigma_n^2 = \sigma_{\mathrm{AWGN}}^2 + \sigma_{\mathrm{ICI}}^2$. The iterative AOFDM mode assignment commences by calculating in the first step for each sub-band and for all four modulation modes the expected overall sub-band BER by means of averaging the estimated individual subcarrier BERs [2]. Throughout the second step of the algorithm - commencing

---

[14]Note that in order to support the process of decision-directed channel prediction, BPSK modulated random sequences - known to the receiver - were also assigned to those sub-bands to which zero bits were mapped by the modulation mode adaptation procedure.

with the lowest-throughput but most robust modulation mode in all sub-bands - in each iteration the number of bits/subcarrier of that particular sub-band is increased, which provides the
best compromise in terms of increasing the number of expected bit errors and the number of
additional data bits accommodated. This process continues, until the target number of bits to
be transmitted by the OFDM symbol is reached. This algorithm originates from the philosophy of the Hughes-Harthogs algorithm [483]. As a result of intensive research in the area
recently several computationally efficient versions of the algorithm have emerged [2]. Again,
the computed AOFDM mode assignment is explicitly signalled to the remote transmitter A
of Figure 15.17 on the next uplink OFDM symbol transmitted by transmitter B and it is also
stored locally in receiver B for employment during the forthcoming downlink time slot. In
the next section we will embark on the performance assessment of the proposed system.

15.4.2   BER Performance

In this section the performance of the decision-directed channel prediction assisted sub-band
adaptive OFDM transceiver will be assessed in the context of the indoor WATM system-
and channel environment of Section 14.3.1. The structure of this section is as follows. In
Section 15.4.2.1 we motivate the employment of channel prediction techniques in the context
of AOFDM systems by highlighting the impact of the channel's decorrelation as a function of
time on the accuracy of the modulation mode assignment. In Section 15.4.2.2 we characterise
the performance of the AOFDM system without channel coding in the context of decision-
directed channel prediction. By contrast, in Section 15.4.2.3 the performance of the decision-
directed channel prediction-assisted AOFDM system is assessed in conjunction with turbo
coding.

15.4.2.1   Motivation of Channel Transfer Function Prediction-Assisted AOFDM

It has been demonstrated in various publications [2,219] that the employment of the constant-
throughput adaptive modulation technique is a viable approach for exploiting the different
sub-band qualities imposed by a wideband channel. In the context of OFDM it would be best
to adapt the modulation mode assignment on a subcarrier-by-subcarrier basis, but as outlined
in Section 15.4.1.1, the associated signalling overhead would be prohibitive. Hence, sets
of adjacent similar quality subcarriers are usually grouped into sub-bands, and an identical
modulation mode is assigned to each sub-band by the adaptation procedure. The maximum
bandwidth of each sub-band is related to the channel's coherence bandwidth, namely to the
bandwidth over which the fading of adjacent subcarriers can be considered correlated, as it
was argued in Section 14.2.

   In order to highlight the influence of the sub-band width on the performance of the
AOFDM modem and to motivate our specific choice of the sub-band width employed in the
context of the simulation results presented throughout this section at the left-hand side of Figure 15.18 we have plotted the BER performance of the AOFDM modem as a function of the
SNR encountered at the reception antenna, parametrised with the number of sub-bands per
OFDM symbol. These results were generated upon invoking the indoor WATM system- and
channel model of Section 14.3.1. The throughput of the AOFDM modem was assumed to be
identical to that of an OFDM modem employing QPSK modulation, which was 1024 bits per
OFDM symbol (BPOS), when using the indoor WATM system's parameters. The modulation

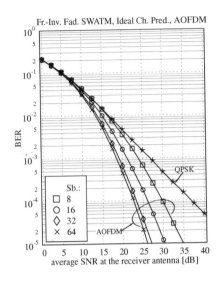
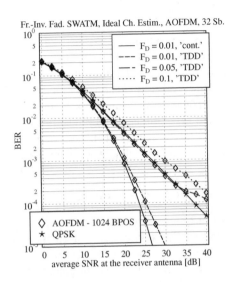

Figure 15.18: **(Left):** **Bit Error Ratio (BER)** exhibited by an AOFDM system employing perfect channel prediction as a function of the SNR at the reception antenna and further parametrised with the number of sub-bands invoked in the modulation mode adaptation. **(Right):** **Bit Error Ratio (BER)** exhibited by a 32 sub-band AOFDM system employing perfect channel estimation as a function of the SNR at the reception antenna and further parametrised with the OFDM symbol-normalised Doppler frequency of the channel. The simulations were conducted in the **"frame-invariant"** fading **indoor WATM channel** environment of Section 14.3.1. The throughput of the AOFDM modem was equivalent to that of an AOFDM system employing fixed QPSK modulation, namely 1024 Bit per OFDM Symbol (BPOS).

mode adaptation performed during the n-th OFDM symbol period for the $(n + 1)$-th OFDM symbol period was based on perfect knowledge of the channel transfer function experienced during the $(n + 1)$-th OFDM symbol period. We observe that by increasing the number of sub-bands per OFDM symbol beyond 32, no significant further performance improvement is attained in the context of the specific indoor WATM channel model. Compared to the BER performance of the fixed-mode QPSK modulation assisted OFDM modem, which has also been plotted in Figure 15.18, the BER of the AOFDM modem is substantially improved.

Our previous investigations were conducted under the idealistic assumption of employing perfect knowledge of the channel transfer function experienced by the OFDM symbol during the $(n + 1)$-th OFDM symbol period, which was used for the AOFDM modulation mode adaptation performed in the n-th OFDM symbol period. To be more realistic and in order to motivate the employment of channel transfer function prediction techniques in the AOFDM modem, let us now stipulate the availability of perfect channel transfer function knowledge only for the n-th OFDM symbol period, which is again invoked in the adaptation of the modulation mode assignment to be used during the $(n + 1)$-th OFDM symbol period.

The associated BER simulation results are portrayed on the right-hand side of Figure 15.18. Again, we have plotted both the performance of the fixed-mode QPSK modulation assisted OFDM system, as well as that of the AOFDM modem of the same throughput, namely 1024 bits per OFDM symbol (BPOS). A total of 32 sub-bands per OFDM symbol was used. As described in Section 15.4.1, the modulation mode assignment for the $(n + 1)$-th downlink timeslot was transmitted to the remote transmitter A of Figure 15.17 on the previous uplink timeslot, assuming a slot-by-slot uplink/downlink TDD scenario. Furthermore, we have also considered the rather idealistic scenario of a continuous transmission of downlink time slots only, where the AOFDM modulation mode assignment for the $(n+1)$-th downlink time slot is instantaneously signalled to the remote transmitter A of Figure 15.17 after the n-th downlink timeslot. Note that in the context of the above TDD scenario the effective Doppler frequency with respect to the channel's decorrelation incurred between two consecutive downlink times-lots is actually doubled compared to the scenario of a continuous transmission. In the graph on the right-hand side of Figure 15.18 we observe that as a result of the channel's decorre-lation experienced at the relatively high OFDM symbol normalised Doppler frequencies of $F_D = 0.05$ and $F_D = 0.1$, the AOFDM modem performs in fact worse, than the fixed-mode QPSK modulation assisted OFDM system. This is because the receiver is expected to per-form the modulation mode adaptation virtually based on near-uncorrelated channel estimates. As it will be demonstrated in the following paragraphs, this problem can be mitigated upon invoking channel transfer function prediction techniques.

15.4.2.2   BER Performance of the Uncoded System

While in the context of our investigations presented in Section 15.4.2.1 we capitalised on the idealistic assumption of benefitting from perfect channel transfer function knowledge, here we will investigate the more realistic environment where the channel transfer function estimates for the $(n + 1)$-th OFDM symbol period are generated during the n-th OFDM symbol period with the aid of a decision-directed channel predictor. Both the demodulation of the transmitted subcarrier symbols performed during the $(n+1)$-th OFDM symbol period, as well as the accuracy of the AOFDM modulation mode assignment computed during the n-th OFDM symbol period for use during the $(n+1)$-th OFDM symbol period will benefit from the more accurate estimates provided by the channel transfer function predictor of Section 15.2.4 instead of the estimator of Section 15.2.3 originally proposed in [68].

Furthermore, we have dropped the idealistic assumption of encountering a frame-invariant fading channel in favour of the more realistic frame-variant fading channel sce-nario, with the consequence of incurring an ICI-induced limitation of the effective SNR in the demodulation process, regardless of the SNR at the reception antenna.

The corresponding BER simulation results are portrayed in Figure 15.19 at the left-hand side for an OFDM symbol normalised Doppler frequency of $F_D = 0.05$, while at the right-hand side for $F_D = 0.01$. Simulation results are provided both for the case of a sliced potentially error-contaminated reference, where the received symbols are demodulated, sliced and remodulated, as well as for an idealistic error-free reference due to stipulating perfect knowledge of the transmitted symbols at the receiver. In the scenario of employing a sliced reference a relatively modest pilot-based training overhead due to transmitting one dedicated training OFDM symbol in every block of 32 OFDM symbols was imposed. Furthermore, the BER curves corresponding to the fixed-mode BPSK- or QPSK modulation assisted scenarios

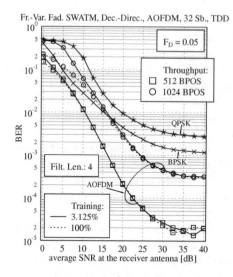
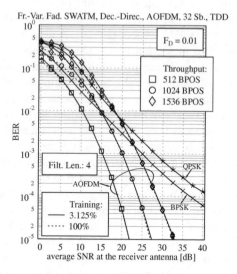

Figure 15.19: **Bit Error Ratio (BER)** exhibited by an AOFDM system employing **robust decision-directed *a priori* channel prediction** using a prediction filter length of 4 taps as a function of the SNR at the reception antenna, and parametrised with the throughput in terms of the number of bits transmitted per OFDM symbol. Both an ideal error-free and a sliced potentially error-contaminated reference were employed in the process of decision-directed channel prediction. The **"frame-variant"** fading **indoor WATM channel** environment of Figure 14.4 was encountered at OFDM symbol-normalised Doppler frequencies F_D of (**left:**) 0.05 and (**right:**) 0.01. The BER performance curves associated with fixed BPSK and QPSK modulation are also plotted as a reference. The CIR window size was $K_0 = 12$ taps.

have been plotted as a reference. Again, a CIR-related tap prediction filter length of four taps was invoked in our simulations.

Note that for the more rapidly fading scenario of $F_D = 0.05$ no BER curve is provided for a throughput of 1536 bits per OFDM symbol (BPOS). This is, because as outlined in Section 15.3.4.2 in the context of the fixed modulation mode based scenarios, the channel predictor tends to become unstable due to error propagation effects, when employing the relatively vulnerable 16QAM modulation scheme in rapidly fading channel environments.

We observe that the AOFDM scheme is capable of significantly outperforming the fixed modulation assisted OFDM scheme. More specifically, for an OFDM symbol normalised Doppler frequency of $F_D = 0.01$ the channel SNR gain due to employing adaptive modulation is in excess of 15dB at a BER of 10^{-4} for both a BPSK- and a QPSK based fixed-mode equivalent throughput system. Similar advantages are observed under the more rapidly fading channel conditions of $F_D = 0.05$, where a comparison to the performance results of Figure 15.18 reveals that the employment of adaptive modulation is facilitated by the four-tap channel prediction filter. In contrast to the case of $F_D = 0.01$ a BER floor is observed at higher SNRs, which is due to the effects of the ICI. At these SNRs the BER reduction attained with the advent of adaptive modulation compared to fixed modulation was about two orders of magnitude in the context of a throughput of 512 bits per OFDM symbol. At the higher throughput of 1024 bits per OFDM symbol the performance advantage of adaptive modulation was still about an order of magnitude.[15] We also note that not only the modulation mode adaptation benefitted from the enhanced channel transfer function accuracy due to employing prediction filtering in the process of decision-directed channel estimation, but also the process of decision-directed channel estimation itself gained from the more reliable symbol decisions delivered by adaptive modulation.

In conclusion, our discussions in this section demonstrated that with the aid of channel prediction filtering constant bitrate adaptive modulation techniques can be supported even at relatively high OFDM symbol-normalised Doppler frequencies.

15.4.2.3   BER Performance of the Turbo-Coded System

Having portrayed the BER versus SNR performance of the uncoded decision-directed channel prediction assisted AOFDM system in Section 15.4.2.2, we will now investigate the achievable performance with the aid of turbo-coding. The associated turbo coding specific parameters were summarised in Table 15.6. Again, we employ a sliced reference, generated upon performing hard decisions to the received subcarrier symbols, followed by remodulation. We demonstrated earlier in Section 15.3.5.4 that employing a turbo-decoded reference based on the "source"-related soft-output bits of the turbo-decoder provides a relatively modest performance advantage compared to a sliced reference, in conjunction with QPSK modulation, while in the context of BPSK modulation and 16QAM no performance advantage was observed.

The simulation results are portrayed in Figure 15.20, at the left-hand side for an OFDM symbol normalised Doppler frequency of $F_D = 0.05$, while at the right-hand side for $F_D = 0.01$. In contrast to the system using no turbo-coding, which was characterised in terms of its BER in Section 15.4.2.2, the performance advantage of the AOFDM system in comparison to a system having the same throughput, but using a fixed modulation mode is significantly reduced. A slight advantage in favour of the AOFDM system is observed at relatively low SNRs, where the decision-directed channel predictor is sensitive to the effects of error propagation. This advantage is further eroded upon invoking a potentially error-contaminated decoded reference signal in the process of decision-directed channel prediction. The reason for the similar performance exhibited by the turbo-coded AOFDM- and fixed-mode BPSK- or QPSK modulation assisted OFDM systems is that both adaptive modulation and turbo coding attempt to mitigate the same channel impairments, namely the time and frequency-dependent channel quality offered by the wideband channel.

---

[15]Note that here we have not taken into account the AOFDM mode signalling overhead imposed by explicitly signalling the modulation mode assignment to the remote transmitter. This is necessary, unless blind modulation mode detection techniques [484] are employed, which are known to be relatively inaccurate at low SNRs. Hence, for the sake of a fairer comparison to the fixed modulation mode assisted system, the number of sub-bands available to the procedure of modulation mode adaptation for data transmissions would have to be reduced appropriately. To be more explicit, in case of 32 sub-bands per OFDM symbol and four available modulation modes per sub-band to be encoded in groups of two bits, the total number of signalling bits per OFDM symbol would be 64. Furthermore, upon invoking halfrate channel coding for providing error protection to the signalling bits, the total overhead is increased to 128 bits per OFDM symbol.

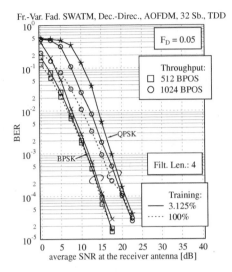
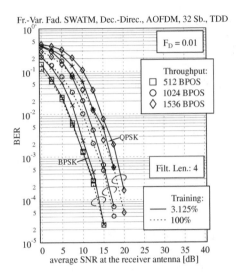

Figure 15.20: Bit Error Ratio (BER) exhibited by a turbo-coded AOFDM system employing **robust decision-directed *a priori* channel prediction** using a prediction filter length of 4 taps as a function of the SNR encountered at the reception antenna, and parametrised with the throughput in terms of the number of bits transmitted per OFDM symbol. The turbo coding parameters were given in Table 15.6. Both an ideal and a sliced reference were employed in the process of decision-directed channel prediction. The **"frame-variant"** fading **indoor WATM channel** environment of Figure 14.4 was encountered at OFDM symbol-normalised Doppler frequencies of **(left:)** $F_D = 0.05$ and **(right:)** $F_D = 0.01$. The BER performance curves associated with fixed mode BPSK and QPSK modulation are also plotted as a reference. The CIR window size was $K_0 = 12$ taps.

We concluded that although we were unable to demonstrate a significant advantage of turbo-coded AOFDM compared to turbo coded fixed-mode BPSK- or QPSK modulation assisted OFDM at high Doppler frequencies, we have shown that adaptive modulation assisted OFDM can in fact be invoked even in channel scenarios having relatively high OFDM symbol normalised Doppler frequencies with the advent of capitalising on FIR Wiener filter prediction techniques.

15.4.3 Conclusion on Robust Decision-Directed Channel Prediction Assisted AOFDM

In Section 15.4 we highlighted that with the aid of channel transfer function prediction techniques constant throughput AOFDM systems can be operated in the context of relatively rapidly fluctuating channel conditions, characterised by high OFDM symbol-normalised Doppler frequencies. In our simulations AOFDM transmissions were successfully demonstrated for Doppler frequencies of $F_D = 0.01$ and 0.05. For the more rapidly fading channel associated with $F_D = 0.05$ the range of throughputs supported was restricted to values below 1536 bits per OFDM symbol. At higher throughputs the decision-directed channel predictor became unstable due to the excessive activation of the relatively vulnerable 16QAM modulation mode. A similar behaviour was also observed in Section 15.3.4.2 for fixed-mode 16QAM assisted OFDM transmissions. In the context of an uncoded system not only the modulation mode adaptation conducted during the n-th OFDM symbol period for the transmission of the OFDM symbol associated with the $(n + 1)$-th downlink period benefitted from the accurate channel estimates provided by the predictor, but also the decision-directed channel predictor gained from the less error-contaminated reference with the advent of the modulation mode adaptation. Furthermore, a turbo-coded AOFDM system was also considered. A slight performance advantage was observed for the turbo-coded AOFDM scheme at a given throughput compared to a turbo-coded BPSK or QPSK modulation assisted OFDM system. In the turbo-coded cases the main advantage of adaptive modulation remains its flexibility with respect to the desired throughput, since fixed-mode transmission is only capable of conveying 512, 1024 or 2048 bits per OFDM symbol.

15.5 Chapter Summary and Conclusion

15.5.1 Description

In Chapter 15 decision-directed channel estimation (DDCE) and prediction (DDCP) has been investigated in the context of single-user single-transmit antenna based OFDM systems. The corresponding OFDM receiver's associated structure was shown in Figure 15.1.

Our specific discussions commenced in Section 15.2.1 with an outline of the procedure of generating *a posteriori* least-squares channel transfer factor estimates based on the received subcarrier signals $x[n, k]$ and based on the classified symbols $\check{s}[n, k]$ that are most likely to have been transmitted in the different subcarriers $k = 0, \ldots, K - 1$. Furthermore, in Section 15.2.2 the 1D-MMSE channel transfer function estimator proposed by Edfors *et al.* [62, 69, 157] and Sandell [158] to inferr improved *a posteriori* channel transfer factor estimates from the initial least-squares channel transfer factor estimates of Section 15.2.1 was described. Based on these ideas Li *et al.* [68] proposed a 2D-MMSE channel transfer function estimator, which additionally capitalises on the least-squares channel transfer factor estimates of past OFDM symbols. This approach was outlined in Section 15.2.3 with an emphasis on the aspects of "robustness" as advocated by Li *et al.* [68].

As a result of the potentially excessive decorrelation of the channel transfer function experienced by consecutive OFDM symbols as demonstrated in Section 15.2.3.3, we found that it is more beneficial to employ a CIR-related tap prediction filter, which operates in the time direction instead of the CIR-related tap estimation filter as in Li's contribution [68]. More explicitly, instead of generating improved *a posteriori* channel transfer factor estimates for the n-th OFDM symbol period, which are then used - upon neglecting the channel's decorrelation - as *a priori* channel transfer factor estimates for demodulation during the $(n + 1)$-th OFDM symbol period, we directly generate *a priori* channel transfer factor predictions for the $(n + 1)$-th OFDM symbol period during the n-th OFDM symbol period. The specific structure of the MMSE CIR-related tap prediction filter was discussed in Section 15.2.4. This included the derivation of an expression for the subcarrier-averaged *a priori* prediction MSE in Section 15.2.4.6.

During our performance study provided in Section 15.3 two approaches were considered for evaluating the CIR-related tap predictor coefficients based on the Wiener solution of Equation 15.38. These were the robust approach advocated by Li *et al.* [68], which was also discussed in Section 15.2.3 and the block-adaptive approach proposed by Al-Susa and Ormondroyd [72], which is based on the Burg algorithm [85].

Specifically, in the context of the robust approach of Section 15.2.3, a uniform, ideally support-limited Doppler power spectrum was employed, which is related to a sinc-shaped spaced-time correlation function $\tilde{r}_H^{[f]}[\Delta n]$, as given by Equation 15.55. Furthermore, the different CIR-related tap variances were assumed to be identical, which is related to imposing a uniform, ideally support-limited multipath intensity profile in the design of the estimator. Recall from our discussions in Section 15.2.3.2 that the estimator's MSE can be rendered insensitive with respect to the specific shape of the channel's scattering function, provided that we can capitalise on an infinite number of previous OFDM symbols in the time direction and on an infinite number of subcarriers in the frequency direction. However, there is an exceptional case, when perfect robustness can be achieved with respect to the channel scattering function's associated multipath intensity profile, namely in the context of a sample-spaced CIR.

In contrast to the robust approach, the block-adaptive approach employs estimates of the CIR-related taps' correlation in the time direction, which are periodically generated from blocks of consecutive OFDM symbols. Note, however, that these estimates do not become explicitly available, since the Burg algorithm of [85] directly produces the optimum predictor weights. In order to avoid the inconvenience of storing a potentially large number of previous OFDM symbols, which may exceed the $N_{tap}^{[t]}$ number of OFDM symbols employed in the actual prediction process, an attractive alternative to the block-adaptive approach is an OFDM symbol-by-symbol adaptive approach, which could be based on the RLS algorithm. This will be demonstrated in Section 16.6 in the context of our investigations on DDCP designed for multi-user OFDM systems.

15.5.2 Performance Assessment

The performance investigations conducted in Section 15.3 commenced with a portrayal of both the robust and that of the adaptive decision-directed channel predictor's MSE in Sections 15.3.1 and 15.3.2, respectively, upon assuming error-free symbol decisions. We found that as a result of the CIR-related tap-by-tap adaptation of the predictor coefficients, the *a priori* estimation MSE performance achieved with the aid of the block-adaptive predictor - assuming the availability of a sufficiently high number of previous OFDM symbols' CIR-related *a posteriori* tap estimates - is potentially lower than that of the robust predictor. However, in the specific case of encountering a uniform multipath intensity profile, both predictors are expected to exhibit the same MSE performance. Our detailed conclusions for these investigations were offered in Sections 15.3.1.5 and 15.3.2.4, respectively. Furthermore, in Section 15.3.3 we assessed the robust predictor's MSE in the context of an uncoded system, based on Monte Carlo simulations. Again, the related detailed conclusions were offered in Section 15.3.3.4. The uncoded system's corresponding BER performance in conjunction with various modulation schemes, namely BPSK, QPSK and 16QAM was studied in Section 15.3.4 and our detailed conclusions were provided in Section 15.3.4.3. In order to even further reduce the DDCE assisted system's BER, additionally turbo-coding was invoked, which was the topic of Section 15.3.5. Our detailed conclusions were offered in Section 15.3.5.5.

In order to summarise, from the investigations of Section 15.3 we conclude that upon invoking 2D-MMSE channel transfer function prediction filtering, decision-directed channel estimation is rendered attractive even in the context of high-mobility channel scenarios associated with OFDM symbol-normalised Doppler frequencies ranging up to $F_D = 0.1$. We found that already in conjunction with a relatively short linear prediction filter length of four taps, most of the channel's time direction correlation can be exploited, while a further increase of the predictor's length results in a further reduction of the channel noise. At higher values of F_D, the BER performance was rather limited due to the effects of ICI. Even in conjunction with turbo coding for example, only the most robust BPSK modulation was capable of facilitating near error-free OFDM transmissions in the worst-case scenario of $F_D = 0.1$. For the less error-resilient QPSK modulation scheme the $F_D = 0.05$ scenario was found to be the highest-speed environment, where reliable operation was feasible and for 16QAM even less time-variant channel conditions were required. This was a result of the instability observed for the robust predictor – specifically at higher OFDM symbol-normalised Doppler frequencies – which was mainly due to the increased amount of ICI. This caused an increased number of subcarrier symbol decisions, which were then propagated to the following OFDM symbols with the effects of a "noise amplification problem" due to the robust estimator's inflexibility to react to the degraded channel conditions caused by the subcarrier symbol errors. In this context the employment of the adaptive predictor would be advantageous, with the disadvantage of having a higher complexity as a result of the coefficient adaptation.

15.5.3   Adaptive OFDM Transceiver

Our further investigations conducted in Section 15.4 concentrated on employing robust decision-directed channel prediction in the context of a constant-throughput AOFDM TDD system. After a description of the transceiver's structure in Section 15.4.1 the associated performance assessments of Section 15.4.2 considered both an uncoded as well as a turbo-coded scenario. Our more detailed conclusions concerning these investigations were offered in Section 15.4.3. In order to summarise, we found that with the aid of channel transfer function prediction techniques constant throughput AOFDM TDD systems can be reliably supported in the context of even relatively rapidly fluctuating channel conditions. In our simulations AOFDM transmissions were successfully demonstrated for Doppler frequencies of $F_D = 0.01$ and 0.05. For the more rapidly fading channel associated with $F_D = 0.05$ the range of throughputs supported was restricted to values below 1536 bits per OFDM symbol. Having characterised the family of single-user decision-directed channel estimation- and prediction techniques in this chapter, in the next chapter we will study multi-user channel estimation techniques.

15.6   List of Symbols Used in Chapter 15

$c_{\{est/pre\}}[\acute{n}, l]$:   Estimator/predictor coefficient associated with the l-th CIR-related tap of the OFDM symbol having an offset of $-\acute{n}$, where $\acute{n} \in \{0, \ldots, N_{tap}^{[t]} - 1\}$, relative to the current OFDM symbol.

$\mathbf{c}_{\{est/pre\}}[l]$:   Vector of estimator/predictor coefficients $c_{\{est/pre\}}[\acute{n}, l]$, $\acute{n} = 0, \ldots, N_{tap}^{[t]} - 1$ associated with the l-th CIR-related tap: $\mathbf{c}_{\{est/pre\}}[l] \in \mathbb{C}^{N_{tap}^{[t]} \times 1}$.

$\mathbf{C}_{est}[\acute{n}]$:   Diagonal matrix of CIR-related tap estimator coefficients associated with the OFDM symbol having an offset of $-\acute{n}$, where $\acute{n} \in \{0, \ldots, N_{tap}^{[t]} - 1\}$, relative to the current OFDM symbol: $\mathbf{C}_{est}[\acute{n}] \in \mathbb{C}^{K \times K}$.

$e_{pre}[n+1, l]$:   Prediction error of the l-th CIR-related tap observed during the $(n+1)$-th OFDM symbol period, when using the prediction generated during the n-th OFDM symbol period.

$\mathbf{e}_{pre}[n+1]|_{K_0}$:   Vector of the K CIR-related taps' prediction errors observed during the $(n+1)$-th OFDM symbol period when using the predictions generated during the n-th OFDM symbol period, upon assuming a rank-K_0 predictor: $\mathbf{e}_{pre}[n+1]|_{K_0} \in \mathbb{C}^{K \times 1}$.

$\mathbf{E}_{pre}[n+1]|_{K_0}$:   Frequency-domain representation of the vector $\mathbf{e}_{pre}[n+1]|_{K_0}$ of the K CIR-related taps' prediction errors observed during the $(n+1)$-th OFDM symbol period when using the predictions generated during the n-th OFDM symbol period, upon assuming a rank-K_0 predictor: $\mathbf{E}_{pre}[n+1]|_{K_0} = \tilde{\mathbf{U}}^{[f]} \mathbf{e}_{pre}[n+1]|_{K_0} \in \mathbb{C}^{K \times 1}$.

$h_{expo}[l]$:   Sample-spaced negative exponentially decaying CIR: $h_{expo}[l] = \beta_{expo} e^{-\alpha_{expo} l}$, $l = 0, \ldots, K_0 - 1$.

$h[n, l]$: l-th CIR-related tap associated with the n-th OFDM symbol period.

$\tilde{h}_{apt}[n, l]$:   CIR-related *a posteriori* least-squares tap estimate.

$\hat{h}_{apt}[n, l]$:   Improved CIR-related *a posteriori* tap estimate.

$\hat{h}_{apr}[n+1, l]$:   CIR-related *a priori* tap estimate generated during the n-th OFDM symbol period for employment during the $(n+1)$-th OFDM symbol period.

$\mathbf{h}[n]$:   Vector of the K different CIR-related taps $h[n, l]$, $l = 0, \ldots, K - 1$: $\mathbf{h}[n] = \tilde{\mathbf{U}}^{[f]H} \mathbf{H}[n] \in \mathbb{C}^{K \times 1}$. [16]

$\tilde{\mathbf{h}}_{apt}[n]$:   Vector of the K different CIR-related *a posteriori* least-squares tap estimates $\tilde{h}_{apt}[n, l]$, $l = 0, \ldots, K - 1$: $\tilde{\mathbf{h}}_{apt}[n] = \tilde{\mathbf{U}}^{[f]H} \tilde{\mathbf{H}}_{apt}[n] \in \mathbb{C}^{K \times 1}$. [17]

---

[16]Note that the choice of $\tilde{\mathbf{U}}^{[f]}$ is arbitrary, as long as the matrix is unitary. However, the optimum choice is $\tilde{\mathbf{U}}^{[f]} = \mathbf{U}^{[f]}$. In this case the CIR-related taps are uncorrelated.

[17]See Footnote 16.

$\hat{\mathbf{h}}_{apt}[n]$:

Vector of the K different improved CIR-related *a posteriori* tap estimates, namely $\hat{h}_{apt}[n,l]$, $l = 0, \ldots, K-1$: $\hat{\mathbf{h}}_{apt}[n] = \tilde{\mathbf{U}}^{[f]H}\hat{\mathbf{H}}_{apt}[n] \in \mathbb{C}^{K \times 1}$. ([18])

$\mathbf{h}[n,l]$:

Vector of the l-th CIR-related tap's current plus the $N_{tap}^{[t]} - 1$ number of past values, namely $h[n + \acute{n}, l]$, $\acute{n} = -N_{tap}^{[t]} + 1, \ldots, 0$ each associated with either the current or one of the past $N_{tap}^{[t]} - 1$ number of OFDM symbols: $\mathbf{h}[n,l] \in \mathbb{C}^{N_{tap}^{[t]} \times 1}$.

$\tilde{\mathbf{h}}_{apt}[n,l]$:

Vector of the l-th CIR-related tap's current plus the $N_{tap}^{[t]} - 1$ number of past OFDM symbols' *a posteriori* least-squares estimates, namely $\tilde{h}_{apt}[n + \acute{n}, l]$, $\acute{n} = -N_{tap}^{[t]} + 1, \ldots, 0$ each associated with either the current or one of the $N_{tap}^{[t]} - 1$ number of past OFDM symbols: $\tilde{\mathbf{h}}_{apt}[n,l] \in \mathbb{C}^{N_{tap}^{[t]} \times 1}$.

$H[n,k]$:

Channel transfer factor associated with the k-th subcarrier of the n-th OFDM symbol.

$\tilde{H}_{apt}[n,k]$:

A posteriori least-squares channel transfer factor estimate.

$\hat{H}_{apt}[n,k]$:

Improved *a posteriori* channel transfer factor estimate.

$\hat{H}_{apr}[n+1,k]$:

A priori channel transfer factor estimate generated during the n-th OFDM symbol period for the $(n + 1)$-th OFDM symbol duration.

$\mathbf{H}[n]$:

Vector of the K different subcarriers' channel transfer factors $H[n,k]$, $k = 0, \ldots, K - 1$: $\mathbf{H}[n] \in \mathbb{C}^{K \times 1}$.

$\tilde{\mathbf{H}}_{apt}[n]$:

Vector of the K different subcarriers' *a posteriori* least-squares channel transfer factor estimates $\tilde{H}_{apt}[n,k]$, $k = 0, \ldots, K - 1$: $\tilde{\mathbf{H}}_{apt}[n] \in \mathbb{C}^{K \times 1}$.

$\hat{\mathbf{H}}_{apt}[n]$:

Vector of the K different subcarriers' improved *a posteriori* channel transfer factor estimates $\hat{H}_{apt}[n,k]$, $k = 0, \ldots, K - 1$: $\hat{\mathbf{H}}_{apt}[n] \in \mathbb{C}^{K \times 1}$.

$\hat{\mathbf{H}}_{apr}[n+1]$:

Vector of the K different subcarriers' *a priori* channel transfer factor estimates $\hat{H}_{apr}[n + 1, k]$, $k = 0, \ldots, K - 1$: $\hat{\mathbf{H}}_{apr}[n + 1] \in \mathbb{C}^{K \times 1}$.

$J_0(x)$:

Zero-order Bessel function of the first kind.

$J(\mathbf{c}_{pre}[l])$:

MSE-related cost-function used for deriving the l-th CIR-related tap's vector of optimum predictor coefficients.

K_0:

Number of significant CIR-related taps.

---

[18]See Footnote 16.

$\mathcal{M}_c$:

Set of M_c number of constellation points associated with the modulation scheme employed.

M_s:

Number of OFDM symbols contained in the input block of the Burg algorithm-assisted predictor. Note that we have $M_s \geq N_{tap}^{[t]}$.

$\text{MSE}_{dec}[n, k]$:

Subcarrier-based MSE imposed by the channel's time-variance in the absence of prediction filtering, when using the channel transfer factors of the current OFDM symbol for the equalisation conducted during the next OFDM symbol period. In the context of Jakes' Doppler spectrum the decorrelation-induced MSE is denoted as $\text{MSE}_{dec,J}[n, k]$.

$\text{MSE}_{apr}[l]\big|_{(\mathbf{c}_{pre}[l])}$: MSE associated with the linear prediction of the l-th CIR-related tap in the context of using the arbitrary vector $\mathbf{c}_{pre}[l]$ of predictor coefficients.

$\text{MMSE}_{apr}[l]$:

Minimum MSE associated with the linear prediction of the l-th CIR-related tap in the context of using the vector $\mathbf{c}_{pre}[l]\big|_{opt}$ of optimum predictor coefficients.

$\overline{\text{MSE}}_{apr}\big|_{K_0}$:

Subcarrier-averaged *a priori* estimation MSE of the rank-K_0 2D-MMSE based channel predictor in the context of arbitrary estimator coefficients.

$\overline{\text{MSE}}_{apt}\big|_{K_0}^{N_{tap}^{[t]}=1}$:

Subcarrier-averaged *a posteriori* estimation MSE of the rank-K_0 1D-MMSE based channel estimator in the context of arbitrary estimator coefficients.

$\overline{\text{MMSE}}_{apt}\big|^{N_{tap}^{[t]}\to\infty}$: Subcarrier-averaged *a posteriori* estimation MSE of the 2D-MMSE based channel estimator employing no rank-reduction in the context of the optimum estimator coefficients, when using infinite-length CIR-related tap estimation filters.

$\overline{\text{MMSE}}_{apt}\big|_{rob}^{N_{tap}^{[t]}\to\infty}$: Subcarrier-averaged *a posteriori* estimation MSE of the 2D-MMSE based channel estimator employing no rank-reduction in the context of the sub-optimum estimator coefficients based on the assumptions of 'robustness' with respect to the channel's multipath intensity profile, when using infinite-length CIR-related tap estimation filters.

$n[n, k]$:

AWGN contribution having a variance of σ_n^2.

$N_{tap}^{[t]}$:

Number of filter taps associated with the CIR-related tap estimator- or predictor.

r_{expo}:

Amplitude ratio of the sample-spaced exponential CIR: $r_{expo} = \frac{h_{expo}[K_0-1]}{h_{expo}[0]}$.

$r_H^{[f]}[\Delta k]$:

Spaced-frequency correlation function in the context of the OFDM system: $r_H^{[f]}[\Delta k] = r_H^{[f]}(\Delta k \Delta F) = r_H(\Delta k \Delta F)$.

$r_H^{[t]}[\Delta n]$:

Spaced-time correlation function in the context of the OFDM system: $r_H^{[t]}[\Delta n] = r_H^{[t]}(\Delta n T_f) = r_H(\Delta n T_f)$. Upon assuming Jakes' Doppler spectrum whose associated correlation function is given by the zero-order Bessel function of the first kind, we have $r_{H,J}^{[t]}[\Delta n] = J_0(\Delta n \cdot \omega_D)$.

$\mathbf{R}^{[f]}$:

Spaced-frequency correlation matrix: $\mathbf{R}^{[f]} = E\{\mathbf{H}[n]\mathbf{H}^H[n]\} \in \mathbb{C}^{K \times K}$. The Eigen-Value Decomposition (EVD) of the correlation matrix $\mathbf{R}^{[f]}$ is given by $\mathbf{R}^{[f]} = \mathbf{U}^{[f]}\Lambda^{[f]}\mathbf{U}^{[f]H}$.

$\mathbf{r}_h^{[t]}[l]$:

Spaced-time cross-correlation vector of the l-th CIR-related taps, which is given by: $\mathbf{r}_h^{[t]}[l] = E\{h^*[n+1,l]\mathbf{h}[n,l]\} \in \mathbb{C}^{N_{tap}^{[t]} \times 1}$.

$\mathbf{r}^{[t]}[l]$:

Normalized spaced-time cross-correlation vector of the l-th CIR-related taps: $\mathbf{r}^{[t]}[l] = \frac{1}{\lambda_l^{[f]}}\mathbf{r}_h^{[t]}[l] \in \mathbb{C}^{N_{tap}^{[t]} \times 1}$.

$\mathbf{r}_{apt}^{[t]}[l]$:

Spaced-time cross-correlation vector of the l-th CIR-related *a posteriori* least-squares tap estimates: $\mathbf{r}_{apt}^{[t]}[l] = E\{h^*[n+1,l]\tilde{\mathbf{h}}_{apt}[n,l]\} = \mathbf{r}_h^{[t]}[l] \in \mathbb{C}^{N_{tap}^{[t]} \times 1}$.

$\mathbf{R}_h^{[t]}[l]$:

Spaced-time auto-correlation matrix of the l-th CIR-related taps: $\mathbf{R}_h^{[t]}[l] = E\{\mathbf{h}[n,l]\mathbf{h}^H[n,l]\} \in \mathbb{C}^{N_{tap}^{[t]} \times N_{tap}^{[t]}}$.

$\mathbf{R}^{[t]}[l]$:

Normalized spaced-time auto-correlation matrix of the l-th CIR-related taps: $\mathbf{R}^{[t]}[l] = \frac{1}{\lambda_l^{[f]}}\mathbf{R}_h^{[t]}[l] \in \mathbb{C}^{N_{tap}^{[t]} \times N_{tap}^{[t]}}$.

$\mathbf{R}_{apt}^{[t]}[l]$:

Spaced-time auto-correlation matrix of the l-th CIR-related *a posteriori* least-squares tap estimates: $\mathbf{R}_{apt}^{[t]}[l] = E\{\tilde{\mathbf{h}}_{apt}[n,l]\tilde{\mathbf{h}}_{apt}^H[n,l]\} = \mathbf{R}_h^{[t]}[l] + \rho\mathbf{I} \in \mathbb{C}^{N_{tap}^{[t]} \times N_{tap}^{[t]}}$.

$s[n,k]$:

Transmitted subcarrier symbol having a variance of σ_s^2.

$\hat{s}[n,k]$:

Linear estimate of the transmitted subcarrier symbol.

$\check{s}[n,k]$:

Classified transmitted subcarrier symbol.

$\hat{\text{SNR}}_{apr}[n+1,k]$:

SNR predicted for the k-th subcarrier of the $(n+1)$-th OFDM symbol to be used in the modulation mode assignment during the n-th OFDM symbol period: $\hat{\text{SNR}}_{apr}[n+1,k] = |\hat{H}_{apr}[n+1,k]|^2\frac{\sigma_s^2}{\sigma_n^2}$.

$\mathbf{U}^{[f]}$:

Unitary KLT matrix associated with the EVD of the spaced-frequency correlation matrix $\mathbf{R}^{[f]}$: $\mathbf{U}^{[f]} \in \mathbb{C}^{K \times K}$.

$x[n,k]$:

Received subcarrier signal.

α:

Modulation-noise enhancement factor $\alpha = E\{\frac{1}{|s[n,k]|^2}\}$ [69,477], where for the specific case of using M-PSK modulation we have $\alpha_{\text{M-PSK}} = 1$, while for 16-QAM we have $\alpha_{\text{16-QAM}} = \frac{17}{9}$ [69].

α_{expo}:   Decay factor of the sample-spaced exponential CIR.

β_{expo}:   Amplitude scaling factor of the sample-spaced exponential CIR.

$\lambda_l^{[f]}$: l-th eigenvalue associated with the EVD of the spaced-frequency correlation matrix $\mathbf{R}^{[f]}$: $\lambda_l^{[f]} = (\mathbf{U}^{[f]H}\mathbf{R}^{[f]}\mathbf{U}^{[f]})_{[l,l]}$.

$v_l^{[f]}$: l-th eigenvalue associated with the decomposition of the spaced-frequency correlation matrix $\mathbf{R}^{[f]}$ with respect to the unitary matrix $\tilde{\mathbf{U}}^{[f]}$ employed in the estimator design: $v_l^{[f]} = (\tilde{\mathbf{U}}^{[f]H}\mathbf{R}^{[f]}\tilde{\mathbf{U}}^{[f]})_{[l,l]}$. Note that for $\tilde{\mathbf{U}}^{[f]} = \mathbf{U}^{[f]}$ we have $v_l^{[f]} = \lambda_l^{[f]}$.

$\Lambda^{[f]}$:   Diagonal matrix of eigenvalues $\lambda_l^{[f]}$, $l = 0, \ldots, K - 1$ associated with the EVD of the spaced-frequency correlation matrix $\mathbf{R}^{[f]}$: $\Lambda^{[f]} \in \mathbb{R}^{K \times K}$.

$()|_{\tilde{\mathbf{U}}^{[f]}}$:   Index used in conjunction with the CIR-related taps' spaced-time cross-correlation vector and auto-correlation matrix in order to indicate that the transform of the channel transfer factors to the CIR-related domain was carried out with the aid of the unitary matrix $\tilde{\mathbf{U}}^{[f]}$ associated with the design of the estimator, where potentially $\tilde{\mathbf{U}}^{[f]} = \mathbf{U}^{[f]}$.

Chapter 16

Channel Transfer Function Estimation for Multi-User OFDM[1]

16.1 Motivation

The topic of decision-directed channel estimation has been addressed in a variety of contributions, notably for example in the detailed discussions by Li *et al.* [68] in the context of *single-user single-transmit antenna* OFDM environments. As outlined in great detail in Chapter 15, the basic idea is to equalise the channel transfer function experienced by an OFDM symbol during the current transmission period by capitalising on that encountered during the previous OFDM symbol period. This implies assuming quasi-invariance of the channel's transfer function between the two consecutive OFDM symbols' transmission intervals. An improved channel transfer function estimate can then be obtained for detecting the most recently received OFDM symbol upon dividing the complex symbol received in each subcarrier by the sliced and remodulated information symbol hosted by a subcarrier. The updated channel estimate is then employed again as an initial channel estimate during the next OFDM symbol's transmission period.

By contrast, in the multi-user OFDM scenario to be outlined in Section 16.2 the signal received by each antenna is constituted by the superposition of the signal contributions associated with the different users or transmit antennas. Note that in terms of the multiple-input multiple-output (MIMO) structure of the channel, the multi-user single-transmit antenna scenario may be viewed as a relative of a single-user space-time coded (STC) scenario using multiple transmit antennas for improving the transmission integrity or to a MIMO system aiming for achieving an increased bitrate for a single user by employing multiple antennas. For the latter a Least-Squares (LS) error channel estimator was proposed by Li *et al.* [91], which aims at recovering the different transmit antennas' channel transfer functions on the basis of the output signal of a specific reception antenna element and by also capitalising on

---

[1]*OFDM and MC-CDMA for Broadband Multi-user Communications WLANS and Broadcasting.*
L.Hanzo, Münster, T. Keller and B.J. Choi,
©2003 John Wiley & Sons, Ltd. ISBN 0-470-85879-6

Year	Author	Contribution
'99	Li, Seshadri and Ariyavisitakul [91]	The LS-assisted DDCE proposed exploits the independence of the transmitted subcarrier symbol sequences.
'00	Jeon, Paik and Cho [92]	Frequency domain PIC-assisted DDCE is studied, which exploits the channel's slow variation versus time.
	Li [93]	Time domain PIC-assisted DDCE is investigated as a simplification of the LS-assisted DDCE of [91]. Optimum training sequences are proposed for the LS-assisted DDCE of [91].
'01	Mody and Stüber [94]	Channel transfer factor estimation designed for frequency domain PSAM based on CIR-related domain filtering is studied.
	Gong and Letaief [95]	MMSE-assisted DDCE is advocated which represents an extension of the LS-assisted DDCE of [95]. The MMSE-assisted DDCE is shown to be practical in the context of transmitting consecutive training blocks. Additionally, a low-rank approximation of the MMSE-assisted DDCE is considered.
	Jeon, Paik and Cho [96]	2D MMSE-based channel estimation is proposed for frequency domain PSAM.
	Vook and Thomas [97]	2D MMSE-based channel estimation is invoked for frequency domain PSAM. A complexity reduction is achieved by CIR-related domain based processing.
	Xie and Georghiades [98]	Expectation maximisation (EM) based channel transfer factor estimation approach for DDCE.
'02	Li [99]	A more detailed discussion on time domain PIC-assisted DDCE is provided and optimum training sequences are proposed [93].
	Bölcskei, Heath Jr. and Paulray [100]	Blind channel identification and equalisation using second-order cyclostationary statistics as well as antenna precoding were studied.
	Minn, Kim and Bhargava [101]	A reduced-complexity version of the LS-assisted DDCE of [91] is introduced, based on exploiting the channel's correlation in the frequency direction, as opposed to invoking the simplified scheme of [99], which exploits the channel's correlation in the time direction. A similar approach was suggested by Slimane [102] for the specific case of two transmit antennas.
	Komninakis, Fragouli, Sayed and Wesel [103]	Fading channel tracking and equalisation were proposed for employment in MIMO systems assisted by Kalman estimation and channel prediction.

Table 16.1: Contributions on channel transfer factor estimation for multiple-transmit antenna-assisted OFDM.

the remodulated received symbols associated with the different users. The performance of this estimator was found to be limited in terms of the mean-square estimation error in scenarios, where the product of the number of transmit antennas and the number of CIR taps to be estimated per transmit antenna approaches the total number of subcarriers hosted by an OFDM symbol.

In [92] a DDCE was proposed by Jeon *et al.* for a space-time coded OFDM scenario of two transmit antennas and two receive antennas. Specifically, the channel transfer function[2] associated with each transmit-receive antenna pair was estimated on the basis of the output signal of the specific receive antenna upon *subtracting* the interfering signal contributions associated with the remaining transmit antennas. These interference contributions were estimated by capitalising on the knowledge of the channel transfer functions of all interfering transmit antennas predicted during the $(n-1)$-th OFDM symbol period for the n-th OFDM symbol, also invoking the corresponding remodulated symbols associated with the n-th OFDM symbol. To elaborate further, the difference between the subtraction-based channel transfer function estimator of [92] and the LS estimator proposed by Li *et al.* in [91] is that in the former the channel transfer functions predicted during the previous, i.e. the $(n-1)$-th OFDM symbol period for the current, i.e. the n-th OFDM symbol, are employed for both symbol detection *as well as* for obtaining an updated channel estimate for employment during the $(n+1)$-th OFDM symbol period. In the approach advocated in [92] the subtraction of the different transmit antennas' interfering signals is performed in the frequency domain.

By contrast, in [93] a similar technique was proposed by Li with the aim of simplifying the DDCE approach of [91], which operates in the time domain. A prerequisite for the operation of this parallel interference cancellation (PIC)-assisted DDCE is the availability of a reliable estimate of the various channel transfer functions for the current OFDM symbol, which are employed in the cancellation process in order to obtain updated channel transfer function estimates for the demodulation of the next OFDM symbol. In order to compensate for the channel's variation as a function of the OFDM symbol index, linear prediction techniques can be employed, as was also proposed, for example, in [93]. However, due to the estimator's recursive structure, determining the optimum predictor coefficients is not as straightforward as for the transversal FIR filter-assisted predictor as described in Section 15.2.4 for the single-user DDCE.

A comprehensive overview of further publications on channel transfer factor estimation for OFDM systems supported by multiple transmit antennas is provided in Table 16.1.

Our further discourse in Chapter 16 evolves as follows. In Section 16.2 we portray the signal model associated with the SDMA uplink transmission scenario. Note again that in terms of the MIMO structure of the channel this SDMA system may be viewed as a relative of a single-user STC scenario employing multiple transmit antennas for the sake of achieving a diversity gain. Hence, the algorithms discussed here are amenable to a wide range of applications involving multiple transmit antennas. Our discussion of specific channel estimation techniques commences in Section 16.4 with a portrayal of the least-squares error assisted estimator proposed by Li *et al.* [91]. In this section we will derive a necessary condition for channel identification and the equations describing the estimator's MSE in the context of both sample-spaced as well as non-sample-spaced CIRs. The estimator's potentially high complexity provides a further motivation for devising alternative approaches. Specifically, in

---

[2]In the context of the OFDM system the set of K different subcarriers' channel transfer factors is referred to as the channel transfer function, or simply as the channel.

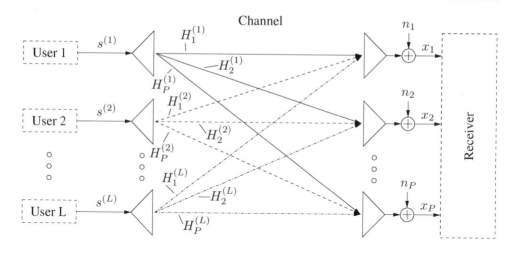

Figure 16.1: Schematic of an SDMA *uplink* scenario as observed on an OFDM subcarrier basis, where each of the L users is equipped with a single transmit antenna and the base station's receiver is assisted by a P-element antenna front-end. For comparison, in an STC scenario the L transmit antennas are used to provide L-th order transmit diversity for a single user.

Section 16.5 we will focus on the far less complex PIC-assisted DDCE employing prediction filtering along the time-axis in the CIR-related domain. Then expressions are derived for the estimator's MSE and an iteration-based novel approach is devised for evaluating the optimum CIR-related tap predictor coefficients. This is followed by an extensive performance assessment under both sample-spaced and non-sample-spaced CIR conditions. In order to avoid the off-line optimisation of the predictor coefficients based on certain assumptions about the channel's statistics, the CIR-related tap predictors can be rendered adaptive with the aid of the Recursive Least Squares (RLS) algorithm. This has the potential of significantly further improving the performance, while maintaining the system's stability under time-variant channel conditions. Our conclusions will be offered in Section 16.7.

16.2   The SDMA Signal Model on a Subcarrier Basis

In Figure 16.1 we have portrayed an SDMA uplink transmission scenario, where each of the L simultaneous users is equipped with a single transmission antenna, while the receiver capitalises on a P-element antenna front-end. The set of complex signals, $x_p[n, k], p = 1, \ldots, P$ received by the P-element antenna array in the k-th subcarrier of the n-th OFDM symbol is constituted by the superposition of the independently faded frequency domain signals associated with the L users sharing the same space-frequency resource. The received signal was corrupted by the Gaussian noise at the array elements. Regarding the statistical properties of the different signal components depicted in Figure 16.1, we assume that the complex data signal $s^{(l)}$ transmitted by the l-th user has zero-mean and a variance of σ_l^2. The AWGN noise process n_p at any antenna array element p exhibits also zero-mean and a variance of

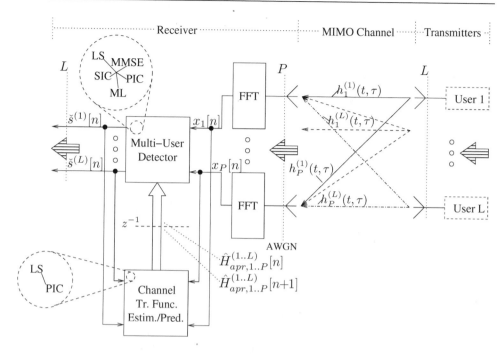

Figure 16.2: Simplified illustration of the SDMA-OFDM scenario. The L different users, each equipped with a single transmit antenna are linked to the BS receiver, which employs P number of antenna elements, via the $L \times P$-dimensional MIMO channel. The output signal of each receiver antenna element is separately subjected to FFT-based demodulation, which results in the subcarrier signals $x_p[n, k]$, $p = 1, \ldots, P$, $k = 0, \ldots, K - 1$. This is followed by subcarrier-based multi-user detection with the aim of obtaining the sliced symbols $\check{s}^{(l)}[n, k]$, $l = 1, \ldots, L$, $k = 0, \ldots, K - 1$. Based on these sliced symbols and on the received subcarrier signals, *a priori* estimates $\hat{H}_{apr,p}^{(l)}[n, k]$ of the channel transfer factors encountered during the next OFDM symbol period are generated.

σ_n^2, which is identical for all array elements. The frequency domain channel transfer factors $H_p^{(l)}$ of the different array elements $p = 1, \ldots, P$ or users $l = 1, \ldots, L$ are independent, stationary, complex Gaussian distributed random variables with zero-mean and unit variance.

16.3 Multi-User Multiple Reception Antenna Aided OFDM

In Figure 14.1 we outlined the basic structure of an OFDM scheme explicitly showing the channel estimation block in the context of a *single-user single-reception antenna-assisted scenario*. In Figure 16.1 we introduced the basic concepts of SDMA assisted multi-user OFDM systems. More specifically, the L different users' transmitted signals are separated at the receiver with the aid of their spatial signature, namely based on the knowledge of their unique set of channel impulse responses or transfer functions associated with each user's transmit antenna and with the P different BS receiver antenna elements. This SDMA-MIMO channel scenario is shown on the right-hand side of Figure 16.2.

While in the multi-user SDMA-OFDM scenario the same OFDM transmitter design can be employed as in the single-user scenario, the SDMA-OFDM receiver shown on the left-hand side of Figure 16.2 exhibits important changes compared to the single-user, single-reception antenna based receiver of Figure 14.1. First of all, we have simplified the multi-user schematic by removing the time domain HF down-conversion-, analogue-to-digital conversion-, synchronisation-, serial-to-parallel and parallel-to-serial conversion- as well as the demodulation and decoding blocks explicitly shown in the single-user receiver of Figure 14.1. Furthermore, we have substituted the single FFT block by separate FFT blocks for the different receiver branches. The different reception branches' frequency domain subcarrier signals $x_p[n, k]$, $p = 1, \ldots, P$, $k = 0, \ldots, K - 1$ become available at the FFT blocks' outputs.

Furthermore, the channel equaliser and demodulator blocks of Figure 14.1 have been substituted by the multi-user detector block as seen in Figure 16.2. At the output of the multi-user detector each of, the L different users' sliced symbols $\check{s}^{(l)}[n, k]$, $l = 1, \ldots, L$, $k = 0, \ldots, K - 1$ become available. As it will be shown in Chapter 17 for the case of the so-called linear detectors, namely the Least-Squares (LS)- and the Minimum Mean-Square (MMSE) based detectors, the detection process constituted by the signal combining- and classification process is sequentially performed, which is equivalent to the channel equalisation and demodulation processes performed in the context of the single-user receiver of Figure 14.1. By contrast, in the context of the more effective detectors, namely Successive Interference Cancellation (SIC), Parallel Interference Cancellation (PIC) and Maximum Likelihood (ML) detection, *a priori* knowledge of the other users' transmitted symbols is required at each subcarrier position for the identification of a specific user's transmitted subcarrier symbol.

The multi-user detection process is based on estimates $\hat{H}_p^{(l)}[n, k]$ of the channel transfer factors $H_p^{(l)}[n, k]$, $l = 1, \ldots, L$, $p = 1, \ldots, P$, $k = 0, \ldots, K - 1$, associated with the K different subcarriers of the $L \cdot P$ number of MIMO channels. Hence, the channel estimator of Figure 14.1 designed to estimate a single set of K number of channel transfer factors has to be substituted by a multi-channel estimator, as shown in Figure 16.2. While in general both pilot-assisted as well as decision-directed approaches could be employed as in the single-user scenario of Figure 14.1, in Figure 16.2 we have portrayed the scenario of decision-directed channel estimation. More specifically, the channel transfer factor estimates generated during the current OFDM symbol period – after detection of the different users' symbols – are employed in the detection stage during the next OFDM symbol period. Two of the most promising estimation approaches are the LS-assisted DDCE and the PIC-assisted DDCE, which will be detailed in Chapter 16.

16.4 Least-Squares Error-Assisted Decision-Directed Channel Transfer Function Estimation

In [91] the Least-Squares Error (LSE) approach was proposed by Li *et al.* for estimating the vector of channel transfer factors associated with each specific transmit-receive antenna pair in the context of the multi-user scenario outlined in Section 16.1. In contrast to [91] our derivation is based on a more compact matrix notation. The outline of Section 16.4 is as follows. In Section 16.4.1 the estimator's specific structure is derived, while in Section 16.4.2 the equations describing its MSE are developed for channel scenarios associated with both sample-spaced and *non-sample-spaced* CIRs. The method of further enhancing the estimator's MSE performance with the aid of CIR-related tap prediction filtering is briefly addressed in Section 16.4.3. Furthermore, in Section 16.4.4 Li's simplified – parallel interference cancellation assisted – approach [93] is outlined, which is then further developed in Section 16.5. The analysis of the estimator's complexity is provided in Section 16.4.5. Our conclusions will be offered in Section 16.4.6.

16.4.1 Derivation of the LS-Estimator

Let us commence by providing a brief outline of Section 16.4.1. In Section 16.4.1.1 the signal recorded for a specific receiver antenna element is described using vector notations, while in Section 16.4.1.2 we embark on the characterisation of the LS estimator based on a simplified, noise-free version of the received signal's model outlined in Section 16.4.1.1.

16.4.1.1 The SDMA Signal Model on a Receiver Antenna Basis

Recall from Section 16.2 that the complex signal $x_p[n, k]$ associated with the p-th receiver antenna element in the k-th subcarrier of the n-th OFDM symbol is given as the superposition of the different users' channel-impaired received signal contributions plus the AWGN, which is expressed as:

$$x_p[n, k] = \sum_{i=1}^{L} H_p^{(i)}[n, k] s^{(i)}[n, k] + n_p[n, k], \tag{16.1}$$

where again, the different variables have been defined in Section 16.2. Upon invoking vector notations, the set of equations constituted by Equation 16.1 for $k = 0, \ldots, K - 1$ can be rewritten as:

$$\mathbf{x}_p[n] = \sum_{i=1}^{L} \mathbf{S}^{(i)}[n] \mathbf{H}_p^{(i)}[n] + \mathbf{n}_p[n] \tag{16.2}$$

$$= \mathbf{S}^T[n] \mathbf{H}_p[n] + \mathbf{n}_p[n], \tag{16.3}$$

where in Equation 16.2 $\mathbf{x}_p[n] \in \mathbb{C}^{K \times 1}$, $\mathbf{H}_p^{(i)}[n] \in \mathbb{C}^{K \times 1}$ and $\mathbf{n}_p[n] \in \mathbb{C}^{K \times 1}$ are column vectors hosting the subcarrier-related variables $x_p[n, k]$, $H_p^{(i)}[n, k]$ and $n_p[n, k]$, respectively, and $\mathbf{S}^{(i)}[n] \in \mathbb{C}^{K \times K}$ is a diagonal matrix with elements given by $s^{(i)}[n, k]$, where $k = 0, \ldots, K - 1$. Furthermore, for the inner-product based representation of Equation 16.3 we have defined $\mathbf{S}[n] \in \mathbb{C}^{LK \times K}$ and $\mathbf{H}_p[n] \in \mathbb{C}^{LK \times 1}$:[3]

$$\mathbf{S}[n] = \left(\mathbf{S}^{(1)}[n] \ldots \mathbf{S}^{(L)}[n] \right)^T, \quad \mathbf{H}_p[n] = \left(\mathbf{H}_p^{(1)T}[n] \ldots \mathbf{H}_p^{(L)T}[n] \right)^T. \tag{16.4}$$

---

[3]In order to avoid conflicts with the previous definition of $\mathbf{H}[n]$ cast in the context of single-user OFDM systems it would be useful to add a superscript of $()^{[L]}$ to this variable. However, this superscript has been omitted here for reasons of notational simplicity.

From now on we will omit the receiver antenna's index p. The dimension of the estimation task, namely that of determining the L users' vectors of channel transfer factors $\mathbf{H}_p^{(i)}[n], i = 1, \ldots, L$ containing K entries each – separately for each receiver antenna element $p = 1, \ldots, P$ – can be significantly reduced with the aid of a sub-space approach, which is the topic of the next section.

16.4.1.2 Sub-Space-Based Approach

The key idea of sub-space-based techniques is to project the received signal on the basis of which the desired signal is to be estimated, onto the desired signal's vector sub-space, which is potentially spanned by only a fraction of the number of basis vectors which span the received signal's vector space. Thus the contributions outside the desired signal's sub-space, which should ideally only be the undesired AWGN, are removed. Unfortunately, however, in the context of non-sample-spaced CIRs the desired signal's space has potentially the same dimension as the received signal's space. Hence, the projection of the received signal onto the lower-dimensional- or lower-rank sub-space implies an imperfect representation of the desired signal.

The further structure of this section is as follows. The low-rank approximation of the i-th user's vector $\mathbf{H}^{(i)}[n]$ of channel transfer factors across the K different subcarriers is outlined in Section 16.4.1.2.1 and the associated coefficients- or CIR-related taps are then determined in Section 16.4.1.2.2. A necessary condition for the unambiguous identification of the different user's CIR-related taps will be outlined in Section 16.4.1.2.3. Our discussions in Section 16.4.1.2 will be concluded in Section 16.4.1.2.4 upon portraying an alternative approach capitalising on the QR matrix factorisation [90] for evaluating the vector $\tilde{\mathbf{h}}_{apt,K_0}[n]$ of CIR-related tap estimates instead of directly performing the inversion of matrix $\mathbf{Q}[n]$ as suggested by Equation 16.22.

16.4.1.2.1 Low-Rank Approximation of the i-th User's Vector of Different Subcarriers' Channel Transfer Factors   Let us commence our detailed discourse by stating that Li *et al.* [91] assumed that upon neglecting the effects of leakage[4] due to the potentially non-sample-spaced nature of the "true" CIR of practical channels, the i-th user's vector $\mathbf{H}^{(i)}[n]$ of channel transfer factors can be approximated by the Discrete Fourier Transform (DFT) vector $\hat{\mathbf{H}}^{(i)}[n] \in \mathbb{C}^{K \times 1}$ of the vector $\hat{\mathbf{h}}_{K_0}^{(i)}[n] \in \mathbb{C}^{K_0 \times 1}$ of $K_0 < K$ number of significant CIR-related *a posteriori* tap estimates as follows:

$$\hat{\mathbf{H}}^{(i)}[n] \;=\; \mathbf{W}\mathbf{J}_{K_0}^{(i)}\hat{\mathbf{h}}_{K_0}^{(i)}[n] \tag{16.5}$$

$$\;=\; \mathbf{W}_J^{(i)}\hat{\mathbf{h}}_{K_0}^{(i)}[n], \tag{16.6}$$

where $\mathbf{W}$ denotes the unitary DFT matrix [90] and $\mathbf{J}_{K_0}^{(i)} \in \mathbb{C}^{K \times K_0}$ has the role of mapping the K_0 CIR-related *a posteriori* tap estimates contained in the vector $\hat{\mathbf{h}}_{K_0}^{(i)}[n]$ to their "true" positions in terms of an OFDM time domain sample-spaced delay, as it will be discussed

---

[4]Consider the case of a non-sample-spaced CIR when the set of K frequency domain channel transfer factor samples is transferred to the CIR-related domain with the aid of a unitary transform, which could be e.g. the DFT matrix $\mathbf{W}$. Although the original CIR might consist of a single non-sample-spaced tap only, its effects are potentially spread across all the K taps in the output of the unitary transform. This effect is known as leakage.

below. The product of these matrices is denoted here as $\mathbf{W}_J^{(i)} \in \mathbb{C}^{K \times K_0}$, namely:

$$\mathbf{W}_J^{(i)} = \mathbf{W}\mathbf{J}_{K_0}^{(i)}. \tag{16.7}$$

More specifically, if the $\acute{m}$-th component of the estimated CIR-related tap vector $\hat{\mathbf{h}}_{K_0}^{(i)}[n]$ is associated with the $\acute{n}$-th integer delay tap raster position, then matrix $\mathbf{J}_{K_0}^{(i)}$ will have a numerical value of unity at the position given by the $\acute{n}$-th row and $\acute{m}$-th column. Since the mapping has to be unambiguous, we note that each column of $\mathbf{J}_{K_0}^{(i)}$ contains one and only one unity entry, while the remaining $K - 1$ entries are zero. Hence, the total number of unity entries contained in $\mathbf{J}_{K_0}^{(i)}$ is equal to K_0. A more compact form of Equation 16.6 for the different users $i = 1, \dots, L$ is given by:

$$\hat{\mathbf{H}}[n] = \mathbf{W}_J \hat{\mathbf{h}}_{K_0}[n], \tag{16.8}$$

where $\hat{\mathbf{H}}[n] \in \mathbb{C}^{LK \times 1}$ and $\hat{\mathbf{h}}_{K_0}[n] \in \mathbb{C}^{LK_0 \times 1}$ are defined as:

$$\hat{\mathbf{H}}[n] = \left(\hat{\mathbf{H}}^{(1)T}[n] \dots \hat{\mathbf{H}}^{(L)T}[n]\right)^T, \quad \hat{\mathbf{h}}_{K_0}[n] = \left(\hat{\mathbf{h}}_{K_0}^{(1)T}[n] \dots \hat{\mathbf{h}}_{K_0}^{(L)T}[n]\right)^T, \tag{16.9}$$

and the block-diagonal matrix $\mathbf{W}_J \in \mathbb{C}^{LK \times LK_0}$ is defined as:

$$\mathbf{W}_J = diag(\mathbf{W}_J^{(1)} \dots \mathbf{W}_J^{(L)}). \tag{16.10}$$

Then following the philosophy of Equation 16.2 the vector $\mathbf{x}[n]$ of received subcarrier signals can be approximated by:

$$\hat{\mathbf{x}}[n] = \sum_{i=1}^{L} \check{\mathbf{S}}^{(i)}[n]\hat{\mathbf{H}}^{(i)}[n] \tag{16.11}$$

$$= \check{\mathbf{S}}^T[n]\hat{\mathbf{H}}[n] \tag{16.12}$$

$$= \mathbf{A}[n]\hat{\mathbf{h}}_{K_0}[n], \tag{16.13}$$

where Equation 16.13 was obtained upon substituting Equation 16.8 into Equation 16.12 and by introducing the short-hand of $\mathbf{A}[n] \in \mathbb{C}^{K \times LK_0}$, namely:

$$\mathbf{A}[n] = \check{\mathbf{S}}^T[n]\mathbf{W}_J. \tag{16.14}$$

Hence, the initial estimation task is reduced to that of determining the L users' associated vectors $\hat{\mathbf{h}}_{K_0}^{(i)}[n]$, $i = 1, \dots, L$, constituting $\hat{\mathbf{h}}_{K_0}[n]$, each hosting K_0 number of significant CIR-related taps. Note that in Equation 16.11 $\check{\mathbf{S}}^{(i)}[n] \in \mathbb{C}^{K \times K}$ denotes the i-th user's matrix of sliced symbol decisions, while $\check{\mathbf{S}}[n] \in \mathbb{C}^{LK \times LK}$ is the L different users' matrix of sliced symbol decisions, which is defined similarly to the matrix $\mathbf{S}[n]$ of transmitted symbols given by Equation 16.4.

16.4.1.2.2 Determination of the LS-DDCE Coefficients Using the Gradient Approach

The model-mismatch related error vector $\Delta\hat{\mathbf{x}}[n] \in \mathbb{C}^{K \times 1}$ between the vector $\mathbf{x}$ of received

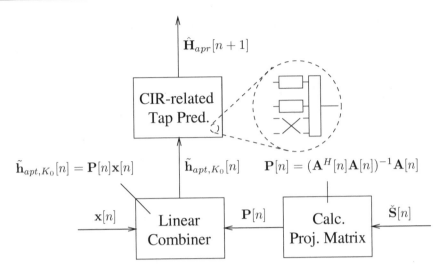

Figure 16.3: Stylised illustration of the LS-assisted DDCE as associated with the p-th receiver antenna. Based on the matrix $\check{\mathbf{S}}[n]$ of tentative symbol decisions, which follows the formulation of $\mathbf{S}[n]$ in Equation 16.4, the projection matrix $\mathbf{P}[n]$ is calculated as shown in Equation 16.24. By multiplication with the vector $\mathbf{x}[n]$ of received subcarrier signals during the combiner stage, the vector $\tilde{\mathbf{h}}_{apt,K_0}[n] \in \mathbb{C}^{LK_0 \times 1}$ of the L different users' CIR-related least-squares *a posteriori* tap estimates is obtained. In the simplest case the vector $\hat{\mathbf{H}}_{apr}[n+1] \in \mathbb{C}^{LK \times 1}$ of channel transfer factor estimates is generated upon invoking Equation 16.8 and by assuming that we have $\hat{\mathbf{H}}_{apr}[n+1] = \hat{\mathbf{H}}_{apt}[n]$. However, a further reduction of the estimation MSE can be achieved with the aid of the CIR-related tap prediction filtering techniques discussed in Section 15.2.4 for single-user scenarios.

channel-impaired signals of the K different subcarriers, and the vector $\hat{\mathbf{x}}$ of estimated subcarrier signals given by Equation 16.13 can be defined as:

$$\Delta\hat{\mathbf{x}}[n] = \mathbf{x}[n] - \hat{\mathbf{x}}[n] \tag{16.15}$$

$$= \mathbf{x}[n] - \mathbf{A}\hat{\mathbf{h}}_{K_0}[n]. \tag{16.16}$$

Furthermore, the total squared error of all the subcarriers of an OFDM symbol is given by the inner product of the vector of subcarrier errors as stated by Equation 16.16, yielding:

$$\|\Delta\hat{\mathbf{x}}[n]\|_2^2 = \Delta\hat{\mathbf{x}}^H[n]\Delta\hat{\mathbf{x}}[n] \tag{16.17}$$

$$= \mathbf{x}^H[n]\mathbf{x}[n] - 2\Re\left(\hat{\mathbf{h}}_{K_0}^H[n]\mathbf{p}[n]\right) +$$

$$+ \hat{\mathbf{h}}_{K_0}^H[n]\mathbf{Q}[n]\hat{\mathbf{h}}_{K_0}[n], \tag{16.18}$$

where the "cross-correlation" vector $\mathbf{p}[n] \in \mathbb{C}^{LK_0 \times 1}$ is given by:

$$\mathbf{p}[n] = \mathbf{A}^H[n]\mathbf{x}[n], \tag{16.19}$$

and the "auto-correlation" matrix[5] $\mathbf{Q}[n] \in \mathbb{C}^{LK_0 \times LK_0}$ is given by:

$$\mathbf{Q}[n] = \mathbf{A}^H[n]\mathbf{A}[n]. \tag{16.20}$$

Following the philosophy of the least-squares (LS) error criterion, a standard approach for determining the optimum vector $\tilde{\mathbf{h}}_{apt,K_0}[n] \in \mathbb{C}^{LK_0 \times 1}$ of CIR-related *a posteriori* tap estimates[6] is to minimiSe the model-mismatch related squared error given by Equation 16.18. Note that in the optimum point the conjugate gradient $\nabla^* = \frac{\partial}{\partial \hat{\mathbf{h}}^*_{K_0}[n]}$ of the error term of Equation 16.18 $||\Delta\hat{\mathbf{x}}[n]||_2^2$ with respect to the vector $\hat{\mathbf{h}}_{K_0}[n]$ of CIR-related tap estimates is zero [91], which can be formulated as:

$$||\Delta\hat{\mathbf{x}}[n]||_2^2 = \text{Min.} \quad \Longleftrightarrow \quad \nabla^*||\Delta\hat{\mathbf{x}}[n]||_2^2 \overset{!}{=} \mathbf{0}. \tag{16.21}$$

Upon substituting Equation 16.18 into Equation 16.21 and using a number of mathematical manipulations, we obtain the following expression [91]:

$$\mathbf{Q}[n]\tilde{\mathbf{h}}_{apt,K_0}[n] = \mathbf{p}[n] \quad \Longleftrightarrow \quad \tilde{\mathbf{h}}_{apt,K_0}[n] = \mathbf{Q}^{-1}[n]\mathbf{p}[n]. \tag{16.22}$$

Here we have specifically exploited that the gradient with respect to the constant first term of Equation 16.18 is equal to zero, while the gradient with respect to the second term was evaluated by capitalising on $\frac{\partial}{\partial \mathbf{z}^*}\mathfrak{Re}(\mathbf{z}^H\mathbf{a}) = \frac{1}{2}\mathbf{a}$ [90], which is valid for complex valued column vectors $\mathbf{z}$ and $\mathbf{a}$. The gradient of the last term in Equation 16.18 was evaluated upon exploiting that $\frac{\partial}{\partial \mathbf{z}^*}\mathbf{z}^H\mathbf{R}\mathbf{z} = \mathbf{R}\mathbf{z}$ [90] for a Hermitian matrix $\mathbf{R}$.

Note that we could have also directly obtained the solution for this CIR-related tap estimation problem by recognizing the characteristic shape of Equation 16.16, for which the least-squares solution is given by [90]:

$$\tilde{\mathbf{h}}_{apt,K_0}[n] = \mathbf{P}[n]\mathbf{x}[n], \tag{16.23}$$

where:

$$\mathbf{P}[n] = (\mathbf{A}^H[n]\mathbf{A}[n])^{-1}\mathbf{A}^H[n]. \tag{16.24}$$

The matrix $\mathbf{P}[n] \in \mathbb{C}^{LK_0 \times K}$ of Equation 16.24 is also known from the literature [90] as a matrix, which projects – when multiplied from the right with $\mathbf{x}[n]$ – on the column vectors of $\mathbf{A}[n]$. These span the signal's sub-space [90]. Alternatively, $\mathbf{P}[n]$ can be interpreted as the pseudo-inverse- or so-called Moore-Penrose inverse [90] of $\mathbf{A}[n]$, denoted by $\mathbf{A}^\dagger[n]$. The representation of the LS-assisted DDCE by means of Equations 16.23 and 16.24 is further illustrated in Figure 16.3. Note that based on Equation 16.8 the vector $\tilde{\mathbf{H}}_{apt}[n] \in \mathbb{C}^{LK \times 1}$ of *a posteriori* channel transfer factor estimates is given by:

$$\tilde{\mathbf{H}}_{apt}[n] = \mathbf{W}_J\tilde{\mathbf{h}}_{apt,K_0}[n]. \tag{16.25}$$

---

[5]Note that according to standard definition evaluating the auto-correlation or cross-correlation involves using the expectation operator, which is not the case in the context of the LS-approach. However, these expressions have still been used in inverted commas, in order to highlight the similarities to a potential MMSE solution.

[6]In accordance with the notation used for DDCE in single-user scenarios, which was discussed in Section 15.2, the least-squares solution is denoted by a superscript of $\tilde{()}$.

Upon resubstituting the vector $\tilde{\mathbf{h}}_{apt,K_0}[n]$ of LS-optimum CIR-related tap estimates given by Equations 16.22 or 16.23 into the model-mismatch related cost-function of Equation 16.18 we obtain:

$$(\|\Delta\hat{\mathbf{x}}[n]\|_2^2)|_{\tilde{\mathbf{h}}_{apt,K_0}[n]} = \mathbf{x}^H[n]\mathbf{x}[n] - \mathbf{p}^H[n]\tilde{\mathbf{h}}_{apt,K_0}[n], \qquad (16.26)$$

which is the minimum mismatch error.

16.4.1.2.3 Necessary Condition for Identification of the LS-DDCE Coefficients

As mentioned in Section 16.4.1.2.2, the solution of the estimation problem according to Equation 16.23 requires computation of the Moore-Penrose- or left-inverse of the matrix $\mathbf{A}[n] \in \mathbb{C}^{K \times LK_0}$ given in Equation 16.24. A necessary and sufficient condition for its existence is that the LK_0 number of columns of the matrix $\mathbf{A}$ are linearly independent [90]. This implies that we have $rank(\mathbf{A}[n]) = LK_0$, where a necessary condition for this to be the case is that the number of rows of $\mathbf{A}[n]$, namely K is higher than the number of columns, namely LK_0, which is formulated as:

$$LK_0 \leq K \quad \Longleftrightarrow \quad L \leq \frac{K}{K_0}. \qquad (16.27)$$

Equation 16.27 imposes a constraint with respect to the L number of simultaneous users- or the number of transmit antennas. Avoiding this limitation is a further motivation for devising alternative channel transfer function estimation approaches.

16.4.1.2.4 Implementation by QR Decomposition

It is well known from the literature that the matrix $\mathbf{Q}[n] = \mathbf{A}^H[n]\mathbf{A}[n]$ associated with the projection matrix $\mathbf{P}[n]$ of Equation 16.24, which is also known as the Moore-Penrose- or pseudo-inverse $\mathbf{A}^\dagger[n]$ of $\mathbf{A}[n]$ [90] exhibits a potentially high condition number $\kappa(\mathbf{A}[n])$, which reflects the degree of ill-conditioning or proximity to rank-deficiency [90]. Hence, a direct inversion of $\mathbf{Q}[n]$ is likely to be inaccurate, specifically for $L \lesssim \frac{K}{K_0}$, as already mentioned in Section 16.4.1.2.3. These effects can be mitigated with the aid of the procedure of QR matrix factorisation or decomposition, as demonstrated for example in [90]. The QR factorisation implies that the matrix $\mathbf{A}[n]$ can be factored as:

$$\mathbf{A}[n] = \mathbf{Q}\mathbf{R} = \mathbf{Q}\begin{pmatrix} \mathbf{R}_1 \\ \mathbf{0} \end{pmatrix}, \qquad (16.28)$$

where $\mathbf{Q} \in \mathbb{C}^{K \times K}$ is a unitary matrix which intentionally lacks the index $[n]$ in order to distinguish it from $\mathbf{Q}[n]$ of Equation 16.20 and $\mathbf{R} \in \mathbb{C}^{K \times LK_0}$ is an upper triangular matrix which can be expressed as $\mathbf{R} = (\mathbf{R}_1 \, \mathbf{0})^T$ since $K \geq LK_0$, with $\mathbf{R}_1 \in \mathbb{C}^{LK_0 \times LK_0}$ and $\mathbf{0}$ being a $(K - LK_0) \times LK_0$-dimensional zero-matrix. Upon substituting Equation 16.28 into Equation 16.16 and by substituting the result into Equation 16.17 we obtain for $\|\Delta\hat{\mathbf{x}}[n]\|_2^2$

[90]:

$$||\Delta\hat{\mathbf{x}}[n]||_2^2 = ||\mathbf{x}[n] - \mathbf{QR}\hat{\mathbf{h}}[n]||_2^2 \qquad (16.29)$$

$$= ||\mathbf{Q}(\mathbf{Q}^H\mathbf{x}[n] - \mathbf{R}\hat{\mathbf{h}}[n])||_2^2 \qquad (16.30)$$

$$= ||\mathbf{Q}^H\mathbf{x}[n] - \mathbf{R}\hat{\mathbf{h}}[n])||_2^2 \qquad (16.31)$$

$$= \left\|\begin{pmatrix} \mathbf{c} \\ \mathbf{d} \end{pmatrix} - \begin{pmatrix} \mathbf{R}_1 \\ \mathbf{0} \end{pmatrix}\hat{\mathbf{h}}[n]\right\|_2^2 \qquad (16.32)$$

$$= ||\mathbf{c} - \mathbf{R}_1\hat{\mathbf{h}}[n]||_2^2 + ||\mathbf{d}||_2^2, \qquad (16.33)$$

where all variables are defined below. The solution in the sense of minimising $||\Delta\hat{\mathbf{x}}[n]||_2^2$ is given by [90]:

$$\hat{\mathbf{R}}_1\tilde{\mathbf{h}}[n] = \mathbf{c}, \qquad (16.34)$$

which can be directly solved for the vector $\tilde{\mathbf{h}}[n]$ of the L different users' CIR-related tap estimates by back-substitution, since $\mathbf{R}_1[n]$ is an upper triangular matrix. Note that in the context of Equation 16.30 we have exploited the unitarity nature of the matrix $\mathbf{Q}$, namely, that we have $\mathbf{QQ}^H = \mathbf{I}$, while in the context of Equation 16.31 that the Euclidean norm of a matrix or vector remains unchanged, when subjected to a unitary transform [90]. Furthermore, we have exploited in Equation 16.32 that $\mathbf{Q}^H\mathbf{x}[n] = (\mathbf{c}\ \mathbf{d})^T$, where $\mathbf{c} \in \mathbb{C}^{LK_0 \times LK_0}$ and $\mathbf{d} \in \mathbb{C}^{(K-LK_0) \times (K-LK_0)}$. Note that other matrix factorisations or decompositions can also be applied to the LS estimation problem, but the solution acquired with the aid of the QR factorisation is the most prominent one [90].

16.4.2 Least-Squares Channel Estimation MSE in the Context of Both Sample-Spaced and Non-Sample-Spaced CIRs

Our analysis commences in Section 16.4.2.1 with the derivation of the channel transfer factor estimation errors' auto-correlation matrix, where more specifically, the average of its main-diagonal elements determines the average channel estimation MSE. In this section we impose no specific assumptions on the CIRs or the transmitted subcarrier symbol sequences. By contrast, in Section 16.4.2.2 we will focus our attention on the idealistic case of sample-spaced CIRs. However, in order to render our investigations more realistic, non-sample-spaced CIRs will be considered in Section 16.4.2.3. In characterising the estimator's performance, we will capitalise on the properties of optimum training sequences, as proposed by Li [93].

16.4.2.1 Correlation Matrix of the Channel Transfer Factor Estimates

Upon substituting Equation 16.3 into Equation 16.23 and by further substituting the result into Equation 16.8 the vector $\tilde{\mathbf{H}}_{apt}[n]$[7] of least-squares channel transfer factor estimates is obtained as a function of the vector $\mathbf{H}[n]$ of "true" channel transfer factors associated with the specific channel encountered and as a function of the vector of different subcarrier noise

---

[7]Note that here we have again omitted the receiver antenna index p.

processes $\mathbf{n}[n]$, which is formulated as:[8]

$$\tilde{\mathbf{H}}_{apt}[n] = \mathbf{W}_J \mathbf{P}[n](\mathbf{S}^T[n]\mathbf{H}[n] + \mathbf{n}[n]). \tag{16.35}$$

Furthermore, the vector $\Delta\tilde{\mathbf{H}}_{apt}[n] \in \mathbb{C}^{LK \times 1}$ of the L different users' channel transfer factor estimation errors is defined as:

$$\Delta\tilde{\mathbf{H}}_{apt}[n] = \mathbf{H}[n] - \tilde{\mathbf{H}}_{apt}[n], \tag{16.36}$$

while its auto-correlation matrix $\mathbf{R}_{\Delta\tilde{\mathbf{H}}_{apt}[n]} \in \mathbb{C}^{LK \times LK}$ is given by:

$$\mathbf{R}_{\Delta\tilde{\mathbf{H}}_{apt}[n]} = E\{\Delta\tilde{\mathbf{H}}_{apt}[n]\Delta\tilde{\mathbf{H}}_{apt}^H[n]\} \tag{16.37}$$

$$= \mathbf{R}^{[f]} - 2\mathfrak{Re}\left\{\mathbf{R}_c^{[f]}\right\} + \mathbf{R}_a^{[f]}. \tag{16.38}$$

More specifically, the first term in Equation 16.38, namely the block-diagonal matrix $\mathbf{R}^{[f]} \in \mathbb{C}^{LK \times LK}$ hosting the different users' spaced-frequency correlation matrices $\mathbf{R}^{[f](i)}, i = 1, \ldots, L$ is given by:

$$\mathbf{R}^{[f]} = E\{\mathbf{H}[n]\mathbf{H}^H[n]\} \tag{16.39}$$

$$= diag(\mathbf{R}^{[f](1)} \ldots \mathbf{R}^{[f](L)}). \tag{16.40}$$

Here we have exploited that the channel transfer functions associated with different transmit antennas are uncorrelated. Again, in Equation 16.40 $\mathbf{R}^{[f](i)} \in \mathbb{C}^{K \times K}$ denotes the i-th user's channel's spaced-frequency correlation matrix, which is defined as:

$$\mathbf{R}^{[f](i)} = E\{\mathbf{H}^{(i)}[n]\mathbf{H}^{(i)H}[n]\}. \tag{16.41}$$

Similarly, for the second term of Equation 16.38, namely the channel transfer factor estimates' cross-correlation matrix $\mathbf{R}_c^{[f]} \in \mathbb{C}^{LK \times LK}$ we obtain:

$$\mathbf{R}_c^{[f]} = E\{\mathbf{H}[n]\tilde{\mathbf{H}}_{apt}^H[n]\} \tag{16.42}$$

$$= \mathbf{R}^{[f]}\mathbf{S}^*[n]\mathbf{P}^H[n]\mathbf{W}_J^H, \tag{16.43}$$

while the last term of Equation 16.38, namely the channel transfer factor estimates' auto-correlation matrix $\mathbf{R}_a^{[f]} \in \mathbb{C}^{LK \times LK}$ is given by:

$$\mathbf{R}_a^{[f]} = E\{\tilde{\mathbf{H}}_{apt}[n]\tilde{\mathbf{H}}_{apt}^H[n]\} \tag{16.44}$$

$$= \mathbf{W}_J\mathbf{P}[n](\mathbf{S}^T[n]\mathbf{R}^{[f]}\mathbf{S}^*[n] + \sigma_n^2\mathbf{I})\mathbf{P}^H[n]\mathbf{W}_J^H. \tag{16.45}$$

---

[8]In the context of a sample-spaced CIR it can be shown that we have $\mathbf{W}_J\mathbf{P}[n]\mathbf{S}^T[n] = \mathbf{I}$, conditioned on an appropriate selection of $\mathbf{J}_{K_0}^{(i)}$, $i = 1, \ldots, L$, such that all of the non-zero taps' energy is retained. In this case $\tilde{\mathbf{H}}_{apt}[n]$ is an unbiased estimate [90] of $\mathbf{H}[n]$, which implies that we have $\tilde{\mathbf{H}}[n] = \mathbf{H}[n] + \mathbf{W}_J\mathbf{P}[n]\mathbf{n}[n]$.

Note that $\mathbf{R}_a^{[f]}$ can be split into a channel- and a noise-related matrix contribution $\mathbf{R}_{a,\text{Channel}}^{[f]} \in \mathbb{C}^{LK \times LK}$ and $\mathbf{R}_{a,\text{AWGN}}^{[f]} \in \mathbb{C}^{LK \times LK}$, respectively, yielding:

$$\mathbf{R}_a^{[f]} = \mathbf{R}_{a,\text{Channel}}^{[f]} + \mathbf{R}_{a,\text{AWGN}}^{[f]}, \tag{16.46}$$

where the two components are given by:

$$\mathbf{R}_{a,\text{Channel}}^{[f]} = \mathbf{W}_J \mathbf{P}[n] \mathbf{S}^T[n] \mathbf{R}^{[f]} \mathbf{S}^*[n] \mathbf{P}^H[n] \mathbf{W}_J^H \tag{16.47}$$

$$\mathbf{R}_{a,\text{AWGN}}^{[f]} = \sigma_n^2 \mathbf{W}_J \mathbf{P}[n] \mathbf{P}^H[n] \mathbf{W}_J^H. \tag{16.48}$$

While in Section 16.4.2.2 we will further elaborate on the specific structure of the auto-correlation matrix $\mathbf{R}_{\Delta \tilde{\mathbf{H}}_{apt}[n]}$ in the context of the idealistic scenario of a sample-spaced CIR, in Section 16.4.2.3 we will consider its specific structure in the context of a more realistic scenario of a non-sample-spaced CIR, with the ultimate aim of determining the *a posteriori* estimation MSEs.

16.4.2.2 Sample-Spaced CIRs

The discussions in this section are structured as follows. The specific effects of a sample-spaced CIR on the channel transfer factor estimation error's auto-correlation matrix are discussed in Section 16.4.2.2.1. In order to further characterise the estimator's MSE performance, the properties of the optimum training sequences proposed by Li [93] are reviewed in Section 16.4.2.2.2, while their impact on the estimator's performance is elaborated on in Section 16.4.2.2.3.

16.4.2.2.1 Auto-Correlation Matrix of the Channel Transfer Factor Estimation Errors
The decomposition of the i-th user's channel's spaced-frequency correlation matrix $\mathbf{R}^{[f](i)}$, $i = 1, \ldots, L$ with respect to the DFT matrix $\mathbf{W}$ can be expressed as:

$$\mathbf{R}^{[f](i)} = \mathbf{W} \Lambda_W^{[f](i)} \mathbf{W}^H, \tag{16.49}$$

where the matrix $\Lambda_W^{[f](i)} \in \mathbb{C}^{K \times K}$ has a diagonal structure with real-valued elements only in the context of a sample-spaced CIR. This is, because in this specific case the Karhunen-Loeve Transform (KLT) matrix [90] of $\mathbf{R}^{[f](i)}$ is identical to the DFT matrix $\mathbf{W}$. Hence the diagonal elements of $\Lambda_W^{[f](i)}$ are the real-valued eigenvalues of the matrix $\mathbf{R}^{[f](i)}$, while the eigenvectors of $\mathbf{R}^{[f](i)}$ are identical to the column vectors of the DFT matrix $\mathbf{W}$. Upon retaining all of the CIR-related taps having a non-zero energy, which is achieved by appropriately designing the matrix $\mathbf{J}_{K_0}^{(i)}$ defined in the context of Equation 16.7, Equation 16.49 can also be formulated as:

$$\mathbf{R}^{[f](i)} = \mathbf{W} \left(\mathbf{J}_{K_0}^{(i)} \mathbf{J}_{K_0}^{(i)H} \right) \Lambda_W^{[f](i)} \left(\mathbf{J}_{K_0}^{(i)} \mathbf{J}_{K_0}^{(i)H} \right) \mathbf{W}^H \tag{16.50}$$

$$= \mathbf{W}_J^{(i)} \Lambda_{W,J}^{[f](i)} \mathbf{W}_J^{(i)H}, \tag{16.51}$$

where $\mathbf{W}_J^{(i)}$ was defined in Equation 16.7 and $\Lambda_{W,J}^{[f](i)} = \mathbf{J}_{K_0}^{(i)H} \Lambda_W^{[f](i)} \mathbf{J}_{K_0}^{(i)} \in \mathbb{C}^{K_0 \times K_0}$. Recall the definition of the matrix $\mathbf{A}[n]$ from Equation 16.14, namely that we have $\mathbf{A}[n] =$

$\mathbf{S}^T[n]\mathbf{W}_J$ and that the Hermitian transpose of $\mathbf{A}[n]$ is given by $\mathbf{A}^H[n] = \mathbf{W}_J^H\mathbf{S}^*[n]$. Then, upon substituting Equation 16.51 into the formulae of the channel transfer factor estimates' cross-correlation matrix- and auto-correlation matrix given in Equations 16.43 and 16.45, respectively, and upon resubstituting the results into Equation 16.38 we obtain the following expression for the auto-correlation matrix of the channel transfer factor estimation errors:

$$\mathbf{R}_{\Delta\tilde{\mathbf{H}}_{apt}[n]} = \mathbf{R}_{a,\text{AWGN}}^{[f]} = \sigma_n^2 \mathbf{W}_J \mathbf{P}[n]\mathbf{P}^H[n]\mathbf{W}_J^H, \qquad (16.52)$$

which results from the noise-related contribution of the matrix $\mathbf{R}_a^{[f]}$ in Equation 16.46, namely from its second term. Upon recalling from Equation 16.24 that the projection matrix was defined as $\mathbf{P}[n] = (\mathbf{A}^H[n]\mathbf{A}[n])^{-1}\mathbf{A}^H[n]$, Equation 16.52 is transformed into:

$$\mathbf{R}_{\Delta\tilde{\mathbf{H}}_{apt}[n]} = \sigma_n^2 \mathbf{W}_J \mathbf{Q}^{-1}[n]\mathbf{W}_J^H, \qquad (16.53)$$

where according to Equation 16.20 we have $\mathbf{Q}[n] = \mathbf{A}^H[n]\mathbf{A}[n]$. Note from Equation 16.53, that since $\mathbf{A}[n] = \mathbf{S}^T[n]\mathbf{W}_J$ according to Equation 16.14, the matrix $\mathbf{R}_{\Delta\tilde{\mathbf{H}}_{apt}[n]}$ and consequently also the average estimation MSE depends on the specific subcarrier symbols transmitted.

16.4.2.2.2 Properties of Optimum Training Sequences

In order to draw further conclusions with respect to the estimator's MSE, let us consider the choice of channel sounding or training subcarrier symbol sequences, as proposed by Li [93], which are given for the i-th user or transmit antenna by:

$$t^{(i)}[n, k] = \sigma_i t_p[n, k] W_K^{-\overline{K}_0(i-1)k}, \qquad (16.54)$$

where $\sigma_i = \sqrt{\sigma_i^2}$ is the i-th user's signal standard deviation, $t_p[n, k]$, $k = 0, \ldots, K-1$ denotes an arbitrary training sequence having potentially complex subcarrier symbols of unit variance, $W_K = e^{-j\frac{2\pi}{K}}$ is the complex Fourier kernel, while $\overline{K}_0$ is defined as $\overline{K}_0 = \lfloor \frac{K}{L} \rfloor$. In [93] it was argued that in the context of these training sequences for identical matrices of $\mathbf{J}_{K_0} = \mathbf{J}_{K_0}^{(1)} = \ldots = \mathbf{J}_{K_0}^{(L)}$, where the K_0 number of unity elements in $\mathbf{J}_{K_0}$ are arranged on a diagonal, such that K_0 number of significant adjacent taps are extracted in the CIR-related domain, the matrix $\mathbf{Q}[n]$ is of diagonal shape. This can be demonstrated by recalling the definition of $\mathbf{Q}[n]$ in Equation 16.20, and substituting $\mathbf{A}[n]$ of Equation 16.14 into Equation 16.20 yields:

$$\begin{aligned}
\mathbf{Q}[n] &= \mathbf{A}^H[n]\mathbf{A}[n] & (16.55) \\
&= \mathbf{W}_J^H\mathbf{S}^*[n]\mathbf{S}^T[n]\mathbf{W}_J. & (16.56)
\end{aligned}$$

By further capitalising on the expressions of $\mathbf{W}_J$ and $\mathbf{S}[n]$ given in Equations 16.10 and 16.4, respectively, the sub-matrix $\mathbf{Q}^{(i,j)}[n] \in \mathbb{C}^{K_0 \times K_0}$ of the block-matrix $\mathbf{Q}[n] \in \mathbb{C}^{LK_0 \times LK_0}$, which is associated with the i-th "row" and j-th "column" of $\mathbf{Q}[n]$, where $i, j \in \{1, \ldots, L\}$

is given by:

$$\mathbf{Q}^{(i,j)}[n] = \mathbf{W}_J^{(i)H}\mathbf{S}^{(i)*}[n]\mathbf{S}^{(j)}[n]\mathbf{W}_J^{(j)} \tag{16.57}$$

$$= \mathbf{J}_{K_0}^{(i)H}\acute{\mathbf{Q}}^{(i,j)}[n]\mathbf{J}_{K_0}^{(j)}, \tag{16.58}$$

where $\acute{\mathbf{Q}}^{(i,j)}[n] \in \mathbb{C}^{K \times K}$ is defined as:

$$\acute{\mathbf{Q}}^{(i,j)}[n] = \mathbf{W}^H\mathbf{S}^{(i)*}[n]\mathbf{S}^{(j)}[n]\mathbf{W}. \tag{16.59}$$

More explicitly, Equation 16.58 was obtained by substituting $\mathbf{W}_J^{(i)}$ given in Equation 16.7 into Equation 16.57. When employing channel sounding or training sequences obeying Equation 16.54 as proposed in [93], the elements of the matrix $\mathbf{S}^{(i)*}[n]\mathbf{S}^{(j)T}[n]$, which is the key component of Equation 16.59, are given by:

$$t^{(i)*}[n,k]t^{(j)}[n,k] = \sigma_i\sigma_j W_K^{\overline{K}_0(i-j)k}. \tag{16.60}$$

By further noting that the elements of the DFT matrix $\mathbf{W}$ are given by $\frac{1}{\sqrt{K}}W_K^{i_1j_1}$ and that of its Hermitian transpose by $\frac{1}{\sqrt{K}}W_K^{-i_1j_1}$, where $i_1, j_1 \in \{0,\dots,K-1\}$, it can readily be shown that the element (i_1, j_1) of the product matrix $\acute{\mathbf{Q}}^{(i,j)}[n]$ defined in Equation 16.59 is given by:

$$\acute{\mathbf{Q}}^{(i,j)}[n]|_{i_1,j_1} = \frac{1}{K}\sum_{k=0}^{K-1}\sigma_i\sigma_j e^{j\frac{2\pi}{K}k[(i_1-j_1)-\overline{K}_0(i-j)]} \tag{16.61}$$

$$= \sigma_i\sigma_j\delta\left[(i_1-j_1)-\overline{K}_0(i-j)-xK\right], \tag{16.62}$$

where:

$$x = \begin{cases} -1 & \wedge \quad (i > j \wedge i_1 < j_1) \\ 0 & \wedge \quad (i > j \wedge i_1 > j_1) \vee (i < j \wedge i_1 < j_1) \\ 1 & \wedge \quad (i < j \wedge i_1 > j_1) \end{cases} . \tag{16.63}$$

In the derivation of Equation 16.62 we have exploited that the summation of the complex exponentials in Equation 16.61 yields a non-zero contribution, if the exponential argument in square brackets is equal to an integer multiple of K, namely, if we have $(i_1 - j_1) - \overline{K}_0(i - j) = xK$. When taking into account the constraint of $i_1, j_1 \in \{0,\dots,K-1\}$, we have $x \in \{-1, 0, 1\}$. Observe furthermore from Equation 16.62 that when we have $(i = j)$, the matrix $\acute{\mathbf{Q}}^{(i,j)}[n]$ is a diagonal matrix having identical elements of σ_i^2, while for $(i > j)$ the matrix is a diagonal matrix, cyclically shifted to the left. Finally, for $(i < j)$ the diagonal matrix $\acute{\mathbf{Q}}^{(i,j)}[n]$ is *cyclically* shifted to the right, exhibiting in both cases the form of:

$$\acute{\mathbf{Q}}^{(i,j)}[n] = \sigma_i\sigma_j\mathbf{I}_{cyc}^{(i,j)}, \quad \text{where } \mathbf{I}_{cyc}^{(i,j)} = \begin{pmatrix} \mathbf{0} & \mathbf{I}_B \\ \mathbf{I}_A & \mathbf{0} \end{pmatrix} \in \mathbb{C}^{K \times K}, \tag{16.64}$$

and where $\mathbf{I}_A \in \mathbb{R}^{K_A \times K_A}$ with $K_A = [K - \overline{K}_0(i-j)] \bmod K$, while $\mathbf{I}_B \in \mathbb{R}^{K_B \times K_B}$ with $K_B = [K - \overline{K}_0(j-i)] \bmod K$. Note that the matrix $\acute{\mathbf{Q}}^{(i,j)}[n]$ multiplied from the right with a column vector performs a cyclic rotation of the column vector's elements.

Upon also taking into account the "masking" effects imposed by the matrix $\mathbf{J}_{K_0} = \mathbf{J}_{K_0}^{(1)} = \ldots = \mathbf{J}_{K_0}^{(L)}$ as seen in Equation 16.58, it follows that for CIRs having K_0 adjacent significant CIR-related taps we have $\mathbf{Q}^{(i,i)}[n] = \sigma_i^2\mathbf{I}$ and $\mathbf{Q}^{(i,j)}[n] = \mathbf{0}$, $i \neq j$.

16.4.2.2.3 *A Posteriori* Estimation MSE Using Optimum Training Sequences

By capitalising on the specific properties of $\mathbf{Q}[n]$ in the context of optimum training sequences, namely that $\mathbf{Q}^{(i,i)}[n] = \sigma_i^2\mathbf{I}$ and $\mathbf{Q}^{(i,j)}[n] = \mathbf{0}, i \neq j$, Equation 16.53 can be rewritten as:

$$\mathbf{R}_{\Delta\tilde{\mathbf{H}}_{apt}[n]} = \sigma_n^2 diag(\frac{1}{\sigma_1^2}\mathbf{T}_{W,K_0}^{(1)} \cdots \frac{1}{\sigma_L^2}\mathbf{T}_{W,K_0}^{(L)}), \tag{16.65}$$

where the matrix $\mathbf{T}_{W,K_0}^{(i)} \in \mathbb{C}^{K \times K}$ is given by:

$$\mathbf{T}_{W,K_0}^{(i)} = \mathbf{W}_J^{(i)}\mathbf{W}_J^{(i)H} \tag{16.66}$$

$$= \mathbf{W}\mathbf{J}_{K_0}^{(i)}\mathbf{J}_{K_0}^{(i)H}\mathbf{W}^H \tag{16.67}$$

$$= \mathbf{W}\mathbf{I}_{K_0}^{(i)}\mathbf{W}^H, \tag{16.68}$$

with $\mathbf{W}_J^{(i)}$ according to Equation 16.10 and $\mathbf{I}_{K_0}^{(i)} = \mathbf{J}_{K_0}^{(i)}\mathbf{J}_{K_0}^{(i)H} \in \mathbb{R}^{K \times K}$. The latter formula will specifically draw our attention in the context of PIC-assisted DDCE in Section 16.5. The estimation MSE averaged over the subcarriers of the i-th user is then given by [91]:

$$\overline{\mathrm{MSE}}_{apt}^{(i)}[n] = \frac{1}{K}\mathrm{Trace}(\mathbf{R}_{\Delta\tilde{\mathbf{H}}_{apt}[n]}^{(i,i)}) \tag{16.69}$$

$$= \frac{K_0}{K}\frac{\sigma_n^2}{\sigma_i^2}, \tag{16.70}$$

while by further averaging over the L different users the total average MSE becomes:

$$\overline{\mathrm{MSE}}_{apt}[n] = \frac{1}{L}\sum_{i=1}^{L}\overline{\mathrm{MSE}}_{apt}^{(i)}[n] \tag{16.71}$$

$$= \frac{K_0}{LK}(\sum_{i=1}^{L}\frac{1}{\sigma_i^2})\sigma_n^2. \tag{16.72}$$

Note that the i-th user's *a posteriori* estimation MSE given by Equation 16.70 is equivalent to that of the single-user scenario, when performing CIR-related tap windowing of the initial least-squares *a posteriori* channel transfer factor estimates $\tilde{H}_{apt}[n, k], k = 0, \ldots, K - 1$, which were defined in Equation 15.2. Note however that when transmitting random symbols the estimation MSE is potentially degraded compared to the scenario of transmitting optimum training sequences. This is since the different channels' CIRs were imperfectly separated from each other.

In the next section we will analyse the estimator's performance in the context of the more realistic environment of a non-sample-spaced CIR.

16.4.2.3 Non-Sample-Spaced CIRs

In contrast to the sample-spaced CIR of the previous section, for a non-sample-spaced CIR the decomposition $\Lambda_W^{[f](i)}$ seen in Equation 16.49 for the i-th user's channel's spaced-frequency correlation matrix $\mathbf{R}^{[f](i)}$ with respect to the DFT matrix $\mathbf{W}$ is not a diagonal matrix and does not have a reduced number of $K_0 \ll K$ non-zero elements. In fact, potentially all of the decomposition's taps may become non-zero.

In order to further characterise the estimator's MSE performance, we will commence our elaborations again by considering the channel transfer factor estimation errors' auto-correlation matrix $\mathbf{R}_{\Delta\tilde{\mathbf{H}}_{apt}[n]}$, which was given by Equation 16.38. We note furthermore that the associated components of $\mathbf{R}_{\Delta\tilde{\mathbf{H}}_{apt}[n]}$, namely the block-diagonal matrix $\mathbf{R}^{[f]}$ of the different users' channels' spaced-frequency correlation matrices, as well as the channel transfer factor estimates' cross-correlation as well as auto-correlation matrix, namely $\mathbf{R}_c^{[f]}$ and $\mathbf{R}_a^{[f]}$ were formulated in Equations 16.40, 16.43 and 16.45. In order to simplify $\mathbf{R}_c^{[f]}$ and $\mathbf{R}_a^{[f]}$ of Equations 16.43 and 16.45, we will again capitalise on the properties of the optimum training sequences proposed by Li [93], which also constituted the basis of our derivations in Section 16.4.2.2.

Hence, our further proceedings are as follows. In Section 16.4.2.3.1 the channel transfer factor estimates' cross-correlation matrix is derived, while the estimates' auto-correlation matrix is derived in Section 16.4.2.3.2, both in the context of the optimum training sequences [93] described in Section 16.4.2.2.2. With the aid of these matrices the average channel transfer factor estimation MSE is then derived in Section 16.4.2.3.3, based on the averaged sum of the main diagonal elements associated with the subcarrier estimation errors' auto-correlation matrix of Equation 16.38.

16.4.2.3.1 Cross-Correlation Matrix of the Channel Transfer Factor Estimates in the Context of Optimum Training Sequences
Let us commence by studying the estimates' cross-correlation matrix, which was given by Equation 16.43. Upon substituting $\mathbf{P}[n]$ of Equation 16.24 and $\mathbf{A}[n]$ of Equation 16.14 into Equation 16.43 we arrive at:

$$\mathbf{R}_c^{[f]} = diag(\mathbf{R}^{[f](1)} \ldots \mathbf{R}^{[f](L)}) \cdot \mathbf{S}^*[n]\mathbf{S}^T[n] \cdot diag(\frac{1}{\sigma_1^2}\mathbf{T}_{W,K_0}^{(1)} \cdots \frac{1}{\sigma_L^2}\mathbf{T}_{W,K_0}^{(L)}), \quad (16.73)$$

where the diagonal matrix on the right-hand side of the product in Equation 16.73 was obtained upon exploiting the specific properties of the training sequences [91] outlined in Section 16.4.2.2.2, implying that $\mathbf{Q}^{(i,i)}[n] = \sigma_i^2\mathbf{I}$ and $\mathbf{Q}^{(i,j)}[n] = \mathbf{0}$, $i \neq j$, as argued in the context of deriving Equation 16.65. From Equation 16.73 we infer that the sub-matrix $\mathbf{R}_c^{[f](i,i)} \in \mathbb{C}^{K \times K}$, located on the diagonal of the matrix $\mathbf{R}_c^{[f]}$, which is associated with the i-th user is given by:

$$\mathbf{R}_c^{[f](i,i)} = \mathbf{R}^{[f](i)}\mathbf{T}_{W,K_0}^{(i)}, \quad (16.74)$$

where we have also exploited that the i-th sub-matrix on the diagonal of $\mathbf{S}^*[n]\mathbf{S}^T[n]$ is given by $\sigma_i^2\mathbf{I}$.

16.4.2.3.2 Auto-Correlation Matrix of the Channel Transfer Factor Estimates in the Context of Optimum Training Sequences

The AWGN-related contribution of the estimates' auto-correlation matrix $\mathbf{R}_a^{[f]}$ given by Equation 16.43 was elaborated on in Section 16.4.2.2, and the same derivation is also valid for the case of the non-sample-spaced CIR discussed here. Specifically, from Equations 16.52 and 16.65 we recall that:

$$\mathbf{R}_{a,\text{AWGN}}^{[f]} = \sigma_n^2 diag(\frac{1}{\sigma_1^2}\mathbf{T}_{W,K_0}^{(1)} \cdots \frac{1}{\sigma_L^2}\mathbf{T}_{W,K_0}^{(L)}), \tag{16.75}$$

and hence, the i-th user's associated sub-matrix in the matrix of Equation 16.75 is given by:

$$\mathbf{R}_{a,\text{AWGN}}^{[f](i,i)} = \frac{\sigma_n^2}{\sigma_i^2}\mathbf{T}_{W,K_0}^{(i)}. \tag{16.76}$$

By contrast, the channel-related component $\mathbf{R}_{a,\text{Channel}}^{[f]}$ of $\mathbf{R}_a^{[f]}$ defined in Equation 16.46, requires some further elaborations. Specifically, from Equation 16.47 we infer that:

$$\mathbf{R}_{a,\text{Channel}}^{[f]} = \mathbf{W}_J\mathbf{P}[n]\mathbf{S}^T[n]\mathbf{R}^{[f]}\mathbf{S}^*[n]\mathbf{P}^H[n]\mathbf{W}_J^H, \tag{16.77}$$

which can be transformed into:

$$\begin{aligned}
\mathbf{R}_{a,\text{Channel}}^{[f]} = {}& diag(\frac{1}{\sigma_1^2}\mathbf{T}_{W,K_0}^{(1)} \cdots \frac{1}{\sigma_L^2}\mathbf{T}_{W,K_0}^{(L)}) \cdot \\
& \cdot \mathbf{S}^*[n]\mathbf{S}^T[n] \cdot diag(\mathbf{R}^{[f](1)} \ldots \mathbf{R}^{[f](L)}) \cdot \mathbf{S}^*[n]\mathbf{S}^T[n] \cdot \\
& \cdot diag(\frac{1}{\sigma_1^2}\mathbf{T}_{W,K_0}^{(1)} \cdots \frac{1}{\sigma_L^2}\mathbf{T}_{W,K_0}^{(L)})
\end{aligned} \tag{16.78}$$

by substituting $\mathbf{P}[n]$ of Equation 16.24, $\mathbf{A}[n]$ of Equation 16.14 and $\mathbf{R}^{[f]}$ of Equation 16.40 into Equation 16.77. Here we have exploited again the effects of the optimum training sequences on the specific structure of the matrix $\mathbf{Q}[n] = \mathbf{A}^H[n]\mathbf{A}[n]$, similar to the procedure employed in the context of deriving Equation 16.73. From Equation 16.78 we infer that the sub-matrix $\mathbf{R}_{a,\text{Channel}}^{[f](i,i)} \in \mathbb{C}^{K \times K}$, located on the diagonal of matrix $\mathbf{R}_a^{[f]}$, which is associated with the i-th user is given by:

$$\mathbf{R}_{a,\text{Channel}}^{[f](i,i)} = \frac{1}{\sigma_i^4}\mathbf{T}_{W,K_0}^{(i)} \left(\sum_{j=1}^{L} \mathbf{S}^{(i)*}\mathbf{S}^{(j)}\mathbf{R}^{[f](j)}\mathbf{S}^{(j)*}\mathbf{S}^{(i)} \right) \mathbf{T}_{W,K_0}^{(i)}. \tag{16.79}$$

In order to proceed further let us again recall the decomposition of the i-th user's channel's spaced-frequency correlation matrix $\mathbf{R}^{[f](i)}$ with respect to the DFT matrix $\mathbf{W}$, which was formulated in Equation 16.49. Recall furthermore the decomposition of $\mathbf{T}_{W,K_0}^{(i)}$ from Equation 16.68. Then Equation 16.79 can be reformulated as:

$$\mathbf{R}_{a,\text{Channel}}^{[f](i,i)} = \frac{1}{\sigma_i^4}\mathbf{W}\mathbf{I}_{K_0}^{(i)} \left(\sum_{j=1}^{L} \acute{\mathbf{Q}}^{(i,j)}[n]\Lambda_W^{[f](j)}\acute{\mathbf{Q}}^{(i,j)H}[n] \right) \mathbf{I}_{K_0}^{(i)}\mathbf{W}^H, \tag{16.80}$$

where $\acute{\mathbf{Q}}^{(i,j)}[n]$ was defined in Equation 16.59. Recall also that the specific structure of $\acute{\mathbf{Q}}^{(i,j)}[n]$ in the context of employing optimum training sequences was outlined in Section 16.4.2.3. While for $(i = j)$ we have $\acute{\mathbf{Q}}^{(i,j)}[n] = \sigma_i^2 \mathbf{I}$, for $(i \neq j)$ we obtain $\acute{\mathbf{Q}}^{(i,j)}[n] = \sigma_i \sigma_j \mathbf{I}_{cyc}^{(i,j)}$, which was further detailed in Equation 16.64. Upon taking into account these specific properties in Equation 16.80, we obtain:

$$\mathbf{R}_{a,\text{Channel}}^{[f](i,i)} = \frac{1}{\sigma_i^2} \mathbf{W} \mathbf{I}_{K_0}^{(i)} \left(\sigma_i^2 \Lambda_W^{[f](i)} + \sum_{\substack{j=1 \\ j \neq i}}^{L} \sigma_j^2 \mathbf{I}_{cyc}^{(i,j)} \Lambda_W^{[f](j)} \mathbf{I}_{cyc}^{(i,j)T} \right) \mathbf{I}_{K_0}^{(i)} \mathbf{W}^H. \tag{16.81}$$

Note in Equation 16.81 that the matrix $\mathbf{I}_{cyc}^{(i,j)}$ performs a rotation of the matrix $\Lambda_W^{[f](i)}$ with respect to its rows, while its transpose, namely $\mathbf{I}_{cyc}^{(i,j)T}$, performs a rotation of the matrix $\Lambda_W^{[f](i)}$ with respect to its columns. The effect of the two rotations within the sum of Equation 16.81 is that different parts of the decomposition $\Lambda_W^{[f](i)}$ become visible to the "masking window" defined by the matrix $\mathbf{I}_{K_0}^{(i)}$.

16.4.2.3.3 Channel Estimation MSE in the Context of Optimum Training Sequences

Hence by following the philosophy of Equation 16.38 the i-th user's associated sub-matrix $\mathbf{R}_{\Delta \tilde{\mathbf{H}}_{apt}[n]}^{(i,i)}$ of the channel transfer factor estimation error correlation matrix $\mathbf{R}_{\Delta \tilde{\mathbf{H}}_{apt}[n]}$ is given by:

$$\mathbf{R}_{\Delta \tilde{\mathbf{H}}_{apt}[n]}^{(i,i)} = \mathbf{R}^{[f](i)} - 2\Re\{\mathbf{R}_c^{[f](i,i)}\} + \mathbf{R}_a^{[f](i,i)}, \tag{16.82}$$

where according to Equation 16.46 we have $\mathbf{R}_a^{[f](i,i)} = \mathbf{R}_{a,\text{Channel}}^{[f](i,i)} + \mathbf{R}_{a,\text{AWGN}}^{[f](i,i)}$ and $\mathbf{R}_c^{[f](i,i)}$, $\mathbf{R}_{a,\text{Channel}}^{[f](i,i)}$ and $\mathbf{R}_{a,\text{AWGN}}^{[f](i,i)}$ were given by Equations 16.74, 16.81 and 16.76. The *a posteriori* estimation MSE averaged over all subcarriers of the i-th user is then given by:

$$\begin{aligned} \text{MSE}_{apt}^{(i)}[n] &= \frac{1}{K} \text{Trace}(\mathbf{R}_{\Delta \tilde{\mathbf{H}}_{apt}[n]}^{(i,i)}) \tag{16.83} \\ &= \sigma_H^2 - \frac{1}{K} \text{Trace}(\Lambda_W^{[f](i)} \mathbf{I}_{K_0}^{(i)}) + \frac{K_0}{K} \frac{\sigma_n^2}{\sigma_i^2} + \\ &\quad + \sum_{\substack{j=1 \\ j \neq i}}^{L} \frac{\sigma_j^2}{\sigma_i^2} \frac{1}{K} \text{Trace}(\mathbf{I}_{cyc}^{(i,j)} \Lambda_W^{[f](j)} \mathbf{I}_{cyc}^{(i,j)T} \mathbf{I}_{K_0}^{(i)}), \tag{16.84} \end{aligned}$$

where we have exploited that $\text{Trace}(\mathbf{U}\mathbf{A}\mathbf{U}^H) = \text{Trace}(\mathbf{A})$ for a unitary matrix $\mathbf{U}$. Note that in Equation 16.84 the sum of the first three terms is identical to the MSE, which would be achieved in a single-user scenario in the context of a non-sample-spaced CIR. However, as a result of the multi-user interference experienced in the CIR-related domain, the MSE is further degraded in the multi-user scenario by the contribution constituted by the last term of Equation 16.84. Furthermore, we note again that the MSE of Equation 16.84 is only achieved upon employing optimum channel sounding or training sequences [93] of Section 16.4.2.2.2, while for random subcarrier symbol sequences the MSE is potentially further degraded.

16.4.3 *A Priori* Channel Transfer Function Estimation MSE Enhancement by Linear Prediction of the *A Posteriori* CIR-Related Tap Estimates

The vector of CIR-related tap estimates obtained with the aid of Equation 16.22 or 16.23 and its associated vector of channel transfer factor estimates given by Equation 16.8 are actually the *a posteriori* estimates generated for the current OFDM symbol after the detection of the transmitted subcarrier symbols. These estimates could for example be employed again following an iterative approach for obtaining potentially enhanced symbol decisions, since the channel transfer factor estimates employed during the initial symbol detection were imperfect. In the conventional decision-directed channel estimator, however, these *a posteriori* estimates are employed as *a priori* estimates for the demodulation of the next OFDM symbol, assuming that the channel's transfer function remained constant. However, as proposed by Li *et al.* [93], enhanced *a priori* channel transfer function estimates can be obtained for the next OFDM symbol period upon invoking linear prediction techniques, which could operate for example in the context of the estimator structures discussed here, on a time direction tap-by-tap basis in the CIR-related domain. This was alluded to by the CIR-related tap prediction block shown at the top of Figure 16.3. Recall that linear CIR-related tap prediction was also used in the context of single-user channel estimation techniques in Section 15.2.4. We will make extensive use of these techniques in the context of PIC-assisted DDCE discussed in Section 16.5. Note that Li's LS-assisted DDCE [91] allows for a purely transversal FIR filter-related implementation of the linear predictor operating along the time direction. Thus, the optimum predictor coefficients can be determined with the aid of the Wiener equation. We note, however, that this approach is different from incorporating a channel transfer function predictor in the context of the PIC-assisted DDCE, a technique, which will be outlined in Section 16.4.4. The evaluation of the associated predictor coefficients is not as straightforward as for the LS-assisted DDCE and will be the topic of our discussions in Section 16.5.

16.4.4 Simplified Approach to LS-Assisted DDCE

A disadvantage of the LS estimation approach discussed in Section 16.4.1 is the potentially significant computational complexity imposed by the calculation of the inverse of the correlation matrix $\mathbf{Q}[n]$, which is required for computing the different users' CIR-related tap estimates, as shown in Equation 16.22. As a potential means of reducing the complexity, Li [93] suggested performing a parallel cancellation of the users' CIR-related contributions. More explicitly, upon invoking Equation 16.22 and by exploiting its specific block structure, the j-th user's vector of CIR-related *a posteriori* tap estimates can be expressed as a function of the $(L-1)$ remaining users' vectors of CIR-related *a posteriori* tap estimates by "removing" their effect with the aid of the Parallel Interference Cancellation (PIC) step constituted by the second bracketed term in:

$$\tilde{\mathbf{h}}_{apt,K_0}^{(j)}[n] = \mathbf{Q}^{(j,j)-1}[n] \left(\mathbf{p}^{(j)}[n] - \sum_{\substack{i=1 \\ i \neq j}}^{L} \mathbf{Q}^{(j,i)}[n] \hat{\mathbf{h}}_{apr,K_0}^{(i)}[n] \right), \qquad (16.85)$$

K_0	number of significant CIR-related taps
K	number of subcarriers
L	number of simultaneous users
P	number of receiver antennas

Table 16.2: Summary of the parameters influencing the LS-assisted DDCE's complexity.

where $\mathbf{p}^{(j)}[n] \in \mathbb{C}^{K_0 \times 1}$ is the j-th sub-vector of the vector $\mathbf{p}[n]$ of cross-correlations defined in Equation 16.19:

$$\mathbf{p}^{(j)}[n] = \mathbf{W}_J^{(j)H} \mathbf{S}^{(i)*}[n]\mathbf{x}[n], \tag{16.86}$$

and $\mathbf{Q}^{(j,i)}[n] \in \mathbb{C}^{LK_0 \times LK_0}$ is the sub-matrix associated with the j-th "row" and i-th "column" of the auto-correlation matrix $\mathbf{Q}[n]$ defined in Equation 16.20, which is expressed as:

$$\mathbf{Q}^{(j,i)}[n] = \mathbf{W}_J^{(j)H} \mathbf{S}^{(j)*}[n] \mathbf{S}^{(i)} \mathbf{W}_J^{(i)}. \tag{16.87}$$

In the simplest case tentative estimates of the $(L-1)$ remaining users' vectors of CIR-related taps, which are required for the cancellation process outlined in Equation 16.85, could be provided upon assuming that $\hat{\mathbf{h}}_{apr,K_0}^{(j)}[n] \approx \tilde{\mathbf{h}}_{apt,K_0}^{(j)}[n-1]$ for reasonably slowly varying channels [93]. A further improvement of the estimator's MSE can be achieved with the aid of an $N_{tap}^{[t]}$-tap prediction filter, which operates individually on each CIR-related tap in the time direction. This solution will be the topic of Section 16.5, where the optimum predictor coefficients are determined.

Before concluding this section, we emphasise again, that the difference between the scheme alluded to here and that to be presented in Section 16.5 is essentially the employment of the parallel interference cancellation, which is once conducted in the CIR-related domain and then once again in the frequency domain. However, it can be readily demonstrated that the scheme proposed in this section and that of Section 16.5 produce identical results.

16.4.5 Complexity Analysis of the Original- and Simplified LS-Assisted DDCE

In this section we will analyse the computational complexity imposed by the LS-assisted DDCE. The relevant system parameters are summarised in Table 16.2. More specifically in Section 16.4.5.1 we will characterise the complexity of the original approach of Section 16.4.1.2 capitalising on the inversion of the correlation matrix $\mathbf{Q}[n]$, while in Section 16.4.5.2 the complexity of the simplified approach of Section 16.4.4 will be assessed.

16.4.5.1 Complexity of the Original LS-Assisted DDCE

The complexity of the original LS-assisted DDCE technique of [91], which was described in Section 16.4.1, is attributed to three different operations. The first operation is the calculation of the correlation matrix $\mathbf{Q}[n]$ defined in Equation 16.20, which will be considered in

Section 16.4.5.1.1. This operation is independent of the specific antenna element considered. The second operation is that of evaluating $\mathbf{p}[n]$ of Equation 16.19, which has to be carried out separately for each of the P receiver antenna elements. The associated complexity will be quantified in Section 16.4.5.1.2. Finally, the system of LS-related equations has to be solved for the vector $\tilde{\mathbf{h}}_{apt}[n]$ of the different users' CIR-related *a posteriori* tap estimates according to Equation 16.22, which is transferred to the frequency domain with the aid of the FFT in order to obtain the channel transfer factor estimates. These steps will be further analysed in terms of their computational complexity in Section 16.4.5.1.3. Based on the analysis of the individual components a formula will then be presented in Section 16.4.5.1.4 for the LS-assisted DDCE's total complexity.

16.4.5.1.1 Complexity Associated with Assembling Matrix $\mathbf{Q}[n]$

More specifically, for the calculation of the sub-matrix $\mathbf{Q}^{(i,j)}[n]$ of the correlation matrix $\mathbf{Q}[n]$ we have relied on the mathematical discourse of [91][9] and found that the normalised number of complex multiplications and additions is given by:

$$C_{\mathbf{Q}^{(i,j)}[n]}^{(\mathbb{C}*\mathbb{C})}\Big|_{\text{norm}} = \frac{1}{LP}(\frac{1}{2}\log_2 K + 1) \tag{16.88}$$

$$C_{\mathbf{Q}^{(ij)}[n]}^{(\mathbb{C}+\mathbb{C})}\Big|_{\text{norm}} = \frac{1}{LP}\log_2 K, \tag{16.89}$$

where the number of operations was normalised to the K number of subcarriers, L number of users and P number of receiver antenna elements. Note that the additive unity contribution enclosed in brackets seen in the expression of the $C_{\mathbf{Q}^{(i,j)}[n]}^{(\mathbb{C}*\mathbb{C})}\Big|_{\text{norm}}$ number of complex multiplications in Equation 16.88 is due to calculating the correlation between the i-th and the j-th user's transmitted symbols on a subcarrier basis, which is explicitly visible for example in the representation of the correlation matrix $\mathbf{Q}^{(i,j)}[n]$ given by Equation 16.87. The entire correlation matrix $\mathbf{Q}[n]$ consists of L^2 sub-matrices and correspondingly the total computational complexity imposed is increased by the same factor of L^2 compared to the complexity of calculating a single sub-matrix. A reduction of this factor of L^2 can be achieved by exploiting the Hermitian structure of the correlation matrix $\mathbf{Q}[n]$, which implies that $\mathbf{Q}^{(i,j)}[n] = \mathbf{Q}^{(j,i)H}[n]$. Hence, the number of off-diagonal sub-matrices to be evaluated is $\binom{L}{2} = \frac{1}{2}L(L-1)$. Furthermore in the context of MPSK modulation schemes the sub-matrices $\mathbf{Q}^{(i,i)}[n]$, $i = 1, \ldots, L$ along the main diagonal of the matrix $\mathbf{Q}[n]$ are weighted unity matrices and therefore no additional complexity is imposed by their computation. Upon incorporating the factor of $\binom{L}{2}$ into Equations 16.88 and 16.89, the total normalised complexity associated with assembling $\mathbf{Q}[n]$ becomes:

$$C_{\mathbf{Q}[n]}^{(\mathbb{C}*\mathbb{C})}\Big|_{\text{norm}} = \frac{1}{2P}(L-1)(\frac{1}{2}\log_2 K + 1) \tag{16.90}$$

$$C_{\mathbf{Q}[n]}^{(\mathbb{C}+\mathbb{C})}\Big|_{\text{norm}} = \frac{1}{2P}(L-1)\log_2 K. \tag{16.91}$$

---

[9]Note, however, that in terms of the scheme's actual implementation it is not advisable to directly employ the representation suggested by Equations 16.19 and 16.20. It is more efficient to implement the associated multiplication involving the DFT matrix $\mathbf{W}$, with the aid of an FFT, which requires $\frac{K}{2}\log_2 K$ number of complex multiplications and twice the number of complex additions.

16.4.5.1.2   Complexity Associated with Assembling Vector p[n]   The computational complexity imposed by evaluating $\mathbf{p}_p[n]$ of Equation 16.19 for $p = 1, \ldots, P$, follows similar considerations, which results in a normalised number of complex multiplications and additions as given by:

$$C_{\mathbf{p}[n]}^{(\mathbb{C}*\mathbb{C})}\big|_{\text{norm}} = \frac{1}{2}\log_2 K + 1 \tag{16.92}$$

$$C_{\mathbf{p}[n]}^{(\mathbb{C}+\mathbb{C})}\big|_{\text{norm}} = \log_2 K. \tag{16.93}$$

Here we have taken into account that $\mathbf{p}[n]$ has to be calculated separately for each antenna element. Again, the unity contribution observed in the expression of the $C_{\mathbf{p}[n]}^{(\mathbb{C}*\mathbb{C})}\big|_{\text{norm}}$ number of complex multiplications is required, because evaluating $\mathbf{p}[n]$ involves calculating the correlation between the received signal and the transmitted symbols, which are assumed to have been correctly detected, on a subcarrier basis, as observed with the i-th user's block-matrix of $\mathbf{p}[n]$, shown in Equation 16.86.

16.4.5.1.3   Complexity Associated with Solving the LS System Equations for the Vector of CIR-Related Tap Estimates $\tilde{\mathbf{h}}_{apt,K_0}[n]$ Once the auto-correlation matrix $\mathbf{Q}[n]$ of Equation 16.20 and the different reception antennas' associated cross-correlation vectors $\mathbf{p}_p[n]$, $p = 1, \ldots, P$ are available, the vector $\tilde{\mathbf{h}}_{apt,K_0,p}[n]$ can be evaluated for each reception antenna separately upon solving Equation 16.22. However, the associated processing can be significantly accelerated by noting that only the right-hand side of Equation 16.22 is changed for the different reception antenna elements. Specifically, we can employ a decomposition-based solution of Equation 16.22 with respect to the matrix $\mathbf{Q}[n]$, such as the LU decomposition [90], for example, or the QR decomposition as described in Section 16.4.1.2.4. This decomposition has to be performed only once, requiring approximately $\frac{1}{3}(LK_0)^3$ complex multiplications and the same number of additions [89]. Then the solution of Equation 16.22 can be carried out by forward and backward substitutions as outlined in [90]. These operations impose a computational complexity of approximately $(LK_0)^2$ complex multiplications and additions per reception antenna element. Hence, we have a total normalised complexity of:

$$C_{\tilde{\mathbf{h}}_{apt,K_0}[n]}^{(\mathbb{C}*\mathbb{C})}\big|_{\text{norm}} = C_{\tilde{\mathbf{h}}_{apt,K_0}[n]}^{(\mathbb{C}+\mathbb{C})}\big|_{\text{norm}} = \frac{(LK_0)^2}{KL}\left[\frac{1}{3}(LK_0) + 1\right]. \tag{16.94}$$

16.4.5.1.4   Total Complexity   Finally, we have to account for the complexity imposed by transforming the different receiver antennas' and users' CIR-related tap estimates to the frequency domain with the aid of the DFT matrix $\mathbf{W}$. This operation can also be implemented with the computationally more efficient FFT. It can be shown that the corresponding nor-

	$K_0 = 8/K = 64$	$K_0 = 8/K = 512$	$K_0 = 64/K = 512$
$C^{(\mathbb{C}*\mathbb{C})}_{\text{MU-CE,LS}}\big|_{\text{norm}}$	$8.50 + 11.6\dot{6}$	$12.06 + 1.4\dot{6}$	$12.06 + 690.6\dot{6}$
$C^{(\mathbb{C}+\mathbb{C})}_{\text{MU-CE,LS}}\big|_{\text{norm}}$	$14.25 + 11.6\dot{6}$	$21.38 + 1.4\dot{6}$	$21.38 + 690.6\dot{6}$

Table 16.3: Computational complexity of the LS-assisted DDCE [91] in terms of the number of complex multiplications and additions normalised to the K number of subcarriers, L number of users and P number of receiver antennas. Here we have assumed that $L = P = 4$.

malised number of complex multiplications and additions is given by:

$$C^{(\mathbb{C}*\mathbb{C})}_{\tilde{\mathbf{h}}_{apt,K_0}[n]->\tilde{\mathbf{H}}_{apt}[n]}\bigg|_{\text{norm}} = \frac{1}{2}\log_2 K \qquad (16.95)$$

$$C^{(\mathbb{C}+\mathbb{C})}_{\tilde{\mathbf{h}}_{apt,K_0}[n]->\tilde{\mathbf{H}}_{apt}[n]}\bigg|_{\text{norm}} = \log_2 K. \qquad (16.96)$$

Hence, the total normalised number of complex multiplications and additions is given by:

$$C^{()}_{\text{MU-CE,LS}}\big|_{\text{norm}} = C^{()}_{\mathbf{Q}[n]}\big|_{\text{norm}} + C^{()}_{\mathbf{p}[n]}\big|_{\text{norm}} + \\
+ C^{()}_{\tilde{\mathbf{h}}_{apt,K_0}[n]}\big|_{\text{norm}} + C^{()}_{\tilde{\mathbf{h}}_{apt,K_0}[n]->\tilde{\mathbf{H}}_{apt}[n]}\big|_{\text{norm}}. \qquad (16.97)$$

Specifically, upon assuming that $L = P$, we obtain:

$$C^{(\mathbb{C}*\mathbb{C})}_{\text{MU-CE,LS}}\big|_{\text{norm}} = \frac{1}{4}\left(5 - \frac{1}{L}\right)\log_2 K + \frac{1}{2}\left(3 - \frac{1}{L}\right) + \frac{K_0^2}{K}\left[\frac{1}{3}(LK_0) + 1\right] \quad (16.98)$$

$$C^{(\mathbb{C}+\mathbb{C})}_{\text{MU-CE,LS}}\big|_{\text{norm}} = \frac{1}{2}\left(5 - \frac{1}{L}\right)\log_2 K + \frac{K_0^2}{K}\left[\frac{1}{3}(LK_0) + 1\right], \qquad (16.99)$$

where a normalisation to $KLP = KL^2$ has been performed. Observe that upon increasing the number of simultaneous users L, the normalised complexity increases linearly. We have evaluated the normalised complexity for a number of configurations and the corresponding results are listed in Table 16.3. Note that the second additive term associated with the complexity entries of Table 16.3 is related to the last additive term in Equations 16.98 and 16.99, which originates from the solution of Equation 16.22 for the vectors $\tilde{\mathbf{h}}_{apt,K_0,p}[n]$, $p = 1, \ldots, P$ of CIR-related taps using the LU decomposition. Observe, for example, that for $K_0 = 64$ CIR-related taps, the contribution due to the matrix inversion is excessive.

16.4.5.2 Complexity of the Simplified LS-Assisted DDCE

For the sake of comparison, in this section let us consider the normalised computational complexity associated with the simplified DDCE approach of Section 16.4.4. Recall from Section 16.4.4 that the difference with respect to the original DDCE approach of Section 16.4.1.2 resides in performing the parallel cancellation of the interfering users' CIR-related tap contributions according to Equation 16.85 instead of carrying out matrix multiplication with the inverse of the correlation matrix $\mathbf{Q}[n]$ of Equation 16.20 as suggested by Equation 16.22. It can be readily shown that the normalised computational complexity of the PIC operations of

	$K_0 = 8/K = 64$	$K_0 = 8/K = 512$	$K_0 = 64/K = 512$
$C^{(\mathbb{C}*\mathbb{C})}_{\text{MU-CE,PIC-LS}}\big|_{\text{norm}}$	$8.50 + 3.0$	$12.06 + 0.38$	$12.06 + 24.0$
$C^{(\mathbb{C}+\mathbb{C})}_{\text{MU-CE,PIC-LS}}\big|_{\text{norm}}$	$14.25 + 3.0$	$21.38 + 0.38$	$21.38 + 24.0$

Table 16.4: Computational complexity of the simplified LS-assisted DDCE [93] in terms of the number of complex multiplications and additions normalised to the K number of subcarriers, L number of users and P number of receiver antennas. Here we have assumed that $L = P = 4$.

Equation 16.85 for $p = 1, \dots, P$ is given by:

$$C^{(\mathbb{C}*\mathbb{C})}_{\tilde{\mathbf{h}}_{apt,K_0},\text{PIC}[n]}\bigg|_{\text{norm}} = C^{(\mathbb{C}+\mathbb{C})}_{\tilde{\mathbf{h}}_{apt,K_0},\text{PIC}[n]}\bigg|_{\text{norm}} = \frac{1}{K}(L-1)K_0^2. \tag{16.100}$$

Hence, following the philosophy of Equation 16.97, the total normalised complexity becomes:

$$\begin{aligned} C^{()}_{\text{MU-CE,PIC-LS}}\big|_{\text{norm}} &= C^{()}_{\mathbf{Q}[n]}\big|_{\text{norm}} + C^{()}_{\mathbf{P}[n]}\big|_{\text{norm}} + \\ &+ C^{()}_{\tilde{\mathbf{h}}_{apt,K_0},\text{PIC}[n]}\big|_{\text{norm}} + C^{()}_{\tilde{\mathbf{h}}_{apt,K_0}[n]->\tilde{\mathbf{H}}_{apt}[n]}\big|_{\text{norm}}. \end{aligned} \tag{16.101}$$

With the aid of Equation 16.101 in the specific scenario of $L = P$ we arrive at the following expressions:

$$C^{(\mathbb{C}*\mathbb{C})}_{\text{MU-CE,PIC-LS}}\big|_{\text{norm}} = \frac{1}{4}\left(5 - \frac{1}{L}\right)\log_2 K + \frac{1}{2}\left(3 - \frac{1}{L}\right) + \frac{1}{K}(L-1)K_0^2 \tag{16.102}$$

$$C^{(\mathbb{C}+\mathbb{C})}_{\text{MU-CE,PIC-LS}}\big|_{\text{norm}} = \frac{1}{2}\left(5 - \frac{1}{L}\right)\log_2 K + \frac{1}{K}(L-1)K_0^2, \tag{16.103}$$

where again a normalisation to $KLP = KL^2$ has been performed. Note in Equations 16.102 and 16.103 that the complexity is a function of the second power of the number of significant CIR-related taps K_0. This has to be contrasted with the cubical dependency observed in Equations 16.98 and 16.99 in the context of the original approach, as discussed in Section 16.4.5.1. In Table 16.4 we have exemplified the complexity of the simplified LS-assisted DDCE, again, for the specific parameters of $L = P = 4$. We observe a significant complexity reduction compared to that of the original estimator, which was characterised for the same parameters in Table 16.3.

16.4.6   Conclusion on the Original and Simplified LS-Assisted DDCE

As an introduction to the topic of channel transfer factor estimation for multi-user OFDM systems in Section 16.4 we have portrayed Li's sub-space based estimation approach [91]. The corresponding equations, specifically that of the vector of optimum CIR-related tap estimates of Equation 16.22 were derived in Section 16.4.1 using a more compact matrix notation rather than the original notation of Li *et al.* [91]. The matrix notation had the advantage that the standard form of this solution, namely that of minimising the squared estimation error as formulated in Equation 16.17 became visible. We also highlighted in Equation 16.27 that a necessary condition for the identification of the L different users' CIR-related taps – where each user has a total of K_0 significant taps – is that $L \leq \frac{K}{K_0}$, which is an additional motivation for identifying alternative channel transfer function estimation approaches. In an effort to further characterise the estimation approach proposed by Li *et al.* [91], its associated channel transfer factor estimation MSE was derived in Section 16.4.2, in the context of both sample-spaced and non-sample-spaced CIRs. Our discussions commenced by deriving the channel transfer factor estimation errors' auto-correlation matrix in Section 16.4.2.1. It was evident that the estimation errors' correlation matrix of Equation 16.38 and hence also the estimation MSE are dependent on the transmitted subcarrier symbol sequences. Li demonstrated in [91] that the LS-DDCE's MSE is minimised, if the sub-matrices on the "side-diagonals" of the correlation matrix $\mathbf{Q}[n]$ of Equation 16.22 are zero-matrices and thus no interference occurs between the different users' CIR-related tap estimation processes. For this to be the case, the constraints proposed by Li in [93] have to be imposed on the subcarrier symbol sequences transmitted by the different antennas of the different users, which is only applicable during the transmission of the training OFDM symbols. However, in order to simplify our analysis, in Section 16.4.2 we have exploited the specific properties of the training sequences. Hence, Equations 16.70 and 16.84 derived in Sections 16.4.2.2 and 16.4.2.3 describing the LS-DDCE's MSE in the context of both sample-spaced- and non-sample-spaced CIRs, respectively, represent the lowest possible estimation MSE achieved in the presence of random subcarrier symbol sequences. Furthermore, in Section 16.4.3 the strategy of an estimation MSE enhancement by transversal linear prediction was alluded to, while in Section 16.4.4 a simplified approach to DDCE based on parallel interference cancellation in the CIR-related domain, as suggested by Li [93] was outlined, which will be the basis of our further detailed discussions of PIC-assisted DDCE in Section 16.5. In this context it was also argued that as a result of the linear signal processing operations applied prior to extracting the most significant CIR-related taps, the PIC can be conducted both in the CIR-related domain upon invoking the correlation matrix $\mathbf{Q}[n]$ as shown in Equation 16.85, or in the frequency domain by direct subtraction of the subcarriers' channel transfer factors after weighting them by the transmitted symbols, as will be demonstrated in Section 16.5. Our complexity analysis conducted in Section 16.4.5 revealed that in the context of the original DDCE approach of Section 16.4.1 the computational complexity imposed by solving Equation 16.22 for generating the different reception antennas' associated vectors of CIR-related tap estimates is potentially excessive due to its cubical dependence on the product of the number of users L and the number of significant CIR-related taps K_0. This is a further motivation for identifying alternative ways of performing the estimation of the different channels' transfer functions.

16.5 Frequency Domain Parallel Interference Cancellation Aided Decision-Directed Channel Estimation

In this section a PIC-assisted DDCE scheme will be introduced and characterised. Specifically, its analytical description is provided in Section 16.5.1, while in Section 16.5.2 its performance is assessed in terms of the achievable *a priori* channel estimation[10] MSE as well

---

[10]Unless otherwise stated, the channel transfer function estimator is simply referred to as the channel estimator.

as the system's BER in the context of both sample-spaced- and non-sample-spaced channel scenarios. The computational complexity of the PIC-assisted DDCE will be analysed in Section 16.5.3. Finally conclusions will be offered in Section 16.5.4. Let us commence our discussions in the next section by considering the recursive estimator's structure.

16.5.1 The Recursive Channel Estimator

The specific structure of Section 16.5.1 is as follows. Our portrayal of the frequency domain PIC-assisted DDCE commences in Section 16.5.1.1, where we provide expressions both for the *a posteriori* channel transfer factor estimates arrived at after the parallel interference cancellation as well as for the *a priori* channel transfer factor estimates upon taking into account the effects of the CIR-related tap prediction filter. The specific structure of the predictor arrangement is detailed in Section 16.5.1.2. Furthermore, in Section 16.5.1.3 we derived an expression for the average *a priori* channel estimation MSE, while in Section 16.5.1.4 an expression of the average *a posteriori* channel estimation MSE for the current OFDM symbol as a function of the corresponding estimation MSEs associated with the previous $N_{tap}^{[t]}$ number of OFDM symbols is shown. After an analysis of the estimator's stability conditions in Section 16.5.1.5, the expression derived for the *a priori* estimation MSE is then employed in Section 16.5.1.6 – under the assumption that the system is in its steady-state condition – to generate the different users' vectors of optimum predictor coefficients, again, as a function of the predictor coefficient-dependent *a priori* estimation MSEs. Since the recursive structure of the channel transfer function estimator does not allow for an algebraic solution to be generated for the desired predictor coefficients, an iterative approach is applied, which exploits the contractive properties of the system equations. This approach was proposed earlier by Rashid-Farrokhi *et al.* [485] in the context of simultaneously optimising the transmit power allocation and base station antenna array weights in wireless networks. Since normally the exact knowledge of the channel's statistics in the form of the spaced-time spaced-frequency correlation function is not available, in Section 16.5.1.7 we discuss potential strategies for providing estimates of the statistics required.

16.5.1.1 *A Priori* and *A Posteriori* Channel Estimates

Recall from Equation 16.1 that the complex output signal $x_p[n, k]$ of the p-th receiver antenna element in the k-th subcarrier of the n-th OFDM symbol is given by:

$$x_p[n, k] = \sum_{i=1}^{L} H_p^{(i)}[n, k] s^{(i)}[n, k] + n_p[n, k], \qquad (16.104)$$

where the different variables have been defined in Section 16.2. Upon invoking vector notation, Equation 16.104 can be rewritten as:

$$\mathbf{x}_p[n] = \sum_{i=1}^{L} \mathbf{S}^{(i)}[n] \mathbf{H}_p^{(i)}[n] + \mathbf{n}_p[n], \qquad (16.105)$$

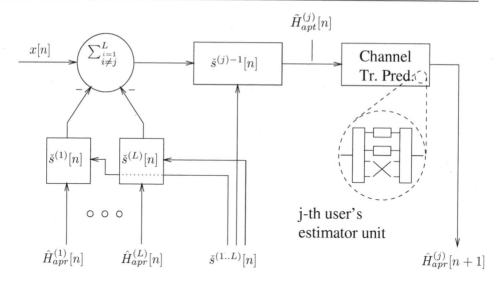

Figure 16.4: Illustration of the PIC-assisted channel transfer function estimation- or prediction block, associated with the j-th user and any of the P receiver antenna elements. The PIC process is described by Equation 16.106. The structure of the channel transfer function predictor follows that described in Section 15.2 for the single-user scenario.

where $\mathbf{x}_p[n] \in \mathbb{C}^{K \times 1}$, $\mathbf{H}_p^{(i)}[n] \in \mathbb{C}^{K \times 1}$ and $\mathbf{n}_p[n] \in \mathbb{C}^{K \times 1}$ are column vectors hosting the subcarrier-related variables $x_p[n,k]$, $H_p^{(i)}[n,k]$ and $n_p[n,k]$, respectively, and $\mathbf{S}^{(i)}[n] \in \mathbb{C}^{K \times K}$ is a diagonal matrix having elements given by $s^{(i)}[n,k]$, where $k = 0, \dots, K - 1$. An *a posteriori* (**apt**) estimate $\tilde{\mathbf{H}}_{apt}^{(j)}[n] \in \mathbb{C}^{K \times 1}$ of the vector $\mathbf{H}^{(j)}[n]$ of "true" channel transfer factors between the j-th user's single transmit antenna and the p-th receiver antenna can be obtained by subtracting all the $(L - 1)$ vectors of interfering users' estimated signal contributions from the vector $\mathbf{x}_p[n]$ of composite received signals of the L users, followed by normalisation with the j-th user's diagonal matrix of detected complex symbols $\check{\mathbf{S}}^{(j)}[n]$, yielding:

$$\tilde{\mathbf{H}}_{apt}^{(j)}[n] = \check{\mathbf{S}}^{(j)-1}[n] \left(\mathbf{x}[n] - \sum_{\substack{i=1 \\ i \neq j}}^{L} \check{\mathbf{S}}^{(i)}[n] \hat{\mathbf{H}}_{apr}^{(i)}[n] \right), \qquad (16.106)$$

where for simplicity's sake we have omitted the receiver antenna's index p. The PIC process based on Equation 16.106, has been further illustrated in Figure 16.4. In Equation 16.106, $\hat{\mathbf{H}}_{apr}^{(i)}[n] \in \mathbb{C}^{K \times 1}$ denotes the i-th user's vector of complex *a priori* (**apr**) channel transfer factor estimates predicted during the $(n - 1)$-th OFDM symbol period for the n-th OFDM symbol, as a function of the vectors of *a posteriori* channel transfer factor estimates $\tilde{\mathbf{H}}_{apt}^{(i)}[n -$

$\acute{n}]$ associated with the previous $N_{tap}^{[t]}$ number of OFDM symbols, which is formulated as:

$$\hat{\mathbf{H}}_{apr}^{(i)}[n] = f\left(\tilde{\mathbf{H}}_{apt}^{(i)}[n-1], \ldots, \tilde{\mathbf{H}}_{apt}^{(i)}[n - N_{tap}^{[t]}]\right). \tag{16.107}$$

We will further elaborate on the specific structure of the predictor in the next section.

16.5.1.2 *A Priori* Channel Prediction Filtering

The channel transfer function prediction along the time direction follows the philosophy of the 2D-MMSE channel transfer function estimation approach proposed by Li *et al.* [68], which in turn is based on the rank-reduction assisted 1D-MMSE channel estimation scheme proposed by Edfors and Sandell *et al.* [62,69]. These schemes were described in Section 15.2 in the context of single-user OFDM systems. Specifically recall Figure 15.3, which illustrated the process of 2D-MMSE based channel estimation- or prediction.

- **In a first step**, in order to obtain the i-th user's vector of *a priori* channel transfer factor estimates for the n-th OFDM symbol period during the $(n-1)$-th OFDM symbol period, which is denoted by $\hat{\mathbf{H}}_{apr}^{(i)}[n]$, the vector of *a posteriori* channel transfer factor estimates $\tilde{\mathbf{H}}_{apt}^{(i)}[n-1]$ is subjected to a unitary linear inverse transform $\tilde{\mathbf{U}}^{[f](i)H} \in \mathbb{C}^{K \times K}$, yielding the vector $\tilde{\mathbf{h}}_{apt}^{(i)}[n-1] \in \mathbb{C}^{K \times 1}$ of CIR-related *a posteriori* tap values:

$$\tilde{\mathbf{h}}_{apt}^{(i)}[n-1] = \tilde{\mathbf{U}}^{[f](i)H}\tilde{\mathbf{H}}_{apt}^{(i)}[n-1]. \tag{16.108}$$

From a statistical point of view the optimum unitary transform to be employed is the Karhunen-Loeve Transform (KLT) [69, 158] with respect to the Hermitian spaced-frequency correlation matrix of *a posteriori* channel transfer factor estimates, which is given by $\mathbf{R}_{apt}^{[f](i)} = E\{\tilde{\mathbf{H}}_{apt}^{(i)}\tilde{\mathbf{H}}_{apt}^{(i)H}\}$, when assuming the wide-sense stationarity of $\tilde{\mathbf{H}}_{apt}^{(i)}[n]$. The matrix $\mathbf{R}_{apt}^{[f](i)} \in \mathbb{C}^{K \times K}$ can be decomposed as $\mathbf{R}_{apt}^{[f](i)} = \mathbf{U}_{apt}^{[f](i)}\Lambda_{apt}^{[f](i)}\mathbf{U}_{apt}^{[f](i)H}$, where $\mathbf{U}_{apt}^{[f](i)} \in \mathbb{C}^{K \times K}$ is the unitary KLT matrix of *eigenvectors*, and $\Lambda_{apt}^{[f](i)} \in \mathbb{R}^{K \times K}$ exhibits the diagonal form of $\Lambda_{apt}^{[f](i)} = diag(\lambda_{apt,0}^{[f](i)}, \ldots, \lambda_{apt,K-1}^{[f](i)})$. The diagonal elements of $\Lambda_{apt}^{[f](i)}$ are referred to as the *eigenvalues* of $\mathbf{R}_{apt}^{[f](i)}$ [90]. Similarly, the desired channel's "true" spaced-frequency correlation matrix $\mathbf{R}^{[f](i)} = E\{\mathbf{H}^{[i]}\mathbf{H}^{[i]H}\}$ can be decomposed as $\mathbf{R}^{[f](i)} = \mathbf{U}^{[f](i)}\Lambda^{[f](i)}\mathbf{U}^{[f](i)H}$. At this stage we note that the error components contaminating the vector $\tilde{\mathbf{H}}_{apt}^{(i)}[n-1]$ estimating the vector $\mathbf{H}^{(i)}[n-1]$ of "true" channel transfer factors are uncorrelated due to the statistical independence of the AWGN and that of the modulated symbols transmitted in the different subcarriers. Hence both $\mathbf{R}_{apt}^{[f](i)}$ and $\mathbf{R}^{[f](i)}$ share the same eigenvectors [158], which implies that we have $\mathbf{U}_{apt}^{[f](i)} = \mathbf{U}^{[f](i)}$. In reality, however, the explicit knowledge of the channel's spaced-frequency correlation matrix $\mathbf{R}^{[f](i)}$ and that of its unitary KLT matrix $\mathbf{U}^{[f](i)}$ is typically unavailable. Instead, an estimate $\tilde{\mathbf{R}}^{[f](i)}$ and its associated unitary KLT matrix $\tilde{\mathbf{U}}^{[f](i)}$ have to be employed, which – in contrast to the optimum KLT matrix $\mathbf{U}^{[f](i)}$ – results in an imperfect decorrelation of the *a posteriori* channel transfer factor estimates.

- **In a second step** linear $N_{tap}^{[t]}$-tap filtering is performed in the time direction separately for those K_0 number of CIR-related components of $\tilde{\mathbf{h}}_{apt}^{(i)}$, for which the variance is significant. This is achieved by capitalising on the current vector $\tilde{\mathbf{h}}_{apt}^{(i)}[n-1]$ and the vectors $\tilde{\mathbf{h}}_{apt}^{(i)}[n-\acute{n}]$, $\acute{n} = 2, \ldots, N_{tap}^{[t]}$ of the previous $(N_{tap}^{[t]} - 1)$ number of OFDM symbols. As a result, in the case of estimation filtering [68] an improved estimate $\hat{\mathbf{h}}_{apt}^{(i)}[n-1]$ of $\mathbf{h}^{(i)}[n-1]$ is obtained, although this technique was not employed here. By contrast, in case of the prediction filtering employed here, an *a priori* estimate $\hat{\mathbf{h}}_{apr}^{(i)}[n] \in \mathbb{C}^{K \times 1}$ of $\mathbf{h}^{(i)}[n]$ is obtained. In mathematical terms this can be formulated as:

$$\hat{\mathbf{h}}_{apr}^{(i)}[n] = \mathbf{I}_{K_0}^{(i)} \sum_{\acute{n}=1}^{N_{tap}^{[t]}} \tilde{c}_{pre}^{(i)}[\acute{n} - 1]\tilde{\mathbf{h}}_{apt}^{(i)}[n - \acute{n}], \qquad (16.109)$$

where $\mathbf{I}_{K_0}^{(i)} \in \mathbb{C}^{K \times K}$ denotes a sparse unity matrix having unity entries only at those K_0 number of diagonal positions, for which the variance of the associated components of $\tilde{\mathbf{h}}_{apt}^{(i)}$ is significant. Furthermore, in Equation 16.109 the variable $\tilde{c}_{pre}^{(i)}[\acute{n} - 1] \in \mathbb{C}$ denotes the $(\acute{n} - 1)$-th CIR-related tap prediction filter coefficient. Note that for simplicity here we employ the same coefficient $\tilde{c}_{pre}^{(i)}[\acute{n} - 1]$ for filtering each of the different K_0 number of taps of the specific $\acute{n}$-th CIR-related vector $\tilde{\mathbf{h}}_{apt}^{(i)}[n - \acute{n}]$, which follows the concepts of robust channel estimation advocated by Li *et al.* [68]. These concepts were outlined in Section 15.2.3.1.

- **In a final step** the vector of CIR-related *a priori* tap estimates $\hat{\mathbf{h}}_{apr}^{(i)}[n]$ is transformed back to the OFDM frequency domain with the aid of the unitary KLT matrix $\tilde{\mathbf{U}}^{[f](i)}$, yielding the vector of *a priori* channel transfer factor estimates $\hat{\mathbf{H}}_{apr}^{(i)}[n]$ for the n-th OFDM symbol period:

$$\hat{\mathbf{H}}_{apr}^{(i)}[n] = \tilde{\mathbf{U}}^{[f](i)}\hat{\mathbf{h}}_{apr}^{(i)}[n]. \qquad (16.110)$$

This vector of *a priori* channel transfer factor estimates is in turn employed in the detection stage during the n-th OFDM symbol period. Upon substituting Equation 16.108 into Equation 16.109 and by substituting the result into Equation 16.110 we obtain the following relation between the vector of *a priori* channel transfer factor estimates derived for the n-th OFDM symbol and the vectors of *a posteriori* channel transfer factor estimates of the past $N_{tap}^{[t]}$ number of OFDM symbols:

$$\hat{\mathbf{H}}_{apr}^{(i)}[n] = \mathbf{T}_{K_0}^{(i)} \sum_{\acute{n}=1}^{N_{tap}^{[t]}} \tilde{c}_{pre}^{(i)}[\acute{n} - 1]\tilde{\mathbf{H}}_{apt}^{(i)}[n - \acute{n}], \qquad (16.111)$$

where $\mathbf{T}_{K_0,p}^{(i)} \in \mathbb{C}^{K \times K}$ is given by:

$$\mathbf{T}_{K_0}^{(i)} = \tilde{\mathbf{U}}^{[f](i)}\mathbf{I}_{K_0}^{(i)}\tilde{\mathbf{U}}^{[f](i)H}. \qquad (16.112)$$

After having described the process of generating the vectors of *a posteriori* and *a priori* channel transfer factor estimates in Sections 16.5.1.1 and 16.5.1.2, we will embark in Section 16.5.1.3 on an evaluation of the associated *a priori* estimation MSE.

16.5.1.3   *A Priori* Channel Estimation MSE

Let us commence our discussions in this section by developing an expression for the vector of *a priori* channel transfer factor estimation errors associated with the j-th user during the n-th OFDM symbol period as a function of the vectors of *a priori* channel transfer factor estimation errors of the $(L-1)$ remaining users during the $N_{tap}^{[t]}$ number of previous OFDM symbol periods. Assuming error-free symbol decisions we have $\check{\mathbf{S}}^{(j)}[n-\acute{n}] = \mathbf{S}^{(j)}[n-\acute{n}]$, $j = 1,\dots,L$, $\acute{n} = 1,\dots,N_{tap}^{[t]}$. Upon substituting Equation 16.105 into Equation 16.106 and then substituting the result into Equation 16.111, yields an expression for the vector of channel transfer factor estimation errors $\Delta\hat{\mathbf{H}}_{apr}^{(j)}[n] \in \mathbb{C}^{K\times 1}$ in the following form:

$$
\begin{aligned}
\Delta\hat{\mathbf{H}}_{apr}^{(j)}[n] \;=\; & -\mathbf{T}_{K_0}^{(j)} \sum_{\acute{n}=1}^{N_{tap}^{[t]}} \tilde{c}_{pre}^{(j)}[\acute{n}-1]\mathbf{S}^{(j)-1}[n-\acute{n}] \sum_{\substack{i=1\\i\neq j}}^{L} \mathbf{S}^{(i)}[n-\acute{n}]\Delta\hat{\mathbf{H}}_{apr}^{(i)}[n-\acute{n}] - \\[4pt]
& -\mathbf{T}_{K_0}^{(j)} \sum_{\acute{n}=1}^{N_{tap}^{[t]}} \tilde{c}_{pre}^{(j)}[\acute{n}-1]\mathbf{S}^{(j)-1}[n-\acute{n}]\mathbf{n}[n-\acute{n}] + \\[4pt]
& + \mathbf{H}^{(j)}[n] - \mathbf{T}_{K_0}^{(j)} \sum_{\acute{n}=1}^{N_{tap}^{[t]}} \tilde{c}_{pre}^{(j)}[\acute{n}-1]\mathbf{H}^{(j)}[n-\acute{n}],
\end{aligned}
\tag{16.113}
$$

where

$$
\Delta\hat{\mathbf{H}}_{apr}^{(j)}[n] = \mathbf{H}^{(j)}[n] - \hat{\mathbf{H}}_{apr}^{(j)}[n].
\tag{16.114}
$$

Please observe that for the sake of avoiding notational confusion the variable i of Equation 16.111 has been substituted by the variable j. The vector of *a priori* channel transfer factor estimation errors given by Equation 16.113 is constituted by three components. Specifically, the first term of Equation 16.113 is due to the effects of the *a priori* prediction errors of the $N_{tap}^{[t]}$ number of past OFDM symbols, the second term is attributed to the contaminating effect of the AWGN and the third term is due to the lack of "perfect predictability" of the channel transfer factors by the $(N_{tap}^{[t]}-1)$-order predictor. In other words, the last term is due to the channel transfer function's decorrelation with time.

The average variance of the j-th user's vector of *a priori* channel transfer factor estimation errors or in other words the average mean-square *a priori* estimation error can be expressed in mathematical terms as:

$$
\overline{\text{MSE}}_{apr}^{(j)}[n] = \frac{1}{K}\text{Trace}(\mathbf{R}_{\Delta\hat{\mathbf{H}}_{apr}^{(j)}}[n]),
\tag{16.115}
$$

where $\mathbf{R}_{\Delta\hat{\mathbf{H}}_{apr}^{(j)}}[n] \in \mathbb{C}^{K\times K}$ denotes the auto-correlation matrix of the vector $\Delta\hat{\mathbf{H}}_{apr}^{(j)}[n]$ of *a priori* channel transfer factor estimation errors. The computation of $\overline{\text{MSE}}_{apr}^{(j)}[n]$ of the j-th user's vector of *a priori* estimation errors associated with the n-th OFDM symbol period as given by Equation 16.113 will be carried out in two steps.

First, let us evaluate the auto-correlation matrix $\mathbf{R}_{\Delta\hat{\mathbf{H}}_{apr}^{(j)}}[n]$. This is achieved by substituting Equation 16.113 into:

$$\mathbf{R}_{\Delta\hat{\mathbf{H}}_{apr}^{(j)}}[n] \;=\; E\left\{\Delta\hat{\mathbf{H}}_{apr}^{(j)}[n]\Delta\hat{\mathbf{H}}_{apr}^{(j)H}[n]\right\} \tag{16.116}$$

$$=\; \frac{\alpha_j}{\sigma_j^2}\mathbf{T}_{K_0}^{(j)}\left(\sum_{\acute{n}=1}^{N_{tap}^{[t]}}|\tilde{c}_{pre}^{(j)}[\acute{n}-1]|^2\sum_{\substack{i=1\\i\neq j}}^{L}\sigma_i^2\mathbf{R}_{\Delta\hat{\mathbf{H}}_{apr}^{(i)}}[n-\acute{n}]|_{Diag}\right)\mathbf{T}_{K_0}^{(j)H} \;+$$

$$+\; \frac{\alpha_j}{\sigma_j^2}\sigma_n^2\sum_{\acute{n}=1}^{N_{tap}^{[t]}}|\tilde{c}_{pre}^{(j)}[\acute{n}-1]|^2\mathbf{T}_{K_0}^{(j)}\mathbf{T}_{K_0}^{(j)H} \;+\; \mathbf{R}_{\mathbf{H}_{dec}^{(j)}}, \tag{16.117}$$

where we introduced a new definition, namely that of the channel transfer function decorrelation-related matrix $\mathbf{R}_{\mathbf{H}_{dec}^{(j)}} \in \mathbb{C}^{K\times K}$, which is given by:

$$\mathbf{R}_{\mathbf{H}_{dec}^{(j)}} \;=\; \mathbf{R}^{[f](j)} - \mathbf{T}_{K_0}^j\mathbf{R}^{[f](j)H}\cdot(\tilde{\mathbf{c}}_{pre}^{(j)H}\mathbf{r}^{[t](j)}) - \mathbf{R}^{[f](j)}\mathbf{T}_{K_0}^{(j)H}\cdot(\tilde{\mathbf{c}}_{pre}^{(j)T}\mathbf{r}^{[t](j)*}) \;+$$

$$+\; \mathbf{T}_{K_0}^{(j)}\mathbf{R}^{[f](j)}\mathbf{T}_{K_0}^{(j)H}\cdot(\tilde{\mathbf{c}}_{pre}^{(j)H}\mathbf{R}^{[t](j)}\tilde{\mathbf{c}}_{pre}^{(j)}). \tag{16.118}$$

In the context of Equation 16.117 we have exploited that the three additive components of the vector $\Delta\hat{\mathbf{H}}_{apr}^{(j)}[n]$ of *a priori* channel transfer factor estimation errors in Equation 16.113 are uncorrelated. The uncorrelated nature of these three terms accrues from the statistical independence of the complex AWGN process and that of the complex valued process describing the channel transfer function's evolution versus frequency and time. We have also exploited that the complex symbols transmitted in different subcarriers of a specific user's signal during a specific OFDM symbol period, as well as the symbols transmitted by the same user in different OFDM symbol periods and the symbols transmitted by different users are statistically independent, which also implies that they are uncorrelated. Still considering Equation 16.117, the variable α_j denotes the so-called "modulation noise enhancement factor" [69, 477], defined as $\alpha_j = E\{|s^{(j)}[n,k]|^2\}E\{|1/s^{(j)}[n,k]|^2\}$. For M-ary Phase Shift Keying (MPSK) based modulation schemes, such as, for example, QPSK we have $\alpha = 1$, while for higher-order Quadrature Amplitude Modulation (QAM) schemes we have $\alpha > 1$ [69, 477]. Note that here we have implicitly assumed that the same modulation scheme is employed on different subcarriers of a specific user's transmitted signal. To elaborate further, the variables to be defined in Equation 16.118 are the spaced-time correlation function related auto-correlation vector $\mathbf{r}^{[t](j)} \in \mathbb{C}^{N_{tap}^{[t]}\times 1}$, of the channel transfer function, where the $\acute{n}$-th element is given by $\mathbf{r}^{[t](j)}|_{\acute{n}} = E\{H^{(j)*}[n,k]H^{(j)}[n-\acute{n},k]\}$, and the spaced-time correlation function related auto-correlation matrix $\mathbf{R}^{[t](j)} \in \mathbb{C}^{N_{tap}^{[t]}\times N_{tap}^{[t]}}$, of the channel transfer function, with the element $(\acute{n}_1,\acute{n}_2)$ given by $\mathbf{R}^{[t](j)}|_{\acute{n}_1,\acute{n}_2} = E\{H^{(j)}[n-\acute{n}_1,k]H^{(j)*}[n-\acute{n}_2,k]\}$. Furthermore, $\tilde{\mathbf{c}}_{pre}^{(j)} \in \mathbb{C}^{N_{tap}^{[t]}\times 1}$ is the vector of conjugate complex CIR-related tap prediction filter coefficients with its $\acute{n}$-th element given by $\tilde{\mathbf{c}}_{pre}^{(j)}|_{\acute{n}} = \tilde{c}_{pre}^{(j)*}[\acute{n}]$. The channel's spaced-frequency correlation matrix $\mathbf{R}^{[f](j)}$ was defined earlier in Section 16.5.1.2. Let us now return to our original objective, namely that of developing an expression for the average *a priori* channel transfer factor estimation MSE during the n-th OFDM symbol period.

Second, Equation 16.117 is invoked in conjunction with Equation 16.115 for obtaining an expression for the j-th user's average *a priori* channel transfer factor estimation MSE as a function of the remaining users' *a priori* estimation MSEs associated with the $N_{tap}^{[t]}$ number of previous OFDM symbol periods:

$$\overline{\mathrm{MSE}}_{apr}^{(j)}[n] = \frac{K_0}{K}\frac{\alpha_j}{\sigma_j^2}\sum_{\acute{n}=1}^{N_{tap}^{[t]}}|\tilde{c}_{pre}^{(j)}[\acute{n}-1]|^2\sum_{\substack{i=1\\i\neq j}}^{L}\sigma_i^2\overline{\mathrm{MSE}}_{apr}^{(i)}[n-\acute{n}]+$$

$$+\frac{K_0}{K}\frac{\alpha_j}{\sigma_j^2}\sigma_n^2\tilde{\mathbf{c}}_{pre}^{(j)H}\tilde{\mathbf{c}}_{pre}^{(j)}+\overline{\mathrm{MSE}}_{dec}^{(j)}, \tag{16.119}$$

where we have:

$$\overline{\mathrm{MSE}}_{dec}^{(j)} = \frac{1}{K}\mathrm{Trace}(\mathbf{R}_{\mathbf{H}_{dec}^{(j)}}) \tag{16.120}$$

$$= \sigma_H^2 - \frac{1}{K}\mathrm{Trace}(\Upsilon^{[f](j)}\mathbf{I}_{K_0}^{(j)})\cdot$$

$$\cdot\left(\tilde{\mathbf{c}}_{pre}^{(j)H}\mathbf{r}^{[t](j)}+\tilde{\mathbf{c}}_{pre}^{(j)T}\mathbf{r}^{[t](j)*}-\tilde{\mathbf{c}}_{pre}^{(j)H}\mathbf{R}^{[t](j)}\tilde{\mathbf{c}}_{pre}^{(j)}\right). \tag{16.121}$$

In the context of deriving Equation 16.119 we have capitalised on the relations $\mathrm{Trace}(\mathbf{A} + \mathbf{B}) = \mathrm{Trace}(\mathbf{A}) + \mathrm{Trace}(\mathbf{B})$, as well as on $\mathrm{Trace}(\mathbf{U}\mathbf{A}\mathbf{U}^H) = \mathrm{Trace}(\mathbf{A})$, which are valid for a unitary matrix $\mathbf{U}$ [158, 478]. Furthermore, in the context of deriving the first additive term in Equation 16.119 we exploited that $\frac{1}{K}\mathrm{Trace}(\mathbf{T}_{K_0}^{(j)}\mathbf{R}_{\Delta\hat{\mathbf{H}}_{apr}^{(i)}}[n-\acute{n}]|_{Diag}\mathbf{T}_{K_0}^{(j)H}) = \frac{K_0}{K}\overline{\mathrm{MSE}}_{apr}^{(i)}[n-\acute{n}]$, which is only valid for a unitary transform matrix $\tilde{\mathbf{U}}^{[f](j)}$ having elements of unity magnitude. This is the case for example, when employing the DFT matrix $\mathbf{W}$ as the unitary transform matrix. The second additive term in Equation 16.119 is based on exploiting the relationship of $\frac{1}{K}\mathrm{Trace}(\mathbf{T}_{K_0}^{(j)}\mathbf{T}_{K_0}^{(j)H}) = \frac{K_0}{K}$. We also note in this context that $\mathbf{T}_{K_0}^{(j)H} = \mathbf{T}_{K_0}^{(j)}$ and that $\mathbf{T}_{K_0}^{(j)}\mathbf{T}_{K_0}^{(j)H} = \mathbf{T}_{K_0}^{(j)}$.

Furthermore, in Equation 16.121, the matrix $\Upsilon^{[f](j)}\in\mathbb{C}^{K\times K}$ denotes the decomposition of the j-th user's channel's spaced-frequency correlation matrix $\mathbf{R}^{[f](j)}$ with respect to the unitary transform matrix $\tilde{\mathbf{U}}^{[f](j)}$, which is expressed as $\Upsilon^{[f](j)} = \tilde{\mathbf{U}}^{[f](j)H}\mathbf{R}^{[f](j)}\tilde{\mathbf{U}}^{[f](j)}$. Note that in contrast to $\Lambda^{[f](j)}$ associated with the decomposition of $\mathbf{R}^{[f](j)}$ with respect to $\mathbf{U}^{[f](j)}$, the matrix $\Upsilon^{[f](j)}$ is not necessarily of diagonal shape constrained to having real-valued elements only.

16.5.1.4 *A Posteriori* Channel Estimation MSE

Following the philosophy of Section 16.5.1.3 related to our derivation of an expression describing the j-th user's average *a priori* channel estimation MSE as a function of the remaining $(L-1)$ users' *a priori* channel estimation MSEs associated with the previous $N_{tap}^{[t]}$ number of OFDM symbol periods, in this section a similar expression is derived for the average *a posteriori* channel transfer factor estimation MSE. This is achieved in a first step upon substituting Equations 16.105 and 16.111 into Equation 16.106. Similar to the definition of the vector of *a priori* estimation errors in Equation 16.114, the j-th user's vector of *a posteriori* estimation errors $\Delta\tilde{\mathbf{H}}_{apt}^{(j)}[n] \in \mathbb{C}^{K \times 1}$ can be defined as:

$$\Delta\tilde{\mathbf{H}}_{apt}^{(j)}[n] = \mathbf{H}^{(j)}[n] - \tilde{\mathbf{H}}_{apt}^{(j)}[n]. \tag{16.122}$$

In accordance with the definition of the average *a priori* channel transfer factor estimation MSE in Equation 16.115, we can also define the average *a posteriori* estimation MSE as:

$$\overline{\mathrm{MSE}}_{apt}^{(j)}[n] = \frac{1}{K}\mathrm{Trace}(\mathbf{R}_{\Delta\tilde{\mathbf{H}}_{apt}^{(j)}}[n]), \tag{16.123}$$

where $\mathbf{R}_{\Delta\tilde{\mathbf{H}}_{apt}^{(j)}}[n] \in \mathbb{C}^{K \times K}$ denotes the auto-correlation matrix of the vector $\Delta\tilde{\mathbf{H}}_{apt}^{(j)}[n]$ of *a posteriori* estimation errors. Our further mathematical manipulations, which are not detailed here for reasons of space yield the following expression for the j-th user's average *a posteriori* estimation MSE during the n-th OFDM symbol period:

$$\overline{\mathrm{MSE}}_{apt}^{(j)}[n] = \frac{K_0}{K}\frac{\alpha_j}{\sigma_j^2}\sum_{\substack{i=1 \\ i \neq j}}^{L}\sigma_i^2\sum_{\acute{n}=1}^{N_{tap}^{[t]}}|\tilde{c}_{pre}^{(i)}[\acute{n}-1]|^2\overline{\mathrm{MSE}}_{apt}^{(i)}[n-\acute{n}] +$$

$$+ \frac{\alpha_j}{\sigma_j^2}\sigma_n^2 + \frac{\alpha_j}{\sigma_j^2}\sum_{\substack{i=1 \\ i \neq j}}^{L}\sigma_i^2\overline{\mathrm{MSE}}_{dec}^{(i)}. \tag{16.124}$$

Finally, the channel decorrelation-related MSE, namely $\overline{\mathrm{MSE}}_{dec}^{(i)}$ of Equation 16.124 is given by Equation 16.121, which is identical for the *a priori* and *a posteriori* estimates.

16.5.1.5 Stability Analysis of the Recursive Channel Estimator

In the steady-state condition we can assume that the specific user's *a priori* and *a posteriori* estimation MSEs are identical for different OFDM symbols, which are expressed as:

$$\overline{\mathrm{MSE}}_{ap(t/r)}^{(i)} = \overline{\mathrm{MSE}}_{ap(t/r)}^{(i)}[n-\acute{n}], \tag{16.125}$$

where $i = 1, \ldots, L$ and $\acute{n} = 0, \ldots, N_{tap}^{[t]}$. Hence, Equation 16.119 simplifies to:

$$\overline{\mathrm{MSE}}_{apr}^{(j)} = \frac{K_0}{K}\frac{\alpha_j}{\sigma_j^2}\tilde{\mathbf{c}}_{pre}^{(j)H}\tilde{\mathbf{c}}_{pre}^{(j)}\sum_{\substack{i=1 \\ i \neq j}}^{L}\sigma_i^2\overline{\mathrm{MSE}}_{apr}^{(i)} +$$

$$+ \frac{K_0}{K}\frac{\alpha_j}{\sigma_j^2}\tilde{\mathbf{c}}_{pre}^{(j)H}\tilde{\mathbf{c}}_{pre}^{(j)}\sigma_n^2 + \overline{\mathrm{MSE}}_{dec}^{(j)}. \tag{16.126}$$

Note that Equation 16.126 can be viewed as a system of equations for different values of $j = 1, \ldots, L$, namely for the different users. It can be shown that Equation 16.126 can be represented in a compact vectorial notation as:

$$\mathbf{MSE}_{apr} = \mathbf{C}_{pre} \cdot \mathbf{P}_s^{-1} \cdot \mathbf{F} \cdot \mathbf{P}_s \cdot \mathbf{MSE}_{apr} + \mathbf{C}_{pre} \cdot \mathbf{P}_s^{-1} \cdot \mathbf{p}_n + \mathbf{MSE}_{dec}, \qquad (16.127)$$

where $\mathbf{MSE}_{apr} \in \mathbb{R}^{L \times 1}$ hosts the different users' *a priori* estimation MSEs denoted by $\overline{\mathrm{MSE}}_{apr}^{(j)}$, $j = 1, \ldots, L$, and the diagonal matrix $\mathbf{C}_{pre} \in \mathbb{R}^{L \times L}$ hosts the different users' CIR-related tap prediction coefficient related terms of $\frac{K_0}{K} \alpha_j \tilde{\mathbf{c}}_{pre}^{(j)H} \tilde{\mathbf{c}}_{pre}^{(j)}$, $j = 1, \ldots, L$. A characteristic component is the feedback matrix $\mathbf{F} \in \mathbb{R}^{L \times L}$, which exhibits the following structure:

$$\mathbf{F} = \begin{pmatrix} 0 & 1 & \cdots & 1 \\ 1 & \ddots & \ddots & \vdots \\ \vdots & \ddots & \ddots & 1 \\ 1 & \cdots & 1 & 0 \end{pmatrix}, \qquad (16.128)$$

where the elements on the side diagonals are of unit value except for the main diagonal, whose elements are zero. The relation to the PIC process is that for the estimation of the j-th user's channel transfer function, the co-channel interference imposed by the $L - 1$ remaining users has to be removed. Note in this context that the j-th row of matrix $\mathbf{F}$ is associated with the estimation process of the j-th user's channel. Furthermore, the diagonal matrix $\mathbf{P}_s \in \mathbb{R}^{L \times L}$ hosts the different users' signal variances σ_i^2, $i = 1, \ldots, L$, while the vector $\mathbf{p}_n \in \mathbb{R}^{L \times 1}$ exhibits identical elements equal to the AWGN noise variance σ_n^2. Finally, the matrix $\mathbf{MSE}_{dec} \in \mathbb{R}^{L \times 1}$ hosts the different users' residual channel decorrelation-related MSEs values, given by $\frac{1}{K} \mathrm{Trace}(\mathbf{R}_{\mathbf{H}_{dec}^{(j)}})$, $j = 1, \ldots, L$, which are also a function of the individual users' CIR-related tap predictor coefficients, as evidenced by Equation 16.121. In order to proceed further, Equation 16.127 can be solved for the vector of *a priori* estimation MSEs, conditioned on the knowledge of the vectors $\tilde{\mathbf{c}}_{pre}^{(j)}$, $j = 1, \ldots, L$ of predictor coefficients, yielding:

$$\mathbf{MSE}_{apr} = \left(\mathbf{I} - \mathbf{C}_{pre} \cdot \mathbf{P}_s^{-1} \cdot \mathbf{F} \cdot \mathbf{P}_s \right)^{-1} \cdot \left(\mathbf{C}_{pre} \cdot \mathbf{P}_s^{-1} \cdot \mathbf{p}_n + \mathbf{MSE}_{dec} \right). \qquad (16.129)$$

By definition, the elements of $\mathbf{MSE}_{apr}$ or, equivalently, the different users' *a priori* estimation MSEs must have a finite positive value. This is coupled to the following conditions:

1) existence of $(\mathbf{I} - \acute{\mathbf{F}})^{-1}$, where $\acute{\mathbf{F}} = \mathbf{C}_{pre} \cdot \mathbf{P}_s^{-1} \cdot \mathbf{F} \cdot \mathbf{P}_s$

2) all elements of $(\mathbf{I} - \acute{\mathbf{F}})^{-1}$ must be positive.

It can be demonstrated that these two conditions are fulfilled, if the spectral radius $\rho(\acute{\mathbf{F}})$[11] of the matrix $\acute{\mathbf{F}} = \mathbf{C}_{pre} \cdot \mathbf{P}_s^{-1} \cdot \mathbf{F} \cdot \mathbf{P}_s$ is less than unity [486]. An *upper-bound* estimate of the spectral radius is given by the largest Euclidean distance measured from the origin in $\mathbb{C}$, exhibited by a point contained in the union $G(\acute{\mathbf{F}})$ of Gershgorin disks[12] of $\acute{\mathbf{F}}$. Hence,

---

[11]Recall that the spectral radius of a matrix is the smallest radius of a circle centred around the origin of $\mathbb{C}$ that contains all the matrix's eigenvalues [90].

[12]With the aid of the Gershgorin circle theorem [90] explicit bounds can be placed on the regions in $\mathbb{C}$, which host the eigenvalues of a matrix $\mathbf{A} \in \mathbb{C}^{m \times m}$. The i-th Gershgorin disk is defined as: $R_i(\mathbf{A}) = \{x \in \mathbb{C} :$

provided that we have:

$$\max_{j=1...L} \left(\frac{K_0}{K} \alpha_j \tilde{\mathbf{c}}_{pre}^{H(j)} \tilde{\mathbf{c}}_{pre}^{(j)} \sum_{\substack{i=1 \\ i \neq j}}^{L} \frac{\sigma_i^2}{\sigma_j^2} \right) < 1, \tag{16.130}$$

it can be shown that $\acute{\mathbf{F}}$ is invertible. By contrast, if this condition is not fulfilled, no immediate conclusion can be drawn with respect to the invertibility of $\acute{\mathbf{F}}$.

A further criterion for the existence of the matrix inverse $(\mathbf{I} - \acute{\mathbf{F}})^{-1}$ is coupled to the condition that the determinant of $(\mathbf{I} - \acute{\mathbf{F}})$ is non-zero, namely that we have $\det(\mathbf{I} - \acute{\mathbf{F}}) \neq 0$. Furthermore, it can be shown that for all elements of this specific matrix inverse to be positive as stipulated in (2), we have to satisfy the condition of $\det(\mathbf{I} - \acute{\mathbf{F}}) > 0$. It can be shown that $\det(\mathbf{I} - \acute{\mathbf{F}}) = \det(\mathbf{I} - \mathbf{C}_{pre} \cdot \mathbf{F})$, which implies that the channel estimator's stability is only a function of the estimator coefficients to be determined. Even if the channel conditions are subjected to variations, the estimator remains stable for a "stable" set of coefficients – provided that correct symbol decisions are performed.

16.5.1.6   Iterative Calculation of the CIR-Related Tap Predictor Coefficients

Upon invoking Equation 16.126 the j-th user's vector of CIR-related tap predictor coefficients $\tilde{\mathbf{c}}_{pre}^{(j)}$ can be evaluated conditional on the remaining $(L - 1)$ number of users' *a priori* estimation MSEs, namely on $\overline{\mathrm{MSE}}_{apr}^{(i)}$, $i = 1, \ldots, L$, $i \neq j$, which ensues by calculating the gradient of $\overline{\mathrm{MSE}}_{apr}^{(j)}$ with respect to the j-th user's coefficients, yielding:

$$\nabla^{(j)} \overline{\mathrm{MSE}}_{apr}^{(j)} = \frac{K_0}{K} \frac{\alpha_j}{\sigma_j^2} \tilde{\mathbf{c}}_{pre}^{(j)} \left(\sum_{\substack{i=1 \\ i \neq j}}^{L} \sigma_i^2 \overline{\mathrm{MSE}}_{apr}^{(i)} + \sigma_n^2 \right) +$$
$$- \frac{1}{K} \mathrm{Trace}(\Upsilon^{[f](j)} \mathbf{I}_{K_0}^{(j)}) \cdot (\mathbf{r}^{[t](j)} - \mathbf{R}^{[t](j)} \tilde{\mathbf{c}}_{pre}^{(j)}), \tag{16.131}$$

where $\mathbf{R}^{[t](j)}$ and $\mathbf{r}^{[t](j)}$ were defined in the context of Equation 16.118. The gradient vector with respect to the j-th user's coefficients is defined here as $\nabla^{(j)} = \frac{\partial}{\partial \tilde{c}_{pre}^{(j)*}}$, with individual components given by the Wirtinger calculus $\frac{\partial}{\partial \tilde{c}_{pre}^{(j)*}} = \frac{1}{2} \left(\frac{\partial}{\partial \tilde{c}_r^{(j)}} + j \frac{\partial}{\partial \tilde{c}_i^{(j)}} \right)$ [487], where $\tilde{c}_r^{(j)}$ and $\tilde{c}_i^{(j)}$ are the real and imaginary parts of $\tilde{c}_{pre}^{(j)}$. In the context of Equation 16.131 we have exploited that $\nabla^{(j)} \tilde{\mathbf{c}}_{pre}^{(j)H} = \mathbf{I}$, as well as that $\nabla^{(j)} \tilde{\mathbf{c}}_{pre}^{(j)T} = \mathbf{0}$ and $\nabla^{(j)} (\tilde{\mathbf{c}}_{pre}^{(j)H} \tilde{\mathbf{c}}_{pre}^{(j)}) = \tilde{\mathbf{c}}_{pre}^{(j)}$.

In the optimum point of operation we have $\nabla^{(j)} \overline{\mathrm{MSE}}_{apr}^{(j)} = \mathbf{0}$ and hence Equation 16.131 can be solved for the j-th user's vector of predictor coefficients, resulting in the Wiener filter-

---

$|x - a_{ii}| \leq \sum_{\substack{j=1 \\ j \neq i}}^{m} |a_{ij}|\}$, where a_{ij} is the element of the matrix $\mathbf{A}$ associated with its i-th row and j-th column.
The eigenvalues of the matrix $\mathbf{A}$ reside within the union of Gershgorin disks of $\mathbf{A}$, which is formulated in a compact form as $\lambda(\mathbf{A}) \subset \bigcup_{i=1}^{m} R_i(\mathbf{A}) = G(\mathbf{A})$ [90].

related solution of:

$$\tilde{\mathbf{c}}_{pre}^{(j)}|_{opt} = \left[\mathbf{R}^{[t](j)} + \frac{K_0}{\text{Trace}(\Upsilon^{[f](j)}\mathbf{I}_{K_0}^{(j)})}\frac{\alpha_j}{\sigma_j^2}\left(\sum_{\substack{i=1\\i\neq j}}^{L}\sigma_i^2\overline{\text{MSE}}_{apr}^{(i)} + \sigma_n^2\right)\mathbf{I}\right]^{-1} \cdot \mathbf{r}^{[t](j)}.$$

(16.132)

Based on Equations 16.126 and 16.132 a fixed-point iteration algorithm [90] can be devised to obtain the different users' vectors of predictor coefficients under the constraint of minimising the sum of the different users' *a priori* estimation MSEs. This approach was proposed earlier by Rashid-Farrokhi *et al.* [485] in the context of simultaneously optimising both the transmit power allocation and the base station antenna array weights in wireless networks, leading to formulae similar to Equations 16.126 and 16.132. In our forthcoming discourse we will briefly present the steps of the algorithm with respect to our specific optimisation problem, but for a formal proof of the algorithm's convergence and that of the uniqueness of the solution, we refer to [485]. Note that in the context of our description of the algorithm, the iteration index – and not the OFDM symbol index – is given in the square brackets.

1) Initialise the different users' *a priori* estimation MSEs, for example by setting $\overline{\text{MSE}}_{apr}^{(j)}[0] = 0$ for $j = 1, \ldots, L$.

2) *For the n-th iteration*: Conditioned on the *a priori* estimation MSE values obtained during the $(n-1)$-th iteration, namely on $\overline{\text{MSE}}_{apr}^{(j)}[n-1]$, $j = 1, \ldots, L$ calculate the different users' vectors of optimum predictor coefficients for the n-th iteration, namely $\tilde{\mathbf{c}}_{pre}^{(j)}[n]|_{opt}$, $j = 1, \ldots, L$, with the aid of Equation 16.132.

3) Conditioned on the n-th iteration's predictor coefficient vectors $\tilde{\mathbf{c}}_{pre}^{(j)}[n]|_{opt}$, $j = 1, \ldots, L$ obtained in step (2) and also conditioned on the $(n-1)$-th iteration's *a priori* estimation MSE values, namely on $\overline{\text{MSE}}_{apr}^{(j)}[n-1]$, $j = 1, \ldots, L$ calculate the n-th iteration's *a priori* estimation MSE values of $\overline{\text{MSE}}_{apr}^{(j)}[n]$, $j = 1, \ldots, L$ with the aid of Equation 16.126.

4) Start a new iteration by returning to step (2).

Note that instead of invoking Equation 16.126 separately for each user, the different users' *a priori* estimation MSEs can also be calculated in parallel with the aid of Equation 16.129, as a result of which an even faster convergence is achieved. The price to be paid is a higher computational complexity, since an explicit matrix inversion is required in Equation 16.129.

16.5.1.6.1 Simplified Approach for Identical User Statistics
A simplification of Equations 16.126 and 16.132, which we will sometimes invoke during our performance assessment in Section 16.5.2 is achieved by imposing a number of assumptions. Specifically, we will assume perfect power control, implying that $\sigma_s^2 = \sigma_1^2 = \ldots = \sigma_L^2$ and that the same modulation mode is employed by all the L users, yielding $\alpha = \alpha_1 = \ldots = \alpha_L$. Additionally, identical spaced-time correlation functions and hence identical auto-correlation matrices $\mathbf{R}^{[t]} = \mathbf{R}^{[t](1)} = \ldots = \mathbf{R}^{[t](L)}$ and auto-correlation vectors $\mathbf{r}^{[t]} = \mathbf{r}^{[t](1)} = \ldots = \mathbf{r}^{[t](L)}$

are associated with the different users. As a result of these assumptions the same *a priori* estimation MSE, namely $\overline{\text{MSE}}_{apr} = \overline{\text{MSE}}_{apr}^{(1)} = \ldots = \overline{\text{MSE}}_{apr}^{(L)}$ and the same CIR-related tap predictor coefficient vector $\tilde{\mathbf{c}}_{pre} = \tilde{\mathbf{c}}_{pre}^{(1)} = \ldots = \tilde{\mathbf{c}}_{pre}^{(L)}$ are associated with the different users. Hence, Equation 16.126 can be directly solved for $\overline{\text{MSE}}_{apr}$ conditioned on a specific vector $\tilde{\mathbf{c}}_{pre}$ of predictor coefficients, yielding:

$$\overline{\text{MSE}}_{apr}\Big|^{\text{SIMPLE}} = \frac{\chi \frac{\sigma_n^2}{\sigma_s^2} \tilde{\mathbf{c}}_{pre}^H \tilde{\mathbf{c}}_{pre} + \frac{1}{K}\text{Trace}(\mathbf{R}_{\mathbf{H}_{dec}})}{1 - \chi(L-1)\tilde{\mathbf{c}}_{pre}^H \tilde{\mathbf{c}}_{pre}}, \tag{16.133}$$

where $\chi = \frac{K_0}{K}\alpha$. The denominator of the fraction on the right-hand side of Equation 16.133 suggests a particularly simple form of the steady-state stability condition, namely that of:

$$\chi(L-1)\tilde{\mathbf{c}}_{pre}^H \tilde{\mathbf{c}}_{pre} < 1, \tag{16.134}$$

which follows from Equation 16.130 upon invoking the above assumptions. Furthermore, Equation 16.132, which delivers the vector of optimum predictor coefficients as a function of the *a priori* estimation MSE simplifies to:

$$\tilde{\mathbf{c}}_{pre}\Big|_{opt}^{\text{SIMPLE}} = \left[\mathbf{R}^{[t]} + \frac{K_0}{\text{Trace}(\Upsilon^{[f]}\mathbf{I}_{K_0})}\alpha\left((L-1)\overline{\text{MSE}}_{apr}\Big|^{\text{SIMPLE}} + \frac{\sigma_n^2}{\sigma_s^2}\right)\mathbf{I}\right]^{-1} \cdot \mathbf{r}^{[t]}. \tag{16.135}$$

Note that upon removing the $(L-1)$ number of contributions in Equations 16.133 and 16.135, which are related to the PIC process, we obtain the expressions for the estimation MSE and the vector of coefficients associated with a transversal predictor, which can be expressed as:

$$\tilde{\mathbf{c}}_{pre,\text{FIR}}\Big|_{opt}^{\text{SIMPLE}} = \left[\mathbf{R}^{[t]} + \frac{K_0}{\text{Trace}(\Upsilon^{[f]}\mathbf{I}_{K_0})}\alpha\frac{\sigma_n^2}{\sigma_s^2}\mathbf{I}\right]^{-1} \cdot \mathbf{r}^{[t]}. \tag{16.136}$$

By stipulating a sample-spaced CIR hosting K_0 number of non-zero taps and upon appropriately designing the matrix $\mathbf{I}_{K_0}$, namely by assigning a numerical value of unity to those diagonal elements, which are related to the different CIR taps' sample-spaced delays, we obtain $\text{Trace}(\Upsilon^{[f]}\mathbf{I}_{K_0}) = K\sigma_H^2$. This further simplifies Equation 16.136.

In the next section a closed-form solution is presented for the optimum "predictor" coefficient in the context of one-tap *a priori* channel estimation.

16.5.1.6.2 Closed Form Solution for Identical User Statistics and One-Tap CIR-Related Tap Prediction Filtering

For the case of simple zero-order CIR-related tap prediction a closed form solution can be derived from Equations 16.136 and 16.133 for the optimum predictor coefficient in the context of a sample-spaced CIR, which is given by:

$$\tilde{c}_{pre}[0]\Big|_{opt}^{\text{SIMPLE}} = -\sqrt{\nu^2 - \frac{1}{\chi \cdot (L-1)}} + \nu, \tag{16.137}$$

where

$$\nu = \frac{\chi \cdot \frac{\sigma_n^2}{\sigma_s^2} + \chi \cdot (L-1) + 1}{2 \cdot \chi \cdot (L-1) \cdot r_H^{[t]}[1]}.\qquad(16.138)$$

By contrast, the optimum predictor coefficient for the case of a transversal one-tap predictor is given by:

$$\tilde{c}_{pre,\text{FIR}}[0]\Big|_{opt}^{\text{SIMPLE}} = \frac{r_H^{[t]}[1]}{1 + \chi \frac{\sigma_n^2}{\sigma_s^2}},\qquad(16.139)$$

which directly follows from Equation 16.136. In Equations 16.138 and 16.139, $r_H^{[t]}[1]$ denotes the channel transfer factor correlation coefficient for a time-lag of one OFDM symbol. Physically this equation simply states that if the channel transfer function varies slowly, $r_H^{[t]}[1] \approx 1$ and hence the predictor coefficient has a high value. By contrast, if the channel correlation is low, the predictor coefficient has to be low.

In the next section we will address the problem of a potential lack of knowledge about the channel's exact statistics namely that of the spaced-time spaced-frequency correlation function.

16.5.1.7 Channel Statistics

As it was observed in Equations 16.126 and 16.132, a prerequisite for determining the different users' vectors of optimum CIR-related tap predictor coefficients is the knowledge of the users' spaced-time channel transfer factor correlation functions $r_H^{[t](j)}[\Delta n]$, $j = 1, \ldots, L$, defined by:

$$r_H^{[t](j)}[\Delta n] = E\{H^{(j)}[n,k] \cdot H^{(j)*}[n - \Delta n, k]\}.\qquad(16.140)$$

These are required for evaluating the auto-correlation matrices $\mathbf{R}^{[t](j)}$ and cross-correlation vectors $\mathbf{r}^{[t](j)}$ for $j = 1, \ldots, L$. Assuming Jakes' fading model [474] for example, the channel correlation along the time direction is given by [68]:

$$r_{H,J}^{[t](j)}[\Delta n] = J_0(\Delta n \cdot \omega_D^{(j)}),\qquad(16.141)$$

$$\approx 1 - \frac{1}{4}(\Delta n \cdot \omega_D^{(j)})^2, \quad \Delta n \cdot \omega_D^{(j)} \ll 1,\qquad(16.142)$$

where $J_0()$ denotes the zero-order Bessel function of the first kind and $\omega_D^{(j)} = 2\pi T_f f_D^{(j)}$, and T_f being the OFDM symbol duration including the guard period time, while $f_D^{(j)}$ denotes the channel's Doppler frequency. Since usually the exact Doppler frequency $f_D^{(j)}$ is not known, it was demonstrated in [68] in the context of a transversal-type estimator, that the MSE performance degradation incurred due to a mismatch of the channel statistics is only marginal, if a uniform, ideally support-limited Doppler power spectrum associated with $\tilde{f}_D^{(j)} \geq f_D^{(j)}$ is assumed for the calculation of the correlation coefficients of Equation 16.140. The associated spaced-time correlation function is given as the inverse Fourier Transform (FT) of the uniform Doppler power spectrum, which leads to:

$$\tilde{r}_{H,\text{unif}}^{[t](j)}[\Delta n] = \frac{sin(\Delta n \cdot \tilde{\omega}_D^{(j)})}{\Delta n \cdot \tilde{\omega}_D^{(j)}}. \tag{16.143}$$

Furthermore, the calculation of the vectors of CIR-related tap predictor coefficients according to Equation 16.132 also requires the evaluation of the expression $\text{Trace}(\Upsilon^{[f](j)}\mathbf{I}_{K_0}^{(j)})$. More explicitly we recall from Section 16.5.1.2 that $\Upsilon^{[f](j)}$ is the decomposition of the j-th user's channel's spaced-frequency correlation matrix $\mathbf{R}^{[f](j)}$ with respect to the unitary transform matrix $\tilde{\mathbf{U}}^{[f](j)}$, which is formulated as $\Upsilon^{[f](j)} = \tilde{\mathbf{U}}^{[f](j)H}\mathbf{R}^{[f](j)}\tilde{\mathbf{U}}^{[f](j)}$, and $\mathbf{I}_{K_0}^{(j)}$ is a sparse identity matrix having unity entries only at those K_0 number of positions, which are associated with a significant value of $\Upsilon^{[f](j)}$. Hence, we note that the evaluation of $\text{Trace}(\Upsilon^{[f](j)}\mathbf{I}_{K_0}^{(j)})$ requires the knowledge of $\mathbf{R}^{[f](j)}$, which is not directly available in practice. Below we have listed several potential approaches which can be pursued for addressing this problem:

1) Periodically estimate the diagonal elements of $\Upsilon^{[f](j)}$ upon evaluating the variance of the components associated with the vector $\tilde{\mathbf{h}}_{apt}^{(j)}[n]$, as it was formulated in Equation 16.108. For the l-th component the variance can be approximated by averaging the component's squared magnitude over the past N_{apt} number of OFDM symbols' contributions, which is expressed as:

$$\hat{\sigma}_{\tilde{h}_{apt}^{(j)}[n,l]}^2 = \frac{1}{N_{apt}} \sum_{\acute{n}=0}^{N_{apt}-1} |\tilde{h}_{apt}^{(j)}[n-\acute{n},l]|^2. \tag{16.144}$$

In a second step the sum of those K_0 number of CIR-related tap estimates is calculated, which exhibit the highest variance. This value constitutes the desired estimate of $\text{Trace}(\Upsilon^{[f](j)}\mathbf{I}_{K_0}^{(j)})$. Note that the corresponding positions in the matrix $\mathbf{I}_{K_0}^{(j)}$ associated with the K_0 highest values are set to unity, in order to perform the masking based activation of significant CIR-related taps during the following OFDM symbol periods. A similar strategy of selecting the significant CIR-related taps was also advocated by Li *et al.* [91] with the aim of reducing the complexity of a transversal filter-type channel estimator employed in Space-Time Coded (STC) OFDM systems.

2) The second alternative is that of obtaining an "average" value of $\text{Trace}(\Upsilon^{[f](j)}\mathbf{I}_{K_0}^{(j)})$ by employing the spaced-frequency correlation matrix $\tilde{\mathbf{R}}^{[f](j)}$ based on the spaced-frequency correlation function associated with a uniform ideally support limited multipath intensity profile. The sparse identity matrix $\mathbf{I}_{K_0}^{(j)}$ could be designed for retaining the *first* K_0 CIR-related coefficients of $\Upsilon^{[f](j)}$ – rather than the K_0 largest one – or alternatively, to retain the first K_0^I and the last K_0^{II} CIR-related coefficients of $\Upsilon^{[f](j)}$, where $K_0 = K_0^I + K_0^{II}$. This was suggested by van de Beek *et al.* [61] in the context of DFT-based channel transfer function estimation employed for single-user OFDM systems.

In the next section we will embark on the performance assessment of the various channel estimation techniques studied.

16.5.2   Performance Assessment

With the exception of the results to be presented in Section 16.5.2.2.3, our investigations were conducted for an SDMA uplink scenario supporting four simultaneous OFDM users each equipped with one transmit antenna. At the base station (BS) four reception antennas were assumed. Unless otherwise stated, we impose an OFDM symbol normalised Doppler frequency of $F_D = 0.007$, which corresponds to a vehicular speed of 50km/h, or equivalently, 31.25mph in the context of the indoor WATM system's parameters, as outlined in Section 14.3.1.

Our assessment commences in Section 16.5.2.2 with a study of the average *a priori* channel transfer factor estimation MSE evaluated in the context of the idealistic scenario of a sample-spaced CIR and upon assuming error-free symbol decisions. Specifically, in the simplified scenario of a sample-spaced CIR the exact shape of the multipath intensity profile does not influence the performance of the PIC-assisted DDCE when employing the same vector of predictor coefficients for the different CIR-related taps. This is as outlined in Section 16.5.1.6.1 the case when the channel's entire energy is conveyed by the K_0 number of significant CIR-related taps. By contrast, in Section 16.5.2.3 in the context of our more realistic investigations of the *a priori* estimation MSE involving non-sample-spaced CIRs, three types of multipath intensity profiles, namely the sparse profiles, uniform profiles and exponential profiles are invoked. The analytical MSE performance results of Sections 16.5.2.2 and 16.5.2.3 were generated upon invoking the iterative approach described in Section 16.5.1.6 for simultaneously optimising the CIR-related tap predictor coefficients and evaluating the different users' *a priori* estimation MSEs.

By contrast, at a later stage, namely in Section 16.5.2.4 we will consider the more realistic case of encountering imperfect, error-contaminated symbol decisions at the detection stage. Both uncoded as well as turbo-coded arrangements will be studied in terms of their *a priori* estimation MSE and system BER. These performance results were generated with the aid of Monte Carlo simulations upon invoking the indoor WATM system- and channel parameters of Section 14.3.1. Since the indoor WATM channel's CIR is composed of sample-spaced taps the performance curves presented in this context characterise the system's best possible performance.

Let us commence our discussions in the next section by studying the effects of the specific choice of the CIR-related tap predictor coefficients on the average *a priori* channel transfer factor estimation MSE.

16.5.2.1   Evolution of the *A Priori* Channel Estimation MSE in a Simplified 2-Tap CIR-Related Tap Prediction Scenario

In Figure 16.5 we have exemplified the evolution of the average *a priori* channel transfer factor estimation MSE according to Equation 16.133 as a function of the CIR-related tap predictor coefficients' associated values, where we employed a 2-tap predictor, since for a higher number of predictor taps a visualisation is less convenient. Also note that the predictor coefficients are real-valued due to employing the real-valued spaced-time channel correlation function of Equation 16.143. In our particular example the *a priori* channel estimation MSE evaluated from Equation 16.133 is minimised for a coefficient vector of $\tilde{\mathbf{c}}_{pre}|_{opt} \approx (1,771, -0.898)^T$. By contrast, for coefficient pairs outside the circle having a

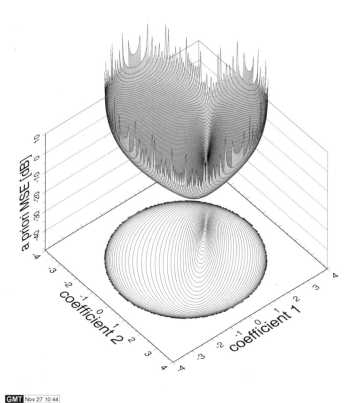

GMT Nov 27 10:44

Figure 16.5: Evolution of the *a priori* channel estimation MSE according to Equation 16.133 associated
with the simplified scenario of Section 16.5.1.6.1 as a function of the real-valued coefficients of the 2-tap CIR-related tap predictor employed in this particular example. The
number of subcarriers was $K = 512$, while the number of significant CIR-related taps
was $K_0 = 16$ in the context of a sample-spaced CIR. Furthermore, the number of users
was $L = 4$ and the OFDM symbol normalised Doppler frequency was $F_D = 0.1$. The
spaced-time channel correlation function of Equation 16.143, associated with a uniform,
ideally support limited Doppler power spectrum was invoked. The SNR at the reception
antenna was equal to 20dB.

radius of $\sqrt{\frac{K}{K_0\alpha(L-1)}} \approx 3.27$, centred around the origin of the $\mathbb{R}^2$ space, the channel estimator is unstable, which is evidenced by the excessive MSEs.

16.5.2.2 *A Priori* Channel Estimation MSE in the Context of Ideal, Error-Free Symbol Decisions Assuming a Sample-Spaced CIR

The simulation results to be presented in this section characterise the average *a priori* channel
transfer factor estimation MSE, which is a by-product of the iterative optimisation of the CIR-

related tap predictor coefficients with the aid of the algorithm outlined in Section 16.5.1.6.

In order to reduce the number of different parameter combinations to be investigated, we stipulate here a simplified scenario, as described in Section 16.5.1.6.1, where identical transmit powers, modulation modes and channel statistics are associated with the different users. More specifically, all users are assumed to employ MPSK modulation which renders the so-called modulation-mode enhancement factor equal to $\alpha_{\text{M-PSK}} = 1$. Unless otherwise stated, the specific channel statistics invoked were that of the channel's spaced-time correlation function provided by the Jakes' model, as given by Equation 16.141.

Furthermore, we considered "frame-invariant" fading, where the fading envelope of each CIR-related tap has been kept constant during each OFDM symbol's transmission period. This avoided the obfuscating effects of inter-subcarrier interference and hence enabled us to study the various channel transfer function estimation effects in isolation.

The further structure of Section 16.5.2.2 is as follows. In Sections 16.5.2.2.1 and 16.5.2.2.2 a comparison in terms of the *a priori* estimation MSE between employing the optimum recursive predictor coefficients and the sub-optimum transversal predictor coefficients – obtained upon setting the $(L - 1)$ number of feedback signals in Equations 16.133 and 16.135 equal to zero. In Section 16.5.2.2.1 we will use a one-tap predictor while in Section 16.5.2.2.2 higher-order predictors. While furthermore in Section 16.5.2.2.3 the influence of the number of simultaneous users on the *a priori* estimation MSE is investigated, in Sections 16.5.2.2.4 and 16.5.2.2.5 the influence of the OFDM symbol normalised Doppler frequency under matched and mismatched channel conditions is portrayed, respectively. Finally, in Section 16.5.2.2.6 a performance comparison to Li's LS-assisted DDCE of Section 16.4 is conducted.

16.5.2.2.1   Optimum Recursive versus Sub-Optimum Transversal CIR-Related Tap Predictor Coefficients - One Tap

In the left-hand graph of Figure 16.6 we have portrayed the average *a priori* channel transfer factor estimation MSE exhibited by the PIC-assisted DDCE in the context of one-tap CIR-related tap prediction filtering. The single predictor coefficient was calculated with the aid of the – in this context sub-optimum transversal filter-related Wiener solution of Equation 16.139 – while the associated MSE was evaluated with the aid of Monte Carlo simulations (labelled as "Simulation") as well as by direct evaluation of Equation 16.133 (labelled as "Formula"). We observe that both our analytical evaluations, as well as the simulations result in a similar MSE performance, thus supporting the validity of our derivations. When increasing the K_0 number of significant CIR taps, the *a priori* estimation MSE degrades, since the effects of the AWGN and of the additional noise due to channel variations are less mitigated. Note from Equation 16.139 that for sufficiently high SNRs the sub-optimum transversal filter-related coefficient approaches the value of the channel's spaced-time correlation function for a unity time-lag, which is in turn close to unity for the relatively slowly fading channels of $f_D = 0.007$ considered here. Hence, following from Equation 16.134, the maximum value of K_0, which guarantees a stable operation in the absence of symbol errors in the context of a scenario having $K = 512$ MPSK modulated ($\alpha = 1$) subcarriers and $L = 4$ simultaneous users each equipped with a single transmit antenna is $K_0 = 170$. This is also reflected by the curves in the left-hand graph of Figure 16.6.

By contrast, in the right-hand graph of Figure 16.6 we have compared – for the same scenario – the average *a priori* channel transfer factor estimation MSE achieved with the aid of the optimum CIR-related tap predictor coefficient given by Equation 16.137, which takes

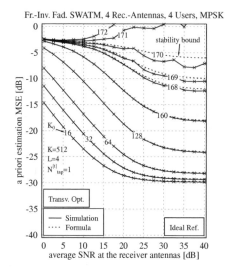

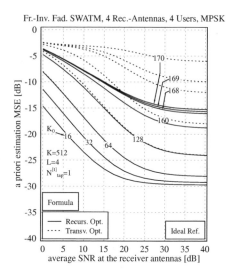

Figure 16.6: (a: Left) Comparison between the *a priori* channel estimation MSE versus SNR performance, generated by an analytical evaluation of Equation 16.133 as well as that obtained by Monte Carlo simulations for the PIC-assisted DDCE of Figure 16.4 in the context of one-tap CIR-related tap prediction filtering invoking the *sub-optimum transversal predictor coefficient* according to Equation 16.139.

(b: Right) Comparison between the *a priori* channel estimation MSE versus SNR performance exhibited by the PIC-assisted DDCE in the context of one-tap prediction filtering, again, upon invoking the sub-optimum transversal predictor coefficient according to Equation 16.139 or alternatively the *optimum recursive predictor coefficient* according to Equation 16.137. The curves are parametrised with the number of significant CIR-related taps K_0. Each of the SDMA scenario's independently faded channels is characterised by the indoor WATM channel parameters of Section 14.3.1.

into account the recursive structure of the PIC-assisted DDCE, against the *a priori* estimation MSE offered by a system employing the suboptimum transversal filter based predictor coefficient of Equation 16.139. Both sets of curves have been obtained by direct evaluation of Equation 16.133 ("Formula"), upon assuming error-free symbol decisions. The results suggest that upon employing the optimum recursive predictor coefficient, the system's stability is increased for relatively high values of $\chi \cdot (L - 1)$ in the sense of Equation 16.134.

16.5.2.2.2 Optimum Recursive- versus Sub-Optimum Transversal CIR-related Tap Predictor Coefficients - Higher Order Following the approach of Figure 16.6, in Figure 16.7 we have compared the average *a priori* channel transfer factor estimation MSE achieved using the optimum recursive CIR-related tap predictor coefficients which were evaluated with the aid of the iterative approach of Section 16.5.1.6 upon capitalising on the simplified Equations of Section 16.5.1.6.1 against the – in this context suboptimum –

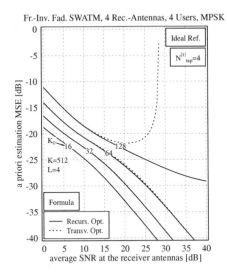
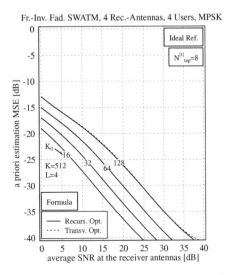

Figure 16.7: *A priori* channel estimation MSE versus SNR performance exhibited by the PIC-assisted DDCE of Figure 16.4 in the context of **(a: Left)** four-tap- and **(b: Right)** eight-tap CIR-related tap prediction filtering upon invoking the optimum recursive predictor coefficients evaluated with the aid of the iterative approach of Section 16.5.1.6 on the basis of the simplified equations of Section 16.5.1.6.1. Again, we have plotted the *a priori* channel estimation MSE performance achieved with the aid of the – in this case – sub-optimum transversal predictor coefficients of Equation 16.136 as a reference. The curves are parametrised with the number of significant CIR-related taps K_0. Each of the SDMA scenario's independently faded channels is characterised by the indoor WATM channel parameters of Section 14.3.1.

transversal filter coefficients provided by Equation 16.136. Two different predictor lengths, namely four- and eight taps are employed and the curves are additionally parametrised with the number of significant CIR-related taps. Again, we observe that the optimum recursive predictor coefficients allow us to avoid the problem of instability, which was potentially incurred in conjunction with the sub-optimum transversal predictor coefficients.

16.5.2.2.3 Influence of the Number of Simultaneous Users in the Context of the Optimum Recursive CIR-Related Tap Predictor Coefficients

So far we considered the scenario of four simultaneous users, each equipped with a single transmit antenna. From the perspective of the number of channels to be estimated, this is equivalent to a space-time coded (STC) scenario of two simultaneous users each employing two transmit antennas or to an STC scenario of a single user employing four transmit antennas. Based on this analogy, it is worth investigating whether the PIC-assisted DDCE approach advocated here is capable of supporting scenarios of a higher complexity in terms of the $L \times P$ number of channels involved. In Figure 16.8 we have plotted the average *a priori* channel transfer factor estimation

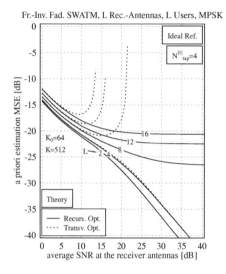
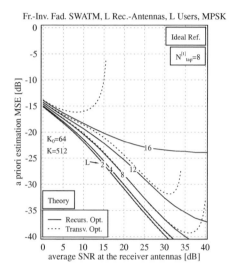

Figure 16.8: *A priori* channel estimation MSE versus SNR performance exhibited by the PIC-assisted DDCE of Figure 16.4 in the context of **(a: Left)** four-tap- and **(b: Right)** eight-tap CIR-related tap prediction filtering upon invoking the optimum recursive predictor coefficients evaluated with the aid of the iterative approach of Section 16.5.1.6 on the basis of the simplified equations of Section 16.5.1.6.1. Again, we have plotted the *a priori* channel estimation MSE performance achieved with the aid of the – in this case – sub-optimum transversal predictor coefficients of Equation 16.136 as a reference. The curves are parametrised with the L number of simultaneous users, while the K_0 number of significant CIR taps was kept constant. Each of the SDMA scenario's independently faded channels is characterised by the indoor WATM channel parameters of Section 14.3.1.

MSE as a function of the L number of simultaneous users, assuming a four- or eight-tap CIR-related tap prediction filter and a fixed number of $K_0 = 64$ significant CIR-related taps. This associated CIR duration corresponds to 12.5% of the duration of a 512-subcarrier OFDM symbol's time domain representation. This may be viewed as the relative upper bound of the CIR length in a well-designed OFDM system. Here we capitalised again on the idealistic assumption of error-free symbol decisions. We observe in Figure 16.8 that the *a priori* channel estimation MSE performance is degraded upon increasing the number of users supported. This is, because more multi-user interference-related noise is inflicted by the *a posteriori* channel estimates during the PIC process, which is then injected into the *a priori* channel estimates' prediction process. As a comparison between the MSE curves corresponding to the four-tap and eight-tap prediction arrangements suggests, these effects can be mitigated by increasing the predictor's range. Again, we observe that in the context of the sub-optimum transversal predictor coefficients the PIC-assisted DDCE tends to become unstable at higher SNRs.

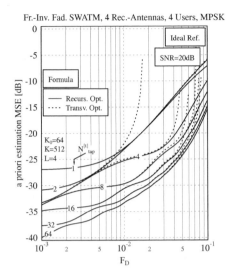
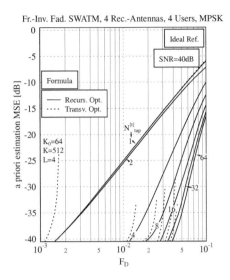

Figure 16.9: *A priori* channel estimation MSE versus OFDM symbol normalised Doppler frequency performance exhibited by the PIC-assisted DDCE of Figure 16.4 in the context of an SNR of **(a: Left)** 20dB and **(b: Right)** 40dB at the reception antennas upon invoking the optimum recursive CIR-related tap predictor coefficients evaluated with the aid of the iterative approach of Section 16.5.1.6 on the basis of the simplified equations of Section 16.5.1.6.1. Again, we have plotted the *a priori* channel estimation MSE performance achieved with the aid of the – in this case – sub-optimum transversal predictor coefficients of Equation 16.136 as a reference. The curves are parametrised with the $N_{tap}^{[t]}$ number of predictor taps, while the K_0 number of significant CIR-related taps was kept constant. Each of the SDMA scenario's independently faded channels is characterised by the indoor WATM channel parameters of Section 14.3.1.

16.5.2.2.4 Influence of the OFDM Symbol Normalised Doppler Frequency
In the context of our previous assessment a fixed OFDM symbol-normalised Doppler frequency of $F_D = 0.007$ was assumed. By contrast, in this section we will investigate the influence of the OFDM symbol normalised Doppler frequency on the average *a priori* channel transfer factor estimation MSE for different CIR-related tap predictor lengths of $N_{tap}^{[t]}$. The results of our analytical evaluations are portrayed in Figure 16.9, both for an SNR of 20dB measured at the reception antennas and for an SNR of 40dB. For both SNRs we observe that the *a priori* channel estimation MSE recorded at a given OFDM symbol normalised Doppler frequency improves only marginally upon increasing the predictor length beyond a value of about 32 taps, for which most of the channel's correlation is exploited. Furthermore, we infer that the predictor's length should be at least four taps, in order to achieve a significant reduction of the *a priori* channel estimation MSE compared to the case of one-tap filtering. Note furthermore in Figure 16.9 that in the higher-SNR scenario of 40dB the MSE reduction due to employing a higher number of CIR-related predictor taps is even more dramatic than in the lower SNR

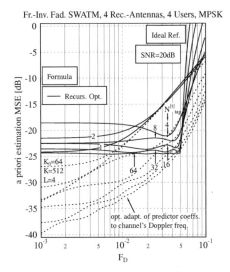 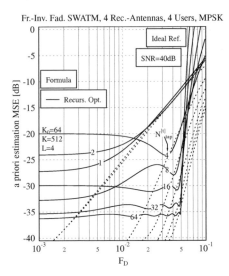

Figure 16.10: *A priori* channel estimation MSE versus OFDM symbol normalised Doppler frequency performance exhibited by the PIC-assisted DDCE of Figure 16.4 in the context of an SNR of **(a: Left)** 20dB and **(b: Right)** 40dB at the reception antennas upon invoking a fixed vector of recursive CIR-related tap predictor coefficients optimised for an OFDM symbol normalised Doppler frequency of $\tilde{F}_D = 0.05$ with the aid of the iterative approach of Section 16.5.1.6 on the basis of the simplified equations of Section 16.5.1.6.1. As in previous graphs, a Jakes' spectrum-related spaced-time correlation function obeying Equation 16.141 was associated with the channel. The predictor coefficients were calculated on the basis of the spaced-time correlation function of Equation 16.143 associated with a uniform, ideally support-limited Doppler power spectrum. Furthermore, we have also plotted the MSE curves corresponding to predictor coefficients optimised for the channel's Doppler frequency as a reference. The MSE curves are parametrised with the $N_{tap}^{[t]}$ number of prediction filter taps, while K_0, namely the number of significant CIR-related taps was fixed. Each of the SDMA scenario's independently faded channels is characterised by the indoor WATM channel parameters of Section 14.3.1.

scenario. This is because the MMSE predictor strikes a trade-off between the mitigation of the AWGN and the exploitation of the channel's correlation between the channel transfer functions experienced by successive OFDM symbols.

So far we have assumed a perfect matching between the channel statistics invoked in the calculation of the predictor coefficients and that of the channel. By contrast, in the next section we investigate the effects of a mismatch with respect to the maximum Doppler frequencies.

16.5.2.2.5 Influence of a Mismatch of the OFDM Symbol Normalised Doppler Frequency

In our previous investigations we employed a Doppler power spectrum according to Jakes' model [474] which is related to a spaced-time channel correlation function given by the zero-order Bessel function of its first kind, as shown in Equation 16.141. Jakes' model was employed both for the derivation of the CIR-related tap predictor coefficients as well as in the "simulated" channel used for evaluating the *a priori* channel estimation MSE. The maximum Doppler frequency used for the calculation of the predictor coefficients as well as for the simulated channel was identical.

In order to render a transversal filter-assisted channel estimator insensitive to the variations of the maximum OFDM symbol normalised Doppler frequency associated with the channel, compared to that imposed in the calculation of the filter coefficients, it was proposed in [67,68] to invoke a uniform, ideally support-limited Doppler power spectrum in the calculation of the filter coefficients, having a maximum OFDM symbol normalised Doppler frequency higher than that of the channel. Recall that the same "robust" DDCE was also used in Section 15.3 for single-user channel estimation. It was argued in [68] that in this case, regardless of the specific shape of the "true" Doppler power spectrum associated with the channel, the channel estimator would exhibit the same channel estimation MSE performance, as if a uniform, ideally support-limited Doppler spectrum was assumed *both* in the evaluation of the filter coefficients, as well as for the channel encountered. Also note that in this case the channel estimator's MSE performance represented the worst-case MSE performance that might be encountered when communicating over channels having non-uniform Doppler power spectra, but using optimally adapted filter coefficients. Furthermore, note from [68] that the MSE performance difference observed between the scenario, where Jakes' Doppler spectrum is associated both with the channel encountered as well as with the computation of the filter coefficients and the case, where a uniform, ideally support-limited Doppler power spectrum is associated with both the channel encountered and with the computation of the filter coefficients, is marginal.

As will be highlighted in the context of our forthcoming discussions, the channel estimator's insensitivity with respect to the channel's Doppler power spectral shape is valid only upon employing an infinite number of predictor taps. Naturally, in practical situations this is not the case.

In Figure 16.10 we have portrayed the average *a priori* channel transfer factor estimation MSE versus OFDM symbol normalised Doppler frequency performance of the recursive estimator of Figure 16.4 in the context of employing a uniform, ideally support-limited Doppler power spectrum having a spaced-time correlation function obeying Equation 16.143 in the calculation of the CIR-related tap predictor coefficients. Furthermore, as in our previous investigations of Section 16.5.2.2 a Doppler power spectrum obeying Jakes' model [474] and having a spaced-time correlation function given by Equation 16.141 was associated with the channel. In our particular example the predictor coefficients were calculated upon invoking once again, the iterative approach of Section 16.5.1.6 for an OFDM symbol normalised Doppler frequency of $\tilde{F}_D = 0.05$. Furthermore, as a reference we have also plotted the *a priori* channel estimation MSE performance in the context of predictor coefficients, which were optimised for the channel's specific Doppler frequency. As reported in [68] and also observed in Figure 16.10, upon increasing the number of predictor taps the *a priori* channel estimation MSE is rendered quasi-invariant for OFDM symbol normalised Doppler frequencies on the channel, which are lower than that assumed in the calculation of the CIR-related tap predictor coefficients, namely $\tilde{F}_D = 0.05$. By contrast, for higher Doppler frequencies a rapid degradation of the MSE is observed in Figure 16.10. This "robustness" is achieved at the cost of a

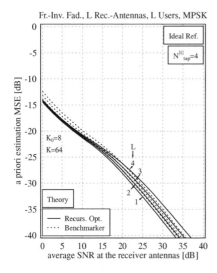
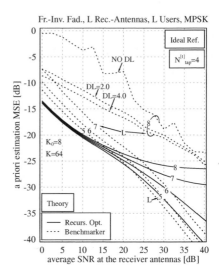

Figure 16.11: *A priori* channel estimation MSE versus SNR at the reception antennas exhibited by the LS-assisted DDCE [91] of Figure 16.3 outlined in Section 16.4 and by the PIC-assisted DDCE of Figure 16.4 seen in Section 16.5, for a number of users L ranging between **(a: Left)** one to four and **(b: Right)** five to eight, in the context of $N_{tap}^{[t]} = 4$-tap *a priori* CIR-related tap prediction filtering for an OFDM symbol normalised Doppler frequency of $F_D = 0.007$. A Jakes' spectrum-related spaced-time correlation function was associated with both the channel encountered as well as with the calculation of the predictor coefficients. Note that in contrast to our previous discussions we have assumed here $K = 64$ and $K_0 = 8$. "DL" indicates Diagonal Loading of the matrix $\mathbf{Q}[n]$ defined in Equation 16.20 in order to support its invertibility.

potentially significant loss in performance compared to the case of optimally adapted predictor coefficients. To give an example, for an SNR of 20dB and for 64 predictor coefficients the *a priori* channel estimation MSE performance loss is as high as 10dB at an OFDM symbol normalised Doppler frequency of $F_D = 0.007$, when the predictor coefficients were designed for $\tilde{F}_D = 0.05$.

16.5.2.2.6 Performance Comparison to Li's LS-Assisted DDCE In this section we will evaluate the *a priori* channel estimation MSE exhibited by the LS-assisted DDCE of Figure 16.3 proposed by Li *et al.* [91] capitalising on a direct inversion of the correlation matrix $\mathbf{Q}[n]$ of the different users' transmitted subcarrier symbol sequences' as outlined in Section 16.4. We will then compare the performance of Li's LS-assisted DDCE against the MSE performance achieved by the PIC-assisted DDCE of Figure 16.4. In both cases we have employed a $N_{tap}^{[t]} = 4$-tap CIR-related tap predictor, where in the context of the LS-assisted DDCE of Figure 16.3 the optimum predictor coefficients were obtained by direct solution of the Wiener equation, while for the PIC-assisted DDCE of Figure 16.4 the predictor coefficients were optimised with the aid of the iterative approach portrayed in Section 16.5.1.6. Note that in contrast to our performance results presented in previous sections for $K = 512$, here we have assumed $K = 64$ subcarriers and $K_0 = 8$ significant CIR-related taps. It can be argued that from the perspective of an AWGN reduction in the context of the sample-spaced CIRs considered here, this system is equivalent to the scenario of $K = 512$ and $K_0 = 64$. However, in the latter case the correlation matrix $\mathbf{Q}[n]$ of Equation 16.20 employed in the LS-assisted DDCE would be of potentially excessive dimension, depending on the number of users L. Hence $K = 512$ and $K_0 = 64$ would result in a complexity, which would be impractical in terms of our simulations. More specifically, our aim in this section is also to demonstrate the deficiencies associated with the direct inversion of the correlation matrix $\mathbf{Q}[n]$, namely that error amplification may be encountered, when the matrix becomes rank deficient.[13] This is observed upon increasing the L number of simultaneous users supported in conjunction with a fixed value of K_0.

Our corresponding simulation results are portrayed in Figure 16.11. On the left-hand side we presented results for a number of simultaneous users ranging between one and four while on the right-hand side for five to eight users. For a lower number of users we observe in the left illustration of Figure 16.11 that up to SNRs of about 15dB measured at the reception antennas the PIC-assisted DDCE of Figure 16.4 performs slightly better, while at higher SNRs the LS-assisted DDCE of Figure 16.3 exhibits a slight performance advantage. The reason for these performance trends is that at lower SNRs the multi-user interference (MUI) imposed, when the matrix $\mathbf{Q}[n]$ has non-zero off-diagonal elements results in a degradation of the MSE of the LS-assisted DDCE shown in Figure 16.3, while at higher SNRs the effects of imperfect CIR-related tap prediction yield an increased MUI in the PIC-assisted DDCE, again, as seen in Figure 16.11. This effect becomes more obvious upon further increasing the number of simultaneous users L, as shown in the right-hand side of Figure 16.11. Note that the less attractive performance exhibited by the LS-assisted DDCE is due to two effects. First of all, the effects of MUI are more pronounced due to the higher number of users supported and second, the imperfections of the direct numerical inversion of the correlation matrix $\mathbf{Q}[n]$ of Equation 16.20 carried out here in the context of our simulations contributes to the significant MSE degradation. This is, because matrices of the form $\mathbf{Q}[n] = \mathbf{A}^H[n]\mathbf{A}[n]$ as encountered in the context of calculating the Moore-Penrose inverse [90] $\mathbf{A}^\dagger[n]$ of $\mathbf{A}[n]$ exhibit a potentially high condition number $\kappa(\mathbf{A}[n])$, where the condition number indicates the degree to which a matrix is ill-conditioned or close to rank-deficiency [90]. Note again that specifically in the context of least-squares estimation problems the direct inversion of the correlation matrix $\mathbf{Q}[n] = \mathbf{A}^H[n]\mathbf{A}[n]$ and the associated deficiencies can be mitigated with the aid of the QR matrix factorisation [90].

16.5.2.3 Effects of a Non-Sample Spaced CIR in the Context of Ideal, Error-Free Symbol Decisions

In Equations 16.126 and 16.132 as well as correspondingly also in Equations 16.133 and 16.135 associated with the simplified scenario of identical transmit powers and identical channel statistics, we observe that the evaluation of the optimum CIR-related tap predic-

---

[13] A square matrix is referred to as rank deficient, if the number of linearly independent rows or columns is lower than the number of rows or columns.

tor coefficients with the aid of the iterative approach described in Section 16.5.1.6 relies on the *a priori* knowledge of the term $\text{Trace}(\Upsilon^{[f](j)}\mathbf{I}_{K_0}^{(j)})$ associated with the j-th user, where $j = 1, \ldots, L$.

Our analytical evaluations in Section 16.5.2 were conducted so far under the assumption of a sample-spaced CIR. Using a sample-spaced CIR facilitates the recovery of almost all the energy of the channel's output, upon invoking a finite number of K_0 significant taps, despite the fact that K_0 is potentially smaller than the total number of K available CIR-related taps. In this case we have $\text{Trace}(\Upsilon^{[f](j)}\mathbf{I}_{K_0}^{(j)}) = K\sigma_H^2$. By contrast, in the context of the more realistic scenario of a non-sample-spaced CIR the energy conveyed by the channel is distributed over a higher number of CIR-related taps, i.e. it potentially "leaks" to all CIR-related taps. This is particularly the case when employing the DFT matrix $\mathbf{W}$ as the unitary transform matrix. While $\text{Trace}(\Upsilon^{[f](j)}) = K\sigma_H^2$ still holds, we potentially incur $\text{Trace}(\Upsilon^{[f](j)}\mathbf{I}_{K_0}^{(j)}) < K\sigma_H^2$ for $K_0 < K$. The motivation for choosing $K_0 < K$ is twofold, namely that of reducing the predictor's complexity, but also to further reduce the noise in the context of the simplified predictor design of Section 16.5.1.6.1 employed here, relying on identical predictor coefficients for each of the different CIR-related taps.

Our further efforts in characterising the *a priori* channel estimation MSE performance in the context of non-sample-spaced CIRs will concentrate on three different types of power-delay profiles, namely, what we refer to as sparse profiles, uniform profiles and exponential profiles. These profiles will be outlined in Sections 16.5.2.3.1, 16.5.2.3.2 and 16.5.2.3.3. On the basis of these profiles we will then assess the PIC-assisted DDCE's average – as well as the subcarrier based – *a priori* channel transfer factor estimation MSE in Sections 16.5.2.3.4 and Section 16.5.2.3.5, respectively.

16.5.2.3.1 Sparse Profiles Here we define as a sparse multipath intensity profile a finite number of Dirac impulses, each characterised by its delay τ_i and variance $\sigma_{h,i}^2$, which is formulated as:

$$r_{h,\text{spar}}(\tau) = \sum_i \sigma_{h,i}^2 \delta(\tau - \tau_i). \tag{16.145}$$

We impose the condition that the sum of the multipath intensity profile's different tap variances is unity:

$$\sum_i \sigma_{h,i}^2 = 1, \tag{16.146}$$

which implies that no energy is lost or gained during the signal's transmission over the channel. By applying the Fourier transform to Equation 16.145, the spaced-frequency correlation function of the associated multipath intensity profile is obtained:

$$r_{H,\text{spar}}(\Delta f) = \sum_i \sigma_{h,i}^2 e^{-j2\pi\tau_i\Delta f}. \tag{16.147}$$

In the context of employing the DFT matrix $\mathbf{W}$ as the unitary transform matrix $\tilde{\mathbf{U}}^{[f]}$ it can be shown that the element $[n_1, n_2]$ of the matrix $\Upsilon_{\text{spar}}^{[f]}$, which is the result of the decomposition

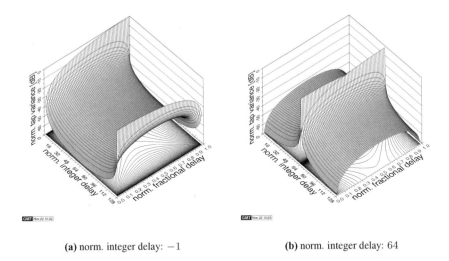

(a) norm. integer delay: -1 (b) norm. integer delay: 64

Figure 16.12: Illustration of the normalised leakage across the diagonal elements of the decomposition $\Upsilon_{\text{spar}}^{[f]}$ of $\mathbf{R}_{\text{spar}}^{[f]}$ with respect to $\tilde{\mathbf{U}}^{[f]} = \mathbf{W}$ in the context of a CIR exhibiting a single tap of unit variance, characterised by its normalised integer- and fractional delay. The normalisation of the diagonal elements of $\Upsilon_{\text{spar}}^{[f]}$ was carried out with respect to the $K = 128$ number of subcarriers, while the normalisation of the single CIR tap's integer and fractional delay was performed with respect to the sampling period duration T_s.

of the spaced-frequency correlation matrix $\mathbf{R}_{\text{spar}}^{[f]}$ with respect to $\tilde{\mathbf{U}}^{[f]}$, which is formulated as $\Upsilon_{\text{spar}}^{[f]} = \tilde{\mathbf{U}}^{[f]H} \mathbf{R}_{\text{spar}}^{[f]} \tilde{\mathbf{U}}^{[f]}$, is given by [61]:

$$\Upsilon_{\text{spar}}^{[f]}[n_1, n_2] = \frac{1}{K} \sum_i \sigma_{h,i}^2 e^{-j\frac{K-1}{K}\pi(n_1-n_2)} \frac{\sin(\pi\acute{n}_1) \cdot \sin(\pi\acute{n}_2)}{\sin(\pi\frac{\acute{n}_1}{K}) \cdot \sin(\pi\frac{\acute{n}_2}{K})}, \qquad (16.148)$$

where $\acute{n}_1 = \frac{\tau_i}{T_s} - n_1$ and $\acute{n}_2 = \frac{\tau_i}{T_s} - n_2$ and T_s is the sampling period duration. For the diagonal elements of $\Upsilon_{\text{spar}}^{[f]}$, identified by the condition of $n_1 = n_2$, Equation 16.148 simplifies to:

$$\Upsilon_{\text{spar}}^{[f]}[n_1, n_1] = \frac{1}{K} \sum_i \sigma_{h,i}^2 \frac{\sin^2(\pi\acute{n}_1)}{\sin^2(\pi\frac{\acute{n}_1}{K})}. \qquad (16.149)$$

With the aid of Equation 16.149 we have evaluated the influence of the fractional component of a single CIR tap's delay with respect to the sampling period duration T_s on the leakage[14] experienced by neighbouring integer delay taps in the context of employing the DFT matrix

---

[14] Although as in our experiments the CIR might consist of a single non-sample-spaced tap only, after transforming the associated set of K different subcarriers' channel transfer factors to the CIR-related domain with the aid of the unitary transform $\tilde{\mathbf{U}}^{[f]}$ as shown in Equation 16.108, all of the transform's output taps are potentially non-zero. This effect is referred to as leakage.

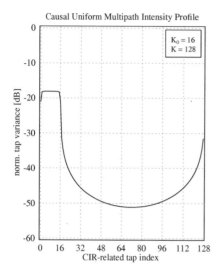
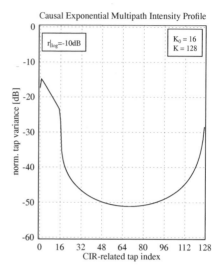

Figure 16.13: Illustration of the normalised leakage among the diagonal elements of the decomposition $\Upsilon^{[f]}$ of $\mathbf{R}^{[f]}$ with respect to $\tilde{\mathbf{U}}^{[f]} = \mathbf{W}$ in the context of a **(a: Left)** uniform multipath intensity profile or **(b: Right)** exponential multipath intensity profile. Both profiles exhibited the same total sampling period duration normalised multipath spread of $T_m/T_s = 16$. The normalisation of the diagonal elements of $\Upsilon^{[f]}$ was carried out with respect to the $K = 128$ number of subcarriers.

$\mathbf{W}$ as the unitary transform matrix $\tilde{\mathbf{U}}^{[f]}$ in the sense of Equation 16.108. Recall that Equation 16.149 reflects the n_1-th CIR-related tap's variance, where $n_1 = 0, \ldots, K - 1$. The results – normalised to the K number of subcarriers – are portrayed in Figure 16.12 on the left-hand side for a normalised integer delay of -1 tap or, equivalently, 127 taps and on the right-hand side for a normalised integer delay of 64 taps. Let us concentrate here on the second figure, namely that of a normalised integer delay of 64 taps associated with the single CIR tap employed. For an additional fractional delay of 0 all of the single CIR tap's energy is projected onto the diagonal element of the matrix $\Upsilon^{[f]}_{\mathrm{spar}}$ at index 64, or on the equivalent tap in the output of the unitary inverse linear transform following the philosophy of Equation 16.108. Upon increasing the additional normalised fractional delay, more of the CIR-tap's energy is conveyed to the tap at the index 65 and the other surrounding taps. Specifically, for a normalised fractional delay of 0.5 both contributions, namely that of the 64-th and 65-th tap are identical. Upon further increasing the normalised fractional delay to unity, all of the single CIR tap's energy is projected onto the 65-th CIR-related tap. Also note the cyclic symmetry of the figures with respect to the integer delay.

16.5.2.3.2 Uniform Profiles

While in the previous section we considered sparse profiles having discrete delay taps, let us now focus on a continuous uniform power delay profile, which is characterised by the following equation:

$$r_{h,\text{unif}}(\tau) = \frac{1}{T_m}\text{rect}\left(\frac{\tau - \tau_{shift}}{T_m}\right),$$
(16.150)

where T_m is the associated delay spread of the channel and τ_{shift} denotes the average delay with respect to the origin of the time-axis. Normalisation to the multipath spread T_m ensures that the energy transfer factor of the CIR, which is the integral of the multipath intensity profile across its region of support, where it exhibits non-zero values, is equal to unity. This implies that no energy is gained or lost during transmission over the channel. Also note that for an average delay of $\tau_{shift} = T_m/2$ an "ideally causal" delay profile is obtained. The spaced-frequency correlation function is given as the Fourier Transform (FT) of the power delay profile, yielding the expected sinc-shaped function of:

$$r_{H,\text{unif}}(\Delta f) = si(\pi T_m \Delta f) \cdot e^{-j2\pi\tau_{shift}\Delta f}.$$
(16.151)

Observe in Equation 16.151 that the higher the delay-spread, the more rapid the frequency domain fading envelope fluctuation and hence the correlation function decays more rapidly, as a function of the frequency spacing. Similar to our investigations of the leakage effects due to the CIR's non-sample-spaced nature in the context of a sparse profile, which were conducted in Section 16.5.2.3.1, at the left-hand side of Figure 16.13 we have plotted the normalised variance of the diagonal elements of the decomposition of $\mathbf{R}_{\text{unif}}^{[f]}$ based on $r_{H,\text{unif}}(\Delta f)$ with respect to employing the DFT matrix $\mathbf{W}$ as the unitary transform matrix $\tilde{\mathbf{U}}^{[f]}$, which is expressed mathematically as $\Upsilon_{\text{unif}}^{[f]} = \tilde{\mathbf{U}}^{[f]H}\mathbf{R}_{\text{unif}}^{[f]}\tilde{\mathbf{U}}^{[f]}$. The normalised multipath spread was assumed to be $T_m/T_s = 16$, which corresponds to 16 time domain OFDM sample durations and the number of subcarriers was $K = 128$. The "u"-shaped evolution of the tap variances seen on the left-hand side of Figure 16.13 for tap indices in excess of $T_m/T_s = 16$ is again a result of the leakage incurred. Note that in contrast to the leakage floor observed in conjunction with the sparse multipath intensity profile hosting in the particular example of Figure 16.12 a single tap only, the leakage floor incurred here is reduced by more than 10dB. This is, because as demonstrated in Figure 16.12 of Section 16.5.2.3.1 for a single Dirac impulse-like CIR, the maximum leakage is observed at a normalised fractional delay of 0.5. By contrast, in the context of the continuous delay profiles considered here we encounter fractional delay components ranging between the values of zero and one. Note that Equation 16.149 suggests an alternative way of obtaining the diagonal elements of $\Upsilon^{[f]}$ for different power delay profiles, namely that of directly integrating the right-hand side of Equation 16.149 with respect to the specific power delay profile, instead of performing the decomposition of the spaced-frequency correlation matrix.

16.5.2.3.3 Exponential Profiles

The exponential multipath intensity delay profile is characterised by:

$$r_{h,\text{expo}}(\tau) = \begin{cases} \beta_{\text{expo}} e^{-\alpha_{\text{expo}}\tau} & \text{for } \tau = 0 \le \tau \le T_m \\ 0 & \text{otherwise} \end{cases},$$
(16.152)

where the decay factor α_{expo} is determined on the basis of the value of the quotient of $r_{h,\text{expo}}(\tau)$ at a delay of $\tau = 0$ and at $\tau = T_m$, namely:

$$r_{\text{expo}} = \frac{r_{h,\text{expo}}(T_m)}{r_{h,\text{expo}}(0)}, \qquad (16.153)$$

yielding:

$$\alpha_{\text{expo}} = -\frac{1}{T_m}\ln(r_{\text{expo}}). \qquad (16.154)$$

Furthermore, the amplitude scaling factor β_{expo} can be determined as a function of α_{expo}, again, based on the condition that the integral of the multipath intensity profile across its region of support is unity, resulting in:

$$\beta_{\text{expo}} = \frac{\alpha_{\text{expo}}}{1 - e^{-\alpha_{\text{expo}}T_m}}. \qquad (16.155)$$

The exponential multipath intensity profile's spaced-frequency correlation function is given by the Fourier transform of the multipath intensity profile of Equation 16.152, yielding:

$$r_{H,\text{expo}}(\Delta f) = \frac{\beta_{\text{expo}}}{\alpha_{\text{expo}} + j2\pi\Delta f}\left[1 - e^{-(\alpha_{\text{expo}}+j2\pi\Delta f)T_m}\right]. \qquad (16.156)$$

Once again, on the right of Figure 16.13 we have plotted the normalised diagonal elements of the decomposition $\Upsilon_{\text{expo}}^{[f]} = \tilde{\mathbf{U}}^{[f]H}\mathbf{R}_{\text{expo}}^{[f]}\tilde{\mathbf{U}}^{[f]}$ of the spaced-frequency correlation matrix $\mathbf{R}_{\text{expo}}^{[f]}$ based on $r_{H,\text{expo}}(\Delta f)$ with respect to $\tilde{\mathbf{U}}^{[f]} = \mathbf{W}$. The normalisation was performed with respect to the $K = 128$ number of subcarriers. Furthermore, as in our previous example of a uniform multipath intensity profile discussed in Section 16.5.2.3.2, the normalised delay spread was $T_m/T_s = 16$. Again, the result of our evaluations is portrayed on the right-hand side of Figure 16.13. For a higher number of subcarriers than $K = 128$, the leakage floor observed in Figure 16.13 is expected to be found at lower variance values, which can be demonstrated with the aid of Equation 16.149.

Having demonstrated the effects of non-sample-spaced multipath intensity profiles on the decomposition $\Upsilon^{[f]} = \tilde{\mathbf{U}}^{[f]H}\mathbf{R}^{[f]}\tilde{\mathbf{U}}^{[f]}$ of the associated spaced-frequency correlation matrix $\mathbf{R}^{[f]}$ with respect to the unitary transform matrix $\tilde{\mathbf{U}}^{[f]} = \mathbf{W}$, we will now embark in the next section on a performance assessment of the PIC-assisted DDCE in the context of non-sample-spaced multipath intensity profiles.

16.5.2.3.4 *A Priori* Channel Estimation MSE for a Non-Sample Spaced CIR

The PIC-assisted DDCE's average *a priori* channel transfer factor estimation MSE performance was evaluated with the aid of the iterative approach proposed in Section 16.5.1.6. The spaced-frequency correlation matrix $\mathbf{R}^{[f]}$ was calculated on the basis of the spaced-frequency correlation function of Equation 16.147 associated with a sparse multipath intensity profile consisting of only a single non-sample-spaced tap having a normalised integer delay of zero and a normalised fractional delay of 0.5. Following from our discussions in Section 16.5.2.3.1, this particular choice of the CIR ensures that the maximum amount of leakage is generated and hence our performance results constitute the worst-case performance. Alternatively, a

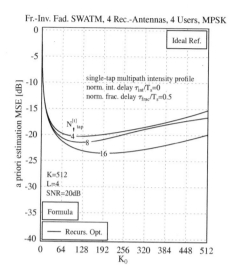
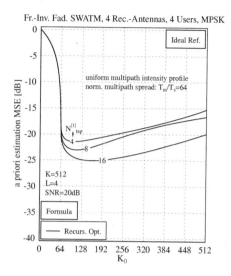

Figure 16.14: *A priori* channel estimation MSE performance of the PIC-assisted DDCE of Figure 16.4 versus the K_0 number of significant CIR-related taps retained in the context of **(a: Left)** a single-tap multipath intensity profile having a normalised integer delay of $\tau_{int}/T_s = 0$ and a normalised fractional delay of $\tau_{frac}/T_s = 0.5$ and **(b: Right)** a uniform multipath intensity profile having a normalised multipath spread of $T_m/T_s = 64$. The curves are furthermore parametrised with the $N_{tap}^{[t]}$ number of CIR predictor taps. The number of subcarriers was equal to $K = 512$ and the SNR at the reception antennas was assumed to be 20dB.

spaced-frequency correlation function obeying Equation 16.151, associated with a uniform multipath intensity profile was employed. The normalised multipath spread was set equal to one eighth of the $K = 512$ subcarriers assumed here, namely to $T_m/T_s = 64$.

The knowledge of the factor $\dfrac{K_0}{\text{Trace}(\Upsilon^{[f](j)}\mathbf{I}_{K_0}^{(j)})}$ is a prerequisite for determining both the optimum CIR-related tap predictor coefficients and the *a priori* channel estimation MSE as outlined in Section 16.5.1.6. The factor $\dfrac{K_0}{\text{Trace}(\Upsilon^{[f](j)}\mathbf{I}_{K_0}^{(j)})}$ was evaluated upon selecting the K_0 largest tap variances from the decomposition $\Upsilon^{[f]} = \tilde{\mathbf{U}}^{[f]H}\mathbf{R}^{[f]}\tilde{\mathbf{U}}^{[f]}$ of the specific spaced-frequency correlation matrices of the channel with respect to $\tilde{\mathbf{U}}^{[f](j)} = \mathbf{W}$.

The corresponding *a priori* channel estimation MSE curves are plotted in Figure 16.14 as a function of the K_0 number of significant CIR-related taps. The curves are also parametrised with the $N_{tap}^{[t]}$ number of predictor taps. For both the single-tap and for the uniform multipath intensity profile channel scenarios a rapid improvement of the estimator's MSE is observed upon increasing the K_0 number of significant CIR-related taps up to a certain optimum K_0 value, which is a consequence of retaining more of the channel's energy. At the same time more of the undesired noise is retained since a gradually decreasing fraction of the CIR-related taps are discarded. Upon increasing the K_0 number of significant taps beyond the optimum point seen in Figure 16.14, the opposite behaviour is observed, namely that the MSE is degraded again. This is, because for these taps the benefit of extracting more of the channel's energy is lower than the penalty incurred due to retaining more of the undesired noise. Note that this behaviour is a result of employing the same set of predictor coefficients for filtering each of the different CIR-related taps. By contrast, in the context of a predictor arrangement employing individually optimised sets of coefficients for the prediction of each of the different CIR-related taps, a "levelling out" of the *a priori* estimation MSE performance would be observed, instead of the explicit degradation seen in Figure 16.14. This is, because for the low-energy CIR-related taps suffering from a low channel-related signal component-to-noise ratio the noise would be more mitigated. As expected, in the context of the channel associated with a uniform multipath intensity profile on the right-hand side of Figure 16.14 an improved MSE performance is observed compared to the case of a single-tap multipath intensity profile, where the tap's normalised fractional delay had been intentionally adjusted to 0.5 in order to maximise the leakage.

Our previous investigations only delivered the *a priori* channel transfer factor estimation MSE averaged over the K subcarriers hosted by each OFDM symbol. By contrast, for the process of multi-user detection the quality of the channel transfer factor estimates recorded on a subcarrier-by-subcarrier basis is of relevance. Hence, our further aim is to characterise the *a priori* channel transfer factor estimation MSE distribution across the various subcarriers, which is the topic of the next section.

16.5.2.3.5   *A Priori* Channel Transfer Factor Estimation MSE for a Non-Sample Spaced CIR on a Subcarrier Basis

The specific distribution of the *a priori* channel transfer factor estimation MSE across the different subcarriers can also be obtained using the approach outlined in Section 16.5.1.6 to jointly optimise the average *a priori* channel estimation MSE and the predictor coefficients. This procedure involved updating the average *a priori* channel estimation MSE of the j-th user, where $j = 1, \ldots, L$, based on the average *a priori* channel estimation MSEs of the remaining users, employing the j-th user's specific vector of predictor coefficients determined with the aid of Equation 16.126. In a second step the j-th user's vector of predictor coefficients was then recomputed based on the updated *a priori* channel estimation MSEs of the remaining users with the aid of Equation 16.132.

By contrast, here we are interested in the exact distribution of the *a priori* channel transfer factor estimation MSE across the different subcarriers. This can be obtained upon invoking Equation 16.117 instead of Equation 16.126 in the algorithm outlined above. Again, in the context of a stable operation as defined in Section 16.5.1.5 we assume that the estimator's statistics recorded in form of the *a priori* channel transfer factor estimation errors' correlation matrix $\mathbf{R}_{\Delta\hat{\mathbf{H}}_{apr}^{(j)}}[n] = \mathbf{R}_{\Delta\hat{\mathbf{H}}_{apr}^{(j)}}[n - \acute{n}]$ is invariant for $\acute{n} = 1, \ldots, N_{tap}^{[t]}$, yielding:

$$\mathbf{R}_{\Delta\hat{\mathbf{H}}_{apr}^{(j)}}[n] = \frac{\alpha_j}{\sigma_j^2}\tilde{\mathbf{c}}_{pre}^{(j)H}\tilde{\mathbf{c}}_{pre}^{(j)}\mathbf{T}_{K_0}^{(j)}\left(\sum_{\substack{i=1\\i\neq j}}^{L}\sigma_i^2\mathbf{R}_{\Delta\hat{\mathbf{H}}_{apr}^{(i)}}[n]|_{Diag}\right)\mathbf{T}_{K_0}^{(j)H} +$$

$$+ \frac{\alpha_j}{\sigma_j^2}\sigma_n^2\tilde{\mathbf{c}}_{pre}^{(j)H}\tilde{\mathbf{c}}_{pre}^{(j)}\mathbf{T}_{K_0}^{(j)}\mathbf{T}_{K_0}^{(j)H} + \mathbf{R}_{\mathbf{H}_{dec}^{(j)}}. \tag{16.157}$$

Recall that the desired subcarrier-based *a priori* channel transfer factor estimation MSE vari-

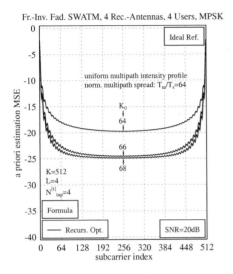
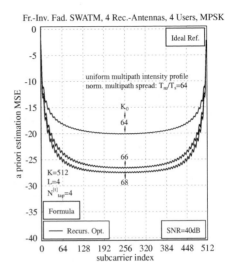

Figure 16.15: *A priori* channel estimation MSE performance versus the subcarrier index exhibited by the PIC-assisted DDCE of Figure 16.4 in the context of a uniform multipath intensity profile with a normalised multipath spread of $T_m/T_s = 64$, at an SNR of **(a: Left)** 20dB and **(b: Right)** 40dB. The curves are further parametrised with the K_0 number of significant CIR-related taps retained. The number of predictor taps employed was $N_{tap}^{[t]} = 4$ and the number of subcarriers was equal to $K = 512$.

ances are found on the main diagonal of the matrix $\mathbf{R}_{\Delta\hat{\mathbf{H}}_{apr}^{(j)}}[n]$ of Equation 16.157. The iteration commences with an initial assignment for the matrices $\mathbf{R}_{\Delta\hat{\mathbf{H}}_{apr}^{(j)}}[n]$, $j = 1, \ldots, L$, potentially constrained by the condition that the matrices are supposed to be Hermitian. The j-th user's *a priori* channel transfer factor estimation error correlation matrix is then updated with the aid of Equation 16.157 on the basis of the remaining users' error correlation matrices' diagonals denoted by $\mathbf{R}_{\Delta\hat{\mathbf{H}}_{apr}^{(i)}}[n]|_{Diag}$, employing the remaining users' associated current vectors of predictor coefficients. After updating all users' error correlation matrices, the vectors of predictor coefficients are updated with the aid of Equation 16.132. This involves evaluating first the average *a priori* channel transfer factor estimation MSEs with the aid of Equation 16.115 on the basis of the updated error correlation matrices. The iteration continues by updating the error correlation matrices again upon invoking the updated vectors of predictor coefficients.

Our analytical performance evaluations have been carried out for the uniform multipath intensity profile, again, in conjunction with a normalised multipath spread of $T_m/T_s = 64$ and for $K = 512$ subcarriers. The number of predictor taps was $N_{tap}^{[t]} = 4$. Our simulation results are portrayed in Figure 16.15, on the left-hand side for an SNR of 20dB at the reception antenna and on the right-hand side for 40dB. The curves are further parametrised with the K_0 number of significant CIR-related taps. As also evidenced by the simulation results of Figure 16.14 the number K_0 should be in excess of $T_m/T_s = 64$ in order to be able to extract all the significant taps and hence to prevent an excessive degradation of the MSE. The most important observation drawn from Figure 16.15 is that as a result of the effects of leakage imposed by the uniform multipath intensity profile the estimation MSE is substantially degraded for the subcarriers near the beginning and the end of the frequency domain OFDM symbol. Estimation MSEs as high as -5dB are observed. Furthermore, upon increasing the SNR measured at the reception antenna to 40dB, the MSE remains relatively high. This is, because the variance of the leakage as defined in Section 16.5.2.3 is independent from the SNR measured at the reception antenna. The somewhat lower MSE observed on the right-hand side of Figure 16.15 for the higher SNR of 40dB is achieved, because the Wiener filter-based CIR-related tap predictor becomes capable of more efficiently exploiting the channel's correlation. Note that at relatively high SNRs, such as, for example, 40dB, the MSE could be further improved upon employing a more beneficial, smooth window function, such as for example the Hamming window, instead of a rectangular window as employed here for windowing the *a posteriori* channel transfer factor estimates. Naturally, this would have to be appropriately considered in the calculation of the predictor coefficients. Based on the relatively high MSE associated with the outer subcarriers, we also expect for these subcarriers a significantly deteriorated BER performance, compared to the subcarriers at the centre of the OFDM symbol.

16.5.2.4 *A Priori* Channel Estimation MSE and System BER in the Context of Imperfect, Error-Contaminated Symbol Decisions Assuming a Sample-Spaced CIR

So far in this section we have capitalised on the idealistic assumption of error-free symbol decisions. By contrast, in a realistic scenario the channel estimation process is impaired by erroneous symbol decisions. These effects will be further highlighted during our forthcoming discussions. While a qualitative description of the associated effects is given in Section 16.5.2.4.1 a more quantitative analysis will be provided in Section 16.5.2.4.2, where the *a priori* channel estimation MSE performance and the system's BER performance will be assessed for the uncoded case. Furthermore, in Section 16.5.2.4.3 the impact of employing turbo coding will be highlighted.

16.5.2.4.1 Effects of Error-Contaminated Symbol Decisions The effect of erroneous symbol decisions can be viewed as an additional source of noise associated with statistical properties that are different from those of the AWGN. As expected, the variance of the impairment induced by erroneous symbol decisions in the DDCE process is a function of the channel SNR. If any of the multi-user channel transfer functions encountered during the current OFDM symbol exhibits a deep fade, potentially causing multiple symbol errors in the corresponding subcarriers, the quality of the associated *a priori* channel transfer factor estimates derived for the next OFDM symbol will be degraded. This will in turn increase the subcarrier symbol error probability at the output of the demodulator during the next OFDM symbol period. In this context the extraction of the K_0 number of significant *a posteriori* CIR-related taps has an adverse effect. This is because the time domain multiplication of the rectangular window-based extraction mask defined by the sparse diagonal matrix $\mathbf{I}_{K_0}$ exhibiting K_0 non-zero entries with the output of the unitary inverse transform $\tilde{\mathbf{U}}^{[f]H}$ given by Equation 16.108 corresponds to the cyclic convolution of the associated transfer functions

in the subcarrier domain. Thus, after time domain CIR-related tap windowing even a single subcarrier symbol error will affect the *a priori* channel transfer factor estimates of multiple adjacent subcarriers. As a result, an error-propagation effect may be observed, where the channel estimation quality is gradually degraded over a period of several consecutive OFDM symbols, also depending on the depth and width of the channel fades. In order to curtail these error propagation effects, a standard technique is to periodically transmit dedicated training OFDM symbols [68].

The effect of symbol errors can potentially be further mitigated upon increasing the *a priori* channel predictor's range, thus reducing the relative influence of each individual OFDM symbol's *a posteriori* estimated channel on the *a priori* channel predictor's output. On the other hand, as a result of using an increased number of OFDM symbols, at the same time the probability of incurring erroneous subcarrier symbol decisions during the estimation process is also increased. A less obvious effect related to the role of erroneous symbol decisions yet acting as an additional source of noise is that the system can be rendered unstable. This is because correct symbol decisions were assumed in the calculation of the predictor coefficients as outlined in Section 16.5.1, and hence a noise amplification problem classically known from the behaviour of conventional zero-forcing channel equalisers might occur. Since it is difficult to quantify the symbol-error-induced noise contribution, the best strategy for avoiding these effects is to employ an *a priori* CIR-related tap predictor having a sufficient range such as for example four or eight taps. The probability of incurring the adverse effects of erroneous symbol decision is of course also reduced upon employing more effective symbol detection schemes at the receiver, such as for example Successive Interference Cancellation (SIC) [104, 109, 113, 117, 119, 124, 126, 128, 135, 137] instead of MMSE detection [105–108, 110, 113, 117, 121, 135–137] and additionally by capitalising on powerful channel coding techniques such as turbo coding [480, 481]. The essence of the SIC and MMSE detection techniques was highlighted in Sections 17.2.3 and 17.3.1.

16.5.2.4.2 MSE and BER Performance in an Uncoded Scenario

In Figure 16.16 we have plotted on the left-hand side the average *a priori* channel transfer factor estimation MSE, while on the right-hand side the system's BER as a function of the SNR measured at the receiver antennas upon invoking both MMSE detection and the more effective, but also more complex M-SIC (M=2) detection technique, which are discussed in Sections 17.2.3 and 17.3.1, respectively. The curves are further parametrised with the number of CIR-related predictor taps and the fraction of training overhead incorporated. While a training overhead of 6.25% corresponds to transmitting one dedicated training OFDM symbol in every block of 16 OFDM symbols, a training overhead of 100% indicates here the scenario, where an error-free reference was made available to the DDCE for benchmarking. In terms of the *a priori* channel estimation MSE we observe on the left of Figure 16.16 that as a result of the M-SIC detector's lower error probability – compared to a system employing MMSE detection – the remodulated reference employed in the PIC-assisted DDCE is of better quality and hence the DDCE's MSE is significantly improved. By comparing the corresponding dashed and continuous curves in Figure 16.16 we observe that for SNRs up to about 7.5dB an MSE degradation is observed with respect to the scenario benefitting from an error-free reference. An interesting phenomenon is observed in the context of MMSE detection, when using two predictor taps. Due to an "excessive" number of erroneous subcarrier symbol decisions encountered in a specific OFDM symbol, which may be potentially induced by a

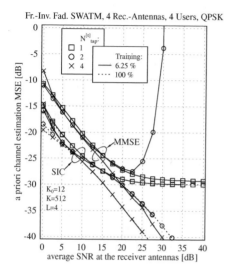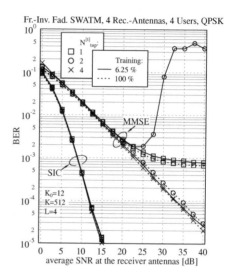

Figure 16.16: (**a: Left**) *A priori* channel estimation MSE versus SNR performance and (**b: Right**) BER versus SNR performance of an uncoded system employing the PIC-assisted DDCE of Figure 16.4 in conjunction with both MMSE and M-SIC (M=2) based detection at the receiver. The curves are further parametrised with the number of predictor taps $N_{tap}^{[t]}$ – ranging from one to four – and with the fraction of training overhead imposed, where 6.25% overhead corresponds to transmitting one dedicated training OFDM symbol per every block of 16 OFDM symbols, and 100% overhead denotes the idealistic case of an error-free reference. QPSK was employed as the modulation scheme.

deep fade on one of the channels of the multiple users, the *a priori* channel estimation MSE encountered during the next OFDM symbol is severely degraded, which in turn may trigger an avalanche of errors, which may lead to the system's instability. In the context of one-tap CIR-related tap prediction filtering error-propagation, events exceeding the length of a training period duration are prevented by periodically transmitting dedicated training OFDM symbols. By contrast, in the case of a higher number of predictor taps the OFDM training block length should ideally be identical to the number of predictor taps in order to eliminate the possibility of error propagation across the training OFDM symbols. Note furthermore that these effects are not observed for the four-tap predictor – at least not in the range of SNRs considered – since the effects of errors imposed by a single OFDM symbol are more efficiently mitigated. The graph on the right-hand side of Figure 16.16 is again evidence of the SIC combiner's more powerful detection capability.

16.5.2.4.3 BER Performance in the Turbo-Coded Scenario

In addition to our discussions of Section 16.5.2.4.2, which were cast in the context of uncoded systems, here we

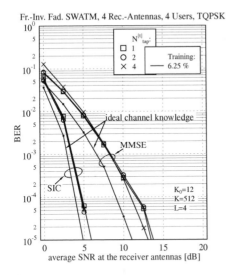
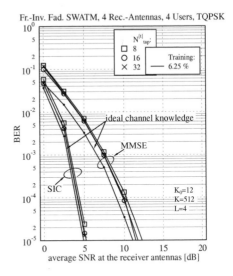

Figure 16.17: BER versus SNR performance of a turbo-coded system employing the PIC-assisted DDCE of Figure 16.4 in conjunction with MMSE or M-SIC (M = 2) detection at the receiver. The curves are parametrised with the number of CIR predictor taps (**a: Left**) ranging from one to four and (**b: Right**) ranging from eight to 32. A training overhead of 6.25% was incorporated by transmitting one dedicated training OFDM symbol in every block of 16 OFDM symbols. As a reference, we have also plotted the BER performance curves associated with the case of ideal channel parameter knowledge. QPSK was employed as the modulation scheme.

consider a turbo-coded system.[15] Instead of plotting the performance of a system capitalising on an error-free reference in the context of PIC-assisted DDCE, as a benchmark, the case of ideal channel parameter knowledge was considered for our further comparisons. Our BER performance results are portrayed on the left-hand side of Figure 16.17 for various numbers of CIR-related predictor taps ranging from one to four, while on the right-hand side the number of predictor taps was between eight and 32. Apart from noticing the M-SIC detector's more effective operation compared to MMSE detection, we observe that upon increasing the number of predictor taps to a sufficiently high value, the system's BER is within a fraction of a dB in comparison to that exhibited by a system capitalising on perfect channel knowledge. By contrast, for the MMSE detector of Section 17.2.3 – even in conjunction with the predictor spanning the highest $N_{tap}^{[t]}$ number of CIR-related taps – a significant performance difference is observed in comparison to the perfect-reference scenario. This phenomenon is attributed to the imperfect error-contaminated remodulated reference.

---

[15]The SIC-related soft-bits were generated with the aid of the simplified method of Section 17.3.1.3.1 rather than with the weighted soft-bit method of Section 17.3.1.3.2. Hence there is still some potential for a performance improvement.

K_0	number of significant CIR-related taps
K	number of subcarriers
L	number of simultaneous users
$N_{tap}^{[t]}$	number of CIR predictor taps
P	number of receiver antennas

Table 16.5: Summary of the parameters influencing the PIC-assisted DDCE's complexity.

Note that in the context of the simulations conducted in this section we have employed an "undecoded" reference, where the output signal of the combiner was sliced and remodulated. As a more complex design alternative, one could also employ a channel decoded reference, where the "source"-related soft-output bits of the turbo-decoder are sliced, re-encoded and remodulated. A further improvement can be achieved by generating the reference based on slicing the "source- plus parity"-related soft-output bits of the turbo-decoder.

16.5.3   Computational Complexity

In this section we will estimate the complexity of the proposed PIC-assisted DDCE scheme of Figure 16.4 in terms of the number of complex multiplications and additions required during the estimation phase. We will consider two specific cases. In the first scenario PIC was carried out in the frequency domain, which was based on the concept of Section 16.5, while in the second case PIC was invoked in the time domain, as proposed by Li [93]. The latter approach was outlined in Section 16.4.4. Note again that these schemes are equivalent in terms of their system equations and performance. Hence the decision, as to which domain to perform the PIC in, should be made on the basis of the computational complexity imposed.

In Table 16.5 we have once again summarised the system parameters, which will influence the achievable performance. Our further discourse is divided into two parts. While in Section 16.5.3.1 we will focus on the complexity imposed by the frequency domain *a posteriori* channel estimation, in Section 16.5.3.2 we will focus on the estimation of the complexity associated with the *a priori* channel estimation conducted in the CIR-related domain.

16.5.3.1   *A Posteriori* Channel Estimation Complexity

Following our discussions outlined in Section 16.5.1.1 an *a posteriori* channel transfer function estimate can be obtained for the j-th user and the p-th receiver antenna element during the n-th OFDM symbol period upon subtracting all the $(L-1)$ remaining users' estimated signal contributions from the frequency domain representation of the signal received by the p-th antenna element. The schematic of this arrangement was shown in Figure 16.4. According to Equation 16.106 this implies multiplying the *a priori* channel transfer factor estimates of the $(L-1)$ users with the users' associated sliced and remodulated subcarrier symbols, followed by the subtraction of these components from the received signal and by its normalisation with respect to the desired user's sliced and remodulated subcarrier symbols. Upon neglecting the specific properties of the modulated signal constellations employed, such as for example the constant modulus property of the various MPSK modulation schemes, the complexity imposed in terms of the number of complex multiplications and additions – normalised to the K number of subcarriers, L number of users and P number of receiver antenna elements is given by: [16]

$$C_{apt}^{(\mathbb{C}*\mathbb{C})}\big|_{\text{norm}} = 2 \tag{16.158}$$

$$C_{apt}^{(\mathbb{C}+\mathbb{C})}\big|_{\text{norm}} = L - 1, \tag{16.159}$$

where the computational complexity of the division-based normalisation has been accounted for as a complex multiplication. Note in this context that $1/\check{s}^{(j)}[n,k] = \check{s}^{(j)*}[n,k]/|\check{s}^{(j)}[n,k]|^2$, where $|\check{s}^{(j)}[n,k]|^2 = \sigma_j^2$ for constant-modulus MPSK modulation schemes.

16.5.3.2 *A Priori* Channel Estimation Complexity

Calculating the i-th user's *a priori* channel transfer factor estimates for the next OFDM symbol period involves transforming the user's current set of *a posteriori* channel transfer factor estimates to the CIR-related domain with the aid of the inverse unitary linear transform $\tilde{U}^{[f](i)H}$ according to Equation 16.108. In our specific case this unitary transform is implemented with the aid of the inverse DFT matrix $\mathbf{W}^H$. Instead of directly multiplying the vector of *a posteriori* channel transfer factor estimates with the inverse DFT matrix $\mathbf{W}^H$ according to Equation 16.108, we rather employ here the IFFT. Employing the IFFT requires a $\frac{K}{2}\log_2 K$ number of complex multiplications and twice the number of additions. Hence, the corresponding normalised complexity contribution in the sense of Equations 16.158 and 16.159 is given by:

$$C_{\text{IFFT}}^{(\mathbb{C}*\mathbb{C})}\big|_{\text{norm}} = \frac{1}{2}\log_2 K \tag{16.160}$$

$$C_{\text{IFFT}}^{(\mathbb{C}+\mathbb{C})}\big|_{\text{norm}} = \log_2 K. \tag{16.161}$$

Furthermore, filtering each of the K_0 number of most significant CIR-related taps along the time direction according to Equation 16.109 using an N-tap prediction filter results in a normalised complexity contribution of:

$$C_{apr}^{(\mathbb{C}*\mathbb{C})}\big|_{\text{norm}} = \frac{K_0}{K}N_{tap}^{[t]} \tag{16.162}$$

$$C_{apr}^{(\mathbb{C}+\mathbb{C})}\big|_{\text{norm}} = \frac{K_0}{K}N_{tap}^{[t]}. \tag{16.163}$$

Upon invoking Equation 16.110, the predicted CIR-related taps are transformed back to the frequency domain, where again, for the specific case of $\tilde{U}^{[f](i)} = \mathbf{W}$ the FFT is employed. The FFT requires the same number of operations, as the IFFT, namely those given by Equations 16.160 and 16.161.

Hence, the total number of complex multiplications and additions, normalised to the product of the K number of subcarriers, L number of users and P number of transmit antennas is

---

[16]For reasons of space economy in the rest of this section we refrain from spelling out always that normalisation was carried out with respect to K, L and P. For the sake of brevity we will simply refer to "normalised" complexity.

	$K_0 = 8/K = 64$	$K_0 = 8/K = 512$	$K_0 = 64/K = 512$
$C_{\text{MU-CE,PIC}}^{(\mathbb{C}*\mathbb{C})}\big|_{\text{norm}}$	8.13	11.02	11.13
$C_{\text{MU-CE,PIC}}^{(\mathbb{C}+\mathbb{C})}\big|_{\text{norm}}$	15.13	21.02	21.13

Table 16.6: Computational complexity of the frequency domain PIC-assisted DDCE of Figure 16.4 in terms of the number of complex multiplications and additions normalised to the K number of subcarriers, L number of users and P number of receiver antennas. Here we have assumed that $L = P = 4$ and furthermore we had $N_{tap}^{[t]} = 1$.

given by:

$$C_{\text{MU-CE,PIC}}^{(\mathbb{C}*\mathbb{C})}\big|_{\text{norm}} = 2 + \log_2 K + \frac{K_0}{K} N_{tap}^{[t]} \qquad (16.164)$$

$$C_{\text{MU-CE,PIC}}^{(\mathbb{C}+\mathbb{C})}\big|_{\text{norm}} = L - 1 + 2\log_2 K + \frac{K_0}{K} N_{tap}^{[t]}. \qquad (16.165)$$

We conclude that in most scenarios the computational complexity imposed will be dominated by the contribution associated with performing the IFFT and FFT operations. In order to provide an illustrative example, we have evaluated the computational complexity of the frequency domain (FD) PIC-assisted DDCE of Figure 16.4 again, for the standard configurations as employed in the context of our rudimentary complexity analysis of Li's original approach and for the simplified, time domain (TD)-PIC-assisted DDCE in Tables 16.3 and 16.4. Specifically, compared to the TD-PIC-assisted DDCE's complexity summarised in Table 16.4, we observe that the FD-PIC's complexity is lower in terms of the number of complex multiplications and additions required. While this applies to the number of complex multiplications regardless of the specific choice of the system parameters, the number of complex additions imposed has to be considered in more detail. Upon comparing Equations 16.99 and 16.165, which reflect the complexity of the TD-PIC and FD-PIC in terms of the number of complex additions, we conclude that if we have $K \geq 2^{2(L-1)}$, then the TD-PIC is more complex for any choice of K_0. By contrast, for the case of $K < 2^{2(L-1)}$, it depends on the specific choice of K_0 whether the TD- or the FD-PIC is more complex. It can readily be shown that the "complexity cross-over point" is given by $\acute{K}_0 = \sqrt{K(1 - \frac{1}{2(L-1)} \log_2 K)}$. More explicitly, for $K_0 < \acute{K}_0$ the TD-PIC is less complex, while for $K_0 > \acute{K}_0$ the FD-PIC is less complex. Note that in formulating these considerations we have assumed in Equation 16.99 that we have $\frac{1}{L} \approx 0$ and furthermore that the term of $\frac{K_0}{K} N_{tap}^{[t]}$ was neglected in Equation 16.165. In order to provide an example, for $L = 8$ and $K = 64$ we have $\acute{K}_0 = 6.04$, while for $K = 512$ we obtain $\acute{K}_0 = 13.52$.

16.5.4   Summary and Conclusions

In summary, in Section 16.5 we have discussed frequency domain PIC-assisted DDCE employed in the context of multi-user OFDM systems- or phrased in more general terms, used in OFDM systems relying on multiple transmit antennas. The outline of this section is as follows. In Section 16.5.4.1 a summary and conclusion will be provided for the specific structure of the PIC-assisted DDCE. Furthermore, in Section 16.5.4.2 our summary and conclusion will be offered for the performance assessment of the PIC-assisted DDCE. Finally, in Section 16.5.4.3 the results of our complexity analysis conducted with respect to the PIC-assisted DDCE will be summarised.

16.5.4.1 Summary and Conclusions on the PIC-Assisted DDCE's Structure

After an introduction of the vectors of *a priori* and *a posteriori* channel transfer factor estimates and their associated expressions in Section 16.5.1.1, the employment of *a priori* CIR-related tap prediction filtering was discussed in Section 16.5.1.2 for the sake of obtaining improved channel transfer factor estimates for the next OFDM symbol period. On the basis of the associated system equations an expression, namely, Equation 16.113 was derived for the vector of different subcarriers' *a priori* channel transfer factor estimation errors associated with the j-th user and p-th receiver antenna element during the current OFDM symbol period as a function of the vectors of *a priori* channel transfer factor estimation errors of the remaining $(L-1)$ users during the past $N_{tap}^{[t]}$ number of OFDM symbol periods, where $N_{tap}^{[t]}$ denotes the predictor's range. This expression was then employed to derive the j-th user's *a priori* channel transfer factor estimation error correlation matrix of Equation 16.117. Furthermore, the j-th user's average channel transfer factor- or simply channel estimation MSE given by Equation 16.119 was derived, again, as a function of the remaining users' corresponding magnitudes associated with the past $N_{tap}^{[t]}$ number of OFDM symbol periods. A similar expression, namely, Equation 16.124 was also derived for the j-th user's *a posteriori* channel estimation MSE. Our discussions continued in Section 16.5.1.5 upon further considering the channel estimator's operation in the steady-state condition, implying that the estimator's MSE has reached its steady-state value and hence it was time-invariant for the consecutive OFDM symbols. Upon employing a more compact matrix notation a closed-form solution, namely Equation 16.129 was derived for the vector of the different users' *a priori* channel estimation MSEs as a function of the different users' transmit powers as well as that of the AWGN noise variance, that of the CIR-related predictor coefficients, that of the channel statistics expressed in form of the channels' spaced-time spaced-frequency correlation functions, and finally that of the specific unitary transforms employed. The criteria to be satisfied for the existence of the equation's solution were provided, which also ensure the system's stability. Furthermore, in Section 16.5.1.6 an expression, namely Equation 16.132 was derived, for the j-th user's vector of CIR-related predictor coefficients as a function of the $(L-1)$ remaining users' *a priori* channel estimation MSEs. This equation was then employed in conjunction with Equation 16.126 in the context of a fixed-point iteration based approach for jointly optimising the expected *a priori* channel estimation MSE and the predictor coefficients, following a similar strategy to that proposed by Rashid-Farrokhi *et al.* [485] to jointly optimise the transmit power allocation and base station antenna array weights in wireless networks. Furthermore, in Section 16.5.1.6.1 simplified expressions were presented for the *a priori* channel estimation MSE, the stability condition and for the vector of optimum predictor coefficients, namely, which were given by Equations 16.133, 16.134 and 16.135, respectively. These equations were valid in the context of a scenario of identical transmit powers- and channel statistics associated with the different users. For this specific scenario, while employing single-tap prediction filtering a closed form solution was presented in Section 16.5.1.6.2 for the predictor's optimum coefficient, which was given by Equation 16.137. Finally, in Section 16.5.1.7

various strategies were discussed for providing estimates of the channel statistics, which constitute the prerequisites for obtaining the optimum predictor coefficients, as was highlighted in Section 16.5.1.6.

16.5.4.2 Summary and Conclusions on the Performance Assessment of the PIC-Assisted DDCE

The further structure of this section is as follows. The PIC-assisted DDCE's performance in the context of sample-spaced- as well as non-sample-spaced CIRs and error-free symbol decisions is summarised in Sections 16.5.4.2.1 and 16.5.4.2.2, respectively. In contrast, in Section 16.5.4.2.3 our findings for the scenario of a sample-spaced CIR and imperfect, potentially error-contaminated symbol decisions are summarised.

16.5.4.2.1 Performance of the PIC-Assisted DDCE in the Context of Sample-Spaced CIRs and Error-Free Symbol Decisions

Our performance assessment of the various techniques studied commenced with Figure 16.5, portraying the *a priori* channel estimation MSE's evolution as a function of the CIR-related tap predictor coefficients in the context of two-tap CIR prediction filtering. Figure 16.5 highlighted the estimator's sensitivity with respect to the choice of the predictor coefficients. Our further investigations in Section 16.5.2.2.1 then concentrated on the average *a priori* channel estimation MSE in the idealistic scenario of a sample-spaced CIR and error-free symbol decisions, which provided us with a useful benchmark. Specifically the investigations conducted in Sections 16.5.2.2.1 and 16.5.2.2.2 demonstrated that without exploiting the system's recursive structure characterised by our derivations presented in Sections 16.5.1.3, 16.5.1.4 and 16.5.1.5 the system may potentially become unstable as observed in Figures 16.6 and 16.7. This is the case, for example, when calculating the predictor coefficients associated with a specific user with the aid of the conventional Wiener solution, while neglecting the remaining users' *a priori* channel estimation MSEs. Our further investigations in Sections 16.5.2.2.1, 16.5.2.2.2 as well as 16.5.2.2.3, 16.5.2.2.4 and 16.5.2.2.5, respectively, which were associated with Figures 16.6, 16.7 as well as 16.8, 16.9 and 16.10, respectively, concentrated on portraying the influence of the various system parameters on the *a priori* channel estimation MSE. More explicitly, we studied the effects of the $N_{tap}^{[t]}$ number of predictor coefficients, that of the K_0 number of significant CIR-related taps, that of the L number of simultaneous users, that of the impact of the OFDM symbol normalised Doppler frequency denoted by F_D, and finally, that of the potential mismatch between the channel statistics assumed in the derivation of the predictor coefficients and that of the channel encountered. Specifically, the investigations conducted in Section 16.5.2.2.5 with respect to the mismatch of the channel statistics supported the arguments of Li *et al.* [68] that if an ideally support-limited Doppler power spectrum having a maximum OFDM symbol normalised Doppler frequency of $\tilde{F}_D$ is assumed in the calculation of the predictor coefficients, then for channels obeying $F_D \leq \tilde{F}_D$ no further *a priori* channel estimation MSE performance degradation is observed. Figure 16.10 demonstrated that this is only true in the strict sense, when increasing the CIR-related tap predictor's order towards infinity. In more practical terms, however, a number of 64 predictor taps appeared to be sufficient for rendering the evolution of the MSE almost flat as a function of the channel's OFDM symbol normalised Doppler frequency. On the other hand, a moderate number of predictor taps, which was as low as four, was found sufficient for exploiting much of the channel transfer function's correlation in the time direction. Furthermore, we observed that if the difference between the target OFDM symbol normalised Doppler frequency for which the predictor coefficients were designed, and that of the channel encountered is too high, then the channel's correlation cannot be optimally exploited. The conclusion that transpired was that of rendering the channel transfer function estimator adaptive, which will be the topic of our discussions in Section 16.6. Our investigations conducted in the context of a sample-spaced CIR and error-free symbol decisions were concluded in Section 16.5.2.2.6 by comparing the *a priori* channel estimation MSE performance of PIC-assisted DDCE to that of the LS-assisted DDCE proposed by Li *et al.* [91], which was characterised in Section 16.4. As shown in Figure 16.11 the general tendency is that for lower SNRs, namely for SNR values up to about 15dB, the PIC-assisted DDCE outperforms the LS-assisted DDCE, while for higher SNRs the opposite trend is observed. The reason is that the LS-assisted DDCE suffers from the MUI imposed due to the imperfect cross-correlation properties of the different users' transmitted subcarrier symbol sequences. By contrast, the PIC-assisted DDCE of Figure 16.4 suffers from MUI due to the imperfections of the *a priori* channel estimates imposed on the PIC process, which becomes more obvious for higher SNRs. It was observed furthermore in conjunction with the LS-assisted DDCE of Figure 16.3 that when the L number of simultaneous SDMA users approaches the maximum tolerable number of users given by the ratio $\frac{K}{K_0}$, as outlined in Section 16.4.1.2.3, then the *a priori* channel estimation MSE is further degraded. This was explained by the increased MUI due to supporting more users, but also due to the numerical imperfections associated with a potentially excessive condition number of the subcarrier symbol sequences' auto-correlation matrix $\mathbf{Q}[n]$, which was defined earlier in Equation 16.20. These imperfections were mitigated for example with the aid of the QR matrix factorisation based approach outlined in Section 16.4.1.2.4.

16.5.4.2.2 Performance of the PIC-Assisted DDCE in the Context of Non-Sample-Spaced CIRs and Error-Free Symbol Decisions

In order to render our investigations more realistic, in Section 16.5.2.3 we considered the estimator's MSE performance in the context of non-sample-spaced CIRs. More specifically, we introduced three multipath intensity profiles, namely the sparse-, the uniform- and the exponential multipath intensity profile. Their associated spaced-frequency channel correlation functions were reviewed in Sections 16.5.2.3.1, 16.5.2.3.2 and 16.5.2.3.3, respectively. Further investigations were conducted with respect to the leakage effects incurred upon decomposing the associated spaced-frequency correlation matrices with respect to the DFT matrix $\mathbf{W}$. Specifically, in case of a sparse multipath intensity profile, using a single tap in the most basic scenario, it was found in Figure 16.12 that the leakage was maximal for a fractional delay of 0.5. The effects of leakage observed when using the uniform- and exponential multipath intensity profiles were portrayed in Figure 16.13. Our more specific investigations of the *a priori* channel estimation MSE exhibited by the PIC-assisted DDCE in the context of non-sample-spaced CIRs focussed further on the specific influence of the K_0 number of significant CIR-related taps. The corresponding simulation results were presented in Figure 16.14 for a single-tap- and for a uniform multipath intensity profile. In both scenarios the estimation MSE exhibited the same tendencies, namely that of having an MSE floor between -20- and -25dB, even for an optimum choice of the number K_0 of significant CIR-related taps. This was attributed to the leakage effects associated with employing the DFT matrix $\mathbf{W}$ instead of the optimum KLT matrix for transforming the *a posteriori* channel transfer factor estimates to the CIR-related domain in the sense of Equation 16.108. Upon further increasing the number of significant CIR-related taps K_0 beyond its optimum point, the *a priori* channel estimation MSE was degraded. This was related to employing the same set of tap predictor coefficients in the different CIR-related taps and hence the AWGN was not optimally suppressed. Our more detailed investigations conducted in Section 16.5.2.3.5, which were quantified in Figure 16.15, focussed on the distribution of the *a priori* channel transfer factor estimation MSE across the various subcarriers of the OFDM symbol. For evaluating the MSE distribution an iterative algorithm following the philosophy of that in Section 16.5.1.6 was devised. We observed in Figure 16.15 that the *a priori* channel transfer factor estimation MSEs at the boundaries of the OFDM symbol interpreted in the sense of the DFT-index based notation or at the centre of the OFDM symbol with respect to the frequency may potentially become excessive, again, as a result of the effects of leakage. Hence, in order to optimise the entire system's performance, the value of the *a priori* channel estimation MSE to be expected in the different subcarriers should be passed on to the combining- or detection stage.

16.5.4.2.3 Performance of the PIC-Assisted DDCE in the Context of Sample-Spaced CIRs and Imperfect, Error-Contaminated Symbol Decisions

Our performance assessment was concluded in Section 16.5.2.4 upon considering sample-spaced CIRs and potentially erroneous symbol decisions, as encountered in practical physical implementations. While a qualitative assessment of the associated effects was provided in Section 16.5.2.4.1, the corresponding MSE- and BER simulation results recorded for the uncoded scenario were portrayed in Figure 16.16 of Section 16.5.2.4.2. Two MUD detection schemes, namely the MMSE and SIC detection schemes of Sections 17.2.3 and 17.3.1 were compared against each other. It was found on the basis of the MSE curves of Figure 16.16 that the impact of erroneous symbol decisions was distinctively visible. From the further evolution of the different MMSE detection-related performance curves of Figure 16.16 it can be concluded that a relatively powerful MUD scheme, such as SIC or M-SIC is necessary for avoiding the PIC-assisted DDCE's instability induced as a result of erroneous symbol decisions. Upon employing a further enhanced turbo-decoding assisted system, the associated BER performance curves provided in Figure 16.17, suggested that the entire system's BER performance was significantly enhanced. More specifically, it was observed in Figure 16.17 that in the context of SIC assisted detection, while employing CIR predictor lengths in excess of four taps, the system's BER performance may approach that exhibited in the context of perfect channel transfer function knowledge.

16.5.4.3 Summary and Conclusions on the PIC-Assisted DDCE's Computational Complexity

Our discussions in Section 16.5 were concluded by a complexity study. It was argued on the basis of the associated complexity under which conditions should the PIC be performed in the CIR-related domain and in the frequency domain, respectively, while noting that both schemes exhibit the same MSE performance. As our comparisons revealed earlier in Section 16.4, the PIC-assisted DDCE – regardless of whether the PIC is performed in the time or in the frequency domain – is significantly less complex, than the LS-assisted DDCE, particularly for a high number of significant CIR-related taps.

The advantages in favour of PIC-assisted DDCE compared to LS-assisted DDCE, namely that of a significantly reduced complexity and the potential support of a higher number of simultaneous users or transmit antennas have to be considered in the light of the disadvantage of requiring a less straightforward procedure for evaluating the CIR-related tap predictor coefficients, while also suffering from a slight performance disadvantage at higher SNRs. Furthermore, common to both approaches was the MSE floor observed at higher SNRs in the context of non-sample-spaced CIRs, which was attributed to the imperfections of the DFT matrix $\mathbf{W}$ in the context of transforming the *a posteriori* channel transfer factor estimates to the CIR-related domain. From these observations we conclude that an improved channel transfer function estimation scheme should perform all filtering operations in the frequency domain, and thus potentially avoiding the effects of transform-related leakage. Furthermore the filtering process should be adaptive with respect to the filters' coefficients. Alternatively, the previous design could be improved upon employing an adaptive unitary transform – such as the KLT – which takes into account the channel's statistics, and again to render the estimator adaptive with respect to the CIR-related tap prediction coefficients. In the next section we will focus on rendering the PIC-assisted DDCE of Figure 16.4 adaptive with respect to the channel's statistics, which is achieved with the aid of employing the Recursive Least Squares (RLS) algorithm.

16.6 RLS-Adaptive Parallel Interference Cancellation-Assisted Decision-Directed Channel Estimation

In an effort to further improve the PIC-assisted DDCE's MSE performance under time-varying conditions with respect to the channel's specific multipath intensity profile and Doppler power spectrum, a viable approach is to adaptively adjust the CIR-related tap predictors' coefficients. A number of rudimentary approaches for rendering the predictor adaptive were outlined in Section 16.5.1.7. A common feature of these techniques was that the adaptation was assumed to be performed on a training block-by-block basis, possibly during training OFDM symbol periods. However, from a computational perspective it might be more advantageous to perform the adaptation on an OFDM symbol-by-symbol basis, once the new *a posteriori* channel transfer factor samples or CIR-related tap estimates become available. Furthermore, it was assumed in Section 16.5.1.2 that in order to simplify the task of determining the predictor coefficients, the same set of predictor coefficients would be employed for each of the K_0 number of different CIR-related taps, regardless of the specific tap's variance. This followed the philosophy of "robust" channel transfer function estimation, which was originally proposed by Li *et al.* [68] with respect to the specific shape of the channel's associated multipath intensity profile. This robustness was achieved at the cost of a sub-optimum MSE performance. As a result, the predictor's complexity was found to be relatively low, which was attributed to the off-line optimisation of the predictor coefficients.

An approach for providing block-adaptivity for the CIR-related tap predictor's coefficients was outlined in Section 15.2.5.2 in form of the Burg algorithms-assisted predictor proposed by Al-Susa and Ormondroyd [72] for single-user OFDM systems. Although the Burg algorithm, which has been initially proposed in the context of speech processing [85] is known to achieve a low MSE due to its strategy of simultaneously minimising both the forward and backward prediction errors, its main disadvantage follows from the necessity of storing a potentially large number of past OFDM symbols' channel transfer factor estimates beyond the prediction filter's range. This can be avoided by performing the CIR-related tap prediction coefficient adaptation on an OFDM symbol by symbol. Here we opted for employing the RLS algorithm [90], which is known to converge relatively rapidly.

The further structure of Section 16.6 is as follows. In Section 16.6.1 the application of the RLS algorithm to the problem of CIR-related tap prediction is discussed, including the assessment of the *a priori* channel estimation MSE and the estimation of the computational complexity imposed. Without any modifications, we then employ the RLS-based CIR-related tap predictor in the context of the PIC-assisted DDCE for multi-user OFDM communications, which is the topic of Section 16.6.2. We also characterise the BER and MSE performance of the RLS-based PIC-assisted DDCE in the context of sample-spaced CIRs. Our conclusion will then be offered in Section 16.6.3.

16.6.1   Single-User RLS-Adaptive CIR-Related Tap Prediction

As an introduction, we will consider in this section the application of RLS-adaptive CIR-related tap prediction filtering in the context of a single-user scenario. While in Section 16.6.1.1 the standard RLS algorithm is reviewed with respect to its potential application in CIR-related tap prediction filtering, a simplified scheme based on ensemble averaging is outlined in Section 16.6.1.2. Furthermore, in Section 16.6.1.3 a rudimentary MSE performance assessment is provided concerning the influence of the RLS-specific forgetting factor α_{RLS} and of the OFDM symbol normalised Doppler frequency F_D on the *a priori* channel estimation MSE. Finally, in Section 16.6.1.4 we will evaluate the complexity imposed by the RLS-assisted "on-line" adaptation of the CIR-related tap predictor coefficients.

16.6.1.1   Review of the RLS Algorithm

For the single-user scenario we recall from Section 15.2.4.4, more specifically from Equation 15.38, that the l-th CIR-related tap's vector $\tilde{\mathbf{c}}_{pre}[n, l]|_{opt} \in \mathbb{C}^{N_{tap}^{[t]} \times 1}$ of optimum predictor coefficients[17] is determined by the Wiener equation [90], namely:

$$\tilde{\mathbf{c}}_{pre}[n, l]|_{opt} = \tilde{\mathbf{R}}_{apt}^{[t]-1}[n, l]\tilde{\mathbf{r}}_{apt}^{[t]}[n, l], \qquad (16.166)$$

where $\tilde{\mathbf{R}}_{apt}^{[t]}[n, l] \in \mathbb{C}^{N_{tap}^{[t]} \times N_{tap}^{[t]}}$ is the l-th CIR-related tap's estimated auto-correlation matrix and $\tilde{\mathbf{r}}_{apt}^{[t]}[n, l] \in \mathbb{C}^{N_{tap}^{[t]} \times 1}$ is the estimated cross-correlation vector, both of which are valid for the n-th OFDM symbol period. The estimate $\tilde{\mathbf{R}}_{apt}^{[t]}[n, l]$ for the n-th OFDM symbol period could be obtained on the basis of the estimate $\tilde{\mathbf{R}}_{apt}^{[t]}[n - 1, l]$ associated with the $(n - 1)$-th OFDM symbol period by evaluating [90]:

$$\tilde{\mathbf{R}}_{apt}^{[t]}[n, l] = \alpha_{\text{RLS}}\tilde{\mathbf{R}}_{apt}^{[t]}[n - 1, l] + (1 - \alpha_{\text{RLS}})\tilde{\mathbf{h}}_{apt}[n - 1, l]\tilde{\mathbf{h}}_{apt}^{H}[n - 1, l], \qquad (16.167)$$

where $\tilde{\mathbf{h}}_{apt}[n - 1, l]$ is defined in correspondence with Equation 15.23 as the vector of $N_{tap}^{[t]}$ number of past CIR-related tap estimates starting with the tap index $(n - 1)$. Furthermore, the "update-term" of Equation 16.167 is identical to that of Equation 15.25 upon removing

---

[17]Note that in contrast to our simplified analysis of PIC-assisted DDCE in Section 16.5 different CIR-related taps are potentially associated with individual vectors of predictor coefficients.

the expectation operator. Furthermore, in Equation 16.167 the variable $\alpha_{\text{RLS}} \in \mathbb{R}$ denotes the so-called forgetting factor [90]. Similarly, the estimate $\tilde{\mathbf{r}}_{apt}^{[t]}[n, l]$ for the n-th OFDM symbol period can be obtained following the philosophy of Equation 16.167, yielding [90]:

$$\tilde{\mathbf{r}}_{apt}^{[t]}[n, l] = \alpha_{\text{RLS}}\tilde{\mathbf{r}}^{[t]}[n-1, l] + (1 - \alpha_{\text{RLS}})\tilde{h}_{apt}^{*}[n, l]\tilde{\mathbf{h}}_{apt}[n-1, l], \tag{16.168}$$

where again, the "update-term" of Equation 16.168 is identical to that of Equation 15.28 upon removing the expectation operator. Instead of explicitly inverting the estimated auto-correlation matrix $\tilde{\mathbf{R}}_{apt}^{[t]}[n]$ associated with the n-th OFDM symbol period, an iterative update strategy based on the matrix inversion lemma – also known as the Sherman-Morrison formula – or Woodbury's identity [90] can be invoked, which is known from the literature as the RLS algorithm [90]. In the context of our specific CIR-related tap prediction problem the RLS-algorithm is summarised below. Specifically, the so-called Kalman gain vector $\mathbf{k}[n, l] \in \mathbb{C}^{N_{tap}^{[t]} \times 1}$ for the n-th OFDM symbol period is given by [90]:

$$\mathbf{k}[n, l] = \frac{(1 - \alpha_{\text{RLS}})\tilde{\mathbf{R}}_{apt}^{[t]-1}[n-1, l]\tilde{\mathbf{h}}_{apt}[n-1, l]}{\alpha_{\text{RLS}} + (1 - \alpha_{\text{RLS}})\tilde{\mathbf{h}}_{apt}^{H}[n-1, l]\tilde{\mathbf{R}}_{apt}^{[t]-1}[n-1]\tilde{\mathbf{h}}_{apt}[n-1, l]}, \tag{16.169}$$

which is then employed in the process of updating the inverse of the CIR-related taps' auto-correlation matrix, namely [90]:

$$\tilde{\mathbf{R}}_{apt}^{[t]-1}[n, l] = \frac{1}{\alpha_{\text{RLS}}} \left[\tilde{\mathbf{R}}_{apt}^{[t]-1}[n-1, l] - \mathbf{k}[n, l]\tilde{\mathbf{h}}_{apt}^{H}[n-1, l]\tilde{\mathbf{R}}_{apt}^{[t]-1}[n-1, l] \right]. \tag{16.170}$$

Furthermore, the CIR-related tap predictor coefficient vector for the n-th OFDM symbol period is given by [90]:

$$\tilde{\mathbf{c}}_{pre}[n, l]|_{opt} = \tilde{\mathbf{c}}_{pre}[n-1, l]|_{opt} + \mathbf{k}[n, l] \left[\tilde{h}_{apt}[n, l] - \tilde{\mathbf{c}}_{pre}^{H}[n-1, l]\tilde{\mathbf{h}}_{apt}[n-1, l] \right]^{*}, \tag{16.171}$$

where the term in brackets denotes the prediction error associated with the n-th OFDM symbol period. A standard approach for initialising the RLS algorithm [90] is that of assuming an inverse correlation matrix having a diagonal shape defined as:

$$\tilde{\mathbf{R}}_{apt}^{[t]-1}[0, l] = \frac{1}{\varepsilon_{\text{RLS},0}}\mathbf{I}, \tag{16.172}$$

where the specific choice of $\varepsilon_{\text{RLS},0} \in \mathbb{R}$ is less critical in our application, than the specific value of the forgetting-factor α_{RLS}. A plausible choice for $\tilde{\mathbf{c}}_{pre}[0, l]$ is for example $(1, 0, \ldots, 0)^{T}$, which corresponds to the case of zero-forcing based one-tap prediction.

16.6.1.2 Potential Simplification by Ensemble Averaging

According to the formulae presented in the previous section, the RLS adaptation has to be performed separately for each of the K_0 number of significant CIR-related taps. A simplification in terms of the computational complexity can potentially be achieved upon invoking the concepts of "robustness" with respect to the channel's specific spaced-frequency correlation function, as proposed by Li *et al.* [68]. These were discussed in Section 15.2.3 in the context of a single-user OFDM scenario. An implication was that the same set of CIR-related tap predictor coefficients was invoked in the process of predicting the different CIR-related tap values - at the cost of an MSE performance degradation. More specifically, in each OFDM symbol period the auto-correlation matrix of the CIR-related *a posteriori* tap estimates is updated according to Equation 16.167 but upon substituting the tap-specific update term by its ensemble average:

$$\tilde{\mathbf{R}}_{apt}^{[t]}[n] = \alpha_{\text{RLS}}\tilde{\mathbf{R}}_{apt}^{[t]}[n-1] + (1 - \alpha_{\text{RLS}})\frac{1}{K_0}\sum_{l \in S_{tap}} \tilde{\mathbf{h}}_{apt}[n-1,l]\tilde{\mathbf{h}}_{apt}^{H}[n-1,l], \quad (16.173)$$

where S_{tap} denotes the set of indices associated with the K_0 number of significant CIR-related taps. Following the same concepts instead of Equation 16.168 we obtain the following cross-correlation vector update expression:

$$\tilde{\mathbf{r}}_{apt}^{[t]}[n] = \alpha_{\text{RLS}}\tilde{\mathbf{r}}^{[t]}[n-1] + (1 - \alpha_{\text{RLS}})\frac{1}{K_0}\sum_{l \in S_{tap}} \tilde{h}_{apt}^{*}[n,l]\tilde{\mathbf{h}}_{apt}[n-1,l]. \quad (16.174)$$

Note that as a result of the ensemble averaging, as seen in Equation 16.173 the innovation-related term of the auto-correlation matrix $\tilde{\mathbf{R}}_{apt}^{[t]}[n]$ may exhibit a rank, which is potentially higher than unity. This precludes the application of the Sherman-Morrison formula [90]. Consequently, the vector of CIR-related tap predictor coefficients has to be evaluated with the aid of Equation 16.166, namely:

$$\tilde{\mathbf{c}}_{pre}[n]|_{opt} = \tilde{\mathbf{R}}_{apt}^{[t]-1}[n]\tilde{\mathbf{r}}_{apt}^{[t]}[n]. \quad (16.175)$$

Note that in terms of the computational complexity imposed the direct matrix inversion based solution of Equation 16.175 is proportional to the cube of the $N_{tap}^{[t]}$ number of predictor taps.

16.6.1.3 MSE Performance Assessment

In order to demonstrate the applicability of the RLS algorithm to the problem of CIR-related tap prediction in the context of DDCE we have portrayed in Figure 16.18 the evolution of the *a priori* channel estimation MSE versus the OFDM symbol index for an arbitrary time segment commencing with an initial vector of prediction coefficients given by $\tilde{\mathbf{c}}_{pre}[0,l] = (1,0,\dots,0)^T$, where $l = 0,\dots,K_0-1$. Here we have employed the sample-spaced indoor WATM channel model of Section 14.3.1, where the highest CIR tap delay is given by $11T_s$. Hence, the number of significant CIR-related taps was chosen as $K_0 = 12$. Furthermore, the CIR-related tap predictor's range was equal to $N_{tap}^{[t]} = 4$. Note that for different time segments the specific MSE evolution is potentially different from that of Figure 16.18, but obeys the same general trend. Here we have investigated the influence of the Kalman forgetting factor α_{RLS} and of the OFDM symbol normalised Doppler frequency on the *a priori* channel estimation MSE performance. Specifically, on the left-hand side of Figure 16.18 the OFDM symbol normalised Doppler frequency was set to $F_D = 0.007$, while the forgetting factor α_{RLS} was varied. We observe in Figure 16.18 that for lower values of α_{RLS}

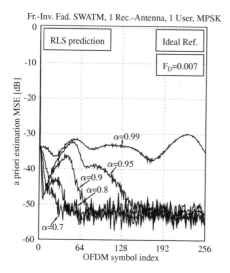
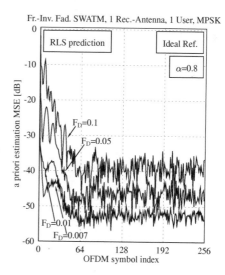

Figure 16.18: Evolution of the *a priori* channel estimation MSE observed with the aid of the RLS prediction-assisted DDCE in the single-reception antenna based single-user scenario for a specific time segment associated with the sample-spaced indoor WATM channel of Section 14.3.1, as a function of the OFDM symbol index; **(a: Left)** parametrised with the forgetting factor $\alpha = \alpha_{\text{RLS}}$, for a fixed OFDM symbol normalised Doppler frequency of $F_D = 0.007$; **(b: Right)** parametrised with the OFDM symbol normalised Doppler frequency F_D and for a fixed forgetting factor of $\alpha_{\text{RLS}} = 0.8$; in both cases the RLS predictor's start-up constant was $\varepsilon_{\text{RLS},0} = 0.1$, the $N_{tap}^{[t]}$ number of CIR predictor taps was equal to four and the SNR at the reception antennas was assumed to be 40dB; furthermore the number of significant CIR-related taps was $K_0 = 12$ and the number of subcarriers was $K = 512$.

a faster adaptation is achieved, while the residual error after adaptation is potentially higher, than that achieved with the aid of a forgetting factor of a higher value, although the latter effect is not explicitly visible in Figure 16.18 due to the limited time span. By contrast, on the right-hand side of Figure 16.18 we have plotted the *a priori* channel estimation MSE for various OFDM symbol normalised Doppler frequencies, F_D, while keeping the forgetting factor α_{RLS} constant. As expected, the *a priori* estimation MSE is increased in scenarios having a higher OFDM symbol normalised Doppler frequency, while the speed of adaptation was almost identical for the different scenarios. Note that for values of α_{RLS} that are significantly lower than those employed in Figure 16.18, the RLS predictor may potentially become unstable. Specifically in our experiments a value of $\alpha_{\text{RLS}} = 0.77$ was just acceptable, and yielded the highest speed of convergence while at the same time potentially the largest residual error after adaptation.

Curves similar to those seen in Figure 16.18 can also be generated for the lower-complexity ensemble averaging assisted adaptive DDCE, which was outlined in Sec-

	$\mathbf{k}[n,l]$	$\tilde{\mathbf{R}}_{apt}^{[t]-1}[n,l]$	$\tilde{\mathbf{c}}_{pre}[n,l]$	Σ
$C_{\mathrm{RLS},tap}^{(\mathbb{C}*\mathbb{C})}$	$N_{tap}^{[t]2} + N_{tap}^{[t]}$	$2N_{tap}^{[t]2}$	$N_{tap}^{[t]}$	$3N_{tap}^{[t]2} + 2N_{tap}^{[t]}$
$C_{\mathrm{RLS},tap}^{(\mathbb{C}+\mathbb{C})}$	$N_{tap}^{[t]2} + N_{tap}^{[t]}$	$N_{tap}^{[t]2} + N_{tap}^{[t]}$	$2N_{tap}^{[t]}$	$2N_{tap}^{[t]2} + 4N_{tap}^{[t]}$
$C_{\mathrm{RLS},tap}^{(\mathbb{R}*\mathbb{C})}$	$N_{tap}^{[t]}$	$N_{tap}^{[t]2}$	$N_{tap}^{[t]}$	$N_{tap}^{[t]2} + 2N_{tap}^{[t]}$

Table 16.7: Computational complexity per CIR-related tap in terms of the $C_{\mathrm{RLS},tap}^{(\mathbb{C}*\mathbb{C})}$ number of complex multiplications, the $C_{\mathrm{RLS},tap}^{(\mathbb{C}+\mathbb{C})}$ number of complex additions and the $C_{\mathrm{RLS},tap}^{(\mathbb{R}*\mathbb{C})}$ number of "mixed" multiplications (real/complex) associated with the different components of the RLS-based adaptation of the vector of predictor coefficients.

	$\tilde{\mathbf{R}}_{apt}^{[t]}[n]$	$\tilde{\mathbf{r}}_{apt}[n]$	$\tilde{\mathbf{c}}_{pre}[n]$	Σ
$C_{\mathrm{RLS}}^{(\mathbb{C}*\mathbb{C})}$	$K_0 N_{tap}^{[t]2}$	$K_0 N_{tap}^{[t]}$	$\frac{1}{3}N_{tap}^{[t]3}$	$\frac{1}{3}N_{tap}^{[t]3} + K_0 N_{tap}^{[t]2} + K_0 N_{tap}^{[t]}$
$C_{\mathrm{RLS}}^{(\mathbb{C}+\mathbb{C})}$	$K_0 N_{tap}^{[t]2}$	$K_0 N_{tap}^{[t]}$	$N_{tap}^{[t]2}$	$(K_0 + 1)N_{tap}^{[t]2} + K_0 N_{tap}^{[t]}$
$C_{\mathrm{RLS}}^{(\mathbb{R}*\mathbb{C})}$	$2N_{tap}^{[t]2}$	$2N_{tap}^{[t]}$	—	$2N_{tap}^{[t]2} + 2N_{tap}^{[t]}$

Table 16.8: Total computational complexity in terms of the $C_{\mathrm{RLS}}^{(\mathbb{C}*\mathbb{C})}$ number of complex multiplications, the $C_{\mathrm{RLS}}^{(\mathbb{C}+\mathbb{C})}$ number of complex additions and the $C_{\mathrm{RLS}}^{(\mathbb{R}*\mathbb{C})}$ number of "mixed" multiplications (real/complex) associated with the different components of the ensemble-averaging assisted RLS-based adaptation of the CIR-related taps' single vector of predictor coefficients.

tion 16.6.1.2. Our simulation results, which are not explicitly portrayed here for reasons of space revealed that the MSE recorded after adaptation is potentially higher than that of the RLS-assisted predictor, which adjusts the predictor coefficients on a CIR-related tap-by-tap basis. However, as will be demonstrated in Section 16.6.1.4, one advantage of the ensemble averaging assisted adaptive DDCE is its potentially reduced complexity.

16.6.1.4 Complexity Study

In Table 16.7 we summarised the implementational complexity associated with evaluating the Kalman gain vector $\mathbf{k}[n,l]$ according to Equation 16.169, that of updating the inverse auto-correlation matrix according to Equation 16.170 and that of updating the CIR-related tap predictor coefficient vector according to Equation 16.171. We observe that the complexity of this operation is proportional to the square of the $N_{tap}^{[t]}$ number of predictor coefficients. Note that the computational complexity could possibly be further reduced by exploiting that in the vector $\tilde{\mathbf{h}}_{apt}[n,l]$ associated with the n-th OFDM symbol period the last $(N_{tap}^{[t]} - 1)$ elements are constituted by the first $(N_{tap}^{[t]} - 1)$ elements of $\tilde{\mathbf{h}}_{apt}[n-1,l]$, while the first element is constituted by the estimated tap of $\tilde{h}_{apt}[n,l]$. For comparison we have summarised in Table 16.8 the computational complexity associated with updating the single vector of predictor coefficients based on the method outlined in Section 16.6.1.2, namely that of ensemble-averaging, rather than updating all the K_0 number of coefficients individually. Note that in Table 16.8 we listed the total complexity imposed by updating K_0 number of CIR-related tap predictor coefficient vectors, while in Table 16.7 we listed the complexity associated with the processing of a specific CIR-related tap. We emphasise that the standard RLS approach of Section 16.6.1.1 always imposes a higher complexity, than the ensemble-averaging assisted approach of Section 16.6.1.2 in terms of both the $C^{(\mathbb{C}+\mathbb{C})}$ number of complex additions and the $C^{(\mathbb{R}*\mathbb{C})}$ number of mixed multiplications.

By contrast, the $C^{(\mathbb{C}*\mathbb{C})}$ number of complex multiplications depends on the parameters K_0 and $N_{tap}^{[t]}$, which jointly determine, as to which of the above-mentioned two approaches exhibits a lower complexity. However, for a realistic choice of these parameters, for example, for $K_0 \geq 3$ and $N_{tap}^{[t]} = \{4, 8, 16\}$, the ensemble-averaging assisted approach of Section 16.6.1.2 is always of a lower complexity. Again, this complexity reduction is achieved at the cost of an MSE performance degradation.

While in Section 16.6.1 the application of the RLS algorithm was discussed in the context of prediction-assisted DDCE for employment in single-user OFDM systems, in Section 16.6.1 we will portray its benefits in the context of the PIC-assisted DDCE for multi-user OFDM systems.

16.6.2 RLS-Adaptive PIC-Assisted DDCE for Multi-User OFDM

After having reviewed the theory of RLS-adaptive CIR-related tap-by-tap prediction for single-user OFDM systems in Section 16.6.1.1 we will directly embark here on an assessment of RLS-adaptive prediction based PIC-assisted DDCE for multi-user OFDM, or more generally, for OFDM systems which support multiple transmit antennas. Here we will concentrate on the idealistic scenario of a sample-spaced CIR, while investigations which take into account the specific properties of the more realistic non sample-spaced channels remain a potential part of our future work.

16.6.2.1 MSE Performance Assessment

Throughout our investigations in this section we will focus again on our standard multi-user OFDM scenario of four simultaneous users, each equipped with one transmit antenna, while at the base station (BS) a four-element antenna array is employed. The channel between each transmit-receive antenna pair, characterised in terms of its sample-spaced impulse response, and the OFDM parameters are fixed to those used by the indoor WATM system of Section 14.3.1. Two detection techniques are invoked in our study, namely, the MMSE and M-SIC detection techniques of Sections 17.2.3 and 17.3.1. The corresponding BER and *a priori* channel estimation MSE simulation results – after the initial adaptation[18] of the predictor coefficients – are portrayed on the left- and right-hand side of Figure 16.19, respectively.

In the context of the BER performance assessment shown on the left-hand side of Figure 16.19 we observe that with the aid of the imperfect channel estimates produced by the RLS-adaptive PIC-assisted DDCE almost the same performance is achieved as when employing ideal channel knowledge. This is particularly true for the powerful M-SIC detection algorithm, which produces relatively reliable symbol decisions and hence also a reliable remodulated reference for the RLS-adaptive PIC-assisted DDCE. In contrast, a slight BER degradation is observed when using the less powerful MMSE detection scheme particularly

---

[18]The initial adaptation of the predictor coefficients was observed e.g. in Figure 16.18. After the adaptation the *a priori* channel estimation MSE fluctuates around its specific mean value.

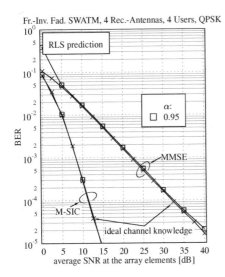
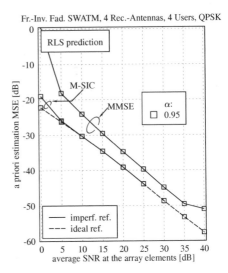

Figure 16.19: **(a: Left)** BER versus SNR performance and **(b: Right)** *a priori* channel estimation MSE versus SNR performance associated with the CIR-related tap-by-tap based RLS-adaptive PIC-assisted DDCE of Figure 16.4 in the context of a scenario of four receiver antennas at the BS and four simultaneous users, each equipped with one transmit antenna. Note that the PIC-assisted DDCE of Section 16.5 and the RLS-adaptive PIC-assisted DDCE of Section 16.6.2 exhibit the same macroscopic structure. The difference resides in the CIR-related tap prediction filter seen on the right-hand side of Figure 16.4. The channel between each transmitter-receiver antenna pair is characterised in terms of its sample-spaced impulse response and OFDM parameters by the indoor WATM channel- and system parameters of Section 14.3.1. The OFDM symbol normalised Doppler frequency was $F_D = 0.007$. Furthermore, the number of significant CIR-related taps was $K_0 = 12$ and the number of subcarriers was $K = 512$; MMSE- as well as M-SIC (M=2) detection was employed at the receiver and both an ideal, error-free reference and an imperfect, error-contaminated reference was invoked in the DDCE; the RLS-specific forgetting factor was set to $\alpha = \alpha_{RLS} = 0.95$.

for the range of SNRs up to 5dB, while for higher SNRs the BER performance is also almost identical to that when using perfect channel estimates.

The benefits of a more reliable remodulated reference used in the RLS-adaptive PIC-assisted DDCE become even more evident from the MSE performance results shown on the right-hand side of Figure 16.19. Here we observe a significant MSE reduction when employing the M-SIC assisted generation of the remodulated reference rather than that of the MMSE detector. In our specific example, which employs the sample-spaced three-path indoor WATM CIR, all the channel's energy is concentrated in three CIR-related taps, namely those at zero, six and eleven sampling period delays, while at all other tap positions within the CIR window of the first $K_0 = 12$ taps, the RLS-based adaptive predictor succeeds in effectively reducing the noise without setting these taps by "brute force" to zero. Hence the maximum noise reduction factor is about $3/512$. Note, however, again that in the more realistic scenario of a non-sample-spaced CIR in conjunction with employing the unitary DFT matrix $\mathbf{W}$ for transforming the least-squares channel transfer factor estimates from the frequency domain to the CIR-related domain, the noise reduction is more moderate due to the effects of spectral leakage as discussed in Section 16.5.2.3. More explicitly, the energy conveyed by the channel is rather spread across the different CIR-related taps.

Finally, let us comment on the specific choice of the forgetting factor $\alpha_{\text{RLS}} = 0.95$. As suggested during our investigations of RLS-based adaptive prediction-assisted DDCE employed in single-user OFDM systems in Section 16.6.1.3, for small values of α_{RLS}, the predictor may become unstable, as a result of which an excessive estimation MSE is observed. Our experiments conducted in the context of the PIC-assisted DDCE of Figure 16.4 further underlined that the appropriate range of α_{RLS} values has to be re-optimised when invoking an imperfect, potentially error-contaminated reference. Again, relatively small values of α_{RLS} yield a fast convergence, but also a high sensitivity to erroneous symbol decisions, while for higher values of α_{RLS} the opposite is true. A choice of $\alpha_{\text{RLS}} = 0.95$ was deemed reasonable in our application. However, for the robust RLS prediction approach outlined in Section 16.6.1.2, our results not included here for reasons of space economy suggested that the predictor's stability was less dependent on the specific choice of α_{RLS} due to the ensemble averaging carried out across the different CIR-related taps.

16.6.3 Conclusion

One drawback of the PIC-assisted DDCE of Figure 16.4 discussed in Section 16.5 was the relatively cumbersome procedure of the off-line optimisation of the predictor coefficients by means of the iterative approach discussed in Section 16.5.1.6. As a prerequisite for its application, assumptions had to be made about the channel's spaced-time spaced-frequency correlation function, potentially inflicting a performance loss. In order to avoid these problems – but at the complexity-related disadvantage of an on-line optimisation during the reception of the OFDM symbols – it was proposed in Section 16.6 to render the CIR-related tap predictors adaptive by means of the RLS algorithm. This was motivated by the observation in the context of the iterative off-line optimisation proposed in Section 16.5.1.6, that although in each iteration the adaptation of the different predictors' coefficients is performed independently, the average channel estimation MSE associated with the different users' channels converges to its minimum. In order to argue further, each iteration of the off-line optimisation has its analogy in the predictor coefficient adjustment conducted by the RLS algorithm after the reception of each OFDM symbol.

Our more specific discussions commenced in Section 16.6.1 by considering a single-user scenario in Section 16.6.1.1. The philosophy of RLS-assisted CIR-related tap prediction was introduced, where the adaptation of the CIR-related tap predictor coefficients was carried out separately for each CIR-related tap. By contrast, in Section 16.6.1.2 a simplification was achieved in terms of the computational complexity by invoking ensemble averaging of the different CIR-related taps' auto-correlation matrices and cross-correlation vectors, according to Equations 16.173 and 16.174, respectively. However, the ensemble averaging prohibited the application of the Sherman-Morrison formula [90], which was invoked in the context of the standard RLS algorithm in an attempt to avoid an explicit solution of Equation 16.166

for the vector of optimum CIR-related tap predictor coefficients. Although the complexity of the direct solution of Equation 16.175 is proportional to the cube of the number of predictor coefficients, its application is in most cases significantly less complex than that of the RLS-assisted CIR-related tap-by-tap adaptation procedure of Section 16.6.1.1. This argument was supported by our complexity analysis provided in Section 16.6.1.4. The complexity reduction is achieved at the cost of an MSE performance degradation in conjunction with the ensemble-averaging assisted approach of Section 16.6.1.2, since the same vector of predictor coefficients is employed for the prediction of each of the different CIR-related taps, potentially exhibiting different variances. Our rudimentary *a priori* estimation MSE performance assessment in Section 16.6.1.3 focussed on the influence of the RLS-specific forgetting factor α_{RLS} and on that of the OFDM symbol normalised Doppler frequency F_D. For lower values of α_{RLS} the speed of convergence was faster, while at the same time the residual MSE after adaptation remained higher, than that recorded for higher values of α_{RLS}.

By contrast to our discussions in Section 16.6.1, in Section 16.6.2 we concentrated on the performance assessment of the RLS-adaptive PIC-assisted DDCE employed in the context of multi-user OFDM. From the BER- and MSE performance curves of Figure 16.19 presented in Section 16.6.2.1 the system's convergence was confirmed. In the context of an imperfect, potentially error-contaminated DDCE-related reference, we found that the RLS-related forgetting factor α_{RLS}, for which a stable operation is maintained even in the presence of sporadic subcarrier symbol errors, was required to be closer to unity, which implies slower loss of memory. A value of $\alpha_{\mathrm{RLS}} = 0.95$ was deemed acceptable for this application. Simulation results generated for the ensemble-averaging based RLS-adaptive PIC-assisted DDCE were not presented here for reasons of space, but our experiments revealed that as a positive side-effect of the ensemble averaging the specific choice of the forgetting factor α_{RLS} was deemed less critical. Part of our future research will be the assessment of the RLS-adaptive PIC-assisted DDCE in the context of channels exhibiting non-sample-spaced CIRs.

16.7   Chapter Summary and Conclusion

In Chapter 16 we have portrayed a range of decision-directed channel estimation (DDCE) approaches designed for multi-user OFDM scenarios, or more generally, for OFDM scenarios supporting multiple transmit antennas. While in Section 16.1 the motivation of the different channel estimation approaches was detailed, and in Section 16.2 the SDMA-MIMO channel scenario was outlined, our more specific discussions commenced in Section 16.4 with the portrayal of Li's least-squares error assisted estimator. Our detailed conclusions with respect to this approach were provided in Section 16.4.6. In the context of our discussions in Section 16.4 we provided a more general mathematical description of the estimator proposed by Li *et al.* [91], while at the same time providing expressions for the *a posteriori* channel transfer function estimation MSE in the context of both sample-spaced- as well as non-sample-spaced CIRs. Furthermore, the estimator's restriction in terms of supporting a maximum of $L = \frac{K}{K_0}$ number of simultaneous users was highlighted, where K is the number of subcarriers and K_0 is the number of significant CIR-related taps. Our discussions of the LS-assisted DDCE were concluded by a detailed complexity analysis, which motivated our quest for alternative channel transfer function estimation approaches. More specifically, the explicit solution for the vector of optimum CIR-related taps according to Equation 16.22

imposes a complexity, which is partially a cubical function of the product LK_0. Hence, depending on the specific choice of the L number of transmit antennas and the K_0 number of significant CIR-related taps, the complexity might become excessive.

In order to reduce this complexity it was proposed by Li [93] to perform a parallel cancellation of the interfering CIR contributions with the aim of avoiding a direct solution of the equation system seen in Equation 16.22. This was achieved on the basis of CIR-related *a posteriori* tap estimates generated during the previous OFDM symbol period by potentially invoking a linear CIR-related tap prediction filter. It was argued furthermore that from a mathematical point of view performing both the parallel interference cancellation and the CIR-related tap prediction in the CIR-related domain is equivalent to performing the PIC in the frequency domain and the tap-prediction in the CIR-related domain, which was the topic of our in-depth discussions in Section 16.5. For detailed conclusions on this topic we refer to Section 16.5.4. As a result of our efforts in Section 16.5.1, expressions were derived for both the *a priori* and *a posteriori* channel transfer function estimation MSE – taking into account the recursive nature of the estimator – and the conditions for its stability were presented. Furthermore, an iterative approach was proposed for the off-line optimisation of the CIR-related tap predictor coefficients. Our performance investigations in Section 16.5.2 were conducted with respect to both the estimator's MSE- and the system's BER performance, as a function of the various system parameters, namely that of the L number of simultaneous users, that of the K_0 number of significant CIR-related taps, the K number of subcarriers, the $N_{tap}^{[t]}$ number of predictor coefficients, and also that as a function of the OFDM symbol normalised Doppler frequency. Furthermore, the effects of non-sample-spaced CIRs were described. Our discussions were concluded by a complexity study in Section 16.5.3 by highlighting under which conditions it was more beneficial to perform the PIC in the frequency or in the CIR-related domain, in order to minimise the associated complexity. It was found furthermore that the PIC-assisted DDCE's complexity is significantly lower, than that of the LS-assisted DDCE described in Section 16.4.

One disadvantage associated with the iterative off-line optimisation of the CIR-related tap predictor coefficients is the requirement of having *a priori* knowledge concerning the channel's spaced-time spaced-frequency correlation function. Since this is normally not available, we argued in Section 16.5.1.7 that a standard procedure is to assume a uniform, ideally support-limited channel scattering function and its associated spaced-time spaced-frequency correlation function with the consequence of a concomitant performance loss in comparison to the perfect knowledge of these parameters. In order to alleviate this problem, a viable approach is to render the PIC-assisted DDCE's prediction filter adaptive, which can be achieved with the aid of the RLS algorithm. This was the topic of Section 16.6. Again, for our detailed conclusions we refer to Section 16.6.3. Our strategy in this section was first to demonstrate the applicability of the RLS algorithm to the problem of adjusting the CIR-related tap predictor's coefficients in a DDCE-assisted single-user OFDM system, and then to demonstrate its applicability to the PIC-assisted DDCE of Figure 16.4 for multi-user OFDM. In order to further reduce the associated computational complexity of RLS-adaptive CIR-related tap-by-tap prediction it was proposed to employ the same vector of predictor coefficients for the prediction of each of the different CIR-related taps and hence to base its calculation on the ensemble average of the different CIR-related taps' auto-correlation matrices and cross-correlation vectors. As a side-effect, we expect a slight performance degradation compared to the higher-complexity case of CIR-related tap-by-tap prediction filtering having individual vectors of predictor coefficients. As our experiments showed, the choice of the RLS-specific forgetting factor α_{RLS} was also less critical than for the standard RLS algorithm. Again, for more detailed conclusions on this topic we refer the reader to Section 16.6.3.

16.8 List of Symbols Used in Chapter 16

Special Symbols - Common

$H_p^{(l)}[n, k]$:

Channel transfer factor associated with the channel encountered between the l-th user's transmit antenna and the p-th receiver antenna-element in the k-th subcarrier of the n-th OFDM symbol period.

$\mathbf{H}_p^{(i)}[n]$:

Vector of channel transfer factors $H_p^{(i)}[n, k]$, $k = 0, \dots, K - 1$ associated with the channel encountered between the i-th user's transmit antenna and the p-th receiver antenna element: $\mathbf{H}_p^{(i)}[n] \in \mathbb{C}^{K \times 1}$.

$\mathbf{I}_{K_0,p}^{(i)}$:

Diagonal masking-matrix used for retaining the significant CIR-related taps: $\mathbf{I}_{K_0,p}^{(i)} = \mathbf{J}_{K_0,p}^{(i)} \mathbf{J}_{K_0,p}^{(i)H} \in \mathbb{R}^{K \times K}$.

$\mathbf{J}_{K_0,p}^{(i)}$:

Matrix invoked for mapping the K_0 significant CIR-related taps' estimates contained in the vector $\hat{\mathbf{h}}_{K_0,p}^{(i)}[n]$ to their 'true' integer positions within the K-tap FFT window: $\mathbf{J}_{K_0,p}^{(i)} \in \mathbb{C}^{K \times K_0}$.

K_0:

Number of significant CIR-related taps.

L:

Number of simultaneous SDMA users, each equipped with a single transmit antenna.

$n_p[n, k]$:

AWGN signal contribution of variance σ_n^2 associated with the p-th receiver antenna element.

$\mathbf{n}_p[n]$:

Vector of AWGN signal contributions $n_p[n, k]$, $k = 0, \dots, K - 1$, associated with the p-th receiver antenna element in each of the K subcarriers: $\mathbf{n}_p[n] \in \mathbb{C}^{K \times 1}$.

P:

Number of BS receiver antenna elements.

$s^{(l)}[n, k]$:

Symbol transmitted by the l-th user.

$\mathbf{S}^{(i)}[n]$:

Diagonal matrix of the i-th user's transmitted subcarrier symbols $s^{(i)}[n, k]$, $k = 0, \dots, K - 1$: $\mathbf{S}^{(i)}[n] \in \mathbb{C}^{K \times K}$.

$\check{\mathbf{S}}^{(i)}[n]$:

Diagonal matrix of the i-th user's sliced subcarrier symbols $\check{s}^{(i)}[n, k]$, $k = 0, \dots, K - 1$: $\check{\mathbf{S}}^{(i)}[n] \in \mathbb{C}^{K \times K}$.

W_K:

Complex Fourier kernel: $W_K = e^{-j\frac{2\pi}{K}}$.

$\mathbf{W}$:

DFT matrix hosting the complex exponentials $\mathbf{W}|_{[i,j]} = \frac{1}{\sqrt{K}} W_K^{ij}$: $\mathbf{W} \in \mathbb{C}^{K \times K}$.

$x_p[n, k]$:

Signal received by the p-th receiver antenna element.

$\mathbf{x}_p[n]$:

Vector of signals $x_p[n, k]$, $k = 0, \dots, K - 1$ associated with the p-th receiver antenna element in the K subcarriers: $\mathbf{x}_p[n] \in \mathbb{C}^{K \times 1}$.

σ_s^2:   Variance of the l-th user's transmitted symbols.

$()[n, k]$:   Index employed for indicating that the signal in round brackets is associated with the k-th subcarrier of the n-th OFDM symbol period.

$()^{(i,j)}$:   Sub-matrix associated with the i-th 'row' and j-th 'column' of the block matrix in brackets.

$()|_{[i,j]}$:   Element associated with the i-th row and j-th column of the matrix in brackets.

Special Symbols - LS-Assisted DDCE

$\mathbf{A}_p[n]$:

Shorthand: $\mathbf{A}_p[n] = \check{\mathbf{S}}^T[n]\mathbf{W}_{J,p} \in \mathbb{C}^{K \times LK_0}$.

$\mathbb{C}^{(\mathbb{C}\{*,+\}\mathbb{C})}\big|_{\text{norm}}$:

Normalized computational complexity quantified in terms of the number of complex multiplications or additions[19].

$\mathbb{C}^{(\mathbb{C}\{*,+\}\mathbb{C})}_{\text{MU-CE,LS}}\big|_{\text{norm}}$:

Normalized total computational complexity quantified in terms of the number of complex multiplications or additions, associated with the LS-assisted DDCE designed for multi-user OFDM[20].

$\mathbb{C}^{(\mathbb{C}\{*,+\}\mathbb{C})}_{\text{MU-CE,PIC-LS}}\big|_{\text{norm}}$:

Normalized total computational complexity in terms of the number of complex multiplications or additions, associated with the PIC-based LS-assisted DDCE designed for multi-user OFDM[21].

$\hat{\mathbf{h}}^{(i)}_{K_0,p}[n]$:

Vector of K_0 significant CIR-related tap estimates associated with the channel encountered between the i-th user's transmit antenna and the p-th receiver antenna element: $\hat{\mathbf{h}}^{(i)}_{K_0,p}[n] \in \mathbb{C}^{K_0 \times 1}$.

$\hat{\mathbf{h}}_{K_0,p}[n]$:

Block-vector hosting the L different users' vectors $\hat{\mathbf{h}}^{(i)}_{K_0,p}[n]$, $i = 1, \ldots, L$ of K_0 significant CIR-related tap estimates, associated with the p-th receiver antenna element: $\hat{\mathbf{h}}_{K_0,p}[n] = (\hat{h}^{(1)T}_{K_0,p}[n], \ldots, \hat{h}^{(L)T}_{K_0,p}[n])^T \in \mathbb{C}^{LK_0 \times 1}$.

$\tilde{\mathbf{h}}_{apt,K_0,p}[n]$:

Block-vector hosting the L different users' vectors of least-squares error optimized significant CIR-related tap estimates, associated with the p-th receiver antenna element: $\tilde{\mathbf{h}}_{apt,K_0,p}[n] = \hat{\mathbf{h}}_{K_0,p}[n]\big|_{opt} \in \mathbb{C}^{LK_0 \times 1}$.

$\hat{\mathbf{H}}^{(i)}_p[n]$:

Vector of channel transfer factor estimates associated with the channel encountered between the i-th user's transmit antenna and the p-th receiver antenna element, based on the vector $\hat{\mathbf{h}}^{(i)}_{K_0,p}[n]$ of significant CIR-related tap estimates: $\hat{\mathbf{H}}^{(i)}_p[n] = \mathbf{W}\mathbf{J}^{(i)}_{K_0,p}\hat{\mathbf{h}}^{(i)}_{K_0,p}[n] \in \mathbb{C}^{K \times 1}$.

$\mathbf{H}_p[n]$:

Block-vector hosting the L different users' vectors $\mathbf{H}^{(i)}_p[n]$, $i = 1, \ldots, L$ of channel transfer factors, associated with the p-th receiver antenna element, which is given by: $\mathbf{H}_p[n] = (H^{(1)T}_p[n], \ldots, H^{(L)T}_p[n])^T \in \mathbb{C}^{LK \times 1}$.

$\hat{\mathbf{H}}_p[n]$:

Block-vector hosting the L different users' vectors $\hat{\mathbf{H}}^{(i)}_p[n]$, $i = 1, \ldots, L$ of channel transfer factor estimates, associated with the p-th receiver antenna element: $\hat{\mathbf{H}}_p[n] = (\hat{\mathbf{H}}^{(1)T}_p[n], \ldots, \hat{\mathbf{H}}^{(L)T}_p[n])^T \in \mathbb{C}^{LK \times 1}$.

$\tilde{\mathbf{H}}_{apt,p}[n]$:

Block-vector of the L different users' vectors $\tilde{\mathbf{H}}^{(i)}_{apt,p}[n]$, $i = 1, \ldots, L$ of least-squares error optimized channel transfer factor estimates, associated with the p-th receiver antenna element: $\tilde{\mathbf{H}}_{apt,p}[n] \in \mathbb{C}^{LK \times 1}$.

---

[19]The normalization was carried out with respect to the number of subcarriers K, number of users L and number of receiver antenna elements P.

[20]See Footnote 19.

[21]See Footnote 19.

$\mathbf{I}_A$:   Unity matrix: $\mathbf{I}_A \in \mathbb{C}^{K_A \times K_A}$.

$\mathbf{I}_B$:   Unity matrix: $\mathbf{I}_B \in \mathbb{C}^{K_B \times K_B}$.

$\mathbf{I}_{cyc}^{(i,j)}$:   Matrix which performs the cyclic rotation of a matrix' rows, when multiplied from the left- or rotation of the matrix' columns, when multiplied from the right: $\mathbf{I}_{cyc}^{(i,j)} = \begin{pmatrix} 0 & \mathbf{I}_B \\ \mathbf{I}_A & 0 \end{pmatrix} \in \mathbb{C}^{K_0 \times K_0}$.

K_A: Short-hand: $K_A = [K - \overline{K}_0(i - j)] \bmod K$.

K_B: Short-hand: $K_B = [K - \overline{K}_0(j - i)] \bmod K$.

$\overline{K}_0$:   Short-hand: $\overline{K}_0 = \lfloor \frac{K}{L} \rfloor$.

$\overline{\mathrm{MSE}}_{apt,p}^{(i)}[n]$:   Subcarrier-averaged channel transfer factor estimation MSE associated with the channel encountered between the i-th user's single transmit antenna and the p-th receiver antenna element.

$\overline{\mathrm{MSE}}_{apt,p}[n]$:   Subcarrier- and user-averaged estimation MSE associated with the channels encountered between the L users' single transmit antennas and the p-th receiver antenna element.

$\mathbf{p}_p[n]$:   LS-related 'cross-correlation' vector: $\mathbf{p}_p[n] = \mathbf{A}_p^H[n]\mathbf{x}[n] \in \mathbb{C}^{LK_0 \times 1}$.

$\mathbf{P}_p[n]$:   LS-related projection matrix- or Moore-Penrose pseudo-inverse of the matrix $\mathbf{A}_p[n]$: $\mathbf{P}_p[n] \in \mathbb{C}^{LK_0 \times K}$.

$\mathbf{Q}_p[n]$:   LS-related 'auto-correlation' matrix: $\mathbf{Q}_p[n] = \mathbf{A}_p^H[n]\mathbf{A}_p[n] \in \mathbb{C}^{LK_0 \times LK_0}$.

$\acute{\mathbf{Q}}_p^{(i,j)}[n]$:   Short-hand: $\acute{\mathbf{Q}}_p^{(i,j)}[n] = \mathbf{W}^H \mathbf{S}^{(i)*}[n]\mathbf{S}^{(j)}[n]\mathbf{W}$.

$\mathbf{R}_p^{[f](i)}$:   Spaced-frequency correlation matrix associated with the channel encountered between the i-th user's transmit antenna and the p-th receiver antenna element: $\mathbf{R}_p^{[f](i)} = E\{\mathbf{H}_p^{(i)}[n]\mathbf{H}_p^{(i)H}[n]\} \in \mathbb{C}^{K \times K}$.

$\mathbf{R}_p^{[f]}$:   Block-diagonal auto-correlation matrix hosting the L spaced-frequency correlation matrices $\mathbf{R}_p^{[f](i)}, i = 1, \ldots, L$ associated with the channels encountered between the L users' single transmit antennas and the p-th receiver antenna element: $\mathbf{R}_p^{[f]} = E\{\mathbf{H}_p[n]\mathbf{H}_p^H[n]\} = diag(\mathbf{R}_p^{[f]1} \ldots \mathbf{R}_p^{[f]L}) \in \mathbb{C}^{LK \times LK}$.

$\mathbf{R}_{a,p}^{[f]}$:   Auto-correlation matrix of the multi-user vector $\tilde{\mathbf{H}}_{apt,p}[n]$ hosting K subcarrier channel transfer factor estimates per user: $\mathbf{R}_{a,p}^{[f]} = E\{\tilde{\mathbf{H}}_{apt,p}[n]\tilde{\mathbf{H}}_{apt,p}^H[n]\} = \mathbf{R}_{a,\mathrm{Channel,p}}^{[f]} + \mathbf{R}_{a,\mathrm{AWGN,p}}^{[f]} \in \mathbb{C}^{LK \times LK}$.

$\mathbf{R}_{a,\mathrm{AWGN},p}^{[f]}$:   AWGN-related component of the auto-correlation matrix $\mathbf{R}_{a,p}^{[f]}$: $\mathbf{R}_{a,\mathrm{AWGN},p}^{[f]} \in \mathbb{C}^{LK \times LK}$.

$\mathbf{R}_{a,\text{Channel},p}^{[f]}$:　　　Channel-related component of the auto-correlation matrix: $\mathbf{R}_{a,p}^{[f]}$: $\mathbf{R}_{a,\text{Channel},p}^{[f]} \in \mathbb{C}^{LK \times LK}$.

$\mathbf{R}_{c,p}^{[f]}$:　　　Cross-correlation matrix between the L-user vector $\mathbf{H}_p[n]$ of the K subcarriers' 'true' channel transfer factors and the L-user vector $\tilde{\mathbf{H}}_{apt,p}[n]$ of the K subcarriers' transfer factor estimates: $\mathbf{R}_{c,p}^{[f]} = E\{\mathbf{H}_p[n]\tilde{\mathbf{H}}_{apt,p}^H[n]\} \in \mathbb{C}^{LK \times LK}$.

$\mathbf{R}_{\Delta\tilde{\mathbf{H}}_{apt,p}[n]}$:　　　Auto-correlation matrix of the vector $\Delta\tilde{\mathbf{H}}_{apt,p}[n]$ of channel transfer factor estimation errors: $\mathbf{R}_{\Delta\tilde{\mathbf{H}}_{apt,p}[n]} = E\{\Delta\tilde{\mathbf{H}}_{apt,p}[n]\Delta\tilde{\mathbf{H}}_{apt,p}^H[n]\} = \mathbf{R}_p^{[f]} - 2\mathfrak{Re}\{\mathbf{R}_{c,p}^{[f]}\} + \mathbf{R}_{a,p}^{[f]} \in \mathbb{C}^{LK \times LK}$.

$\mathbf{S}[n]$:　　　Block-matrix hosting the L different users' diagonal matrices $\mathbf{S}^{(i)}[n]$, $i = 1, \ldots, L$ of transmitted subcarrier symbols: $\mathbf{S}[n] = (\mathbf{S}^{(1)T}[n], \ldots, \mathbf{S}^{(L)T}[n])^T \in \mathbb{C}^{LK \times K}$.

$\check{\mathbf{S}}[n]$:　　　Block-matrix hosting the L different users' diagonal matrices $\check{\mathbf{S}}^{(i)}[n]$, $i = 1, \ldots, L$ of sliced subcarrier symbols: $\check{\mathbf{S}}[n] \in \mathbb{C}^{LK \times K}$.

$t^{(i)}[n,k]$:　　　k-th element of the optimum training subcarrier symbol sequence associated with the i-th user as proposed by Li [93]: $t^{(i)}[n,k] = \sigma_i t_p[n,k]W_K^{-\overline{K}_0(i-1)K}$

$t_p[n,k]$:　　　k-th element of an arbitrary training subcarrier symbol sequence $t_p[n,k]$, $k = 0, \ldots, K - 1$ having unit-variance elements.

$\mathbf{T}_{W,K_0,p}^{(i)}$:　　　Short-hand: $\mathbf{T}_{W,K_0,p}^{(i)} = \mathbf{W}_{J,p}^{(i)}\mathbf{W}_{J,p}^{(i)H} \in \mathbb{C}^{K \times K}$.

$\mathbf{W}_{J,p}^{(i)}$:　　　Combined DFT- and mapping matrix for transforming the vector $\hat{\mathbf{h}}_{K_0,p}^{(i)}[n]$ of K_0 significant CIR-related tap estimates to the vector $\hat{\mathbf{H}}_p^{(i)}[n]$ of K channel transfer factor estimates: $\mathbf{W}_{J,p}^{(i)} = \mathbf{W}\mathbf{J}_{K_0,p}^{(i)} \in \mathbb{C}^{K \times K_0}$.

$\mathbf{W}_{J,p}$:　　　Block-diagonal matrix of the L different users' combined DFT- and mapping matrices $\mathbf{W}_{J,p}^{(i)}$, $i = 1, \ldots, K$ associated with the p-th receiver antenna element: $\mathbf{W}_{J,p} = diag(\mathbf{W}_{J,p}^{(1)}, \ldots, \mathbf{W}_{J,p}^{(L)}) \in \mathbb{C}^{LK \times LK_0}$.

$\hat{\mathbf{x}}_p[n]$:　　　Vector of estimated signals $\hat{x}_p[n,k]$, $k = 0, \ldots, K - 1$ associated with the p-th receiver antenna element in each of the K subcarriers: $\hat{\mathbf{x}}_p[n] \in \mathbb{C}^{K \times 1}$.

$\mathbf{\Lambda}_{W,p}^{[f](i)}$:　　　Decomposition of the spaced-frequency correlation matrix $\mathbf{R}_p^{[f](i)}$ associated with the channel between the i-th user's transmit antenna and the p-th receiver antenna element, with respect to the DFT matrix $\mathbf{W}$: $\mathbf{\Lambda}_{W,p}^{[f](i)} = (\mathbf{W}^H\mathbf{R}_p^{[f](i)}\mathbf{W}) \in \mathbb{C}^{K \times K}$.

$\mathbf{\Lambda}_{W,J,p}^{[f](i)}$:　　　Short-hand: $\mathbf{\Lambda}_{W,J,p}^{[f](i)} = \mathbf{J}_{K_0,p}^{(i)H}\mathbf{\Lambda}_{W,p}^{[f](i)}\mathbf{J}_{K_0,p}^{(i)} \in \mathbb{C}^{K_0 \times K_0}$.

$\Delta\tilde{\mathbf{H}}_{apt,p}[n]$:   Vector of channel transfer factor estimation errors associated with the K subcarriers of each of the L users, with respect to the p-th receiver antenna element: $\Delta\tilde{\mathbf{H}}_{apt,p}[n] = \mathbf{H}_p[n] - \tilde{\mathbf{H}}_{apt,p}[n] \in \mathbb{C}^{LK \times 1}$.

$\Delta\hat{\mathbf{x}}[n]$:   Vector of estimation errors between the received vector $\mathbf{x}[n]$- and the synthesized vector $\hat{\mathbf{x}}[n]$ of subcarrier signals: $\Delta\hat{\mathbf{x}}[n] = \mathbf{x}[n] - \hat{\mathbf{x}}[n] \in \mathbb{C}^{K \times 1}$.

Special Symbols - Decision-Directed Channel Estimation for Multi-User OFDM - PIC-assisted DDCE

$\acute{c}_{pre,p}^{(i)}[\acute{n}]$:   Predictor coefficient employed in the context of the CIR prediction process associated with the channel encountered between the i-th user's transmit antenna and the p-th receiver antenna element. Note that the $\acute{n}$-th predictor coefficient, where $\acute{n} \in \{0, \ldots, N_{tap}^{[t]} - 1\}$, is associated with the OFDM symbol having an offset of $-\acute{n}$ relative to the current OFDM symbol.

$\tilde{\mathbf{c}}_{pre,p}^{(i)}$:   Vector of predictor coefficients employed in the context of the CIR prediction process associated with the channel encountered between the i-th user's transmit antenna and the p-th receiver antenna element: $\tilde{\mathbf{c}}_{pre,p}^{(i)} \in \mathbb{C}^{N_{tap}^{[t]} \times 1}$.

$\tilde{\mathbf{c}}_{pre,p}^{(j)}|_{opt}$:   Vector of optimum predictor coefficients, based on the knowledge of the remaining users' average *a priori* estimation MSEs, namely on $\overline{\mathrm{MSE}}_{apr}^{(i)}$, $i = 1, \ldots, L$, $i \neq j$: $\tilde{\mathbf{c}}_{pre,p}^{(j)}|_{opt} \in \mathbb{C}^{N_{tap}^{[t]} \times 1}$.

$\mathbf{C}_{pre}$:   Diagonal matrix of the L different users' CIR-related tap prediction coefficient terms of $\frac{K_0}{K}\alpha_j \tilde{\mathbf{c}}_{pre,p}^{(j)H}\tilde{\mathbf{c}}_{pre,p}^{(j)}$, $j = 1, \ldots, L$: $\mathbf{C}_{pre} \in \mathbb{C}^{L \times L}$.

$\mathbb{C}_{\mathrm{MU\text{-}CE,PIC}}^{(\mathbb{C}\{*,+\}\mathbb{C})}|_{\mathrm{norm}}$:   Normalized total computational complexity expressed in terms of the number of complex multiplications or additions, associated with the PIC-assisted DDCE designed for multi-user OFDM[22].

$\mathbb{C}_{\mathrm{RLS}}^{\mathbb{C}\{*,+\}\mathbb{C}}$:   Computational complexity of the CIR-related ensemble-averaging assisted RLS-based tap predictor, expressed in terms of the number of complex multiplications and additions, inflicted per channel and OFDM symbol.

$\mathbb{C}_{\mathrm{RLS},tap}^{\mathbb{C}\{*,+\}\mathbb{C}}$:   Computational complexity of the CIR-related RLS-based tap predictor, expressed in terms of the number of complex multiplications and additions, inflicted per CIR-related tap and OFDM symbol.

$\mathbf{F}$:   Feedback matrix characterizing the PIC process. While the main-diagonal elements are zero, the side-diagonal elements are unity: $\mathbf{F} \in \mathbb{R}^{L \times L}$.

$\acute{\mathbf{F}}$:   Short-hand: $\acute{\mathbf{F}} = \mathbf{C}_{pre} \cdot \mathbf{P}_s^{-1} \cdot \mathbf{F} \cdot \mathbf{P}_s \in \mathbb{R}^{L \times L}$.

$G(\acute{\mathbf{F}})$:   Union of Gershgorin disks of $\acute{\mathbf{F}}$. With the aid of the Gershgorin circle theorem explicit bounds can be imposed on the regions in $\mathbb{C}$, which host the eigenvalues of the matrix $\acute{\mathbf{F}}$ [90].

$\hat{\mathbf{h}}_{apr,p}^{(i)}[n]$:   Vector of the K CIR-related taps' *a priori* estimates $\hat{h}_{apr,p}^{(i)}[n,l]$, $l = 0, \ldots, K - 1$: $\hat{\mathbf{h}}_{apr,p}^{(i)}[n] \in \mathbb{C}^{K \times 1}$.

---

[22]See Footnote 19.

$\tilde{\mathbf{h}}_{apt,p}^{(i)}[n]$:　Vector of the K CIR-related taps' *a posteriori* estimates, namely $\tilde{h}_{apt,p}^{(i)}[n,l]$, $l = 0, \ldots, K-1$, which is given by: $\tilde{\mathbf{h}}_{apt,p}^{(i)}[n] \in \mathbb{C}^{K \times 1}$.

$\hat{\mathbf{H}}_{apr,p}^{(i)}[n]$:　Vector of the K different subcarriers' *a posteriori*[23] channel transfer factor estimates $\hat{H}_{apr,p}^{(i)}[n,k]$, $k = 0, \ldots, K-1$: $\hat{\mathbf{H}}_{apr,p}^{(i)}[n] \in \mathbb{C}^{K \times 1}$.

$\tilde{\mathbf{H}}_{apt,p}^{(i)}[n]$:　Vector of the K different subcarriers' *a priori* channel transfer factor estimates $\tilde{H}_{apt,p}^{(i)}[n,k]$, $k = 0, \ldots, K-1$: $\tilde{\mathbf{H}}_{apt,p}^{(i)}[n] \in \mathbb{C}^{K \times 1}$.

$\mathbf{k}_p^{(i)}[n,l]$:　Kalman gain vector associated with the RLS-assisted prediction of the l-th CIR-related tap: $\mathbf{k}_p^{(i)}[n,l] \in \mathbb{C}^{N_{tap}^{[t]} \times 1}$.

$\overline{\mathrm{MSE}}_{apr,p}^{(j)}[n]$:　Subcarrier-averaged *a priori* channel transfer factor estimation MSE, which is expressed as: $\overline{\mathrm{MSE}}_{apr,p}^{(j)}[n] = \frac{1}{K}\mathrm{Trace}(\mathbf{R}_{\Delta\hat{\mathbf{H}}_{apr,p}^{(j)}}[n])$.

$\overline{\mathrm{MSE}}_{apt,p}^{(j)}[n]$:　Subcarrier-averaged *a posteriori* channel transfer factor estimation MSE, which is expressed as: $\overline{\mathrm{MSE}}_{apt,p}^{(j)}[n] = \frac{1}{K}\mathrm{Trace}(\mathbf{R}_{\Delta\tilde{\mathbf{H}}_{apt,p}^{(j)}}[n])$.

$\overline{\mathrm{MSE}}_{dec,p}^{(j)}[n]$:　Subcarrier-averaged channel-decorrelation related MSE, which is expressed as: $\overline{\mathrm{MSE}}_{dec,p}^{(j)}[n] = \frac{1}{K}\mathrm{Trace}(\mathbf{R}_{\Delta\mathbf{H}_{dec,p}^{(j)}})$.

$\mathbf{MSE}_{apr,p}$:　Vector of average *a priori* channel transfer factor estimation MSEs, namely $\overline{\mathrm{MSE}}_{apr,p}^{(j)}[n]$, $j = 1, \ldots, L$: $\mathbf{MSE}_{apr,p}[n] \in \mathbb{R}^{L \times 1}$.

$\mathbf{MSE}_{dec,p}$:　Vector of average channel decorrelation related MSEs, namely $\overline{\mathrm{MSE}}_{dec,p}^{(j)}[n]$, $j = 1, \ldots, L$: $\mathbf{MSE}_{dec,p}[n] \in \mathbb{R}^{L \times 1}$.

$N_{tap}^{[t]}$:　Number of filter taps associated with the CIR-related tap estimator- or predictor.

$\mathbf{P}_s$:　Diagonal-matrix hosting the different users' signal variances σ_l^2, $l = 1, \ldots, L$: $\mathbf{P}_s \in \mathbb{R}^{L \times L}$.

r_{expo}:　Amplitude quotient associated with the exponential multipath intensity profile $r_{h,\mathrm{expo}}(\tau)$: $r_{\mathrm{expo}} = \frac{r_{h,\mathrm{expo}}(T_m)}{r_{h,\mathrm{expo}}(0)}$.

$\mathbf{r}_p^{[t](j)}$:　Spaced-time cross-correlation vector, whose elements are given by $\mathbf{r}_p^{[t](j)}|_{\acute{n}} = E\{H^{(j)*}[n+1,k]H^{(j)}[n-\acute{n},k]\}$: $\mathbf{r}_p^{[t](j)} \in \mathbb{C}^{N_{tap}^{[t]} \times 1}$.

$\tilde{\mathbf{r}}_{apt,p}^{[t](j)}[n,l]$:　l-th CIR-related tap's RLS-related spaced-time sample-cross-correlation vector associated with the n-th OFDM symbol period: $\tilde{\mathbf{r}}_{apt,p}^{[t](j)}[n,l] \in \mathbb{C}^{N_{tap}^{[t]} \times 1}$.

---

[23]Note that the *a posteriori* channel transfer factor estimates for the n-th OFDM symbol were generated during the $(n-1)$-th OFDM symbol period by linear $N_{tap}^{[t]}$ filtering.

$\mathbf{R}_p^{[f](i)}$:

Spaced-frequency correlation matrix: $\mathbf{R}_p^{[f](i)} = E\{\mathbf{H}_p^{(i)}[n]\mathbf{H}_p^{(i)H}[n]\} \in \mathbb{C}^{K \times K}$. The EigenValue Decomposition (EVD) of the matrix $\mathbf{R}_p^{[f](i)}$ is given by: $\mathbf{R}_p^{[f](i)} = \mathbf{U}_p^{[f](i)}\Lambda_p^{[f](i)}\mathbf{U}_p^{[f](i)H}$.

$\tilde{\mathbf{R}}_p^{[f](i)}$:

Spaced-frequency correlation matrix invoked in the design of the predictor: $\tilde{\mathbf{R}}_p^{[f](i)} \in \mathbb{C}^{K \times K}$. The EigenValue Decomposition (EVD) of the matrix $\tilde{\mathbf{R}}_p^{[f](i)}$ is given by: $\tilde{\mathbf{R}}_p^{[f](i)} = \tilde{\mathbf{U}}_p^{[f](i)}\tilde{\Lambda}_p^{[f](i)}\tilde{\mathbf{U}}_p^{[f](i)H}$.

$\mathbf{R}_{apt,p}^{[f](i)}$:

Spaced-frequency correlation matrix associated with the *a posteriori* transfer factor estimates of the channel: $\mathbf{R}_{apt,p}^{[f](i)} = E\{\tilde{\mathbf{H}}_{apt,p}^{(i)}[n]\tilde{\mathbf{H}}_{apt,p}^{(i)H}[n]\} \in \mathbb{C}^{K \times K}$. The EigenValue Decomposition (EVD) of the matrix $\mathbf{R}_{apt,p}^{[f](i)}$ is given by: $\mathbf{R}_{apt,p}^{[f](i)} = \mathbf{U}_{apt,p}^{[f](i)}\Lambda_{apt,p}^{[f](i)}\mathbf{U}_{apt,p}^{[f](i)H}$.

$\mathbf{R}_p^{[t](i)}$:

Spaced-time correlation matrix, whose elements are given by $\mathbf{R}^{[t](j)}|_{\acute{n}_1,\acute{n}_2} = E\{H^{(j)}[n-\acute{n}_1,k]H^{(j)*}[n-\acute{n}_2,k]\}$: $\mathbf{R}_p^{[t](i)} \in \mathbb{C}^{N_{tap}^{[t]} \times N_{tap}^{[t]}}$.

$\tilde{\mathbf{R}}_{apt,p}^{[t](j)}[n,l]$:

l-th CIR-related tap's RLS-related spaced-time sample-correlation matrix associated with the n-th OFDM symbol period: $\tilde{\mathbf{R}}_{apt,p}^{[t](j)}[n,l] \in \mathbb{C}^{N_{tap}^{[t]} \times N_{tap}^{[t]}}$.

$\mathbf{R}_{\mathbf{H}_{dec,p}^{(j)}}$:

Channel transfer function decorrelation related matrix: $\mathbf{R}_{\mathbf{H}_{dec,p}^{(j)}} \in \mathbb{C}^{K \times K}$.

$\mathbf{R}_{\Delta\hat{\mathbf{H}}_{apr,p}^{(j)}}[n]$:

Auto-correlation matrix of the vector $\Delta\hat{\mathbf{H}}_{apr,p}^{(j)}[n]$ of *a priori* channel transfer factor estimation errors: $\mathbf{R}_{\Delta\hat{\mathbf{H}}_{apr,p}^{(j)}}[n] = E\{\Delta\hat{\mathbf{H}}_{apr,p}^{(j)}[n]\Delta\hat{\mathbf{H}}_{apr,p}^{(j)H}[n]\} \in \mathbb{C}^{K \times K}$.

$\mathbf{R}_{\Delta\tilde{\mathbf{H}}_{apt,p}^{(j)}}[n]$:

Auto-correlation matrix of the vector $\Delta\tilde{\mathbf{H}}_{apt,p}^{(j)}[n]$ of *a posteriori* channel transfer factor estimation errors: $\mathbf{R}_{\Delta\tilde{\mathbf{H}}_{apt,p}^{(j)}}[n] = E\{\Delta\tilde{\mathbf{H}}_{apt,p}^{(j)}[n]\Delta\tilde{\mathbf{H}}_{apt,p}^{(j)H}[n]\} \in \mathbb{C}^{K \times K}$.

$\mathbf{T}_{K_0,p}^{(i)}$:

Short-hand: $\mathbf{T}_{K_0,p}^{(i)} = \tilde{\mathbf{U}}_p^{[f](i)}\mathbf{I}_{K_0,p}^{(i)}\tilde{\mathbf{U}}_p^{[f](i)H} \in \mathbb{C}^{K \times K}$.

$\mathbf{U}_p^{[f](i)}$:

Unitary KLT matrix associated with the EVD of the spaced-frequency correlation matrix $\mathbf{R}_p^{[f](i)}$: $\mathbf{U}_p^{[f](i)} \in \mathbb{C}^{K \times K}$.

$\mathbf{U}_{apt,p}^{[f](i)}$:

Unitary KLT matrix associated with the EVD of the *a posteriori* channel transfer factor estimates' spaced-frequency correlation matrix $\mathbf{R}_{apt,p}^{[f](i)}$: $\mathbf{U}_{apt,p}^{[f](i)} \in \mathbb{C}^{K \times K}$.

$\tilde{\mathbf{U}}_p^{[f](i)}$:

Unitary KLT matrix associated with the EVD of the spaced-frequency correlation matrix $\tilde{\mathbf{R}}_p^{[f](i)}$ invoked in the design of the predictor: $\tilde{\mathbf{U}}_p^{[f](i)} \in \mathbb{C}^{K \times K}$.

α_{expo}: Decay factor associated with the exponentially decaying multipath intensity profile $r_{h,\text{expo}}(\tau)$.

α_j: Modulation-noise enhancement factor $\alpha_j = E\{\frac{1}{|s^{(j)}[n,k]|^2}\}$ associated with the j-th user's transmitted subcarrier symbols.

α_{RLS}: Forgetting factor employed in the context of RLS-assisted CIR-related tap prediction filtering: $\alpha_{\text{RLS}} \in \mathbb{R}$.

β_{expo}: Amplitude scaling factor associated with the exponentially decaying multipath intensity profile $r_{h,\text{expo}}(\tau)$.

$\varepsilon_{\text{RLS},0}$: Startup-constant employed in the context of RLS-assisted CIR-related tap prediction filtering: $\varepsilon_{\text{RLS},0} \in \mathbb{R}$.

$\lambda_{p,l}^{[f](i)}$: l-th eigenvalue associated with the EVD of the spaced-frequency correlation matrix $\mathbf{R}_p^{[f](i)}$: $\lambda_{p,l}^{[f](i)} = (\mathbf{U}_p^{[f](i)H}\mathbf{R}_p^{[f](i)}\mathbf{U}_p^{[f](i)})_{[l,l]}$.

$\lambda_{apt,p,l}^{[f](i)}$: l-th eigenvalue associated with the EVD of the *a posteriori* channel transfer factor estimates' spaced-frequency correlation matrix $\mathbf{R}_{apt,p}^{[f](i)}$, which is expressed as: $\lambda_{apt,p,l}^{[f](i)} = (\mathbf{U}_{apt,p}^{[f](i)H}\mathbf{R}_{apt,p}^{[f](i)}\mathbf{U}_{apt,p}^{[f](i)})_{[l,l]}$.

ν: Short-hand employed in the context of calculating the single coefficient of a CIR-related one-tap predictor.

$\rho(\acute{\mathbf{F}})$: Spectral radius of the matrix $\acute{\mathbf{F}}$. The spectral radius is the smallest radius of a circle centered around the origin of $\mathbb{C}$ that contains all the matrix' eigenvalues [90].

$\hat{\sigma}^2_{\tilde{h}^{(j)}_{apt}[n,l]}$: Estimated variance associated with the l-th CIR-related *a posteriori* tap estimate based upon the current and the $N_{tap} - 1$ number of previous CIR-related *a posteriori* tap estimates.

τ_{int}: Integer part of the delay τ normalized to the sampling period duration T_s, where we have: $\tau_{int} = \lfloor \frac{\tau}{T_s} \rfloor$.

τ_{frac}: Fractional part of the delay τ normalized to the sampling period duration T_s: $\tau_{frac} = \frac{\tau}{T_s} - \tau_{int}$.

$\upsilon_{p,l}^{[f](i)}$: l-th diagonal element associated with the decomposition of the spaced-frequency correlation matrix $\mathbf{R}_p^{[f](i)}$ with respect to the channel predictor's unitary transform matrix $\tilde{\mathbf{U}}_p^{[f](i)}$, where we have: $\upsilon_{p,l}^{[f](i)} = (\tilde{\mathbf{U}}_p^{[f](i)H}\mathbf{R}_p^{[f](i)}\tilde{\mathbf{U}}^{[f](i)})|_{[l,l]}$.

χ: Short-hand: $\chi = \frac{K_0}{K}\alpha$.

$\Lambda_p^{[f](i)}$:
Diagonal matrix of eigenvalues $\lambda_{p,l}^{[f](i)}$, $l = 0, \ldots, K-1$ associated with the EVD of the spaced-frequency correlation matrix $\mathbf{R}_p^{[f](i)}$: $\Lambda_p^{[f](i)} \in \mathbb{R}^{K \times K}$.

$\Lambda_{apt,p}^{[f](i)}$:
Diagonal matrix of eigenvalues $\lambda_{apt,p,l}^{[f](i)}$, $l = 0, \ldots, K-1$ associated with the EVD of the *a posteriori* channel transfer factor estimates' spaced-frequency correlation matrix $\mathbf{R}_{apt,p}^{[f](i)}$: $\Lambda_{apt,p}^{[f](i)} \in \mathbb{R}^{K \times K}$.

$\tilde{\Lambda}_p^{[f](i)}$:
Diagonal matrix of eigenvalues $\tilde{\lambda}_{p,l}^{[f](i)}$, $l = 0, \ldots, K-1$ associated with the EVD of the spaced-frequency correlation matrix $\tilde{\mathbf{R}}_p^{[f](i)}$ invoked in the design of the channel predictor, where we have: $\tilde{\Lambda}_p^{[f](i)} \in \mathbb{R}^{K \times K}$.

$\Upsilon_p^{[f](i)}$:
Decomposition of the spaced-frequency correlation matrix $\mathbf{R}_p^{[f](i)}$, with respect to the channel predictor's unitary transform matrix $\tilde{\mathbf{U}}_p^{[f](i)}$, which is given by: $\Upsilon_p^{[f](i)} = \tilde{\mathbf{U}}_p^{[f](i)H} \mathbf{R}_p^{[f](i)} \tilde{\mathbf{U}}_p^{[f](i)} \in \mathbb{C}^{K \times K}$.

$\Delta \hat{\mathbf{H}}_{apr,p}^{(i)}[n]$:
Vector of the K different subcarriers' *a priori* channel transfer factor estimation errors: $\Delta \hat{\mathbf{H}}_{apr,p}^{(i)}[n,k] = \mathbf{H}_p^{(i)}[n] - \hat{\mathbf{H}}_{apr,p}^{(i)}[n] \in \mathbb{C}^{K \times 1}$.

$\Delta \tilde{\mathbf{H}}_{apt,p}^{(i)}[n]$:
Vector of the K different subcarriers' *a posteriori* channel transfer factor estimation errors: $\Delta \tilde{\mathbf{H}}_{apt,p}^{(i)}[n,k] = \mathbf{H}_p^{(i)}[n] - \tilde{\mathbf{H}}_{apt,p}^{(i)}[n] \in \mathbb{C}^{K \times 1}$.

$\nabla_p^{(j)}$:
Nabla operator with respect to the vector of predictor coefficients $\tilde{\mathbf{c}}_{pre,p}^{(j)*}$, which is expressed as: $\nabla_p^{(j)} = \frac{\partial}{\partial \tilde{\mathbf{c}}_{pre,p}^{(j)*}}$.

$()_p^{(j)}$:
Notation used for indicating that the variable in round brackets is associated with the channel encountered between the j-th user's single transmit antenna and the p-th receiver antenna element.

$()|^{\text{SIMPLE}}$:
Notation used for indicating that the variable in round brackets, which could either be the vector of optimum predictor coefficients, or the associated estimation MSE is employed in the context of the simplified scenario of different users having identical transmit powers, as well as identical modulation-mode noise enhancement factors and identical channel statistics.

$()_{\text{FIR}}$:
Notation used in conjunction with the vector of optimum predictor coefficients in order to indicate that the $(L-1)$ number of interfering users' MSE contributions were neglected in the coefficients' calculation.

$()_{\text{spar}}$:
Notation used for indicating a sparse multipath intensity profile $r_{h,\text{spar}}(\tau)$, a spaced-frequency correlation function $r_{H,\text{spar}}(\Delta f)$, a spaced-frequency correlation matrix $\mathbf{R}_{\text{spar}}^{[f]}$ and a decomposition $\Upsilon_{\text{spar}}^{[f]}$ of the spaced-frequency correlation matrix with respect to the DFT matrix $\mathbf{W}$.

$()_{\text{unif}}$:   Notation used for specifying a uniform multipath intensity profile $r_{h,\text{unif}}(\tau)$, a spaced-frequency correlation function $r_{H,\text{unif}}(\Delta f)$, a spaced-frequency correlation matrix $\mathbf{R}_{\text{unif}}^{[f]}$ and a decomposition $\Upsilon_{\text{unif}}^{[f]}$ of the spaced-frequency correlation matrix with respect to the DFT matrix $\mathbf{W}$.

$()_{\text{expo}}$:   Notation used for a negative exponentially decaying multipath intensity profile $r_{h,\text{expo}}(\tau)$, a spaced-frequency correlation function $r_{H,\text{expo}}(\Delta f)$, a spaced-frequency correlation matrix $\mathbf{R}_{\text{expo}}^{[f]}$ and a decomposition $\Upsilon_{\text{expo}}^{[f]}$ of the spaced-frequency correlation matrix with respect to the DFT matrix $\mathbf{W}$.

Chapter **17**

Uplink Detection Techniques for Multi-User SDMA-OFDM[1]

17.1 Introduction

17.1.1 Classification of Smart Antennas

In recent years various smart antenna designs have emerged, which have found application in diverse scenarios, as seen in Table 17.3. The main objective of employing smart antennas is that of combating the effects of multipath fading on the desired signal and suppressing interfering signals, thereby increasing both the performance and capacity of wireless systems [488]. Specifically, in smart antenna assisted systems multiple antennas may be invoked at the transmitter and/or the receiver, where the antennas may be arranged for achieving spatial diversity, directional beamforming or for attaining both diversity and beamforming. In smart antenna systems the achievable performance improvements are usually a function of the antenna spacing and that of the algorithms invoked for processing the signals received by the antenna elements.

In beamforming arrangements [209] typically $\lambda/2$-spaced antenna elements are used for the sake of creating a spatially selective transmitter/receiver beam. Smart antennas using beamforming have widely been employed for mitigating the effects of various interfering signals and for providing beamforming gain. Furthermore, the beamforming arrangement is capable of suppressing co-channel interference, which allows the system to support multiple users within the same bandwidth and/or same time-slot by separating them spatially. This spatial separation becomes however only feasible, if the corresponding users are separable in terms of the angle of arrival of their beams. These beamforming schemes, which employ appropriately phased antenna array elements that are spaced at distances of $\lambda/2$ typically result in an improved SINR distribution and enhanced network capacity [209].

---

[1]*OFDM and MC-CDMA for Broadband Multi-user Communications WLANS and Broadcasting.* L.Hanzo, Münster, T. Keller and B.J. Choi,
©2003 John Wiley & Sons, Ltd. ISBN 0-470-85879-6

Year	Author	Contribution
'96	Foschini [104]	The concept of the BLAST architecture was introduced.
'98	Vook and Baum [105]	SMI-assisted MMSE combining was invoked on an OFDM subcarrier basis.
	Wang and Poor [106]	Robust sub-space-based weight vector calculation and tracking were employed for co-channel interference suppression, as an improvement of the SMI-algorithm.
	Wong, Cheng, Letaief and Murch [107]	Optimisation of an OFDM system was reported in the context of multiple transmit and receive antennas upon invoking the maximum SINR criterion. The computational complexity was reduced by exploiting the channel's correlation in the frequency direction.
	Li and Sollenberger [108]	Tracking of the channel correlation matrix' entries was suggested in the context of SMI-assisted MMSE combining for multiple receiver antenna-assisted OFDM, by capitalising on the principles of [68].
'99	Golden, Foschini, Valenzuela and Wolniansky [109]	The SIC detection assisted V-BLAST algorithm was introduced.
	Li and Sollenberger [110]	The system introduced in [108] was further detailed.
	Vandenameele, Van Der Perre, Engels and H. D. Man [111]	A comparative study of different SDMA detection techniques, namely that of MMSE, SIC and ML detection was provided. Further improvements of SIC detection were suggested by adaptively tracking multiple symbol decisions at each detection node.
	Speth and Senst [112]	Soft-bit generation techniques were proposed for MLSE in the context of a coded SDMA-OFDM system.
'00	Sweatman, Thompson, Mulgrew and Grant [113]	Comparisons of various detection algorithms including LS, MMSE, D-BLAST and V-BLAST (SIC detection) were carried out.
	van Nee, van Zelst and Awater [114–116]	The evaluation of ML detection in the context of a Space-Division Multiplexing (SDM) system was provided, considering various simplified ML detection techniques.
	Vandenameele, Van Der Perre, Engels, Gyselinckx and De Man [117]	More detailed discussions were provided on the topics of [111].

Table 17.1: Contributions on detection techniques for MIMO systems and more specifically multiple transmit antenna-assisted OFDM systems.

Year	Author	Contribution
'00	Li, Huang, Lozano and Foschini [118]	Reduced complexity ML detection was proposed for multiple transmit antenna systems employing adaptive antenna grouping and multi-step reduced-complexity detection.
'01	Degen, Walke, Lecomte and Rembold [119]	An overview of various adaptive MIMO techniques was provided. Specifically, pre-distortion was employed at the transmitter, as well as LS- or BLAST detection were used at the receiver or balanced equalisation was invoked at both the transmitter and receiver.
	Zhu and Murch [120]	A tight upper bound on the SER performance of ML detection was derived.
	Li, Letaief, Cheng and Cao [121]	Joint adaptive power control and detection were investigated in the context of an OFDM/SDMA system, based on the approach of Farrokhi *et al.* [122].
	van Zelst, van Nee and Awater [123]	Iterative decoding was proposed for the BLAST system following the turbo principle.
	Benjebbour, Murata and Yoshida [124]	The performance of V-BLAST or SIC detection was studied in the context of a backward iterative cancellation scheme employed after the conventional forward cancellation stage.
	Sellathurai and Haykin [125]	A simplified D-BLAST was proposed, which used iterative PIC capitalising on the extrinsic soft-bit information provided by the FEC scheme employed.
	Bhargave, Figueiredo and Eltoft [126]	A detection algorithm was suggested, which followed the concepts of V-BLAST or SIC. However, multiple symbols states are tracked from each detection stage, where – in contrast to [117] – an intermediate decision is made at intermediate detection stages.
	Thoen, Deneire, Van Der Perre and Engels [127]	A constrained LS detector was proposed for OFDM/SDMA, which was based on exploiting the constant modulus property of PSK signals.
'02	Li and Luo [128]	The block error probability of optimally ordered V-BLAST was studied. Furthermore, the block error probability was also investigated for the case of tracking multiple parallel symbol decisions from the first detection stage, following an approach similar to that of [117].

Table 17.2: Contributions on detection techniques for MIMO systems and for multiple transmit antenna-assisted OFDM systems.

Beamforming [209]	Typically $\lambda/2$-spaced antenna elements are used for the sake of creating a spatially selective transmitter/receiver beam. Smart antennas using beamforming have been employed for mitigating the effects of cochannel interfering signals and for providing beamforming gain.
Spatial Diversity [208] and Space-Time Spreading	In contrast to the $\lambda/2$-spaced phased array elements, in spatial diversity schemes, such as space-time block or trellis codes [208] the multiple antennas are positioned as far apart as possible, so that the transmitted signals of the different antennas experience independent fading, resulting in the maximum achievable diversity gain.
Space Division Multiple Access	SDMA exploits the unique, user-specific "spatial signature" of the individual users for differentiating amongst them. This allows the system to support multiple users within the same frequency band and/or time slot.
Multiple Input Multiple Output Systems [104]	MIMO systems also employ multiple antennas, but in contrast to SDMA arrangements, not for the sake of supporting multiple users. Instead, they aim for increasing the throughput of a wireless system in terms of the number of bits per symbol that can be transmitted by a given user in a given bandwidth at a given integrity.

Table 17.3: Applications of multiple antennas in wireless communications

In contrast to the $\lambda/2$-spaced phased array elements, **in spatial diversity schemes**, such as space-time coding [208] aided transmit diversity arrangements, the multiple antennas are positioned as far apart as possible. A typical antenna element spacing of 10λ [488] may be used, so that the transmitted signals of the different antennas experience independent fading, when they reach the receiver. This is because the maximum diversity gain can be achieved, when the received signal replicas experience independent fading. Although spatial diversity can be achieved by employing multiple antennas at either the base station, mobile station, or both, it is more cost effective and practical to employ multiple transmit antennas at the base station. A system having multiple receiver antennas has the potential of achieving receiver diversity, while that employing multiple transmit antennas exhibits transmit diversity. Recently, the family of transmit diversity schemes based on space-time coding, either space-time block codes or space-time trellis codes, has received wide attention and has been invoked in the 3rd-generation systems [209,489]. The aim of using spatial diversity is to provide both transmit as well as receive diversity and hence enhance the system's integrity/robustness. This typically results in a better physical-layer performance and hence a better network-layer performance, hence space-time codes indirectly increase not only the transmission integrity, but also the achievable spectral efficiency.

A third application of smart antennas is often referred to as **Space Division Multiple Access** (SDMA), which exploits the unique, user-specific "spatial signature" of the individual users for differentiating amongst them. In simple conceptual terms one could argue that both a conventional CDMA spreading code and the Channel Impulse Response (CIR) affect the transmitted signal similarly - they are namely convolved with it. Hence, provided that the CIR is accurately estimated, it becomes known and certainly unique, although - as opposed to orthogonal Walsh-Hadamad spreading codes, for example - not orthogonal to the other CIRs. Nonetheless, it may be used for uniquely identifying users after channel estimation and hence for supporting several users within the same bandwidth. Provided that a powerful multiuser detector is available, one can support even more users than the number of antennas. Hence this method enhances the achievable spectral efficiency directly.

Finally, Multiple Input Multiple Output (MIMO) systems [104, 490–493] also employ multiple antennas, but in contrast to SDMA arrangements, not for the sake of supporting multiple users. Instead, they aim for increasing the throughput of a wireless system in terms of the number of bits per symbol that can be transmitted by a single user in a given bandwidth at a given integrity.

17.1.2   Introduction to Space-Division-Multiple-Access

Space-Division-Multiple-Access (SDMA) communication systems have recently drawn wide interests. In these systems the L different users' transmitted signals are separated at the base-station (BS) with the aid of their unique, user-specific spatial signature, which is constituted by the P-element vector of channel transfer factors between the users' single transmit antenna and the P different receiver antenna elements at the BS, upon assuming flat-fading channel conditions such as in each of the OFDM subcarriers. This will be further detailed during our portrayal of the SDMA-MIMO channel model in Section 17.1.5.

A whole host of multi-user detection (MUD) techniques known from Code-Division-Multiple-Access (CDMA) communications lend themselves also to an application in the context of SDMA-OFDM on a per-subcarrier basis. Some of these techniques are the Least-Squares (LS) [113, 119, 127, 135], Minimum Mean-Square Error (MMSE) [105–108, 110, 113, 117, 121, 135–137], Successive Interference Cancellation (SIC) [104, 109, 113, 117, 119, 124, 126, 128, 135, 137], Parallel Interference Cancellation (PIC) [125, 135] and Maximum Likelihood (ML) detection [112, 114–118, 120, 123, 135, 137]. A comprehensive overview of recent publications on MUD techniques for MIMO systems is given in Tables 17.1 and 17.2.

In Section 17.1.3 a detailed classification of the different MUD techniques is provided. By contrast, a more simple classification is employed in Section 17.1.4, which reflects the structure of this chapter. Before we embark on the discussion of linear MUD techniques in Section 17.2, the SDMA-MIMO channel model, to be used in our further discussions will be reviewed in Section 17.1.5

17.1.3   Classification of Multi-User Detection Techniques

The above techniques have been classified in Figure 17.1. Among the different techniques, the ML detection principle shown on the left-hand side of Figure 17.1 is known to exhibit the optimum performance, but also imposes the highest complexity. This is, because in ML detection the number of M_c^L trial symbol combinations, which are constituted by all possible combinations of the L different users' transmitted symbols belonging to an M_c-ary constellation, has to be evaluated in terms of the Euclidean distance between the vector of signals actually received by the P different antenna elements and the vector of trial signals, which are generated from all legitimate transmitted symbols, impaired according to the estimated

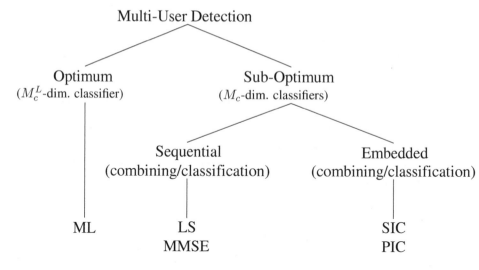

Figure 17.1: Classification of the various MUD techniques discussed in Chapter 17.

channel.

In order to avoid the potentially excessive complexity of the optimum ML detection, a range of sub-optimum detection techniques have been devised, which are summarised on the right-hand side of Figure 17.1. The philosophy of the suboptimum detectors is to reduce the dimensionality of the classification problem associated with selecting the specific constellation point, which is most likely to have been transmitted by each user.

Specifically, in the context of the LS- and MMSE detection techniques to be detailed in Sections 17.2.2 and 17.2.3, first linear estimates of the different users' transmitted signals are provided with the aid of the weighted combining of the signals received by the different antenna elements at the BS. This is followed by separately demodulating each of the L different users' combiner output signals. Hence, the original M_c^L-dimensional classification problem associated with the optimum ML detection has been reduced to L number of individual classification steps, each having a dimensionality of M_c. This is achieved at the cost of a BER degradation, which is associated with ignoring the residual interference contaminating the linear combiner's output signals.

We note again that in the context of LS- and MMSE detection the linear combining and classification steps are invoked in a sequential manner. However, a significant BER performance improvement can be achieved by embedding the classification process, which is a non-linear operation, into the linear combining process. Two of the most prominent representatives of this family of techniques are the SIC- and the PIC based detectors, which will be the topic of Sections 17.3.1 and 17.3.2, respectively.

17.1.4   Outline of Chapter 17

A possible classification of the various MUD techniques to be discussed in this chapter was presented in Figure 17.1. However, an alternative classification, which will serve as a guideline for our forthcoming discussions is portrayed in Figure 17.2. Here we have introduced

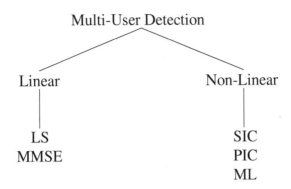

Figure 17.2: Classification of the various MUD techniques

the classes of linear- and non-linear detection techniques. The rationale of this classification is that in the context of linear detection techniques, such as LS- and MMSE detection, no *a priori* knowledge of the remaining users' transmitted symbols is required for the detection of a specific user. However, in the case of SIC, PIC and ML detection, *a priori* knowledge is involved, which must be provided by the non-linear classification operation involved in the demodulation process.

As shown in Figure 17.2, the further structure of this Chapter 17 is as follows. In Section 17.2 the most salient linear detection techniques, namely LS- and MMSE detection will be discussed. These discussions include their MSE- and BER performance analysis in the context of both uncoded and turbo-coded scenarios, as well as the analysis of the computational complexity.

In Section 17.3 we will then embark on a detailed analysis of the family of non-linear detection techniques, namely that of SIC, PIC and ML detection, again, with respect to their BER performance in both uncoded and turbo-coded scenarios. Furthermore, a complexity analysis will be carried out. Specifically, in the context of our analysis of SIC detection and its derivatives, namely of M-SIC and partial M-SIC we will focus our efforts on the effects of error propagation across the different detection stages. These investigations motivated the introduction of a weighted soft-bit metric to be employed in the context of turbo-decoding. Furthermore, in the context of PIC detection we proposed to embed turbo-decoding into the detection process, with the aim of increasing the reliability of the *a priori* symbol estimates employed in the PIC process. A final comparison between the different linear- and non-linear detection techniques in terms of their BER performance and computational complexity will be conducted at the end of Section 17.3.

In an effort to further enhance the performance of the different detection techniques without reducing their effective throughput, as in the case of turbo-coding, the applicability of adaptive modulation and Walsh-Hadamard Transform (WHT) based spreading will be investigated in Section 17.4. Our conclusions for this chapter will then be offered in Section 17.5.

However, before we embark on the investigation of linear detection techniques in Section 17.2, the SDMA-MIMO channel model will be introduced in the next section.

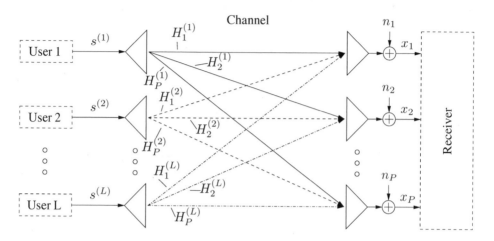

Figure 17.3: Schematic of an SDMA uplink MIMO channel scenario, where each of the L users is equipped with a single transmit antenna and the receiver is assisted by a P-element antenna front-end.

17.1.5　SDMA-MIMO Channel Model

In Figure 17.3 we have portrayed a Space-Division-Multiple-Access (SDMA) uplink transmission scenario, where each of the L simultaneous users is equipped with a single transmission antenna, while the receiver capitalises on a P-element antenna front-end. The vector of complex signals, $\mathbf{x}[n,k]$, received by the P-element antenna array in the k-th subcarrier of the n-th OFDM symbol is constituted by the superposition of the independently faded signals associated with the L users sharing the same space-frequency resource. The received signal was corrupted by the Gaussian noise at the array elements. The indices $[n,k]$ have been omitted for notational convenience during our forthcoming discourse, yielding:

$$\mathbf{x} \;=\; \mathbf{Hs} + \mathbf{n}, \tag{17.1}$$

where the vector $\mathbf{x} \in \mathbb{C}^{P \times 1}$ of received signals, the vector $\mathbf{s} \in \mathbb{C}^{L \times 1}$ of transmitted signals and the array noise vector $\mathbf{n} \in \mathbb{C}^{P \times 1}$, respectively, are given by:

$$\mathbf{x} \;=\; (x_1, x_2, \ldots, x_P)^T, \tag{17.2}$$

$$\mathbf{s} \;=\; (s^{(1)}, s^{(2)}, \ldots, s^{(L)})^T, \tag{17.3}$$

$$\mathbf{n} \;=\; (n_1, n_2, \ldots, n_P)^T. \tag{17.4}$$

The frequency domain channel transfer factor matrix $\mathbf{H} \in \mathbb{C}^{P \times L}$ is constituted by the set of channel transfer factor vectors $\mathbf{H}^{(l)} \in \mathbb{C}^{P \times 1}$, $l = 1, \ldots, L$ of the L users:

$$\mathbf{H} \;=\; (\mathbf{H}^{(1)}, \mathbf{H}^{(2)}, \ldots, \mathbf{H}^{(L)}), \tag{17.5}$$

each of which hosts the frequency domain channel transfer factors between the single transmitter antenna associated with a particular user l and the reception antenna elements

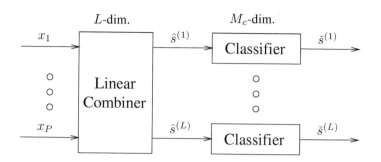

Figure 17.4: Illustration of the main signal paths associated with a linear detector such as LS or MMSE. The signals $x_p, p = 1, \ldots, P$ received by the different antenna elements are fed into the linear combiner, whose associated output vector $\hat{\mathbf{s}}$ of linear signal estimates with elements $\hat{s}^{(l)}, l = 1, \ldots, L$ is defined by Equation 17.7, and the LS- or MMSE-specific weight matrices are given by Equations 17.39 or 17.64 in its right-inverse related form and 17.68 in its left-inverse related form. The l-th user's signal, where $l = 1, \ldots, L$, is then conveyed to a separate classifier or demodulator, at the output of which the amplified constellation point $\check{s}^{(l)}$ most likely transmitted by the l-th user becomes available. The demodulator is described by Equation 17.94. Note that here we have omitted the signal paths associated with the channel transfer factor estimates required by the linear combiner.

$p = 1, \ldots, P$:

$$\mathbf{H}^{(l)} = (H_1^{(l)}, H_2^{(l)}, \ldots, H_P^{(l)})^T, \qquad (17.6)$$

with $l \in \{1, \ldots, L\}$. Regarding the statistical properties of the components associated with the vectors involved in Equation 17.1, we assume that the complex data signal $s^{(l)}$ transmitted by the l-th user has zero-mean and a variance of σ_l^2. The AWGN noise process n_p at any antenna array element p exhibits also zero-mean and a variance of σ_n^2. The frequency domain channel transfer factors $H_p^{(l)}$ of the different array elements $p \in \{1, \ldots, P\}$ or users $l \in \{1, \ldots, L\}$ are independent, stationary, complex Gaussian distributed processes with zero-mean and unit variance.

17.2   Linear Detection Techniques

The first class of detectors portrayed in this chapter belong to the family of the so-called linear detectors. Their employment is motivated by the observation that in the context of the optimum ML detector to be discussed in Section 17.3.3 a potentially excessive number of M_c^L trial symbol combinations has to be tested in terms of their associated trial signals' Euclidean distance measured from the vector of signals received by the different antenna elements. Recall that L represents the number of simultaneously transmitting users and M_c represents the number of legitimate transmitted symbols. Depending on the L number of simultaneous users supported and the M_c number of constellation points, a practical implementation of the ML detector may become unrealistic, as we will show during our complexity comparison in Section 17.3.4.2. A more practical approach is hence to generate estimates of the different users' transmitted signals with the aid of a linear combiner. These signal estimates would then be demodulated separately for each of the L users upon neglecting the residual interference caused by the remaining users in a specific user's combiner output signal. Hence the dimensionality of the receiver's classification task during demodulation is reduced from evaluating the multi-user Euclidean distance metric M_c^L times to the evaluation of the single-user Euclidean distance metric L times for all the M_c symbols. This reduces the total complexity to evaluating the Euclidean metric LM_c times. A simplified block diagram of the linear detector is also shown in Figure 17.4.

Our discussions commence in Section 17.2.1 with the characterisation of the linear combiner's output signal and its components. By contrast, in Sections 17.2.2 and 17.2.3 we will focus our attention on two specific linear combiners, namely on the LS combiner and on the MMSE combiner, respectively.[2] These combiners constitute the basis for our discussions on non-linear detection techniques, such as SIC, PIC and transform-based ML detection in Sections 17.3.1, 17.3.2 and 17.3.3. Furthermore, in Section 17.2.4 the process of symbol classification or demodulation - as seen on the right-hand side of Figure 17.4 - is described while in Section 17.2.5 the generation of soft-bit values for turbo-decoding is outlined. The LS and MMSE detectors are characterised in terms of their associated combiner's MSE and SINR as well as the detector's BER performance in Section 17.2.6. This is followed by a detailed complexity analysis in Section 17.2.7. Finally, our conclusions on linear detection techniques will be offered in Section 17.2.8.

17.2.1 Characterisation of the Linear Combiner's Output Signal

As the terminology suggests, an estimate $\hat{\mathbf{s}} \in \mathbb{C}^{L \times 1}$ of the vector of transmitted signals $\mathbf{s}$ of the L simultaneous users is generated by linearly combining the signals received by the P different receiver antenna elements with the aid of the weight matrix $\mathbf{W} \in \mathbb{C}^{P \times L}$, resulting in:

$$\hat{\mathbf{s}} = \mathbf{W}^H \mathbf{x}. \tag{17.7}$$

In order to gain further insight into the specific structure of the combiner's output signal on a component basis, let us substitute Equation 17.1 into Equation 17.7 and consider the l-th user's associated vector component:

$$\hat{s}^{(l)} = \mathbf{w}^{(l)H} \mathbf{x} \tag{17.8}$$

$$= \mathbf{w}^{(l)H} (\mathbf{Hs} + \mathbf{n}) \tag{17.9}$$

$$= \mathbf{w}^{(l)H} \mathbf{H}^{(l)} s^{(l)} + \mathbf{w}^{(l)H} \sum_{\substack{i=1 \\ i \neq l}}^{L} \mathbf{H}^{(i)} s^{(i)} + \mathbf{w}^{(l)H} \mathbf{n}, \tag{17.10}$$

where the weight vector $\mathbf{w}^{(l)} \in \mathbb{C}^{P \times 1}$ is the l-th column vector of the weight matrix $\mathbf{W}$. While in Section 17.2.1.1 we briefly characterise the different additive components of the combiner's output signal, their statistical properties recorded in terms of the different contributions' variances are highlighted in Section 17.2.1.2. On the basis of these, the three most

---

[2]Note that in the following the specific linear detector is referred to as the LS- or MMSE-detector, depending on whether LS- or MMSE combining is employed.

prominent performance measures used for assessing the quality of the combiner's output signal namely, the SINR, the SIR and the SNR will be introduced in Section 17.2.1.3.

17.2.1.1 Description of the Different Signal Components

We observe from Equation 17.10 that the combiner's output signal is constituted by three additive components. More specifically, in Equation 17.10 the first term, namely:

$$\hat{s}_S^{(l)} = \mathbf{w}^{(l)H} \mathbf{H}^{(l)} s^{(l)} \tag{17.11}$$

denotes the desired user's associated contribution, while the second term, namely:

$$\hat{s}_I^{(l)} = \mathbf{w}^{(l)H} \sum_{\substack{i=1 \\ i \neq l}}^{L} \mathbf{H}^{(i)} s^{(i)} \tag{17.12}$$

denotes the interfering users' residual contribution. Finally, the last term, namely:

$$\hat{s}_N^{(l)} = \mathbf{w}^{(l)H} \mathbf{n} \tag{17.13}$$

is related to the AWGN. These components can be further characterised in terms of their variances, which will be further elaborated on in the next section.

17.2.1.2 Statistical Characterisation

Specifically, the variance of the desired user's detected signal is given by:

$$
\begin{align}
\sigma_S^{(l)2} &= E\{\hat{s}_S^{(l)H} \hat{s}_S^{(l)}\} \tag{17.14} \\
&= \mathbf{w}^{(l)H} \mathbf{R}_{a,S}^{(l)} \mathbf{w}^{(l)}, \quad \text{where} \tag{17.15} \\
\mathbf{R}_{a,S}^{(l)} &= \sigma_l^2 \mathbf{H}^{(l)} \mathbf{H}^{(l)H} \in \mathbb{C}^{P \times P} \tag{17.16}
\end{align}
$$

is the auto-correlation matrix of the desired user's signal. Following similar calculations, the variance of the interfering users' contribution is given as:

$$
\begin{align}
\sigma_I^{(l)2} &= E\{\hat{s}_I^{(l)H} \hat{s}_I^{(l)}\} \tag{17.17} \\
&= \mathbf{w}^{(l)H} \mathbf{R}_{a,I}^{(l)} \mathbf{w}^{(l)}, \quad \text{where} \tag{17.18} \\
\mathbf{R}_{a,I}^{(l)} &= \sum_{\substack{i=1 \\ i \neq l}}^{L} \sigma_i^2 \mathbf{H}^{(i)} \mathbf{H}^{(i)H} \in \mathbb{C}^{P \times P} \tag{17.19}
\end{align}
$$

is the auto-correlation matrix of the interfering users' signals. Finally, the residual AWGN related variance can be expressed as:

$$
\sigma_N^{(l)2} = E\{\hat{s}_N^{(l)H}\hat{s}_N^{(l)}\} \tag{17.20}
$$

$$
= \mathbf{w}^{(l)H}\mathbf{R}_{a,N}\mathbf{w}^{(l)}, \quad \text{where} \tag{17.21}
$$

$$
\mathbf{R}_{a,N} = \sigma_n^2\mathbf{I} \in \mathbb{C}^{P \times P} \tag{17.22}
$$

is the diagonal noise correlation matrix. Specifically, in the context of Equation 17.22 we have exploited again that the AWGN observed at different elements of the receiver antenna array is uncorrelated. For employment at a later stage we will additionally define here the undesired signal's auto-correlation matrix, which is related to the sum of the residual interference plus the AWGN expressed as:

$$
\mathbf{R}_{a,I+N}^{(l)} = \mathbf{R}_{a,I}^{(l)} + \mathbf{R}_{a,N}, \tag{17.23}
$$

where the matrices $\mathbf{R}_{a,I}^{(l)}$ and $\mathbf{R}_{a,N}$ were given by Equations 17.19 and 17.22, respectively.

17.2.1.3   Performance Measures

Three different performance measures can be defined on the basis of the desired signal's variance $\sigma_S^{(l)2}$, the interfering signal's variance $\sigma_I^{(l)2}$ and the noise variance $\sigma_N^{(l)2}$, which were given by Equations 17.14, 17.17 and 17.20. These measures can be employed to characterise the quality of the linear combiner's output signal. These are the Signal-to-Interference-plus-Noise Ratio (SINR) at the combiner's output, defined as [136]:

$$
\text{SINR}^{(l)} = \frac{\sigma_S^{(l)2}}{\sigma_I^{(l)2} + \sigma_N^{(l)2}} = \frac{\mathbf{w}^{(l)H}\mathbf{R}_{a,S}^{(l)}\mathbf{w}^{(l)}}{\mathbf{w}^{(l)H}\mathbf{R}_{a,I+N}^{(l)}\mathbf{w}^{(l)}}, \tag{17.24}
$$

the Signal-to-Interference Ratio (SIR), defined as [136]:

$$
\text{SIR}^{(l)} = \frac{\sigma_S^{(l)2}}{\sigma_I^{(l)2}} = \frac{\mathbf{w}^{(l)H}\mathbf{R}_{a,S}^{(l)}\mathbf{w}^{(l)}}{\mathbf{w}^{(l)H}\mathbf{R}_{a,I}^{(l)}\mathbf{w}^{(l)}}, \tag{17.25}
$$

and the Signal-to-Noise Ratio (SNR) given by [136]:

$$
\text{SNR}^{(l)} = \frac{\sigma_S^{(l)2}}{\sigma_N^{(l)2}} = \frac{\mathbf{w}^{(l)H}\mathbf{R}_{a,S}^{(l)}\mathbf{w}^{(l)}}{\mathbf{w}^{(l)H}\mathbf{R}_{a,N}^{(l)}\mathbf{w}^{(l)}}. \tag{17.26}
$$

In the next section we will embark on the portrayal of least-squares error detection, in an effort to compare a number of different criteria that can be invoked to adjust the detector's associated combiner weight matrix $\mathbf{W}$ introduced in Equation 17.7.

17.2.2 Least-Squares Error Detector

With reference to Figure 17.4, in this section we will derive the Least-Squares (LS) error- or Zero-Forcing (ZF) combiner [113, 119, 127, 135], which attempts to recover the vector $s[n, k]$ of signals transmitted by the L different users in the k-th subcarrier of the n-th OFDM symbol period, regardless of the signal quality quantified in terms of the SNR at the reception antennas. For simplicity, we will again omit the index $[n, k]$ throughout our forthcoming discourse. Our description of the LS combiner is structured as follows. In Section 17.2.2.1 a simplified model $\hat{\mathbf{x}}$ of the vector $\mathbf{x}$ of signals received by the P different antenna elements is introduced as a function of the estimate $\hat{\mathbf{s}}$ of the vector of L number of transmitted signals. On the basis of this simplified model a cost-function is established in Section 17.2.2.2, which follows the philosophy of the squared model mismatch error. The estimate $\hat{\mathbf{s}}$ of the L different users' transmitted symbols is then determined in Section 17.2.2.3 with the aid of the conjugate-gradient method [90]. Alternatively, the so-called orthogonality principle [90] could be invoked. Furthermore, in Section 17.2.2.4 a condition is provided, which has to be satisfied in order to be able to identify the estimate $\hat{\mathbf{s}}$, while in Sections 17.2.2.5 and 17.2.2.6 expressions are presented for both the squared error measured in the received signal's domain and for the mean-square error evaluated in the transmitted signal's domain.

17.2.2.1 Simplified Model of the Received Signal

Upon assuming perfect knowledge of the channel transfer factor matrix $\mathbf{H}$ an estimate $\hat{\mathbf{x}} \in \mathbb{C}^{P \times 1}$ of the vector of signals received by the P different antenna elements in a specific subcarrier is given in teh same way as Equation 17.1 by:

$$\hat{\mathbf{x}} = \mathbf{H}\hat{\mathbf{s}}, \tag{17.27}$$

where $\hat{\mathbf{s}} \in \mathbb{C}^{L \times 1}$ is the estimate of the vector of signals transmitted by the L different users, which we are attempting to recover.

17.2.2.2 Least-Squares Error Cost-Function

The estimation error $\Delta\hat{\mathbf{x}} \in \mathbb{C}^{P \times 1}$ in the received signal's domain can hence be expressed as:

$$\Delta\hat{\mathbf{x}} = \mathbf{x} - \hat{\mathbf{x}} \tag{17.28}$$
$$= \mathbf{x} - \mathbf{H}\hat{\mathbf{s}}. \tag{17.29}$$

Correspondingly, the squared error $||\Delta\hat{\mathbf{x}}||_2^2 \in \mathbb{R}$ is given as the inner product of the vector of LS estimation errors formulated in Equation 17.28, namely as:

$$||\Delta\hat{\mathbf{x}}||_2^2 = \Delta\hat{\mathbf{x}}^H \Delta\hat{\mathbf{x}} \tag{17.30}$$
$$= \mathbf{x}^H\mathbf{x} - 2\Re(\hat{\mathbf{s}}^H \mathbf{p}_{LS}) + \hat{\mathbf{s}}^H \mathbf{Q}_{LS}\hat{\mathbf{s}}, \tag{17.31}$$

where the "cross-correlation" vector $\mathbf{p}_{LS} \in \mathbb{C}^{L \times 1}$ is given by:

$$\mathbf{p}_{LS} = \mathbf{H}^H\mathbf{x}, \tag{17.32}$$

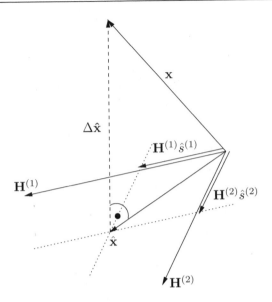

Figure 17.5: Illustration of the principle of LS detection for a scenario of $L = 2$ simultaneous users. The vector $\mathbf{x}$ of received signals is projected onto the vector space spanned by the column vectors $\mathbf{H}^{(1)}$, $\mathbf{H}^{(2)}$ hosted by the channel matrix $\mathbf{H}$.

while the "auto-correlation" matrix $\mathbf{Q}_{LS} \in \mathbb{C}^{L \times L}$ is given by:[3]

$$\mathbf{Q}_{LS} = \mathbf{H}^H \mathbf{H}. \tag{17.33}$$

17.2.2.3 Recovery of the Transmitted Signals by the Gradient Approach

A standard approach designed to determine the desired vector $\hat{\mathbf{s}}$ representing the estimated transmitted signals of the L users is to minimise the squared error given by Equation 17.30. This can be achieved by noting that in the optimum point of operation, associated with the weight matrix having the optimum weights, the conjugate gradient $\nabla^* = \frac{\partial}{\partial \hat{\mathbf{s}}^*} \in \mathbb{C}^{L \times 1}$ of $||\Delta \hat{\mathbf{x}}||_2^2$ with respect to $\hat{\mathbf{s}}$ is equal to zero, which can be expressed as:

$$||\Delta \hat{\mathbf{x}}||_2^2 = \text{Min.} \quad \Longleftrightarrow \quad \nabla^* ||\Delta \hat{\mathbf{x}}||_2^2 \overset{!}{=} 0. \tag{17.34}$$

Upon substituting Equation 17.31 into Equation 17.34, after some mathematical manipulations we obtain:

$$\mathbf{Q}_{LS} \hat{\mathbf{s}}_{LS} = \mathbf{p}_{LS} \quad \Longleftrightarrow \quad \hat{\mathbf{s}}_{LS} = \mathbf{Q}_{LS}^{-1} \mathbf{p}_{LS}. \tag{17.35}$$

---

[3]Note that $\mathbf{p}_{LS}$ and $\mathbf{Q}_{LS}$ have been intentionally termed "cross-correlation" vector and "auto-correlation" matrices, in order to highlight the similarities to the corresponding matrices in the context of MMSE combining, although the expectation operator is not invoked here.

Here we have specifically exploited that the gradient with respect to the first term of Equation 17.31 is equal to zero, while the gradient with respect to the second term was evaluated by capitalising on $\frac{\partial}{\partial \mathbf{z}^*}\mathfrak{Re}(\mathbf{z}^H\mathbf{a}) = \frac{1}{2}\mathbf{a}$ [90] for complex valued vectors $\mathbf{z}$ and $\mathbf{a}$. Furthermore, in the context of evaluating the gradient of the last term in Equation 17.31 we have exploited that we have $\frac{\partial}{\partial \mathbf{z}^*}\mathbf{z}^H\mathbf{R}\mathbf{z} = \mathbf{R}\mathbf{z}$ [90] in conjunction with a Hermitian matrix $\mathbf{R}$.

Upon substituting Equations 17.32 and 17.33 into Equation 17.35 we obtain the following expression for the vector $\hat{\mathbf{s}}_{LS}$ of estimated transmitted signals of the L simultaneous users:

$$\hat{\mathbf{s}}_{LS} = \mathbf{P}_{LS}\mathbf{x}, \tag{17.36}$$

where the projection matrix $\mathbf{P}_{LS} \in \mathbb{C}^{L \times P}$ is given by:

$$\mathbf{P}_{LS} = (\mathbf{H}^H\mathbf{H})^{-1}\mathbf{H}^H. \tag{17.37}$$

More specifically, the matrix $\mathbf{P}_{LS}$ projects the vector $\mathbf{x}$ of the P different antenna elements' received signals onto the column space of the channel matrix $\mathbf{H}$ [90]. These principles are further illustrated in Figure 17.5. As a comparison of Equations 17.7 and 17.36 reveals, the least-squares estimation based weight matrix $\mathbf{W}_{LS} \in \mathbb{C}^{P \times L}$ is hence given by:

$$\mathbf{W}_{LS} = \mathbf{P}_{LS}^H \tag{17.38}$$
$$= \mathbf{H}(\mathbf{H}^H\mathbf{H})^{-1}, \tag{17.39}$$

while the l-th user's associated weight vector $\mathbf{w}_{LS}^{(l)}$, which is the l-th column vector of matrix $\mathbf{W}_{LS}$ can be expressed as:

$$\mathbf{w}_{LS}^{(l)} = \mathbf{H} \cdot \mathbf{col}^{(l)}\{(\mathbf{H}^H\mathbf{H})^{-1}\}, \tag{17.40}$$

where $\mathbf{col}^{(l)}\{\}$ denotes the l-th column vector of the matrix enclosed in curly brackets.

17.2.2.4 Condition for Identification

From the literature [90], the projection matrix $\mathbf{P}_{LS}$ of Equation 17.37 is also known as the Moore-Penrose pseudo-inverse or left-inverse of $\mathbf{H}$ [90], which is denoted by $\mathbf{H}^\dagger$. A sufficient condition for its existence is that the L number of columns of the matrix $\mathbf{H}$ are linearly independent, which implies that we have $rank(\mathbf{H}) = L$. A necessary condition for this is that the P number of rows of $\mathbf{H}$ is equal to or larger than its L number of columns, namely that $P \geq L$. This implies that the maximum number of simultaneous users or transmit antennas supported by the LS combiner must be lower or equal to the P number of receiver antennas.

17.2.2.5 Squared Estimation Error in the Received Signals' Domain

The squared estimation error formulated in the sense of Equation 17.30 and associated with the vector $\hat{\mathbf{s}}_{LS}$ of least-squares signal estimates is given upon substituting Equation 17.35 into Equation 17.31, resulting in:

$$(||\Delta\hat{\mathbf{x}}||_2^2)_{LS} = \mathbf{x}^H\mathbf{x} - \mathbf{p}_{LS}^H\hat{\mathbf{s}}_{LS}. \tag{17.41}$$

17.2.2.6   Mean-Square Estimation Error in the Transmitted Signals' Domain

By substituting the received signal's model of Equation 17.1 into Equation 17.36 we obtain:

$$\hat{\mathbf{s}}_{LS} = \mathbf{s} + \mathbf{P}_{LS}\mathbf{n}, \tag{17.42}$$

which indicates that the LS-estimate $\hat{\mathbf{s}}_{LS}$ of the transmitted signal vector $\mathbf{s}$ of the L simultaneous users is based on a noise-contaminated version of $\mathbf{s}$. We note that the vector $\mathbf{s}$ of transmitted signals is restored regardless of the potential noise amplification incurred, which coined the term Zero-Forcing (ZF) combiner. Since $E\{\hat{\mathbf{s}}_{LS}\} = \mathbf{s}$, the vector $\hat{\mathbf{s}}_{LS}$ is also called an unbiased estimate of $\mathbf{s}$ [90]. Furthermore, from the literature [90] the vector $\hat{\mathbf{s}}_{LS}$ is also known as the Maximum Likelihood (ML) estimate of $\mathbf{s}$. More specifically, the vector $\hat{\mathbf{s}}_{LS}$ is a sample of an L-dimensional multi-variate complex Gaussian distribution, namely, $\hat{\mathbf{s}}_{LS} \sim \mathcal{CN}(\mathbf{s}, \mathbf{R}_{\Delta\hat{\mathbf{s}}_{LS}})$, [4], with the mean vector $\mathbf{s}$ and the covariance matrix $\mathbf{R}_{\Delta\hat{\mathbf{s}}_{LS}} \in \mathbb{C}^{L \times L}$.[5] given by:

$$\mathbf{R}_{\Delta\hat{\mathbf{s}}_{LS}} = E\{(\mathbf{P}_{LS}\mathbf{n})(\mathbf{P}_{LS}\mathbf{n})^H\} \tag{17.43}$$

$$= \sigma_n^2(\mathbf{H}^H\mathbf{H})^{-1}, \tag{17.44}$$

and where Equation 17.44 has been obtained by substituting Equation 17.37 into Equation 17.43, and by exploiting that $E\{\mathbf{n}\mathbf{n}^H\} = \sigma_n^2\mathbf{I}$. The average estimation Mean-Square Error (MSE) evaluated in the transmitted signals' domain is hence given by:

$$\overline{MSE}_{LS} = \frac{1}{L}\mathrm{Trace}(\mathbf{R}_{\Delta\hat{\mathbf{s}}_{LS}}) \tag{17.45}$$

$$= \frac{1}{L}\sigma_n^2\mathrm{Trace}((\mathbf{H}^H\mathbf{H})^{-1}), \tag{17.46}$$

while the l-th user's associated minimum MSE is given as the l-th diagonal element of the matrix $\mathbf{R}_{\Delta\hat{\mathbf{s}}_{LS}}$ in Equation 17.43. Hence, according to Equations 17.43 and 17.44 we have:

$$MSE_{LS}^{(l)} = \sigma_n^2\mathbf{w}_{LS}^{(l)H}\mathbf{w}_{LS}^{(l)} \tag{17.47}$$

$$= \sigma_n^2((\mathbf{H}^H\mathbf{H})^{-1})_{[l,l]}. \tag{17.48}$$

In the next section the potentially more effective MMSE detection approach will be discussed.

17.2.3   Minimum Mean-Square Error Detector

With reference to Figure 17.4, in contrast to the LS combiner of Section 17.2.2 the Minimum Mean-Square Error (MMSE) detector's associated MMSE combiner [105–108, 110, 113, 117, 121, 135–137] exploits the available statistical knowledge concerning the signals transmitted by the different users, as well as that related to the AWGN at the receiver antenna elements.

---

[4]The complex Gaussian distribution function is denoted here as $\mathcal{CN}()$, in order to distinguish it from the Gaussian distribution function defined for real-valued random variables.

[5]Note that $\Delta\mathbf{s}_{LS} = \mathbf{s} - \hat{\mathbf{s}}_{LS}$.

In Section 17.2.3.1 the mean-square error related cost-function is introduced, which is then employed in Section 17.2.3.2 to derive the optimum weight matrix with the aid of the conjugate-gradient approach [90]. Furthermore, in Section 17.2.3.3 expressions are provided for the average MSE and the user-specific MSE, respectively, in the context of employing the optimum weight matrix. Our discussions of the MMSE combiner are concluded in Section 17.2.3.4 by reducing the expression derived for the optimum weight matrix to a standard form, which will be shown later in Section 17.2.3.4 to differ only by a scalar factor from the corresponding expressions associated with the Minimum Variance (MV) combiner and also from that of the maximum SINR combiner.

17.2.3.1 Mean-Square Error Cost-Function

In contrast to the derivation of the LS combiner in Section 17.2.2, the cost-function employed here directly reflects the quality of the combiner weights in the transmitted signals' domain. In order to elaborate further, the vector $\Delta\hat{s} \in \mathbb{C}^{L \times 1}$ of the L simultaneous users' estimation errors evaluated in the transmitted signals' domain can be defined as:

$$\Delta\hat{s} = \mathbf{s} - \hat{\mathbf{s}} \tag{17.49}$$

$$= \mathbf{s} - \mathbf{W}^H\mathbf{x}, \tag{17.50}$$

where Equation 17.50 has been obtained by substituting Equation 17.7 into Equation 17.49. Furthermore, the estimation error's auto-correlation matrix $\mathbf{R}_{\Delta\hat{s}} \in \mathbb{C}^{L \times L}$ is given by:

$$\mathbf{R}_{\Delta\hat{s}} = E\{\Delta\hat{s}\Delta\hat{s}^H\} \tag{17.51}$$

$$= \mathbf{P} - \mathbf{R}_c^H\mathbf{W} - \mathbf{W}^H\mathbf{R}_c + \mathbf{W}^H\mathbf{R}_a\mathbf{W}, \tag{17.52}$$

where the cross-correlation matrix $\mathbf{R}_c \in \mathbb{C}^{P \times L}$ of the received and transmitted signals is defined as:

$$\mathbf{R}_c = E\{\mathbf{x}\mathbf{s}^H\} \tag{17.53}$$

$$= \mathbf{H}\mathbf{P}. \tag{17.54}$$

Similarly, for the auto-correlation matrix $\mathbf{R}_a \in \mathbb{C}^{P \times P}$ of the received signals we obtain:

$$\mathbf{R}_a = E\{\mathbf{x}\mathbf{x}^H\} \tag{17.55}$$

$$= \mathbf{H}\mathbf{P}\mathbf{H}^H + \sigma_n^2\mathbf{I} \tag{17.56}$$

$$= \sum_{l=1}^{L}\sigma_l^2\mathbf{H}^{(l)}\mathbf{H}^{(l)H} + \sigma_n^2\mathbf{I}. \tag{17.57}$$

Note that clearly the sum of the auto-correlation matrices $\mathbf{R}_{a,S}$, $\mathbf{R}_{a,I}$ and $\mathbf{R}_{a,N}$, given by Equations 17.16, 17.19 and 17.22, respectively, constitutes the auto-correlation matrix $\mathbf{R}_a \in \mathbb{C}^{P \times P}$ of the different reception antennas' associated signals, which can be expressed as:

$$\mathbf{R}_a = \mathbf{R}_{a,S}^{(l)} + \mathbf{R}_{a,I+N}^{(l)}, \tag{17.58}$$

where $\mathbf{R}^{(l)}_{a,I+N} = \mathbf{R}^{(l)}_{a,I} + \mathbf{R}_{a,N}$ from Equation 17.23. Furthermore, in Equations 17.54 and 17.56 the matrix $\mathbf{P} \in \mathbb{R}^{L \times L}$ is the diagonal matrix of the different users' associated transmit powers or signal variances, given by:

$$\mathbf{P} = Diag(\sigma_1^2, \sigma_2^2, \dots, \sigma_L^2). \tag{17.59}$$

In the context of deriving Equations 17.54 and 17.56 we have also exploited that $E\{\mathbf{ss}^H\} = \mathbf{P}$, as well as that $E\{\mathbf{nn}^H\} = \sigma_n^2 \mathbf{I}$ and that $E\{\mathbf{sn}^H\} = 0$. On the basis of Equation 17.52, the total mean-square estimation error $E\{||\Delta\hat{s}||_2^2\} \in \mathbb{R}$ accumulated for the different users is given by:

$$
\begin{aligned}
E\{||\Delta\hat{s}||_2^2\} &= \mathrm{Trace}(\mathbf{R}_{\Delta\hat{s}}) \tag{17.60} \\
&= \mathrm{Trace}(\mathbf{P}) - \mathrm{Trace}(\mathbf{R}_c^H \mathbf{W}) - \mathrm{Trace}(\mathbf{W}^H \mathbf{R}_c) + \\
&\quad + \mathrm{Trace}(\mathbf{W}^H \mathbf{R}_a \mathbf{W}). \tag{17.61}
\end{aligned}
$$

This equation will be employed in the next section in order to optimally adjust the matrix $\mathbf{W}$ of combiner coefficients.

17.2.3.2 Recovery of the Transmitted Signals by the Gradient Approach

Determining the weight matrix on the basis of evaluating the gradient with respect to the different users' total mean-square estimation error given by Equation 17.61 results in the standard form of the MMSE combiner, which is related to the right-inverse of the channel matrix $\mathbf{H}$. This will be further elaborated on in Section 17.2.3.2.1. Alternatively, the weight matrix can be represented in a form related to the left-inverse of the channel matrix $\mathbf{H}$, which had attracted our interest earlier in Section 17.2.2 in the context of the LS combiner characterised by Equations 17.36 and 17.37. The left-inverse related form of the MMSE combiner will briefly be addressed in Section 17.2.3.2.2.

17.2.3.2.1 Right-Inverse Related Form of the MMSE Combiner   Similarly to our proceedings in Section 17.2.2.3, the matrix $\mathbf{W}$ of optimum weights can be determined by noting that when $E\{||\Delta\hat{s}||_2^2\}$ of Equation 17.60 is minimised, its conjugate gradient evaluated with respect to the weight matrix $\mathbf{W}$ is identical to the zero-matrix. Hence, we obtain the following equation for the matrix $\mathbf{W}_{\mathrm{MMSE}} \in \mathbb{C}^{P \times L}$ of optimum weights:

$$\mathbf{R}_a \mathbf{W}_{\mathrm{MMSE}} = \mathbf{R}_c \quad \Longleftrightarrow \quad \mathbf{W}_{\mathrm{MMSE}} = \mathbf{R}_a^{-1} \mathbf{R}_c. \tag{17.62}$$

More specifically, upon substituting Equations 17.54 and 17.56 into Equation 17.62 we have:

$$\mathbf{W}_{\mathrm{MMSE}} = (\mathbf{HPH}^H + \sigma_n^2 \mathbf{I})^{-1} \mathbf{HP}. \tag{17.63}$$

In the context of deriving the conjugate gradient of Equation 17.61 with respect to the weight matrix $\mathbf{W}$ we have exploited that the constant first term yields a zero contribution, while the remaining terms were evaluated upon noting that $\frac{\partial \mathrm{Trace}(\mathbf{AXB})}{\partial \overline{\mathbf{X}}^*} = 0$, as well as that $\frac{\partial \mathrm{Trace}(\mathbf{AX}^H \mathbf{B})}{\partial \overline{\mathbf{X}}^*} = \mathbf{BA}$ and $\frac{\partial \mathrm{Trace}(\mathbf{X}^H \mathbf{AXB})}{\partial \overline{\mathbf{X}}^*} = \mathbf{AXB}$ for the complex matrices $\mathbf{A}$, $\mathbf{B}$ and

$\mathbf{X}$ [90]. Note that Equation 17.63 can be rewritten as:

$$\mathbf{W}_{\text{MMSE}} = (\mathbf{HP}_{\text{SNR}}\mathbf{H}^H + \mathbf{I})^{-1}\mathbf{HP}_{\text{SNR}}, \tag{17.64}$$

where in the same way as Equation 17.59, the matrix $\mathbf{P}_{\text{SNR}} \in \mathbb{R}^{L \times L}$ is the diagonal matrix of the different users' associated SNRs at the receiver antennas, which can be written as:

$$\mathbf{P}_{\text{SNR}} = Diag(\text{SNR}^{(1)}, \text{SNR}^{(2)}, \dots, \text{SNR}^{(L)}), \tag{17.65}$$

and where the l-th user's SNR is given by $\text{SNR}^{(l)} = \frac{\sigma_l^2}{\sigma_n^2}$. Furthermore, note from Equations 17.62 and 17.63 that the l-th user's associated weight vector $\mathbf{w}_{\text{MMSE}}^{(l)} \in \mathbb{C}^{P \times 1}$ is given by:

$$\begin{aligned} \mathbf{w}_{\text{MMSE}}^{(l)} &= \mathbf{R}_a^{-1}\mathbf{H}^{(l)}\sigma_l^2 &(17.66)\\ &= (\mathbf{HPH}^H + \sigma_n^2\mathbf{I})^{-1}\mathbf{H}^{(l)}\sigma_l^2. &(17.67) \end{aligned}$$

17.2.3.2.2 Left-Inverse Related Form of the MMSE Combiner

Recall that as demonstrated in Equation 17.22, the auto-correlation matrix of the AWGN is represented by a scaled identity matrix. Hence, it can be shown that an alternative expression with respect to Equation 17.63 for the MMSE combiner's weight matrix is given by:

$$\mathbf{W}_{\text{MMSE}} = \mathbf{HP}_{\text{SNR}}(\mathbf{H}^H\mathbf{HP}_{\text{SNR}} + \mathbf{I})^{-1}. \tag{17.68}$$

Upon substituting $\mathbf{R}_c = \mathbf{HP}$ as defined in Equation 17.54[6] as well as by substituting $\mathbf{R}_{\overline{a}} \in \mathbb{C}^{L \times L}$[7] defined as:

$$\mathbf{R}_{\overline{a}} = \mathbf{H}^H\mathbf{HP} + \sigma_n^2\mathbf{I}, \tag{17.69}$$

in the same way as Equation 17.62 we obtain the following relation:

$$\mathbf{R}_{\overline{a}}^H\mathbf{W}_{\text{MMSE}}^H = \mathbf{R}_c^H \iff \mathbf{W}_{\text{MMSE}}^H = \mathbf{R}_{\overline{a}}^{H-1}\mathbf{R}_c^H. \tag{17.70}$$

In terms of the required numerical accuracy, calculating the weight matrix $\mathbf{W}_{\text{MMSE}}$ by solving the system of equations as shown on the left-hand side of Equation 17.70 is more attractive, than a solution by the direct inversion of the auto-correlation matrix $\mathbf{R}_{\overline{a}}$, as shown on the right-hand side of Equation 17.70.

Note that in contrast to the auto-correlation matrix $\mathbf{R}_a$ defined in Equation 17.56, which is the core element of the right-inverse related representation of the weight matrix according to Equation 17.62, the correlation matrix $\mathbf{R}_{\overline{a}}$ defined in Equation 17.69, which was associated with the left-inverse related representation of the weight matrix in Equation 17.70 is not Hermitian. As a result, unfortunately the same computationally efficient methods which can

---

[6]Recall from Equations 17.59 and 17.65 that $\mathbf{P} = \sigma_n^2\mathbf{P}_{\text{SNR}}$.

[7]The left-inverse related form of the auto-correlation matrix is denoted here as $\mathbf{R}_{\overline{a}}$ in order to distinguish it from its right-inverse related form, namely $\mathbf{R}_a$, defined in Equation 17.56.

be invoked to solve the system of equations associated with the right-inverse related representation of the weight matrix, namely the auto-correlation matrix's Toeplitz structure, are not applicable here. However, a computational advantage is potentially achievable with the advent of the lower dimensionality of the matrix $\mathbf{R}_{\overline{a}} \in \mathbb{C}^{L \times L}$ compared to that of matrix $\mathbf{R}_a \in \mathbb{C}^{P \times P}$, provided that we have $P > L$. This property renders the left-inverse related form of the MMSE combiner particularly attractive for its repeated application in each cancellation stage of the SIC detector, which will be discussed in Section 17.3.1.

17.2.3.3   Mean-Square Estimation Error in the Transmitted Signals' Domain

Upon substituting the weight matrix $\mathbf{W}_{\text{MMSE}}$ defined in Equation 17.62 into Equation 17.52 we obtain for the auto-correlation matrix of the estimation errors associated with the different users' transmitted signals the following expression:

$$
\begin{aligned}
\mathbf{R}_{\Delta\hat{s}_{\text{MMSE}}} &= \mathbf{P} - \mathbf{R}_c^H \mathbf{R}_a^{-1} \mathbf{R}_c & (17.71) \\
&= \mathbf{P} - \mathbf{R}_c^H \mathbf{W}_{\text{MMSE}}. & (17.72)
\end{aligned}
$$

Hence, following the philosophy of Equation 17.60, the average minimum MSE (MMSE) of the L simultaneous users is given by:

$$
\overline{\text{MMSE}}_{\text{MMSE}} = \frac{1}{L}\text{Trace}(\mathbf{R}_{\Delta\hat{s}_{\text{MMSE}}}), \tag{17.73}
$$

while the l-th user's MMSE is given as the l-th diagonal element of the estimation errors' auto-correlation matrix $\mathbf{R}_{\Delta\hat{s}_{\text{MMSE}}}$ defined in Equation 17.71, namely as:

$$
\begin{aligned}
\text{MMSE}_{\text{MMSE}}^{(l)} &= \sigma_l^2(1 - \mathbf{H}^{(l)H}\mathbf{R}_a^{-1}\mathbf{H}^{(l)}\sigma_l^2) & (17.74) \\
&= \sigma_l^2(1 - \mathbf{H}^{(l)H}\mathbf{w}_{\text{MMSE}}^{(l)}). & (17.75)
\end{aligned}
$$

17.2.3.4   Optimum Weight Vector in Standard Form

We recall from Equation 17.58 that the received signals' auto-correlation matrix $\mathbf{R}_a$ can be expressed as the sum of the desired and undesired signals' contributions, namely as:

$$
\mathbf{R}_a = \mathbf{R}_{a,S}^{(l)} + \mathbf{R}_{a,I+N}^{(l)}. \tag{17.76}
$$

Upon invoking the well-known matrix-inversion lemma or Sherman-Morrison formula [90], the inverse of the auto-correlation matrix, namely $\mathbf{R}_a^{-1}$, can be rewritten as:

$$
\mathbf{R}_a^{-1} = \mathbf{R}_{a,I+N}^{(l)-1} - \frac{\mathbf{R}_{a,I+N}^{(l)-1}\sigma_l^2\mathbf{H}^{(l)}\mathbf{H}^{(l)H}\mathbf{R}_{a,I+N}^{(l)-1}}{1 + \sigma_l^2\mathbf{H}^{(l)H}\mathbf{R}_{a,I+N}^{(l)-1}\mathbf{H}^{(l)}}. \tag{17.77}
$$

Upon further substituting Equation 17.77 into Equation 17.66 we obtain:

$$
\mathbf{w}_{\text{MMSE}}^{(l)} = \beta_{\text{MMSE}}\mathbf{R}_{a,I+N}^{(l)-1}\mathbf{H}^{(l)}, \tag{17.78}
$$

where:

$$\beta_{\text{MMSE}} = \frac{\sigma_l^2}{1 + \text{SINR}^{(l)}}, \tag{17.79}$$

and where the achievable SINR is given by:

$$\text{SINR}^{(l)} = \sigma_l^2 \mathbf{H}^{(l)H} \mathbf{R}_{a,I+N}^{(l)-1} \mathbf{H}^{(l)}. \tag{17.80}$$

This is immediately seen by substituting Equation 17.78 into Equation 17.24. It is interesting to note that the SINR given by Equation 17.80 is independent from the factor β_{MMSE} related to the MMSE-criterion. As a consequence, different combining approaches result in achieving the same SINR at the combiner's output [136], provided that their weight vectors can be expressed in the form of Equation 17.78, despite having a constant β, which is potentially different from β_{MMSE} of the MMSE combiner. It was demonstrated in [136] that the most prominent combiners, which obey Equation 17.78, are the MMSE-, the Minimum Variance (MV) and the Maximum SINR combiners. As argued before, all three of these techniques exhibit the same SINR at the combiner's output, although they have different MSEs.

17.2.3.5   Relation between MMSE and MV Combining

In order to motivate the employment of the Minimum Variance (MV) combiner let us recall from Section 17.2.2 that the LS combiner's philosophy was to fully recover the original signal transmitted – as illustrated by Equation 17.42 – without relying on any information concerning the AWGN process, which corrupts the signal received by the different antenna elements. By contrast, the philosophy of the MMSE combiner portrayed in Section 17.2.3 was to strike a balance between the recovery of the signals transmitted and the suppression of the AWGN. An attractive compromise is constituted by the MV approach, which aims to recover the original signals transmitted whilst ensuring a partial suppression of the AWGN based on the knowledge of its statistics. In other words, the l-th user's associated weight vector $\mathbf{w}^{(l)}$ has to be adjusted such that its transfer factor, which is seen from Equation 17.11 to be equal to $\mathbf{w}^{(l)H}\mathbf{H}^{(l)}$, assumes a specific predefined value of $g = \mathbf{w}^{(l)H}\mathbf{H}^{(l)}$. The corresponding interference and noise variances of $\sigma_I^{(l)2}$ and $\sigma_N^{(l)2}$ are given by Equations 17.17 and 17.20, respectively.

Usually the MV combiner is derived by minimising a Lagrangian cost-function, which incorporates both a constraint on the desired user's effective transfer factor, as well as the undesired signal's variance [90, 136]. However, as argued in the previous section, the different combiners' associated weight vectors, namely those of the MMSE, MV and Maximum SINR combiners, differ only by a scalar multiplier. Hence, the MV-related weight vector $\mathbf{w}_{\text{MV}}^{(l)}$ of the l-th user can be directly inferred from the MMSE-related weight vector $\mathbf{w}_{\text{MMSE}}^{(l)}$ by simple normalisation according to:

$$\mathbf{w}_{\text{MV}}^{(l)} = \frac{g}{\mathbf{w}_{\text{MMSE}}^{(l)H}\mathbf{H}^{(l)}} \mathbf{w}_{\text{MMSE}}^{(l)}. \tag{17.81}$$

Here the term in the nominator denotes the l-th user's gain factor valid in the context of MMSE combining. Upon substituting the MMSE-specific weight vector given by Equation 17.78 into Equation 17.81 we obtain:

$$\mathbf{w}_{\text{MV}}^{(l)} = \beta_{\text{MV}} \mathbf{R}_{a,I+N}^{(l)-1} \mathbf{H}^{(l)}, \tag{17.82}$$

where:

$$\beta_{\text{MV}} = \frac{g}{\mathbf{H}^{(l)H} \mathbf{R}_{a,I+N}^{(l)} \mathbf{H}^{(l)}}. \tag{17.83}$$

Specifically, for $g = 1$ this "normalised MMSE combiner" is also known as the Minimum Variance Distortionless Response (MVDR) combiner.

17.2.4 Demodulation of the Different Users' Combiner Output Signals

As observed on the left-hand side of Figure 17.4, the linear detector is constituted by the linear combiner, which produces estimates of the signals transmitted by the L different users. Based on these linear estimates the task of the classifiers seen on the right-hand side of Figure 17.4 is to determine the complex symbols- or constellation points that are most likely to have been transmitted by the different users.

In Section 17.2.4.1 each user's combiner output signal is approximated as a sample of a complex Gaussian distribution function. This representation is then employed in Section 17.2.4.2 to determine the complex symbol or constellation point that is most likely to have been transmitted by a specific user.

17.2.4.1 Approximation of a Specific User's Combiner Output Signal as a Sample of a Complex Gaussian Distribution

In Sections 17.2.2, 17.2.3 and 17.2.3.5 various methods of detecting the different users' transmitted signals were discussed, namely, the LS- MMSE- and MV techniques, respectively. Common to these techniques was their linear structure, which was conveniently illustrated by Equation 17.10. Specifically, the l-th user's combiner output signal $x_{\text{eff}}^{(l)} = \hat{s}^{(l)}$ is constituted by a superposition of the desired user's signal $H_{\text{eff}}^{(l)} s^{(l)}$ and of the undesired signal $n_{\text{eff}}^{(l)}$, which is expressed as:

$$x_{\text{eff}}^{(l)} = H_{\text{eff}}^{(l)} s^{(l)} + n_{\text{eff}}^{(l)}. \tag{17.84}$$

Comparing Equation 17.84 with Equation 17.10 reveals that the desired user's effective transfer factor $H_{\text{eff}}^{(l)}$ is given by:

$$H_{\text{eff}}^{(l)} = \mathbf{w}^{(l)H} \mathbf{H}^{(l)}, \tag{17.85}$$

while the effective undesired signal $n_{\text{eff}}^{(l)}$, namely the sum of the $L - 1$ interfering users' residual signals plus the residual AWGN, is given by:

$$n_{\text{eff}}^{(l)} = \hat{s}_I^{(l)} + \hat{s}_N^{(l)}. \tag{17.86}$$

The individual components were defined in Equations 17.12 and 17.13, while their associated variances $\sigma_I^{(l)2}$ and $\sigma_N^{(l)2}$ were given by Equations 17.18 and 17.21. The l-th user's combiner output signal can therefore be approximately modelled, as a sample of a complex Gaussian distribution having a mean value of $H_{\text{eff}}^{(l)} s^{(l)}$ and a variance of $\sigma_{n_{\text{eff}}}^{(l)2} = \sigma_I^{(l)2} + \sigma_N^{(l)2}$ which is formulated as, $x_{\text{eff}}^{(l)} \sim \mathcal{CN}(H_{\text{eff}}^{(l)} s^{(l)}, \sigma_{n_{\text{eff}}}^{(l)2})$. [8] We note, however, that this relationship is only exactly true for an infinite number of interferers, as a result of the Central-Limit Theorem [177]. This complex Gaussian distribution can be expressed as [494]:

$$f\left(x_{\text{eff}}^{(l)} \big| s^{(l)}, H_{\text{eff}}^{(l)}\right) = \frac{1}{\pi \sigma_{n_{\text{eff}}}^{(l)2}} \exp\left(-\frac{1}{\sigma_{n_{\text{eff}}}^{(l)2}} \left| x_{\text{eff}}^{(l)} - H_{\text{eff}}^{(l)} s^{(l)} \right|^2\right). \tag{17.87}$$

More explicitly, $P(x_{\text{eff}}^{(l)} | s^{(l)}, H_{\text{eff}}^{(l)}) = f(x_{\text{eff}}^{(l)} | s^{(l)}, H_{\text{eff}}^{(l)})$ denotes the *a priori* probability that $x_{\text{eff}}^{(l)}$ is observed at the l-th user's combiner output under the condition that the symbol $s^{(l)}$ is transmitted over a channel characterised by the effective transfer factor $H_{\text{eff}}^{(l)}$ of Equation 17.85.

17.2.4.2 Determination of a Specific User's Transmitted Symbol by Maximising the *A Posteriori* Probability

The complex symbol $\check{s}_{\text{ML}\approx}^{(l)}$ [9] that is most likely to have been transmitted by the l-th user can be determined upon maximising the *a posteriori* probability $P(\check{s}^{(l)} | x_{\text{eff}}^{(l)}, H_{\text{eff}}^{(l)})$, that the complex symbol $\check{s}$ was transmitted under the condition that the signal $x_{\text{eff}}^{(l)}$ is observed at the combiner output, for all symbols contained in the trial-set $\mathcal{M}^{(l)}$ given by:

$$\mathcal{M}^{(l)} = \left\{ \check{s}^{(l)} \bigg| \frac{\check{s}^{(l)}}{\sigma_l} \in \mathcal{M}_c \right\}. \tag{17.88}$$

In Equation 17.88 $\mathcal{M}_c$ denotes the set of constellation points associated with the specific modulation scheme employed. In mathematical terms this can be formulated as:

$$\check{s}_{\text{ML}\approx}^{(l)} = arg \max_{\check{s}^{(l)} \in \mathcal{M}^{(l)}} P\left(\check{s}^{(l)} | x_{\text{eff}}^{(l)}, H_{\text{eff}}^{(l)}\right). \tag{17.89}$$

Upon invoking the definition of the conditional probability, the *a posteriori* probability $P(\check{s}^{(l)} | x_{\text{eff}}^{(l)}, H_{\text{eff}}^{(l)})$ seen in Equation 17.89 can be rewritten as:

$$P(\check{s}^{(l)} | x_{\text{eff}}^{(l)}, H_{\text{eff}}^{(l)}) = P(x_{\text{eff}}^{(l)} | \check{s}^{(l)}, H_{\text{eff}}^{(l)}) \frac{P(\check{s}^{(l)})}{P(x_{\text{eff}}^{(l)})}, \tag{17.90}$$

---

[8] The complex Gaussian distribution function is denoted here as $\mathcal{CN}()$, in order to distinguish it from the Gaussian distribution function $\mathcal{N}()$ defined for real-valued random variables.

[9] Here we have denoted the most likely transmitted symbol as $\check{s}_{\text{ML}\approx}^{(l)}$, in order to emphasise that the Gaussian approximation was used to model the residual interference contaminating the combiner's output signal.

where the total probability $P(x_{\text{eff}}^{(l)})$ follows from the condition that:

$$\sum_{\breve{s}^{(l)} \in \mathcal{M}^{(l)}} P(\breve{s}^{(l)}|x_{\text{eff}}^{(l)}, H_{\text{eff}}^{(l)}) \overset{!}{=} 1, \tag{17.91}$$

which yields:

$$P(x_{\text{eff}}^{(l)}) = \sum_{\breve{s}^{(l)} \in \mathcal{M}^{(l)}} P(x_{\text{eff}}^{(l)}|\breve{s}^{(l)}, H_{\text{eff}}^{(l)})P(\breve{s}^{(l)}). \tag{17.92}$$

Note that Equation 17.90 in conjunction with Equation 17.92 is also known as Bayes' theorem [90].

Upon substituting Equation 17.90 into Equation 17.89 and by noting again that the *a priori* probability $P(x_{\text{eff}}^{(l)}|\breve{s}^{(l)}, H_{\text{eff}}^{(l)})$ is given by the complex Gaussian distribution function of Equation 17.87, namely by $f(x_{\text{eff}}^{(l)}|\breve{s}^{(l)}, H_{\text{eff}}^{(l)})$, we obtain for the ML symbol estimate $\breve{s}_{\text{ML}\approx}^{(l)}$ the following expression:

$$\breve{s}_{\text{ML}\approx}^{(l)} = arg \max_{\breve{s}^{(l)} \in \mathcal{M}^{(l)}} P\left(x_{\text{eff}}^{(l)}|\breve{s}^{(l)}, H_{\text{eff}}^{(l)}\right) \iff \breve{s}_{\text{ML}\approx}^{(l)} = arg \min_{\breve{s}^{(l)} \in \mathcal{M}^{(l)}} \left|x_{\text{eff}}^{(l)} - H_{\text{eff}}^{(l)}\breve{s}^{(l)}\right|^2. \tag{17.93}$$

Note from Equation 17.93 that determining the ML symbol estimate implies minimising the Euclidean distance in the argument of the exponential term associated with the Gaussian distribution function of Equation 17.87. In the context of our derivation we have also exploited that we have $P(\breve{s}^{(l)}) = \frac{1}{M_c} = const.$, as well as that we have $P(x_{\text{eff}}^{(l)}) = const.$ as seen in Equation 17.91. Hence, these terms are irrelevant in the context of the minimisation required by Equation 17.93.

In order to avoid the multiplication of each trial-symbol $\breve{s}^{(l)}$ with the effective transfer factor $H_{\text{eff}}^{(l)}$, as required by Equation 17.93, it is legitimate to evaluate:

$$\breve{s}_{\text{ML}\approx}^{(l)} = arg \min_{\breve{s}^{(l)} \in \mathcal{M}^{(l)}} \left|\frac{1}{H_{\text{eff}}^{(l)}}x_{\text{eff}}^{(l)} - \breve{s}^{(l)}\right|^2, \tag{17.94}$$

instead. Note again that if the estimate $x_{\text{eff}}^{(l)}$ at the l-th user's combiner output is generated with the aid of the MMSE criterion, then the normalised estimate $x_{\text{eff}}^{(l)}/H_{\text{eff}}^{(l)}$ is actually the complex symbol, which would be observed at the output of the MVDR combiner, as described in Section 17.2.3.5. Note, however, that in the context of MPSK modulation schemes the normalisation by the real-valued factor of $H_{\text{eff}}^{(l)}$ is not necessary, since only the signal's phase is of importance to the detection process.

17.2.5   Generation of Soft-Bit Information for Turbo-Decoding

Employing turbo decoding at the receiver is a powerful means of further enhancing the system's BER. Naturally, this is achieved at the cost of a reduction of the system's effective throughput. A prerequisite for the employment of turbo codes is the availability of soft-bit information at the detector's output, whose generation will be discussed in this section.

Our discussions will be based on Equation 17.84, which described the l-th user's combiner output signal $x_{\text{eff}}^{(l)}$ as the superposition of the desired user's signal contribution $s^{(l)}$, which has a gain of $H_{\text{eff}}^{(l)}$, plus the effective noise contribution $n_{\text{eff}}^{(l)}$, which comprises the $L - 1$ remaining users' residual interference and the residual AWGN. The residual interference was approximated by a Gaussian process and hence the total variance of the effective noise became $\sigma_{n_{\text{eff}}}^{(l)2} = \sigma_I^{(l)2} + \sigma_N^{(l)2}$.

With respect to Equation 17.84 the soft-bit value or log-likelihood ratio $L_{m\approx}^{(l)}$ associated with the l-th user at the m-th bit-position is given by [90]:

$$L_{m\approx}^{(l)} = \ln \frac{P(b_m^{(l)} = 1 | x_{\text{eff}}^{(l)}, H_{\text{eff}}^{(l)})}{P(b_m^{(l)} = 0 | x_{\text{eff}}^{(l)}, H_{\text{eff}}^{(l)})}, \tag{17.95}$$

which is the natural logarithm of the quotient of *a posteriori* probabilities that the m-th bit transmitted by the l-th user in the k-th subcarrier is associated with a logical value of $b_m^{(l)} = 1$ or $b_m^{(l)} = 0$. Note that here we have again omitted the index $[n, k]$ for the k-th subcarrier of the n-th OFDM symbol. Equation 17.95 can be further expanded by noting that the *a posteriori* probability that a bit of $b_m^{(l)} = 1$ was transmitted is given by the sum of the *a posteriori* probabilities of those symbols, which are associated with a bit value of $b_m^{(l)} = 1$, again, at the m-th bit position. The *a posteriori* probability that a bit value of $b_m^{(l)} = 0$ was transmitted can be represented equivalently. Hence we obtain:

$$L_{m\approx}^{(l)} = \ln \frac{\sum_{(\check{s}^{(l)}/\sigma_l) \in \mathcal{M}_{cm}^1} P(\check{s}^{(l)} | x_{\text{eff}}^{(l)}, H_{\text{eff}}^{(l)})}{\sum_{(\check{s}^{(l)}/\sigma_l) \in \mathcal{M}_{cm}^0} P(\check{s}^{(l)} | x_{\text{eff}}^{(l)}, H_{\text{eff}}^{(l)})}, \tag{17.96}$$

where $\mathcal{M}_{cm}^b$ denotes the specific subset of the set $\mathcal{M}_c$ of constellation points of the modulation scheme employed, which are associated with a bit value of $b \in \{0, 1\}$ at the m-th bit position. For notational convenience we can define the l-th user's associated set of trial vectors employed for determining the probability that the m-th transmitted bit exhibits a value of $b \in \{0, 1\}$ as follows:

$$\mathcal{M}_m^{b(l)} = \left\{ \check{s}^{(l)} \left| \frac{\check{s}^{(l)}}{\sigma_l} \in \mathcal{M}_{cm}^b \right. \right\}. \tag{17.97}$$

Substituting Bayes theorem of Equation 17.90 into Equation 17.96 then yields for the l-th user's soft-bit value at the m-th bit position the following expression:

$$L_{m\approx}^{(l)} = \ln \frac{\sum_{\check{s}^{(l)} \in \mathcal{M}_m^{1(l)}} P(x_{\text{eff}}^{(l)} | \check{s}^{(l)}, H_{\text{eff}}^{(l)})}{\sum_{\check{s}^{(l)} \in \mathcal{M}_m^{0(l)}} P(x_{\text{eff}}^{(l)} | \check{s}^{(l)}, H_{\text{eff}}^{(l)})}. \tag{17.98}$$

Here we have exploited that the different trial symbols $\check{s}^{(l)}$ have the same probability, namely $P(\check{s}^{(l)}) = const.$, $\check{s}^{(l)} \in \mathcal{M}^{(l)}$, where $\mathcal{M}^{(l)} = \mathcal{M}_m^{(0)(l)} \bigcup \mathcal{M}_m^{(1)(l)}$. Upon recalling from Section 17.2.4 that the *a priori* probability $P(x_{\text{eff}}^{(l)} | \check{s}^{(l)}, H_{\text{eff}}^{(l)})$ is given by the complex Gaussian distribution function $f(x_{\text{eff}}^{(l)} | \check{s}^{(l)}, H_{\text{eff}}^{(l)})$ defined in Equation 17.87, we obtain that:

$$L_{m\approx}^{(l)} = \ln \frac{\sum_{\check{s}^{(l)} \in \mathcal{M}_m^{1(l)}} \exp\left(-\frac{1}{\sigma_{n_{\text{eff}}}^2} |x_{\text{eff}}^{(l)} - H_{\text{eff}}^{(l)} \check{s}^{(l)}|^2\right)}{\sum_{\check{s}^{(l)} \in \mathcal{M}_m^{0(l)}} \exp\left(-\frac{1}{\sigma_{n_{\text{eff}}}^2} |x_{\text{eff}}^{(l)} - H_{\text{eff}}^{(l)} \check{s}^{(l)}|^2\right)}. \tag{17.99}$$

Observe that evaluating the l-th user's soft-bit value at the m-th bit position with the aid of Equation 17.99 involves the exponential function, which is computationally demanding.

17.2.5.1 Simplification by Maximum Approximation

In order to avoid the explicit evaluation of the exponential function, a common approach is constituted by the so-called maximum approximation, which implies that only that specific additive term is retained in the calculation of the numerator and nominator of Equation 17.99, which yields the maximum contribution. It can be readily shown that as a result of this simplification we obtain instead of Equation 17.99 the following expression:

$$L_{m\approx}^{(l)} \approx \frac{1}{\sigma_{n_{\text{eff}}}^{(l)2}} \left[|x_{\text{eff}}^{(l)} - H_{\text{eff}}^{(l)} \check{s}_{m\approx}^{0(l)}|^2 - |x_{\text{eff}}^{(l)} - H_{\text{eff}}^{(l)} \check{s}_{m\approx}^{1(l)}|^2 \right], \tag{17.100}$$

where

$$\check{s}_{m\approx}^{b(l)} = arg \min_{\check{s}^{(l)} \in \mathcal{M}_m^{b(l)}} |x_{\text{eff}}^{(l)} - H_{\text{eff}}^{(l)} \check{s}^{(l)}|^2, \quad b \in \{0, 1\}, \tag{17.101}$$

while the set $\mathcal{M}_m^{b(l)}$ was defined in Equation 17.97. We note that for each soft-bit to be determined, Equation 17.101 has to be invoked twice, namely, once for a bit value of $b = 1$ and once for $b = 0$.

A significant complexity reduction can be achieved by exploiting that $\mathcal{M}^{(l)} = \mathcal{M}_m^{0(l)} \bigcup \mathcal{M}_m^{1(l)}$. Hence, the calculation of the Euclidean distance metric $|x_{\text{eff}}^{(l)} - H_{\text{eff}}^{(l)} \check{s}^{(l)}|^2$ only has to be performed once for the different trial symbols $\check{s}^{(l)} \in \mathcal{M}^{(l)}$, followed by an appropriate selection in the context of the soft-bit generation assisted by Equation 17.101. Specifically half of the symbols $\check{s}_{m\approx}^{b(l)}$ for the different bit polarities $b \in \{0, 1\}$ and bit positions m can readily be inferred by conducting an initial search of the entire set $\mathcal{M}^{(l)}$. This results in the ML estimate $\check{s}_{\text{ML}\approx}^{(l)}$ of the transmitted symbol according to Equation 17.94. The initial ML symbol estimate is constituted by a specific bit representation. The minimisation obeying Equation 17.101 has to be conducted over the set of specific symbols, which contain the inverted versions of the bits identified during the previously mentioned initial ML symbol search.

17.2.6 Performance Analysis

In the context of our simulations the frame-invariant fading indoor WATM channel and system model described in Section 14.3.1 will be employed. Furthermore, perfect knowledge of the channel transfer functions associated with the different transmit-receive antenna pairs will be assumed. Note that as a result of performing the detection of the different users' transmitted symbols independently on an OFDM subcarrier-by-subcarrier basis, the performance

system parameters	choice
CIR model	3-path indoor WATM of Section 14.3.1
CIR tap fading	OFDM symbol invariant
system model	indoor WATM of Section 14.3.1
channel estimation	ideal
transmit antennas per user	1

Table 17.4: Summary of the system set-up; also note that the fading was assumed to be uncorrelated for the different CIR taps associated with the channel between a specific transmitter-receiver antenna pair, as well as uncorrelated for the same CIR tap of different transmitter-receiver antenna pairs.

turbo coding parameters	choice
coding rate R_c	$1/2$
constraint length K_c	3
generator polynomial	$(7,5)_8$
number of iterations	4

Table 17.5: Summary of the turbo-coding parameters.

results presented here for the uncoded system are independent of the indoor WATM channel's specific multipath intensity profile. The advantage of employing the idealistic model of an OFDM symbol invariant fading channel is that the performance results are not impaired by the obfuscating effects of Inter-subCarrier Interference (ICI). Again the general system set-up has been summarised in Table 17.4, while in Table 17.5 we have summarised the turbo-coding parameters to be employed in the context of our investigation of turbo-coded systems.

In Section 17.2.6.1 the different detectors, namely LS, MMSE and MV are compared to each other in terms of the achievable MSE at the associated combiner's output, as well as in terms of the BER. By contrast, in Sections 17.2.6.2 and 17.2.6.3 we concentrate on the assessment of the specific MMSE detector's performance in terms of the distribution of the SINR measured at the associated combiner's output, as well as that of the detector's BER performance, respectively. These investigations are conducted as a function of the number of users L and that of the number of reception antennas P. Our investigations are concluded upon evaluating the BER performance of turbo-coded MMSE detection-assisted SDMA-OFDM in Section 17.2.6.4.

17.2.6.1 MSE and BER Performance Comparison of LS, MMSE and MVDR Detection

On the left-hand side of Figure 17.6 we have portrayed the average MSE performance recorded at the different detectors' combiner outputs, as a function of the SNR at the reception antennas. More specifically, the MSE was evaluated as the squared error between the signal transmitted by a specific user and that observed at its associated combiner output, normalised to the user's signal variance. Here we have considered the scenario of two reception antennas and two simultaneous SDMA users, each equipped with one transmit an-

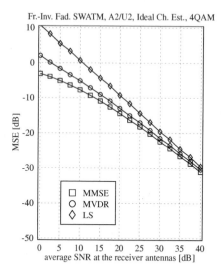
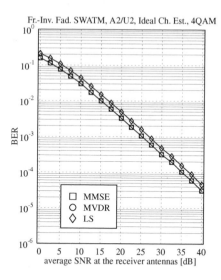

Figure 17.6: Comparison of the different linear detectors, namely, the LS, MMSE and MVDR techniques as a function of the SNR at the reception antennas, with respect to **(a: Left)** the MSE at the combiner's output and **(b: Right)** the detector's BER; here we have employed the configuration of $L = 2$ simultaneous users and $P = 2$ reception antennas at the BS $(A2/U2)$; for the basic simulation parameters we refer to Table 17.4

tenna, which we denoted as (A2/U2). As expected, the best MSE performance is exhibited by the MMSE combiner, closely followed by the MVDR combiner, as seen in Figure 17.6. The worst MSE performance was exhibited by the LS combiner, which is also widely known as the zero-forcing combiner. Furthermore, observe that upon increasing the users' SNRs towards infinity, the different combiners' MSE curves merge. This is, because when increasing the SNR, a Wiener filter-based combiner effectively operates as an LS combiner, aiming to minimise purely the interfering signals' variances, rather than that of the joint noise and interference contributions. Note that in the context of our simulations the LS combiner's correlation matrix given by Equation 17.33 was regularised [90] upon adding a value of 10^{-6} to its main diagonal elements. This contributed towards mitigating the problems associated with its inversion. By contrast, on the right-hand side of Figure 17.6 we have portrayed the system's BER associated with the different detectors in the context of 4QAM modulation. We observe that both Wiener filter-based detectors, namely, the MMSE and MVDR schemes achieve the same BER performance, as a result of their identical SINR performance as highlighted in Sections 17.2.3.4 and 17.2.3.5. Note, however, that in the context of higher-order QAM modulation schemes, such as 16QAM, for example, where also the constellation points' amplitude conveys information, a slight BER performance advantage was observed for the MVDR detector although the corresponding results are not included here for reasons

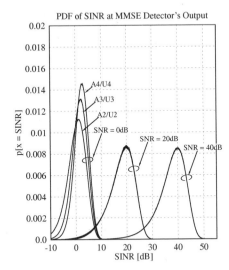
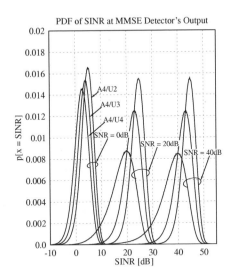

Figure 17.7: Characterisation of the MMSE detector in terms of its associated SINR-PDF observed at the combiner's output; **(a: Left)** for each configuration the L number of users is identical to the P number of reception antennas, using the configurations of $A2/U2$, $A3/U3$ and $A4/U4$; **(b: Right)** the P number of reception antennas is equal to four, while the L number of users is varied, where we have the configurations of $A4/U2$, $A4/U3$ and $A4/U4$; for the basic simulation parameters we refer to Table 17.4.

of space. Furthermore, similar to our observations with respect to the different detectors' associated combiner MSE performance, the LS detector performs significantly worse, than the MMSE and the MVDR detectors also in terms of the BER. Hence, in our following discussions we will focus on the MMSE detector, which will also be employed as the core element of SIC and PIC detectors to be discussed in Sections 17.3.1 and 17.3.2, respectively.

17.2.6.2 SINR Performance of MMSE Detection for Different Numbers of Users and Reception Antennas

In order to further characterise the MMSE detector, in Figure 17.7 we have plotted the Probability Density Function (PDF) of the SINR at the combiner's output for different combinations of the number of simultaneous users L and the number of reception antennas P, as well as for SNRs of 0dB, 20dB and 40dB recorded at the reception antennas. Specifically, in the graph seen on the left-hand side of Figure 17.7 we have compared those PDFs against each other, which are associated with the particular configurations of $P = L \in \{2, 3, 4\}$. We observe that at sufficiently high SNRs the SINR distributions become almost identical, which is because the different arrangements have the same diversity order. Here we emphasise the expression "almost identical", since at higher SNRs – although visually they appear

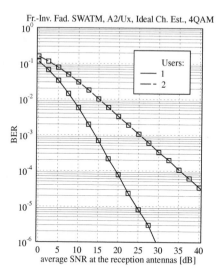
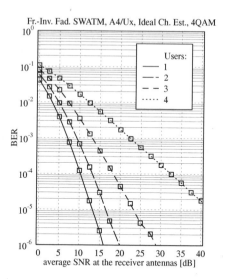

Figure 17.8: BER performance of 4QAM-modulated, MMSE detection-assisted SDMA-OFDM as a function of the SNR at the reception antennas; the curves are further parametrised with the number of simultaneous users L and the number of reception antennas P, where more specifically **(a: Left)** two reception antennas, **(b: Right)** four reception antennas were employed; for the basic simulation parameters we refer to Table 17.4.

identical in Figure 17.7 – perceivable differences were found in terms of the corresponding average BER performance for the configurations of A2/U2, A3/U3 as well as A4/U4. More explicitly, upon increasing the MIMO system's order, the BER performance is improved.

By contrast, on the right-hand side of Figure 17.7 we have considered configurations of $P = 4$, $L \in \{2, 3, 4\}$. Here we observe that upon increasing the diversity order, namely by decreasing the L number of SDMA users, while keeping the P number of reception antennas constant, the probability of incurring higher SINRs is increased. The effects of these SINR improvements on the system's BER performance will be further investigated in the next section,

17.2.6.3 BER Performance of MMSE Detection for Different Numbers of Users and Reception Antennas

In Figure 17.8 we have portrayed the BER performance of a 4QAM-modulated MMSE detection-assisted SDMA-OFDM scheme as a function of the SNR at the reception antennas. The curves are further parametrised with the number of simultaneous users L and the number of reception antennas P. Upon decreasing the number of users L while keeping the number of reception antennas P constant, we observe that the BER performance is dramatically improved. This is because the system's diversity order is increased and hence the detector

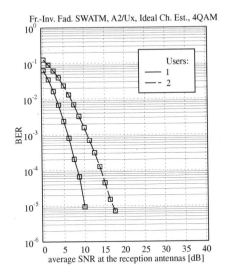

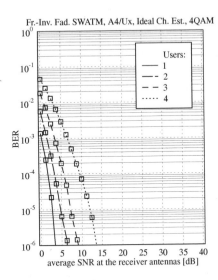

Figure 17.9: BER performance of turbo-coded, 4QAM-modulated, MMSE detection-assisted SDMA-OFDM as a function of the SNR at the reception antennas; the curves are further parametrised with the number of simultaneous users L and the number of reception antennas P, where **(a: Left)** two reception antennas, and **(b: Right)** four reception antennas were employed, respectively; for the basic simulation parameters and for the turbo-coding parameters we refer to Tables 17.4 and 17.5, respectively.

benefits from a degree of freedom to adjust its associated combiner weights in favour of a better exploitation of the channel's diversity, in favour of a reduction of the AWGN, rather than aiming for the mitigation of the remaining users' interference.

17.2.6.4 BER Performance of Turbo-Coded MMSE Detection-Assisted SDMA-OFDM

Turbo-decoding at the receiver is a powerful means of further enhancing the system's BER performance. This is achieved at the cost of reducing the system's effective throughput and by investing additional computational complexity. The turbo coding parameters were summarised in Table 17.5, but for the reader's convenience they are repeated here. Namely, the coding rate was $R_c = \frac{1}{2}$, the constraint length was $K_c = 3$, the octally represented generator polynomials of $(7, 5)_8$ were used and four iterations were performed. The generation of the soft-bits required for turbo-decoding was discussed earlier in Section 17.2.5.

Our BER simulation results are portrayed in Figure 17.9, on the left-hand side for $P = 2$ reception antennas, while on the right-hand side for $P = 4$ reception antennas, when supporting up to $L = P$ number of users. We observe that compared to the uncoded scenario, whose associated simulation results were shown in Figure 17.8 the BER is significantly reduced. To provide an example, for a so-called "fully loaded" system associated with $L = P = 4$ the

SNR at the reception antennas required for a BER of 10^{-5} was around 42dB, while in the context of turbo-decoding the same BER was reached at an SNR of around 13dB. Again, this performance improvement is achieved at the cost of halving the system's throughput and at an additional computational complexity imposed by the turbo-decoder. Furthermore, similar to the uncoded scenario, upon removing one user from the "fully loaded" system results in a significant reduction of the BER. This is because the MMSE combiner has a higher degree of freedom in terms of the choice of the optimum weight matrix, with the beneficial effect of a better suppression of the undesired AWGN.

17.2.7 Complexity Analysis

In this section we will analyse the computational complexity inflicted per subcarrier when evaluating the vector of estimated transmitted signals, followed by hard-decision based demodulation carried out with the aid of minimising the Euclidean distance metric of Equation 17.94. While in Section 17.2.7.1 the LS combiner's complexity is quantified, in Section 17.2.7.2 we will concentrate on the portrayal of the MMSE combiner's complexity. Finally, the computational complexity related to the demodulation of the linear combiners' output signals will then be analysed in Section 17.2.7.3.

17.2.7.1 LS Combining

An expression for the LS combiner's associated weight vector was provided in Equation 17.35 for the case of $P \geq L$. First of all, the calculation of the "auto-correlation" matrix $\mathbf{Q}_{\mathrm{LS}}$ defined in Equation 17.33 requires $L^2 P$ number of complex multiplications and the same number of additions. By contrast, the evaluation of the "cross-correlation" vector $\mathbf{p}_{\mathrm{LS}}$ requires LP number of complex multiplications and additions. Due to the potentially high condition number [90] of the "auto-correlation" matrix $\mathbf{Q}_{\mathrm{LS}}$ it is disadvantageous to directly invert it, since it requires a high numerical accuracy. Recall that this direct solution for the vector $\hat{\mathbf{s}}_{\mathrm{LS}}$ was shown on the right-hand side of Equation 17.35.

17.2.7.1.1 LS Combining without Generating the Weight Matrix   In order to circumvent this problem, the preferred method is that of solving the equation system shown on the left-hand side of Equation 17.35. In order to ensure numerical stability, a matrix decomposition based approach such as the Cholesky, LU or QR decomposition [90] could be invoked. In the context of the LS solution the most prominent matrix factorisation technique is the QR decomposition. However, here we assume that the LU decomposition technique outlined in [90] is employed. Hence, in a first step the matrix $\mathbf{Q}_{\mathrm{LS}}$ is LU decomposed, imposing a computational complexity of $\frac{1}{3}L^3$ complex multiplications and additions. Then, in a second step the desired vector $\hat{\mathbf{s}}_{\mathrm{LS}}$ is determined with the aid of forward and backward substitutions using the procedure outlined in [90], which imposes a complexity of L^2 complex multiplications and additions. Hence, the total computational complexity of solving Equation 17.35 per subcarrier is given by:

$$C_{\mathrm{LS,direct}}^{(\mathbb{C}*\mathbb{C})} = C_{\mathrm{LS,direct}}^{(\mathbb{C}+\mathbb{C})} = PL + (P+1)L^2 + \frac{1}{3}L^3. \tag{17.102}$$

Note, however, that as a result of this procedure the weight matrix $\mathbf{W}_{\mathrm{LS}}$ defined in Equation 17.38 does not become explicitly available, although it might be required to determine the estimation MSE, the SNR or SINR on a subcarrier basis.

17.2.7.1.2 LS Combining Generating the Weight Matrix

As an example, in the context of the SIC detection[10] procedure to be discussed in Section 17.3.1, explicit knowledge of the weight matrix is required to calculate the subcarrier-based S(I)NR values employed for selecting the most dominant user to be cancelled next in a specific detection stage. A possible solution for determining $\mathbf{W}_{\mathrm{LS}}$ is first to solve Equation 17.37 to obtain the projection matrix $\mathbf{P}_{\mathrm{LS}}$, which is related to the weight matrix $\mathbf{W}_{\mathrm{LS}}$ by the Hermitian transpose as seen in Equation 17.38. This step is then followed by appropriately combining the output signals of the array elements according to Equation 17.36. As a consequence, the associated total computational complexity would be:

$$C_{\mathrm{LS,W+cmb}}^{(\mathbb{C}*\mathbb{C})} = C_{\mathrm{LS,W+cmb}}^{(\mathbb{C}+\mathbb{C})} = PL + 2PL^2 + \frac{1}{3}L^3. \qquad (17.103)$$

The concomitant increase in computational complexity compared to that quantified by Equation 17.102 is because in the process of evaluating $\mathbf{W}_{\mathrm{LS}}$, the forward and backward substitutions as outlined in [90] would have to be carried out for P different matrix right-hand sides.

Note that since the computational complexity is dominated in both cases by the third order as a function of the number of simultaneous users L, increasing this parameter will dramatically increase the associated complexity.

17.2.7.2 MMSE Combining

The second combiner which we will analyse in terms of its computational complexity is the MMSE combiner of Section 17.2.3. Here the relevant equations are the general combiner's formula, namely Equation 17.7, as well as the MMSE-specific expression that has to be evaluated for the determination of the optimum weight matrix, namely, Equation 17.64 or 17.68. Recall that both forms are equivalent to each other. Nonetheless, as argued in Section 17.2.3.2, there is a difference in the dimension of the auto-correlation matrices. Specifically, in the context of the left-inverse related form of the weight matrix given by Equation 17.68 an auto-correlation matrix $\mathbf{R}_{\overline{a}}$ of dimension $L \times L$ has to be inverted, while in conjunction with the right-inverse related form of Equation 17.64 the auto-correlation matrix $\mathbf{R}_a$ to be inverted is of dimension $P \times P$.

17.2.7.2.1 Left-Inverse Related Form of MMSE Combining without Generating the Weight Matrix

As shown earlier in Section 17.2.7.1 for the LS combiner, if the weight matrix $\mathbf{W}_{\mathrm{MMSE}}$ is not explicitly required, a complexity reduction can be achieved in conjunction with the left-inverse related representation of the MMSE combiner by directly solving Equation 17.70 for the vector of the transmitted symbols' estimates. The solution of Equa-

---

[10]The LS detector, or alternatively the MMSE detector could be employed as a baseline detector in each SIC detection stage.

tion 17.70 imposes a complexity of:

$$C_{\text{MMSE,direct}}^{(\mathbb{C}*\mathbb{C})} \quad = \quad C_{\text{MMSE,direct}}^{(\mathbb{C}+\mathbb{C})} = PL + (P+1)L^2 + \frac{1}{3}L^3 \tag{17.104}$$

$$C_{\text{MMSE,direct}}^{(\mathbb{R}*\mathbb{C})} \quad = \quad PL \tag{17.105}$$

$$C_{\text{MMSE,direct}}^{(\mathbb{R}+\mathbb{C})} \quad = \quad L. \tag{17.106}$$

which implies a complexity reduction by a factor of $(P-1)L^2$ in terms of the number of complex multiplications and additions.

17.2.7.2.2 Left-Inverse Related Form of MMSE Combining Generating the Weight Matrix

Upon following similar steps, as in the context of our analysis of the LS combiner's complexity, which was considered for the scenario of $P \geq L$ in Section 17.2.7.1, we found that the complexity of the MMSE combiner, as represented by Equation 17.68 in its left-inverse related form can be related to the complexity formula originally derived for the LS combiner namely, to Equation 17.103, which is repeated here for the reader's convenience:

$$C_{\text{MMSE,W+cmb}}^{(\mathbb{C}*\mathbb{C})} = C_{\text{MMSE,W+cmb}}^{(\mathbb{C}+\mathbb{C})} = PL + 2PL^2 + \frac{1}{3}L^3. \tag{17.107}$$

More explicitly, this LS combining related formula quantifies the number of complex multiplications and additions required by Equations 17.37 and 17.7, respectively. However, the MMSE combiner is somewhat more complex, since in Equation 17.68 an additional complexity contribution of mixed real-complex multiplications and additions is incurred due to incorporating the SNR matrix of $\mathbf{P}_{\text{SNR}}$. The number of these operations is given by:

$$C_{\text{MMSE,W+cmb}}^{(\mathbb{R}*\mathbb{C})} \quad = \quad PL \tag{17.108}$$

$$C_{\text{MMSE,W+cmb}}^{(\mathbb{R}+\mathbb{C})} \quad = \quad L. \tag{17.109}$$

17.2.7.3 Demodulation of the Linear Combiner's Output Signal

In addition to the computational complexity associated with the process of linear combining, we also have to account for the complexity imposed by demodulating the different users' associated combiner output signals with the aid of Equation 17.94. More specifically, from Equation 17.94 we infer that evaluating the Euclidean distance metric for a single M_c-ary trial-symbol $\check{s}$ of a subcarrier requires one complex addition, as well as "half" a complex multiplication, which is related to the operation of actually calculating the Euclidean norm[11] of the complex-valued difference between the received signal and the trial-symbol. Hence, in the context of M_c number of symbols per trial-set and for L number of simultaneous users to be demodulated, the total computational complexity related to the demodulation of

---

[11]$|a_x + ja_y|^2 = (a_x + ja_y) \cdot (a_x + ja_y)^* = a_x^2 + a_y^2.$

Equation 17.94 is given by:

$$C_{\text{lin,dem}}^{(\mathbb{C}*\mathbb{C})} = \frac{1}{2}LM_c \tag{17.110}$$

$$C_{\text{lin,dem}}^{(\mathbb{C}+\mathbb{C})} = C_{\text{lin,dem}}^{(\mathbb{R}\lessgtr\mathbb{R})} = LM_c, \tag{17.111}$$

where we have introduced the number of real-valued comparisons $C_{\text{lin,dem}}^{(\mathbb{R}\lessgtr\mathbb{R})}$ between the Euclidean distance metric outcomes as a further index of complexity.

17.2.7.4 Simplified Complexity Formulae to be Used in the Comparison of the Different Detectors

In Sections 17.2.7.1, 17.2.7.2 and 17.2.7.3 we elaborated on the individual computational complexity exhibited by the LS and MMSE combiners described in Sections 17.2.2 and 17.2.3, respectively, as well as by the process of demodulating the combiner's output signal as outlined in Section 17.2.4. By contrast, in this section, we will present simplified complexity formulae for the LS and MMSE detectors, which will be employed in our final comparison of the different detectors' complexities in Section 17.3.4.2. Our aim was to give a more compact representation of the complexity, in terms of the number of complex multiplications and additions, as well as real-valued comparisons. Specifically, in the context of the MMSE detector the number of mixed real-complex multiplications and additions has been expressed in terms of the number of complex multiplications and additions[12] upon weighting them by a factor of $\frac{1}{2}$. Here we assume that the weight matrix is not determined, which allows for a lower-complexity implementation, as argued in the previous sections. Hence, for the LS detector of Equation 17.35 and the associated process of demodulation in Equation 17.94 we obtain the following simplified complexity formulae:

$$C_{\text{LS}}^{\mathbb{C}*\mathbb{C}} = \frac{1}{2}LM_c + PL + (P+1)L^2 + \frac{1}{3}L^3 \tag{17.112}$$

$$C_{\text{LS}}^{\mathbb{C}+\mathbb{C}} = LM_c + PL + (P+1)L^2 + \frac{1}{3}L^3 \tag{17.113}$$

$$C_{\text{LS}}^{\mathbb{R}\lessgtr\mathbb{R}} = LM_c. \tag{17.114}$$

By contrast, for the left-inverse related form of the MMSE detector in Equation 17.70 plus for the associated demodulation procedure of Equation 17.94 we obtain:

$$C_{\text{MMSE}}^{\mathbb{C}*\mathbb{C}} = \frac{1}{2}LM_c + \frac{3}{2}PL + (P+1)L^2 + \frac{1}{3}L^3 \tag{17.115}$$

$$C_{\text{MMSE}}^{\mathbb{C}+\mathbb{C}} = LM_c + \left(P+\frac{1}{2}\right)L + (P+1)L^2 + \frac{1}{3}L^3 \tag{17.116}$$

$$C_{\text{MMSE}}^{\mathbb{R}\lessgtr\mathbb{R}} = LM_c. \tag{17.117}$$

Again, these simplified formulae will be employed in the context of Section 17.3.4.2 to compare the different detectors' complexities.

---

[12]Here we have neglected the real-valued additions required to evaluate the product of two complex numbers and hence our complexity formulae provide an upper bound estimate.

17.2.8   Conclusion on Linear Detection Techniques

In Section 17.2 we concentrated on the mathematical portrayal, performance- and complexity comparison of the most prominent linear detection techniques, namely on the LS and MMSE procedures of Sections 17.2.2 and 17.2.3, respectively. Our discussions commenced in Section 17.2.1 with the characterisation of a linear combiner's output signal and its components, while our more specific discussions in Section 17.2.2 focussed on the LS detector, also known as the ZF detector or decorrelating detector. Its associated combiner weight matrix was shown in Equation 17.38 to be given as the Hermitian transpose of the Moore-Penrose pseudo-inverse or left-inverse of the channel transfer factor matrix $\mathbf{H}$. In contrast to the calculation of the MMSE related weight matrix, its calculation outlined in Equation 17.38 does not require any statistical information. Although, as shown in Equation 17.42 the transmitted signal is recovered with unit-gain, it is contaminated by the residual AWGN, which is potentially boosted due to the effects of the actual channel matrix. In order to achieve a lower average MSE than that of Equation 17.46, derived to characterise the LS combiner's output, the MMSE combiner of Section 17.2.3 can be invoked.

As suggested by the terminology, from the set of all linear combiners the MMSE combiner exhibits the lowest MSE at the output. As shown in Section 17.2.3, this is achieved upon incorporating statistical information concerning the transmitted signals' variances and the AWGN variance into the detection process, resulting in a Wiener filter-related weight matrix as evidenced by Equation 17.63 or 17.64. This representation of the weight matrix $\mathbf{W}_{\text{MMSE}}$ is related to the right-inverse of the channel matrix. If the auto-correlation matrix associated with the different antenna elements' AWGN is a scaled unity matrix, then an alternative representation, namely Equation 17.68, can be obtained, which is related to the channel matrix's left-inverse rather than its right-inverse. Depending on the dimensions of the channel transfer factor matrix $\mathbf{H}$ in terms of the L number of users and the P number of reception antenna elements, the left- or right-inverse related form may be preferred in terms of the computational complexity imposed. In the context of these discussions the relation between the MMSE- and MVDR combiner was briefly addressed. Similar to the LS combiner of Section 17.2.2, the desired signal is recovered with unity gain, while at the same time suppressing the AWGN, again, based upon knowledge of the different users' SNRs encountered at the reception antennas. Specifically, from Equations 17.78 and 17.82 we recall that both the l-th user's MMSE weight vector $\mathbf{w}_{\text{MMSE}}^{(l)}$ and the MVDR combiner's weight vector $\mathbf{w}_{\text{MV}}^{(l)}$ can be represented as the scaled product of the inverse interference-plus-noise correlation matrix $\mathbf{R}_{a,I+N}^{-1}$ and the desired user's channel vector $\mathbf{H}^{(l)}$. The difference between the two solutions resides in the choice of the scalar factor β, which for the specific case of MMSE detection was given by Equation 17.79, while for MV detection by Equation 17.83. As further argued in Section 17.2.3.4, both solutions exhibit the same SINR, which in turn is identical to that of the maximum SINR combiner [90, 136] not detailed here. The maximum SINR combiner can also be represented in a similar form as Equations 17.79 and 17.83, but with a different value of β. Hence, the Wiener filter-related linear combiners, namely the MMSE, MVDR and maximum SINR combiners maximise the SINR. Furthermore, in Section 17.2.5 the generation of soft-bit information for turbo-decoding was demonstrated.

Our performance assessment with respect to the MSE and SINR at the combiner's output, as well as with respect to the detector's BER was carried out in Section 17.2.6. The curves, which were shown on the left-hand side of Figure 17.6 supported the theory that the MSE at the combiner's output is minimised by the MMSE weight matrix, while a slight MSE degradation was observed for the MVDR weight matrix of Equation 17.81 as a consequence of the requirement to recover the desired user's transmitted signal with a specific gain, which was assumed to be unity in our case. As expected, the worst MSE was exhibited by the LS combiner of Section 17.2.2. In terms of the system's 4QAM-related BER on the right-hand side of Figure 17.6 we observed an identical performance for both Wiener filter-related detectors,[13] namely for the MMSE and MV detector. Again, as argued in Section 17.2.3.5, their performance was identical, because the different Wiener filter-related detectors achieve the same SINR. However, for the LS detector of Section 17.2.2 a significant performance degradation was also observed in terms of the BER as shown on the right-hand side of Figure 17.6. Our further investigations concentrated on portraying the influence of the relation between the number of users L and the number of reception antennas P on the MMSE detector's SINR and on the associated BER performance. Specifically, from Figures 17.7 and 17.8 we observed that upon decreasing the number of users L, while keeping the number of reception antennas P constant, the PDF of the SINR shifted towards higher SINRs, while at the same time the system's BER is significantly improved. This is also a motivation for the employment of the successive interference cancellation approach, which will be discussed in Section 17.3.1.

Finally, the BER performance of MMSE detection-assisted SDMA-OFDM was analysed in a turbo-coded scenario. The associated BER versus SNR performance curves were portrayed in Figure 17.9. Compared to the uncoded scenario the BER was significantly improved – although at the cost of halving the system's effective throughput. Our estimates of the computational complexity associated with the different linear detectors, namely LS and MMSE were presented in Section 17.2.7. We found that the computational complexity is of third order, namely $\mathcal{O}(3)$ with respect to the number of users L, upon assuming the weight matrix's representation in its left-inverse related form.

17.3   Non-Linear Detection Techniques

In Section 17.2 the family of linear detection techniques was discussed. These detectors aimed at reducing the potentially excessive M_c^L number of evaluations of the multi-user Euclidean distance metric associated with the optimum ML detector to a significantly lower LM_c number of evaluations of the single-user Euclidean distance metric of Equation 17.94. This implies a substantial complexity reduction. Recall that the variable L represents the number of users supported, while M_c is the number of constellation points for the specific modulation scheme employed. As portrayed in the linear detector's block diagram shown in Figure 17.4, the strategy is first to provide linear estimates of the different users' transmitted signals and then to perform the non-linear classification or demodulation separately for each user. This philosophy was based on the assumption that the different users' associated linear combiner output signals are corrupted only by the residual AWGN, which is, however, only an approximation. In fact the linear combiners' output signals in Figure 17.4 also contain residual interference, which is not Gaussian distributed and hence represents an important source of further information.

Instead of sequentially performing the operations of linear combining and classification-

---

[13]Recall that the linear detector is constituted by the concatenation of the linear combiner and the classifier.

or demodulation as in the linear detector's case of Figure 17.4, a more effective strategy is to embed the demodulation into the process of linear combining, which is known from the family of classic channel equalisers as decision-feedback. As a result, the residual multi-user interference observed at the classifier's inputs is reduced. Hence, the classifier's accuracy due to neglecting the residual interference is less impaired. Two of the most prominent multi-user detection techniques known from CDMA communications, which incorporate these ideas, are the SIC and PIC detection techniques. These techniques are also applicable in the context of communicating over flat-fading channels as observed for example on an OFDM subcarrier basis. In the context of our portrayal of SIC detection in Section 17.3.1, apart from discussing various techniques for its improvement, a detailed analysis of the effects of error propagation occurring at the different detection stages will be provided. This error propagation analysis motivated the employment of weighted soft-bit metric assisted turbo-decoding. Furthermore, our discussions on PIC detection will be presented in Section 17.3.2. We will demonstrate that a significant enhancement of the PIC detector's performance can be achieved by embedding turbo-decoding into the PIC detection process, instead of simply serially concatenating the PIC detector with the turbo-decoder. Finally, the optimum ML detector will be analysed in Section 17.3.3.

17.3.1 SIC Detection

The philosophy of the Successive Interference Cancellation (SIC) assisted detector [104, 109, 113, 117, 119, 124, 126, 128, 135, 137] is motivated by two observations. First of all, we note that for a specific subcarrier the MSE and SINR at the output of the LS or MMSE combiner might substantially differ for the different users, depending on their spatial signatures. Second, we recall from our investigations in Section 17.2.6.3 that upon increasing the MIMO system's diversity order, e.g. by decreasing the number of simultaneous users L while keeping the number of reception antennas P constant, the MSE performance of the LS or MMSE combiner and correspondingly the system's BER performance are improved as a consequence of assigning a higher grade of diversity to mitigate the effects of fading. This was illustrated in Figure 17.8. Hence, an attractive strategy, which has recently attracted a great deal of interest is to detect only the specific user having the highest SINR, SIR or SNR in each iteration at the output of the LS or MMSE combiner. Having detected this user's signal, the corresponding remodulated signal is subtracted from the composite signal received by the different antenna elements. Furthermore, the channel transfer factor matrix – and the SNR matrix formulated in the context of the MMSE combiner characterised by Equation 17.68 in its left-inverse related form - are updated accordingly.

In Section 17.3.1.1 the standard SIC algorithm is portrayed, which allows only the most likely symbol decision to be retained in each detection stage. This section also includes a detailed analysis of the effects of error propagation, which occurs across the different detection stages. By contrast, in Section 17.3.1.2 M-SIC and its derivatives are discussed, where potentially the M most likely tentative symbol decisions are retained at each node of the detection process – as will be further explained in Section 17.3.1.2.1 – rather than retaining only the most likely symbol decision. Note that each detection node is associated with a specific appropriately updated array output vector of the SIC-aided detection process. Furthermore, in Section 17.3.1.3 the various techniques of soft-bit generation will be discussed and a weighted soft-bit metric will be proposed for employment in turbo-decoding, which

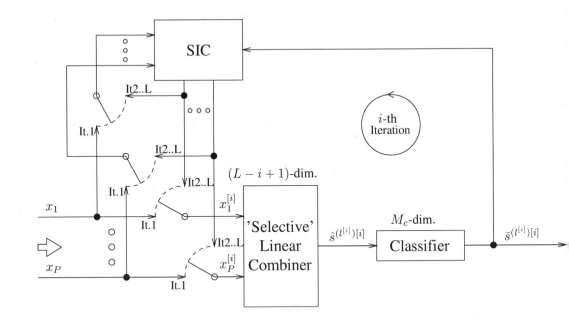

Figure 17.10: Illustration of the main signal paths associated with the standard SIC detector. During the first iteration the signals $x_p, p = 1, \ldots, P$ received by the different antenna elements are directly fed into the "selective" linear combiner, where we have $\mathbf{x}^{[1]} = \mathbf{x}$ at the detection stage or iteration of $i = 1$. The task of the "selective" linear combiner is to identify the most dominant remaining user in terms of its SINR at the combiner output - from the set of $(L - i + 1)$ remaining users during the i-th detection stage or iteration - and to provide its signal estimate $\hat{s}^{(l^{[i]})[i]}$ at the combiner's output. This is described by Equations 17.123 and 17.124. The selected $l^{[i]}$-th user's linear signal estimate $\hat{s}^{(l^{[i]})[i]}$ is then classified or demodulated according to Equation 17.125, yielding the amplified constellation point $\breve{s}^{(l^{[i]})[i]}$ that is most likely to have been transmitted by the $l^{[i]}$-th user. Now the corresponding modulated signal can be regenerated. The influence of the $l^{[i]}$-th user's modulated signal is then removed from the vector $\mathbf{x}^{[1]}$ of signals received by the different antenna elements with the aid of the SIC module. This cancellation operation is described by Equation 17.127. The first iteration $(i = 1)$ is deemed to have been completed, when the decontaminated signal appears at the output of the SIC stage. Hence, beginning with the second SIC iteration the "selective" linear combiner's input, namely the "decontaminated" vector $\mathbf{x}^{[i]}$ of signals received by the different antenna elements, which contains only the influence of the $(L - i + 1)$ remaining users, is constituted by the output of the SIC module, provided that correct symbol decisions were conducted in the previous detection stages. Note that for the sake of visual clarity here we have omitted the signal paths associated with the channel transfer factor estimates required by the linear combiner and by the SIC module. The role of the switches is to indicate that at the first detection stage the SIC is directly fed with the signals received by the different array elements, while during the remaining iterations of $i = 2, \ldots, L$ with the partially "decontaminated" composite signal of the remaining $(L - i + 1)$ users.

is capable of substantially enhancing the performance of turbo-coded SIC detection-assisted SDMA-OFDM systems. Finally, a detailed performance analysis of the standard SIC and that of the M-SIC is offered in terms of the associated system's BER and SER performance both in the context of uncoded and turbo-coded scenarios in Section 17.3.1.4. The complexity of the different detection schemes will be analysed in Section 17.3.1.5. Finally, the summary of Section 17.3.1 will be offered in Section 17.3.1.6 along with our conclusions.

17.3.1.1   Standard SIC

From now on we assume that the MMSE combiner in its specific left-inverse related form as given by Equation 17.68 is assumed to perform the detection of the most dominant user in each cancellation stage. For the reader's convenience we have repeated here the formula describing the combiner's operation from Equation 17.70:

$$\mathbf{R}_{\overline{a}}^{H}\mathbf{W}_{\text{MMSE}}^{H} = \mathbf{R}_{c}^{H} \quad \Longleftrightarrow \quad \mathbf{W}_{\text{MMSE}}^{H} = \mathbf{R}_{\overline{a}}^{H-1}\mathbf{R}_{c}^{H}, \tag{17.118}$$

where the received signals' auto-correlation matrix $\mathbf{R}_{\overline{a}}$ was defined in Equation 17.69 as:

$$\mathbf{R}_{\overline{a}} = \mathbf{H}^{H}\mathbf{H}\mathbf{P} + \sigma_{n}^{2}\mathbf{I}, \tag{17.119}$$

while the cross-correlation matrix of the transmitted and received signals was given in Equation 17.54 namely:

$$\mathbf{R}_{c} = \mathbf{H}\mathbf{P}. \tag{17.120}$$

It is computationally efficient to refrain from recalculating the correlation matrix $\mathbf{R}_{\overline{a}}$ of Equation 17.119 and the cross-correlation matrix $\mathbf{R}_{c}$ of Equation 17.120 at each cancellation stage. This complexity reduction can be achieved by updating these matrices based on the specific index of the most recently detected user. Nonetheless, these matrices have to be calculated once at the beginning of the SIC detection procedure. The more detailed structure of the SIC detector will be portrayed below.

1) *Initialisation*: Initialise the detector upon setting $\mathbf{x}^{[1]} = \mathbf{x} \in \mathbb{C}^{P \times 1}$, as well as upon evaluating $\mathbf{R}_{\overline{a}}^{[1]H} = \mathbf{P}\mathbf{H}^{H}\mathbf{H} + \sigma_{n}^{2}\mathbf{I}$ and $\mathbf{R}_{c}^{[1]H} = \mathbf{P}\mathbf{H}^{H}$. Here the index in the superscript, namely $()^{[i]}$ indicates the detection stage index, which is initially set to $i = 1$.

2) *i-th Detection Stage*: At the beginning of the i-th SIC detection stage, given correct symbol decisions in the previous detection stages, the updated vector $\mathbf{x}^{[i]}$ of received signals only contains the remaining $L^{[i]} = L - i + 1$ users' signal contributions plus the AWGN since the remodulated signals of the previously detected $(i - 1)$ users have been deducted from the originally received composite signal of $\mathbf{x}$. Furthermore, the dimension of the auto-correlation matrix $\mathbf{R}_{\overline{a}}^{[i]H}$ – represented here in its Hermitian transposed form – has been reduced to $\mathbf{R}_{\overline{a}}^{[i]H} \in \mathbb{C}^{L^{[i]} \times L^{[i]}}$, while the dimension of the cross-correlation matrix $\mathbf{R}_{c}^{[1]H}$ also represented in its Hermitian transposed form has been reduced to $\mathbf{R}_{c}^{[i]H} \in \mathbb{C}^{L^{[i]} \times P}$ upon removing the previously cancelled users' associated entries. This matrix dimension reduction potentially facilitates the reduction of the system's overall complexity. Then the specific steps at the i-th detection stage are as follows:

- *Calculation of the Remaining Users' Weight Matrix*: Generate the $L^{[i]}$ number of remaining users' associated weight matrix upon invoking the MMSE approach, which is represented in its left-inverse related form by:

$$\mathbf{R}_{\overline{a}}^{[i]H}\mathbf{W}_{\text{MMSE}}^{[i]H} = \mathbf{R}_{c}^{[i]H} \quad \Longleftrightarrow \quad \mathbf{W}_{\text{MMSE}}^{[i]H} = \mathbf{R}_{\overline{a}}^{[i]H-1}\mathbf{R}_{c}^{[i]H}. \quad (17.121)$$

Observe in Equation 17.121 that in contrast to Equation 17.119 and 17.120 we have substituted the matrices $\mathbf{R}_{\overline{a}}^{H}$ and $\mathbf{R}_{c}^{H}$ by their reduced-dimensional counterparts associated with the i-th detection stage, namely by $\mathbf{R}_{\overline{a}}^{[i]H}$ and $\mathbf{R}_{c}^{[i]H}$. The set of $L^{[i]}$ number of remaining users at the i-th detection stage is denoted here by $\mathcal{L}^{[i]}$.

- *Selection of the Most Dominant User*: Calculate the objective function, which could be the SINR, SIR or SNR at the MMSE combiner's output according to Equations 17.24, 17.25 or 17.26, respectively employing the different users' weight vectors. As an example, here we employ the SNR of Equation 17.26, since its calculation is significantly less complex than that of the SINR or SIR given by Equations 17.25 and 17.26. Based on Equation 17.26, the l-th user's associated SNR at the MMSE combiner's output during the i-th detection stage is given by:

$$\text{SNR}^{(l)[i]} = \frac{\mathbf{w}^{(l)[i]H}\mathbf{R}_{a,S}^{(l)}\mathbf{w}^{(l)[i]}}{\mathbf{w}^{(l)[i]H}\mathbf{R}_{a,N}\mathbf{w}^{(l)[i]}}, \quad (17.122)$$

where the auto-correlation matrix $\mathbf{R}_{a,S}^{(l)}$ of the l-th user's channel transfer factors was defined in Equation 17.16, while the noise correlation matrix $\mathbf{R}_{a,N}$ recorded in case of encountering uncorrelated AWGN at the different reception antenna elements of the BS was given in Equation 17.22. Furthermore, the l-th user's weight vector $\mathbf{w}^{(l)[i]}$ is given here in form of the corresponding column vector of the weight matrix $\mathbf{W}_{\text{MMSE}}^{[i]}$, which has been obtained upon solving Equation 17.121. The selection of the most dominant user, which is assumed here to be the $l^{[i]}$-th user, can then be expressed as:

$$l^{[i]} = \underset{l \in \mathcal{L}^{[i]}}{argmax}(\text{SNR}^{(l)[i]}). \quad (17.123)$$

- *Detection of the Most Dominant User*: Under the assumption that the $l^{[i]}$-th user has been found to be the most dominant one among the $L^{[i]}$ remaining users at the i-th detection stage, detect the user's transmitted signal upon invoking Equation 17.8, namely:

$$\hat{\mathbf{s}}^{(l^{[i]})[i]} = \mathbf{w}^{(l^{[i]})[i]H}\mathbf{x}^{[i]}. \quad (17.124)$$

- *Demodulation of the Most Dominant User*: Carry out the demodulation by mapping the detected signal $\hat{\mathbf{s}}^{(l^{[i]})[i]}$[14] to one of the M_c number of constellation points

---

[14]Note that while the superscript in round brackets denotes the user index, the superscript in squared brackets denotes the detection stage or iteration index.

contained in the set $\mathcal{M}_c$ associated with a particular modulation scheme. As shown in Equation 17.94, this involves minimising the Euclidean distance metric, namely:

$$\check{s}^{(l^{[i]})[i]} = arg \min_{\check{s}/\sigma_{l^{[i]}} \in \mathcal{M}_c} \left| \frac{1}{H_{\text{eff}}^{(l^{[i]})[i]}} \hat{s}^{(l^{[i]})[i]} - \check{s} \right|^2 , \qquad (17.125)$$

where the detected user's transfer factor $H_{\text{eff}}^{(l^{[i]})[i]}$ is given by:

$$H_{\text{eff}}^{(l^{[i]})[i]} = \mathbf{w}^{(l^{[i]})[i]H} \mathbf{H}^{(l^{[i]})}. \qquad (17.126)$$

Note, however, that for MPSK modulation schemes the normalisation to $H_{\text{eff}}^{(l^{[i]})[i]}$ is not necessary, because the information transmitted is incorporated into the signal's phase. Furthermore, in the context of Equation 17.125 the variance of the $l^{[i]}$-th user's $\mathcal{M}_c$ number of legitimate trial symbols is given by $\sigma_{l^{[i]}}^2$. Alternatively the MMSE combiner's output signal $\hat{s}^{(l^{[i]})[i]}$ can be normalised by $\sigma_{l^{[i]}} = \sqrt{\sigma_{l^{[i]}}^2}$ instead of amplifying the individual constellation points contained in the set $\mathcal{M}_c$.

- *Detector Update by Removing the Most Dominant User's Contribution*: Based on the demodulated signal $\check{s}^{(l^{[i]})[i]}$, the $l^{[i]}$-th user's remodulated contribution is removed from the current vector of composite received signals, yielding:

$$\mathbf{x}^{[i+1]} = \mathbf{x}^{[i]} - \mathbf{H}^{(l^{[i]})} \check{s}^{(l^{[i]})[i]}. \qquad (17.127)$$

Furthermore, the influence of the $l^{[i]}$-th user's associated channel transfer factor vector $\mathbf{H}^{(l^{[i]})}$ is eliminated from the auto-correlation matrix $\mathbf{R}_{\bar{a}}^{[i]H}$, yielding the reduced dimensional matrix of:

$$\mathbf{R}_{\bar{a}}^{[i]H} \longrightarrow \mathbf{R}_{\bar{a}}^{[i+1]H} \in \mathbb{C}^{(L^{[i]}-1)\times(L^{[i]}-1)}, \qquad (17.128)$$

as well as from the cross-correlation matrix $\mathbf{R}_c^{[i]H}$, yielding the reduced dimensional matrix of:

$$\mathbf{R}_c^{[i]H} \longrightarrow \mathbf{R}_c^{[i+1]H} \in \mathbb{C}^{(L^{[i]}-1)\times P}. \qquad (17.129)$$

More specifically, this is achieved by removing the $\acute{l}^{[i]}$-th row and column from the matrix $\mathbf{R}_{\bar{a}}^{[i]H}$ as well as by eliminating the $\acute{l}^{[i]}$-th row from the matrix $\mathbf{R}_c^{[i]H}$, where the index $\acute{l}^{[i]}$ denotes the position of the column vector $\mathbf{H}^{(l^{[i]})}$ in the hypothetic reduced-size channel transfer factor matrix $\mathbf{H}^{[i]}$, which is associated with the i-th detection stage.

3) Commence the $(i+1)$-th iteration by returning to Step (2). Iterate, until all the L users have been detected.

We have summarised the standard SIC algorithm once again in Table 17.6. Furthermore, a simplified block diagram of the SIC detector was portrayed in Figure 17.10. Note that here we

Description	Instruction
Initialisation	$\mathbf{x}^{[1]} = \mathbf{x}; \mathbf{R}_{\overline{a}}^{[1]H} = \mathbf{P}\mathbf{H}^H\mathbf{H} + \sigma_n^2\mathbf{I}; \mathbf{R}_c^{[1]H} = \mathbf{P}\mathbf{H}^H; L^{[1]} = L$
i-th iteration:	
Weight calc.	$\mathbf{R}_{\overline{a}}^{[i]H}\mathbf{W}_{\mathrm{MMSE}}^{[i]H} = \mathbf{R}_c^{[i]H} \iff \mathbf{W}_{\mathrm{MMSE}}^{[i]H} = \mathbf{R}_{\overline{a}}^{[i]H-1}\mathbf{R}_c^{[i]H}$
Selection	$\mathrm{SNR}^{(l)[i]} = \dfrac{\mathbf{w}^{(l)[i]H}\mathbf{R}_{a,S}^{(l)}\mathbf{w}^{(l)[i]}}{\mathbf{w}^{(l)[i]H}\mathbf{R}_{a,N}\mathbf{w}^{(l)[i]}}, \ l \in \mathcal{L}^{[i]}, l^{[i]} = \underset{l\in\mathcal{L}^{[i]}}{argmax}(\mathrm{SNR}^{(l)[i]})$
Combining	$\hat{\mathbf{s}}^{(l^{[i]})[i]} = \mathbf{w}^{(l^{[i]})[i]H}\mathbf{x}^{[i]}$
Demodulation	$\check{\mathbf{s}}^{(l^{[i]})[i]} = arg\ \underset{\check{\mathbf{s}}/\sigma_{l^{[i]}}\in\mathcal{M}_c}{min}\left\|\dfrac{1}{H_{\mathrm{eff}}^{(l^{[i]})[i]}}\hat{\mathbf{s}}^{(l^{[i]})[i]} - \check{\mathbf{s}}\right\|^2, \ H_{\mathrm{eff}}^{(l^{[i]})[i]} = \mathbf{w}^{(l^{[i]})[i]H}\mathbf{H}^{(l^{[i]})}$
Updating	$\mathbf{x}^{[i+1]} = \mathbf{x}^{[i]} - \mathbf{H}^{(l^{[i]})}\check{\mathbf{s}}^{(l^{[i]})[i]}$
	$\mathbf{R}_{\overline{a}}^{[i]H} \longrightarrow \mathbf{R}_{\overline{a}}^{[i+1]H} \in \mathbb{C}^{(L^{[i]}-1)\times(L^{[i]}-1)}$
	$\mathbf{R}_c^{[i]H} \longrightarrow \mathbf{R}_c^{[i+1]H} \in \mathbb{C}^{(L^{[i]}-1)\times P}$
	$L^{[i+1]} = L^{[i]} - 1$
Return	Start $(i+1)$-th iteration

Table 17.6: Summary of the standard SIC detector's operation in the context of employing MMSE combining and the SNR as an objective function for the selection of the most dominant user at each detection stage.

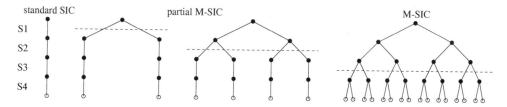

Figure 17.11: Illustration of the **(a: Left)** standard SIC, **(b: Right)** partial M-SIC retaining multiple branches per detection node up to the first- and second detection stage, (c) M-SIC with multiple tentative symbol decisions per detection node at all detection stages; here we have assumed that $M = 2$; in the graph each detection node represents an updated vector of signals received by the different antenna elements, for which also the linear combining but not the array weight calculation has to be performed separately; furthermore, each branch represents a tentative symbol decision made at a given detection stage.

have omitted the signal paths associated with supplying the channel transfer factor estimates to the linear combiner and to the SIC module. In the next section a higher-complexity strategy is proposed to further enhance the standard SIC detector's performance.

17.3.1.2 M-SIC and its Derivatives

As we will highlight in Section 17.3.1.4, the standard SIC detector's performance is impaired as a result of the error-propagation occurring between the different consecutive detection stages. Hence efficient countermeasures capable of significantly reducing these effects will be the topic of this section. Specifically in Section 17.3.1.2.1 M-SIC will be investigated, while partial M-SIC in Section 17.3.1.2.2 and Selective Decision Insertion (SDI) M-SIC in Section 17.3.1.2.3.

17.3.1.2.1   M-SIC   A viable strategy to reduce the error propagation effects is to track from each detection stage not only the single most likely symbol decision, but an increased number of $M \leq M_c$ most likely tentative symbol decisions, where M_c denotes the number of constellation points associated with a specific modulation scheme. To provide an example, for $M = 2$ in the first detection stage we have a total of $M = 2$ possible symbol decisions, while in the second detection stage $M^2 = 4$ tentative symbol decisions and correspondingly, in the i-th detection stage we encounter M^i possible tentative symbol decisions. Following our description of the standard SIC detector in Table 17.6, associated with each tentative symbol decision there is a specific updated vector of signals, generated by cancelling the effects of the most dominant $L - i + 1$ number of users from the P-dimensional vector of signals received by the P number of different antenna elements. Hence, in the following detection stage the MMSE combining has to be performed separately for the different updated P-dimensional vectors of received signals. Correspondingly, the number of parallel tentative symbol decisions to be tracked is increased by the factor of M compared to that of the current detection stage. This process can conveniently be portrayed with the aid of a tree-structure, as shown on the right-hand side of Figure 17.11, where again, we have assumed that $M = 2$ was used. Specifically, each detection node represents an updated P-dimensional vector of signals received by the P different antenna elements, while the branches are associated with the various tentative symbol decisions at the $i = 1, \ldots, L$ detection stages. Note that the first detection node at the top of the figure is associated with the original P-dimensional vector of signals received by the different antenna elements. In the final detection stage, after the subtraction of the least dominant user's estimated P-dimensional signal contribution, a decision must be made concerning which specific combination of L number of symbols – represented by the branches connecting the different detection nodes – has most likely been transmitted by the L different users in the specific subcarrier considered. A suitable criterion for performing this decision is given by the Euclidean distance between the original P-dimensional vector of signals received by the P different antenna elements and the estimated P-dimensional vector of received signals based on the tentative symbol decisions and upon taking into account the effects of the channel. The same decision metric is employed also by the ML detector, which will be discussed in Section 17.3.3. Note, furthermore, that this distance measure is identical to the Euclidean norm of the P-dimensional vector of residual signals after the subtraction of the last detected user's P-dimensional signal contribution vector.

The performance improvement potentially observed for the M-SIC scheme compared to the standard SIC arrangement is achieved at the cost of a significantly increased computational complexity. This is since the number of parallel tentative symbol decisions associated with a specific detection stage is a factor of M higher than that of the previous detection stage, and hence in the last detection stage we potentially have to consider M^L number of different tentative symbol decisions. Again, this implies that the approach of the M-SIC scheme resembles that of the ML detector to be discussed in Section 17.3.3.

17.3.1.2.2   Partial M-SIC   A viable approach to further reduce the associated computational complexity is motivated by the observation that for sufficiently high SNRs the standard

SIC detector's performance is predetermined by the bit or symbol-error probabilities incurred during the first detection stage. This is, because if the most dominant user's associated symbol decision is erroneous, its effects potentially spread to all other users' decisions conducted in the following detection stages. Furthermore, as observed previously in the context of our investigations of the MMSE detector's performance in Section 17.2.6.3 as a function of the number of simultaneous users L and the number of reception antennas P, the highest performance gain in terms of the achievable SNR reduction at the reception antennas, whilst maintaining a specific BER, is observed upon removing the first user from a fully loaded system. An example of such a fully loaded system is that supporting four simultaneous users with the aid of four reception antennas. Hence we conclude that the symbol error probability specifically of the first detection stage should be as low as possible, while the tentative symbol decisions carried out at later detection stages become automatically more reliable as a result of the system's increased diversity order due to removing the previously detected users.

Hence, our suggestion is to retain $M > 1$ number of tentative symbol decisions at each detection node, characterised by its associated updated P-dimensional vector of received signals only up to the specific $L_{\text{pM-SIC}}$-th stage in the detection process. By contrast, at later detection stages only one symbol decision is retained at each detection node, as in the standard SIC scheme. This philosophy is further highlighted with the aid of the two graphs at the centre of Figure 17.11. Specifically, in the illustration second from the left of Figure 17.11 we have portrayed the case of retaining two tentative symbol decisions per detection node only in the first detection stage, while in the illustration second from the right-hand side of Figure 17.11 two tentative symbol decisions per detection node are retained in both of the first two detection stages.

17.3.1.2.3 Selective-Decision-Insertion Aided M-SIC

In order to reduce even further the computational complexity, an improved strategy termed Selective-Decision-Insertion (SDI) can be applied, which was initially proposed in [117, 137]. The philosophy of the SDI technique is that of tracking additional tentative symbol decisions only in those $N_{\text{SDI}}^{[1]}$ number of subcarriers, which exhibit the lowest SINR during the first detection stage, since these are most likely to cause symbol errors.

17.3.1.3 Generation of Soft-Bit Information for Turbo-Decoding

In Section 17.2.5 we elaborated on the process of soft-bit generation supporting the employment of turbo-decoding in the context of linear detection techniques, such as the LS and MMSE schemes. Specifically, we capitalised on the assumption that the residual interference at the combiner's output is Gaussian, which enabled us to employ the same strategies for generating the soft-bit values, as in a single-user scenario. In Section 17.3.1.3.1 we will demonstrate that the soft-bit generation process designed for the non-linear SIC detector can be based on that of the linear detection schemes, although, as shown in Section 17.3.1.3.2, a further performance enhancement can be achieved upon accounting for the effects of error propagation, which occur through the different detection stages, as will be demonstrated in Section 17.3.1.4.2.

17.3.1.3.1 Generation of Rudimentary Soft-Bits

As highlighted in Section 17.3.1.1, at each stage of the standard SIC-related detection process we generate estimates of the remaining users' transmitted signals with the aid of a linear combiner. Hence, a feasible approach employed to generate soft-bit values is to invoke the linear combiner's output signals of the most dominant user as it was demonstrated in Section 17.2.5. More specifically, the $l^{[i]}$-th user's soft-bit values – where the superscript i in $()^{[i]}$ indicates that this particular user was found to be the most dominant remaining user during the i-th detection stage – can be generated upon invoking the associated combiner output signal $x_{\text{eff}}^{(l^{[i]})[i]} = \hat{s}^{(l^{[i]})[i]}$ defined in the context of Equation 17.95 namely:

$$L_{m\approx}^{(l^{[i]})[i]} = \ln \frac{P(b_m^{(l^{[i]})[i]} = 1 | x_{\text{eff}}^{(l^{[i]})[i]}, H_{\text{eff}}^{(l^{[i]})[i]})}{P(b_m^{(l^{[i]})[i]} = 0 | x_{\text{eff}}^{(l^{[i]})[i]}, H_{\text{eff}}^{(l^{[i]})[i]})}, \tag{17.130}$$

where $H_{\text{eff}}^{(l^{[i]})[i]} = \mathbf{w}^{(l^{[i]})[i]H} \mathbf{H}^{(l^{[i]})[i]}$, as given by Equation 17.126 is the l-th detected user's effective channel transfer factor.

However, generating the soft-bit values of the $l^{[i]}$-th user – whose associated signal is linearly detected during the i-th detection stage – with the aid of Equation 17.130 inherently assumes that the signal components of those users, which have already been detected and demodulated during the previous SIC detection stages, have been correctly removed from the P-dimensional vector $\mathbf{x}^{[1]} = \mathbf{x}$ of signals received by the P different antenna elements for the successful employment of this principle. A necessary condition is that the associated symbol decisions were free of errors, namely that we had $s^{(l^{[j]})[j]} = \check{s}^{(l^{[j]})[j]}$, $j = 1, \ldots, i - 1$. Naturally, this assumption only holds with a certain probability. In the following our aim will be to estimate this probability and draw our further conclusions.

17.3.1.3.2 Generation of Weighted Soft-Bits To elaborate a little further, during the first detection stage, the probability that the $l^{[1]}$-th user, which was found to be the most dominant one, has been correctly demodulated is given by the *a posteriori* probability of $P(\check{s}^{(l^{[1]})[1]} | x_{\text{eff}}^{(l^{[1]})[1]}, H_{\text{eff}}^{(l^{[1]})[1]})$. Hence, by contrast during the second detection stage the probability that the $l^{[2]}$-th user, which was found to be the most dominant one among the $L - 1$ remaining users, has been correctly demodulated is given by the *a posteriori* probability of $P(\check{s}^{(l^{[2]})[2]} | x_{\text{eff}}^{(l^{[2]})[2]}, H_{\text{eff}}^{(l^{[2]})[2]})$, conditioned on a correct symbol decision during the first detection stage.

Furthermore, since we have $P(x_{\text{eff}}^{(l^{[2]})[2]} | x_{\text{eff}}^{(l^{[1]})[1]}) = P(s^{(l^{[1]})[1]} | x_{\text{eff}}^{(l^{[1]})[1]})$, where we have exploited that $P(H_{\text{eff}}^{(l^{[1]})[1]}) = P(H_{\text{eff}}^{(l^{[2]})[2]}) = 1$, we obtain an estimate for the probability of the joint event that $\check{s}^{(l^{[1]})[1]}$ and $\check{s}^{(l^{[2]})[2]}$ are the correct symbol decisions at the first and second detection stages, respectively, which can be expressed as:

$$P(\check{s}^{(l^{[2]})[2]}, \check{s}^{(l^{[1]})[1]} | x_{\text{eff}}^{(l^{[1]})[1]}) = P(\check{s}^{(l^{[2]})[2]} | x_{\text{eff}}^{(l^{[2]})[2]}) P(\check{s}^{(l^{[1]})[1]} | x_{\text{eff}}^{(l^{[1]})[1]}). \tag{17.131}$$

More generally, for the demodulated symbols of the first $i - 1$ number of detection stages we have the joint probability:

$$P_{\text{joint}}^{[i-1]} = P(\check{s}^{(l^{[i-1]})[i-1]}, \ldots, \check{s}^{(l^{[1]})[1]} | x_{\text{eff}}^{(l^{[1]})[1]}) = \prod_{j=1}^{i-1} P(\check{s}^{(l^{[j]})[j]} | x_{\text{eff}}^{(l^{[j]})[j]}). \tag{17.132}$$

Note, however, that this is only an estimate of the true joint probability, since for a finite number of users the residual interference at a specific stage's combiner output is potentially non-Gaussian, which is particularly the case, if an error has occurred in one of the detection stages.

The estimated joint probability $P_{\text{joint}}^{[i-1]}$ of correct symbol decisions during the first $i-1$ number of detection stages, which was given by Equation 17.132, can be invoked as a measure of confidence for the soft-bit values generated during the i-th detection stage. Specifically, it is expected that if an error has occurred during one of the detection stages, then the soft-bit values produced with the aid of Equation 17.130 for the following detection stages will be relatively unreliable. Hence, a viable approach of mitigating these effects is the employment of weighting, namely by weighting of the *a posteriori* probabilities that a bit having a polarity of $b \in \{0,1\}$ has been transmitted, as seen in the numerator and denominator of Equation 17.130, yielding:

$$L_{m\approx}^{(l^{[i]})[i]}\Big|_{\text{weight}} = \ln \frac{P(b_m^{(l^{[i]})[i]} = 1|x_{\text{eff}}^{(l^{[i]})[i]}, H_{\text{eff}}^{(l^{[i]})[i]}) \cdot P_{\text{joint}}^{[i-1]} + \frac{1}{2}(1 - P_{\text{joint}}^{[i-1]})}{P(b_m^{(l^{[i]})[i]} = 0|x_{\text{eff}}^{(l^{[i]})[i]}, H_{\text{eff}}^{(l^{[i]})[i]}) \cdot P_{\text{joint}}^{[i-1]} + \frac{1}{2}(1 - P_{\text{joint}}^{[i-1]})}. \quad (17.133)$$

We observe that if $P_{\text{joint}}^{[i-1]}$ approaches unity, which reflects a high confidence in having no symbol errors during the previous $(i-1)$ number of detection stages, then the expression of Equation 17.133 transforms into that of Equation 17.130, which was based on the assumption of benefitting from the perfect removal of the previously detected users' signal contributions. By contrast, if $P_{\text{joint}}^{[i-1]}$ tends towards zero, which indicates a high probability of encountering symbol errors in the SIC process, then the bit probabilities $P(b_m^{(l^{[i]})[i]}|x_{\text{eff}}^{(l^{[i]})[i]}, H_{\text{eff}}^{(l^{[i]})[i]})$, $b_m^{(l^{[i]})[i]} \in \{0,1\}$ are potentially unreliable and hence should be de-weighted. This has the effect that the L-value in Equation 17.133 tends to zero.

The advantage of weighted soft-bits will be demonstrated in the context of our investigations on turbo-coded SIC schemes in Section 17.3.1.4.7.

17.3.1.4 Performance Analysis

In this section the standard SIC algorithm and its derivative, namely the M-SIC scheme will be investigated in terms of their achievable Bit Error-Ratio (BER) and Symbol Error-Ratio (SER)[15] performance. Again, the frame-invariant fading indoor WATM channel model and its associated OFDM system model described in Section 14.3.1 were invoked and ideal knowledge of the channel transfer functions associated with the different transmit-receive antenna pairs was assumed. The aim of stipulating a frame-invariant fading channel was that of avoiding the obfuscating effects of Inter-subCarrier Interference (ICI). Note that as a result of this assumption in the uncoded scenario the different detectors' BER and SER performance curves are independent from the indoor SWATM channel's specific multipath intensity profile. For a summary of the basic simulation set-up we refer to Table 17.4.

In Section 17.3.1.4.1 standard SIC and M-SIC are characterised in terms of their BER and SER performance for different numbers of communicating users and receiver antennas. Furthermore, in an effort to illustrate the effects of error-propagation across the different detection stages, in Section 17.3.1.4.2 more detailed investigations are conducted with respect

---

[15]Note that here we refer to the symbol transmitted on an OFDM subcarrier basis.

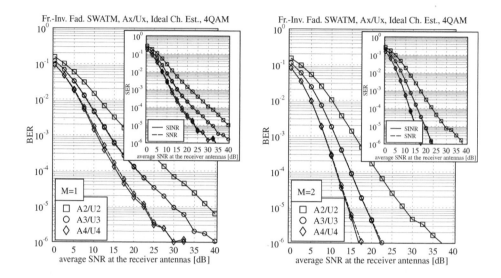

Figure 17.12: BER and SER performance of 4QAM-modulated **(a: Left)** standard SIC and **(b: Right)** M-SIC ($M = 2$) detection-assisted SDMA-OFDM as a function of the SNR recorded at the reception antennas and parametrised for the different system configurations in terms of the number of users L and the number of reception antennas P; two different measures, namely the SINR and the SNR at the combiner's output were employed for performing the selection of the most dominant user at each detection stage; for the basic simulation parameters we refer to Table 17.4.

to the associated system's SER performance, which is evaluated on a detection stage-by-stage basis. These investigations are further extended in Section 17.3.1.4.3 to the scenario, where an error-free remodulated reference signal is employed in the context of updating the vector of signals received by the different antenna elements in each detection stage. In order to further augment our understanding of the effects of error-propagation, in Section 17.3.1.4.4 the detection stages' symbol-error event probabilities are analysed. More specifically, each error-event is captured as the unique combination of the presence of a symbol error ("1") - or the absence of a symbol error ("0") at the different detection stages. Furthermore, in Section 17.3.1.4.5 the SER performance of the partial M-SIC scheme is analysed with respect to the detection stage $L_{\text{pM-SIC}}$ up to which the M most likely symbol decisions are retained at each detection node of the detection process. Finally, in Section 17.3.1.4.6 the technique of SDI-M-SIC is characterised briefly. Explicitly we will quantify the effect of the $N_{\text{SDI}}^{[1]}$ number of subcarriers for which $M = 2$ tentative symbol decisions are made during the first SIC detection stage. Our performance assessments will be concluded in Section 17.3.1.4.7 with the analysis of turbo-decoded, standard SIC detection-assisted SDMA-OFDM systems.

17.3.1.4.1 BER and SER Performance of Standard SIC and M-SIC for Different Numbers of Users and Receiver Antennas

In Figure 17.12 we have portrayed the BER as well as the SER performance as a function of the SNR at the reception antennas. Specifically, on the left-hand side of Figure 17.12 we characterised the standard SIC, while on the right-hand side of Figure 17.12 the M-SIC scheme. The results of Figure 17.12 are also parametrised for the different configurations by the number of users L and by the number of reception antennas P, which were assumed here to be identical. Furthermore, both the SINR as well as the SNR recorded at the MMSE combiner's output are considered potential alternatives for performing the selection of the most dominant user at each detection stage.

For both detectors, namely for the standard SIC and for the M-SIC scheme the general trend is that by increasing the MIMO system's order upon employing, for example, four reception antennas to support four simultaneous users (A4/U4) instead of two reception antennas to support two simultaneous users (A2/U2), the system's performance evaluated in terms of the achievable BER and SER is significantly improved. This is because in the context of encountering correct symbol decisions at the different stages of the detection process, the associated MIMO system's diversity order, namely the ratio between the number of transmit and receive antennas, is increased. As argued in Sections 17.2.6.2 and 17.2.6.3, this has the effect of providing a higher degree of freedom at the MMSE combiner of each detection stage for adjusting the reception antennas' weights, which results in a more efficient suppression of the AWGN. Furthermore, for a MIMO system having a higher order, at a specific detection stage there is also the choice between a larger number of users to be selected as the most dominant user to be detected next, which also implies additional diversity. We also observe in Figure 17.12 that compared to standard SIC the M-SIC scheme retaining $M = 2$ tentative 4QAM symbol decisions out of the $M_c = 4$ legitimate symbols exhibits a significant performance advantage, which is achieved at the cost of an increased computational complexity. Hence, potential complexity reduction strategies, namely partial M-SIC and SDI-M-SIC will be characterised in Section 17.3.1.4.5 and 17.3.1.4.6.

With respect to the different objective measures, namely the SNR and SINR, employed to perform the selection of the most dominant user at each stage of the detection process, a slight advantage is observed in favour of the SINR measure, although only for the system configurations of a higher order. In the context of these configurations the residual interference was apparently higher than for the system configurations supporting a lower number of users. In the next section we will focus our attention on the effects of error propagation between the different detection stages.

17.3.1.4.2 SER Performance of Standard SIC and M-SIC on a Per-Detection Stage Basis

In this section we will further investigate the effects of error propagation across the different detection stages. Hence on the left-hand side of Figure 17.13 we have portrayed the SER associated with the different detection stages for standard SIC, while on the right-hand side of Figure 17.13 for the M-SIC scheme. In the context of the simulation results presented in each of the larger graphs, the SINR was invoked as the metric used for performing the selection of the most dominant user at each stage of the detection process. By contrast, in the smaller-sized sub-graphs of Figure 17.13, we have compared the SINR and SNR criteria against each other, at the first detection stage.

More specifically, for the standard SIC we observe on the left-hand side of Figure 17.13 that upon traversing through the different detection stages the SER is significantly increased

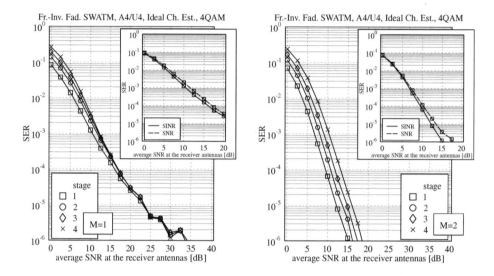

Figure 17.13: SER performance observed in different detection stages in the context of 4QAM-modulated **(a: Left)** standard SIC and **(b: Right)** M-SIC ($M = 2$) detection-assisted SDMA-OFDM as a function of the SNR at the reception antennas; here a system configuration of $L = 4$ simultaneous users and $P = 4$ reception antennas is considered; in the smaller-sized sub-figures two measures, namely the SINR and the SNR evaluated at the combiner's output, are compared against each other in terms of performing the selection of the most dominant user at each detection stage; for the basic simulation parameters we refer to Table 17.4.

at lower SNRs, while at higher SNRs the different detection stages' SERs become virtually identical. As will be supported by our forthcoming analysis of the standard SIC detector's SER recorded in the context of an error-free reference signal, a viable explanation of this phenomenon is that the majority of symbol errors caused by the early detection stages propagates to the later stages. In other words, only a comparably small additional error contribution is caused by the later detection stages, which is a consequence of the increased diversity order associated with the gradually decreasing number of undetected user symbols. For the M-SIC scheme on the right-hand side of Figure 17.13 a similar behaviour is observed, although, for higher SNRs the different detection stages' SER performance curves do not merge. This is, because at each detection stage the SER contribution induced by the error propagation from previous detection stages is of a similar significance as the additional contribution due to the AWGN.

17.3.1.4.3   SER Performance of Standard SIC and M-SIC on a Per-Detection Stage Basis for an Error-Free Reference

In order to further highlight the associated error propagation effects, let us stipulate the availability of an ideal, error-free remodulated reference dur-

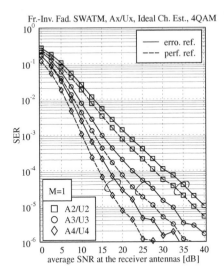
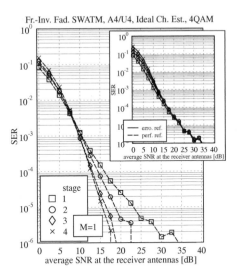

Figure 17.14: (a: Left) SER performance of 4QAM-modulated standard SIC detection-assisted SDMA-OFDM as a function of the SNR at the reception antennas, parametrised with the number of simultaneous users L and the number of reception antennas P, where $L = P$; **(b: Right)** SER performance of standard SIC recorded separately for each detection stage in the context of a scenario of $L = P = 4$; an ideal, error-free reference is employed in the subtractive interference cancellation process associated with each detection stage, while the curves associated with the more realistic error-contaminated reference have been plotted as a benchmark; the SINR was invoked as the metric for the selection of the most dominant user at each detection stage; for the basic simulation parameters we refer to Table 17.4.

ing the subtraction of the most recently detected user's contribution from the P-dimensional vector of signals received by the P different antenna elements. Corresponding to this scenario on the left-hand side of Figure 17.14 we have plotted the SER versus SNR performance of various system configurations parametrised again with the number of simultaneous users L, each equipped with one transmit antenna, which was assumed for each configuration to be identical to the number of reception antennas P, namely we had $L = P$. In contrast to the curves presented in Figure 17.12, which are repeated here as a benchmark, the error propagation between the different detection stages was prevented due to the employment of an error-free remodulated reference. Again, the standard SIC detector with $M = 1$ is considered here. As a result of the idealistic nature of the reference, the system's SER performance is significantly improved compared to the more realistic case of an imperfect, potentially error-contaminated remodulated reference. The corresponding SER performance curves recorded during the different detection stages are shown on the right-hand side of Figure 17.14, where again, in the reduced-sized sub-figure we have plotted the curves associated with an im-

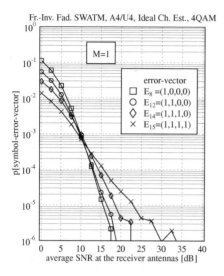
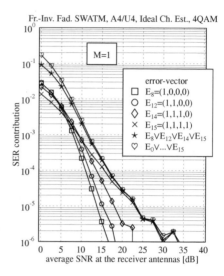

Figure 17.15: (a: Left) Probability- and **(b: Right)** SER contribution of the different significant error-vectors as a function of the SNR measured at the reception antennas; here 4QAM-modulated standard SIC detection-assisted SDMA-OFDM is considered in the context of a scenario of $L = 4$ simultaneous users and $P = 4$ reception antennas (A4/U4); the SINR was invoked as the metric for the selection of the most dominant user at each detection stage; for the basic simulation parameters we refer to Table 17.4.

perfect remodulated reference. These curves were originally shown on the left-hand side of Figure 17.13. For the error-free remodulated reference related curves we observe at the higher-index detection stages that for SNRs up to 7.5dB only a relatively modest SER degradation is encountered, when compared to the previous detection stages. This phenomenon is in contrast to the more distinct degradation observed in conjunction with the original curves generated in the context of a potentially error-contaminated remodulated reference. This was a result of the effects of error propagation. The residual SER degradation observed in Figure 17.14 for the SNR range up to 7.5dB is attributed to the process of ranking the remaining users' SINRs, where the most dominant remaining user is detected and consequently during the last detection stages only the weaker users are still to be detected. By contrast, at higher SNRs this trend is reversed. Specifically, the SER achieved during the last detection stages is far lower than that of the first detection stages, which is a result of the implicit increase of the system's diversity order. Again, as also supported by the BER curves of Figure 17.8, for a lower number of users the MMSE combiner employed at each detection stage is less restricted in terms of the specific choice of the receiver antenna weights, which results in a more efficient suppression of the AWGN.

17.3.1.4.4   Evaluation of the Error-Propagation-Related Event Probabilities   In order
to obtain further insight into the effects of error propagation across the different detection
stages, we have measured the probabilities of the various symbol error events. More specifically, the error-event associated with index j can be defined as the following vector:

$$E_j = \mathbf{e}_j = (e_{L-1}, \dots, e_i, \dots, e_0), \quad e_i \in \{0,1\}, \tag{17.134}$$

where the index j is given by interpreting the vector $\mathbf{e}_j$ as a binary number, yielding:

$$j = \sum_{i=0}^{L-1} \begin{cases} 2^i & \wedge\, e_i = 1 \\ 0 & \wedge\, e_i = 0 \end{cases}. \tag{17.135}$$

Explicitly, in case of an error at the i-th detection stage we have $e_i = 1$. In the context of
our evaluations we have focussed on four specific symbol error events, which are suitable for
further demonstrating the effects of error propagation through the different detection stages.
These error events are $E_8 = (1,0,0,0)$, $E_{12} = (1,1,0,0)$, as well as $E_{14} = (1,1,1,0)$ and
$E_{15} = (1,1,1,1)$, where E_8 indicates encountering an error event during the last detection
stage. The corresponding probabilities of these events as a function of the SNR recorded at
the reception antennas are portrayed on the left-hand side of Figure 17.15 for the standard SIC
scheme. Here we observe that for SNRs below the cross-over point near 10dB the specific
error propagation events, which extend from the first detection stage to the last stage are less
likely than those events which commence in one of the last stages of the detection process.

This phenomenon can be explained as follows. Even for SNRs as low as 0dB, the probability of an AWGN-induced symbol error at a specific detection stage is far lower than the
probability of incurring an error-free detection. Hence, the event having the highest probability is that of no symbol errors at all detection stages, which is denoted as E_0. The next
most likely class of error events – assuming the independence of encountering decision errors
at the different detection stages, as we also assumed in the case of an error-free reference –
is constituted by those events which host a symbol error only in one of the detection stages.
As highlighted earlier in the context of our description of the graph on the right-hand side
of Figure 17.14, for relatively low SNRs it is more likely to incur symbol errors during the
higher-index detection stages than in the lower-index stages, which was attributed to the
process of ranking the users at the different detection stages. This effect is even further augmented here as a result of the associated error-propagation phenomenon. More explicitly,
the probability of incurring an error event during the first detection stage only, which is indicated by $E_1 = (0,0,0,1)$ is likely to be lower than the probability of encountering the error
event of $E_{15} = (1,1,1,1)$ due to the error-propagation effects, although in case of independent errors the probability of the latter would be expected to be significantly lower. Hence,
following these argumentations, among the four different error events considered here, the
event of $E_8 = (1,0,0,0)$, which is associated with incurring a symbol error only in the
last detection stage, appears with the highest probability at low to medium SNRs of up to
10dB. By contrast, for higher SNRs the reverse behaviour is observed, namely that the event
of incurring symbol errors in all detection stages occurs with the highest probability. This
is because upon increasing the SNR, the probability of incurring a symbol error in the last
detection stage decreases far more rapidly than the probability of incurring a symbol error
in the first detection stage due to its interference-contaminated nature, despite detecting the

highest-power user first. As argued earlier, this is also a consequence of the higher diversity order available for the MMSE combiner during the last detection stages, which is due to removing the interference imposed by other users during the previous detection stages.

On the right-hand side of Figure 17.15 we have related the different symbol error events to their SER contribution. For this purpose we have weighted the event probabilities presented on the left-hand side of Figure 17.15 by their relative contribution to the average SER. More specifically, for errors events which host a symbol error in a single detection stage only, the weighting factor is $1/L$, while correspondingly for two symbol errors we have $2/L$ and so on. Additionally, we have plotted here the joint contribution of the four most significant error events, as well as the total average SER, which is the joint contribution of all error events. While for lower SNRs some difference is observed between the SER predicted with the aid of the four most significant error events, namely $E_8 \vee E_{12} \vee E_{14} \vee E_{15}$ and the actual SER curves associated with $E_1 \vee \ldots \vee E_{15}$, at higher SNRs the former events of $E_8 \vee E_{12} \vee E_{14} \vee E_{15}$ closely predict the actual SER. A plausible interpretation of this phenomenon is that at higher SNRs the SER is constituted by error propagation events, which is even further augmented, since the MMSE combiner does not take into account the extra non-Gaussian "noise" caused by symbol errors encountered in previous detection stages. This problem results in a noise amplification, a phenomenon, which is also known from zero-forcing combiners.

17.3.1.4.5 SER Performance of the Partial M-SIC As observed in the previous sections, the M-SIC scheme described in Section 17.3.1.2.1 is capable of significantly outperforming the standard SIC arrangement, which was found to suffer from the consequences of error-propagation between the different detection stages. Furthermore, compared to the optimum ML detection scheme the performance degradation was observed to be less than 0.5dB in terms of the SNR at the receiver antennas required at a specific SER. This impressive performance was achieved at the cost of a significantly increased computational complexity compared to standard SIC. In order to reduce this potentially excessive complexity, it was proposed in Section 17.3.1.2.2 to retain $M > 1$ symbol decisions at each detection node, characterised by its associated updated vector of received signals up to the specific $L_{\text{pM-SIC}}$-th stage in the detection process, where the updating process implied cancelling the effects of the remodulated tentatively demodulated symbols from the composite multi-user signal. By contrast, at higher-index detection stages only one symbol decision would be tracked from each detection node, as in the standard SIC scheme. These principles were illustrated in Figure 17.11.

The benefits of employing this strategy have again been verified here with the aid of simulations, as shown on the left-hand side of Figure 17.16. Again, we have plotted the performance of standard SIC ($M = 1$) as a benchmark. Also note that the case of tracking multiple decisions per detection node in the first three detection stages is equivalent to the original M-SIC and hence it does not yield a complexity reduction. However, employing multiple tentative symbol decisions only at the first two detection stages will be shown to yield a significant complexity reduction in Section 17.3.1.5.2, which is achieved at the modest cost of increasing the SNR required for maintaining an SER of 10^{-5} by only about 1dB.

17.3.1.4.6 SER Performance of Selective-Decision-Insertion Aided M-SIC The philosophy of the SDI technique [117, 137] of Section 17.3.1.2 is that of tracking additional ten-

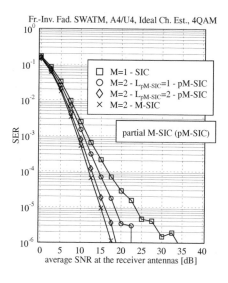

Fr.-Inv. Fad. SWATM, A4/U4, Ideal Ch. Est., 4QAM

□ M=1 - SIC
○ M=2 - $L_{pM\text{-}SIC}$=1 - pM-SIC
◇ M=2 - $L_{pM\text{-}SIC}$=2 - pM-SIC
× M=2 - M-SIC

partial M-SIC (pM-SIC)

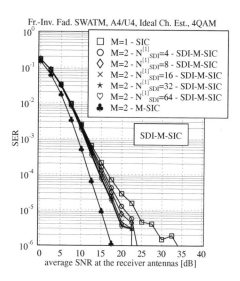

Fr.-Inv. Fad. SWATM, A4/U4, Ideal Ch. Est., 4QAM

□ M=1 - SIC
○ M=2 - $N^{[1]}_{SDI}$=4 - SDI-M-SIC
◇ M=2 - $N^{[1]}_{SDI}$=8 - SDI-M-SIC
× M=2 - $N^{[1]}_{SDI}$=16 - SDI-M-SIC
★ M=2 - $N^{[1]}_{SDI}$=32 - SDI-M-SIC
♡ M=2 - $N^{[1]}_{SDI}$=64 - SDI-M-SIC
♣ M=2 - M-SIC

SDI-M-SIC

Figure 17.16: SER performance as a function of the SNR recorded at the reception antennas of 4QAM-modulated **(a: Left)** partial M-SIC ($M = 2$) detection-assisted SDMA-OFDM, parametrised with the index of the detection stage up to which multiple tentative symbol decisions are retained at each detection node of the detection process and **(b: Right)** SDI-M-SIC ($M = 2$) detection-assisted SDMA-OFDM, parametrised with the $N^{[1]}_{SDI}$ number of subcarriers associated with multiple tentative symbol decisions during the first detection stage; here a system configuration of $L = 4$ simultaneous users and $P = 4$ reception antennas (A4/U4) is considered; the SINR was employed as the metric for the selection of the most dominant user at each detection stage; for the basic simulation parameters we refer to Table 17.4.

tative symbol decisions only in those $N^{[1]}_{SDI}$ number of subcarriers, which exhibit the lowest SINR during the first detection stage, since these are most likely to cause symbol errors. As argued in the context of Figure 17.15, of this approach is motivated by the observation that at higher SNRs symbol errors are mainly caused during the first detection stage because of the higher number of interfering users than at later stages. Again, we have evaluated the performance of this strategy with the aid of computer simulations. The associated SER performance results as a function of the SNR measured at the reception antennas and further parametrised with the number of low-quality subcarriers N_{SDI} where $M = 2$ tentative symbol decisions are retained during the first detection stage are shown on the right-hand side of Figure 17.16. By contrast, in all other subcarriers only one symbol decision is retained at each detection node for the sake of maintaining a low complexity. As a reference, on the right-hand side of Figure 17.16 we have again plotted the SER performance curves associated with the partial M-SIC, as well as that of the original M-SIC, as shown on the left-hand side of Figure 17.16. We observe that when the $N^{[1]}_{SDI} = 64$ number of lowest-quality subcarriers classified in terms

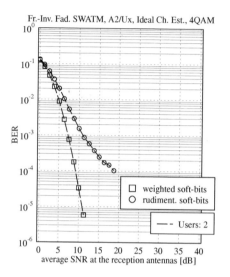
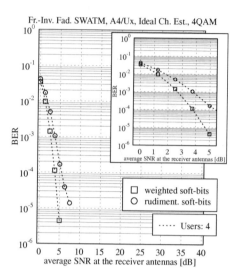

Figure 17.17: BER performance of turbo-coded, 4QAM-modulated, standard SIC detection-assisted SDMA-OFDM as a function of the SNR recorded at the reception antennas for the rudimentary soft-bit generation approach as suggested by Equation 17.130 and for the weighted soft-bit generation approach of Equation 17.133; the curves are further parametrised with the number of simultaneous users and reception antennas using $L = P$, where more specifically **(a: Left)** two reception antennas, and **(b: Right)** four reception antennas were employed; in the smaller-sized sub-figure we have magnified the range of SNRs between 0dB and 5dB; for the basic simulation parameters and the turbo-coding parameters we refer to Tables 17.4 and 17.5, respectively.

of the SINR experienced during the first detection stage are associated with $M = 2$ tentative symbol decisions during this specific detection stage, then a similar SER performance is observed, to the significantly more complex scenario, where all subcarriers are associated with $M = 2$ tentative symbol decisions during the first detection stage. This is because by retaining multiple tentative symbol decisions exclusively during the first detection stage, only the probability of those error propagation events is reduced, which are caused during the first detection stage and these are likely to be those associated with the $N_{\mathrm{SDI}}^{[1]}$ number of lowest-SINR subcarriers.

However, compared to the SER performance of the M-SIC, the performance degradation incurred by the first-stage SDI-M-SIC characterised on the right-hand side of Figure 17.16 is substantial. In order to achieve further performance improvements, additional tentative symbol decisions could also be retained during the higher-index detection stages although naturally at the cost of a higher complexity.

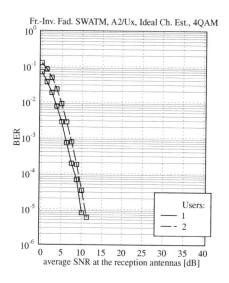
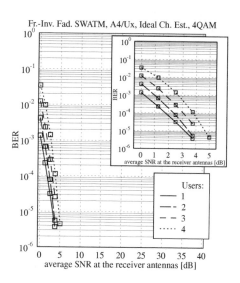

Figure 17.18: BER performance of turbo-coded, 4QAM-modulated, standard SIC detection-assisted SDMA-OFDM as a function of the SNR recorded at the reception antennas; the curves are further parametrised with the number of simultaneous users L and the number of reception antennas P, where more specifically **(a: Left)** two reception antennas, and **(b: Right)** four reception antennas were employed; in the smaller-sized sub-figure we have magnified the range of SNRs between 0dB and 5dB; for the basic simulation parameters and the turbo-coding parameters we refer to Tables 17.4 and 17.5, respectively.

17.3.1.4.7 BER Performance of Turbo-Coded SIC Detection-Assisted SDMA-OFDM

As demonstrated in Section 17.2.6.4 in the context of MMSE detection-assisted SDMA-OFDM employing turbo-decoding at the receiver is a powerful means of further enhancing the system's BER performance. Naturally, this is achieved at the cost of reducing the system's effective throughput and by investing an additional amount of computational complexity. The relevant turbo coding parameters were summarised in Table 17.5, but for the reader's convenience they are repeated here: the coding rate was $R_c = \frac{1}{2}$, the constraint length was $K_c = 3$, the octally represented generator polynomials of $(7, 5)_8$ were used and four turbo decoding iterations were performed.

Our discussions will first of all concentrate on a comparison between the various methods of soft-bit generation, namely on the rudimentary approach of Equation 17.130 and on the weighted approach of Equation 17.133. The associated BER versus SNR simulation results valid for a scenario of $L = P = 2$ and $L = P = 4$ number of simultaneous users and reception antennas are portrayed on the left- and right-hand sides of Figure 17.17, respectively. In both scenarios a significant performance advantage is observed for the weighted soft-bit generation approach, which attempts to take into account the effects of error propagation through the different SIC detection stages upon de-weighting potentially unreliable soft-bit values. For the lower-order SDMA scenario of $L = P = 2$ the performance improvement achieved by the weighted soft-bit generation is even more dramatic than for the scenario of $L = P = 4$. This is, because for the lower-order scenario the probability that a symbol error has incurred during the first detection stage is higher than for the higher-order scenario, since the latter system benefits more substantially from the increased grade of diversity experienced in conjunction with the higher number of users, supported by a higher number of antennas. Hence, weighting of the soft-bits is more effective in scenarios supporting a lower number of users with the aid of less antennas.

Having found that the weighted soft-bit metric of Equation 17.133 is the advantageous one in terms of the system's associated BER performance, further investigations were conducted with respect to different scenarios in terms of the number of simultaneous users L and the number of reception antennas P. The corresponding simulation results are shown in Figure 17.18 again, on the left-hand side for $P = 2$ reception antennas while on the right-hand side for $P = 4$ reception antennas. Similar to our observations discussed in the context of turbo-coded MMSE detection-assisted SDMA-OFDM in Section 17.2.6.4, the system's BER performance is improved upon decreasing the number of simultaneous users L. Again, this is because the associated MMSE combiner of each detection stage is less constrained with respect to the specific choice of the antenna array weights and hence more of the channel's diversity potential can be dedicated to mitigating the effects of AWGN, rather than to suppressing the undesired co-channel users.

17.3.1.5 Complexity Analysis

In this section we will analyse the computational complexity exhibited by the various forms of the SIC algorithm on a per-subcarrier basis. In Section 17.3.1.5.1 the computational complexity of the standard SIC algorithm will be investigated, while in Sections 17.3.1.5.2 and 17.3.1.5.3 the complexity of the M-SIC and partial M-SIC schemes will be analysed, respectively.

17.3.1.5.1 Complexity of Standard SIC   Here we assume that at each stage of the detection process the MMSE combiner is employed in its specific left-inverse related representation as described in Section 17.2.3.2.2. As highlighted in Section 17.3.1.1, a prerequisite for the calculation of the weight matrix $\mathbf{W}_{\mathrm{MMSE}}^{[i]H}$ according to Equation 17.118 in the i-th detection stage, is the availability of the left-inverse related correlation matrix $\mathbf{R}_a^{[i]H}$ defined in Equation 17.119, and the knowledge of the cross-correlation matrix $\mathbf{R}_c^{[i]H}$ defined in Equation 17.120, which were employed in the context of our description of the SIC detector in Section 17.3.1.1 in their Hermitian transposed form. It was highlighted furthermore in the same section that instead of entirely recalculating these matrices at the beginning of each detection stage on the basis of the current reduced-size channel transfer factor matrix $\mathbf{H}^{[i]}$, it is computationally more advantageous to calculate them only once at the beginning of the detection process and then eliminate the most recently detected user's influence at the end of each detection stage. This is achieved by removing the $\tilde{l}^{[i]}$-th row- and column from the matrix $\mathbf{R}_a^{[i]H}$ and by eliminating the $\tilde{l}^{[i]}$-th column from the matrix $\mathbf{R}_c^{[i]H}$, respectively, followed by appropriately rearranging these matrices with the result of a reduced dimensionality. Note that $\tilde{l}^{[i]}$ is the column vector index of the reduced-size channel transfer factor matrix $\mathbf{H}^{[i]}$, which is associated with the most recently detected user, namely the $l^{[i]}$-th user. Let us now consider the individual steps one by one.

- *Initialisation*: Specifically, the computational overhead incurred in terms of the number of complex, as well as mixed real-complex multiplications and additions at the beginning of the detection process required for the computation of the auto-correlation matrices $\mathbf{R}_{\overline{a}}^{[1]H}$ and $\mathbf{R}_{c}^{[1]H}$ is given by:

$$C_{\text{SIC,stp}}^{(\mathbb{C}*\mathbb{C})} = C_{\text{SIC,stp}}^{(\mathbb{C}+\mathbb{C})} = PL^2 \tag{17.136}$$

$$C_{\text{SIC,stp}}^{(\mathbb{R}*\mathbb{C})} = PL \tag{17.137}$$

$$C_{\text{SIC,stp}}^{(\mathbb{R}+\mathbb{C})} = L. \tag{17.138}$$

- *Weight Calculation*: Furthermore, during the i-th detection stage Equation 17.118 has to be solved for the weight matrix $\mathbf{W}_{\text{MMSE}}^{[i]H}$ with the aid of the LU-decomposition [90]. More specifically, the LU decomposition implies factorising the reduced-size correlation matrix $\mathbf{R}_{\overline{a}}^{[i]H} \in \mathbb{C}^{L^{[i]} \times L^{[i]}}$ into a lower (L) and upper (U) triangular matrix, followed by forward- and backward substitutions as outlined in [90]. The associated computational complexity becomes:

$$C_{\text{SIC,w}}^{(\mathbb{C}*\mathbb{C})[i]} = C_{\text{SIC,w}}^{(\mathbb{C}+\mathbb{C})[i]} = PL^{[i]2} + \frac{1}{3}L^{[i]3}. \tag{17.139}$$

Upon taking into account the reduction of the number of users to be detected and the corresponding reduction of the channel matrix' dimension across the different detection stages, the total complexity imposed by calculating the weight matrices defined in Equation 17.121 is given by:

$$C_{\text{SIC,w}}^{(\mathbb{C}*\mathbb{C})} = C_{\text{SIC,w}}^{(\mathbb{C}+\mathbb{C})} = P \sum_{L^{[i]}=1}^{L} (L^{[i]})^2 + \frac{1}{3} \sum_{L^{[i]}=1}^{L} (L^{[i]})^3 \tag{17.140}$$

$$= \frac{1}{3}P(2L+1)\alpha_{\text{SIC}} + \frac{1}{3}\alpha_{\text{SIC}}^2, \tag{17.141}$$

where the factor α_{SIC} is defined as:

$$\alpha_{\text{SIC}} = \sum_{i=1}^{L} i = \frac{1}{2}L(L+1). \tag{17.142}$$

Here we have exploited that [473] $\sum_{i=1}^{n} i = n(n+1)/2$ and $\sum_{i=1}^{n} i^2 = n(n+1)(2n+1)/6$ as well as $\sum_{i=1}^{n} i^3 = n^2(n+1)^2/4$.

- *Selection*: As the metric employed for the selection of the most dominant user we opted for the SNR at the combiner's output, as given by Equation 17.26. The rationale of this choice was that we demonstrated in Section 17.3.1.4.1 that the employment of the SNR only imposes a marginal performance loss compared to employing the SINR as the dominant user ranking metric. Here we have found that the number of complex multiplications and additions, as well as the number of real divisions and multiplications and the number of comparisons required are given by:

$$C_{\text{SIC,obj}}^{(\mathbb{C}*\mathbb{C})} = (\alpha_{\text{SIC}} - 1)(2P + 1) \tag{17.143}$$

$$C_{\text{SIC,obj}}^{(\mathbb{C}+\mathbb{C})} = (\alpha_{\text{SIC}} - 1)(2P) \tag{17.144}$$

$$C_{\text{SIC,obj}}^{(\mathbb{R}/\mathbb{R})} = C_{\text{SIC,obj}}^{(\mathbb{R}*\mathbb{R})} = C_{\text{SIC,obj}}^{(\mathbb{R}\lessgtr\mathbb{R})} = (\alpha_{\text{SIC}} - 1). \tag{17.145}$$

In this context we implicitly assumed that the SNR calculation associated with the selection of the most dominant user is not required in the last detection stage, since only a single user remains to be detected.

- *Combining*: Furthermore, the linear combining at each stage of the successive cancellation process – carried out with the aid of Equation 17.124 – necessitates the following number of complex multiplications and additions as given by:

$$C_{\text{SIC,cmb}}^{(\mathbb{C}*\mathbb{C})} = C_{\text{SIC,cmb}}^{(\mathbb{C}+\mathbb{C})} = LP. \tag{17.146}$$

- *Demodulation*: In addition, the process of MPSK-related demodulation,[16] which is realised with the aid of Equation 17.125, exhibits a total complexity of:

$$C_{\text{SIC,dem}}^{(\mathbb{C}*\mathbb{C})} = \frac{1}{2}LM_c \tag{17.147}$$

$$C_{\text{SIC,dem}}^{(\mathbb{C}+\mathbb{C})} = C_{\text{SIC,dem}}^{(\mathbb{R}\lessgtr\mathbb{R})} = LM_c, \tag{17.148}$$

where again, we have included the number of comparisons between the outcomes of the Euclidean distance metric as an additional factor increasing the complexity.

- *Updating*: Finally, the process of updating the vector of signals received by the different antenna elements according to Equation 17.127 by cancelling the effects of the remodulated signal of a specific user from the composite multi-user signal imposes a total complexity of:

$$C_{\text{SIC,upd}}^{(\mathbb{C}*\mathbb{C})} = C_{\text{SIC,upd}}^{(\mathbb{C}+\mathbb{C})} = (L - 1)P, \tag{17.149}$$

where again, no updating or remodulated signal cancellation is required during the last detection stage. Here we have also taken into account the average number of explicit data transfers, which was quantified as:

$$C_{\text{SIC,upd}}^{\Rightarrow} = L \left[(L + 1) \left[\frac{1}{9}(2L + 1) - \frac{3}{4} \right] + \frac{5}{6} \right] + \frac{1}{4}PL(L - 1). \tag{17.150}$$

More explicitly, the first term corresponds to rearranging the auto-correlation matrix $\mathbf{R}_{\bar{a}}^{[i]H}$, while the second term is associated with rearranging the cross-correlation matrix $\mathbf{R}_c^{[i]H}$ at the end of each detection stage, $i = 1, \ldots, L$.

---

[16]In the context of the MPSK modulation scheme the normalisation of the MMSE combiner's output signal is avoided.

Upon combining the different contributions given in the previous equations, we obtain for the total complexity of the standard SIC detector as a function of the number of users L, the number of reception antennas P and the number of constellation points M_c the following expressions:

$$C_{\text{SIC}}^{(\mathbb{C}*\mathbb{C})} = \frac{1}{2}LM_c + [P(L-3)-1] + \frac{1}{3}[3 + P(2L+13)]\alpha_{\text{SIC}} + \frac{1}{3}\alpha_{\text{SIC}}^2 \quad (17.151)$$

$$C_{\text{SIC}}^{(\mathbb{C}+\mathbb{C})} = LM_c + [P(L-3)-1] + \frac{1}{3}[3 + P(2L+13)]\alpha_{\text{SIC}} + \frac{1}{3}\alpha_{\text{SIC}}^2 \quad (17.152)$$

$$C_{\text{SIC}}^{(\mathbb{R}\lessgtr\mathbb{R})} = (\alpha_{\text{SIC}} - 1) + LM_c \quad (17.153)$$

$$C_{\text{SIC}}^{(\mathbb{R}*\mathbb{C})} = PL \quad (17.154)$$

$$C_{\text{SIC}}^{(\mathbb{R}+\mathbb{C})} = L \quad (17.155)$$

$$C_{\text{SIC}}^{(\mathbb{R}/\mathbb{R})} = C_{\text{SIC}}^{(\mathbb{R}*\mathbb{R})} = (\alpha_{\text{SIC}} - 1) \quad (17.156)$$

where again, $\alpha_{\text{SIC}} = \frac{1}{2}L(L+1)$ was defined in Equation 17.142. In order to further simplify the complexity analysis, it is reasonable to assume that mixed real-complex multiplications[17] and additions are only half as complex as those performed on two complex variables, while the multiplication of two real-valued variables exhibits only one quarter of the complexity. Hence, a slightly simplified characterisation of the standard SIC detector's complexity is given by:

$$C_{\text{SIC}}^{(\mathbb{C}*\mathbb{C})} = \frac{1}{2}LM_c + \left[P\left(\frac{3}{2}L-3\right) - \frac{5}{4}\right] + \frac{1}{3}\left[\frac{15}{4} + P(2L+13)\right]\alpha_{\text{SIC}} + \frac{1}{3}\alpha_{\text{SIC}}^2 \quad (17.157)$$

$$C_{\text{SIC}}^{(\mathbb{C}+\mathbb{C})} = L\left(M_c + \frac{1}{2}\right) + P(L-3) + \frac{1}{3}P(2L+13)\alpha_{\text{SIC}} + \frac{1}{3}\alpha_{\text{SIC}}^2 \quad (17.158)$$

$$C_{\text{SIC}}^{(\mathbb{R}\lessgtr\mathbb{R})} = (\alpha_{\text{SIC}} - 1) + LM_c \quad (17.159)$$

$$C_{\text{SIC}}^{(\mathbb{R}/\mathbb{R})} = (\alpha_{\text{SIC}} - 1). \quad (17.160)$$

17.3.1.5.2 Complexity of M-SIC As argued in Section 17.3.1.4.2, the standard SIC algorithm suffers from error propagation, especially in those subcarriers, where the S(I)NR observed at the selected user's associated combiner output during the first detection stage is relatively low. A viable strategy to improve the achievable performance and to further reduce the BER/SER is that of allowing not only the single most likely constellation point to be selected from the constellation during a specific detection stage's demodulation process, but rather to retain the $M > 1$ number of most likely constellation points. As a result, in the i-th stage $M^{(i-1)}$ number of detection and demodulation operations has to be performed. This also requires M^i number of updating or interference cancellation operations with respect to the local vector $\mathbf{x}^{[i]}$ of received signals. The complexity contribution required for evaluating the weight matrices and SNR estimates, namely "Initialisation", "Weight Calculation" and

---

[17]Here we neglect that the multiplication of two complex numbers involves real-valued additions as well. Hence our simplified formulae are to be understood as best-case estimates of the complexity.

"Selection" of Section 17.3.1.5.1 remains unchanged.

- *Combining*: More specifically, the computational complexity associated with the operation of combining hence becomes:

$$C_{\text{M-SIC,cmb}}^{(\mathbb{C}*\mathbb{C})} = C_{\text{M-SIC,cmb}}^{(\mathbb{C}+\mathbb{C})} = \beta_{\text{M-SIC}}P, \tag{17.161}$$

where the factor $\beta_{\text{M-SIC}}$, which corresponds to the total number of different detection nodes in the detection process is given by:

$$\beta_{\text{M-SIC}} = \sum_{i=0}^{L-1} M^i = \begin{cases} L & \wedge\, M = 1 \\ \frac{M^L - 1}{M-1} & \wedge\, M > 1. \end{cases} \tag{17.162}$$

This is the formula characterising a geometrical series [473].

- *Demodulation*: For the MPSK-related demodulation operation we have:

$$C_{\text{M-SIC,dem}}^{(\mathbb{C}*\mathbb{C})} = \frac{1}{2}\beta_{\text{M-SIC}}M_c \tag{17.163}$$

$$C_{\text{M-SIC,dem}}^{(\mathbb{C}+\mathbb{C})} = \beta_{\text{M-SIC}}M_c, \tag{17.164}$$

where the factor $\beta_{\text{M-SIC}}$ was given by Equation 17.162. Additionally, it is also useful to take into account the number of comparisons to be carried out in the context of the demodulation process invoked at the different detection stages. Here we have to differentiate between the first $L-1$ detection stages and the last detection stage. Specifically, during the first $L-1$ stages of Figure 17.11 there are:

$$\gamma_{\text{M-SIC,i}} = \sum_{i=0}^{L-2} M^i = \begin{cases} L-1 & \wedge\, M = 1 \\ \frac{M^{(L-1)}-1}{M-1} & \wedge\, M > 1, \end{cases} \tag{17.165}$$

number of different detection nodes, each of which is associated with selecting the M number of most likely tentative symbol decisions out of the M_c number of possible symbol decisions associated with the specific modulation scheme employed. The number of comparisons per detection node required by this operation is given by:

$$\delta_{\text{M-SIC,i}} = \sum_{i=M_c-(M-1)}^{M_c} i = \frac{1}{2}M(1 + 2M_c - M), \tag{17.166}$$

which could potentially be further reduced with the aid of more effective binary tree-based search methods.[18] By contrast, during the last detection stage we incur:

$$\gamma_{\text{M-SIC,ii}} = \begin{cases} 1 & \wedge\, M = 1 \\ M^{(L-1)} & \wedge\, M > 1 \end{cases} \tag{17.167}$$

number of different detection nodes, each of which is associated with the selection of

---

[18]From the literature it is well known that recursive search algorithms, such as "quick-sort" require a potentially lower number of comparisons than those based on the principles of "sorting by selection".

the single most likely symbol decision. This imposes a total number of comparisons per detection node, which is given by:

$$\delta_{\text{M-SIC,ii}} = M_c. \tag{17.168}$$

The final operation in the M-SIC detection process is to select the specific symbol vector from the various possible symbol vectors defined by the different tentative symbol decisions retained at each detection node, whose associated Euclidean distance metric is the lowest during the last detection stage. This requires an additional $M^{(L-1)} - 1$ number of comparisons. Hence, the total number of demodulation-related comparisons during the entire detection process is given by:

$$C_{\text{M-SIC,dem}}^{(\mathbb{R} \lessgtr \mathbb{R})} = \epsilon_{\text{M-SIC}}, \tag{17.169}$$

where:

$$\epsilon_{\text{M-SIC}} = \gamma_{\text{M-SIC,i}} \delta_{\text{M-SIC,i}} + \gamma_{\text{M-SIC,ii}} \delta_{\text{M-SIC,ii}} + M^{(L-1)} - 1. \tag{17.170}$$

- *Updating*: Furthermore, for the total number of updating operations with respect to the vector of signals received by the different antenna elements which implies cancelling the effects of the remodulated signal from the composite multi-user signal as seen in Equation 17.127 we obtain:

$$C_{\text{M-SIC,upd}}^{(\mathbb{C}*\mathbb{C})} = C_{\text{M-SIC,upd}}^{(\mathbb{C}+\mathbb{C})} = \zeta_{\text{M-SIC}} P, \tag{17.171}$$

where $\zeta_{\text{M-SIC}}$ is defined here as:

$$\zeta_{\text{M-SIC}} = \sum_{i=1}^{L-1} M^i = \begin{cases} L-1 & \wedge M = 1 \\ M \frac{M^{(L-1)} - 1}{M-1} & \wedge M > 1. \end{cases} \tag{17.172}$$

Upon additionally recalling the complexity-related contributions corresponding to the initialisation, weight calculation and selection from Section 17.3.1.5.1, the M-SIC detector's complexity is characterised by the following equations:

$$C_{\text{M-SIC}}^{(\mathbb{C}*\mathbb{C})} = \frac{1}{2}\beta_{\text{M-SIC}} M_c + \left[P\left(\beta_{\text{M-SIC}} + \zeta_{\text{M-SIC}} - \frac{1}{2}L - 2\right) - \frac{5}{4} \right] +$$

$$+ \frac{1}{3}\left[\frac{15}{4} + P\left(2L + 13\right) \right] \alpha_{\text{SIC}} + \frac{1}{3}\alpha_{\text{SIC}}^2 \tag{17.173}$$

$$C_{\text{M-SIC}}^{(\mathbb{C}+\mathbb{C})} = \beta_{\text{M-SIC}} M_c + L\left(\frac{1}{2} - P\right) + P(\beta_{\text{M-SIC}} + \zeta_{\text{M-SIC}} - 2) +$$

$$+ \frac{1}{3}P(2L + 13)\alpha_{\text{SIC}} + \frac{1}{3}\alpha_{\text{SIC}}^2 \tag{17.174}$$

$$C_{\text{M-SIC}}^{(\mathbb{R}/\mathbb{R})} = (\alpha_{\text{SIC}} - 1) \tag{17.175}$$

$$C_{\text{M-SIC}}^{(\mathbb{R} \lessgtr \mathbb{R})} = (\alpha_{\text{SIC}} - 1) + \epsilon_{\text{M-SIC}}, \tag{17.176}$$

where again, α_{SIC}, $\beta_{\text{M-SIC}}$, as well as $\gamma_{\text{M-SIC,i}}$, $\delta_{\text{M-SIC,i}}$ and $\gamma_{\text{M-SIC,ii}}$, $\delta_{\text{M-SIC,ii}}$ and additionally $\epsilon_{\text{M-SIC}}$, $\zeta_{\text{M-SIC}}$ were given by Equations 17.142, 17.162, as well as 17.165, 17.166 and 17.167, 17.168 and also Equations 17.170, 17.172, respectively. In the context of these equations we have once again taken into account that mixed real-complex multiplications and additions are only half as complex as the corresponding operations associated with two complex numbers.[19]

17.3.1.5.3 Complexity of Partial M-SIC

In this section we will briefly assess the computational complexity of partial M-SIC. We recall from Section 17.3.1.2.2 that in contrast to the M-SIC described in Section 17.3.1.2.1, the M most likely symbol decisions per detection node are only retained up to the specific $L' = L_{\text{pM-SIC}}$-th detection stage. By contrast, for the higher-index detection stages only the single most likely symbol decision per detection node is retained, as in the case of the standard SIC scheme, which was described in Section 17.3.1.1. This implies that for $L' = 1$ the partial M-SIC detector is identical to the standard SIC detector, while for $L' = L$ the M-SIC detector is obtained. The analysis of the partial M-SIC's complexity follows a similar procedure to that applied in Section 17.3.1.5.2 in the case of the M-SIC scheme. The difference resides in the specific composition of the factors $\beta_{\text{M-SIC}}$ and $\gamma_{\text{pM-SIC,i}}$, $\gamma_{\text{pM-SIC,ii}}$, as well as $\epsilon_{\text{pM-SIC}}$ and $\zeta_{\text{pM-SIC}}$, which were defined for M-SIC in Equations 17.162 and 17.165, 17.167 as well as 17.170, 17.172. By contrast, here we have:

$$\beta_{\text{pM-SIC}} = \sum_{i=0}^{L'-1} M^i + \sum_{i=L'}^{L-1} M^{L'-1} = \begin{cases} L & \wedge M = 1 \\ \frac{M^{L'}-1}{M-1} + (L-L')M^{L'-1} & \wedge M > 1, \end{cases} \quad (17.177)$$

as well as:

$$\gamma_{\text{pM-SIC,i}} = \sum_{i=0}^{L'-2} M^i = \begin{cases} L'-1 & \wedge M = 1 \\ \frac{M^{L'-1}-1}{M-1} & \wedge M > 1, \end{cases} \quad (17.178)$$

and:

$$\gamma_{\text{pM-SIC,ii}} = \sum_{i=L'-1}^{L-1} M^{(L'-1)} = (L-L'+1)M^{(L'-1)}. \quad (17.179)$$

Hence, similarly to Equation 17.169 the total number of demodulation-related comparisons during the entire detection process is given by:

$$C_{\text{pM-SIC,dem}}^{(\mathbb{R}\lessgtr\mathbb{R})} = \epsilon_{\text{pM-SIC}}, \quad (17.180)$$

where we have:

$$\epsilon_{\text{pM-SIC}} = \gamma_{\text{pM-SIC,i}}\delta_{\text{pM-SIC,i}} + \gamma_{\text{pM-SIC,ii}}\delta_{\text{pM-SIC,ii}} + M^{(L'-1)} - 1, \quad (17.181)$$

---

[19]Again, this neglects that the multiplication of two complex numbers also involves real-valued additions.

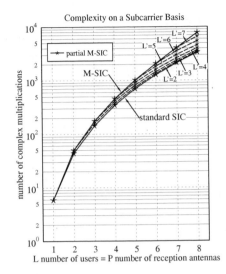
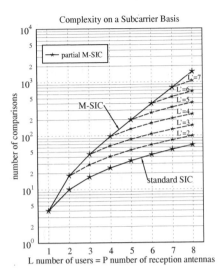

Figure 17.19: Complexity of standard SIC, M-SIC and partial M-SIC in terms of **(a: Left)** the number of complex multiplications $C^{(\mathbb{C}*\mathbb{C})}$ and **(b: Right)** the number of real-valued comparisons $C^{(\mathbb{R}\lessgtr\mathbb{R})}$ recorded on a subcarrier basis, as a function of the number of simultaneous users L, which was assumed to be equal to the number of reception antennas P; the curves related to partial M-SIC are further parametrised with the index of the detection stage namely, $L' = L_{\text{pM-SIC}}$, up to which M number of tentative symbol decisions per detection node were retained; specifically for M-SIC and partial M-SIC the number of retained tentative symbol decisions per detection node was equal to $M = 2$, while in all scenarios $M_c = 4$ constellation points were assumed, which is for example the case in the context of 4QAM modulation.

and $\delta_{\text{pM-SIC,i}} = \delta_{\text{M-SIC,i}}$ as well as $\delta_{\text{pM-SIC,ii}} = \delta_{\text{M-SIC,ii}}$. Furthermore, the interference-cancellation based updating-related factor $\zeta_{\text{pM-SIC}}$ is given by:

$$\zeta_{\text{pM-SIC}} = \sum_{i=1}^{L'-1} M^i + \sum_{i=L'}^{L-1} M^{L'-1} = \begin{cases} L-1 & \wedge M = 1 \\ M\frac{M^{L'-1}-1}{M-1} + (L-L')M^{(L'-1)} & \wedge M > 1. \end{cases}$$

$$(17.182)$$

In order to obtain the expressions for the total complexity associated with partial M-SIC, the variables $\beta_{\text{M-SIC}}$, $\gamma_{\text{M-SIC,i}}$, $\gamma_{\text{M-SIC,ii}}$, $\epsilon_{\text{M-SIC}}$ and $\zeta_{\text{pM-SIC}}$ employed in the context of Equations 17.173, 17.174, 17.175, 17.176 have to be substituted against the corresponding partial M-SIC-specific expressions defined in this section.

17.3.1.5.4 Complexity Comparison of the Different SIC Detectors In this section the different SIC-related detection techniques, namely standard SIC, M-SIC and partial M-SIC will be compared against each other on the basis of the complexity formulae developed in the previous sections. In the context of our evaluations, the M_c number of constellation points associated with the modulation scheme employed was assumed to be $M_c = 4$, which is the case, for example, for 4QAM modulation. The M-SIC retained $M = 2$ number of symbol decisions at each detection node. In the context of the 4QAM modulation scheme we found that further increasing the M number of tentative symbol decisions per detection node retained does not yield a significant additional performance improvement.

Specifically, on the left-hand side of Figure 17.19 we have compared standard SIC against the M-SIC scheme as well as to the partial M-SIC arrangement in terms of the number of complex multiplications imposed on a subcarrier basis, as a function of the number of simultaneous users L, which was assumed here to be equal to the number of reception antennas P. Furthermore, the curves associated with the partial M-SIC scheme are additionally parametrised with the index of the detection stage, namely $L' = L_{\text{pM-SIC}}$, up to which the M most likely tentative symbol decisions per detection node were retained. By contrast, for the higher-index detection stages only one symbol decision per detection node was retained. The curves associated with the different detection techniques in Figure 17.19 were generated with the aid of the complexity formulae given by Equations 17.157 as well as 17.173. For the partial M-SIC scheme furthermore the specific variables have to be substituted into Equation 17.173. The number of complex additions associated with the different detectors has not been illustrated here, since these values were found only to differ slightly from those of the number of complex multiplications. As expected, it is evidenced by Figure 17.19 that the M-SIC detector may become significantly more complex than standard SIC, depending on the number of simultaneous users L. While for a scenario of $L = P = 4$ the complexity of the M-SIC scheme is observed in Figure 17.19 to be higher than that of standard SIC by about 32%, for a scenario of $L = P = 8$ the M-SIC is by a factor of about 2.32 or equivalently, by about 132% more complex. This is a result of the exponential growth of the number of demodulation and updating operations associated with the M-SIC scheme, when increasing the number of users L as seen in Figure 17.19. As expected, the complexity of partial M-SIC is shown in Figure 17.19 to be between that of standard SIC and M-SIC for $1 < L' < L$.

Similar observations can also be inferred for the computational complexity in terms of the number of comparisons between real-valued numbers, as portrayed on the right-hand side of Figure 17.19. Here the complexity difference between standard SIC and M-SIC is even more dramatic. While for a scenario of $L = P = 4$ the M-SIC is a factor 3.88 more complex than standard SIC, for a scenario of $L = P = 8$ this factor is as high as 23.32. Again, the partial M-SIC is capable of significantly reducing the complexity – also depending on the cut-off level L' – although this is achieved at the cost of a performance degradation, as was observed in Section 17.3.1.4.5.

In order to further characterise the complexity of the standard SIC and that of the M-SIC detector, in Table 17.7 we have listed the complexity-related contribution of the different components involved in the detection process for a scenario of $L = P = 4$. Note that the process of initialisation, weight calculation and selection – as described in Section 17.3.1.1 – is identical for both the standard and the M-SIC detector, which is indicated by "%" in the corresponding entries of Table 17.7. However, as argued earlier in this section, the computational complexity related to the operations of demodulation, combining and updating is

description	$C_{\text{SIC}}^{(\mathbb{C}*\mathbb{C})}$	$C_{\text{SIC}}^{(\mathbb{C}+\mathbb{C})}$	$C_{\text{SIC}}^{(\mathbb{R}\lessgtr\mathbb{R})}$	$C_{\text{M-SIC}}^{(\mathbb{C}*\mathbb{C})}$	$C_{\text{M-SIC}}^{(\mathbb{C}+\mathbb{C})}$	$C_{\text{M-SIC}}^{(\mathbb{R}\lessgtr\mathbb{R})}$
initialisation:	72	66	-	%	%	%
weight calc.	153.33	153.33	-	%	%	%
selection	83.25	72	9	%	%	%
combining	16	16	-	60	60	-
demodulation	8	16	16	30	60	88
updating	12	12	-	56	56	-
→ total	345	335.33	25	454.58	467.33	97

Table 17.7: Computational complexity of the different processing steps involved in standard SIC (columns 1...3) and M-SIC (columns 4...6) for a scenario of $L = P = 4$ simultaneous users- and reception antennas; specifically for M-SIC the number of tentative symbol decisions per detection node was equal to $M = 2$, while in all scenarios $M_c = 4$ constellation points were assumed, which is, for example, the case with 4QAM modulation.

significantly increased for the M-SIC scheme compared to standard SIC due to the higher number of detection nodes.

17.3.1.6 Summary and Conclusions on SIC Detection Techniques

Our discussions commenced in Section 17.3.1 with a portrayal and characterisation of standard SIC-based detection and its derivatives, namely the M-SIC and partial M-SIC detection schemes in the context of both uncoded and turbo-coded scenarios. More specifically, standard SIC detection was detailed in Section 17.3.1.1. Its associated block diagram was portrayed in Figure 17.10 and the most significant equations were summarised in Table 17.6. As a result of the SIC detector's strategy of detecting only the most dominant user having the highest SINR, SIR or SNR at its associated linear combiner's output, the dimensionality of the associated symbol classification or demodulation was reduced to evaluating the single-user Euclidean distance metric of Equation 17.125 LM_c times in contrast to calculating the multi-user Euclidean distance metric of Equation 17.223 M_c^L-times, as in the case of joint optimum ML detection.

The performance of the SIC detector critically relies on correct symbol decisions at the different detection stages, otherwise potentially catastrophic error propagation is encountered. An attractive strategy of reducing these effects is that of tracking multiple symbols decisions from each detection node. The philosophy of this technique was addressed in Section 17.3.1.2. Specifically in Section 17.3.1.2.1 we considered M-SIC, where the M number of most likely tentative symbol decisions were tracked from all detection nodes in the detection "tree" of Figure 17.11.

In an effort to reduce the potentially high computational complexity of M-SIC compared to that of standard SIC and motivated by the observation that the highest symbol error probabilities are associated with the "early" or low-index detection stages, in Section 17.3.1.2.2 the partial M-SIC was briefly characterised. Recall that at higher SNRs for the detection stages encountered towards the end of the detection process the symbol error probability is lower than for the detection process at the beginning, provided that error-free symbol decisions were encountered in the previous detection stages. This is because towards the end of the detection process the MIMO system's effective diversity order is increased. More specifically, the principle of partial M-SIC was to track the $M > 1$ number of most likely tentative symbol decisions per detection node only during the first few detection stages. By contrast, for the later detection stages only one symbol decision per detection node was made, as in the case of standard SIC. The philosophy of these techniques was further proved with the aid of Figure 17.11, which illustrates the associated detection trees for the specific scenario of $L = 4$ users and for retaining $M = 2$ number of tentative symbol decisions per detection node in case of M-SIC and partial M-SIC. Specifically, the graph on the left-hand side of Figure 17.11 was associated with standard SIC, while that on the right-hand side of Figure 17.11 with M-SIC. Finally, in the centre of Figure 17.11 the various partial M-SIC related detection "trees" were portrayed. A further reduction of the computational complexity could potentially be achieved with the aid of the SDI-M-SIC technique [117, 137], which was briefly addressed in Section 17.3.1.2.3.

Our further deliberations in Section 17.3.1.3 then addressed the problem of SIC-specific soft-bit generation required for turbo-decoding. Based on our observation of the effects of error propagation through the different SIC detection stages the weighted soft-bit metric of Equation 17.133 was proposed. It was demonstrated later in Section 17.3.1.4.7 that the employment of this weighted soft-bit metric resulted in a significant BER performance improvement in scenarios, where the number of users L is of similar value to the number of BS reception antenna elements P.

Our performance analysis of the various SIC-related detection techniques was conducted in Section 17.3.1.4. Specifically, in Section 17.3.1.4.1 our discussions commenced with the analysis of the BER and SER performance of standard SIC and M-SIC, parametrised with the number of users L and the number of reception antennas P. A significant performance improvement was observed in Figure 17.12 upon increasing the SDMA-MIMO system's order under the constraint of $L = P$. This was a result of the higher "diversity" of users in terms of their different received signal quality observed at each detection stage. Furthermore, we found that using the SINR instead of the SNR at the linear combiner's output for selecting the most dominant user from the set of remaining users at each SIC detection stage had a modest, but noticeable beneficial effect at low BERs in conjunction with more than three users.

Our further investigations conducted in Section 17.3.1.4.2 then focused on the analysis of the SER encountered at each detection stage, again for both the standard SIC and the M-SIC. As was shown for the standard SIC scheme on the left-hand side of Figure 17.13, the SER monotonously increases upon approaching the last detection stage. Our explanation of this phenomenon was that the symbol error probability encountered at a specific detection stage is composed of the symbol error probability of the previous stage "plus" an "additional" error probability, which is related to the effects of the residual AWGN at the specific stage considered under the assumption that no symbol errors have occurred in the previous detection stages. For higher SNRs we observed that the different detection stages' SER curves merge, since the "AWGN"-related symbol error contribution is decreased for the higher-index detection stages, which is a result of the system's increased grade of diversity. For M-SIC similar SER curves were shown on the right-hand side of Figure 17.13.

In order to further highlight the effects of error propagation through the different SIC detection stages, in Section 17.3.1.4.3 the standard SIC detector's SER performance was portrayed in the context employing error-free symbol decisions in the SIC module of Figure 17.10. Specifically on the left-hand side of Figure 17.14 the SER results were averaged over the different detection stages, while on the right-hand side of Figure 17.14 the SER results were portrayed on a per-detection stage basis. Compared to the more realistic case of employing an imperfect, potentially error-contaminated remodulated reference the SER improvement is substantial, which provided a further motivation for mitigating the effects of error-propagation for example with the aid of the M-SIC scheme of Section 17.3.1.2.

In an effort to further characterise the effects of error-propagation, in Section 17.3.1.4.4 we analysed the probability of the various symbol error events, namely that a symbol error occurred in the first detection stage, while the symbol decisions carried out during the higher-index detection stages were error-free. The associated probabilities of the various error-events at the reception antennas were shown on the left-hand side of Figure 17.15 as a function of the SNR, while on the right-hand side their contribution to the total SER was portrayed. From these curves we inferred that for higher SNRs the SER is governed by specific error-events, which originated from the first detection-stage, followed by those events, which commenced in the second stage and so on.

This was the main motivation for employing partial M-SIC, which was characterised in terms of its SER in Section 17.3.1.4.5, specifically on the left-hand side of Figure 17.16. Here we observed that in a scenario of $L = P = 4$ users and reception antennas as well as at an SER of 10^{-5} achieved by the partial M-SIC scheme up to and including the second detection stage, the SNR must be only a modest 1dB higher than that required by the M-SIC arrangement, while at the same time halving the complexity.

Furthermore, in Section 17.3.1.4.6 SDI-M-SIC was discussed and compared to both standard SIC and M-SIC on the right-hand side of Figure 17.16. The philosophy of SDI-M-SIC was to allow the employment of M-SIC or partial M-SIC only in a limited number of low-quality OFDM subcarriers, namely in those which exhibit the lowest SINR during the first detection stage, while using standard SIC in all other subcarriers. In our specific example we employed partial M-SIC, where multiple tentative symbol decisions per detection node were permitted only during the first detection stage. In the context of the indoor WATM channel model of Section 14.3.1 it was shown on the right-hand side of Figure 17.16 that a modest number of $N_{\mathrm{SDI}}^{[1]} = 64$ decision-insertion related subcarriers is sufficient for closely approximating the performance of the partial M-SIC scheme. This corresponded to $1/8$-th of the total subcarriers.

Finally, in Section 17.3.1.4.7 the BER performance of turbo-coded standard SIC detection-assisted SDMA-OFDM was analysed. Specifically, in the context of Figure 17.17 we compared the benefits of employing the rudimentary soft-bit metric of Equation 17.130 against those of the improved soft-bit metric of Equation 17.133, which accounted for the effects of error-propagation through the different detection stages. The employment of the weighted soft-bit metric of Equation 17.133 was found to be particularly beneficial in the context of a "fully-loaded" SDMA-OFDM system, where the number of users L equals the number of reception antennas P. Further BER performance results related to the weighted soft-bit metric were presented in Figure 17.18, which characterised the influence of the number of users L on the SDMA-OFDM system's BER performance. Here we observed that the BER curves associated with different numbers of users were within an SNR range of 2dB at a BER of 10^{-5}. This is an indication of the higher quality of the soft-bit estimates.

Our discussions in Section 17.3.1 were concluded in Section 17.3.1.5 with the aid of a complexity analysis of standard SIC, M-SIC and partial M-SIC. Specifically, the associated complexity formulae of standard SIC, which reflect the number of complex multiplications and additions as well as real-valued comparisons and divisions were given by Equations 17.157, 17.158, 17.159 and 17.160, respectively. Furthermore, the corresponding complexity formulae of M-SIC were given by Equations 17.173, 17.174, 17.175 and 17.176, respectively. Based on these equations the different SIC detectors' complexities were graphically compared in Figure 17.19. Specifically on the left-hand side of Figure 17.19 we plotted the number of complex multiplications, while on the right-hand side the number of real-valued comparisons required. We observed that standard SIC exhibits the lowest complexity, while M-SIC the highest complexity. A compromise in terms of both the achievable BER performance and the complexity imposed is achieved with the aid of the partial M-SIC scheme. In order to support our analysis, the corresponding numerical complexity values were provided in Table 17.6 for the standard SIC and the M-SIC.

17.3.2 PIC Detection

One of the key justifications for proposing SIC was that upon decreasing the number of users during the successive detection stages a higher grade of antenna array diversity potential can be dedicated by the MMSE combiner to the mitigation of the serious channel transfer factor fades, rather than to suppressing the interfering signal sources. Hence the highest array noise mitigation is achieved during the last SIC iteration, when after correct detection and subtraction of all the co-channel users' remodulated signals, the interference-free array output vector is constituted by the transmitted signal of the least dominant user plus an array noise contribution. The above interference cancellation principle – which was portrayed in the context of SIC in Section 17.3.1 – can be invoked also in form of a PIC scheme [125, 135].

Our discussions commence in Section 17.3.2.1 with the description of the PIC detector's structure in the context of an uncoded scenario, while in Section 17.3.2.2 the principles of turbo-coded PIC are discussed. This is followed in Section 17.3.2.3 by the analysis of its performance in both an uncoded- and a turbo-coded scenario, while in Section 17.3.2.4 the analysis of the detector's complexity is carried out. A summary and conclusions will be offered in Section 17.3.2.5.

17.3.2.1 Uncoded PIC

In this section we will highlight the PIC detector's structure, which is depicted in Figure 17.20. Let us commence our discussions upon recalling from Equation 17.1 the specific structure of the vector $\mathbf{x} \in \mathbb{C}^{P \times 1}$ of signals received by the different antenna elements, namely that we have:

$$\mathbf{x} = \mathbf{Hs} + \mathbf{n} \tag{17.183}$$

$$= \mathbf{H}^{(l)} s^{(l)} + \sum_{\substack{i=1 \\ i \neq l}}^{L} \mathbf{H}^{(i)} s^{(i)} + \mathbf{n}, \tag{17.184}$$

where again, $\mathbf{H} \in \mathbb{C}^{P \times L}$ is the channel transfer factor matrix, $\mathbf{s} \in \mathbb{C}^{L \times 1}$ is the composite multi-user vector of signals transmitted by the L different users and $\mathbf{n} \in \mathbb{C}^{P \times 1}$ is the vector of AWGN contributions encountered at the P different antenna elements. Specifi-

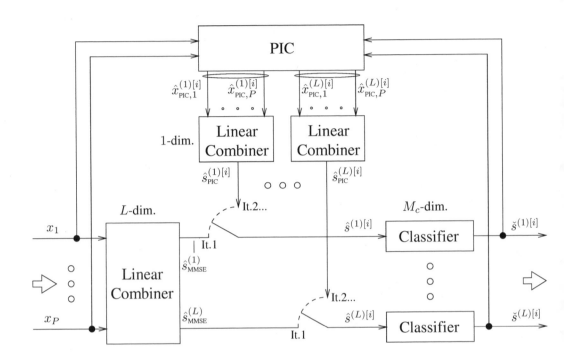

Figure 17.20: Illustration of the main signal paths associated with the hard-decision based PIC detector. The signals $x_p, p = 1, \ldots, P$ received by the different antenna elements are fed into the MMSE linear combiner described by Equations 17.185 and 17.186, which is active only during the first PIC detection stage. Furthermore, the signals $x_p, p = 1, \ldots, P$ are fed into the PIC module described by Equation 17.189, which is active for all PIC iterations associated with $i \geq 2$. The outputs $\breve{s}^{(l)[i]}, l = 1, \ldots, L$ of the bank of M_c-ary symbol classifiers or demodulators obeying Equations 17.187 and 17.193 are then fed back to the PIC module seen at the top of the figure. According to Equation 17.189, P different signals $\hat{x}_{\mathrm{PIC},p}^{(l)[i]}, p = 1, \ldots, P$, namely the potentially interference-free i.e. "decontaminated" antenna output signals are available for the l-th user at the output of the PIC module, which are again linearly combined with the aid of Equations 17.191 and 17.192, in order to form the estimate $\hat{s}_{\mathrm{PIC}}^{(l)[i]}$ of the signal transmitted by the l-th user, where $l = 1, \ldots, L$. Note that for the sake of visual clarity here we have omitted the signal paths associated with the channel transfer factor estimates required by the linear combiners and the PIC module. Also note than in our specific case the linear combiner employed is the MMSE combiner of Section 17.2.3

cally, from the component representation given by Equation 17.184 we observe that the array output vector $\mathbf{x}$ is composed of the l-th user's signal contribution vector and the $L - 1$ interfering users' signal contribution vectors plus the AWGN vector. Hence, if initial estimates $\check{s}^{(i)}, i \in \{1, \dots, L\} \setminus \{l\}$ of the interfering users' transmitted signals were available, a noisy estimate $\hat{\mathbf{x}}^{(l)}$ of the l-th user's signal contribution could be obtained upon removing the $L - 1$ interfering users' estimated signal contributions given by $\mathbf{H}^{(i)}\check{s}^{(i)}, i \in \{1, \dots, L\} \setminus \{l\}$ from the vector $\mathbf{x}$ of signals received by the different antenna elements. An estimate $\hat{s}^{(l)}$ of the l-th user's transmitted signal could then be inferred by linear antenna diversity combining. The more specific processing steps of the PIC detector proposed here will be further detailed in the following sections with reference to Figure 17.20.

First-Stage - MMSE Detection

- *Combining*: During the first PIC iteration seen in Figure 17.20 each user is detected by means of the MMSE combiner, where the linear combiner's output vector $\hat{\mathbf{s}}_{\text{MMSE}} \in \mathbb{C}^{L \times 1}$ is given according to Equation 17.50 by:

$$\hat{\mathbf{s}}_{\text{MMSE}} = \mathbf{W}_{\text{MMSE}}^{[1]H}\mathbf{x}, \tag{17.185}$$

and the weight matrix $\mathbf{W}_{\text{MMSE}}^{[1]} \in \mathbb{C}^{L \times L}$ is given in its left-inverse related form according to Equation 17.68 as:

$$\mathbf{W}_{\text{MMSE}}^{[1]} = \mathbf{HP}_{\text{SNR}}(\mathbf{H}^H\mathbf{HP}_{\text{SNR}} + \mathbf{I})^{-1}, \tag{17.186}$$

where $\mathbf{P}_{\text{SNR}} \in \mathbb{C}^{L \times L}$ is the diagonal-shaped SNR matrix.

- *Classification/Demodulation*: Then the linear combiner's output vector $\hat{\mathbf{s}}^{[1]} = \hat{\mathbf{s}}_{\text{MMSE}}$ is demodulated, with the aid of the blocks seen at the bottom right-hand corner of Figure 17.20, resulting in the vector $\check{\mathbf{s}}^{[1]} \in \mathbb{C}^{L \times 1}$ of symbols that are most likely to have been transmitted by the L different users. More specifically, as shown in Section 17.2.4, the demodulation is carried out upon evaluating Equation 17.94, namely:

$$\check{s}^{(l)[1]} = arg \min_{\check{s}/\sigma_l \in \mathcal{M}_c} \left| \frac{1}{H_{\text{eff}}^{(l)[1]}} \hat{s}^{(l)[1]} - \check{s} \right|^2, \quad l = 1, \dots, L, \tag{17.187}$$

where the l-th user's effective channel transfer factor $H_{\text{eff}}^{(l)[1]}$ is given by:

$$H_{\text{eff}}^{(l)[1]} = \mathbf{w}_{\text{MMSE}}^{(l)[1]}\mathbf{H}^{(l)}, \tag{17.188}$$

and the l-th user's weight vector $\mathbf{w}_{\text{MMSE}}^{(l)[1]}$ is the l-th column vector of the weight matrix $\mathbf{W}_{\text{MMSE}}^{[1]}$. Equation 17.187 implies calculating the Euclidean distance between each of the normalised elements[20] of the combiner's output vector $\hat{\mathbf{s}}^{[i]}$ namely, $\hat{s}^{(l)[i]}, l = 1, \dots, L$, and all the legitimate trial-symbols, which are the amplified constellation points contained in the set $\mathcal{M}_c$, associated with the specific modulation scheme em-

---

[20]The normalisation is not necessary in the context of employing MPSK modulation schemes in the absence of turbo-decoding.

ployed. According to the ML principle that specific trial-symbol is retained as the most likely transmitted one for the l-th user, which exhibits the smallest Euclidean distance from the combiner's normalised output signal $\hat{s}^{(l)}$. However, as argued in Section 17.2.4, this decision principle is based on the assumption, that the residual interference contaminating the combiner's output signal is also Gaussian, which in general is not the case. Hence the demodulation principle formulated according to Equation 17.187 is sub-optimum.

i-th Stage: PIC Detection

- *Parallel Interference Cancellation*: During the i-th PIC iteration seen in Figure 17.20, where $i \geq 2$ a potentially improved estimate $\hat{s}_{\mathrm{PIC}}^{(l)[i]}$ of the complex symbol $s^{(l)}$ transmitted by the l-th user is obtained upon subtracting in a first step the $L-1$ interfering users' estimated signal contributions, from the original vector $\mathbf{x}$ of signals received by the different antenna elements, which can be expressed as:

$$\hat{\mathbf{x}}_{\mathrm{PIC}}^{(l)[i]} = \mathbf{x} - \sum_{\substack{j=1 \\ j \neq l}}^{L} \mathbf{H}^{(j)} \breve{s}^{(j)[i-1]}. \tag{17.189}$$

This operation takes place within the PIC block shown at the top of Figure 17.20. Provided that correct tentative symbol decisions were made during the previous detection stage, namely we have $\breve{s}^{(j)[i-1]} = s^{(j)}, j \in \{1, \ldots, L\} \setminus \{l\}$ for the $L-1$ interfering users, the estimated array output vector $\hat{\mathbf{x}}_{\mathrm{PIC}}^{(l)[i]} \in \mathbb{C}^{P \times 1}$ will only consist of the l-th user's namely the desired user's signal contribution vector $\mathbf{H}^{(l)} s^{(l)}$ plus the AWGN vector $\mathbf{n}$, which is expressed as:

$$\hat{\mathbf{x}}_{\mathrm{PIC}}^{(l)[i]} = \mathbf{H}^{(l)} s^{(l)} + \mathbf{n} \quad \wedge \breve{s}^{(j)[i-1]} = s^{(j)}, j \in \{1, \ldots, L\} \setminus \{l\}. \tag{17.190}$$

- *Combining*: The final task is hence to extract an estimate $\hat{s}_{\mathrm{PIC}}^{(l)[i]}$ of the signal $s^{(l)}$ transmitted by the l-th user from the l-th user's PIC-related array output vector $\hat{\mathbf{x}}_{\mathrm{PIC}}^{(l)[i]}$. This can be achieved upon invoking once again the left-inverse related MMSE combiner, seen below the PIC block at the top of Figure 17.20, whose associated weight matrix was given by Equation 17.186 for the more general case of detecting L users. As observed in Equation 17.190, the signal vector $\hat{\mathbf{x}}_{\mathrm{PIC}}^{(l)[i]}$ at the output of the PIC block at the top of Figure 17.20 is now potentially free of interference. Hence the channel transfer factor matrix $\mathbf{H}$ and the SNR matrix $\mathbf{P}_{\mathrm{SNR}}$ defined in Equation 17.65, which are integral parts of the MMSE-related weight matrix according to Equation 17.186, have to be substituted by the l-th user's related components namely, by $\mathbf{H}^{(l)}$ and $\mathrm{SNR}^{(l)} = \frac{\sigma_l^2}{\sigma_n^2}$. This results in the weight vector $\mathbf{w}_{\mathrm{MMSE}}^{(l)[i]}$, given by:

$$\mathbf{w}_{\mathrm{MMSE}}^{(l)[i]} = \frac{\mathbf{H}^{(l)}}{\|\mathbf{H}^{(l)}\|_2^2 + \frac{1}{\mathrm{SNR}^{(l)}}}. \tag{17.191}$$

With the aid of the weight vector of Equation 17.191 an estimate $\hat{s}^{(l)[i]} = \hat{s}_{\mathrm{PIC}}^{(l)[i]}$ of the l-th user's transmitted signal $s^{(l)}$ can then be extracted from the vector $\hat{\mathbf{x}}_{\mathrm{PIC}}^{(l)[i]}$ seen at the output of the linear MMSE combiner in the centre of Figure 17.20 - similar to Equation 17.185 - as follows:

$$\hat{s}_{\text{PIC}}^{(l)[i]} = \mathbf{w}_{\text{MMSE}}^{(l)[i]H} \hat{\mathbf{x}}_{\text{PIC}}^{(l)[i]}. \tag{17.192}$$

- *Classification/Demodulation*: The above PIC and MMSE-combining steps are again followed by the classification, demodulation stage seen on the right of Figure 17.20, which obeys:

$$\check{s}^{(l)[i]} = arg \min_{\check{s}/\sigma_l \in \mathcal{M}_c} \left| \frac{1}{H_{\text{eff}}^{(l)[i]}} \hat{s}^{(l)[i]} - \check{s} \right|^2, \; l = 1, \dots, L, \tag{17.193}$$

where the *l*-th user's effective channel transfer factor $H_{\text{eff}}^{(l)[i]}$ is given by:

$$H_{\text{eff}}^{(l)[i]} = \mathbf{w}_{\text{MMSE}}^{(l)[i]} \mathbf{H}^{(l)}. \tag{17.194}$$

In other words, Equation 17.193 delivers the symbol $\check{s}^{(l)[i]}$ that is most likely to have been transmitted by the *l*-th user. The *i*-th PIC iteration described above potentially has to be performed for all the different SDMA users namely, for $l = 1, \dots, L$.

We have summarised the steps of the PIC algorithm once again in Table 17.8, while the schematic of the PIC detector was provided in Figure 17.20. Note, however, that for the sake of visual clarity in the context of this simplified schematic we have omitted the signal paths associated with the channel transfer factor estimates required by the linear combiners and within the PIC module.

17.3.2.2 Turbo-Coded PIC

In the context of our investigations concerning uncoded PIC detection-assisted SDMA-OFDM in Section 17.3.2.3.1 we will highlight that the system's relatively poor performance compared to that of uncoded SIC detection-assisted SDMA-OFDM, discussed in Section 17.3.1.4.1, is related to the effects of "error propagation" between the different users' symbol estimates during the second PIC detection stage. This is, because if amongst the L different users' tentative symbol decisions made during the first PIC stage there is a specific subcarrier, which has an unreliable tentative symbol decision while all the $L - 1$ remaining users' tentative symbol decisions are relatively reliable, then after the second PIC detection stage all of these $L - 1$ users' symbol decisions will become potentially unreliable. By contrast, the single user's tentative symbol decision, which was unreliable after the first detection stage is expected to become more reliable.

In Sections 17.2.6.4 and 17.3.1.4.7 we demonstrated in the context of both MMSE- and SIC detection-assisted SDMA-OFDM that turbo-decoding at the receiver is a powerful means of further enhancing the system's BER performance. Specifically, the turbo-decoder was incorporated into the system by simply feeding the demodulator's soft-bit output into the turbo-decoder. Recall that the generation of the soft-bits required for turbo-decoding was discussed in Section 17.2.5 and 17.3.1.3, respectively. In the context of the associated simulations four turbo decoding iterations were performed by the turbo-decoder, followed by slicing or hard-

Description	Operations
First-Stage - MMSE Det.	
Calc. MMSE weight matrix	$\mathbf{W}_{\mathrm{MMSE}}^{[1]} = \mathbf{H}\mathbf{P}_{\mathrm{SNR}}(\mathbf{H}^H\mathbf{H}\mathbf{P}_{\mathrm{SNR}} + \mathbf{I})^{-1} \in \mathbb{C}^{P \times L}$
Detection	$\hat{\mathbf{s}}_{\mathrm{MMSE}} = \mathbf{W}_{\mathrm{MMSE}}^{[1]H}\mathbf{x} \in \mathbb{C}^{L \times 1}, \quad \hat{s}^{(l)[1]} = \hat{s}_{\mathrm{MMSE}}, \ l = 1, \ldots, L$
Demodulation, $l = 1, \ldots, L$	$\breve{s}^{(l)[1]} = arg \min_{\breve{s}/\sigma_l \in \mathcal{M}_c} \left|\frac{1}{H_{\mathrm{eff}}^{(l)[1]}}\hat{s}^{(l)[1]} - \breve{s}\right|^2, H_{\mathrm{eff}}^{(l)[1]} = \mathbf{w}_{\mathrm{MMSE}}^{(l)[1]}\mathbf{H}^{(l)}$
i-th Stage - PIC $(l = 1, .., L)$	
Subtraction	$\hat{\mathbf{x}}_{\mathrm{PIC}}^{(l)[i]} = \mathbf{x} - \sum_{\substack{j=1 \\ j \neq l}}^{L} \mathbf{H}^{(j)}\breve{s}^{(j)[i-1]} \in \mathbb{C}^{P \times 1}$
Calc. MMSE weight vectors	$\mathbf{w}_{\mathrm{MMSE}}^{(l)[i]} = \frac{\mathbf{H}^{(l)}}{\|\mathbf{H}^{(l)}\|_2^2 + \frac{1}{\mathrm{SNR}^{(l)}}} \in \mathbb{C}^{P \times 1}$
Detection	$\hat{s}_{\mathrm{PIC}}^{(l)[i]} = \mathbf{w}_{\mathrm{MMSE}}^{(l)[i]H}\hat{\mathbf{x}}_{\mathrm{PIC}}^{(l)[i]}, \quad \hat{s}^{(l)[i]} = \hat{s}_{\mathrm{PIC}}^{(l)[i]}$
Demodulation	$\breve{s}^{(l)[i]} = arg \min_{\breve{s}/\sigma_l \in \mathcal{M}_c} \left|\frac{1}{H_{\mathrm{eff}}^{(l)[i]}}\hat{s}^{(l)[i]} - \breve{s}\right|^2, H_{\mathrm{eff}}^{(l)[i]} = \mathbf{w}_{\mathrm{MMSE}}^{(l)[i]H}\mathbf{H}^{(l)}$

Table 17.8: Summary of the standard hard-decision-based PIC detector.

decision of the turbo-decoder's "source"-related soft-output bits. Here we note explicitly that the turbo-iterations were entirely performed within the turbo-decoder, without invoking again the system's remaining components. The turbo-coding parameters were summarised in Table 17.5.

Similar to the MMSE and SIC schemes, in order to enhance the BER performance of PIC detection-assisted SDMA-OFDM a trivial approach would be to feed the turbo-decoder with the demodulator's soft-bit output after the PIC detection process and again, to perform a number of iterations within the turbo-decoder, followed by slicing the turbo-decoder's "source"-related soft-output bits. However, recall our observation that for some users the reliability of the second PIC detection stage's symbol decisions might be degraded compared to that of the first PIC detection stage due to feeding potentially unreliable tentative symbol decisions into the second PIC detection stage. Hence it is potentially beneficial to embed the turbo-decoding into the PIC detection process. To be more specific, after the first detection stage, based on the different users' soft-bit values derived from the associated MMSE combiner's output signals, as it was demonstrated in Section 17.2.5, only a fraction of the total number of turbo-decoding iterations are performed. A reference signal is then generated upon slicing and remodulating the turbo-decoder's "source- plus parity"-related *a posteriori* soft-output bits for the following second PIC detection stage. Our experiments revealed that it is less effective to slice, *reencode* and remodulate only the "source"-related *a posteriori* soft-output bits. After the second PIC detection stage, again, soft-bits would again be generated, which are fed into the turbo-decoder, followed by a number of turbo-iterations and a final slicing of the "source"-related soft-output bits. These constitute the PIC receiver's output bits. Alternatively, further PIC iterations could be performed, but in the context of the hard-decision- or slicing based PIC scheme employed here no additional performance gain was observed.

A further BER performance improvement can potentially be achieved upon feeding soft-bit values, rather than sliced bits, into the second PIC detection stage following the concepts of turbo-equalisation [125] instead of the hard-decision based remodulated reference signals.

17.3.2.3 Performance Analysis

In this section the PIC algorithm will be investigated with respect to its BER performance in both a scenario without channel coding, as well as a scenario where turbo-coding is employed. Specifically, the interference cancellation process carried out during the PIC detector's second stage will be shown to benefit from the less error-contaminated first stage tentative symbol decisions. Again, the frame-invariant fading indoor WATM channel model and its associated OFDM system model described in Section 14.3.1 were invoked and ideal knowledge of the channel transfer functions associated with the different transmit-receive antenna pairs was assumed. For a summary of the basic simulation set-up we refer again to Table 17.4.

While in Section 17.3.2.3.1 the PIC detection assisted SDMA-OFDM system's BER performance is considered without employing channel coding, our simulation results for the turbo-coded scenario will be discussed in Section 17.3.2.3.2.

17.3.2.3.1 BER Performance of Uncoded PIC Detection-Assisted SDMA-OFDM for Different Numbers of Users and Receiver Antennas

In Figure 17.21 we have portrayed the BER as well as the SER performance of 4QAM-modulated PIC detection-assisted SDMA-OFDM as a function of the SNR encountered at the reception antennas. Specifically, in the context of the curves shown on the left-hand side of Figure 17.21 $P = 2$ reception antennas and up to $L = 2$ simultaneous users were assumed. By contrast, the curves at the right-hand side of Figure 17.21 characterise the scenario of $P = 4$ reception antennas supporting up to $L = 4$ simultaneous users. The MMSE detection-related BER performance curves, which were shown earlier in Figure 17.8 have again been plotted as a reference. We observe that upon increasing the SNR, the PIC detection-assisted system's BER performance exhibits the same trends as that of the MMSE detector, although a specific BER performance is achieved at consistently lower SNRs. The highest SNR gain achievable with the advent of employing PIC detection compared to MMSE detection is approximately 3.03dB at a BER of 10^{-4}, which was observed here for the basic SDMA scenario of $P = 2$ reception antennas and $L = 2$ simultaneous users (A2/U2). By contrast, for the higher-order SDMA scenario of $P = 4$ reception antennas the highest SNR gain of 1.61dB is observed for $L = 2$ simultaneous users namely, while for $L = 3$ and $L = 4$ users the SNR gains are approximately 1.43dB and 1.25dB, again, at a BER of 10^{-4}. The reason for the reduction of the SNR gain along with increasing the number of simultaneous users L is, because upon supporting more users, the probability of an erroneous first-stage tentative symbol decision among one of the users is increased. Hence also the users benefitting from correct tentative symbol decisions during the first detection stage are more likely to be corrupted. As mentioned before, in an effort to render the first-stage tentative symbol decisions more reliable, channel coding can be employed. Hence, in the next section we will concentrate our attention on characterising the PIC detection-assisted system's BER performance in a turbo-coded scenario.

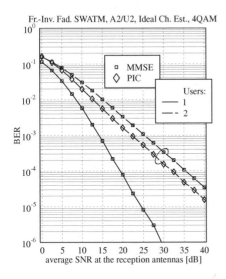 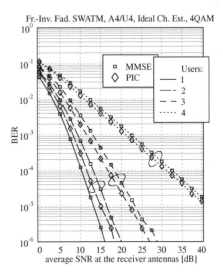

Figure 17.21: BER performance of 4QAM-modulated, PIC detection-assisted SDMA-OFDM as a function of the SNR encountered at the reception antennas for **(a: Left)** $P = 2$ reception antennas and up to $L = 2$ simultaneous users and **(b: Right)** $P = 4$ reception antennas and up to $L = 4$ simultaneous users; additionally, we have plotted the BER performance of MMSE detection-assisted SDMA-OFDM; for the basic simulation parameters we refer to Table 17.4.

17.3.2.3.2 BER Performance of Turbo-Coded PIC Detection-Assisted SDMA-OFDM for Different Numbers of Users and Receiver Antennas
As demonstrated in Sections 17.2.6.4 and 17.3.1.4.7 in the context of turbo-coded MMSE- and SIC detection-assisted SDMA-OFDM, employing turbo-decoding at the receiver is a powerful means of further enhancing the system's BER performance. Again, this is achieved at the cost of reducing the system's effective throughput and by imposing additional computational complexity.

In the context of PIC detection-assisted SDMA-OFDM we conjectured in Section 17.3.2.2 that instead of simply concatenating PIC detection and turbo-decoding as we did in the case of turbo-coded MMSE- or SIC detection-assisted SDMA-OFDM, it is potentially more beneficial to embed the turbo-decoding in the PIC detection process. As a result, a set of more accurate remodulated reference signals can be obtained for the second PIC detection stage.

More explicitly, a fraction of the total number of affordable turbo-decoding iterations would be performed after the first PIC detection stage, while the rest of the turbo iterations are carried out during the second PIC detection stage. More explicitly, in order to render the associated BER performance results comparable to those of turbo-coded MMSE-, SIC and ML detection-assisted SDMA-OFDM presented in Sections 17.2.6.4, 17.3.1.4.7 and 17.3.3.4.2, respectively, it is useful to split the total number of turbo iterations available between the

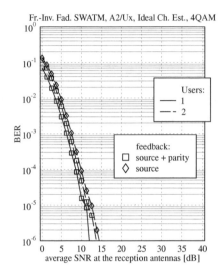

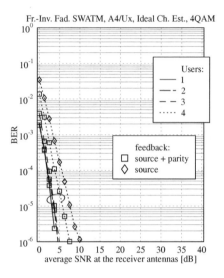

Figure 17.22: BER performance of turbo-coded, 4QAM-modulated, PIC detection-assisted SDMA-OFDM as a function of the SNR recorded at the reception antennas for **(a: Left)** $P = 2$ reception antennas and up to $L = 2$ simultaneous users and **(b: Right)** $P = 4$ reception antennas and up to $L = 4$ simultaneous users. Two methods of generating the remodulated reference signal were employed by the PIC module of Figure 17.20, namely a "source"- related *a posteriori* soft-output based reference and a "source- plus parity"-related *a posteriori* soft-output based reference. For the list of basic simulation parameters we refer to Table 17.4.

turbo-decoding conducted after the first and the second PIC detection stage. Since four turbo iterations were employed in our previous investigations, a plausible choice is to assign two iterations to both the first and the second PIC detection stage. An equal splitting of the number of turbo iterations is also motivated by the observation that the most significant BER improvement due to turbo decoding is achieved during the first few turbo-decoding iterations.

The other relevant turbo coding parameters were summarised in Table 17.5, but for the reader's convenience they are repeated: the coding rate was $R_c = \frac{1}{2}$, the constraint length was $K_c = 3$, and octally represented generator polynomials of $(7, 5)_8$ were used. Again, a total of four turbo iterations was performed.

Our BER simulation results are portrayed in Figure 17.22 on the left-hand side for $P = 2$ reception antennas, while on the right-hand side for $P = 4$ reception antennas, when supporting up to $L = P$ number of users. Two methods of generating the remodulated reference- or reconstructed received signal used in the PIC module of Figure 17.20 are compared against each other. Specifically in the first case we used slicing, reencoding, interleaving and remodulating only for the "source"-related *a posteriori* soft-output bits, while in the second scenario we employed slicing, interleaving and remodulating for the "source- plus parity"-

related *a posteriori* soft-output bits of the turbo-decoder. As expected, compared to the un-coded scenario, whose associated simulation results were shown in Figure 17.21, the BER is significantly reduced for both methods of generating the remodulated reference signal. How-ever, for the "source- plus parity" related remodulated reference signal of the second scenario a performance advantage of about 1.8dB was observed in Figure 17.22 at a BER of 10^{-5} compared to the "source"-related reference based scenario. In general, note the significant BER performance difference in favour of the scenario of four reception antennas and three simultaneous users, compared to supporting four simultaneous users. The explanation of this phenomenon is that in the scenario supporting a lower number of users the tentative sym-bol decisions provided by the first PIC detection stage are more reliable. This is a result of the higher relative diversity order encountered by the MMSE combiner, which constitutes the first detection stage of the PIC process, as argued in Section 17.3.2.1. Hence the effects of "error propagation" between the different users' signals during the second PIC detection stage are reduced.

17.3.2.4 Complexity Analysis

In this section we will analyse the complexity exhibited by the PIC detector described in Sec-tion 17.3.2.1, which was also summarised in Table 17.8. We will consider each processing step of Section 17.3.2.1 and Table 17.8 in terms of the associated complexity.

First-Stage - MMSE Detection:

- *Combining*: In the standard PIC algorithm of Section 17.3.2.1 the MMSE-related weight matrix $\mathbf{W}_{\text{MMSE}}^{[1]}$ to be employed for detection during the first PIC iteration does not explicitly have to be made available. Hence, according to Equations 17.104, 17.105 and 17.106 an initial estimate $\hat{\mathbf{s}}^{[1]}$ of the vector $\mathbf{s}$ of signals transmitted by the different users can be obtained, which imposes a computational complexity quantified in terms of the number of complex multiplications and additions, as follows:

$$C_{\text{PIC,direct}}^{[1](\mathbb{C}*\mathbb{C})} = C_{\text{PIC,direct}}^{[1](\mathbb{C}+\mathbb{C})} = PL + (P+1)L^2 + \frac{1}{3}L^3 \qquad (17.195)$$

$$C_{\text{PIC,direct}}^{[1](\mathbb{R}*\mathbb{C})} = PL \qquad (17.196)$$

$$C_{\text{PIC,direct}}^{[1](\mathbb{R}+\mathbb{C})} = L. \qquad (17.197)$$

- *Demodulation*: Furthermore, the demodulation operation carried out during the first PIC iteration, which follows the philosophy of Equation 17.187, imposes a computa-tional complexity of:

$$C_{\text{PIC,dem}}^{[1](\mathbb{C}*\mathbb{C})} = \frac{1}{2}LM_c \qquad (17.198)$$

$$C_{\text{PIC,dem}}^{[1](\mathbb{C}+\mathbb{C})} = C_{\text{PIC,dem}}^{[1](\mathbb{R}\lessgtr\mathbb{R})} = LM_c, \qquad (17.199)$$

where M_c is the number of constellation points contained in the set $\mathcal{M}_c$. Note that the number of complex multiplications has been weighted by a factor of $\frac{1}{2}$ in order to account for the reduced complexity associated with calculating the product between a complex number and its conjugate complex value in the context of the Euclidean distance metric evaluation of Equation 17.187.

Second-Stage - PIC Detection:

- *Parallel Interference Cancellation*: As observed in Table 17.8 the second PIC iteration commences by the operation of parallel interference cancellation, described by Equation 17.189, which is associated with a complexity of:

$$C_{\text{PIC,sub}}^{[2](\mathbb{C}*\mathbb{C})} = LP \tag{17.200}$$

$$C_{\text{PIC,sub}}^{[2](\mathbb{C}+\mathbb{C})} = L(\log_2 L)P. \tag{17.201}$$

Here we have assumed that the parallel subtraction based interference cancellation is organized in form of a binary tree. This reduces the complexity $C_{\text{PIC,sub}}^{[2](\mathbb{C}+\mathbb{C})}$ from originally $L(L-1)P$ complex additions associated with a linear implementation to that of $L(\log_2 L)P$.

- *Combining*: The operation of combining was described by Equations 17.191 and 17.192. A complexity reduction is achieved, when employing constant-modulus M-PSK modulation schemes. Then it is sufficient to perform the combining of the different users' associated signals by multiplying them with the Hermitian transpose of their associated channel vectors, which results in:

$$C_{\text{PIC,cmb-MPSK}}^{[2](\mathbb{C}*\mathbb{C})} = C_{\text{PIC,cmb-MPSK}}^{[2](\mathbb{C}+\mathbb{C})} = PL. \tag{17.202}$$

By contrast, in the more general case of QAM modulation schemes it can be shown that we have:

$$C_{\text{PIC,cmb-QAM}}^{[2](\mathbb{C}*\mathbb{C})} = \frac{3}{2}LP \tag{17.203}$$

$$C_{\text{PIC,cmb-QAM}}^{[2](\mathbb{C}+\mathbb{C})} = 2LP \tag{17.204}$$

$$C_{\text{PIC,cmb-QAM}}^{[2](\mathbb{R}/\mathbb{C})} = C_{\text{PIC,cmb-QAM}}^{[2](\mathbb{R}+\mathbb{R})} = L, \tag{17.205}$$

where the number of complex multiplications inflicted by calculating the Euclidean norm $||\mathbf{H}^{(l)}||_2^2$ has, again, been weighted by a factor of $\frac{1}{2}$ in order to account for the reduced complexity of calculating the product of a complex number with its conjugate complex value.

- *Demodulation*: In a final step, the different users' estimated signals are again demodulated according to Equation 17.193, which requires a computational complexity of:

$$C_{\text{PIC,dem}}^{[2](\mathbb{C}*\mathbb{C})} = \frac{1}{2}LM_c \tag{17.206}$$

$$C_{\text{PIC,dem}}^{[2](\mathbb{C}+\mathbb{C})} = C_{\text{PIC,dem}}^{[2](\mathbb{R}\lessgtr\mathbb{R})} = LM_c. \tag{17.207}$$

Hence, upon combining the different implementational complexity contributions we obtain the following expression for the total computational complexity:

$$C_{\text{PIC}}^{\mathbb{C}*\mathbb{C}} = LM_c + 5PL + (P+1)L^2 + \frac{1}{3}L^3 \qquad (17.208)$$

$$C_{\text{PIC}}^{\mathbb{C}+\mathbb{C}} = 2L(M_c + \frac{1}{2}) + (4 + \log_2 L)PL + (P+1)L^2 + \frac{1}{3}L^3 \qquad (17.209)$$

$$C_{\text{PIC}}^{\mathbb{R}\lessgtr\mathbb{R}} = 2LM_c \qquad (17.210)$$

$$C_{\text{PIC}}^{\mathbb{R}/\mathbb{C}} = L, \qquad (17.211)$$

where again, mixed real-complex multiplications and additions as well as real additions were assumed to have half the complexity of those, which involve complex numbers.[21]

17.3.2.5 Summary and Conclusions on PIC Detection

In Section 17.3.2.5 PIC assisted detection of SDMA-OFDM was introduced and characterised with respect to its BER performance in the context of both uncoded and turbo-coded scenarios. Furthermore, its complexity was analysed.

The employment of PIC detection was motivated by two observations. In Section 17.3.1.6 we found that SIC-based detection is capable of significantly outperforming MMSE detection in terms of the system's BER performance. This was a result of increasing the system's diversity order by successively removing the already detected users' remodulated signal contributions from the vector x of composite multi-user signals received by the different BS antenna elements. A substantial computational complexity was associated with the calculation of the linear combiner's weight matrix at each detection stage according to Equation 17.121, and with the calculation of the remaining users' SINRs according to Equation 17.122, which was followed by the selection of the most dominant user according to Equation 17.123. Hence, in our quest for alternative, potentially less complex detection techniques, PIC detection was considered.

As shown in the PIC detector's block diagram of Figure 17.20 and as described in Section 17.3.2.1, during the first PIC iteration linear MMSE estimates of the different users' transmitted signals were generated with the aid of a linear combiner, obeying Equations 17.185 and 17.186. These linear estimates seen in Figure 17.20 were demodulated, as shown in Equation 17.187, and employed in the context of the PIC module seen at the bottom left corner of Figure 17.20 as a reference for reconstructing the different users' transmitted signal contributions. A potentially more accurate estimate of the l-th user's transmitted signal, where $l = 1, \ldots, L$, was then generated during the next PIC iteration by subtracting the $L - 1$ remaining users' reconstructed signal contributions from the vector x of received composite multi-user signals as suggested by Equation 17.189. These operations were followed by diversity combining, obeying Equations 17.191 and 17.192 and demodulation, described by Equation 17.193, in order to obtain the specific constellation point that is most likely to have been transmitted. Provided that correct tentative symbol decisions were made for all the $L - 1$ remaining users in the previous PIC iteration, an improved linear MMSE estimate of the l-th user's transmitted signal would become available at the associated demodulator's

---

[21] Here we have again neglected that no real-valued additions are required in the context of real-complex multiplications and hence our calculation produces an upper-bound estimate of the complexity imposed.

input. This procedure has to be invoked for all the L different users. These processing steps were also summarised in Table 17.8 and in Figure 17.20.

Motivated by the PIC detector's relatively limited BER improvement compared to MMSE detection in the context of an uncoded scenario, in Section 17.3.2.2 we proposed to embed turbo-decoding into the PIC iteration, instead of simply concatenating the PIC detector and the turbo-decoder. As a result of this embedded turbo-decoding operation we expected to reduce the effects of "error-propagation" in the PIC module of Figure 17.20. In these experiments soft-bits were generated for turbo-decoding as it was described in Section 17.2.5 in the context of MMSE detection. This procedure was similar to the single-user scenario. Hence, there still remains some potential for further further improvement. In the context of turbo-decoding a sliced reference was generated for employment in the PIC module, which was the result of performing hard-decisions on the turbo-decoder's "source"-related soft-output bits followed by reencoding. An alternative strategy was that of performing hard-decisions on both the 'source'- and the 'parity'-related soft-output bits, which was shown in our performance investigations in Section 17.3.2.3.2 to be advantageous.

Our BER performance studies were conducted in Section 17.3.2.3. Specifically, Figure 17.21 characterised an uncoded scenario, while Figure 17.22 was recorded in the context of a turbo-coded scenario, again upon portraying the influence of the number of users L and the number of reception antennas P. For the uncoded scenario the SNR performance improvement compared to a system employing MMSE detection was at most 3.03dB, in a scenario of $L = P = 2$, while for a scenario of $L = P = 4$ the corresponding gain was at least 1.25dB. For a turbo-coded scenario the SNR improvement was more substantial. This will be further detailed in the context of our final comparison of all the different detectors in Section 17.3.4.1.

In order to conclude our discussions, the computational complexity imposed by PIC detection was analysed in Section 17.3.2.4, which resulted in Equations 17.208, 17.209 and 17.210, describing the number of complex multiplications and additions, as well as real-valued comparisons. These equations were invoked in our graphical portrayal of the different detectors' implementational complexities in Section 17.3.4.2.

17.3.3 ML Detection

In this section we will outline the philosophy of the Maximum Likelihood (ML) detector [112, 114–118, 120, 123, 135, 137], which is optimum from a statistical point of view. An associated disadvantage is its potentially excessive computational complexity, which results from the strategy of jointly detecting the L different users. This implies assessing the M_c^L possible combinations of symbols transmitted by the L different users by evaluating their Euclidean distance from the received signal, upon taking into account the effects of the channel. The stylised ML detector has once again been portrayed in Figure 17.23.

In Section 17.3.3.1 the philosophy of standard ML detection will be portrayed. In scenarios, where the number of users L is lower than the number of reception antennas P, a complexity reduction can be achieved by transforming the vector of composite multi-user signals received by the different antenna elements first to the so-called "trial-space" with the aid of a linear transform, as will be discussed in Section 17.3.3.2. In an attempt to further enhance the system's BER performance turbo-coding can be invoked. The generation of the soft-bit information required will be discussed in Section 17.3.3.3, where a further reduction

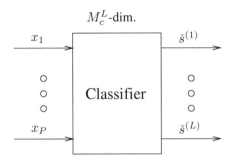

Figure 17.23: Representation of the optimum ML detector. In contrast to the sub-optimum linear- and non-linear detectors, namely LS and MMSE discussed in Sections 17.2.2 and 17.2.3, respectively, as well as SIC and PIC discussed in Sections 17.3.1 and 17.3.2, respectively, the L different users' complex symbols most likely transmitted are jointly detected. This is achieved upon evaluating the M_c^L number of trial-symbols in terms of their multi-user Euclidean distance metric with respect to the vector $\mathbf{x}$ of the signals received by the different antenna elements, namely Equation 17.223. A disadvantage is the associated computational complexity, which might be excessive.

of the computational complexity can be achieved by applying the well-known maximum approximation. Our BER performance investigation will be conducted in Section 17.3.3.4 for both uncoded as well as turbo-coded scenarios. Furthermore, estimates of the computational complexity imposed will be provided in Section 17.3.3.5. A summary and conclusions will be offered in Section 17.3.3.6.

17.3.3.1 Standard ML Detection

The structure of our discussions on the ML detector is as follows. In Section 17.3.3.1.1 the vector $\mathbf{x}$ of received signals is interpreted as a manifestation of a multi-variate complex Gaussian distribution function $f(\mathbf{x}|\breve{\mathbf{s}}, \mathbf{H})$, which reflects the *a priori* probability that the vector $\mathbf{x}$ was received. Furthermore, in Section 17.3.3.1.2 it will be shown that determining the desired vector of symbols $\breve{\mathbf{s}}_{\text{ML}}$ that is most likely to have been transmitted by the L different users is equivalent to maximising the transmitted symbols' *a posteriori* probability. However, it will be demonstrated with the aid of Bayes' theorem [90] that this is equivalent to maximising the *a priori* probability, which again, is available in terms of the Gaussian distribution function.

17.3.3.1.1 Representation of the Vector of Received Signals as a Sample of a Multi-Variate Complex Gaussian Distribution Function In order to commence our discussions, let us recall from Equation 17.1 the definition of the vector $\mathbf{x}$ of signals received by the P different antenna elements, namely that we have:

$$\mathbf{x} = \mathbf{Hs} + \mathbf{n}, \tag{17.212}$$

where again, we have omitted the index $[n, k]$, which denotes the k-th subcarrier of the n-th OFDM symbol. We observe that $\mathbf{x} \sim \mathcal{CN}(\mathbf{Hs}, \mathbf{R_n})$, namely $\mathbf{x}$ is a sample of an L-

dimensional multi-variate complex Gaussian distribution, having a vector of mean values given by $\mathbf{Hs}$ and a covariance matrix of $\mathbf{R_n} \in \mathbb{C}^{P \times P}$, where the latter is given by:

$$\mathbf{R_n} = E\{\mathbf{nn}^H\} \tag{17.213}$$

$$= \sigma_n^2 \mathbf{I}, \tag{17.214}$$

implying that the different noise contributions are assumed to be uncorrelated. This multi-variate complex Gaussian distribution function can be expressed as [494]:

$$f(\mathbf{x}|\mathbf{s}, \mathbf{H}) = \frac{1}{\pi^P |\mathbf{R_n}|} \exp\left(-(\mathbf{x} - \mathbf{Hs})^H \mathbf{R_n}^{-1}(\mathbf{x} - \mathbf{Hs})\right) \tag{17.215}$$

$$= \frac{1}{\pi^P (\sigma_n^2)^P} \exp\left(-\frac{1}{\sigma_n^2}||\mathbf{x} - \mathbf{Hs}||_2^2\right), \tag{17.216}$$

where Equation 17.216 was obtained by substituting Equation 17.214 into Equation 17.215. The representation of the complex Gaussian distribution function is legitimate, since again, the noise at the different receiver antenna elements is assumed to be uncorrelated. More explicitly, $P(\mathbf{x}|\mathbf{s}, \mathbf{H}) = f(\mathbf{x}|\mathbf{s}, \mathbf{H})$ denotes the *a priori* probability that the vector $\mathbf{x}$ has been received by the BS antenna elements under the condition that the vector $\mathbf{s}$ was transmitted by the different users over a channel characterised by the matrix $\mathbf{H}$.

17.3.3.1.2 Determination of the Vector of Transmitted Symbols by Maximising the *A Posteriori* Probability

In simple verbal terms the ML detector finds the specific L-dimensional vector of M_c-ary symbols, which is most likely to have been transmitted. In more formal terms ML detection is based on the idea of maximising the *a posteriori* probability $P(\check{\mathbf{s}}|\mathbf{x}, \mathbf{H})$ that the specific vector $\check{\mathbf{s}} \in \mathbb{C}^{L \times 1}$ of the different users' symbols – which is an element of the set $\mathcal{M}^L$ of trial-vectors – was transmitted over the SDMA-MIMO channel characterised by the channel transfer factor matrix $\mathbf{H} \in \mathbb{C}^{P \times L}$ under the condition that the vector $\mathbf{x} \in \mathbb{C}^{P \times 1}$ was received by the different BS receiver antenna elements. This maximisation procedure can be expressed as:

$$\check{\mathbf{s}}_{\mathrm{ML}} = arg \max_{\check{\mathbf{s}} \in \mathcal{M}^L} P(\check{\mathbf{s}}|\mathbf{x}, \mathbf{H}), \tag{17.217}$$

where the set $\mathcal{M}^L$ of trial-vectors is given by:

$$\mathcal{M}^L = \left\{ \check{\mathbf{s}} = \begin{pmatrix} \check{s}^{(1)} \\ \vdots \\ \check{s}^{(L)} \end{pmatrix} \middle| \frac{\check{s}^{(1)}}{\sigma_1}, \dots, \frac{\check{s}^{(L)}}{\sigma_L} \in \mathcal{M}_c \right\}, \tag{17.218}$$

and where $\sigma_l = \sqrt{\sigma_l^2}$ denotes the l-th user's standard deviation, while $\mathcal{M}_c$ denotes the set of complex constellations points associated with the specific modulation scheme employed.

The maximisation procedure obeying Equation 17.217 involves knowledge of the *a posteriori* probabilities $P(\check{\mathbf{s}}|\mathbf{x}, \mathbf{H}), \check{\mathbf{s}} \in \mathcal{M}^L$, which can be obtained from the *a priori* probabilities

$P(\mathbf{x}|\check{s}, \mathbf{H})$ with the aid of Bayes' theorem [90], namely:

$$P(\check{s}|\mathbf{x}, \mathbf{H}) = P(\mathbf{x}|\check{s}, \mathbf{H})\frac{P(\check{s})}{P(\mathbf{x})}, \qquad (17.219)$$

where all symbol vector probabilities are assumed to be identical, i.e. we have $P(\check{s}) = \frac{1}{M_c^L} = const.$, and for the total probability $P(\mathbf{x})$ we have:

$$P(\mathbf{x}) = \sum_{\check{s} \in \mathcal{M}^L} P(\mathbf{x}|\check{s}, \mathbf{H})P(\check{s}) = const., \qquad (17.220)$$

which follows from the simple fact that all probabilities have to sum to unity, i.e. that:

$$\sum_{\check{s} \in \mathcal{M}^L} P(\check{s}|\mathbf{x}, \mathbf{H}) \overset{!}{=} 1. \qquad (17.221)$$

Hence, upon substituting Equation 17.219 into Equation 17.217 and exploiting that we have $P(\check{s}) = const.$ as well as that $P(\mathbf{x}) = const.$ for all $\check{s} \in \mathcal{M}^L$, we obtain:

$$\check{s}_{\text{ML}} = arg \max_{\check{s} \in \mathcal{M}^L} P(\check{s}|\mathbf{x}, \mathbf{H}) \quad \Longleftrightarrow \quad \check{s}_{\text{ML}} = arg \max_{\check{s} \in \mathcal{M}^L} P(\mathbf{x}|\check{s}, \mathbf{H}), \qquad (17.222)$$

where $P(\mathbf{x}|\check{s}, \mathbf{H}) = f(\mathbf{x}|\check{s}, \mathbf{H})$ was given by Equation 17.216. Note from Equation 17.216 that maximising $f(\mathbf{x}|\check{s}, \mathbf{H})$ is equivalent to minimising the Euclidean distance metric $||\mathbf{x} - \mathbf{H}\check{s}||_2^2 \; \forall \; \check{s} \in \mathcal{M}^L$, and hence we have:

$$\check{s}_{\text{ML}} = arg \max_{\check{s} \in \mathcal{M}^L} P(\check{s}|\mathbf{x}, \mathbf{H}) \quad \Longleftrightarrow \quad \check{s}_{\text{ML}} = arg \min_{\check{s} \in \mathcal{M}^L} ||\mathbf{x} - \mathbf{H}\check{s}||_2^2. \qquad (17.223)$$

Note, however, that the complexity associated with evaluating Equation 17.223 might potentially be excessive, depending on the M_c^L number of vectors contained in the trial-set $\mathcal{M}^L$. An attractive strategy of reducing the complexity in the context of scenarios, where the L number of users is lower than the P number of BS reception antenna elements will be outlined in the next section.

17.3.3.2 Transform-Based ML Detection

As observed in Equation 17.223, determining the ML symbol estimate requires comparing the Euclidean distance between the vector $\mathbf{x}$ of signals actually received by the different antenna elements and the vector $\mathbf{H}\check{s}$ of signals, which would be received in the absence of AWGN, for all the different vectors $\check{s}$ of symbol combinations contained in the set $\mathcal{M}^L$. In order to potentially reduce the computational complexity in a specific scenario, where the L number of users is lower than the P number of reception antenna elements, it was proposed in [112] to transform the vector $\mathbf{x}$ of received signals first to the trial-domain[22] with the aid of a linear transform and then to perform the calculation of the Euclidean distance directly in the trial-domain. More explicitly applying this transform-based approach results in a potentially lower complexity than that associated with evaluating Equation 17.223. A linear transform

---

[22]The trial-domain is equal to the transmitted signal's domain.

which yields a particularly simple form of the Euclidean distance metric to be evaluated is based on the left-inverse- or Moore-Penrose pseudo-inverse of the channel matrix $\mathbf{H}$, which was discussed in Section 17.2.2. The resultant vector is also known as the LS estimate or ML estimate of the vector s of transmitted signals. More specifically, the LS estimate $\hat{\mathbf{s}}_{LS}$ was given by Equation 17.36, namely by:

$$\hat{\mathbf{s}}_{LS} = \mathbf{P}_{LS}\mathbf{x}, \tag{17.224}$$

with the associated projection matrix $\mathbf{P}_{LS}$ defined in Equation 17.37 as:

$$\mathbf{P}_{LS} = (\mathbf{H}^H\mathbf{H})^{-1}\mathbf{H}^H. \tag{17.225}$$

Furthermore, it was shown in Equation 17.42 that the LS combiner's output vector $\mathbf{s}_{LS}$ is composed of the vector s of transmitted signals plus an additional contribution due to the AWGN encountered at the array elements, which is formulated as:

$$\hat{\mathbf{s}}_{LS} = \mathbf{s} + \mathbf{P}_{LS}\mathbf{n}. \tag{17.226}$$

Recall from Equation 17.44 that the noise at the LS combiner's output is correlated, having a covariance matrix $\mathbf{R}_{\Delta\hat{\mathbf{s}}_{LS}}$ expressed in the form of:

$$\mathbf{R}_{\Delta\hat{\mathbf{s}}_{LS}} = \frac{1}{\sigma_n^2}(\mathbf{H}^H\mathbf{H})^{-1}. \tag{17.227}$$

Hence, in equivalence to Equation 17.215, the multi-variate complex Gaussian distribution function, which reflects the probability that the vector $\hat{\mathbf{s}}_{LS}$ is observed at the output of the linear combiner under the condition that the vector s was transmitted over a channel characterised by the channel transfer factor matrix $\mathbf{H}$ is given by:

$$f(\hat{\mathbf{s}}_{LS}|\mathbf{s}, \mathbf{H}) = \frac{1}{\pi^P|\mathbf{R}_{\Delta\hat{\mathbf{s}}_{LS}}|} \exp\left(-(\hat{\mathbf{s}}_{LS} - \mathbf{s})^H\mathbf{R}_{\Delta\hat{\mathbf{s}}_{LS}}^{-1}(\hat{\mathbf{s}}_{LS} - \mathbf{s})\right). \tag{17.228}$$

Upon following the steps outlined in the context of Section 17.3.3.1, the vector $\check{\mathbf{s}}_{ML}$ of symbols that is most likely to have been transmitted is then given by:

$$\check{\mathbf{s}}_{ML} = \underset{\check{\mathbf{s}}\in\mathcal{M}^L}{arg\,max}\, P(\check{\mathbf{s}}|\mathbf{x}, \mathbf{H}) \quad\Longleftrightarrow\quad \check{\mathbf{s}}_{ML} = \underset{\check{\mathbf{s}}\in\mathcal{M}^L}{arg\,min}\, (\hat{\mathbf{s}}_{LS} - \check{\mathbf{s}})^H\mathbf{R}_{\Delta\hat{\mathbf{s}}_{LS}}^{-1}(\hat{\mathbf{s}}_{LS} - \check{\mathbf{s}}). \tag{17.229}$$

As a result of the linear properties of the transform applied in Equation 17.224 to the vector x of signals received by the different antenna elements, the same symbol detection error probability is achieved as with the aid of the standard approach of Equation 17.223, however, at a potentially lower complexity. Note, however, that a necessary condition for the existence of the projection matrix $\mathbf{P}_{LS}$ is that the number of users L must be lower than or equal to the number of reception antenna elements P, which imposes a limitation compared to the standard ML detector of Section 17.3.3.1.

17.3.3.3 ML-Assisted Soft-Bit Generation for Turbo-Decoding

Turbo coding-based error protection of the different subcarriers hosted by an OFDM symbol is a powerful means of further enhancing the system's BER performance. This is achieved at the cost of reducing the system's effective throughput and increasing the system's complexity. A prerequisite for performing turbo decoding at the receiver is the availability of soft-bit information. As suggested in [112], it is desirable in terms of keeping the computational complexity as low as possible to perform the turbo trellis-decoding separately for the different users. This is, because the joint trellis-decoding of the different users' transmitted signals would potentially require an excessive number of trellis decisions and hence impose a high complexity.

17.3.3.3.1 Standard ML-Assisted Soft-Bit Generation

Following the concepts of Equation 17.95 the soft-bit value or log-likelihood ratio- or LLR-value associated with the l-th user at the m-th bit-position is given by:

$$L_m^{(l)} = \ln \frac{P(b_m^{(l)} = 1 | \mathbf{x}, \mathbf{H})}{P(b_m^{(l)} = 0 | \mathbf{x}, \mathbf{H})}, \tag{17.230}$$

which is the natural logarithm of the quotient of probabilities that the bit considered assumes either a value of $b_m^{(l)} = 1$ or $b_m^{(l)} = 0$. Note that here we have again omitted the index $[n, k]$ for the k-th subcarrier of the n-th OFDM symbol, which is associated with the different variables. Equation 17.230 can be further expanded by noting that the probability that a binary bit value of $b_m^{(l)} = 1$ was transmitted at the m-th bit position associated with the l-th user in the k-th subcarrier is given by the sum of the probabilities of those symbol combinations, where the l-th user's transmitted symbol is associated with a bit value of $b_m^{(l)} = 1$. The probability that a bit value of $b_m^{(l)} = 0$ was transmitted can be expanded equivalently. Hence we obtain:

$$L_m^{(l)} = \ln \frac{\sum_{\check{s}^{(1)}/\sigma_1 \in \mathcal{M}_c} \cdots \sum_{\check{s}^{(l)}/\sigma_l \in \mathcal{M}_{cm}^1} \cdots \sum_{\check{s}^{(L)}/\sigma_L \in \mathcal{M}_c} P(\check{\mathbf{s}} | \mathbf{x}, \mathbf{H})}{\sum_{\check{s}^{(1)}/\sigma_1 \in \mathcal{M}_c} \cdots \sum_{\check{s}^{(l)}/\sigma_l \in \mathcal{M}_{cm}^0} \cdots \sum_{\check{s}^{(L)}/\sigma_L \in \mathcal{M}_c} P(\check{\mathbf{s}} | \mathbf{x}, \mathbf{H})}, \tag{17.231}$$

where $\mathcal{M}_{cm}^b$ denotes the specific subset of the set $\mathcal{M}_c$ of constellation points of the modulation scheme employed, which are associated with a bit value of $b \in \{0, 1\}$ at the m-th bit position. For notational convenience we can define the l-th user's associated set of trial-vectors employed for determining the probability that the m-th transmitted bit exhibits a value of $b \in \{0, 1\}$ as follows:

$$\mathcal{M}_m^{b(l)L} = \left\{ \check{\mathbf{s}} = \begin{pmatrix} \check{s}^{(1)} \\ \vdots \\ \check{s}^{(L)} \end{pmatrix} \middle| \frac{\check{s}^{(1)}}{\sigma_1} \in \mathcal{M}_c, \ldots, \frac{\check{s}^{(l)}}{\sigma_l} \in \mathcal{M}_{cm}^b, \ldots, \frac{\check{s}^{(L)}}{\sigma_L} \in \mathcal{M}_c \right\}. \tag{17.232}$$

Upon invoking again Bayes' theorem given by Equation 17.219, namely that:

$$P(\check{s}|\mathbf{x}, \mathbf{H}) = P(\mathbf{x}|\check{s}, \mathbf{H})\frac{P(\check{s})}{P(\mathbf{x})}, \tag{17.233}$$

and re-substituting Equation 17.233 into Equation 17.231 we obtain the following expression for the l-th user's soft-bit value at the m-th bit position:

$$L_m^{(l)} = \ln \frac{\sum_{\check{s} \in \mathcal{M}_m^{1(l)L}} P(\mathbf{x}|\check{s}, \mathbf{H})}{\sum_{\check{s} \in \mathcal{M}_m^{0(l)L}} P(\mathbf{x}|\check{s}, \mathbf{H})}. \tag{17.234}$$

Here we have exploited that the different symbol combination vectors $\check{s}$ have the same probability namely that $P(\check{s}) = const.$, $\check{s} \in \mathcal{M}^L$. Upon recalling from Section 17.3.3.1 that the probability $P(\mathbf{x}|\check{s}, \mathbf{H})$ is given by the multi-variate complex Gaussian distribution function $f(\mathbf{x}|\check{s}, \mathbf{H})$ defined in Equation 17.216, we obtain:

$$L_m^{(l)} = \ln \frac{\sum_{\check{s} \in \mathcal{M}_m^{1(l)L}} \exp\left(-\frac{1}{\sigma_n^2}||\mathbf{x} - \mathbf{H}\check{s}||_2^2\right)}{\sum_{\check{s} \in \mathcal{M}_m^{0(l)L}} \exp\left(-\frac{1}{\sigma_n^2}||\mathbf{x} - \mathbf{H}\check{s}||_2^2\right)}. \tag{17.235}$$

Observe that evaluating the l-th user's soft-bit value based on its LLR at the m-th bit position with the aid of Equation 17.235 involves the exponential function, which might be computationally expensive.

17.3.3.3.2 Simplification by Maximum Approximation In order to avoid the explicit evaluation of the exponential function in Equation 17.235, a common approach is the employment of the so-called maximum-approximation, which implies that only that specific additive term is retained in the calculation of the numerator and nominator of Equation 17.235, which yields the maximum contribution. It can be readily shown that as a result of this simplification we obtain instead of Equation 17.235 the following expression:

$$L_m^{(l)} \approx \frac{1}{\sigma_n^2}\left[||\mathbf{x} - \mathbf{H}\check{s}_m^{0(l)}||_2^2 - ||\mathbf{x} - \mathbf{H}\check{s}_m^{1(l)}||_2^2\right], \tag{17.236}$$

where

$$\check{s}_m^{b(l)} = arg \min_{\check{s} \in \mathcal{M}_m^{b(l)L}} ||\mathbf{x} - \mathbf{H}\check{s}||_2^2, \quad b \in \{0, 1\}, \tag{17.237}$$

while the set $\mathcal{M}_m^{b(l)L}$ was defined in Equation 17.232. We note that for each soft-bit to be determined, Equation 17.237 has to be invoked twice, once for a bit status of $b = 1$ and once for $b = 0$.

Observe, however, that a significant complexity reduction can be achieved by exploiting that the union of the two subspaces associated with the binary bit values of 0 and 1 of the L users at bit position m constitutes the entire trial space, namely that $\mathcal{M}^L = \mathcal{M}_m^{0(l)L} \bigcup \mathcal{M}_m^{1(l)L}$. Hence, the calculation of the Euclidean distance metric $||\mathbf{x} - \mathbf{H}\check{s}||_2^2$ has to be performed only once for the different trial vectors $\check{s} \in \mathcal{M}^L$, followed by an appropriate selection in the context of the soft-bit generation assisted by Equation 17.237. Specifically, in a first step the Euclidean distance metric can be determined for half of the vectors $\breve{s}_m^{b(l)}$ associated with the different bit polarities $b \in \{0, 1\}$ and bit positions m by searching the entire set $\mathcal{M}^L$, which results in the ML estimate $\breve{s}_{\text{ML}}$ of the vectors of transmitted symbols according to Equation 17.223. This initial L-dimensional, M_c-ary ML symbol estimate is given by a specific bit vector. The inverse of this ML bit vector contains the specific bit polarities, for which the further minimisation according to Equation 17.237 still has to be conducted.

17.3.3.4 Performance Analysis

In this section the BER performance of ML detection-assisted SDMA-OFDM will be investigated in both a lower-complexity, higher effective throughput scenario using no channel coding, as well as in a higher-complexity, lower throughput scenario where turbo-coding is employed. Again, the frame-invariant fading indoor WATM channel model and its associated OFDM system model described in Section 14.3.1 were invoked and ideal knowledge of the channel transfer functions associated with the different transmit-receive antenna pairs was assumed. For a summary of the basic simulation setup we refer again to Table 17.4. While in Section 17.3.3.4.1 the ML detection-assisted SDMA-OFDM system's BER performance is considered in the uncoded scenario, our simulation results characterising the turbo-coded scheme will be discussed in Section 17.3.3.4.2.

17.3.3.4.1 BER Performance of ML Detection-Assisted SDMA-OFDM for Different Numbers of Users and Reception Antennas In Figure 17.24 we have portrayed the BER performance of ML detection-assisted OFDM as a function of the SNR encountered at the reception antennas. Specifically, the curves on the left-hand side of Figure 17.24 are parametrised with both the number of users L and the number of reception antennas P, where only scenarios associated with $L \leq P$ are considered. We observe that upon increasing the MIMO system's order, namely by considering a system of four reception antennas and four simultaneous users compared to a system of two reception antennas and two simultaneous users, the system's BER performance is significantly improved. This is in contrast to the behaviour observed for the MMSE detector in Figure 17.8, where a more modest improvement was observed, but it follows similar trends to those exhibited by the SIC detector characterised in Figure 17.12. More specifically, the ML detector benefits from the higher grade of diversity provided by a higher-order MIMO system. Also observe that by increasing the number of users L at a fixed number of reception antennas P, the system's performance degrades gracefully. More explicitly, the performance difference between the lowest-complexity system supporting one user and that of a "fully loaded" system associated with $L = P$ users is less than 2dB, which is in contrast to the significant degradation observed for the MMSE detector in Figure 17.8.

Our further investigations were conducted with respect to supporting L number of users, which was higher than the P number of reception antennas. By contrast, employing a configuration, where $L > P$ was prohibited in the context of the linear detectors, such as the LS or MMSE as well as the MMSE-based SIC and PIC schemes. The associated simulation results are portrayed on the right-hand side of Figure 17.24 for a scenario of $P = 4$ reception antennas, supporting up to $L = 6$ simultaneous users. Here we observe again that the performance degradation incurred upon increasing the number of users L beyond the num-

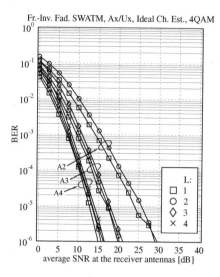
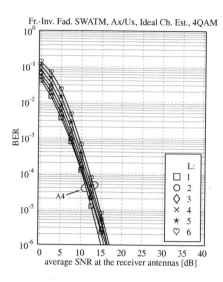

Figure 17.24: BER performance of 4QAM-modulated, uncoded and ML detection-assisted SDMA-OFDM as a function of the SNR encountered at the reception antennas; **(a: Left)** the curves are further parametrised with the number of simultaneous users L and the number of reception antennas P, using the configurations of A2/U1..2, A3/U1..3 and A4/U1..4; **(b: Right)** the curves are parametrised with the L number of users for a fixed number of $P = 4$ reception antennas, namely using configurations of A4/U1..6; for the basic simulation parameters we refer to Table 17.4.

ber of reception antennas P is gradual. This is in contrast to the more abrupt degradation, which would potentially be observed in conjunction with MMSE-based detection schemes, when allowing for example five simultaneous users instead of four users in a scenario of four reception antennas.

17.3.3.4.2 BER Performance of Turbo-Coded ML Detection-Assisted SDMA-OFDM for Different Numbers of Users and Reception Antennas

As was shown in Sections 17.2.6.4, 17.3.1.4.7 and 17.3.2.3.2 for MMSE, SIC and PIC detection-assisted SDMA-OFDM systems, respectively, the employment of turbo-decoding at the receiver is a powerful means of further enhancing the system's BER performance. This is achieved at the cost of a reduction of the system's effective throughput and by investing additional computational complexity. The associated turbo coding parameters were summarised in Table 17.5, but for the reader's convenience they will be repeated here: the coding rate was $R_c = \frac{1}{2}$, the constraint length was $K_c = 3$, the octally represented generator polynomials of $(7, 5)_8$ were used and 4 iterations were performed. The generation of the soft-bits required for turbo-decoding in the context of ML detection was discussed earlier in Section 17.3.3.3.

Our BER simulation results are portrayed in Figure 17.25, on the left-hand side for $P = 2$

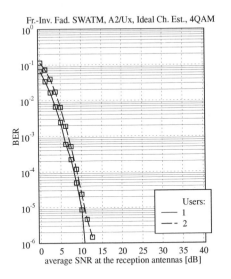

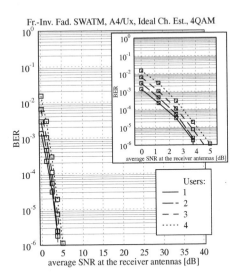

Figure 17.25: BER performance of turbo-coded, 4QAM-modulated, ML detection-assisted SDMA-OFDM as a function of the SNR recorded at the reception antennas; the curves are further parametrised with the number of simultaneous users L and the number of reception antennas P, where more specifically **(a: Left)** two reception antennas and **(b: Right)** four reception antennas were employed; for the basic simulation and turbo-coding parameters we refer to Tables 17.4 and 17.5, respectively.

reception antennas, while on the right-hand side for $P = 4$ reception antennas and up to $L = P$ number of simultaneous users. Again, we observe that compared to the uncoded scenario, whose associated BER simulation results were shown in Figure 17.24, the SNR at the reception antennas required for maintaining a specific BER is significantly reduced. In order to provide an example, in the context of a scenario associated with $L = P = 2$ and in the absence of channel coding an SNR of about 25.7dB is required to maintain a target BER of 10^{-5}, while with the assistance of turbo-coding this target BER is reached at a reduced SNR of about 10.6dB. Similarly, for a scenario of $L = P = 4$ the corresponding SNRs for the uncoded and coded case are given by 13.9dB and 4dB, respectively.

17.3.3.5 Complexity Analysis

The structure of our complexity analysis of the ML-related detection techniques is as follows. While in Section 17.3.3.5.1 the complexity of standard ML detection is discussed, we will focus our attention in Section 17.3.3.5.2 on the analysis of transform-based ML detection. Our discussions will be concluded in Section 17.3.3.5.3 by elaborating on the complexity associated with the generation of soft-bits to be used in the context of turbo-coded ML detection-assisted SDMA-OFDM.

17.3.3.5.1   Complexity of Standard ML Detection   As observed in Equation 17.223, an M_c^L number of symbol combinations has to be compared in terms of the Euclidean distance metric for the detection of the different users' transmitted symbols in a specific OFDM subcarrier. This imposes a computational complexity quantified in terms of the number of complex multiplications, additions as well as real-valued comparisons, which is given by:

$$C_{\mathrm{ML}}^{(\mathbb{C}*\mathbb{C})} \;\; = \;\; M_c^L P\left(\frac{1}{2}+L\right) \tag{17.238}$$

$$C_{\mathrm{ML}}^{(\mathbb{C}+\mathbb{C})} \;\; = \;\; M_c^L P\left(\frac{3}{2}+L\right) \tag{17.239}$$

$$C_{\mathrm{ML}}^{(\mathbb{R}\lessgtr\mathbb{R})} \;\; = \;\; M_c^L, \tag{17.240}$$

where again, the number of complex multiplications and additions involved in actually calculating the Euclidean norm $||.||_2^2$ has been weighted with a factor of $\frac{1}{2}$, because the multiplication of a complex number with its conjugate complex value inflicts two real-valued multiplications.[23] Note, however, that the number of multiplications $M_c^L PL$ required in the context of evaluating the term $\mathbf{H}\check{\mathbf{s}}$ in Equation 17.223 for the M_c^L number of different trial vectors can be reduced to $M_c PL$ by evaluating $\mathbf{H}^{(l)}\check{s}^{(l)}$, $\check{s}^{(l)} \in \mathcal{M}_c$, $l = 1, \dots, L$ and storing the resultant vectors in a lookup table. A similar technique could also be applied to reduce the number of complex multiplications in the context of the transform-based ML detection technique of Section 17.3.3.2, as will be discussed in the next section.

17.3.3.5.2   Complexity of Transform-Based ML Detection   The transform-based ML detection technique of Equation 17.229 capitalised on the LS estimate $\hat{s}_{\mathrm{LS}}$ of the vector $\mathbf{x}$ of signals received by the different antenna elements. This estimate was provided with the aid of Equation 17.224 which employs the projection matrix $\mathbf{P}_{\mathrm{LS}}$ given by Equation 17.225. Furthermore, in the transform-based ML detection process obeying Equation 17.229 explicit knowledge of the error or noise covariance matrix $\mathbf{R}_{\Delta\hat{s}_{\mathrm{LS}}}$ given by Equation 17.213, was required for describing the statistical properties of the LS combiner's vector of output signals.

Hence, it is a reasonable strategy to determine the error covariance matrix $\mathbf{R}_{\Delta\hat{s}_{\mathrm{LS}}}$ first with the aid of Equation 17.213. This imposes a computational complexity given by:

$$C_{\mathrm{ML\text{-}trf,err\text{-}cov}}^{(\mathbb{C}*\mathbb{C})} = C_{\mathrm{ML\text{-}trf,err\text{-}cov}}^{(\mathbb{C}+\mathbb{C})} = PL^2 + \frac{4}{3}L^3, \tag{17.241}$$

where the second term accounts for performing the direct inversion in Equation 17.213 with the aid of the LU decomposition [90]. Specifically, the LU decomposition [90] itself requires a complexity of $\frac{1}{3}L^3$ complex multiplications and additions, while the ensuing forward- and backward substitutions as outlined in [90] contribute another $L \cdot L^2$ number of complex multiplications and additions. The LU decomposition technique is only used here for a baseline comparison – more efficient techniques of performing the matrix inversion are known from the literature. Common to most of these matrix inversion techniques is that they are associated with a complexity order of $\mathcal{O}(m^3)$, where m is the dimension of the square matrix to be inverted.

---

[23]$(a_x + ja_y) \cdot (a_x - ja_y) = a_x^2 + a_y^2.$

Generating the LS estimate $\hat{\mathbf{s}}_{\text{LS}}$ of the vector **s** of transmitted signals with the aid of Equations 17.224 and 17.225, where the latter can be simplified upon substituting the expression of Equation 17.213 derived for the error covariance matrix $\mathbf{R}_{\Delta\hat{\mathbf{s}}_{\text{LS}}}$ imposes an additional complexity of:

$$C_{\text{ML-trf,LS-est}}^{(\mathbb{C}*\mathbb{C})} = C_{\text{ML-trf,LS-est}}^{(\mathbb{C}+\mathbb{C})} = PL + L^2. \tag{17.242}$$

The major part of the computational complexity, however, is imposed – similar to standard ML detection – by evaluating the M_c^L number of possible trial vectors $\check{\mathbf{s}} \in \mathcal{M}^L$ with the aid of Equation 17.229. This inflicts an additional number of operations given by:

$$C_{\text{ML-trf,trial}}^{(\mathbb{C}*\mathbb{C})} = M_c^L L \left(\frac{1}{2} + L\right) \tag{17.243}$$

$$C_{\text{ML-trf,trial}}^{(\mathbb{C}+\mathbb{C})} = M_c^L L \left(\frac{3}{2} + L\right) \tag{17.244}$$

$$C_{\text{ML-trf,trial}}^{(\mathbb{R}\lessgtr\mathbb{R})} = M_c^L. \tag{17.245}$$

Upon combining the different contributions quantified in this section, the total complexity of the transform-based ML detector becomes:

$$C_{\text{ML-trf}}^{(\mathbb{C}*\mathbb{C})} = PL + (P+1)L^2 + \frac{4}{3}L^3 + M_c^L L \left(\frac{1}{2} + L\right) \tag{17.246}$$

$$C_{\text{ML-trf}}^{(\mathbb{C}+\mathbb{C})} = PL + (P+1)L^2 + \frac{4}{3}L^3 + M_c^L L \left(\frac{3}{2} + L\right) \tag{17.247}$$

$$C_{\text{ML-trf}}^{(\mathbb{R}\lessgtr\mathbb{R})} = M_c^L. \tag{17.248}$$

A comparison between the equations associated with the standard ML detector and the transform-based ML detector reveals that employing the latter is only recommended, when the number of users L is smaller than the number of reception antenna elements P, but even then it still depends on the particular scenario, whether the latter is really advantageous.

In conclusion, particularly in the context of the higher-order QAM modulation schemes, such as, for example, 16QAM and in conjunction with a relatively high number of simultaneous users the computational complexity might become prohibitive for the application of the ML detector.

17.3.3.5.3 Complexity of ML-Assisted Maximum Approximation Based Soft-Bit Generation

In this section the complexity of maximum approximation based soft-bit generation will be analysed which was discussed in Section 17.3.3.3.2. We argued that the soft-bit generation procedure commences by evaluating the Euclidean distance metric, which is part of Equations 17.223 and 17.237, for all the different vectors of symbols contained in the set $\mathcal{M}^L$. The computational complexity of this processing step expressed in terms of the number of complex multiplications and additions was already quantified in Equations 17.238 and 17.239. Additionally, a substantial number of comparisons has to be carried out between the real-valued metric values, which constitute an integral part of the search across the different subsets of symbols denoted by $\mathcal{M}_m^{b(l)L}$ according to Equation 17.237. As argued in

Section 17.3.3.3.2, this complexity can be halved by determining in a first step the ML symbol estimate and its associated bit vector with the aid of Equation 17.223. This complexity is quantified in terms of the number of comparisons between real-valued metric values, as given by Equation 17.240, namely M_c^L. Furthermore, in the second step a $L \log_2 M_c$ number of search steps – where $\log_2 M_c$ is the number of bits per symbol – has to be conducted across a set of dimension $M_c^L/2$ each. Hence the total complexity is given by:

$$C_{\text{ML-soft}}^{(\mathbb{C}*\mathbb{C})} = C_{\text{ML}}^{(\mathbb{C}*\mathbb{C})} = M_c^L P \left(\frac{1}{2} + L \right) \tag{17.249}$$

$$C_{\text{ML-soft}}^{(\mathbb{C}+\mathbb{C})} = C_{\text{ML}}^{(\mathbb{C}+\mathbb{C})} = M_c^L P \left(\frac{3}{2} + L \right) \tag{17.250}$$

$$C_{\text{ML-soft}}^{(\mathbb{R} \lessgtr \mathbb{R})} = C_{\text{ML}}^{(\mathbb{R} \lessgtr \mathbb{R})} (1 + \frac{1}{2} L \log_2 M_c) \tag{17.251}$$

$$= M_c^L (1 + \frac{1}{2} L \log_2 M_c) \tag{17.252}$$

$$C_{\text{ML-soft}}^{(\mathbb{R}+\mathbb{R})} = L \log_2 M_c, \tag{17.253}$$

where the contribution $C_{\text{ML-soft}}^{(\mathbb{R}+\mathbb{R})}$ accounts for performing the subtraction of the metric values as shown in Equation 17.236.

17.3.3.6   Summary and Conclusions on ML Detection

In Section 17.3.3 the ML detector was discussed, which is optimum from a statistical point of view. That specific vector $\check{s}_{\text{ML}}$ of the different users' symbols is deemed to be optimum at the output of the ML detector, which exhibits the highest *a posteriori* probability of $P(\check{s}_{\text{ML}}|\mathbf{x}, \mathbf{H})$ amongst the M_c^L number of trial-vectors contained in the set $\mathcal{M}^L$ defined by Equation 17.218. Recall from Section 17.3.3.1.2 that the *a posteriori* probability $P(\check{s}_{\text{ML}}|\mathbf{x}, \mathbf{H})$ can be expressed with the aid of the Bayesian theorem of Equation 17.219, in terms of the *a priori* probability $P(\mathbf{x}|\check{s}_{\text{ML}}, \mathbf{H})$, which is given by the multi-variate complex Gaussian distribution function, of Equation 17.216. Identifying the optimum trial-vector $\check{s}_{\text{ML}}$ is equivalent to minimising the Euclidean distance between the vector $\mathbf{x}$ of received signals and the trial-vector $\check{s}$ transmitted over the MIMO channel described by the channel transfer factor matrix $\mathbf{H}$ for all trial-vectors contained in the set $\mathcal{M}^L$ as it was highlighted in Equation 17.223.

The computational complexity associated with this minimisation process is potentially excessive, since all the number of M_c^L trial-vectors contained in the set $\mathcal{M}^L$ have to be compared to each other in terms of the Euclidean distance metric of Equation 17.223. Provided that the number of users L is lower than the number of BS reception antenna elements P, a reduction of the complexity can be achieved by transforming the vector $\mathbf{x}$ of received signals first to the trial-space with the aid of the linear transform of Equation 17.224, instead of transforming each trial-vector $\check{s} \in \mathcal{M}^L$ separately to the received signal's space upon multiplication with the channel transfer factor matrix $\mathbf{H}$. These discussions were conducted in the context of Section 17.3.3.2.

Furthermore, in Section 17.3.3.3 the principles of ML detection-assisted soft-bit generation for employment in turbo-decoding were discussed, where a complexity reduction was achieved by applying the maximum approximation.

The ML detector's associated BER performance evaluated in the context of both a low-complexity, higher effective throughput uncoded and a higher-complexity, lower-throughput turbo-coded scenario was the topic of Section 17.3.3.4. Specifically, in Section 17.3.3.4.1 the influence of the number of users L and the number of BS reception antennas P on the BER performance of 4QAM-modulated ML detection-assisted SDMA-OFDM was analysed. We found that as shown on the left-hand side of Figure 17.24 – regardless of the number of users L – upon increasing the number of reception antennas P, the ML detector's associated BER performance was significantly improved. This was a result of the increased grade of channel diversity available. Furthermore, as shown on the right-hand side of Figure 17.24 the ML detector also exhibited a high resilience against the increase of the number of users L. Specifically, for a scenario of $P = 4$ reception antennas the BER curves of one to six users were confined to a narrow interval of about 2.5dB. This performance trend is in contrast to that observed for the linear combining based detectors, such as the LS and MMSE as well as SIC and PIC schemes, where a necessary condition of high integrity detection is that $L \leq P$. Furthermore, our BER performance results recorded for the turbo-coding based system were presented in Figure 17.25, as part of Section 17.3.3.4.2.

Finally our analysis of the ML detector's computational complexity was conducted in Section 17.3.3.5. Specifically, in Section 17.3.3.5.1 we analysed the complexity of standard ML detection, where the number of complex multiplications and additions as well as real-valued comparisons was given by Equations 17.238, 17.239 and 17.240, respectively. As expected, these complexities were proportional to the number of vectors M_c^L contained in the trial-set $\mathcal{M}^L$. Furthermore, in Section 17.3.3.5.2 the complexity of transform-based ML detection was evaluated. The associated complexity formulae were given by Equations 17.246, 17.247 and 17.248. It was clear from these equations that the transform based ML detection may only be preferred against standard ML detection, if the number of users L is lower than the number of reception antennas P. Our analysis of the complexity of ML-assisted maximum approximation-based soft-bit generation in Section 17.3.3.5.3 revealed that compared to standard ML detection the complexity quantified in terms of the number of real-valued comparisons due to comparing the values of the Euclidean distance metric across subsets of the set $\mathcal{M}^L$ of trial-vectors is significantly increased.

17.3.4   Final Comparison of the Different Detection Techniques

In this section a final comparison of the different linear and non-linear detection techniques namely, that of the MMSE, standard SIC, M-SIC, PIC and ML schemes will be carried out, which were described and characterised in Sections 17.2.3, 17.3.1.1, 17.3.1.2, 17.3.2 and 17.3.3, respectively. Again, as in previous sections, our comparison will focus on the system's BER performance in both an uncoded- and a turbo-coded scenario, which will be the topic of Section 17.3.4.1. By contrast in Section 17.3.4.2 a comparison between the different detectors' complexities will be carried out.

17.3.4.1   BER Performance Comparison of the Different Detection Techniques in Uncoded and Turbo-Coded Scenarios

In Figure 17.26 we have compared the different detectors' SDMA-OFDM related BER performance, on the left-hand side for the uncoded scenario and on the right-hand side for the turbo-coded scenario.

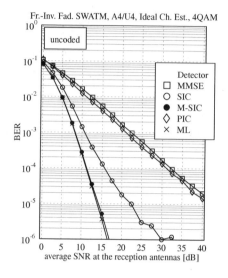
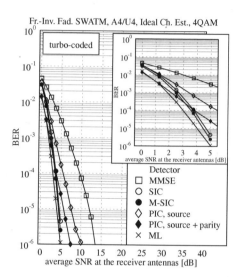

Figure 17.26: BER performance of **(a: Left)** uncoded and **(b: Right)** turbo-coded, 4QAM-modulated, MMSE, standard SIC, M-SIC (M=2), PIC and ML detection-assisted SDMA-OFDM as a function of the SNR at the reception antennas; in the context of the turbo-coded scenario both a PIC scheme capitalising on a "source"-related reference and a "source-plus parity"-related reference are compared against each other; $L = P = 4$ simultaneous users and reception antennas were employed; for the list of basic simulation- and turbo-coding parameters we refer to Tables 17.4 and 17.5, respectively.

Let us first of all focus our attention on the uncoded scenario. As expected, the best performance is exhibited by the most complex ML detector, closely followed by the M-SIC scheme, where $M = 2$. By contrast, a significant BER degradation is observed for the standard SIC scheme potentially as a result of the effects of error propagation through the different detection stages, a phenomenon which was analysed in Section 17.3.1.4.2. The second worst performance is exhibited by the PIC arrangement, while a further degradation by about 1.25dB is incurred upon employing the rudimentary MMSE detection, as argued earlier in Section 17.3.2.3.1. Specifically, the PIC detector's performance was impaired by the lower-power users, potentially propagating errors to those users, which benefitted from a relatively high SNR at the first-stage combiner output. An attractive approach to significantly improving the PIC detector's performance in the context of an SDMA-OFDM system is that of employing channel decoding after the first detection stage, as suggested in Section 17.3.2.2.

Let us now summarise our observations inferred in the context of the turbo-coded scenario, portrayed on the right-hand side of Figure 17.26. Again, the SDMA-OFDM system exhibits the best BER performance in the context of employing soft-bit values, which are generated with the aid of the ML-related metric of Section 17.3.3.3. Note that here we employed the simplified, maximum-approximation based formula of Equation 17.236. By contrast, a modest SNR degradation of only around 0.6dB is observed for the M-SIC (M = 2) and of about 0.8dB for the standard SIC scheme, both recorded at a BER of 10^{-4}. Again, the SIC detector's soft-bit values were generated with the aid of the weighted soft-bit metric of Equation 17.133. These performance trends are closely followed by the BER performance evaluated in the context of PIC-aided soft-bit generation, where we have compared two different approaches of generating the PIC-related remodulated reference or feedback signals against each other. Recall from Section 17.3.2.3.2 that the "source"-related reference generation implied slicing, reencoding, interleaving and remodulating the "source"-related *a posteriori* soft-output bits of the turbo-decoder. By contrast, the "source- plus parity"-related remodulated reference implied slicing, interleaving and remodulating the "source- plus parity"-related *a posteriori* soft-output bits of the turbo-decoder. The performance degradation of the "source- plus parity"-related reference -assisted PIC scheme compared to standard SIC is observed in Figure 17.26 to be about 1dB, while for the "source"-related reference-assisted PIC scheme an SNR degradation of 2.8dB is observed compared to standard SIC. Again, for MMSE detection-assisted SDMA-OFDM the worst performance is observed, namely an additional SNR degradation of about 4.3dB, compared to the "source"-related reference assisted PIC arrangement.

17.3.4.2 Complexity Comparison of the Different Detection Techniques

Having compared the various detection techniques, namely MMSE, SIC, M-SIC, PIC and ML in terms of the associated system's BER performance, in this section we will compare them with respect to their computational complexity. Here we will concentrate on the previously introduced two measures of complexity namely, the number of complex multiplications, as well as the number of comparisons between real-valued variables, which occur in the process of demodulation and during the selection of the most dominant user in each of the SIC's detection stages. In the context of our evaluations the number of constellation points associated with the modulation scheme employed was assumed to be $M_c = 4$, which is the case, for example, in 4QAM modulation, while for the M-SIC the number of symbols retained at each detection node was $M = 2$.

Specifically, on the left-hand side of Figure 17.19 we have compared the MMSE,[24] standard SIC, M-SIC, PIC and the ML detection schemes in terms of the number of complex multiplications $C^{\mathbb{C}*\mathbb{C}}$ incurred on a subcarrier basis, as a function of the number of simultaneous users L, which was assumed here to be equal to the number of reception antennas P. The curves associated with the different detection techniques were generated with the aid of the complexity formulae given by Equations 17.115, 17.157, 17.173, 17.208 and 17.238, respectively. As expected, the lowest computational complexity expressed in terms of the number of multiplications is exhibited by the MMSE detector, followed by PIC, standard SIC and M-SIC, while the highest complexity is exhibited by the optimum ML detector.

Similar observations can also be made for the computational complexity quantified in terms of the number of comparisons between real-valued numbers $C^{(\mathbb{R}\lessgtr\mathbb{R})}$, as portrayed on the right-hand side of Figure 17.27. Here the associated complexity formulae were given for the MMSE, standard SIC, M-SIC, PIC and ML detection schemes by Equations 17.117, 17.159, 17.176, 17.210 and 17.240, respectively. We observe a similar ranking of the different

---

[24]The LS detector's complexity has not been portrayed here explicitly, since it is only marginally less complex than the MMSE detector.

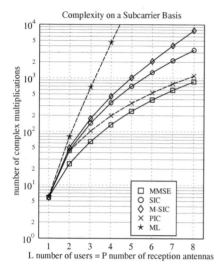
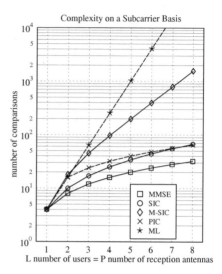

Figure 17.27: Complexity of the MMSE, standard SIC, M-SIC, PIC and ML detection schemes in terms of **(a: Left)** the $C^{(\mathbb{C}*\mathbb{C})}$ number of complex multiplications and **(b: Right)** the $C^{(\mathbb{R}\lesseqgtr\mathbb{R})}$ number of real-valued comparisons on a subcarrier basis, as a function of the number of simultaneous users L, which was assumed here to be equal to the number of reception antennas P; specifically for M-SIC the number of tentative symbol decisions per detection node was equal to $M = 2$, while in all scenarios $M_c = 4$ constellation points were assumed, which is for example the case in the context of 4QAM modulation.

detectors in terms of their associated complexity, as previously seen on the left-hand side of Figure 17.27 in terms of the associated performance. An exception is given by the number of comparisons associated with the PIC detector, which was found to be higher in the context of lower-order SDMA scenarios than for the standard SIC detector. The reason for this trend is that the PIC detector's complexity is increased compared to that of the MMSE detector, since the demodulation of each user's signal has to be performed twice, namely during the first- and the second detection stage. By contrast, the complexity of the standard SIC is increased compared to MMSE, since in each detection stage the most dominant user has to be selected from the set of remaining users.

The number of complex additions associated with the different detectors has not been illustrated here, since these values were found only to differ slightly from those characterising the number of complex multiplications.

In order to further support our comparison of the different detectors in terms of their associated computational complexity, we have summarised in Table 17.9 the number of complex multiplications and additions, as well as real-valued comparisons imposed in a scenario of $L = 4$ simultaneous users and $P = 4$ reception antenna elements.

	MMSE	std. SIC	M-SIC	PIC	ML
$C^{\mathbb{C}*\mathbb{C}}$	133.3̄3̄	344.58	454.58	197.3̄3̄	4608
$C^{\mathbb{C}+\mathbb{C}}$	135.3̄3̄	335.3̄3̄	467.3̄3̄	233.3̄3̄	5632
$C^{\mathbb{C}\lessgtr\mathbb{C}}$	16	25	97	32	256

Table 17.9: Computational complexity of the different detection schemes, namely MMSE, standard SIC, M-SIC, PIC and ML detection quantified in terms of the number of complex multiplications and additions $C^{\mathbb{C}*\mathbb{C}}$, $C^{\mathbb{C}+\mathbb{C}}$ as well as the number of real-valued comparisons $C^{\mathbb{R}\lessgtr\mathbb{R}}$ for a scenario of $L = P = 4$ simultaneous users- and reception antennas; specifically for M-SIC the number of tentative symbol decisions per detection node was equal to $M = 2$, while in all scenarios $M_c = 4$ constellation points were assumed, which is, for example, the case in conjunction with 4QAM modulation.

17.4 Performance Enhancement

The BER reduction observed in the context of turbo-coded SDMA-OFDM in conjunction with various detection techniques, namely MMSE, SIC, PIC and ML detection in Sections 17.2.6.4, 17.3.1.4.7, 17.3.2.3.2 and 17.3.3.4.2 was achieved at the cost of a substantial reduction of the system's effective throughput, namely by 50%, upon employing half-rate turbo-coding. This loss in throughput however, could have, been compensated upon employing a higher-order modulation scheme, namely 16QAM instead of 4QAM, thus further increasing the computational complexity. Obviously there is a trade-off between the BER performance, the throughput and the computational complexity.

In this section we will study potential techniques for further enhancing the BER performance of SDMA-OFDM on the uplink channel to the base station, without reducing the system's effective throughput. The techniques envisaged are constant throughput adaptive modulation as well as Walsh-Hadamard Transform (WHT) spreading across the different subcarriers. Both of these techniques have been recognised as being effective for exploiting the diversity offered by a wideband channel. Specifically adaptive modulation has widely been discussed in the context of single-user OFDM systems, namely in [2], and furthermore it was also successfully employed in conjunction with decision-directed channel prediction in Section 15.4. On the other hand, spreading the transmitted signal by means of orthogonal codes has been extensively discussed in the context of single- and multi-carrier CDMA systems, potentially supporting multiple simultaneous users. In our contribution, however, spreading is employed to further exploit the channel's diversity potential, while multiple users are supported with the aid of the multiple BS receiver antennas.

In Section 17.4.1 adaptive modulation assisted SDMA-OFDM or in short form SDMA-AOFDM will be discussed. We will then embark in Section 17.4.2 on a discussion of WHT spreading-assisted SDMA-OFDM or again, in short form SDMA-WHTS-OFDM.

17.4.1 Adaptive Modulation-Assisted SDMA-OFDM

In order to commence our discussions, let us briefly review in the next section the concepts of adaptive modulation as employed in the context of a single-user OFDM scenario.

17.4.1.1 Outline of the Adaptive Single-User Receiver

Adaptive modulation employed in single-user OFDM systems has previously been discussed in Section 15.4 in the context of our assessment of channel transfer function prediction techniques. Recall that invoking adaptive modulation was motivated by the observation that the BER performance of an OFDM modem, which employs a fixed-mode modulation scheme is severely degraded due to the deep frequency domain channel transfer function fades experienced. This deficiency of the fixed-mode modems may be mitigated by assigning a more robust, but lower throughput modulation mode to those subcarriers, which are severely affected by the deep fades. By contrast, a potentially less robust, but higher throughput modulation mode may be assigned to the higher quality subcarriers.

A prerequisite of performing the modulation mode assignment during the n-th uplink[25] OFDM symbol period for employment during the $(n + 1)$-th OFDM symbol period is the availability of a reliable estimate of the channel transfer function to be experienced by the OFDM symbol received during $(n + 1)$-th OFDM symbol period. The simplest approach to subcarrier channel quality estimation would be to employ the pilot-based or decision-directed channel estimate[26] for the n-th OFDM symbol period as an *a priori* estimate of the channel experienced during the $(n + 1)$-th OFDM symbol period. However, as shown in Section 15.4 depending on the OFDM symbol normalised Doppler frequency of the channel this *a priori* estimate may result in an inaccurate assignment of the modulation modes to the different subcarriers, which is a consequence of the channel variations incurred between the two OFDM symbol periods.

A significant improvement leading to a more accurate *a priori* channel estimate for the $(n + 1)$-th OFDM symbol period could, however, be achieved with the aid of the decision-directed Wiener filter-based channel prediction techniques discussed in Section 15. As a result of the modulation mode adaptation portrayed the BER performance of the AOFDM modem was observed in Figure 15.19 to be significantly improved compared to that of an OFDM modem having the same throughput, but using a fixed modulation mode. At the same time, relatively rapidly varying channels having a high OFDM symbol normalised Doppler frequency could be supported.

Motivated by the successful employment of constant throughput adaptive modulation techniques in the context of single-user OFDM systems, we will now investigate their potential for employment in multi-user SDMA-OFDM systems. We will focus our attention on employing a linear detector at the receiver, which exhibits the highest potential of achieving a significant BER performance improvement with the aid of adaptive modulation techniques.

Our discussions commence in Section 17.4.1.2 with the outline of the adaptive multi-user receiver's structure. This is followed in Section 17.4.1.3 by an assessment of the system's BER performance. Our summary and conclusions will be offered in Section 17.4.1.4.

17.4.1.2 Outline of the Adaptive Multi-User SDMA-OFDM Receiver

In Figure 17.28 we have portrayed the basic block diagram of the adaptive multi-user SDMA-OFDM receiver employed at the BS. During the n-th OFDM symbol period, after removing the cyclic OFDM prefix- or guard interval, which is not shown here, the complex time domain

---

[25]Unless otherwise stated, we refer to "uplink OFDM symbol period" simply as "OFDM symbol period".

[26]Again, we synonymously use the expressions "channel estimate" and "channel transfer function estimate".

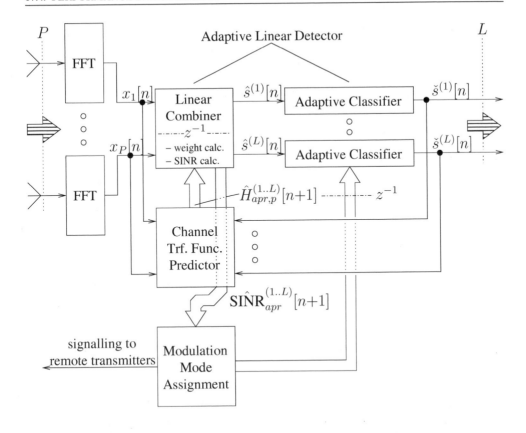

Figure 17.28: Block diagram of the adaptive multi-user SDMA-OFDM receiver, which is supported
by a P-element antenna front-end in order to facilitate the separation of the L number
of simultaneous users' associated signals at the BS's receiver. The subcarrier index k,
where $k = 0, \ldots, K - 1$ has been omitted for reasons of simplicity.

signals received by the P different BS antenna elements are independently subjected to the
FFT, which delivers the frequency domain or subcarrier-based representation of the signals
received, namely $x_p[n, k]$, where $p = 1, \ldots, P$ and $k = 0, \ldots, K - 1$. Note that for
notational convenience the subcarrier index k has been omitted in Figure 17.28.

The various signals $x_p[n, k]$ are then conveyed to the linear combiner, represented by
Equation 17.7, which linear estimates $\hat{s}^{(l)}[n, k]$ of the signals transmitted by the L
different users, namely $l = 1, \ldots, L$, separately for each subcarrier. The combiner weights
were already generated during the $(n - 1)$-th OFDM symbol period for use during the n-th
OFDM symbol period.

The linear signal estimates $\hat{s}^{(l)}[n, k]$ are then conveyed to the adaptive classifiers, which
deliver the sliced symbols $\check{s}^{(l)}[n, k]$ that are most likely to have been transmitted according
to the Euclidean distance metric of Equation 17.94. Again, this classification takes place
separately for the different users $l = 1, \ldots, L$ and subcarriers $k = 0, \ldots, K - 1$. Note
that the classifiers of the adaptive OFDM modem require side information concerning the

subcarrier or sub-band modulation mode assignment employed, which was generated during the previous OFDM symbol period and was locally stored at the receiver. The sliced symbols are then demapped to their bit-representation not shown in Figure 17.28, in order to obtain the bits transmitted.

Furthermore, the sliced symbols $\check{s}^{(l)}[n, k]$ of Figure 17.28 are conveyed together with the received subcarrier symbols $x_p[n, k]$ to the channel transfer function predictor, which generates the *a priori* estimates $\hat{H}_{apr,p}^{(l)}[n + 1, k]$ of the channel transfer factors $H_p^{(l)}[n + 1, k]$, associated with the $L \cdot P$ number of SDMA-MIMO channels portrayed in Figure 17.3 during the $(n + 1)$-th OFDM symbol period. However, these *a priori* channel transfer factor estimates $\hat{H}_{apr,p}^{(l)}[n+1, k]$ have already been employed during the n-th OFDM symbol period for generating the matrices $\hat{\mathbf{W}}_{apr}[n + 1, k]$, $k = 0, \ldots, K - 1$ of the combiner weights associated with the $(n + 1)$-th OFDM symbol period, upon invoking Equations 17.64 or 17.68.[27]

Furthermore, the combiner weights are then employed in conjunction with Equation 17.24 to obtain *a priori* estimates of the subcarrier-based SINRs, namely of $\mathrm{SÎNR}_{apr}^{(l)}[n + 1, k]$, potentially observed by the L different users at the linear combiner's output during the $(n+1)$-th OFDM symbol period. These *a priori* subcarrier SINR estimates are required for computing the different users' modulation mode assignments to be employed during the $(n+1)$-th uplink OFDM symbol period. The algorithm used for performing the modem mode assignment was summarised earlier in Section 15.4.1.1. Note that the updated modulation mode assignment is conveyed to the remote transmitters during the next downlink OFDM symbol period.

17.4.1.3 Performance Assessment

In this section we will briefly assess the BER performance of MMSE detection-assisted SDMA-AOFDM. Again, we employed the indoor WATM system and channel model of Section 14.3.1, where the fading was OFDM symbol invariant in order to avoid the obfuscating effects of inter-subcarrier interference.

Furthermore, perfect channel prediction was invoked, namely perfect knowledge of the channel transfer functions experienced during $(n + 1)$-th OFDM symbol period was made available during the n-th OFDM symbol period for calculating the MMSE combiner's weights to be employed during the $(n + 1)$-th OFDM symbol period. The number of BS receiver antennas P was equal to four. A total of 32 sub-bands each hosting 16 subcarriers was employed in the context of AOFDM, which capitalised on four modulation modes, namely "no transmission", BPSK, 4QAM and 16QAM.

Our simulation results are portrayed in Figure 17.29. On the left-hand side of Figure 17.29 we have compared SDMA-OFDM using fixed 4QAM modulation against 32 Sub-band (Sb)-SDMA-AOFDM having the same throughput, namely 1024 Bit per OFDM Symbol (BPOS). Note that here we have neglected the signalling overhead required for transmitting side information related to the modulation mode assignment to be employed during the next downlink OFDM symbol period. The BER curves are parametrised with the number of users L. The highest beneficial impact of adaptive modulation is observed for a "fully loaded" SDMA-

---

[27]Here the combiner weight matrix for a specific subcarrier is represented by $\hat{\mathbf{W}}_{apr}$ instead of $\mathbf{W}$ as in Equations 17.64 and 17.68, in order to indicate that its calculation is based on the imperfect estimates delivered by the *a priori* channel transfer function predictor.

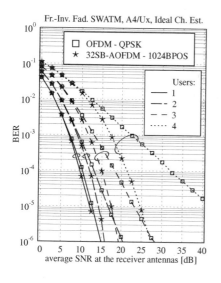
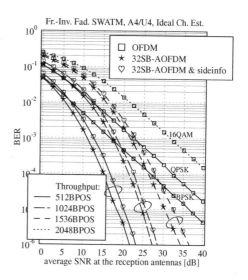

Figure 17.29: BER performance of MMSE detection-assisted SDMA-AOFDM as a function of the SNR recorded at the reception antennas. The curves are further parametrised with the number of simultaneous users L while the number of reception antennas P was fixed to four. Specifically, **(a: Left)** we have compared SDMA-OFDM using fixed 4QAM modulation to 32 Sub-band (Sb) AOFDM having the same throughput. Furthermore, **(b: Right)** we have compared SDMA-OFDM using fixed BPSK, 4QAM or 16QAM modulation to 32Sb-AOFDM having equivalent throughputs, upon once neglecting ("32Sb-AOFDM")- and once incorporating ("32Sb-AOFDM & sideinfo") the additional overhead required for transmitting side-information for the reverse link. Note that in the context of the 512-subcarrier indoor WATM system model of Section 14.3.1 employed here, fixed BPSK-, 4QAM- or 16QAM modulation-assisted OFDM is associated with throughputs of 512, 1024 or 2048 Bit per OFDM Symbol (BPOS). Perfect channel transfer function prediction was employed.

AOFDM system, where the number of users equals the number of BS receiver antennas. The is because for a lower number of users the effective channel – namely the SDMA-MIMO channel concatenated with the linear combiner – experienced by the different users across the various subcarriers fluctuates less dramatically. The justification of this observation is that the linear combiner is capable of dedicating more of the channel's diversity potential to mitigating the serious frequency domain channel fades, rather than to suppressing the interference imposed by the undesired co-channel users. As a result, the benefits of adaptive modulation are eroded.

Having found that adaptive modulation is rendered attractive only in the context of an almost fully loaded SDMA-OFDM scenario, we will now focus our attention on the BER performance of 32Sb-SDMA-AOFDM in the context of a scenario of four simultaneous users and reception antennas. Here we have considered both cases, namely that where no side-information is transmitted as in the context of the results presented on the left-hand side of Figure 17.29, and that, where explicit side-information related to the modulation mode assignment to be used on the next downlink OFDM symbol period is received from the remote transmitters. Upon assuming that the AOFDM modem supports four modulation modes, namely "no transmission", BPSK, 4QAM and 16QAM, a total number of two bits per sub-band are required. This number is increased to four bits upon assuming the employment of half-rate error-correction coding. Hence, in the context of 32Sb-SDMA-AOFDM the transmission overhead required for signalling the modulation mode assignment to be used on the reverse link is equal to 128 bit per OFDM symbol and user. Hence, for the effective system throughputs of 512, 1024 and 1536 BPOS the total target throughputs of the AOFDM modem are 640, 1152 and 1664 BPOS, respectively. From the BER curves shown on the right-hand side of Figure 17.29 we infer that in the context of the more realistic arrangement of transmitting explicit side-information the SNR required for attaining a specific BER is increased by a maximum of about 2dB, compared to the rather idealistic scenario, which neglects the transmission of side-information. Note furthermore that compared to the SDMA-OFDM schemes using fixed BPSK and 4QAM modulation modes, which support throughputs of 512 and 1024 BPOS, the BER reduction achieved by AOFDM at a fixed SNR or equivalently, the SNR reduction attained at a specific BER, is substantial. To provide an example, for a throughput of 1024 BPOS the SNR reduction due to employing SDMA-AOFDM compared to SDMA-OFDM using fixed 4QAM modulation is around 16dB at a BER of 10^{-5}, upon considering explicit side-information in the AOFDM transmissions.

17.4.1.4   Summary and Conclusions

In summary, in Section 17.4.1 we have described and characterised adaptive modulation assisted SDMA-OFDM. More specifically, in Section 17.4.1.1 adaptive modulation employed in single-user scenarios such as those described in Section 15.4 was briefly revisited. Furthermore, in Section 17.4.1.2 the architecture of multi-user SDMA-AOFDM receiver was detailed in the context of employing linear detection techniques, such as MMSE. Its simplified block diagram was shown in Figure 17.28. Our BER performance assessment was then conducted in Section 17.4.1.3. We found that the employment of adaptive modulation in SDMA-OFDM is useful only in the context of an almost fully-loaded SDMA-OFDM scenario, where the number of users L approaches the number of receiver antennas P. Using the indoor WATM system- and channel parameters as described in Section 14.3.1, the SNR advantage owing to employing 32Sb-SDMA-AOFDM having an effective throughput of 1024 BPOS compared to SDMA-OFDM using fixed 4QAM modulation was around 16dB at a BER of 10^{-5} in the context of assuming perfect channel transfer function prediction.

In our further experiments, which are not explicitly described here for reasons of space, we found that PIC detection, which was discussed in Section 17.3.2 is also amenable to employment in conjunction with adaptive modulation techniques, resulting in a similar BER improvement as recorded in the context of MMSE detection. The modulation mode assignment to be used would be based on the SNR or SINR observed at the output of the linear combiner, which constitutes the first PIC stage, as shown in Figure 17.20. By contrast, in the context of the SIC detection scheme investigated in Section 17.3.1 the employment of adaptive modulation techniques turned out to be less attractive. This is because the effects of deep channel transfer function fades experienced by some of the subcarriers have already been mitigated by detecting in each SIC stage only the most dominant remaining user. Similarly, the advantages of adaptive modulation are also expected to erode in the context of ML detection. In the next section we consider an alternative frequency domain fading counter measure, namely that of averaging the effects of fading, rather than accommodating them.

17.4.2   Walsh-Hadamard Transform Spreading Assisted SDMA-OFDM

Spreading the information symbols to be transmitted with the aid of orthogonal codes is the basis of supporting multiple-access capabilities in the context of single- and multicarrier CDMA (MC-CDMA) systems [495]. Instead of transmitting each complex symbol delivered by the modulator separately on a specific subcarrier in the context of multi-carrier OFDM modems, its influence is spread over several subcarriers with the aid of orthogonal multi-chip spreading codes. The advantage of employing orthogonal codes for performing the spreading is related to the resultant simple receiver design. A prominent class of orthogonal codes, which have been often used in CDMA systems is constituted by the family of orthogonal Walsh codes [495], which are particularly attractive, since the operation of spreading with the aid of these codes can be implemented in form of a "fast" transform, which takes advantage of the codes' recursive structure, similarly to the FFT.

Note, however, that in the context of our discussions presented in this section, we are more interested in spreading as a means of exploiting the wideband channel's diversity potential, rather than in its ability to support multiple users, since multiple users are supported in the context of the SDMA-OFDM receiver with the aid of the P-element antenna array and the associated detection techniques.

Due to the operations of spreading and despreading combined with MMSE-based frequency domain equalisation, the adverse effects of the low-SNR subcarriers on the average BER performance are potentially improved. This is a direct consequence of spreading, because even if the signal corresponding to a specific chip is obliterated by a deep frequency domain channel fade, after despreading its effects are spread over the Walsh-Hadamard Transform (WHT) length. Hence there is a high chance of still recovering all the partially affected subcarrier symbols without errors.

In Section 17.4.2.1 the structure of the WHT spreading-assisted single-user OFDM receiver is outlined. Specifically, we will demonstrate the separability of the operations of frequency domain channel transfer factor equalisation and despreading in the context of orthogonal codes. In Section 17.4.2.2 we will then describe the WHT spreading (WHTS) assisted multi-user SDMA-OFDM (SDMA-WHTS-OFDM) receiver's specific structure. The BER performance assessment of SDMA-WHTS-OFDM cast in the context of employing either MMSE- or PIC detection at the receiver will then be conducted in Section 17.4.2.3. Our conclusions will be offered in Section 17.4.2.4.

17.4.2.1   Outline of WHTS-Assisted Single User OFDM Receiver

In Figure 17.30 we have portrayed the simplified block diagram of the single-user WHTS-OFDM transmission scenario. More specifically, at the top of Figure 17.30 the schematic of the WHTS-OFDM transmitter is shown, which consists of WHT-assisted spreading, followed by the OFDM-related IFFT-based modulator. The IFFT-assisted modulator's output

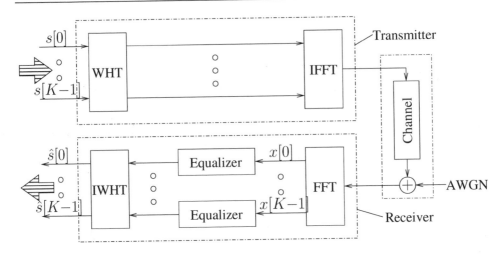

Figure 17.30: Simplified block diagram of the single-user WHTS-OFDM scenario. For reasons of notational simplicity the OFDM symbol index n and the static index $()_{MC}$ have been removed from the different variables.

samples are then conveyed – upon neglecting here the further OFDM transmitter processing steps – through the stylised AWGN contaminated transmission channel, as shown on the right-hand side of Figure 17.30. The WHTS-OFDM receiver shown at the bottom of Figure 17.30 then performs the OFDM-related FFT-aided demodulation of the incoming signal samples, followed by subcarrier-based channel transfer factor equalisation and despreading with the aid of the inverse WHT (IWHT). As we will show in the context of the forthcoming derivations, the separation of the WHTS-OFDM receiver into subcarrier-based equalisation and despreading is a consequence of the orthogonality of the Walsh codes employed.

Our discussions commence in Section 17.4.2.1.1 with an outline of the WHT matrix's specific properties, namely its unitary nature. Furthermore, in Section 17.4.2.1.2 the WHTS-OFDM receiver's design will be outlined. We will highlight that due to the WHT matrix's unitary nature the processes of channel transfer function equalisation and despreading can be conducted separately.

17.4.2.1.1 Properties of the Walsh-Hadamard Transform The lowest-dimensional WHT from which the higher dimensional WHTs can be recursively derived, is given by the WHT$_2$ transform, which is described by the following unitary matrix:

$$\mathbf{U}_{\text{WHT}_2} = \frac{1}{\sqrt{2}} \begin{pmatrix} 1 & 1 \\ 1 & -1 \end{pmatrix}. \tag{17.254}$$

More generally, the N-th order WHT is given by the following recursive expression:

$$\mathbf{U}_{\text{WHT}_N} = \frac{1}{\sqrt{2}} \begin{pmatrix} 1 \cdot \mathbf{U}_{\text{WHT}_{N-1}} & 1 \cdot \mathbf{U}_{\text{WHT}_{N-1}} \\ 1 \cdot \mathbf{U}_{\text{WHT}_{N-1}} & -1 \cdot \mathbf{U}_{\text{WHT}_{N-1}} \end{pmatrix}, \tag{17.255}$$

as a function of the $(N-1)$-th order WHT matrix, namely $\mathbf{U}_{\text{WHT}_{N-1}}$. Note that the column vectors of matrix $\mathbf{U}_{\text{WHT}_N}$ represent the orthogonal Walsh code vectors, for which we have:

$$col_i\{\mathbf{U}_{\text{WHT}_N}\} \cdot col_j\{\mathbf{U}_{\text{WHT}_N}\} = \begin{cases} 1 & i = j \\ 0 & i \neq j \end{cases}. \qquad (17.256)$$

This implies that $\mathbf{U}_{\text{WHT}_N}$ is an orthogonal matrix, namely that we have [90]:

$$\mathbf{U}_{\text{WHT}_N}^T \mathbf{U}_{\text{WHT}_N} = \mathbf{I}, \qquad (17.257)$$

which therefore allows us to conclude that $\mathbf{U}_{\text{WHT}_N}$ is also unitary, satisfying [90]:

$$\mathbf{U}_{\text{WHT}_N}^H \mathbf{U}_{\text{WHT}_N} = \mathbf{I}. \qquad (17.258)$$

The unitary property allows us to separate the signal processing at the receiver into the operations of subcarrier-based channel transfer factor equalisation followed by despreading, as it will be demonstrated in the next section.

17.4.2.1.2 Receiver Design The vector $\mathbf{x}[n] \in \mathbb{C}^{K \times 1}$ of complex signals observed in the K different subcarriers at the output of the receiver's FFT-based demodulation is given for the WHT-OFDM system portrayed in Figure 17.30 by:

$$\mathbf{x}[n] = \mathbf{H}[n]\mathbf{U}_{\text{WHT}}\mathbf{s}[n] + \mathbf{n}[n], \qquad (17.259)$$

where $\mathbf{H}[n] \in \mathbb{C}^{K \times K}$ is the diagonal matrix of subcarrier channel transfer factors, namely:

$$\mathbf{H}[n] = diag\left(H[0], H[1], \ldots, H[K-1]\right), \qquad (17.260)$$

and $\mathbf{U}_{\text{WHT}} \in \mathbb{C}^{K \times K}$ is the unitary WHT matrix[28] of K-th order, which was defined by Equation 17.255. Furthermore, in Equation 17.259 $\mathbf{s}[n] \in \mathbb{C}^{K \times 1}$ denotes the vector of transmitted subcarrier symbols, namely:

$$\mathbf{s}[n] = \left(s[0], s[1], \ldots, s[K-1]\right)^T, \qquad (17.261)$$

and $\mathbf{n}[n] \in \mathbb{C}^{K \times 1}$ is the vector of subcarrier-related AWGN samples, namely:

$$\mathbf{n}[n] = \left(n[0], n[1], \ldots, n[K-1]\right)^T. \qquad (17.262)$$

Note that in the context of the above definitions we have omitted the OFDM symbol index $[n]$ for reasons of notational simplicity.

Equation 17.259 can be transferred into standard form, namely to:

$$\mathbf{x}[n] = \mathbf{H}_{\text{WHT}}[n]\mathbf{s}[n] + \mathbf{n}[n], \qquad (17.263)$$

by considering the product of the diagonal channel matrix $\mathbf{H}[n]$ and the WHT matrix $\mathbf{U}_{\text{WHT}}$

---

[28]Note that here we have omitted the lower case index, which indicates the order of the WHT.

as the effective channel matrix $\mathbf{H}_{\text{WHT}}[n] \in \mathbb{C}^{K \times K}$, namely by introducing:

$$\mathbf{H}_{\text{WHT}}[n] = \mathbf{H}[n]\mathbf{U}_{\text{WHT}}. \tag{17.264}$$

Note that Equation 17.263 exhibits the same structure as Equation 17.1, describing the SDMA-MIMO channel scenario on a subcarrier basis. Hence the same techniques can be invoked for recovering the vector $s[n]$ of symbols transmitted over the K different OFDM subcarriers. These detection techniques were investigated in Sections 17.2 and 17.3 to recover the symbols transmitted by the L different users on a subcarrier basis in the context of SDMA-OFDM.

Here we will focus our attention on the case of linear equalisation, namely where an estimate $\hat{s}[n] \in \mathbb{C}^{K \times 1}$ of the vector of transmitted subcarrier symbols $s[n]$ is obtained by linearly combining the complex signals received in the different subcarriers, which are represented by the vector $\mathbf{x}[n]$. The combining can be achieved with the aid of the weight matrix $\mathbf{W}[n] \in \mathbb{C}^{K \times K}$, as shown in Equation 17.7, namely:

$$\hat{s}[n] = \mathbf{W}^H[n]\mathbf{x}[n]. \tag{17.265}$$

In the context of the MMSE criterion we obtain – as demonstrated earlier in Equation 17.63 – the weight matrix $\mathbf{W}_{\text{MMSE}}[n] \in \mathbb{C}^{K \times K}$, which is given in its right-inverse related form as follows:

$$\mathbf{W}_{\text{MMSE}}[n] = (\mathbf{H}_{\text{WHT}}[n]\mathbf{P}_{\text{MC}}\mathbf{H}_{\text{WHT}}^H[n] + \sigma_n^2\mathbf{I})^{-1}\mathbf{H}_{\text{WHT}}[n]\mathbf{P}_{\text{MC}}. \tag{17.266}$$

In Equation 17.266 the diagonal matrix $\mathbf{P}_{\text{MC}} \in \mathbb{R}^{K \times K}$ of transmit powers associated with the different subcarriers is given for an equal power allocation by:

$$\mathbf{P}_{\text{MC}} = \sigma_s^2\mathbf{I}, \tag{17.267}$$

where σ_s^2 denotes the signal variance, and σ_n^2 is the AWGN variance. Upon substituting Equations 17.264 and 17.267 into Equation 17.266 we obtain the following equation for the weight matrix $\mathbf{W}_{\text{MMSE}}[n]$:

$$\mathbf{W}_{\text{MMSE}}[n] = \mathbf{E}_{\text{MMSE}}[n]\mathbf{U}_{\text{WHT}}, \tag{17.268}$$

where the channel-related equaliser matrix $\mathbf{E}_{\text{MMSE}}[n] \in \mathbb{C}^{K \times K}$ is given by:

$$\mathbf{E}_{\text{MMSE}}[n] = (\mathbf{H}[n]\mathbf{H}^H[n] + \frac{\sigma_n^2}{\sigma_s^2}\mathbf{I})^{-1}\mathbf{H}[n]. \tag{17.269}$$

This matrix describes the operation of the equaliser seen in Figure 17.30. Here we have capitalised on the unitary nature of the matrix $\mathbf{U}_{\text{WHT}}$, as reflected by Equation 17.258. Note that $\mathbf{E}_{\text{MMSE}}[n]$ given by Equation 17.269 is a diagonal matrix, where the k-th diagonal element is given by:

$$\mathbf{E}_{\text{MMSE}}[n]|_{(k,k)} = \frac{H[n,k]}{|H[n,k]|^2 + \frac{\sigma_n^2}{\sigma_s^2}}, \tag{17.270}$$

and where $H[n, k]$ is the k-th subcarrier's channel transfer factor. Upon substituting Equation 17.268 into Equation 17.265 the MMSE combining related vector $\hat{s}_{\text{MMSE}}[n] \in \mathbb{C}^{K \times 1}$ of the transmitted subcarrier symbols' estimates is given by:

$$\hat{s}_{\text{MMSE}}[n] = \mathbf{U}_{\text{WHT}}^{H} \mathbf{E}_{\text{MMSE}}^{H}[n] \mathbf{x}[n]. \tag{17.271}$$

Note in Equation 17.271 that the receiver's operation is separated into two steps. The first step is the subcarrier-based one-tap equalisation, which is carried out by multiplying the FFT-based OFDM demodulator's output vector $\mathbf{x}[n]$ in Figure 17.30 by the Hermitian transpose of the diagonal matrix $\mathbf{E}_{\text{MMSE}}[n]$ of Equation 17.269. The second step is the IWHT based despreading carried out by multiplying with the unitary matrix $\mathbf{U}_{\text{WHT}}^{H} = \mathbf{U}_{\text{WHT}}$, which was also shown in Figure 17.30.

Following the philosophy of Equation 17.73 it can be demonstrated furthermore that the signal estimation MSE averaged over the different subcarrier-related components of $\hat{s}_{\text{MMSE}}[n]$ is given by:

$$\overline{\text{MMSE}}_{\text{MMSE}}[n] = \sigma_s^2 \left(1 - \frac{1}{K} \sum_{k=0}^{K-1} H_{\text{MMSE}}[n, k] \right), \tag{17.272}$$

where the k-th subcarrier's effective "channel" transfer factor $H_{\text{MMSE}}[n, k]$, includes both the effects of the channel and that of the one-tap equalisation at the receiver, namely:

$$H_{\text{MMSE}}[n, k] = \frac{|H[n, k]|^2}{|H[n, k]|^2 + \frac{\sigma_n^2}{\sigma_s^2}}. \tag{17.273}$$

In the context of deriving Equation 17.272 we have exploited only that $\text{Trace}(\mathbf{U}^H \mathbf{A} \mathbf{U}) = \text{Trace}(\mathbf{A})$ for a unitary matrix $\mathbf{U}$ and for an arbitrary matrix $\mathbf{A}$ [478]. Note, however, that in the specific case of employing the WHT as the unitary transform, an estimation MSE identical to that averaged over an OFDM symbol, namely that quantified by Equation 17.272 is also observed for each individual subcarrier. This could be shown by following the philosophy of Equation 17.74. Furthermore, it can be demonstrated that after IWHT assisted despreading, as seen in Figure 17.30 the subcarrier based SINR is identical for all the different subcarriers, which is given by:

$$\text{SINR}_{\text{WHT,MMSE}}[n] = \frac{\sigma_{S,\text{WHT,MMSE}}^2[n]}{\sigma_{I,\text{WHT,MMSE}}^2[n] + \sigma_{N,\text{WHT,MMSE}}^2[n]}, \tag{17.274}$$

where we have:

$$\sigma^2_{S,\text{WHT,MMSE}}[n] = \frac{1}{K^2}\left(\sum_{k=1}^{K} H_{\text{MMSE}}[n,k]\right)^2 \sigma^2_s \tag{17.275}$$

$$\sigma^2_{I,\text{WHT,MMSE}}[n] = \frac{1}{K^2}\left(\sum_{k=0}^{K-1}\left[(K-1)H_{\text{MMSE}}[n,k] - \sum_{\substack{k'=0 \\ k'\neq k}}^{K-1} H_{\text{MMSE}}[n,k']\right]H_{\text{MMSE}}[n,k]\right)\sigma^2_s \tag{17.276}$$

$$\sigma^2_{N,\text{WHT,MMSE}}[n] = \frac{1}{K}\left(\sum_{k=1}^{K}\frac{H_{\text{MMSE}}[n,k]}{\left(|H[n,k]|^2 + \frac{\sigma^2_n}{\sigma^2_s}\right)^2}\right)\sigma^2_n. \tag{17.277}$$

In the next section we will embark on describing the multi-user SDMA-WHTS-OFDM receiver.

17.4.2.2 Outline of the WHTS Assisted Multi-User SDMA-OFDM Receiver

In the previous section we have demonstrated in the context of a single-user WHTS-OFDM receiver that the operations of linear frequency domain channel transfer factor equalisation and despreading can be sequentially performed. Similar derivations can also be conducted in the context of the multi-user SDMA-WHTS-OFDM scenario, resulting in the receiver design shown in Figure 17.31. Following the design concepts of the multi-user SDMA-AOFDM receiver, shown in Figure 17.28, we have included a decision-directed channel transfer function predictor in Figure 17.31 to provide the channel estimates required by the linear combiner.

Again, as seen in Figure 17.31, blocks of K consecutive samples of the signals received by the P number of BS antenna elements are independently subjected to a K-point FFT, which yields the signal samples' frequency domain representation, namely $x_p[n,k]$, $p = 1,\ldots,P$, $k = 0,\ldots,K-1$. Following this step linear combining is performed on a per subcarrier basis with the aid of Equation 17.7 in order obtain estimates $\hat{s}^{(l)}[n,k]$ of the signals $s^{(l)}[n,k]$ transmitted by the L different users, where $l = 1,\ldots,L$. Recall that the combiner matrix associated with the MMSE criterion was given in its right-inverse related form by Equation 17.63. Following the design concepts of the single-user WHTS-OFDM receiver shown in Figure 17.30, the linear signal estimates associated with the L different users are independently subjected to despreading by means of a K-point IWHT, which results in the despread signal estimates $\hat{s}^{(l)}_{\text{IWHT}}[n,k]$ of Figure 17.31. These are classified separately for each subcarrier and each user with the aid of Equation 17.94 in order to obtain the complex symbols $\check{s}^{(l)}[n,k]$ that are most likely to have been transmitted. In the context of the receiver design proposed here and depicted in Figure 17.31, *a priori* estimates of the channel transfer factors $H_p^{(l)}[n,k]$ employed in the calculation of the different subcarriers' weight matrices according to Equation 17.63 are again generated upon feeding back the current OFDM symbol's subcarrier symbol decisions, which are subjected to WHT-based spreading in order to regenerate the complex symbols transmitted by the L different users in each subcarrier.

Since the employment of WHTs having a high transform length, such as, for example, 512, as required in case of the indoor WATM system model employed in our investigations

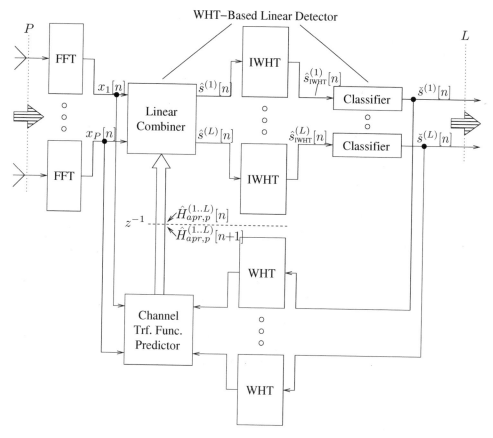

Figure 17.31: Block diagram of the linear combining assisted multi-user SDMA-WHTS-OFDM receiver, which is supported by a P-element antenna front-end in order to facilitate the separation of the L number of simultaneous users' associated signals at the receiver. Decision-directed channel transfer function prediction is performed in order to facilitate the separation of the different users' transmitted signals with the aid of the linear combiner. The subcarrier index k, where $k = 0, \ldots, K - 1$ has been omitted for reasons of notational simplicity.

in Section 17.4.2.3 would impose a high computational complexity, we divide the OFDM symbol into several WHTs as seen in Figure 17.32. This is also justified by the observation that most of the channel's frequency domain diversity potential can be exploited with the aid of a relatively short spreading length, as illustrated in Figure 17.33. Furthermore, depending on the particular power delay profile of the channel, the OFDM symbol bandwidth of K subcarriers can be divided into K/M_{WHT} interleaved blocks of size M_{WHT} each, which are separately subjected to the WHT. More specifically, the i-th WHT block of an OFDM symbol

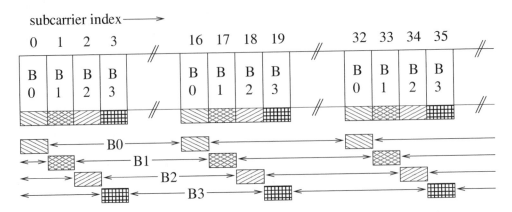

Figure 17.32: Illustration of WHT-based subcarrier spreading using a limited WHT blocksize M_{WHT}, which is typically a fraction of the total number of subcarriers K hosted by the OFDM symbol for the sake of low implementational complexity.

contains subcarriers having indices j given by:

$$j = i + r\frac{K}{M_{\text{WHT}}}, \qquad 0 \le r \le M_{\text{WHT}} - 1, \tag{17.278}$$

where according to our definition both the first WHT block and the first OFDM subcarrier are represented by an index of zero. In Figure 17.32 we have further illustrated the operation of WHT-based spreading applied to blocks of an identical size, where each block hosts only a fraction of the total number of subcarriers K associated with the OFDM symbol. More specifically, in this particular example the OFDM symbol is composed of 16 interleaved WHT blocks and the specific subcarriers, which are 16 frequency positions apart from each other belong to the same WHT block.

17.4.2.3 Performance Assessment

Our performance investigations of WHTS-OFDM are conducted separately for single- and multi-user OFDM scenarios. Specifically in Section 17.4.2.3.1 we will demonstrate the influence of the spreading code length on the WHTS-OFDM system's performance. Furthermore, a comparison between WHTS-OFDM and AOFDM is also carried out. Our investigations of multi-user SDMA-WHTS-OFDM cast in the context of MMSE- and PIC detection will then be conducted in Section 17.4.2.3.2.

17.4.2.3.1　Single-User WHTS-OFDM

Simulation results have been obtained for the indoor WATM system- and channel model of Section 14.3.1. We commenced our investigations by assessing the impact of Walsh-Hadamard spreading using different spreading code lengths on the BER performance of a 4QAM single reception antenna, single user OFDM system in the indoor WATM channel environment. MMSE-based frequency domain channel equalisation, as described in Section 17.4.2.1.2, was performed at the receiver. The corresponding

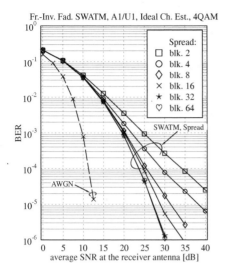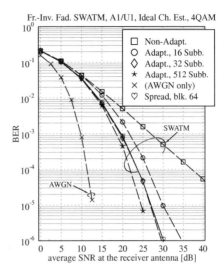

Figure 17.33: **(a: Left)** BER performance of a WHT spreading-assisted 4QAM modulated single reception antenna, single user OFDM system parametrised with the spreading block size. **(b: Right)** BER performance comparison between a 4QAM modulated single reception antenna, single user standard OFDM system, an OFDM system employing four-mode ("no transmission", BPSK, 4QAM, 16QAM), constant-throughput, zero-delay signalling based adaptive modulation using 16 or 512 sub-bands, and a Walsh-Hadamard spreading assisted OFDM system. The simulations were conducted in the context of the "frame-invariant" fading indoor WATM system- and channel model of Section 14.3.1. Ideal channel transfer function knowledge was assumed. The BER performance of 4QAM signalling over an AWGN channel has been plotted as a reference.

results are portrayed on the left-hand side of Figure 17.33. We observe that as a consequence of the residual '"multiple-access" interference imposed by the spread signals of the different subcarriers hosted by each WHT block, the BER performance is not particularly sensitive to the WHT block length, provided that it is in excess of 32 subcarriers for the SNRs of our interest. It should be noted that the benefit of spreading is directly related to the frequency domain diversity potential offered by a specific dispersive channel. More specifically, the higher the channel's delay spread, the less separated are the frequency domain fades, hence tolerating a higher WHT length, while achieving as high a randomisation of the frequency domain fading effects as possible.

In our next investigation we compared the BER performance of the WHT-assisted OFDM system to a non-spread OFDM system and to an OFDM system employing adaptive modulation [2], under the constraint of having a target throughput equivalent to that of the fixed mode 4QAM modulated OFDM system. The modulation mode adaptation regime employed a total of four modes, namely "no transmission", BPSK, 4QAM and 16QAM transmission. In order to reduce the signalling overhead required, the modulation modes were assigned on a sub-band basis, where each sub-band hosted either one or a number of subcarriers. Specifically, using one subcarrier per sub-band allowed us to determine the upperbound performance of the system. Furthermore, the best-case scenario of perfect channel transfer function knowledge was invoked in the process of determining the optimum modulation mode assignment. The corresponding simulation results are illustrated on the right-hand side of Figure 17.33. We observe that in the specific indoor WATM channel environment, assuming the separation of the total bandwidth into 32 equal-sized sub-bands, each hosting 16 subcarriers, the OFDM system employing adaptive modulation exhibits a similar performance to that of the spread OFDM system for a WHT blocksize of 64 subcarriers. By contrast, a hypothetic system assigning the best-matching individual modulation mode to each subcarrier outperformed the WHT OFDM scheme by about 2dB in terms of the required SNR. Hence, taking also into account the signalling overhead required by the adaptive modulation scheme, as well as its limited applicability restricted to relatively slowly varying channels in the absence of channel transfer function prediction techniques, we conclude that subcarrier spreading is a more convenient approach to exploiting the wideband channel's diversity potential.

17.4.2.3.2 Multi-User SDMA-WHTS-OFDM   Our further aim was to investigate the applicability of WHT-based spreading in the context of an SDMA-OFDM system, where the signals of L simultaneous users, each equipped with one transmission antenna, are separated at the BS with the aid of a P-element antenna array. The design of the corresponding receiver was outlined in Section 17.4.2.2. In our investigations we invoked the MMSE and PIC-based multi-user detection approaches of Sections 17.2.3 and 17.3.2, while SIC described in Section 17.3.1 was not directly applicable to a spread OFDM system. This is because in a specific subcarrier or sub-band in each iteration the highest-power user is detected first, followed by the subtraction of its sliced and remodulated signal from the residual composite multi-user signal received by each antenna. Since the WHT-based spreading is performed across subcarriers spaced apart from each other as far as possible for the sake of maximising the achievable frequency domain diversity effect, these subcarriers would potentially require a different SIC detection order. Hence not all the symbols of a specific user contained in a WHT block are available at the same time for demodulation. Simulation results have been obtained for a two reception antenna, two user SDMA scenario. The results are portrayed on the left-hand side of Figure 17.34. We observe that both the MMSE and the PIC detector using WHT based spreading outperform the M-SIC detector in the non-spread case, which tracked $M = 1$ or $M = 2$ tentative symbol decisions from each detection node as shown in Figure 17.11. Note that the least complex multi-user receiver, namely the MMSE detector performed about 2.5dB worse on average, than PIC, which was also observed in the non-spread scenario. Again, by contrast, in case of a four reception antenna, four user SDMA scenario characterised on the right-hand side of Figure 17.34, the $(M = 2)$-SIC detector applied in the non-spread scenario outperforms both spreading assisted arrangements, namely those employing MMSE and PIC based detection. However, compared to the standard $(M = 1)$-SIC detector, at sufficiently high SNRs both the MMSE and PIC detection assisted SDMA-OFDM systems exhibit a better BER performance. This is achieved at a significantly lower complexity than that of the SIC scheme, upon assuming that a "fast" implementation of the WHT is employed for performing the spreading.

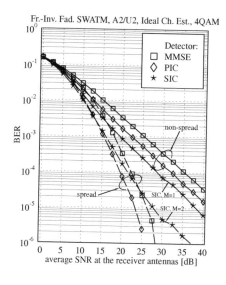
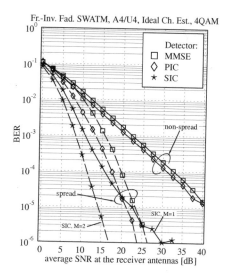

Figure 17.34: BER performance comparison between 4QAM modulated MMSE, PIC or SIC detection-assisted SDMA-OFDM systems as a function of the SNR at the reception antennas, **(a: Left)** for a $L = P = 2$ number of simultaneous users and reception antennas, while **(b: Right)** for a $L = P = 4$ number of simultaneous users and reception antennas. The systems employing MMSE and PIC multi-user detection were further assisted by Walsh-Hadamard spreading using a block size of 16 subcarriers. Ideal channel transfer function knowledge was assumed. The simulations were conducted in the context of the frame-invariant fading indoor WATM system- and channel model described in Section 14.3.1.

17.4.2.4 Summary and Conclusions

In Section 17.4.2 WHT spreading-assisted OFDM was discussed in the context of both single- and multi-user OFDM scenarios. Our discussions commenced in Section 17.4.2.1 with an outline of the WHTS-OFDM receiver's structure in the context of the single-user scenario. Specifically, in Section 17.4.2.1.1 the WHT's properties, namely its recursive structure, as reflected by Equation 17.255, which allows for a "fast" implementation, similar to the FFT, were described. Of further interest was the WHT matrix' unitary nature, as described by Equation 17.258, which follows from the orthogonality of the different Walsh code vectors. Next we highlighted in Section 17.4.2.1.2 that the unitary nature of the WHT matrix facilitates a particularly simple receiver design in the context of linear channel transfer function equalisation. The block diagram of the single-user WHTS-OFDM receiver was shown at the bottom of Figure 17.30. Specifically, the IWHT-based despreading was shown to be decoupled from the channel transfer factor equalisation. This was also demonstrated in the context of the formal derivation of a linear estimate $\hat{s}_{MC,MMSE}$ of the vector s_{MC} of signals transmitted in the K different subcarriers, which was formulated in Equation 17.271. Furthermore, we found that the signal estimation MSE at the receiver's output is identical for the different sub-carriers, which was given by Equation 17.272. We also found that the SINR at the receiver's output, which is given by Equation 17.274, is identical for the different subcarriers.

In Section 17.4.2.2 we then embarked on the description of the multi-user SDMA-WHTS-OFDM receiver's structure shown in Figure 17.31, which followed the same design concepts as the single-user WHTS-OFDM receiver. Specifically the multi-user SDMA scheme employed a sequential equalisation or combining scheme, followed by WHT despreading, implemented separately for the different users. In order to reduce the computational complexity, we found that it was attractive to perform the spreading separately for K/M_{WHT} number of "interleaved" blocks of size M_{WHT} subcarriers each, as it was shown in Figure 17.32, instead of implementing it for a single larger-size block of K subcarriers, without incurring a noticeable increase of the BER.

The evaluation of the single-user WHTS-OFDM and multi-user SDMA-WHTS-OFDM systems' BER performance was then carried out in Section 17.4.2.3. Specifically, in Figure 17.33 we found that already for relatively small spreading block lengths, namely 64 sub-carriers per block employed in the context of the indoor WATM channel of Figure 14.3.1, the wideband channel's diversity can be exploited, while a further increase of the spreading length did not yield a significant BER improvement, as shown on the left-hand side of Figure 17.33. This was because by increasing the spreading block length, the channel transfer factors associated with neighbouring subcarriers contained in a block became more similar. Furthermore, our BER comparisons portrayed on the right-hand side of Figure 17.33 revealed that WHTS-OFDM is also more attractive than AOFDM. More explicitly, AOFDM is capable of slightly outperforming WHTS-OFDM, but only for an unrealistically high number of sub-bands, namely in excess of 64, when neglecting the transmission of side-information and by assuming perfect channel transfer function knowledge in the modulation mode assignment, the AOFDM is capable of slightly outperforming WHTS-AOFDM.

Our BER performance assessment of multi-user WHTS-OFDM was then conducted in the context of Figure 17.34 for scenarios of two BS reception antennas and two simultaneous users as well as for four reception antennas and four simultaneous users, as shown on the left-hand and right-hand sides of Figure 17.34, respectively. Various detection techniques were compared against each other, namely MMSE, PIC and SIC detection, noting that in the context of the latter WHTS was not directly applicable. We found that in the lower-dimensional SDMA-OFDM scenario of two users both PIC as well as MMSE detection-assisted SDMA-WHTS-OFDM are capable of outperforming $(M = 1)$- and $(M = 2)$-aided SIC detection-assisted SDMA-OFDM, while in the higher-dimensional SDMA-OFDM scenario $(M = 2)$-SIC detection-assisted SDMA-OFDM performed best, while at the same time exhibiting by far the highest complexity. By contrast, at sufficiently high SNRs $(M = 1)$-SIC detection-assisted SDMA-OFDM was outperformed by both PIC as well as MMSE detection-assisted SDMA-WHTS-OFDM, while at the same time exhibiting a potentially far lower computational complexity, than the SIC detection-assisted systems.

17.5 Chapter Summary and Conclusion

In this section our summary and conclusions will be provided for this chapter, where we discussed a range of uplink detection techniques designed for multi-user SDMA-OFDM. In

Section 17.5.1 the motivation of employing multiple reception antenna assisted SDMA receivers is reviewed, which is followed in Section 17.5.2 by a summary of the family of linear detection schemes. Our summary and conclusions related to the set of non-linear detection schemes will be offered in Section 17.5.3. Finally, our overall comparison between the different detection schemes investigated will then be offered in Section 17.5.3.4. This is followed in Section 17.5.3.5 by our conclusions on the suite of performance enhancement techniques studied.

17.5.1 Review of the Motivation for Multiple Reception Antenna SDMA Receivers

During the past few decades a variety of Multiple Access (MA) techniques, such as Time Division Multiple Access (TDMA), Frequency Division Multiple Access (FDMA), Code Division Multiple Access (CDMA) have found favour in the various wireless communications systems. More recently Space Division Multiple Access (SDMA) has been proposed to support multiple users in OFDM-based communications system. In the context of the more conventional techniques, namely TDMA, FDMA and CDMA, both the mobiles as well as the base station are typically equipped with a single transmit and receive antenna, respectively. The access of the different users to the finite capacity transmission channel is then facilitated in TDMA by allowing each user to access the channel's entire bandwidth for a finite time duration, namely for the duration of a time slot. By contrast, in FDMA we assign each user a frequency slot. Finally, in CDMA all users share the same frequency band and we facilitate the separation of the different users' transmitted signals with the aid of unique, user-specific spreading codes.

Alternatively, all users could potentially share the same frequency band, as in CDMA, and we could exploit for their separation that the different users' transmitted signals experience different channel transfer functions. The separation of the different users based on their unique channel transfer function constitutes the principle of a minimalistic SDMA scheme. More specifically, in the context of each of the flat-fading OFDM subcarriers, the channel matrix $\mathbf{H}$ associated with the scenario of $L > 1$ different users and $P = 1$ receiver antenna is of unity rank. Hence, only the ML detector is capable of separating the different users' transmitted signals, upon inflicting a potentially large complexity.

In order to render SDMA amenable to employment of lower complexity linear combining-based multi-user detection techniques, such as the LS, MMSE, SIC and PIC detection arrangements discussed in Sections 17.2.2, 17.2.3, 17.3.1 and 17.3.2, respectively, a viable strategy is to increase the P number of receiver antennas, such that we have $P \geq L$. Hence the channel matrix $\mathbf{H}$ is of potentially "full rank" with respect to the number of users L. In other words, we have $L = rank(\mathbf{H})$, which facilitates the linear separation of the different users' transmitted signals based on their spatial signature. More specifically, the spatial signature of a particular user is constituted by the set of channel transfer factors between the user's single transmit antenna and the P number of different receiver antennas, again, upon assuming flat-fading channel conditions for each of the OFDM subcarriers. This Multiple-Input Multiple-Output (MIMO) channel scenario was further detailed in Section 17.1.5, where for simplicity we assumed that the fading experienced by each of the MIMO sub-channels associated with a specific transmitter-receiver antenna pair is independent from that of the other MIMO sub-channels.

Our more specific discussions of multi-user detection techniques applicable to SDMA-OFDM were separated into the subclasses of linear- and non-linear detection techniques of Sections 17.2 and 17.3, respectively. As argued in Sections 17.1.3 and 17.1.4 the rationale of this classification was that in the context of linear detection techniques, such as the LS and MMSE detection schemes discussed in Section 17.2, no *a priori* knowledge of the remaining users' transmitted symbols is required for the detection of a specific user. However, in the case of the SIC, PIC and ML detection techniques discussed in Section 17.3, *a priori* knowledge of the likely values of the symbol is involved, which must be provided by the non-linear classification or decision operation involved in the demodulation process.

17.5.2 Summary and Conclusions Related to Linear Detectors

The typical structure of a linear detector was highlighted in Figure 17.4. More specifically, in a first step linear estimates of the L different users' transmitted signals are obtained by appropriately combining the signals received by the P different antenna elements. In a next step these signal estimates are classified or demodulated in order to determine the complex symbol or constellation point that is most likely to have been transmitted by each user. As a result of the user signal separation facilitated with the aid of the linear combiner, the process of classification is substantially simplified compared to that of the multi-user ML detector discussed in Section 17.3.3. Instead of evaluating the multi-user Euclidean distance metric associated with the multi-user ML detector M_c^L number of times, in case of the linear detector the single-user Euclidean distance metric has to be evaluated LM_c number of times, which constitutes a complexity reduction. However, this complexity reduction is achieved at the cost of a significant performance degradation in the context of the linear detector compared to that of the optimum ML detector, since the effects of the residual interference contaminating the linear combiner's output signals are neglected by the former.

The linear combiner's associated weight matrix can be adjusted according to a number of different criteria. Explicitly, the Least-Squares (LS) error- and the Minimum Mean-Square Error (MMSE) criteria were investigated in Sections 17.2.2 and 17.2.3, respectively. More specifically, as detailed in Section 17.2.2, the LS detector's associated LS combiner generates linear estimates of the signals transmitted by the different users based solely on the knowledge of the channel's frequency domain transfer factors. In contrast to the LS combiner, the MMSE detector's associated MMSE combiner, which belongs to the class of Wiener-filter related combiners, as argued in Section 17.2.3.4, additionally capitalises on statistical knowledge of the AWGN process, which contaminates the signals received by the P number of different antenna elements. As the terminology suggests, the MMSE combiner achieves the minimum signal estimation MSE. However, the transmitted signals to be estimated cannot be recovered with a unity gain, which is in contrast to the LS combiner. As argued in Section 17.2.3.5, this non-unity gain may be compensated by normalising the MMSE combiner's output signals, at the cost of increasing the estimation MSE. The resultant normalised combiner weight vectors were identical to those of the MV combiner.

Our analysis of the LS- and MMSE detector's MSE and BER performance characterised in Section 17.2.6 underlined the MMSE detector's advantage compared to the LS detector in terms of achieving a lower signal estimation MSE and BER, as shown in Figure 17.6. Furthermore, we found that upon decreasing the number of simultaneous SDMA users L, while keeping the number of reception antennas P constant, the MMSE detector's performance quantified in terms of the SINR and BER was significantly improved, as shown in Figures 17.7 and 17.8. A further BER reduction was achieved with the aid of turbo-coding, as shown in Figure 17.9, although as usual, this was achieved at the cost of reducing each SDMA user's effective throughput, while also imposing further additional computational complexity.

Our detailed analysis of the different linear detectors' computational complexity provided in Section 17.2.7 revealed that the MMSE detector is slightly more complex than the LS detector. However, the general trend is that the complexity is proportional to the cube of the number of users L, as in case of LS detection or MMSE detection implemented in its left-inverse related form. A similar cubically proportional complexity dependence is valid also with respect to the number of reception antennas P, as in case of the MMSE detection implemented in its right-inverse related form.

For a more detailed summary and conclusions related to the family of linear detection techniques we refer the reader to Section 17.2.8.

17.5.3   Summary and Conclusions Related to Non-Linear Detectors

Our summary and conclusions on non-linear detectors are separated into Sections 17.5.3.1, 17.5.3.2 and 17.5.3.3, considering SIC, PIC and ML detection, respectively.

17.5.3.1   SIC Detection

The employment of SIC detection was motivated earlier in the context of our performance analysis of MMSE detection in Section 17.2.6.3 by the specific observation that upon decreasing the number of simultaneous users L, while keeping the number of reception antennas P constant, the MMSE detector's BER performance was improved. This was because for a lower number of SDMA users the associated MMSE combiner was less constrained with respect to the specific choice of the combiner weights optimised to suppress the interfering users' signal contributions. This allowed for a more effective noise mitigation. The same principle can be invoked in the context of an iterative detector, namely the standard SIC detector of Figure 17.10, where in each iteration or cancellation stage only the most dominant user having, for example, the highest SNR, SIR or SINR at the linear combiner's output was detected. The detected user's influence is eliminated from the partially decontaminated vector of signals received by the different antenna elements, upon invoking the detected user's remodulated signal. This principle was further detailed in Section 17.3.1.1.

In the context of the BER and SER performance results of Section 17.3.1.4 we found that the standard SIC detector suffers from the effects of error propagation across the different detection stages. In fact, it was observed that if a symbol error occurred in one of the lower-index detection stages, then there was a relatively high probability that symbol errors would occur in the higher-index detection stages. By contrast, if correct symbol decisions were made in the lower-index detection stages, then the probability that an error would occurr in one of the higher-index detection stages was lower than for the lower-index detection stages. This was a consequence of the MIMO channel's increased diversity order in the context of the higher-index detection stages, following the removal of the signal contributions of those users, which had already been detected. In order to mitigate the effects of error propagation across the different detection stages, the standard SIC detector was appropriately modified.

More specifically, in the context of the M-SIC scheme discussed in Section 17.3.1.2, an

$M > 1$ number of tentative symbol decisions are tracked from each detection node. Consequently, after the detection of the last user a decision has to be made as to which of the $M^{(i-1)}$ number of vectors of the different users' tentative symbols is most likely to have been transmitted. The significant performance advantage offered by M-SIC compared to standard SIC is achieved at the cost of an increased computational complexity, which is related to the additional number of "decontamination" and demodulation operations associated with the increased number of detection nodes involved.

Based on the observation that symbol error propagation events are predominantly triggered by the lower-index detection stages, as evidenced by Figure 17.16, a viable strategy of reducing the complexity of M-SIC is to restrict the employment of $M > 1$ number of tentative symbol decision per detection node to the lower-index detection stages, while at the higher-index detection stages employing $M = 1$. This strategy was discussed in Section 17.3.1.2.2 and it was termed partial M-SIC- or pM-SIC.

A further complexity reduction was achieved in Section 17.3.1.2.3 by restricting the employment of M-SIC- or partial M-SIC to those specific OFDM subcarriers, which exhibited a relatively low SINR during the first detection stage, while using standard SIC in conjunction with $M = 1$ in the subcarriers exhibiting a higher SINR. This strategy, which was initially proposed in [117, 137] was termed as Selective-Decision-Insertion M-SIC (SDI-M-SIC).

Our further discussions presented in Section 17.3.1.3 addressed the task of soft-bit generation in the context of standard SIC. While our rudimentary approach in Section 17.3.1.3.1 followed the philosophy of soft-bit generation contrived for the MMSE detection technique, as discussed in Section 17.2.5, the improved "weighted" soft-bit metric of Section 17.3.1.3.2 additionally accounted for the effects of error-propagation across the different SIC stages. More specifically, in case of relatively unreliable symbol decisions generated during the previous SIC detection stages a viable strategy is to de-emphasise the soft-bits generated for the current detection stage by appropriately decreasing their value and thus indicating a low associated confidence.

The assessment of the BER and SER performance exhibited by the standard SIC, M-SIC, pM-SIC and SDI-M-SIC schemes was conducted in Section 17.3.1.4. Specifically, we found in Figure 17.12 that the BER and SER performance of SIC- and M-SIC detection was significantly improved upon increasing the $L = P$ number of users and reception antennas of the "fully loaded" SDMA-OFDM system. This is, because for a higher number of users, the SIC detector benefits from selecting the most dominant user from a larger "pool" of different users at a specific detection stage, with the desirable effect of reducing the probability of incurring a low-SINR user as the most dominant user. Furthermore, in Figure 17.12 we found that upon employing M-SIC instead of standard SIC, a further substantial reduction of the BER or SER can be achieved. Using the SINR instead of the SNR recorded at the linear combiner's output in each detection stage for identifying the most dominant user yielded a noticeable BER or SER reduction, although only for SDMA scenarios, where the number of users and reception antennas was in excess of four. The effects of error propagation were detailed in Sections 17.3.1.4.2, 17.3.1.4.3 and 17.3.1.4.4. Furthermore, the SER performance of both pM-SIC and SDI-M-SIC was evaluated in Sections 17.3.1.4.5 and 17.3.1.4.6 with associated Figure 17.16. These schemes were employed to reduce the potentially substantial computational complexity associated with M-SIC. Specifically, in the context of an SDMA scenario supporting $L = 4$ simultaneous users with the aid of $P = 4$ reception antennas we found that employing $M = 2$ tentative symbol decisions per detection node during the first two detection stages ($L_{\text{pM-SIC}} = 2$) of pM-SIC, which was reduced to $M = 1$ symbol decision per detection node during the higher-index detection stages, results in an SNR degradation of approximately 1dB at an SER of 10^{-5}. This 1dB SNR degradation was the price of halving the computational complexity quantified in terms of the number of comparisons to be conducted, as it was shown with the aid of Figure 17.19. Our performance assessment of SDI-pM-SIC was the topic of Section 17.16, which demonstrated that in the context of the specific indoor WATM channel model of Section 14.3.1, SDI-pM-SIC yields the same SER performance as pM-SIC ($L_{\text{pM-SIC}} = 1$),[29] provided that, in the context of SDI-pM-SIC, pM-SIC is employed in the $N_{\text{SDI}}^{[1]} = 64$ number of lowest-SINR subcarriers recorded during the first detection stage, while using standard SIC in the remaining subcarriers. The performance assessment of the various SIC schemes was concluded with the evaluation of the BER performance exhibited by turbo-coded standard SIC detection-assisted SDMA-OFDM in Figures 17.17 and 17.18, which conveniently highlighted the benefits of the weighted soft-bit metric in comparison to the standard soft-bit metric, both of which were outlined in Section 17.3.1.3.

Finally, an analysis of the computational complexity exhibited by the various SIC schemes was carried out in Section 17.3.1.5. Specifically, in Figure 17.19 it was demonstrated that amongst the various successive interference cancellation based detectors the standard SIC detector is the least complex one, while M-SIC exhibits the highest complexity. A compromise between performance and complexity is constituted by the partial M-SIC scheme. For a more detailed summary and conclusions we refer to Section 17.3.1.6.

17.5.3.2 PIC Detection

The employment of the PIC detection scheme discussed in Section 17.3.2 was partially motivated by the SIC detector's potentially high complexity, which is related to the requirement of identifying the most dominant user – as well as recalculating the selected user's weight vector – in each detection stage, as outlined in Section 17.3.1. Furthermore, we found in the context of SIC detection that the highest AWGN mitigation was achieved by the linear combiner employed in each of the SIC detection stages, during its last detection stage, following the successful removal of all interfering co-channel users' contributions.

As a consequence, the PIC scheme portrayed in Figure 17.20 was investigated. In the context of this arrangement tentative estimates of the different users' transmitted signals were generated with the aid of a linear combiner, which was the MMSE combiner of Section 17.2.3 in our specific case. These signal estimates were then demodulated in order to obtain tentative symbol decisions, which were remodulated and subtracted from the vector of signals received by the different antenna elements, upon taking into account the effects of the channel. As a result, a potentially interference-free vector of received signals was obtained for each user, provided that correct symbol decisions were made for the remaining users. Hence, the MMSE combiners, which were employed to obtain improved signal estimates from the decontaminated array output vectors became then capable of more effectively suppressing the AWGN. This principle was further detailed in Section 17.3.2.1.

However, in the context of our performance study provided in Section 17.3.2.3 we found that in the absence of channel coding, the detector performs only slightly better than the MMSE detector. Specifically, in the context of the "fully loaded" SDMA scenario of four re-

---

[29]This implied using $M = 2$ tentative symbol decisions at the first detection stage, which was reduced to employing $M = 1$ symbol decision during the remaining detection stages.

ception antennas supporting four simultaneous SDMA users, as characterised in Figure 17.21, the SNR advantage of PIC detection over MMSE detection when aiming for maintaining a BER of 10^{-4} was as low as 1.25dB, while for a "minimalistic" SDMA scenario of two reception antennas and two simultaneous users an SNR advantage of 3.03dB was observed at the same BER. The relatively modest SNR improvement of 1.25dB was related to the effect that if the symbol decisions obtained during the first detection stage were erroneous even for a single user, then during the PIC process of Figure 17.20 the remaining users' received signals were imperfectly decontaminated. This phenomenon had the effect of potentially incurring erroneous symbol decisions for all the other users as well, during the following demodulation process.

In order to combat these effects it was proposed in Section 17.3.2.2 to combine the PIC detection scheme with turbo-decoding, incorporated into the classification module of Figure 17.20. A remodulated reference signal to be used in the PIC process may be generated based on the original information bit-positions or "source"-related soft-output bits of the turbo-decoder, requiring the slicing, re-encoding and remodulation of these bits. Alternatively, the "source- plus parity"-related soft-output bits may be sliced and remodulated, which exhibited a slight advantage in Figure 17.22 in terms of the system's BER performance. In order to render the associated BER simulation results presented in Figure 17.22 comparable to those of the other detectors we decided to equally split the total affordable number of PIC iterations into those employed during the first- and the second PIC stage. Compared to turbo-coded MMSE detection-assisted SDMA-OFDM a dramatic performance advantage was achieved with the aid of this arrangement, as shown in our final performance comparison of the different detection schemes portrayed in Figure 17.26.

A detailed complexity analysis of the PIC detector was conducted in Section 17.3.2.4 and based on these equations it became obvious in our final comparison of the different detectors' complexities quantified in Figure 17.27 that the PIC detector constitutes an attractive design compromise between the MMSE- and standard SIC detectors. For a more detailed summary and conclusions on PIC detection we refer the reader to Section 17.3.2.5.

17.5.3.3 ML Detection

In Section 17.3.3 the optimum ML detector was described and characterised. As argued in Section 17.3.3.1, ML detection is based on the strategy of maximising the *a posteriori* probability $P(\check{s}|x, H)$ that a hypothetic 'L-user' vector of symbols $\check{s}$ was composed of the individual symbols transmitted by the L different users over a channel characterised by the matrix H defined in Equation 17.212, conditioned on the vector x of signals received by the P different antenna elements. The maximisation of the likelihood metric was carried out over the entire set $\mathcal{M}^L$ of M_c^L number of hypothetic "L-user" symbol vectors constituted by the L different users' M_c-ary constellations. With the aid of Bayes' theorem [90] and upon exploiting that the different symbol combination vectors were transmitted with equal probability, it was furthermore shown that maximising the *a posteriori* probability $P(\check{s}|x, H)$ is equivalent to maximising the *a priori* probability $P(x|\check{s}, H)$, which is the probability that the signal vector x was received by the different antenna elements, conditioned on transmitting the hypothetic "L-user" symbol vector $\check{s}$. It was furthermore shown that the *a priori* probability $P(x|\check{s}, H)$ is given by the multi-variate complex Gaussian distribution function $f(x|\check{s}, H)$, which is defined by its vector of mean values and by its covariance matrix. Hence, it was argued that maximisation of the *a priori* probability $P(\mathbf{x}|\check{\mathbf{s}}, \mathbf{H})$ is equivalent to minimising the argument of the exponential function of Equation 17.216 constituting the multi-variate complex Gaussian distribution function. This involved minimising the Euclidean distance between the vector $\mathbf{x}$ of received signals and the hypothetic "L-user" vector of transmitted signals $\check{\mathbf{s}}$, upon taking into account the effects of the MIMO channel described by the channel matrix $\mathbf{H}$ of Equation 17.212, again for all trial-vectors contained in the set $\mathcal{M}^L$ of M_c^L number of "L-user" symbol vectors.

As part of minimising the multi-user Euclidean distance metric, each of the M_c^L different hypothetic trial-vectors $\check{\mathbf{s}} \in \mathcal{M}^L$ has to be transformed to the received signal's space upon multiplication with the channel matrix $\mathbf{H}$ of Equation 17.212. It was demonstrated in Section 17.3.3.1 that if the number of simultaneous users L is significantly lower than the number of receiver antennas P, the associated complexity can potentially be reduced upon transforming each trial-vector first to the transmitted signal's space with the aid of a linear transform, followed by evaluating a modified Euclidean distance metric. It was shown in Section 17.3.3.2 that a particularly suitable transform is the LS-related transform matrix of Equation 17.225, which delivers a noise-contaminated unity-gain estimate of the "L-user" vector of transmitted signals, simplifying the Euclidean distance metric employed.

Furthermore, the generation of soft-bit values for turbo-decoding at the receiver was alluded to in Section 17.3.3.3 based on the assumption of employing a separate trellis decoding of the different users' signals.

The BER performance of both the uncoded and turbo-coded scenarios was then characterised in Section 17.3.3.4. Specifically, in Figure 17.24 we found for the uncoded scenario that the ML detector's performance is relatively insensitive to the number of users L, given a fixed number of reception antennas P. In contrast to the linear combining-based detectors discussed in this chapter, even when increasing the number of users beyond the number of reception antennas, the performance degradation is graceful. Furthermore, we found that similar to the SIC detector's behaviour recorded when increasing the number of reception antennas, the ML detector's BER performance was significantly improved owing to the higher degree of diversity provided by a MIMO system of a higher order. Again, for the turbo-coded scenario we observed in Figure 17.25 a substantially improved BER performance compared to that of the uncoded scenario. For example, for the "fully loaded" system of four reception antennas supporting four simultaneous users the BER at an SNR of 5dB was as low as 10^{-6}.

Our complexity analysis documented in Section 17.3.3.5 revealed that the ML detector's complexity is proportional to the M_c^L number of symbol combinations constituted by the L different users' M_c-ary trial symbols. For a more detailed summary and conclusions were refer the reader to Section 17.3.3.6.

17.5.3.4 Overall Comparison of the Different Detection Techniques

Our final comparison of the different linear- and non-linear detection techniques in both uncoded- and turbo-coded scenarios was documented in Section 17.3.4. Specifically the achievable BER performance was documented in Figure 17.26, while the associated computational complexity, in Figure 17.27.

The essence of this comparison was that in all investigated scenarios – as expected – the ML detector constituted the best performing, but highest complexity solution, while the MMSE detector was the worst-performing, lowest-complexity solution.

A compromise in terms of performance and complexity was provided by the class of SIC detectors and its derivatives as documented in Section 17.3.1. While in the uncoded scenario of Figure 17.26 the M-SIC (M = 2) scheme performed almost identically to the ML detector, a substantial performance degradation was observed in Figure 17.26 for the lower-complexity standard SIC detector. A trade-off between the performance and the complexity associated with the standard SIC and M-SIC schemes of Sections 17.3.1.1 and 17.3.1.2.1 was achievable with the aid of the partial M-SIC or SDI-M-SIC arrangements of Sections 17.3.1.2.2 and 17.3.1.2.3. The associated performance results were, however, not repeated in Figure 17.26. By contrast, in the turbo-coded scenario, both detectors, namely the standard SIC and M-SIC (M = 2) schemes of Sections 17.3.1.1 and 17.3.1.2.1 performed within a range of 1dB in excess of the SNR required by turbo-coded ML detection-assisted SDMA-OFDM, when maintaining a specific BER.

While PIC detection was unattractive in the uncoded scenario, owing to its relatively modest BER improvement compared to MMSE detection, in the turbo-coded scenario a significant BER improvement was achieved. However, for the range of BERs of our interest, namely below 10^{-4}, the PIC detector's performance was worse than that of turbo-coded standard SIC. This was related to the imperfections of the soft-bit estimates employed and hence there is still some potential for its improvement. Note that this performance improvement was achieved, while exhibiting a lower complexity than that of standard SIC.

17.5.3.5 Summary and Conclusions Related to Performance Enhancement Techniques

In order to render the less complex, but also less powerful detection techniques, such as MMSE and PIC also more attractive for employment in uncoded scenarios, the performance enhancement techniques of Section 17.4.1 may be invoked, which are well known from the field of single reception antenna-based communications systems. Specifically, constant throughput adaptive modulation as well as Walsh-Hadamard spreading using orthogonal spreading codes may be employed. While AOFDM capitalises on the difference of the different subcarriers' channel quality offered by the wide-band channel, spreading aims for averaging the subcarriers' quality differences.

17.5.3.5.1 Adaptive Modulation Assisted SDMA-OFDM   Adaptive modulation employed in the context of OFDM, which we termed as AOFDM, is based on the idea of assigning a more robust, lower-throughput modulation mode to those subcarriers, which are likely to cause symbol errors in the context of a fixed-mode transceiver as a result of their associated low SNR. By contrast, a less robust, higher throughput modulation mode is assigned to the higher-quality subcarriers. These concepts were further elaborated on in Section 17.4.1. In the context of our associated BER performance investigations conducted in Section 17.4.1.3 we found that, as shown in Figure 17.29, the employment of adaptive modulation in an MMSE detection-assisted SDMA-OFDM system is only advantageous in terms of further reducing the BER in specific scenarios, where the number of simultaneous users L approaches the number of reception antennas P. This is because in this specific scenario the effective channel transfer function experienced by the different users is "sufficiently non-flat" across the different subcarriers, with the result of exhibiting sufficient difference in terms of the associated channel quality for AOFDM to excel.

However, there are two aspects of adaptive modulation, which render its employment less convenient. First of all, it is necessary to provide an estimate of the subcarrier channel quality for the next transmission time slot, which could be generated on the basis of the channel estimates available for the current transmission time slot, upon assuming time-invariance of the channel. However, under time-variant channel conditions, the more elaborate, potentially decision-directed channel prediction techniques as seen in Figure 17.28 have to be invoked, in order to obtain accurate estimates of the channel transfer function for the next transmission time slot. Second, the employment of adaptive modulation requires the signalling of the requested modulation mode assignment to be used during the next transmission time slot to the remote users, which requires the existence of a reverse-link and hence its applicability is mainly confined to Time-Division Duplexing (TDD) systems, where every uplink transmission time slot is followed by a down-link time slot and vice versa. Furthermore, the transmission of channel-quality related side-information reduces the AOFDM modem's effective throughput, which has to be compensated for by appropriately increasing the target throughput of the AOFDM modem.

17.5.3.5.2   Walsh-Hadamard Transform Spreading Assisted SDMA-OFDM   The above-mentioned deficiencies of AOFDM constituted the motivation for the alternative technique of employing spreading with the aid of orthogonal spreading codes across the various subcarriers, which was the topic of Section 17.4.2. We argued that as a result of the spreading codes' orthogonality the operations to be carried out at the receiver - also in the multi-user SDMA-WHTS-OFDM scenario - can be separated into that of channel transfer function equalisation and WHT despreading, which substantially simplifies the receiver's design. This was illustrated in Figures 17.30 and 17.31 for the single- and multi-user OFDM scenarios, respectively.

The initial BER performance results portrayed in Figure 17.32 in the context of the single user WHTS-OFDM scenario demonstrated that with the aid of spreading almost the same BER performance can be achieved, as in conjunction with employing adaptive modulation capitalising on perfect channel transfer function predictions used for estimating the subcarrier channel quality during the modulation mode assignment. Furthermore, in contrast to AOFDM, there is no need to transmit side-information to the remote transmitters. Hence, the employment of WHTS-OFDM is amenable to a wider range of transmission scenarios. As shown in [496] the WHT spreading employed at the transmitter can be efficiently combined with the OFDM-related IFFT. As for the receiver, we can argue that the additional complexity imposed by performing the despreading operation is marginal, compared to that owing to multi-user detection.

However, the main essence of our investigations was that as portrayed in Figure 17.34, with the aid of spreading the performance of the MMSE- or PIC detection-assisted SDMA-OFDM systems can be significantly improved, again, provided that the effective transmission channel is strongly frequency selective. As outlined in the previous section, this is the case in almost fully-loaded SDMA-OFDM scenarios, where the number of simultaneous users L approaches the number of reception antennas P. For sufficiently high SNRs the MMSE- or PIC detection-assisted WHTS-OFDM systems of Section 17.4.2.2 were capable of outperforming standard SIC detection-assisted SDMA-OFDM of Section 17.3.1.1.

17.6   List of Symbols Used in Chapter 17

$b_m^{(l)}[n, k]$:   Bit polarity associated with the l-th user at the m-th bit position.

$\mathcal{C}_{\text{lin,dem}}$:   Computational complexity associated with the operation of received symbol classification or synonymously demodulation in the context of the linear detectors.

$\mathcal{C}_{\text{LS,direct}}$:   Computational complexity associated with the LS solution $\hat{s}_{\text{LS}}[n, k]$ found for a specific subcarrier, without explicitly generating the weight matrix.

$\mathcal{C}_{\text{LS,W+cmb}}$:   Computational complexity associated with the LS solution $\hat{s}_{\text{LS}}[n, k]$ found for a specific subcarrier, upon explicitly generating the weight matrix, followed by combining the different antennas' received signals.

$\mathcal{C}_{\text{LS}}$:   Total computational complexity associated with LS detection invoked for a specific subcarrier, without explicitly generating the weight matrix.

$\mathcal{C}_{\text{ML}}$:   Computational complexity associated with the joint ML detection of the different users' transmitted symbols.

$\mathcal{C}_{\text{ML-trf,x}}$:   Computational complexity associated with the transform-based joint ML detection-related (x = err-cov:) error-covariance matrix calculation, (x = LS-est:) calculation of the LS estimates and (x = trial:) evaluation of the trial-symbol-vectors.

$\mathcal{C}_{\text{ML-trf}}$:   Total computational complex associated with the transform-based joint ML detection of the different users' transmitted symbols for a specific subcarrier.

$\mathcal{C}_{\text{MLSE}}$:   Computational complexity associated with the joint ML detection assisted soft-bit generation for a specific subcarrier.

$\mathcal{C}_{\text{MMSE,direct}}$:   Computational complexity associated with the MMSE solution $\hat{s}_{\text{MMSE}}[n, k]$ for a specific subcarrier, without explicitly generating the weight matrix.

$\mathcal{C}_{\text{MMSE,W+cmb}}$:   Computational complexity associated with the MMSE solution for a specific subcarrier, upon explicitly generating the weight matrix, followed by combining the different antennas received signals.

$\mathcal{C}_{\text{MMSE}}$:   Total computational complexity associated with the MMSE detection for a specific subcarrier, without explicitly generating the weight matrix.

$\mathcal{C}_{\text{PIC,x}}^{[1]}$:   Computational complexity associated with the PIC-related (x = MMSE, direct:) and MMSE-assisted generation of linear signal estimates and (x = dem:) with the demodulation process during the first PIC detection stage for a specific subcarrier.

$\mathcal{C}_{\text{PIC,x}}^{[2]}$:   Computational complexity associated with the PIC-related (x = sub:) subtraction, (x = cmb:) diversity combining and (x = dem:) demodulation during the second PIC detection stage for a specific subcarrier.

$\mathcal{C}_{\text{PIC}}$:   Total computational complexity associated with the PIC detection for a specific subcarrier.

$\mathcal{C}_{\text{SIC,x}}$:   Computational complexity associated with the SIC-related (x = stp:) startup, (x = W:) weight vector calculation, (x = obj:) objective function calculation, (x = cmb:) combining, (x = dem:) demodulation and (x = upd:) updating for a specific subcarrier.

$\mathcal{C}_{\text{SIC}}$:   Total computational complexity of standard SIC detection for a specific subcarrier.

$\mathcal{C}_{\text{p/M-SIC,x}}$:   Computational complexity associated with the M-SIC- or pM-SIC -related (x = cmb:) combining, (x = dem) demodulation and (x = upd:) updating for a specific subcarrier.

$\mathcal{C}_{\text{p/M-SIC}}$:   Total computational complexity of M-SIC- or pM-SIC detection for a subcarrier.

E_j: j-th error-event employed for characterising the effects of SIC-related error propagation. The index j is generated by interpreting the different detection stages conditions of an error-free detection ('0') or an erroneous detection ('1') as a binary number.

$\mathbf{E}_{\text{MMSE}}[n]$:   Diagonal-shaped MMSE-based equaliser matrix employed in the context of the single-user WHTS-OFDM system: $\mathbf{E}_{\text{MMSE}}[n] \in \mathbb{C}^{K \times K}$.

$f(x_{\text{eff}}^{(l)}|s^{(l)}, H_{\text{eff}}^{(l)})$:   Complex Gaussian distribution function of the signal $x_{\text{eff}}^{(l)}[n, k]$ observed at the l-th combiner's output, conditioned on transmitting the symbol $s^{(l)}[n, k]$ over a channel characterised by the effective transfer factor $H_{\text{eff}}^{(l)}[n, k]$.

$f(\mathbf{x}|\mathbf{s}, \mathbf{H})$:   Multi-variate complex Gaussian distribution function of the vector $\mathbf{x}[n, k]$ of signals observed at the different BS receiver antenna elements conditioned on transmitting the vector $\mathbf{s}[n, k]$ of symbols over a channel characterised by the matrix $\mathbf{H}[n, k]$.

$H_{\text{MMSE}}[n, k]$:   Effective joint transfer factor of the equaliser and the channel in the context of single-user WHTS-OFDM.

$H_{\text{eff}}^{(l)}[n, k]$:   Effective transfer factor associated with the l-th user's signal contribution to the l-th user's linear combiner output signal $x_{\text{eff}}^{(l)}[n, k]$, which is given by: $H_{\text{eff}}^{(l)}[n, k] = \mathbf{w}^{(l)H}[n, k]\mathbf{H}^{(l)}[n, k]$.

$H_p^{(l)}[n, k]$:   Channel transfer factor associated with the channel encountered between the l-th user's single transmit antenna and the p-th receiver antenna element in the k-th subcarrier of the n-th OFDM symbol period: $H_p^{(l)}[n, k] \in \mathbb{C}$.

$\mathbf{H}^{(l)}[n, k]$:   Vector of channel transfer factors $H_p^{(l)}[n, k]$, $p = 1, \dots, P$ encountered between the l-th user's transmit antenna and the P receiver antenna elements: $\mathbf{H}^{(l)}[n, k] \in \mathbb{C}^{P \times 1}$.

$\mathbf{H}[n, k]$:   Matrix constituted by the vectors of channel transfer factors $\mathbf{H}^{(l)}[n, k]$, $l = 1, \dots, L$, each hosting the channel transfer factors between a specific user's single transmit antenna and the L receiver antenna elements: $\mathbf{H}[n, k] \in \mathbb{C}^{P \times L}$.

$\mathbf{H}_{\text{WHT}}$:   Short-hand: $\mathbf{H}_{\text{WHT}}[n] = \mathbf{H}[n] \mathbf{U}_{\text{WHT}}$.

$l^{[i]}$:   Index of the user selected during the i-th SIC detection stage from the set of remaining users with indices contained in $\mathcal{L}^{[i]}$.

L: Number of simultaneous users.

$\mathcal{L}^{[i]}$:   Set of indices associated with the $L - i + 1$ number of remaining users during the i-th SIC detection stage.

$L_m^{(l)}[n, k]$:   Soft-bit value or log-likelihood ratio associated with the l-th user at the m-th bit position in the context of the joint ML detection of the different users' symbols.

$L_{m, \approx}^{(l)}[n, k]$:   Soft-bit value or log-likelihood ratio associated with the l-th user at the m-th bit position. The index $()_\approx$ serves to distinguish the soft-bit value from the optimum soft-bit value generated in the context of the joint ML detection of the different users' symbols.

$L_{\text{pM-SIC}}$:   Index of the detection stage up to which $M > 1$ number of tentative symbol decisions are tracked in the context of pM-SIC.

M: Number of tentative symbol decisions per detection node in the context of M-SIC, where we have $M \leq M_c$.

M_c: Number of constellation points associated with the modulation scheme employed.

M_{WHT}:   Number of subcarriers contained in a WHT block.

$\text{MSE}^{(l)}[n, k]$:   MSE of the l-th user's linear symbol estimate in the context of linear detection techniques.

$\overline{\text{MSE}}[n, k]$:   Average estimation MSE of the L different users' transmitted symbols, which is given by: $\overline{\text{MSE}}[n, k] = \frac{1}{L} \text{Trace}(\mathbf{R}_{\Delta \hat{\mathbf{s}}}[n, k])$.

$\overline{\text{MMSE}}_{\text{MMSE}}[n]$:   Estimation MSE averaged over the different subcarriers, which is identical for all linear subcarrier symbol estimates $\hat{s}[n, k]$, $k = 0, \dots, K - 1$ after equalisation and despreading in the context of the single-user WHTS-OFDM system.

$\mathcal{M}_c$: 　Set of the M_c number of constellation points associated with the modulation scheme employed.

$\mathcal{M}_{cm}^b$: 　Subset of the set $\mathcal{M}_c$ of constellation points, associated with a bit polarity of $b \in \{0, 1\}$ at the m-th bit position.

$\mathcal{M}^{(l)}$: 　Set of the M_c number of trial-symbols associated with the modulation scheme employed by the l-th user. The trial-symbols are the appropriately amplified constellation points.

$\mathcal{M}_{cm}^{b(l)}$: 　Subset of the set $\mathcal{M}^{(l)}$ of the l-th users trial symbols, associated with a bit polarity of $b \in \{0, 1\}$ at the m-th bit position.

$\mathcal{M}^L$: 　Set of the M_c^L number of trial-symbol-vectors š.

$\mathcal{M}_m^{b(l)L}$: 　Subset of the set $\mathcal{M}^L$ of the L users' trial-symbol-vectors, associated with a bit polarity of $b \in \{0, 1\}$ at the l-th user's m-th bit position.

$n_{\text{eff}}^{(l)}[n, k]$: 　Effective interference- plus noise contribution associated with the l-th user's linear combiner output signal $x_{\text{eff}}^{(l)}[n, k] = \hat{s}^{(l)}[n, k]$, which is given by: $n_{\text{eff}}^{(l)}[n, k] = \hat{s}_I^{(l)}[n, k] + \hat{s}_N^{(l)}[n, k]$.

$n_p[n, k]$: 　AWGN signal contribution associated with the p-th receiver antenna element.

$\mathbf{n}[n, k]$: 　Vector of AWGN signal contributions $n_p[n, k]$, $p = 1, \ldots, P$ associated with the P different receiver antenna elements: $\mathbf{n}[n, k] \in \mathbb{C}^{P \times 1}$.

$N_{\text{SDI}}^{[1]}$: 　Number of subcarriers for which more than one symbol decision is tracked from the first detection stage in the context of SDI-M-SIC.

$\mathbf{p}_{\text{LS}}[n, k]$: 　LS-related 'cross-correlation' vector: $\mathbf{p}_{\text{LS}}[n, k] = \mathbf{H}^H[n, k]\mathbf{x}[n, k] \in \mathbb{C}^{L \times 1}$.

P: 　Number of BS receiver antenna elements.

$P(x_{\text{eff}}^{(l)}[n, k])$: 　Probability of the signal $x_{\text{eff}}^{(l)}[n, k]$, which is given by the total probability, namely by $P(x_{\text{eff}}^{(l)}) = \sum_{\check{s} \in \mathcal{M}^{(l)}} P(x_{\text{eff}}^{(l)}|\check{s}, H_{\text{eff}}^{(l)}) \cdot P(\check{s})$.

$P(x_{\text{eff}}^{(l)}|s^{(l)}, H_{\text{eff}}^{(l)})$: 　*A priori* probability that $x_{\text{eff}}^{(l)}[n, k]$ is observed at the l-th user's combiner output under the condition, that $s^{(l)}[n, k]$ was transmitted, which is given by: $P(x_{\text{eff}}^{(l)}|s^{(l)}, H_{\text{eff}}^{(l)}) = f(x_{\text{eff}}^{(l)}|s^{(l)}, H_{\text{eff}}^{(l)})$.

$P(\check{s}^{(l)}[n, k])$: 　Probability of the symbol $\check{s}^{(l)}[n, k]$. When no *a priori* probability is available about the transmitted symbols, it is assumed that all symbols $\check{s}^{(l)}[n, k] \in \mathcal{M}^{(l)}$ appear with the same probability $P(\check{s}^{(l)}[n, k]) = \frac{1}{M_c}$.

$P(\check{s}^{(l)}|x_{\text{eff}}^{(l)}, H_{\text{eff}}^{(l)})$: *A posteriori* probability that the symbol $\check{s}^{(l)}[n,k]$ was transmitted by the l-th user under the condition that the signal $x_{\text{eff}}^{(l)}[n,k]$ has been received, which can be expressed with the aid of Bayes' rule as:
$$P(\check{s}^{(l)}|x_{\text{eff}}^{(l)}, H_{\text{eff}}^{(l)}) = P(x_{\text{eff}}^{(l)}|\check{s}^{(l)}, H_{\text{eff}}^{(l)})\frac{P(\check{s}^{(l)})}{P(x_{\text{eff}}^{(l)})}.$$

$P(b_m^{(l)}|x_{\text{eff}}^{(l)}, H_{\text{eff}}^{(l)})$: *A posteriori* probability that a bit having a polarity of $b_m^{(l)}[n,k]$ was transmitted by the l-th user at the m-th bit position, under the condition that the signal $x_{\text{eff}}^{(l)}[n,k]$ has been received.

$P_{\text{joint}}^{[i]}[n,k]$: Joint probability of the symbol decisions up to- and including the i-th SIC detection stage, employed in the context of the weighted soft-bit generation for the $(i+1)$-th detection stage.

$P(\check{s}[n,k])$: Probability of the vector $\check{s}[n,k]$ of symbols. When no *a priori* probability is available about the transmitted symbols, it is assumed that all symbols $\check{s}[n,k] \in \mathcal{M}^L$ appear with the same probability $P(\check{s}[n,k]) = \frac{1}{M_c^L}$.

$P(\mathbf{x}[n,k])$: Probability of the vector $\mathbf{x}[n,k]$ of signals, which is given as the total probability, namely: $P(\mathbf{x}) = \sum_{\check{s}\in\mathcal{M}^L} P(\mathbf{x}|\check{s}, \mathbf{H})P(\check{s})$.

$P(\mathbf{x}|\mathbf{s}, \mathbf{H})$: *A priori* probability that the vector $\mathbf{x}[n,k]$ of signals is observed at the BS receiver antenna elements under the condition that the vector $\mathbf{s}[n,k]$ of symbols has been transmitted: $P(\mathbf{x}|\mathbf{s}, \mathbf{H}) = f(\mathbf{x}|\mathbf{s}, \mathbf{H})$.

$P(\check{s}|\mathbf{x}, \mathbf{H})$: *A posteriori* probability that the vector $\check{s}[n,k]$ of symbols was transmitted under the condition that the vector $\mathbf{x}[n,k]$ of signals has been received, which can be expressed according to Bayes' rule as: $P(\check{s}|\mathbf{x}, \mathbf{H}) = P(\mathbf{x}|\mathbf{s}, \mathbf{H})\frac{P(\check{s})}{P(\mathbf{x})}$.

$\mathbf{P}$: Diagonal matrix of the L different users' transmit powers σ_l^2, $l = 1,\dots,L$: $\mathbf{P} \in \mathbb{C}^{L\times L}$.

$\mathbf{P}_{\text{MC}}$: Diagonal matrix of the K subcarriers' transmit powers in the context of the single-user WHTS-OFDM system: $\mathbf{P}_{\text{MC}} = \sigma_s^2\mathbf{I} \in \mathbb{C}^{K\times K}$.

$\mathbf{P}_{\text{SNR}}$: Diagonal matrix of the L different users' SNR values given by $\text{SNR}^{(l)} = \frac{\sigma_l^2}{\sigma_n^2}$, $l = 1,\dots,L$ at the receiver antenna elements: $\mathbf{P}_{\text{SNR}} \in \mathbb{C}^{L\times L}$.

$\mathbf{P}_{\text{LS}}[n,k]$: LS-related projection matrix: $\mathbf{P}_{\text{LS}}[n,k] = (\mathbf{H}^H[n,k]\mathbf{H}[n,k])^{-1}\mathbf{H}[n,k] \in \mathbb{C}^{L\times P}$.

$\mathbf{Q}_{\text{LS}}[n,k]$: LS-related 'auto-correlation' matrix: $\mathbf{Q}_{\text{LS}}[n,k] = \mathbf{H}^H[n,k]\mathbf{H}[n,k] \in \mathbb{C}^{L\times L}$.

$\mathbf{R}_{a,S}^{(l)}[n,k]$: Auto-correlation matrix of the l-th user's signal contribution to the vector $\mathbf{x}[n]$ of received signals: $\mathbf{R}_{a,S}^{(l)}[n,k] \in \mathbb{C}^{P\times P}$.

$\mathbf{R}_{a,I}^{(l)}[n,k]$: Auto-correlation matrix of the $(L-1)$ interfering users' signal contributions to the vector $\mathbf{x}[n]$ of received signals, when regarding the l-th user as the desired user: $\mathbf{R}_{a,I}^{(l)}[n,k] \in \mathbb{C}^{P \times P}$.

$\mathbf{R}_{a,N}[n,k]$: Auto-correlation matrix of the AWGN signal contribution to the vector $\mathbf{x}[n]$ of received signals: $\mathbf{R}_{a,N}[n,k] \in \mathbb{C}^{P \times P}$.

$\mathbf{R}_{a,I+N}^{(l)}[n,k]$: Short-hand: $\mathbf{R}_{a,I+N}^{(l)}[n,k] = \mathbf{R}_{a,I}^{(l)}[n,k] + \mathbf{R}_{a,N}[n,k] \in \mathbb{C}^{P \times P}$.

$\mathbf{R}_a[n,k]$: Auto-correlation matrix of the vector of received signals $\mathbf{x}[n,k]$: $\mathbf{R}_a[n,k] = E\{\mathbf{x}[n,k]\mathbf{x}^H[n,k]\} = \mathbf{HPH}^H + \sigma_n^2\mathbf{I} \in \mathbb{C}^{P \times P}$.

$\mathbf{R}_{\overline{a}}[n,k]$: Left-inverse related form of the auto-correlation matrix: $\mathbf{R}_{\overline{a}}[n,k] = \mathbf{H}^H\mathbf{HP} + \sigma_n^2\mathbf{I} \in \mathbb{C}^{L \times L}$.

$\mathbf{R}_c[n,k]$: Cross-correlation matrix of the vectors of received signals $\mathbf{x}[n,k]$ and transmitted symbols $\mathbf{s}[n,k]$: $\mathbf{R}_c[n,k] = E\{\mathbf{x}[n,k]\mathbf{s}^H[n,k]\} = \mathbf{H}[n,k]\mathbf{P} \in \mathbb{C}^{P \times L}$.

$\mathbf{R}_{\Delta\hat{\mathbf{s}}}[n,k]$: Auto-correlation matrix of the vector $\Delta\hat{\mathbf{s}}[n,k]$ constituted by the transmitted symbols' estimation errors: $\mathbf{R}_{\Delta\hat{\mathbf{s}}}[n,k] = E\{\Delta\hat{\mathbf{s}}[n,k]\Delta\hat{\mathbf{s}}^H[n,k]\} \in \mathbb{C}^{L \times L}$.

$\mathbf{R_n}$: Auto-correlation matrix of the AWGN at the receiver antenna elements: $\mathbf{R_n} = \sigma_n^2\mathbf{I} \in \mathbb{C}^{P \times P}$.

$s^{(l)}[n,k]$: Symbol transmitted by the l-th user.

$\hat{s}^{(l)}[n,k]$: Linear estimate of the symbol transmitted by the l-th user, which is given by: $\hat{s}^{(l)}[n,k] = \mathbf{w}^{(l)H}[n,k]\mathbf{x}[n,k]$.

$\hat{s}_S^{(l)}[n,k]$: Desired l-th user's contribution to the linear estimate $\hat{s}^{(l)}[n,k]$ of the symbol transmitted by the l-th user, based on using the weight vector $\mathbf{w}^{(l)}[n,k]$.

$\hat{s}_I^{(l)}[n,k]$: Undesired interfering users' contribution to the linear estimate $\hat{s}^{(l)}[n,k]$ of the symbol transmitted by the l-th user, based on using the weight vector $\mathbf{w}^{(l)}[n,k]$.

$\hat{s}_N^{(l)}[n,k]$: Undesired noise-related contribution to the linear estimate $\hat{s}^{(l)}[n,k]$ of the symbol transmitted by the l-th user, based on using the weight vector $\mathbf{w}^{(l)}[n,k]$.

$\hat{s}^{(l)}[n,k]$: Linear estimate of the symbol transmitted by the l-th user after despreading in the context of the multi-user WHTS-OFDM system.

$\check{s}^{(l)}[n,k]$: Classified linear estimate of the symbol transmitted by the l-th user.

$\check{s}_{\mathrm{ML}\approx}^{(l)}[n,k]$: The symbol that is most likely to have been transmitted by the l-th user, found upon maximising the *a posteriori* probability $P(\check{s}^{(l)}|x_{\mathrm{eff}}^{(l)},H_{\mathrm{eff}}^{(l)})$ across the set of trial-symbols $\mathcal{M}^{(l)}$. The index $()_\approx$ is employed in order to distinguish this solution from the optimum solution $\check{s}_{\mathrm{ML}}[n,k]$ found by joint ML detection of the L different users' transmitted symbols.

$\check{s}_{m\approx}^{b(l)}[n,k]$: The symbol that is most likely to have been transmitted by the l-th user, found upon maximising the *a posteriori* probability $P(\check{s}^{(l)}|x_{\mathrm{eff}}^{(l)},H_{\mathrm{eff}}^{(l)})$ across the subset of trial-symbols $\mathcal{M}_m^{b(l)}$.

$\hat{\mathbf{s}}[n]$: Vector of the K subcarriers' linear symbol estimates in the context of the single-user WHTS-OFDM system: $\hat{\mathbf{s}}[n] \in \mathbb{C}^{K\times 1}$.

$\hat{\mathbf{s}}_{\mathrm{MMSE}}[n]$: Vector of the K subcarriers' MMSE-based linear symbol estimates in the context of the single-user WHTS-OFDM system: $\hat{\mathbf{s}}_{\mathrm{MMSE}}[n] \in \mathbb{C}^{K\times 1}$.

$\mathbf{s}[n,k]$: Vector of symbols $s^{(l)}[n,k]$, $l = 1, \ldots, L$ transmitted by the L different users: $\mathbf{s}[n,k] \in \mathbb{C}^{L\times 1}$.

$\hat{\mathbf{s}}[n,k]$: Vector of linear symbol estimates $\hat{s}^{(l)}[n,k]$, $l = 1, \ldots, L$, which is given by: $\hat{\mathbf{s}}[n,k] = \mathbf{W}^H[n,k]\mathbf{x}[n,k] \in \mathbb{C}^{L\times 1}$.

$\hat{\mathbf{s}}_{\mathrm{LS}}[n,k]$: LS-assisted vector of linear symbol estimates: $\hat{\mathbf{s}}_{\mathrm{LS}}[n,k] = \mathbf{Q}_{\mathrm{LS}}^{-1}[n,k]\mathbf{p}_{\mathrm{LS}}[n,k] = \mathbf{P}_{\mathrm{LS}}[n,k]\mathbf{x}[n,k] = \mathbf{W}_{\mathrm{LS}}^H[n,k]\mathbf{x}[n,k] \in \mathbb{C}^{L\times 1}$.

$\check{\mathbf{s}}[n,k]$: Vector of classified linear symbol estimates $\check{s}^{(l)}[n,k]$, $l = 1, \ldots, L$, which is given by: $\check{\mathbf{s}}[n,k] \in \mathbb{C}^{L\times 1}$.

$\check{\mathbf{s}}_{\mathrm{ML}}[n,k]$: Vector of symbols that are most likely to have been transmitted by the L different users, found upon maximising the *a posteriori* probability $P(\check{\mathbf{s}}|\mathbf{x},\mathbf{H})$ across the set of trial-symbol-vectors $\mathcal{M}^L$.

$\check{\mathbf{s}}_m^{b(l)}$: Vector of symbols that are most likely to have been transmitted by the L different users found upon maximising the *a posteriori* probability $P(\check{\mathbf{s}}|\mathbf{x},\mathbf{H})$ across the subset of trial-symbol-vectors $\mathcal{M}_m^{b(l)L}$.

$\mathrm{SINR}_{\mathrm{WHT,MMSE}}[n]$: Subcarrier-averaged SINR associated with the linear subcarrier symbol estimates $\hat{s}[n,k]$, $k = 0, \ldots, K - 1$ in the context of the single-user WHTS-OFDM system: $\mathrm{SINR}_{\mathrm{WHT,MMSE}}[n] = \frac{\sigma_{S,\mathrm{WHT,MMSE}}^2}{\sigma_{I,\mathrm{WHT,MMSE}}^2 + \sigma_{N,\mathrm{WHT,MMSE}}^2}$.

$\mathrm{SINR}^{(l)}[n,k]$: SINR at the l-th user's associated combiner output, which is expressed as:
$\mathrm{SINR}^{(l)}[n,k] = \frac{\sigma_S^{(l)2}[n,k]}{\sigma_I^{(l)2}[n,k] + \sigma_N^{(l)2}[n,k]}$.

$\mathrm{SIR}^{(l)}[n,k]$: SIR at the l-th user's associated combiner output: $\mathrm{SIR}^{(l)}[n,k] = \frac{\sigma_S^{(l)2}[n,k]}{\sigma_I^{(l)2}[n,k]}$.

$\text{SNR}^{(l)}[n,k]$:   SNR at the l-th user's associated combiner output: $\text{SNR}^{(l)}[n,k] = \frac{\sigma_S^{(l)2}[n,k]}{\sigma_N^{(l)2}[n,k]}$.

$\mathbf{U}_{\text{WHT}_N}$: N-th order Walsh-Hadamard Transform (WHT) matrix.

$\mathbf{w}^{(l)}[n,k]$:   Weight vector used for estimating the l-th user's transmitted symbol: $\mathbf{w}^{(l)}[n,k] \in \mathbb{C}^{P \times 1}$.

$\mathbf{W}[n,k]$:   Matrix of weight vectors $\mathbf{w}^{(l)}[n,k]$, $l = 1, \dots, L$ used for estimating the L different users' transmitted symbols: $\mathbf{W}[n,k] \in \mathbb{C}^{P \times L}$.

$\mathbf{W}_{\text{LS}}[n,k]$:   LS-optimised weight matrix: $\mathbf{W}_{\text{LS}}[n,k] = \mathbf{P}_{\text{LS}}^H \in \mathbb{C}^{P \times L}$.

$\mathbf{W}_{\text{MMSE}}[n,k]$:   MMSE-optimised weight matrix: $\mathbf{W}_{\text{MMSE}}[n,k] = \mathbf{R}_a^{-1}[n,k]\mathbf{R}_c[n,k] \in \mathbb{C}^{P \times L}$ in its right-inverse related form, or $\mathbf{W}_{\text{MMSE}}[n,k] = \mathbf{R}_c[n,k]\mathbf{R}_{\overline{a}}^{-1}[n,k]$ in its left-inverse related form.

$\mathbf{W}[n]$:   Weight matrix in the context of the single-user WHTS-OFDM system: $\mathbf{W}[n] \in \mathbb{C}^{K \times K}$.

$\mathbf{W}_{\text{MMSE}}[n]$:   MMSE-based weight matrix employed in the context of the single-user WHTS-OFDM system: $\mathbf{W}_{\text{MMSE}}[n] = \mathbf{E}_{\text{MMSE}}[n]\mathbf{U}_{\text{WHT}} \in \mathbb{C}^{K \times K}$.

$x_{\text{eff}}^{(l)}[n,k]$:   Model of the l-th user's combiner output signal: $x_{\text{eff}}^{(l)}[n,k] = H_{\text{eff}}^{(l)}[n,k]s^{(l)}[n,k] + n_{\text{eff}}^{(l)}[n,k] = \hat{s}^{(l)}[n,k]$.

$x_p[n,k]$:   Signal recorded at the p-th receiver antenna element.

$\mathbf{x}[n,k]$:   Vector of signals $x_p[n,k]$, $p = 1, \dots, P$ recorded at the P receiver antenna elements: $\mathbf{x}[n,k] \in \mathbb{C}^{P \times 1}$.

$\hat{\mathbf{x}}[n,k]$:   Vector of linear estimates of the signals $x_p[n,k]$, $p = 1, \dots, P$ recorded at the P receiver antenna elements: $\hat{\mathbf{x}}[n,k] \in \mathbb{C}^{P \times 1}$.

α_{SIC}:   Constant employed in the context of quantifying the SIC detector's computational complexity: $\alpha_{\text{SIC}} = \sum_{i=1}^{L} i = \frac{1}{2}L(L+1)$.

β_{MMSE}:   Proportionality constant associated with the MMSE combiner' weight vector in the context of its representation in standard form.

β_{MV}:   Proportionality constant associated with the MV combiner's weight vector in the context of its representation in standard form.

$\beta_{\text{p/M-SIC}}$:   Constants employed in the context of quantifying the M-SIC- or pM-SIC detector's computational complexity.

$\gamma_{\text{p/M-SIC},\{i/ii\}}$:   Constants employed in the context of quantifying the M-SIC- or pM-SIC detector's computational complexity.

$\delta_{\text{p/M-SIC},\{i/ii\}}$:   Constants employed in the context of quantifying the M-SIC- or pM-SIC detector's computational complexity.

$\epsilon_{\text{p/M-SIC}}$:   Constant employed in the context of quantifying the M-SIC- or pM-SIC detector's computational complexity.

ζ_{PIC}:   Constant employed in the context of quantifying the M-SIC- or pM-SIC detector's computational complexity.

σ_l^2:   Variance of the l-th user's transmitted subcarrier symbols.

$\sigma_{x,\text{WHT,MMSE}}^2[n]$:   Average (x = S:) desired signal variance, (x = I:) undesired interference signal variance and (x = N:) undesired AWGN variance associated with the linear subcarrier symbols estimates $\hat{s}[n,k]$, $k = 0, \ldots, K-1$ after equalisation and despreading in the context of single-user WHTS-OFDM.

$\sigma_x^{(l)2}[n,k]$:   Variance of the signal $\hat{s}_x^{(l)}[n,k]$, where (x = S:) desired signal, (x=I:) undesired interference signal, (x = N:) AWGN noise.

σ_n^2:   Variance of the AWGN.

$\sigma_{\text{eff}}^{(l)2}[n,k]$:   Variance of the effective interference- plus noise contribution $n_{\text{eff}}^{(l)}[n,k]$, which is given by: $\sigma_{\text{eff}}^{(l)2}[n,k] = \sigma_I^{(l)2} + \sigma_N^{(l)2}$.

$\Delta\hat{s}[n,k]$:   Vector of the transmitted symbols' estimation errors: $\Delta\hat{s}[n,k] = s[n,k] - \hat{s}[n,k] \in \mathbb{C}^{L \times 1}$.

$\Delta\hat{x}[n,k]$:   Vector of the received signals' estimation errors: $\Delta\hat{x}[n,k] = x[n,k] - \hat{x}[n,k] \in \mathbb{C}^{P \times 1}$.

$()[n,k]$:   Signal associated with the k-th subcarrier of the n-th OFDM symbol.

$()_{\text{LS}}$:   Variables associated with the LS combiner.

$()_{\text{MMSE}}$:   Variables associated with the MMSE combiner.

$()_{\text{MV}}$:   Variables associated with the MV combiner.

$()_{\text{p/M-/SIC}}$:   Variables associated with the standard SIC, M-SIC- or pM-SIC detectors.

$()_{\text{PIC}}$:   Variables associated with the PIC detector.

$()^{[i]}$:   Notation used in the context of SIC detection in order to indicate that the variable enclosed in round brackets is associated with the i-th detection stage, where $i = 1, \ldots, L$.

$()^{(l^{[i]})[i]}$:   Notation used in the context of SIC detection in order to indicate that the variable enclosed in round brackets is associated with the i-th detection stage, conditioned on regarding the $l^{[i]}$-th user as the strongest user to be detected during the i-th stage.

$()^{(l)[i]}$:   Notation used in the context of PIC detection in order to indicate that the variable enclosed in round brackets is associated with the l-th user during the i-th detection stage.

Chapter 18

OFDM-Based Wireless Video System Design [1]

P.J. Cherriman, L. Hanzo, T. Keller and M. Münster

In this chapter we provide a few application examples based on powerful turbo-coded OFDM systems invoked for the wireless transmission of video telephone signals. Here we refrain from detailing the associated video compression aspects and refer the interested reader for example to [7] for further details on video communications.

18.1 Sub-band-Adaptive Turbo-Coded OFDM-Based Interactive Videotelephony [2]

18.1.1 Motivation and Background

In this section, burst-by-burst adaptive OFDM is proposed and investigated in the context of interactive videotelephony.

As mentioned earlier, burst-by-burst adaptive quadrature amplitude modulation [164] (AQAM) was advocated, for example, by Steele and Webb [164, 198] in order to enable the transceiver to cope with the time-variant channel quality of narrowband fading channels. The intricacies of adaptive transceivers were detailed in Chapters 6 and 12. Further related research was also conducted at the University of Osaka by Sampei and his colleagues, investigating variable coding rate concatenated coded schemes [199]; at the University of Stanford by Goldsmith and her team, studying the effects of variable-rate, variable-power arrangements [200]; and at the University of Southampton in the United Kingdom, investigating a variety of practical aspects of AQAM [203, 204, 207]. The channel's quality is estimated

[1] *OFDM and MC-CDMA for Broadband Multi-user Communications WLANS and Broadcasting.*
L. Hanzo, Münster, T. Keller and B.J. Choi,
©2003 John Wiley & Sons, Ltd. ISBN 0-470-85879-6
[2] This section is based on P.J. Cherriman, T. Keller, and L. Hanzo: "Sub-band-adaptive Turbo-coded OFDM-based Interactive Video Telephony", *IEEE Transactions on Circuits and Systems for Video Technology*, Oct. 2002, Vol. 12, No. 10, pp. 829–840 ©2002 IEEE.

on a burst-by-burst basis, and the most appropriate modulation mode is selected in order to maintain the required target bit error rate (BER) performance, while maximising the system's Bit Per Symbol (BPS) throughput. Though use of this reconfiguration regime, the distribution of channel errors becomes typically less bursty, than in conjunction with nonadaptive modems, which potentially increases the channel coding gains [247]. Furthermore, the soft-decision channel codec metrics can also be invoked in estimating the instantaneous channel quality [247], regardless of the type of channel impairments.

A range of coded AQAM schemes was analysed by Matsuoka *et al.* [199], Lau *et al.* [202] and Goldsmith *et al.* [497]. For data transmission systems, which do not necessarily require a low transmission delay, variable-throughput adaptive schemes can be devised, which operate efficiently in conjunction with powerful error correction codecs, such as long block length turbo codes [215]. However, the acceptable turbo interleaving delay is rather low in the context of low-delay interactive speech. Video communications systems typically require a higher bit rate than speech systems, and hence they can afford a higher interleaving delay.

The above principles – which were typically investigated in the context of narrowband modems – were further advanced in conjunction with wideband modems, employing powerful block turbo-coded, wideband Decision Feedback Equaliser (DFE) assisted AQAM transceivers [247, 250]. A neural-network Radial Basis Function (RBF) DFE-based AQAM modem design was proposed in [498], where the RBF DFE provided the channel-quality estimates for the modem mode switching regime. This modem was capable of removing the residual BER of conventional DFEs, when linearly non-separable received phasor constellations were encountered.

These burst-by-burst adaptive principles can also be extended to Adaptive Orthogonal Frequency Division Multiplexing (AOFDM) schemes [219] and to adaptive joint-detection-based Code Division Multiple Access (JD-ACDMA) arrangements [499]. The associated AQAM principles were invoked in the context of parallel AOFDM modems by Czylwik *et al.* [205], Fischer [223], and Chow *et al.* [206]. Adaptive subcarrier selection has also been advocated by Rohling *et al.* [500] in order to achieve BER performance improvements. Due to lack of space without completeness, further significant advances over benign, slowly varying dispersive Gaussian fixed links – rather than over hostile wireless links – are due to Chow, Cioffi, and Bingham [206] from the United States, rendering OFDM the dominant solution for asymmetric digital subscriber loop (ADSL) applications, potentially up to bit rates of 54 Mbps. In Europe OFDM has been favoured for both Digital Audio Broadcasting (DAB) and Digital Video Broadcasting [47, 190] (DVB) as well as for high-rate Wireless Asynchronous Transfer Mode (WATM) systems due to its ability to combat the effects of highly dispersive channels [150]. The idea of "water-filling" – as allocating different modem modes to different subcarriers was referred to – was proposed for OFDM by Kalet [19] and later further advanced by Chow *et al.* [206]. This approach was adapted later in the context of time-variant mobile channels for duplex wireless links, for example, in [219]. Finally, various OFDM-based speech and video systems were proposed in [501, 502], while the co-channel interference sensitivity of OFDM can be mitigated with the aid of adaptive beam-forming [105, 129] in multi-user scenarios.

Section 18.1.1 outlines the architecture of the proposed video transceiver, while Section 18.1.5 quantifies the performance benefits of AOFDM transceivers in comparison to conventional fixed transceivers. Section 18.1.6 endeavuors to highlight the effects of more "aggressive" loading of the subcarriers in both BER and video quality terms, while Section 18.1.7 proposed time-variant rather than constant rate AOFDM as a means of more accurately matching the transceiver to the time-variant channel quality fluctuations, before concluding in Section 18.3.

18.1.2 AOFDM Modem Mode Adaptation and Signalling

The proposed duplex AOFDM scheme operates on the following basis:

- *Channel quality estimation* is invoked upon receiving an AOFDM symbol in order to select the modem mode allocation of the next AOFDM symbol.

- *The decision concerning the modem modes for the next AOFDM symbol* is based on the prediction of the expected channel conditions. Then the transmitter has to select the appropriate modem modes for the groups or sub-bands of OFDM subcarriers, where the subcarriers were grouped into sub-bands of identical modem modes in order to reduce the required number of signaling bits.

- *Explicit signalling or blind detection of the modem modes* is used to inform the receiver as to what type of demodulation to invoke.

If the channel quality of the uplink and downlink can be considered similar, then the channel-quality estimate for the uplink can be extracted from the downlink and vice versa. We refer to this regime as open-loop adaptation. In this case, the transmitter has to convey the modem modes to the receiver, or the receiver can attempt blind detection of the transmission parameters employed. By contrast, if the channel cannot be considered reciprocal, then the channel-quality estimation has to be performed at the receiver, and the receiver has to instruct the transmitter as to what modem modes have to be used at the transmitter, in order to satisfy the target integrity requirements of the receiver. We refer to this mode as closed-loop adaptation. Blind modem mode recognition was invoked, for example, in [219] – a technique that results in bit rate savings due to refraining from dedicating bits to explicit modem mode signaling at the cost of increased complexity. Let us address the issues of channel quality estimation on a sub-band-by-sub-band basis in the next subsection.

18.1.3 AOFDM Sub-band BER Estimation

A reliable channel-quality metric can be devised by calculating the expected overall bit error probability for all available modulation schemes M_n in each sub-band, which is denoted by $\bar{p}_e(n) = 1/N_s \sum_j p_e(\gamma_j, M_n)$. For each AOFDM sub-band, the modem mode having the highest throughput, while exhibiting an estimated BER below the target value is then chosen. Although the adaptation granularity is limited to the sub-band width, the channel-quality estimation is quite reliable, even in interference-impaired environments.

Against this background in our forthcoming discussions, the design trade-offs of turbo-coded Adaptive Orthogonal Frequency Division Multiplex (AOFDM) wideband video transceivers are presented. We will demonstrate that AOFDM provides a convenient framework for adjusting the required target integrity and throughput both with and without turbo channel coding and lends itself to attractive video system construction, provided that a near-instantaneously programmable rate video codec – such as the H.263 scheme highlighted in the next section – can be invoked.

18.1.4　Video Compression and Transmission Aspects

In this study we investigate the transmission of 704 x 576 pixel Four-times Common Intermediate Format (4CIF) high-resolution video sequences at 30 frames/s using sub-band-adaptive turbo-coded Orthogonal Frequency Division Multiplex (AOFDM) transceivers. The transceiver can modulate 1, 2, or 4 bits onto each AOFDM subcarrier, or simply disable transmissions for subcarriers that exhibit a high attenuation or phase distortion due to channel effects.

The H.263 video codec [7] exhibits an impressive compression ratio, although this is achieved at the cost of a high vulnerability to transmission errors, since a run-length coded bit stream is rendered undecodable by a single bit error. In order to mitigate this problem, when the channel codec protecting the video stream is overwhelmed by the transmission errors, we refrain from decoding the corrupted video packet in order to prevent error propagation through the reconstructed video frame buffer [503]. We found that it was more beneficial in video-quality terms if these corrupted video packets were dropped and the reconstructed frame buffer was not updated, until the next video packet replenishing the specific video frame area was received. The associated video performance degradation was found to be perceptually unobjectionable for packet dropping or transmission frame error rates (FER) below about 5%. These packet dropping events were signalled to the remote video decoder by superimposing a strongly protected one-bit packet acknowledgment flag on the reverse-direction packet, as outlined in [503]. Turbo error correction codes [215] were used.

18.1.5　Comparison of Sub-band-Adaptive OFDM
and Fixed Mode OFDM Transceivers

In order to show the benefits of the proposed sub-band-adaptive OFDM transceiver, we compare its performance to that of a fixed modulation mode transceiver under identical propagation conditions, while having the same transmission bit rate. The sub-band-adaptive modem is capable of achieving a low bit error ratio (BER), since it can disable transmissions over low-quality subcarriers and compensate for the lost throughput by invoking a higher modulation mode than that of the fixed-mode transceiver over the high-quality subcarriers.

Table 18.1 shows the system parameters for the fixed BPSK and QPSK transceivers, as well as for the corresponding sub-band-adaptive OFDM (AOFDM) transceivers. The system employs constraint length three, half-rate turbo coding, using octal generator polynomials of 5 and 7 as well as random turbo interleavers. Therefore, the unprotected bit rate is approximately half the channel-coded bit rate. The protected to unprotected video bit rate ratio is not exactly half, since two tailing bits are required to reset the convolutional encoders' memory to their default state in each transmission burst. In both modes, a 16-bit Cyclic Redundancy Checking (CRC) is used for error detection, and 9 bits are used to encode the reverse link feedback acknowledgment information by simple repetition coding. The feedback flag decoding ensues using majority logic decisions. The packetisation requires a small amount of header information added to each transmitted packet, which is 11 and 12 bits per packet for BPSK and QPSK, respectively. The effective or useful video bit rates for the BPSK and QPSK modes are then 3.4 and 7.0 Mbps.

The fixed-mode BPSK and QPSK transceivers are limited to 1 and 2 bits per symbol, respectively. By contrast, the proposed AOFDM transceivers operate at the same bit rate, as

	BPSK mode	QPSK mode
Packet rate	4687.5 packets/s	
FFT length	512	
OFDM symbols/packet	3	
OFDM symbol duration	$2.6667 \mu s$	
OFDM time frame	80 time slots = 213 μs	
Normalised Doppler frequency, f'_d	1.235×10^{-4}	
OFDM symbol normalised Doppler frequency, F_D	7.41×10^{-2}	
FEC coded bits/packet	1536	3072
FEC-coded video bit rate	7.2 Mbps	14.4 Mbps
Unprotected bits/packet	766	1534
Unprotected bit rate	3.6 Mbps	7.2 Mbps
Error detection CRC (bits)	16	16
Feedback error flag bits	9	9
Packet header bits/packet	11	12
Effective video bits/packet	730	1497
Effective video bit rate	3.4 Mbps	7.0 Mbps

Table 18.1: System Parameters for the Fixed QPSK and BPSK Transceivers, as well as for the Corresponding Sub-band-adaptive OFDM (AOFDM) Transceivers for Wireless Local Area Networks (WLANs)

their corresponding fixed modem mode counterparts, although they can vary their modulation mode on a subcarrier-by-subcarrier basis between 0, 1, 2, and 4 bits per symbol. Zero bits per symbol implies that transmissions are disabled for the subcarrier concerned.

The "micro-adaptive" nature of the sub-band-adaptive modem is characterised by Figure 18.1, portraying at the top a contour plot of the channel signal-to-noise ratio (SNR) for each subcarrier versus time. At the centre and bottom of the figure, the modulation mode chosen for each 32-subcarrier sub-band is shown versus time for the 3.4 and 7.0 Mbps target-rate sub-band-adaptive modems, respectively. The channel SNR variation versus both time and frequency is also shown in three-dimensional form in Figure 18.2, which may be more convenient to visualise. This was recorded for the channel impulse response of Figure 18.3. It can be seen that when the channel is of high quality – as for example, at about frame 1080 – the sub-band-adaptive modem used the same modulation mode, as the equivalent fixed-rate modem in all subcarriers. When the channel is hostile – for example, around frame 1060 – the sub-band-adaptive modem used a lower-order modulation mode in some sub-bands than the equivalent fixed-mode scheme, or in extreme cases disabled transmission for that sub-band. In order to compensate for the loss of throughput in this sub-band, a higher-order modulation mode was used in the higher quality sub-bands.

One video packet is transmitted per OFDM symbol; therefore, the video packet-loss ratio is the same as the OFDM symbol error ratio. The video packet-loss ratio is plotted against the channel SNR in Figure 18.4. It is shown in the graph that the sub-band-adaptive transceivers – or synonymously termed as microscopic-adaptive (μAOFDM), in contrast to OFDM symbol-by-symbol adaptive transceivers – have a lower packet-loss ratio (PLR) at the same SNR compared to the fixed modulation mode transceiver. Note in Figure 18.4 that

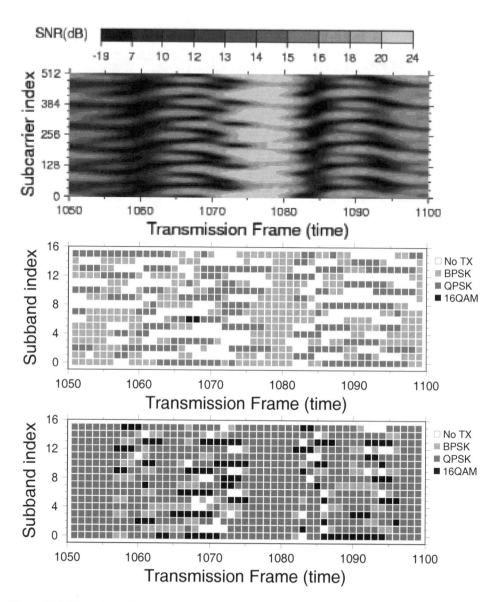

Figure 18.1: The micro-adaptive nature of the sub-band-adaptive OFDM modem. The top graph is a contour plot of the channel SNR for all 512 subcarriers versus time. The bottom two graphs show the modulation modes chosen for all 16 32-subcarrier sub-bands for the same period of time. The middle graph shows the performance of the 3.4 Mbps sub-band-adaptive modem, which operates at the same bit rate as a fixed BPSK modem. The bottom graph represents the 7.0 Mbps sub-band-adaptive modem, which operated at the same bit rate as a fixed QPSK modem. The average channel SNR was 16 dB.

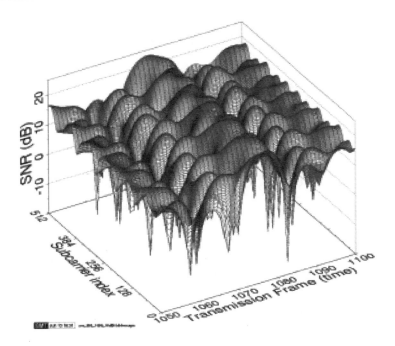

Figure 18.2: Instantaneous channel SNR for all 512 subcarriers versus time, for an average channel SNR of 16 dB over the channel characterised by the channel impulse response (CIR) of Figure 18.3.

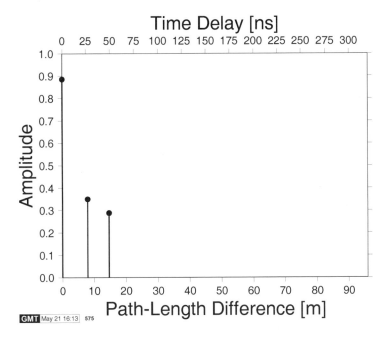

Figure 18.3: Indoor three-path WATM channel impulse response.

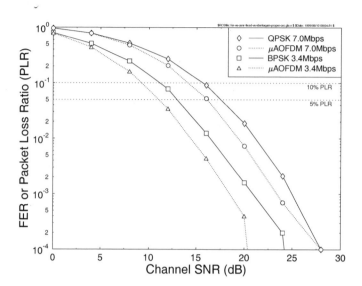

Figure 18.4: Frame Error Rate (FER) or video packet-loss ratio (PLR) versus channel SNR for the BPSK and QPSK fixed modulation mode OFDM transceivers and for the corresponding sub-band-adaptive μAOFDM transceiver, operating at identical effective video bit rates, namely, at 3.4 and 7.0 Mbps, over the channel model of Figure 18.3 at a normalised Doppler frequency of $F_D = 7.41 \times 10^{-2}$.

the sub-band-adaptive transceivers can operate at lower channel SNRs than the fixed modem mode transceivers, while maintaining the same required video packet-loss ratio. Again, the figure labels the sub-band-adaptive OFDM transceivers as μAOFDM, implying that the adaption is not noticeable from the upper layers of the system. A macro-adaption could be applied in addition to the microscopic adaption by switching between different target bit rates, as the longer-term channel quality improves and degrades. This issue is the subject of Section 18.1.7.

Having shown that the sub-band-adaptive OFDM transceiver achieved a reduced video packet loss in comparison to fixed modulation mode transceivers under identical channel conditions, we now compare the effective throughput bit rate of the fixed and adaptive OFDM transceivers in Figure 18.5. The figure shows that when the channel quality is high, the throughput bit rates of the fixed and adaptive transceivers are identical. However, as the channel degrades, the loss of packets results in a lower throughput bit rate. The lower packet-loss ratio of the sub-band-adaptive transceiver results in a higher throughput bit rate than that of the fixed modulation mode transceiver.

The throughput bit rate performance results translate to the decoded video-quality performance results evaluated in terms of PSNR in Figure 18.6. Again, for high-channel SNRs the performance of the fixed and adaptive OFDM transceivers is identical. However, as the channel quality degrades, the video quality of the sub-band-adaptive transceiver degrades less dramatically than that of the corresponding fixed modulation mode transceiver.

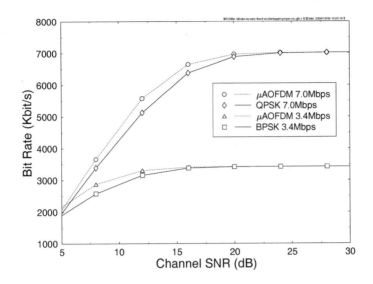

Figure 18.5: Effective throughput bit rate versus channel SNR for the BPSK and QPSK fixed modulation mode OFDM transceivers and that of the corresponding sub-band-adaptive or μAOFDM transceiver operating at identical effective video bit rates of 3.4 and 7.0 Mbps, over the channel of Figure 18.3 at a normalised Doppler frequency of $F_D = 7.41 \times 10^{-2}$.

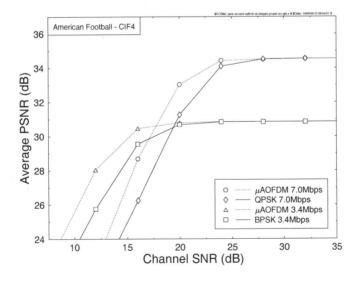

Figure 18.6: Average video quality expressed in PSNR versus channel SNR for the BPSK and QPSK fixed modulation mode OFDM transceivers and for the corresponding μAOFDM transceiver operating at identical channel SNRs over the channel model of Figure 18.3 at a normalised Doppler frequency of $F_D = 7.41 \times 10^{-2}$.

18.1.6   Sub-band-Adaptive OFDM Transceivers Having Different Target Bit Rates

As mentioned earlier, the sub-band-adaptive modems employ different modulation modes for different subcarriers in order to meet the target bit rate requirement at the lowest possible channel SNR. This is achieved by using a more robust modulation mode or eventually by disabling transmissions over subcarriers having a low channel quality. By contrast, the adaptive system can invoke less robust, but higher throughput, modulation modes over subcarriers exhibiting a high-channel quality. In the examples we have previously considered, we chose the AOFDM target bit rate to be identical to that of a fixed modulation mode transceiver. In this section, we comparatively study the performance of various μAOFDM systems having different target bit rates.

The previously described μAOFDM transceiver of Table 18.1 exhibited a FEC-coded bit rate of 7.2 Mbps, which provided an effective video bit rate of 3.4 Mbps. If the video target bit rate is lower than 3.4 Mbps, then the system can disable transmission in more of the subcarriers, where the channel quality is low. Such a transceiver would have a lower bit error rate than the previous BPSK-equivalent μAOFDM transceiver and therefore could be used at lower average channel SNRs, while maintaining the same bit error ratio target. By contrast, as the target bit rate is increased, the system has to employ higher-order modulation modes in more subcarriers at the cost of an increased bit error ratio. Therefore, high target bit-rate μAOFDM transceivers can only perform within the required bit error ratio constraints at high-channel SNRs, while low target bit-rate μAOFDM systems can operate at low-channel SNRs without causing excessive BERs. Therefore, a system that can adjust its target bit rate as the channel SNR changes would operate over a wide range of channel SNRs, providing the maximum possible average throughput bit rate, while maintaining the required bit error ratio.

Hence, below we provide a performance comparison of various μAOFDM transceivers that have four different target bit rates, of which two are equivalent to that of the BPSK and QPSK fixed modulation mode transceivers of Table 18.1. The system parameters for all four different bit-rate modes are summarised in Table 18.2. The modes having effective video bit rates of 3.4 and 7.0 Mbps are equivalent to the bit rates of a fixed BPSK and QPSK mode transceiver, respectively.

Figure 18.7 shows the Frame Error Rate (FER) or video packet-loss ratio (PLR) performance versus channel SNR for the four different target bit rates of Table 18.2, demonstrating, as expected, that the higher target bit-rate modes require higher channel SNRs in order to operate within given PLR constraints. For example, the mode having an effective video bit rate of 9.8 Mbps can only operate for channel SNRs in excess of 19 dB under the constraint of a maximum PLR of 5%. However, the mode that has an effective video bit rate of 3.4 Mbps can operate at channel SNRs of 11 dB and above, while maintaining the same 5% PLR constraint, albeit at about half the throughput bit rate and so at a lower video quality.

The trade-offs between video quality and channel SNR for the various target bit rates can be judged from Figure 18.8, suggesting, as expected, that the higher target bit rates result in a higher video quality, provided that channel conditions are favourable. However, as the channel quality degrades, the video packet-loss ratio increases, thereby reducing the throughput bit rate and hence the associated video quality. The lower target bit-rate transceivers operate at an inherently lower video quality, but they are more robust to the prevailing channel condi-

Packet rate	4687.5 Packets/s			
FFT length	512			
OFDM symbols/packet	3			
OFDM symbol duration	$2.6667\mu s$			
OFDM time frame	80 time-slots $= 213\mu s$			
Normalised Doppler frequency, f'_d	1.235×10^{-4}			
OFDM symbol normalised Doppler frequency, F_D	7.41×10^{-2}			
FEC-coded bits/packet	858	1536	3072	4272
FEC-coded video bit rate	4.0 Mbps	7.2 Mbps	14.4 Mbps	20.0 Mbps
No. of unprotected bits/packet	427	766	1534	2134
Unprotected bit rate	2.0 Mbps	3.6 Mbps	7.2 Mbps	10.0 Mbps
No. of CRC bits	16	16	16	16
No. of feedback error flag bits	9	9	9	9
No. of packet header bits/packet	10	11	12	13
Effective video bits/packet	392	730	1497	2096
Effective video bit rate	1.8 Mbps	3.4 Mbps	7.0 Mbps	9.8 Mbps
Equivalent modulation mode		BPSK	QPSK	
Minimum channel SNR for 5% PLR (dB)	8.8	11.0	16.1	19.2
Minimum channel SNR for 10% PLR (dB)	7.1	9.2	14.1	17.3

Table 18.2: System parameters for the four different target bit rates of the various sub-band-adaptive OFDM (μAOFDM) transceivers

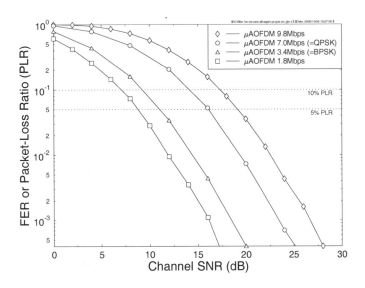

Figure 18.7: FER or video packet-loss ratio (PLR) versus channel SNR for the sub-band adaptive OFDM transceivers of Table 18.2 operating at four different target bit rates, over the channel model of Figure 18.3 at a normalised Doppler frequency of $F_D = 7.41 \times 10^{-2}$.

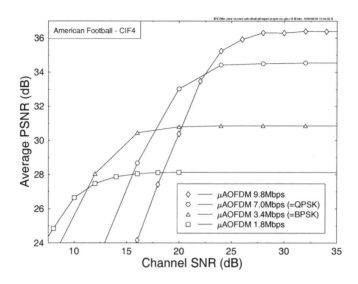

Figure 18.8: Average video quality expressed in PSNR versus channel SNR for the sub-band-adaptive OFDM transceivers of Table 18.2, operating at four different target bit rates, over the channel model of Figure 18.3 at a normalised Doppler frequency of $F_D = 7.41 \times 10^{-2}$.

tions and thus can operate at lower channel SNRs, while guaranteeing a video quality, which is essentially unaffected by channel errors. It was found that the perceived video quality became impaired for packet-loss ratios in excess of about 5%.

The trade-offs between video quality, packet-loss ratio, and target bit rate are further developed with reference to Figure 18.9. The figure shows the video quality measured in PSNR versus video frame index at a channel SNR of 16 dB as well as for an error-free situation. At the bottom of each graph, the packet-loss ratio per video frame is shown. The three figures indicate the trade-offs to be made in choosing the target bit rate for the specific channel conditions experienced – in this specific example for a channel SNR of 16dB. Note that under error-free conditions the video quality improved upon increasing the bit rate.

Specifically, video PSNRs of about 40, 41.5, and 43 dB were observed for the effective video bit rates of 1.8, 3.4, and 7.0 Mbps. The figure shows that for the target bit rate of 1.8 Mbps, the system has a high grade of freedom in choosing which subcarriers to invoke. Therefore, it is capable of reducing the number of packets that are lost. The packet-loss ratio remains low, and the video quality remains similar to that of the error-free situation. The two instances where the PSNR is significantly different from the error-free performance correspond to video frames in which video packets were lost. However, in both instances the system recovers in the following video frame.

As the target bit rate of the sub-band-adaptive OFDM transceiver is increased to 3.4 Mbps, the sub-band modulation mode selection process has to be more "aggressive", resulting in increased video packet loss. Observe in the figure that the transceiver having an effective video bit rate of 3.4 Mbps exhibits increased packet loss. In one frame as much as 5% of the packets transmitted for that video frame were lost, although the average PLR was only

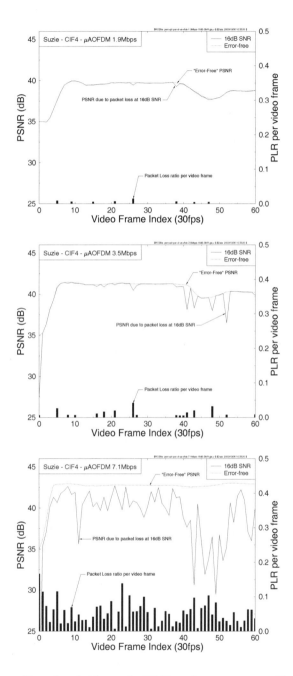

Figure 18.9: Video-quality and packet-loss ratio (PLR) performance versus video-frame index (time) comparison of sub-band-adaptive OFDM transceivers having target bit rates of 1.8, 3.4, and 7.0 Mbps, under the same channel conditions, at 16 dB SNR over the channel of Figure 18.3 at a normalised Doppler frequency of $F_D = 7.41 \times 10^{-2}$.

0.4%. Because of the increased packet loss, the video PSNR curve diverges from the error-free performance curve more often. However, in almost all cases the effects of the packet losses are masked in the next video frame, indicated by the re-merging PSNR curves in the figure, maintaining a close to error-free PSNR. The subjective effect of this level of packet loss is almost imperceivable.

When the target bit rate is further increased to 7.0 Mbps, the average PLR is about 5% under the same channel conditions, and the effects of this packet-loss ratio are becoming objectionable in perceived video-quality terms. At this target bit rate, there are several video frames where at least 10% of the video packets have been lost. The video quality measured in PSNR terms rarely reaches its error-free level, because every video frame contains at least one lost packet. The perceived video quality remains virtually unimpaired until the head movement in the "Suzie" video sequence around frames 40–50, where the effect of lost packets becomes obvious, and the PSNR drops to about 30 dB.

18.1.7 Time-Variant Target Bit Rate OFDM Transceivers

By using a high target bit rate, when the channel quality is high, and employing a reduced target bit rate, when the channel quality is poor, an adaptive system is capable of maximising the average throughput bit rate over a wide range of channel SNRs, while satisfying a given quality constraint. This quality constraint for our video system could be a maximum packet-loss ratio.

Because a substantial processing delay is associated with evaluating the packet-loss information, modem mode switching based on this metric is less efficient due to this latency. Therefore, we decided to invoke an estimate of the bit error ratio (BER) for mode switching, as follows. Since the noise energy in each subcarrier is independent of the channel's frequency domain transfer function H_n, the local signal-to-noise ratio (SNR) in subcarrier n can be expressed as

$$\gamma_n = |H_n|^2 \cdot \gamma, \tag{18.1}$$

where γ is the overall SNR. If no signal degradation due to Inter–Subcarrier Interference (ISI) or interference from other sources appears, then the value of γ_n determines the bit error probability for the transmission of data symbols over the subcarrier n. Given γ_j across the N_s subcarriers in the jth sub-band, the expected overall BER for all available modulation schemes M_n in each sub-band can be estimated, which is denoted by $\bar{p}_e(n) = 1/N_s \sum_j p_e(\gamma_j, M_n)$. For each sub-band, the scheme with the highest throughput, whose estimated BER is lower than a given threshold, is then chosen.

We decided to use a quadruple-mode switched sub-band-adaptive modem using the four target bit rates of Table 18.2. The channel estimator can then estimate the expected bit error ratio of the four possible modem modes. Our switching scheme opted for the modem mode, whose estimated BER was below the required threshold. This threshold could be varied in order to tune the behaviour of the switched sub-band-adaptive modem for a high or a low throughput. The advantage of a higher throughput was a higher error-free video quality at the expense of increased video packet losses, which could reduce the perceived video quality.

Figure 18.10 demonstrates how the switching algorithm operates for a 1% estimated BER threshold. Specifically, the figure portrays the estimate of the bit error ratio for the four

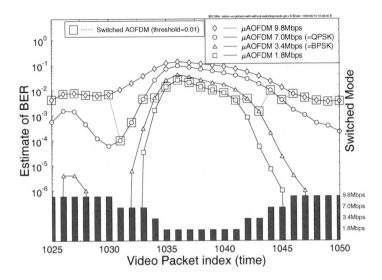

Figure 18.10: Illustration of mode switching for the switched sub-band adaptive modem. The figure shows the estimate of the bit error ratio for the four possible modes. The large square and the dotted line indicate the modem mode chosen for each time interval by the mode switching algorithm. At the bottom of the graph, the bar chart specifies the bit rate of the switched sub-band adaptive modem on the right-hand axis versus time when using the channel model of Figure 18.3 at a normalised Doppler frequency of $F_D = 7.41 \times 10^{-2}$.

possible modem modes versus time. The large square and the dotted line indicate the mode chosen for each time interval by the mode switching algorithm. The algorithm attempts to use the highest bit-rate mode, whose BER estimate is less than the target threshold, namely, 1% in this case. However, if all the four modes' estimate of the BER is above the 1% threshold, then the lowest bit-rate mode is chosen, since this will be the most robust to channel errors. An example of this is shown around frames 1035–1040. At the bottom of the graph a bar chart specifies the bit rate of the switched sub-band adaptive modem versus time in order to emphasise when the switching occurs.

An example of the algorithm, when switching among the target bit rates of 1.8, 3.4, 7, and 9.8 Mbps, is shown in Figure 18.11. The upper part of the figure portrays the contour plot of the channel SNR for each subcarrier versus time. The lower part of the figure displays the modulation mode chosen for each 32-subcarrier sub-band versus time for the time-variant target bit-rate (TVTBR) sub-band adaptive modem. It can be seen at frames 1051–1055 that all the sub-bands employ QPSK modulation. Therefore, the TVTBR-AOFDM modem has an instantaneous target bit rate of 7 Mbps. As the channel degrades around frame 1060, the modem has switched to the more robust 1.8 Mbps mode. When the channel quality is high around frames 1074–1081, the highest bit-rate 9.8 Mbps mode is used. This demonstrates that the TVTBR-AOFDM modem can reduce the number of lost video packets by using reduced bit-rate but more robust modulation modes, when the channel quality is poor. However, this is at the expense of a slightly reduced average throughput bit rate. Usually, a higher throughput bit rate results in a higher video quality. However, a high bit rate is also associated with a

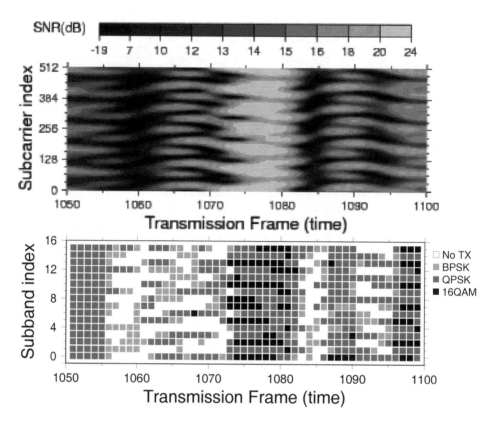

Figure 18.11: The micro-adaptive nature of the time-variant target bit-rate sub-band adaptive (TVTBR-AOFDM) modem. The top graph is a contour plot of the channel SNR for all 512 subcarriers versus time. The bottom graph shows the modulation mode chosen for all 16 subbands for the same period of time. Each sub-band is composed of 32 subcarriers. The TVTBR AOFDM modem switches between target bit rates of 2, 3.4, 7, and 9.8 Mbps, while attempting to maintain an estimated BER of 0.1% before channel coding. Average channel SNR is 16 dB over the channel of Figure 18.3 at a normalised Doppler frequency of $F_D = 7.41 \times 10^{-2}$.

high packet-loss ratio, which is usually less attractive in terms of perceived video quality than a lower bit-rate, lower packet-loss ratio mode.

Having highlighted how the time domain mode switching algorithm operates, we will now characterise its performance for a range of different BER switching thresholds. A low BER switching threshold implies that the switching algorithm is cautious about switching to the higher bit-rate modes. Therefore the system performance is characterised by a low video packet-loss ratio and a low throughput bit rate. A high BER switching threshold results in the switching algorithm attempting to use the highest bit-rate modes in all but the worst channel conditions. This results in a higher video packet-loss ratio. However, if the packet-loss ratio is not excessively high, a higher video throughput is achieved.

Figure 18.12 portrays the video packet-loss ratio or FER performance of the TVTBR-

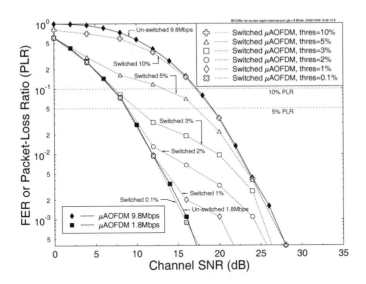

Figure 18.12: FER or video packet-loss ratio versus channel SNR for the TVTBR-AOFDM modem for a variety of BER switching thresholds. The switched modem uses four modes, with target bit rates of 1.8, 3.4, 7, and 9.8 Mbps. The unswitched 1.8 and 9.8 Mbps results are also shown in the graph as solid markers using the channel model of Figure 18.3 at a normalised Doppler frequency of $F_D = 7.41 \times 10^{-2}$.

AOFDM modem for a variety of BER thresholds, compared to the minimum and maximum rate unswitched modes. For a conservative BER switching threshold of 0.1%, the time-variant target bit-rate sub-band adaptive (TVTBR-AOFDM) modem has a similar packet-loss ratio performance to that of the 1.8 Mbps nonswitched or constant target bit-rate (CTBR) sub-band adaptive modem. However, as we will show, the throughput of the switched modem is always better than or equal to that of the unswitched modem and becomes far superior, as the channel quality improves. Observe in the figure that the "aggressive" switching threshold of 10% has a similar packet-loss ratio performance to that of the 9.8 Mbps CTBR-AOFDM modem. We found that in order to maintain a packet-loss ratio of below 5%, the BER switching thresholds of 2 and 3% offered the best overall performance, since the packet-loss ratio was fairly low, while the throughput bit rate was higher than that of an unswitched CTBR-AOFDM modem.

A high BER switching threshold results in the switched sub-band adaptive modem transmitting at a high average bit rate. However, we have shown in Figure 18.12 how the packet-loss ratio increases as the BER switching threshold increases. Therefore, the overall useful or effective throughput bit rate – that is, the bit rate excluding lost packets – may in fact be reduced in conjunction with high BER switching thresholds. Figure 18.13 demonstrates how the transmitted bit rate of the switched TVTBR-AOFDM modem increases with higher BER switching thresholds. However, when this is compared to the effective throughput bit rate, where the effects of packet loss are taken into account, the trade-off between the BER switching threshold and the effective bit rate is less obvious. Figure 18.14 portrays the corresponding effective throughput bit rate versus channel SNR for a range of BER switching thresholds. The figure demonstrates that for a BER switching threshold of 10% the effec-

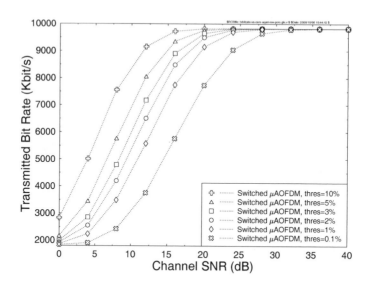

Figure 18.13: Transmitted bit rate of the switched TVTBR-AOFDM modem for a variety of BER switching thresholds. The switched modem uses four modes, having target bit rates of 1.8, 3.4, 7, and 9.8 Mbps, over the channel model of Figure 18.3 at a normalised Doppler frequency of $F_D = 7.41 \times 10^{-2}$.

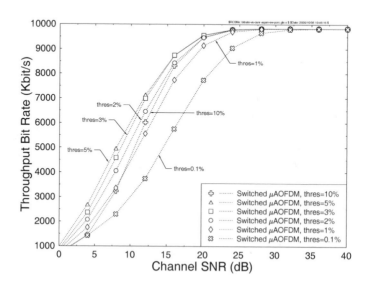

Figure 18.14: Effective throughput bit rate of the switched TVTBR-AOFDM modem for a variety of BER switching thresholds. The switched modem uses four modes, with target bit rates of 1.8, 3.4, 7, and 9.8 Mbps. The channel model of Figure 18.3 is used at a normalised Doppler frequency of $F_D = 7.41 \times 10^{-2}$.

tive throughput bit-rate performance was reduced in comparison to some of the lower BER switching threshold scenarios. Therefore, the BER = 10% switching threshold is obviously too aggressive, resulting in a high packet-loss ratio and a reduced effective throughput bit rate. For the switching thresholds considered, the BER = 5% threshold achieved the highest effective throughput bit rate. However, even though the BER = 5% switching threshold produces the highest effective throughput bit rate, this is at the expense of a relatively high video packet-loss ratio, which, as we will show, has a detrimental effect on the perceived video quality.

We will now demonstrate the effects associated with different BER switching thresholds on the video quality represented by the peak-signal-to-noise ratio (PSNR). Figure 18.15 portrays the PSNR and packet-loss performance versus time for a range of BER switching thresholds. The top graph in the figure indicates that for a BER switching threshold of 1% the PSNR performance is very similar to the corresponding error-free video quality. However, the PSNR performance diverges from the error-free curve when video packets are lost, although the highest PSNR degradation is limited to 2 dB. Furthermore, the PSNR curve typically reverts to the error-free PSNR performance curve in the next frame. In this example, about 80% of the video frames have no video packet loss. When the BER switching threshold is increased to 2%, as shown in the centre graph of Figure 18.15, the video-packet loss ratio has increased, such that now only 41% of video frames have no packet loss. The result of the increased packet loss is a PSNR curve, which diverges from the error-free PSNR performance curve more regularly, with PSNR degradations of up to 7 dB. When there are video frames with no packet losses, the PSNR typically recovers, achieving a similar PSNR performance to the error-free case. When the BER switching threshold was further increased to 3% – which is not shown in the figure – the maximum PSNR degradation increased to 10.5 dB, and the number of video frames without packet losses was reduced to 6%.

The bottom graph of Figure 18.15 depicts the PSNR and packet loss performance for a BER switching threshold of 5%. The PSNR degradation in this case ranges from 1.8 to 13 dB and all video frames contain at least one lost video packet. Even though the BER = 5% switching threshold provides the highest effective throughput bit rate, the associated video quality is poor. The PSNR degradation in most video frames is about 10 dB. Clearly, the highest effective throughput bit rate does not guarantee the best video quality. We will now demonstrate that the switching threshold of BER = 1% provides the best video quality, when using the average PSNR as our performance metric.

Figure 18.16(a) compares the average PSNR versus channel SNR performance for a range of switched (TVTBR) and unswitched (CTBR) AOFDM modems. The figure compares the four unswitched (i.e., CTBR sub-band adaptive modems) with switching (i.e., TVTBR sub-band adaptive modems), which switch between the four fixed-rate modes, depending on the BER switching threshold. The figure indicates that the switched TVTBR sub-band adaptive modem having a switching threshold of BER = 10% results in similar PSNR performance to the unswitched CTBR 9.8 Mbps sub-band adaptive modem. When the switching threshold is reduced to BER = 3%, the switched TVTBR AOFDM modem outperforms all of the unswitched CTBR AOFDM modems. A switching threshold of BER = 5% achieves a PSNR performance, which is better than the unswitched 9.8 Mbps CTBR AOFDM modem, but worse than that of the unswitched 7.0 Mbps modem, at low- and medium-channel SNRs.

A comparison of the switched TVTBR AOFDM modem employing all six switching thresholds that we have used previously is shown in Figure 18.16(b). This figure suggests

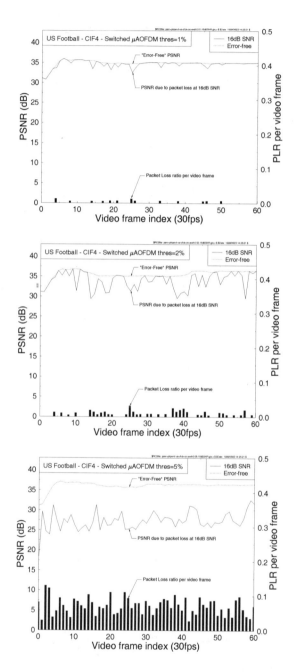

Figure 18.15: Video-quality and packet-loss ratio performance versus video-frame index (time) comparison between switched TVTBR-AOFDM transceivers with different BER switching thresholds, at an average of 16dB SNR, using the channel model of Figure 18.3 at a normalised Doppler frequency of $F_D = 7.41 \times 10^{-2}$.

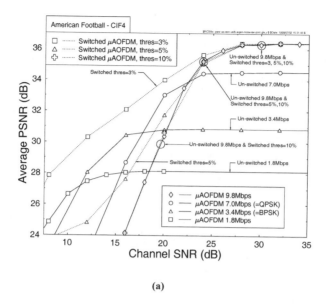

(a)

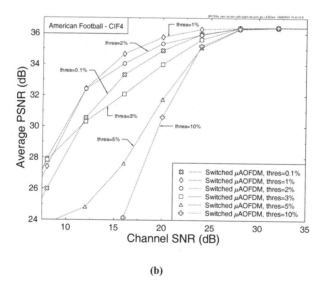

(b)

Figure 18.16: Average PSNR versus channel SNR performance for switched and unswitched sub-band adaptive modems. Figure (a) compares the four unswitched CTBR sub-band adaptive modems with switched TVTBR sub-band adaptive modems (using the same four modem modes) for switching thresholds of BER = 3, 5, and 10%. Figure (b) compares the switched TVTBR sub-band adaptive modems for switching thresholds of BER = 0.1, 1, 2, 3, 5, and 10%.

that switching thresholds of BER = 0.1, 1, and 2% perform better than the BER = 3% threshold, which outperformed all of the unswitched CTBR sub-band adaptive modems. The best average PSNR performance was achieved by a switching threshold of BER = 1%. The more conservative BER = 0.1% switching threshold results in a lower PSNR performance, since its throughput bit rate was significantly reduced. Therefore, the best trade-off in terms of PSNR, throughput bit rate, and video packet-loss ratio was achieved with a switching threshold of about BER = 1%.

18.2 Multi-user Detection-Assisted Turbo-Coded OFDM/H.263 Interactive Videotelephony in HIPERLAN-like Systems[3]

18.2.1 Overview

Recently a substantial amount of research has been conducted on combining Orthogonal Frequency Division Multiplexing (OFDM) [2] with multiple-antenna reception assisted co-channel interference suppression. Since unique user-specific transmit antennas, i.e. spatial signatures, are used, these arrangements are also referred to as Space-Division Multiple Access (SDMA) techniques. One of the key issues is the separation of different users, which can be performed based on their spatial signature, assuming the knowledge of the channel parameters. Numerous algorithms have been proposed for performing the task of user separation. The so-called Sample-Matrix Inversion (SMI) [130, 134] algorithm's roots date back to the early years of development in this research area. Other related algorithms, which have recently attracted interest are, for example, the Successive Interference Cancellation (SIC) technique originally proposed in the context of Lucent's V-BLAST system advocated by Foschini et al. in [104, 109]. These techniques were further considered by Vandenameele and Münster in [111] and [163], respectively. Furthermore, a Parallel Interference Cancellation (PIC) algorithm has been considered in [163], while Maximum-Likelihood (ML) combining and potential complexity reduction strategies were discussed by Awater et al. in [115]. A similar contribution was presented by Speth et al. in [112], where in contrast to [115], the focus is on the problem of maximum likelihood soft-bit generation. A comparative study of these detection schemes was also presented in [163], which concluded that MMSE combining has the lowest-complexity and hence it is the least effective detection scheme. The best performance was exhibited by the ML combiner at the cost of a substantially increased complexity, particularly in the context of higher-order modulation schemes, such as 16QAM. The complexity of PIC and SIC was observed to be within these two boundaries, with an advantage in favour of SIC, specifically in high-complexity scenarios, such as for example in conjunction with four receiver antennas, potentially supporting four simultaneous users. Hence, we opted for focussing our further investigations on the MMSE antenna-array output combining and the ML combining schemes, more precisely on the MLSE soft-bit generation approach proposed by Speth et al., which was found to be more amenable to our turbo-coded transmission scenario.

---

[3]This section is based on P.J. Cherriman, T. Keller, and L. Hanzo: "Sub-band-adaptive HIPERLAN-like Turbo Coded Multi-user Detection Assisted OFDM Based Interactive Video Telephony", VTC 2001 Fall, Atlantic City, 7-10 October 2001, USA, pp. 87-91 ©2002 IEEE.

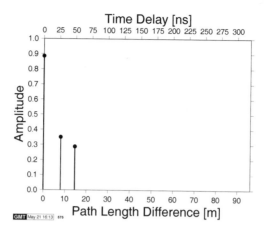

Figure 18.17: Short WATM channel impulse response [2].

The description of our OFDM-based video system is given in Section 18.2.2, which also comprises a discussion of the specific transmission packet structure employed. The OFDM signal model is outlined in Section 18.2.3, while the MMSE and ML antenna array output combining strategies are briefly reviewed in Section 18.2.4. Modem and video performance results are provided in Section 18.2.5 for the three-path fading indoor Wireless Asynchronous Transfer Mode (WATM) channel model of Figure 18.17 [2, 219].

18.2.2 The Video System

18.2.2.1 OFDM Transceiver

We assume an OFDM-SDMA uplink environment, where the number of co-channel users is upper-bound by the number of antenna elements associated with the receiver array at the base station. By contrast, each user is equipped with a single transmission antenna. In our OFDM scheme 512 subcarriers and a 64-sample guard-interval was used, in order to avoid inter-symbol interference. Five consecutive OFDM symbols were hosted by a packet, which was transmitted at specific intervals, in order to fulfil the bitrate requirements of the video codec. The packet structure is portrayed in Figure 18.18. Specifically, the first OFDM symbol of a packet is a dedicated training symbol, hosting exclusively pilot tones, while the following four OFDM symbols host exclusively data symbols. More specifically, the training OFDM symbol exhibits the structure illustrated in Figure 18.19, potentially supporting four simultaneous users.

There are two typical types of pilot arrangements, which can be differentiated on the basis, whether the different users are assigned separate dedicated pilot positions. Specifically, in case of separate dedicated pilot positions the l-th user for example would transmit pilot tones at the subcarrier positions indicated by R_{l-1} $l = 1 \ldots 4$ in Figure 18.19, while transmitting no symbols at the other subcarrier frequencies. At the receiver MMSE lowpass interpolation is performed separately for generating a channel transfer function estimate, in order to equalise the frequency domain signals received by the different antenna array elements. Again, this is carried out separately between all pilot symbols of the same relative index $l - 1$ associated

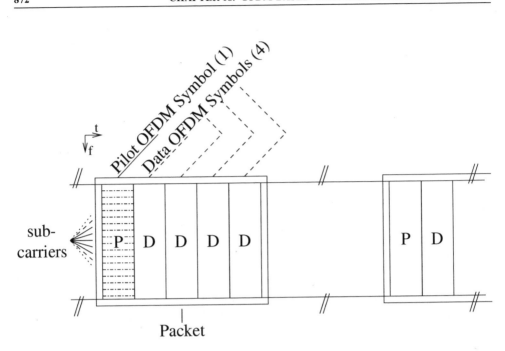

Figure 18.18: Packet structure of the proposed multi-user video transmission arrangement. A dedicated pilot OFDM symbol is transmitted at the beginning of the five-symbol burst, followed by a set of four OFDM symbols hosting exclusively data subcarriers.

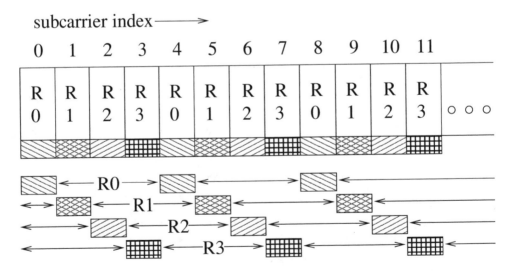

Figure 18.19: Pilot arrangement hosted by the first OFDM symbol in the video packet to simultaneously estimate the channel transfer functions between the different users' transmit antennas and the multiple array elements at the Base Station's (BS) receiver.

with the *l*-th user, as seen in Figure 18.19. This procedure allows us to generate a subcarrier-based estimate of the channel transfer function between the single transmit antenna of the *l*-th user and the reception antenna array element considered.

The second pilot arrangement allows different users to share the same pilot positions. This is achieved upon invoking the principles of code-division known from MC-CDMA. Specifically, spreading can be implemented by means of assigning an individual orthogonal code to each user. The *l*-th user would hence spread each single pilot tone across several subcarriers, as shown in Figure 18.19, in all consecutive blocks hosting indices R_0-R_3. At the receiver, again, interpolation is carried out separately across subcarriers having the same relative index R_i, followed by cross-correlation with the user's individual spreading codes, in order to extract the different channel transfer functions.

Both pilot-based channel estimation methods are capable of achieving the same estimation performance, assuming that in the case of separate pilot tones each user is allowed to boost its transmit power by a factor equal to the number of simultaneous users. This power-boosting measure can be invoked, since – as outlined before – no symbols are transmitted on subcarriers between the pilot positions. The expected difference between the above-mentioned two methods is likely to be their different Peak-to-Average Power Ratio (PAPR) at the output of the OFDM modulator – an issue, which requires further investigations.

Throughout the simulations conducted in Section 18.2.5 we assumed "packet-invariant" fading, where again, a packet was constituted by five OFDM symbols. Accordingly, the fading magnitude and phase was considered constant during the packet hosting a total of five OFDM symbols. This assumption is justified by the low OFDM symbol-normalised Doppler frequency of $F_d = 1.532 \cdot 10^{-3}$ encountered in the indoor WATM channel scenario of [2] at the maximum pedestrian speed of $3m/s$, which was assumed here.

Furthermore, for our simulations we assumed that the fading experienced by consecutive video packets was uncorrelated, which is justified by the relatively high time domain separation between two consecutive video packets at the bitrates envisaged. In addition, a turbo convolutional codec has been incorporated into the system, which resulted in a substantial Bit Error Ratio (BER) reduction. The turbo coding parameters can be summarised as follows. The coding rate was $R = 1/2$, the constraint length was $K = 3$, the octally represented generator-polynomials of $\{7, 5\}$ were used, and four turbo-decoding iterations were performed.

18.2.2.2 Video Transmission Regime

In this contribution we transmitted 176x144 pixel Quarter Common Intermediate Format (QCIF) video sequences at 30 frames/s using the above OFDM transceiver, which can be configured as a 1, 2 or 4 bit/symbol scheme. The H.263 video codec was used [7], which extensively employs variable-length compression techniques and hence achieves a high compression ratio. However, as all entropy- and variable-length coded bit streams, its bits are extremely sensitive to transmission errors.

This error sensitivity was counteracted in our system by invoking an adaptive packetisation and packet dropping regime, when the channel codec protecting the video stream became incapable of removing all channel errors. Specifically, we refrained from decoding the corrupted video packets in order to prevent error propagation through the reconstructed video frame buffer [503]. Hence these corrupted video packets were dropped at both the transmit-

Features	Multi-rate system		
Mode	BPSK	4QAM	16QAM
Coded data bits/packet	254	510	1020
Packet Rate (packets/s)	976.560000		
Transmission data bitrate (kbit/s)	248.0	498.0	996.1
Video packet CRC (bits)	16		
Feedback protection (bits)	15		
Video packet header (bits)	9	10	11
Video bits/packet	214	469	978
Effective Video-rate (kbit/s)	209.0	458.0	955.1
Video framerate (Hz)	30		

Table 18.3: Video system parameters

ter and receiver and the reconstructed frame buffer was not updated, until the next error-free video packet replenishing the specific video frame area was received. This required a low-delay, strongly protected video packet acknowledgement flag, which was superimposed on the transmitted payload packets [503]. The associated video performance degradation was found to be perceptually unobjectionable for Packet Loss Rates (PLR) below about 5%, although this issue will be detailed in more depth during our further discourse. Due to lack of space some of the implementation details of the video packetisation and transmission regime are omitted, which can be found in [503].

More explicitly, the H.263 decoder discards all the corrupted video packets. A 16 bit Cyclic Redundancy Check (CRC) is added to every video packet for detecting corrupted packets. In addition to the CRC, each video packet has a header, containing information which allows the packet disassembly block to appropriately combine the macroblocks' video information, which has been mapped to and transmitted in several consecutive video transmission packets. The video encoder receives feedback information from the decoder, informing it of the success or failure of the previous packet. In case of corruption the packet disassembly block effectively conceals the packet loss by replacing the affected area of the picture with the corresponding area from the previous video frame.

In addition to rendering the video information more robust the packet assembly and disassembly block is capable of assembling video packets containing more bits, when higher order modulation modes are used in case of a high instantaneous channel quality. The bitrate control machanism of the video encoder is closely linked to the packet assembly block, in order to compensate for the rapid changes in the QAM transmission mode and the resultant packet size. The video system's parameters are summarised in Table 18.3.

18.2.3   The OFDM Signal Model

In Figure 18.20 we have illustrated a typical OFDM multi-user uplink scenario. More specifically, the $(P \times 1)$-dimensional vector of complex signals, $\mathbf{x}[n, k]$, received by the P-element antenna array in the k-th subcarrier of the n-th OFDM symbol is constituted by the superposition of the independently faded signals associated with the L users sharing the same space-frequency resource. The received signal was corrupted by the Gaussian noise at the

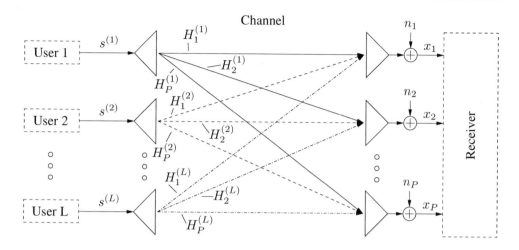

Figure 18.20: Multi-User OFDM Uplink Scenario as observed in the frequency domain on a subcarrier basis; in this model the transmitted symbols $s^{(l)}[n,k]$ are multiplied with uncorrelated channel transfer factors $h_m^{(l)}[n,k]$ for each user $l \in \{1,\ldots,L\}$ and for each antenna element $p \in \{1,\ldots,P\}$; the symbol and subcarrier indices $[n,k]$ have been omitted for notational simplicity.

array elements. The indices $[n,k]$ have been omitted for notational convenience during our forthcoming discourse, yielding:

$$\mathbf{x} = \mathbf{H}\mathbf{s} + \mathbf{n}, \tag{18.2}$$

where the $(P \times 1)$-dimensional vector of received signals $\mathbf{x}$, the vector of transmitted signals $\mathbf{s}$ and the array noise vector $\mathbf{n}$, respectively, are given by:

$$\mathbf{x} = (x_1, x_2, \ldots, x_P)^T, \tag{18.3}$$

$$\mathbf{s} = (s^{(1)}, s^{(2)}, \ldots, s^{(L)})^T, \tag{18.4}$$

$$\mathbf{n} = (n_1, n_2, \ldots, n_P)^T. \tag{18.5}$$

The frequency domain channel transfer function matrix $\mathbf{H}$ of dimension $L \times P$ is constituted by the set of frequency domain channel transfer functions of the L users:

$$\mathbf{H} = (\mathbf{h}^{(1)}, \mathbf{h}^{(2)}, \ldots, \mathbf{h}^{(L)}), \tag{18.6}$$

between the single transmitter antenna associated with a particular user l and the reception array elements $p \in \{1,\ldots,P\}$ in the form of:

$$\mathbf{h}^{(l)} = (h_1^{(l)}, h_2^{(l)}, \ldots, h_P^{(l)})^T, \tag{18.7}$$

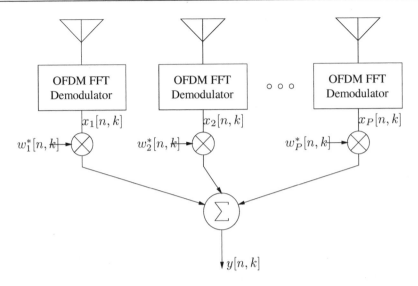

Figure 18.21: Array output combiner in an OFDM system

with $l \in \{1, \dots, L\}$. Regarding the statistical properties of the vector components in Equations 18.3 and 18.7, we assume that the complex data signal $s^{(l)}$ transmitted by the l-th user has zero-mean and unit variance. The AWGN noise process n_p at any antenna array element p exhibits also zero-mean and a variance of σ^2. The frequency domain channel transfer functions $h_p^{(l)}$ for the different array elements $p \in \{1, \dots, P\}$ or users $l \in \{1, \dots, L\}$ are independent, stationary, complex Gaussian distributed processes with zero-mean and different variances σ_l^2 [130].

18.2.4   Co-Channel Interference Cancellation Techniques

In this section we will provide a brief overview of the two antenna array output combining strategies considered here, specifically that of MMSE combining and ML combining.

The SMI-Algorithm The SMI algorithm [130, 134] generates the optimum antenna array weights in the sense of the Minimum Mean Square Error (MMSE) between the associated combiner's output signal and a reference signal known to the receiver. The comniner's output vector $\mathbf{y}$, hosting the L different users' signals is hence given by:

$$
\begin{aligned}
\mathbf{y} &= (y^{(1)}, y^{(2)}, \dots, y^{(L)})^T \\
&= \mathbf{W}_{\text{MMSE}}^H \mathbf{x},
\end{aligned} \tag{18.8}
$$

which is further illustrated in Figure 18.21. The $(P \times L)$-dimensional antenna weight matrix $\mathbf{W}_{\text{MMSE}}$ is given by [130, 134]:

$$
\begin{aligned}
\mathbf{W}_{\text{MMSE}} &= \left(\mathbf{w}_{\text{MMSE}}^{(1)}, \mathbf{w}_{\text{MMSE}}^{(2)}, \dots, \mathbf{w}_{\text{MMSE}}^{(L)} \right) \\
&= (\mathbf{H}\mathbf{H}^H + \sigma^2 \mathbf{I})^{-1} \mathbf{H},
\end{aligned} \tag{18.9}
$$

as the direct solution of the Wiener-equation in the context of a multi-user scenario. Capitalising on Equations 18.2, 18.8 and 18.9, the SNR at the combiner's output for the l-th user is given by [136]:

$$SNR^{(l)} = \left(\mathbf{w}_{\mathrm{MMSE}}^{(l)H} \mathbf{h}^{(l)} \mathbf{h}^{(l)H} \mathbf{w}_{\mathrm{MMSE}}^{(l)}\right) \cdot \left(\sigma^2 \mathbf{w}_{\mathrm{MMSE}}^{(l)H} \mathbf{w}_{\mathrm{MMSE}}^{(l)}\right)^{-1},$$

(18.10)

where $l \in \{1, \dots, L\}$. Equation 18.10 will be invoked in conjunction with turbo-coding in order to generate the soft-input information for the turbo decoder in Section 18.2.5. In the next paragraph we will briefly review the MLSE algorithm.

The MLSE-Algorithm The second algorithm studied is the MLSE soft-bit generation algorithm proposed by Speth *et al.* in [112]. This technique constitutes a derivative of the ML combiner, which can be employed in both coded and uncoded systems after applying hard decisions to the so-called Log-Likelihood Ratios (LLR) of the bits. This approach is based on the idea of detecting one out of L users at a time, so as to avoid the complex joint detection of all users by a Viterbi-like tree-search based detector. In order to avoid the explicit evaluation of the exponential terms which constitute the LLR-values, Speth *et al.* [112] proposed transforming the received signal vector $\mathbf{x}[n, k]$ of the P-element array into a so-called trial-vector using the ML estimate $\hat{\mathbf{s}}$ of the L-dimensional transmitted signal vectors $\mathbf{s}$ of the L users, which is given similarly to Equation 18.8 by:

$$\hat{\mathbf{s}} = \mathbf{W}_{\mathrm{ML}}^{H}\mathbf{x}, \quad \text{with} \tag{18.11}$$

$$\mathbf{W}_{\mathrm{ML}} = (\mathbf{H}\mathbf{H}^{H})^{-1}\mathbf{H}, \tag{18.12}$$

once again, invoking the signal model outlined in Section 18.2.3. It should be noted that the ML estimate of $\mathbf{s}$, namely $\hat{\mathbf{s}}$ of Equation 18.11 maximises the SIR at the combiner's output. This is in contrast to the MMSE combiner defined by Equations 18.8 and 18.9, which seeks to maximise the joint effects of both the noise and interference expressed in terms of the Signal-to-Interference plus Noise Ratio (SINR). Capitalising on the so-called maximum approximation of the exponential LLR terms [112], Speth *et al.* applied the following log-likelihood value for the first user [112]:

$$L_{1,k} \approx (\hat{\mathbf{s}} - \check{\mathbf{s}}_k^{0x})^{H}\mathbf{R}_{\epsilon}^{-1}(\hat{\mathbf{s}} - \check{\mathbf{s}}_k^{0x}) - (\hat{\mathbf{s}} - \check{\mathbf{s}}_k^{1x})^{H}\mathbf{R}_{\epsilon}^{-1}(\hat{\mathbf{s}} - \check{\mathbf{s}}_k^{1x}),$$

(18.13)

where $\mathbf{R}_{\epsilon}$ denotes the covariance matrix of the estimation error $\epsilon = \hat{\mathbf{s}} - \mathbf{s}$, which is given by [112]:

$$\mathbf{R}_{\epsilon} = \sigma^2(\mathbf{H}^{H}\mathbf{H})^{-1}, \tag{18.14}$$

and $\check{\mathbf{s}} = \check{\mathbf{s}}_k^{bx}$ represents the optimum trial-vector defined in [112] as:

$$\arg\min_{\check{\mathbf{s}} \in \mathcal{M}_k^{bx}} |(\hat{\mathbf{s}} - \check{\mathbf{s}})^{H}\mathbf{R}_{\epsilon}^{-1}(\hat{\mathbf{s}} - \check{\mathbf{s}})|^2, \tag{18.15}$$

where the minimisation is performed over all two-user trial-vectors of the form $\check{\mathbf{s}} = (\check{s}_1, \check{s}_2)^{T}$

and the trial-vectors are contained in the set:

$$\mathcal{M}_k^{bx} = \left\{ \begin{pmatrix} \check{s}_1 \\ \check{s}_2 \end{pmatrix} | \check{s}_1 \in \mathcal{M}_k^b, \check{s}_2 \in \mathcal{M} \right\}. \qquad (18.16)$$

In Equation 18.16, $\mathcal{M}$ denotes the set of constellation points associated with a particular modulation scheme. For example, we have $M = 4$ elements in $\mathcal{M}$ for 4-QAM, and $\mathcal{M}_k^{bx}$ is the specific reduced-size subset of the M^L possible symbol combinations of the L users, comprising only those constellation points, which have an associated bit combination that carries a binary value b at bit position k. We have now determined the soft-bit value $L_{1,k}$ of user 1 at bit-position k. In order to obtain the soft-bit values of user 2, the trial set given by Equation 18.16 has to be modified correspondingly, by exchanging the appropriate indices. The computational complexity associated with evaluating Equation 18.15 strongly depends on the size M^L of the "search-set". Each additional user increases the size of the trial-set by one "dimension".

18.2.5   Video System Performance Results

Again, the video system's parameters are summarised in Table 18.3. The two-antenna, two-user system's BER versus channel SNR performance is shown in Figure 18.22, indicating that the MLSE algorithm consistently outperforms the MMSE algorithm. The corresponding video Packet Loss Ratio (PLR) results are portrayed in Figure 18.23, where again, a video packet was constituted by four consecutive OFDM symbols. The PLR curves suggest similar performance trends to the BER curves of Figure 18.22. When the number of users and antennas is increased to four, the BER performance results of Figure 18.24 are obtained.

For the sake of convenience, Figure 18.25 compares the BER versus channel SNR performance of the MMSE algorithm for the two-antenna, two-user scenario and for the four-antenna, four-user system. Observe that the four-antenna scenario outperforms the two-antenna scenario. This is a consequence of having four independent diversity channels for each of the four users, rather than only two.

Observe furthermore that the performance difference of the two- and four-user scenarios is more pronounced for the lower-order modulation schemes. Specifically, a 2-3 dB SNR difference can be observed for BPSK, which is reduced to about 1-1.5 dB for 16QAM. The same performance trends become also explicit in the PLR versus channel SNR performance graphs displayed in Figure 18.26.

Here the effective video throughput bitrate is defined as the video bitrate provided by the correctly received packets, since the erroneously received packets are discarded. Therefore, the effective thoughput bitrate is that of the decoded error-free video stream.

The effective video throughput bitrate is shown in Figure 18.27 as a function of the channel SNR. Since the MLSE scheme outperformed the MMSE algorithm in Figures 18.22 and 18.23 in terms of the achievable BER and PLR, respectively, similar trends are observed also in the context of the throughput. However, the performance difference is less apparent in terms of the achievable throughput.

The average Peak-Signal-to-Noise-Ratio (PSNR) objective video quality metric was invoked in Figure 18.28 in order to demonstrate that the average video quality degrades, as the PLR increases. Our informal subjective video quality investigations indicated that video PSNR degradations of 1-2dB are unobjectionable, suggesting that the system is capable of

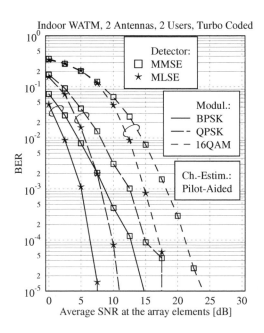

Figure 18.22: BER versus channel SNR performance of the turbo coded OFDM system using MMSE or ML array output combining at the two-element antenna array-assisted receiver supporting two simultaneous users of equal signal power, for 1, 2 and 4 bit/symbol modulation modes when using pilot-aided channel estimation according to Figures 18.18 and 18.19 over the WATM channel of Figure 18.17.

operating at PLRs of up to about 5%. When the PLR monitored by the system exceeds the above threshold value, the transceiver is instructed by the system controller to switch to a more robust modulation mode, before the subjective video quality becomes noticably degraded.

Figure 18.28 illustrates the robustness of the video transmission scheme, since for example in the BPSK mode the average PSNR only drops by about 5dB, when 30% of the packets are corrupted and discarded. The PSNR versus PLR performance of the MMSE and MLSE algoritm is fairly similar for a given PLR, although at a given channel SNR the MMSE algorithm always exhibits a higher PLR. Furthermore, the performance of the four-antenna, four-user system is almost identical to that of the two-antenna, two-user scheme shown in Figure 18.28. This is because the video performance is closely related to the PLR, while other modem performance metrics have a limited effect on the entire system's video performance.

Figures 18.29 and 18.30, characterise the average PSNR video quality versus channel SNR performance. As we have seen before, the MLSE algorithm outperforms the MMSE

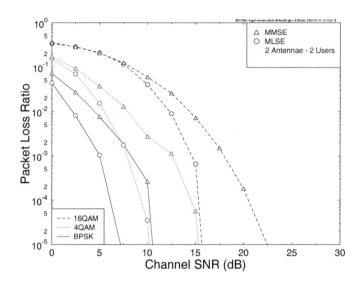

Figure 18.23: PLR versus channel SNR for the two-antenna, two-user system for transmission over the WATM channel of Figure 18.17.

algorithm. The performance difference is more noticable than that seen in the context of the video throughput bitrate. The MLSE algorithm allows the video system to operate at about 5dB lower channel SNRs, than the MMSE algorithm, although this is achieved at a significantly higher complexity. Furthermore, the four-antenna, four-user scenario requires a lower channel SNR, than the two-antenna, two-user scenario. In order to study the worst-case scenario, whilst maintaining a lower complexity, in our further investigations the MMSE algorithm was used.

In order to provide a further perspective on the MMSE system's performance, Figure 18.31 shows the effective video throughput bitrate versus channel SNR performance for both the two- and four-user scenarios. Again, the four-antenna, four-user system exhibits a better performance. However, as before, the throughput bitrate difference between the two- and four-user systems is modest, except when the bitrate starts to degrade, as the channel SNR degrades.

Recall from Figures 18.25 and 18.26 that the performance difference between the two- and four-antenna scenarios is higher for the lower-order modulation modes and the performance difference decreases, as the number of bits per modulated symbol is increased. The corresponding average PSNR or video quality versus channel SNR graph is shown in Figure 18.32, quantifying the video performance difference of the two- and four-antenna scenarios. Again, the performance difference reduces, as the modulation order is increased.

18.3   Chapter Summary and Conclusion

In this chapter we considered a few application examples in the context of OFDM-based wireless video telephony. Specifically, in Section 18.1 a range of AOFDM video transceivers

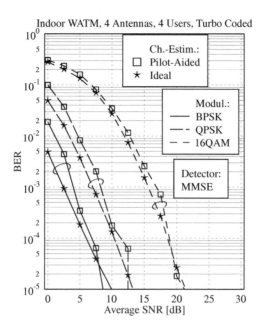

Figure 18.24: BER versus channel SNR performance of the turbo coded OFDM system using MMSE or ML array output combining at the four-element antenna array-assisted receiver supporting four simultaneous users of equal signal power, for 1, 2 and 4 bit/symbol modulation modes when using pilot-aided channel estimation according to Figures 18.18 and 18.19 over the WATM channel of Figure 18.17.

has been proposed for robust, flexible and low-delay interactive videotelephony. In order to minimise the amount of signalling required, we divided the OFDM subcarriers into sub-bands and controlled the modulation modes on a sub-band-by-sub-band basis. The proposed constant target bit-rate AOFDM modems provided a lower BER than the corresponding conventional OFDM modems. The slightly more complex switched TVTBR-AOFDM modems can provide a balanced video-quality performance, across a wider range of channel SNRs than the other schemes investigated.

In Section 18.2 multi-user detection-assisted, multiple transmit antenna based OFDM schemes were studied in the context of HIPRLAN 2-like systems. It was demonstrated that the system's user capacity can be improved with the aid of unique spatial user signatures. The MLSE detection algorithm outperformed the MMSE scheme by about 5dB in terms of the required SNR and this performance gain manifested itself also in terms of the system's video performance.

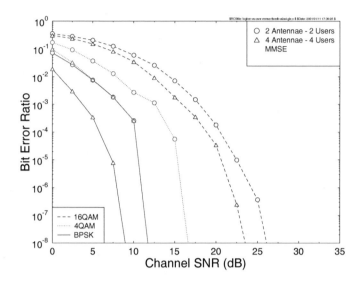

Figure 18.25: BER versus Channel SNR performance of the MMSE OFDM scheme for a two-antenna, two-user scenario and a four-antenna, four-user scenario for transmission over the WATM channel of Figure 18.17.

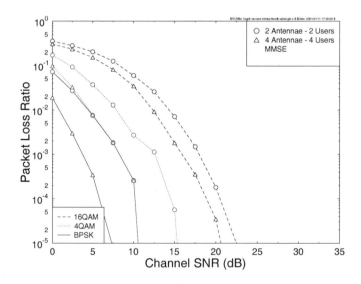

Figure 18.26: PLR versus Channel SNR performance of the MMSE OFDM scheme for a two-antenna, two-user scenario and a four-antenna, four-user scenario for transmission over the WATM channel of Figure 18.17.

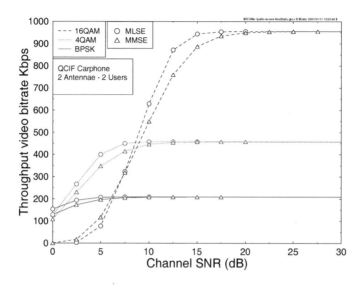

Figure 18.27: Effective video throughput bitrate versus channel SNR for the two-antenna, two-user OFDM system for transmission over the WATM channel of Figure 18.17.

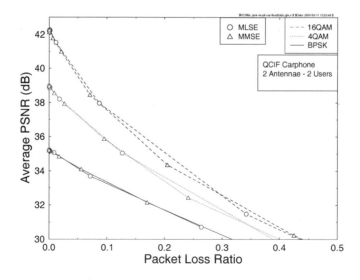

Figure 18.28: Average PSNR versus PLR of the two-antenna, two-user OFDM scheme for transmission over the WATM channel of Figure 18.17.

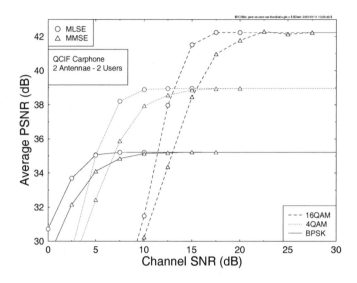

Figure 18.29: Average PSNR versus channel SNR performance of the two-antenna, two-user OFDM system for transmission over the WATM channel of Figure 18.17.

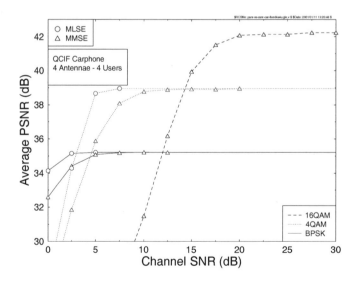

Figure 18.30: Average PSNR versus channel SNR performance of the four-antenna, four-user OFDM system for transmission over the WATM channel of Figure 18.17.

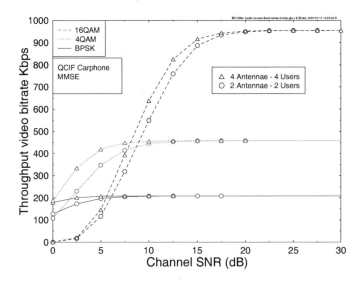

Figure 18.31: Effective video throughput bitrate versus channel SNR using the MMSE OFDM scheme for a two-antenna, two-user scenario and for a four-antenna, four-user scenario for transmission over the WATM channel of Figure 18.17.

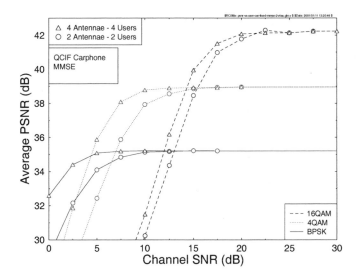

Figure 18.32: Average PSNR versus channel SNR performance of the MMSE OFDM scheme for two-antenna, two-user and for four-antenna, four-user scenarios for transmission over the WATM channel of Figure 18.17.

Chapter **19**

Conclusion and Further Research Problems [1]

19.1 Summary of Part I

19.1.1 Summary of Part I

In Chapters 2 – 7 we discussed the basic implementational, algorithmic and performance aspects of orthogonal frequency division multiplexing in predominantly duplex mobile communications environments. Specifically, following a rudimentary introduction to OFDM in Chapter 2 in Chapter 3 we have further studied the structure of an OFDM modem and we have investigated the problem of the high peak-to-mean power ratio observed for OFDM signals, and that of clipping amplification caused by insufficient amplifier back-off. We have investigated the BER performance and the spectrum of the OFDM signal in the presence of clipping, and we have seen that for an amplifier back-off of 6 dB the BER performance was indistinguishable from the perfectly amplified case. We have investigated the effects of quantisation of the time domain OFDM signal. The effects of phase noise on the OFDM transmission were studied, and two-phase noise models were suggested. One model was based on white phase noise, only relying on the integrated phase jitter, while a second model used coloured noise, which was generated from the phase noise mask.

In Chapter 4 we studied OFDM transmissions over time dispersive channels. The spectrum of the transmitted frequency domain symbols is multiplied with the channel's frequency domain channel transfer function, hence the amplitude and phase of the received subcarriers are distorted. If the channel is varying significantly during each OFDM symbol's duration, then additional inter-subcarrier interference occurs, affecting the modem's performance. We have seen the importance of channel estimation on the performance of coherently detected OFDM, and we have studied two simple pilot-based channel estimation schemes. Differen-

[1]*OFDM and MC-CDMA for Broadband Multi-user Communications WLANS and Broadcasting.*
L.Hanzo, Münster, T. Keller and B.J. Choi,
©2003 John Wiley & Sons, Ltd. ISBN 0-470-85879-6

tially detected modulation can operate without channel estimation, but exhibits lower BER performance than coherent detection. We have seen that the signal-to-noise ratio is not constant across the OFDM symbol's subcarriers, and that this translates into a varying bit error probability across the different subcarriers.

The effects of timing and frequency errors between transmitter and receiver were studied in Chapter 5. We saw that a timing error results in a phase rotation of the frequency domain symbols, and possibly inter-OFDM-symbol interference, while a carrier frequency error leads to inter-subcarrier-interference. We have suggested a cyclic postamble, in order to suppress inter-OFDM-symbol interference for small timing errors, but we have seen that frequency errors of more than 5% of the subcarrier distance lead to severe performance losses. In order to combat this, we have investigated a set of frequency- and timing-error estimation algorithms. We suggested a time domain-based joint time and frequency error acquisition algorithm, and studied the performance of the resulting system over fading time dispersive channels.

Based on the findings of Chapter 4 we investigated adaptive modulation techniques to exploit the frequency diversity of the channel. Specifically, in Chapter 6, three adaptive modulation algorithms were proposed and their performance was investigated. The issue of signalling was discussed, and we saw that adaptive OFDM systems require a significantly higher amount of signalling information than adaptive serial systems. In order to limit the amount of signalling overhead, a sub-band adaptive scheme was suggested, and the performance trade-offs against a subcarrier-by-subcarrier adaptive scheme were discussed. Blind modulation mode detection schemes were investigated, and combined with an error correction decoder. We saw that by combining adaptive modulation techniques with a strong convolutional turbo channel codec significant system throughput improvements were achieved for low SNR values. Finally, frequency domain pre-distortion techniques were investigated in order to pre-equalise the time-dispersive channel's transfer function. We saw that by incorporating pre-distortion in adaptive modulation, significant throughput performance gains were achieved compared to adaptive modems without pre-equalisation.

We saw in Chapter 6 that although channel coding significantly improves the achievable throughput for low SNR values in adaptive modems, with increasing average channel quality the coding overhead limits the system's throughput. In order to combat this problem, in Chapter 7 we investigated coding schemes which offer readily adjustable code rates. RRNS [208] and turbo BCH codes [208] were employed for implementing an adaptive coding scheme allowing us to adjust the code rate across the subcarriers of an OFDM symbol. Combinations of adaptive coding and the adaptive modulation techniques introduced in Chapter 7 were studied, and we saw that a good compromise between the coded and uncoded transmission characteristics can be found.

19.1.2   Conclusions of Part I

(1) Based on the implementation-oriented characterisation of OFDM modems, leading to a real-time testbed implementation and demonstration at 34 Mbps we concluded that OFDM is amenable to the implementation of high bit rate wireless ATM networks, which is underlined by the recent ratification of the HIPERLAN II standard.

(2) The range of proposed joint time and frequency synchronisation algorithms efficiently supported the operation of OFDM modems in a variety of propagation environments,

resulting in virtually no BER degradation in comparison to the perfectly synchronised modems. For implementation in the above-mentioned 34 Mbps real-time testbed simplified versions of these algorithms were invoked.

(3) Symbol-by-symbol adaptive OFDM substantially increases the BPS throughput of the system at the cost of moderately increased complexity. It was demonstrated in the context of an adaptive real-time audio system that this increased modem throughput can be translated into improved audio quality at a given channel quality.

(4) The proposed blind symbol-by-symbol adaptive OFDM modem mode detection algorithms were shown to be robust against channel impairments in conjunction with twin-mode AOFDM. However, it was necessary to combine it with higher-complexity channel coding based mode detection techniques, in order to maintain sufficient robustness, when using quadruple-mode AOFDM.

(5) The combination of frequency domain pre-equalisation with AOFDM resulted in further performance benefits at the cost of a moderate increase in the peak-to-mean envelope flluctuation and system complexity.

(6) The combination of adaptive RRNS and turbo BCH FEC coding with AOFDM provided further flexibility for the system, in order to cope with hostile channel conditions, in particular in the low SNR region.

19.2 Summary and Conclusions of Part II

19.2.1 Summary of Part II

Since their initial introduction in 1993 [52, 55, 277, 278], multi-carrier spread-spectrum systems have attracted significant research interest. Existing advanced techniques originally developed for DS-CDMA and OFDM have also been applied to MC-CDMA, while a range of new unique techniques have been proposed for solving various problems specific to multi-carrier CDMA systems. The first two chapters of Part II, namely Chapters 8 and 9, reviewed the basic concepts of MC-CDMA and the various spreading sequences applicable to MC-CDMA transmissions. Chapter 10 characterised the achievable performance of MC-CDMA schemes employing various detectors. Then, three further topics closely related to MC-CDMA based communications were investigated in depth.

Specifically, the peak factor reduction techniques were presented in Chapter 11. While analysing the signal envelope of MC-CDMA in Section 11.5.1, it was found that the aperiodic correlations of the spreading sequences play an important role in determining the associated peak factor. By investigating several orthogonal sequences, in Section 11.5.2.2 it was shown that a set of orthogonal complementary sequences applied to MC-CDMA has the advantageous property of limiting the peak factor of the MC-CDMA signal to 2 or 3dB, when the number of simultaneously transmitting users is less than or equal to four, regardless of the spreading factor. When the MC-CDMA system was "fully-loaded", i.e. supported the highest possible number of users, Walsh codes exhibited the lowest worst-case peak-factor as well as the lowest median peak-factor. The latter property was exploited in terms of limiting the peak factor of "fully-loaded" MC-CDMA systems having a spreading factor of 15 to a maximum peak factor of 2.5, which is comparable to the peak factor of filtered single carrier signals. This was achieved by employing a peak-factor limiting block code, as detailed in Section 11.5.2.3. However, our simulation study of MC-CDMA systems employing realistic power amplifier models revealed in Section 11.5.3 that Zadoff-Chu spreading-based MC-CDMA outperforms Walsh spreading-based MC-CDMA, when transmitting over a Rayleigh fading channel. Investigating this unexpected result further, it was found that Zadoff-Chu spreading-based MC-CDMA better exploits the frequency diversity in comparison to Walsh spreading-assisted MC-CDMA, since the combined symbol after Zadoff-Chu spreading over each sub-carrier is less likely to have a low value compared to Walsh spread and combined symbols. This suggests that the effects of the frequency domain peak-factor, in other words, the dynamic range of the frequency domain symbols on the achievable diversity gain has to be further investigated. A preliminary idea related to this idea was presented in Section 11.5.4.

Chapter 12 was devoted to adaptive modulation techniques. After the introduction of a general model of various adaptive modulation schemes in Section 12.3.1, the existing techniques proposed for determining the transceiver mode-switching levels were reviewed in Section 12.4.1, 12.4.2 and 12.4.3, since the performance of adaptive schemes is predetermined by the switching levels employed as well as by the average SNR per symbol. In search of the globally optimum switching levels, the Lagrangian optimisation technique was invoked in Section 12.4.4 and the relationship amongst the different mode-switching levels was found to be independent of the underlying channel scenarios such the Multi Path Intensity (MIP) profile or the fading magnitude distribution. Having established the technique for determining the optimum switching levels, a comprehensive set of performance results was presented for adaptive PSK schemes, adaptive Star QAM schemes as well as for adaptive Square QAM schemes, when they were communicating over a flat Nakagami m fading channel. Since adaptive Square QAM exhibited the highest BPS throughput among the three adaptive schemes investigated, the performance of adaptive Square QAM schemes was further studied, employing two-dimensional Rake receivers in Section 12.5.2 and applying concatenated space-time coding in conjunction with turbo convolution coding in Section 12.5.5. As expected, in Section 12.5.1, 12.5.2 and 12.5.5, it was found that the SNR gain of the adaptive schemes over fixed-mode schemes decreased, as the fading became less severe. Hence, the adaptive schemes exhibited only a modest performance gain, when (a) the Nakagami fading parameter m was higher than four, (b) the number of diversity antennas was higher than four when MRC diversity combining was used over independent Rayleigh fading channels, or (c) the total number of transmit antennas plus receive antennas was higher than four when space-time coding was used. Specifically, in Section 12.5.5 it was found that ST block coding reduces the relative performance advantage of adaptive modulation as summarised in Figure 12.49(b), since it increases the diversity order and eventually reduces the channel quality variations. When the adaptive modulation schemes were operating over wide-band channels, their SNR gain eroded even in conjunction with a low number of antennas, since the fading depth of wideband channels is typically less deep than that of narrow-band channels. For transmission over correlated fading channels it is expected that the SNR gain of adaptive schemes over their fixed-mode counterparts erodes less, as the number of antenna increased. However, further investigations are required to quantify the effects of correlated fading. Even though some channel-coding based results were presented in Section 12.5.5, the investigation of coded adaptive modulation schemes is far from complete. Since our AQAM mode switching levels were optimised for uncoded systems, the performance results presented for the coded adaptive schemes are not optimum. The technique similar to that developed in Section 12.4.4 may be applied, provided that a closed-form mode-specific average BER expression becomes available for channel coded fixed modulation schemes. available. In addition, the "spread of BER" has to be quantified by some measure, such as its PDF. This concept arises, since the quality of service experienced by a user may differ for fixed-mode schemes and for adaptive modulation schemes, even if their average BERs are identical. The variance or in other words fluctuation of the instantaneous BER distribution may serve as a useful quality measure. The PDF of this instantaneous BER may be obtained by applying Jacobian transformation [453] to the PDF of the instantaneous SNR per symbol. It is expected that there may be a direct relationship between this measure of "spread of BER" and the required interleaver length of the channel codec associated. More explicitly, it is expected that the distribution of the bit errors becomes less bursty in conjunction with AQAM schemes, since they are capable of near-instantaneously involving a more robust but reduced-throughput AQAM mode, when the channel quality degrades. Hence AQAM schemes may require shorter channel interleavers for randomising the position of channel errors.

In Chapter 13, a Successive Partial Detection (SPD) scheme was introduced. We analysed the BEP performance and the implementational complexity of three proposed SPD detectors, when communicating over the AWGN contaminated by an impulse noise in the time domain or a narrow band interference / a tone jamming signal in the frequency domain, assuming that one transform-domain symbol was obliterated due to a severe corruption. We found that the Type II SPD detector exhibits a lower BEP and requires a lower complexity than those of the conventional full-despreading based detector. This work has to be expanded for transmission over wideband channels. The application of the SPD scheme in the context of a joint-detector is also an interesting future research issue, since it will reduce the complexity of the joint detector.

Part II of the book concentrated on investigating the MC-CDMA [52, 53, 55] scheme, which is one of the family of three different multi-carrier CDMA techniques [326]. This technique was advocated, because MC-CDMA results in the lowest BER among the three schemes investigated in a similar scenario [326]. Our investigations concentrated on the downlink, because in the uplink stringent synchronisation of the mobile terminals has to be met. Future research should extend the results of Chapter 11 and 12 to both multi-carrier DS-CDMA [277] and to multi-tone (MT) CDMA [278], as well as to the family of more sophisticated adaptive MC-CDMA schemes [314].

19.2.2   Conclusions of Part II

The main contributions and conclusions of Part II of the book are in three specific areas closely related to Multi-Carrier Code Division Multiple Access (MC-CDMA), which are summarised in this subsection.

Crest Factor of Multi-Code MC-CDMA Signal

One of the main drawbacks of Orthogonal Frequency Division Multiplexing (OFDM) [2] is that the envelope power of the transmitted signal fluctuates widely, requiring a highly linear RF power amplifier. Finding a solution to this problem has stimulated intensive research. The accrueing advances were critically reviewed in Sections 11.3 and 11.4. Since

MC-CDMA [52, 55] schemes spread a message symbol across the frequency domain and employ an OFDM transmitter for conveying each spread bit, their transmitted signal also exhibits a high crest factor (CF), which is defined as the ratio of the peak amplitude of the modulated time domain signal to its root mean square (RMS) value. Hence, the characteristics of the crest factor of the MC-CDMA signal were studied in Section 11.5, with a view to reduce the associated power envelope variations.

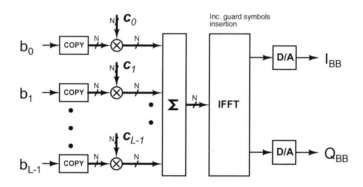

Figure 19.1: MC-CDMA transmitter model: b_l and $\mathbf{C}_l$ are the l-th message symbol and spreading sequence, respectively.

The simplified transmitter structure of a multi-code MC-CDMA system is portrayed in Figure 19.1, where L M-ary Phase Shift Keying (MPSK) modulated symbols $\{b_l \mid 0 \leq l < L\}$ are transmitted simultaneously using L orthogonal spreading sequences $\{\mathbf{C}_l \mid 0 \leq l < L\}$. Each of the spreading sequences is constituted by N chips according to $\mathbf{C}_l = \{c_l[n] \mid 0 \leq n < N\}$, where the complex spreading sequence c_l has a unit magnitude of $|c_l[n]| = 1$. Then, the normalised complex envelope $s(t)$ of an MPSK modulated multi-code MC-CDMA signal is represented for the duration of a symbol period T as:

$$s(t) = \frac{1}{\sqrt{N}} \sum_{l=0}^{L-1} \sum_{n=0}^{N-1} b_l \, c_l[n] \, e^{j2\pi F n \frac{t}{T}} .$$

The power envelope $|s(t)|^2$ of this MPSK modulated multi-code MC-CDMA signal was analysed in Section 11.5.1 and it was expressed as [57]:

$$|s(t)|^2 = \frac{2}{N} \mathrm{Re} \left[\sum_{n=0}^{N-1} (A[n] + X[n]) \, e^{j2\pi n \frac{t}{T}} \right] ,$$

where $A[n]$ is the collective aperiodic auto-correlation of the spreading sequences $\{\mathbf{C}_l\}$ defined in the context of Equation 11.38, while $X[n]$ is the collective aperiodic cross-correlation defined in 11.39.

Having observed that the power envelope of the MPSK modulated multi-code MC-CDMA signal is completely characterised by the collective auto-correlations and cross-correlations of the spreading sequence employed, a set of various spreading sequences was investigated in quest of the sequences yielding low CF or Peak Factor (PF), where the PF was defined as the square of the CF, namely $PF \triangleq CF^2$. **It was shown by Theorem 11.5 in Section 11.5.2.2 that a specific version of Sivaswamy's complementary set [389] results in a low CF, bounded by 3dB, in the context of BPSK modulated four-code MC-CDMA systems, regardless of the spreading factor employed, as seen in Figure 19.2.**

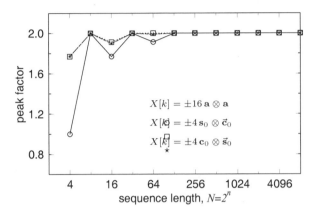

Figure 19.2: The peak-factors of Sivaswamy's complementary set of sequences. The number of simultaneously used sequences is $L = 4$ and the sequence length of $N = 2^n$ is between $2^2 = 4$ and $2^{13} = 8192$.

Since a similar family of spreading sequences yielding such a low CF could not be found for code sets having more than four codes, i.e. for $L > 4$, in Section 11.5.2.3 a crest factor reduction coding scheme [31] was applied instead. We found in Section 11.5.2.3 that the peak factors of BPSK modulated MC-CDMA employing an appropriate family of spreading codes could be reduced below 2.5, when using the crest factor reduction code proposed in [57], as seen in Figure 19.3(a). In order to investigate the effects of the modulated signal peak clipping by the non-linear power amplifier, the Bit Error Ratio (BER) performance was studied in Section 11.5.3.3. It was expected that the Walsh spreading based scheme would result in the lowest BER, since it resulted in the lowest worst-case peak factor for 'fully-loaded $(L = N)$' BPSK modulated MC-CDMA employing a Minimum Mean Square Error (MMSE) Block-Decision Feedback Equaliser (BDFE) based Joint Detector (JD) [323], when experiencing independent Rayleigh fading for each subcarriers. However, the Zadoff-Chu based scheme performed better than Walsh based scheme [57]. We found that the Zadoff-Chu spreading based MC-CDMA scheme inflicted less Multiple User Interference (MUI), than the Walsh-spreading based scheme and it utilised the available diversity more efficiently, than the Walsh-spreading based scheme.

Adaptive Modulation

The second field, where novel contributions were made, was in realms of adaptive modulation schemes. Adaptive modulation [2, 198, 395, 397] attempts to provide the highest possible throughput given the current near-instantaneous channel quality, while maintaining the required data transmission integrity. We analysed in Section 12.3.3.1 the performance of adaptive modulation schemes and derived a closed form expression for the average BER and for

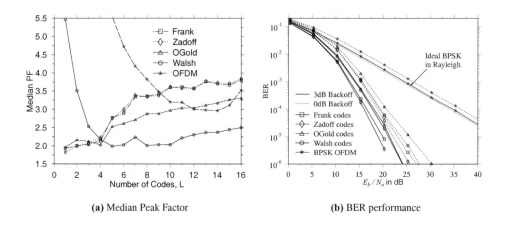

(a) Median Peak Factor

(b) BER performance

Figure 19.3: Effects of the $(L-1)/L$-rate crest factor reduction coding for the spreading factor of $N = 16$. (a) The worst-case peak factor when using various spreading codes for BPSK modulated MC-CDMA. (b) BER versus E_b/N_o performance of BPSK modulated MMSE-JD MC-CDMA and that of BPSK modulated OFDM, when transmitting over independent Rayleigh channels for each subcarriers.

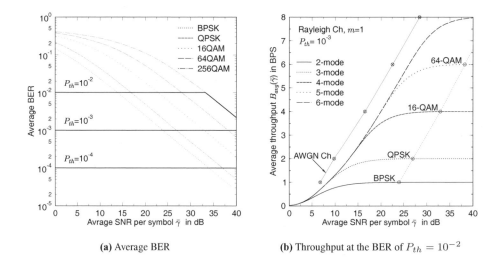

(a) Average BER

(b) Throughput at the BER of $P_{th} = 10^{-2}$

Figure 19.4: The average BER and average throughput performance of a six-mode adaptive Square QAM scheme for communicating over a flat Rayleigh channel ($m = 1$). (a) The constant target average BER is maintained over the entire range of the average SNR values up the avalanche SNR. (b) The markers '$\otimes$' and '$\odot$' represent the required SNR of the corresponding fixed-mode Square QAM schemes achieving the same target BER as the adaptive schemes, operating over an AWGN channel and a Rayleigh channel, respectively.

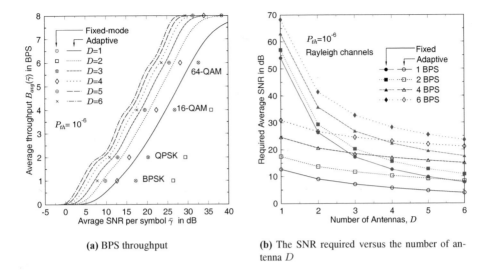

(a) BPS throughput

(b) The SNR required versus the number of antenna D

Figure 19.5: The performance of the MRC-aided antenna-diversity assisted AQAM scheme operating over independent Rayleigh fading channels at the target average BEP of $P_{th} = 10^{-6}$. (a) The markers represent the corresponding fixed-mode QAM performance. (b) The required SNR achieving the various throughput for the AQAM and fixed-mode QAM schemes.

the average Bits Per Symbol (BPS) throughput [504]. Since the modulation mode switching levels predetermine the average BER as well as the average BPS, it is important to optimise these switching levels for achieving the maximum possible average throughput. Having reviewed the existing techniques of determining the switching levels [214, 397], a per-SNR optimisation technique was proposed in [415], which was based on Powell's multi-dimensional optimisation [334], yielding a constant target BER and the maximum BPS throughput. Since this optimisation procedure was often trapped in local optima, rather than reaching the global optimum, the Lagrangian-based optimisation technique was developed [505] and was subsequently used in Section 12.5 in order to investigate the performance of adaptive modulation schemes employing various modulation modes, namely PSK, Star QAM and Square QAM phasor constellations.

Since our modulation mode switching levels were optimised in order to achieve the highest throughput, while maintaining the target BER, the average BER of our six-mode AQAM scheme remained constant over the entire range of the average SNR values up the avalanche SNR, beyond which it followed the BER of the highest throughput mode, namely that of 256-QAM, as seen in Figure 19.4(a). On the other hand, the BPS throughput increased steadily, as the average SNR increased, as seen in Figure 19.4(b). Our optimised AQAM operating over a Rayleigh fading channel required 4dB more SNR, in comparison to fixed-mode QAM operating over an AWGN channel.

Since in Section 12.5.2 receiver antenna diversity has been used for combating the effects of fading, we investigated the performance of Maximal Ratio Combining (MRC) antenna diversity assisted AQAM schemes [504]. As it is seen in Figure 19.5, the average SNRs required for achieving the target BEP of $P_{th} = 10^{-6}$ for the fixed-mode schemes and that for

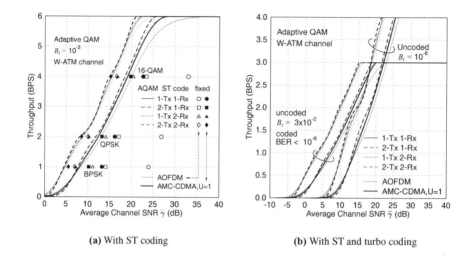

(a) With ST coding

(b) With ST and turbo coding

Figure 19.6: The BPS throughput performance of five-mode AOFDM and AMC-CDMA for communicating over the W-ATM channel [2, pp.474]. (a) The SNR gain of the adaptive modems decreases, as the diversity of the ST coding increases. The BPS curves appear in pairs, corresponding to AOFDM and AMC-CDMA - indicated by the thin and thick lines, respectively - for each of the four different ST code configurations. The markers represent the SNRs required by the fixed-mode OFDM and MC-CDMA schemes for maintaining the target BER of 10^{-3} in conjunction with the four ST-coded schemes considered. (b) The turbo convolutional coding assisted adaptive modems have SNR gains up to 7dB compared to their uncoded counterparts achieving a comparable average BER.

the adaptive schemes decrease, as the antenna diversity order increases. However, the differences between the required SNRs of the adaptive schemes and their fixed-mode counterparts also decrease, as the antenna diversity order increases.

In Section 12.5.5, the performance of Space-Time (ST) block coded constant-power adaptive multi-carrier modems employing optimum SNR-dependent modem mode switching levels were investigated [423, 506]. As expected, it was found that ST block coding reduces the relative performance advantage of adaptive modulation, since it increases the diversity order and eventually reduces the channel quality variations, as it can be observed in Figure 19.6(a). **Having observed that 1-Tx aided AOFDM and 2-Tx ST coding aided fixed-mode MC-CDMA resulted in a similar BPS throughput performance, we concluded that fixed-mode MC-CDMA in conjunction with 2-Tx ST coding could be employed, provided that we could afford the associated complexity. By contrast, AOFDM could be a low complexity alternative of counteracting the near-instantaneous channel quality variations.** When turbo convolutional coding was concatenated to the ST block codes, near-error-free transmission was achieved at the expense of halving the average throughput, as seen in Figure 19.6(b) . Compared to the uncoded system, the turbo coded system was capable of achieving a higher throughput in the low SNR region at the cost of a higher complexity. Our study of the relationship between the uncoded BER and the corresponding coded BER showed that adaptive modems obtain higher coding gains, than that of fixed modems.

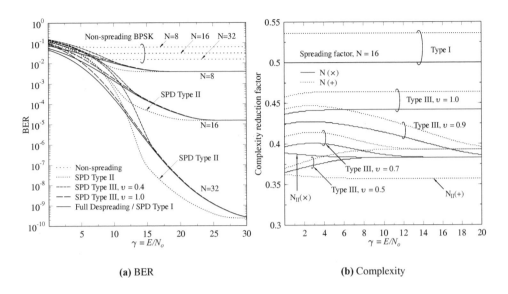

(a) BER   (b) Complexity

Figure 19.7: The bit error rates and complexities of the various SPD detectors and the conventional despreading scheme, where N is the spreading factor. (a) The non-spread BPSK scheme has a high irreducible BER of $0.5/N$, while the spread systems have a significantly lower error floor of $1/2^N$. The Type II SPD Detector shows a superior performance in the high SNR region in comparison to the other schemes. (b) The complexity of three types of SPD detectors quantified in terms of the required number of multiplications and additions, normalised to those of the conventional full-despreading based detector associated with the spreading factor of $N = 16$. The Type II SPD detector requires less than 40% of the operations necessitated by the conventional scheme.

This was due to the fact that the adaptive modem avoids burst errors even in deep channel fades by reducing the number of bits per modulated symbol eventually to zero.

Successive Partial Despreading

In Chapter 13, a Successive Partial Detection (SPD) scheme was introduced in an effort to reduce the computational efforts involved in despreading in the context of a 'fully loaded' CDMA downlink scenario. We analysed the bit error rate performance and the implementational complexity of three different types of SPD detectors, when communicating over the AWGN channel, contaminated by time domain impulse noise or frequency domain tone-jamming. It was found that the Type II SPD scheme resulted in a lower BER, than the conventional despreading scheme in the SNR range of $\gamma > 13dB$, while requiring significantly reduced computational efforts, namely less than 40% in comparison to the conventional scheme, when quantified in terms of the required number of multiplications and additions, as it can be observed in Figure 19.7.

19.3 Summary and Conclusions of Part III

19.3.1 Pilot-Assisted Channel Estimation for Single-User OFDM

In the context of our portrayal of single-user OFDM in Chapter 14 we highlighted that a prerequisite for performing coherent detection of the different subcarriers' symbols at the receiver is the availability of an estimate of the subcarriers' channel transfer factors.

An initial *a posteriori* estimate of the channel transfer factor associated with a specific subcarrier can be generated upon dividing the received subcarrier signal by the remodulated symbol decision obtained for this subcarrier. The initial *a posteriori* channel transfer factor estimates can then be further enhanced with the aid of generating smoothed estimates using filtering between neighbouring estimates based on exploiting the channel transfer factors' correlation in both the frequency and the time direction.

Depending on the origin of the sliced and remodulated symbol employed as a reference for deriving the initial *a posteriori* channel transfer factor estimates, the channel estimators can be divided into two categories, namely pilot-assisted and decision-directed approaches.

More specifically, in the context of pilot-assisted channel estimation the initial *a posteriori* channel transfer factor estimates are calculated only for a number of so-called pilot subcarriers, for which the complex transmitted symbol is known *a priori* at the receiver. This is followed by an interpolation between the different pilot subcarriers' *a posteriori* channel transfer factor estimates, in order to obtain channel transfer factor estimates for all subcarriers. While early publications on pilot-assisted channel estimation considered periodic one-dimensional (1D) pilot patterns, more recent publications considered the employment of two-dimensional (2D) pilot patterns, based on exploiting the channel transfer factors' correlation not only in the frequency-direction, but also in the time direction. The most often used pilot patterns are of rectangular shape, but other geometries have also been investigated in the literature. As a result of employing 2D pilot patterns, the pilot overhead imposed can be reduced, while potentially maintaining the same channel transfer factor estimation MSE as in case of a 1D pattern. One disadvantage of using 2D pilot patterns is, however, that a block-based processing has to be used at the receiver, where each channel estimation block typically contains several OFDM symbols. This increases the entire system's delay, which may become prohibitively high in voice-based communications. Furthermore, extra storage requirements are imposed due to the necessity of retaining a number of past OFDM symbols. However, the block-based processing could be avoided by rendering the channel transfer function interpolation filter causal, namely by employing only the *a posteriori* channel transfer factor estimates of the current- and the past OFDM symbols in the filtering process but no "future" OFDM symbols. A consequence would be the associated degradation of the channel estimation MSE.

Two of the most prominent methods reported in the literature for performing the interpolation between the pilot subcarriers' *a posteriori* channel transfer factor estimates, are polynomial and Wiener filter-based approaches. Polynomial interpolation attempts to approximate the channel transfer function's evolution based on minimising the squared error between the curve or surface spanned by the polynomial and the *a posteriori* channel transfer factor estimates at the pilot positions. In contrast, Wiener filtering is based on minimising the expected squared error between the desired channel transfer factor of a *specific subcarrier* and a linear combination of the *a posteriori* channel transfer factor estimates associated with the pilot positions. In contrast to the polynomial approach the Wiener filter requires knowledge of the channel's statistics in form of the spaced-time spaced-frequency correlation function or the related scattering function.

Due to the fundamental importance of Wiener filtering both in the context of channel transfer factor predicton as well as in multi-user detection, we concentrated in our review of pilot-assisted channel estimation techniques originally proposed in Chapter 14 on the 2D-FIR- and cascaded 1D-FIR Wiener filter-based channel estimation as proposed by Höher *et al.* [58, 66, 67, 462]. Although 2D-FIR Wiener filtering provides the optimum solution for an unlimited number of filter taps, for a given moderate complexity the cascaded 1D-FIR Wiener filtering becomes more attractive in terms of achieving a lower estimation MSE. In terms of the pilot grid's dimensions, a two-times oversampling in both the time and frequency direction was found to be sufficient for almost completely eliminating the residual estimation MSE observed at higher SNRs. Our further investigations were cast in the context of maintaining "robustness" for the channel estimator. This was achieved by assuming a uniform, ideally support-limited scattering function and its associated 2D-sinc-shaped spaced-time spaced-frequency correlation function in the calculation of the estimator's coefficients. As long as the support region of the channel's actual scattering function – potentially shifted along the multipath delay axis as a result of synchronisation errors – was contained in the support region of the "robust" estimator's associated uniform scattering function, no significant MSE degradation was observed compared to the case, when the channel's encountered scattering function was uniform as well.

For a given pilot grid the MSE performance of pilot-assisted Wiener filter-aided channel estimation can be potentially improved upon increasing the number of filter taps. However, a fundamental performance limitation is imposed, because the *a posteriori* channel transfer factor estimates, which are associated with pilot subcarriers spaced further apart from the specific position, for which the channel transfer factor is to be estimated, are less correlated. This effect can be addressed upon increasing the pilot grid density. In the limit, every subcarrier could potentially be employed for pilot transmissions. This strategy is exploited in the context of decision-directed channel estimation based on the premise that the demodulated symbols are error-free or exhibit a low detection-error probability.

19.3.2 Decision-Directed Channel Estimation for Single-User OFDM

Similar to the scenario addressed in Chapter 14 dealing with pilot-assisted chaannel estimation, in Chapter 15 we also argued that a prerequisite for performing coherent detection of the different subcarriers' symbols at the receiver is the availability of a reliable estimate of the subcarriers' channel transfer factors.

An initial *a posteriori* estimate of the channel transfer factor associated with a specific subcarrier can be generated upon dividing the received subcarrier signal by the remodulated symbol decision obtained for this subcarrier. The initial *a posteriori* channel transfer factor estimates can then be further enhanced with the aid of generating smoothed estimates using filtering between neighbouring estimates based on exploiting the channel transfer factors' correlation in both the frequency and the time direction.

Depending on the origin of the sliced and remodulated symbol employed as a reference for deriving the initial *a posteriori* channel transfer factor estimates, the channel estimators can be divided into two categories, namely pilot-assisted and decision-directed approaches.

More specifically, in the context of pilot-assisted channel estimation the initial *a posteriori* channel transfer factor estimates are calculated only for a number of so-called pilot subcarriers, for which the complex transmitted symbol is known *a priori* at the receiver. This is followed by an interpolation between the different pilot subcarriers' *a posteriori* channel transfer factor estimates, in order to obtain channel transfer factor estimates for all subcarriers.

By contrast, the philosophy of the decision-directed channel estimation discussed in Chapter 15 was based on the idea of employing the remodulated subcarrier symbol decisions as "pilots", or in other words as a reference signal for generating the set of K initial *a posteriori* channel transfer factor estimates for the current OFDM symbol. These initial estimates could then be further enhanced by MMSE-based 1D-FIR Wiener filtering across the K subcarriers exploiting also the channel transfer factors' correlation in the frequency-direction, as suggested by Edfors *et al.* [62, 69, 157]. An further MSE enhancement was achieved by Li *et al.* [68] with the aid of MMSE-based 2D-FIR Wiener filtering upon also exploiting the channel's correlation in the time direction, namely with the aid of employing also a number of previous OFDM symbols' initial *a posteriori* channel transfer factor estimates in the filtering process. These enhanced *a posteriori* channel transfer factor estimates derived for the current OFDM symbol would then be employed as *a priori* channel transfer factor estimates for frequency domain equalisation conducted during the next OFDM symbol period upon neglecting the channel's decorrelation between the two OFDM symbol periods.

19.3.2.1 Complexity Reduction by CIR-Related Domain Filtering

However, the complexity imposed upon performing the filtering across all the K different subcarriers in the frequency domain may potentially become excessive. To be more specific, a computational complexity of K^2 number of complex multiplications and the same number of complex additions would be inflicted. In order to reduce the associated computational complexity, it was suggested by Edfors *et al.* [62, 69, 157] to transform the initial *a posteriori* channel transfer factor estimates to the CIR-related domain with the aid of the Karhunen-Loeve Transform (KLT), followed by CIR-related one-tap filtering of only the first K_0 number of uncorrelated CIR-related taps, which are assumed to be the most significant taps in terms of their variance. Finally, the remaining filtered CIR-related taps are transformed back to the frequency domain. In this case the computational complexity would be quantified in terms of $2K_0K$ complex multiplications and additions. While the KLT achieves a perfect decorrelation and hence the best possible energy compaction in the CIR-related domain, one disadvantage is that for its calculation explicit knowledge of the channel's statistics, namely the spaced frequency correlation matrix is required, which is not known *a priori*. Hence, as suggested by van de Beek *et al.* [61], Edfors *et al.* [69] and Li *et al.* [68], the channel-independent DFT matrix could be employed instead of the optimum KLT matrix for transforming the initial *a posteriori* channel transfer factor estimates to the CIR-related domain. In the context of a sample-spaced CIR the DFT matrix is identical to the KLT matrix and hence the CIR-related taps are uncorrelated. By contrast, in the context of a non-sample-spaced CIR the DFT matrix is sub-optimum in the sense that the CIR-related taps are not perfectly decorrelated and hence the energy compaction becomes sub-optimum. As a result of windowing, the CIR-related taps generated upon retaining only a limited number of $K_0 \ll K$ taps with the aim of reducing the estimator's complexity, the DFT based estimator's MSE is significantly degraded compared to that of the KLT-based estimator. This is because significant signal components are removed. Note that windowing is particularly effective in terms of complexity reduction, when the CIR-related taps are also filtered in the time direction, as in case of the 2D-MMSE based estimator proposed by Li *et al.* [68]. An attractive compromise between the optimum KLT and the channel-independent DFT was found by Li and Sollenberger [80] upon employing the unitary matrix, which is related to the KLT of the uniform multipath intensity profile's spaced-frequency correlation matrix as the transform matrix required to transform the initial *a posteriori* channel transfer factor estimates to the CIR-related domain. Thus attractive energy compaction properties, similar to those of the optimum KLT, are achieved without having exact knowledge of the channel's statistics. As a result, only a slight MSE degradation compared to that of the optimum KLT-based estimator is incurred.

19.3.2.2 Compensation of the Channel's Time-Variance by CIR-Related Tap Prediction Filtering

Our more detailed investigations of Chapter 15 concentrated on the effects of the channel's decorrelation incurred between consecutive OFDM symbols. Li *et al.* [68] proposed employing 2D-MMSE-based estimation filtering for deriving improved *a posteriori* channel transfer factor estimates for the current OFDM symbol based on the current and a number of previous OFDM symbols' initial *a posteriori* channel transfer factor estimates. These channel estimates could then be employed as *a priori* channel transfer factor estimates for demodulation during the next OFDM symbol period. However, in the context of rapidly time-variant channels, associated with a potentially high OFDM symbol normalised Doppler frequency, it is more effective to directly predict the channel transfer factor for the next OFDM symbol period. This can be achieved by substituting the Wiener filter-based CIR-related tap estimation filters of Li's 2D-MMSE based channel estimator design [68] by Wiener filter-based CIR-related tap prediction filters. Our investigations demonstrated that with the aid of CIR-related tap prediction filtering even channel scenarios having OFDM symbol normalised Doppler frequency as high as $F_D = 0.1$ can be supported, while capitalising on relatively short prediction filters. Employing four predictor taps seemed to be sufficient for compensating most of the channel's variation, while further increasing the predictor's length resulted in additional modest MSE reduction due to averaging over a higher number of noisy samples. However, our experiments demonstrated that at lower SNRs and for higher-order modulation schemes the channel estimation MSE is potentially high, which is the result of employing erroneous subcarrier symbol decisions in the DDCE process. As a consequence, error propagation effects occur, which have to be curtailed by regularly transmitting training OFDM symbols. In order to further reduce the system's BER, the employment of turbo-coding was considered as a viable option. However, generating the DDCE's reference signal by slicing, reencoding, interleaving and remodulating the turbo-decoder's "source"-related soft-output bits only exhibited no significant advantage compared to slicing and remodulating the turbo-decoder's soft-input bits. By contrast, generating the DDCE's reference from the turbo-decoder's "source- plus parity"-related soft-output bits proved to be more effective.

19.3.2.3   Subject for Future Research: Successive Adaptivity of KLT and CIR-Related Tap Prediction Filtering

In the context of the 2D-MMSE based channel prediction the concepts of "robustness" with respect to the channel's true scattering function can be applied, as it was originally proposed by Li *et al.* [68] for the 2D-MMSE based channel estimator. However, in order to further improve the 2D-MMSE based channel predictor's MSE performance, the channel predictor could be rendered adaptive with respect to two components, namely the transform, which conveys the initial *a posteriori* channel transfer factor estimates to the CIR-related domain and, second with respect to the CIR-related tap predictors.

Recall that the optimum transform is known to be the Karhunen-Loeve transform with respect to the channel's spaced-frequency correlation matrix. An estimate of the channel's spaced-frequency correlation matrix is given by the auto-correlation matrix of the initial *a posteriori* channel transfer factor estimates. Although this matrix differs from the channel's spaced-frequency correlation matrix by an additive weighted identity matrix, which is associated with the noise contributions, the eigenvectors of both matrices are identical [90, 157]. Also recall that these eigenvectors constitute the optimum transform in terms of achieving a perfect decorrelation of the frequency domain channel transfer factors. The auto-correlation matrix of the initial *a posteriori* channel transfer factor estimates can be estimated with the aid of the sample-correlation method [90]. Based on the current OFDM symbol's *a posteriori* channel transfer factor estimates the sample-correlation matrix can be regularly updated. Instead of entirely recomputing the KLT matrix in every OFDM symbol period, an iterative update- or tracking method proposed by Davila [507] as well as Rezayee and Gazor [508] in the context of speech coding is expected to be computationally more effective.

Apart from the above-mentioned adaptive transform, the second component of the 2D-MMSE based channel predictor, which can be rendered adaptive, are the CIR-related tap predictors. In the context of our investigations we compared the block-based Burg algorithm-assisted CIR-related tap prediction approach proposed by Al-Susa and Ormondroyd [72] against the non-adaptive robust approach proposed by Li *et al.* [68]. Although these investigations demonstrated that an adaptive predictor is capable of outperforming the robust predictor, one disadvantage of the block-based adaptation was constituted by the extra storage requirements imposed. In order to avoid these disadvantages, an OFDM symbol-by-symbol adaptation technique relying on the LMS- or RLS algorithms is expected to be more attractive. We have further illustrated the concepts of the fully adaptive 2D-MMSE based channel predictor in Figure 19.8. Besides the potentially reduced estimation MSE, an additional advantage of the adaptive predictor compared to a robust predictor is its ability to better compensate for the effects of erroneous subcarrier symbol decisions, which manifest themselves similarly to the effects of impulsive noise in the DDCE process. However, the advantages of the adaptive predictor have to be viewed in the light of the disadvantage of a higher complexity compared to the robust predictor.

Our further investigations in Chapter 15 were cast in the context of employing 2D-MMSE based channel prediction in an OFDM system, which employs adaptive modulation. This was motivated by the observation that as a result of the channel's variation versus time, the modulation mode assignment computed during the current OFDM symbol period – based on the current OFDM symbol's channel transfer factor estimates – for application during the next OFDM symbol period becomes inaccurate. Hence the AOFDM modem's performance

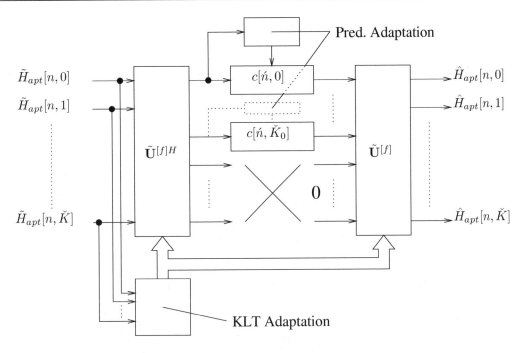

Figure 19.8: Stylised illustration of a fully adaptive 2D-MMSE based channel predictor. The KLT co-efficients are adapted for the sake of transforming the initial *a posteriori* channel transfer factor estimates to the CIR-related domain and for transforming the filtered CIR-related tap predictions back to the frequency domain. Furthermore, also the CIR-related tap predictor coefficients are subjected to adaptation, potentially invoking the RLS algorithm. Note that we have defined $\check{K} = K - 1$.

is limited. A significant performance improvement was hence achieved by computing the modulation mode assignment based on the *a priori* channel transfer factor estimates generated for the following OFDM symbol period with the aid of the 2D-MMSE based channel predictor. Note that the benefits of combining decision-directed channel prediction and adaptive modulation are twofold. On the one hand, the modulation mode assignment profits from the accurate channel predictions and on the other hand the channel predictions benefit from the more reliable remodulated reference signal invoked in the DDCE process in the light of employing adaptive modulation.

19.3.3 Channel Estimation for Multi-User SDMA-OFDM

In the context of a single-user, single-transmit antenna-assisted scenario, the task of acquiring a set of K initial *a posteriori* channel transfer factor estimates was accomplished by simply dividing the signal received in each subcarrier by the subcarrier's complex transmitted symbol. These initial *a posteriori* channel transfer factor estimates were then employed for deriving improved *a posteriori* channel transfer factor estimates for the current OFDM symbol period or for generating *a priori* channel transfer factor estimates for use during the next OFDM symbol period upon invoking CIR-related tap estimation- or prediction filtering, respectively.

By contrast, in the context of a multi-user SDMA-OFDM scenario, the signal received by each antenna element at the base station is constituted by the superposition of the L different users' transmitted and independently faded signals.

19.3.3.1 LS-Assisted DDCE

Based on the observation that the channel's multipath spread T_m – normalised to the OFDM sampling period duration T_s – is typically only a fraction of the number of subcarriers K, a subspace-based approach was proposed by Li *et al.* [91] to recover the L sets of K-subcarrier channel transfer factors associated with a specific BS receiver antenna. Based on the Least-Squares (LS) error criterion the vector of received subcarrier signals is projected onto the sub-space spanned by the first K_0 number of column vectors associated with the K-th order DFT matrix upon taking into account the different users' unique transmitted subcarrier symbol sequences. As a result, for each of the L users a set of K_0 CIR-related tap estimates is obtained, which are then subjected to the DFT in order to obtain the desired L number of sets of channel transfer factor estimates.

Our mathematical portrayal of this estimation problem capitalised on a more compact matrix notation than that of [91], which further motivated the employment of the LS error criterion as proposed by Li *et al.* [91]. Based on this notation a necessary condition was provided for the identification of the different users' CIR-related taps. More explicitly, the product of the L number of users and the K_0 number of CIR-related taps to be estimated was required to be lower than or equal to the K number of OFDM subcarriers, namely, we required that the condition of $LK_0 \leq K$ was satisfied. While this constitutes a necessary condition, it was observed that if the L number of users was close to the tolerable limit, the estimation MSE was potentially degraded depending on the specific subcarrier symbol sequences transmitted by the different users. This provided an additional motivation for de-vising alternative multi-user channel estimation approaches.

Furthermore, the computational complexity of the LS-assisted DDCE was found to be substantial, because a system of equations associated with a left-hand matrix of dimension $LK_0 \times LK_0$ and a right-hand matrix of dimension $LK_0 \times P$ had to be solved for estimating the CIR-related taps of the MIMO system's channels between the L users' single transmit an-tennas and the base-station's P receiver antennas. However, an advantage of the LS-assisted DDCE is that its MSE can be further improved by invoking pure transversal CIR-related tap filtering.

19.3.3.2 PIC-Assisted DDCE

In order to address the LS-assisted DDCE's deficiency of supporting only a limited number of users and imposing a potentially excessive computational complexity, the idea of Parallel Interference Cancellation (PIC)-assisted DDCE, which was originally suggested for the CIR-related domain by Li [93] and for the frequency domain by Jeon *et al.* [92], was significantly further developed. We argued above that since in a multi-user SDMA-OFDM scenario the signal received by a specific antenna element is given by the superposition of the different users' transmitted signals, the low-complexity single-user techniques for deriving initial *a posteriori* channel transfer factor estimates cannot be directly applied. However, a viable approach for the estimation of a specific user's channel transfer factors is constituted by first removing the remaining users' interference from the received composite multi-user signal, and then performing the channel estimation with the aid of the same techniques as in the single transmit-antenna assisted scenario. This involves generating the initial *a posteriori* channel transfer factor estimates, followed by CIR-related domain filtering. We found that from a mathematical point of view, performing the PIC in the frequency domain or in the time domain are identical. However, performing the PIC in the frequency domain, while the filtering in the CIR-related domain seems to be the least complex solution. Note that in the context of the PIC process the remaining users' interfering signal components are reconstructed based on the sliced symbols generated at the multi-user detector's output, and upon invoking furthermore the associated *a priori* channel transfer factor estimates generated during the previous OFDM symbol period for the current OFDM symbol period.

However, compared to the single-user single-transmit antenna-assisted scenario, the prediction filters potentially to be employed in the CIR-related domain for further enhancing the estimator's MSE were shown not to be transversal, but recursive. In the context of our discussions, mathematical expressions were derived for the current OFDM symbol's *a posteriori* channel transfer factor estimates' MSE and for the next OFDM symbol's predicted *a priori* channel transfer factor estimates' MSE. Furthermore, conditions for the estimator's stability were provided. Based on the system equations' contractive properties [90] an iterative algorithm was devised for the off-line optimisation of the CIR-related tap predictor's coefficients with the aim of minimising the *a priori* estimator's average MSE. Our simulation results demonstrated that the number of users each having a single transmit antenna is not limited by K/K_0, as it is the case for the LS-assisted DDCE. Furthermore we found that also the principles of "robustness" with respect to the channel's actual scattering function as invoked in the single-user scenario were also applicable for the multi-user scenario, which renders the off-line optimisation of the CIR-related tap predictor coefficients an attractive option. In the context of our investigations we also illustrated the effects of a non-sampled spaced CIR on the estimator's MSE, based on using the DFT matrix as the unitary transform matrix for conveying the inital *a posteriori* channel transfer factor estimates to the CIR-related domain. We found that as a result of retaining only the K_0 most significant CIR-related taps with the aim of reducing the estimator's complexity, in the frequency domain the channel transfer factors' estimation MSE significantly increased towards the edges of the OFDM symbol. This problem can be addressed by using the "robust" transform basis, as proposed by Li and Sollenberger [80] instead of the unitary DFT matrix.

In order to potentially further improve the estimator's MSE and to render the estimator capable of appropriately reacting to impulsive noise as caused for example by erroneous subcarrier symbol decisions, we demonstrated that the adaptation of the CIR-related tap predictor coefficients could also be performed with the aid of the RLS algorithm. It is interesting to note that although the different predictors associated with the CIR-related taps of a specific channel or with different channels perform their coefficient adaptation independently, the estimators' average MSE is minimised.

19.3.4   Uplink Detection Techniques for SDMA-OFDM

Finally, in Chapter 17 we discussed a range of uplink multi-user detection techniques, which
– in addition to multi-user channel estimation – constitute one of the pivotal components of
the SDMA-OFDM receiver. The detection techniques investigated were separated into the
sub-classes of linear- and non-linear detection. Specifically, in the context of linear detection
techniques, such as the Least-Squares (LS) and Minimum Mean-Square Error (MMSE) approaches no *a priori* knowledge about any of the other users' transmitted symbols is required
for the detection of a specific user's transmitted symbol. This in contrast to the family of
non-linear detectors, namely to the Successive Interference Cancellation (SIC), Parallel Interference Cancellation (PIC) and Maximum Likelihood (ML) detection based approaches.
Explicitly, in the context of these schemes *a priori* knowledge about one or more of the
remaining users' transmitted symbols is required for the detection of a specific user's transmitted symbol. An exception is the first cancellation stage of the SIC detector.

For the different multi-user detectors investigated, a mathematical analysis as well as a
performance and complexity analysis was conducted. We found that the linear detectors,
which perform the operations of linear combining and classification sequentially, constitute
the lowest complexity, but also the least powerful solutions in terms of the achievable BER.
By contrast, the ML detector is associated with the highest computational complexity due
to evaluating M_c^L number of L-dimensional trial-vectors in each subcarrier in terms of their
Euclidean distance from the vector of received signals upon taking into account the effects
of the channel. As a benefit, the ML detector's BER performance is optimum. Recall in this
context that L denotes the number of simultaneous SDMA users and M_c is the number of
constellation points associated with the specific modulation scheme employed. A compromise between the achievable performance and complexity imposed is given by the SIC and
PIC detectors.

19.3.4.1   SIC Detection

The philosophy of the SIC scheme is based on linearly detecting and cancelling successively
in each stage of the detection process the strongest remaining user in terms of an objective
measure, which could either be the SNR, SIR or SINR at the linear combiner's output of each
detection stage. Thus, the minimum occurring value of the objective function is maximised.
The potential of various schemes was assessed for further enhancing the SIC detector's performance, namely that of the M-SIC, partial M-SIC and SDI-M-SIC schemes, which were
contrived based on the idea of tracking multiple tentative symbol decisions from each detection stage, while performing a decision as regards to which symbols were transmitted, after
the cancellation of the last user. Our analysis of SIC also considered the effects of error
propagation potentially occurring between the different detection stages. Based on these observations an improved metric was developed for soft-bit generation to be employed in the
context of a system using turbo-decoding at the receiver. While the SIC and M-SIC schemes
potentially perform close to ML detection, this is achieved at the cost of a significantly increased computational complexity compared to the MMSE detection.

19.3.4.2 PIC Detection

We found that a further enhancement of the performance versus complexity trade-offs is constituted by the PIC detector. Recall that the signal received by the different BS antenna elements is constituted by the superposition of the different users' transmitted signals. Hence, a linear estimate of a specific user's transmitted signal can be generated upon removing the remaining users' transmitted signals, followed by single-user diversity combining. In the context of PIC detection, initial symbol decisions are generated with the aid of an MMSE detector for reconstructing the channel-impaired transmitted signals to be used in the actual cancellation process. While in the context of an uncoded scenario the PIC detector was found to perform only marginally better, than the MMSE detector, in the turbo-coded scenario a significant performance improvement was observed when using the sliced, interleaved and remodulated "source- plus parity"-related soft-output bits of the turbo-decoder obtained during the first detection stage as a reference for the PIC process of the second stage. A further performance improvement can potentially be achieved with the aid of a soft-bit-based PIC process instead of a hard-decision-based PIC.

19.3.4.3 Improvement of MMSE and PIC Detection by Adaptive Modulation or WHT Spreading

In order to render the employment of MMSE and PIC detection more attractive also in an uncoded scenario, investigations were conducted with respect to additional employing either adaptive modulation or Walsh-Hadamard Transform (WHT) based spreading in the multi-user SDMA-OFDM scenario. We found that these techniques are most effective in the context of a fully-loaded system, namely where the number of simultaneous users – and hence the number of transmit antennas – equals the number of BS receiver antenna elements. Given the restriction of employing AOFDM only in Time Division Duplexing (TDD) scenarios and the requirement of employing prediction filtering for compensating for the channel's variations with time, a more straightforward solution is constituted by WHT spreading across the different subcarriers. While at the transmitters the Walsh Hadamard Transform can be combined with the OFDM-related IFFT, as proposed in [496], at the receiver the computational complexity is increased, since the despreading has to be performed separately for each user. In the context of our investigations we highlighted that both in the single-user and in the multi-user SDMA-OFDM system, the equalisation or linear combining at the receiver and the despreading can be performed sequentially as a result of the WHT matrix's unitary nature. Our simulation results demonstrated that with the aid of adaptive modulation or WHT spreading the performance of the MMSE or PIC detection-assisted systems can be significantly improved. Hence these schemes constitute an attractive compromise in comparison to the significantly more complex SIC-assisted SDMA-OFDM schemes. For further reading on related topics we refer the reader to [509–517].

19.3.5 OFDM-Based Wireless Video System Design

Finally, in Chapter 18 we considered a few application examples in the context of OFDM-based wireless video telephony. Specifically, in Section 18.1 a range of AOFDM video transceivers was proposed for robust, flexible and low-delay interactive videotelephony. In order to minimise the amount of signalling required, we divided the OFDM subcarriers into sub-bands and controlled the modulation modes on a sub-band-by-sub-band basis. The proposed constant target bit-rate AOFDM modems provided a lower BER than the corresponding conventional OFDM modems. The slightly more complex switched TVTBR-AOFDM modems can provide a balanced video-quality performance, across a wider range of channel SNRs than the other schemes investigated.

In Section 18.2 multi-user detection-assisted, multiple transmit antenna-based OFDM schemes were studied in the context of HIPRLAN 2-like systems. It was demonstrated that the system's user capacity can be improved with the aid of unique spatial user signatures. The MLSE detection algorithm outperformed the MMSE scheme by about 5dB in terms of the required SNR and this performance gain manifested itself also in terms of the system's video performance.

19.4   Closing Remarks

This monograph considered a range of OFDM and MC-CDMA-related topics applicable to both single-user and multi-user communications. However, a whole host of further recent advances in the field of communications research are applicable also to OFDM. Specifically, the family of classification and learning-based neural network-assisted receivers investigated in the context of conventional single-carrier systems provides a rich set of further research topics. Partial response modulation techniques also have the promise of performance advantages in OFDM schemes. The joint optimisation of adaptive subcarrier bit-allocation and crestfactor reduction techniques constitutes a further research challange in the context of multi-user OFDM and MC-CDMA systems. All the above-mentioned techniques have the potential of improving the complexity versus performance balance of the system. The design of joint coding and modulation schemes is particularly promising in the context of OFDM and MC-CDMA. Finally, the use of OFDM in ultra-wide band systems invoking various frequency-hopping and multiple access techniques is likely to grow in popularity as an exciting research area.

These enabling techniques along with those detailed in the book are expected to find their way into future standards, such as the successors of the 802.11, the High Performance Local Area Network standard known as HiPerLAN, the European Digital Audio Broadcast (DAB) and Digital Video Broadcast (DVB) arrangements and their descendants. They are also likely to be adopted by the standardisation bodies in future generations of personal communications systems.

It is expected that wireless systems of the near future are likely to witness the co-existence of space-time-coded transmit diversity arrangements and near-instantaneously adaptive OFDM as well as MC-CDMA schemes for years to come. Intelligent learning algorithms will configure the transceivers in the appropriate mode that ultimately provides the best trade-off in terms of satisfying the user's preference in the context of the service requested [7, 207, 209].

A further advantage of the near-instantaneously adaptive OFDM and MC-CDMA transceivers is that they allow the system to instantaneously drop its transmission rate, when the channel quality is reduced, for example, as a consequence of the instantaneously peaking co-channel interference. By contrast, a conventional fixed-mode transceiver would drop the call and hence degrade both the quality of service and the network's teletraffic capacity. The achievable teletraffic performance of adaptive CDMA systems was documented in depth in conjunction with adaptive antenna-assisted dynamic channel allocation schemes in [209].[2]

Throughout this monograph we have endeavoured to picture the range of contradictory system design trade-offs associated with the conception of OFDM and MC-CDMA systems. We intended to present the material in an unbiased fashion and sufficiently richly illustrated in terms of the associated design trade-offs so that readers will be able to find recipes and examples for solving their own particular wireless communications problems. In this rapidly evolving field it is a challenge to complete a timely, yet self-contained treatise, since new advances are discovered at an accelerating pace, which the authors would like to report on. Our sincere hope is that you, the readers, have found the book a useful source of information, but above all a catalyst for further research.

---

[2]A range of related research papers and book chapters can be found at http://www-mobile.ecs.soton.ac.uk.

Glossary

ACF	Auto-correlation Function
ACTS	Advanced Communications Technologies and Services - a European research programme
ADSL	Asynchronous Digital Subscriber Loop
AOFDM	Adaptive Orthogonal Frequency Division Multiplexing
APR	A Priori
APT	A Posteriori
AWGN	Additive White Gaussian Noise
BER	Bit-Error Ratio
BLAST	Bell Labs Space-Time architecture
BPOS	Bit per OFDM Symbol
BPSK	Binary Phase-Shift Keying
BS	Basestation
CDF	Cumulative Distribution Function
CDMA	Code-Division Multiple Access
CE	Channel Estimation
CIR	Channel Impulse Response
DAB	Digital Audio Broadcasting
DDCE	Decision-Directed Channel Estimation

DDCP	Decision-Directed Channel Prediction
DFT	Discrete Fourier Transform
DMUX	Demultiplexer
DTTB	Digital Terrestrial Television Broadcast
D-BLAST	Diagonal BLAST
EM	Expectation Maximization
EVD	EigenValue Decomposition
FDM	Frequency Division Multiplexing
FDMA	Frequency Division Multiple Access
FEC	Forward Error Correction
FFT	Fast Fourier Transform
FIR	Finite Impulse Response
HF	High-Frequency
ICI	Inter-subCarrier Interference
IDFT	Inverse Discrete Fourier Transform
IFFT	Inverse Fast Fourier Transform
IIR	Infinite Impulse Response
ISI	Inter-Symbol Interference
IWHT	Inverse Walsh Hadamard Transform
KLT	Karhunen-Loeve Transform
LLR	Log-Likelihood Ratio
LS	Least-Squares
LSE	Least-Squares Error
MA	Multiple Access
MC	Multi-Carrier
MIMO	Multiple-Input Multiple-Output
ML	Maximum Likelihood

MLSE	Maximum Likelihood Sequence Estimation
MMSE	Minimum Mean-Square Error
MSE	Mean-Square Error
MU	Multi-User
MUD	Multi-User Detection
MUI	Multi-User Interference
MUX	Multiplexer
MV	Minimum Variance
MVDR	Minimum Variance Distortionless Response
OFDM	Orthogonal Frequency Division Multiplexing
PAPR	Peak-to-Average Power Ratio
PDF	Probability Density Function
PIC	Parallel Interference Cancellation
PSAM	Pilot Symbol Aided Modulation
PSD	Power Spectral Density
PSK	Phase-Shift Keying
QAM	Quadrature Amplitude Modulation
QPSK	Quadrature Phase-Shift Keying
RLS	Recursive Least-Squares
RNS	Residue Number System
SB	Subband
SDM	Space-Division Multiplexing
SDMA	Space-Division Multiple Access
SDI	Selective Decision Insertion
SER	Symbol Error Ratio
SIC	Successive Interference Cancellation
SINR	Signal-to-Interference-plus-Noise Ratio

SIR	Signal-to-Interference Ratio
SMI	Sample Matrix Inversion
SNR	Signal-to-Noise Ratio
STC	Space-Time Coding
SVD	Singular-Value Decomposition
TCM	Trellis-Coded Modulation
TDD	Time-Division Duplexing
TDMA	Time-Division Multiple Access
TTCM	Turbo-Trellis Coded Modulation
V-BLAST	Vertical BLAST
WATM	Wireless Asynchronous Transfer Mode
WHT	Walsh-Hadamard Transform
WHTS	Walsh-Hadamard Transform Spreading
ZF	Zero-Forcing
1D	One-Dimensional
2D	Two-Dimensional

Bibliography

[1] R. W. Chang, "Synthesis of band-limited orthogonal signals for multichannel data transmission," *Bell Systems Technical Journal*, vol. 46, pp. 1775–1796, December 1966.

[2] L. Hanzo, W. Webb, and T. Keller, *Single- and Multi-carrier Quadrature Amplitude Modulation*. New York, USA: IEEE Press-John Wiley, April 2000.

[3] W. Webb and R. Steele, "Variable rate QAM for mobile radio," *IEEE Transactions on Communications*, vol. 43, pp. 2223–2230, July 1995.

[4] L. Hanzo, "Bandwidth-efficient wireless multimedia communications," *Proceedings of the IEEE*, vol. 86, pp. 1342–1382, July 1998.

[5] S. Nanda, K. Balachandran, and S. Kumar, "Adaptation techniques in wireless packet data services," *IEEE Communications Magazine*, vol. 38, pp. 54–64, January 2000.

[6] L. Hanzo, F. Somerville, and J. Woodard, *Voice Compression and Communications: Principles and Applications for Fixed and Wireless Channels*. IEEE Press and John Wiley, 2001. (For detailed contents and sample chapters please refer to http://www-mobile.ecs.soton.ac.uk.).

[7] L. Hanzo, P. Cherriman, and J. Streit, *Wireless Video Communications: From Second to Third Generation Systems, WLANs and Beyond*. IEEE Press and John Wiley, 2001. (For detailed contents please refer to http://www-mobile.ecs.soton.ac.uk.).

[8] M. Zimmermann and A. Kirsch, "The AN/GSC-10/KATHRYN/variable rate data modem for HF radio," *IEEE Transactions on Communication Technology*, vol. CCM–15, pp. 197–205, April 1967.

[9] S. B. Weinstein and P. M. Ebert, "Data transmission by frequency division multiplexing using the discrete fourier transform," *IEEE Transactions on Communication Technology*, vol. COM–19, pp. 628–634, October 1971.

[10] L. Cimini, "Analysis and simulation of a digital mobile channel using orthogonal frequency division multiplexing," *IEEE Transactions on Communications*, vol. 33, pp. 665–675, July 1985.

[11] M. Alard and R. Lassalle, "Principles of modulation and channel coding for digital broadcasting for mobile receivers," *EBU Review, Technical No. 224*, pp. 47–69, August 1987.

[12] *Proceedings of 1st International Symposium,DAB,* (Montreux, Switzerland), June 1992.

[13] A. Peled and A. Ruiz, "Frequency domain data transmission using reduced computational complexity algorithms," in *Proceedings of International Conference on Acoustics, Speech, and Signal Processing, ICASSP'80,* vol. 3, (Denver, CO, USA), pp. 964–967, IEEE, 9–11 April 1980.

[14] B. Hirosaki, "An orthogonally multiplexed QAM system using the discrete fourier transform," *IEEE Transactions on Communications,* vol. COM-29, pp. 983–989, July 1981.

[15] H. Kolb, "Untersuchungen über ein digitales mehrfrequenzverfahren zur datenübertragung," in *Ausgewählte Arbeiten über Nachrichtensysteme,* no. 50, Universität Erlangen-Nürnberg, 1982.

[16] H. Schüssler, "Ein digitales Mehrfrequenzverfahren zur Datenübertragung," in *Professoren-Konferenz, Stand und Entwicklungsaussichten der Daten und Telekommunikation,* (Darmstadt, Germany), pp. 179–196, 1983.

[17] K. Preuss, "Ein Parallelverfahren zur schnellen Datenübertragung Im Ortsnetz," in *Ausgewählte Arbeiten über Nachrichtensysteme,* no. 56, Universität Erlangen-Nürnberg, 1984.

[18] R. Rückriem, "Realisierung und messtechnische Untersuchung an einem digitalen Parallelverfahren zur Datenübertragung im Fernsprechkanal," in *Ausgewählte Arbeiten über Nachrichtensysteme,* no. 59, Universität Erlangen-Nürnberg, 1985.

[19] I. Kalet, "The multitone channel," *IEEE Transactions on Communications,* vol. 37, pp. 119–124, February 1989.

[20] B. Hirosaki, "An analysis of automatic equalizers for orthogonally multiplexed QAM systems," *IEEE Transactions on Communications,* vol. COM-28, pp. 73–83, January 1980.

[21] P. Bello, "Selective fading limitations of the KATHRYN modem and some system design considerations," *IEEE Trabsactions on Communications Technology,* vol. COM-13, pp. 320–333, September 1965.

[22] E. Powers and M. Zimmermann, "A digital implementation of a multichannel data modem," in *Proceedings of the IEEE International Conference on Communications,* (Philadelphia, USA), 1968.

[23] R. Chang and R. Gibby, "A theoretical study of performance of an orthogonal multiplexing data transmission scheme," *IEEE Transactions on Communication Technology,* vol. COM–16, pp. 529–540, August 1968.

[24] B. R. Saltzberg, "Performance of an efficient parallel data transmission system," *IEEE Transactions on Communication Technology,* pp. 805–813, December 1967.

[25] K. Fazel and G. Fettweis, eds., *Multi-Carrier Spread-Spectrum.* Dordrecht: Kluwer, 1997. ISBN 0-7923-9973-0.

[26] F. Classen and H. Meyr, "Synchronisation algorithms for an OFDM system for mobile communications," in *Codierung für Quelle, Kanal und Übertragung,* no. 130 in ITG Fachbericht, (Berlin), pp. 105–113, VDE–Verlag, 1994.

[27] F. Classen and H. Meyr, "Frequency synchronisation algorithms for OFDM systems suitable for communication over frequency selective fading channels," in *Proceedings of IEEE VTC '94*, (Stockholm, Sweden), pp. 1655–1659, IEEE, 8–10 June 1994.

[28] R. van Nee and R. Prasad, *OFDM for wireless multimedia communications*. London: Artech House Publishers, 2000.

[29] P. Vandenameele, L. van der Perre, and M. Engels, *Space division multiple access for wireless local area networks*. Kluwer, 2001.

[30] S. Shepherd, P. van Eetvelt, C. Wyatt-Millington, and S. Barton, "Simple coding scheme to reduce peak factor in QPSK multicarrier modulation," *Electronics Letters*, vol. 31, pp. 1131–1132, July 1995.

[31] A. E. Jones, T. A. Wilkinson, and S. K. Barton, "Block coding scheme for reduction of peak to mean envelope power ratio of multicarrier transmission schemes," *Electronics Letters*, vol. 30, pp. 2098–2099, December 1994.

[32] D. Wulich, "Reduction of peak to mean ratio of multicarrier modulation by cyclic coding," *Electronics Letters*, vol. 32, pp. 432–433, 1996.

[33] S. Müller and J. Huber, "Vergleich von OFDM–Verfahren mit reduzierter Spitzenleistung," in *2. OFDM–Fachgespräch in Braunschweig*, 1997.

[34] M. Pauli and H.-P. Kuchenbecker, "Neue Aspekte zur Reduzierung der durch Nichtlinearitäten hervorgerufenen Außerbandstrahlung eines OFDM–Signals," in *2. OFDM–Fachgespräch in Braunschweig*, 1997.

[35] T. May and H. Rohling, "Reduktion von Nachbarkanalstörungen in OFDM–Funkübertragungssystemen," in *2. OFDM–Fachgespräch in Braunschweig*, 1997.

[36] D. Wulich, "Peak factor in orthogonal multicarrier modulation with variable levels," *Electronics Letters*, vol. 32, no. 20, pp. 1859–1861, 1996.

[37] H. Schmidt and K. Kammeyer, "Adaptive Subträgerselektion zur Reduktion des Crest faktors bei OFDM," in *3. OFDM Fachgespräch in Braunschweig*, 1998.

[38] R. Dinis and A. Gusmao, "Performance evaluation of OFDM transmission with conventional and 2-branch combining power amplification schemes," in *Proceeding of IEEE Global Telecommunications Conference, Globecom 96*, (London, UK), pp. 734–739, IEEE, 18–22 November 1996.

[39] R. Dinis, P. Montezuma, and A. Gusmao, "Performance trade-offs with quasi-linearly amplified OFDM through a 2-branch combining technique," in *Proceedings of IEEE VTC'96*, (Atlanta, GA, USA), pp. 899–903, IEEE, 28 April–1 May 1996.

[40] R. Dinis, A. Gusmao, and J. Fernandes, "Adaptive transmission techniques for the mobile broadband system," in *Proceeding of ACTS Mobile Communication Summit '97*, (Aalborg, Denmark), pp. 757–762, ACTS, 7–10 October 1997.

[41] B. Daneshrad, L. Cimini Jr., and M. Carloni, "Clustered-OFDM transmitter implementation," in *Proceedings of IEEE International Symposium on Personal, Indoor, and Mobile Radio Communications (PIMRC'96)*, (Taipei, Taiwan), pp. 1064–1068, IEEE, 15–18 October 1996.

[42] M. Okada, H. Nishijima, and S. Komaki, "A maximum likelihood decision based nonlinear distortion compensator for multi-carrier modulated signals," *IEICE Transactions on Communications*, vol. E81B, no. 4, pp. 737–744, 1998.

[43] R. Dinis and A. Gusmao, "Performance evaluation of a multicarrier modulation technique allowing strongly nonlinear amplification," in *Proceedings of ICC 1998*, pp. 791–796, IEEE, 1998.

[44] T. Pollet, M. van Bladel, and M. Moeneclaey, "BER sensitivity of OFDM systems to carrier frequency offset and wiener phase noise," *IEEE Transactions on Communications*, vol. 43, pp. 191–193, February/March/April 1995.

[45] H. Nikookar and R. Prasad, "On the sensitivity of multicarrier transmission over multipath channels to phase noise and frequency offset," in *Proceedings of IEEE International Symposium on Personal, Indoor, and Mobile Radio Communications (PIMRC'96)*, (Taipei, Taiwan), pp. 68–72, IEEE, 15–18 October 1996.

[46] W. Warner and C. Leung, "OFDM/FM frame synchronization for mobile radio data communication," *IEEE Transactions on Vehicular Technology*, vol. 42, pp. 302–313, August 1993.

[47] H. Sari, G. Karam, and I. Jeanclaude, "Transmission techniques for digital terrestrial TV broadcasting," *IEEE Communications Magazine*, pp. 100–109, February 1995.

[48] P. Moose, "A technique for orthogonal frequency division multiplexing frequency offset correction," *IEEE Transactions on Communications*, vol. 42, pp. 2908–2914, October 1994.

[49] K. Brüninghaus and H. Rohling, "Verfahren zur Rahmensynchronisation in einem OFDM-System," in *3. OFDM Fachgespräch in Braunschweig*, 1998.

[50] F. Daffara and O. Adami, "A new frequency detector for orthogonal multicarrier transmission techniques," in *Proceedings of IEEE Vehicular Technology Conference (VTC'95)*, (Chicago, USA), pp. 804–809, IEEE, 15–28 July 1995.

[51] M. Sandell, J.-J. van de Beek, and P. Börjesson, "Timing and frequency synchronisation in OFDM systems using the cyclic prefix," in *Proceedings of International Symposium on Synchronisation*, (Essen, Germany), pp. 16–19, 14–15 December 1995.

[52] N. Yee, J.-P. Linnartz, and G. Fettweis, "Multicarrier CDMA in indoor wireless radio networks," in *PIMRC'93*, pp. 109–113, 1993.

[53] A. Chouly, A. Brajal, and S. Jourdan, "Orthogonal multicarrier techniques applied to direct sequence spread spectrum CDMA systems," in *Proceedings of the IEEE Global Telecommunications Conference 1993*, (Houston, TX, USA), pp. 1723–1728, 29 November – 2 December 1993.

[54] G. Fettweis, A. Bahai, and K. Anvari, "On multi-carrier code division multiple access (MC-CDMA) modem design," in *Proceedings of IEEE VTC '94*, (Stockholm, Sweden), pp. 1670–1674, IEEE, 8–10 June 1994.

[55] K. Fazel and L. Papke, "On the performance of convolutionally-coded CDMA/OFDM for mobile communication system," in *PIMRC'93*, pp. 468–472, 1993.

[56] R. Prasad and S. Hara, "Overview of multicarrier CDMA," *IEEE Communications Magazine*, pp. 126–133, December 1997.

[57] B.-J. Choi, E.-L. Kuan, and L. Hanzo, "Crest–factor study of MC-CDMA and OFDM," in *Proceeding of VTC'99 (Fall)*, vol. 1, (Amsterdam, Netherlands), pp. 233–237, IEEE, 19–22 September 1999.

[58] P. Höher, "TCM on frequency-selective land-mobile fading channels," in *International Workshop on Digital Communications*, (Tirrenia, Italy), pp. 317–328, September 1991.

[59] J. Chow, J. Cioffi, and J. Bingham, "Equalizer training algorithms for multicarrier modulation systems.," in *International Conference on Communications*, (Geneva, Switzerland), pp. 761–765, IEEE, May 1993.

[60] S. Wilson, R. E. Khayata, and J. Cioffi, "16QAM Modulation with Orthogonal Frequency Division Multiplexing in a Rayleigh-Fading Environment," in *Vehicular Technology Conference*, vol. 3, (Stockholm, Sweden), pp. 1660–1664, IEEE, June 1994.

[61] J.-J. van de Beek, O. Edfors, M. Sandell, S. Wilson, and P. Börjesson, "On channel estimation in OFDM systems," in *Proceedings of Vehicular Technology Conference*, vol. 2, (Chicago, IL USA), pp. 815–819, IEEE, July 1995.

[62] O. Edfors, M. Sandell, J. van den Beek, S. K. Wilson, and P. Börjesson, "OFDM Channel Estimation by Singular Value Decomposition," in *Proceedings of Vehicular Technology Conference*, vol. 2, (Atlanta, GA USA), pp. 923–927, IEEE, April 28 - May 1 1996.

[63] P. Frenger and A. Svensson, "A Decision Directed Coherent Detector for OFDM," in *Proceedings of Vehicular Technology Conference*, vol. 3, (Atlanta, GA USA), pp. 1584–1588, IEEE, Apr 28 - May 1 1996.

[64] V. Mignone and A. Morello, "CD3-OFDM: A Novel Demodulation Scheme for Fixed and Mobile Receivers," *IEEE Transactions on Communications*, vol. 44, pp. 1144–1151, September 1996.

[65] F. Tufvesson and T. Maseng, "Pilot Assisted Channel Estimation for OFDM in Mobile Cellular Systems," in *Proceedings of Vehicular Technology Conference*, vol. 3, (Phoenix, Arizona), pp. 1639–1643, IEEE, May 4-7 1997.

[66] P. Höher, S. Kaiser, and P. Robertson, "Two-dimensional pilot-symbol-aided channel estimation by Wiener filtering," in *International Conference on Acoustics, Speech and Signal Processing*, (Munich, Germany), pp. 1845–1848, IEEE, April 1997.

[67] P. Höher, S. Kaiser, and P. Robertson, "Pilot–symbol–aided channel estimation in time and frequency," in *Proceedings of Global Telecommunications Conference: The Mini–Conf.*, (Phoenix, AZ), pp. 90–96, IEEE, November 1997.

[68] Y. Li, L. Cimini, and N. Sollenberger, "Robust Channel Estimation for OFDM Systems with Rapid Dispersive Fading Channels," *IEEE Transactions on Communications*, vol. 46, pp. 902–915, April 1998.

[69] O. Edfors, M. Sandell, J.-J. van den Beek, S. Wilson, and P. Börjesson, "OFDM Channel Estimation by Singular Value Decomposition," *IEEE Transactions on Communications*, vol. 46, pp. 931–939, April 1998.

[70] F. Tufvesson, M. Faulkner, and T. Maseng, "Pre-Compensation for Rayleigh Fading Channels in Time Division Duplex OFDM Systems," in *Proceedings of 6th International Workshop on Intelligent Signal Processing and Communications Systems*, (Melbourne, Australia), pp. 57–33, IEEE, November 5-6 1998.

[71] M. Itami, M. Kuwabara, M. Yamashita, H. Ohta, and K. Itoh, "Equalization of Orthogonal Frequency Division Multiplexed Signal by Pilot Symbol Assisted Multipath

Estimation," in *Proceedings of Global Telecommunications Conference*, vol. 1, (Sydney, Australia), pp. 225–230, IEEE, November 8-12 1998.

[72] E. Al-Susa and R. Ormondroyd, "A Predictor-Based Decision Feedback Channel Estimation Method for COFDM with High Resilience to Rapid Time-Variations," in *Proceedings of Vehicular Technology Conference*, vol. 1, (Amsterdam, Netherlands), pp. 273–278, IEEE, September 19-22 1999.

[73] B. Yang, K. Letaief, R. Cheng, and Z. Cao, "Robust and Improved Channel Estimation for OFDM Systems in Frequency Selective Fading Channels," in *Proceedings of Global Telecommunications Conference*, vol. 5, (Rio de Janeiro, Brazil), pp. 2499–2503, IEEE, December 5-9 1999.

[74] Y. Li, "Pilot-Symbol-Aided Channel Estimation for OFDM in Wireless Systems," *IEEE Transactions on Vehicular Technology*, vol. 49, pp. 1207–1215, July 2000.

[75] B. Yang, K. Letaief, R. Cheng, and Z. Cao, "Channel Estimation for OFDM Transmission in Multipath Fading Channels Based on Parametric Channel Modeling," *IEEE Transactions on Communications*, vol. 49, pp. 467–479, March 2001.

[76] S. Zhou and G. Giannakis, "Finite-Alphabet Based Channel Estimation for OFDM and Related Multicarrier Systems," *IEEE Transactions on Communications*, vol. 49, pp. 1402–1414, August 2001.

[77] X. Wang and K. Liu, "OFDM Channel Estimation Based on Time-Frequency Polynomial Model of Fading Multipath Channel," in *Proceedings of Vehicular Technology Conference*, vol. 1, (Atlantic City, NJ USA), pp. 460–464, IEEE, October 7-11 2001.

[78] B. Yang, Z. Cao, and K. Letaief, "Analysis of Low-Complexity Windowed DFT-Based MMSE Channel Estimator for OFDM Systems," *IEEE Transactions on Communications*, vol. 49, pp. 1977–1987, November 2001.

[79] B. Lu and X. Wang, "Bayesian Blind Turbo Receiver for Coded OFDM Systems with Frequency Offset and Frequency-Selective Fading," *IEEE Journal on Selected Areas in Communications*, vol. 19, pp. 2516–2527, December 2001.

[80] Y. Li and N. Sollenberger, "Clustered OFDM with Channel Estimation for High Rate Wireless Data," *IEEE Transactions on Communications*, vol. 49, pp. 2071–2076, December 2001.

[81] M. Morelli and U. Mengali, "A Comparison of Pilot-Aided Channel Estimation Methods for OFDM Systems," *IEEE Transactions on Signal Processing*, vol. 49, pp. 3065–3073, December 2001.

[82] M.-X. Chang and Y. Su, "Model-Based Channel Estimation for OFDM Signals in Rayleigh Fading," *IEEE Transactions on Communications*, vol. 50, pp. 540–544, April 2002.

[83] M. Necker and G. Stüber, "Totally Blind Channel Estimation for OFDM over Fast Varying Mobile Channels," in *Proceedings of International Conference on Communications*, (New York, NY USA), IEEE, April 28 - May 2 2002.

[84] B. Yang, Z. Cao, and K. Letaief, "Low Complexity Channel Estimator Based on Windowed DFT and Scalar Wiener Filter for OFDM Systems," in *Proceedings of International Conference on Communications*, vol. 6, (Helsinki, Finnland), pp. 1643–1647, IEEE, June 11-14 2001.

[85] J. Deller, J. Proakis, and J. Hansen, *Discrete-Time Processing of Speech Signals.* Macmillan Publishing Company, 1993.

[86] L. Hanzo, F. Somerville, and J. Woodard, *Voice Compression and Communications.* IEEE Press Wiley Inter-Science, 2001.

[87] A. Duel-Hallen, S. Hu, and H. Hallen, "Long Range Prediction of Fading Signals," *IEEE Signal Processing Magazine*, vol. 17, pp. 62–75, May 2000.

[88] F. Tufvesson, *Design of Wireless Communication Systems - Issues on Synchronization, Channel Estimation and Multi-Carrier Systems.* Department of Applied Electronics, Lund University, Sweden, 2000.

[89] W. Press, S. Teukolshy, W. Vetterling, and B. Flannery, *Numerical Recipes in C.* Cambridge University Press, 1992.

[90] T. Moon and W. Stirling, *Mathematical Methods and Algorithms for Signal Processing.* Prentice Hall, 2000.

[91] Y. Li, N. Seshadri, and S. Ariyavisitakul, "Channel Estimation for OFDM Systems with Transmitter Diversity in Mobile Wireless Channels," *IEEE Journal on Selected Areas in Communications*, vol. 17, pp. 461–471, March 1999.

[92] W. Jeon, K. Paik, and Y. Cho, "An Efficient Channel Estimation Technique for OFDM Systems with Transmitter Diversity," in *Proceedings of International Symposium on Personal, Indoor and Mobile Radio Communications*, vol. 2, (Hilton London Metropole Hotel, London, UK), pp. 1246–1250, IEEE, September 18-21 2000.

[93] Y. Li, "Optimum Training Sequences for OFDM Systems with Multiple Transmit Antennas," in *Proc. of Global Telecommunications Conference*, vol. 3, (San Francisco, United States), pp. 1478–1482, IEEE, November 27 - December 1 2000.

[94] A. Mody and G. Stüber, "Parameter Estimation for OFDM with Transmit Receive Diversity," in *Proceedings of Vehicular Technology Conference*, vol. 2, (Rhodes, Greece), pp. 820–824, IEEE, May 6-9 2001.

[95] Y. Gong and K. Letaief, "Low Rank Channel Estimation for Space-Time Coded Wideband OFDM Systems," in *Proceedings of Vehicular Technology Conference*, vol. 2, (Atlantic City Convention Center, Atlantic City, NJ USA), pp. 772–776, IEEE, October 7-11 2001.

[96] W. Jeon, K. Paik, and Y. Cho, "Two-Dimensional MMSE Channel Estimation for OFDM Systems with Transmitter Diversity," in *Proceedings of Vehicular Technology Conference*, vol. 3, (Atlantic City Convention Center, Atlantic City, NJ USA), pp. 1682–1685, IEEE, October 7-11 2001.

[97] F. Vook and T. Thomas, "MMSE Multi-User Channel Estimation for Broadband Wireless Communications," in *Proceedings of Global Telecommunications Conference*, vol. 1, (San Antonio, Texas, USA), pp. 470–474, IEEE, November 25-29 2001.

[98] Y. Xie and C. Georghiades, "An EM-based Channel Estimation Algorithm for OFDM with Transmitter Diversity," in *Proceedings of Global Telecommunications Conference*, vol. 2, (San Antonio, Texas, USA), pp. 871–875, IEEE, November 25-29 2001.

[99] Y. Li, "Simplified Channel Estimation for OFDM Systems with Multiple Transmit Antennas," *IEEE Transactions on Wireless Communications*, vol. 1, pp. 67–75, January 2002.

[100] H. Bölcskei, R. Heath, and A. Paulraj, "Blind Channel Identification and Equalization in OFDM-Based Multi-Antenna Systems," *IEEE Transactions on Signal Processing*, vol. 50, pp. 96–109, January 2002.

[101] H. Minn, D. Kim, and V. Bhargava, "A Reduced Complexity Channel Estimation for OFDM Systems with Transmit Diversity in Mobile Wireless Channels," *IEEE Transactions on Wireless Communications*, vol. 50, pp. 799–807, May 2002.

[102] S. Slimane, "Channel Estimation for HIPERLAN/2 with Transmitter Diversity," in *International Conference on Communications*, (New York, NY USA), IEEE, April 28 - May 2 2002.

[103] C. Komninakis, C. Fragouli, A. Sayed, and R. Wesel, "Multi-Input Multi-Output Fading Channel Tracking and Equalization Using Kalman Estimation," *IEEE Transactions on Signal Processing*, vol. 50, pp. 1065–1076, May 2002.

[104] G. Foschini, "Layered Space-Time Architecture for Wireless Communication in a Fading Environment when using Multi-Element Antennas"," *Bell Labs Technical Journal*, vol. Autumn, pp. 41–59, 1996.

[105] F. Vook and K. Baum, "Adaptive antennas for OFDM," in *Proceedings of IEEE Vehicular Technology Conference (VTC'98)*, vol. 2, (Ottawa, Canada), pp. 608–610, IEEE, 18–21 May 1998.

[106] X. Wang and H. Poor, "Robust Adaptive Array for Wireless Communications," *IEEE Transactions on Communications*, vol. 16, pp. 1352–1366, October 1998.

[107] K.-K. Wong, R.-K. Cheng, K. Letaief, and R. Murch, "Adaptive Antennas at the Mobile and Base Station in an OFDM/TDMA System," in *Proceedings of Global Telecommunications Conference*, vol. 1, (Sydney, Australia), pp. 183–190, IEEE, November 8-12 1998.

[108] Y. Li and N. Sollenberger, "Interference Suppression in OFDM Systems using Adaptive Antenna Arrays," in *Proceedings of Global Telecommunications Conference*, vol. 1, (Sydney, Australia), pp. 213–218, IEEE, November 8-12 1998.

[109] G. Golden, G. Foschini, R. Valenzuela, and P. Wolniansky, "Detection Algorithms and Initial Laboratory Results using V-BLAST Space-Time Communication Architecture," *IEE Electronics Letters*, vol. 35, pp. 14–16, January 1999.

[110] Y. Li and N. Sollenberger, "Adaptive Antenna Arrays for OFDM Systems with Cochannel Interference," *IEEE Transactions on Communications*, vol. 47, pp. 217–229, February 1999.

[111] P. Vandenameele, L. Van der Perre, M. Engels, and H. Man, "A novel class of uplink OFDM/SDMA algorithms for WLAN," in *Proceedings of Global Telecommunications Conference — Globecom'99*, vol. 1, (Rio de Janeiro, Brazil), pp. 6–10, IEEE, 5–9 December 1999.

[112] M. Speth, A. Senst, and H. Meyr, "Low complexity space-frequency MLSE for multi-user COFDM," in *Proceedings of Global Telecommunications Conference — Globecom'99*, vol. 1, (Rio de Janeiro, Brazil), pp. 2395–2399, IEEE, 5–9 December 1999.

[113] C. H. Sweatman, J. Thompson, B. Mulgrew, and P. Grant, "A Comparison of Detection Algorithms including BLAST for Wireless Communication using Multiple Antennas,"

in *Proceedings of International Symposium on Personal, Indoor and Mobile Radio Communications*, vol. 1, (Hilton London Metropole Hotel, London, UK), pp. 698–703, IEEE, September 18-21 2000.

[114] R. van Nee, A. van Zelst, and G. Awater, "Maximum Likelihood Decoding in a Space-Division Multiplexing System," in *Proceedings of Vehicular Technology Conference*, vol. 1, (Tokyo, Japan), pp. 6–10, IEEE, May 15-18 2000.

[115] G. Awater, A. van Zelst, and R. van Nee, "Reduced Complexity Space Division Multiplexing Receivers," in *Proceedings of Vehicular Technology Conference*, vol. 1, (Tokyo, Japan), pp. 11–15, IEEE, May 15-18 2000.

[116] A. van Zelst, R. van Nee, and G. Awater, "Space Division Multiplexing (SDM) for OFDM systems," in *Proceedings of Vehicular Technology Conference*, vol. 2, (Tokyo, Japan), pp. 1070–1074, IEEE, May 15-18 2000.

[117] P. Vandenameele, L. V. D. Perre, M. Engels, B. Gyselinckx, and H. D. Man, "A Combined OFDM/SDMA Approach," *IEEE Journal on Selected Areas in Communications*, vol. 18, pp. 2312–2321, November 2000.

[118] X. Li, H. Huang, A. Lozano, and G. Foschini, "Reduced-Complexity Detection Algorithms for Systems Using Multi-Element Arrays," in *Proc. of Global Telecommunications Conference*, vol. 2, (San Francisco, United States), pp. 1072–1076, IEEE, November 27 - December 1 2000.

[119] C. Degen, C. Walke, A. Lecomte, and B. Rembold, "Adaptive MIMO Techniques for the UTRA-TDD Mode," in *Proceedings of Vehicular Technology Conference*, vol. 1, (Rhodes, Greece), pp. 108–112, IEEE, May 6-9 2001.

[120] X. Zhu and R. Murch, "Multi-Input Multi-Output Maximum Likelihood Detection for a Wireless System," in *Proceedings of Vehicular Technology Conference*, vol. 1, (Rhodes, Greece), pp. 137–141, IEEE, May 6-9 2001.

[121] J. Li, K. Letaief, R. Cheng, and Z. Cao, "Joint Adaptive Power Control and Detection in OFDM/SDMA Wireless LANs," in *Proceedings of Vehicular Technology Conference*, vol. 1, (Rhodes, Greece), pp. 746–750, IEEE, May 6-9 2001.

[122] F. Rashid-Farrokhi, K. Liu, and L. Tassiulas, "Transmit Beamforming and Power Control for Cellular Wireless Systems," *IEEE Journal on Selected Areas in Communications*, vol. 16, pp. 1437–1450, October 1998.

[123] A. van Zelst, R. van Nee, and G. Awater, "Turbo-BLAST and its Performance," in *Proceedings of Vehicular Technology Conference*, vol. 2, (Rhodes, Greece), pp. 1282–1286, IEEE, May 6-9 2001.

[124] A. Benjebbour, H. Murata, and S. Yoshida, "Performance of Iterative Successive Detection Algorithm with Space-Time Transmission," in *Proceedings of Vehicular Technology Conference*, vol. 2, (Rhodes, Greece), pp. 1287–1291, IEEE, May 6-9 2001.

[125] M. Sellathurai and S. Haykin, "A Simplified Diagonal BLAST Architecture with Iterative Parallel-Interference Cancellation Receivers," in *Proceedings of International Conference on Communications*, vol. 10, (Helsinki, Finnland), pp. 3067–3071, IEEE, June 11-14 2001.

[126] A. Bhargave, R. Figueiredo, and T. Eltoft, "A Detection Algorithm for the V-BLAST System," in *Proceedings of Global Telecommunications Conference*, vol. 1, (San Antonio, Texas, USA), pp. 494–498, IEEE, November 25-29 2001.

[127] S. Thoen, L. Deneire, L. V. D. Perre, and M. Engels, "Constrained Least Squares Detector for OFDM/SDMA-based Wireless Networks," in *Proceedings of Global Telecommunications Conference*, vol. 2, (San Antonio, Texas, USA), pp. 866–870, IEEE, November 25-29 2001.

[128] Y. Li and Z.-Q. Luo, "Parallel Detection for V-BLAST System," in *Proceedings of International Conference on Communications*, (New York, NY USA), IEEE, April 28 - May 2 2002.

[129] Y. Li and N. Sollenberger, "Interference suppression in OFDM systems using adaptive antenna arrays," in *Proceeding of Globecom'98*, (Sydney, Australia), pp. 213–218, IEEE, 8–12 November 1998.

[130] Y. Li and N. Sollenberger, "Adaptive antenna arrays for OFDM systems with cochannel interference," *IEEE Transactions on Communications*, vol. 47, pp. 217–229, February 1999.

[131] Y. Li, L. Cimini, and N. Sollenberger, "Robust channel estimation for OFDM systems with rapid dispersive fading channels," *IEEE Transactions on Communications*, vol. 46, pp. 902–915, April 1998.

[132] C. Kim, S. Choi, and Y. Cho, "Adaptive beamforming for an OFDM sytem," in *Proceeding of VTC'99 (Spring)*, (Houston, TX, USA), IEEE, 16–20 May 1999.

[133] L. Lin, L. Cimini Jr., and J.-I. Chuang, "Turbo codes for OFDM with antenna diversity," in *Proceeding of VTC'99 (Spring)*, (Houston, TX, USA), IEEE, 16–20 May 1999.

[134] M. Münster, T. Keller, and L. Hanzo, "Co–channel interference suppression assisted adaptive OFDM in interference limited environments," in *Proceeding of VTC'99 (Fall)*, vol. 1, (Amsterdam, Netherlands), pp. 284–288, IEEE, 19–22 September 1999.

[135] S. Verdu, *Multiuser Detection*. Cambridge University Press, 1998.

[136] J. Litva and T.-Y. Lo, *Digital Beamforming in Wireless Communications*. London: Artech House Publishers, 1996.

[137] P. Vandenameele, L. Van der Perre, M. Engels, B. Gyselinckx, and H. Man, "A novel class of uplink OFDM/SDMA algorithms: A statistical performance analysis," in *Proceedings of Vehicular Technology Conference*, vol. 1, (Amsterdam, Netherlands), pp. 324–328, IEEE, 19–22 September 1999.

[138] F. Mueller-Roemer, "Directions in audio broadcasting," *Journal Audio Engineering Society*, vol. 41, pp. 158–173, March 1993.

[139] G. Plenge, "DAB — a new radio broadcasting system — state of development and ways for its introduction," *Rundfunktech. Mitt.*, vol. 35, no. 2, 1991.

[140] ETSI, *Digital Audio Broadcasting (DAB)*, 2nd ed., May 1997. ETS 300 401.

[141] ETSI, *Digital Video Broadcasting (DVB); Framing structure, channel coding and modulation for digital terrestrial television*, August 1997. EN 300 744 V1.1.2.

[142] P. Chow, J. Tu, and J. Cioffi, "A discrete multitone transceiver system for HDSL applications," *IEEE journal on selected areas in communications*, vol. 9, pp. 895–908, August 1991.

[143] P. Chow, J. Tu, and J. Cioffi, "Performance evaluation of a multichannel transceiver system for ADSL and VHDSL services," *IEEE journal on selected areas in communications*, vol. 9, pp. 909–919, August 1991.

[144] K. Sistanizadeh, P. Chow, and J. Cioffi, "Multi-tone transmission for asymmetric digital subscriber lines (ADSL)," in *Proceedings of ICC'93*, pp. 756–760, IEEE, 1993.

[145] ANSI, *ANSI/T1E1.4/94-007, Asymmetric Digital Subscriber Line (ADSL) Metallic Interface.*, August 1997.

[146] A. Burr and P. Brown, "Application of OFDM to powerline telecommunications," in *3rd International Symposium On Power-Line Communications*, (Lancaster, UK), 30 March – 1 April 1999.

[147] M. Deinzer and M. Stoger, "Integrated PLC-modem based on OFDM," in *3rd International Symposium On Power-Line Communications*, (Lancaster, UK), 30 March – 1 April 1999.

[148] R. Prasad and H. Harada, "A novel OFDM based wireless ATM system for future broadband multimedia communications," in *Proceeding of ACTS Mobile Communication Summit '97*, (Aalborg, Denmark), pp. 757–762, ACTS, 7–10 October 1997.

[149] C. Ciotti and J. Borowski, "The AC006 MEDIAN project — overview and state–of–the–art," in *Proc. ACTS Summit '96*, (Granada, Spain), pp. 362–367, 27–29 November 1996.

[150] J. Borowski, S. Zeisberg, J. Hübner, K. Koora, E. Bogenfeld, and B. Kull, "Performance of OFDM and comparable single carrier system in MEDIAN demonstrator 60GHz channel," in *Proceeding of ACTS Mobile Communication Summit '97*, (Aalborg, Denmark), pp. 653–658, ACTS, 7–10 October 1997.

[151] M. D. Benedetto, P. Mandarini, and L. Piazzo, "Effects of a mismatch in the in–phase and in–quadrature paths, and of phase noise, in QDCPSK-OFDM modems," in *Proceeding of ACTS Mobile Communication Summit '97*, (Aalborg, Denmark), pp. 769–774, ACTS, 7–10 October 1997.

[152] T. Rautio, M. Pietikainen, J. Niemi, J. Rautio, K. Rautiola, and A. Mammela, "Architecture and implementation of the 150 Mbit/s OFDM modem (invited paper)," in *IEEE Benelux Joint Chapter on Communications and Vehicular Technology, 6th Symposium on Vehicular Technology and Communications*, (Helsinki, Finland), p. 11, 12–13 October 1998.

[153] J. Ala-Laurila and G. Awater, "The magic WAND — wireless ATM network demondtrator system," in *Proceeding of ACTS Mobile Communication Summit '97*, (Aalborg, Denmark), pp. 356–362, ACTS, 7–10 October 1997.

[154] J. Aldis, E. Busking, T. Kleijne, R. Kopmeiners, R. van Nee, R. Mann-Pelz, and T. Mark, "Magic into reality, building the WAND modem," in *Proceeding of ACTS Mobile Communication Summit '97*, (Aalborg, Denmark), pp. 775–780, ACTS, 7–10 October 1997.

[155] E. Hallmann and H. Rohling, "OFDM-Vorschläge für UMTS," in *3. OFDM Fachge-spräch in Braunschweig*, 1998.

[156] "Universal mobile telecommunications system (UMTS); UMTS terrestrial radio access (UTRA); concept evaluation," tech. rep., ETSI, 1997. TR 101 146.

[157] O. Edfors, *Low-Complexity Algorithms in Digital Receivers*. Lulea University of Technology, 1996.

[158] M. Sandell, *Design and analysis of estimators for multicarrier modulation and ultrasonic imaging*. Lulea University of Technology, 1996.

[159] M. Münster and L. Hanzo, "MMSE Channel Prediction For Symbol-by-symbol Adaptive OFDM Systems," in 5th *International OFDM-Workshop 2000*, pp. 35/1–35/6, Technische Universität Hamburg-Harburg, September 2000.

[160] M. Münster and L. Hanzo, "MMSE Channel Prediction Assisted Symbol-by-Symbol Adaptive OFDM," in *Proceedings of International Conference on Communications*, (Birmingham, Alabama), IEEE, April 28 - May 2 2002.

[161] M. Münster and L. Hanzo, "Parallel Interference Cancellation Assisted Decision-Directed Channel Estimation for Multi-User OFDM," in 6th *International OFDM-Workshop 2000*, (Hotel Hafen Hamburg, Germany), pp. 35/1–35/5, Technische Universität Hamburg-Harburg, September 18-19 2001.

[162] M. Münster and L. Hanzo, "PIC-Assisted Decision-Directed Channel Estimation for Multi-User OFDM Environments," in *accepted for publication in Proceedings of Vehicular Technology Conference*, IEEE, September 2002.

[163] M. Münster and L. Hanzo, "Co-Channel Interference Cancellation Techniques for Antenna Array Assisted Multiuser OFDM Systems," in *Proceedings of 3G-'2000 Conference*, vol. 1, (London, Great Britain), pp. 256–260, IEE, IEE, March 27-29 2000.

[164] L. Hanzo, W. Webb, and T. Keller, *Single- and Multi-Carrier Quadrature Amplitude Modulation: Principles and Applications for Personal Communications, WLANs and Broadcasting*. John Wiley and IEEE Press, 2000.

[165] H. Kolb Private Communications.

[166] J. Lindner Private Communications.

[167] W. Webb and L. Hanzo, *Modern Quadrature Amplitude Modulation: Principles and Applications for Wireless Communications*. IEEE Press-Pentech Pressm, 1994.

[168] D. Schnidman, "A generalized nyquist criterion and an optimum linear receiver for a pulse modulation system," *Bell Systems Technical Journal*, pp. 2163–2177, November 1967.

[169] W. V. Etten, "An optimum linear receiver for multiple channel digital transmission systems," *IEEE Transactions on Communications*, vol. COM-23, pp. 828–834, August 1975.

[170] A. Kaye and D. George, "Transmission of multiplexed PAM signals over multiple channel and diversity systems," *IEEE Tranactions on Communications Technology*, vol. COM-18, pp. 520–525, October 1970.

[171] M. Aaron and D. Tufts, "Intersymbol interference and error probability," *IEEE Transactions on Information Theory*, vol. IT-12, pp. 26–34, January 1966.

[172] D. Tufts, "Nyquist's problem: The joint optimization of transmitter and receiver in pulse amplitude modulation," *Proceedings of IEEE*, vol. 53, pp. 248–259, March 1965.

[173] H. Schüssler, *Digitale Systeme zur Signalverarbeitung*. Berlin, Heidelberg, and New York: Springer Verlag, 1974.

[174] J. K. Cavers, "An Analysis of Pilot Symbol Assisted Modulation for Rayleigh Fading Channels," *IEEE Transactions on Vehicular Technology*, vol. 40, pp. 686–693, November 1991.

[175] H. Harmuth, *Transmission of Information by Orthogonal Time Functions*. Berlin: Springer Verlag, 1969.

[176] H. Harmuth, "On the transmission of information by orthogonal time functions," *AIEE*, July 1960.

[177] J. Proakis, *Digital communications*. McGraw-Hill, 1995.

[178] R. O'Neill and L. Lopes, "Performance of amplitude limited multitone signals," in *Proceedings of IEEE VTC '94*, (Stockholm, Sweden), IEEE, 8–10 June 1994.

[179] X. Li and L. Cimini, "Effects of clipping and filtering on the performance of OFDM," in *Proceedings of IEEE VTC'97*, (Phoenix, AZ, USA), pp. 1634–1638, IEEE, 4–7 May 1997.

[180] A. Garcia and M. Calvo, "Phase noise and sub–carrier spacing effects on the performance of an OFDM communications system," *IEEE Communications Letters*, vol. 2, pp. 11–13, January 1998.

[181] W. Robins, *Phase Noise in signal sources*, vol. 9 of *IEE Telecommunication series*. Peter Peregrinus Ltd., 1982.

[182] C. Tellambura, Y. Guo, and S. Barton, "Equaliser performance for HIPERLAN in indoor channels," *Wireless Personal Communications*, vol. 3, no. 4, pp. 397–410, 1996.

[183] T. Ojanperä, M. Gudmundson, P. Jung, J. Sköld, R. Pirhonen, G. Kramer, and A. Toskala, "FRAMES: - hybrid multiple access technology," in *Proceedings of IEEE ISSSTA'96*, (Mainz, Germany), pp. 334–338, IEEE, September 1996.

[184] M. Failli, "Digital land mobile radio communications COST 207," tech. rep., European Commission, 1989.

[185] J. Torrance and L. Hanzo, "Comparative study of pilot symbol assisted modem schemes," in *Proceedings of IEE Conference on Radio Receivers and Associated Systems (RRAS'95)*, (Bath, UK), pp. 36–41, IEE, 26–28 September 1995.

[186] R. Hasholzner, C. Drewes, and A. Hutter, "Untersuchungen zur linearen ICI-Kompensation bei OFDM (in German)," in *3.OFDM Fachgespräch*, Technische Universität Braunschweig, Germany, Sept. 1998.

[187] W. Jeon, K. Chang, and Y. Cho, "An Equalization Technique for Orthogonal Frequency-Division Multiplexing Systems in Time-Variant Multipath Channels," *IEEE Transactions on Communications*, vol. 47, pp. 27–32, January 1999.

[188] A. Hutter and R. Hasholzner, "Determination of Intercarrier Interference Covariance Matrices and their Application to Advanced Equalization for Mobile OFDM," in 5th *International OFDM-Workshop 2000*, pp. 33/1–33/5, Technische Universität Hamburg-Harburg, Sept. 2000.

[189] M. Ruessel and G. Stüber, "Terrestrial digital video broadcasting for mobile reception using OFDM," *Wireless Pers. Commun.*, vol. 2, no. 3, pp. 45–66, 1995.

[190] K. Fazel, S. Kaiser, P. Robertson, and M. Ruf, "A concept of digital terrestrial television broadcasting," *Wireless Personal Communications*, vol. 2, pp. 9–27, 1995.

[191] J. Kuronen, V.-P. Kaasila, and A. Mammela, "An all-digital symbol tracking algorithm in an OFDM system by using the cyclic prefix," in *Proc. ACTS Summit '96*, (Granada, Spain), pp. 340–345, 27–29 November 1996.

[192] M. Kiviranta and A. Mammela, "Coarse frame synchronization structures in OFDM," in *Proc. ACTS Summit '96*, (Granada, Spain), pp. 464–470, 27–29 November 1996.

[193] Z. Li and A. Mammela, "An all digital frequency synchronization scheme for OFDM systems," in *Proceedings of the IEEE International Symposium on Personal, Indoor and Mobile Radio Communications (PIMRC)*, (Helsinki, Finland), pp. 327–331, 1–4 September 1997.

[194] J. Bingham, "Method and apparatus for correcting for clock and carrier frequency offset, and phase jitter in multicarrier modems." U.S. Patent No. 5206886, 27 April 1993.

[195] T. de Couasnon, R. Monnier, and J. Rault, "OFDM for digital TV broadcasting," *Signal Processing*, vol. 39, pp. 1–32, 1994.

[196] P. Mandarini and A. Falaschi, "SYNC proposals." MEDIAN Design Note, January 1996.

[197] T. Keller and L. Hanzo, "Orthogonal frequency division multiplex synchronisation techniques for wireless local area networks," in *Proceedings of IEEE International Symposium on Personal, Indoor, and Mobile Radio Communications (PIMRC'96)*, vol. 3, (Taipei, Taiwan), pp. 963–967, IEEE, 15–18 October 1996.

[198] R. Steele and W. Webb, "Variable rate QAM for data transmission over Rayleigh fading channels," in *Proceeedings of Wireless '91*, (Calgary, Alberta), pp. 1–14, IEEE, 1991.

[199] H. Matsuoka, S. Sampei, N. Morinaga, and Y. Kamio, "Adaptive modulation system with variable coding rate concatenated code for high quality multi-media communications systems," in *Proceedings of IEEE VTC'96*, vol. 1, (Atlanta, GA, USA), pp. 487–491, IEEE, 28 April–1 May 1996.

[200] S.-G. Chua and A. Goldsmith, "Variable-rate variable-power mQAM for fading channels," in *Proceedings of IEEE VTC'96*, (Atlanta, GA, USA), pp. 815–819, IEEE, 28 April–1 May 1996.

[201] D. Pearce, A. Burr, and T. Tozer, "Comparison of counter-measures against slow Rayleigh fading for TDMA systems," in *IEE Colloquium on Advanced TDMA Techniques and Applications*, (London, UK), pp. 9/1–9/6, IEE, 28 October 1996. digest 1996/234.

[202] V. Lau and M. Macleod, "Variable rate adaptive trellis coded QAM for high bandwidth efficiency applications in Rayleigh fading channels," in *Proceedings of IEEE Vehicular Technology Conference (VTC'98)*, (Ottawa, Canada), pp. 348–352, IEEE, 18–21 May 1998.

[203] J. Torrance and L. Hanzo, "Latency and networking aspects of adaptive modems over slow indoors Rayleigh fading channels," *IEEE Transactions on Vehicular Technology*, vol. 48, no. 4, pp. 1237–1251, 1998.

[204] J. Torrance, L. Hanzo, and T. Keller, "Interference aspects of adaptive modems over slow Rayleigh fading channels," *IEEE Transactions on Vehicular Technology*, vol. 48, pp. 1527–1545, September 1999.

[205] A. Czylwik, "Adaptive OFDM for wideband radio channels," in *Proceeding of IEEE Global Telecommunications Conference, Globecom 96*, (London, UK), pp. 713–718, IEEE, 18–22 November 1996.

[206] P. Chow, J. Cioffi, and J. Bingham, "A practical discrete multitone transceiver loading algorithm for data transmission over spectrally shaped channels," *IEEE Transactions on Communications*, vol. 48, pp. 772–775, 1995.

[207] L. Hanzo, C. Wong, and M. Yee, *Adaptive Wireless Transceivers*. John Wiley, IEEE Press, 2002. (For detailed contents, please refer to http://www-mobile.ecs.soton.ac.uk.).

[208] L. Hanzo, T. Liew, and B. Yeap, *Turbo Coding, Turbo Equalisation and Space-Time Coding*. John Wiley, IEEE Press, 2002. (For detailed contents, please refer to http://www-mobile.ecs.soton.ac.uk.).

[209] J. Blogh and L. Hanzo, *3G Systems and Intelligent Networking*. John Wiley and IEEE Press, 2002. (For detailed contents, please refer to http://www-mobile.ecs.soton.ac.uk.).

[210] J. Torrance, *Adaptive Full Response Digital Modulation for Wireless Communications Systems*. PhD thesis, Department of Electronics and Computer Science, University of Southampton, UK, 1997.

[211] K. Miya, O. Kato, K. Homma, T. Kitade, M. Hayashi, and T. Ue, "Wideband CDMA systems in TDD-mode operation for IMT-2000," *IEICE Transactions on Communications*, vol. E81-B, pp. 1317–1326, July 1998.

[212] O. Kato, K. Miya, K. Homma, T. Kitade, M. Hayashi, and M. Watanabe, "Experimental performance results of coherent wideband DS-CDMA with TDD scheme," *IEICE Transactions on Communications.*, vol. E81-B, pp. 1337–1344, July 1998.

[213] T. Keller and L. Hanzo, "Blind-detection assisted sub-band adaptive turbo-coded OFDM schemes," in *Proceeding of VTC'99 (Spring)*, (Houston, TX, USA), pp. 489–493, IEEE, 16–20 May 1999.

[214] J. Torrance and L. Hanzo, "Optimisation of switching levels for adaptive modulation in a slow Rayleigh fading channel," *Electronics Letters*, vol. 32, pp. 1167–1169, 20 June 1996.

[215] C. Berrou, A. Glavieux, and P. Thitimajshima, "Near shannon limit error-correcting coding and decoding: Turbo codes," in *Proceedings of the International Conference on Communications*, (Geneva, Switzerland), pp. 1064–1070, May 1993.

[216] L. Bahl, J. Cocke, F. Jelinek, and J. Raviv, "Optimal decoding of linear codes for minimising symbol error rate," *IEEE Transactions on Information Theory*, vol. 20, pp. 284–287, March 1974.

[217] T. Keller, M. Muenster, and L. Hanzo, "A burst–by–burst adaptive OFDM wideband speech transceiver." submitted to IEEE JSAC, 1999.

[218] T. Keller, J. Woodard, and L. Hanzo, "Turbo-coded parallel modem techniques for personal communications," in *Proceedings of IEEE VTC'97*, (Phoenix, AZ, USA), pp. 2158–2162, IEEE, 4–7 May 1997.

[219] T. Keller and L. Hanzo, "Adaptive orthogonal frequency division multiplexing schemes," in *Proceeding of ACTS Mobile Communication Summit '98*, (Rhodes, Greece), pp. 794–799, ACTS, 8–11 June 1998.

[220] C. E. Shannon, "Communication in the presence of noise," *Proceedings of the I.R.E.*, vol. 37, pp. 10–22, January 1949.

[221] L. Piazzo, "A fast algorithm for near-optimum power and bit allocation in OFDM systems." to appear in Electronics Letters, December 1999.

[222] T. Willink and P. Wittke, "Optimization and performance evaluation of multicarrier transmission," *IEEE Transactions on Information Theory*, vol. 43, pp. 426–440, March 1997.

[223] R. Fischer and J. Huber, "A new loading algorithm for discrete multitone transmission," in *Proceeding of IEEE Global Telecommunications Conference, Globecom 96*, (London, UK), pp. 713–718, IEEE, 18–22 November 1996.

[224] S. Lai, R. Cheng, K. Letaief, and R. Murch, "Adaptive trellis coded mqam and power optimization for ofdm transmission," in *Proceedings of VTC'99 (Spring)*, (Houston, TX, USA), IEEE, 16–20 May 1999.

[225] D. Hughes-Hartogs, "Ensemble modem structure for imperfect transmission media." U.S Patents Nos. 4,679,227 (July 1988) 4,731,816 (March 1988) and 4,833,796 (May 1989).

[226] J. Bingham, "Multicarrier modulation for data transmission: an idea whose time has come," *IEEE Communications Magazine*, pp. 5–14, May 1990.

[227] L. Godara, "Applications of antenna arrays to mobile communications, part II: Beamforming and direction-of-arrival considerations," *Proceedings of the IEEE*, vol. 85, pp. 1193–1245, August 1997.

[228] Y. Li, "Pilot-symbol-aided channel estimation for OFDM in wireless systems," in *Proceedings of VTC'99 (Spring)*, (Houston, TX, USA), IEEE, 16–20 May 1999.

[229] N. Szabo and R. Tanaka, *Residue Arithmetic and Its Applications to Computer Technology*. New York, USA: McGraw-Hill, 1967.

[230] R. Watson and C. Hastings, "Self-checked computation using residue arithmetic," *Proceedings of the IEEE*, vol. 54, pp. 1920–1931, December 1966.

[231] R. Pyndiah, "Iterative decoding of product codes: Block turbo codes," in *Proceedings of the International Symposium on Turbo Codes & Related Topics*, (Brest, France), pp. 71–79, 3–5 September 1997.

[232] P. Adde, R. Pyndiah, O. Raoul, and J.-R. Inisan, "Block turbo decoder design," in *Proceedings of the International Symposium on Turbo Codes & Related Topics*, (Brest, France), pp. 166–169, 3–5 September 1997.

[233] W. Jenkins and B. Leon, "The use of residue number system in the design of finite impulse response filters," *IEEE Transactions on Circuits Systems*, vol. CAS-24, pp. 191–201, April 1977.

[234] M. Soderstrand, "A high-speed, low-cost, recursive digital filter using residue number arithmetic," *Proceedings of IEEE*, vol. 65, pp. 1065–1067, July 1977.

[235] M. Soderstrand and E. Fields, "Multipliers for residue number arithmetic digital filters," *Electronics Letters*, vol. 13, pp. 164–166, March 1977.

[236] M. Soderstrand, W. Jenkins, and G. Jullien, *Residue Number System Arithmetic: Modern Applications in Digital Signal Processing*. New York, USA: IEEE Press, 1986.

[237] E. Claudio, G. Orlandi, and F. Piazza, "A Systolic Redundant Residue Arithmetic Error Correction Circuit," *IEEE Transactions on Computers*, vol. 42, pp. 427–432, April 1993.

[238] H. Krishna, K.-Y. Lin, and J.-D. Sun, "A coding theory approach to error control in redundant residue number systems - part I: theory and single error correction," *IEEE Transactions on Circuits Systems*, vol. 39, pp. 8–17, January 1992.

[239] J.-D. Sun and H. Krishna, "A coding theory approach to error control in redundant residue number systems — part II: multiple error detection and correction," *IEEE Transactions on Circuits Systems*, vol. 39, pp. 18–34, January 1992.

[240] T. Liew, L.-L. Yang, and L. Hanzo, "Soft-decision redundant residue number system based error correction coding," in *Proceeding of VTC'99 (Fall)*, (Amsterdam, Netherlands), pp. 2974–2978, IEEE, 19–22 September 1999.

[241] L.-L. Yang and L. Hanzo, "Residue number system arithmetic assisted m-ary modulation," *IEEE Communications Letters*, vol. 3, pp. 28–30, February 1999.

[242] L.-L. Yang and L. Hanzo, "Performance of residue number system based DS-CDMA over multipath fading channels using orthogonal sequences," *ETT*, vol. 9, pp. 525–536, November–December 1998.

[243] H. Krishna and J.-D. Sun, "On theory and fast algorithms for error correction in residue number system product codes," *IEEE Transactions on Comput.*, vol. 42, pp. 840–852, July 1993.

[244] D. Chase, "A class of algorithms for decoding block codes with channel measurement information," *IEEE Transactions on Information Theory*, vol. IT-18, pp. 170–182, January 1972.

[245] J. Hagenauer, E. Offer, and L. Papke, "Iterative decoding of binary block and convolutional codes," *IEEE Transactions on Information Theory*, vol. 42, pp. 429–445, March 1996.

[246] H. Nickl, J. Hagenauer, and F. Burkett, "Approaching shannon's capacity limit by 0.27 dB using simple hamming codes," *IEEE Communications Letters*, vol. 1, pp. 130–132, September 1997.

[247] T. Liew, C. Wong, and L. Hanzo, "Block turbo coded burst-by-burst adaptive modems," in *Proceedings of Microcoll'99, Budapest, Hungary*, pp. 59–62, 21–24 March 1999.

[248] B. Yeap, T. Liew, J. Hamorsky, and L. Hanzo, "Comparative study of turbo equalisers using convolutional codes and block-based turbo-codes for GMSK modulation," in *Proceedings of VTC 1999 Fall*, (Amsterdam, Holland), pp. 2974–2978, 19-22 September 1999.

[249] M. Yee, T. Liew, and L. Hanzo, "Radial basis function decision feedback equalisation assisted block turbo burst-by-burst adaptive modems," in *Proceedings of VTC '99 Fall*, (Amsterdam, Holland), pp. 1600–1604, 19-22 September 1999.

[250] C. Wong, T. Liew, and L. Hanzo, "Turbo coded burst by burst adaptive wideband modulation with blind modem mode detection," in *Proceeding of ACTS Mobile Communication Summit '99*, (Sorrento, Italy), pp. 303–308, ACTS, 8–11 June 1999.

[251] A. Viterbi, *CDMA: Principles of Spread Spectrum Communication*. Reading MA, USA: Addison-Wesley, June 1995. ISBN 0201633744.

[252] L. Miller and J. Lee, *CDMA Systems Engineering Handbook*. London, UK: Artech House, 1998.

[253] R. A. Scholtz, "The origins of spread spectrum communications," *IEEE Transactions on Communications*, vol. 30, no. 5, pp. 822–854, 1982.

[254] G. R. Cooper and R. W. Nettleton, "A spread-spectrum technique for high-capacity mobile communications," *IEEE Transactions on Vehicular Technology*, vol. 27, pp. 264–275, November 1978.

[255] O. C. Yue, "Spread spectrum mobile radio, 1977-1982," *IEEE Transactions on Vehicular Technology*, vol. 32, pp. 98–105, February 1983.

[256] M. Simon, J. Omura, R. Scholtz, and B. Levitt, *Spread Spectrum Communications Handbook*. New York, USA: McGraw-Hill, 1994.

[257] Y. Yoon, R. Kohno, and H. Imai, "A SSMA system with cochannel interference cancellation with multipath fading channels," *IEEE Journal on Selected Areas in Communications*, vol. 11, pp. 1067–1075, September 1993.

[258] K. Gilhousen, I. Jacobs, R. Padovani, A. Viterbi, L. Weaver Jr., and C. Wheatley III, "On the capacity of a cellular CDMA system," *IEEE Transactions on Vehicular Technology*, vol. 40, pp. 303–312, May 1991.

[259] P. Patel and J. Holtzman, "Analysis of a simple successive interference cancellation scheme in a DS/CDMA system," *IEEE Journal on Selected Areas in Communications*, vol. 12, pp. 796–807, June 1994.

[260] R. Lupas and S. Verdú, "Linear multiuser detectors for synchronous code divison multiple access channels," *IEEE Transactions on Information Theory*, vol. 35, pp. 123–136, January 1989.

[261] P. Jung and J. Blanz, "Joint detection with coherent receiver antenna diversity in CDMA mobile radio systems," *IEEE Transactions on Vehicular Technology*, vol. 44, pp. 76–88, February 1995.

[262] J. Thompson, P. Grant, and B. Mulgrew, "Smart antenna arrays for CDMA systems," *IEEE Personal Communications Magazine*, vol. 3, pp. 16–25, October 1996.

[263] R. Steele and L. Hanzo, eds., *Mobile Radio Communications*. New York, USA: IEEE Press - John Wiley & Sons, 2nd ed., 1999.

[264] K. Feher, *Wireless digital communications: Modulation and spread spectrum.* Englewood Cliffs, NJ: Prentice-Hall, 1995.

[265] R. Kohno, R. Meidan, and L. Milstein, "Spread spectrum access methods for wireless communication," *IEEE Communications Magazine*, vol. 33, pp. 58–67, January 1995.

[266] R. Price and E. Green Jr., "A communication technique for multipath channels," *Proceedings of the IRE*, vol. 46, pp. 555–570, March 1958.

[267] U. Grob, A. L. Welti, E. Zollinger, R. Kung, and H. Kaufmann, "Microcellular direct-sequence spread-spectrum radio system using n-path rake receiver," *IEEE Journal on Selected Areas in Communications*, vol. 8, no. 5, pp. 772–780, 1990.

[268] P. W. Baier, "A critical review of CDMA," in *Proceedings of the IEEE Vehicular Technology Conference (VTC)*, (Atlanta, GA, USA), pp. 6–10, 28 April–1 May 1996.

[269] Telcomm. Industry Association (TIA), Washington, DC, USA, *Mobile station — Base station compatibility standard for dual-mode wideband spread spectrum cellular system, EIA/TIA Interim Standard IS-95*, July 1993.

[270] A. J. Viterbi, A. M. Viterbi, and E. Zehavi, "Performance of power-controlled wideband terrestrial digital communication," *IEEE Transactions on Communications*, vol. 41, no. 4, pp. 559–569, 1993.

[271] F. Simpson and J. Holtzman, "Direct sequence CDMA power control, interleaving, and coding," *IEEE Journal on Selected Areas in Communications*, vol. 11, pp. 1085–1095, September 1993.

[272] S. Ariyavisitakul and L. Chang, "Signal and interference statistics of a CDMA system with feedback power control," *IEEE Transactions on Communications*, vol. 41, pp. 1626–1634, November 1993.

[273] S. Ariyavisitakul, "Signal and interference statistics of a CDMA system with feedback power control - Part II," *IEEE Transactions on Communications*, vol. 42, no. 2/3/4, pp. 597–605, 1994.

[274] J. G. Proakis, *Digital Communications.* Mc-Graw Hill International Editions, 3rd ed., 1995.

[275] R. W. Chang and R. A. Gibby, "A theoretical study of performance of an orthogonal multiplexing data transmission scheme," *IEEE Transactions on Communication Technology*, vol. 16, no. 4, pp. 529–540, 1968.

[276] P. V. Eetvelt, S. J. Shepherd, and S. K. Barton, "The distribution of peak factor in QPSK multi-carrier modulation," *Wireless Personal Communications*, vol. 2, pp. 87–96, 1995.

[277] V. M. DaSilva and E. S. Sousa, "Performance of orthogonal CDMA codes for quasi-synchronous communication systems," in *Proceedings of IEEE ICUPC 1993*, (Ottawa, Canada), pp. 995–999, October 1993.

[278] L. Vandendorpe, "Multitone direct sequence CDMA system in an indoor wireless environment," in *Proceedings of IEEE SCVT 1993*, (Delft, The Netherlands), pp. 4.1:1–8, October 1993.

[279] R. Prasad and S. Hara, "Overview of multi-carrier CDMA," in *Proceedings of the IEEE International Symposium on Spread Spectrum Techniques and Applications (ISSSTA)*, (Mainz, Germany), pp. 107–114, 22–25 September 1996.

[280] D. Scott, P. Grant, S. McLaughlin, G. Povey, and D. Cruickshank, "Research in recon-figurable terminal design for mobile and personal communications," tech. rep., Department of Electrical Engineering, The University of Edinburgh, March 1997.

[281] N. Yee and J. P. Linnartz, "MICRO 93-101: Multi-carrier CDMA in an indoor wireless radio channel," tech. rep., University of California at Berkeley, 1994.

[282] L.-L. Yang and L. Hanzo, "Performance of generalized multicarrier DS-CDMA over Nakagami-m fading channels," *IEEE Transactions on Communications, (http://www-mobile.ecs.soton.ac.uk/lly)*, vol. 50, pp. 956–966, June 2002.

[283] L.-L. Yang and L. Hanzo, "Slow frequency-hopping multicarrier DS-CDMAfor transmission over Nakagami multipath fading channels," *IEEE Journal on Selected Areas in Communications*, vol. 19, no. 7, pp. 1211–1221, 2001.

[284] E. A. Sourour and M. Nakagawa, "Performance of orthogonal multicarrier CDMA in a multipath fading channel," *IEEE Transactions on Communications*, vol. 44, pp. 356–367, March 1996.

[285] L.-L. Yang and L. Hanzo, "Slow frequency-hopping multicarrier DS-CDMA," in *International Symposium on Wireless Personal Multimedia Communications (WPMC'99)*, (Amsterdam, The Netherlands), pp. 224–229, September:21–23 1999.

[286] L.-L. Yang and L. Hanzo, "Blind soft-detection assisted frequency-hopping multicarrier DS-CDMA systems," in *Proceedings of IEEE GLOBECOM'99*, (Rio de Janeiro, Brazil), pp. 842–846, December:5-9 1999.

[287] S. Slimane, "MC-CDMA with quadrature spreading for wireless communication systems," *European Transactions on Telecommunications*, vol. 9, pp. 371–378, July–August 1998.

[288] I. Kalet, "The multitone channel," *IEEE Transactions on Communications*, vol. 37, pp. 119–124, February 1989.

[289] R.-Y. Li and G. Stette, "Time-limited orthogonal multicarrier modulation schemes," *IEEE Transactions on Communications*, vol. 43, pp. 1269–1272, February/March/April 1995.

[290] L. Goldfeld and D. Wulich, "Multicarrier modulation system with erasures-correcting decoding for nakagami fading channels," *European Trans. on Telecommunications*, vol. 8, pp. 591–595, November–December 1997.

[291] E. Sousa, "Performance of a direct sequence spread spectrum multiple access system utilizing unequal carrier frequencies," *IEICE Transactions on Communications*, vol. E76-B, pp. 906–912, August 1993.

[292] B. Saltzberg, "Performance of an efficient parallel data transmission system," *IEEE Transactions on Communication Technology*, vol. 15, pp. 805–811, December 1967.

[293] C. Baum and K. Conner, "A multicarrier transmission scheme for wireless local communications," *IEEE Journal on Selected Areas in Communications*, vol. 14, pp. 512–529, April 1996.

[294] V. Dasilva and E. Sousa, "Multicarrier orthogonal CDMA signals for quasi-synchronous communication systems," *IEEE Journal on Selected Areas in Communications*, vol. 12, pp. 842–852, June 1994.

[295] L. Vandendorpe and O. V. de Wiel, "MIMO DEF equalization for multitone DS/SS systems over multipath channels," *IEEE Journal on Selected Areas in Communications*, vol. 14, pp. 502–511, April 1996.

[296] N. Al-Dhahir and J. Cioffi, "A bandwidth-optimized reduced-complexity equalized multicarrier transceiver," *IEEE Transactions on Communications*, vol. 45, pp. 948–956, August 1997.

[297] P. Jung, F. Berens, and J. Plechinger, "A generalized view on multicarrier CDMA mobile radio systems with joint detection (Part i)," *FREQUENZ*, vol. 51, pp. 174–184, July–August 1997.

[298] S. Hara and R. Prasad, "Design and performance of multicarrier CDMA system in frequency-selective Rayleigh fading channels," *IEEE Transactions on Vehicular Technology*, vol. 48, pp. 1584–1595, September 1999.

[299] V. Tarokh and H. Jafarkhani, "On the computation and reduction of the peak-to-average power ratio in multicarrier communications," *IEEE Transactions on Communications*, vol. 48, pp. 37–44, January 2000.

[300] D. Wulich and L. Goldfied, "Reduction of peak factor in orthogonal multicarrier modulation by amplitude limiting and coding," *IEEE Transactions on Communications*, vol. 47, pp. 18–21, January 1999.

[301] H.-W. Kang, Y.-S. Cho, and D.-H. Youn, "On compensating nonlinear distortions of an OFDM system using an efficient adaptive predistorter," *IEEE Transactions on Communications*, vol. 47, pp. 522–526, April 1999.

[302] Y.-H. Kim, I. Song, Seokho, and S.-R. Park, "A multicarrier CDMA system with adaptive subchannel allocation for forward links," *IEEE Transactions on Vehicular Technology*, vol. 48, pp. 1428–1436, September 1999.

[303] X. Gui and T.-S. Ng, "Performance of asynchronous orthogonal multicarrier CDMA system in frequency selective fading channel," *IEEE Transactions on Communications*, vol. 47, pp. 1084–1091, July 1999.

[304] T.-M. Lok, T.-F. Wong, and J. Lehnert, "Blind adaptive signal reception for MC-CDMA systems in Rayleigh fading channels," *IEEE Transactions on Communications*, vol. 47, pp. 464–471, March 1999.

[305] B. Rainbolt and S. Miller, "Multicarrier CDMA for cellular overlay systems," *IEEE Journal on Selected Areas in Communications*, vol. 17, pp. 1807–1814, October 1999.

[306] S.-M. Tseng and M. Bell, "Asynchronous multicarrier DS-CDMA using mutually orthogonal complementary sets of sequences," *IEEE Transactions on Communications*, vol. 48, pp. 53–59, January 2000.

[307] D. Rowitch and L. Milstein, "Convolutionally coded multicarrier DS-CDMA systems in a multipath fading channel – Part I: Performance analysis," *IEEE Transactions on Communications*, vol. 47, pp. 1570–1582, October 1999.

[308] D. Rowitch and L. Milstein, "Convolutionally coded multicarrier DS-CDMA systems in a multipath fading channel – Part II: Narrow-band interference suppression," *IEEE Transactions on Communications*, vol. 47, pp. 1729–1736, November 1999.

[309] D.-W. Lee and L. Milstein, "Comparison of multicarrier DS-CDMA broadcast systems in a multipath fading channel," *IEEE Transactions on Communications*, vol. 47, pp. 1897–1904, December 1999.

[310] N. Yee, J.-P. Linnartz, and G. Fettweis, "Multi-carrier CDMA in indoor wireless radio network," *IEICE Transactions on Communications*, vol. E77-B, pp. 900–904, July 1994.

[311] S. Kondo and L. Milstein, "On the use of multicarrier direct sequence spread spectrum systems," in *Proceedings of IEEE MILCOM'93*, (Boston, MA), pp. 52–56, Oct. 1993.

[312] V. M. DaSilva and E. S. Sousa, "Performance of orthogonal CDMA codes for quasi-synchronous communication systems," in *Proceedings of IEEE ICUPC'93*, (Ottawa, Canada), pp. 995–999, Oct. 1993.

[313] L. Vandendorpe, "Multitone direct sequence CDMA system in an indoor wireless environment," in *Proceedings of IEEE First Symposium of Communications and Vehicular Technology in the Benelux, Delft, The Netherlands*, pp. 4.1-1–4.1-8, Oct. 1993.

[314] L. L. Yang and L. Hanzo, "Blind joint soft-detection assisted slow frequency-hopping multi-carrier DS-CDMA," *IEEE Transactions on Communication*, vol. 48, no. 9, pp. 1520–1529, 2000.

[315] D.-W. Lee and L. Milstein, "Analysis of a multicarrier DS-CDMA code-acquisition system," *IEEE Transactions on Communications*, vol. 47, pp. 1233–1244, August 1999.

[316] B. Steiner, "Time domain uplink channel estimation in multicarrier-CDMA mobile radio system concepts," in *Multi-Carrier Spread-Spectrum* (K. Fazel and G. Fettweis, eds.), pp. 153–160, Kluwer Academic Publishers, 1997.

[317] K. W. Yip and T. S. Ng, "Tight error bounds for asynchronous multicarrier cdma and their application," *IEEE Communications Letters*, vol. 2, pp. 295–297, November 1998.

[318] S. Kondo and L. Milstein, "Performance of multicarrier DS CDMA systems," *IEEE Transactions on Communications*, vol. 44, pp. 238–246, February 1996.

[319] B. M. Popovic, "Spreading sequences for multicarrier CDMA systems," *IEEE Transactions on Comminications*, vol. 47, no. 6, pp. 918–926, 1999.

[320] P. Jung, P. Berens, and J. Plechinger, "Uplink spectral efficiency of multicarrier joint detection code division multiple access based cellular radio systems," *Electronics Letters*, vol. 33, no. 8, pp. 664–665, 1997.

[321] D.-W. Lee, H. Lee, and J.-S. Kim, "Performance of a modified multicarrier direct sequence CDMA system," *Electronics and Telecommunications Research Institute Journal*, vol. 19, pp. 1–11, April 1997.

[322] L. Rasmussen and T. Lim, "Detection techniques for direct sequence and multicarrier variable rate for broadband CDMA," in *Proceedings of the ICCS/ISPACS '96*, pp. 1526–1530, 1996.

[323] P. Jung, F. Berens, and J. Plechinger, "Joint detection for multicarrier CDMA mobile radio systems - Part II: Detection techniques," in *Proceedings of IEEE ISSSTA 1996*, vol. 3, (Mainz, Germany), pp. 996–1000, September 1996.

[324] Y. Sanada and M. Nakagawa, "A multiuser interference cancellation technique utilizing convolutional codes and multicarrier modulation for wireless indoor communications," *IEEE Journal on Selected Areas in Communications*, vol. 14, pp. 1500–1509, October 1996.

[325] Q. Chen, E. S. Sousa, and S. Pasupathy, "Multicarrier CDMA with adaptive frequency hopping for mobile radio systems," *IEEE Journal on Selected Areas in Communications*, vol. 14, pp. 1852–1857, December 1996.

[326] T. Ojanperä and R. Prasad, "An overview of air interface multiple access for IMT-2000/UMTS," *IEEE Communications Magazine*, vol. 36, pp. 82–95, September 1998.

[327] R. L. Peterson, R. E. Ziemer, and D. E. Borth, *Introduction to Spread Spectrum Communications*. Prentice Hall International Editions, 1995.

[328] S. W. Golomb, *Shift Register Sequences*. Holden-Day, 1967.

[329] F. Adachi, K. Ohno, A. Higashi, T. Dohi, and Y. Okumura, "Coherent multicode DS-CDMA mobile Radio Access," *IEICE Transactions on Communications*, vol. E79-B, pp. 1316–1324, September 1996.

[330] S. C. Bang, "ETRI wideband CDMA system for IMT-2000," in *Presentation material in 1st IMT-2000 Workshop*, (Seoul, Korea), pp. II–1–1 – II–1–21, KIET, August 1997.

[331] H. Rohling, K. Brüninghaus, and R. Grünheid, "Comparison of multiple access schemes for an OFDM downlink system," in *Multi-Carrier Spread-Spectrum* (K. Fazel and G. P. Fettweis, eds.), pp. 23–30, Kluwer Academic Publishers, 1997.

[332] R. Prasad, *CDMA for Wireless Personal Communications*. London: Artech House, May 1996. ISBN 0890065713.

[333] M. Schnell and S. Kaiser, "Diversity considerations forMC-CDMA systems in mobile communications," in *Proceedings of IEEE ISSSTA 1996*, pp. 131–135, 1996.

[334] W. H. Press, S. A. Teukolsky, W. T. Vetterling, and B. P. Flannery, *Numerical Recipies in C*. Cambridge University Press, 1992.

[335] S. Verdú, *Multiuser Detection*. Cambridge, UK: Cambridge University Press, 1998.

[336] E. Kuan and L. Hanzo, "Burst-by-burst adaptive multiuser detection cdma: A framework for existing and future wireless standards," *Proceedings of the IEEE*, December 2002.

[337] R. Prasad, *Universal Wireless Personal Communications*. London, UK: Artech House Publishers, 1998.

[338] S. Glisic and B. Vucetic, *Spread Spectrum CDMA Systems for Wireless Communications*. London, UK: Artech House, April 1997. ISBN 0890068585.

[339] G. Woodward and B. Vucetic, "Adaptive detection for DS-CDMA," *Proceedings of the IEEE*, vol. 86, pp. 1413–1434, July 1998.

[340] S. Moshavi, "Multi-user detection for DS-CDMA communications," *IEEE Communications Magazine*, vol. 34, pp. 124–136, October 1996.

[341] A. Duel-Hallen, J. Holtzman, and Z. Zvonar, "Multiuser detection for cdma systems," *IEEE Personal Communications*, vol. 2, pp. 46–58, April 1995.

[342] J. Laster and J. Reed, "Interference rejection in digital wireless communications," *IEEE Signal Processing Magazine*, vol. 14, pp. 37–62, May 1997.

[343] J. Thompson, P. Grant, and B. Mulgrew, "Performance of antenna array receiver algorithms for CDMA," in *Proceedings of the IEEE Global Telecommunications Conference (GLOBECOM)*, (London, UK), pp. 570–574, 18–22 November 1996.

[344] A. Naguib and A. Paulraj, "Performance of wireless CDMA with m-ary orthogonal modulation and cell site antenna arrays," *IEEE Journal on Selected Areas in Communications*, vol. 14, pp. 1770–1783, December 1996.

[345] L. Godara, "Applications of antenna arrays to mobile communications, part I: Performance improvement, feasibility, and system considerations," *Proceedings of the IEEE*, vol. 85, pp. 1029–1060, July 1997.

[346] R. Kohno, H. Imai, M. Hatori, and S. Pasupathy, "Combination of adaptive array antenna and a canceller of interference for direct-sequence spread-specturm multiple-access system," *IEEE Journal on Selected Areas in Communications*, vol. 8, pp. 675–681, May 1998.

[347] M. R. Schroeder, "Synthesis of low-peak-factor signals and binary sequences with low autocorrelation," *IEEE Transactions on Information Theory*, pp. 85–89, 1970.

[348] E. V. der Ouderaa, J. Schoukens, and J. Renneboog, "Peak factor minimization using a time-frequency domain swapping algorithm," *IEEE Transactions on Instrumentation and Measurement*, vol. 37, pp. 145–147, March 1988.

[349] L. J. Greenstein and P. J. Fitzgerald, "Phasing multitone signals to minimize peak factors," *IEEE Transactions on Communications*, vol. 29, no. 7, pp. 1072–1074, 1981.

[350] R. W. Bäuml, R. F. H. Fischer, and J. B. Huber, "Reducing the peak-to-average power ratio of multicarrier modulation by selected mapping," *Electronics Letters*, vol. 32, pp. 2056–2057, October 1996.

[351] X. Li and J. A. Ritcey, "M-sequences for OFDM peak-to-average power ratio reduction and error correction," *Electronics Letters*, vol. 33, pp. 554–555, March 1997.

[352] J. Jedwab, "Comment : M-sequences for OFDM peak-to-average power ratio reduction and error correction," *Electronics Letters*, vol. 33, pp. 1293–1294, July 1997.

[353] C. Tellambura, "Use of m-sequences for OFDM peak-to-average power ratio reduction," *Electronics Letters*, vol. 33, pp. 1300–1301, July 1997.

[354] S. Narahashi and T. Nojima, "New phasing scheme of N-multiple carriers for reducing peak-to-average power ratio," *Electronics Letters*, vol. 30, pp. 1382–1383, August 1994.

[355] M. Friese, "Multicarrier modulation with low peak-to-average power ratio," *Electronics Letters*, vol. 32, pp. 713–714, April 1996.

[356] P. V. Eetvelt, G. Wade, and M. Tomlinson, "Peak to average power reduction for OFDM schemes by selective scrambling," *Electronics Letters*, vol. 32, pp. 1963–1964, October 1996.

[357] J. A. Davis and J. Jedwab, "Peak-to-mean power control and error correction for OFDM transmission using Golay sequences and Reed-Muller codes," *Electronics Letters*, vol. 33, no. 4, pp. 267–268, 1997.

[358] S. Boyd, "Multitone signals with low crest factor," *IEEE Transactions on Circuits and Systems*, vol. 33, pp. 1018–1022, October 1986.

[359] W. Rudin, "Some theorems on Fourier coefficients," *Proceedings of American Mathematics Society*, vol. 10, pp. 855–859, December 1959.

[360] D. J. Newman, "An L^1 extrimal problem for polynomials," *Proceedings of American Mathematics Society*, vol. 16, pp. 1287–1290, December 1965.

[361] B. M. Popovic, "Synthesis of power efficient multitone signals with flat amplitude spectrum," *IEEE Transactions on Communications*, vol. 39, no. 7, pp. 1031–1033, 1991.

[362] H. S. Shapiro, "Extremal problems for polynomials and power series," Master's thesis, Massachusetts Institute of Technology, 1951.

[363] M. Golay, "Complementary series,," *IRE Transactions on Information Theory*, vol. IT-7, pp. 82–87, 1961.

[364] C. Tellambura, "Upper bound on peak factor of n-multiple carriers," *Electronics Letters*, vol. 33, pp. 1608–1609, September 1997.

[365] C.-C. Tseng and C. L. Liu, "Complementary sets of sequences," *IEEE Transactions on Information Theory*, vol. 18, pp. 644–652, September 1972.

[366] R. Sivaswamy, "Multiphase complementarty codes," *IEEE Transactions on Information Theory*, vol. 24, pp. 546–552, September 1978.

[367] R. L. Frank, "Polyphase complementary codes," *IEEE Transactions on Information Theory*, vol. 26, pp. 641–647, November 1980.

[368] D. Wulich and L. Goldfeld, "Reduction of peak factor in orthogonal multcarrier modulation by amplitude limiting and coding," *IEEE Transactions on Communications*, vol. 47, no. 1, pp. 18–21, 1999.

[369] R. H. Barker, "Group synchronizing of binary digital systems," *in Communication Theory*, pp. 273–287, 1953.

[370] R. H. Pettit, "Pulse sequences with good autocorrelation properties," *The Microwave Journal*, pp. 63–67, February 1967.

[371] R. Turyn and J. Storer, "On binary sequences," *Proceedings of American Mathematics Society*, vol. 12, pp. 394–399, June 1961.

[372] S. W. Golomb and R. A. Scholtz, "Generalized Barker sequences," *IEEE Transactions on Information Theory*, vol. 4, pp. 533–537, October 1965.

[373] D. J. G. Mestdagh and P. M. P. Spruyt, "A method to reduce the probability of clipping in DMT-based transceivers," *IEEE Transactions on Communications*, vol. 44, no. 10, pp. 1234–1238, 1996.

[374] D. J. G. Mestdagh, P. M. P. Spruyt, and B. Biran, "Effect of amplitude clipping in DMT-ADSL transceivers," *Electronics Letters*, vol. 29, pp. 1354–1355, July 1993.

[375] S. J. Shepherd, J. Oriss, and S. K. Barton, "Asymptotic limits in peak envelope power reduction by redundant coding in orthogonal frequency-division multiplex modulation," *IEEE Transactions on Communications*, vol. 46, no. 1, pp. 5–10, 1998.

[376] M. Breiling, S. H. Müller-Weinfurtner, and J. B. Huber, "Peak-power reduction in OFDM without explicit side information," in *Proceedings of International OFDM Workshop 2000*, (Hamburg, Germany), September 2000.

[377] S. H. Müller and J. B. Huber, "OFDM with reduced peak-to-mean power ratio by optimum combination of partial transmit sequences," *Electronics Letters*, vol. 33, pp. 368–369, February 1997.

[378] C. Tellambura, "Phase optimisation criterion for reducing peak-to-average power ratio in OFDM," *Electronics Letters*, vol. 34, pp. 169–170, January 1998.

[379] J. Kreyszig, *Advanced engineering mathematics*. Wiley, 7th edition ed., 1993.

[380] R. C. Heimiller, "Phase shift pulse codes with good periodic correlation properties," *IRE Transactions on Information Theory*, vol. 7, pp. 254–257, October 1961.

[381] R. L. Frank, "Phase shift pulse codes with good periodic correlation properties," *IRE Transactions on Information Theory*, vol. 8, pp. 381–382, October 1962.

[382] R. L. Frank, "Polyphase codes with good nonperiodic correlation properties," *IEEE Transactions on Information Theory*, vol. 9, pp. 43–45, 1963.

[383] D. C. Chu, "Polyphase codes with good periodic correlation properties," *IEEE Transactions on Information Theory*, vol. 18, pp. 531–532, July 1972.

[384] R. L. Frank, "Comments on polyphase codes with good correlation properties," *IEEE Transactions on Information Theory*, vol. 19, p. 244, March 1973.

[385] B. M. Popović, "Comment: Merit factor of Chu and Frank sequences," *Electronics Letters*, vol. 27, pp. 776–777, April 1991.

[386] M. Antweiler and L. Bömer, "Reply: Merit factor of Chu and Frank sequences," *Electronics Letters*, vol. 27, pp. 777–778, April 1991.

[387] M. Antweiler and L. Bömer, "Merit factor of Chu and Frank sequences," *Electronics Letters*, vol. 26, pp. 2068–2070, December 1990.

[388] R. Gold, "Optimal binary sequences for spread spectrum multiplexing," *IEEE Transactions on Information Theory*, vol. 13, pp. 619–621, October 1967.

[389] R. Sivaswamy, "Digital and analog subcomplementary sequences for pulse compression," *IEEE Transactions on Aerospace and Electronic Systems*, vol. 14, pp. 343–350, March 1978.

[390] E. Costa, M. Midrio, and S. Pupolin, "Impact of amplifier nonlinearities on OFDM transmission system performance," *IEEE Communications Letters*, vol. 3, pp. 37–39, February 1999.

[391] G. Santella and F. Mazzanga, "A model for performance evaluation in M-QAM-OFDM schemes in presence of non-linear distortions," in *Proceedings of IEEE VTC 1995*, pp. 830–834, July 1995.

[392] C. Rapp, "Effects of HPA-nonlinearity on 4-DPSK-OFDM-signal for digital sound broadcasting system," in *Proceedings of 2nd European Conference on Satellite Communications*, (Liege, Belgium), ESA-SP-332, October 1991.

[393] P802.11a/D6.0, *Draft supplement to standard for tellecommunications and information exchange between systems - LAN/MAN specific requirements - Part 11: Wireless MAC and PHY specifications: High Speed Physical Layer in the 5 GHz Band*, May 1999.

[394] D. V. Sarwate, "Bounds on crosscorrelation and autocorrelation of sequences," *IEEE Transactions on Information Theory*, vol. 25, pp. 720–724, November 1979.

[395] J. F. Hayes, "Adaptive feedback communications," *IEEE Transactions on Communication Technology*, vol. 16, no. 1, pp. 29–34, 1968.

[396] J. K. Cavers, "Variable rate transmission for rayleigh fading channels," *IEEE Transactions on Communications Technology*, vol. COM-20, pp. 15–22, February 1972.

[397] W. T. Webb and R. Steele, "Variable rate QAM for mobile radio," *IEEE Transactions on Communications*, vol. 43, no. 7, pp. 2223–2230, 1995.

[398] M. Moher and J. Lodge, "TCMP—a modulation and coding strategy for rician fading channels," *IEEE Journal on Selected Areas in Communications*, vol. 7, pp. 1347–1355, December 1989.

[399] S. Sampei and T. Sunaga, "Rayleigh fading compensation for QAM in land mobile radio communications," *IEEE Transactions on Vehicular Technology*, vol. 42, pp. 137–147, May 1993.

[400] S. Otsuki, S. Sampei, and N. Morinaga, "Square QAM adaptive modulation/TDMA/TDD systems using modulation level estimation with Walsh function," *Electronics Letters*, vol. 31, pp. 169–171, February 1995.

[401] W. Lee, "Estimate of channel capacity in Rayleigh fading environment," *IEEE Transactions on Vehicular Technology*, vol. 39, pp. 187–189, August 1990.

[402] A. Goldsmith and P. Varaiya, "Capacity of fading channels with channel side information," *IEEE Transactions on Information Theory*, vol. 43, pp. 1986–1992, November 1997.

[403] M. S. Alouini and A. J. Goldsmith, "Capacity of Rayleigh fading channels under different adaptive transmission and diversity-combining technique," *IEEE Transactions on Vehicular Technology*, vol. 48, pp. 1165–1181, July 1999.

[404] A. Goldsmith and S. Chua, "Variable rate variable power MQAM for fading channels," *IEEE Transactions on Communications*, vol. 45, pp. 1218–1230, October 1997.

[405] M. S. Alouini, X. Tand, and A. J. Goldsmith, "An adaptive modulation scheme for simultaneous voice and data transmission over fading channels," *IEEE Journal on Selected Areas in Communications*, vol. 17, pp. 837–850, May 1999.

[406] D. Yoon, K. Cho, and J. Lee, "Bit error probability of M-ary Quadrature Amplitude Modulation," in *Proc. IEEE VTC 2000-Fall*, vol. 5, pp. 2422–2427, IEEE, September 2000.

[407] C. Wong and L. Hanzo, "Upper-bound performance of a wideband burst-by-burst adaptive modem," *IEEE Transactions on Communications*, vol. 48, pp. 367–369, March 2000.

[408] E. L. Kuan, C. H. Wong, and L. Hanzo, "Burst-by-burst adaptive joint-detection CDMA," in *Proc. of IEEE VTC'99 Fall*, vol. 2, (Amsterdam, Netherland), pp. 1628–1632, September 1999.

[409] K. J. Hole, H. Holm, and G. E. Oien, "Adaptive multidimensional coded modulation over flat fading channels," *IEEE Journal on Selected Areas in Communications*, vol. 18, pp. 1153–1158, July 2000.

[410] J. Torrance and L. Hanzo, "Upper bound performance of adaptive modulation in a slow Rayleigh fading channel," *Electronics Letters*, vol. 32, pp. 718–719, 11 April 1996.

[411] M. Nakagami, "The m-distribution - A general formula of intensity distribution of rapid fading," in *Statistical Methods in Radio Wave Propagation* (W. C. Hoffman, ed.), pp. 3–36, Pergamon Press, 1960.

[412] I. S. Gradshteyn and I. M. Ryzhik, *Table of Integrals, Series and Products*. New York, USA: Academic Press, 1980.

[413] J. Lu, K. B. Letaief, C. I. J. Chuang, and M. L. Lio, "M-PSK and M-QAM BER computation using signal-space concepts," *IEEE Transactions on Communications*, vol. 47, no. 2, pp. 181–184, 1999.

[414] T. Keller and L. Hanzo, "Adaptive modulation technique for duplex OFDM transmission," *IEEE Transactions on Vehicular Technology*, vol. 49, pp. 1893–1906, September 2000.

[415] B. J. Choi, M. Münster, L. L. Yang, and L. Hanzo, "Performance of Rake receiver assisted adaptive-modulation based CDMA over frequency selective slow Rayleigh fading channel," *Electronics Letters*, vol. 37, pp. 247–249, February 2001.

[416] G. S. G. Beveridge and R. S. Schechter, *Optimization: Theory and Practice*. McGraw-Hill, 1970.

[417] W. Jakes Jr., ed., *Microwave Mobile Communications*. New York, USA: John Wiley & Sons, 1974.

[418] "COST 207 : Digital land mobile radio communications, final report," tech. rep., Luxembourg, 1989.

[419] M. K. Simon and M. S. Alouini, *Digital Communication over Fading Channels: A Unified Approach to Performance Analysis*. John Wiley & Sons, Inc., 2000. ISBN 0471317799.

[420] C. Y. Wong, R. S. Cheng, K. B. Letaief, and R. D. Murch, "Multiuser OFDM with adaptive subcarrier, bit, and power allocation," *IEEE Journal on Selected Areas in Communications*, vol. 17, pp. 1747–1758, October 1999.

[421] T. Keller and L. Hanzo, "Adaptive multicarrier modulation: A convenient framework for time-frequency processing in wireless communications," *Proceedings of the IEEE*, vol. 88, pp. 611–642, May 2000.

[422] A. Klein, G. Kaleh, and P. Baier, "Zero forcing and minimum mean square error equalization for multiuser detection in code division multiple access channels," *IEEE Transactions on Vehicular Technology*, vol. 45, pp. 276–287, May 1996.

[423] B. J. Choi, T. H. Liew, and L. Hanzo, "Concatenated space-time block coded and turbo coded symbol-by-symbol adaptive OFDM and multi-carrier CDMA systems," in *Proceedings of IEEE VTC 2001-Spring*, p. P.528, IEEE, May 2001.

[424] B. Vucetic, "An adaptive coding scheme for time-varying channels," *IEEE Transactions on Communications*, vol. 39, no. 5, pp. 653–663, 1991.

[425] H. Imai and S. Hirakawa, "A new multi-level coding method using error correcting codes," *IEEE Transactions on Information Theory*, vol. 23, pp. 371–377, May 1977.

[426] G. Ungerboeck, "Channel coding with multilevel/phase signals," *IEEE Transactions on Information Theory*, vol. IT-28, pp. 55–67, January 1982.

[427] S. M. Alamouti and S. Kallel, "Adaptive trellis-coded multiple-phased-shift keying Rayleigh fading channels," *IEEE Transactions on Communications*, vol. 42, pp. 2305–2341, June 1994.

[428] S. Chua and A. Goldsmith, "Adaptive coded modulation for fading channels," *IEEE Transactions on Communications*, vol. 46, pp. 595–602, May 1998.

[429] T. Keller, T. Liew, and L. Hanzo, "Adaptive rate RRNS coded OFDM transmission for mobile communication channels," in *Proceedings of VTC 2000 Spring*, (Tokyo, Japan), pp. 230–234, 15-18 May 2000.

[430] T. Keller, T. H. Liew, and L. Hanzo, "Adaptive redundant residue number system coded multicarrier modulation," *IEEE Journal on Selected Areas in Communications*, vol. 18, pp. 1292–2301, November 2000.

[431] T. Liew, C. Wong, and L. Hanzo, "Block turbo coded burst-by-burst adaptive modems," in *Proceedings of Microcoll'99*, (Budapest, Hungary), pp. 59–62, 21-24 March 1999.

[432] C. Wong, T. Liew, and L. Hanzo, "Turbo coded burst by burst adaptive wideband modulation with blind modem mode detection," in *ACTS Mobile Communications Summit*, (Sorrento, Italy), pp. 303–308, 8-11 June 1999.

[433] V. Tarokh, N. Seshadri, and A. R. Calderbank, "Space-Time Codes for High Data Rate Wireless Communication: Performance Criterion and Code Construction," *IEEE Transactions on Information Theory*, vol. 44, pp. 744–765, March 1998.

[434] S. M. Alamouti, "A simple transmit diversity technique for wireless communications," *IEEE Journal on Selected Areas in Communications*, vol. 16, pp. 1451–1458, October 1998.

[435] V. Tarokh, H. Jafarkhani, and A. R. Calderbank, "Space-time block coding for wireless communications: Performance results," *IEEE Journal on Selected Areas in Communications*, vol. 17, pp. 451–460, March 1999.

[436] C. Berrou and A. Glavieux, "Near optimum error correcting coding and decoding: Turbo codes," *IEEE Transactions on Communications*, vol. 44, pp. 1261–1271, October 1996.

[437] W. T. Webb and L. Hanzo, *Modern Quadrature Amplitude Modulation : Principles and Applications for Fixed and Wireless Channels*. London: IEEE Press, and John Wiley & Sons, 1994.

[438] Y. Chow, A. Nix, and J. McGeehan, "Analysis of 16-APSK modulation in AWGN and rayleigh fading channel," *Electronics Letters*, vol. 28, pp. 1608–1610, November 1992.

[439] F. Adachi and M. Sawahashi, "Performance analysis of various 16 level modulation schemes under Rrayleigh fading," *Electronics Letters*, vol. 28, pp. 1579–1581, November 1992.

[440] F. Adachi, "Error rate analysis of differentially encoded and detected 16APSK under rician fading," *IEEE Transactions on Vehicular Technology*, vol. 45, pp. 1–12, February 1996.

[441] Y. C. Chow, A. R. Nix, and J. P. McGeehan, "Diversity improvement for 16-DAPSK in Rayleigh fading channel," *Electronics Letters*, vol. 29, pp. 387–389, February 1993.

[442] Y. C. Chow, A. R. Nix, and J. P. McGeehan, "Error analysis for circular 16-DAPSK in frquency-selective Rayleigh fading channels with diversity reception," *Electronics Letters*, vol. 30, pp. 2006–2007, November 1994.

[443] C. M. Lo and W. H. Lam, "Performance analysis of bandwidth efficient coherent modulation schems with L-fold MRC and SC in Nakagami-m fading channels," in *Proceedings of IEEE PIMRC 2000*, vol. 1, pp. 572–576, September 2000.

[444] S. Benedetto, E. Biglierri, and V. Castellani, *Digital Transmission Theory*. Prentice-Hall, 1987.

[445] Z. Dlugaszewski and K. Wesolowski, "WHT/OFDM - an imporved OFDM transmission method for selective fading channels," in *Proceedings of IEEE SCVT 2000*, (Leuven, Belgium), pp. 71–74, 19 October 2000.

[446] Z. Dlugaszewski and K. Wesolowski, "Performance of several OFDM transceivers in the indoor radio channels in 17 ghz band," in *Proceedings of IEEE VTC 2001 Spring*, p. P.644, IEEE, May 2001.

[447] S. Verdú, "Minimum probability of error for asynchronous Gaussian multiple-access channel," *IEEE Transactions on Communications*, vol. 32, pp. 85–96, January 1986.

[448] K. M. Wong, J. Wu, T. N. Davidson, Q. Jin, and P. C. Ching, "Performance of wavelet packet-division multiplexing in impulsive and Gaussian noise," *IEEE Transactions on Communication*, vol. 48, no. 7, pp. 1083–1086, 2000.

[449] M. Ghosh, "Analysis of the effect of impulse noise on multicarrier and single carrier QAM systems," *IEEE Transactions on Communications*, vol. 44, no. 2, pp. 145–147, 1996.

[450] Q. Wang, T. A. Gulliver, L. J. Mason, and V. K. Bhargava, "Performance of SFH/MDPSK in tone interference and Gaussian noise," *IEEE Transactions on Communications*, vol. 42, no. 2/3/4, pp. 1450–1454, 1994.

[451] G. E. Atkin and I. F. Blake, "Performance of multitone FFH/MFSK systems in the presence of jamming," *IEEE Transactions on Communications*, vol. 35, no. 2, pp. 428–435, 1989.

[452] L. B. Milstein and P. K. Das, "An analysis of a real-time transform domain filtering digital communication system - Part i: Narrow-band interference rejection," *IEEE Transactions on Communications*, vol. 28, no. 6, pp. 816–824, 1980.

[453] A. Papoulis, *Probability, Random Variables, and Stochastic Processes*. McGraw-Hill, 3 ed., 1991.

[454] M. R. Spiegel, *Mathematical Handbook of Formulas and Tables*. McGraw-Hill Book Company, 1968.

[455] W. X. Zheng, "Design of adaptive envelope-constrained filters in the presence of impulsive noise," *IEEE Transactions on Circuits and Systems II*, vol. 48, pp. 188–192, February 2001.

[456] Y. Zhang and R. S. Blum, "Iterative multiuser detection for turbo-coded synchronous cdma in gaussian and non-gaussian impulsive noise," *IEEE Transactions on Communication Technology*, vol. 49, no. 3, pp. 397–400, 2001.

[457] M. Moeneclaey, M. V. Bladel, and H. Sari, "Sensitivity of multiple-access techniques to narrow-band interference," *IEEE Transactions on Communication Technology*, vol. 49, no. 3, pp. 497–505, 2001.

[458] K. Cheun and T. Jung, "Performance of asynchronous fhss-ma networks under rayleigh fading and tone jamming," *IEEE Transactions on Communications*, vol. 49, no. 3, pp. 405–408, 2001.

[459] Y. Li, "Pilot-symbol-aided channel estimation for OFDM in wireless systems," in *Proc. of Vehicular Technology Conference*, (Houston, Texas), p. 1999, IEEE, May 1999.

[460] R. Mersereau and T. Speake, "The Processing of Periodically Sampled Multidimensional Signals," *IEEE Transactions on Acoustics, Speech and Signal Processing*, vol. ASSP-31, pp. 188–194, February 1983.

[461] D. Dudgeon and R. Mersereau, *Multidimensional Digital Signal Processing*. Prentice Hall, Inc., 1984.

[462] P. Höher, S. Kaiser, and P. Robertson, "Pilot-symbol-aided Channel Estimation in Time and Frequency," in *Proceedings of International Workshop on Multi-Carrier Spread-Spectrum*, pp. 169–178, K. Fazel and G.P. Fettweis (Eds.), Kluver Academic Publishers, April 1997.

[463] H. Lüke, *Signalübertragung*. Springer-Verlag Berlin Heidelberg New York, 1995.

[464] R.Steele and L. Hanzo, *Mobile Radio Communications*. John Wiley & Sons, LTD, Chichester, 1999.

[465] P. Bello, "Time-frequency duality," *IEEE Transactions on Information Theory*, vol. 10, pp. 18–33, January 1964.

[466] T. Keller, *Adaptive OFDM techniques for personal communications systems and local area networks*. PhD Thesis, Department of Electronics and Computer Science, University of Southampton, United Kingdom, 1999.

[467] T. Chen and P. Vaidyanathan, "The Role of Integer Matrices in Multidimensional Multirate Systems," *IEEE Transactions on Signal Processing*, vol. 41, pp. 1035–1047, March 1993.

[468] C. M. Duffee, *The Theory of Matrices*. Verlag Julius Springer, 1933.

[469] M. Newman, *Integral Matrices*. Academic Press, Inc., 1972.

[470] S. Kaiser, *Multi-Carrier CDMA Mobile Radio Systems - Analysis and Optimization of Detection, Decoding and Channel Estimation*. VDI Verlag GmbH, 1998.

[471] S. Weiss and R. Stewart, *On Adaptive Filtering in Oversampled Subbands*. Shaker Verlag GmbH, Aachen, Germany, 1998.

[472] S. Haykin, *Adaptive Filter Theory*. Prentice Hall, Inc., 1996.

[473] G. Grosche, V. Ziegler, and D. Ziegler, *Taschenbuch der Mathematik*. BSB B. G. Teubner Verlagsgesellschaft, Leipzig, 1989.

[474] W. Jakes, *Mobile Microwave Communications*. John Wiley & Sons, New York, 1974.

[475] T. Petermann, S. Vogeler, K.-D. Kammeyer, and D. Boss, "Blind turbo channel estimation in ofdm receivers," *Conference Record of the Thirty-Fifth Asilomar Conference on Signals, Systems and Computers*, vol. 2, pp. 1489–1493, 2001.

[476] M. Münster and L. Hanzo, "First-Order Channel Parameter Estimation Assisted Cancellation of Channel Variation-Induced Inter-Subcarrier Interference in OFDM Systems," in *Proceedings of EUROCON '2001*, vol. 1, (Bratislava, Czech), pp. 1–5, IEEE, July 2001.

[477] A. Chini, Y. Wu, M. El-Tanany, and S. Mahmoud, "Filtered Decision Feedback Channel Estimation for OFDM Based DTV Terrestrial Broadcasting System," *IEEE Transactions on Broadcasting*, vol. 44, pp. 2–11, Mar 1998.

[478] G. Strang, *Linear Algebra and Its Applications, 2nd edition*. Academic Press, 1980.

[479] "EN 300 744, V1.1.2, Digital Video Broadcasting (DVB); Framing structure, channel coding and modulation for digital terrestrial television," in *European Standard (Telecommunications Series)*, pp. 26–28, ETSI, ETSI, 1997.

[480] C. Berrou, A. Glavieux, and P. Thitimajshima, "Near shannon limit error-correcting coding and decoding: Turbo codes," in *Proc. of International Conference on Communications*, (Geneva, Switzerland), pp. 1064–1070, IEEE, May 1993.

[481] C. Berrou and A. Glavieux, "Near optimum error correcting coding and decoding: turbo codes," *IEEE Transactions on Communications*, vol. 44, pp. 1261–1271, October 1996.

[482] H. Rohling and R. Gruenheid, "Adaptive Coding and Modulation in an OFDM-TDMA Communication System," in *Proceedings of Vehicular Technology Conference*, vol. 2, (Ottawa, Canada), pp. 773–776, IEEE, May 18-21 1998.

[483] D. Hughes-Hartogs, "Ensemble modem structure for imperfect transmission media," *U.S. Patents Nos. 4,679,222 (July 1987), 4,731,816 (March 1988) and 4,833,796 (May 1989)*.

[484] T. Keller and L. Hanzo, " Blind-detection Assisted Sub-band Adaptive Turbo-Coded OFDM Schemes," in *Proc. of Vehicular Technology Conference*, (Houston, USA), pp. 489–493, IEEE, May 1999.

[485] F. Rashid-Farrokhi, L. Tassiulas, and K. R. Liu, "Joint Optimal Power Control and Beamforming in Wireless Networks Using Antenna Arrays," *IEEE Transactions on Communications*, vol. 46, pp. 1313–1324, October 1998.

[486] F. Gantmacher, *The Theory of Matrices*, vol. 2. New York: Chelsea, 1990.

[487] W. Fischer and I. Lieb, *Funktionentheorie*. Vieweg Verlag, Braunschweig, Germany, 1983.

[488] J. Winters, "Smart antennas for wireless systems," *IEEE Personal Communications*, vol. 5, pp. 23–27, February 1998.

[489] R. Derryberry, S. Gray, D. Ionescu, G. Mandyam, and B. Raghothaman, "Transmit diversity in 3g cdma systems," *IEEE Communications Magazine*, vol. 40, pp. 68–75, April 2002.

[490] A. Molisch, M. Win, and J. Winters, "Space-time-frequency (stf) coding for mimo-ofdm systems," *IEEE Communications Letters*, vol. 6, pp. 370–372, September 2002.

[491] A. Molisch, M. Steinbauer, M. Toeltsch, E. Bonek, and R. Thoma, "Capacity of mimo systems based on measured wireless channels," *IEEE Journal on Selected Areas in Communications*, vol. 20, pp. 561–569, April 2002.

[492] D. Gesbert, M. Shafi, D.-S. Shiu, P. Smith, and A. Naguib, "From theory to practice: an overview of mimo space-time coded wireless systems," *IEEE Journal on Selected Areas in Communications*, vol. 21, pp. 281–302, April 2003.

[493] M. Shafi, D. Gesbert, D.-S. Shiu, P. Smith, and W. Tranter, "Guest editorial: Mimo systems and applications," *IEEE Journal on Selected Areas in Communications*, vol. 21, pp. 277–280, April 2003.

[494] S. Kay, *Fundamentals of Statistical Signal Processing, Estimation Theory*. Prentice Hall, New Jersey, 1993.

[495] K. Fazel and G. Fettweis, *Multi-carrier Spread Spectrum & Related Topics, ISBN 0-7923-7740-0*. Kluwer, 2000.

[496] H. Bogucka, "Application of the New Joint Complex Hadamard Transform - Inverse Fourier Transform in a OFDM/CDMA Wireless Communication System," in *Proceedings of Vehicular Technology Conference*, vol. 5, (Amsterdam, Netherlands), pp. 2929–2933, IEEE, September 19-22 1999.

[497] A. Goldsmith and S. Chua, "Variable-rate variable-power MQAM for fading channels," *IEEE Transactions on Communications*, vol. 45, pp. 1218–1230, October 1997.

[498] M. Yee and L. Hanzo, "Upper-bound performance of radial basis function decision feedback equalised burst-by-burst adaptive modulation," in *Proceedings of ECMCS'99, CD-ROM*, (Krakow, Poland), 24–26 June 1999.

[499] E. Kuan, C. Wong, and L. Hanzo, "Burst-by-burst adaptive joint detection CDMA," in *Proceeding of VTC'99 (Spring)*, (Houston, TX, USA), IEEE, 16–20 May 1999.

[500] H. Rohling and R. Grünheid, "Peformance of an OFDM-TDMA mobile communication system," in *Proceeding of IEEE Global Telecommunications Conference, Globecom 96*, (London, UK), pp. 1589–1593, IEEE, 18–22 November 1996.

[501] P. Cherriman, T. Keller, and L. Hanzo, "Constant-rate turbo-coded and block-coded orthogonal frequency division multiplex videophony over UMTS," in *Proceeding of Globecom'98*, vol. 5, (Sydney, Australia), pp. 2848–2852, IEEE, 8–12 November 1998.

[502] J. Woodard, T. Keller, and L. Hanzo, "Turbo-coded orthogonal frequency division multiplex transmission of 8 kbps encoded speech," in *Proceeding of ACTS Mobile Communication Summit '97*, (Aalborg, Denmark), pp. 894–899, ACTS, 7–10 October 1997.

[503] P. Cherriman and L. Hanzo, "Programmable H.263-based wireless video transceivers for interference-limited environments," *IEEE Transactions on Circuits and Systems for Video Technology*, vol. 8, pp. 275–286, June 1998.

[504] B. J. Choi and L. Hanzo, "Performance analysis of Rake receiver assisted adaptive QAM using antenna diversity over frequency selective Rayleigh channels," *Submitted to IEEE Journal on Selected Areas in Communications*, January 2001.

[505] B. J. Choi and L. Hanzo, "Optimum mode-switching levels for adaptive modulation systems," in *Submitted to IEEE GLOBECOM 2001*, 2001.

[506] T. H. Liew, B. J. Choi, and L. Hanzo, "Space-time block coded and space-time trellis coded OFDM," in *Proceedings of IEEE VTC 2001-Spring*, p. P.533, IEEE, May 2001.

[507] C. Davila, "Blind Adaptive Estimation of KLT Basis Vectors," *IEEE Transactions on Signal Processing*, vol. 49, pp. 1364–1369, July 2001.

[508] A. Rezayee and S. Gazor, "An Adaptive KLT Approach for Speech Enhancement," *IEEE Transactions on Speech and Audio Processing*, vol. 9, pp. 87–95, February 2001.

[509] Y. Li, J. Chuang, and N. Sollenberger, "Transmitter Diversity for OFDM Systems and Its Impact on High-Rate Data Wireless Networks," *IEEE Journal on Selected Areas in Communications*, vol. 17, pp. 1233–1243, July 1999.

[510] B. Lu and X. Wang, "Iterative Receivers for Multi-User Space-Time Coding Systems," *IEEE Journal on Selected Areas in Communications*, vol. 18, pp. 2322–2335, November 2000.

[511] M. Honig and W. Xiao, "Performance of Reduced-Rank Linear Interference Cancellation," *IEEE Transactions on Information Theory*, vol. 47, pp. 1928–1946, July 2001.

[512] D. Brown, M. Motani, V. Veeravalli, H. Poor, and C. Johnson, "On the Performance of Linear Parallel Interference Cancellation," *IEEE Transactions on Information Theory*, vol. 47, pp. 1957–1970, July 2001.

[513] A. Grant and C. Schlegel, "Convergence of Linear Interference Cancellation Multiuser Receivers," *IEEE Transactions on Communications*, vol. 49, pp. 1824–1834, October 2001.

[514] H. Dai and H. Poor, "Turbo Multiuser Detection for Coded DMT VDSL Systems," *IEEE Journal on Selected Areas in Communications*, vol. 20, pp. 351–362, February 2002.

[515] K.-W. Cheong, W.-J. Choi, and J. Cioffi, "Multiuser Soft Interference Canceler via Iterative Decoding for DSL Applications," *IEEE Journal on Selected Areas in Communications*, vol. 20, pp. 363–371, February 2002.

[516] A. Lampe, R. Schober, W. Gerstacker, and J. Huber, "A Novel Iterative Multiuser Detector for Complex Modulation Schemes," *IEEE Journal on Selected Areas in Communications*, vol. 20, pp. 339–350, February 2002.

[517] B. Lu, X. Wang, and Y. Li, "Iterative Receivers for Space-Time Block-Coded OFDM Systems in Dispersive Fading Channels," *IEEE Transactions on Wireless Communications*, vol. 1, pp. 213–225, April 2002.

Subject Index

Author Index